Some Important Organic Functional Groups

	Functional Group*	Example	IUPAC Name
Acid anhydride	$-\!C(\!=\!O)\!-\!O\!-\!C(\!=\!O)\!-$	CH_3COCCH_3 (di-keto)	Ethanoic anhydride (Acetic anhydride)
Acid chloride	$-\!C(\!=\!O)\!-\!Cl$	CH_3CCl (keto)	Ethanoyl chloride (Acetyl chloride)
Alcohol	$-\!OH$	CH_3CH_2OH	Ethanol (Ethyl alcohol)
Aldehyde	$-\!C(\!=\!O)\!-\!H$	CH_3CH (keto)	Ethanal (Acetaldehyde)
Alkane	(none)	CH_3CH_3	Ethane
Alkene	$C\!=\!C$	$CH_2\!=\!CH_2$	Ethene (Ethylene)
Alkyne	$-\!C\!\equiv\!C\!-$	$HC\!\equiv\!CH$	Ethyne (Acetylene)
Amide	$-\!C(\!=\!O)\!-\!N\!-$	CH_3CNH_2 (keto)	Ethanamide (Acetamide)
Amine, primary	$-\!NH_2$	$CH_3CH_2NH_2$	Ethylamine
Amine, secondary	$-\!NH\!-$	$(CH_3CH_2)_2NH$	Diethylamine
Amine, tertiary	$-\!N\!-$	$(CH_3CH_2)_3N$	Triethylamine
Carboxylic acid	$-\!C(\!=\!O)\!-\!O\!-\!H$	CH_3COH (keto)	Ethanoic acid (Acetic acid)
Disulfide	$-\!S\!-\!S\!-$	CH_3SSCH_3	Dimethyl disulfide
Epoxide	epoxide ring (C–C with bridging O)	$H_2C\!-\!CH_2$ (O bridge)	Oxirane (Ethylene oxide)
Ester	$-\!C(\!=\!O)\!-\!O\!-\!C\!-$	CH_3COCH_3 (keto)	Methyl ethanoate (Methyl acetate)
Ether	$-\!O\!-$	CH_3COCH_3	Diethyl ether
Haloalkane	$-\!X$, $X = F,\ Cl,\ Br,\ I$	CH_3CH_2Cl	Chloroethane (Ethyl chloride)
Ketone	$-\!C(\!=\!O)\!-$	CH_3CCH_3 (keto)	Propanone (Acetone)
Nitrile	$-\!C\!\equiv\!N$	$CH_3\!-\!C\!\equiv\!N$	Ethanenitrile (Acetonitrile)
Nitro	$-\!N^{+}(\!=\!O)\!-\!O^{-}$	CH_3NO_2	Nitromethane
Phenol	aromatic ring with $-\!OH$	phenol (C_6H_5OH)	Phenol
Sulfide	$-\!S\!-$	CH_3SCH_3	Dimethyl sulfide
Thiol	$-\!S\!-\!H$	CH_3CH_2SH	Ethanethiol (Ethyl mercaptan)

* Where bonds to an atom are not specified, the atom is assumed to be bonded to one or more carbon or hydrogen atoms in the rest of the molecule.

Organic Chemistry

Third Edition

William H. Brown

Beloit College

Christopher S. Foote

University of California, Los Angeles

Harcourt College Publishers

Forth Worth   Philadelphia   San Diego   New York   Orlando   Austin
San Antonio   Toronto   Montreal   London   Sydney   Tokyo

i

Publisher: Emily Barrosse
Acquisitions Editor: John Vondeling
Marketing Strategist: Pauline Mula
Developmental Editor: Sandra Kiselica
Project Editor: Ellen Sklar
Production Manager: Charlene Squibb
Art Director: Jonel Sofian

Cover Credit: Yellow crysanthemums, by Alan Pitcairn, of Grant Heilman Photography. Red mum, by Christi Carter, of Grant Heilman Photography.

Organic Chemistry, Third Edition
ISBN: 0-03-033497-7
Library of Congress Catalog Card Number: 2001086006

Address for domestic orders:
Harcourt College Publishers, 6277 Sea Harbor Drive, Orlando, FL 32887-6777
1-800-782-4479
e-mail   collegesales@harcourt.com

Address for international orders:
International Customer Service, Harcourt, Inc.
6277 Sea Harbor Drive, Orlando FL 32887-6777
(407) 345-3800
Fax (407) 345-4060
e-mail   hbintl@harcourt.com

Address for editorial correspondence:
Harcourt College Publishers, Public Ledger Building, Suite 1250, 150 S. Independence Mall West, Philadelphia, PA 19106-3412

Web Site Address
http://www.harcourtcollege.com

Printed in the United States of America
1234567890 032 10 987654321

For John Vondeling
Mentor and Friend

BIOGRAPHIES

William H. Brown William H. Brown is Professor of Chemistry at Beloit College, where he has twice been named Teacher of the Year. He is also the author of the college textbook *Introduction to Organic Chemistry 2/e*, published in 2000. His regular teaching responsibilities include organic chemistry, advanced organic chemistry, and, more recently, special topics in pharmacology and drug synthesis. He received his PhD from Columbia University under the direction of Gilbert Stork and did postdoctoral work at California Institute of Technology and the University of Arizona.

Bill Brown and his wife Carolyn enjoy hiking in the Southwest and the study of petroglyphs and pictographs. Twice he has been the Director of Beloit College's World Outlook Seminar, a program coordinated with the University of Glasgow in Scotland.

Bill Brown in Capitol Reef National Park, Utah *(Carolyn S. Simonton)*

Christopher S. Foote Christopher S. Foote is Professor of Chemistry at the University of California, Los Angeles. He received his BS degree from Yale University and his PhD in Organic Chemistry from Harvard University. In 1995, he received the Tolman Award of the ACS Southern California Section for his contributions to chemistry. Foote's research has focused on the chemistry of oxygen in organic and biological systems and on the chemistry of fullerenes. Other awards he has received include the Yale Science and Engineering Award for the Advancement of Basic and Applied Science, the ACS North Jersey Sections' Leo Hendrick Baekeland Award, and the 2000 American Society of Photo Biology Distinguished Research Award. He was also an ACS Cope Scholar in 1994 and is the author of more than 250 research papers.

Christopher S. Foote

PREFACE

The Audience

This book provides an introduction to organic chemistry for students majoring in chemistry and in related disciplines, especially the health and biological sciences. Fundamental to our approach to the material of this course is the fact that organic chemistry has an underlying rationale, namely the mechanistic themes that unify it. With an appreciation for the power of mechanisms to provide an underpinning for its information content, organic chemistry can become an exciting area for exploration in its own right. We hope students will see organic chemistry as a dynamic and ever-expanding area of science waiting openly for those who are prepared, both by training and inquisitive nature, to ask questions and to explore.

New to the Third Edition

In this edition, we have made major changes to unify the approach, to add new material that is at the forefront of organic synthesis, and to make the treatment of energy and mechanisms more complete. For almost every reaction for which a mechanism can be written, we present it. The treatment of reaction energetics has been expanded. Most importantly, we have incorporated feedback from students and users throughout this revision.

We now use SI units. Almost all elementary chemistry texts and most biochemistry texts now prefer kJ to kcal, and it seems time to begin the transition for organic chemistry texts as well. However, many users of the book will have learned to think in kcal. For this reason, we give energies as kJ (kcal)/mol throughout.

We have removed stereoviews from this edition and replaced them with computer models generated by CambridgeSoft's Chem3D. All models are also available on the bundled CD-ROM. Computed electrostatic potential maps and orbitals generated with Wavefunction's MacSpartan computer program are used in the text where relevant.

Chapters 3 and 4 have been switched to put stereochemistry/chirality immediately following the introduction to structure. Chapter 4 introduces acids and bases, followed by the alkenes chapter, which now uses the acid/base concepts more effectively.

The spectroscopy chapters (12–14) have been expanded somewhat, and have been made modular in that material for all functional groups has been introduced. In the present form, IR and NMR could be used as early as after Chapter 2, but could also be introduced at any convenient point in a course, including much later. Topicity has been added to the NMR chapter.

A separate chapter on organometallics (15) has been added. This chapter alerts students to the enormous advances in this area of chemistry. The Heck, Simmons-Smith, and alkene methathesis reactions are used as examples of important organometallic reactions. Synthesis and concepts of stereocontrol have been greatly expanded and stereospecific or enantiospecific syntheses are used in many places.

A new interchapter on medicinal chemistry unifies and reinforces the reactions introduced previously in the text by using them in problems on synthesis of important medicinal compounds. Explanatory material with these problems introduces more material on medicinal chemistry. Many of these problems can be introduced earlier to deepen the synthesis component of the course.

Major changes have been made in the detailed treatment of mechanisms in the carboxyl and enolate chapters (18 and 19). The chapter on conjugated systems (23) has been expanded, and limited coverage of UV/Vis spectroscopy has been moved to this chapter. There is expanded coverage of the Diels-Alder reaction. The chapter on proteins (27) has been updated with the use of mass spectra in sequencing, and a section on the genome decoding has been added to the DNA chapter (28).

Chapter-by-Chapter Overview

Chapter 1 begins with a review of the electronic structure of atoms and molecules, the Lewis model of bonding, and use of the VSEPR model to predict shapes of molecules and ions. Within this discussion, we introduce the functional groups encountered most frequently in the text along with the theory of resonance and the use of curved arrows and electron pushing. A knowledge of resonance theory combined with a facility for moving electrons gives students two powerful tools for writing reaction mechanisms and understanding chemical reactivity. Chapter 1 concludes with an introduction to quantum mechanics and the molecular orbital theory of covalent bonding.

Chapter 2 opens with a description of the structure, nomenclature, and conformational analysis of alkanes and cycloalkanes. Beginning here and continuing throughout the text, a clear distinction is made between IUPAC and common names. Where names are introduced, IUPAC names are given first and common names, where appropriate, follow in parentheses. The IUPAC system is introduced in Section 2.3A through the naming of alkanes, and in Section 2.5, it is presented as a general system of nomenclature. The concept of stereoisomerism is introduced in this chapter with a discussion of cis,trans isomerism in cycloalkanes.

(Charles D. Winters)

Chapter 3 begins with a review of constitutional, conformational, and cis,trans/E,Z isomerism and then introduces the concepts of chirality, enantiomerism, and diastereomerism.

Chapter 4 contains a general introduction to acid-base chemistry with emphasis on both qualitative and quantitative determination of the position of equilibrium in acid-base reactions. We include an in-depth discussion of relationships between molecular structure and acidity. With an understanding of the structural basis for these relationships, students can then deal with questions such as "Why is a carboxylic acid a stronger acid than an alcohol", and "Why is acetylene a stronger acid than ethane?"

Chapters 5 and 6 cover the chemistry of alkenes. Their structure and physical properties are presented in Chapter 5. The chapter concludes with the structure of terpenes and an introduction to one theme in the molecular logic of biomolecules. The focus in Chapter 6 is on the chemical reactivity of alkenes. It opens with an introduction to the concept of a reaction mechanism, energy diagrams, Gibbs free energy, activation energy, and reactive intermediates. Reactions of alkenes are organized in the following order: electrophilic additions, hydroboration, oxidation, and reduction. The twin concepts of regiospecificity/selectivity and stereospecificity/selectivity are introduced in the context of electrophilic additions. This chapter builds on the concepts of acid-base reactions introduced in Chapter 4 and develop the concepts of carbocation stability and rearrangement.

Chapter 7   introduces the haloalkanes and their structure and nomenclature. It has as its central theme the radical halogenation of alkanes and provides an introduction to the mechanistic concepts of chain initiation, propagation, and termination. Regioselectivity of radical bromination compared with radical chlorination is interpreted in terms of Hammond's postulate. Free radical autoxidation is introduced.

Chapter 8   presents what, in our experience, is one of the most formidable and anxiety-producing aspects of introductory organic chemistry, namely S_N1, S_N2, E1, E2 mechanisms and the attendant concepts of stereochemistry, kinetics, and relationships between structure and chemical reactivity. The difficulty does not lie in any single part of this material, but in the number of concepts to be assimilated at one time. This introduction is made much simpler because the key concepts of carbocation stability and rearrangements have already been introduced in Chapter 6. By this stage in the course, students have a good grounding in the structure of organic molecules, the theory of resonance, electron pushing, and reaction mechanisms. Nucleophilic substitution and β-elimination then become a vehicle for integration of previously covered chemistry into a larger pattern.

Chapter 9   continues the theme of the relationships between structure and reactivity by considering the chemistry of alcohols. A significant body of the chemistry of alcohols can be understood using the concepts of nucleophilic substitution, β-elimination, and the relative stability of carbocations, all of which have been developed in previous chapters.

Chapter 10   opens with the structure and nomenclature of alkynes followed by the acidity of terminal alkynes. It then emphasizes the usefulness of alkynes as building blocks in organic synthesis through alkylation of acetylide anions, reduction, hydroboration-oxidation, and electrophilic additions. The chapter concludes with an introduction to the strategy of organic synthesis, namely retrosynthetic analysis.

(Charles D. Winters)

Chapter 11   is a logical extension of nucleophilic substitutions as applied to the synthesis and reactions of ethers and epoxides. The value of epoxides in organic synthesis is stressed, including the regio- and stereochemistry of their reactions with a variety of nucleophiles. Sharpless stereospecific epoxidation is introduced.

Chapters 12–14   examine the molecular spectroscopy of the most common functional groups presented in the course. First is infrared spectroscopy in Chapter 12, followed by ^{1}H-NMR and ^{13}C-NMR spectroscopy in Chapter 13, and then mass spectrometry in Chapter 14. While this material is presented as a cluster of chapters midway through the text, the chapters have been made free-standing and can be used in different order as appropriate to a particular course.

Chapter 15 new in this edition, begins with the preparation and structure of Grignard and organolithium reagents, followed by their basicity, reaction with proton acids, and reaction with oxiranes. Next we show the preparation of lithium diorganocopper (Gilman) reagents and their coupling with alkyl and alkenyl halides to form new carbon-carbon bonds. This chapter introduces carbenes and carbenoids, including the Simmons-Smith reaction. In keeping with our desire to give even greater emphasis to the chemistry of organometallic compounds, we introduce both the Heck and olefin metathesis reactions and demonstrate their usefulness in forming new carbon-carbon bonds.

(Omikron/Photo Researchers Inc.)

Chapters 16–19 concentrate on the chemistry of carbonyl-containing compounds. First is the chemistry of aldehydes and ketones in Chapter 16 followed by the chemistry of carboxylic acids and their functional derivatives in Chapters 17 and 18. Collected in Chapter 19 are various carbonyl condensation reactions, including the aldol reaction, and the Claisen and Dieckmann condensations as well as alkylation and acylation reactions of acetoacetic esters, malonic esters, and enamines. Mechanistic treatment of these reactions is enhanced.

Chapters 20 and 21 present the chemistry of aromatic compounds. The first of these chapters concentrates on structure and nomenclature of aromatic compounds, the concept of aromaticity, and the structure and acid-base properties of phenols. The second is devoted to aromatic substitution reactions.

Chapter 22 presents the chemistry of aliphatic and aromatic amines.

Chapter 23 completes the introduction to organic functional groups with the particular chemistry of conjugated dienes, including both 1,2- and 1,4-addition and the Diels-Alder reaction. This discussion is followed by an overview of pericyclic reactions and the concept of transition state aromaticity. The chapter concludes with an introduction to UV-Vis spectroscopy.

Problems in Medicinal Chemistry With completion of Chapter 23, we have covered the chemistry of the major functional groups and, in this interchapter section, we present a set of problems in organic synthesis. In recognition of the fact that many students taking introductory organic chemistry are interested in careers in the health and biological sciences, we have chosen problems entirely from the area of medicinal chemistry. Problems are grouped by their therapeutic category, and for several we include information on drug design and discovery. Many of them may be introduced earlier to deepen the synthesis component of the course.

Chapter 24 is a systematic introduction to organic polymer chemistry. New to this edition is a discussion of ring-opening metathesis polymerization (ROMP). Given the importance of organic polymers in the world around us, we have made this chapter more extensive than that found in most other organic textbooks.

Chapters 25–28 present an introduction to the chemistry to carbohydrates, lipids, amino acids and proteins, and nucleic acids. Emphasis in this chapter is on the structure of these biomolecules. These chapters have been updated to include cutting-edge topics related to proteomics and genomics.

Chapter 29 presents a discussion of two key metabolic pathways, namely glycolysis and β-oxidation of fatty acids. It is our purpose in this chapter to show that the reactions of these pathways are biochemical equivalents of organic functional group reactions we have already studied in detail.

Special Features

- **Full-Color Art Program** One of the most distinctive features of this text is its visual impact. The text's extensive full-color art program includes over 250 pieces of art by professional artists John and Bette Woolsey. A large number of molecular models have been generated and energy minimized in CambridgeSoft's Chem3D, and then rendered by these artists to provide easily visualized pictures of molecular structures. All of these models are available on the bundled CD-ROM. Computed electrostatic potential maps generated by Wavefunction's MacSpartan are provided at appropriate places throughout the text to illustrate the important concepts of resonance, electronegativity, and nucleophilicity.

- **Photo Art** Photos, conceived and developed for this text, show organic chemistry as it occurs in the laboratory and in everyday life and depict the natural sources of many organic compounds.

- **Using ChemOffice Web with Organic Chemistry, Third Edition** Packaged with the text is a CD-ROM prepared by the authors in conjunction with CambridgeSoft Corporation and containing over 300 models rendered in Chem3D. Their purpose is to assist students in visualizing organic molecules as three-dimensional objects. With the plug-in supplied on the CD-ROM, students can rotate each model, change from ball-and-stick to space-filling, measure bond angles and interatomic distances, and invert configuration at a stereocenter in a cyclic molecule. With the plug-in, students can also build models in ChemDraw, assign R,S configuration to each stereocenter, and import ChemDraw structures into other documents. Icons in the text alert students to use the CD.

- **Bioorganic Chemistry** Bioorganic chemistry is emphasized throughout the text, in the Chemistry in Action boxes, in end-of-chapter problems, and in the 50 medical chemistry problems collected in the interchapter section following Chapter 23. Merck Index references are from *The Merck Index* (Susan Budavari, editor, 12th edition, Merck Research Laboratories, 1996).

- **Chemistry in Action Boxes** These boxes illustrate applications of organic chemistry to everyday settings. Topics range from "Radical Autoxidation" to "Drugs That Lower Plasma Levels of Cholesterol."

- **In-Chapter Examples** There are an abundance of in-chapter examples, all with detailed solutions. Following each in-chapter example is a comparable in-chapter problem designed to give students the opportunity to solve a problem related to the example.

- **End-of-Chapter Summaries and Summaries of Key Reactions** End-of-chapter summaries highlight all important new terms found in a chapter. In addition, each reaction is annotated and keyed to the section where it is discussed.

- **End-of-Chapter Problems** There are plentiful end-of-chapter problems. The majority of problems are categorized by topics. A problem number set in red indicates a more challenging problem.

- **Glossary of Key Terms** Throughout the book we place definitions for new terms in the margin. In addition, all definitions are collected in a glossary at the end of the text. Each glossary listing is keyed to the section of the text where the term is introduced.

- **Color** Color is used to highlight parts of molecules and to follow the course of reactions. The following graphic shows some of the colors used consistently in the artwork in this book.

- **Interviews** Four interviews with prominent scientists describe how each of these people became interested in chemistry, then insights into their careers as educators and research professionals. Their enthusiasm for their work is evident, and they invite students to pursue similar interests in the sciences.

(Photo Asscociates)

Support Package

For the Student

- **Student Study Guide** by Brent and Sheila Iverson of the University of Texas, Austin. Contains detailed solutions to all in-text and end-of-chapter problems.

- **Pushing Electrons: A Guide for Students of Organic Chemistry, third edition** by Daniel P. Weeks, Northwestern University. A paperback workbook designed to help students learn techniques of electron pushing. Its programmed approach emphasizes repetition and active participation.

- **Problem Book for Organic Chemistry** by Andrew Ternay (University of Denver) contains more than 800 additional problems with explanations about the important concepts in organic chemistry. Keyed to the textbook.

- **Harcourt Interactive Organic Chemistry CD-ROM** by William Vining and Vincent Rotello, University of Massachusetts. This dual-platform tutorial CD-ROM includes six modules to help students learn organic chemistry in an interactive fashion. The modules include Mechanisms, Nomenclature, Reactivity Explorer, Multistep Synthesis,

  Spectroscopy, and Supporting Concepts. The text is keyed by an icon shown in

  the margin. An accompanying **workbook** by Steven Hardinger, University of California, Los Angeles, contains questions to guide students through the CD-ROM modules.

- **CSC ChemOffice Limited, version 4.5** includes ChemDraw, Chem3D, and ChemFinder and is available at a very reasonable price from the publisher.

- **Harcourt Organic Chemistry Web Site** (http://www.harcourtcollege.com) contains tutorials, animations, molecular modeling, and practice problems. **PowerPoint**™ lecture slides by William H. Brown can be found on the Harcourt Web site for organic chemistry.

For the Instructor

- **Test Bank** A multiple-choice bank of over 1000 problems for instructors to use for tests, quizzes, or homework assignments. Also available in computerized form for Windows™ and Macintosh® platforms.

- **PowerPoint™** lecture slides by William H. Brown can be found on the Harcourt Web site for organic chemistry.

- **Instructors' Resource CD-ROM** contains all images and tables from the text for instructors to incorporate into lectures.

- **Overhead Transparency Acetates** A selection of 150 full-color figures from the text.

To the Student

ChemOffice Web — CD-ROM

How to Use the CD-ROM

The CD-ROM enclosed in your text is your exploration to molecular structure. In the text you will find CD-ROM icons 💿, which direct you to specific models and exercises on the CD-ROM. Icons are located next to particular models, examples, practice problems, or additional problems at the end of each chapter.

When you find an icon in the text, mount the CD-ROM, locate the menu of chapters on the left side of your screen, and open (click on) the appropriate chapter. You will then see a menu of models for that chapter along with a brief descriptive title of each model. Select (click on) the appropriate title. The screen that appears shows you the model you have selected and, under it, one or more exercises related to that model. In addition, there are molecular modeling exercises at the end of many of the in-text chapters.

The purpose of the CD-ROM, along with its models and exercises, is to help you appreciate that organic molecules are three-dimensional objects. With the CD-ROM, you can measure bond lengths, bond angles, and examine molecular sizes and shapes. You can rotate the models in space and view them from any perspective you choose. You can display the models as ball-and-stick, cylindrical bond, wire frame, or space-filling.

Tutorials

Included on the CD-ROM is a set of tutorials. We suggest that you work through these tutorials first before you try any of the interactive exercises on the CD-ROM. The tutorials will introduce you quickly to the tools.

Tutorial Contents

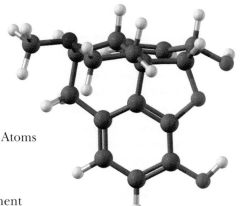

A. Measure Bond Lengths
B. Measure Bond Angles
C. Determine Bond Order
D. Rotate a Molecule
E. Change Among Model Types
F. Resize a Model or Models
G. Resize and Center a Model or Models
H. Rotate About a Bond
I. Measure Distance Between Nonbonded Atoms
J. Hide/Show all Hydrogens (H's)
K. Create Mirror Images
L. Invert Configuration
M. Paste a Chem3D Structure into a Document
N. Play a Movie
O. Assign Formal Charges

We do not provide solutions in the accompanying Study Guide for the Interchapter problems. We do, however, provide solutions on our Web site to instructors who have the option of supplying them to the students. Select Organic Chemistry from our Web site at http://www.harcourtcollege.com/chem

Acknowledgments

While one or a few persons are listed as "author" of any textbook, the book is in fact the product of collaboration of many individuals, some obvious, others not so obvious. It is with gratitude that we herein acknowledge the contributions of the many. It is only fitting to begin with John Vondeling, Vice President and Publisher of Saunders College Publishing. John's contribution began with the faith that we could do this book and then marshaling the support systems necessary to bring it from rough manuscript to bound book form, assembling the elements of the supplemental materials, and finally bringing to bear his keen sense of the marketplace.

Sandi Kiselica has been a rock of support as Senior Developmental Editor. We so appreciate her ability to set challenging but manageable schedules for us and her constant encouragement as we worked to meet those deadlines. She was also an invaluable resource person with whom we could discuss everything from pedagogy to details of art work. We also want to acknowledge others at Saunders who contributed to this project, in particular, Ellen Sklar, Project Editor; Jonel Sofian, Art Director; Pauline Mula, Senior Marketing Strategist; and Charlene Squibb, Senior Production Manager.

We are also indebted to the many reviewers of our manuscript who helped shape its contents. With their guidance we have revised this text to better meet the needs of their students. A special thanks goes to William Vining (University of Massachusettes) who keyed our textbook by icon to the Harourt Interactive organic chemistry CD-ROM. Also, we are grateful to Gary Lyon (Louisianna State University) who read all the galleys and page proofs for accuracy.

Reviewers for the Third Edition:

Eric Anslyn, *University of Texas*
James Canary, *New York University*
Dorothy Goldish, *California State University, Long Beach*
Steven Hardinger, *University of California, Los Angeles*
Ian Hunt, *University of Calgary*
Francis Klein, *Creighton University*
Robert Loeschen, *California State University, Long Beach*
Marco Lopez, *California State University, Long Beach*
Gary Lyon, *Louisiana State University*
Andrew MacMillan, *University of Toronto*
Aaron Odom, *Michigan State University*
Anne Padias, *University of Arizona*

Brian Pagenkopf, *University of Texas*
Michael Rathke, *Michigan State University*
Carmelo Rizzo, *Vanderbilt University*
K. C. Russell, *University of Miami*
John Scheffer, *University of British Columbia*
William Shay, *University of North Dakota*
Valerie Sheares, *Iowa State University*
J. William Suggs, *Brown University*
Edward Waali, *University of Montana*
Theodore Widlanski, *Indiana University*
Jeffrey Winkler, *University of Pennsylvania*

Reviewers for the Second Edition of *Organic Chemistry:*

William Bailey, *University of Connecticut*
John Benbow, *Lehigh University*
Thomas Bryson, *University of South Carolina*
Mary Campbell, *Mount Holyoke College*
Lyle Castle, *Idaho State University*
Claire Castro, *University of San Francisco*
Clair Cheer, *San Jose State University*
Barry Coddens, *Northwestern University*
Mark DeCamp, *University of Michigan, Dearborn*
Dale Drueckhammer, *Stanford University*
William Epstein, *University of Utah*
Morris Fishman, *New York University*
Jack Gilbert, *University of Texas, Austin*
Stanley I. Goldberg, *University of New Orleans*
Scott Gronert, *San Francisco State University*
Dan Harvey, *University of California, San Diego*
Gene Hiegel, *California State University, Fullerton*
John Landgrebe, *University of Kansas*

Norman Lebel, *Wayne State University*
Richard Luibrand, *California State University, Hayward*
Eugene Mash, *University of Arizona*
Dominic McGrath, *University of Connecticut*
Kirk McMichael, *Washington State University*
Richard Morrison, *West Virginia University*
James Mulvaney, *University of Arizona*
Kathy Nabona, *Austin Community College*
Gary Newton, *University of Georgia*
Bruce Norcross, *State University of New York, Binghamton*
Steven Pedersen, *University of California, Berkeley*
Michael Rathke, *Michigan State University*
Carmelo Rizzo, *Vanderbilt University*
Alan Rosan, *Drew University*
Daniel Singleton, *Texas A&M University*
Robert Stern, *Oakland University*
J. William Suggs, *Brown University*
Michelle Sulikowski, *Texas A&M University*

Reviewers for the First Edition of *Organic Chemistry:*

Neil T. Allison, *University of Arkansas*
Rodney Badger, *Southern Oregon State College*
Shelton Bank, *State University of New York, Albany*
Nancy Barta, graduate student, *Michigan State University*
John L. Belletiri, *University of Cincinnati*
Edwin Bryant, graduate student, *Michigan State University*
Edward M. Burgess, *Georgia Institute of Technology*
Robert G. Carlson, *University of Kansas*
Dana S. Chatellier, *University of Delaware*
William D. Closson, *State University of New York, Albany*
David Crich, *University of Illinois, Chicago*
Dennis D. Davis, *New Mexico State University*
James A. Deyrup, *University of Florida*
Thomas A. Dix, *University of California, Irvine*
Paul Dowd, *University of Pittsburgh*
Michael B. East, *Florida Institute of Technology*
Raymond C. Fort, Jr., *University of Maine*
Warren Giering, *Boston University*
Leland Harris, *University of Arizona*
John Helling, *University of Florida*
John L. Hogg, *Texas A&M University*
John W. Huffman, *Clemson University*
Brent Iverson, *University of Texas, Austin*
Ronald Kluger, *University of Toronto*
Joseph B. Lambert, *Northwestern University*
Allan K. Lazarus, *Trenton State University*
Jerry March, *Adelphi University*
Kenneth L. Marsi, *California State University, Long Beach*

David M. McKinnon, *University of Manitoba*
James Mulvaney, *University of Arizona*
Walter Ott, *Emory University*
E. Paul Papadopoulos, *University of New Mexico, Albuquerque*
Russell C. Petter, *Sandoz Research Institute*
Joseph M. Prokipcak, *University of Guelph*
William A. Pryor, *Louisiana State University*
Michael Rathke, *Michigan State University*
Charles B. Rose, *University of Nevada, Reno*
James Schreck, *University of Northern Colorado*
Jonathan Sessler, *University of Texas*
Martin Sobczak, graduate student, *Michigan State University*
Steve Steffke, graduate student, *Michigan State University*
John Stille, *Michigan State University*
J. William Suggs, *Brown University*
Peter Trumper, *Bowdoin College*
Ken Turnbull, *Write State University*
George Wahl, *North Carolina State University*
Michael Waldo, graduate student, *Michigan State University*
Daniel Weeks, *Northwestern University*
David F. Weimer, *University of Iowa*
Desmond Wheeler, *University of Nebraska*
Darrell J. Woodman, *University of Washington*
Ali Zand, graduate student, *Michigan State University*

We have enjoyed writing this text, and hope that instructors and students alike find in it a measure of the excitement we feel for organic chemistry.

WILLIAM H. BROWN
Beloit College
Beloit, WI
brownwh@beloit.edu

CHRISTOPHER S. FOOTE
University of California, Los Angeles
Los Angeles, CA
foote@chem.ucla.edu

April, 2001

CONTENTS OVERVIEW

CONTENTS

(Charles D. Winters)

(Earl Roberge/Photo Researchers, Inc.)

(T. J. Florian, Rainbow)

(Paul Shambroom/Science Source/Photo Researchers)

(Herb Charles Ohlmeyer/Fran Heyl Associates)

(Charles D. Winters)

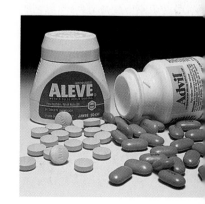

(Charles D. Winters)

(Frank Orel/Stone)

(Charles D. Winters)

(Hans Strand/Stone)

CHEMISTRY IN ACTION ESSAYS

LIST OF
MECHANISMS

Most mechanisms are set off in a Mechanism box. A few mechanisms are discussed in the text, but not boxed.

COVALENT BONDING AND SHAPES OF MOLECULES

A ccording to the simplest definition, **organic chemistry** is the study of the compounds of carbon. Perhaps its most remarkable feature is that most organic compounds consist of carbon and only a few other elements—chiefly, hydrogen, oxygen, and nitrogen. Chemists have discovered or made well over ten million compounds composed of carbon and these three other elements. Organic compounds are everywhere around us—in our foods, flavors, and fragrances;

■ A model of the structure of diamond, one form of pure carbon. Each carbon in a diamond atom is bonded to four other carbons at the corners of a tetrahedron. *(Charles D. Winters)* Inset: A model of Buckyball. See the box "Buckyball—A New Form of Carbon."

1

in our medicines, toiletries, and cosmetics; in our plastics, films, fibers, and resins; in our paints and varnishes; in our glues and adhesives; and, of course, in our bodies and those of all living things.

But organic chemistry is more than the chemistry of carbon and the few other elements listed here. It is also the chemistry of compounds containing carbon-metal bonds, among them bonds to Li, Na, K, Mg, B, Cu, Al, Si, Fe, Ni, Pd, Pt, Ru, Rh, Ti, Zn, and Hg. The chemistry of organometallic compounds is an extremely large area of study today and one of the major interfaces between organic chemistry, inorganic chemistry, and biochemistry. We introduce organometallics in Chapter 15 and continue their study in several following chapters.

Let us begin our study of organic chemistry with a review of how the elements of C, H, O, and N combine by sharing electron pairs to form molecules. There is a great deal of material in this chapter, but much of it should be familiar from your previous chemistry courses. However, because all subsequent chapters in this book use this material, it is essential that you understand it and can use it fluently.

1.1   Electronic Structure of Atoms

You should already be familiar with the fundamentals of the electronic structure of atoms. Briefly, an atom contains a small, dense nucleus made of neutrons and positively charged protons. Most of the mass of an atom is contained in its nucleus. The nucleus is surrounded by a much larger extranuclear space containing negatively charged electrons. The nucleus of an atom has a diameter of 10^{-14} to 10^{-15} meters (m). The extranuclear space where its electrons are found is a much larger volume with a diameter of approximately 10^{-10} m (Figure 1.1).

Electrons do not move freely in the space around the nucleus but are confined to regions of space called **principal energy levels** or, more simply, **shells.** Electron shells are identified by the principal quantum numbers 1, 2, 3, and so on. Each shell can contain up to $2n^2$ **electrons,** where n is the number of the shell. Thus, the first shell can contain 2 electrons, the second 8 electrons, the third 18 electrons, the fourth 32 electrons, and so on (Table 1.1). Electrons in the first shell are nearest to the positively charged nucleus and are held most strongly by it; these electrons are said to be lowest in energy. To say that they are lowest in energy also means that it takes a greater energy to remove an electron from the first shell of an atom than from any other shell. Electrons in higher numbered shells are farther from the positively charged nucleus. They are held less strongly, it takes less energy to remove them from the atom, and, accordingly, they are said to be higher in energy.

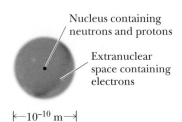

Nucleus containing neutrons and protons

Extranuclear space containing electrons

⊢—10^{-10} m—⊣

Figure 1.1

A schematic view of an atom. Most of the mass of an atom is concentrated in its small, dense nucleus.

Shell   A region of space around a nucleus in which electrons are found.

Table 1.1	Distribution of Electrons in Shells	
Shell	Number of Electrons Shell Can Hold	Relative Energies of Electrons in These Shells
4	32	higher
3	18	↑
2	8	
1	2	lower

Table 1.2	Distribution of Orbitals Within Shells
Shell	**Orbitals Contained in That Shell**
3	$3s$, $3p_x$, $3p_y$, $3p_z$, plus five $3d$ orbitals
2	$2s$, $2p_x$, $2p_y$, $2p_z$
1	$1s$

Shells are divided into subshells designated by the letters s, p, d, and f, and, within these subshells, electrons are grouped in orbitals (Table 1.2). An **orbital** is a region of space that can hold two electrons. The first shell contains a single orbital called a $1s$ orbital. The second shell contains one s orbital and three p orbitals. The three p orbitals are directed along the x, y, and z axes and are designated $2p_x$, $2p_y$, and $2p_z$. The third shell contains one $3s$ orbital, three $3p$ orbitals, and five $3d$ orbitals.

One way to visualize the electron density associated with a particular orbital is to draw a boundary surface around the region of space that contains some arbitrary percentage of the negative charge associated with that orbital. Most commonly, we draw the boundary surface at 95%. In this course, we concentrate on s and p orbitals, boundary surface shapes for which are shown in Section 1.7B.

Orbital A region of space that can hold two electrons.

A. Electron Configuration of Atoms

The electron configuration of an atom is a description of the orbitals its electrons occupy. Every atom has an infinite number of possible electron configurations. At this stage, we are concerned primarily with the **ground-state electron configuration**—the electron configuration of lowest energy. We determine the ground-state electron configuration of an atom by using the following three rules.

Ground-state electron configuration The electron configuration of lowest energy for an atom, molecule, or ion.

Rule 1: The Aufbau ("Build-Up") Principle. Orbitals fill in order of increasing energy, from lowest to highest. In this course, we are concerned primarily with the elements of the first, second, and third periods of the Periodic Table. Orbitals of these elements fill in the order $1s$, $2s$, $2p$, $3s$, $3p$, and so on.

Aufbau principle Orbitals fill in order of increasing energy, from lowest to highest.

Rule 2: The Pauli Exclusion Principle. No more than two electrons may be present in an orbital, and their spins must be paired. To understand the reason for spin pairing, recall from general chemistry that electrons behave as if they have a spin, just as the earth has a spin. And, just as the earth has magnetic north (N) and south (S) poles, so also do spinning electrons. We know from our everyday experiences that identical magnetic poles (N-N or S-S) repel each other and opposite magnetic poles (N-S) attract each other. The Pauli exclusion principle states that only two electrons can occupy an orbital and that their spins (magnetic fields) must be opposite (N to S and S to N). For example, with four electrons, the $1s$ and $2s$ orbitals are filled and are written $1s^2 2s^2$. With an additional six electrons, the set of three $2p$ orbitals is filled and is written $2p_x{}^2 2p_y{}^2 2p_z{}^2$. Alternatively, a filled set of three $2p$ orbitals may be written $2p^6$.

Pauli exclusion principle No more than two electrons may be present in an orbital. If two electrons are present, their spins must be paired.

Rule 3: Hund's Rule. When orbitals of equivalent energy are available but there are not enough electrons to fill all of them completely, then one electron is added to each orbital before a second electron is added to any one of them. After the $1s$ and $2s$ orbitals are filled with four electrons, a fifth electron is added to the $2p_x$

Hund's rule When orbitals of equivalent energy are available but there are not enough electrons to fill all of them completely, then one electron is added to each equivalent orbital before a second electron is added to any one of them.

Table 1.3	Ground-State Electron Configurations for Elements 1–18		
First Period*	**Second Period**		**Third Period**
H 1 $1s^1$	Li 3 [He] $2s^1$		Na 11 [Ne] $3s^1$
He 2 $1s^2$	Be 4 [He] $2s^2$		Mg 12 [Ne] $3s^2$
	B 5 [He] $2s^2 2p^1$		Al 13 [Ne] $3s^2 3p^1$
	C 6 [He] $2s^2 2p^2$		Si 14 [Ne] $3s^2 3p^2$
	N 7 [He] $2s^2 2p^3$		P 15 [Ne] $3s^2 3p^3$
	O 8 [He] $2s^2 2p^4$		S 16 [Ne] $3s^2 3p^4$
	F 9 [He] $2s^2 2p^5$		Cl 17 [Ne] $3s^2 3p^5$
	Ne 10 [He] $2s^2 2p^6$		Ar 18 [Ne] $3s^2 3p^6$

*Elements are listed by symbol, atomic number, and simplified ground-state electron configuration.

orbital, a sixth to the $2p_y$ orbital, and a seventh to the $2p_z$ orbital. Only after each $2p$ orbital contains one electron is a second electron added to the $2p_x$ orbital. Carbon, for example, has six electrons, and its ground-state electron configuration is $1s^2 2s^2 2p_x^1 2p_y^1 2p_z^0$. Alternatively, it may be simplified to $1s^2 2s^2 2p^2$. Table 1.3 shows ground-state electron configurations in the first 18 elements of the Periodic Table.

Example 1.1

Write the ground-state electron configuration for each element showing the occupancy of each p orbital.

(a) Lithium   **(b)** Oxygen   **(c)** Chlorine

Solution

(a) Lithium (atomic number 3): $1s^2 2s^1$
(b) Oxygen (atomic number 8): $1s^2 2s^2 2p_x^2 2p_y^1 2p_z^1$
(c) Chlorine (atomic number 17): $1s^2 2s^2 2p_x^2 2p_y^2 2p_z^2 3s^2 3p_x^2 3p_y^2 3p_z^1$

Problem 1.1

Write and compare the ground-state electron configurations for each pair of elements.

(a) Carbon and silicon   **(b)** Oxygen and sulfur   **(c)** Nitrogen and phosphorus

Valence electrons Electrons in the valence (outermost) shell of an atom.
Valence shell The outermost electron shell of an atom.
Lewis structure of an atom The symbol of an element surrounded by a number of dots equal to the number of electrons in the valence shell of the atom.

B.   Lewis Structures

When discussing the physical and chemical properties of an element, chemists often focus on the electrons in the outermost shell of its atoms because these electrons are involved in the formation of chemical bonds and in chemical reactions. Carbon, for example, with the ground-state electron configuration $1s^2 2s^2 2p^2$, has four outer-shell electrons. Outer-shell electrons are called **valence electrons,** and the energy level in which they are found is called the **valence shell.** To show the outermost electrons of an atom, we commonly use a representation called a **Lewis structure,** after the Ameri-

Table 1.4	**Lewis Structures for Elements 1–18***						
1A	2A	3A	4A	5A	6A	7A	8A
H·							He:
Li·	Be:	B̈:	·C̈:	·N̈:	:Ö:	:F̈:	:N̈e:
Na·	Mg:	Ȧl:	·S̈i:	·P̈:	:S̈:	:C̈l:	:Är:

*Electron dots are arranged in this table in accordance with Hund's rule.

can chemist Gilbert N. Lewis (1875–1946) who devised this notation. A Lewis structure shows the symbol of the element surrounded by a number of dots equal to the number of electrons in the outer shell of an atom of that element. In Lewis structures, the atomic symbol represents the "core," that is, the nucleus and all inner shells. Table 1.4 shows Lewis structures for the first 18 elements of the Periodic Table.

The noble gases helium and neon have filled valence shells. The valence shell of helium is filled with two electrons; that of neon is filled with eight electrons. Neon and argon have in common an electron configuration in which the s and p orbitals of their valence shells are filled with eight electrons. The valence shells of all other elements shown in Table 1.4 contain fewer than eight electrons.

Compare the Lewis structures given in Table 1.4 with the ground-state electron configurations given in Table 1.3. The Lewis structure of boron (B), for example, is shown in Table 1.4 with three valence electrons; these are the paired $2s$ electrons and the single $2p_x$ electron shown in Table 1.3. The Lewis structure of carbon (C) is shown in Table 1.4 with four valence electrons; these are the two paired $2s$ electrons and the single $2p_x$ and $2p_y$ electrons shown in Table 1.3.

For C, N, O, and F in period 2 of the Periodic Table, the valence electrons belong to the second shell. With eight electrons, this shell is completely filled. For Si, P, S, and Cl in period 3 of the Periodic Table, the valence electrons belong to the third shell. This shell is only partially filled with eight electrons; the $3s$ and $3p$ orbitals are fully occupied, but the five $3d$ orbitals can accommodate an additional ten electrons. Because of the differences in number and kind of valence-shell orbitals available to elements of the second and third periods, significant differences exist in the covalent bonding of oxygen and sulfur and of nitrogen and phosphorus (Section 1.2F). For example, although oxygen and nitrogen can accommodate no more than 8 electrons in their valence shells, many phosphorus-containing compounds have 10 electrons in the valence shell of phosphorus, and many sulfur-containing compounds have 10 and even 12 electrons in the valence shell of sulfur.

1.2 Lewis Model of Bonding

A. Formation of Ions

In 1916, Lewis devised a beautifully simple model that unified many of the observations about chemical bonding and reactions of the elements. He pointed out that the chemical inertness of the noble gases indicates a high degree of stability of the electron configurations of these elements: helium with a valence shell of two electrons

Gilbert N. Lewis (1875–1946) introduced the theory of the electron pair that extended our understanding of covalent bonding and of the concept of acids and bases. It is in his honor that we often refer to an "electron" dot structure as a Lewis structure. *(Corbis)*

Octet rule Group 1A–7A elements react to achieve an outer shell of eight valence electrons.

$(1s^2)$, neon with a valence shell of eight electrons $(2s^22p^6)$, and argon with a valence shell of eight electrons $(3s^23p^6)$. The tendency of atoms to react in ways that achieve an outer shell of eight valence electrons is particularly common among elements of Groups 1A–8A (the main-group elements) and is given the special name **octet rule.**

Example 1.2

Show how sodium follows the octet rule in forming Na^+.

Solution

The ground-state electron configurations for Na and Na^+ are

$$Na \text{ (11 electrons): } 1s^22s^22p^63s^1$$

$$Na^+ \text{ (10 electrons): } 1s^22s^22p^6$$

Thus, Na^+ has a complete octet of electrons in its outermost (valence) shell and has the same electron configuration as neon, the noble gas nearest it in atomic number.

Problem 1.2

Show that the following obey the octet rule.

(a) Sulfur forms S^{2-}. **(b)** Magnesium forms Mg^{2+}.

B. Formation of Chemical Bonds

According to Lewis' model, atoms bond together in such a way that each participating atom acquires a completed outer-shell electron configuration resembling that of the noble gas nearest to it in atomic number. Atoms acquire completed valence shells in two ways.

Anion An atom or group of atoms bearing a negative charge.
Cation An atom or group of atoms bearing a positive charge.
Ionic bond A chemical bond resulting from the electrostatic attraction of an anion and a cation.
Covalent bond A chemical bond formed between two atoms by sharing one or more pairs of electrons.

1. An atom may lose or gain enough electrons to acquire a completely filled valence shell. An atom that gains electrons becomes an **anion** (a negatively charged ion), and an atom that loses electrons becomes a **cation** (a positively charged ion). A chemical bond between a positively charged ion and a negatively charged ion is called an **ionic bond.**
2. An atom may share electrons with one or more other atoms to complete its valence shell. A chemical bond formed by sharing electrons is called a **covalent bond.**

C. Electronegativity and Chemical Bonds

Electronegativity A measure of the force of an atom's attraction for electrons it shares with another atom in a chemical bond.

How do we estimate the degree of ionic or covalent character in a chemical bond? One way is to compare the electronegativities of the atoms involved. **Electronegativity** is a measure of an atom's attraction for electrons that it shares in a chemical bond with another atom. The most widely used scale of electronegativities (Table 1.5) was devised by Linus Pauling in the 1930s and is based on bond dissociation energies.

On the Pauling scale, fluorine, the most electronegative element, is assigned an electronegativity of 4.0, and all other elements are assigned values in relation to fluorine. As you study the electronegativity values in this table, note that they increase

Table 1.5 Electronegativity Values for Some Atoms (Pauling Scale)

												H 2.1							

1A	2A							8B			1B	2B	3A	4A	5A	6A	7A
Li 1.0	Be 1.5												B 2.0	C 2.5	N 3.0	O 3.5	F 4.0
Na 0.9	Mg 1.2	3B	4B	5B	6B	7B					1B	2B	Al 1.5	Si 1.8	P 2.1	S 2.5	Cl 3.0
K 0.8	Ca 1.0	Sc 1.3	Ti 1.5	V 1.6	Cr 1.6	Mn 1.5	Fe 1.8	Co 1.8	Ni 1.8	Cu 1.9	Zn 1.6	Ga 1.6	Ge 1.8	As 2.0	Se 2.4	Br 2.8
Rb 0.8	Sr 1.0	Y 1.2	Zr 1.4	Nb 1.6	Mo 1.8	Tc 1.9	Ru 2.2	Rh 2.2	Pd 2.2	Ag 1.9	Cd 1.7	In 1.7	Sn 1.8	Sb 1.9	Te 2.1	I 2.5
Cs 0.7	Ba 0.9	La 1.1	Hf 1.3	Ta 1.5	W 1.7	Re 1.9	Os 2.2	Ir 2.2	Pt 2.2	Au 2.4	Hg 1.9	Tl 1.8	Pb 1.8	Bi 1.9	Po 2.0	At 2.2

■ <1.0   □ 1.5 – 1.9   □ 2.5 – 2.9
□ 1.0 – 1.4   ■ 2.0 – 2.4   ■ 3.0 – 4.0

from left to right within a period of the Periodic Table and decrease from top to bottom within a group. Values increase from left to right because the increasing positive charge on the nucleus results in a greater force of attraction for the atom's valence electrons. Electronegativity decreases from top to bottom because the increasing distance of the valence electrons from the nucleus results in a lower attraction of the nucleus for them.

Example 1.3

Judging from their relative positions in the Periodic Table, which element in each set is the more electronegative?

(a) Lithium or carbon   **(b)** Nitrogen or oxygen   **(c)** Carbon or oxygen

Solution

The elements in these sets are all in the second period of the Periodic Table. Electronegativity in this period increases from left to right.

(a) C > Li   **(b)** O > N   **(c)** O > C

Problem 1.3

Judging from their relative positions in the Periodic Table, which element in each set is the more electronegative?

(a) Lithium or potassium   **(b)** Nitrogen or phosphorus   **(c)** Carbon or silicon

Ionic Bonds

An ionic bond is a chemical bond formed between two atoms by the attractive force between positive and negative ions. An ionic bond is formed by the transfer of electrons from the valence shell of an atom of lower electronegativity to the valence shell

Linus Pauling (1901–1994) was the first person ever to receive two unshared Nobel Prizes. He received the 1954 Nobel Prize for chemistry for his contributions to our understanding of chemical bonding. He received the 1962 Peace Prize for his efforts on behalf of international control of nuclear weapons testing. (*Corbis*)

of an atom of higher electronegativity. The more electronegative atom gains one or more valence electrons and becomes an anion; the less electronegative atom loses one or more valence electrons and becomes a cation. As a rough guideline, we say that a chemical bond is ionic if the difference in electronegativity between the bonded atoms is 1.9 or greater. An example of an ionic bond is that formed between sodium (electronegativity 0.9) and fluorine (electronegativity 4.0). In the following equation, we use a single-headed (barbed) arrow to show the transfer of one electron from sodium to fluorine.

$$\text{Na} \cdot + \overset{\cdot\cdot}{\underset{\cdot\cdot}{\text{F}}} : \longrightarrow \text{Na}^+ : \overset{\cdot\cdot}{\underset{\cdot\cdot}{\text{F}}} :^-$$

In forming Na^+F^-, the single $3s$ valence electron of sodium is transferred to the partially filled valence shell of fluorine:

$$\text{Na}(1s^2 2s^2 2p^6 3s^1) + \text{F}(1s^2 2s^2 2p^5) \longrightarrow \text{Na}^+(1s^2 2s^2 2p^6) + \text{F}^-(1s^2 2s^2 2p^6)$$

As a result of this transfer of one electron, both sodium and fluorine form ions that have the same electron configuration as neon, the noble gas nearest each in atomic number.

Covalent Bonds

A covalent bond is a chemical bond formed between atoms by the sharing of one or more pairs of electrons. The simplest example of a covalent bond occurs in the hydrogen molecule. When two hydrogen atoms bond, the single electrons from each combine to form an electron pair. This shared pair completes the valence shell of each hydrogen. According to the Lewis model, a pair of electrons in a covalent bond functions in two ways simultaneously; it is shared by two atoms and at the same time fills the outer (valence) shell of each. We use a line between the two hydrogens to symbolize the covalent bond formed by the sharing of a pair of electrons.

$$\text{H} \cdot + \cdot \text{H} \longrightarrow \text{H—H} \qquad \Delta\text{H}° = -435 \text{ kJ } (-104 \text{ kcal})/\text{mol}$$

The Lewis model accounts for the stability of two covalently bonded atoms in the following way. In forming a covalent bond, an electron pair occupies the region between two nuclei and serves to shield one positively charged nucleus from the repulsive force of the other. At the same time, the electron pair attracts both nuclei. In other words, an electron pair in the space between two nuclei bonds them together and fixes the internuclear distance to within very narrow limits. The distance between nuclei participating in a chemical bond is called the **bond length.** Every covalent bond has a characteristic bond length. In H—H, it is 74 pm (picometer; 1 pm $= 10^{-12}$ m). We use SI units of picometers; formerly, chemists used Å (Ångstroms); 1 pm $= 0.01$ Å.

Polar Covalent Bonds

Although all covalent bonds involve the sharing of electrons, they differ widely in the degree of sharing. We divide covalent bonds arbitrarily into two categories, polar and nonpolar, depending on the difference in electronegativity between bonded atoms (Table 1.6).

A covalent bond between carbon and hydrogen, for example, is classified as **nonpolar covalent** because the difference in electronegativity between these two atoms is $2.5 - 2.1 = 0.4$ Pauling unit. An example of a **polar covalent bond** is that of H—Cl. The difference in electronegativity between chlorine and hydrogen is $3.0 - 2.1 = 0.9$ Pauling unit. An important consequence of the unequal sharing of electrons in a po-

Bond length The distance between atoms in a covalent bond. Now most commonly given in picometers (pm; 1 pm $= 10^{-12}$ m).

Nonpolar covalent bond A covalent bond between atoms whose difference in electronegativity is less than approximately 0.5 Pauling unit.

Polar covalent bonds A covalent bond between atoms whose difference in electronegativity is between approximately 0.5 and 1.9 Pauling units.

Table 1.6 **Classification of Chemical Bonds**	
Difference in Electronegativity Between Bonded Atoms	Type of Bond
Less than 0.5	Nonpolar covalent
0.5 to 1.9	Polar covalent
Greater than 1.9	Ionic

lar covalent bond is that the more electronegative atom gains a greater fraction of the shared electrons and acquires a partial negative charge, indicated by the symbol $\delta-$. The less electronegative atom has a smaller fraction of the shared electrons and acquires a partial positive charge, indicated by the symbol $\delta+$. Alternatively, we show the direction of bond polarity by an arrow with the arrowhead pointing toward the negative end and a plus sign on the tail of the arrow at the positive end.

$$\overset{\delta+}{H} - \overset{\delta-}{Cl} \qquad \overset{\longrightarrow}{H-Cl}$$

Example 1.4

Classify these bonds as nonpolar covalent, polar covalent, or ionic.

(a) O—H   **(b)** N—H   **(c)** K—Br   **(d)** C—Mg

Solution

Based on differences in electronegativity between the bonded atoms, three of these bonds are polar covalent and one is ionic.

Bond	Difference in Electronegativity	Type of Bond
(a) O—H	$3.5 - 2.1 = 1.4$	Polar covalent
(b) N—H	$3.0 - 2.1 = 0.9$	Polar covalent
(c) K—Br	$2.8 - 0.8 = 2.0$	Ionic
(d) C—Mg	$2.5 - 1.2 = 1.3$	Polar covalent

Problem 1.4

Classify these bonds as nonpolar covalent, polar covalent, or ionic.

(a) S—H   **(b)** P—H   **(c)** C—F   **(d)** C—Cl

We must point out that electronegativity varies somewhat depending on the chemical environment and oxidation state of an atom; therefore, these rules are only guidelines and must be used with caution. Lithium iodide, for example, has a high melting point (449°C) and boiling point (1180°C) characteristic of ionic compounds. Yet, based on the difference in electronegativity between these two elements of $2.5 - 1.0 = 1.5$ Pauling units, we would classify LiI as a polar covalent compound.

Example 1.5

Using the symbols $\delta-$ and $\delta+$, indicate the direction of polarity in these polar covalent bonds.

(a) C—O   **(b)** N—H   **(c)** C—Mg

Solution

Electronegativity is given beneath each atom. The atom with the greater electronegativity has the partial negative charge; the atom with the lesser electronegativity has the partial positive charge.

$$\begin{array}{ccc} \delta+ \ \delta- & \delta- \ \delta+ & \delta- \ \ \ \delta+ \\ \textbf{(a) } \text{C—O} & \textbf{(b) } \text{N—H} & \textbf{(c) } \text{C—Mg} \\ 2.5 \ \ 3.5 & 3.0 \ \ 2.1 & 2.5 \ \ 1.2 \end{array}$$

Problem 1.5

Using the symbols $\delta-$ and $\delta+$, indicate the direction of polarity in these polar covalent bonds.

(a) C—N   **(b)** N—O   **(c)** C—Cl

Bond dipole moment (μ)   A measure of the polarity of a covalent bond. The product of the charge on either atom of a polar bond times the distance between the atoms.

The polarity of a covalent bond is measured by a quantity called a **bond dipole moment** and is given the symbol **μ** (Greek mu). Bond dipole moment is defined as the product of the charge, e (either the $\delta+$ or $\delta-$ because each is the same in absolute magnitude), on one of its atoms times the distance, d, separating the two atoms. The SI unit for a dipole moment is the coulomb·meter, but it is commonly reported instead in a derived unit called the debye (D: 1 D = 3.34×10^{-30} C·m). Table 1.7 lists bond dipole moments for the types of covalent bonds we deal with most frequently in this course.

D.   Drawing Lewis Structures for Molecules and Polyatomic Ions

The ability to write Lewis structures for molecules and polyatomic ions is a fundamental skill for the study of organic chemistry. The following guidelines will help you do this. As you study these guidelines, look at the examples in Table 1.8.

Table 1.7 **Average Bond Dipole Moments of Selected Covalent Bonds**					
Bond	**Bond Dipole (D)**	**Bond**	**Bond Dipole (D)**	**Bond**	**Bond Dipole (D)**
H—C	0.3	C—F	1.4	C—O	0.7
H—N	1.3	C—Cl	1.5	C=O	2.3
H—O	1.5	C—Br	1.4	C—N	0.2
H—S	0.7	C—I	1.2	C≡N	3.5

Table 1.8 Lewis Structures for Several Compounds*

H—Ö—H	H—N̈—H with H below	H—C—H with H above and H below	H—Cl̈:
H_2O (8)	NH_3 (8)	CH_4 (8)	HCl (8)
Water	Ammonia	Methane	Hydrogen chloride

H₂C=CH₂ (drawn with H's)	H—C≡C—H	H₂C=Ö: (drawn with H's)	Carbonic acid structure
C_2H_4 (12)	C_2H_2 (10)	CH_2O (12)	H_2CO_3 (24)
Ethylene	Acetylene	Formaldehyde	Carbonic acid

*The number of valence electrons is shown in parentheses.

1. Determine the number of valence electrons in the molecule or ion. To do this, add the number of valence electrons contributed by each atom. For ions, add one electron for each negative charge on the ion, and subtract one electron for each positive charge on the ion. For example, the Lewis structure for a water molecule, H_2O, must show eight valence electrons: one from each hydrogen and six from oxygen. The Lewis structure for the hydroxide ion, OH^-, must also show eight valence electrons: one from hydrogen, six from oxygen, plus one for the negative charge on the ion.

2. Determine the connectivity (arrangement) of atoms in the molecule or ion. Except for the simplest molecules and ions, this connectivity must be determined experimentally. For some molecules and ions given as examples in the text, you are asked to propose an arrangement of atoms. For most, however, you are given the experimentally determined arrangement.

3. Connect the atoms with single bonds. Then arrange the remaining electrons in pairs so that each atom in the molecule or ion has a complete outer shell. Each hydrogen atom must be surrounded by two electrons. Each atom of carbon, oxygen, nitrogen, and halogen must be surrounded by eight electrons (per the octet rule). There are exceptions to this generalization. One example is the perchlorate ion, ClO_4^-, in which chlorine is bonded to 4 oxygen atoms and has 14 electrons in its valence shell.

4. A pair of electrons involved in a covalent bond **(bonding electrons)** is shown as a single bond; an unshared pair of electrons **(nonbonding electrons)** is shown as a pair of dots.

5. In a **single bond** two atoms share one pair of electrons. In a **double bond** they share two pairs of electrons, and in a **triple bond** they share three pairs of electrons.

Table 1.8 shows Lewis structures, molecular formulas, and names for several compounds. The number of valence electrons each molecule contains is shown in parentheses. Notice that, in these molecules, each hydrogen is surrounded by two valence electrons, and each carbon, nitrogen, oxygen, and chlorine is surrounded by eight valence electrons. Furthermore, each carbon has four bonds, nitrogen has three

Bonding electrons Valence electrons involved in forming a covalent bond (i.e., shared electrons).
Nonbonding electrons Valence electrons not involved in forming covalent bonds. Also called unshared pairs or lone pairs.

bonds and one unshared pair of electrons, oxygen has two bonds and two unshared pairs of electrons (often called **lone pairs**), and chlorine (and other halogens as well) has one bond and three unshared pairs of electrons.

Example 1.6

Draw Lewis structures, showing all valence electrons, for these molecules.

(a) CO_2 **(b)** CH_3OH **(c)** CH_3Cl

Solution

Under the Lewis structure for each molecule is the number of valence electrons it contains.

(a) $\ddot{O}=C=\ddot{O}$ **(b)**   H—C—O—H   **(c)**   H—C—Cl:
                                                        | | |
                                                        H   H   H

Carbon dioxide | Methanol | Chloromethane
(16 valence electrons) | (14 valence electrons) | (14 valence electrons)

Problem 1.6

Draw Lewis structures, showing all valence electrons, for these molecules.

(a) C_2H_6 **(b)** CS_2 **(c)** HCN

E. Formal Charge

Supporting Concepts; Formal Charge

Formal charge The charge on an atom in a polyatomic ion or molecule.

Throughout this course we deal not only with molecules but also with polyatomic cations and anions. Examples of polyatomic cations are the hydronium ion, H_3O^+, and the ammonium ion, NH_4^+. An example of a polyatomic anion is the bicarbonate ion, HCO_3^-. It is important that you be able to determine which atom or atoms in a molecule or polyatomic ion bear the positive or negative charge. The charge on an atom in a molecule or polyatomic ion is called its **formal charge.** To derive a formal charge:

1. Write a correct Lewis structure for the molecule or ion.
2. Assign to each atom all its unshared (nonbonding) electrons and one half its shared (bonding) electrons.
3. Compare this number with the number of valence electrons in the neutral, un-bonded atom. If the number of electrons assigned to a bonded atom is less than that assigned to the unbonded atom, then there are more positive charges in the nucleus than counterbalancing negative charges outside the nucleus, and the atom has a positive formal charge. Conversely, if the number of electrons assigned to a bonded atom is greater than that assigned to the unbonded atom, the atom has a negative formal charge.

$$\text{Formal charge} = \begin{pmatrix} \text{Number of valence} \\ \text{electrons in the neutral,} \\ \text{unbonded atom} \end{pmatrix} - \begin{pmatrix} \text{All unshared} \\ \text{electrons} \end{pmatrix} + \begin{pmatrix} \text{One half of all} \\ \text{shared electrons} \end{pmatrix}$$

Example 1.7

Draw Lewis structures for these ions, and show which atom in each bears the formal charge.

(a) H_3O^+ **(b)** HCO_3^-

Solution

(a) The Lewis structure for the hydronium ion must show 8 valence electrons: 3 from the three hydrogens, 6 from oxygen, minus 1 for the single positive charge. An oxygen atom has 6 valence electrons. The oxygen atom in H_3O^+ is assigned 2 unshared electrons and 1 from each shared pair of electrons, giving it a formal charge of $6 - (2 + 3) = +1$.

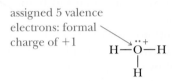

assigned 5 valence electrons: formal charge of +1

(b) The Lewis structure for the bicarbonate ion must show 24 valence electrons: 4 from carbon, 18 from the three oxygens, 1 from hydrogen, plus 1 for the single negative charge. Loss of a hydrogen ion from carbonic acid (Table 1.8) gives the bicarbonate ion. Carbon is assigned 1 electron from each shared pair and has no formal charge $(4 - 4 = 0)$. Two oxygens are assigned 6 valence electrons each and have no formal charges $(6 - 6 = 0)$. The third oxygen is assigned 7 valence electrons and has a formal charge of $6 - (6 + 1) = -1$.

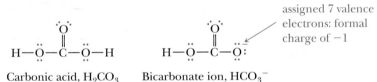

assigned 7 valence electrons: formal charge of −1

Carbonic acid, H_2CO_3 Bicarbonate ion, HCO_3^-

Problem 1.7

Draw Lewis structures for these ions, and show which atom in each bears the formal charge.

(a) $CH_3NH_3^+$ **(b)** CO_3^{2-} **(c)** OH^-

When writing Lewis structures for molecules and ions, you must remember that elements of the second period, including carbon, nitrogen, oxygen, and fluorine, can accommodate no more than eight electrons in the four orbitals ($2s$, $2p_x$, $2p_y$, and $2p_z$) of their valence shells. Following are two Lewis structures for nitric acid, HNO_3, each with the correct number of valence electrons (24):

10 electrons in the valence shell of nitrogen

An acceptable   Not an acceptable
Lewis structure   Lewis structure

The structure on the left is an acceptable Lewis structure. It shows the required 24 valence electrons, and each oxygen and nitrogen has a completed valence shell of 8 electrons. Further, it shows a positive formal charge on nitrogen and a negative formal charge on one of the oxygens. Note that the sum of the formal charges on the acceptable Lewis structure for HNO_3 is zero. The structure on the right is not an acceptable Lewis structure. Although it shows the correct number of valence electrons, it places 10 electrons in the valence shell of nitrogen.

F.   Exceptions to the Octet Rule

The Lewis model of covalent bonding focuses on valence electrons and the necessity for each atom other than H participating in a covalent bond to have a completed valence shell of eight electrons. Although most molecules formed by main-group elements (Groups 1A–7A) have structures that satisfy the octet rule, there are two important exceptions to this rule.

The first group of exceptions consists of molecules containing atoms of Group 3A elements. Following is a Lewis structure for BF_3. In this uncharged covalent compound, boron is surrounded by only six valence electrons. Aluminum chloride is an example of a compound in which aluminum, the element immediately below boron in Group 3A, has an incomplete valence shell. Because their valence shells are only partially filled, trivalent compounds of boron and aluminum are highly reactive.

Boron trifluoride   Aluminum chloride

A second group of exceptions to the octet rule consists of molecules and ions that contain an atom with more than eight electrons in its valence shell. Atoms of second-period elements use $2s$ and $2p$ orbitals for bonding, and these orbitals can contain only eight valence electrons, hence the octet rule. Atoms of third-period elements have $3d$ orbitals and may expand their valence shells to accommodate more than eight electrons. Following are Lewis structures for trimethylphosphine, phosphorus pentachloride, and phosphoric acid. The first compound has eight electrons in the valence shell of phosphorus; the second and third compounds have ten electrons in the valence shell of phosphorus.

Trimethylphosphine   Phosphorus pentachloride   Phosphoric acid

Sulfur, another third-period element, forms compounds in which its valence shell contains 8, 10, or 12 electrons.

Hydrogen sulfide   Dimethyl sulfoxide   Sulfuric acid

C H E M I S T R Y   I N   A C T I O N

The Octet Rule

Because the octet rule of G. N. Lewis gives us a powerful and simple model for understanding bonding in organic compounds, experiments designed to prepare molecules with ten electrons in the valence shell of carbon or nitrogen atoms might seem pointless. However, because chemistry is an experimental science, the truth of concepts such as the octet rule can be established only by experiment. No matter how many molecules obey the octet rule, a single exception would result in major modifications to the rule, or even in its replacement.

In 1949, the German chemist Georg Wittig attempted to prepare pentamethylnitrogen, a compound with ten electrons in nitrogen's valence shell, by the following reaction:

$$CH_3 \underset{\underset{CH_3}{|}}{\overset{\overset{CH_3}{|}}{-\overset{+}{N}-}} CH_3 \ Br^- \ + \ Li^+ : CH_3^- \ \xrightarrow{\ ?\ }$$

Tetramethylammonium   Methyllithium
bromide

$$\underset{H_3C}{\overset{H_3C}{>}}\overset{\overset{CH_3}{|}}{\underset{\underset{CH_3}{|}}{N}}{-}CH_3 \ + \ Li^+ \ Br^-$$

Pentamethylnitrogen

Instead, an acid-base reaction took place with one of the C—H bonds in the tetramethylammonium ion to give an unstable compound that has a positive charge on nitrogen and a negative charge on carbon.

$$CH_3 \underset{\underset{CH_3}{|}}{\overset{\overset{CH_3}{|}}{-\overset{+}{N}-}} \overset{..}{C}H_2^-$$

A nitrogen ylide

This novel type of molecule had not been made before. Wittig gave this class of molecules the name ylide. Nitrogen ylides cannot be isolated as stable compounds, but they can be used as intermediates in other reactions.

Reasoning that phosphorus (just below nitrogen in the Periodic Table) is capable of expanding its octet and might form stable ylides, Wittig carried out an analogous reaction with phosphorus. Once again, an acid-base reaction took place, this time to form a phosphorus ylide. Phosphorus ylides, Wittig discovered, can be isolated as stable compounds, which illustrates a difference between the chemistry of nitrogen and phosphorus.

$$CH_3 \underset{\underset{CH_3}{|}}{\overset{\overset{CH_3}{|}}{-\overset{+}{P}-}} CH_3 \ Br^- \ + \ Li^+ : CH_3^- \ \longrightarrow$$

Tetramethylphosphonium   Methyllithium
bromide

$$CH_3 \underset{\underset{CH_3}{|}}{\overset{\overset{CH_3}{|}}{-\overset{+}{P}-}} \overset{..}{C}H_2^- \ + \ Li^+ Br^- + CH_4$$

A phosphorus ylide

This ylide is shown with eight electrons in the valence shell of phosphorus. It can also be written with the valence shell of phosphorus expanded to accommodate ten electrons.

$$CH_3 \underset{\underset{CH_3}{|}}{\overset{\overset{CH_3}{|}}{-P=}} CH_2$$

An alternative prepresentation for a phosphorus ylide

Wittig soon abandoned his attempts to make compounds with five bonds to nitrogen and instead studied the chemistry of the newly discovered phosphorus ylides (Section 16.7). He found that these ylides are extraordinarily useful reagents for preparing complex organic molecules, including such important compounds as vitamin A. Professor Wittig shared the 1979 Nobel Prize for chemistry for his discovery and work with phosphorus ylides. See G. Wittig, *Science,* **210:** 600 (1980).

1.3   Functional Groups

Functional group   An atom or group of atoms within a molecule that shows a characteristic set of physical and chemical properties.

Carbon combines with other atoms (e.g., H, N, O, S, halogens) to form structural units called **functional groups.** Functional groups are important for three reasons. First, they are the units by which we divide organic compounds into classes. Second, they are sites of characteristic chemical reactions; a particular functional group, in whatever compound it is found, undergoes the same types of chemical reactions. Third, functional groups serve as a basis for naming organic compounds.

Introduced here are several of the functional groups we encounter early in this course. At this point, our concern is only with pattern recognition. We shall have more to say about the structure and properties of these functional groups in following chapters. A complete list of the major functional groups we study in this text is presented on the inside front cover of the text.

A.   Alcohols

Hydroxyl group   An —OH group.
Alcohol   A compound containing an —OH (hydroxyl) group bonded to an sp^3 hybridized carbon.

The functional group of an **alcohol** is an **—OH (hydroxyl)** group bonded to a tetrahedral carbon atom (a carbon having bonds to four other atoms).

—C—Ö—H   H—C—C—O—H

Functional   An alcohol
group   (Ethanol)

We can also represent this alcohol in a more abbreviated form called a condensed structural formula. In a **condensed structural formula,** CH_3 indicates a carbon with three attached hydrogens, CH_2 indicates a carbon with two attached hydrogens, and CH indicates a carbon with one attached hydrogen. Unshared pairs of electrons are generally not shown in a condensed structural formula. Thus, the condensed structural formula for the alcohol of molecular formula C_2H_6O is CH_3—CH_2—OH. It is also common to write these formulas in an even more condensed manner, by omitting all single bonds: CH_3CH_2OH.

Alcohols are classified as **primary (1°), secondary (2°),** or **tertiary (3°)** depending on the number of carbon atoms bonded to the carbon bearing the —OH group.

Primary (1°) alcohol   An alcohol in which the —OH group is bonded to a carbon that is bonded to one carbon.
Secondary (2°) alcohol   An alcohol in which the —OH group is bonded to a carbon that is bonded to two carbons.
Tertiary (3°) alcohol   An alcohol in which the —OH group is bonded to a carbon that is bonded to three carbons.

CH₃—C—OH   CH₃—C—OH   CH₃—C—OH

A 1° alcohol   A 2° alcohol   A 3° alcohol

Example 1.8

Draw Lewis structures and condensed structural formulas for the two alcohols of molecular formula C_3H_8O. Classify each as primary, secondary, or tertiary.

Solution

Begin by drawing the three carbon atoms in a chain. The oxygen atom of the hydroxyl group may be bonded to the carbon chain in two ways: either to an end carbon or to the middle carbon.

$$C-C-C \qquad C-C-C-OH \qquad \overset{\displaystyle OH}{\underset{\displaystyle |}{C}}-\overset{|}{C}-C$$

The carbon chain   The two locations for the OH group

Finally, add seven more hydrogens for a total of eight shown in the molecular formula. Show unshared electron pairs on the Lewis structures but not on the condensed structural formulas.

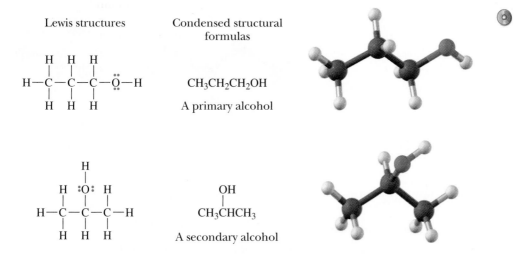

Lewis structures

Condensed structural
formulas

$CH_3CH_2CH_2OH$

A primary alcohol

OH
|
CH_3CHCH_3

A secondary alcohol

The secondary alcohol, whose common name is isopropyl alcohol, is the cooling, soothing component in rubbing alcohol.

Problem 1.8

Draw Lewis structures and condensed structural formulas for the four alcohols of molecular formula $C_4H_{10}O$. Classify each alcohol as primary, secondary, or tertiary.

B. Amines

The functional group of an amine is an **amino group;** a nitrogen atom bonded to one, two, or three carbon atoms. In a **primary (1°) amine,** nitrogen is bonded to one carbon group. In a **secondary (2°) amine,** it is bonded to two carbon groups, and in a **tertiary (3°) amine,** it is bonded to three carbon groups.

Amino group   A compound containing an sp^3 hybridized nitrogen atom bonded to one, two, or three carbon atoms.
Primary (1°) amine   An amine in which nitrogen is bonded to only one carbon atom.
Secondary (2°) amine   An amine in which nitrogen is bonded to two carbon atoms.
Tertiary (3°) amine   An amine in which nitrogen is bonded to three carbon atoms.

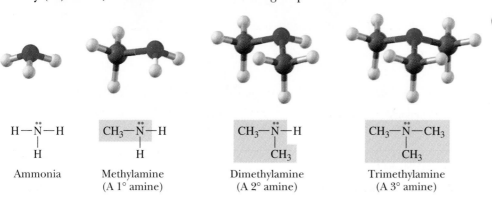

Ammonia

Methylamine
(A 1° amine)

Dimethylamine
(A 2° amine)

Trimethylamine
(A 3° amine)

Example 1.9

Draw condensed structural formulas for the two primary amines of molecular formula C_3H_9N.

Solution

For a primary amine, draw a nitrogen atom bonded to two hydrogens and one carbon.

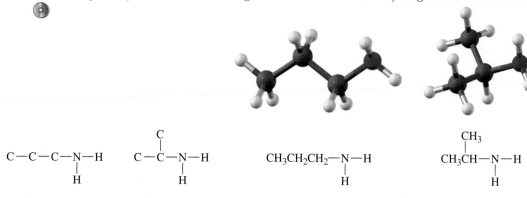

$$C-C-C-\underset{\underset{H}{|}}{N}-H \qquad C-\underset{|}{\overset{\overset{C}{|}}{C}}-\underset{\underset{H}{|}}{N}-H \qquad CH_3CH_2CH_2-\underset{\underset{H}{|}}{N}-H \qquad \overset{\overset{CH_3}{|}}{CH_3CH}-\underset{\underset{H}{|}}{N}-H$$

The three carbons may be bonded to nitrogen in two ways.

Add seven hydrogens to give each carbon four bonds and give the correct molecular formula.

Problem 1.9

Draw structural formulas for the three secondary amines of molecular formula $C_4H_{11}N$.

C. Aldehydes and Ketones

The functional group of both aldehydes and ketones is the **C=O (carbonyl)** group. In formaldehyde, CH_2O, the simplest **aldehyde,** the carbonyl carbon is bonded to two hydrogens. In all other aldehydes, it is bonded to one hydrogen and one carbon. In a condensed structural formula, the aldehyde group may be written showing the carbon-oxygen double bond as —CH=O; alternatively, it may be written —CHO. In a **ketone,** the carbonyl carbon is bonded to two carbon atoms.

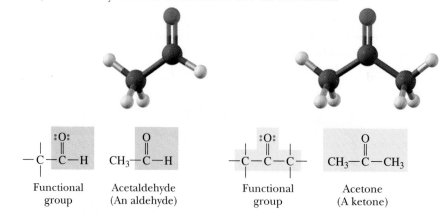

Functional group   Acetaldehyde (An aldehyde)   Functional group   Acetone (A ketone)

Example 1.10

Draw condensed structural formulas for the two aldehydes of molecular formula C_4H_8O.

Solution

First, draw the functional group of an aldehyde and then add the remaining carbons. These may be attached in two ways. Then, add seven hydrogens to complete the four bonds of each carbon.

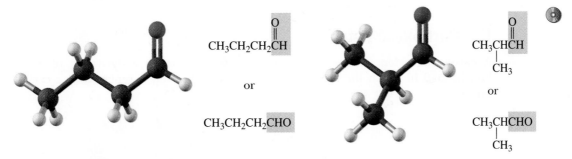

$$CH_3CH_2CH_2CH$$ with O double bonded to the CH carbon

or

$$CH_3CH_2CH_2CHO$$

$$CH_3CHCH$$ with O double bonded to the CH carbon and CH_3 below

or

$$CH_3CHCHO$$ with CH_3 below

Problem 1.10

Draw condensed structural formulas for the three ketones of molecular formula $C_5H_{10}O$.

D. Carboxylic Acids

The functional group of a **carboxylic acid** is a —COOH (**carboxyl:** *carb*onyl + hy- *droxyl*) group. In a condensed structural formula, a carboxyl group may also be writ- ten —CO₂H.

Carboxyl group   A —COOH group.
Carboxylic acid   A compound containing a carboxyl (—COOH) group.

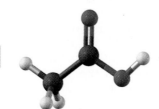

Functional group   Acetic acid (A carboxylic acid)

Example 1.11

Draw a condensed structural formula for the single carboxylic acid of molecular for- mula $C_3H_6O_2$.

Solution

The only way the carbon atoms can be written is three in a chain, and the —COOH group must be on an end carbon of the chain.

$$CH_3—CH_2—C—O—H$$ with O double bonded to C   or $$CH_3CH_2COOH$$

Problem 1.11

Draw condensed structural formulas for the two carboxylic acids of molecular formula $C_4H_8O_2$.

E.   Carboxylic Esters

Carboxylic ester   A derivative of a carboxylic acid in which H of the carboxyl group is replaced by a carbon group.

A **carboxylic ester,** commonly referred to as an **ester,** is a derivative of a carboxylic acid in which the hydrogen of the carboxyl group is replaced by a carbon group.

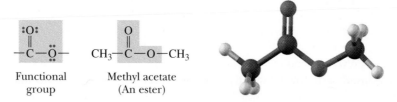

Functional group

Methyl acetate (An ester)

Example 1.12

The molecular formula of methyl acetate is $C_3H_6O_2$. Draw the structural formula of another ester of this same molecular formula.

Solution

There is only one other ester of this molecular formula. Its structural formula is

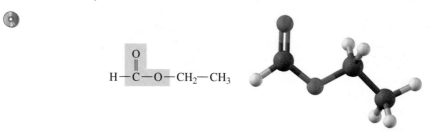

Problem 1.12

Draw structural formulas for the four esters of molecular formula $C_4H_8O_2$.

1.4   Bond Angles and Shapes of Molecules

In Section 1.2, we used a shared pair of electrons as the fundamental unit of a covalent bond and drew Lewis structures for several molecules and ions containing various combinations of single, double, and triple bonds. We can predict bond angles in these and other molecules and ions in a very straightforward way using the

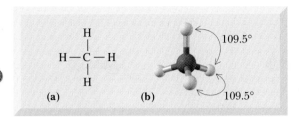

Figure 1.2

A methane molecule, CH_4. (a) Lewis structure and (b) shape. The hydrogens occupy the four corners of a regular tetrahedron, and all H—C—H bond angles are 109.5°.

valence-shell electron-pair repulsion (VSEPR) model. According to the VSEPR model, an atom is surrounded by an outer shell of valence electrons. These valence electrons may be involved in the formation of single, double, or triple bonds, or they may be unshared. Each combination creates a negatively charged region of space, and, because like charges repel each other, the various regions of electron density around an atom will spread out so that each is as far away from the others as possible.

We use the VSEPR model in the following way to predict the shape of a methane molecule, CH_4. The Lewis structure for CH_4 shows a carbon atom surrounded by four regions of electron density, each of which contains a pair of electrons forming a bond to a hydrogen atom. According to the VSEPR model, the four regions radiate from carbon so that they are as far away from each other as possible. This occurs when the angle between any two pairs of electrons is 109.5°. Therefore, we predict all H—C—H bond angles to be 109.5°, and the shape of the molecule to be **tetrahedral** (Figure 1.2). The H—C—H bond angles in methane have been measured experimentally and found to be 109.5°, identical to those predicted.

We predict the shape of an ammonia molecule, NH_3, in exactly the same manner. The Lewis structure of NH_3 shows nitrogen surrounded by four regions of electron density. Three regions contain single pairs of electrons forming covalent bonds with hydrogen atoms. The fourth region contains an unshared pair of electrons (Figure 1.3). Using the VSEPR model, we predict that the four regions of electron density around nitrogen are arranged in a tetrahedral manner, that all H—N—H bond angles are 109.5°, and that the shape of the molecule is **pyramidal** (like a pyramid). The observed bond angles are 107.3°. This small difference between the predicted and observed angles can be explained by proposing that the unshared pair of electrons on nitrogen repels adjacent electron pairs more strongly than do bonding pairs.

Figure 1.4 shows a Lewis structure and a ball-and-stick model of a water molecule. In H_2O, oxygen is surrounded by four regions of electron density. Two of these regions contain pairs of electrons used to form single covalent bonds to the two hydrogens; the remaining two contain unshared electron pairs. Using the VSEPR model, we predict that the four regions of electron density around oxygen repel each other and are arranged in a tetrahedral manner. The predicted H—O—H bond angle is 109.5°.

Experimental measurements show that the actual bond angle is 104.5°, a value smaller than that predicted. This difference between the predicted and observed bond angles can be explained by proposing, as we did for NH_3, that unshared pairs of electrons repel adjacent pairs more strongly than do bonding pairs. Note that the distortion from 109.5° is greater in H_2O, which has two unshared pairs of electrons, than it is in NH_3, which has only one unshared pair.

A general prediction emerges from this discussion of the shapes of CH_4, NH_3, and H_2O molecules. If a Lewis structure shows four regions of electron density

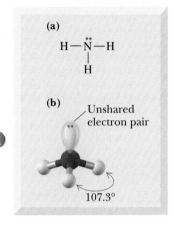

Figure 1.3

An ammonia molecule, NH_3. (a) Lewis structure and (b) shape.

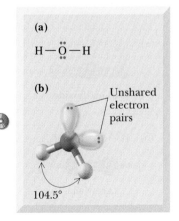

Figure 1.4

A water molecule, H_2O. (a) Lewis structure and (b) shape.

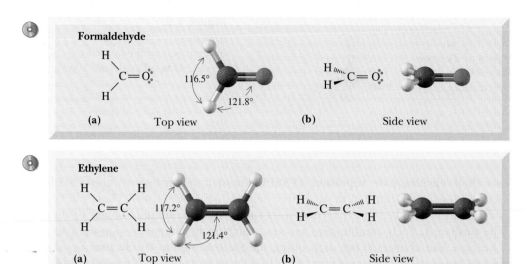

Figure 1.5
Shapes of formaldehyde, CH_2O, and ethylene, C_2H_4, molecules shown from (a) side view and (b) top view.

around a central atom, the VSEPR model predicts a tetrahedral distribution of electron density and bond angles of approximately 109.5°.

In many of the molecules we shall encounter, an atom is surrounded by three regions of electron density. Shown in Figure 1.5 are Lewis structures for formaldehyde, CH_2O, and ethylene, C_2H_4.

According to the VSEPR model, a double bond is treated as a single region of electron density. In formaldehyde, carbon is surrounded by three regions of electron density: two regions contain single pairs of electrons forming single bonds to hydrogen atoms; the third region contains two pairs of electrons forming a double bond to oxygen. In ethylene, each carbon atom is also surrounded by three regions of electron density: two contain single pairs of electrons, and the third contains two pairs of electrons.

Three regions of electron density about an atom are farthest apart when they are coplanar (in the same plane) and make angles of 120° with each other. Thus, the predicted H—C—H and H—C—O bond angles in formaldehyde and the predicted H—C—H and H—C—C bond angles in ethylene are all 120°. Further, all bonds in each compound lie in a plane. Both formaldehyde and ethylene are **planar** molecules.

In still other types of molecules, a central atom is surrounded by only two regions of electron density. Shown in Figure 1.6 are Lewis structures and ball-and-stick models of carbon dioxide, CO_2, and acetylene, C_2H_2.

In carbon dioxide, carbon is surrounded by two regions of electron density: each contains two pairs of electrons and forms a double bond to an oxygen atom. In acetylene, each carbon is also surrounded by two regions of electron density. One contains a single pair of electrons and forms a single bond to a hydrogen atom, and the other contains three pairs of electrons and forms a triple bond to a carbon atom. In each case, the two regions of electron density are farthest apart if they form a straight line through the central atom and create an angle of 180°. Both carbon dioxide and acetylene are **linear** molecules.

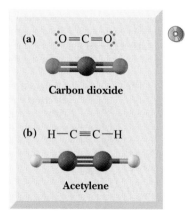

Figure 1.6
Shapes of (a) carbon dioxide, CO_2, and (b) acetylene, C_2H_2, molecules.

Table 1.9   Predicted Molecular Shapes (VSEPR Model)

Regions of Electron Density Around Central Atom	Predicted Distribution of Electron Density	Predicted Bond Angles	Examples
4	Tetrahedral	109.5°	
3	Trigonal planar	120°	
2	Linear	180°	

Predictions of the VSEPR model are summarized in Table 1.9. In these three-dimensional drawings, a solid line indicates a bond in the plane of the paper. A solid wedge indicates a bond projecting out of the plane toward the reader, and a broken wedge indicates a bond projecting behind the plane away from the reader.

Example 1.13

Predict all bond angles in these molecules.

(a) CH_3Cl **(b)** $CH_2{=}CHCl$

Solution

(a) The Lewis structure for CH_3Cl shows carbon surrounded by four regions of electron density. Therefore, we predict the distribution of electron pairs about carbon to be tetrahedral, all bond angles to be 109.5°, and the shape of CH_3Cl to be tetrahedral. The actual H—C—Cl bond angle is 108°.

(b) The Lewis structure for $CH_2{=}CHCl$ shows each carbon surrounded by three regions of electron density. Therefore, we predict all bond angles to be 120°. The actual C—C—Cl bond angle is 122.4°.

Problem 1.13

Predict all bond angles for these molecules.

 (a) CH$_3$OH   **(b)** PF$_3$ **(c)** H$_2$CO$_3$

C H E M I S T R Y   I N   A C T I O N

Buckyball — A New Form of Carbon

A favorite chemistry examination question is: what are the elemental forms of carbon? The usual answer is that pure carbon is found in two forms: graphite and diamond. These forms have been known for centuries, and it was generally believed that they are the only forms of carbon having extended networks of C atoms in well-defined structures. But that is not so! The scientific world was startled in 1985 when Richard Smalley of Rice University and Harry W. Kroto of the University of Sussex, UK, and their co-workers announced that they had detected a new form of carbon with the molecular formula C$_{60}$. They suggested that the molecule has a structure that resembles a soccer ball; it has 12 five-membered rings and 20 six-membered rings arranged such that each five-membered ring is surrounded by five six-membered rings. This structure reminded its discoverers of a geodesic dome, a structure invented by the innovative American engineer and philosopher R. Buckminster Fuller. Therefore, the official name of this allotrope of carbon has become buckminsterfullerene or more simply, "fullerene," and many chemists refer to C$_{60}$ simply as buckyball. Kroto, Smalley, and Robert F. Curl were awarded the 1996 Nobel Prize for chemistry for this work. Many higher fullerenes, such as C$_{70}$ and C$_{84}$, have also been isolated and studied.

Fullerenes have a rich chemistry. They behave as if they have electron-deficient double bonds. Many different fullerene adducts with a variety of structures have been prepared. Cationic derivatives of these adducts, for example, bind tightly to DNA and can be used to visualize it by electron microscopy.

An astonishing recent development in this field has been the preparation of single-wall "nanotubes,"

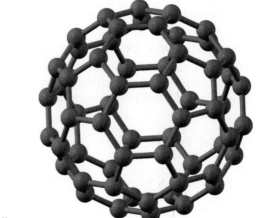

A buckyball.

which are based on C$_{60}$ or higher fullerenes and are extended for a very long distance to make long molecules that are hundreds of times stronger than steel and can act as molecular wires. Nanotubes can be used as the probe (a very sharp tip) in atomic force microscopes, which can detect single molecules. The nanotubes make the sharpest possible tips because they are of molecular dimensions.

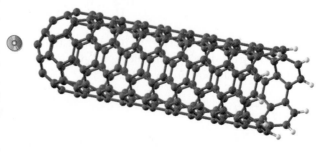

A nanotube. (Yves Rubin, UCLA)

1.5 Polar and Nonpolar Molecules

We can now combine our understanding of bond dipole moments (Section 1.2C) and molecular geometry (Section 1.4) to predict the polarity of polyatomic molecules. As we shall see, to be polar, a molecule must have one or more polar bonds. But, as we shall also see, not every molecule with polar bonds is polar.

To predict whether a molecule is polar, we need to determine (1) if the molecule has polar bonds and (2) the arrangement of its atoms in space. The **dipole moment** (μ) of a molecule is the vector sum of its individual bond dipoles. In carbon dioxide, for example, each C—O bond is polar with oxygen, the more electronegative atom, bearing a partial negative charge and with carbon bearing a partial positive charge. Because carbon dioxide is a linear molecule, the vector sum of its two bond dipoles is zero; therefore, the dipole moment of a CO_2 molecule is zero. Boron trifluoride is planar with bond angles of 120°. Although each B—F bond is polar, the vector sum of its bond dipoles is zero, and BF_3 has no dipole moment. Carbon tetrachloride is tetrahedral with bond angles of 109.5°. Although it has four polar C—Cl bonds, the vector sum of its bond dipoles is zero, and CCl_4 also has no dipole moment.

Dipole moment (μ) The vector sum of individual bond dipole moments in a molecule. Reported in a unit called the debye (D).

Carbon dioxide
$\mu = 0$ D

Boron trifluoride
$\mu = 0$ D

Carbon tetrachloride
$\mu = 0$ D

Other molecules, such as water and ammonia, have polar bonds and dipole moments greater than zero; they are polar molecules. Each O—H bond in a water molecule and each N—H bond in ammonia is polar with oxygen and nitrogen, the more electronegative atoms, bearing a partial negative charge and each hydrogen bearing a partial positive charge. Also shown are electrostatic potential plots that display the computed electronic charge density in water and ammonia. In these models, red represents negative charge and blue represents positive charge. In agreement with the dipole moment diagram and our expectations, the more electronegative atom has substantial negative charge in both molecules. For more information on how to interpret these plots, see Appendix 9.

An electrostatic potential plot of an ammonia molecule

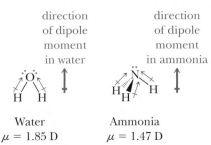

direction of dipole moment in water

direction of dipole moment in ammonia

Water
$\mu = 1.85$ D

Ammonia
$\mu = 1.47$ D

An electrostatic potential plot of a water molecule

Example 1.14

Which of these molecules are polar? For each that is, specify the direction of its dipole moment.

(a) CH_3Cl **(b)** CH_2O **(c)** C_2H_2

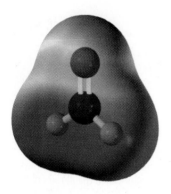

An electrostatic potential plot of a formaldehyde molecule

Solution

Both chloromethane, CH_3Cl, and formaldehyde, CH_2O, have polar bonds and, because of their geometry, are polar molecules. Because of its linear geometry, acetylene, C_2H_2, has no dipole moment.

(a) Chloromethane $\mu = 1.87$ D

(b) Formaldehyde $\mu = 2.33$ D

(c) H—C≡C—H

Acetylene $\mu = 0$ D

Problem 1.14

Which molecules are polar? For each that is, specify the direction of its dipole moment.

(a) CH_2Cl_2 (b) HCN   (c) H_2O_2

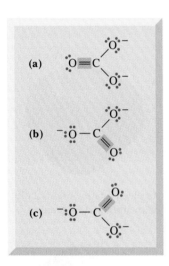

Figure 1.7
(a to c) Three Lewis structures for the carbonate ion.

Contributing structures Representations of a molecule, ion, or radical that differ only in the distribution of valence electrons.
Resonance hybrid A molecule, ion, or radical described as a composite of a number of contributing structures.
Double-headed arrow A symbol used to connect resonance contributing structures.

1.6   Resonance

Supporting Concepts; Resonance

As chemists developed more understanding of covalent bonding in organic compounds, it became obvious that, for a great many molecules and ions, no single Lewis structure provides a truly accurate representation. For example, Figure 1.7 shows three Lewis structures for the carbonate ion, CO_3^{2-}, each of which shows carbon bonded to three oxygen atoms by a combination of one double bond and two single bonds. Each Lewis structure implies that one carbon-oxygen bond is different from the other two. However, this is not the case. All three carbon-oxygen bonds are identical. The problem for chemists, then, is how to describe the structure of molecules and ions for which no single Lewis structure was adequate and yet still retain Lewis structures. As an answer to this problem, Linus Pauling proposed the theory of resonance.

A.   Theory of Resonance

The **theory of resonance** was developed primarily by Pauling in the 1930s. According to this theory, many molecules and ions are best described by writing two or more Lewis structures and considering the real molecule or ion to be a composite of these structures. Individual Lewis structures are called **contributing structures.** We show that the real molecule or ion is a **resonance hybrid** of the various contributing structures by interconnecting them with **double-headed arrows.** Do not confuse the double-headed arrow with the double arrow used to show chemical equilibrium. As we explain shortly, resonance structures are not in equilibrium with each other.

Three contributing structures for the carbonate ion are shown in Figure 1.8. These three contributing structures are said to be equivalent; they have identical patterns of covalent bonding and are of equal energy.

The use of the term "resonance" for this theory of covalent bonding might suggest to you that bonds and electron pairs are constantly changing back and forth from one position to another over time. This notion is not at all correct. The carbonate ion, for example, has one and only one real structure. The problem is ours—how

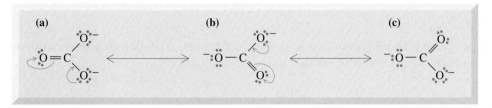

Figure 1.8
(a to c) The carbonate ion represented as a resonance hybrid of three equivalent contributing structures. Curved arrows show the redistribution of valence electrons between one contributing structure and the next.

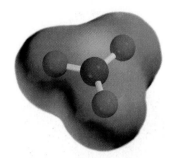

An electrostatic potential plot of a carbonate ion shows that the negative charge is distributed equally among the three oxygens

do we draw that one real structure? The resonance method is a way to describe the real structure and at the same time retain Lewis structures with electron-pair bonds. Thus, although we realize that the carbonate ion is not accurately represented by any one contributing structure shown in Figure 1.8, we continue to represent it by one of these for convenience. We understand, of course, that what is intended is the resonance hybrid.

B. Curved Arrows and Electron Pushing

Notice in Figure 1.8 that the only difference among contributing structures (a), (b), and (c) is the position of valence electrons. To show how this repositioning occurs, chemists use a symbol called a **curved arrow.**

A curved arrow is nothing more than a bookkeeping symbol for keeping track of electron pairs, or, as some call it, **electron pushing.** Do not be misled by its simplicity. Electron pushing will help you see the relationship among contributing structures. Later in the course, it will help you follow bond-breaking and bond-forming steps in organic reactions. Stated directly, electron pushing is a survival skill in organic chemistry.

Following are contributing structures for the nitrite and acetate ions. Curved arrows show how the contributing structures are interconverted. For each ion, the contributing structures are equivalent.

Curved arrow   A symbol used to show the redistribution of valence electrons in resonance contributing structures or reactions.

Nitrite ion
(equivalent contributing structures)

Acetate ion
(equivalent contributing structures)

Following are contributing structures for the resonance hybrid of acetone. These contributing structures are nonequivalent; they have different patterns of covalent bonding and different energies.

Acetone
(nonequivalent contributing structures)

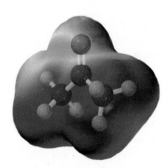

An electrostatic potential plot of an acetone molecule

Example 1.15

Draw the contributing structure indicated by the curved arrows. Be certain to show all valence electrons and all formal charges.

(a) $CH_3-\overset{\overset{\displaystyle \cdot\cdot}{O}}{\underset{\displaystyle ||}{C}}-H \longleftrightarrow$ (b) $H-\overset{\displaystyle C}{\underset{\displaystyle H}{|}}-\overset{\overset{\displaystyle \cdot\cdot}{O}}{\underset{\displaystyle ||}{C}}-H \longleftrightarrow$ (c) $CH_3-\overset{\displaystyle \cdot\cdot}{O}-\overset{\displaystyle +}{\underset{\displaystyle H}{C}}-H \longleftrightarrow$

Solution

(a) $CH_3-\overset{\overset{\displaystyle \cdot\cdot}{\ddot{O}:^-}}{\underset{\displaystyle +}{C}}-H$ (b) $H-\overset{\displaystyle C}{\underset{\displaystyle H}{|}}=\overset{\overset{\displaystyle \cdot\cdot}{\ddot{O}:^-}}{C}-H$ (c) $CH_3-\overset{\displaystyle +}{\overset{\displaystyle \cdot\cdot}{O}}=\overset{\displaystyle C}{\underset{\displaystyle H}{|}}-H$

Problem 1.15

Draw the contributing structure indicated by the curved arrows. Be certain to show all valence electrons and all formal charges.

(a) $H-\overset{\overset{\displaystyle \cdot\cdot}{O}}{\underset{\displaystyle ||}{C}}-\overset{\displaystyle \cdot\cdot}{O}:^- \longleftrightarrow$ (b) $H-\overset{\overset{\displaystyle \cdot\cdot}{O}}{\underset{\displaystyle ||}{C}}-\overset{\displaystyle \cdot\cdot}{O}:^- \longleftrightarrow$ (c) $CH_3-\overset{\overset{\displaystyle \cdot\cdot}{O}}{\underset{\displaystyle ||}{C}}-\overset{\displaystyle \cdot\cdot}{O}-CH_3 \longleftrightarrow$

C. Rules for Writing Acceptable Contributing Structures

Certain rules must be followed in writing acceptable contributing structures:

1. All contributing structures must have the same number of valence electrons.
2. All contributing structures must obey the rules of covalent bonding; no contributing structure may have more than two electrons in the valence shell of hydrogen or more than eight electrons in the valence shell of a second-period element. Third-period elements, such as phosphorus and sulfur, may have up to 12 electrons in their valence shells.
3. The positions of all nuclei must be the same; that is, contributing structures differ only in the distribution of valence electrons.
4. All contributing structures must have the same number of paired and unpaired electrons.

Example 1.16

Which sets are pairs of contributing structures?

(a) $CH_3-\overset{\overset{\displaystyle \cdot\cdot}{O}}{\underset{\displaystyle ||}{C}}-CH_3$ and $CH_3-\overset{\overset{\displaystyle \cdot\cdot}{\ddot{O}:^-}}{\underset{\displaystyle +}{C}}-CH_3$ (b) $CH_3-\overset{\overset{\displaystyle \cdot\cdot}{O}}{\underset{\displaystyle ||}{C}}-CH_3$ and $CH_2=\overset{\overset{\displaystyle \cdot\cdot}{O}-H}{\underset{\displaystyle |}{C}}-CH_3$

Solution

(a) They are contributing structures. They differ only in the distribution of valence electrons.

(b) They are not contributing structures. They differ in the arrangement of their atoms.

Problem 1.16

Which sets are pairs of contributing structures?

(a) CH_3-C and CH_3-C^+ (b) CH_3-C and $CH_3-\overset{-}{C}$

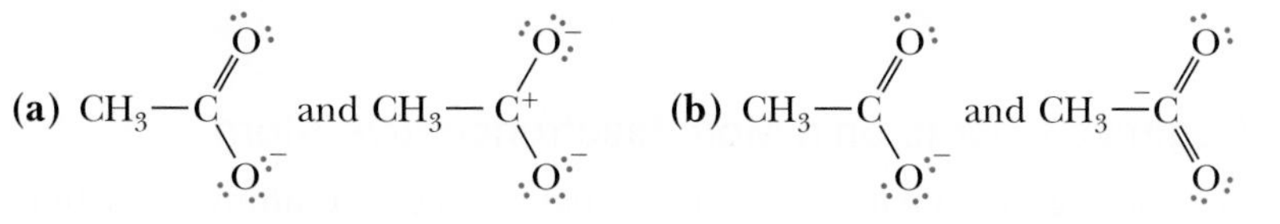

D. Estimating the Relative Importance of Contributing Structures

Not all structures contribute equally to a hybrid. The following preferences will help you to estimate the relative importance of various contributing structures. In fact, structures can be ranked by the number of preferences they follow. Those that follow the most preferences contribute most to the hybrid. Any structure that violates all four of these preferences (or number 4 alone) can be ignored and never written.

Preference 1: Filled Valence Shells

Structures in which all atoms have filled valence shells (completed octets) contribute more than those in which one or more valence shells are unfilled. For example, the following are the contributing structures for Example 1.15(c) and its solution.

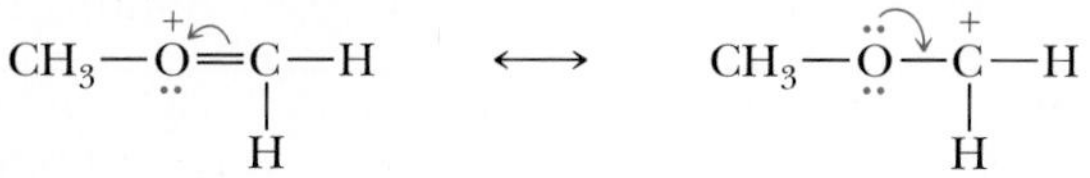

$$CH_3-\overset{+}{O}=C-H \quad \longleftrightarrow \quad CH_3-\overset{..}{O}-\overset{+}{C}-H$$

Greater contribution: Lesser contribution:
both carbon and oxygen have carbon has only six electrons
complete valence shells in its valence shell

Preference 2: Maximum Number of Covalent Bonds

Structures with a greater number of covalent bonds contribute more than those with fewer covalent bonds. In the illustration for preference 1, the structure on the left has eight covalent bonds and makes the greater contribution to the hybrid. The structure on the right has only seven covalent bonds.

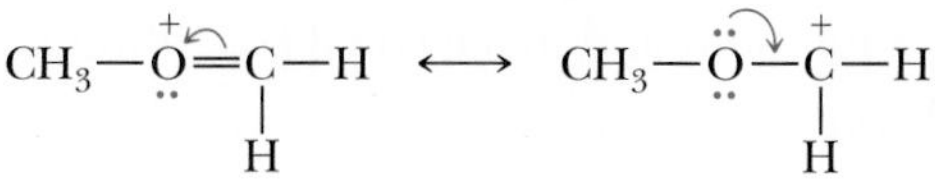

$$CH_3-\overset{+}{O}=C-H \quad \longleftrightarrow \quad CH_3-\overset{..}{O}-\overset{+}{C}-H$$

Greater contribution: Lesser contribution:
eight covalent bonds seven covalent bonds

Preference 3: Least Separation of Unlike Charges

Structures involving separation of unlike charges contribute less than those that do not involve charge separation.

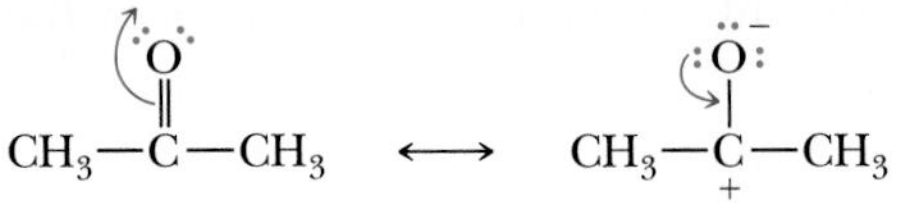

Greater contribution: no separation of unlike charges

Lesser contribution: separation of unlike charges

Preference 4: Negative Charge on a More Electronegative Atom

Structures that carry a negative charge on a more electronegative atom contribute more than those with the negative charge on a less electronegative atom. Conversely, structures that carry a positive charge on a less electronegative atom contribute more than those that carry the positive charge on a more electronegative atom. Following are three contributing structures for acetone:

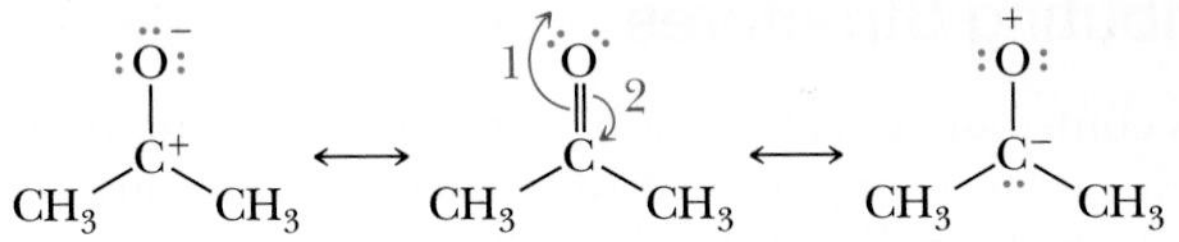

(a)
Lesser
contribution

(b)
Greater
contribution

(c)
Should not
be drawn

Structure (b) makes the largest contribution to the hybrid. Structure (a) contributes less because it involves separation of unlike charges; thus, arrow 1 on structure (b) is preferred. Structure (c) violates all four preference rules and should not be drawn, and arrow 2 on structure (b) can be ignored.

Example 1.17

Estimate the relative contribution of the members in each set of contributing structures.

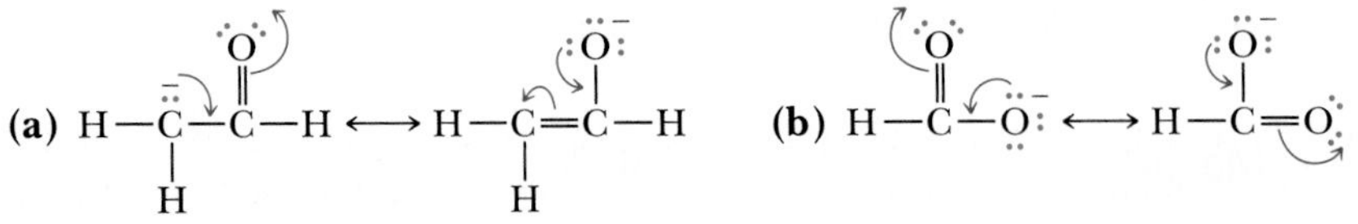

Solution

(a) The structure on the right makes a greater contribution to the hybrid because it places the negative charge on oxygen, the more electronegative atom.

(b) The structures are equivalent and make equal contributions to the hybrid.

Problem 1.17

Estimate the relative contribution of the members in each set of contributing structures.

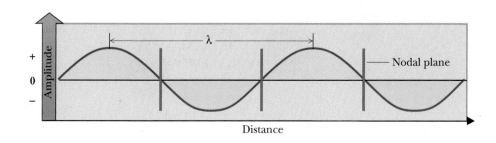

(a) and (b) resonance structures with curved arrows

1.7 Quantum or Wave Mechanics

Thus far in this chapter, we have concentrated on the Lewis model of bonding and on the VSEPR model. The Lewis model deals primarily with the coordination numbers of atoms (the number of bonds a given atom can form), and the VSEPR model deals primarily with bond angles and molecular geometries. Although each model is useful in its own way, neither gives us any means of accounting in a quantitative or even semiquantitative way for the reasons atoms combine in the first place to form covalent bonds with the liberation of energy. At this point, we need to study an entirely new approach to the theory of covalent bonding, one that provides a means of understanding not only the coordination numbers of atoms and molecular geometries but also the energetics of chemical bonding.

A. Moving Particles Exhibit the Properties of a Wave

The beginning of this new approach to the theory of chemical bonding was provided by Albert Einstein (1879–1955), a German-born American physicist. In 1905, Einstein postulated that light consists of photons of electromagnetic radiation. The energy, E, of a photon is proportional to the frequency, ν (Greek nu), of the light. The proportionality constant in this equation is Planck's constant, h.

$$E = h\nu$$

In 1923, the French physicist Louis de Broglie followed Einstein's lead and advanced the revolutionary idea that if light exhibits properties of particles in motion, then a particle in motion should exhibit the properties of a wave. He proposed that a particle of mass m and speed v has an associated wavelength λ (Greek lambda), given by the equation

$$\lambda = \frac{h}{mv} \qquad \text{(the de Broglie relationship)}$$

Illustrated in Figure 1.9 is a wave such as might result from plucking a guitar string. The mathematical equation that describes this wave is called a wave equation.

Einstein received a PhD in 1905, and in that same year, he published a paper in which he used his photon concept to explain the photoelectric effect. For this work, he was awarded the 1921 Nobel Prize for physics.

Figure 1.9

Characteristics of a wave associated with a moving particle. Wavelength is designated by the symbol λ.

Amplitude / + / 0 / − / λ / Nodal plane / Distance

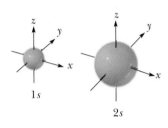

Node Any point in space where the value of a wave function is zero.

Quantum mechanics The branch of science that studies particles and their associated waves.

Wave function A solution to a set of equations that defines the energy of an electron in an atom.

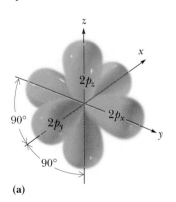

1s

2s

Figure 1.10
Probability distribution for 1s and 2s atomic orbitals.

Figure 1.11
Three-dimensional shapes of $2p_x$, $2p_y$, and $2p_z$ atomic orbitals and their orientation in space relative to one another.

The numerical value(s) of the solution(s) of a wave equation may be positive (corresponding to a wave crest), negative (corresponding to a wave trough), or zero. A **node** is any point where the value of a solution of a wave equation is zero. A **nodal plane** is any plane perpendicular to the direction of propagation that runs through a node. Shown in Figure 1.9 are three nodal planes.

Erwin Schrödinger built on the idea of de Broglie and in 1926 proposed an equation that could be used to describe the wave properties associated with an electron in an atom or a molecule. **Quantum mechanics (wave mechanics)** is the branch of science that studies particles and their associated waves.

Solving the Schrödinger equation gives a set of solutions called **wave functions.** Each wave function ψ (Greek psi) is associated with a unique set of quantum numbers and with a particular atomic or molecular orbital. The value of ψ^2 is proportional to the probability of finding an electron at a given point in space. Looked at in another way, the value of ψ^2 at any point in space is proportional to the electron density at that point. Electron density theoretically reaches to infinity but actually is vanishingly low at long distances from the nucleus.

In this course, we concentrate on wave functions and shapes associated with s and p atomic orbitals because they are the orbitals most often involved in covalent bonding in organic compounds.

B.   Shapes of Atomic s and p Orbitals

All s orbitals have the shape of a sphere, with the center of the sphere at the nucleus. Shown in Figure 1.10 are probability distributions for 1s and 2s orbitals. These orbitals are completely symmetric along all axes. Shown in Figure 1.11 are the three-dimensional shapes of the three 2p orbitals, combined in one diagram to illustrate their relative orientations in space. Each 2p orbital consists of two lobes arranged in a straight line with the nucleus in the middle. The three 2p orbitals are mutually perpendicular and are designated $2p_x$, $2p_y$, and $2p_z$. The sign of the wave function of a 2p orbital is positive in one lobe, zero at the nucleus, and negative in the other lobe. The plus or minus is simply a change in sign of a mathematical function and has no relationship to energy or electron distribution. Because the value of ψ^2 is always positive, the probability of finding an electron in the (+) lobe of a 2p orbital is the same as that of finding it in the (−) lobe.

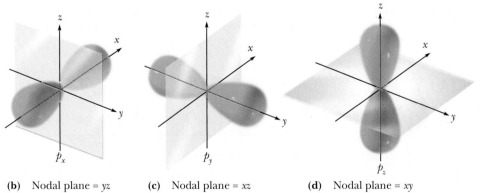

(a)

(b)   Nodal plane = yz   **(c)**   Nodal plane = xz   **(d)**   Nodal plane = xy

Besides providing a way to determine the shapes of atomic orbitals, the Schrödinger equation also provides a way, at least in principle, to quantify the energetics of covalent bond formation. These approximations have taken two forms: (1) the valence bond model and (2) the molecular orbital model. Both models for chemical bonding use the methods of quantum mechanics, but each makes slightly different simplifying assumptions. At sufficiently high levels of theory, both models converge. We concentrate our discussion on the molecular orbital model.

1.8 Molecular Orbital Theory of Covalent Bonding

A. Formation of Molecular Orbitals

Molecular orbital (MO) theory begins with the hypothesis that electrons in atoms exist in atomic orbitals and assumes that electrons in molecules exist in molecular orbitals. Just as the Schrödinger equation can be used to calculate the energies and shapes of atomic orbitals, molecular orbital theory assumes that the Schrödinger equation can also be used to calculate the energies and shapes of molecular orbitals. Following is a summary of the rules used in applying molecular orbital theory to the formation of covalent bonds.

Molecular orbital theory A theory of chemical bonding in which electrons in molecules occupy molecular orbitals formed by the combination of the atomic orbitals that make up the molecule.

1. Combination of n atomic orbitals forms a set of n molecular orbitals; that is, the number of molecular orbitals formed is equal to the number of atomic orbitals combined.
2. Just like atomic orbitals, molecular orbitals are arranged in order of increasing energy. It is possible to calculate reasonably accurate relative energies of a set of molecular orbitals. Experimental measurements such as those derived from molecular spectroscopy can also be used to provide very detailed information about the relative energies of molecular orbitals.
3. Filling of molecular orbitals with electrons is governed by the same principles as the filling of atomic orbitals. Molecular orbitals are filled beginning with the lowest unoccupied molecular orbital (the Aufbau principle). A molecular orbital can accommodate no more than two electrons, and their spins must be paired (the Pauli exclusion principle). When two or more molecular orbitals of the same energy are available, one electron is added to each before any equivalent orbital is filled with two electrons (Hund's rule).

To illustrate the formation of molecular orbitals, consider the shapes and relative energies of the molecular orbitals arising from combination of two $1s$ atomic orbitals. Combination by addition of their wave functions gives the molecular orbital shown in Figure 1.12(a). When electrons occupy this MO, electron density is concentrated in the region between the two positively charged nuclei and serves to offset the repulsive interaction between them. The molecular orbital we have just described is called a sigma bonding molecular orbital and is given the symbol σ_{1s} (pronounced sigma one ess). A **bonding molecular orbital** is one in which electrons have a lower energy than they would in the isolated atomic orbitals. A **sigma (σ) bonding molecular orbital** is one in which electron density lies between the two nuclei, along the axis joining them, and is cylindrically symmetric about the axis.

Combination of two $1s$ atomic orbitals by subtraction of their wave functions gives the molecular orbital shown in Figure 1.12(b). If electrons occupy this MO,

Bonding molecular orbital A molecular orbital in which electrons have a lower energy than they would in isolated atomic orbitals.

Sigma (σ) bonding molecular orbital A molecular orbital in which electron density is concentrated between two nuclei and along the axis joining them.

Figure 1.12

Molecular orbitals derived from combination of two $1s$ atomic orbitals: (a) combination by addition and (b) combination by subtraction. Electrons in the bonding MO spend most of their time in the region between the two nuclei and bond the atoms together. Electrons in the antibonding MO do not contribute to bonding.

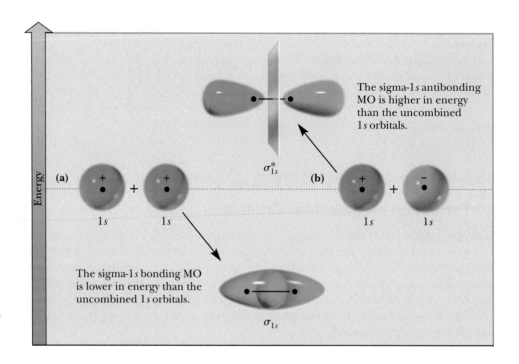

The sigma-$1s$ antibonding MO is higher in energy than the uncombined $1s$ orbitals.

σ_{1s}^*

(a)

$1s$ $+$ $1s$

(b)

$1s$ $+$ $1s$

The sigma-$1s$ bonding MO is lower in energy than the uncombined $1s$ orbitals.

σ_{1s}

Energy

Antibonding molecular orbital
A molecular orbital in which electrons have a higher energy than they would in isolated atomic orbitals.

Pi (π) bonding molecular orbital
A molecular orbital formed by overlap of parallel p orbitals on adjacent atoms; its electron density lies above and below the line connecting the atoms.

electron density is concentrated outside the region between the two nuclei; consequently, there is little or no electron density between the nuclei to offset nuclear repulsion. There is a node, or point of zero electron density, between the atoms. This molecular orbital is called a sigma antibonding molecular orbital and is given the symbol σ_{1s}^* (pronounced sigma star one ess). An **antibonding molecular orbital** is one in which the electrons in it have a higher energy (are more easily removed) than they would in the isolated atomic orbitals. An asterisk shows that a molecular orbital is antibonding.

In representations of orbitals, we use blue to indicate the lobe of a particular orbital in which the sign of the wave function is negative and red where it is positive.

The **ground state** of an atom or molecule is its state of lowest energy. In the ground state of the hydrogen molecule, the two electrons occupy the σ_{1s} MO with paired spins. An **excited state** is any electronic state other than the ground state. In the lowest excited state of the hydrogen molecule, one electron occupies the σ_{1s} MO, and the other occupies the σ_{1s}^* MO. There is no net bonding in this excited state, and dissociation will result from the electrostatic repulsion of the two hydrogen nuclei. Energy-level diagrams of the ground state and the lowest excited state of the hydrogen molecule are shown in Figure 1.13.

Combination of two $2s$ atomic orbitals produces two sigma molecular orbitals, designated σ_{2s} and σ_{2s}^*, that are larger in shape and higher in energy than the σ_{1s} and σ_{1s}^* molecular orbitals illustrated in Figure 1.13.

Next let us consider the molecular orbitals formed by the combination of $2p$ orbitals. According to MO theory, two $2p$ atomic orbitals can overlap end-on to form a sigma bonding (σ_{2p}) and a sigma antibonding (σ_{2p}^*) molecular orbital. We will not encounter these kinds of MOs in this course. What we do encounter is combination of parallel $2p$ atomic orbitals by addition of their wave functions to give a **pi (π) bonding molecular orbital** (π_{2p}) shown in Figure 1.14(a). Combination of their wave functions by subtraction gives the pi antibonding (π_{2p}^*) molecular orbital shown in

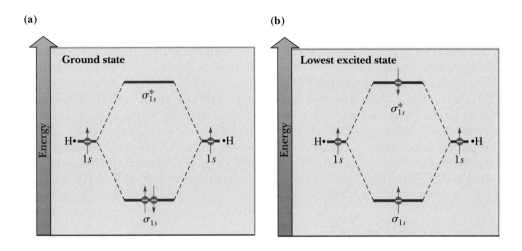

Figure 1.13
A molecular orbital energy diagram for the hydrogen molecule, H_2: (a) ground state and (b) lowest excited state.

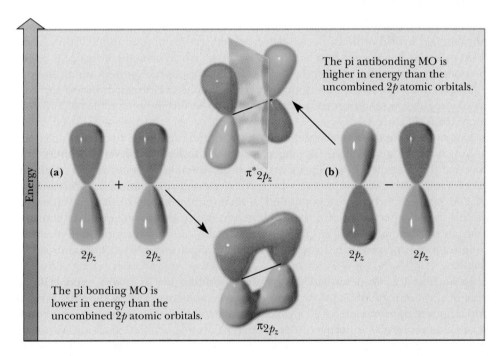

The pi antibonding MO is higher in energy than the uncombined $2p$ atomic orbitals.

$\pi^* 2p_z$

The pi bonding MO is lower in energy than the uncombined $2p$ atomic orbitals.

$\pi 2p_z$

Figure 1.14
Molecular orbitals formed by combination of parallel $2p$ orbitals: (a) combination by addition and (b) combination by subtraction.

Figure 1.14(b). We will have little need to refer to antibonding molecular orbitals throughout the remainder of the course, except for when we treat ultraviolet-visible spectroscopy in Chapter 23. Our concentration will be on bonding molecular orbitals and their participation in chemical reactions.

B.   Hybridization of Atomic Orbitals

Formation of a covalent bond between two hydrogen atoms is straightforward. Formation of covalent bonds between atoms of second-period elements, however, presents a problem. In forming covalent bonds, atoms of carbon, nitrogen, and oxygen (all second-period elements) use $2s$ and $2p$ atomic orbitals. The three $2p$ atomic orbitals are at angles of $90°$ to each other (Figure 1.11), and, if atoms of second-period

Hybrid orbital   An orbital formed by the combination of two or more atomic orbitals.

elements used these orbitals to form covalent bonds, bond angles around each would be approximately 90°. However, we rarely observe bond angles of 90° in organic molecules. What we find instead are bond angles of approximately 109.5° in molecules with only single bonds, 120° in molecules with double bonds, and 180° in molecules with triple bonds, as shown in Table 1.9.

To account for these observed bond angles, Pauling proposed that atomic orbitals combine to form new orbitals, called **hybrid orbitals,** which then interact to form bonds with the angles that we do observe. Hybrid orbitals are formed by combinations of atomic orbitals, a process called hybridization. The number of hybrid orbitals formed is equal to the number of atomic orbitals combined. Elements of the second period form three types of hybrid orbitals, designated sp^3, sp^2, and sp, each of which can contain up to two electrons.

C. sp^3 Hybrid Orbitals—Bond Angles of Approximately 109.5°

sp^3 Hybrid orbital   A hybrid atomic orbital formed by the combination of one *s* atomic orbital and three *p* atomic orbitals.

The combination of one $2s$ atomic orbital and three $2p$ atomic orbitals forms four equivalent *sp^3* **hybrid orbitals.** Because they are derived from four atomic orbitals, sp^3 hybrid orbitals always occur in sets of four. Each sp^3 hybrid orbital consists of a larger lobe pointing in one direction and a smaller lobe of opposite sign pointing in the opposite direction. The axes of the four sp^3 hybrid orbitals are directed toward the corners of a regular tetrahedron, and sp^3 hybridization results in bond angles of approximately 109.5° (Figure 1.15).

You must remember that superscripts in the designation of hybrid orbitals tell you how many atomic orbitals have been combined to form the hybrid orbitals. You know that the designation sp^3 represents a hybrid orbital because it shows a combination of s and p orbitals. The superscripts in this case tell you that *one s* atomic orbital and *three p* atomic orbitals are combined in forming the hybrid orbital. Do not confuse this use of superscripts with that used in writing a ground-state electron configuration, as for example $1s^2 2s^2 2p^5$ for fluorine. In the case of a ground-state electron configuration, superscripts tell you the number of electrons in each orbital or set of orbitals.

In Section 1.2, we described the covalent bonding in CH_4, NH_3, and H_2O in terms of the Lewis model, and in Section 1.4 we used the VSEPR model to predict bond angles of approximately 109.5° in each molecule. Now let us consider the bonding in these molecules in terms of the overlap of atomic orbitals. To bond with four

Figure 1.15

sp^3 Hybrid orbitals. (a) Representation of a single sp^3 hybrid orbital showing two lobes of unequal size. The sign of the wave function is positive in one lobe and negative in the other. (b) Three-dimensional representation of four sp^3 hybrid orbitals directed toward the corners of a regular tetrahedron. The smaller lobe of each sp^3 hybrid orbital is hidden behind the larger lobe. (c) If four balloons of similar size and shape are tied together, they will naturally assume a tetrahedral geometry.

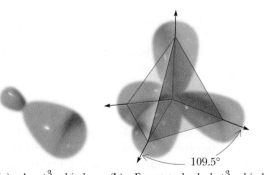

(a)   An sp^3 orbital   **(b)**   Four tetrahedral sp^3 orbitals   **(c)**

109.5°

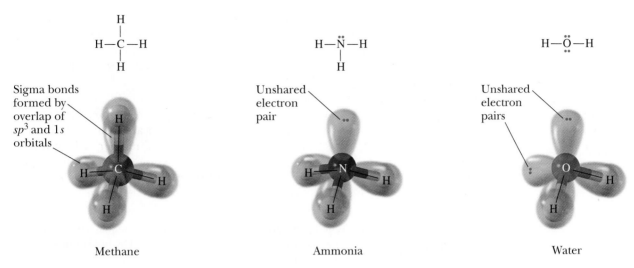

Figure 1.16
Orbital overlap pictures of methane, ammonia, and water.

other atoms with bond angles of 109.5°, carbon uses sp^3 hybrid orbitals. Carbon has four valence electrons, and one electron is placed in each sp^3 hybrid orbital. Each partially filled sp^3 hybrid orbital then overlaps with a partially filled $1s$ atomic orbital of hydrogen to form the four sigma (σ) bonds of methane, and hydrogen atoms occupy the corners of a regular tetrahedron (Figure 1.16).

In bonding with three other atoms, the five valence electrons of nitrogen are distributed so that one sp^3 hybrid orbital is filled with a pair of electrons and the other three sp^3 hybrid orbitals have one electron each. Overlapping of these partially filled sp^3 hybrid orbitals with $1s$ atomic orbitals of three hydrogen atoms produces the NH_3 molecule (Figure 1.16).

In bonding with two other atoms, the six valence electrons of oxygen are distributed so that two sp^3 hybrid orbitals are filled, and the remaining two have one electron each. Each partially filled sp^3 hybrid orbital overlaps with a $1s$ atomic orbital of hydrogen, and hydrogen atoms occupy two corners of a regular tetrahedron. The remaining two sp^3 hybrid orbitals, each occupied by an unshared pair of electrons, are directed toward the other two corners of a regular tetrahedron (Figure 1.16).

D. sp^2 Hybrid Orbitals—Bond Angles of Approximately 120°

The combination of one $2s$ atomic orbital and two $2p$ atomic orbitals forms three equivalent **sp^2 hybrid orbitals.** Because they are derived from three atomic orbitals, sp^2 hybrid orbitals always occur in sets of three. As with sp^3 orbitals, each sp^2 hybrid orbital consists of two lobes, one larger than the other. The axes of the three sp^2 hybrid orbitals lie in a plane and are directed toward the corners of an equilateral triangle; the angle between sp^2 hybrid orbitals is 120°. The third $2p$ atomic orbital (remember $2p_x, 2p_y, 2p_z$) is not involved in hybridization and consists of two lobes lying perpendicular to the plane of the sp^2 hybrid orbitals. Figure 1.17 shows three equivalent sp^2 orbitals along with the remaining unhybridized $2p$ atomic orbital.

sp^2 **Hybrid orbital** A hybrid atomic orbital formed by the combination of one s atomic orbital and two p atomic orbitals.

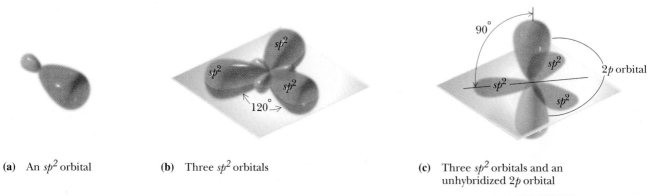

(a) An sp^2 orbital

(b) Three sp^2 orbitals

(c) Three sp^2 orbitals and an unhybridized $2p$ orbital

Figure 1.17

sp^2 Hybrid orbitals. (a) A single sp^2 hybrid orbital showing two lobes of unequal size. (b) The three sp^2 hybrid orbitals with their axes in a plane at angles of 120°. (c) The unhybridized $2p$ orbital perpendicular to the plane created by the three sp^2 hybrid orbitals.

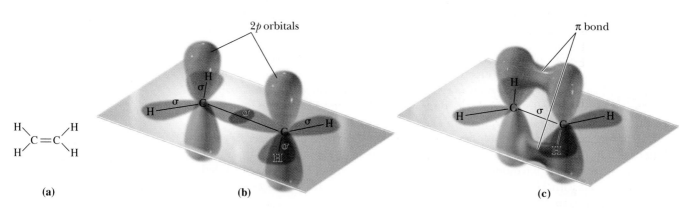

(a)

(b)

(c)

Figure 1.18

Covalent bond formation in ethylene. (a) Lewis structure, (b) overlap of sp^2 hybrid orbitals forms a sigma (σ) bond between the carbon atoms, and (c) overlap of parallel $2p$ orbitals forms a pi (π) bond.

Pi (π) bond A covalent bond formed by the overlap of parallel p orbitals.

Second-period elements use a combination of sp^2 hybrid and $2p$ atomic orbitals to form double bonds. Consider ethylene, C_2H_4, a Lewis structure for which is shown in Figure 1.18(a). A sigma bond between the carbons in ethylene is formed by overlap of sp^2 hybrid orbitals along a common axis as seen in Figure 1.18(b). Each carbon also forms sigma bonds to two hydrogens. The remaining $2p$ orbitals on adjacent carbons lie parallel to each other and overlap to form a pi bond [Figure 1.18(c)]. A **pi (π) bond** is a covalent bond formed by overlap of parallel p orbitals. Because of the lesser degree of overlap of orbitals forming pi bonds compared with those forming sigma bonds, pi bonds are generally weaker than sigma bonds.

Molecular orbital theory describes all double bonds in the same manner we have used to describe carbon-carbon double bonds. In formaldehyde, $CH_2\!=\!\!O$, the simplest

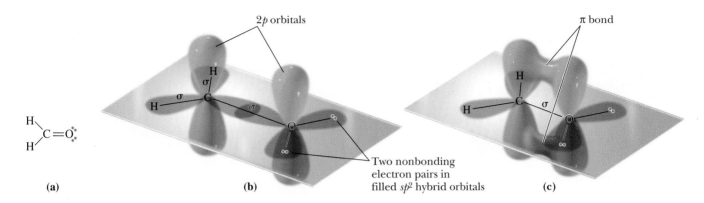

Figure 1.19
A carbon-oxygen double bond. (a) Lewis structure of $CH_2{=}O$, (b) the sigma (σ) bond framework and nonoverlapping parallel $2p$ atomic orbitals, and (c) overlap of parallel $2p$ atomic orbitals to form a pi (π) bond.

organic molecule containing a carbon-oxygen double bond, carbon forms sigma bonds to two hydrogens by overlap of sp^2 orbitals of carbon and $1s$ atomic orbitals of hydrogen. Carbon and oxygen are joined by a sigma bond formed by overlap of sp^2 hybrid orbitals and a pi bond formed by overlap of parallel $2p$ atomic orbitals (Figure 1.19).

E. *sp* Hybrid Orbitals — Bond Angles of Approximately 180°

The combination of one $2s$ atomic orbital and one $2p$ atomic orbital produces two equivalent *sp* **hybrid orbitals.** Because they are derived from two atomic orbitals, *sp* hybrid orbitals always occur in sets of two. The two *sp* hybrid orbitals lie at an angle of 180°. The axes of the unhybridized $2p$ atomic orbitals are perpendicular to each other and to the axis of the two *sp* hybrid orbitals. In Figure 1.20, *sp* hybrid orbitals are shown on the x-axis and unhybridized $2p$ orbitals on the y- and z-axes.

sp **Hybrid orbital**   A hybrid atomic orbital formed by the combination of one s atomic orbital and one p atomic orbital.

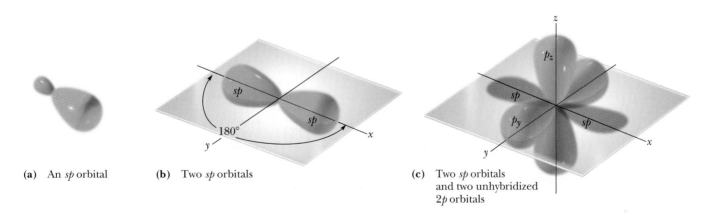

(a) An *sp* orbital   (b) Two *sp* orbitals   (c) Two *sp* orbitals and two unhybridized $2p$ orbitals

Figure 1.20
sp Hybrid orbitals. (a) A single *sp* hybrid orbital consisting of two lobes of unequal size. (b) Two *sp* hybrid orbitals in a linear arrangement. (c) Unhybridized $2p$ atomic orbitals are perpendicular to the line created by the axes of the two *sp* hybrid orbitals.

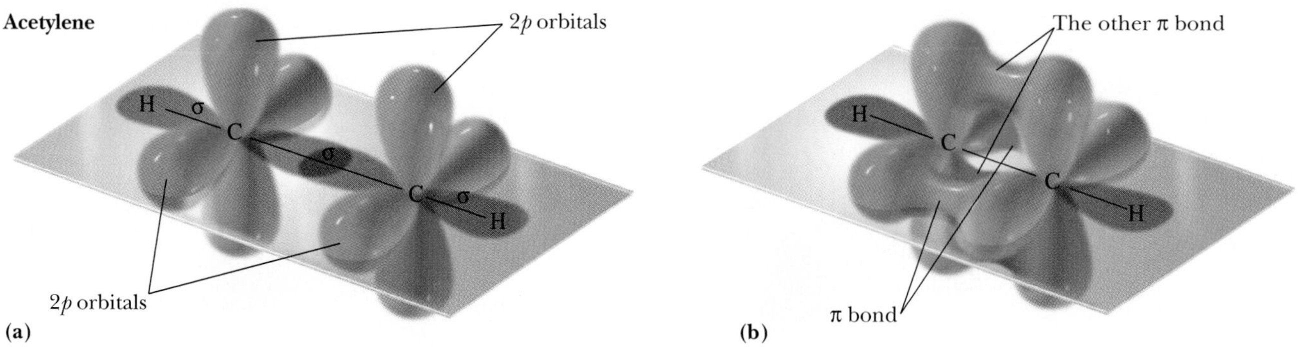

Acetylene

(a)

(b)

Figure 1.21
Covalent bonding in acetylene, C_2H_2. (a) The sigma bond framework shown along with nonoverlapping $2p$ atomic orbitals. (b) Formation of two pi bonds by overlap of two sets of parallel $2p$ atomic orbitals.

Figure 1.21 shows an orbital overlap diagram for acetylene, C_2H_2. A carbon-carbon triple bond consists of one sigma bond formed by overlap of sp hybrid orbitals and two pi bonds. One pi bond is formed by overlap of a pair of parallel $2p$ atomic orbitals. The second pi bond is formed by overlap of another pair of parallel $2p$ atomic orbitals. The relationship among the number of atoms bonded to carbon, orbital hybridization, and types of bonds involved is summarized in Table 1.10.

Table 1.10   Covalent Bonding of Carbon

Groups Bonded to Carbon	Orbital Hybridization	Predicted Bond Angles	Types of Bonds to Carbon	Example	Name
4	sp^3	109.5°	Four sigma bonds	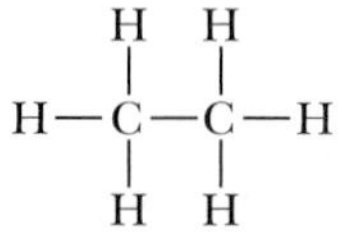 H—C—C—H (with H above and below each C)	Ethane
3	sp^2	120°	Three sigma bonds and one pi bond	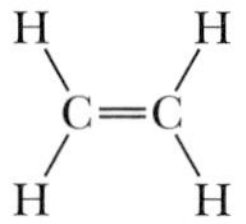 C=C (with H's)	Ethylene
2	sp	180°	Two sigma bonds and two pi bonds	$H—C\equiv C—H$	Acetylene

---

Example 1.18

Describe the bonding in acetic acid, CH_3COOH, in terms of the atomic orbitals involved, and predict all bond angles.

Solution

Following are three identical Lewis structures. Labels on the first Lewis structure point to atoms and show the hybridization of each atom. Labels on the second Lewis structure point to bonds and show the type of bond, either sigma or pi. Labels on the third point to atoms and show bond angles about each atom as predicted by the VSEPR model.

Problem 1.18

Describe the bonding in these molecules in terms of the atomic orbitals involved, and predict all bond angles.

(a) $CH_3CH{=}CH_2$ (b) CH_3NH_2

F. Bond Lengths and Bond Strengths in Ethane, Ethylene, and Acetylene

Values for bond lengths and bond strengths (bond dissociation energies) for ethane, ethylene, and acetylene are given in Table 1.11. As you study this table, note the following points.

1. Carbon-carbon triple bonds are shorter than carbon-carbon double bonds, which in turn are shorter than carbon-carbon single bonds. This order of bond lengths is due to the fact that there are three versus two versus one bond holding the carbon atoms together.
2. The C—H bond in acetylene is shorter than that in ethylene, which in turn is shorter than that in ethane. The relative lengths of these C—H bonds are determined by

Table 1.11 Bond Lengths and Bond Strengths for Ethane, Ethylene, and Acetylene

Name	Formula	Bond	Bond Orbital Overlap	Bond Length (pm)	Bond Strength [kJ (kcal)/mol]
Ethane	H—C—C—H (CH$_3$CH$_3$)	C—C	sp^3–sp^3	153.2	368 (88)
		C—H	sp^3–1s	111.4	410 (98)
Ethylene	C=C (CH$_2$=CH$_2$)	C—C	sp^2–sp^2, 2p–2p	133.9	611 (146)
		C—H	sp^2–1s	110.0	435 (104)
Acetylene	H—C≡C—H	C—C	sp–sp, two 2p–2p	121.2	837 (200)
		C—H	sp–1s	109.0	523 (125)

↳ shortest & strongest C-H bond

the percent *s*-character in the hybrid orbital of carbon forming the sigma bond with hydrogen. The greater the percent *s*-character of a hybrid orbital, the closer to the nucleus electrons in it are held and the shorter the bond. The relative lengths of C—H single bonds correlate with the fact that the percent *s*-character in an *sp* orbital is 50%, in an *sp²* orbital is 33.3%, and in an *sp³* orbital is 25%. Also, because electrons in *s* orbitals are bound more tightly than those in *p* orbitals, the more *s*-character in a bond, the stronger it is (see Appendix 3).

3. There is a correlation between bond length and bond strength; the shorter the bond, the stronger it is. A carbon-carbon triple bond is the shortest C—C bond; it is also the strongest. The carbon-hydrogen bond in acetylene is the shortest; it is also the strongest.

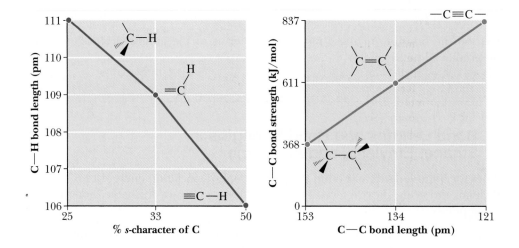

4. Although a C—C double bond is stronger than a C—C single bond, it is not twice as strong. By the same token, a C—C triple bond is stronger than a C—C single bond, but it is not three times as strong. These differences arise because the overlap of orbitals lying on the same axis and forming sigma bonds is more efficient (gives a greater bond strength) than overlap of orbitals lying parallel to each other and forming pi bonds.

Summary

Atoms consist of a small, dense nucleus and electrons distributed about the nucleus in regions of space called **principal energy levels** or **shells** (Section 1.1). Each shell can contain as many as $2n^2$ electrons, where n is the number of the shell. Each principal energy level is subdivided into regions of space called **orbitals**. The **Lewis structure** (Section 1.1B) of an element shows the symbol of the element surrounded by a number of dots equal to the number of electrons in the **valence shell** of the atom. According to the Lewis model of covalent bonding (Section 1.2), atoms bond together in such a way that each atom participating in a chemical bond acquires a completed valence-shell electron configuration resembling that of the noble gas nearest it in atomic number. An **ionic bond** is a chemical bond formed by the attractive force between an anion and a cation. A **covalent bond** is a chemical bond formed by the sharing of electron pairs between atoms. The tendency of main-group elements (those of Groups 1A–7A) to achieve an outer shell of eight valence electrons is called the **octet rule.**

Electronegativity (Section 1.2C) is a measure of the force of attraction by an atom for electrons it shares in a chemical bond with another atom. A **nonpolar covalent bond** (Section 1.2C) is a covalent bond in which the difference in electronegativity of the bonded atoms is less than 0.5 unit. A **polar covalent bond** is a covalent bond in which the difference in electronegativity of the bonded atoms is between 0.5 and 1.9 units. In a polar covalent bond, the more electronegative atom bears a partial negative charge ($\delta-$) and the less electronegative atom bears a partial positive charge ($\delta+$). A polar bond has a bond dipole equal to the product of the absolute value of the partial charge times the distance between the dipolar charges (the bond length). The **dipole moment** of a polyatomic molecule is the vector sum of its bond moments (Section 1.5).

An acceptable **Lewis structure** (Section 1.2D) for a molecule or an ion must show (1) the correct connectivity of atoms, (2) the correct number of valence electrons, (3) no more than two electrons in the outer shell of hydrogen and no more than eight electrons in the outer shell of any second-period element, and (4) all formal charges. **Formal charge** is the charge on an atom in a molecule or polyatomic ion (Section 1.2E).

Functional groups (Section 1.3) are characteristic structural units by which we divide organic compounds into classes and that serve as a basis for nomenclature. They are also sites of chemical reactivity. A particular functional group undergoes the same types of chemical reactions in whatever compound it occurs.

Bond angles of molecules and polyatomic ions can be predicted using Lewis structures and the **valence-shell electron-pair repulsion (VSEPR) model** (Section 1.4). For atoms surrounded by four regions of electron density, the VSEPR model predicts bond angles of 109.5°; for atoms surrounded by three regions of electron density, it predicts bond angles of 120°; and for two regions of electron density, it predicts bond angles of 180°.

According to the **theory of resonance** (Section 1.6A), molecules and ions for which no single Lewis structure is adequate are best described by writing two or more **contributing structures** and considering the real molecule or ion to be a **resonance hybrid** of the various contributing structures.

Contributing structures to the hybrid are interconnected by **double-headed arrows.** The manner in which valence electrons are redistributed from one contributing structure to the next is shown by **curved arrows** (Section 1.6B). Use of curved arrows in this way is commonly referred to as **electron pushing.** The most important contributing structures have (1) filled valence shells, (2) a maximum number of covalent bonds, (3) the least separation of unlike charges, and (4) carry any negative charge on a more electronegative atom and/or any positive charge on a less electronegative atom.

Quantum mechanics (Section 1.7) is the branch of science that studies particles and their associated waves. It provides a way to determine the shapes of atomic orbitals and to quantify the energetics of covalent bond formation.

According to **molecular orbital theory** (Section 1.8), combination of n atomic orbitals gives n molecular orbitals. Molecular orbitals are divided into sigma and pi bonding and antibonding molecular orbitals. These orbitals are arranged in order of increasing energy, and their order of filling with electrons is governed by the same rules as for filling atomic orbitals.

The combination of atomic orbitals is called **hybridization** (Section 1.8B), and the resulting atomic orbitals are called **hybrid orbitals.** Combination of one $2s$ atomic orbital and three $2p$ atomic orbitals produces four equivalent sp^3 **hybrid orbitals,** each directed toward a corner of a regular tetrahedron at angles of 109.5° (Section 1.8C).

The combination of one $2s$ atomic orbital and two $2p$ atomic orbitals produces three equivalent sp^2 **hybrid orbitals,** the axes of which lie in a plane at angles of 120° (Section 1.8D). All C=C, C=O, C=N, N=N, and N=O double bonds are a combination of one sigma (σ) bond formed by the overlap of sp^2 hybrid orbitals and one pi (π) bond formed by overlap of parallel $2p$ orbitals.

The combination of one $2s$ atomic orbital and one $2p$ atomic orbital produces two equivalent sp **hybrid orbitals,** the axes of which lie at an angle of 180° (Section 1.8E). All C≡C and C≡N triple bonds are a combination of one sigma bond formed by the overlap of sp hybrid orbitals and two pi bonds formed by the overlap of two sets of parallel $2p$ orbitals.

Problems

Electronic Structure of Atoms

1.19 Write the ground-state electron configuration for each atom. After each atom is given its atomic number.

  (a) Sodium (11)   **(b)** Magnesium (12)   **(c)** Oxygen (8)   **(d)** Nitrogen (7)

1.20 Identify the atom that has each ground-state electron configuration.

  (a) $1s^2 2s^2 2p^6 3s^2 3p^4$ **(b)** $1s^2 2s^2 2p^4$

1.21 Define valence shell and valence electron.

1.22 How many electrons are in the valence shell of each atom?

(a) Carbon   (b) Nitrogen   (c) Chlorine   (d) Aluminum

Lewis Structures and Formal Charge

1.23 Judging from their relative positions in the Periodic Table, which atom in each set is more electronegative?

(a) Carbon or nitrogen   (b) Chlorine or bromine   (c) Oxygen or sulfur

1.24 Which compounds have nonpolar covalent bonds, which have polar covalent bonds, and which have ionic bonds?

(a) LiF   (b) CH_3F (c) $MgCl_2$ (d) HCl

1.25 Using the symbols $\delta-$ and $\delta+$, indicate the direction of polarity, if any, in each covalent bond.

(a) C—Cl   (b) S—H   (c) C—S   (d) P—H

1.26 Write Lewis structures for these compounds. Show all valence electrons. None of them contains a ring of atoms.

(a) H_2O_2	(b) N_2H_4	(c) CH_3OH
Hydrogen peroxide	Hydrazine	Methanol
(d) CH_3SH	(e) CH_3NH_2	(f) CH_2Cl_2
Methanethiol	Methylamine	Dichloromethane
(g) CH_3OCH_3	(h) H_2CO_3	(i) CH_2O
Dimethyl ether	Carbonic acid	Formaldehyde
(j) CH_3COOH	(k) CH_3COCH_3	(l) HCN
Acetic acid	Acetone	Hydrogen cyanide
(m) HNO_3	(n) HNO_2	(o) HCOOH
Nitric acid	Nitrous acid	Formic acid

1.27 Write Lewis structures for these ions. Show all valence electrons and all formal charges.

(a) NH_2^-	(b) HCO_3^-	(c) CO_3^{2-}
Amide ion	Bicarbonate ion	Carbonate ion
(d) NO_3^-	(e) $HCOO^-$	(f) CH_3COO^-
Nitrate ion	Formate ion	Acetate ion

1.28 Complete these structural formulas by adding enough hydrogens to complete the tetravalence of each carbon. Then write the molecular formula of each compound.

(a)
```
          C
          |
C—C=C—C—C
```
(b)
```
        O
        ||
C—C—C—C—OH
```
(c)
```
      O
      ||
C—C—C—C
```

(d)
```
    O
    ||
C—C—C—H
    |
    C
```
(e)
```
      C
      |
C—C—C—C—NH2
      |
      C
```
(f)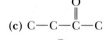
```
      O
      ||
C—C—C—OH
    |
    NH2
```

(g)
```
  OH
  |
C—C—C—C—C
```
(h)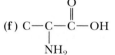
```
  OH   O
  |    ||
C—C—C—C—OH
```
(i) C=C—C—OH

1.29 Some of these structural formulas are incorrect (i.e., they do not represent a real compound) because they have atoms with an incorrect number of bonds. Which structural formulas are incorrect, and which atoms in them have an incorrect number of bonds?

(a)
$$
\begin{array}{cc}
\text{H} & \text{H} \\
| & | \\
\text{H}-\text{C}-\text{C}=\text{O}-\text{H} \\
| & | \\
\text{H} & \text{H}
\end{array}
$$

(b)
$$
\begin{array}{c}
\text{Cl} \\
| \\
\text{H}-\text{C}=\text{C}-\text{H} \\
| \quad | \\
\text{H} \quad \text{H}
\end{array}
$$

(c)
$$
\begin{array}{ccc}
\text{H} & \text{H} \\
| & | \\
\text{H}-\text{N}-\text{C}-\text{C}-\text{O}-\text{H} \\
| & | & | \\
\text{H} & \text{H} & \text{H}
\end{array}
$$

(d)
$$
\begin{array}{c}
\text{H} \\
| \\
\text{H}-\text{C}\equiv\text{C}-\text{C}-\text{H} \\
| \\
\text{H}
\end{array}
$$

(e)
$$
\text{H}-\text{O}-\text{C}-\text{C}-\text{C}-\text{O}-\text{H} \quad (\text{H H O})
$$

(f)
$$
\text{H}-\text{C}-\text{C}-\text{C}-\text{H} \quad (\text{H H O})
$$

(g)
$$
\text{H}-\text{C}-\text{C}=\text{C}=\text{C}-\text{C}-\text{H}
$$

(h)
$$
\text{H}-\text{C}\equiv\text{C}-\text{C}-\text{H}
$$

1.30 Following the rule that each atom of carbon, oxygen, and nitrogen reacts to achieve a complete outer shell of eight valence electrons, add unshared pairs of electrons as necessary to complete the valence shell of each atom in these ions. Then assign formal charges as appropriate.

(a)
$$
\begin{array}{cc}
\text{H} & \text{H} \\
| & | \\
\text{H}-\text{C}-\text{C}-\text{O} \\
| & | \\
\text{H} & \text{H}
\end{array}
$$

(b)
$$
\begin{array}{cc}
\text{H} & \text{H} \\
| & | \\
\text{H}-\text{C}-\text{C} \\
| & | \\
\text{H} & \text{H}
\end{array}
$$

(c)
$$
\begin{array}{ccc}
\text{H} & \text{H} & \text{O} \\
| & | & \diagup\diagdown \\
\text{H}-\text{N}-\text{C}-\text{C} \\
| & | & \diagdown \\
\text{H} & \text{H} & \text{O}
\end{array}
$$

1.31 Following are several Lewis structures showing all valence electrons. Assign formal charges in each structure as appropriate.

(a)
$$
\begin{array}{ccc}
\text{H} & \text{Ö} \\
| & \| \\
\text{H}-\text{C}-\text{C}-\text{C}-\text{H} \\
| & | \\
\text{H} & \text{H}
\end{array}
$$

(b)
$$
\begin{array}{ccc}
& :\text{Ö}: \\
& | \\
\text{H}-\text{N}-\text{C}=\text{C}-\text{H} \\
| & | \\
\text{H} & \text{H}
\end{array}
$$

(c)
$$
\begin{array}{ccc}
& \text{Ö} \\
& \| \\
\text{H}-\text{C}-\text{C}-\text{H} \\
| \\
\text{H}
\end{array}
$$

(d)
$$
\begin{array}{ccc}
\text{H} & :\text{Ö}: \\
| & | \\
\text{H}-\text{C}-\text{C}=\text{C}-\text{H} \\
| & | \\
\text{H} & \text{H}
\end{array}
$$

(e)
$$
\begin{array}{ccc}
\text{H} & \text{H} \\
| & | \\
\text{H}-\text{C}-\text{C}-\text{C}-\text{H} \\
| & | & | \\
\text{H} & \text{H} & \text{H}
\end{array}
$$

(f)
$$
\begin{array}{ccc}
\text{H} \\
| \\
\text{H}-\text{C}-\text{Ö}-\text{H} \\
| \\
\text{H} & \text{H}
\end{array}
$$

Polarity of Covalent Bonds

1.32 Which statements are true about electronegativity?

(a) Electronegativity increases from left to right in a period of the Periodic Table.

(b) Electronegativity increases from top to bottom in a column of the Periodic Table.

(c) Hydrogen, the element with the lowest atomic number, has the smallest electronegativity.

(d) The higher the atomic number of an element, the greater its electronegativity.

1.33 Why does fluorine, the element in the upper right corner of the Periodic Table, have the largest electronegativity of any element?

1.34 Arrange the single covalent bonds within each set in order of increasing polarity.

(a) C—H, O—H, N—H   (b) C—H, B—H, O—H   (c) C—H, C—Cl, C—I

(d) C—S, C—O, C—N   (e) C—Li, C—B, C—Mg

1.35 Using the values of electronegativity given in Table 1.5, predict which indicated bond in each set is the more polar and, using the symbols $\delta+$ and $\delta-$, show the direction of its polarity.

(a) CH_3—OH or CH_3O—OH   (b) H—NH_2 or CH_3—NH_2

(c) CH_3—SH or CH_3S—H   (d) CH_3—F or H—H

1.36 Identify the most polar bond in each molecule.

(a) $HSCH_2CH_2OH$ (b) $CHCl_2F$ (c) $HOCH_2CH_2NH_2$

Bond Angles and Shapes of Molecules

1.37 Use the VSEPR model to predict bond angles about each highlighted atom.

(a) [structure]   (b) [structure]   (c) [structure]

(d) [structure]   (e) [structure]   (f) [structure]

1.38 Use the VSEPR model to predict bond angles about each atom of carbon, nitrogen, and oxygen in these molecules.

(a) CH_3—CH=CH_2 (b) CH_3—N—CH_3 (with CH_3 above)   (c) CH_3—CH_2—C—OH (with O above)

(d) CH_2=C=CH_2 (e) CH_2=C=O   (f) CH_3—CH=N—OH

1.39 Use the VSEPR model to predict the geometry of these ions.

(a) NH_2^- (b) NO_2^- (c) NO_2^+ (d) NO_3^-

(e) CH_3COO^- (f) CH_3^- (g) $AlCl_4^-$

Functional Groups

1.40 Draw Lewis structures for these functional groups. Be certain to show all valence electrons on each.

(a) Carbonyl group   (b) Carboxyl group   (c) Hydroxyl group

1.41 Draw condensed structural formulas for all compounds of molecular formula C_4H_8O that contain

(a) A carbonyl group (there are two aldehydes and one ketone).
(b) A carbon-carbon double bond and a hydroxyl group (there are eight).

1.42 What is the meaning of the term tertiary (3°) when it is used to classify alcohols? Draw a structural formula for the one tertiary (3°) alcohol of molecular formula $C_4H_{10}O$.

1.43 What is the meaning of the term tertiary (3°) when it is used to classify amines? Draw a structural formula for the one tertiary (3°) amine of molecular formula $C_4H_{11}N$.

1.44 Draw structural formulas for

(a) The four primary (1°) amines of molecular formula $C_4H_{11}N$.
(b) The three secondary (2°) amines of molecular formula $C_4H_{11}N$.
(c) The one tertiary (3°) amine of molecular formula $C_4H_{11}N$.

1.45 Draw structural formulas for the three tertiary (3°) amines of molecular formula $C_5H_{13}N$.

1.46 Draw structural formulas for

(a) The eight alcohols of molecular formula $C_5H_{12}O$.
(b) The eight aldehydes of molecular formula $C_6H_{12}O$.

(c) The six ketones of molecular formula $C_6H_{12}O$.

(d) The eight carboxylic acids of molecular formula $C_6H_{12}O_2$.

(e) The nine carboxylic esters of molecular formula $C_5H_{10}O_2$.

1.47 Identify the functional groups in each compound.

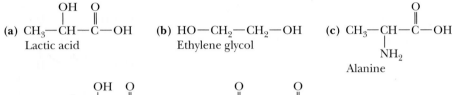

(a)
$$\underset{\text{Lactic acid}}{CH_3-\overset{\overset{\displaystyle OH}{|}}{CH}-\overset{\overset{\displaystyle O}{\|}}{C}-OH}$$
Lactic acid

(b) $HO-CH_2-CH_2-OH$
Ethylene glycol

(c)
$$CH_3-\underset{\underset{\displaystyle NH_2}{|}}{\overset{\overset{\displaystyle O}{\|}}{CH}}-C-OH$$
Alanine

(d)
$$HO-CH_2-\overset{\overset{\displaystyle OH}{|}}{CH}-\overset{\overset{\displaystyle O}{\|}}{C}-H$$
Glyceraldehyde

(e)
$$CH_3-\overset{\overset{\displaystyle O}{\|}}{C}-CH_2-\overset{\overset{\displaystyle O}{\|}}{C}-OH$$
Acetoacetic acid

(f) $H_2NCH_2CH_2CH_2CH_2CH_2CH_2NH_2$
1,6-Hexanediamine

Polar and Nonpolar Molecules

1.48 Draw a three-dimensional representation for each molecule. Indicate which ones have a dipole moment and in what direction it is pointing.

(a) CH_3F (b) CH_2Cl_2 (c) CH_2ClBr

(d) $CFCl_3$ (e) CCl_4 (f) $CH_2{=}CCl_2$

(g) $CH_2{=}CHCl$ (h) $HC{\equiv}C-C{\equiv}CH$ (i) $CH_3C{\equiv}N$

(j) $(CH_3)_2C{=}O$ (k) $BrCH{=}CHBr$ (two answers)

1.49 Tetrafluoroethylene, C_2F_4, is the starting material for the synthesis of the polymer polytetrafluoroethylene (PTFE), one form of which is known as Teflon. Tetrafluoroethylene has a zero dipole moment. Propose a structural formula for this molecule.

Resonance and Contributing Structures

1.50 Which statements are true about resonance contributing structures?

(a) All contributing structures must have the same number of valence electrons.

(b) All contributing structures must have the same arrangement of atoms.

(c) All atoms in a contributing structure must have complete valence shells.

(d) All bond angles in sets of contributing structures must be the same.

1.51 Draw the contributing structure indicated by the curved arrow(s). Assign formal charges as appropriate.

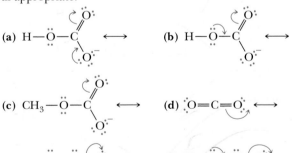

(a) ...

(b) ...

(c) ...

(d) ...

(e) ...

(f) ...

1.52 Using the VSEPR model, predict the bond angles about the carbon and nitrogen atoms in each pair of contributing structures in Problem 1.51. In what way do these bond angles change from one contributing structure to the other?

1.53 In Problem 1.51 you were given one contributing structure and asked to draw another. Label pairs of contributing structures that are equivalent. For those sets in which the contributing structures are not equivalent, label the more important contributing structure.

1.54 Are the structures in each set valid contributing structures?

(a) $\begin{array}{c} H \\ \diagdown \\ \diagup \\ H \end{array} C = \ddot{O} \colon \longleftrightarrow {}^{-}C \equiv \overset{+}{O} \colon$ with H above and below

(b) $H - \ddot{N} = \overset{+}{N} = \ddot{N} \colon^{-} \longleftrightarrow H - \overset{\bar{\cdot\cdot}}{N} - \overset{+}{N} \equiv N \colon$

(c) $H - \underset{\underset{H}{|}}{\overset{\overset{H}{|}}{C}} - \overset{\overset{\cdot\cdot}{O}}{\underset{}{\overset{\|}{C}}} - H \longleftrightarrow H - \underset{\underset{H}{|}}{\overset{\overset{H}{|}}{C}} - \overset{\cdot\cdot}{\underset{\cdot\cdot}{\overset{:\ddot{O}:^{-}}{C}}} - H$

(d) $\begin{array}{c} H \\ \diagdown \\ \diagup \\ H \end{array} C = C \begin{array}{c} :\ddot{O} - H \\ \diagup \\ \diagdown \\ H \end{array} \longleftrightarrow H - \underset{\underset{H}{|}}{\overset{\overset{H}{|}}{C}} - \overset{\overset{\cdot\cdot}{O}}{\overset{\|}{C}} - H$

Molecular Orbital Theory

1.55 State the orbital hybridization of each highlighted atom.

(a) $H - \underset{\underset{H}{|}}{\overset{\overset{H}{|}}{C}} - \underset{\underset{H}{|}}{\overset{\overset{H}{|}}{C}} - H$

(b) $\begin{array}{c} H \\ \diagdown \\ \diagup \\ H \end{array} C = C \begin{array}{c} H \\ \diagup \\ \diagdown \\ H \end{array}$

(c) $H - C \equiv C - H$

(d) $\begin{array}{c} H \\ \diagdown \\ \diagup \\ H \end{array} C = \ddot{O} \colon$

(e) $H - \overset{\overset{\cdot\cdot}{O}}{\overset{\|}{C}} - \ddot{O} - H$

(f) $H - \underset{\underset{H}{|}}{\overset{\overset{H}{|}}{C}} - \ddot{O} - H$

(g) $H - \underset{\underset{H}{|}}{\overset{\overset{H}{|}}{C}} - \ddot{N} - H$

(h) $H - \ddot{O} - \ddot{N} = \ddot{O} \colon$

(i) $CH_2 = C = CH_2$

1.56 Describe each highlighted bond in terms of the overlap of atomic orbitals.

(a) $\begin{array}{c} H \\ \diagdown \\ \diagup \\ H \end{array} C = C \begin{array}{c} H \\ \diagup \\ \diagdown \\ H \end{array}$

(b) $H - C \equiv C - H$

(c) $CH_2 = C = CH_2$

(d) $\begin{array}{c} H \\ \diagdown \\ \diagup \\ H \end{array} C = \ddot{O} \colon$

(e) $H - \overset{\overset{\cdot\cdot}{O}}{\overset{\|}{C}} - \ddot{O} - H$

(f) $H - \underset{\underset{H}{|}}{\overset{\overset{H}{|}}{C}} - \ddot{O} - H$

(g) $H - \underset{\underset{H}{|}}{\overset{\overset{H}{|}}{C}} - \ddot{N} - H$

(h) $H - \underset{\underset{H}{|}}{\overset{\overset{H}{|}}{C}} - \ddot{O} - \overset{\overset{\cdot\cdot}{O}}{\overset{\|}{C}} - H$

(i) $H - \ddot{O} - \ddot{N} = \ddot{O} \colon$

1.57 Following is a structural formula of benzene, C_6H_6.

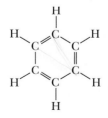

(a) Predict each H—C—C and C—C—C bond angle in benzene.
(b) State the hybridization of each carbon in benzene.
(c) Predict the shape of a benzene molecule.

1.58 Following is a structural formula of the prescription drug famotidine, manufactured by Merck Sharpe & Dohme under the name Pepcid (see *The Merck Index*, 12th ed., #3972). The primary clinical use of Pepcid is for the treatment of active duodenal ulcers and benign gastric ulcers. Pepcid is a competitive inhibitor of histamine H_2 receptors and reduces both gastric acid concentration and the volume of gastric secretions.

(a) Complete the Lewis structure of famotidine showing all valence electrons and any formal positive or negative charges.

(b) Describe each circled bond in terms of the overlap of atomic orbitals.

1.59 Draw a Lewis structure for methyl isocyanate, CH_3NCO, showing all valence electrons. Predict all bond angles in this molecule and the hybridization of each C, N, and O.

Additional Problems

1.60 Why are the following molecular formulas impossible?

(a) CH_5 (b) C_2H_7

1.61 Each compound contains both ionic and covalent bonds. Draw the Lewis structure for each compound and show by dashes which are covalent bonds and by indication of charges which are ionic bonds.

(a) CH_3ONa
Sodium methoxide

(b) NH_4Cl
Ammonium chloride

(c) $NaHCO_3$
Sodium bicarbonate

(d) $NaBH_4$
Sodium borohydride

(e) $LiAlH_4$
Lithium aluminum hydride

1.62 Predict whether the carbon-metal bond in these organometallic compounds is nonpolar covalent, polar covalent, or ionic. For each polar covalent bond, show the direction of its polarity by the symbols $\delta+$ and $\delta-$.

1.63 Silicon is immediately under carbon in the Periodic Table. Predict the geometry of silane, SiH_4.

1.64 Phosphorus is immediately under nitrogen in the Periodic Table. Predict the molecular formula for phosphine, the compound formed by phosphorus and hydrogen. Predict the H—P—H bond angle in phosphine.

1.65 Following are three contributing structures for diazomethane, CH_2N_2.

(a) Using curved arrows, show how each contributing structure is converted to the one on its right.

(b) Which contributing structure makes the largest contribution to the hybrid?

1.66 Draw a Lewis structure for the azide ion, N_3^-. (The order of atom attachment is N—N—N, and they do not form a ring.) How does the resonance model account for the fact that the lengths of the N—N bonds in this ion are identical?

1.67 Draw a Lewis structure for the ozone molecule, O_3. (The order of atom attachment is O—O—O, and they do not form a ring.) How does the resonance model account for the fact that the length of each O—O bond in ozone (128 pm) is shorter than the O—O single bond in hydrogen peroxide (HOOH, 147 pm), but longer than the O—O double bond in the oxygen molecule (123 pm)?

1.68 Cyanic acid, HOCN, and isocyanic acid, HNCO, dissolve in water to yield the same anion on loss of H^+.

(a) Write a Lewis structure for cyanic acid.

(b) Write a Lewis structure for isocyanic acid.

(c) Account for the fact that each acid gives the same anion on loss of H^+.

1.69 In Chapter 6, we study a group of organic cations called carbocations. Following is the structure of one such carbocation, the *tert*-butyl cation.

$$H_3C \diagdown$$
$$\overset{+}{C}-CH_3 \qquad \textit{tert-Butyl cation}$$
$$H_3C \diagup$$

(a) How many electrons are in the valence shell of the carbon bearing the positive charge?

(b) Predict the bond angles about this carbon.

(c) Given the bond angle you predicted in (b), what hybridization do you predict for this carbon?

1.70 Many reactions involve a change in hybridization of one or more atoms in the starting material. In each reaction, identify the atoms in the organic starting material that change hybridization and indicate what the change is. We examine these reactions in more detail later in the course.

(a)
$$\begin{array}{c} H \diagdown H \diagup \\ C = C \\ H \diagup \diagdown H \end{array} + Cl_2 \longrightarrow \begin{array}{c} Cl \quad H \\ | \quad\; | \\ H-C-C-H \\ | \quad\; | \\ H \quad Cl \end{array}$$

(b) $H-C\equiv C-H + Cl_2 \longrightarrow \begin{array}{c} H \diagdown Cl \diagup \\ C = C \\ Cl \diagup \diagdown H \end{array}$

(c) $H-C\equiv C-H + H_2O \longrightarrow \begin{array}{c} H \quad O \\ | \quad\; \| \\ H-C-C-H \\ | \\ H \end{array}$

(d) $\begin{array}{c} O \\ \| \\ H-C-H \end{array} + H_2 \longrightarrow \begin{array}{c} H \\ | \\ H-C-O-H \\ | \\ H \end{array}$

(e) $\begin{array}{c} H \quad\; H \\ | \quad\;\; | \\ H-C-\overset{+}{C}-C-H \\ | \quad\;\; | \quad\;\; | \\ H \quad H \quad H \end{array} + H_2O \longrightarrow \begin{array}{c} H \\ \diagup \\ H \quad O \quad H \\ | \quad\; | \quad\; | \\ H-C-C-C-H \\ | \quad\; | \quad\; | \\ H \quad H \quad H \end{array} + H^+$

(f) $\begin{array}{c} H \quad H \quad\;\; O \quad H \\ | \quad\; | \quad\;\; \| \quad\; | \\ H-C-C-O-C-C-H \\ | \quad\; | \quad\quad\quad | \\ H \quad H \quad\quad\quad H \end{array} \longrightarrow \begin{array}{c} H \diagdown H \diagup \\ C = C \\ H \diagup \diagdown H \end{array} + \begin{array}{c} O \quad H \\ \| \quad\; | \\ H-O-C-C-H \\ | \\ H \end{array}$

ALKANES AND CYCLOALKANES

In this chapter, we begin our study of organic compounds with a discussion of the physical and chemical properties of alkanes and cycloalkanes, both saturated hydrocarbons and the simplest types of organic compounds.

A **hydrocarbon** is a compound that is composed of only carbon and hydrogen. A **saturated hydrocarbon** contains only single bonds. Saturated in this context means that each carbon has the maximum number of hydrogens for the number of carbons it contains.

■ Bunsen burners burn natural gas, which is primarily methane with small amounts of ethane, propane, and butane (Section 2.10A). *(Charles D. Winters)* Inset: A model of methane, the major component of natural gas.

An **alkane** is a saturated hydrocarbon whose carbon atoms are arranged in an open chain. Alkanes are commonly referred to as **aliphatic hydrocarbons** because the physical properties of the higher members of this class resemble those of the long carbon-chain molecules we find in animal fats and plant oils (Greek: *aleiphar,* fat or oil).

The CD-ROM Molecular Models database contains 3D models of many alkanes.

2.1 Structure of Alkanes

Methane, CH_4, and ethane, C_2H_6, are the first two members of the alkane family. Shown in Figure 2.1 are Lewis structures and molecular models for these molecules. The shape of methane is tetrahedral and all H—C—H bond angles are 109.5°. Each of the carbon atoms in ethane is also tetrahedral, and all bond angles are approximately 109.5°.

Although the three-dimensional shapes of larger alkanes are more complex than those of methane and ethane, each carbon atom is still tetrahedral, and all bond angles are approximately 109.5°.

The next members of the alkane family are propane, butane, and pentane. They are shown here as condensed structural formulas and ball-and-stick models. In addition, they are shown in a form called a **line-angle drawing** or stick figure. In this abbreviated way to draw structural formulas, only the carbon framework of the molecule is shown, and you are left to fill in hydrogen atoms as necessary to complete the tetravalence of carbon.

Line-angle drawing An abbreviated way to draw structural formulas in which each angle and line ending represents a carbon and each line represents a bond.

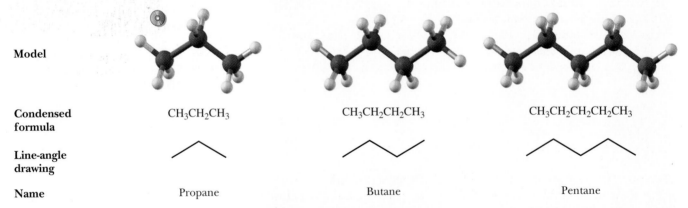

Model			
Condensed formula	$CH_3CH_2CH_3$	$CH_3CH_2CH_2CH_3$	$CH_3CH_2CH_2CH_2CH_3$
Line-angle drawing			
Name	Propane	Butane	Pentane

Condensed structural formulas for alkanes can also be written in an abbreviated form. For example, the structural formula of pentane contains three CH_2 (methylene) groups in the middle of the chain. They can be grouped together and the

Figure 2.1
Methane and ethane.
(a) Lewis structures and
(b) ball-and-stick models.

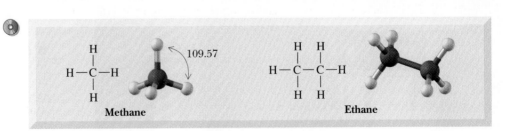

Table 2.1   Names, Molecular Formulas, and Condensed Structural Formulas for the First 20 Alkanes with Unbranched Chains

Name	Molecular Formula	Condensed Structural Formula	Name	Molecular Formula	Condensed Structural Formula
Methane	CH_4	CH_4	Undecane	$C_{11}H_{24}$	$CH_3(CH_2)_9CH_3$
Ethane	C_2H_6	CH_3CH_3	Dodecane	$C_{12}H_{26}$	$CH_3(CH_2)_{10}CH_3$
Propane	C_3H_8	$CH_3CH_2CH_3$	Tridecane	$C_{13}H_{28}$	$CH_3(CH_2)_{11}CH_3$
Butane	C_4H_{10}	$CH_3(CH_2)_2CH_3$	Tetradecane	$C_{14}H_{30}$	$CH_3(CH_2)_{12}CH_3$
Pentane	C_5H_{12}	$CH_3(CH_2)_3CH_3$	Pentadecane	$C_{15}H_{32}$	$CH_3(CH_2)_{13}CH_3$
Hexane	C_6H_{14}	$CH_3(CH_2)_4CH_3$	Hexadecane	$C_{16}H_{34}$	$CH_3(CH_2)_{14}CH_3$
Heptane	C_7H_{16}	$CH_3(CH_2)_5CH_3$	Heptadecane	$C_{17}H_{36}$	$CH_3(CH_2)_{15}CH_3$
Octane	C_8H_{18}	$CH_3(CH_2)_6CH_3$	Octadecane	$C_{18}H_{38}$	$CH_3(CH_2)_{16}CH_3$
Nonane	C_9H_{20}	$CH_3(CH_2)_7CH_3$	Nonadecane	$C_{19}H_{40}$	$CH_3(CH_2)_{17}CH_3$
Decane	$C_{10}H_{22}$	$CH_3(CH_2)_8CH_3$	Eicosane	$C_{20}H_{42}$	$CH_3(CH_2)_{18}CH_3$

abbreviated structural formula written $CH_3(CH_2)_3CH_3$. Names, molecular formulas, and abbreviated structural formulas of the first 20 alkanes are given in Table 2.1. We have more to say about naming alkanes in Section 2.3. Acyclic alkanes (alkanes without rings) have the general molecular formula C_nH_{2n+2}. Thus, given the number of carbon atoms in an alkane, it is easy to determine the number of hydrogens in the molecule and also its molecular formula. For example, decane with 10 carbon atoms must have $(2 \times 10) + 2 = 22$ hydrogen atoms and a molecular formula of $C_{10}H_{22}$.

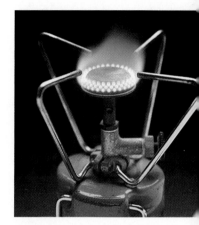

Some camping stoves use butane as a fuel. *(Charles D. Winters)*

2.2 Constitutional Isomerism in Alkanes

Compounds that have the same molecular formula but different structural formulas (different orders of attachment of atoms) are called **constitutional isomers.** In Section 1.3, we encountered several examples of constitutional isomers although we did not call them that at the time. We saw that there are two alcohols of molecular formula C_3H_8O, one ketone and two aldehydes of molecular formula C_4H_8O, and four amines of molecular formula C_3H_9N. In this section, we study constitutional isomers in more detail, including how to recognize them and how to draw structures for them.

For the molecular formulas CH_4, C_2H_6, and C_3H_8, only one order of attachment of atoms is possible. For the molecular formula C_4H_{10}, there are two possible orders of attachment of atoms. In one of these, the four carbons are attached in a chain; in the other, three carbons are attached in a chain with the fourth carbon as a branch on the chain, shown on the next page as condensed structural formulas, line-angle drawings, and ball-and-stick models.

Constitutional isomers   Compounds with the same molecular formula but a different connectivity (order of attachment) of their atoms.

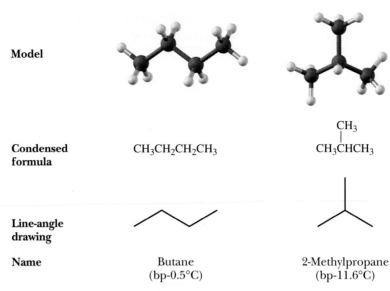

Model		
Condensed formula	$CH_3CH_2CH_2CH_3$	$CH_3\overset{\displaystyle CH_3}{\underset{\displaystyle |}{C}}HCH_3$
Line-angle drawing		
Name	Butane (bp-0.5°C)	2-Methylpropane (bp-11.6°C)

The constitutional isomers of molecular formula C_4H_{10} are named butane and 2-methylpropane. We discuss how to name alkanes in the following section. Butane and 2-methylpropane are different compounds and have different physical and chemical properties.

To determine whether two or more structural formulas represent constitutional isomers, write the molecular formula of each and then compare them. Compounds that have the same molecular formula but different connectivity (the order of attachment of their atoms) are constitutional isomers.

Example 2.1

Do the structural formulas in each set represent the same compound or constitutional isomers?

(a) $CH_3CH_2CH_2CH_2CH_2CH_3$ and $CH_3CH_2CH_2$ (each is C_6H_{14})
                                                          $|$
                                                 $CH_2CH_2CH_3$

(b) $CH_3\overset{\displaystyle CH_3}{\underset{\displaystyle |}{C}}HCH_2\overset{\displaystyle CH_3}{\underset{\displaystyle |}{C}}H$ and $CH_3CH_2\overset{\displaystyle CH_3}{\underset{\displaystyle |}{C}}HCHCH_3$ (each is C_7H_{16})
                                       $|$ $|$
                                      CH_3 CH_3

Solution

(a) Each structural formula has an unbranched chain of six carbons; they are identical and represent the same compound, which is seen most clearly on the accompanying line-angle drawing.

$\overset{1\ \ 2\ \ 3\ \ 4\ \ 5\ \ 6}{CH_3CH_2CH_2CH_2CH_2CH_3}$ and $\overset{1\ \ 2\ \ 3}{CH_3CH_2CH_2}$
                                                                                                                            $\underset{4\ \ 5\ \ 6}{|}$
                                                                                                                            $CH_2CH_2CH_3$

(b) Each structural formula has a chain of five carbons with two CH_3— branches. Although the branches are identical, they are at different locations on the chains.

Therefore, these structural formulas represent constitutional isomers.

$$CH_3CHCH_2CH \quad\quad\quad\quad and \quad\quad CH_3CH_2CHCHCH_3$$

Problem 2.1

Write line-angle drawings for each compound and determine if the formulas in each set represent the same compound or constitutional isomers.

(a) $CH_3CHCHCH_3$ and $CH_3CH_2CHCH_2CHCH_3$

(b) $CH_3CHCHCH_3$ and $CH_3CHCHCH_2CH_3$

Example 2.2

Write condensed structural formulas and line-angle drawings for the five constitutional isomers of molecular formula C_6H_{14}.

Solution

In doing problems of this type, you should devise a strategy and then follow it. Here is one such strategy. First, draw a structural formula for the constitutional isomer with all six carbons in an unbranched chain. Then, draw structural formulas for all constitutional isomers with five carbons in a chain and one carbon as a branch on the chain. Finally, draw structural formulas for all constitutional isomers with four carbons in a chain and two carbons as branches.

6 carbons in an unbranched chain $CH_3CH_2CH_2CH_2CH_2CH_3$

5 carbons in a chain with 1 carbon as a branch $CH_3CHCH_2CH_2CH_3$ $CH_3CH_2CHCH_2CH_3$

4 carbons in a chain; 2 carbons as branches $CH_3CCH_2CH_3$ $CH_3CHCHCH_3$

No constitutional isomers with only three carbons in the longest chain are possible for C_6H_{14}.

Problem 2.2

Write condensed structural formulas and line-angle drawings for the three constitutional isomers of molecular formula C_5H_{12}.

The ability of carbon atoms to form strong, stable bonds with other carbon atoms results in a staggering number of constitutional isomers. As shown in the following table, there are three constitutional isomers of molecular formula C_5H_{12}. For molecular formula $C_{10}H_{22}$, there are 75 constitutional isomers, and for $C_{30}H_{62}$, there are over four billion.

Carbon Atoms	Constitutional Isomers
1	1
5	3
10	75
15	4,347
25	36,797,588
30	4,111,846,763

Thus, for even a small number of carbon and hydrogen atoms, a very large number of constitutional isomers is possible. In fact, the potential for constitutional isomers among organic molecules made from just the basic building blocks of carbon, hydrogen, nitrogen, and oxygen atoms is practically limitless.

Nomenclature; Alkanes

2.3   Nomenclature of Alkanes

A.   The IUPAC System

Ideally, every organic compound should have a name from which a structural formula can be drawn. For this purpose, chemists have adopted a set of rules established by an organization called the International Union of Pure and Applied Chemistry (IUPAC). The IUPAC name of an alkane with an unbranched chain of carbon atoms consists of two parts: (1) a prefix that indicates the number of carbon atoms in the chain and (2) the suffix -ane to show that the compound is a saturated hydrocarbon. Prefixes used to show the presence of 1 to 20 carbon atoms are given in Table 2.2.

The first four prefixes listed in Table 2.2 were chosen by the IUPAC because they were well established in the language of organic chemistry. In fact, they were well established even before there were hints of the structural theory underlying the discipline. For example, the prefix but- appears in the name butyric acid, a compound of four carbon atoms formed by air oxidation of butter (Latin: *butyrum,* butter). Prefixes to show five or more carbons are derived from Greek or Latin numbers. Names, mo-

Table 2.2	Prefixes Used in the IUPAC System to Show the Presence of 1 to 20 Carbon Atoms in an Unbranched Chain		
Prefix	**Number of Carbon Atoms**	**Prefix**	**Number of Carbon Atoms**
meth-	1	undec-	11
eth-	2	dodec-	12
prop-	3	tridec-	13
but-	4	tetradec-	14
pent-	5	pentadec-	15
hex-	6	hexadec-	16
hept-	7	heptadec-	17
oct-	8	octadec-	18
non-	9	nonadec-	19
dec-	10	eicos-	20

lecular formulas, and condensed structural formulas for the first 20 alkanes with un-branched chains were given in Table 2.1.

IUPAC names of alkanes with branched chains consist of a parent name, which indicates the longest chain of carbon atoms in the compound, and substituent names, which indicate the groups attached to the parent chain.

$$CH_3 \longleftarrow \text{substituent}$$
$$| $$
$$CH_3CH_2CH_2CHCH_2CH_2CH_2CH_3 \longleftarrow \begin{array}{l}\text{parent chain of 8}\\ \text{carbon atoms}\end{array}$$

4-Methyloctane

A substituent group derived from an alkane by removal of a hydrogen atom is called an **alkyl group.** The symbol **R—** is commonly used to represent an alkyl group. Alkyl groups are named by dropping the -ane from the name of the parent alkane and adding the suffix -yl. The substituent derived from methane is methyl, CH_3—, for example, and that derived from ethane is ethyl, CH_3CH_2—.

The rules of the IUPAC system for naming alkanes are:

1. The general name for a saturated hydrocarbon with an unbranched chain of carbon atoms consists of a prefix showing the number of carbon atoms in the chain and the ending -ane.
2. For branched-chain alkanes, the longest chain of carbon atoms is taken as the parent chain, and its name becomes the root name.
3. Each substituent is given a name and a number. The number shows the carbon atom of the parent chain to which the substituent is bonded.

$$CH_3$$
$$|$$
$$CH_3CHCH_3$$

2-Methylpropane

Alkyl group A group derived by removing a hydrogen from an alkane.

R— A symbol used to represent an alkyl group.

4. If there is one substituent, number the parent chain from the end that gives it the lower number. The following alkane must be numbered as shown and named 2-methylpentane. Numbering from the other end of the chain gives the incorrect name 4-methylpentane.

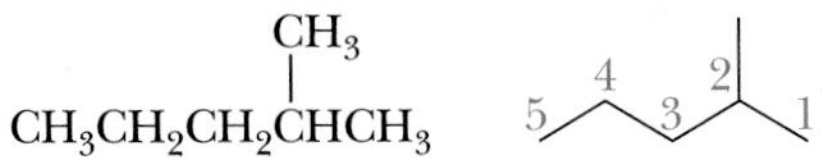

$$CH_3CH_2CH_2CHCH_3$$

2-Methylpentane

5. If the same substituent occurs more than once, number the parent chain from the end that gives the lower number to the substituent encountered first. The number of times the substituent occurs is indicated by a prefix di-, tri-, tetra-, penta-, hexa-, and so on.

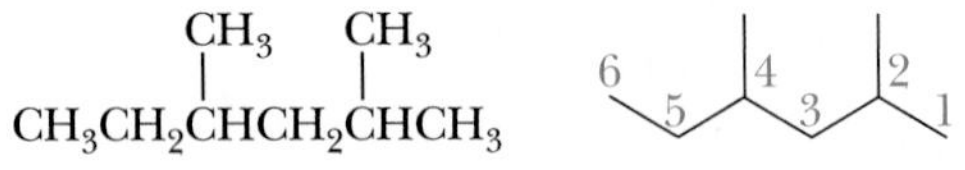

$$CH_3CH_2CHCH_2CHCH_3$$

2,4-Dimethylhexane

6. If there are two or more different substituents, list them in alphabetical order and number the chain from the end that gives the lower number to the substituent encountered first. If there are different substituents in equivalent positions on opposite ends of the parent chain, the substituent of lower alphabetical order is given the lower number.

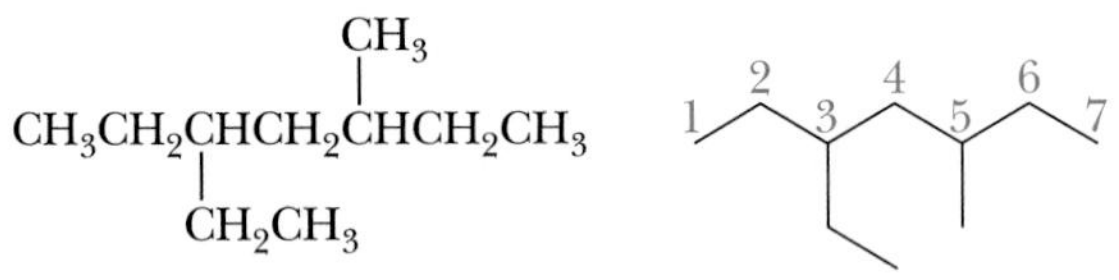

$$CH_3CH_2CHCH_2CHCH_2CH_3$$
$$CH_2CH_3$$

3-Ethyl-5-methylheptane

7. The prefixes di-, tri-, tetra-, and so on are not included in alphabetizing. The names of substituents are alphabetized first, and then these prefixes are inserted. In this example, the alphabetizing parts are ethyl and methyl, not ethyl and di-methyl.

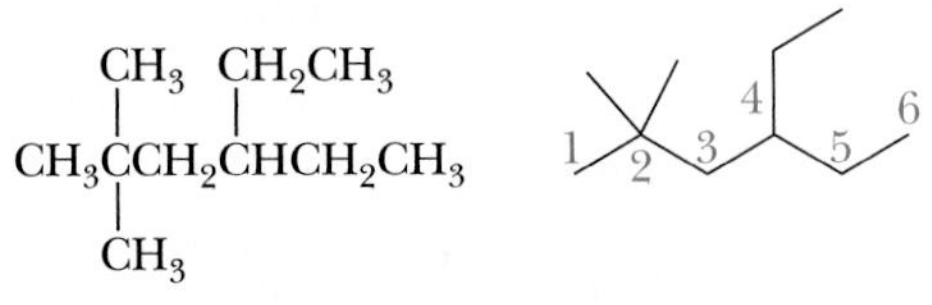

$$CH_3CCH_2CHCH_2CH_3$$
$$CH_3$$

4-Ethyl-2,2-dimethylhexane

Substituents are named following this same set of rules. Those with unbranched chains are named by dropping -ane from the name of the parent alkane and replacing it with -yl. Thus, unbranched alkyl substituents are named methyl, ethyl, propyl, butyl, pentyl, and so forth. Substituents with branched chains are named according to rules 2 and 3. IUPAC names and structural formulas for

Table 2.3 IUPAC and Common Names (in Parentheses) for Alkyl Groups with One to Five Carbons

Name	Condensed Structural Formula	Name	Condensed Structural Formula
Methyl	$-CH_3$		CH_3
Ethyl	$-CH_2CH_3$	1,1-Dimethylethyl (*tert*-butyl)	$-\overset{\displaystyle CH_3}{\underset{\displaystyle CH_3}{C}}CH_3$
Propyl	$-CH_2CH_2CH_3$		
1-Methylethyl (isopropyl)	$-\underset{\displaystyle CH_3}{CH}CH_3$	Pentyl	$-CH_2CH_2CH_2CH_2CH_3$
		3-Methylbutyl (isopentyl)	$-CH_2CH_2\underset{\displaystyle CH_3}{CH}CH_3$
Butyl	$-CH_2CH_2CH_2CH_3$		
2-Methylpropyl (isobutyl)	$-CH_2\underset{\displaystyle CH_3}{CH}CH_3$	2-Methylbutyl	$-CH_2\underset{\displaystyle CH_3}{CH}CH_2CH_3$
1-Methylpropyl (*sec*-butyl)	$-\underset{\displaystyle CH_3}{CH}CH_2CH_3$	2,2-Dimethylpropyl (neopentyl)	$-CH_2\overset{\displaystyle CH_3}{\underset{\displaystyle CH_3}{C}}CH_3$

unbranched and branched alkyl groups containing one to five carbon atoms are given in Table 2.3. Also given in parentheses are common names, which we discuss in Section 2.4B.

Example 2.3

Write the IUPAC name for each alkane.

(a) $CH_3\underset{\displaystyle CH_3}{CH}CH_2CH_3$ (b) $CH_3\underset{\displaystyle CH_3}{CH}CH_2\underset{\displaystyle CH(CH_3)_2}{CH}CH_2CH_2CH_3$

Solution

The longest chain in each is numbered from the end of the chain toward the substituent encountered first (rule 4). The substituents in (b) are listed in alphabetical order (rule 7).

(a)

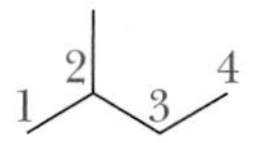

2-Methylbutane

(b)

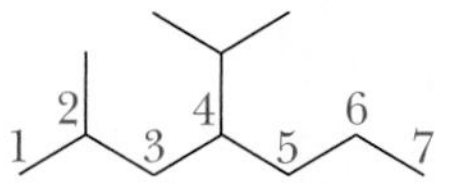

2-Methyl-4-(1-methylethyl)heptane

Problem 2.3

Write the IUPAC name for each alkane.

$$\begin{array}{ccc}
& CH_3 & CH_3 \\
& | & | \\
\text{(a)} & CH_3CHCH_2CH_2CHCHCH_3 \\
& & | \\
& & CH_2CH_2CH_3
\end{array}$$

$$\begin{array}{c}
CH_2CH_2CH_3 \\
| \\
\text{(b)}\quad CH_3CH_2CH_2CCH_2CH_2CH_3 \\
| \\
CH_3CHCH_3
\end{array}$$

B. Common Names

In the older system of common nomenclature, the total number of carbon atoms in an alkane, regardless of their arrangement, determines the name. The first three alkanes are methane, ethane, and propane. All alkanes of formula C_4H_{10} are called butanes, those of formula C_5H_{12} are called pentanes, and those of formula C_6H_{14} are called hexanes. For alkanes beyond propane, iso- is used to indicate that one end of an otherwise unbranched chain terminates in $-CH(CH_3)_2$, and neo- is used to indicate that one end of an otherwise unbranched chain terminates in $-C(CH_3)_3$. Following are examples of common names:

$$\begin{array}{cccc}
& CH_3 & CH_3 & CH_3 \\
& | & | & | \\
CH_3CH_2CH_2CH_2CH_3 & CH_3CHCH_3 & CH_3CH_2CHCH_3 & CH_3CCH_3 \\
& & & | \\
& & & CH_3
\end{array}$$

| Pentane | Isobutane | Isopentane | Neopentane |

As is so often the case in the naming of organic compounds, the alkyl groups were given common names. In this older system, the total number of carbons in the substituent, regardless of their arrangement, determined the name. For one- and two-carbon substituents, there is no ambiguity; they are methyl and ethyl. For a three-carbon substituent, however, there is ambiguity; there are two of them, and they cannot both be named propyl. To solve the problem, one was named propyl and the other isopropyl.

To name other branching patterns, the common system uses a set of prefixes to classify a carbon atom as **primary (1°)**, **secondary (2°)**, **tertiary (3°)**, or **quaternary (4°)**, depending on the number of carbon atoms bonded to it. A carbon bonded to one carbon atom is a primary carbon, a carbon bonded to two carbon atoms is a secondary carbon, and so forth. For example, propane contains two primary carbons and one secondary carbon. 2-Methylpropane contains three primary carbons and one tertiary carbon. 2,2,4-Trimethylpentane contains five primary carbons, one secondary carbon, one tertiary carbon, and one quaternary carbon.

Primary (1°) carbon A carbon bonded to one other carbon.
Secondary (2°) carbon A carbon bonded to two other carbons.
Tertiary (3°) carbon A carbon bonded to three other carbons.
Quaternary (4°) carbon A carbon bonded to four other carbons.

two 1° carbons

$$CH_3-CH_2-CH_3$$

a 2° carbon

Propane

a 3° carbon

$$\begin{array}{c}
CH_3-CH-CH_3 \\
| \\
CH_3
\end{array}$$

2-Methylpropane

a 4° carbon

$$\begin{array}{ccc}
CH_3 & & CH_3 \\
| & & | \\
CH_3-C-CH_2-CH-CH_3 \\
| \\
CH_3
\end{array}$$

2,2,4-Trimethylpentane

Similarly, hydrogens are also classified as primary, secondary, or tertiary, depending on the type of carbon to which each is bonded.

These classifications are applied to the common names of the alkyl group in the following way. For an alkyl substituent of four carbons, there are four possible arrangements of atoms. The unbranched four-carbon substituent is named butyl. The four-carbon substituent with three carbons in a chain that terminates in $-CH(CH_3)_2$ is named isobutyl. In the *sec*-butyl group, the point of attachment of the alkyl group is a secondary carbon. In the *tert*-butyl group, the point of attachment of the group is a tertiary carbon.

When giving alkanes common names, hyphenated prefixes such as *sec-* and *tert-* are not considered when alphabetizing. The prefixes iso- and neo- are not hyphenated and are included when alphabetizing.

This system of common names has no good way of handling other branching patterns; for more complex alkanes, it is necessary to use the more flexible IUPAC system of nomenclature.

In this text, we concentrate on IUPAC names. However, we also use common names, especially when the common name is used almost exclusively in the everyday discussions of chemists. When both IUPAC and common names are given in the text, we always give the IUPAC name first followed by the common name in parentheses. In this way, you should have no doubt about which name is which.

Example 2.4

Write the common name of each alkane in Example 2.3.

Solution

(a) Isopentane   **(b)** 4-Isopropyl-2-methylheptane

Problem 2.4

Write the common name of each alkane in Problem 2.3.

2.4 Cycloalkanes

A hydrocarbon that contains carbon atoms joined to form a ring is called a **cyclic hydrocarbon.** When all carbons of the ring are saturated, the hydrocarbon is called a **cycloalkane.**

Cycloalkane A saturated hydrocarbon that contains carbons joined to form a ring.

A. Structure and Nomenclature

Cycloalkanes of ring sizes ranging from 3·to over 30 are found in nature, and, in principle, there is no limit to ring size. Five-membered rings (cyclopentanes) and six-membered rings (cyclohexanes) are especially common and have, therefore, received special attention.

Figure 2.2 shows structural formulas of cyclopropane, cyclobutane, cyclopentane, and cyclohexane. Chemists commonly represent these and other cycloalkanes by regular polygons having the same number of sides as there are carbons in the ring. For example, cyclopropane is represented by a triangle, and cyclohexane by a hexagon.

Cycloalkanes contain two fewer hydrogen atoms than alkanes of the same number of carbon atoms (the two missing hydrogens come from the ring-closing C—C

H$_2$C $\diagdown$ CH$_2$ or $\triangle$ H$_2$C—CH$_2$ $|$ $|$ H$_2$C—CH$_2$ or $\square$ H$_2$C $\diagup$ CH$_2$ $\diagdown$ CH$_2$ H$_2$C—CH$_2$ or $\pentagon$ H$_2$C $\diagup$ CH$_2$ $\diagdown$ CH$_2$ H$_2$C $\diagdown$ CH$_2$ $\diagup$ CH$_2$ or $\hexagon$

Cyclopropane C$_3$H$_6$ **Cyclobutane** C$_4$H$_8$ **Cyclopentane** C$_5$H$_{10}$ **Cyclohexane** C$_6$H$_{12}$

Figure 2.2
Examples of cycloalkanes.

bond). For example, compare the molecular formulas of cyclohexane, C$_6$H$_{12}$, and hexane, C$_6$H$_{14}$. The general formula of a cycloalkane is **C$_n$H$_{2n}$**.

To name cycloalkanes, prefix the name of the corresponding open-chain alkane with cyclo-, and name each substituent on the ring. If there is only one substituent on the cycloalkane ring, there is no need to give it a number. If there are two substituents, number the ring beginning with the substituent of lower alphabetical order. If there are three or more substituents, number the ring to give them the lowest set of numbers, and list them in alphabetical order.

Example 2.5

Write the molecular formula and IUPAC name for each cycloalkane.

(a) (cyclopentyl isopropyl)   **(b)** (cyclohexyl with ethyl and tert-butyl)   **(c)** (cyclohexyl with ethyl and methyl)

Solution

(a) First replace each angle and line terminus by a carbon and then add hydrogens as necessary to give each carbon four bonds. The molecular formula of this compound is C$_8$H$_{16}$. Because there is only one substituent on the ring, there is no need to number the atoms of the ring. Its IUPAC name is (1-methylethyl)cyclopentane. Its common name is isopropylcyclopentane.

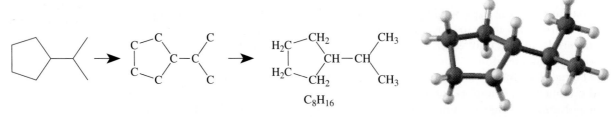

C$_8$H$_{16}$

(b) The two substituents are ethyl and 1,1-dimethylethyl, and the IUPAC name of the cycloalkane is 1-ethyl-4-(1,1-dimethylethyl)cyclohexane. Alternatively, the substituents are named ethyl and *tert*-butyl, giving the cycloalkane the name 1-*tert*-butyl-4-ethylcyclohexane. Its molecular formula is C$_{12}$H$_{24}$.

(c) Number the ring to give the three substituents the lowest set of numbers and then list them in alphabetical order. Its name is 2-ethyl-1,4-dimethylcyclohexane, and its molecular formula is $C_{10}H_{20}$.

Problem 2.5

Write the molecular formula, IUPAC name, and common name for each cycloalkane.

(a) ![structure a] **(b)** ![structure b] **(c)** ![structure c]

B. Bicycloalkanes

An alkane that contains two rings that share two carbon atoms in common is classified as a **bicycloalkane.** The shared carbon atoms are called **bridgehead carbons,** and the carbon chains connecting them are called **bridges.** The general formula of a bicycloalkane is C_nH_{2n-2}. Figure 2.3 shows three examples of bicycloalkanes along with the IUPAC and common name of each.

> **Bicycloalkane** An alkane containing two rings that share two carbons.

Example 2.6

Write the general formula for an alkane, a cycloalkane, and a bicycloalkane. How do these general formulas differ?

Solution

General formulas are alkane, C_nH_{2n+2}; cycloalkane, C_nH_{2n}; and bicycloalkane, C_nH_{2n-2}. Each general formula in this series has two fewer hydrogens than the previous member of the series.

Problem 2.6

Write molecular formulas for each bicycloalkane, given its number of carbon atoms.

(a) Hydrindane (nine carbons)   **(b)** Decalin (ten carbons)
(c) Norbornane (seven carbons)

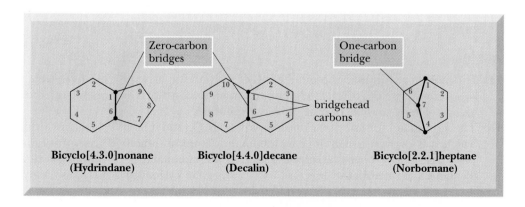

Figure 2.3
Examples of bicycloalkanes.

IUPAC names of bicycloalkanes are derived in the following way:

1. The parent name of a bicycloalkane is that of the hydrocarbon of the same number of carbon atoms as are in the bicyclic ring system. For example, the first compound in Figure 2.3 contains nine carbons and is, therefore, a bicyclononane. The second compound contains ten carbons and is a bicyclodecane.

2. Numbering begins at one bridgehead carbon and proceeds along the longest bridge to the second bridgehead carbon; then it travels along the next longest bridge back to the original bridgehead carbon and so on until all ring carbons are numbered. If there are two bridges of the same length, proceed along the one that gives the lower number of the first encountered substituent. The name and location of substituents are shown by the rules given in Section 2.3A. If there is a choice of numbering patterns, choose the one that gives substituents the lowest possible numbers.

3. Bridge lengths are shown by counting the number of carbons linking the bridgeheads and placing them in decreasing order in brackets between the prefix bicyclo- and the parent name and with periods separating each number. For example, the first compound in Figure 2.3 has two bridgehead carbons. There are four carbons in the first bridge, three in the second, and zero in the third; its name is bicyclo[4.3.0]nonane. Two additional examples follow:

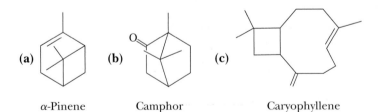

2-Methylbicyclo[4.4.0]decane   1,7,7-Trimethylbicyclo[2.2.1]heptane

---

Example 2.7

Following are structural formulas and common names for three bicyclic compounds. Write the molecular formula of each compound, and name the bicycloalkane from which it is derived.

(a) α-Pinene   (b) Camphor   (c) Caryophyllene

Solution

(a) The molecular formula of α-pinene is $C_{10}H_{16}$, and the bicycloalkane from which it is derived is bicyclo[3.1.1]heptane. α-Pinene is a major component, often as high as 65% by volume, of pine oil and turpentine.

(b) The molecular formula of camphor is $C_{10}H_{16}O$, and the bicycloalkane from which it is derived is bicyclo[2.2.1]heptane. Camphor, obtained from the camphor tree, *Cinnamonium camphora,* is used in the manufacture of certain plastics, lacquers, and varnishes.

(c) The molecular formula of caryophyllene is $C_{15}H_{24}$, and the bicycloalkane from which it is derived is bicyclo[7.2.0]undecane. Caryophyllene is one of the fragrant components of oil of cloves.

Problem 2.7

Draw structural formulas for the following bicycloalkanes.

(a) Bicyclo[3.1.0]hexane   **(b)** Bicyclo[2.2.2]octane
(c) Bicyclo[4.2.0]octane   **(d)** 2,6,6-Trimethylbicyclo[3.1.1]heptane

2.5 The IUPAC System — A General System of Nomenclature

The naming of alkanes and cycloalkanes in Sections 2.3 and 2.4 illustrated the application of the IUPAC system of nomenclature in two specific classes of organic compounds. Now, let us describe the general approach of the IUPAC system. The name assigned to any compound with a chain of carbon atoms consists of three parts: a **prefix,** an **infix** (a modifying element inserted into a word), and a **suffix.** Each part provides specific information about the structure of the compound.

1. The prefix indicates the number of carbon atoms in the parent chain. Prefixes to show the presence of 1 to 20 carbon atoms in an unbranched chain are given in Table 2.2.
2. The infix indicates the nature of the carbon-carbon bonds in the parent chain.

Infix	Nature of Carbon-Carbon Bonds in the Parent Chain
-an-	All single bonds
-en-	One or more double bonds
-yn-	One or more triple bonds

3. The suffix indicates the class of compound to which the substance belongs.

Suffix	Class of Compound
-e	Hydrocarbon
-ol	Alcohol
-al	Aldehyde
-amine	Amine
-one	Ketone
-oic acid	Carboxylic acid

Example 2.8

Following are IUPAC names and structural formulas for four compounds. Divide each name into a prefix, an infix, and a suffix, and specify the information about the structural formula that is contained in each part of the name.

(a) $CH_2{=}CHCH_3$ (b) CH_3CH_2OH (c) $CH_3CH_2CH_2CH_2\overset{\overset{\displaystyle O}{\|}}{C}OH$ (d) $HC{\equiv}CH$

Propene   Ethanol   Pentanoic acid   Ethyne

Solution

(a) prop-en-e ⟶ a carbon-carbon double bond
⟵ a hydrocarbon
three carbon atoms

(b) eth-an-ol ⟶ only carbon-carbon single bonds
⟵ an alcohol
two carbon atoms

(c) pent-an-oic acid ⟶ only carbon-carbon single bonds
⟵ a carboxylic acid
five carbon atoms

(d) eth-yn-e ⟶ a carbon-carbon triple bond
⟵ a hydrocarbon
two carbon atoms

Problem 2.8

Combine the proper prefix, infix, and suffix, and write the IUPAC name for each compound.

(a) $CH_3\overset{\overset{\displaystyle O}{\|}}{C}CH_3$ (b) $CH_3(CH_2)_3\overset{\overset{\displaystyle O}{\|}}{C}H$ (c) ⬠—OH   (d) ⬡

2.6 Conformations of Alkanes and Cycloalkanes

Structural formulas are useful to show the order of attachment of atoms in a molecule. However, they usually do not show three-dimensional shapes. As chemists try to understand more and more about the relationships among structure and the chemical and physical properties of compounds, it becomes increasingly important to know more about the three-dimensional shapes of molecules. In this section, we ask you to look at molecules as three-dimensional objects and to visualize not only bond angles but also distances within molecules between various atoms and groups not bonded to each other. We also describe intramolecular strain, which we divide into three types: torsional strain, nonbonded interaction strain, and angle strain. We very strongly urge you to build models (either physically or by computer using software such as Cambridge-Soft's ChewDraw/Chem3D) of the molecules discussed in this section so that you become comfortable in dealing with them as three-dimensional objects and understand fully the origins of intramolecular strain. Three-dimensional Chem3D models of all these compounds are available on the CD packaged with this text and also on its Web site. You should manipulate these models by rotating parts of each molecule about its single bonds to help you visualize the three-dimensional relationships among its atoms.

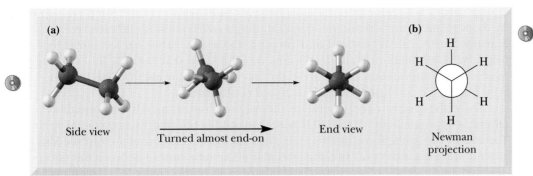

Figure 2.4
A staggered conformation of ethane. (a) Ball-and-stick models and (b) Newman projection.

A. Alkanes

Alkanes of two or more carbons can be transformed into a number of different three-dimensional arrangements of their atoms by rotating parts of the molecule about one or more carbon-carbon bonds. Any three-dimensional arrangement of atoms that results from rotation about single bonds is called a **conformation.** Figure 2.4(a) shows a ball-and-stick model of a **staggered conformation** of ethane. In this conformation, the three C—H bonds on one carbon are as far apart as possible from those bonds on the adjacent carbon. Figure 2.4(b) is a shorthand way to represent this conformation of ethane. It is called a Newman projection. In a **Newman projection,** a molecule is viewed down the axis of a C—C bond. The three atoms or groups of atoms on the carbon nearer your eye are shown on lines extending from the center of the circle at angles of 120°. The three atoms or groups of atoms on the carbon farther from your eye are shown on lines extending from the circumference of the circle, also at angles of 120°. Remember that bond angles about each carbon in ethane are approximately 109.5° and not 120°, as this Newman projection might suggest. The three lines in front represent bonds directed toward you, whereas the three lines in back point away from you.

Figure 2.5 shows a ball-and-stick model and a Newman projection for another conformation of ethane. In this conformation, the three C—H bonds on one carbon are as close as possible to the three C—H bonds on the adjacent carbon. In other words, hydrogen atoms on the back carbon are eclipsed by the hydrogen atoms on the front carbon (as the sun can be eclipsed when the moon passes in front of it). This is called an **eclipsed conformation.**

To discuss energy relationships between conformations, it is convenient to define the term dihedral angle. A **dihedral angle, θ** (Greek letter theta), is the angle created by two intersecting planes, each defined by three atoms. In the Newman projection of the eclipsed conformation of ethane in Figure 2.6(a), two H—C—C planes are shown. The angle at which these planes intersect (the dihedral angle) is 0°. A staggered conformation in which the dihedral angle of the two H—C—C planes is 60° is illustrated in Figure 2.6(b).

In principle, there are an infinite number of conformations of ethane that differ only in the degree of rotation about the carbon-carbon single bond. There is a small

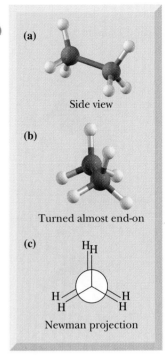

Figure 2.5
An eclipsed conformation of ethane. (a, b) Ball-and-stick models and (c) Newman projection.

Conformation Any three-dimensional arrangement of atoms in a molecule that results by rotation about a single bond.

Staggered conformation A conformation about a carbon-carbon single bond in which the atoms or groups on one carbon are as far apart as possible from atoms or groups on an adjacent carbon.

Newman projection A way to view a molecule by looking along a carbon-carbon single bond.

Eclipsed conformation A conformation about a carbon-carbon single bond in which the atoms or groups on one carbon are as close as possible to the atoms or groups on an adjacent carbon.

Dihedral angle The angle created by two intersecting planes.

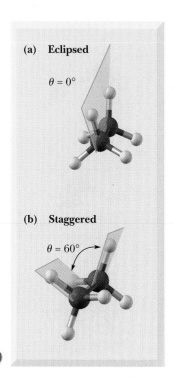

Figure 2.6

Dihedral angles in (a) eclipsed and (b) staggered conformations of ethane.

energy barrier between conformations, however, and rotation is not completely free. At room temperature, ethane molecules undergo collisions with sufficient energy so that the barrier can be crossed and rotation about the carbon-carbon single bond from one conformation to another occurs rapidly. As we shall see, the lowest energy, most stable conformation of ethane is the staggered conformation. The highest energy, least stable conformation is the eclipsed conformation.

The difference in energy between the eclipsed and staggered conformations of ethane is approximately 12.6 kJ (3.0 kcal)/mol and is referred to as torsional strain. **Torsional strain** is the force that opposes the rotation of one part of a molecule about a bond while the other part remains stationary. The relationship between energy and dihedral angle for the conformations of ethane is shown in Figure 2.7. An animation of the energy minimization of an eclipsed conformation of ethane relaxing to a staggered conformation is on the CD packaged with the text.

Energy diagrams in this book use "energy" as a vertical axis. Remember that there are several types of energy—potential energy, free energy, and enthalpy—that are important for various contexts. Diagrams for all these are often nearly indistinguishable, and in most cases "energy" will do for the concepts we are introducing. When it is necessary to be more precise about the type of energy, we will do so and explain why.

The torsional strain in eclipsed ethane is caused by the slight coulombic repulsion of electron pairs of adjacent C—H bonds as they rotate past each other in going from one staggered conformation to another. From inspection of molecular models of staggered and eclipsed ethane, you will see that, in an eclipsed conformation, electron pairs of adjacent C—H bonds are closer to one another than they are in a staggered conformation.

Next let us look at the conformations of butane viewed along the bond between carbons 2 and 3. For butane, there are two types of staggered conformations and two

Figure 2.7

Energy of ethane as a function of dihedral angle. The eclipsed conformations are approximately 12.6 kJ (3 kcal)/mol higher in energy than the staggered conformations.

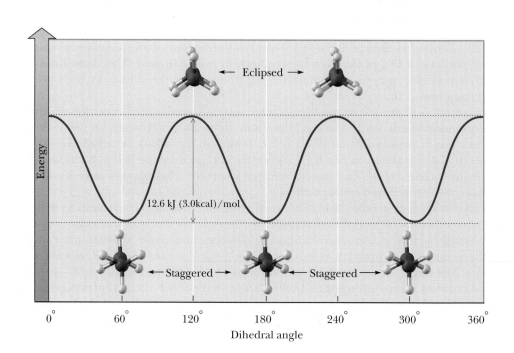

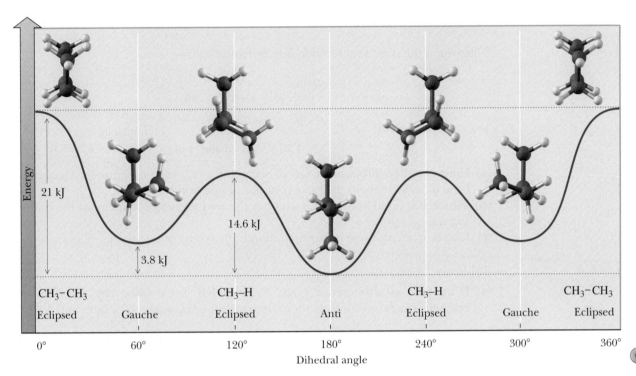

Figure 2.8

Energy of butane as a function of the dihedral angle about the bond between carbons 2 and 3. The lowest energy conformation occurs when the two methyl groups are the maximum distance apart (dihedral angle 180°). The highest energy conformation occurs when the two methyl groups are eclipsed (dihedral angle 0°).

Anti conformation A conformation about a single bond of an alkane in which the groups lie at a dihedral angle of 180°.

Gauche conformation A conformation about a single bond of an alkane in which two groups lie at a dihedral angle of 60°.

Nonbonded interaction strain The strain that arises when atoms not bonded to each other are forced abnormally close to one another.

types of eclipsed conformations. The staggered conformation in which the methyl groups are the maximum distance apart ($\theta = 180°$) is called the **anti conformation;** that in which they are closer together ($\theta = 60°$) is called the **gauche conformation.** In one eclipsed conformation ($\theta = 0°$), methyl is eclipsed by methyl. In the other ($\theta = 120°$), methyl is eclipsed by hydrogen. The energy relationships for rotation from 0° to 360° are illustrated in Figure 2.8. You should view and manipulate the computer models on the CD and measure distances to confirm these relationships.

Note that both gauche and anti conformations of butane are staggered conformations, yet the gauche conformations are approximately 3.8 kJ (0.9 kcal)/mol higher in energy than the anti conformation. The difference in energy between these conformations is due to nonbonded interaction strain. **Nonbonded interaction strain,** or **steric strain** as it is also called, arises when atoms or groups of atoms not bonded to each other are forced into close proximity so that their electrons repel each other. Nonbonded interaction strain in the case of gauche butane arises because the two methyl groups are closer to one another than they are in the anti conformation.

At any given instant, there is a larger number of butane molecules in the anti conformation than in the gauche conformation. The percentage of the anti conformation present at 20°C is about 70%.

You should note that, even though the two staggered conformations with methyl groups gauche (dihedral angles 60° and 300°) have equal energies, they are not identical. They are related by reflection; one is the reflection of the other just as your right hand is the reflection of your left hand. Notice that the conformations with eclipsed —CH_3 and —H groups (dihedral angles of 120° and 240°) are also related by reflection. We shall have more to say about objects and their mirror reflections in Chapter 3.

Example 2.9

Following is the structural formula of 1,2-dichloroethane.

$$\begin{array}{ccc} & H & H \\ & | & | \\ Cl-C- & C-Cl \\ & | & | \\ & H & H \end{array}$$

1,2-Dichloroethane

(a) Draw Newman projections for all staggered conformations formed by rotation from 0° to 360° about the carbon-carbon single bond.
(b) Which staggered conformation(s) has the lowest energy? Which has the highest energy?
(c) Which, if any, of these conformations are related by reflection?

Solution

(a) If we take as a reference point the dihedral angle when the chlorines are eclipsed, staggered conformations occur at dihedral angles 60°, 180°, and 300°.

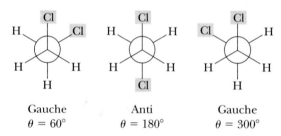

Gauche	Anti	Gauche
$\theta = 60°$	$\theta = 180°$	$\theta = 300°$

(b) We predict that the anti conformation ($\theta = 180°$) has the lowest energy. The two gauche conformations ($\theta = 60°$ and 300°) are of higher but equal energy. We are not given data in the problem to calculate the actual energy differences.
(c) The two gauche conformations are related by reflection.

Problem 2.9

For 1,2-dichloroethane:

(a) Draw Newman projections for all eclipsed conformations formed by rotation from 0° to 360° about the carbon-carbon single bond.
(b) Which eclipsed conformation(s) has the lowest energy? Which has the highest energy?
(c) Which, if any, of these conformations are related by reflection?

B.   Cycloalkanes

Cyclopropane

The observed C—C—C bond angles in cyclopropane are 60° (Figure 2.9), a value considerably smaller than the 109.5° predicted for sp^3 hybridized carbon atoms. Furthermore, hydrogen atoms on adjacent carbons are forced into eclipsed relationships.

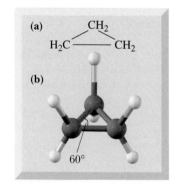

Figure 2.9
Cyclopropane. (a) Structural formula and (b) ball-and-stick model.

Cyclopropane is a strained molecule due to both angle strain and torsional strain. **Angle strain** arises from the creation of abnormal bond angles, in this case compression of C—C—C bond angles from 109.5° to 60°. Torsional strain in cyclopropane arises because of six pairs of eclipsed C—H bonds. The total strain energy in cyclopropane is approximately 116 kJ (27.7 kcal)/mol (see Figure 2.23). Because of its extreme degree of intramolecular strain, cyclopropane and its derivatives undergo several ring-opening reactions not shown by larger cycloalkanes.

> **Angle strain**   The strain that arises when a bond angle is either compressed or expanded compared to its normal value.

Cyclobutane

In all cycloalkanes larger than cyclopropane, nonplanar or puckered conformations are favored. Figure 2.10 shows molecular models of one planar and one puckered conformation of cyclobutane. If cyclobutane were planar, all C—C—C bond angles would be 90°, and there would be eight pairs of eclipsed hydrogen interactions. Puckering of the ring alters the energy in two ways: (1) it decreases the torsional strain associated with eclipsing interactions, but (2) it increases further the angle strain due to compression of C—C—C bond angles. Because the decrease in torsional strain is greater than the increase in angle strain, puckered cyclobutane is more stable than planar cyclobutane. In the conformation of lowest energy, the measured C—C—C bond angles are 88°, and the strain energy in cyclobutane is approximately 110 kJ (26.3 kcal)/mol (see Figure 2.23). An animation of this puckering is on the CD.

Cyclopentane

If cyclopentane were to adopt a planar conformation, all C—C—C bond angles would be 108°. This angle differs only slightly from the tetrahedral angle of 109.5°; consequently, there would be little angle strain in this conformation. There would be, however, ten pairs of fully eclipsed C—H bonds creating a torsional strain of approximately 42 kJ (10 kcal)/mol. To relieve at least a part of this torsional strain, the ring twists into the "envelope" conformation shown in Figure 2.11. In this conformation, four carbon atoms are in a plane, and the fifth is bent out of the plane, rather like an envelope with its flap bent upward. Cyclopentane exists as a dynamic equilibrium of five equivalent envelope conformations in which C—C—C bond angles are reduced (increasing angle strain), but the number of eclipsed C—H interactions is also reduced (decreasing torsional strain). Thus, the molecule is more stable in the envelope conformation than in a planar conformation. The average C—C—C bond

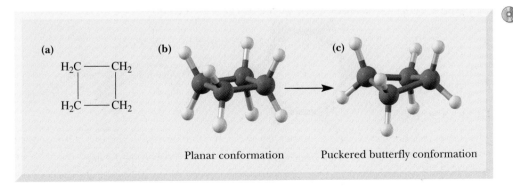

(a)

H₂C —— CH₂

H₂C —— CH₂

(b)

(c)

Planar conformation   Puckered butterfly conformation

Figure 2.10

(a) Cyclobutane. (b) In the planar conformation, there are eight pairs of eclipsed C—H interactions. (c) The energy is a minimum in the puckered butterfly conformation. An animation of this puckering is on the CD.

Figure 2.11
Cyclopentane. (a) Line-angle drawing. (b) In the planar conformation, there are ten pairs of eclipsed C—H interactions. (c) The most stable conformation is a puckered envelope conformation. An animation of the puckering is on the CD.

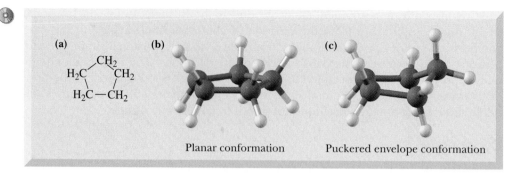

Planar conformation   Puckered envelope conformation

angle in cyclopentane is 105° because of this puckering. The strain energy in cyclopentane is approximately 27 kJ (6.5 kcal)/mol (see Figure 2.23). An animation of this puckering is on the CD.

Cyclohexane

Chair conformation The most stable nonplanar conformation of a cyclohexane ring; all bond angles are approximately 109.5°, and all bonds on adjacent carbons are staggered.

Axial position A position on a chair conformation of a cyclohexane ring that extends from the ring parallel to the imaginary axis of the ring.

Equatorial position A position on a chair conformation of a cyclohexane ring that extends from the ring roughly perpendicular to the imaginary axis of the ring.

Cyclohexane adopts a number of puckered conformations, the most stable of which is the **chair conformation** (Figure 2.12). In it, all C—C—C bond angles are near 109.5° (precluding angle strain), and hydrogens on adjacent carbons are staggered with respect to one another (precluding torsional strain). In addition, no two atoms are close enough to each other for nonbonded interaction strain to exist. Thus, there is no strain of any kind in a chair conformation of cyclohexane. An animation of a planar model of cyclohexane having its energy minimized to give a chair conformation is on the CD.

In a chair conformation of cyclohexane, the C—H bonds are arranged in two different orientations. Six C—H bonds are in the **axial position,** and the other six are in the **equatorial position.** One way to visualize the difference between these two types of bonds is to imagine an axis through the center of the chair, perpendicular to the floor (Figure 2.13). Axial bonds (labeled a) are parallel to this axis. Three axial bonds point straight up; the other three axial bonds point straight down. Notice also that axial bonds alternate, first up and then down as you move from one carbon of

Figure 2.12
Cyclohexane.

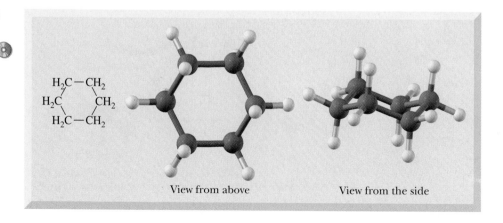

View from above   View from the side

the ring to the next. Equatorial bonds (labeled e) are approximately perpendicular to our imaginary axis and parallel to two carbon-carbon bonds of the ring. Equatorial bonds also alternate, first slightly up and then slightly down as you move from one carbon of the ring to the next. Notice further that if the axial bond on a carbon points upward, then the equatorial bond on that carbon points slightly downward. Conversely, if the axial bond on a particular carbon points downward, then the equatorial bond on that carbon points slightly upward.

There are many other nonplanar conformations of cyclohexane, two of which, a **boat conformation** and a **twist-boat conformation,** are shown in Figure 2.14. You can visualize interconversion of chair and boat conformations by twisting about two carbon-carbon bonds as illustrated in Figure 2.15. To do this with a molecular model, hold five of the carbon atoms of a chair in place and move the sixth upward (or downward, as the case may be). A boat conformation is considerably less stable than a chair conformation because of the torsional strain associated with four pairs of eclipsed (e) C—H interactions and the nonbonded interaction between the two flag-pole hydrogens (f). The difference in energy between chair and boat conformations is approximately 27 kJ (6.5 kcal)/mol.

Some of the strain in a boat conformation can be relieved by a slight twisting of the ring to form a twist-boat conformation. An animation of a boat conformation undergoing energy minimization to a twist-boat is on the CD. It is estimated by

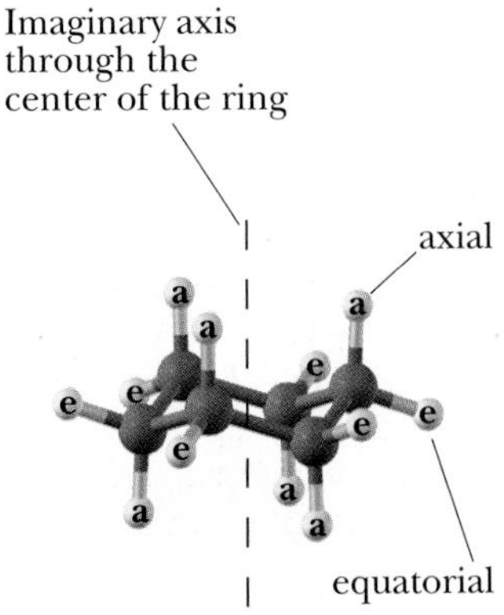

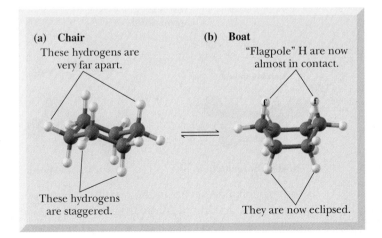
Boat conformation   Twist-boat conformation

Figure 2.13
Chair conformation of cyclohexane, showing axial (a) and equatorial (e) C—H bonds.

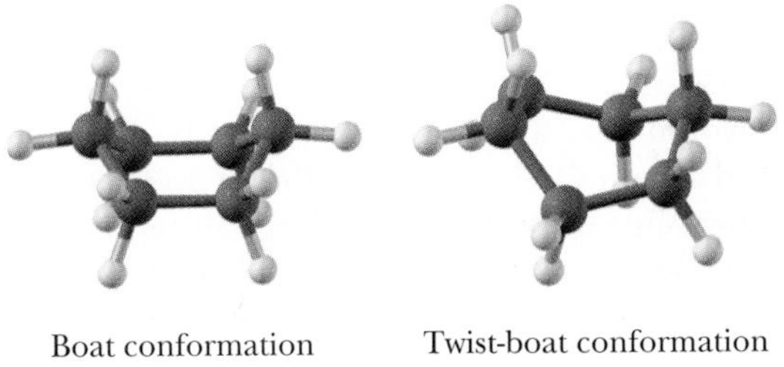

Figure 2.14
Boat and twist-boat conformations of cyclohexane.

Boat conformation A nonplanar conformation of a cyclohexane ring in which carbons 1 and 4 of the ring are bent toward each other.
Twist-boat conformation A nonplanar conformation of a cyclohexane ring that is twisted from and slightly more stable than a boat conformation.

(a) Chair
These hydrogens are very far apart.
These hydrogens are staggered.

(b) Boat
"Flagpole" H are now almost in contact.
They are now eclipsed.

Figure 2.15
Interconversion of (a) a chair conformation to (b) a boat conformation produces one set of nonbonded flagpole interactions and four sets of eclipsed hydrogen interactions.

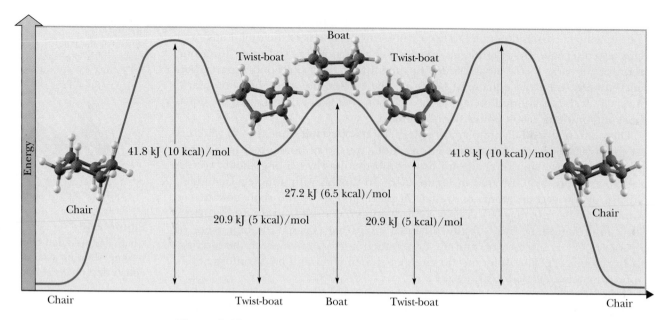

Figure 2.16

Energy diagram for interconversion of chair, twist-boat, and boat conformations of cyclo-hexane. The chair conformation is the most stable because angle, torsional, and nonbonded interaction strain are at a minimum.

computer modeling that a twist-boat is favored over a boat conformation by approximately 6.3 kJ (1.5 kcal)/mol. An energy diagram for interconversion between chair, twist-boat, and boat conformations is shown in Figure 2.16. The large difference in energy between chair and boat or twist-boat conformations means that, at room temperature, molecules in the chair conformation make up more than 99.99% of the equilibrium mixture.

For cyclohexane, the two equivalent chair conformations can be interconverted by twisting one chair first to a boat and then to the other chair.

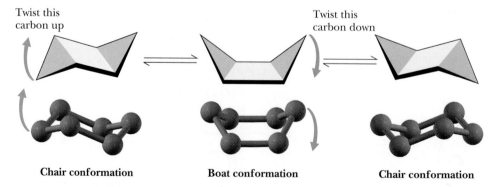

When one chair is converted to the other, a change occurs in the relative orientations in space of the hydrogen atoms attached to each carbon. All hydrogen atoms axial in one chair become equatorial in the other and vice versa (Figure 2.17). The conver-

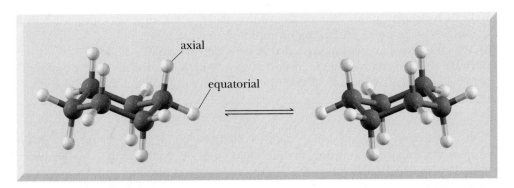

Figure 2.17
Interconversion of chair cyclo-hexanes. All C—H bonds equa-torial in one chair are axial in the alternative chair, and vice versa.

sion of one chair conformation of cyclohexane to the other occurs rapidly at room temperature.

Example 2.10

Following is a chair conformation of cyclohexane showing one methyl group and one hydrogen.

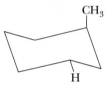

(a) Indicate by a label whether each group is equatorial or axial.
(b) Draw the alternative chair conformation and again label each group equatorial or axial.

Solution

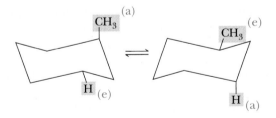

One chair   The alternative chair

Problem 2.10

Following is a chair conformation of cyclohexane with carbon atoms numbered 1 through 6.

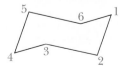

(a) Draw hydrogen atoms that are above the plane of the ring on carbons 1 and 2 and below the plane of the ring on carbon 4.

(b) Which of these hydrogens are equatorial? Which are axial?

(c) Draw the alternative chair conformation. Which hydrogens are equatorial? Which are axial? Which are above the plane of the ring? Which are below it?

If a hydrogen atom of cyclohexane is replaced by an alkyl group, the group occupies an equatorial position in one chair and an axial position in the other chair. This means that the two chairs are no longer equivalent and no longer of equal stability.

A convenient way to describe the relative stabilities of chair conformations with equatorial and axial substituents is in terms of a type of nonbonded interaction strain called **diaxial interaction.** Diaxial interaction refers to the repulsion between an axial substituent and an axial hydrogen (or another group) on the same side of the ring. Because these repulsions are between groups on carbons 1 and 3 of a cyclohexane ring, they are often called 1,3-diaxial interactions. Consider methylcyclohexane (Figure 2.18). When the —CH_3 is axial, it is parallel to the axial C—H bonds on carbons 3 and 5 and rather close to them. Thus, for axial methylcyclohexane, there are two unfavorable methyl-hydrogen diaxial interactions. No such unfavorable interactions exist when the methyl group is in an equatorial position. For methylcyclohexane, the equatorial methyl conformation is favored over the axial methyl conformation by approximately 7.28 kJ (1.74 kcal)/mol.

To understand why axial methylcyclohexane is less stable than equatorial methylcyclohexane, recall our discussion of the conformations of butane. There we saw that gauche butane is less stable than anti butane by approximately 3.8 kJ (0.9 kcal)/mol. Figure 2.19 shows a ball-and-stick model of axial methylcyclohexane turned so that you sight along bonds C_1—C_2 and C_4—C_5 of the ring. As illustrated here, the conformation of the methyl group and an axial hydrogen is identical to a gauche butane conformation. There are actually two gauche butane-like interactions, one with the axial hydrogen on carbon 3 and the other with the axial hydrogen on carbon 5. These two gauche butane-like interactions [approximately $2 \times 3.8 = 7.6$ kJ (1.8 kcal)/mol] account for the intramolecular strain of 7.28 kJ (1.74 kcal)/mol in this conformation compared to equatorial methylcyclohexane. Note that in the model in Figure 2.19 (which was computed to give the minimum energy structure of this conformation), the methyl leans back a bit to minimize the contact!

Let us calculate the amount of axial and equatorial methylcyclohexane conformers in the equilibrium using the thermodynamic data we were given. Given the differ-

Diaxial interactions Nonbonded interactions between atoms or groups in axial positions on the same side of a chair conformation of a cyclohexane ring.

Figure 2.18

Two chair conformations of methylcyclohexane. The two diaxial interactions make the axial methyl conformation less stable by approximately 7.28 kJ (1.74 kcal)/mol.

Diaxial interactions

ence in energy between the axial and equatorial conformations, we can calculate the ratio at equilibrium using the following equation, which relates the change in Gibbs free energy, ΔG^0, for an equilibrium, and the equilibrium constant, K_{eq}.

$$\Delta G^0 = -RT \ln K_{eq}$$

where ΔG^0 is the difference in standard Gibbs free energy in kilojoules per mole between the axial and equatorial conformations of butane, R is the gas constant [8.314 J(1.987 cal) $\cdot$ K^{-1} $\cdot$ mol^{-1}], and T is the temperature in Kelvin. Converting from the natural logarithm to $\log_{10}$ (which requires the factor of 2.303) and solving this equation for $\log K_{eq}$ gives

$$\log K_{eq} = \frac{-\Delta G^0}{2.303\ RT}$$

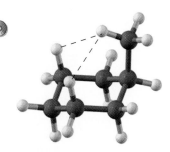

Figure 2.19
Intramolecular strain in axial methylcyclohexane. The ring is oriented to show the two di-axial interactions. The gauche butane-like skeleton is shown in color. Note that the methyl leans back to minimize the contact with the axial hydrogens.

Appendix 1 gives a table of K_{eq} values versus ΔG^0 and some other useful relationships.

Substituting the value of -7.28 kJ/mol (axial methyl $\rightarrow$ equatorial methyl) for ΔG^0 and solving the equation gives a value of 18.9 for the equilibrium constant at room temperature (25°C = 298 K).

$$\log K_{eq} = \frac{-(-7280\,\text{J}\cdot\text{mol}^{-1})}{2.303\ (8.314\,\text{J}\cdot\text{K}^{-1}\cdot\text{mol}^{-1})\ 298\ \text{K}} = 1.276$$

$$K_{eq} = 10^{1.28} = \frac{18.9}{1} = \frac{\text{equatorial}}{\text{axial}}$$

At any given instant, therefore, there is a much larger number of methylcyclohexane molecules in the equatorial conformation than in the axial. The percentage of equatorial is 100 × equatorial/(equatorial + axial), or about 95%.

As the size of the alkyl substituent increases, the preference for conformations with the group equatorial increases. Table 2.4 shows the difference in free energy between axial and equatorial substituents for monosubstituted cyclohexanes. With a

Table 2.4 ΔG^0 (Axial-Equatorial) for Monosubstituted Cyclohexanes at 25°C

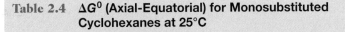

| | $-\Delta G^0$ | | | $-\Delta G^0$ | |
Group	kJ/mol	kcal/mol	Group	kJ/mol	kcal/mol
C≡N	0.8	0.19	NH$_2$	5.9	1.41
F	1.0	0.24	COOH	5.9	1.41
C≡CH	1.7	0.41	CH=CH$_2$	7.1	1.70
I	1.9	0.45	CH$_3$	7.28	1.74
Cl	2.2	0.53	CH$_2$CH$_3$	7.3	1.75
Br	2.4	0.57	CH(CH$_3$)$_2$	9.0	2.15
OH	3.9	0.93	C(CH$_3$)$_3$	21.0	5.00

group as large as *tert*-butyl, the energy of the axial conformer becomes so large that the equatorial conformation is approximately 4000 times more abundant at room temperature than the axial conformation. In fact, a chair with an axial *tert*-butyl group is so unstable that, if a *tert*-butyl group is forced into an axial position, the ring adopts a twist-boat conformation. Note in Table 2.4 that the preference for the equatorial position among the halogens is Br > Cl > I > F. Yet the size of the halogen atoms decreases in the order I > Br > Cl > F. The reason for this anomaly is that the C—I bond is so long that the center of the iodine atom is too far from the axial hydrogen to interact with it.

Example 2.11

Label all methyl-hydrogen diaxial interactions in this chair conformation.

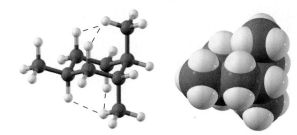

Solution

There are four methyl-hydrogen diaxial interactions in this example; each axial methyl group has two sets of diaxial interactions with parallel hydrogen atoms on the same side of the ring. The equatorial methyl group has no diaxial interactions.

Problem 2.11

Draw the alternative chair conformation for the trisubstituted cyclohexane given in Example 2.11. Label all methyl-hydrogen diaxial interactions in this chair conformation.

Example 2.12

Calculate the ratio of the diequatorial to diaxial conformation of this disubstituted cyclohexane at 25°C.

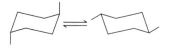

C H E M I S T R Y   I N   A C T I O N

The Poisonous Puffer Fish

Nature is by no means limited to carbon in six-membered rings. Tetrodotoxin, one of the most potent toxins known, is composed of a set of interconnected six-membered rings, each in a chair conformation. All but one of these rings have atoms other than carbon in them. Tetrodotoxin is produced in the liver and ovaries of many species of *Tetraodontidae,* especially the puffer fish, so called because it inflates itself to an almost spherical spiny ball when it is alarmed. It is evidently a species highly preoccupied with defense, but the Japanese are not put off. They regard the puffer, called "fugu" in Japanese, as a delicacy. To serve it in a public restaurant, a chef must be registered as sufficiently skilled in removing the toxic organs so as to make the flesh safe to eat.

A puffer fish with its body inflated. (Tim Rock/Animals Animals)

Symptoms of tetrodotoxin poisoning begin with attacks of severe weakness, progressing to complete paralysis and eventual death. Tetrodotoxin blocks sodium ion channels, which are essential for neurotransmission. This prevents communication between neurons and muscle cells and results in the fatal symptoms described.

Interestingly enough, the ring system in tetrodotoxin is found in a hydrocarbon, adamantane, first isolated from petroleum sources. Adamantane has the same carbon skeleton found infinitely repeated in diamond.

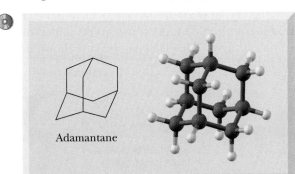

Adamantane

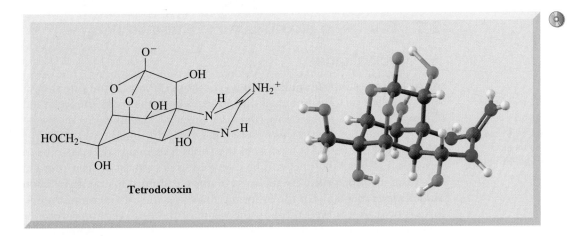

Tetrodotoxin

Solution

For these two chair conformations, ΔG^0 (2 axial $CH_3 \rightarrow$ 2 equatorial CH_3) = -14.6 kJ (3.5 kcal)/mol. Substituting this value in the equation $\Delta G^0 = -2.303 \, RT \log K_{eq}$ gives a ratio 362:1.

$$\log K_{eq} = \frac{-(-14{,}600 \, \text{J/mol})}{2.303 \, (8.314 \, \text{J} \cdot \text{K}^{-1} \cdot \text{mol}^{-1}) \, 298 \, \text{K}} = 2.559$$

$$K_{eq} = 10^{2.559} = 362$$

Problem 2.12

Draw a chair conformation of 1,4-dimethylcyclohexane in which one methyl group is equatorial and the other is axial. Draw the alternative chair conformation and calculate the ratio of the two conformations at 25°C.

Highly Strained Small-Ring Compounds

We have seen that small-ring compounds such as cyclopropane and cyclobutane have a high degree of both angle and torsional strain. It has been a particular challenge to chemists to attempt to synthesize even more highly strained rings, both for the challenge of devising new reaction sequences to make such molecules and to better understand relationships between molecular strain and chemical reactivity. Among the fascinating molecules synthesized in recent years are bicyclo[1.1.0]butane and a variety of propellanes. A **propellane** is a molecule in which two atoms joined by a single bond are also joined by three other bridges. The [1.1.1]propellane shown here is the smallest member of this class of compounds.

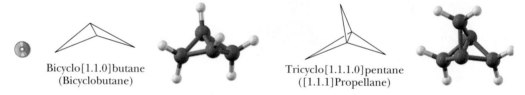

Bicyclo[1.1.0]butane
(Bicyclobutane)

Tricyclo[1.1.1.0]pentane
([1.1.1]Propellane)

2.7   Cis,Trans Isomerism in Cycloalkanes

A.   Cycloalkanes

Cis,trans isomers   Isomers that have the same connectivity but a different arrangement of their atoms in space due to the presence of either a ring or a carbon-carbon double bond.

All cycloalkanes with substituents on two or more carbons of the ring show a type of isomerism called cis,trans isomerism. **Cis,trans isomers** have (1) the same molecular formula, (2) the same connectivity (the same order of attachment of their atoms), but (3) an arrangement of their atoms in space that cannot be interconverted by rotation about single bonds. By way of comparison, the Gibbs free energy difference between conformations is so small that they can be interconverted easily at or near room temperature.

Cis,trans isomerism in cyclic structures can be illustrated by models of 1,2-dimethylcyclopentane. In the following drawings, the cyclopentane ring is shown as a planar pentagon viewed through the plane of the ring. Carbon-carbon bonds of the ring projecting toward the reader are shown as heavy lines. When viewed from this perspective, substituents attached to the ring project above and below the plane of the ring. In one isomer of 1,2-dimethylcyclopentane, the methyl groups are on the same side of the ring; in the other, they are on opposite sides of the ring. The prefix **cis** (Latin: on the same side) is used to indicate that the substituents are on the same side of the ring; the prefix **trans** (Latin: across) is used to indicate that they are on opposite sides of the ring. In each isomer, the configuration of the methyl groups is fixed because of the restricted rotation about the ring carbon-carbon bonds; that is, the cis isomer cannot be converted to the trans isomer and vice versa without breaking and reforming one or more bonds. The cis isomer is approximately 7.1 kJ (1.7 kcal)/mol higher in energy (less stable) than the trans isomer because of the non-bonded interaction of the two methyl groups in the cis isomer.

Cis A prefix meaning on the same side.

Trans A prefix meaning across from.

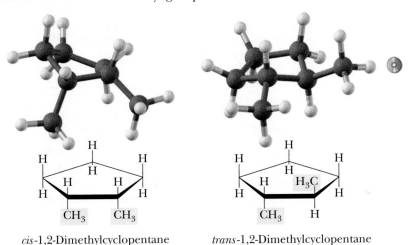

cis-1,2-Dimethylcyclopentane   *trans*-1,2-Dimethylcyclopentane

---

Example 2.13

Which cycloalkanes show cis,trans isomerism? For each that does, draw both isomers.

(a) [structure: cyclopentane with CH₃]

(b) [structure: cyclobutane with two CH₃]

(c) [structure: cyclobutane with H₃C and CH₃]

Solution

(a) Methylcyclopentane does not show cis,trans isomerism. It has only one substituent on the ring.

(b) 1,1-Dimethylcyclobutane does not show cis,trans isomerism. Only one arrangement is possible for the two methyl groups on the ring; they must be trans to each other.

(c) 1,3-Dimethylcyclobutane shows cis,trans isomerism. Note that, in these structures, we show only the hydrogen atoms on carbons bearing the methyl groups.

cis-1,3-Dimethylcyclobutane   *trans*-1,3-Dimethylcyclobutane

Problem 2.13

Which cycloalkanes show cis,trans isomerism? For each that does, draw both isomers.

(a) H_3C———CH_3 (b) ◯—CH_2CH_3 (c) CH_2CH_3 / CH_3

For the purposes of determining the number of cis,trans isomers in substituted cycloalkanes, it is adequate to draw the cycloalkane ring as a planar polygon as is done in the following disubstituted cyclohexane. Two cis,trans isomers are possible for 1,4-dimethylcyclohexane. In these structural formulas, only the hydrogens on methyl-substituted carbons are shown.

H CH_3 H   H

H_3C H H_3C CH_3

trans-1,4-Dimethylcyclohexane   *cis*-1,4-Dimethylcyclohexane

The cis and trans isomers of 1,4-dimethylcyclohexane actually exist as nonplanar chair conformations. When working with alternative chair conformations, it is helpful to remember that all axial groups in one chair become equatorial in the alternative chair, and vice versa.

In one chair conformation of *trans*-1,4-dimethylcyclohexane, the two methyl groups are axial; in the alternative chair conformation, they are equatorial. Of these chair conformations, the one with both methyls equatorial is considerably more stable.

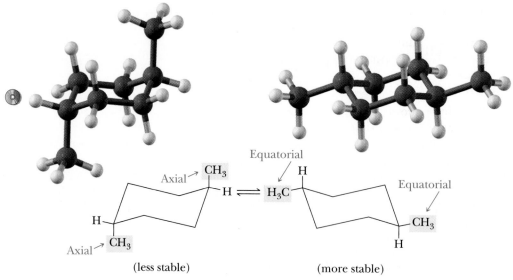

Axial ↗ CH_3 Equatorial ↙ H

Axial → H ⇌ H_3C Equatorial → CH_3

H   H

Axial ↗ CH_3

(less stable)   (more stable)

trans-1,4-Dimethylcyclohexane

The alternative chair conformations of *cis*-1,4-dimethylcyclohexane are of equal energy. In one chair, one methyl group occupies an equatorial position, and the other occupies an axial position. In the other chair, the orientations in space of the methyl groups are reversed.

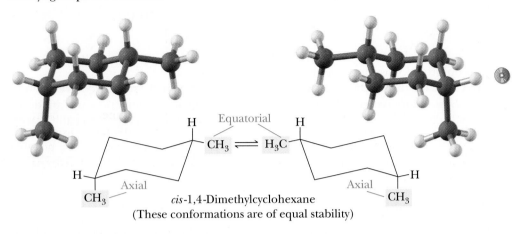

cis-1,4-Dimethylcyclohexane
(These conformations are of equal stability)

Example 2.14

Following is a chair conformation of 1,3-dimethylcyclohexane.

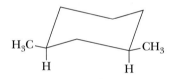

(a) Is this a chair conformation of *cis*-1,3-dimethylcyclohexane or of *trans*-1,3-dimethylcyclohexane?

(b) Draw the alternative chair conformation of this compound. Of the two chair conformations, which is the more stable?

(c) Draw a planar hexagon representation for the isomer shown in this example.

Solution

(a) The isomer shown is *cis*-1,3-dimethylcyclohexane; the two methyl groups are on the same side of the ring.

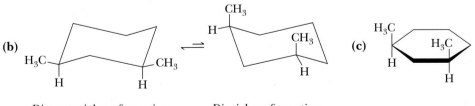

Diequatorial conformation
(more stable)

Diaxial conformation
(less stable)

Problem 2.14

Following is a planar hexagon representation for one isomer of 1,2,4-trimethylcyclohexane.

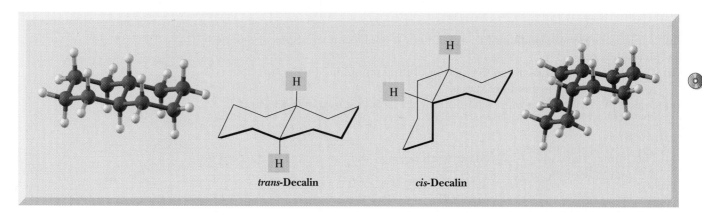

(a) Draw the alternative chair conformations of this compound and state which chair conformation is the more stable.

(b) Calculate the Gibbs free energy difference between these two conformations. (*Hint:* 1,2-*trans* diequatorial substituents have a gauche interaction and are destabilized by about 3.8 kJ (0.9 kcal)/mol. Don't forget to include this factor in your calculation!

B.   Bicycloalkanes

Bicycloalkanes, particularly those with a carbon skeleton like that of bicyclo[4.4.0] decane (decalin), are abundant in the biological world. Figure 2.20 shows structural formulas for *trans*-decalin and *cis*-decalin. In the trans isomer, the two hydrogen atoms on the bridgehead carbons are on opposite sides of the molecule, and in the cis isomer, they are on the same side of the molecule. *Trans*-decalin and *cis*-decalin are different compounds and have different physical and chemical properties. The cis isomer, for example, has a boiling point of 195°C; the trans isomer has a boiling point of 185.5°C.

Figure 2.20
Cis and trans isomers of decalin.

Example 2.15

Compare the relative stabilities of *cis*-decalin and *trans*-decalin in terms of the number of diaxial interactions in the most stable conformation of each isomer.

Solution

There are no diaxial interactions on the all-chair conformation of *trans*-decalin shown in Figure 2.20. In the cis isomer, there are three diaxial methyl-hydrogen (gauche butane) interactions. *Trans*-decalin is more stable than *cis*-decalin by approximately 12.1 kJ (2.9 kcal)/mol, which is consistent with three gauche butane interactions [3 × 3.8 = 11.4 kJ (2.7 kcal)/mol] in the cis isomer.

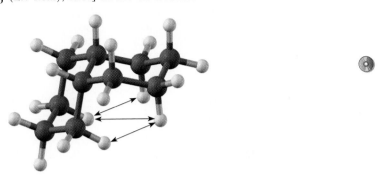

Problem 2.15

Which of the following stereoisomers is the more stable?

(a)   (b)

2.8   Physical Properties of Alkanes and Cycloalkanes

Reactivity Explorer; Alkanes

You are already familiar with the physical properties of some alkanes and cycloalkanes from your everyday experiences. The low-molecular-weight alkanes, such as methane, ethane, propane, and butane, are gases at room temperature. Higher molecular-weight alkanes, such as those in gasoline and kerosene, are liquids. Very-high-molecular-weight alkanes, such as those in paraffin wax, are solids. Melting points, boiling points, and densities of the first ten alkanes are listed in Table 2.5.

Methane is a gas at room temperature and atmospheric pressure. It can be converted to a liquid if cooled to −164°C and to a solid if further cooled to −182°C. The fact that methane (or any other compound, for that matter) can exist as a liquid or solid depends on the existence of intermolecular forces of attraction between the particles of each pure compound. Although the forces of attraction between particles are all electrostatic in nature, they vary widely in their relative strengths. The strongest attractive forces are between ions, for example between Na^+ and Cl^- in NaCl [787 kJ (188 kcal)/mol]. Dipole-dipole interactions and hydrogen bonding [8–42 kJ (2–10 kcal)/mol] are weaker attractive forces. We shall have more to say about these intermolecular attractive forces in Chapter 9 when we discuss the physical properties of alcohols, compounds containing polar O—H groups.

Dispersion forces [0.08–8 kJ (0.02–2 kcal) mol] are the weakest intermolecular attractive forces. The existence of dispersion forces accounts for the fact that low-molecular-weight, nonpolar substances, such as hydrogen, neon, and methane, can be liquefied. To visualize the origin of dispersion forces, it is necessary to think in

Dispersion forces   Very weak intermolecular coulombic forces of attraction.

Table 2.5	Physical Properties of Some Unbranched Alkanes			
Name	Condensed Structural Formula	mp (°C)	bp (°C)	Density of Liquid (g/mL at 0°C)
Methane	CH_4	−182	−164	(a gas)
Ethane	CH_3CH_3	−183	−88	(a gas)
Propane	$CH_3CH_2CH_3$	−190	−42	(a gas)
Butane	$CH_3(CH_2)_2CH_3$	−138	0	(a gas)
Pentane	$CH_3(CH_2)_3CH_3$	−130	36	0.626
Hexane	$CH_3(CH_2)_4CH_3$	−95	69	0.659
Heptane	$CH_3(CH_2)_5CH_3$	−90	98	0.684
Octane	$CH_3(CH_2)_6CH_3$	−57	126	0.703
Nonane	$CH_3(CH_2)_7CH_3$	−51	151	0.718
Decane	$CH_3(CH_2)_8CH_3$	−30	174	0.730

terms of instantaneous distributions of electron density rather than average distributions. Consider neon, for example. Neon is a gas at room temperature and 1.00 atm. It can be liquefied when cooled to −246°C. From the heat of vaporization, it can be calculated that the neon-neon attractive interaction in the liquid state is approximately 0.3 kJ (0.07 kcal)/mol. We account for this intermolecular interaction in the following way. Over time, the distribution of electron density in a neon atom is symmetrical, and there is no dipole moment [Figure 2.21(a)]. However, at any instant, there is a probability that electron density is polarized (shifted) more toward one part of the atom than toward another. This temporary polarization creates a temporary dipole moment, which in turn induces temporary dipole moments in adjacent atoms [Figure 2.21(b)].

The strength of dispersion forces depends on how easily an electron cloud can be polarized. Electrons in smaller atoms and molecules are held closer to their nuclei and, therefore, are not easily polarized. Electrons in larger atoms and molecules are more easily polarized. For this reason, the strength of dispersion forces tends to increase with increasing molecular mass and size. Intermolecular attractive forces between Cl_2 molecules and between Br_2 molecules are estimated to be 2.9 kJ (0.7 kcal)/mol and 4.2 kJ (1.0 kcal)/mol, respectively.

Dispersion forces are inversely proportional to the sixth power of the distance between interacting particles. For them to be important, the interacting particles must be in virtual contact with one another.

Figure 2.21
Dispersion forces. (a) The average distribution of electron density in a neon atom is symmetrical, and there is no polarity. (b) Temporary polarization of one neon atom induces temporary polarization in adjacent atoms. Electrostatic attractions between temporary dipoles are called dispersion forces.

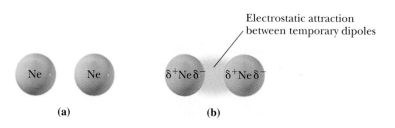

Table 2.6	**Physical Properties of the Isomeric Alkanes of Molecular Formula C_6H_{14}**		
Name	**bp (°C)**	**mp (°C)**	**Density (g/mL)**
Hexane	68.7	−95	0.659
2-Methylpentane	60.3	−154	0.653
3-Methylpentane	63.3	−118	0.664
2,3-Dimethylbutane	58.0	−129	0.661
2,2-Dimethylbutane	49.7	−98	0.649

Because interactions between alkane molecules consist only of these very weak dispersion forces, boiling points of alkanes are lower than those of almost any other type of compound of the same molecular weight. As the number of atoms and the molecular weight of alkanes increase, there is more opportunity for dispersion forces between their molecules and boiling points increase.

Melting points of alkanes also increase with increasing molecular weight. The increase, however, is not as regular as that observed for boiling points because the packing of molecules into ordered patterns of solids changes as molecular size and shape change.

The average density of the alkanes listed in Table 2.5 is about 0.7 g/mL; that of higher molecular-weight alkanes is about 0.8 g/mL. All liquid and solid alkanes are less dense than water (1.0 g/mL) and, therefore, float on it.

Alkanes that are constitutional isomers are different compounds and have different physical and chemical properties. Listed in Table 2.6 are boiling points, melting points, and densities of the five constitutional isomers of molecular formula C_6H_{14}. The boiling point of each of the branched-chain isomers of C_6H_{14} is lower than that of hexane itself, and the more branching there is, the lower the boiling point. These differences in boiling point are related to molecular shape in the following way. The only forces of attraction between alkane molecules are dispersion forces. As branching increases, the shape of an alkane molecule becomes more compact, and its surface area decreases. As surface area decreases, contact among adjacent molecules decreases, the strength of dispersion forces decreases, and boiling points also decrease. Thus, for any group of alkane constitutional isomers, it is usually observed that the least branched isomer has the highest boiling point, and the most branched isomer has the lowest boiling point.

Example 2.16

Arrange the alkanes in each set in order of increasing boiling point.

(a) Butane, decane, and hexane
(b) 2-Methylheptane, octane, and 2,2,4-trimethylpentane

Solution

(a) All compounds are unbranched alkanes. As the number of carbon atoms in the chain increases, dispersion forces between molecules increase and so do boiling points. Decane has the highest boiling point, and butane has the lowest.

$$CH_3CH_2CH_2CH_3 \qquad CH_3CH_2CH_2CH_2CH_2CH_3 \qquad CH_3(CH_2)_8CH_3$$

Butane	Hexane	Decane
bp 0°C	bp 69°C	bp 174°C

(b) These three alkanes are constitutional isomers of molecular formula C_8H_{18}. Their relative boiling points depend on the degree of branching. 2,2,4-Trimethylpentane, the most highly branched isomer, has the smallest surface area and the lowest boiling point. Octane, the unbranched isomer, has the largest surface area and the highest boiling point.

$$\begin{array}{ccc} & CH_3\ \ CH_3 & CH_3 \\ & |\ \ \ \ | & | \\ CH_3CCH_2CHCH_3 & CH_3CHCH_2CH_2CH_2CH_2CH_3 & CH_3(CH_2)_6CH_3 \\ | & & \\ CH_3 & & \end{array}$$

2,2,4-Trimethylpentane	2-Methylheptane	Octane
bp 99°C	bp 118°C	bp 126°C

Problem 2.16

Arrange the alkanes in each set in order of increasing boiling point.

(a) 2-Methylbutane, 2,2-dimethylpropane, and pentane
(b) 3,3-Dimethylheptane, 2,2,4-trimethylhexane, and nonane

Reactivity Explorer; Alkanes

2.9 Reactions of Alkanes

Alkanes and cycloalkanes are quite unreactive toward most reagents, a behavior consistent with the fact that they are nonpolar compounds and contain only strong sigma bonds. They do, however, react under certain conditions with O_2 and with the halogens Cl_2 and Br_2. At this point, we present only their combustion with oxygen. We discuss their reaction with halogens in Chapter 7.

A. Oxidation

Oxidation of alkanes by O_2 to give carbon dioxide and water is by far their most economically important reaction. Oxidation of saturated hydrocarbons is the basis for their use as energy sources for heat [natural gas, liquefied petroleum gas (LPG), and fuel oil] and power (gasoline, diesel fuel, and aviation fuel). Following are balanced equations for the complete oxidation of methane, the major component of natural gas, and propane, the major component of LPG. Also given is the **heat of combustion** for each alkane. Heats of combustion for hydrocarbons are negative; that is, their oxidation is exothermic.

Heat of combustion Standard heat of combustion is the heat released when one mole of a substance in its standard state (gas, liquid, solid) is oxidized completely to carbon dioxide and water and is given the symbol ΔH^0.

$$CH_4 \ + \ 2O_2 \longrightarrow CO_2 \ + \ 2H_2O \qquad \Delta H^0 = -890.4 \text{ kJ} \quad (-212.8 \text{ kcal})/\text{mol}$$

Methane

$$CH_3CH_2CH_3 \ + \ 5O_2 \longrightarrow 3CO_2 \ + \ 4H_2O \qquad \Delta H^0 = -2220 \text{ kJ} \quad (-530.6 \text{ kcal})/\text{mol}$$

Propane

Table 2.7   Heats of Combustion of Four Constitutional Isomers of C_8H_{18}

Hydrocarbon	Structural Formula	ΔH^0 [kJ/mol (kcal/mol)]
Octane		−5470.6 (−1307.5)
2-Methylheptane		−5465.6 (−1306.3)
2,2-Dimethylhexane		−5458.4 (−1304.6)
2,2,3,3-Tetramethylbutane		−5451.8 (−1303.0)

B.   Heats of Combustion and Relative Stability of Alkanes and Cycloalkanes

One important use of heats of combustion is to give information on the relative stabilities of isomeric hydrocarbons. To illustrate, consider the heats of combustion of the four constitutional isomers given in Table 2.7. All four compounds undergo combustion according to this equation.

$$C_8H_{18} + \frac{25}{2}\, O_2 \longrightarrow 8CO_2 + 9H_2O$$

We see that octane has the largest (most negative) heat of combustion. As branching increases, the ΔH^0 decreases (becomes less negative). Of the four isomers shown here, the isomer with four branches has the lowest (least negative) heat of combustion.

Figure 2.22 is a graphical analysis of the data in Table 2.7. Because all four compounds give the same products on oxidation, the only difference between them is

Figure 2.22

Heats of combustion of four isomeric alkanes of molecular formula C_8H_{18}.

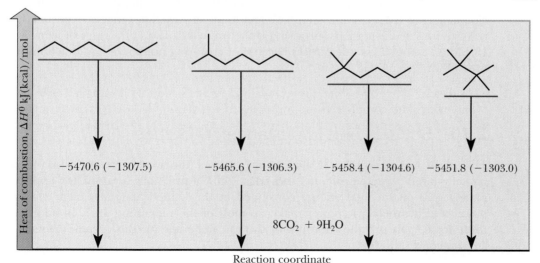

−5470.6 (−1307.5)   −5465.6 (−1306.3)   −5458.4 (−1304.6)   −5451.8 (−1303.0)

$8CO_2 + 9H_2O$

Heat of combustion, ΔH^0 kJ (kcal)/mol

Reaction coordinate

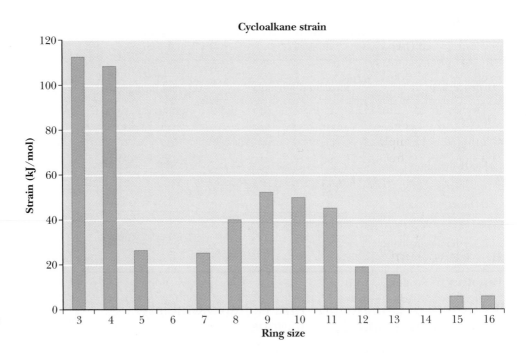

Figure 2.23
Strain energy of cycloalkanes as
a function of ring size.

their relative energies. Therefore, we conclude that branching increases the stability of an alkane.

As we saw in Section 2.4, there is considerable strain in small-ring cycloalkanes. We can measure this strain by measuring the heat of combustion versus ring size. It has been determined by measurement of the heats of combustion of a series of un-branched alkanes that the average heat of combustion per methylene (CH_2) group is 658.7 kJ (157.4 kcal)/mol. Using this value, we can calculate a predicted heat of com-bustion for each cycloalkane. These results are displayed graphically in Figure 2.23. Strain energy is the difference between the actual and predicted heats of combus-tion. We see that cyclopropane has the largest strain energy of any cycloalkane, which is consistent with the extreme compression of its C—C—C bond angles to 60°. Cy-clobutane and cyclopentane each have less strain, and cyclohexane, as expected, has zero strain. What is perhaps surprising is the presence of strain in rings of from 7 to 13 carbon atoms. This strain arises because of cross-ring nonbonded interactions be-tween hydrogen atoms.

2.10   Sources and Importance of Alkanes

The three major sources of alkanes throughout the world are the fossil fuels, namely natural gas, petroleum, and coal. These fossil fuels account for approxi-mately 90% of the total energy consumed in the United States. Nuclear electric power and hydroelectric power make up most of the remaining 10%. In addition, these fossil fuels provide the bulk of the raw materials for the organic chemicals consumed worldwide.

A. Natural Gas

Natural gas consists of approximately 90–95% methane, 5–10% ethane, and a mixture of other relatively low boiling alkanes, chiefly propane, butane, and 2-methylpropane. The current widespread use of ethylene as the organic chemical industry's most important building block is due largely to the ease with which ethane can be separated from natural gas and cracked into ethylene. **Cracking** is a process whereby a hydrocarbon is heated, usually over a catalyst, to convert it into smaller fragments. Ethane, for example, is cracked to give ethylene by heating it in a furnace at 800–900°C for a fraction of a second.

$$CH_3CH_3 \xrightarrow{800-900°C} CH_2{=}CH_2 + H_2$$

Ethane   Ethene
(Ethylene)

A petroleum refinery.
(K. Straiton/Photo Researchers, Inc.)

B. Petroleum

Petroleum is a thick, viscous liquid mixture of literally thousands of compounds, most of them hydrocarbons, formed from the decomposition of ancient marine plants and animals. Petroleum and petroleum-derived products fuel automobiles, aircraft, and trains. They provide most of the greases and lubricants required for the machinery of our highly industrialized society. Furthermore, petroleum, along with natural gas, provides close to 90% of the organic raw materials for the synthesis and manufacture of synthetic fibers, plastics, detergents, drugs, dyes, and a multitude of other products.

It is the task of a petroleum refinery to produce usable products, with a minimum of waste, from the thousands of different hydrocarbons in this liquid mixture. The various physical and chemical processes for this purpose fall into two broad categories: **separation processes,** which separate the complex mixture into various fractions, and **reforming processes,** which alter the molecular structure of the hydrocarbon components themselves.

The fundamental separation process in refining petroleum is fractional distillation (Figure 2.24). Practically all crude oil that enters a refinery goes to distillation units where it is heated gradually to temperatures as high as 370–425°C and separated into fractions. Each fraction contains a mixture of hydrocarbons that boils within a particular range. Following are the common names associated with several of these fractions along with the major uses of each.

1. Gases boiling below 20°C are taken off at the top of the distillation column. This fraction is a mixture of low-molecular-weight hydrocarbons, predominantly propane, butane, and 2-methylpropane, substances that can be liquefied under pressure at room temperature. The liquefied mixture, known as liquefied petroleum gas (LPG), can be stored and shipped in metal tanks and is a convenient gaseous fuel for home heating and cooking.
2. Naphthas, bp 20–200°C, are a mixture of C_5 to C_{12} alkanes and cycloalkanes. The naphthas also contain small amounts of benzene, toluene, xylene, and other aromatic hydrocarbons (Chapter 20). The light naphtha fraction, bp 20–150°C, is the source of straight-run gasoline and averages approximately 25% of crude petroleum. In a sense, naphthas are the most valuable distillation fractions because they are useful not only as fuel but also as sources of raw materials for the organic chemical industry.

Typical octane ratings of commonly used gasolines. *(Charles D. Winters)*

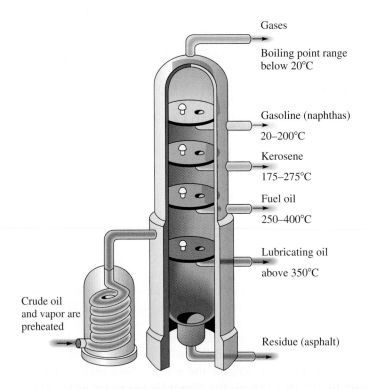

Figure 2.24

Fractional distillation of petroleum. The lighter, more volatile fractions are removed from higher up the column, and the heavier, less volatile fractions are removed from lower down.

CHEMISTRY IN ACTION

Octane Rating — What Those Numbers at the Pump Mean

Gasoline is a complex mixture of C_6 to C_{12} hydrocarbons. The quality of gasoline as a fuel for internal combustion engines is expressed by its octane rating. Engine knocking occurs when a portion of the air-fuel mixture explodes prematurely (usually as a result of heat developed during compression) and independently of ignition by the spark plug. Two compounds were selected as reference fuels. One of these, 2,2,4-trimethylpentane (isooctane), has very good antiknock properties (the air-fuel mixture burns smoothly in the combustion chamber) and was assigned an octane rating of 100. (The name "isooc-

tane" as used here is a common name.) Heptane, the other reference compound, has poor antiknock properties and was assigned an octane rating of 0.

The **octane rating** of a particular gasoline is the percent of isooctane in a mixture of isooctane and heptane that has equivalent antiknock properties. For example, the antiknock properties of 2-methylhexane are the same as those of a mixture of 42% isooctane and 58% heptane; therefore, the octane rating of 2-methylhexane is 42. Octane itself has an octane rating of -20, which means that it produces even more engine knocking than heptane.

$$CH_3(CH_2)_5CH_3$$

$$\underset{\underset{CH_3}{|}}{\overset{\overset{CH_3\ \ CH_3}{|\ \ \ \ |}}{CH_3CCH_2CHCH_3}}$$

Heptane
(octane rating 0)

2,2,4-Trimethylpentane
(octane rating 100)

3. Kerosene, bp 175–275°C, is a mixture of C_9 to C_{15} hydrocarbons. Kerosene is used as jet fuel.

4. Fuel oil, bp 250–400°C, is a mixture of C_{15} to C_{18} hydrocarbons. It is from this fraction that diesel fuel is obtained.

5. Lubricating oil and heavy fuel oil distill from the column at temperatures above 350°C.

6. Asphalt is the black, tarry residue remaining after removal of the other volatile fractions.

The two most common reforming processes are cracking, as illustrated by the thermal conversion of ethane to ethylene (Section 2.10A), and catalytic reforming. Catalytic reforming is illustrated by the conversion of hexane first to cyclohexane and then to benzene.

$$CH_3CH_2CH_2CH_2CH_2CH_3 \xrightarrow[-H_2]{\text{catalyst}} \bigcirc \xrightarrow[-3H_2]{\text{catalyst}} \bigcirc$$

Hexane   Cyclohexane   Benzene

C. Coal

To understand how coal can be used as a raw material for the production of organic compounds, it is necessary to discuss synthesis gas. **Synthesis gas** is a mixture of carbon monoxide and hydrogen in varying proportions depending on the means by which it is manufactured. Synthesis gas is prepared by passing steam over hot coal. It is also prepared by partial oxidation of methane with oxygen.

$$\underset{\text{Coal}}{C} + H_2O \xrightarrow{\text{heat}} CO + H_2$$

$$CH_4 + \tfrac{1}{2} O_2 \xrightarrow{\text{catalyst}} CO + 2H_2$$

Two important organic compounds produced today almost exclusively from carbon monoxide and hydrogen are methanol and acetic acid. In the production of methanol, the ratio of hydrogen to carbon monoxide is adjusted to $2:1$, and the mixture is passed over a catalyst at elevated temperature and pressure.

$$CO + 2H_2 \xrightarrow{\text{catalyst}} \underset{\text{Methanol}}{CH_3OH}$$

Treatment of methanol, in turn, with carbon monoxide over a different catalyst gives acetic acid.

$$\underset{\text{Methanol}}{CH_3OH} + CO \xrightarrow{\text{catalyst}} \underset{\text{Acetic acid}}{CH_3\overset{\displaystyle O}{\overset{\|}{C}}OH}$$

Because the processes for making methanol and acetic acid directly from carbon monoxide are commercially proven, it is likely that the decades ahead will see the development of routes to other organic chemicals from coal via methanol.

Summary

A **hydrocarbon** is a compound composed only of carbon and hydrogen. **Saturated hydrocarbons** (**alkanes** and **cycloalkanes**) contain only single bonds. Alkanes have the general formula C_nH_{2n+2}. **Constitutional isomers** (Section 2.2) have the same molecular formula but a different connectivity (a different order of attachment of their atoms).

Alkanes are named according to a set of rules developed by the International Union of Pure and Applied Chemistry (IUPAC) (Section 2.3A). The IUPAC system is a general system of nomenclature (Section 2.5). The IUPAC name of a compound consists of three parts: (1) a **prefix** that tells the number of carbon atoms in the parent chain, (2) an **infix** that tells the nature of the carbon-carbon bonds in the parent chain, and (3) a **suffix** that tells the class to which the compound belongs. Substituents derived from alkanes are known as **alkyl groups.**

A carbon atom is classified as **primary (1°), secondary (2°), tertiary (3°),** or **quaternary (4°)** depending on the number of alkyl groups bonded to it.

A saturated hydrocarbon that contains carbon atoms bonded to form a ring is called a **cycloalkane** (Section 2.4A). To name a cycloalkane, name and locate each substituent on the ring, and prefix the name of the open-chain alkane by cyclo-. Five-membered rings (cyclopentanes) and six-membered rings (cyclohexanes) are especially abundant in the biological world. A **bicycloalkane** (Section 2.4B) is a hydrocarbon that contains two rings that share two carbon atoms. The shared carbons are called **bridgehead carbons.**

A **conformation** is any three-dimensional arrangement of the atoms of a molecule resulting from rotations about one or more single bonds (Section 2.6). One convention for showing conformations is the **Newman projection. A dihedral angle** is the angle created by two intersecting planes. For ethane, staggered conformations occur at dihedral angles of 60°, 180°, and 300°. Eclipsed conformations occur at dihedral angles of 0°, 120°, and 240°. For butane viewed along the C_2—C_3 bond, the staggered conformation of dihedral angle 180° is called an **anti conformation;** the staggered conformations of dihedral angles of 60° and 300° are called **gauche conformations.** The anti conformation of butane is lower in energy than the gauche conformations by approximately 3.8 kJ (0.9 kcal)/mol.

Intramolecular strain (Section 2.6) is of three types: (1) **torsional strain,** which is the force that opposes the rotation of one part of a molecule about a bond while the other part is held fast; (2) **angle strain,** which arises from creation of either abnormally large or abnormally small bond angles; and (3) **nonbonded interaction strain,** which arises when atoms not bonded to each other are forced abnormally close to one another. The relationship between the change in Gibbs free energy and equilibrium constant in kilojoules (kilocalories) per mole is given by the equation $\Delta G^0 = -2.303\ RT \log K_{eq}$.

In all cycloalkanes larger than cyclopropane, nonplanar conformations are favored. The lowest energy conformation of cyclopentane is an **envelope conformation** (Section 2.6B). The lowest energy conformations of cyclohexane are two interconvertible **chair conformations** (Section 2.6B). In a chair conformation, six bonds are **axial,** and six bonds are **equatorial.** Bonds axial in one chair are equatorial in the alternative chair. **Boat** and **twist-boat conformations** are higher in energy than chair conformations. The more stable conformation of a substituted cyclohexane is the one that minimizes diaxial interactions.

Cis,trans isomers (Section 2.7A) have the same molecular formula and the same order of attachment of atoms, but arrangements of their atoms in space that cannot be interconverted by rotation about single bonds. **Cis** means that substituents are on the same side of the ring; **trans** means that they are on opposite sides of the ring. Most cycloalkanes with substituents on two or more carbons show cis,trans isomerism.

Low-molecular-weight alkanes are gases at room temperature and atmospheric pressure. Higher molecular-weight alkanes are liquids. Very-high-molecular-weight alkanes are solids.

Alkanes are nonpolar compounds, and the only forces of attraction between their molecules are **dispersion forces** (Section 2.8), weak electrostatic interactions between temporary induced dipoles of adjacent atoms or molecules. Among a set of alkane constitutional isomers, the least branched isomer generally has the highest boiling point; the most branched isomer generally has the lowest boiling point.

As determined by **heats of combustion,** strain in cycloalkanes varies with ring size (Section 2.9). Cyclohexane, which has the most common ring size among organic compounds, is strain free.

Natural gas (Section 2.10A) consists of 90–95% methane with lesser amounts of ethane and other low-molecular-weight hydrocarbons. **Petroleum** (Section 2.10B) is a liquid mixture of literally thousands of different hydrocarbons. The most important processes in petroleum refining are fractional distillation, catalytic cracking, and catalytic reforming. **Synthesis gas** (Section 2.10C), a mixture of carbon monoxide and hydrogen, can be derived from natural gas, coal, or petroleum.

Key Reaction

1. Oxidation of Alkanes (Section 2.10)

Oxidation of alkanes to carbon dioxide and water is the basis for their use as energy sources of heat and power.

$$CH_3CH_2CH_3 + 5O_2 \longrightarrow 3CO_2 + 4H_2O \qquad \Delta H^0 = -2220 \text{ kJ } (-530.6 \text{ kcal})/\text{mol}$$

Problems

Constitutional Isomerism

2.17 Which statements are true about constitutional isomers?

    (a) They have the same molecular formula.
    (b) They have the same molecular weight.
    (c) They have the same order of attachment of atoms.
    (d) They have the same physical properties.

2.18 Which structural formulas represent identical compounds and which represent constitutional isomers?

2.19 Indicate whether the compounds in each set are constitutional isomers.

2.20 Draw structural formulas, and write IUPAC names for the nine constitutional isomers of molecular formula C_7H_{16}.

2.21 Draw structural formulas for all the following.

    (a) Alcohols of molecular formula $C_4H_{10}O$
    (b) Aldehydes of molecular formula C_4H_8O
    (c) Ketones of molecular formula $C_5H_{10}O$
    (d) Carboxylic acids of molecular formula $C_5H_{10}O_2$

Nomenclature of Alkanes and Cycloalkanes

2.22 Write IUPAC names for these alkanes and cycloalkanes. Name substituents both by IUPAC names and common names (if common names exist).

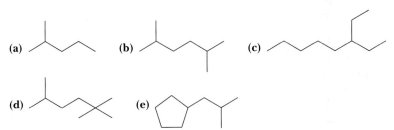

(a)   **(b)**   **(c)**

(d)   **(e)**

2.23 Write structural formulas for these alkanes and cycloalkanes.

  (a) 2,2,4-Trimethylhexane   **(b)** 2,2-Dimethylpropane
  (c) 3-Ethyl-2,4,5-trimethyloctane   **(d)** 5-Butyl-2,2-dimethylnonane
  (e) 4-(1-Methylethyl)octane   **(f)** 3,3-Dimethylpentane
  (g) *trans*-1,3-Dimethylcyclopentane **(h)** *cis*-1,2-Diethylcyclobutane

2.24 Explain why each is an incorrect IUPAC name. Write the correct IUPAC name for the intended compound.

  (a) 1,3-Dimethylbutane   **(b)** 4-Methylpentane
  (c) 2,2-Diethylbutane   **(d)** 2-Ethyl-3-methylpentane
  (e) 2-Propylpentane   **(f)** 2,2-Diethylheptane
  (g) 2,2-Dimethylcyclopropane **(h)** 1-Ethyl-5-methylcyclohexane

The IUPAC System of Nomenclature

2.25 For each IUPAC name, draw the corresponding structural formula.

  (a) Ethanol   **(b)** Butanal   **(c)** Butanoic acid
  (d) Ethanoic acid   **(e)** Heptanoic acid **(f)** Propanoic acid
  (g) Octanal   **(h)** Cyclopentene   **(i)** Cyclopentanol
  (j) Cyclopentanone   **(k)** Cyclohexanol   **(l)** Propanone

2.26 Write the IUPAC name for each compound.

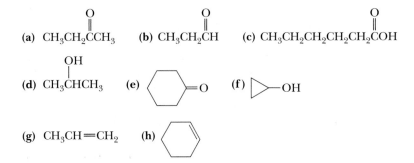

(a) $CH_3CH_2CCH_3$ **(b)** CH_3CH_2CH **(c)** $CH_3CH_2CH_2CH_2CH_2COH$

(d) CH_3CHCH_3 with OH   **(e)** (cyclohexanone) $=O$ **(f)** (cyclopropane)$-OH$

(g) $CH_3CH{=}CH_2$ **(h)** (cyclohexene)

Conformations of Alkanes and Cycloalkanes

2.27 Torsional strain resulting from eclipsed C—H bonds is approximately 4.2 kJ (1.0 kcal)/mol and that for eclipsed C—H and C—CH$_3$ bonds is approximately 6.3 kJ (1.5 kcal)/mol. Given this information, sketch a graph of energy versus dihedral angle for propane.

2.28 How many different staggered conformations are there for 2-methylpropane? How many different eclipsed conformations are there?

2.29 Consider 1-bromopropane, $CH_3CH_2CH_2Br$.

(a) Draw a Newman projection for the conformation in which —CH_3 and —Br are anti (dihedral angle 180°).

(b) Draw Newman projections for the conformations in which —CH_3 and —Br are gauche (dihedral angles 60° and 300°).

(c) Which of these is the lowest energy conformation?

(d) Which of these conformations, if any, are related by reflection?

2.30 Consider 1-bromo-2-methylpropane and draw the following.

(a) The staggered conformation(s) of lowest energy

(b) The staggered conformation(s) of highest energy

2.31 In cyclohexane, an equatorial substituent is equidistant from the axial and the equatorial groups on an adjacent carbon. Show that this is so by examining a molecular model on the CD and measuring these distances.

2.32 *Trans*-1,4-di-*tert*-butylcyclohexane exists in a normal chair conformation. *Cis*-1,4-di-*tert*-butylcyclohexane, however, adopts a twist-boat conformation. Draw both isomers and explain why the cis isomer is more stable in a twist-boat conformation.

2.33 From studies of the dipole moment of 1,2-dichloroethane in the gas phase at room temperature, it is estimated that the ratio of molecules in the anti conformation to gauche conformation is 7.6:1. Calculate the difference in Gibbs free energy between these two conformations.

Cis,Trans Isomerism in Cycloalkanes

2.34 Draw structural formulas for the cis and trans isomers of 1,2-dimethylcyclopropane.

2.35 Name and draw structural formulas for all cycloalkanes of molecular formula C_5H_{10}. Be certain to include cis and trans isomers as well as constitutional isomers.

2.36 Using a planar pentagon representation for the cyclopentane ring, draw structural formulas for the cis and trans isomers of the following.

(a) 1,2-Dimethylcyclopentane   (b) 1,3-Dimethylcyclopentane

2.37 Gibbs free energy differences between axial-substituted and equatorial-substituted chair conformations of cyclohexane were given in Table 2.4.

(a) Calculate the ratio of equatorial to axial *tert*-butylcyclohexane at 25°C.

(b) Explain, by examining the molecular models on the CD, why the conformational equilibria for methyl, ethyl, and isopropyl substituents are comparable but the conformational equilibrium for *tert*-butylcyclohexane lies considerably farther toward the equatorial conformation.

2.38 When cyclohexane is substituted by an ethynyl group, —C≡CH, the energy difference between axial and equatorial conformations is only 1.7 kJ (0.41 kcal)/mol. Compare the conformational equilibrium for methylcyclohexane with that for ethynylcyclohexane and account for the difference between the two.

2.39 Draw the alternative chair conformations for the cis and trans isomers of 1,2-dimethyl-cyclohexane, 1,3-dimethylcyclohexane, and 1,4-dimethylcyclohexane.

(a) Indicate by a label whether each methyl group is axial or equatorial.

(b) For which isomer(s) are the alternative chair conformations of equal stability?

(c) For which isomer(s) is one chair conformation more stable than the other?

2.40 Use your answers from Problem 2.39 to complete the table showing correlations between cis and trans and between axial and equatorial for disubstituted derivatives of cyclohexane.

Position of Substitution	Cis	Trans
1,4-	a,e or e,a	e,e or a,a
1,3-	____ or ____	____ or ____
1,2-	____ or ____	____ or ____

2.41 Calculate the difference in Gibbs free energy in kilojoules per mole between the alternative chair conformations of the following.

(a) *trans*-4-Methylcyclohexanol   (b) *cis*-4-Methylcyclohexanol
(c) *trans*-1,4-Dicyanocyclohexane

2.42 There are four cis,trans isomers of 2-isopropyl-5-methylcyclohexanol.

(a) Using a planar hexagon representation for the cyclohexane ring, draw structural formulas for the four cis,trans isomers.
(b) Draw the more stable chair conformation for each of your answers in part (a).
(c) Of the four cis,trans isomers, which is the most stable? (If you answered this part correctly, you picked the isomer found in nature and given the name menthol.)

2.43 Draw alternative chair conformations for each substituted cyclohexane and state which chair is more stable.

(a) (b) (c) (d)

2.44 Glucose (Section 25.1) contains a six-membered ring. In the more stable chair conformation of this molecule, all substituents on the ring are equatorial. Draw this more stable chair conformation.

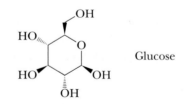

Glucose

2.45 1,2,3,4,5,6-Hexachlorocyclohexane shows cis,trans isomerism. At one time a crude mixture of these isomers was sold as an insecticide. The insecticidal properties of the mixture arise from one isomer, known as lindane, which is *cis*-1,2,4,5-*trans*-3,6-hexachlorocyclohexane (see *The Merck Index*, 12th ed., #5526).

(a) Draw a structural formula for 1,2,3,4,5,6-hexachlorocyclohexane disregarding, for the moment, the existence of cis,trans isomerism. What is the molecular formula of this compound?
(b) Using a planar hexagon representation for the cyclohexane ring, draw a structural formula for lindane.

(c) Draw a chair conformation for lindane, and label which chlorine atoms are axial and which are equatorial.

(d) Draw the alternative chair conformation of lindane, and again label which chlorine atoms are axial and which are equatorial.

(e) Which of the alternative chair conformations of lindane is more stable? Explain.

Physical Properties

2.46 In Problem 2.20, you drew structural formulas for all isomeric alkanes of molecular formula C_7H_{16}. Predict which isomer has the lowest boiling point and which has the highest boiling point.

2.47 What generalization can you make about the densities of alkanes relative to the density of water?

2.48 What unbranched alkane has about the same boiling point as water? (Refer to Table 2.5 on the physical properties of alkanes.) Calculate the molecular weight of this alkane, and compare it with that of water.

Reactions of Alkanes

2.49 Complete and balance the following combustion reactions. Assume that each hydrocarbon is converted completely to carbon dioxide and water.

(a) Propane + $O_2 \rightarrow$ (b) Octane + $O_2 \rightarrow$
(c) Cyclohexane + $O_2 \rightarrow$ (d) 2-Methylpentane + $O_2 \rightarrow$

2.50 Following are heats of combustion per mole for methane, propane, and 2,2,4-trimethylpentane. Each is a major source of energy. On a gram-for-gram basis, which of these hydrocarbons is the best source of heat energy?

Hydrocarbon	Component of	ΔH^0 [kj (kcal)/mol]
CH_4	Natural gas	-891 (-213)
$CH_3CH_2CH_3$	LPG	-2221 (-531)
CH₃CCH₂CHCH₃ with CH₃ groups	Gasoline	-5452 (-1304)

2.51 The structural formulas and heats of combustion of acetaldehyde and ethylene oxide are given here. Which of these compounds is the more stable? Explain.

$$CH_3{-}\overset{\overset{O}{\|}}{C}H \qquad\qquad H_2\overset{\overset{O}{/\backslash}}{C}{-}CH_2$$

Acetaldehyde   Ethylene oxide
-1164 kJ (-278.8 kcal)/mol -1264 kJ (-302.1 kcal)/mol

2.52 Without consulting tables, arrange these compounds in order of decreasing (less negative) heat of combustion: hexane, 2-methylpentane, 2,2-dimethylbutane.

2.53 Which would you predict to have the larger (more negative) heat of combustion, *cis*-1,4-dimethylcyclohexane or *trans*-1,4-dimethylcyclohexane?

Molecular Modeling

2.54 Using the models on the CD, measure the distance between hydrogens on adjacent carbon atoms in the staggered and eclipsed conformations of ethane and estimate the ratio of eclipsed/staggered distance. To measure distance, click on one atom (it will become darkened), and then move the pointer to any other atom.

2.55 Measure all C—C—C bond angles in the envelope conformation of cyclopentane on the CD, and compare them with the value of 105° given in the text. How do you account for the difference in these values?

2.56 As you see from Figure 2.23, the strain energies in cyclopentane and cycloheptane are approximately equal, but there is zero strain in cyclohexane. Examine the model of cycloheptane on the CD, and see if you can determine why the cycloheptane ring is strained. Check for angle strain, and close contacts that result in nonbonded interaction strain.

2.57 Examine the model of a chair cyclohexane on the CD. Rotate it so that you view the chair from above, from the side, from the footpiece to the headpiece, and so on. As you do these rotations, convince yourself that the six axial C—H bonds are parallel and that they alternate up, down, and so on. Also convince yourself that there are six equatorial C—H bonds and that those on opposite carbons (1 and 4, 2 and 5, and 3 and 6) are parallel and trans to each other.

2.58 Examine the molecular model of axial-methylcyclohexane on the CD, and measure the distance between the methyl group and the ring hydrogens on carbons 2, 3, 4, 5, and 6. You should find that the axial methyl group is closer to axial hydrogens on carbons 3 and 5 than to any other ring hydrogens.

2.59 What kinds of conformations do the six-membered rings in adamantane and twistane exhibit? In answering this question, you will find it helpful to examine the models of adamantane and twistane on the CD.

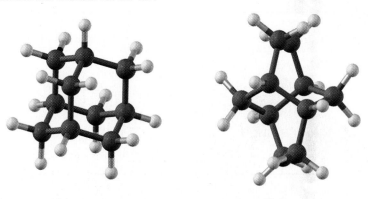

Adamantane   Twistane

2.60 On the CD are models of these three small-ring compounds, and the strain energy of each. Which of the three is the least stable? Which is the most stable? What explanation can you offer for their relative stabilities?

    (a) [1.1.1]Propellane   **(b)** [2.1.1]Propellane   **(c)** [2.2.1]Propellane

2.61 Chem3D provides an easy way to invert the configuration at a tetrahedral carbon atom. Following the directions on the CD, convert

    (a) Equatorial-methylcyclohexane to axial-methylcyclohexane

    (b) *cis*-1,2-Dimethylcyclopentane to *trans*-1,2-dimethylcyclopentane

CHIRALITY

Our goal in this chapter is to expand our awareness of molecules as three-dimensional objects. In particular, we explore the relationships between a three-dimensional object and its mirror image. A **mirror image** is the reflection of an object in a mirror. When you look in a mirror, you see a reflection, or mirror image, of yourself. Now suppose your mirror image became a three-dimensional object. We could then ask, "What is the relationship between

■ Tartaric acid (Section 3.5) is found in grapes and other fruits, both free and as its salts. *(Jerry Alexander/Tony Stone)* Inset: A model of the *R,R* enantiomer of tartaric acid.

101

you and your mirror image?" To clarify what we mean by "relationship," we might instead ask, "Can your reflection be superposed on (placed on top of) the original 'you' in such a way that every detail of the reflection corresponds exactly to the original?" The answer is that you and your mirror image are not superposable. If you have a ring on the little finger of your right hand, for example, your mirror image has the ring on the little finger of its left hand. If you part your hair on your right side, it will be parted on the left side in your reflection. Simply stated, you and your reflection are different objects. You cannot superpose one on the other.

An understanding of spatial relationships of this type is fundamental to an understanding of organic chemistry and biochemistry. In fact, the ability to deal with molecules as three-dimensional objects is a survival skill in organic chemistry and biochemistry.

Supporting Concepts; Stereo Chemistry (contains an interactive flow chart that is used to decide what types of isomers two structures represent)

3.1   Stereoisomerism

Isomers are different compounds with the same molecular formula. Thus far, we have encountered two types of isomers. Constitutional isomers (Section 2.2) have the same molecular formula but a different order of attachment of atoms in their molecules. Examples of pairs of constitutional isomers are pentane and 2-methylbutane, and 1-pentene and cyclopentane.

constitutional isomers:

$$CH_3CH_2CH_2CH_2CH_3 \quad \text{and} \quad \overset{\overset{\displaystyle CH_3}{|}}{CH_3CHCH_2CH_3} \qquad CH_2{=}CHCH_2CH_2CH_3 \quad \text{and}$$

| Pentane | 2-Methylbutane | 1-Pentene | Cyclopentane |
| (C_5H_{12}) | (C_5H_{12}) | (C_5H_{10}) | (C_5H_{10}) |

Stereoisomers   Isomers that have the same molecular formula and the same connectivity but a different orientation of their atoms in space.

A second type of isomerism is stereoisomerism. **Stereoisomers** have the same molecular formula and the same connectivity but different orientations of their atoms in space. The one example of stereoisomerism we have seen thus far is that of cis,trans isomers in cycloalkanes (Section 2.7), which arise because substituents on a ring are locked into an orientation in space with respect to one another by the ring.

stereoisomers:

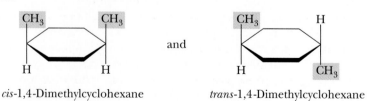

and

cis-1,4-Dimethylcyclohexane   *trans*-1,4-Dimethylcyclohexane

In this chapter, we study another type of stereoisomerism, namely enantiomers and diastereomers (Figure 3.1), and explain the similarities and differences among them.

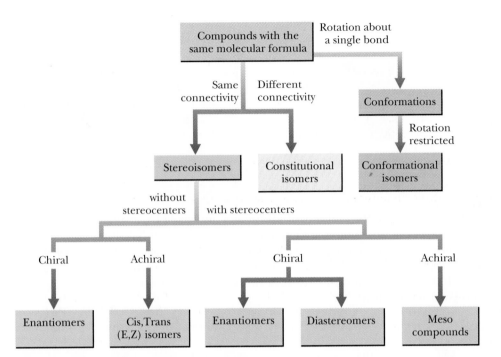

Figure 3.1
Relationships among isomers.

3.2 Chirality

Molecules that are not superposable on their mirror images are said to be **chiral** (pronounced ki-ral, to rhyme with spiral; from the Greek: *cheir,* hand). That is, they show handedness. Chirality is encountered in three-dimensional objects of all sorts. Your left hand is chiral and so is your right hand. A spiral binding on a notebook is chiral. A machine screw with a right-handed thread is chiral. A ship's propeller is chiral. As you examine the objects in the world around you, you will undoubtedly conclude that the vast majority of them are chiral as well.

The contrasting situation to chirality occurs when an object and its mirror image are superposable. An object and its mirror image are superposable if one of them can be oriented in space so that all its features (corners, edges, points, designs, etc.) correspond exactly to those in the other member of the pair. If this can be done, the object and its mirror image are identical; the original object is achiral. An **achiral** object is one that lacks chirality. Examples of objects lacking chirality are an undecorated cup, a regular tetrahedron, a cube, and a perfect sphere.

An object is achiral if it possesses any of certain elements of symmetry, the most important of which in organic compounds are a plane and center of symmetry. As we shall see, any molecule possessing either of these symmetry elements (and one other rarely encountered one) is achiral and can be superposed on its mirror image. A **plane of symmetry** is an imaginary plane passing through an object dividing it such that one half is the reflection of the other half. The cube shown in Figure 3.2 has several planes of symmetry. Both the beaker and the compound bromochloromethane have a single plane of symmetry. A **center of symmetry** is a point so situated that identical components of the object are located equidistant and on opposite sides from the point along any axis passing through that point. The cube shown in Figure 3.2 has a center of symmetry.

The horns of this African gazelle show chirality and are mirror images of each other.
(William H. Brown)

Chiral From the Greek, *cheir,* hand; an object that is not superposable on its mirror image; an object that has handedness.
Achiral An object that lacks chirality; an object that has no handedness.
Plane of symmetry An imaginary plane passing through an object dividing it so that one half is the mirror image of the other half.
Center of symmetry A point so situated that identical components of an object are located on opposite sides and equidistant from that point along any axis passing through it.

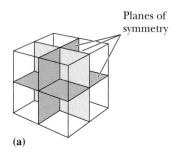

Planes of symmetry

(a)

Plane of symmetry

(b)

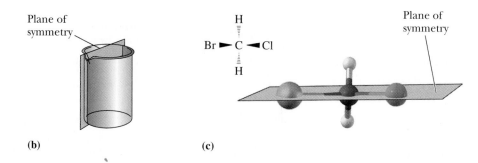

Plane of symmetry

(c)

Figure 3.2

Symmetry in objects. A cube has several planes of symmetry and a center of symmetry. The beaker and CH_2BrCl each have a single plane of symmetry.

We can illustrate the chirality of an organic molecule by considering 2-hydroxypropanoic acid, more commonly named lactic acid. Figure 3.3 shows three-dimensional representations and ball-and-stick models for lactic acid and its mirror image. In these representations, all bond angles about the central carbon atom are approximately 109.5°, and the four bonds from this carbon are directed toward the corners of a regular tetrahedron.

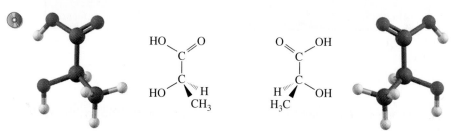

Figure 3.3

Stereorepresentations of lactic acid and its mirror image.

A model of lactic acid can be turned and rotated in any direction in space, but as long as bonds are not broken and rearranged, only two of the four groups attached to the central carbon of one molecule can be made to coincide with those of its mirror image. Because lactic acid and its mirror image are nonsuperposable, they are classified as enantiomers. **Enantiomers** are nonsuperposable mirror images. Note that the terms "chiral" and "achiral" refer to objects; the term "enantiomers" refers to the relationship between objects.

Enantiomers Stereoisomers that are nonsuperposable mirror images of each other; refers to a relationship between pairs of objects.

Stereocenter An atom that has four different atoms or groups of atoms attached to it; also called a stereogenic center.

The most common cause of chirality in organic molecules is a tetrahedral atom, most commonly carbon, bonded to four different groups. A carbon atom with four different groups bonded to it is called a **stereocenter,** or alternatively, a **stereogenic center.** The carbon atom of lactic acid bearing the —OH, —H, —CH_3, and —COOH groups is a stereocenter.

Example 3.1

Each molecule has one stereocenter. Draw stereorepresentations for the enantiomers of each.

(a) $CH_3\overset{\overset{\displaystyle Cl}{|}}{C}HCH_2CH_3$ **(b)**   [structure of cyclohexene with Cl]

Solution

You will find it helpful to view models on the CD of enantiomer pairs from different perspectives as is done in these representations. As you work with these pairs of enantiomers, notice that each has a carbon atom bonded to four different groups, which makes the molecule chiral.

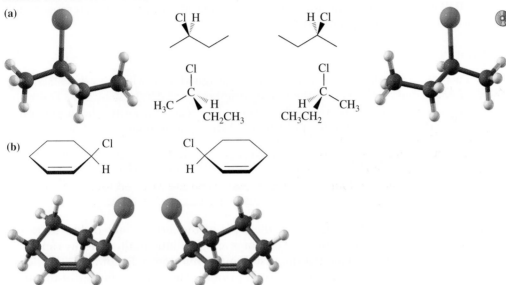

Problem 3.1

Each molecule has one stereocenter. Draw stereorepresentations for the enantiomers of each.

(a) cyclopentyl—CHCH$_3$ with OH   (b) CH$_3$CHCHCH$_3$ with OH and CH$_3$

In all the molecules studied so far, chirality arises because of the presence of a carbon stereocenter. Stereocenters are not limited to carbon. Following are stereorepresentations of a chiral cation in which the stereocenter is nitrogen. We discuss the chirality of nitrogen stereocenters in more detail in Chapter 22.

H$_3$C   CH$_2$CH$_3$ CH$_3$CH$_2$ CH$_3$

A pair of enantiomers

Enantiomers of tetrahedral silicon, phosphorus, and germanium compounds have also been isolated.

3.3 Naming Stereocenters — The *R,S* System

A system for designating the configuration of a stereocenter was devised in the late 1950s by R. S. Cahn and C. K. Ingold in England and V. Prelog in Switzerland and is named after them. The system, also called the **R,S system,** has been incorporated into the IUPAC rules of nomenclature. The orientation of groups about a stereocenter is specified using a set of priority rules.

Priority Rules

1. Each atom bonded to the stereocenter is assigned a priority. Priority is based on atomic number; the higher the atomic number, the higher the priority. Following are several substituents arranged in order of increasing priority. The atomic number of the atom determining priority is shown in parentheses.

$$\underset{-H,}{(1)} \quad \underset{-CH_3,}{(6)} \quad \underset{-NH_2,}{(7)} \quad \underset{-OH,}{(8)} \quad \underset{-SH,}{(16)} \quad \underset{-Cl,}{(17)} \quad \underset{-Br,}{(35)} \quad \underset{-I}{(53)}$$

Increasing priority ⟶

2. If priority cannot be assigned on the basis of the atoms bonded directly to the stereocenter, look at the next set of atoms and continue until a priority can be assigned. Priority is assigned at the first point of difference. Following is a series of groups arranged in order of increasing priority. The atomic number of the atom on which the assignment of priority is based is shown above it.

$$\underset{-CH_2-H}{(1)} \quad \underset{-CH_2-CH_3}{(6)} \quad \underset{-CH_2-NH_2}{(7)} \quad \underset{-CH_2-OH}{(8)} \quad \underset{-CH_2-Cl}{(17)}$$

Increasing priority ⟶

If two carbons have substituents of the same priority, priority is assigned to the carbon that has more of these substituents. Thus, $-CHCl_2 > -CH_2Cl$.

3. Atoms participating in a double or triple bond are considered to be bonded to an equivalent number of similar atoms by single bonds; that is, atoms of the double bond are duplicated and atoms of a triple bond are triplicated.

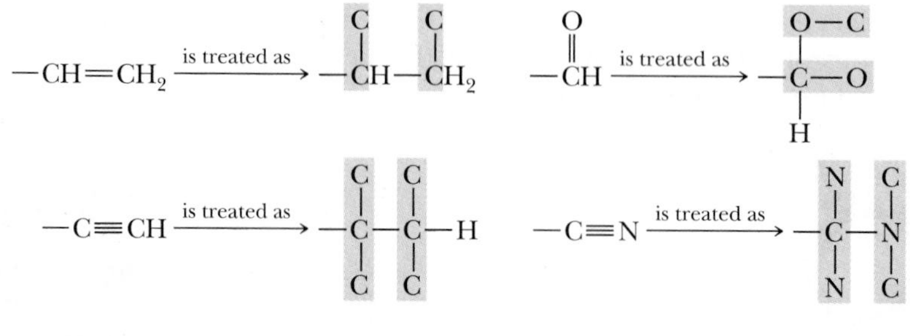

Example 3.2

Assign priorities to the groups in each set.

(a) $-\overset{O}{\overset{\|}{C}}OH$ and $-\overset{O}{\overset{\|}{C}}H$ **(b)** $-CH=CH_2$ and $-CH(CH_3)_2$

Solution

(a) The first point of difference is O of the —OH in the carboxyl group compared to —H in the aldehyde group. The carboxyl group is higher in priority.

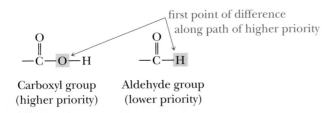

first point of difference
along path of higher priority

Carboxyl group
(higher priority)

Aldehyde group
(lower priority)

(b) Carbon 1 in each group has the same pattern of atoms, namely C(C,C,H). For the vinyl group, bonding at carbon 2 is C(C,H,H). For the isopropyl group, at carbon 2 it is C(H,H,H). The vinyl group is higher in priority than is the isopropyl group.

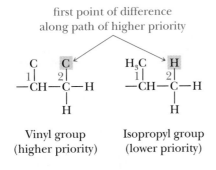

first point of difference
along path of higher priority

Vinyl group
(higher priority)

Isopropyl group
(lower priority)

Problem 3.2

Assign priorities to the groups in each set.

(a) —CH_2OH and   —CH_2CH_2OH **(b)** —CH_2OH and   —CH=CH_2

(c) —CH_2OH and   —$C(CH_3)_3$

To assign *R* or *S* configuration to a stereocenter:

1. Locate the stereocenter, identify its four substituents, and assign a priority from 1 (highest) to 4 (lowest) to each substituent.
2. Orient the molecule in space so that the group of lowest priority (4) is directed away from you as would be, for instance, the steering column of a car. The three groups of higher priority (1–3) then project toward you, as would be the spokes of the steering wheel.
3. Read the three groups projecting toward you in order from highest priority (1) to lowest priority (3).
4. If the groups are read in a clockwise direction, the configuration is designated as **R** (Latin: *rectus,* straight, correct); if they are read in a counterclockwise direction, the configuration is **S** (Latin: *sinister,* left). You can also visualize this as follows: turning the steering wheel to the right (down the order of priority) equals *R;* turning it to the left equals *S*.

R From the Latin, *rectus,* straight, correct; used in the *R,S* convention to show that the order of priority of groups on a stereocenter is clockwise.

S From the Latin, *sinister,* left; used in the *R,S* convention to show that the order of priority of groups on a stereocenter is counterclockwise.

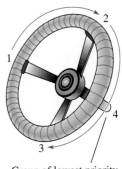

Group of lowest priority points away from you

Example 3.3

Assign an *R* or *S* configuration to the stereocenter in each molecule.

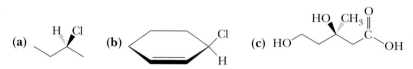

(a)   (b)   (c)

Solution

View each molecule through the stereocenter along the bond from the stereocenter toward the group of lowest priority. In (a), the order of priority is Cl > CH_2CH_3 > CH_3 > H; the configuration is *S*. In (b), the order of priority is Cl > CH=CH > CH_2 > H; the configuration is *R*. In (c), the order of priority is OH > CH_2COOH > CH_2CH_2OH > CH_3; the configuration is *R*.

(a)   (b)

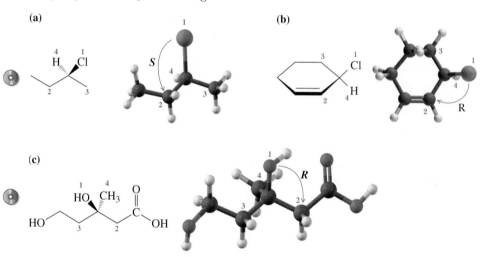

(c)

Problem 3.3

Assign an *R* or *S* configuration to the stereocenter in each molecule.

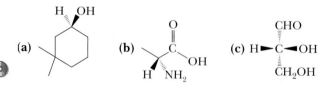

(a)   (b)   (c) H—C—OH

3.4   Acyclic Molecules with Two or More Stereocenters

We have now seen several examples of molecules with one stereocenter and verified that, for each, two stereoisomers (one pair of enantiomers) are possible. Now let us consider molecules with two or more stereocenters. To generalize, for a molecule with *n* stereocenters, the maximum number of stereoisomers possible is 2^n. We have already seen that, for a molecule with one stereocenter, $2^1 = 2$ stereoisomers are possible. For a molecule with two stereocenters, $2^2 = 4$ stereoisomers are possible; for a molecule with three stereocenters, $2^3 = 8$ stereoisomers are possible, and so forth.

A.   Enantiomers and Diastereomers

Let us begin our study of molecules with multiple stereocenters by considering 2,3,4-trihydroxybutanal, a molecule with two stereocenters, shown here in color.

$$HOCH_2-\underset{\overset{|}{OH}}{CH}-\underset{\overset{|}{OH}}{CH}-CHO$$

2,3,4-Trihydroxybutanal

The maximum number of stereoisomers possible for this molecule is $2^2 = 4$, each of which is drawn in Figure 3.4. One of these pairs is called erythrose, and the other, threose.

   Stereoisomers (a) and (b) are nonsuperposable mirror images of each other and are, therefore, a pair of enantiomers. Stereoisomers (c) and (d) are also non-superposable mirror images of each other and are a second pair of enantiomers. One way to describe the four stereoisomers of 2,3,4-trihydroxybutanal is to say that they consist of two pairs of enantiomers. Enantiomers (a) and (b) of 2,3,4-trihydroxybutanal are given the names (2R,3R)-erythrose and (2S,3S)-erythrose; enantiomers (c) and (d) are given the names (2R,3S)-threose and (2S,3R)-threose. Erythrose and threose belong to the class of compounds called carbohydrates, which we discuss in Chapter 25. Erythrose is found in erythrocytes (red blood cells), hence the derivation of its name.

   We have specified the relationship between (a) and (b) and between (c) and (d); each represents a pair of enantiomers. What is the relationship between (a) and (c), between (a) and (d), between (b) and (c), and between (b) and (d)? The answer is that they are called **diastereomers.** As we see in this example, diastereomers require the presence of at least two stereocenters.

Diastereomers   Stereoisomers that are not mirror images of each other; refers to relationships among two or more objects.

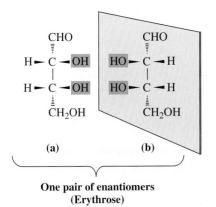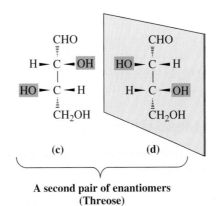

| | | | |
| (a) | (b) | (c) | (d) |

One pair of enantiomers
(Erythrose)

A second pair of enantiomers
(Threose)

Figure 3.4
The four stereoisomers of 2,3,4-trihydroxybutanal, a compound with two stereocenters.

Example 3.4

Following are stereorepresentations for the four stereoisomers of 1,2,3-butanetriol. *R* and *S* configurations are given for the stereocenters in (1) and (4).

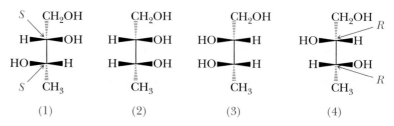

(1) (2) (3) (4)

(a) Write IUPAC names for each molecule showing the *R* or *S* configuration of each stereocenter.
(b) Which compounds are enantiomers?
(c) Which compounds are diastereomers?

Solution

(a) (1) (2*S*,3*S*)-1,2,3-Butanetriol   (2) (2*S*,3*R*)-1,2,3-Butanetriol
  (3) (2*R*,3*S*)-1,2,3-Butanetriol   (4) (2*R*,3*R*)-1,2,3-Butanetriol
(b) Enantiomers are stereoisomers that are nonsuperposable mirror images of each other. As you see from their configurations, compounds (1) and (4) are one pair of enantiomers, and compounds (2) and (3) are a second pair of enantiomers.
(c) Diastereomers are stereoisomers that are not mirror images of each other. Compounds (1) and (2), (1) and (3), (2) and (4), and (3) and (4) are pairs of diastereomers.

Problem 3.4

Following are stereorepresentations for the four stereoisomers of 3-chloro-2-butanol.

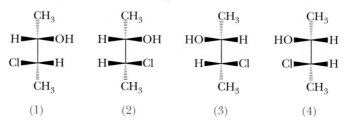

(1) (2) (3) (4)

(a) Assign an *R* or *S* configuration to each stereocenter.
(b) Which compounds are enantiomers?
(c) Which compounds are diastereomers?

B. Meso Compounds

Certain molecules containing two or more stereocenters have special symmetry properties that reduce the number of stereoisomers to fewer than the maximum number that is predicted by the 2^n rule. One such molecule is 2,3-dihydroxybutanedioic acid, more commonly named tartaric acid.

$$HOC—CH—CH—COH$$

2,3-Dihydroxybutanedioic acid
(Tartaric acid)

Tartaric acid is a colorless, crystalline compound. During fermentation of grape juice, potassium bitartrate (one carboxyl group is present as a potassium salt, $—COO^- K^+$) deposits as a crust on the sides of wine casks. When collected and purified, it is sold commercially as cream of tartar (see *The Merck Index,* 12th ed., #7776).

In tartaric acid, carbons 2 and 3 are stereocenters, and, using the 2^n rule, the maximum number of stereoisomers possible is $2^2 = 4$; these stereorepresentations are drawn in Figure 3.5. Structures (a) and (b) are nonsuperposable mirror images and are, therefore, a pair of enantiomers. Structures (c) and (d) are also mirror images, but they are superposable. To see this, imagine that you first rotate (d) by 180° in the plane of the paper, lift it out of the plane of the paper, and place it on top of (c). If you do this mental manipulation correctly, you find that (d) is superposable on (c). Therefore, (c) and (d) are not different molecules; they are the same molecule, just oriented differently. Because (c) and its mirror image are superposable, (c) is achiral.

Another way to determine that (c) is achiral is to see that it has a plane of symmetry that bisects the molecule in such a way that the top half is the reflection of the bottom half. Thus, even though (c) has two stereocenters, it is achiral (Section 3.2).

The stereoisomer of tartaric acid represented by (c) or (d) is called a meso compound. A **meso compound** is an achiral compound that contains two or more stereocenters. We can now answer the question: How many stereoisomers are there of tartaric acid? The answer is three: one meso compound and one pair of enantiomers. Note that the meso compound is a diastereomer of each member of the pair of enantiomers.

From this example, we can make this generalization about meso compounds: they only occur in compounds with two identical stereocenters, and then only if one of the stereocenters is *R* and the other is *S*.

You will notice that the stereoisomers in Figure 3.5 are shown in only one conformation. Because conformational isomers in acyclic systems interconvert extremely rapidly, we need consider only the most symmetric conformers. Other conformers may be rotated to give the most symmetric one for determination of symmetry properties (see Problem 3.5).

Meso compound   An achiral compound possessing two or more stereocenters.

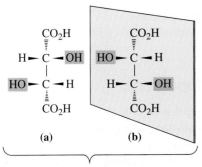

(a)   (b)

A pair of enantiomers

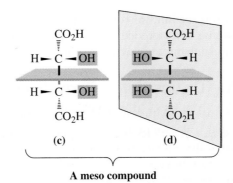

(c)   (d)

A meso compound

Figure 3.5
Stereoisomers of tartaric acid. One pair of enantiomers and one meso compound.

Example 3.5

Following are stereorepresentations for the three stereoisomers of 2,3-butanediol. The carbons are numbered beginning from the left, as shown in 1.

(1)   (2)   (3)

(a) Assign an R or S configuration to each stereocenter.
(b) Which are enantiomers?   (c) Which is the meso compound?
(d) Which are diastereomers?

Solution

(a) (1) (2R,3R)-2,3-Butanediol   (2) (2R,3S)-2,3-Butanediol
    (3) (2S,3S)-Butanediol
(b) Compounds (1) and (3) are enantiomers.
(c) Compound (2) is a meso compound.
(d) (1) and (2) are diastereomers; (2) and (3) are also diastereomers.

Problem 3.5

Following are four Newman projection formulas for tartaric acid.

(1)   (2)   (3)   (4)

(a) Which represent the same compound?   (b) Which represent enantiomers?
(c) Which represent a meso compound?   (d) Which are diastereomers?

3.5   Cyclic Molecules with Two or More Stereocenters

In this section, we concentrate on derivatives of cyclopentane and cyclohexane containing two stereocenters. We can analyze stereoisomerism in cyclic compounds in the same way as in acyclic compounds.

A.   Disubstituted Derivatives of Cyclopentane

Let us start with 2-methylcyclopentanol, a compound with two stereocenters. Using the 2^n rule, we predict a maximum of $2^2 = 4$ stereoisomers. Both the cis isomer and the trans isomer are chiral: the cis isomer exists as one pair of enantiomers, and the trans isomer exists as a second pair of enantiomers. The cis and trans isomers are stereoisomers that are not mirror images of each other, or, more simply, the cis and trans isomers are diastereomers.

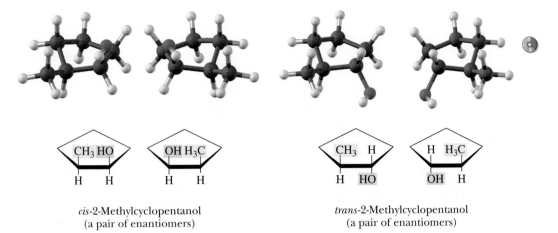

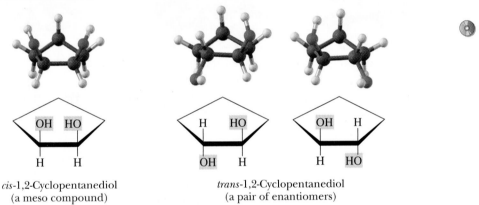

cis-2-Methylcyclopentanol
(a pair of enantiomers)

trans-2-Methylcyclopentanol
(a pair of enantiomers)

1,2-Cyclopentanediol also has two stereocenters; therefore, the 2^n rule predicts a maximum of $2^2 = 4$ stereoisomers. As seen in the following stereodrawings, only three stereoisomers exist for this compound. The cis isomer is achiral (meso) because it and its mirror image are superposable. The trans isomer is chiral and exists as a pair of enantiomers.

cis-1,2-Cyclopentanediol
(a meso compound)

trans-1,2-Cyclopentanediol
(a pair of enantiomers)

Alternatively, the cis isomer is achiral because it possesses a plane of symmetry that bisects the molecule into two mirror halves.

Example 3.6

How many stereoisomers exist for 3-methylcyclopentanol?

Solution

There are four stereoisomers of 3-methylcyclopentanol. The cis isomer exists as one pair of enantiomers; the trans isomer, as a second pair of enantiomers.

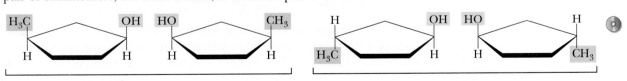

cis-3-Methylcyclopentanol
(a pair of enantiomers)

trans-3-Methylcyclopentanol
(a pair of enantiomers)

Problem 3.6

How many stereoisomers exist for 1,3-cyclopentanediol?

B.   Disubstituted Derivatives of Cyclohexane

As an example of a disubstituted cyclohexane, let us consider the methylcyclohexanols. 4-Methylcyclohexanol can exist as two stereoisomers: a pair of cis,trans isomers. Both the cis and trans isomers are achiral. In each, a plane of symmetry runs through the $—CH_3$ and HO— groups and the carbons attached to them.

3-Methylcyclohexanol has two stereocenters and exists as $2^2 = 4$ stereoisomers. The cis isomer exists as one pair of enantiomers; the trans isomer, as a second pair of enantiomers.

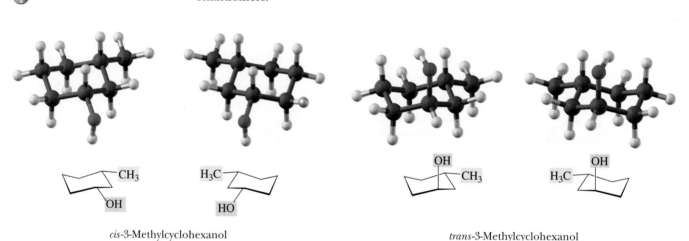

cis-3-Methylcyclohexanol   trans-3-Methylcyclohexanol

Similarly, 2-methylcyclohexanol has two stereocenters and exists as $2^2 = 4$ stereoisomers. The cis isomer exists as one pair of enantiomers; the trans isomer, as a second pair of enantiomers.

Example 3.7

How many stereoisomers exist for 1,3-cyclohexanediol?

Solution

1,3-Cyclohexanediol has two stereocenters and, according to the 2^n rule, has a maximum of $2^2 = 4$ stereoisomers. The trans isomer of this compound exists as a pair of enantiomers. The cis isomer has a plane of symmetry and is a meso compound. Therefore, although the 2^n rule predicts a maximum of four stereoisomers for 1,3-cyclohexanediol, only three exist: one meso compound and one pair of enantiomers.

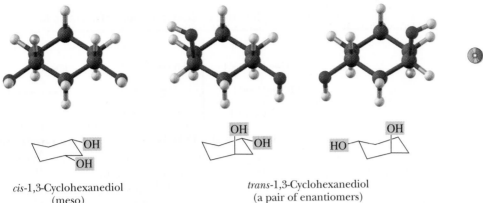

cis-1,3-Cyclohexanediol
(meso)

trans-1,3-Cyclohexanediol
(a pair of enantiomers)

Problem 3.7

How many stereoisomers exist for 1,4-cyclohexanediol?

1,2-Cyclohexanediol has two stereocenters and, according to the 2^n rule, can exist as a maximum of four stereoisomers. The cis isomer exists as one pair of enantiomers, and the trans isomer exists as a second pair of enantiomers.

(1)   (2)   (3)   (4)

cis-1,2-Cyclohexanediol
(a pair of enantiomers)

trans-1,2-Cyclohexanediol
(a pair of enantiomers)

The enantiomers of the cis isomer cannot be separated, however, because each is converted to the other by a rapid chair-to-alternative-chair conversion. As shown in the following structural formulas, the alternative chair of (1) is, in fact, the mirror image of (1). Thus, each enantiomer of the cis isomer interconverts to its mirror image. Therefore, the enantiomers cannot be separated.

(1)

Alternative chair of (1);
also the enantiomer of (1)

3.6   Properties of Stereoisomers

Enantiomers have identical physical and chemical properties in an achiral environment, that is, an environment without chirality. Examples of achiral environments are a chiral compound dissolved in H_2O, CH_3CH_2OH, or CH_2Cl_2.

Table 3.1 Some Physical Properties of the Stereoisomers of Tartaric Acid

	(R,R)-Tartaric Acid	(S,S)-Tartaric Acid	Meso Tartaric Acid
Specific rotation*	+12.7	−12.7	0
Melting point (°C)	171–174	171–174	146–148
Density at 20°C (g/cm³)	1.7598	1.7598	1.660
Solubility in water at 20°C (g/100 mL)	139	139	125
pK_1 (25°C)	2.98	2.98	3.23
pK_2 (25°C)	4.34	4.34	4.82

*Specific rotation is discussed in Section 3.7B.

The enantiomers of tartaric acid (Table 3.1), for example, have the same melting point, the same boiling point, the same solubility in water and other common solvents, the same value of pK_a, and they undergo the same acid-base reactions. The enantiomers of tartaric acid do, however, differ in optical activity (the ability to rotate the plane of plane-polarized light), which we discuss Section 3.7A. Diastereomers have different physical and chemical properties, even in achiral environments. Meso-tartaric acid has different physical properties from those of the enantiomers and can be separated from them by methods such as crystallization.

3.7 Optical Activity — How Chirality Is Detected in the Laboratory

As we have already established, enantiomers are different compounds, and thus we must expect that they differ in some properties. One property that differs between enantiomers is their effect on the rotation of the plane of polarized light. Each member of a pair of enantiomers rotates the plane of polarized light, and for this reason, enantiomers are said to be **optically active.**

The phenomenon of optical activity was discovered in 1815 by the French physicist Jean Baptiste Biot. To understand how it is detected in the laboratory, we must first understand something about plane-polarized light and a polarimeter, the device used to detect optical activity.

Optically active Refers to a compound that rotates the plane of polarized light.

A. Plane-Polarized Light

Ordinary light consists of waves vibrating in all planes perpendicular to its direction of propagation (Figure 3.6). Certain materials such as calcite or Polaroid sheet (a plastic film containing properly oriented crystals of an organic substance embedded

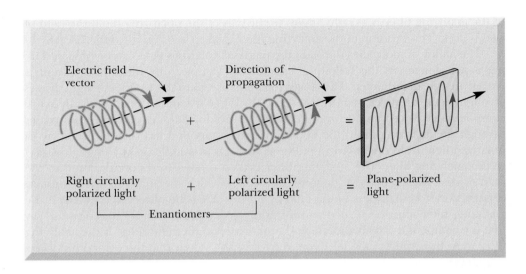

Electric field vector

Direction of propagation

+ =

Right circularly polarized light + Left circularly polarized light = Plane-polarized light

└────── Enantiomers ──────┘

Figure 3.6
Plane-polarized light is a mixture of left and right circularly polarized light.

in it) selectively transmit light waves vibrating in only parallel planes. Electromagnetic radiation vibrating in only parallel planes is said to be **plane polarized.**

Plane-polarized light is the vector sum of left and right circularly polarized light that propagates through space as left- and right-handed helices. These two forms of light are enantiomers, and because of their opposite handedness, each component interacts in an opposite way with chiral molecules. The result of this interaction is that each member of a pair of enantiomers rotates the plane of polarized light in an opposite direction.

Plane-polarized light   Light vibrating in only parallel planes.

B.   Polarimeters

A **polarimeter** consists of a light source, a polarizing filter and an analyzing filter (each made of calcite or Polaroid film), and a sample tube (Figure 3.7). If the sample tube is empty, the intensity of light reaching the detector (your eye) is at its maximum when the polarizing axes of the two filters are parallel. If the analyzing filter is turned either clockwise or counterclockwise, less light is transmitted. When the axis

Polarimeter   An instrument for measuring the ability of a compound to rotate the plane of polarized light.

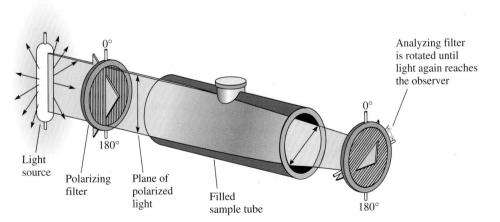

0°

180°

Analyzing filter is rotated until light again reaches the observer

0°

180°

Light source

Polarizing filter

Plane of polarized light

Filled sample tube

Figure 3.7
Schematic diagram of a polarimeter with its sample tube containing a solution of an optically active compound. The analyzing filter has been turned clockwise by α degrees to restore the dark field to the observer.

of the analyzing filter is at right angles to the axis of the polarizing filter, the field of view is dark. This position of the analyzing filter is taken as 0° on the optical scale.

The ability of molecules to rotate the plane of polarized light can be observed using a polarimeter in the following way. First, a sample tube filled with solvent is placed in the polarimeter. The analyzing filter is adjusted so that no light passes through to the observer; that is, it is set to 0°. When a solution of an optically active compound is placed in the sample tube, a certain amount of light now passes through the analyzing filter; the optically active compound has rotated the plane of polarized light from the polarizing filter so that it is now no longer at an angle of 90° to the analyzing filter. The analyzing filter is then rotated to restore darkness in the field of view. The number of degrees, α, through which the analyzing filter must be rotated to restore darkness to the field of view is called the **observed rotation.** If the analyzing filter must be turned to the right (clockwise) to restore darkness, we say that the compound is **dextrorotatory** (Latin: *dexter*, on the right side). If the analyzing filter must be turned to the left (counterclockwise), we say that the compound is **levorotatory** (Latin: *laevus*, on the left side).

Observed rotation The number of degrees through which a compound rotates the plane of polarized light.

Dextrorotatory Refers to a substance that rotates the plane of polarized light to the right.

Levorotatory Refers to a substance that rotates the plane of polarized light to the left.

Specific rotation Observed rotation of the plane of polarized light when a sample is placed in a tube 1.0 dm in length and at a concentration of 1 g/mL.

The magnitude of the observed rotation for a particular compound depends on its concentration, the length of the sample tube, the temperature, the solvent, and the wavelength of the light used. **Specific rotation,** $[\alpha]$, is defined as the observed rotation at a specific cell length and sample concentration.

$$\text{Specific rotation} = [\alpha]_\lambda^T = \frac{\text{Observed rotation (degrees)}}{\text{Length (dm)} \times \text{Concentration (g/mL)}}$$

The standard cell length is 1 decimeter (1 dm or 10 cm). For a pure liquid sample, the concentration is expressed in grams per milliliter (g/mL; density). The concentration of a sample dissolved in a solvent is also usually expressed as grams per milliliter of solution. The temperature (T, in degrees centigrade) and wave length (λ) of light are designated, respectively, as superscript and subscript. The light source most commonly used in polarimetry is the sodium D line ($\lambda = 589$ nm), the line responsible for the yellow color of sodium-vapor lamps.

In reporting either observed or specific rotation, it is common to indicate a dextrorotatory compound with a plus sign in parentheses, $(+)$, and a levorotatory compound with a minus sign in parentheses, $(-)$. For any pair of enantiomers, one enantiomer is dextrorotatory, and the other is levorotatory. For each member of the pair, the value of the specific rotation is exactly the same, but the sign is opposite. Following are the specific rotations of the enantiomers of 2-butanol at 25°C using the D line of sodium.

(S)-(+)-2-Butanol
$[\alpha]_D^{25}$ +13.52

(R)-(−)-2-Butanol
$[\alpha]_D^{25}$ −13.52

Example 3.8

A solution is prepared by dissolving 400 mg of testosterone, a male sex hormone, in 10.0 mL of ethanol and placing it in a sample tube 10.0 cm in length. The observed rotation of this sample at 25°C using the D line of sodium is +4.36°. Calculate the specific rotation of testosterone.

Solution

The concentration of testosterone is 400 mg/10.0 mL = 0.0400 g/mL. The length of the sample tube is 1.00 dm. Inserting these values in the equation for calculating specific rotation gives

$$\text{Specific rotation} = \frac{\text{Observed rotation (degrees)}}{\text{Length (dm)} \times \text{Concentration (g/mL)}} = \frac{+4.36°}{1.00 \times 0.0400} = +109$$

Problem 3.8

The specific rotation of progesterone, a female sex hormone, is +172°. Calculate the observed rotation for a solution prepared by dissolving 300 mg of progesterone in 15.0 mL of dioxane and placing it in a sample tube 10.0 cm long.

A polarimeter is used to measure the rotation of plane-polarized light as it passes through a sample. *(Richard Megna, 1992, Fundamental Photographs)*

C. Racemic Mixtures

An equimolar mixture of two enantiomers is called a **racemic mixture,** a term derived from the name "racemic acid" (Latin: *racemus,* a cluster of grapes). Racemic acid is the name originally given to an equimolar mixture of the enantiomers of tartaric acid (Table 3.1). Because a racemic mixture contains equal numbers of dextrorotatory and levorotatory molecules, its specific rotation is zero. Alternatively, we say that a racemic mixture is optically inactive. A racemic mixture is indicated by adding the prefix (±) to the name of the compound [or sometimes the prefix (D,L)].

Racemic mixture   A mixture of equal amounts of two enantiomers.

D. Optical Purity and Enantiomeric Excess

When dealing with a pair of enantiomers, it is essential to have a means of describing the composition of that mixture and the degree to which one enantiomer is in excess relative to its mirror image. The most common way of describing the composition of a mixture of enantiomers is by its percent **optical purity,** a property that can be observed directly. Optical purity is the specific rotation of a mixture of enantiomers divided by the specific rotation of the enantiomerically pure substance.

Optical purity   The specific rotation of a mixture of enantiomers divided by the specific rotation of the enantiomerically pure substance.

$$\text{Percent optical purity} = \frac{[\alpha]_{\text{sample}}}{[\alpha]_{\text{pure enantiomer}}} \times 100$$

An alternative way to describe the composition of a mixture of enantiomers is by its **enantiomeric excess (ee),** which is the difference in the number of moles of each enantiomer in a mixture compared with the total number of moles of both.

$$\text{Percent optical purity} = \text{Enantiomeric excess (ee)} = \frac{[R] - [S]}{[R] + [S]} \times 100 = \%R - \%S$$

Enantiomeric excess (ee)   The percentage difference in the number of moles of each enantiomer in a mixture compared with the total number of moles of both.

For example, if a mixture consists of 75% of the *R* enantiomer and 25% of the *S* enantiomer, then the enantiomeric excess of the *R* enantiomer is 50%.

Example 3.9

Figure 3.8 presents a scheme for separation of the enantiomers of mandelic acid. The specific rotation of optically pure (S)-(−)-mandelic acid is −158. Suppose that, instead of isolating pure (S)-(−)-mandelic acid from this scheme, the sample is a mixture of enantiomers with a specific rotation of −134. For this sample, calculate the following.

(a) The enantiomeric excess of this sample of (S)-(−)-mandelic acid.
(b) The percentage of (S)-(−)-mandelic acid and of (R)-(+)-mandelic acid in the sample.

Solution

(a) The enantiomeric excess of (S)-(−)-mandelic acid is 84.8%.

$$\text{Enantiomeric excess} = \frac{-134}{-158} \times 100 = 84.8\%$$

(b) This sample is 84.8% (S)-(−)-mandelic acid and 15.2% (R,S)-mandelic acid. The (R,S)-mandelic acid is 7.6% (S)-enantiomer and 7.6% (R)-enantiomer. The sample, therefore, contains 92.4% of the (S)-enantiomer and 7.6% of the (R)-enantiomer. We can check these values by calculating the observed rotation of a mixture containing 92.4% (S)-(−)-mandelic acid and 7.6% (R)-(+)-mandelic acid as follows:

$$\text{Specific rotation} = 0.924 \times (-158) + 0.076 \times (+158) = -146 + 12 = -134$$

which agrees with the experimental specific rotation.

Problem 3.9

One commercial synthesis of naproxen (the active ingredient in Aleve and a score of other over-the-counter and prescription nonsteroidal anti-inflammatory drug preparations) gives the enantiomer shown in 97% enantiomeric excess.

Naproxen
(a nonsteroidal anti-inflammatory drug)

(a) Assign an R or S configuration to this enantiomer of naproxen.
(b) What are the percentages of R and S enantiomers in the mixture?

3.8 Separation of Enantiomers — Resolution

Resolution Separation of a racemic mixture into its enantiomers.

The separation of a racemic mixture into its enantiomers is called **resolution.**

A. Resolution by Means of Diastereomeric Salts

One general scheme for separating enantiomers requires chemical conversion of a pair of enantiomers into two diastereomers with the aid of an enantiomerically pure chiral resolving agent. This chemical resolution is successful because the diastereomers thus formed have different physical properties and often can be separated by physical means (most commonly fractional crystallization or column chromatography) and purified. The final step in this scheme for resolution is chemical conversion of the separated diastereomers back to the individual enantiomers and recovery of the chiral resolving agent.

A reaction that lends itself to chemical resolution is salt formation because it is readily reversible.

$$\text{RCOOH} \quad + \quad \text{:B} \rightleftharpoons \text{RCOO}^- \text{HB}^+$$

$$\text{Carboxylic} \qquad \text{Base} \qquad \text{Salt}$$
$$\text{acid}$$

Several enantiomerically pure bases available from plants have been used as chiral resolving agents for racemic acids. Examples are cinchonine and quinine.

(+)-Cinchonine
$[\alpha]_D^{23} = +228$

(−)-Quinine
$[\alpha]_D^{25} = -165$

The base (+)-cinchonine is found in the bark of most species of *Cinchona,* a genus of evergreen trees or shrubs growing in the tropical valleys of the Andes and now extensively cultivated for its bark in India, Java, and parts of South America. Extracts of its bark have been used for centuries to cure the fevers associated with malaria. The genus was named after the Countess of Cinchon, wife of the viceroy of Peru, who was cured of fever by cinchona bark and later brought a supply of it back to Spain. Also found in cinchona bark is (−)-quinine, an even more potent antimalarial drug than cinchonine.

The resolution of mandelic acid by way of its diastereomeric salts with cinchonine is illustrated in Figure 3.8. Racemic mandelic acid and optically pure (+)-cinchonine (Cin) are dissolved in boiling water, and the solution is allowed to cool, whereupon the less soluble diastereomeric salt crystallizes. This salt is collected and purified by further recrystallization. The filtrates, richer in the more soluble diastereomeric salt, are concentrated to give this salt, which is also purified by further recrystallization. The purified diastereomeric salts are treated with aqueous HCl to precipitate the nearly pure enantiomers of mandelic acid. Cinchonine remains in the aqueous solution as its water-soluble hydrochloride salt.

Optical rotations and melting points of racemic mandelic acid, cinchonine, the purified diastereomeric salts, and the pure enantiomers of mandelic acid are given in Figure 3.8. Note the following two points: (1) The diastereomeric salts have different specific rotations and different melting points. (2) The enantiomers of mandelic acid have identical melting points and have specific rotations that are identical in magnitude but opposite in sign.

Cinchona bark, a source of quinine, one of the first antimalarial drugs. *(Walter H. Hodge/Peter Arnold, Inc.)*

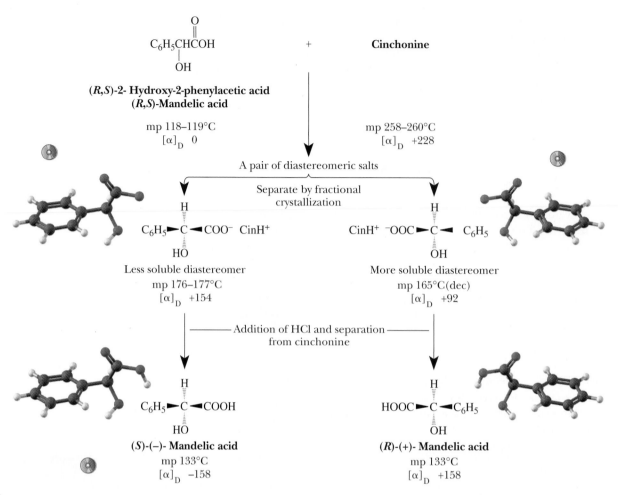

Figure 3.8
Resolution of mandelic acid.

Resolution of a racemic base with a chiral acid is carried out in a similar way. Acids that are commonly used as chiral resolving agents are (+)-tartaric acid, (−)-malic acid, and (+)-camphoric acid (Figure 3.9). These and other naturally occurring chiral resolving agents are produced in plant and animal systems as single enantiomers.

B. Enzymes as Resolving Agents

In their quest for enantiomerically pure compounds, organic chemists have developed several new techniques for chiral synthesis. One approach is to use enzymes as chiral catalysts for the large-scale synthesis of enantiomerically pure substances. A class of enzymes under study is the esterases, which catalyze the hydrolysis of esters to give an alcohol and a carboxylic acid.

The ethyl esters of naproxen crystallize in two enantiomeric crystal forms: one containing the (R)-ester and the other containing the (S)-ester. Each is insoluble in water. Chemists then use an esterase in alkaline solution to hydrolyze selectively the (S)-ester to the (S)-carboxylic acid, which goes into solution as the sodium salt. The

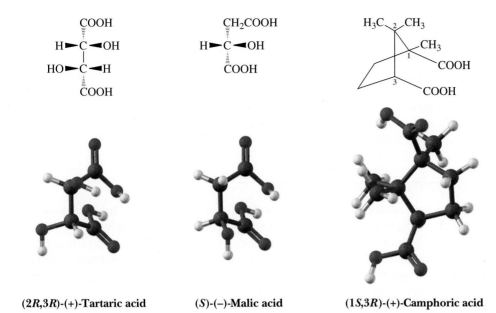

(2R,3R)-(+)-Tartaric acid

$[\alpha]_D^{20}$ +12.5

(S)-(–)-Malic acid

$[\alpha]_D^{20}$ –27

(1S,3R)-(+)-Camphoric acid

$[\alpha]_D^{20}$ +46.5

Figure 3.9
Some carboxylic acids used as chiral resolving agents.

(R)-ester is unaffected by these conditions. Filtering the alkaline solution recovers the crystals of the (R)-ester. After crystals of the (R)-ester are removed, the alkaline solution is acidified to give enantiomerically pure (S)-naproxen. The recovered (R)-ester is racemized (converted to an R,S-mixture) and treated again with the esterase. Thus, by recycling the (R)-ester, all the racemic ester is converted to (S)-naproxen. The sodium salt of (S)-naproxen is the active ingredient in Aleve and several other nonsteroidal anti-inflammatory medications (see *The Merck Index,* 12th ed., #6504).

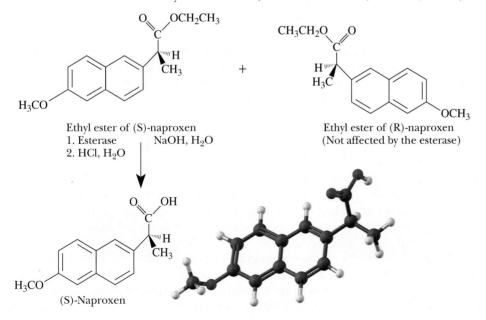

Ethyl ester of (S)-naproxen
1. Esterase   NaOH, H₂O
2. HCl, H₂O

Ethyl ester of (R)-naproxen
(Not affected by the esterase)

(S)-Naproxen

3.9 The Significance of Chirality in the Biological World

Except for inorganic salts and a relatively few low-molecular-weight organic substances, the molecules in living systems, both plant and animal, are chiral. Although these molecules can exist as a number of stereoisomers, almost invariably only one stereoisomer is found in nature. Of course, instances do occur in which more than one stereoisomer is found, but these rarely exist together in the same biological system.

A. Chirality in Biomolecules

Perhaps the most conspicuous examples of chirality among biological molecules are the enzymes, all of which have multiple stereocenters. An illustration is chymotrypsin, an enzyme in the intestines of animals, which catalyzes the hydrolysis of proteins during digestion. Chymotrypsin has 251 stereocenters. The maximum number of stereoisomers possible is 2^{251}, a staggeringly large number, almost beyond comprehension. Fortunately, nature does not squander its precious energy and resources unnecessarily; only one of these stereoisomers is produced and used. Because enzymes are chiral substances, most either produce or react only with substances that match their own chirality.

B. How an Enzyme Distinguishes Between a Molecule and Its Enantiomer

Enzymes are chiral catalysts. Some are completely specific for the catalysis of the reaction of only one particular compound, whereas others are less specific and catalyze similar reactions of a family of compounds. An enzyme catalyzes a biological reaction of molecules by first positioning them at a binding site on its surface. These molecules may be held at the binding site by a combination of hydrogen bonds, electrostatic attractions, dispersion forces, or even covalent bonds.

An enzyme with specific binding sites for three of the four groups on a stereocenter can distinguish between a molecule and its enantiomer or one of its diastereomers. Assume, for example, that an enzyme involved in catalyzing a reaction of glyceraldehyde has three binding sites, one specific for —H, a second specific for

Figure 3.10

A schematic diagram of an enzyme surface capable of interacting with (R)-$(+)$-glyceraldehyde at three binding sites, but with (S)-$(-)$-glyceraldehyde at only two of these sites.

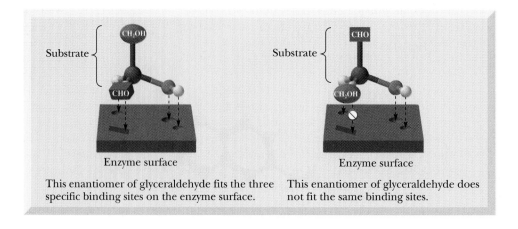

This enantiomer of glyceraldehyde fits the three specific binding sites on the enzyme surface.

This enantiomer of glyceraldehyde does not fit the same binding sites.

—OH, and a third specific for —CHO. Assume further that the three sites are arranged on the enzyme surface as shown in Figure 3.10. The enzyme can distinguish (R)-(+)-glyceraldehyde (the natural or biologically active form) from its enantiomer (S)-(−)-glyceraldehyde because the natural enantiomer can be adsorbed on the en-

C H E M I S T R Y   I N   A C T I O N

Chiral Drugs

Some of the common drugs used in human medicine, for example aspirin (Section 18.6B), are achiral. Others are chiral and sold as single enantiomers. The penicillin and erythromycin classes of antibiotics and the drug captopril are all chiral drugs. Captopril, which is very effective for the treatment of high blood pressure and congestive heart failure, was developed in a research program designed to discover effective inhibitors for angiotensin-converting enzyme (ACE). It is manufactured and sold as the (S,S)-stereoisomer (see *The Merck Index,* 12th ed., #1817). A large number of chiral drugs, however, are sold as racemic mixtures. The popular analgesic ibuprofen (the active ingredient in Motrin, Advil, and many other nonaspirin analgesics) is an example.

Captopril   (S)-Ibuprofen

For racemic drugs, most often only one enantiomer exerts the beneficial effect, whereas the other enantiomer either has no effect or may even exert a detrimental effect. Thus, enantiomerically pure drugs are usually more effective than their racemic counterparts. A case in point is the drug dihydroxyphenylalanine used in the treatment of Parkinson's disease. The active drug is dopamine. Unfortunately, this compound does not cross the blood-brain barrier to the required site of action in the brain. What is administered, instead, is the prodrug, a compound that is not active by itself but is converted in the body to an active drug. 3,4-Dihydroxyphenylalanine (see *The Merck Index,* 12th ed., #3478) is such a prodrug. It crosses the blood-brain barrier and then undergoes

decarboxylation catalyzed by the enzyme dopamine decarboxylase. This enzyme is specific for the S enantiomer, which is commonly known as L-DOPA. It is essential, therefore, to administer the enantiomerically pure prodrug. Were the prodrug to be administered in a racemic form, there could be a dangerous buildup of the R enantiomer, which cannot be metabolized by the enzymes present in the brain.

(S)-(−)-3,4-Dihydroxyphenylalanine
(L-DOPA)
$[\alpha]_D^{13}$ −13.1°

Recently, the U.S. Food and Drug Administration established new guidelines for the testing and marketing of chiral drugs. After reviewing these guidelines, many drug companies have decided to develop only single enantiomers of new chiral drugs. For this reason, there has been a tremendous recent interest in developing stereospecific synthetic methods. In addition to regulatory pressure, there are patent considerations. If a company has patents on a racemic drug, a new patent can often be taken out on one of its enantiomers. Only the S enantiomer of the pain reliever ibuprofen is biologically active. In the case of ibuprofen, however, the body converts the inactive R enantiomer to the active S enantiomer.

zyme surface with three groups interacting with their appropriate binding sites; the other enantiomer can, at best, bind to only two of these sites.

Because interactions between molecules in living systems take place in a chiral environment, it should be no surprise that a molecule and its enantiomer or diastereomers have different physiological properties. The tricarboxylic acid (TCA) cycle, for example, produces and then metabolizes only (S)-(+)-malic acid. Because only one enantiomer is produced, these reactions are said to be **stereospecific.**

$$\text{HOOC} \diagdown \diagup \text{COOH}$$
$$\underset{\text{H}\quad\text{OH}}{\Big|}$$

(S)-(+)-Malic acid

That interactions between molecules in the biological world are highly stereospecific is not surprising, but just how these interactions are accomplished at the molecular level with such precision and efficiency is one of the great puzzles that modern science has only recently begun to unravel.

Summary

Stereoisomers (Section 3.1) have the same order of attachment of atoms in their molecules but a different three-dimensional orientation of their atoms in space. Stereoisomers are divided into enantiomers and diastereomers. **Enantiomers** are stereoisomers that are mirror images of each other. **Diastereomers** are stereoisomers that are not mirror images.

A **mirror image** is the reflection of an object in a mirror. Molecules that are not superposable on their mirror images are said to be **chiral** (Section 3.2). Chirality is a property of an object as a whole, not of a particular atom. An **achiral** object is one that lacks chirality; that is, it is an object that has a superposable mirror image. Almost all achiral objects possess at least one plane or center of symmetry. A **plane of symmetry** is an imaginary plane passing through an object dividing it such that one half is the reflection of the other half. A **center of symmetry** is a point so situated that identical components of the object are located on opposite sides and equidistant from the point along any axis passing through that point.

A **stereocenter** (Section 3.2) is an atom, most commonly carbon, with four different groups bonded to it. Stereocenters are not limited to carbon. Chiral compounds of tetrahedral nitrogen, silicon, phosphorus, and germanium have also been prepared.

The configuration at any stereocenter can be specified by the **R,S system** (Section 3.3). To apply this convention, each atom or group of atoms bonded to the stereocenter is (1) assigned a priority and numbered from highest priority to lowest priority. (2) The molecule is oriented in space so that the group of lowest priority is directed away from the observer, and (3) the remaining three groups are read in order from highest priority to lowest priority. If the order of groups is clockwise,

the configuration is **R** (Latin: *rectus,* right, correct). If the order is counterclockwise, the configuration is **S** (Latin: *sinister,* left).

For a molecule with n stereocenters, the maximum number of stereoisomers possible is 2^n (Section 3.4). Certain molecules have special symmetry properties that reduce the number of stereoisomers to fewer than that predicted by the 2^n rule. A compound is **meso** (Section 3.4B) if it contains two or more stereocenters assembled in such a way that its molecules are achiral.

Enantiomers have identical physical and chemical properties in achiral environments (Section 3.6). They have different properties, however, in chiral environments, as for example, in the presence of plane-polarized light (Section 3.7). They also have different properties in the presence of chiral reagents (Section 3.8A) and enzymes as chiral catalysts (Section 3.8B). **Diastereomers** have different physical and chemical properties even in achiral environments.

Light that vibrates in only parallel planes is said to be **plane polarized** (Section 3.7A). Plane-polarized light contains equal components of left and right circularly polarized light. A **polarimeter** (Section 3.7B) is an instrument used to detect and measure the magnitude of optical activity. A compound is said to be **optically active** if it rotates the plane of polarized light. **Observed rotation** is the number of degrees the plane of polarized light is rotated. **Specific rotation** is the observed rotation measured in a cell 1 dm long and at a sample concentration of 1 g/mL. If the analyzing prism must be turned clockwise to restore the zero point, the compound is **dextrorotatory.** If the analyzing prism must be turned counterclockwise to restore the zero point, the compound is **levorotatory.** Each member of a pair of enantiomers rotates the plane of polarized light an equal number of degrees but opposite in direction (Section 3.7B). A **racemic mixture** (Section 3.7C) is a mixture of equal amounts of two enantiomers and has a specific rotation of zero. Percent **optical purity** (identical to **enantiomeric excess**) is defined as the specific rotation of a mixture of enantiomers divided by the specific rotation of the pure enantiomer times 100 (Section 3.7D).

Resolution (Section 3.8) is the experimental process of separating a mixture of enantiomers into the two pure enantiomers. A common chemical means of resolving organic compounds is to treat the racemic mixture with a chiral resolving agent that converts the mixture of enantiomers into a pair of diastereomers. The diastereomers are separated based on differences in their physical properties; each diastereomer is then converted to a pure stereoisomer, uncontaminated by its enantiomer. Enzymes are also used as resolving agents because of their ability to catalyze a reaction of one enantiomer but not that of its mirror image.

Enzymes catalyze biological reactions by first positioning the molecule or molecules at binding sites and holding them there by a combination of hydrogen bonds, electrostatic attractions, dispersion forces, and covalent bonds. An enzyme with specific binding sites for three of the four groups on a stereocenter can distinguish between a molecule and its enantiomer (Section 3.9B). Almost all enzyme-catalyzed reactions are **stereospecific.**

Problems

Chirality

3.10 Think about the helical coil of a telephone cord or a spiral binding and suppose that you view the spiral from one end and find that it is a left-handed twist. If you view the same spiral from the other end, is it a right-handed or left-handed twist?

3.11 Next time you have the opportunity to view a collection of sea shells that have a helical twist, study the chirality of their twists. Do you find an equal number of left-handed and right-handed spiral shells, or are they mostly all of the same chirality? What about the handedness of different species of spiral shells?

3.12 One reason we can be sure that sp^3-hybridized carbon atoms are tetrahedral is the number of stereoisomers that can exist for different organic compounds.

   (a) How many stereoisomers are possible for $CHCl_3$, CH_2Cl_2, and $CHClBrF$ if the four bonds to carbon have a tetrahedral arrangement?

   (b) How many stereoisomers would be possible for each of these compounds if the four bonds to the carbon had a square planar geometry?

Enantiomers

3.13 Which compounds contain stereocenters?

   (a) 2-Chloropentane   **(b)** 3-Chloropentane
   (c) 3-Chloro-1-pentene   **(d)** 1,2-Dichloropropane

3.14 Using only C, H, and O, write structural formulas for the lowest molecular-weight chiral.

   (a) Alkane   **(b)** Alcohol   **(c)** Aldehyde
   (d) Ketone   **(e)** Carboxylic acid   **(f)** Carboxylic ester

3.15 Draw mirror images for these molecules.

This Atlantic auger shell has a right-handed helical twist.
(*Carolina Biological Supply/Phototake, NYC*)

3.16 Following are several stereorepresentations for lactic acid. Take (a) as a reference structure. Which stereorepresentations are identical with (a), and which are mirror images of (a)?

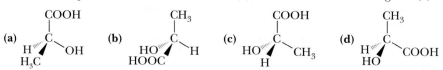

3.17 Mark each stereocenter in the following molecules with an asterisk. How many stereoisomers are possible for each molecule?

$$\text{(a) } CH_3\underset{\underset{OH}{|}}{\overset{\overset{CH_3}{|}}{C}}CH{=}CH_2 \quad \text{(b) } H\underset{\underset{CH_3}{|}}{\overset{\overset{COOH}{|}}{C}}OH \quad \text{(c) } CH_3\underset{\underset{NH_2}{|}}{\overset{\overset{CH_3}{|}}{CH}}CHCOOH \quad \text{(d) } CH_3\overset{\overset{O}{\|}}{C}CH_2CH_3$$

$$\text{(e) } H\underset{\underset{CH_2OH}{|}}{\overset{\overset{CH_2OH}{|}}{C}}OH \quad \text{(f) } CH_3CH_2\underset{\underset{OH}{|}}{CH}CH{=}CH_2 \quad \text{(g) } HO\underset{\underset{CH_2COOH}{|}}{\overset{\overset{CH_2COOH}{|}}{C}}COOH$$

3.18 Show that butane in a gauche conformation is chiral. Do you expect that resolution of butane at room temperature is possible?

Designation of Configuration: The *R,S* System

3.19 Assign priorities to the groups in each set.

(a) —H   —CH₃   —OH   —CH₂OH

(a) $-H$ $\quad-CH_3$ $\quad-OH$ $\quad-CH_2OH$
(b) $-CH_2CH{=}CH_2$ $\quad-CH{=}CH_2$ $\quad-CH_3$ $\quad-CH_2COOH$
(c) $-CH_3$ $\quad-H$ $\quad-COO^-$ $\quad-NH_3{}^+$
(d) $-CH_3$ $\quad-CH_2SH$ $\quad-NH_3{}^+$ $\quad-CHO$

3.20 Following are structural formulas for the enantiomers of carvone. (See *The Merck Index,* 12th ed., #1925.) Each has a distinctive odor characteristic of the source from which it is isolated. Assign an *R* or *S* configuration to the single stereocenter in each enantiomer.

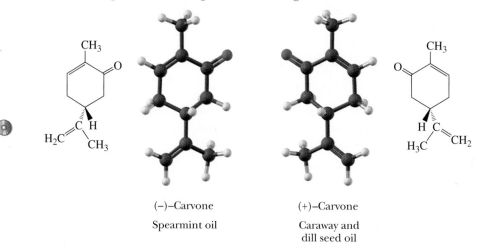

(−)−Carvone

Spearmint oil

(+)−Carvone

Caraway and dill seed oil

3.21 Following is a staggered conformation for one of the enantiomers of 2-butanol.

$$\underset{H}{\overset{H_3C}{\diagdown}}\underset{\diagup}{\overset{}{C}}\underset{CH_3}{\overset{H}{\diagup}}OH$$

(a) Is this (R)-2-butanol or (S)-2-butanol?

(b) Draw a Newman projection for this staggered conformation, viewed along the bond between carbons 2 and 3.

(c) Draw a Newman projection for two more staggered conformations of this molecule. Which of your conformations is the most stable? Assume that —OH and —CH$_3$ are comparable in size.

3.22 For centuries, Chinese herbal medicine has used extracts of *Ephedra sinica* to treat asthma. Phytochemical investigation of this plant resulted in isolation of ephedrine, a very potent dilator of the air passages of the lungs. The naturally occurring stereoisomer is levorotatory and has the following structure (see *The Merck Index,* 12th ed., #3645). Assign an *R* or *S* configuration to each stereocenter.

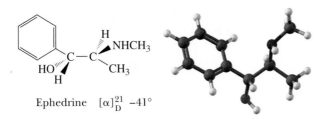

Ephedrine $[\alpha]_D^{21}$ −41°

Ephedra sinica, the source of ephedrine, a potent bronchodilator. *(Paolo Koch/Photo Researchers, Inc.)*

3.23 When oxaloacetic acid and acetyl-coenzyme A (acetyl-CoA) labeled with radioactive carbon-14 in position 2 are incubated with citrate synthase, an enzyme of the tricarboxylic acid cycle, only the following enantiomer of [2-^{14}C]citric acid is formed stereospecifically. Note that citric acid containing only ^{12}C is achiral. Assign an *R* or *S* configuration to this enantiomer of [2-^{14}C]citric acid. *Note:* Carbon-14 has a higher priority than carbon-12.

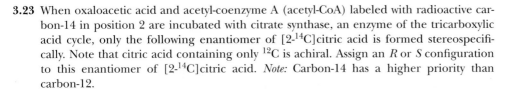

Oxaloacetic acid   Acetyl-CoA   [2-^{14}C]Citric acid

Molecules with Two or More Stereocenters

3.24 Draw stereorepresentations for all stereoisomers of this compound. Label those that are meso compounds and those that are pairs of enantiomers.

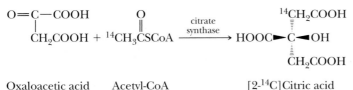

3.25 Mark each stereocenter in the following molecules with an asterisk. How many stereoisomers are possible for each molecule?

(a) CH$_3$CHCHCOOH
        | |
       HO OH

(b) CH$_2$—COOH
        |
       CH—COOH
        |
    HO—CH—COOH

(c) [structure with OH on cyclopentane ring with methyl]

(d) [cyclohexene structure with isopropyl group]

(e) [structure with OH and isopropyl substituents on cyclohexane ring]

(f) [structure with ketone, isopropyl, and cyclohexene ring with COOH]

(g) [tetrahydrofuran ring with two OH groups]

(h) [decalin-type bicyclic structure with ketone O]

3.26 Label the eight stereocenters in cholesterol. How many stereoisomers are possible for a molecule of this many stereocenters?

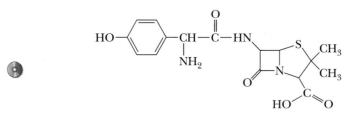

Cholesterol

3.27 Label the four stereocenters in amoxicillin (see *The Merck Index,* 12th ed., #617), which belongs to the family of semisynthetic penicillins.

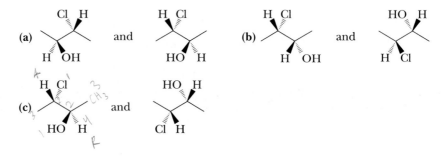

Amoxicillin

3.28 If the optical rotation of a new compound is measured and found to have a specific rotation of +40, how can you tell if the actual rotation is not really +40 plus some multiple of +360? In other words, how can you tell if the rotation is not actually a value such as +400 or +760?

3.29 Are the formulas within each set identical, enantiomers, or diastereomers?

(a) [Cl H / H OH structure] and [H Cl / HO H structure]

(b) [H Cl / H OH structure] and [HO H / H Cl structure]

(c) [H Cl / HO H structure] and [HO H / Cl H structure]

3.30 Which of the following are meso compounds?

(a) 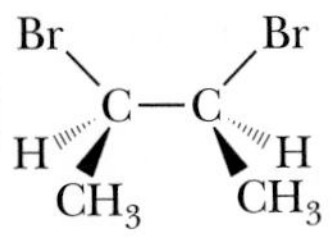
(b) 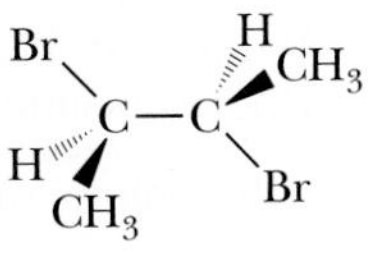
(c) 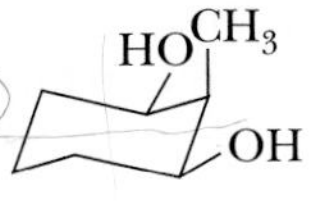

(d) 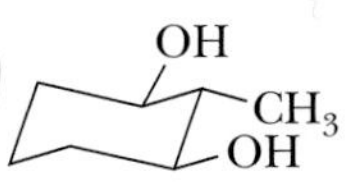
(e) 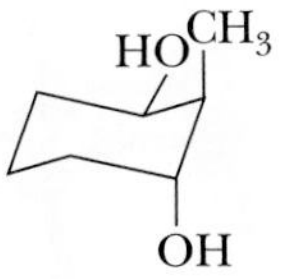
(f) 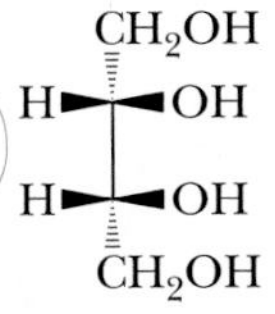

(g) 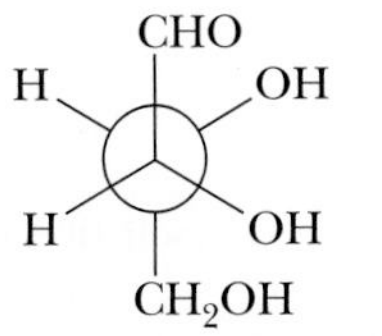
(h) 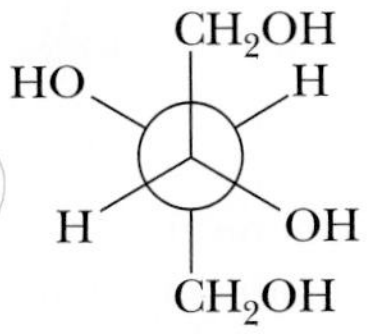
(i) 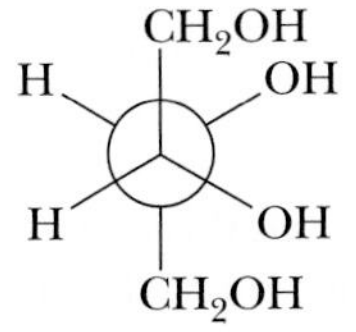

3.31 Vigorous oxidation of the following bicycloalkene breaks the carbon-carbon double bond and converts each carbon of the double bond to a COOH group. Assume that the conditions of oxidation have no effect on the configuration of either the starting bicycloalkene or the resulting dicarboxylic acid. Is the dicarboxylic acid produced from this oxidation one enantiomer, a racemic mixture, or a meso compound?

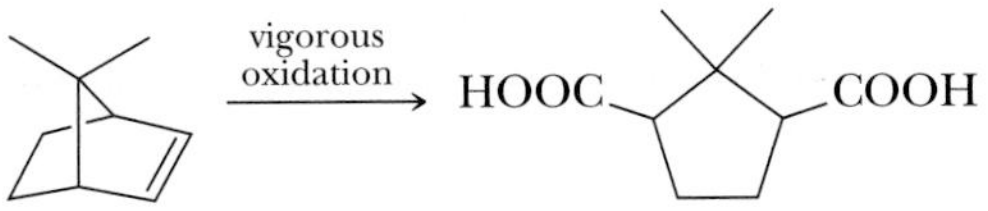

3.32 A long polymer chain, such as polyethylene $(\text{—}CH_2CH_2\text{—})_n$, can potentially exist in solution as a chiral object. Give two examples of chiral structures that a polyethylene chain could adopt.

Molecular Modeling

3.33 ChemDraw provides a very easy way to make mirror images. First, create a stereocenter in ChemDraw, make a copy, and place it adjacent to your original. Second, select the copy and, (3) from the Object menu, select Flip Horizontal. As shown here, this procedure converts an enantiomer to its mirror image.

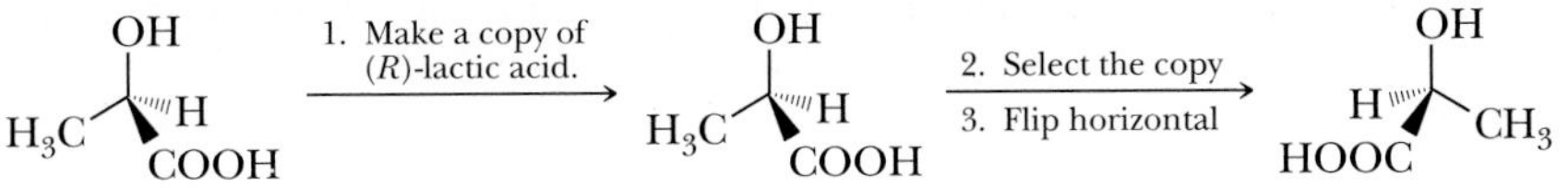

(*R*)-Lactic acid — Copy of (*R*)-lactic acid — (*S*)-Lactic acid

Now try this procedure with these molecules chosen from the text.

(a) 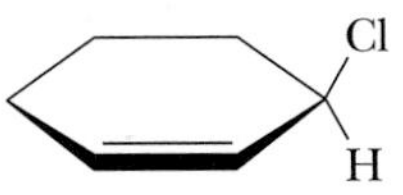
(b) 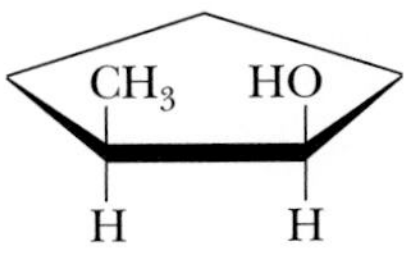
(c) 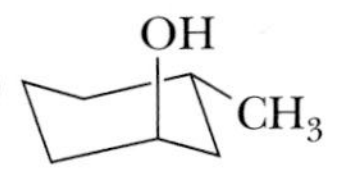
(d) 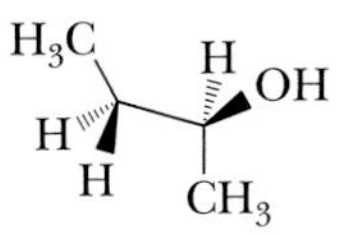

3.34 ChemDraw is able to assign an *R* or *S* configuration to a stereocenter. To do this, use ChemDraw to build a model of a chiral compound. Then, from the Tools pulldown menu, select Show Stereochemistry. As practice, build structures 3.33(a–d) in Chem-Draw, and show the configuration of each stereocenter.

3.35 The following molecule is an attractant pheromone for the olive fly.

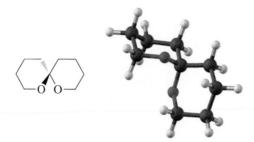

(a) Rotate the three-dimensional model on the CD and convince yourself that each six-membered ring has a strain-free chair conformation.

(b) This molecule has no stereocenter, and yet it is chiral. Examine the three-dimensional model and convince yourself that it has no plane or center of symmetry and that it is, in fact, chiral.

(c) "The presence of a stereocenter in an organic molecule is a sufficient condition for chirality, but it is not a necessary condition." Explain.

3.36 The following molecule belongs to the class of compounds called allenes. The functional group of an allene is two adjacent carbon-carbon double bonds. Disubstituted allenes of this type are chiral. The specific rotation of the enantiomer shown is −314.

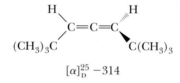

$$[\alpha]_D^{25}\ -314$$

(a) Examine the three-dimensional model on the CD and convince yourself that it has no plane or center of symmetry.

(b) Make its mirror image and convince yourself that the original and the mirror image are nonsuperposable; that is, that they are a pair of enantiomers.

4

ACIDS AND BASES

A great many organic reactions are either acid-base reactions or involve catalysis by an acid or base at some stage. Of the reactions involving acid catalysis, some use proton-donating acids, such as H_3O^+ and $CH_3CH_2OH_2^+$. Others use Lewis acids, such as BF_3 and $AlCl_3$. It is essential, therefore, that you have a good grasp of the fundamentals of acid-base chemistry. In this and following chapters, we study the acid-base properties of the major classes of organic compounds.

■ Citrus fruits are sources of citric acid. Lemon juice, for example, contains 5–8% citric acid. Inset: A model of citric acid. *(Charles D. Winters)*

4.1   Brønsted-Lowry Acids and Bases

Brønsted-Lowry acid   A proton donor.

Brønsted-Lowry base   A proton acceptor.

In 1923, the Danish chemist Johannes Brønsted and the English chemist Thomas Lowry independently proposed the following definitions: an **acid** is a proton (H^+) donor, and a **base** is a proton (H^+) acceptor. In a neutralization reaction, for example, between aqueous H_3O^+ and OH^-, a proton is transferred from H_3O^+, a Brønsted-Lowry acid, to OH^-, a Brønsted-Lowry base. We use curved arrows to show the flow of electrons in acid-base reactions. The curved arrow on the right in the following equation shows an unshared pair of electrons on oxygen forming a new bond with hydrogen; in donating electrons, this oxygen becomes neutral. The curved arrow on the left shows an electron pair from an O—H bond moving onto oxygen; this oxygen gains electrons and becomes neutral.

$$H-\overset{+}{\underset{|}{\overset{..}{O}}}-H + :\overset{-}{\underset{..}{O}}-H \longrightarrow H-\overset{..}{\underset{|}{O}}: + H-\overset{..}{\underset{..}{O}}-H$$

Proton   Proton
donor   acceptor

According to the Brønsted-Lowry theory, any pair of molecules or ions that can be interconverted by loss or gain of a proton is called a conjugate acid-base pair. When an acid transfers a proton to a base, the acid is converted to its **conjugate base;** when a base accepts a proton, the base is converted to its **conjugate acid.** In the reaction between hydronium ion and ammonia, for example, H_3O^+ is converted to its conjugate base, H_2O, and NH_3 is converted to its conjugate acid, NH_4^+.

Conjugate base   The species formed from an acid when it donates a proton to a base.

Conjugate acid   The species formed from a base when it accepts a proton from an acid.

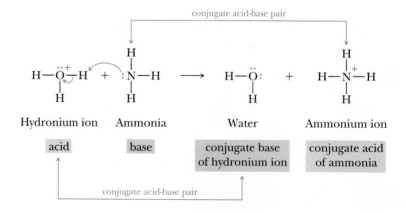

In the acid-base reaction between acetic acid and methylamine, there is transfer of a proton from acetic acid to water to produce acetate ion and methylammonium ion. Acetic acid and acetate ion are a conjugate acid-base pair, and methylamine and methylammonium ion are a second conjugate acid-base pair. These conjugate pairs are in equilibrium.

conjugate acid-base pair

Acetic acid   Methylamine   Acetate ion   Methylammonium
                                                        ion

acid   base   conjugate base   conjugate acid
                    of acetic acid   of methylamine

conjugate acid-base pair

As seen in these examples, there is a reciprocal relationship between any conjugate acid-base pair. The conjugate acid will always have one more proton and an increase in positive charge (or a decrease in negative charge). A conjugate base will always have one less proton and an increase in negative charge (or a decrease in positive charge).

Example 4.1

For each conjugate acid-base pair, identify the first species as an acid or base and the second species as its conjugate base or conjugate acid. In addition, draw Lewis structures for each species, showing all valence electrons and any formal charges.

(a) CH_3OH $CH_3OH_2^+$ **(b)** H_2CO_3 HCO_3^- **(c)** CH_3NH_2 $CH_3NH_3^+$
      base   acid   acid   base   base   acid

Solution

(a) $CH_3-\ddot{O}-H$ $CH_3-\overset{+}{\underset{H}{\ddot{O}}}-H$ **(b)** $H-\ddot{O}-\overset{O}{C}-\ddot{O}-H$ $H-\ddot{O}-\overset{O}{C}-\ddot{O}:^-$

Base   Conjugate   Acid   Conjugate
            acid   base

(c) $CH_3-\underset{H}{\ddot{N}}-H$ $CH_3-\overset{H}{\underset{H}{\overset{+}{N}}}-H$

Base   Conjugate
            acid

Problem 4.1

For each conjugate acid-base pair, identify the first species as an acid or base and the second species as its conjugate acid or conjugate base. In addition, draw Lewis structures for each species, showing all valence electrons and any formal charges.

(a) H_2SO_4 HSO_4^- **(b)** NH_3 NH_2^- **(c)** CH_3OH CH_3O^-

Example 4.2

Write these reactions as proton-transfer reactions. Label which reactant is the acid and which is the base; label which product is the conjugate base of the original acid and which is the conjugate acid of the original base. Write Lewis structures for each reactant and product, and use curved arrows to show the flow of electrons in each reaction.

(a) $H_2O + NH_4^+ \rightarrow H_3O^+ + NH_3$
(b) $CH_3CH_2OH + NH_2^- \rightarrow CH_3CH_2O^- + NH_3$

Solution

(a) Water is the base (proton acceptor), and ammonium ion is the acid (proton donor).

Base   Acid   Conjugate acid of H_2O Conjugate base of NH_4^+

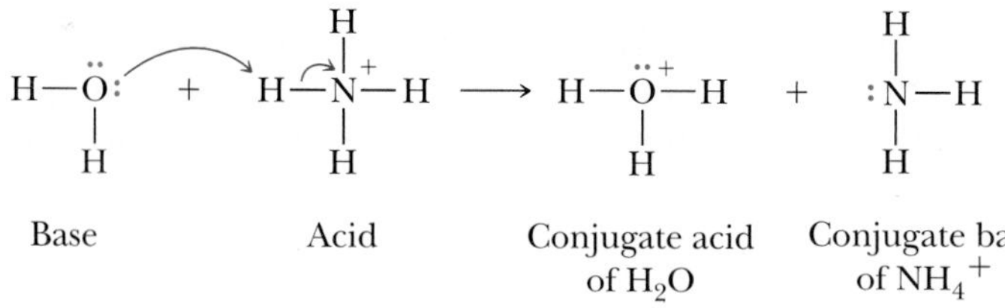

(b) Ethanol is the acid (proton donor), and amide ion is the base (proton acceptor).

Acid   Base   Conjugate base of CH_3CH_2OH Conjugate acid of NH_2^-

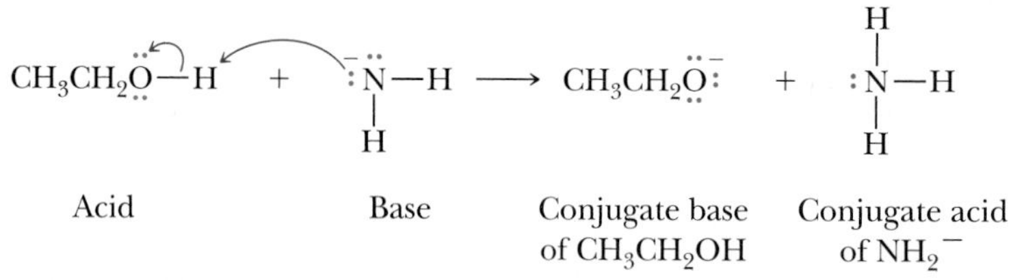

Problem 4.2

Write these reactions as proton-transfer reactions. Label which reactant is the acid and which is the base; label which product is the conjugate base of the original acid and which is the conjugate acid of the original base. Write Lewis structures for each reactant and product, and use curved arrows to show the flow of electrons in each reaction.

(a) $CH_3SH + OH^- \rightarrow CH_3S^- + H_2O$
(b) $CH_2{=}O + HCl \rightarrow CH_2{=}OH^+ + Cl^-$

Thus far, we have dealt with Brønsted-Lowry bases that have only one site that can act as a proton acceptor in an acid-base reaction. Many organic compounds have two or more such sites. In the following discussion, we restrict our discussion to compounds containing a carbonyl group in which the carbonyl carbon is bonded to either an oxygen or a nitrogen atom. The principle we develop here applies to other types of molecules as well, namely that the more stable protonated form is the one in which the positive charge is more delocalized, in this case by resonance (Section 1.6).

Let us consider first the potential sites for proton transfer to an oxygen atom of a carboxylic acid such as acetic acid. Proton transfer to the carbonyl oxygen gives cation A, and proton transfer to the hydroxyl oxygen gives cation B.

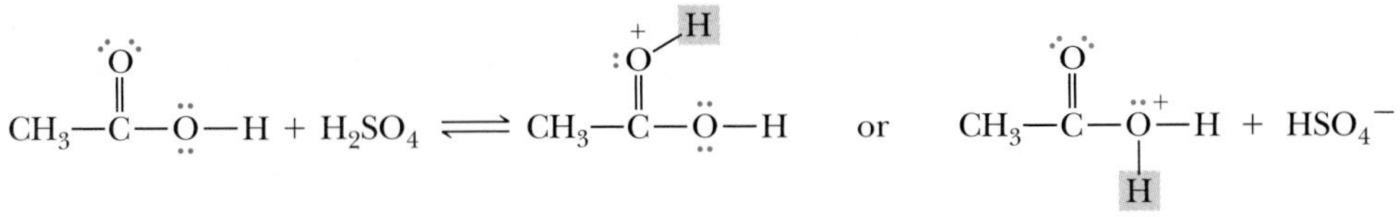

A
(protonation on the carbonyl oxygen)

B
(protonation on the hydroxyl oxygen)

We now examine each cation and determine which is the more stable (lower in energy). For cation A, we can write three resonance structures, two of which place the positive charge on oxygen and one of which places it on carbon. Of these three structures, A-1 and A-3 make the greater contributions to the hybrid; A-2, in which carbon has an incomplete octet, makes a lesser contribution. Thus, on protonation of the carbonyl oxygen, the positive charge is delocalized over three atoms with the greater share of it being on the two equivalent oxygen atoms. (They were not equivalent before protonation, but they are now.)

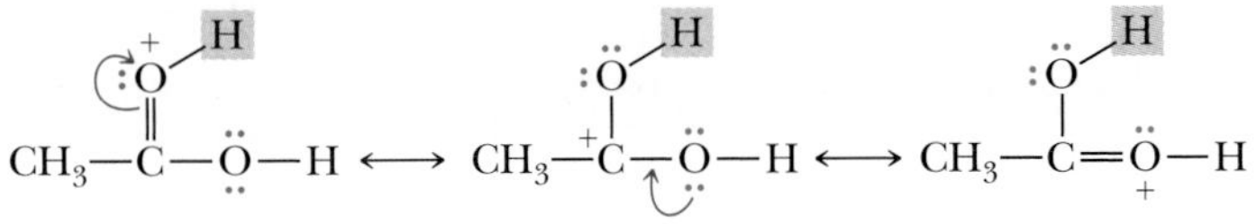

A-1
(C and O have complete octets)

A-2
(C has incomplete octet)

A-3
(C and O have complete octets)

Protonation on the hydroxyl oxygen gives a cation for which we can write two contributing structures. Of these, B-2 makes at best only a minor contribution to the hybrid; therefore, the charge on this cation is, in effect, localized on the hydroxyl oxygen.

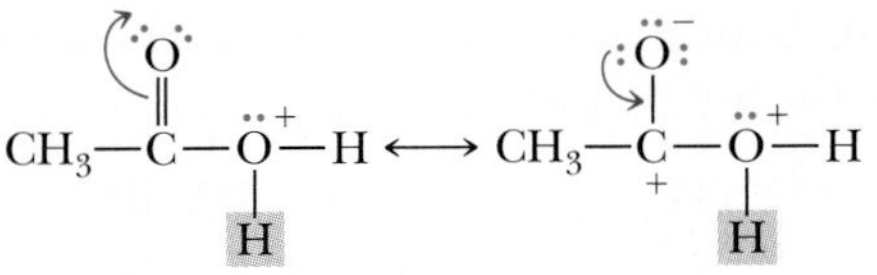

B-1

B-2
(charge separation and adjacent positive charges)

From this analysis of cations A and B, we see that protonation occurs preferentially on the carbonyl oxygen and that cation A is the predominant cation present at equilibrium.

Example 4.3

An amide is a functional group in which the —OH of a carboxyl group is replaced by an NH_2 group. Draw the structural formula of acetamide, which is derived from acetic acid, and determine if proton transfer to the amide group from HCl occurs preferentially on the carbonyl oxygen or the amide nitrogen.

Solution

Following is a Lewis structure for acetamide and its two possible protonated forms.

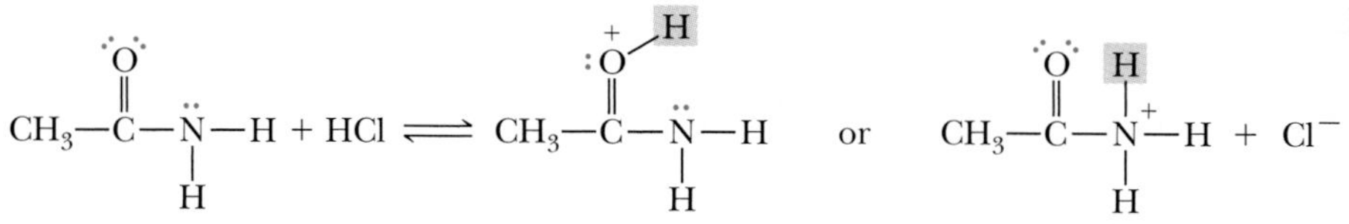

| Acetamide (an amide) | A (protonation on the amide oxygen) | B (protonation on the amide nitrogen) |

Three contributing structures can be drawn for cation A. Structures A-1 and A-3 make the greater contributions to the hybrid; of these, A-3 has the positive charge on the less electronegative atom and, therefore, makes a greater contribution than A-1. The result is that the positive charge in cation A is delocalized over three atoms, the greater share of it being on nitrogen and oxygen.

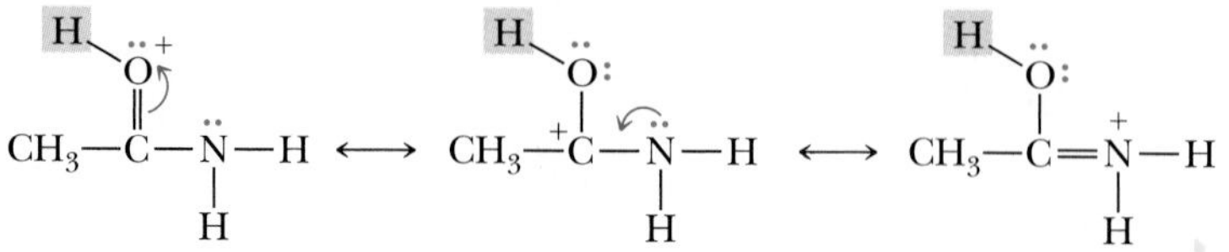

| A-1 (complete octets; positive charge on more electronegative atom) | A-2 (incomplete octet for C) | A-3 (complete octets; positive charge on less electronegative atom) |

Only two contributing structures can be drawn for cation B. Of these, B-2 requires creation and separation of unlike charges and places positive charges on adjacent atoms; it makes little contribution to the hybrid. Thus, the positive charge in cation B is essentially localized on the amide nitrogen.

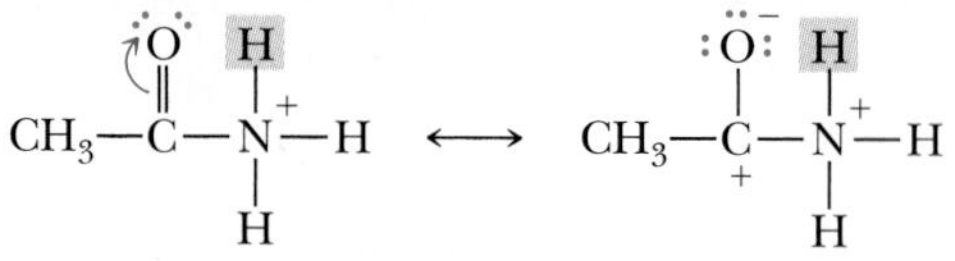

| B-1 (protonation on the amide nitrogen) | B-2 (adjacent plus charges; charge separation) |

From this analysis, we conclude that proton transfer to the carbonyl oxygen of the amide group gives the more stable cation; therefore, cation A is the predominant cation present at equilibrium.

Problem 4.3

Following is a structural formula for guanidine, the form in which migratory birds excrete excess metabolic nitrogen. The hydrochloride salt of this compound is a white crystalline powder, freely soluble in water and ethanol.

(a) Write a Lewis structure for guanidine showing all valence electrons.
(b) Does proton transfer occur preferentially to one of its —NH$_2$ groups or to its =NH group? Explain.

$$\underset{\text{Guanidine}}{H_2N-\overset{\displaystyle \overset{NH}{\|}}{C}-NH_2}$$

Thus far we have considered proton transfer to atoms having a nonbonding pair of electrons. Proton transfer reactions also occur with compounds having pi electrons, as for example, the pi electrons of carbon-carbon double and triple bonds. The pi electrons of the carbon-carbon double bond of 2-butene, for example, react with strong acids such as H$_2$SO$_4$, H$_3$PO$_4$, HCl, HBr, and HI by proton transfer to form a new carbon-hydrogen bond.

$$CH_3-CH=CH-CH_3 + H-\ddot{Br}: \longrightarrow CH_3-\overset{+}{C}-\overset{\overset{H}{|}}{\underset{|}{C}}-CH_3 + :\ddot{Br}:^-$$

| 2-Butene | *sec*-Butyl cation (a 2° carbocation) |

The result of this proton-transfer reaction is formation of a **carbocation,** a species in which one of its carbons has only six electrons in its valence shell and carries a charge of +1. Because the carbon bearing the positive charge in the *sec*-butyl cation has only two other carbons bonded to it, it is classified as a secondary (2°) carbocation.

Example 4.4

The acid-base reaction between 2-methyl-2-butene and HI can in principle form two isomeric carbocations. Write chemical equations for the formation of each carbocation. Use curved arrows to show the proton transfer in each reaction.

$$CH_3-\overset{\overset{\displaystyle CH_3}{|}}{C}=CH-CH_3$$

2-Methyl-2-butene

Solution

Proton transfer to carbon 3 of this alkene gives a tertiary (3°) carbocation. Proton transfer to carbon 2 gives a secondary (2°) carbocation.

$$CH_3-\overset{\overset{\displaystyle |}{C}}{\underset{\overset{\displaystyle |}{CH_3}}{}}=CH-CH_3 + H-\ddot{I}: \longrightarrow H_3C-\overset{+}{C}-\overset{\overset{H}{|}}{\underset{\overset{|}{H}}{C}}-CH_3 + :\ddot{I}:^-$$

A tertiary carbocation

$$CH_3-\underset{\underset{CH_3}{|}}{C}=CH-CH_3 \;+\; \overset{}{H}-\overset{..}{\underset{..}{I}}: \;\longrightarrow\; H_3C-\underset{\underset{H_3C}{|}}{\overset{\overset{H}{|}}{C}}-\overset{+}{\underset{\underset{H}{|}}{C}}-CH_3 \;+\; :\overset{..}{\underset{..}{I}}:^-$$

A secondary carbocation

Problem 4.4

Write an equation to show the proton transfer between each alkene or cycloalkene and HCl. Where isomeric carbocations are possible, show each.

(a) $CH_3CH_2CH=CHCH_3$ **(b)**

2–Pentene

Cyclohexene

4.2 Acid Dissociation Constants, pK_a, and the Relative Strengths of Acids and Bases

Any quantitative measure of the acidity of organic acids or bases involves measuring the equilibrium concentrations of the various components in an acid-base equilibrium. The strength of an acid is then expressed by an equilibrium constant. The dissociation of acetic acid in water is given by the following equation.

$$\underset{\text{Acetic acid}}{CH_3\overset{\overset{O}{\|}}{C}OH} \;+\; \underset{\text{Water}}{H_2O} \;\rightleftharpoons\; \underset{\text{Acetate ion}}{CH_3\overset{\overset{O}{\|}}{C}O^-} \;+\; \underset{\substack{\text{Hydronium}\\ \text{ion}}}{H_3O^+}$$

We can write an equilibrium expression for the dissociation of this or any other acid in a more general form; dissociation of the acid, HA, in water gives an anion, A^-, and the hydronium ion, H_3O^+. The equilibrium constant for this ionization is

$$HA + H_2O \rightleftharpoons A^- + H_3O^+$$

$$K_{eq} = \frac{[H_3O^+][A^-]}{[HA][H_2O]}$$

Because water is the solvent for this reaction and its concentration changes very little when HA is added to it, we can treat the concentration of water as a constant equal to 1000 g/L or approximately 55.6 mol/L. We can then combine these two constants (K_{eq} and the concentration of water) to define a new constant called an **acid dissociation constant**, given the symbol K_a.

$$K_a = K_{eq}[H_2O] = \frac{[H_3O^+][A^-]}{[HA]}$$

The value of the dissociation constant for acetic acid is 1.74×10^{-5}. Because dissociation constants for most acids, including organic acids, are numbers with negative exponents, acid strengths are often expressed as **pK_a** where p$K_a = -\log_{10} K_a$. The pK_a for acetic acid is 4.76. Table 4.1 gives names, molecular formulas, and values of pK_a for some organic and inorganic acids. Note that the larger the value of pK_a, the

$$K_a = 10^{-pK_a}$$

Table 4.1 pK$_a$ Values for Some Organic and Inorganic Acids

	Acid	Formula	pK_a	Conjugate Base	
Weaker acid	Ethane	CH_3CH_3	51	$CH_3CH_2^-$	Stronger base
	Ethylene	$CH_2{=}CH_2$	44	$CH_2{=}CH^-$	
	Ammonia	NH_3	38	NH_2^-	
	Hydrogen	H_2	35	H^-	
	Acetylene	$HC{\equiv}CH$	25	$HC{\equiv}C^-$	
	Ethanol	CH_3CH_2OH	15.9	$CH_3CH_2O^-$	
	Water	H_2O	15.7	HO^-	
	Methylammonium ion	$CH_3NH_3^+$	10.64	CH_3NH_2	
	Bicarbonate ion	HCO_3^-	10.33	CO_3^{2-}	
	Phenol	C_6H_5OH	9.95	$C_6H_5O^-$	
	Ammonium ion	NH_4^+	9.24	NH_3	
	Hydrogen sulfide	H_2S	7.04	HS^-	
	Carbonic acid	H_2CO_3	6.36	HCO_3^-	
	Acetic acid	CH_3COOH	4.76	CH_3COO^-	
	Benzoic acid	C_6H_5COOH	4.19	$C_6H_5COO^-$	
	Phosphoric acid	H_3PO_4	2.1	$H_2PO_4^-$	
	Hydronium ion	H_3O^+	-1.74	H_2O	
	Sulfuric acid	H_2SO_4	-5.2	HSO_4^-	
	Hydrogen chloride	HCl	-7	Cl^-	
Stronger acid	Hydrogen bromide	HBr	-8	Br^-	Weaker base
	Hydrogen iodide	HI	-9	I^-	

weaker the acid. Note also the inverse relationship between the strengths of the conjugate acid-base pairs; the stronger an acid, the weaker its conjugate base, and vice versa.

Example 4.5

For each value of pK_a, calculate the corresponding value of K_a. Which compound is the stronger acid?

(a) Ethanol, p$K_a = 15.9$ **(b)** Carbonic acid, p$K_a = 6.36$

Solution

(a) For ethanol, $K_a = 1.3 \times 10^{-16}$ **(b)** For carbonic acid, $K_a = 4.4 \times 10^{-7}$

Because the value of pK_a for carbonic acid is smaller than that for ethanol, carbonic acid is the stronger acid, and ethanol is the weaker acid.

Problem 4.5

For each value of K_a, calculate the corresponding value of pK_a. Which compound is the stronger acid?

(a) Acetic acid, $K_a = 1.74 \times 10^{-5}$ **(b)** Chloroacetic acid, $K_a = 1.38 \times 10^{-3}$

Values of pK_a in aqueous solution in the range 2 to 12 can be measured quite accurately. Values of pK_a smaller than 2 are less accurate because very strong acids, such as HCl, HBr, and HI, are completely ionized in water and the only acid present in solutions of these acids is H_3O^+. For acids too strong to be measured accurately in water, less basic solvents such as acetic acid or mixtures of water and sulfuric acid are used. Although none of the halogen acids, for example, is completely ionized in acetic acid, HI shows a greater degree of ionization than either HBr or HCl and, therefore, is the strongest acid of the three. Values of pK_a greater than 12 are also less accurate. For bases too strong to be measured in aqueous solution, more basic solvents such as liquid ammonia and dimethyl sulfoxide are used. Because different solvent systems are used to measure relative strengths at either end of the acidity scale, pK_a values smaller than 2 and greater than 12 should be used only in a qualitative way when comparing them with values in the middle of the scale.

4.3   The Position of Equilibrium in Acid-Base Reactions

We know from the value of pK_a for an acid whether an aqueous solution of the acid contains more molecules of the undissociated acid or its anion. For HCl, a strong acid with a pK_a of -7, at equilibrium in aqueous solution there will be no undissociated molecules of HCl present; the major species in solution will be H_3O^+ and Cl^-.

$$HCl + H_2O \longrightarrow H_3O^+ + Cl^- \qquad pK_a = -7$$

For acetic acid on the other hand, which is a weak acid with a pK_a of 4.76, the major species present at equilibrium in aqueous solution are CH_3COOH molecules.

$$\underset{\text{O}}{\overset{\text{O}}{CH_3\overset{\|}{C}OH}} + H_2O \rightleftharpoons \underset{\text{O}}{\overset{\text{O}}{CH_3\overset{\|}{C}O^-}} + H_3O^+ \qquad pK_a = 4.76$$

In these acid-base reactions, water is the base. But what if we have a base other than water as the proton acceptor, or if we have an acid other than hydrogen chloride or acetic acid as the proton donor? How do we determine quantitatively or even qualitatively which species are present at equilibrium; that is, how do we determine where the position of equilibrium lies?

Let us take as an example the acid-base reaction of acetic acid and ammonia to form acetate ion and ammonium ion.

$$\underset{\text{Acetic acid}}{\overset{\text{O}}{CH_3\overset{\|}{C}OH}} + \underset{\text{Ammonia}}{NH_3} \rightleftharpoons \underset{\text{Acetate ion}}{\overset{\text{O}}{CH_3\overset{\|}{C}O^-}} + \underset{\text{Ammonium ion}}{NH_4^+}$$

In this equilibrium there are two acids present, acetic acid and ammonium ion. There are also two bases present, ammonia and acetate ion. A way to analyze this equilibrium is to view it as a competition of the two bases, ammonia and acetate ion, for a proton. Which of these is the stronger base? The information we need to answer this question is in Table 4.1. We first determine which conjugate acid is the stronger acid and couple this with the fact that the stronger the acid, the weaker its conjugate base. From Table 4.1, we see that acetic acid, pK_a 4.76, is the stronger acid, which means that CH_3COO^- is the weaker base. Conversely, ammonium ion, pK_a 9.24, is the weaker acid, which means that NH_3 is the stronger base. We can now label the relative strengths of each acid and base. In an acid-base reaction, the position of equilibrium always favors reaction of the stronger acid and stronger base to form the weaker acid and weaker base. Thus at equilibrium, the major species present are the weaker acid and weaker base. Therefore, in the reaction between acetic acid and ammonia, the equilibrium lies to the right and the major species present are acetate ion and ammon)um ion.

$$
\underset{\substack{\text{Acetic acid} \\ \text{p}K_a\ 4.76 \\ \text{(stronger acid)}}}{CH_3\overset{\overset{\displaystyle O}{\|}}{C}OH} \;+\; \underset{\substack{\text{Ammonia} \\ \text{(stronger base)}}}{NH_3} \;\rightleftharpoons\; \underset{\substack{\text{Acetate ion} \\ \text{(weaker base)}}}{CH_3\overset{\overset{\displaystyle O}{\|}}{C}O^-} \;+\; \underset{\substack{\text{Ammonium ion} \\ \text{p}K_a\ 9.24 \\ \text{(weaker acid)}}}{NH_4{}^+}
$$

Not only can we use values of pK_a for each acid to estimate the position of equilibrium, but we can also use them to calculate an equilibrium constant for the equilibrium. Consider an acid-base reaction represented by the following general equation.

$$
\underset{\text{Acid}}{HA} \;+\; \underset{\text{Base}}{B} \;\rightleftharpoons\; \underset{\substack{\text{Conjugate} \\ \text{base of HA}}}{A^-} \;+\; \underset{\substack{\text{Conjugate} \\ \text{acid of B}}}{HB^+}
$$

The equilibrium constant for this reaction is

$$
K_{eq} = \frac{[A^-][BH^+]}{[HA][B]}
$$

Multiplying the right-hand side of this equation by $[H_3O^+]/[H_3O^+]$ gives a new expression, which, on rearrangement, becomes the K_a of acid HA divided by the K_a of acid BH^+.

$$
K_{eq} = \frac{[A^-][BH^+]}{[HA][B]} \times \frac{[H_3O^+]}{[H_3O^+]} = \frac{[A^-][H_3O^+]}{[HA]} \times \frac{[BH^+]}{[B][H_3O^+]}
$$

$$
= \frac{K_{HA}}{K_{BH^+}}
$$

Alternatively, we can take the logarithm of each side of this equation and arrive at this expression.

$$
pK_{eq} = pK_{HA} - pK_{BH^+}
$$

Thus, if we know the acid dissociation constants of each acid in the equilibrium, we can calculate the position of the acid-base equilibrium.

$$
\underset{\substack{\text{Acetic acid} \\ \text{p}K_a\ 4.76 \\ \text{(stronger acid)}}}{CH_3COOH} \;+\; \underset{\substack{\text{Ammonia} \\ \text{(stronger base)}}}{NH_3} \;\rightleftharpoons\; \underset{\substack{\text{Acetate ion} \\ \text{(weaker base)}}}{CH_3COO^-} \;+\; \underset{\substack{\text{Ammonium ion} \\ \text{p}K_a\ 9.24 \\ \text{(weaker acid)}}}{NH_4{}^+} \qquad \begin{array}{l} pK_{eq} = 4.76 - 9.24 = -4.48 \\ K_{eq} = 3.0 \times 10^4 \end{array}
$$

From the fact that the stronger acid reacts with the stronger base to give the weaker acid and the weaker base, we arrive at the conclusion that equilibrium lies to the right. Using values of pK_a, we calculate that the equilibrium constant for the reaction between acetic acid and ammonia is 3.0×10^4.

Consider now the reaction between aqueous solutions of acetic acid and sodium bicarbonate to give sodium acetate and carbonic acid. In the equation for this equilibrium, we omit the sodium ion, Na^+, because it does not undergo a chemical change in this reaction. Instead, we write the equilibrium as a net ionic equation, which shows only the species undergoing chemical change.

$$\underset{\substack{\text{Acetic acid} \\ pK_a \ 4.76}}{CH_3\overset{\displaystyle O}{\overset{\|}{C}}OH} + \underset{\text{Bicarbonate ion}}{HCO_3^-} \rightleftharpoons \underset{\text{Acetate ion}}{CH_3\overset{\displaystyle O}{\overset{\|}{C}}O^-} + \underset{\substack{\text{Carbonic acid} \\ pK_a \ 6.36}}{H_2CO_3} \qquad \begin{array}{l} pK_{eq} = 4.76 - 6.36 = -1.60 \\ K_{eq} = 10^{1.60} = 40 \end{array}$$

Acetic acid is the stronger acid; therefore, the position of this equilibrium lies to the right. Carbonic acid is formed, which then decomposes to carbon dioxide and water.

Example 4.6

Predict the position of equilibrium and calculate the equilibrium constant, K_{eq}, for each acid-base reaction.

(a) $\underset{\text{Phenol}}{C_6H_5OH} + \underset{\text{Bicarbonate ion}}{HCO_3^-} \rightleftharpoons \underset{\text{Phenoxide ion}}{C_6H_5O^-} + \underset{\text{Carbonic acid}}{H_2CO_3}$

(b) $\underset{\text{Acetylene}}{HC\equiv CH} + \underset{\text{Amide ion}}{NH_2^-} \rightleftharpoons \underset{\text{Acetylide ion}}{HC\equiv C^-} + \underset{\text{Ammonia}}{NH_3}$

Solution

(a) Carbonic acid is the stronger acid; the position of this equilibrium lies to the left. Phenol does not transfer a proton to bicarbonate ion to form carbonic acid.

$$\underset{pK_a \ 9.95}{C_6H_5OH} + HCO_3^- \underset{\longleftarrow}{\rightleftharpoons} C_6H_5O^- + \underset{pK_a \ 6.36}{H_2CO_3} \qquad \begin{array}{l} pK_{eq} = 9.95 - 6.36 = 3.59 \\ K_{eq} = 10^{-3.59} = 2.57 \times 10^{-4} \end{array}$$

(b) Acetylene is the stronger acid; the position of this equilibrium lies to the right.

$$\underset{pK_a \ 25}{HC\equiv CH} + NH_2^- \rightleftharpoons HC\equiv C^- + \underset{pK_a \ 38}{NH_3} \qquad \begin{array}{l} pK_{eq} = 25 - 38 = -13 \\ K_{eq} = 10^{13} \end{array}$$

Problem 4.6

Predict the position of equilibrium and calculate the equilibrium constant, K_{eq}, for these acid-base reactions.

(a) $\underset{\text{Methylamine}}{CH_3NH_2} + \underset{\text{Acetic acid}}{CH_3COOH} \rightleftharpoons \underset{\substack{\text{Methylammonium} \\ \text{ion}}}{CH_3NH_3^+} + \underset{\substack{\text{Acetate} \\ \text{ion}}}{CH_3COO^-}$

(b) $\underset{\text{Ethoxide ion}}{CH_3CH_2O^-} + \underset{\text{Ammonia}}{NH_3} \rightleftharpoons \underset{\text{Ethanol}}{CH_3CH_2OH} + \underset{\text{Amide ion}}{NH_2^-}$

4.4   Molecular Structure and Acidity

Let us now examine in some detail the relationships between molecular structure and acidity. The overriding principle in determining the relative acidities of organic acids is the stability of the anion, A$^-$, resulting from loss of a proton; the more stable the anion, the greater acidity of the acid, HA. As we will see, the relative stabilities of anions can be understood by a combination of effects, including the electronegativity and size of the atom bearing the negative charge, resonance, the inductive effect, and the hybridization of the atom bearing the negative charge.

A.   Electronegativity of the Atom Bearing the Negative Charge

Let us consider the relative acidities of the following series of hydrogen acids, all of which are in the same period of the Periodic Table.

Ethane	Methylamine	Methanol
pK_a 51	pK_a 38	pK_a 16

Increasing acidity

The pK_a value for ethane is given in Table 4.1, but values for methylamine and methanol are not. We can, however, make good guesses about the pK_a values of these acids by reasoning that the nature of the alkyl group bonded to nitrogen or oxygen has only a relatively small effect on the acidity of the hydrogen bonded to the **heteroatom** (in organic chemistry, an atom other than carbon). Therefore, we estimate that the pK_a of methylamine is approximately the same as that of ammonia (pK_a 38), and that the pK_a of methanol is approximately the same as ethanol (pK_a 15.9).

Ethane is the weakest acid in this series, and ethyl anion is the strongest conjugate base. Conversely, methanol is the strongest acid and methoxide ion is the weakest conjugate base.

	Ethyl anion	Methylamide ion	Methoxide ion

Increasing basicity

The relative acidity within a period of the Periodic Table is related to the electronegativity of the atom in the anion that bears the negative charge. The greater the electronegativity of this atom, the more strongly its electrons are held, and the more stable the anion. Conversely, the smaller the electronegativity of this atom, the less tightly its electrons are held and the less stable the anion. Oxygen, the most electronegative of the three atoms compared, has the largest electronegativity (3.5 on the Pauling scale), and methanol forms the most stable anion. Carbon, the least electronegative of the three (2.5) forms the least stable anion. Because methanol forms the most stable anion, it is the strongest acid in this series. Ethane is the weakest acid in the series.

It is essential to understand that this argument based on electronegativity applies only to acids within the same period (row) of the Periodic Table. Anions of atoms within the same period have approximately the same size, and their energies of solvation are approximately the same.

B.   Size of the Atom Bearing the Negative Charge

To illustrate how the acidity of hydrogen acids varies within a group (column) of the Periodic Table, let us compare the acidities of methanol and methanethiol, CH_3SH. We estimated in the previous section that the pK_a of methanol is 16. We can estimate the pK_a of methanethiol in the following way. The pK_a of hydrogen sulfide, H_2S, is given in Table 4.1 as 7.04. If we assume that substitution of a methyl group for a hydrogen makes only a slight change in acidity, we then estimate that the pK_a of methanethiol is approximately 7.0. Thus, methanethiol is the stronger acid, and methanethiolate ion is the weaker conjugate base.

$$CH_3-\overset{..}{\underset{..}{S}}-H \;+\; CH_3-\overset{..}{\underset{..}{O}}:^- \;\rightleftharpoons\; CH_3-\overset{..}{\underset{..}{S}}:^- \;+\; CH_3-\overset{..}{\underset{..}{O}}-H$$

Methanethiol	Methoxide	Methanethiolate	Methanol
$pK_a\,7.0$	ion	ion	$pK_a\,16$
(stronger acid)	(stronger base)	(weaker base)	(weaker acid)

The relative acidity of these two hydrogen acids, and in fact any set of hydrogen acids within a group of the Periodic Table, is related to the size of the atom bearing the negative charge. Size increases from top to bottom within a group because the valence electrons are in increasingly higher principal energy levels. This means that (1) they are farther from the nucleus and (2) they occupy a larger volume of space. Because sulfur is below oxygen in the Periodic Table, it is larger than oxygen. Accordingly, the negative charge on sulfur in methanethiolate is spread over a larger volume of space; therefore, the CH_3S^- anion is more stable. The negative charge on oxygen in methoxide ion is confined to a smaller volume of space; therefore, the CH_3O^- anion is less stable.

CH_3S^- CH_3O^-

We see this same trend in the strength of the halogen acids, HF, HCl, HBr, and HI, which increase in strength from HF (the weakest) to HI (the strongest). Of their anions, iodide ion is the largest; its charge has the largest volume of space, and it is the most stable. HI is, therefore, the strongest acid of the series. Conversely, fluoride is the smallest anion; its charge is the most concentrated, and fluoride ion is the least stable. HF is, therefore, the weakest acid of the series.

C.   Resonance Delocalization of Charge in the Anion

Carboxylic acids are weak acids. Values of pK_a for most unsubstituted carboxylic acids fall within the range 4 to 5. The pK_a for acetic acid, for example, is 4.76. Values of pK_a for most **alcohols,** compounds that also contain an —OH group, fall within the range 15 to 18; the value of pK_a for ethanol, for example, is 15.9. Thus, most alcohols are slightly weaker acids than water (pK_a = 15.7) but much weaker acids than carboxylic acids.

We account for the greater acidity of carboxylic acids compared with alcohols using the resonance model and looking at the relative stabilities of the alkoxide ion and the carboxylate ion. Our guideline is this: the more stable the anion, the farther the position of equilibrium is shifted toward the right, and the more acidic the compound.

Here we take the acid ionization of an alcohol as a reference equilibrium.

$$CH_3CH_2 \!-\! \overset{..}{\underset{..}{O}} \!-\! H \;+\; H_2O \;\rightleftharpoons\; CH_3CH_2 \!-\! \overset{..}{\underset{..}{O}}\!\!\,^{\,-} \;+\; H_3O^+ \qquad pK_a = 15.9$$

An alcohol   An alkoxide ion

In the alkoxide anion, the negative charge is localized on oxygen. In contrast, ionization of a carboxylic acid gives an anion for which we can write two equivalent contributing structures that result in delocalization of the negative charge. Because of this delocalization, a carboxylate anion is more stable than an alkoxide anion. Conversely, a carboxylic acid is a stronger acid than an alcohol.

$$CH_3 \!-\! \overset{\overset{\displaystyle ..O}{\|}}{\underset{\underset{\displaystyle \overset{..}{O} \!-\! H}{}}{C}} \;+\; H_2O \;\rightleftharpoons\; CH_3 \!-\! C \;\longleftrightarrow\; CH_3 \!-\! C \;+\; H_3O^+$$

These contributing structures are equivalent; the carboxylate anion is stabilized by delocalization of the negative charge.

D.   Inductive Effect and Electrostatic Stabilization of the Anion

We see an example of the **inductive effect** in alcohols in the fact that an electronegative substituent adjacent to the carbon bearing the —OH group increases the acidity of the alcohol. Compare, for example, the acidities of ethanol and 2,2,2-trifluoroethanol. The acid dissociation constant for 2,2,2-trifluoroethanol is larger than that of ethanol by more than three orders of magnitude, which means that the 2,2,2-trifluoroethoxide ion is considerably more stable than the ethoxide ion.

$$\underset{\displaystyle \underset{H}{|}}{\overset{\displaystyle \overset{H}{|}}{H \!-\! C \!-\! CH_2O \!-\! H}} \qquad\qquad \underset{\displaystyle \underset{F}{|}}{\overset{\displaystyle \overset{F}{|}}{F \!-\! C \!-\! CH_2O \!-\! H}}$$

Ethanol   2,2,2-Trifluoroethanol
pK_a 15.9 pK_a 12.4

We account for the increased stability of the 2,2,2-trifluoroethoxide ion in the following way. Fluorine is more electronegative than carbon (4.0 versus 2.5); therefore, the

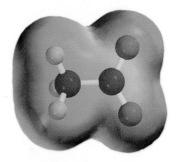

As shown in this electrostatic potential plot, the negative charge of an acetate ion is distributed equally between its two oxygen atoms.

Inductive effect   The polarization of the electron density of a covalent bond due to the electronegativity of a nearby atom.

C—F bond has a significant dipole, indicated in the following figure by directional arrows on the polar bonds as well as symbols to show the partial charges. There is an attractive stabilization by the interaction of the negatively charged oxygen and the partial positive charge on the carbon bearing the fluorines, which results in stabilization of the alkoxide ion.

Electrostatic attraction stabilizes the negative charge on oxygen.

Stabilization by the inductive effect falls off rapidly with increasing distance of the electronegative atoms(s) from the site of the negative charge. Compare, for example, the pK_a values of alcohols substituted with fluorine on carbons 2 versus 3 versus 4. When fluorine atoms are more than two carbons away from the carbon bearing the —OH group, they have almost no effect on acidity.

CF_3—CH_2—OH	CF_3—CH_2—CH_2—OH	CF_3—CH_2—CH_2—CH_2—OH
2,2,2-Trifluoroethanol (pK_a 12.4)	3,3,3-Trifluoro-1-propanol (pK_a 14.6)	4,4,4-Trifluoro-1-butanol (pK_a 15.4)

Inductive effect The polarization of the electron density of a covalent bond due to the electronegativity of a nearby atom.

We also see the operation of the inductive effect in the acidity of halogen-substituted carboxylic acids. The pK_a for chloroacetic acid, for example, is approximately two orders of magnitude greater than that of acetic acid.

Cl—CH_2—C(=O)—OH	CH_3—C(=O)—OH
Chloroacetic acid pK_a 2.86	Acetic acid pK_a 4.76

In the case of chloroacetate anion, the negative charge is stabilized by electrostatic interaction between the oxygens of the anion and the partial positive charge on the carbon bearing the chlorine atom.

The positive charge helps neutralize the negative charge on the oxygens.

In the hybrid, charge is distributed equally between the two oxygens.

As was the case with halogen substitution and the acidity of alcohols, the acid-enhancing effect of halogen substitution in carboxylic acids falls off rapidly with increasing distance between the point of substitution and the carboxyl group.

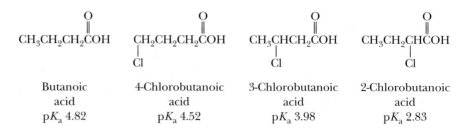

| $CH_3CH_2CH_2COH$ | $CH_2CH_2CH_2COH$ | CH_3CHCH_2COH | CH_3CH_2CHCOH |
	Cl	Cl	Cl
Butanoic acid pK_a 4.82	4-Chlorobutanoic acid pK_a 4.52	3-Chlorobutanoic acid pK_a 3.98	2-Chlorobutanoic acid pK_a 2.83

Table 4.2 Acidity of Alkanes, Alkenes, and Alkynes

	Weak Acid	Conjugate Base	pK_a
Water	HO—H	HO$^-$	15.7
Acetylene	HC≡C—H	HC≡C$^-$	25
Ammonia	H_2N—H	H_2N^-	38
Ethylene	CH_2=CH—H	CH_2=CH$^-$	44
Ethane	CH_3CH_2—H	$CH_3CH_2^-$	51

Increasing acidity →

E. Hybridization and the Percent *s* Character of the Atom Bearing the Negative Charge

To see the effect of hybridization, we consider the case of two or more anions, each with the same charge and each with an atom of the same element bearing the charge. The only difference is the hybridization of the atom bearing the negative charge. Of special importance for us is the acidity of a hydrogen bound to a carbon of an alkane, an alkene, and an alkyne.

One of the major differences between the chemistry of alkynes and alkenes is that a hydrogen attached to a triply bonded carbon atom is sufficiently acidic that it can be removed by a strong base, such as sodium amide or sodium hydride. Shown in Table 4.2 are pK_a values for acetylene, ethylene, and ethane. Also shown for comparison are values for ammonia and water.

We account for the greater acidity of alkynes compared with alkenes and alkanes in the following way. The lone pair of electrons on a carbon anion lies in a hybrid orbital: an *sp* orbital for an alkyne anion, an sp^2 orbital for an alkene anion, and an sp^3 orbital for an alkane anion. An *sp* orbital has 50% *s* character; an sp^2 orbital, 33%; and an sp^3 orbital, 25% (Section 1.8). Electrons in an *s* orbital are lower in energy than those in a *p* orbital; that is, they are held more tightly to the nucleus. Therefore, the more *s* character in a hybrid orbital, the more electronegative the atom will be, and the more acidic a hydrogen bonded to it will be (and the more stable the anion). Of the three types of compounds, the carbon in an alkyne (*sp* hybridized with 50% *s* character) is the most electronegative. Therefore, an alkyne anion is the most stable of the series; an alkyne anion is the least basic, and an alkyne is the strongest acid of the series. By similar reasoning, the alkane carbon (sp^3 hybridized with 25% *s* character) is the least electronegative, an alkane anion is the most basic, and an alkane is the weakest acid of the series. An alkene anion with 33% *s* character is intermediate.

4.5 Lewis Acids and Bases

Gilbert N. Lewis, who proposed that covalent bonds are formed by sharing one or more pairs of electrons (Section 1.2B), further expanded the theory of acids and bases to include a group of substances not included in the Brønsted-Lowry concept. According to the Lewis definition, an **acid** is a species that can form a new covalent bond by accepting a pair of electrons; a **base** is a species that can form a new covalent bond by donating a

Lewis acid Any molecule or ion that can form a new covalent bond by accepting a pair of electrons.
Lewis base Any molecule or ion that can form a new covalent bond by donating a pair of electrons.

pair of electrons. In the following general equation, the Lewis acid, A, accepts a pair of electrons in forming the new covalent bond and acquires a negative formal charge. The Lewis base, :B, donates the pair of electrons and acquires a positive formal charge.

$$ A \quad + \quad :B \quad \rightleftharpoons \quad \bar{A} \!-\! \overset{+}{B} $$

Lewis acid
(electron pair
acceptor)

Lewis base
(electron pair
donor)

new covalent bond
formed in this Lewis
acid-base reaction

CHEMISTRY IN ACTION

The Strongest Acid?

What is the strongest acid? In recent years, organic chemists have prepared mixtures of protic and Lewis acids that have remarkable proton-donating power. Two of the most reactive of these mixtures, termed **superacids,** are HF with SbF_5 and HSO_3F (fluorosulfonic acid) with SbF_5. Either SO_2 or SO_2F_2 is used as a solvent for superacids. SbF_5 is a Lewis acid that reacts with the fluorine of the acid; the net effect is to pull electrons away from hydrogen.

Both theoretical and experimental evidence exists that, in superacids, the electron pairs that make up the carbon-carbon and carbon-hydrogen bonds of hydrocarbons act as Lewis bases and become protonated. The resulting cations have unusual structures. In the case of ethane, for example, the ion $C_2H_7^+$ is produced.

The dashed lines in the structure shown for $C_2H_7^+$ indicate the formation of a three-center, two-electron bond. Notice that because the incoming proton has no electrons of its own, the octet rule is not violated.

Even at low temperatures ($-78°C$), the $C_2H_7^+$ ion is not very stable. One of its decomposition products is methane, CH_4. The observation of methane as a product supports protonation of the carbon-carbon bond electron pair over one of the six carbon-hydrogen electron pairs.

Among those studying the reactions of superacids with alkanes was George Olah, at the University of Southern California, who received the 1994 Nobel Prize for chemistry. Olah's discoveries completely transformed our understanding of the chemistry of hydrocarbon cations.

Although the chemistry of alkanes in superacids may seem esoteric, in fact, these reactions provide a model for one of the most important reactions in industrial organic chemistry, namely the catalytic cracking of petroleum (Section 2.9B). Highly acidic sites on the solid catalysts used in petroleum refining promote protonation of C—C and C—H bonds. Protonation of C—C bonds leads to fragmentation and isomerization of larger hydrocarbons; protonation of C—H bonds leads to the production of hydrogen and alkenes.

$$ H-O-\overset{\overset{O}{\|}}{\underset{\underset{O}{\|}}{S}}-F \; + \; SbF_5 \longrightarrow HO-\overset{\overset{O}{\|}}{\underset{\underset{O}{\|}}{S}}-\overset{+}{F}\overset{-}{SbF_5} $$

Fluorosulfonic
acid

$$ HO-\overset{\overset{O}{\|}}{\underset{\underset{O}{\|}}{S}}-\overset{+}{F}\overset{-}{SbF_5} \; + \; \underset{H}{\overset{H}{>}}C-C\underset{H}{\overset{H}{<}} \longrightarrow \left[\underset{H}{\overset{H}{>}}\overset{H}{\overset{+}{C}}\text{---}C\underset{H}{\overset{H}{<}} \right] \; + \; {}^-O-\overset{\overset{O}{\|}}{\underset{\underset{O}{\|}}{S}}-\overset{+}{F}\overset{-}{SbF_5} $$

(Unstable;
reacts further)

Note that, although we speak of a Lewis base as "donating" a pair of electrons, the term is not fully accurate. Donating in this case does not mean that the electron pair under consideration is removed completely from the valence shell of the base. Rather, donating means that the electron pair becomes shared with another atom to form a covalent bond.

Another example of a Lewis acid-base reaction is that of a carbocation (a Lewis acid) with bromide ion, a Lewis base. The *sec*-butyl cation, for example, reacts with bromide ion to form 2-bromobutane.

sec-Butyl cation	Bromide	2-Bromobutane
(a carbocation)	ion	

The Lewis concept of acids and bases includes proton-transfer reactions; all Brønsted-Lowry bases (proton acceptors) are also Lewis bases, and all Brønsted-Lowry acids (proton donors) are also Lewis acids. The Lewis model, however, is more general in that it is not restricted to proton-transfer reactions.

Consider the reaction that occurs when boron trifluoride gas is dissolved in diethyl ether.

Diethyl ether	Boron trifluoride	A BF_3-ether complex
(a Lewis base)	(a Lewis acid)	

Boron, a Group 3A element, has three electrons in its valence shell, and, after forming single bonds with three fluorine atoms to give BF_3, boron still has only six electrons in its valence shell. Because it has an empty orbital in its valence shell and can accept two electrons into it, boron trifluoride is electron deficient and, therefore, a Lewis acid. In forming the O—B bond, the oxygen atom of diethyl ether (a Lewis base) donates an electron pair, and boron accepts the electron pair. The reaction between diethyl ether and boron trifluoride is classified as an acid-base reaction according to the Lewis model, but because there is no proton transfer involved, it is not classified as an acid-base reaction by the Brønsted-Lowry model. Said another way, all Brønsted-Lowry acids are **protic acids;** Lewis acids may be protic acids or they may be **aprotic acids.**

Protic acid An acid that is a proton donor in an acid-base reaction.

Aprotic acid An acid that is not a proton donor; an acid that is an electron pair acceptor in a Lewis acid-base reaction.

Example 4.7

Write an equation for the reaction between each Lewis acid-base pair, showing electron flow by means of curved arrows.

(a) $BF_3 + NH_3 \rightarrow$ **(b)** $(CH_3)_2CH^+ + Cl^- \rightarrow$

Solution

(a) BF_3 has an empty orbital in the valence shell of boron and is the Lewis acid. NH_3 has an unshared pair of electrons in the valence shell of nitrogen and is the Lewis base. In this example, each of these atoms takes on a formal charge; the resulting structure, however, has no net charge.

Lewis acid   Lewis base

(b) The trivalent carbon atom in the isopropyl cation has an empty orbital in its valence shell and is, therefore, the Lewis acid. Chloride ion is the Lewis base.

Lewis acid   Lewis base

Problem 4.7

Write an equation for the reaction between each Lewis acid-base pair, showing electron flow by means of curved arrows.

(a) $(CH_3CH_2)_3B + OH^- \rightarrow$ **(b)** $CH_3Cl + AlCl_3 \rightarrow$

Summary

A **Brønsted-Lowry acid** is a proton donor, and a **Brønsted-Lowry base** is a proton acceptor (Section 4.1). Neutralization of an acid by a base is a proton-transfer reaction in which the acid is transformed into its **conjugate base,** and the base is transformed into its **conjugate acid.**

A strong acid or strong base is one that is completely ionized in water. A weak acid or weak base is one that is only partially ionized in water (Section 4.2). Among the most common weak organic acids are carboxylic acids, compounds that contain the —COOH (carboxyl) group. The value of K_a (the acid ionization constant) for acetic acid, a representative carboxylic acid, is 1.74×10^{-5}; the value of pK_a (the negative logarithm of K_a) for acetic acid is 4.76. The acidity of hydrogen acids is determined by the stability of the anion formed on deprotonation. Factors that influence the stability of an anion (Section 4.4) are (1) electronegativity of the atom bearing the negative charge, (2) size of the atom bearing the negative charge, (3) delocalization of charge in the anion, (4) the **inductive effect** and (5) the hybridization of the atom bearing the negative charge.

A **Lewis acid** (Section 4.5) is a species that can form a new covalent bond by accepting a pair of electrons; a **Lewis base** is a species that can form a new covalent bond by donating a pair of electrons.

Key Reactions

1. Proton-Transfer Reaction (Section 4.1)

A proton-transfer reaction involves transfer of a proton from a proton donor (a Brønsted-Lowry acid) to a proton acceptor (a Brønsted-Lowry base).

Proton   Proton
donor   acceptor

2. Position of Equilibrium in an Acid-Base Reaction (Section 4.3)

The stronger acid reacts with the stronger base to give a weaker acid and a weaker base. K_{eq} for this equilibrium can be calculated from pK_a values for the two acids.

$$CH_3COH + CN^- \rightleftharpoons CH_3CO^- + HCN \qquad pK_{eq} = 4.76 - 9.31 = -4.55$$

$$pK_a\ 4.76 \qquad\qquad\qquad pK_a\ 9.31 \qquad K_{eq} = 3.55 \times 10^4$$

3. Lewis Acid-Base Reaction (Section 4.5)

A Lewis acid-base reaction involves sharing an electron pair between an electron pair donor (a Lewis base) and an electron pair acceptor (a Lewis acid).

Problems

4.8 For each conjugate acid-base pair, identify the first species as an acid or base and the second species as its conjugate acid or base. In addition, draw Lewis structures for each species, showing all valence electrons and any formal charge.

(a) HCOOH   HCOO$^-$ (b) NH$_4^+$ NH$_3$ (c) CH$_3$CH$_2$O$^-$ CH$_3$CH$_2$OH
(d) HCO$_3^-$ CO$_3^{2-}$ (e) H$_2$PO$_4^-$ HPO$_4^{2-}$ (f) CH$_3$CH$_3$ CH$_3$CH$_2^-$
(g) CH$_3$S$^-$ CH$_3$SH

4.9 Complete a net ionic equation for each proton-transfer reaction using curved arrows to show the flow of electron pairs in each reaction. In addition, write Lewis structures for all starting materials and products. Label the original acid and its conjugate base; also label the original base and its conjugate acid. If you are uncertain about which substance in each equation is the proton donor, refer to Table 4.1 for the relative strengths of proton acids.

(a) NH$_3$ + HCl $\rightarrow$ (b) CH$_3$CH$_2$O$^-$ + HCl $\rightarrow$
(c) HCO$_3^-$ + OH$^-$ $\rightarrow$ (d) CH$_3$COO + NH$_4^+$ $\rightarrow$

4.10 Complete a net ionic equation for each proton-transfer reaction using curved arrows to show the flow of electron pairs in each reaction. Label the original acid and its conjugate base; also label the original base and its conjugate acid.

(a) NH$_4^+$ + OH$^-$ $\rightarrow$ (b) CH$_3$CO$^-$ + CH$_3$NH$_3^+$ $\longrightarrow$
(c) CH$_3$CH$_2$O$^-$ + NH$_4^+$ $\rightarrow$ (d) CH$_3$NH$_3^+$ + OH$^-$ $\rightarrow$

4.11 Each molecule or ion can function as a base. Write a structural formula of the conjugate acid formed by reaction of each with HCl.

$$\overset{O}{\overset{\|}{}}$$

(a) CH_3CH_2OH (b) HCH (c) $(CH_3)_2NH$ (d) HCO_3^-

4.12 In acetic acid, the OH hydrogen is more acidic than the CH_3 hydrogens. Explain.

4.13 As we shall see in Chapter 19, hydrogens on a carbon adjacent to a carbonyl group are far more acidic than those not adjacent to a carbonyl group. The anion derived from acetone, for example, is more stable than is the anion derived from ethane. Account for the greater stability of the anion from acetone.

$$\overset{O}{\overset{\|}{CH_3CCH_2}}-H \qquad CH_3CH_2-H$$

Acetone   Ethane
pK_a 22 pK_a 51

Quantitative Measure of Acid and Base Strength

4.14 Which has the larger numerical value?

(a) The pK_a of a strong acid or the pK_a of a weak acid
(b) The K_a of a strong acid or the K_a of a weak acid

4.15 In each pair, select the stronger acid.

(a) Pyruvic acid (pK_a 2.49) and lactic acid (pK_a 3.85)
(b) Citric acid (pK_{a1} 3.08) and phosphoric acid (pK_{a1} 2.10)

4.16 Arrange the compounds in each set in order of increasing acid strength. Consult Table 4.1 for pK_a values of each acid.

(a) CH_3CH_2OH $HO\overset{O}{\overset{\|}{C}}O^-$ $C_6H_5\overset{O}{\overset{\|}{C}}OH$

Ethanol   Bicarbonate ion   Benzoic acid

(b) $HO\overset{O}{\overset{\|}{C}}OH$ $CH_3\overset{O}{\overset{\|}{C}}OH$ HCl

Carbonic acid   Acetic acid   Hydrogen chloride

4.17 Arrange the compounds in each set in order of increasing base strength. Consult Table 4.1 for pK_a values of the conjugate acid of each base.

(a) NH_3 $HO\overset{O}{\overset{\|}{C}}O^-$ $CH_3CH_2O^-$ (b) OH^- $HO\overset{O}{\overset{\|}{C}}O^-$ $CH_3\overset{O}{\overset{\|}{C}}O^-$

(c) H_2O NH_3 $CH_3\overset{O}{\overset{\|}{C}}O^-$ (d) NH_2^- $CH_3\overset{O}{\overset{\|}{C}}O^-$ OH^-

Position of Equilibrium in Acid-Base Reactions

4.18 Unless under pressure, carbonic acid (Table 1.8) in aqueous solution breaks down into carbon dioxide and water, and carbon dioxide is evolved as bubbles of gas. Write an equation for the conversion of carbonic acid to carbon dioxide and water.

4.19 Will carbon dioxide be evolved when sodium bicarbonate is added to an aqueous solution of these compounds? Explain.

(a) Sulfuric acid   (b) Ethanol   (c) Ammonium chloride

4.20 Acetic acid, CH_3COOH, is a weak organic acid, pK_a 4.76. Write an equation for the equilibrium reactions of acetic acid with each base. Which equilibria lie considerably toward the left? Which lie considerably toward the right?

(a) $NaHCO_3$ (b) NH_3 (c) H_2O (d) $NaOH$

4.21 Alcohols are weak organic acids, pK_a 15–18. The pK_a of ethanol, CH_3CH_2OH, is 15.9. Write equations for the equilibrium reaction of ethanol with each base. Which equilibria lie considerably toward the left? Which lie considerably toward the right?

(a) HCO_3^- (b) OH^- (c) NH_2^- (d) NH_3

4.22 Benzoic acid, C_6H_5COOH, is only slightly soluble in water, but its sodium salt, $C_6H_5COO^-Na^+$, is quite soluble in water. In which solution(s) will benzoic acid dissolve (more readily than in water)?

(a) Aqueous $NaOH$ (b) Aqueous $NaHCO_3$ (c) Aqueous Na_2CO_3

4.23 4-Methylphenol, $CH_3C_6H_4OH$ (pK_a 10.26), is only slightly soluble in water, but its sodium salt, $CH_3C_6H_4O^-Na^+$, is quite soluble in water. In which solution(s) will 4-methylphenol dissolve?

(a) Aqueous $NaOH$ (b) Aqueous $NaHCO_3$ (c) Aqueous Na_2CO_3

4.24 For an acid-base reaction, one way to determine the predominant species at equilibrium is to say that the reaction arrow points to the acid with the higher value of pK_a. For example,

$$NH_4^+ + H_2O \longleftarrow NH_3 + H_3O^+$$

$$pK_a\ 9.24 \qquad\qquad\qquad pK_a\ -1.74$$

$$NH_4^+ + OH^- \longrightarrow NH_3 + H_2O$$

$$pK_a\ 9.24 \qquad\qquad\qquad pK_a\ 15.7$$

Explain why this rule works.

4.25 Will acetylene react with sodium hydride according to the following equation to form a salt and hydrogen? Using pK_a values given in Table 4.1, calculate K_{eq} for this equilibrium.

$$HC\equiv CH + Na^+H^- \longrightarrow HC\equiv C^-Na^+ + H_2$$

Acetylene   Sodium   Sodium   Hydrogen
             hydride   acetylide

4.26 Using pK_a values given in Table 4.1, predict the position of equilibrium in this acid-base reaction and calculate its K_{eq}.

$$H_3PO_4 + CH_3CH_2OH \rightleftharpoons H_2PO_4^- + CH_3CH_2OH_2^+$$

Lewis Acids and Bases

4.27 For each equation, label the Lewis acid and the Lewis base. In addition, use curved arrows to show the flow of electrons in each reaction.

4.28 Complete the equation for the reaction between each Lewis acid-base pair. In each equation, label which starting material is the Lewis acid and which the Lewis base; use curved arrows to show the flow of electrons in each reaction. In doing this problem, it is essential that you show valence electrons for all atoms participating in each reaction.

$$
\text{(a)} \quad CH_3-\underset{\underset{CH_3}{|}}{\overset{\overset{CH_3}{|}}{C}}-Cl + \underset{\underset{Cl}{|}}{Al}-Cl \longrightarrow \qquad \text{(b)} \quad CH_3-\underset{\underset{CH_3}{|}}{\overset{\overset{CH_3}{|}}{C^+}} + H-O-H \longrightarrow
$$

$$
\text{(c)} \quad CH_3-\overset{+}{C}H-CH_3 + Br^- \rightarrow \qquad \text{(d)} \quad CH_3-\overset{+}{C}H-CH_3 + CH_3-O-H \rightarrow
$$

4.29 Each of these reactions can be written as a Lewis acid–Lewis base reaction. Label the Lewis acid and the Lewis base; use curved arrows to show the flow of electrons in each reaction. In doing this problem, it is essential that you show valence electrons for all atoms participating in each reaction.

$$
\text{(a)} \quad CH_3-CH{=}CH_2 + H-Cl \longrightarrow CH_3-\overset{+}{C}H-\overset{\overset{H}{|}}{C}H_2 + Cl^-
$$

$$
\text{(b)} \quad CH_3-\underset{\underset{CH_3}{|}}{C}{=}CH_2 + Br-Br \longrightarrow CH_3-\underset{\underset{CH_3}{|}}{\overset{+}{C}}-CH_2-Br + Br^-
$$

Additional Problems

4.30 2,4-Pentanedione is a considerably stronger acid than acetone. Write a structural formula for the conjugate base of each acid and account for the greater stability of the conjugate base from 2,4-pentanedione.

$$
\underset{\substack{\text{Acetone} \\ pK_a\ 22}}{\overset{\overset{O}{\|}}{CH_3CCH_2}-H} \qquad\qquad \underset{\substack{\text{2,4-Pentanedione} \\ pK_a\ 9}}{\overset{\overset{O\quad O}{\|\ \ \|}}{CH_3C\underset{\underset{H}{|}}{C}HCCH_3}}
$$

4.31 Write an equation for the acid-base reaction between 2,4-pentanedione and sodium ethoxide, and calculate its equilibrium constant, K_{eq}. The pK_a of 2,4-pentanedione is 9; that of ethanol is 15.9.

$$
\underset{\text{2,4-Pentanedione}}{\overset{\overset{O\quad O}{\|\ \ \|}}{CH_3C\underset{\underset{H}{|}}{C}HCCH_3}} \quad + \quad \underset{\text{Sodium ethoxide}}{CH_3CH_2O^-Na^+} \rightleftharpoons
$$

4.32 The *sec*-butyl cation can react as both a Lewis acid and a Brønsted-Lowry acid in the presence of a water-sulfuric acid mixture. In each case, however, the product is different. The two reactions are

$$
\text{(1)} \quad \underset{\text{\textit{sec}-Butyl cation}}{CH_3-\overset{+}{C}H-CH_2-CH_3} + H_2O \longrightarrow CH_3-\underset{\underset{O}{\overset{\overset{H\diagdown\ \overset{+}{\ }\ \diagup H}{}}{|}}}{C}H-CH_2-CH_3
$$

$$
\text{(2)} \quad \underset{\text{\textit{sec}-Butyl cation}}{CH_3-\overset{+}{C}H-CH_2-CH_3} + H_2O \longrightarrow CH_3-CH{=}CH-CH_3 + H_3O^+
$$

(a) In which reaction(s) does this cation react as a Lewis acid? In which does it react as a Brønsted-Lowry acid?

(b) Write Lewis structures for reactants and products and show by the use of curved arrows how each reaction occurs.

4.33 Following is a structural formula for the *tert*-butyl cation.

$$CH_3-\overset{+}{\underset{\underset{CH_3}{|}}{C}}-CH_3$$

tert-Butyl cation
(a carbocation)

(a) Predict all C—C—C bond angles in this cation.

(b) What is the hybridization of the carbon bearing the positive charge?

(c) Write a balanced equation to show its reaction as a Lewis acid with water; to show its reaction as a Brønsted-Lowry acid with water.

4.34 Write equations for the reaction of each compound with H_2SO_4, a strong protic acid.

(a) CH_3OCH_3 (b) $CH_3CH_2SCH_2CH_3$ (c) $CH_3CH_2NHCH_2CH_3$

(d) $CH_3\overset{\overset{\displaystyle CH_3}{|}}{N}CH_3$ (e) $CH_3\overset{\overset{\displaystyle O}{||}}{C}CH_3$ (f) $CH_3\overset{\overset{\displaystyle O}{||}}{C}OCH_3$

4.35 Write equations for the reaction of each compound in Problem 4.34 with BF_3, a Lewis acid.

4.36 Write a structural formula for the conjugate base formed when each compound is treated with one mole of a base stronger than the compound's conjugate base.

(a) $HOCH_2CH_2NH_2$ (b) $HSCH_2CH_2NH_2$ (c) $HOCH_2CH_2C\equiv CH$

(d) $HO\overset{\overset{\displaystyle O}{||}}{C}CH_2CH_2SH$ (e) $CH_3\overset{\overset{\displaystyle HO\ \ O}{\underset{}{|\ \ \ ||}}}{C}HCOH$ (f) $H_3\overset{+}{N}CH_2CH_2\overset{\overset{\displaystyle O}{||}}{C}OH$

(g) $H_3N^+CH_2CH_2\overset{\overset{\displaystyle O}{||}}{C}O^-$ (h) $HSCH_2CH_2OH$

4.37 Explain why the hydronium ion, H_3O^+, is the strongest acid that can exist in aqueous solution. What is the strongest base that can exist in aqueous solution?

4.38 What is the strongest base that can exist in liquid ammonia as a solvent?

4.39 For each pair of molecules or ions, select the stronger base and write its Lewis structure.

(a) CH_3S^- or CH_3O^- (b) CH_3NH^- or CH_3O^-
(c) CH_3COO^- or OH^- (d) $CH_3CH_2O^-$ or H^-
(e) NH_3 or OH^- (f) NH_3 or H_2O
(g) CH_3COO^- or HCO_3^- (h) HSO_4^- or OH^-
(i) OH^- or Br^-

4.40 Account for the fact that nitroacetic acid, O_2NCH_2COOH (pK_a 1.68), is a considerably stronger acid than acetic acid, CH_3COOH (pK_a 4.76).

4.41 Sodium hydride, NaH, is available commercially as a gray-white powder. It melts at 800°C with decomposition. It reacts explosively with water and ignites spontaneously on standing in moist air.

(a) Write a Lewis structure for the hydride ion and for sodium hydride. Is your Lewis structure consistent with the fact that this compound is a high-melting solid? Explain.

(b) When sodium hydride is added very slowly to water, it dissolves with the evolution of a gas. The resulting solution is basic to litmus. What is the gas evolved? Why has the solution become basic?

(c) Write an equation for the reaction between sodium hydride and 1-butyne, $CH_3CH_2C\equiv CH$. Use curved arrows to show the flow of electrons in this reaction.

4.42 An ester is a derivative of a carboxylic acid in which the hydrogen of the carboxyl group is replaced by an alkyl group (Section 1.3E). Draw a structural formula of methyl acetate, which is derived from acetic acid by replacement of the H of its —OH group by a methyl group. Determine if proton transfer to this compound from HCl occurs preferentially on the oxygen of the C=O group or the oxygen of the OCH_3 group.

4.43 Following is a structural formula for imidazole, a building block of the essential amino histidine (Chapter 27). It is also a building block of histamine (see *The Merck Index,* 12th ed., #4756), a compound all too familiar to persons with allergies and takers of antihistamines. When imidazole is dissolved in water, proton transfer to it gives a cation. Is this cation better represented by structure A or B? Explain.

Imidazole + $H_2O \rightleftharpoons$ A or B + OH^-

4.44 Methyl isocyanate, CH_3—N=C=O, is used in the industrial synthesis of a type of pesticide and herbicide known as a carbamate. As a historical note, an industrial accident in Bhopal, India, in 1984 resulted in leakage of an unknown quantity of this chemical into the air. An estimated 200,000 persons were exposed to its vapors, and over 2000 of these people died.

(a) Write a Lewis structure for methyl isocyanate and predict its bond angles. What is the hybridization of its carbonyl carbon? Of its nitrogen atom?

(b) Methyl isocyanate reacts with strong acids, such as sulfuric acid, to form a cation. Will this molecule undergo protonation more readily on its oxygen or nitrogen atom? In considering contributing structures to each hybrid, do not consider structures in which more than one atom has an incomplete octet.

4.45 Offer an explanation for the following observations.

(a) H_3O^+ is a stronger acid than NH_4^+.

(b) Nitric acid, HNO_3, is a stronger acid than nitrous acid, HNO_2.

(c) Ethanol and water have approximately the same acidity.

(d) Trifluoroacetic acid, CF_3COOH, is a stronger acid than trichloroacetic acid, CCl_3COOH.

Methyl isocyanate

5

ALKENES I

An **unsaturated hydrocarbon** contains one or more carbon-carbon double or triple bonds. The term "unsaturated" indicates that there are fewer hydrogens bonded to carbon than in an alkane, C_nH_{2n+2}. The three classes of unsaturated hydrocarbons are alkenes, alkynes, and arenes. Alkenes contain a carbon-carbon double bond and have the general formula C_nH_{2n}. Alkynes contain a carbon-carbon triple bond and have the general formula C_nH_{2n-2}. The simplest alkene is ethylene, and the simplest alkyne is acetylene.

■ Haze of the Blue Ridge Mountains. This haze is caused by light-scattering from the aerosol produced by the photooxidaton of isoprene and other hydrocarbons. See the Chemistry in Action box "Why Plants Emit Isoprene." *(Bob Thomason/Tony Stone Images)* Inset: A model of isoprene.

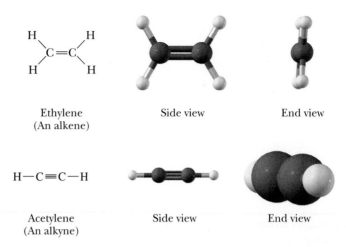

Ethylene
(An alkene)

Side view

End view

Acetylene
(An alkyne)

Side view

End view

In this chapter, we study the structure, nomenclature, and physical properties of alkenes. Alkynes are discussed separately in Chapter 10.

Arene A term used to classify benzene and its derivatives.

Arenes are the third class of unsaturated hydrocarbons. The Lewis structure of benzene, the simplest arene, is

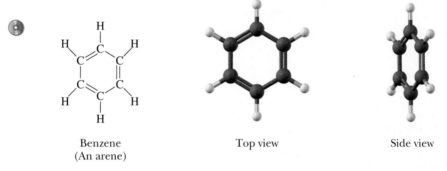

Benzene
(An arene)

Top view

Side view

Aryl group (Ar—) A group derived from an arene by removal of an H.

Phenyl group A group derived by removing an H from benzene; abbreviated C_6H_5— or Ph—.

Just as a group derived by removal of an H from an alkane is called an alkyl group and given the symbol R— (Section 2.3A), a group derived by removal of an H from an arene is called an **aryl group** and given the symbol **Ar—.**

When a benzene ring occurs as a substituent on a parent chain, it is named a **phenyl group.** You might think that when present as a substituent, benzene would become benzyl, just as ethane becomes ethyl. This is not so! "Phene" is a now-obsolete name for benzene and, although this name is no longer used, a derivative has persisted in the name "phenyl." Following is a structural formula for the phenyl group and two alternative representations for it. Throughout this text, we will represent benzene by a hexagon with three inscribed double bonds. It is also common to represent it by a hexagon with an inscribed circle. We explain the reasons for the alternative representations in Chapter 20.

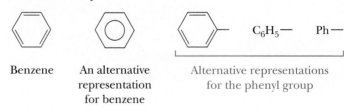

Benzene

An alternative
representation
for benzene

C_6H_5—   Ph—

Alternative representations
for the phenyl group

The chemistry of benzene and its derivatives is quite different from that of alkenes and alkynes, but, even though we do not study the chemistry of arenes until Chapters 20 and 21, we will show structural formulas of compounds containing aryl groups before that time. What you need to remember at this point is that an aryl group is not chemically reactive under any of the conditions we describe in Chapters 6–19.

5.1   Structure of Alkenes

A.   Shapes of Alkenes

Using the valence-shell electron-pair repulsion model (Section 1.4) of a carbon-carbon double bond, we predict a value of 120° for the bond angles about each carbon in a double bond. The observed H—C—C bond angle in ethylene is 121.1°, close to that predicted. In other alkenes, deviations from the predicted angle of 120° may be somewhat larger because of the strain introduced by nonbonded interactions created by groups attached to the carbons of the double bond. The C—C—C bond angle in propene, for example, is 123.9°.

B.   Molecular Orbital Model of a Carbon-Carbon Double Bond

In Section 1.8D, we described the formation of a carbon-carbon double bond in terms of the overlap of atomic orbitals. A carbon-carbon double bond consists of one sigma bond and one pi bond (Figure 5.1). Each carbon of the double bond

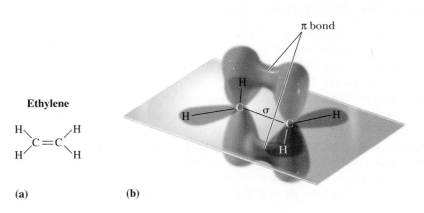

Figure 5.1
Covalent bonding in ethylene. (a) Lewis structure and (b) orbital overlap model showing the sigma and pi bonds.

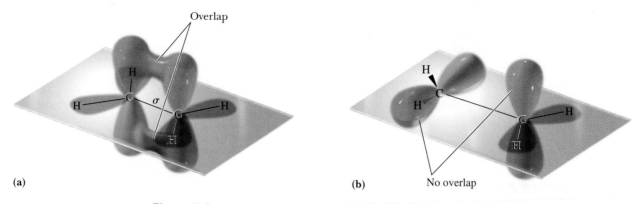

Figure 5.2
Restricted rotation about a carbon-carbon double bond. (a) Orbital overlap model showing the pi bond. (b) The pi bond is broken by rotating the plane of one H—C—H group by 90° with respect to the plane of the other H—C—H group.

uses its three sp^2 hybrid orbitals to form sigma bonds to three atoms. The unhybridized $2p$ atomic orbitals, which lie perpendicular to the plane created by the axes of the three sp^2 hybrid orbitals, combine to form two pi molecular orbitals: one bonding and the other antibonding. We are concerned here with the pi bonding MO only. For the unhybridized $2p$ orbitals to be parallel, thus giving maximum overlap, the two carbon atoms of the double bond and the four attached atoms must lie in a plane.

It takes approximately 264 kJ (63 kcal)/mol to break the pi bond in ethylene, that is, to rotate one carbon by 90° with respect to the other where zero overlap occurs between $2p$ orbitals on adjacent carbons (Figure 5.2). This energy is considerably greater than the thermal energy available at room temperature; consequently, rotation about a carbon-carbon double bond does not occur under normal conditions. You might compare rotation about a carbon-carbon double bond, such as in ethylene, with that about a carbon-carbon single bond, such as in ethane (Section 2.6A). Whereas rotation about the carbon-carbon single bond in ethane is relatively free [rotation barrier approximately 12.5 kJ (3.0 kcal)/mol], rotation about the carbon-carbon double bond in ethylene is severely restricted.

C.   Cis,Trans Isomerism in Alkenes

Because of restricted rotation about a carbon-carbon double bond, any alkene in which each carbon of the double bond has two different groups attached to it shows **cis,trans isomerism.** For example, 2-butene has two stereoisomers. In *cis*-2-butene, the two methyl groups are on one side of the double bond, and the two hydrogens are on the other side. In *trans*-2-butene, the two methyl groups are on opposite sides of the double bond. These two compounds cannot be converted into one another at room temperature because of the restricted rotation about the double bond; they are different compounds, with different physical and chemical properties.

Cis alkenes are less stable than their trans isomers because of nonbonded interaction strain between alkyl substituents on the same side of the double bond, as can be seen in space-filling models of the cis and trans isomers of 2-butene. This is the same

Cis,trans isomers   Isomers that have the same order of attachment of their atoms but a different arrangement of their atoms in space due to the presence of either a ring (Chapter 2) or a carbon-carbon double bond (Chapter 5).

type of strain that results in the preference for equatorial methylcyclohexane over axial methylcyclohexane (Section 2.6B).

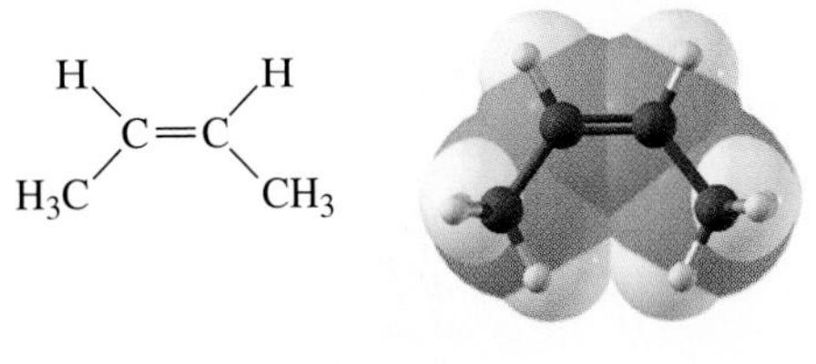

cis-2-Butene
mp −139°C, bp 4°C

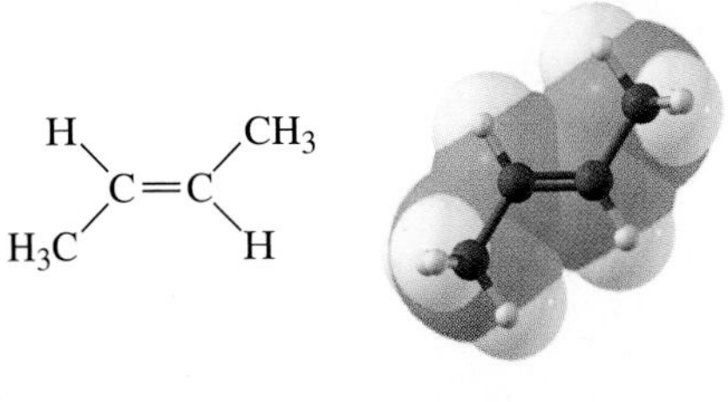

trans-2-Butene
mp −106°C, bp 1°C

5.2 Nomenclature of Alkenes

Alkenes are named using the IUPAC system, but, as we shall see, some are still referred to by their common names.

A. IUPAC Names

To form IUPAC names for alkenes, change the -an- infix of the parent alkane to -en- (Section 2.5). Hence, CH_2=CH_2 is named ethene, and CH_3CH=CH_2 is named propene. In higher alkenes, where isomers exist that differ in location of the double bond, a numbering system must be used. According to the IUPAC system,

1. Number the longest carbon chain that contains the double bond in the direction that gives the carbon atoms of the double bond the lowest possible numbers.
2. Indicate the location of the double bond by the number of its first carbon.
3. Name branched or substituted alkenes in a manner similar to alkanes.
4. Number the carbon atoms, locate and name substituent groups, locate the double bond, and name the main chain.

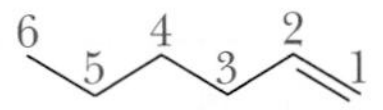

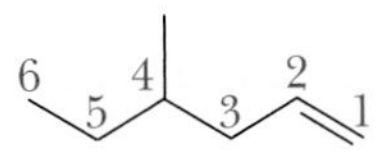

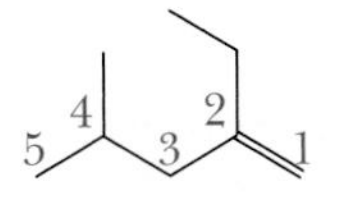

$\underset{CH_3CH_2CH_2CH_2CH=CH_2}{\overset{6\ \ 5\ \ 4\ \ 3\ \ 2\ \ \ 1}{}}$

1-Hexene

$\underset{CH_3CH_2CHCH_2CH=CH_2}{\overset{6\ \ 5\ \ 4\ \ 3\ \ 2\ \ \ 1}{}}$
CH₃

4-Methyl-1-hexene

$\underset{CH_3CHCH_2C=CH_2}{\overset{5\ \ 4\ \ 3\ \ 2\ \ 1}{}}$
CH₃
CH₂CH₃

2-Ethyl-4-methyl-1-pentene

Note that there is a chain of six carbon atoms in 2-ethyl-4-methyl-1-pentene. However, because the longest chain that contains the double bond has only five carbons, the parent hydrocarbon is pentane, and the molecule is named as a disubstituted 1-pentene.

Example 5.1

Write the IUPAC name of each alkene.

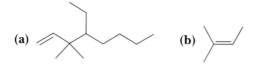

(a)   (b)

Solution

(a) 4-Ethyl-3,3-dimethyl-1-octene   (b) 2-Methyl-2-butene

Problem 5.1

Write the IUPAC name of each alkene.

(a)   (b)

B. Common Names

Despite the precision and universal acceptance of IUPAC nomenclature, some alkenes, particularly those of low molecular weight, are known almost exclusively by their common names, as illustrated by the common names of these alkenes.

$$CH_2{=}CH_2 \qquad CH_3CH{=}CH_2 \qquad \overset{\overset{\displaystyle CH_3}{|}}{CH_3}C{=}CH_2$$

IUPAC name:	Ethene	Propene	2-Methylpropene
Common name:	Ethylene	Propylene	Isobutylene

Furthermore, the common names methylene (a CH_2 group), vinyl, and allyl are often used to show the presence of the following alkenyl groups:

Alkenyl Group	IUPAC Name	Common Name	Example	IUPAC Name (Common Name)
$CH_2{=}$	Methylidene	Methylene	$H_2C{=}$⬠	Methylidenecyclopentane (Methylenecyclopentane)
$CH_2{=}CH{-}$	Ethenyl	Vinyl	$CH_2{=}CH{-}$⬠	Ethenylcyclopentane (Vinylcyclopentane)
$CH_2{=}CHCH_2{-}$	2-Propenyl	Allyl	$CH_2{=}CHCH_2{-}$⬠	2-Propenylcyclopentane (Allylcyclopentane)

C. Systems for Designating Configuration in Alkenes

The Cis,Trans System

The most common method for specifying the configuration in alkenes uses the prefixes cis and trans. There is no doubt whatsoever which isomer is intended by the name *trans*-3-hexene. For more complex alkenes, the orientation of the atoms of the parent chain determines whether the alkene is cis or trans. On the right is a structural formula for the cis isomer of 3,4-dimethyl-2-pentene. In this example, carbon atoms of the main chain (carbons 1 and 4) are on the same side of the double bond; therefore, this alkene is cis.

trans-3-Hexene

cis-3,4-Dimethyl-2-pentene

Example 5.2

Name each alkene and show the configuration about each double bond using the cis,trans system.

(a)

(b)

Solution

(a) The chain contains seven carbon atoms and is numbered from the end that gives the lower number to the first carbon of the double bond. Its name is *trans*-3-heptene.

(b) The longest chain contains seven carbon atoms and is numbered from the right so that the first carbon of the double bond is carbon 3 of the chain. Its name is *cis*-4-methyl-3-heptene.

Problem 5.2

Which alkenes show cis,trans isomerism? For each alkene that does, draw the trans isomer.

(a) 2-Pentene   (b) 2-Methyl-2-pentene   (c) 3-Methyl-2-pentene

---

The *E,Z* System

Because the cis,trans system is not detailed enough to name all alkenes, chemists developed the **E,Z system.** This system uses the priority rules of the *R,S* system (Section 3.3) to assign priority to the substituents on each carbon of the double bond. Using these rules, we decide which group on each carbon has the higher priority. If the groups of higher priority are on the same side of the double bond, the configuration

E,Z system   A system to specify the configuration of groups about a carbon-carbon double bond.

Z From the German, *zusammen*, together. Specifies that groups of higher priority on the carbons of a double bond are on the same side.

E From the German, *entgegen*, opposite. Specifies that groups of higher priority on the carbons of a double bond are on opposite sides.

of the alkene is ***Z*** (German: *zusammen*, together). If they are on opposite sides of the double bond, the alkene is ***E*** (German: *entgegen*, opposite).

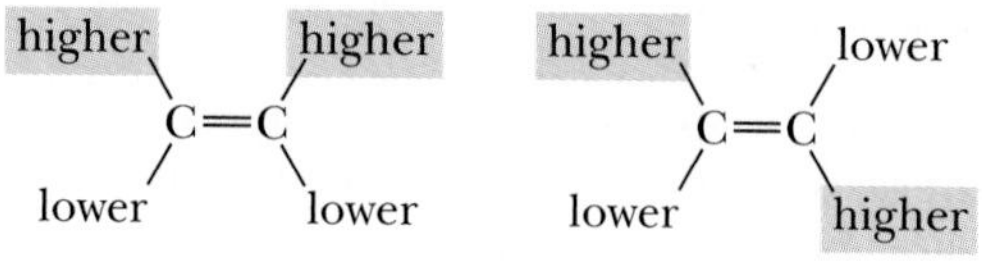

Z (*zusammen*) *E* (*entgegen*)

Throughout this text, we use the cis,trans system for alkenes in which it is clear which is the main carbon chain. We use the *E,Z* system in all other cases.

Example 5.3

Name each alkene and specify its configuration by the *E,Z* system.

(a) 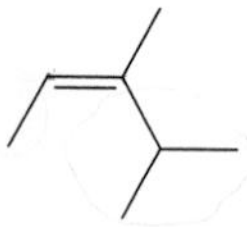 **(b)** 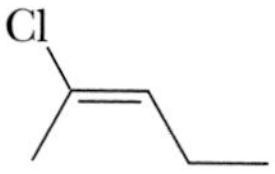

(c) 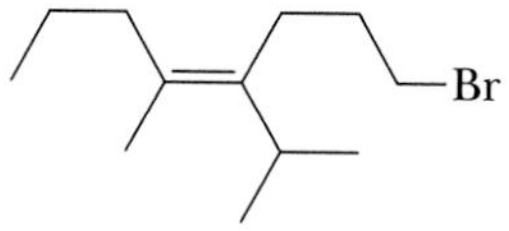

Solution

(a) The group of higher priority on carbon 2 is methyl; that of higher priority on carbon 3 is isopropyl. Because the groups of higher priority are on the same side of the double bond, the alkene has the *Z* configuration. Its name is (*Z*)-3,4-dimethyl-2-pentene. Using the cis,trans system, it is named *cis*-3,4-dimethyl-2-pentene.

(b) Groups of higher priority on carbons 2 and 3 are —Cl and —CH_2CH_3. Because these groups are on opposite sides of the double bond, the configuration of this alkene is *E* and its name is (*E*)-2-chloro-2-pentene. Using the cis,trans system, it is *cis*-2-chloro-2-pentene.

(c) The groups of higher priority are on opposite sides of the double bond; the configuration is *E*. The name of this bromoalkene is (*E*)-1-bromo-4-isopropyl-5-methyl-4-octene. Using the cis,trans system, it is *cis*-1-bromo-4-isopropyl-5-methyl-4-octene.

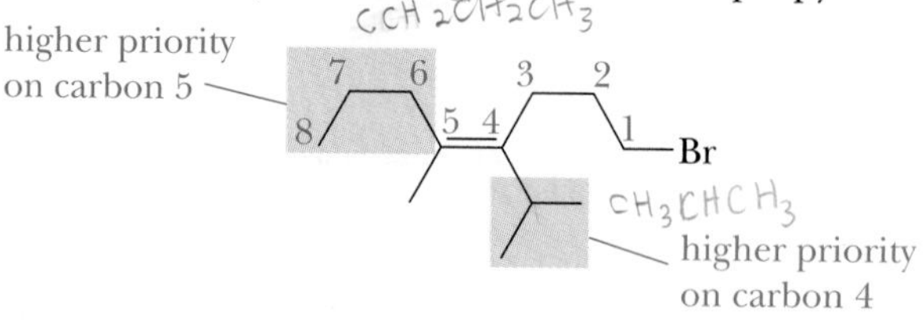

Problem 5.3

Name each alkene and specify its configuration by the *E,Z* system.

(a) 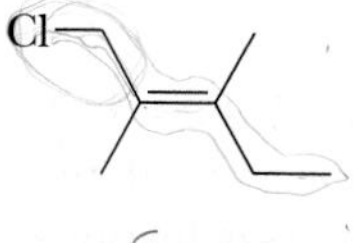

(b) 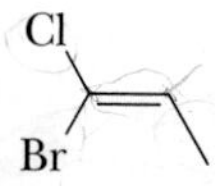

(c) 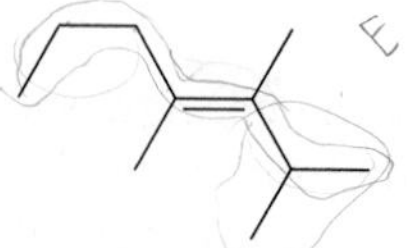

D. Cycloalkenes

In naming cycloalkenes, the carbon atoms of the ring double bond are numbered 1 and 2 in the direction that gives the substituent encountered first the smaller number.

  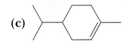

| 3-Methylcyclopentene | 4-Ethyl-1-methylcyclohexene | 1,6-Dimethylcyclohexene |

Example 5.4

Write the IUPAC and common name of each cycloalkene.

 (a)   (b) 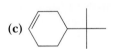 (c)

Solution

(a) 3,3-Dimethylcyclohexene   **(b)** 1,2-Dimethylcyclopentene
(c) 4-(1-methylethyl)-1-methylcyclohexene; 4-isopropyl-1-methylcyclohexene

Problem 5.4

Write the IUPAC and common name of each cycloalkene.

 (a)   (b)   (c)

E. Cis,Trans Isomerism in Cycloalkenes

Following are structural formulas for four cycloalkenes:

| Cyclopentene | Cyclohexene | Cycloheptene | Cyclooctene |

In these representations, the configuration about each double bond is cis. Is it possible to have a trans configuration in these and larger cycloalkenes? To date, *trans*-cyclooctene is the smallest trans cycloalkene that has been prepared in pure form and is stable at room temperature. Yet, even in this trans cycloalkene, there is considerable angle strain; the double bond's *p* orbitals make an angle of 44° to each other. *Cis*-cyclooctene is more stable than its trans isomer by 38 kJ (9.1 kcal)/mol. Note that the trans isomer is chiral even though it has no stereocenter.

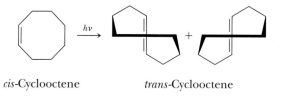

cis-Cyclooctene   *trans*-Cyclooctene

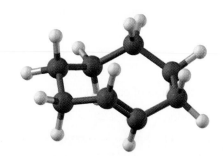

F.   Dienes, Trienes, and Polyenes

For alkenes containing two or more double bonds, the infix -en- is changed to -adien-, -atrien-, and so on. Those that contain several double bonds are also referred to more generally as polyenes (Greek: *poly*, many). Following are examples of three dienes.

CH_2＝$CHCH_2CH$＝CH_2 CH_2＝CCH＝CH_2 ⬠

                                       CH_3 (above)

1,4-Pentadiene   2-Methyl-1,3-butadiene   1,3-Cyclopentadiene
                                (Isoprene)

G.   Cis,Trans Isomerism in Dienes, Trienes, and Polyenes

Thus far we have considered cis,trans isomerism in alkenes containing only one carbon-carbon double bond. For an alkene with one carbon-carbon double bond that can show cis,trans isomerism, two stereoisomers are possible. For an alkene with n carbon-carbon double bonds, each of which can show cis,trans isomerism, 2^n stereoisomers are possible.

Example 5.5

How many stereoisomers are possible for 2,4-heptadiene?

Solution

This molecule has two carbon-carbon double bonds, each of which shows cis,trans isomerism. As shown in this table, there are $2^2 = 4$ stereoisomers. Two of these are drawn on the right.

Double Bond	
C_2—C_3	C_4—C_5
trans	trans
trans	cis
cis	trans
cis	cis

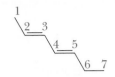

(2E,4E)-2,4-Heptadiene
trans,trans-2,4-Heptadiene

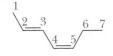

(2E,4Z)-2,4-Heptadiene
trans,cis-2,4-Heptadiene

Problem 5.5

Draw structural formulas for the other two stereoisomers of 2,4-heptadiene.

Example 5.6

How many stereoisomers are possible for 10,12-hexadecadien-1-ol?

$$CH_3(CH_2)_2CH=CHCH=CH(CH_2)_8CH_2OH$$

10,12-Hexadecadien-1-ol

Solution

Cis,trans isomerism is possible about both double bonds. Four stereoisomers are possible.

Problem 5.6

(10*E*,12*Z*)-10,12-hexadecadien-1-ol is a sex pheromone of the silkworm. Draw a structural formula for this compound.

Silk worms spinning cocoons on a loom, silk farm, Japan. *(Paul Chesley/Tony Stone Worldwide)*

An example of a biologically important compound for which a number of cis,trans isomers is possible is vitamin A. There are four carbon-carbon double bonds in the chain of carbon atoms bonded to the substituted cyclohexene ring, and each has the potential for cis,trans isomerism. There are $2^4 = 16$ stereoisomers possible for this structural formula. Vitamin A is the all-*E* (all-trans) isomer.

Vitamin A (retinol)

CHEMISTRY IN ACTION

The Case of the Iowa and New York Strains of the European Corn Borer

Although humans communicate largely by sight and sound, chemicals are the primary means of communication for the vast majority of other species in the animal world. Often, communication within a species is specific for one of two or more configurational isomers. For example, a member of a given species may respond to a cis isomer of a chemical but not to the trans isomer. Or, alternatively, it might respond to a quite precise blend of cis and trans isomers but not to other blends of these same isomers.

Several groups of scientists have studied the components of the sex pheromones of both the Iowa and the New York strains of the European corn borer. Females of these closely related species secrete the sex attractant 11-tetradecenyl acetate. Males of the Iowa strain show maximum response to a mixture containing 96% of the cis isomer and 4% of the trans isomer. When the pure cis isomer is used alone, males are only weakly attracted. Males of the New York strain show an entirely different response. They respond maximally to a mixture containing 3% of the cis isomer and 97% of the trans isomer.

$$CH_3CH_2 \quad\quad (CH_2)_{10}OCCH_3$$
$$\overset{O}{\overset{\|}{}}$$

$$\underset{H}{\overset{}{C}}=\underset{H}{\overset{}{C}}$$

cis-11-Tetradecenyl acetate

$$\underset{CH_3CH_2}{\overset{H}{C}}=\underset{H}{\overset{(CH_2)_{10}OCCH_3}{C}}$$

trans-11-Tetradecenyl acetate

There is evidence that optimum response to a narrow range of stereoisomers as we see here is widespread in nature and that a great many insects maintain species isolation for mating and reproduction by the stereochemistry of their pheromones. See J. A. Klun et al., *Science* **181:** 661 (1973).

The European corn borer, Pyrausta nubilalis (Runk/Schoenberger/Grant Heilman Photography, Inc.)

5.3 Physical Properties of Alkenes

Alkenes are nonpolar compounds, and the only attractive forces between their molecules are dispersion forces (Section 2.8). Their physical properties, therefore, are similar to those of alkanes. Alkenes of two, three, and four carbon atoms are gases at room temperature. Those of five or more carbons are colorless liquids less dense than water. Alkenes are insoluble in water but soluble in one another, in other nonpolar organic liquids, and in ethanol. Table 5.1 lists physical properties of some alkenes.

Table 5.1 **Physical Properties of Some Alkenes**			
Name	**Structural Formula**	**mp (°C)**	**bp (°C)**
Ethylene	$CH_2{=}CH_2$	−169	−104
Propene	$CH_3CH{=}CH_2$	−185	−47
1-Butene	$CH_3CH_2CH{=}CH_2$	−185	−6
1-Pentene	$CH_3CH_2CH_2CH{=}CH_2$	−138	30
cis-2-Pentene	$\begin{array}{c} H_3C \qquad CH_2CH_3 \\ \diagdown \quad \diagup \\ C{=}C \\ \diagup \quad \diagdown \\ H \qquad\quad H \end{array}$	−151	37
trans-2-Pentene	$\begin{array}{c} H_3C \qquad H \\ \diagdown \quad \diagup \\ C{=}C \\ \diagup \quad \diagdown \\ H \qquad CH_2CH_3 \end{array}$	−156	36
2-Methyl-2-butene	$\begin{array}{c} CH_3 \\ | \\ CH_3C{=}CHCH_3 \end{array}$	−134	39

5.4   Naturally Occurring Alkenes — Terpene Hydrocarbons

A **terpene** is a compound whose carbon skeleton can be divided into two or more units that are identical with the carbon skeleton of isoprene. Carbon 1 of an isoprene unit is called the head; carbon 4 is called the tail. Terpenes are almost always formed by bonding the tail of one isoprene unit to the head of another.

Terpene   A compound whose carbon skeleton can be divided into two or more units identical with the carbon skeleton of isoprene.

$$CH_2{=}\overset{\overset{\displaystyle CH_3}{\displaystyle |}}{C}{-}CH{=}CH_2$$

2-Methyl-1,3-butadiene
(Isoprene)

A study of terpenes provides a glimpse of the wondrous diversity that nature can generate from a simple carbon skeleton. Terpenes also illustrate an important principle of the molecular logic of living systems, namely, that in building large molecules, small subunits are bonded together enzymatically by an iterative process and then chemically modified by precise enzyme-catalyzed reactions. Chemists use the same principles in the laboratory, but our methods do not have the precision and selectivity of the enzyme-catalyzed reactions of cellular systems.

Probably the terpenes most familiar to you, at least by odor, are components of the so-called essential oils obtained by steam distillation or ether extraction of various parts of plants. Essential oils contain the relatively low-molecular-weight substances that are in large part responsible for characteristic plant fragrances. Many essential oils, particularly those from flowers, are used in perfumes.

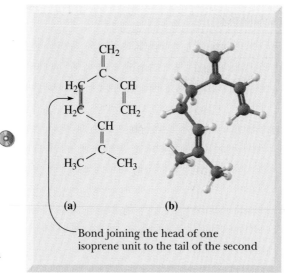

Figure 5.3

Myrcene. (a) Structural formula and (b) ball-and-stick model.

(a) **(b)**

Bond joining the head of one isoprene unit to the tail of the second

One example of a terpene obtained from an essential oil is myrcene, $C_{10}H_{16}$, a component of bayberry wax and oils of bay and verbena. Myrcene is a triene with a parent chain of eight carbon atoms and two one-carbon branches [Figure 5.3(a)].

Head-to-tail bonds between isoprene units are vastly more common in nature than are the alternative head-to-head or tail-to-tail patterns. Figure 5.4 shows structural formulas of five more terpenes, all derived from two isoprene units. Geraniol

Figure 5.4

Five terpenes, each divisible into two isoprene units.

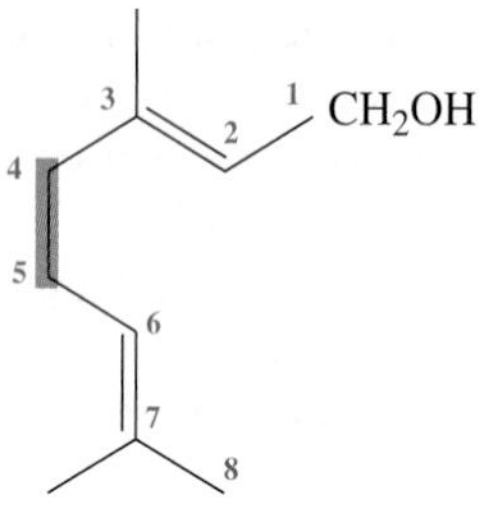

Geraniol
(Rose and other flowers)

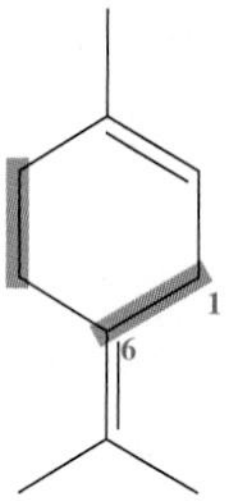

Limonene
(Oil of lemon and orange)

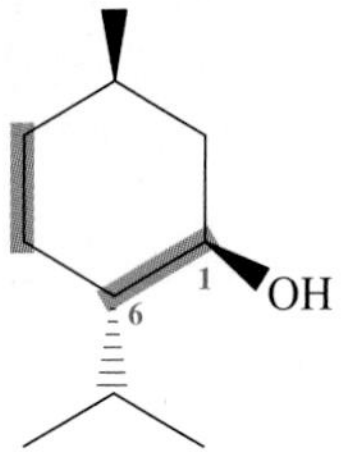

Menthol
(Peppermint)

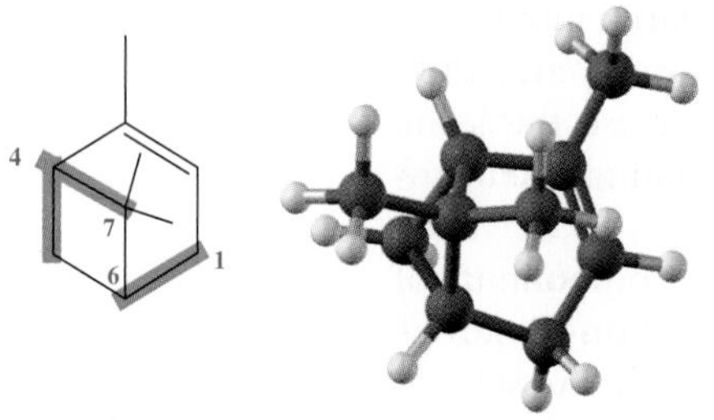

α-Pinene
(Turpentine)

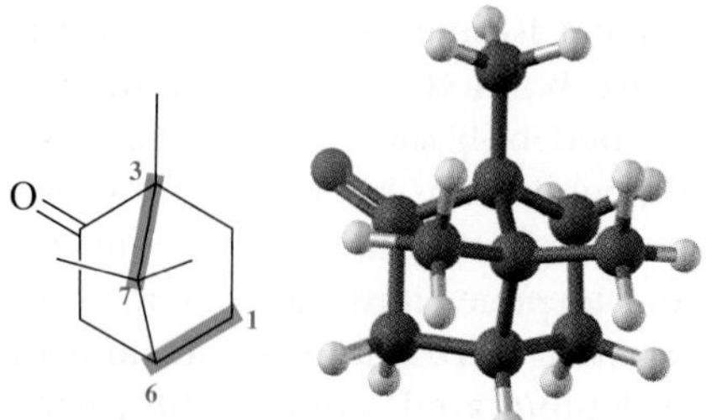

Camphor
(Camphor tree)

CHEMISTRY IN ACTION

Why Plants Emit Isoprene

Names like Virginia's Blue Ridge, Jamaica's Blue Mountain Peak, and Australia's Blue Mountains remind us of the bluish haze that hangs over wooded hills in the summer time. It was discovered in the 1950s that this haze is rich in isoprene, which means that isoprene is far more abundant in the atmosphere than anyone thought. The haze itself is caused by light scattering from an aerosol produced by photooxidation of isoprene and other hydrocarbons. Scientists now estimate that global emission of isoprene by plants is 3×10^{14} g/yr (3×10^{8} tonnes/yr), which represents approximately 2% of all carbon fixed by photosynthesis. A study of hydrocarbon emissions in the Atlanta area showed that plants were by far the largest emitters of hydrocarbons, with plant-derived isoprene accounting for almost 60% of the total.

Why do plants emit so much isoprene into the atmosphere rather than use it for the synthesis of terpenes and other natural products? Tom Sharkey, a University of Wisconsin plant physiologist, found that emission of isoprene is extremely sensitive to temperature. Plants grown at 20°C do not emit isoprene, but they begin to emit it when leaf temperature is increased to 30°C. In certain plants, isoprene emission can increase as much as tenfold for each 10°C increase in leaf temperature. Sharkey studied the relationship between temperature-induced leaf damage and isoprene concentration in leaves of the kudzu [*Pueraria lobata* (Wild.) Ohwi.] plant. He discovered that leaf damage, as measured by chlorophyll destruction, begins to occur at 37.5°C in the absence of isoprene, but not until 45°C in the presence of isoprene. Sharkey speculates that isoprene dissolves in leaf membranes and in some way increases their tolerance to heat stress. Because isoprene is made rapidly and also lost rapidly, its concentration correlates with temperature throughout the day. See "Why Plants Emit Isoprene" by T. D. Sharkey and E. L. Singsass, *Nature* **374:** 769 (1955).

Haze over the Blue Ridge Mountains. (Photo/NATS)

has the same carbon skeleton as myrcene. In the last four terpenes of Figure 5.4, the carbon atoms present in myrcene and geraniol are cross-linked to give cyclic structures. To help you identify the points of cross linkage and ring formation, the carbon atoms of the geraniol skeleton are numbered 1 through 8. This numbering pattern is used in the remaining terpenes to show points of cross linking. In both limonene and menthol, there is a new carbon-carbon bond formed between carbons 1 and 6. In α-pinene, there are new carbon-carbon bonds formed between carbons 1 and 6 and 4 and 7. In camphor, they are formed between carbons 1 and 6 and 3 and 7.

Figure 5.5 shows structural formulas of two terpenes divisible into three isoprene units. For reference, the carbon atoms of the parent chain of farnesol are numbered

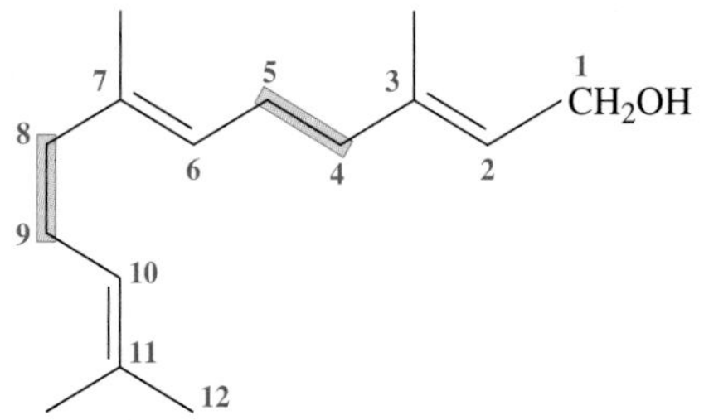

Farnesol
(Lily-of-the-valley)

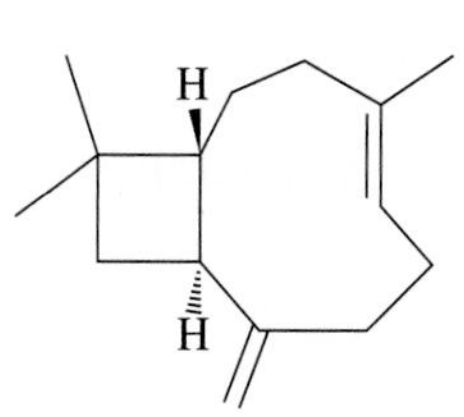

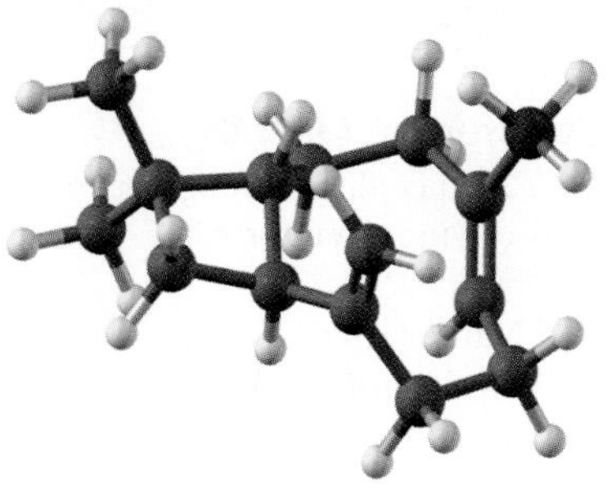

Caryophyllene
(Oil of cloves)

Figure 5.5

Two terpenes, each divisible into three isoprene units.

1 through 12. Try to discover what patterns of cross linking of the carbon skeleton of farnesol gives the carbon skeleton of caryophyllene.

Vitamin A (Section 5.2G), a terpene of molecular formula $C_{20}H_{30}O$, consists of four isoprene units linked head-to-tail and cross-linked at one point to form a six-membered ring.

The synthesis of substances in living systems is a fascinating area of research and one of the links between organic and biochemistry. However tempting it might be to propose that nature synthesizes terpenes by joining together molecules of isoprene, this is not quite the way it is done. We will discuss the synthesis of terpenes in Sections 19.4 and 26.4.

Summary

An alkene is an unsaturated hydrocarbon that contains a carbon-carbon double bond. The general formula of an alkene is C_nH_{2n}. A carbon-carbon double bond consists of one sigma bond formed by the overlap of sp^2 hybrid orbitals and one pi bond formed by the overlap of parallel $2p$ orbitals (Section 5.1B). The strength of the pi bonds in ethylene is approximately 264 kJ (63 kcal)/mol, which is considerably weaker than the sigma bond. The structural feature that makes **cis,trans isomerism** possible in alkenes is lack of rotation about the two carbons of the double bond (Section 5.1C).

According to the IUPAC system (Section 5.2A), the presence of a carbon-carbon double bond is shown by changing the infix of the parent hydrocarbon from -an- to -en-. For compounds containing two or more double bonds, the infix is changed to -adien-, -atrien-, and so on. The names methylene, vinyl, and allyl are commonly used to show the presence of $=CH_2$, $—CH=CH_2$, and $—CH_2CH=CH_2$ groups, respectively.

Whether an alkene is cis or trans is determined by the orientation of the main carbon chain about the double bond

(Section 5.2C). The configuration of a carbon-carbon double bond is specified more precisely by the **E,Z system** using the same set of priority rules used for the R,S system (Section 3.3). If the two groups of higher priority are on the same side of the double bond, the alkene is designated **Z** (German: *zusammen*, together); if they are on opposite sides, the alkene is designated **E** (German: *entgegen*, opposite). To date, *trans*-cyclooctene is the smallest trans cycloalkene that has been prepared in pure form and is stable at room temperature.

Because alkenes are essentially nonpolar compounds and the only attractive forces between their molecules are dispersion forces, their physical properties are similar to those of alkanes (Section 5.3).

The characteristic structural feature of a **terpene** (Section 5.4) is a carbon skeleton that can be divided into two or more isoprene units. Terpenes illustrate an important principle of the molecular logic of living systems, namely that in building large molecules, small subunits are strung together by an iterative process and then chemically modified by precise enzyme-catalyzed reactions.

Problems

Structure of Alkenes

5.7 Predict all bond angles about each highlighted carbon atom. To make these predictions, use the valence-shell electron-pair repulsion (VSEPR) model (Section 1.4).

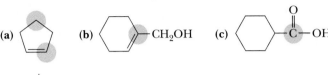

(a)   **(b)** $\quad$—CH$_2$OH $\quad$ **(c)**

(d)   **(e)** HC≡C—CH=CH$_2$

5.8 For each highlighted carbon atom in Problem 5.7, identify which atomic orbitals are used to form each sigma bond and which are used to form each pi bond.

5.9 Following is the structure of 1,2-propadiene (allene).
   (a) Predict all bond angles in this molecule.
   (b) State the orbital hybridization of each carbon.
   (c) Account for the molecular geometry of allene in terms of the orbital overlap model.

5.10 Following are lengths for a series of C—C single bonds. Propose an explanation for the differences in bond lengths.

$$CH_2=C=CH_2$$

**1,2-Propadiene
(Allene)**

Structure	Length of C—C Single Bond (pm)
CH$_3$—CH$_3$	153.7
CH$_2$=CH—CH$_3$	151.0
CH$_2$=CH—CH=CH$_2$	146.5
HC≡C—CH$_3$	145.9

Nomenclature of Alkenes

5.11 Draw structural formulas for these alkenes.

   (a) *trans*-2-Methyl-3-hexene
   (b) 2-Methyl-2-hexene
   (c) 2-Methyl-1-butene
   (d) 3-Ethyl-3-methyl-1-pentene
   (e) 2,3-Dimethyl-2-butene
   (f) *cis*-2-Pentene
   (g) (*Z*)-1-Chloropropene
   (h) 3-Methylcyclohexene
   (i) 1-Isopropyl-4-methylcyclohexene
   (j) (*E*)-2,6-Dimethyl-2,6-octadiene
   (k) 3-Cyclopropyl-1-propene
   (l) Cyclopropylethene
   (m) 2-Chloropropene
   (n) Tetrachloroethylene
   (o) 1-Chlorocyclohexene
   (p) Bicyclo[2.2.1]-2-heptene
   (q) Bicyclo[4.4.0]-1-decene

5.12 Name these alkenes and cycloalkenes.

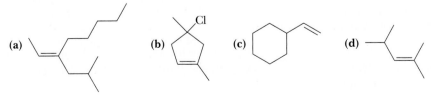

(a)   **(b)**   **(c)**   **(d)**

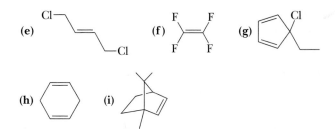

5.13 Arrange the following groups in order of increasing priority.

    (a) —CH_3 —H   —Br   —CH_2CH_3

    (b) —OCH_3 —$CH(CH_3)_2$ —$B(CH_2CH_3)_2$ —H

    (c) —CH_3 —CH_2OH —CH_2NH_2 —CH_2Br

5.14 Assign an *E* or *Z* and a cis or trans configuration to these dicarboxylic acids, each of which is an intermediate in the tricarboxylic acid cycle. Under each is its common name.

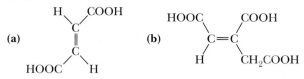

                Fumaric acid   Aconitic acid

5.15 Name and draw structural formulas for all alkenes of molecular formula C_5H_{10}. As you draw these alkenes, remember that cis and trans isomers are different compounds and must be counted separately.

5.16 For each molecule that shows cis,trans isomerism, draw the cis isomer.

5.17 Draw structural formulas for all compounds of molecular formula C_5H_{10} that are:

    (a) Alkenes that do not show cis,trans isomerism

    (b) Alkenes that do show cis,trans isomerism

    (c) Cycloalkanes that do not show cis,trans isomerism

    (d) Cycloalkanes that do show cis,trans isomerism

5.18 β-Ocimene (see *The Merck Index*, 12th ed., #6837), a triene found in the fragrance of cotton blossoms and several essential oils, has the IUPAC name (*Z*)-3,7-dimethyl-1,3,6-octatriene. Draw a structural formula for β-ocimene.

5.19 Draw the structural formula for at least one bromoalkene of molecular formula C_5H_9Br that shows:

    (a) Neither *E,Z* isomerism nor chirality

    (b) *E,Z* isomerism but not chirality

    (c) Chirality but not *E,Z* isomerism

    (d) Both chirality and *E,Z* isomerism

5.20 Following are structural formulas and common names for four molecules that contain both a carbon-carbon double bond and another functional group. Give each an IUPAC name.

(b) Label all stereocenters in santonin. How many stereoisomers are possible for a molecule with this number of stereocenters?

5.30 In many parts of South America, extracts of the leaves and twigs of *Montanoa tomentosa* are brewed with water to make a "tea" used to stimulate menstruation, to facilitate labor, and to act as an abortifacient. Phytochemical investigations of this plant have resulted in isolation of a very potent fertility-regulating compound called zoapatanol (see *The Merck Index*, 12th ed., #10318).

Zoapatanol

(a) Show that the carbon skeleton of zoapatanol can be divided into four isoprene units bonded head-to-tail and then cross-linked in one point along the chain.
(b) Specify the configuration about the carbon-carbon double bond to the seven-membered ring according to the *E,Z* system.
(c) How many stereoisomers are possible for this molecule? In answering this problem, you must consider both *E,Z* isomerism and *R,S* isomerism.

5.31 Pyrethrin II and pyrethrosin are two natural products isolated from plants of the chrysanthemum family (see *The Merck Index*, 12th ed., #8148 and #8149). Pyrethrin II is a natural insecticide and is marketed as such.

Pyrethrin II   Pyrethrosin

(a) Label all stereocenters in each molecule and all carbon-carbon double bonds about which there is the possibility for cis,trans isomerism.
(b) State the number of stereoisomers possible for each molecule.
(c) Show that the bicyclic ring system of pyrethrosin is composed of three isoprene units.

Molecular Modeling

5.32 Measure the CH_3, CH_3 distance in the energy-minimized model of *cis*-2-butene, and the CH_3, H distance in the energy-minimized model of *trans*-2-butene. In which isomer is the nonbonded interaction strain greater?

5.33 Measure the $C=C-C$ bond angles in the energy-minimized models of the cis and trans isomers of 2,2,5,5-tetramethyl-3-hexene. In which case is the deviation from the VSEPR model predictions greater?

5.34 Measure the C—C—C and C—C—H bond angles in the energy-minimized models of cyclohexene and compare them with those predicted by the VSEPR model. Explain any differences.

5.35 Measure the C—C—C and C—C—H bond angles in the energy-minimized models of cis and trans isomers of cyclooctene. Compare these values with those predicted by the VSEPR model. In which isomer are deviations from the VSEPR model predictions greater?

ALKENES II

I n this chapter, we begin our systematic study of one of the most important unifying concepts in organic chemistry: the concept of a **reaction mechanism.** We use the reactions of alkenes as the vehicle by which to introduce this concept.

6.1 Reactions of Alkenes—An Overview

The most characteristic reaction of alkenes is **addition** to a carbon-carbon double bond in such a way that the pi bond is broken, and, in its place, sigma bonds are formed to two new atoms or groups of atoms. Several examples of reactions at a carbon-carbon double bond are shown in Table 6.1 along with the descriptive name(s) associated with each. Some

■ These laboratory squeese bottles are fabricated from polyethylene. *(Charles D. Winters)* Inset: A model of ethylene, the monomer from which polyethylene is derived.

Table 6.1 Characteristic Alkene Addition Reactions

Reaction	Descriptive Name(s)
$\diagdown C = C \diagup$ + HCl (HX) ⟶ $-\overset{\text{H}}{\underset{}{C}}-\overset{}{\underset{\text{Cl (X)}}{C}}-$	Hydrochlorination (hydrohalogenation)
$\diagdown C = C \diagup$ + H_2O ⟶ $-\overset{\text{H}}{\underset{}{C}}-\overset{}{\underset{\text{OH}}{C}}-$	Hydration
$\diagdown C = C \diagup$ + Br_2 (X_2) ⟶ $-\overset{\text{(X) Br}}{\underset{\text{Br (X)}}{C}}-\overset{}{C}-$	Bromination (halogenation)
$\diagdown C = C \diagup$ + Br_2 (X_2) $\xrightarrow{H_2O}$ $-\overset{\text{HO}}{\underset{\text{Br (X)}}{C}}-\overset{}{C}-$	Bromohydrin (halohydrin)
$\diagdown C = C \diagup$ + $Hg(OAc)_2$ $\xrightarrow{H_2O}$ $-\overset{\text{AcOHg}}{\underset{\text{OH}}{C}}-\overset{}{C}-$	Oxymercuration
$\diagdown C = C \diagup$ + BH_3 ⟶ $-\underset{\text{H}}{C}-\underset{\text{BH}_2}{C}-$	Hydroboration
$\diagdown C = C \diagup$ + OsO_4 ⟶ $-\underset{\text{HO}}{C}-\underset{\text{OH}}{C}-$	Diol formation (oxidation)
$\diagdown C = C \diagup$ + H_2 ⟶ $-\underset{\text{H}}{C}-\underset{\text{H}}{C}-$	Hydrogenation (reduction)

of these reactions are treated separately under oxidations (Section 6.5) or reductions (Section 6.6) but are included in this table because they are formally additions. In the following sections, we study these alkene reactions in considerable detail, with particular attention to the mechanism by which each occurs.

A second characteristic reaction of alkenes is the formation of **chain-growth polymers** (Greek: *poly*, many, and *meros*, part). In the presence of certain catalysts called initiators, many alkenes form polymers made by the addition of **monomers** (Greek: *mono*, one, and *meros*, part) to a growing polymer chain as illustrated by the formation of polyethylene from ethylene. In alkene polymers of industrial and commercial importance, *n* is a large number, typically several thousand.

$$n\,CH_2{=}CH_2 \xrightarrow{\text{initiator}} \text{(}CH_2CH_2\text{)}_n$$

We discuss this alkene reaction in Chapter 24.

6.2   Reaction Mechanisms

A **reaction mechanism** describes in detail how a reaction occurs. It describes which bonds are broken and which new ones are formed as well as the order and relative rates of the various bond-breaking and bond-forming steps. If the reaction takes place in solution, it describes the role of the solvent. If the reaction involves a catalyst, it describes the role of the catalyst. A complete reaction mechanism describes the positions of all atoms and the energy of the entire system during each moment of the reaction. This ideal, however, can rarely be approached in practice. Reaction mechanisms are determined by carefully planned experiments. Modern computational methods now allow description of detailed mechanistic pathways, but the results must still be compared with experiment.

A.   Energy Diagrams and Transition States

Think of a chemical bond as a spring. As a spring is stretched from its resting position, its energy is increased. As it returns to its resting position, its energy is decreased. Similarly, during a chemical reaction, bond breaking corresponds to an increase in energy, and bond forming corresponds to a decrease in energy. We use an **energy diagram** to show the changes in energy that occur in going from reactants to products. Energy is measured on the vertical axis, and the change in position of the atoms during the reaction is represented on the horizontal axis, called the **reaction coordinate.** The reaction coordinate corresponds to how far the reaction has progressed (it is not a time axis). Figure 6.1 shows an energy diagram for the reaction of compounds C + A—B to form C—A + B. This reaction occurs in one step, meaning that bond breaking in starting materials and bond forming to give the observed products occur simultaneously.

As we already mentioned (Section 2.6), there are several types of energy that may be important to consider in reactions. In this text, we will be most concerned with changes in Gibbs free energy (ΔG^0) and enthalpy (ΔH^0). **Gibbs free energy** is directly related to reaction rates and equilibria, while **enthalpy** is a parameter that is measurable by determining the heat released or taken up in reactions. The two differ only by the term $T\Delta S^0$:

$$\Delta G^0 = \Delta H^0 - T\Delta S^0$$

Energy diagram   A graph showing the changes in energy that occur during a chemical reaction; energy is plotted on the vertical axis, and reaction progress is plotted on the horizontal axis.
Reaction coordinate   A measure of the change in the positions of atoms during a reaction; plotted on the horizontal axis in a reaction energy diagram.
Gibbs free energy, ΔG^0   A thermodynamic function relating enthalpy, entropy, and temperature, given by the equation $\Delta G^0 = \Delta H^0 - T\Delta S^0$. If $\Delta G < 0$ for a reaction, the reaction is spontaneous. If $\Delta G^0 > 0$, the reaction is nonspontaneous.

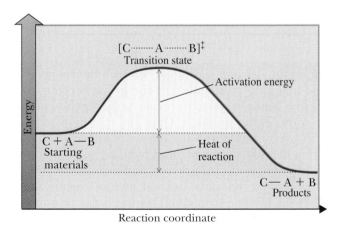

Figure 6.1
An energy diagram for a one-step reaction between C and A—B. The dashed lines indicate that, in the transition state, the new C—A bond is partially formed and the A—B bond is partially broken. The energy of the reactants is higher than that of the products.

where T is the temperature in Kelvins and ΔS^0 is the entropy change. The entropy change in a reaction is determined by the change in the number of degrees of freedom (for example by how many molecules or how many bond rotations there are on each side of a balanced chemical equation or by how many equivalent isomers can be formed in a reaction). If the number of molecules doesn't change in a reaction, ΔS^0 is often very small, in which case $\Delta G^0 \sim \Delta H^0$.

Because the $T\Delta S^0$ term is proportional to temperature, entropy becomes more important at high temperatures. Because $T\Delta S^0$ is negative for reactions in which the number of molecules increases, high-temperature reactions often lead to bond breaking, with an increase in degrees of freedom or number of molecules and a resultant increase in entropy. Cracking of petroleum (Section 2.9) is an example. If we are talking about changes in enthalpy in a reaction, (heat given off or absorbed) we should correctly say it is **exothermic** or **endothermic.** If we are talking about changes in Gibbs free energy, we should say it is **exergonic** or **endergonic.** As noted previously, energy diagrams are useful for conceptualizing reactions and are not very different for free energy and enthalpy.

The difference in enthalpy between the reactants and products (in their standard states) is called the **heat of reaction** (ΔH^0). If the enthalpy of products is lower than that of reactants, heat is released, and the reaction is exothermic. If the enthalpy of products is higher than that of reactants, heat is absorbed, and the reaction is endothermic. The one-step reaction shown in Figure 6.1 is exothermic. Similar relationships hold for Gibbs free energy. If the ΔG^0 for a reaction is negative, the reaction is exergonic and can proceed spontaneously in the forward direction. However, not all exergonic reactions proceed at measurable rates. If the ΔG^0 for a reaction is positive, the reaction is endergonic and does not proceed in the forward reaction. The relationship between ΔG^0, ΔH^0, ΔS^0 and the spontaneity of chemical reactions is summarized in the following table.

Exothermic reaction A reaction in which the enthalpy of the products is lower than that of the reactants; a reaction in which heat is liberated.

Endothermic reaction A reaction in which the enthalpy of the products is higher than the enthalpy of the reactants; a reaction in which heat is absorbed.

Exergonic reaction A reaction in which the Gibbs free energy of the products is lower than that of the reactants.

Endergonic reaction A reaction in which the Gibbs free energy of the products is higher than that of the reactants.

	$\Delta S^0 < 0$	$\Delta S^0 > 0$
$\Delta H^0 > 0$	$\Delta G^0 > 0$ Reaction is never spontaneous.	Reaction is spontaneous at higher temperature where $T\Delta S^0 > \Delta H^0$ and, therefore, $\Delta G^0 < 0$.
$\Delta H^0 < 0$	Reaction is spontaneous at lower temperatures where $T\Delta S^0 < \Delta H^0$ and, therefore, $\Delta G^0 < 0$.	$\Delta G^0 < 0$ Reaction is always spontaneous.

Transition state An unstable species of maximum energy formed during the course of a reaction; a maximum on an energy diagram.

A **transition state** is the point on the reaction coordinate at which the energy is a maximum for a given step. At the transition state, sufficient energy has become concentrated in the proper bonds so that bonds in reactants break. As they break, energy is redistributed, and new bonds form, giving products. Once the transition state is reached, the reaction proceeds to give products with the release of energy. A transition state has a definite geometry, a definite arrangement of bonding and nonbonding electrons, and a definite distribution of electron density and charge. Because a transition state is at an energy maximum, it cannot be isolated, and its structure cannot be determined experimentally. Its lifetime is fleeting, on the order of one picosecond (the duration of a single bond vibration). As we shall see soon, however, even though we usually cannot observe a transition state directly by any experimental means, we can often infer a great deal about its probable structure from other experimental observations.

For the reaction illustrated in Figure 6.1, we use dashed lines to show the partial bonding in the transition state. As C begins to form a new covalent bond with A (shown by the dashed line), the covalent bond between A and B begins to break (also shown by a dashed line). On completion of the reaction, the A—B bond is fully broken, and the C—A bond is fully formed.

The difference in energy between reactants and the transition state is called the **activation energy.** If we are discussing Gibbs free energy, activation energy is given the symbol $\Delta G^{\ddagger}$. (Often this is discussed in terms of the closely related potential energy, and the activation energy is called E_a.) The activation energy is the minimum required for reaction to occur; it can be considered an energy barrier for the reaction. $\Delta G^{\ddagger}$ determines the rate of a reaction, that is, how fast the reaction occurs. If $\Delta G^{\ddagger}$ is large, only a very few molecular collisions occur with sufficient energy to reach the transition state, and the reaction is slow. If $\Delta G^{\ddagger}$ is small, many collisions generate sufficient energy to reach the transition state, and the reaction is fast.

In a reaction that occurs in two or more steps, each step has its own transition state and activation energy. Shown in Figure 6.2 is an energy diagram for conversion of reactants to products in two steps. A **reaction intermediate** corresponds to an energy minimum between two transition states, in this case between transition states 1 and 2. Note that because the energies of the reaction intermediates we describe in this chapter are higher than those of either reactants or products, they are highly reactive, and usually cannot be isolated. However, some have significant lifetimes and can be observed directly.

The slowest step in a multistep reaction, called the **rate-determining step,** is the step that crosses the highest energy barrier. In the two-step sequence shown in Figure 6.2, Step 1 crosses the higher energy barrier and is, therefore, the rate-determining step.

Activation energy The difference in Gibbs free energy between reactants and the transition state.

Intermediate A species, formed between two successive reaction steps, that lies in an energy minimum between the two transition states.

Rate-determining step The step in a multistep reaction sequence that crosses the highest energy barrier.

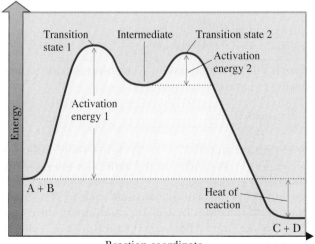

Figure 6.2
Energy diagram for a two-step reaction involving formation of an intermediate. The energy of the reactants is higher than that of the products, and energy is released in the conversion of A + B to C + D.

B. Activation Energy and Rate

The relationship between rate constant, k, and activation energy for chemical reactions is best discussed in terms of Gibbs free energy. It is given by the following equation, which relates the reaction rate constant, k, to the free energy of activation, $(\Delta G^{\ddagger})$.

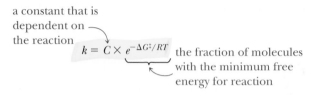

a constant that is dependent on the reaction

$$k = C \times e^{-\Delta G^{\ddagger}/RT}$$

the fraction of molecules with the minimum free energy for reaction

C is a constant (units s^{-1}) that depends on the reaction, and is often about $10^{13}\ s^{-1}$ for unimolecular reactions (a reaction in which only one molecule is involved in the transition state). $\Delta G^{\ddagger}$ is the activation energy in kJ (kcal)·mol^{-1}, R is the gas constant with a value of 8.3145×10^{-3} kJ·$(1.987 \times 10^{-3}$ kcal)·mol^{-1}·deg^{-1}, and T is the temperature in Kelvins.

Most organic reactions we deal with in this course have activation energies in the range of 42–146 kJ (10–35 kcal)/mol. Those with activation energies below 84 kJ (20 kcal)/mol proceed rapidly at room temperature. Those with higher activation energies may require heating or input of some other form of energy, such as light, to provide more molecules with sufficient energy to overcome the activation energy barrier.

Example 6.1

An often-stated generalization is that for many reactions taking place at or near room temperature (25°C), the rate of reaction approximately doubles for every 10°C rise in temperature. This generalization is valid for only a narrow range of activation energies. What is the activation energy for a reaction whose rate at 35°C is two times its rate at 25°C?

Solution

If we assume that the activation energy for a reaction is constant and independent of temperature, then we can determine the ratio of rate constants k_2 and k_1 for a reaction carried out at temperatures T_2 and T_1:

$$\frac{k_2}{k_1} = \frac{C \cdot e^{-\Delta G^{\ddagger}/RT_2}}{C \cdot e^{-\Delta G^{\ddagger}/RT_1}}$$

Taking the natural logarithm of each side of the equation, converting to base 10, and rearranging terms gives the following equation:

$$\log \frac{k_2}{k_1} = \frac{\Delta G^{\ddagger}}{2.303R} \left(\frac{1}{T_1} - \frac{1}{T_2} \right)$$

The reaction is carried out at 25°C (298 K) and 35°C (308 K), respectively. Substituting values of the relative rate constants T_2, T_1, and R in the activation energy equation and solving for the activation energy gives

$$\log = \frac{2}{1} = \frac{\Delta G^{\ddagger}}{2.303 \times 8.315 \times 10^{-3}\ kJ \cdot mol^{-1}} \left(\frac{1}{298} - \frac{1}{308} \right)$$

and

$$\Delta G^{\ddagger} = 52.7\ kJ\ (12.6\ kcal)/mol$$

Problem 6.1

Suppose that the activation energy for a particular chemical reaction is 105.5 kJ (25.2 kcal)/mol. By what factor is the rate of reaction increased when the reaction takes place at 35°C compared with the rate at 25°C?

Example 6.2

If the rate constants for two reactions, each taking place at 25°C and each with the same value of the constant C, differ by a factor of 10, what is the difference in their activation energies in kilojoules per mole?

Solution

The following relationship, derived from the rate equation, relates rate constants k_2 and k_1 of two reactions to their relative activation energies.

$$\Delta G^{\ddagger} = -2.303 RT \log \frac{k_1}{k_2}$$

Substituting the ratio of 10/1 in the equation, we find that their activation energies differ by approximately 5.7 kJ (1.36 kcal)/mol.

$$\Delta G^{\ddagger} = 2.303 \times 8.315 \times 10^{-3} \text{ kJ} \times \text{mol}^{-1} \times 298 \times \log \left(\frac{10}{1} \right) = -5.71 \text{ kJ/mol}$$

Problem 6.2

Complete the first three entries in this table for reactions taking place at 25°C. Given the pattern of these first three entries, estimate the approximate values for the remaining two entries. How many kilojoules per mole in activation energy correspond to a power of 10 in relative rates?

$\Delta G^{\ddagger}$ (kJ/mol)	$\dfrac{k_2}{k_1}$
_____	1
_____	10
_____	100
_____	1,000
_____	10,000

C. Developing a Reaction Mechanism

To develop a reaction mechanism, chemists begin by designing experiments that will reveal details of a particular chemical reaction. Next, through a combination of experience and intuition, they propose several sets of steps or mechanisms, each of which might account for the overall chemical transformation. Finally, each proposed mechanism is tested against the experimental observations to exclude those mechanisms that are not consistent with the facts.

A mechanism becomes generally established by excluding reasonable alternatives and by showing that it is consistent with every test that can be devised. This, of course, does not mean that a generally accepted mechanism is a completely accurate description of the chemical events, but only that it is the best chemists have been able to devise. It is important to keep in mind that, as new experimental evidence is obtained, it may be necessary to modify a generally accepted mechanism or possibly even discard it and start all over again.

Before we go on to consider reactions and reaction mechanisms, we might ask why it is worth the trouble to establish them and worth your time to learn about them. One reason is very practical: mechanisms provide a framework within which to organize a great deal of descriptive chemistry. For example, with insight into how reagents add to particular alkenes, it is possible to make generalizations and then to predict how the same reagents might add to other alkenes. A second reason lies in the intellectual satisfaction derived from constructing models that accurately reflect the behavior of chemical systems. Finally, to a creative scientist, a mechanism is a tool to be used in the search for new information and new understanding. A mechanism consistent with all that is known about a reaction can be used to make predictions about chemical interactions as yet unexplored, and experiments can be designed to test these predictions. Thus, reaction mechanisms provide a way not only to organize knowledge but also to extend it.

Supporting Concepts;
Electrophilic addition
reactions

6.3   Electrophilic Additions

We begin our introduction to the chemistry of alkenes with an examination of five addition reactions: addition of hydrogen halides (HCl, HBr, and HI), water (H_2O), halogens (Cl_2 and Br_2), mercuric acetate [$Hg(OAc)_2$] in the presence of water, and finally Cl_2 and Br_2 in the presence of water. We first study some of the experimental observations about each addition reaction and then its mechanism. Through the study of these particular reactions, we will develop a general understanding of how alkenes undergo addition reactions.

Mechanisms: Alkanes;
Propene hydrochlorination

A.   Addition of Hydrogen Halides

The hydrogen halides HCl, HBr, and HI add to alkenes to give haloalkanes (alkyl halides). These additions may be carried out either with the pure reagents (neat) or in the presence of a polar solvent such as acetic acid. Addition of HCl to ethylene gives chloroethane (ethyl chloride).

$$CH_2\!\!=\!\!CH_2 + HCl \longrightarrow \overset{\overset{\displaystyle H}{|}}{C}H_2\!-\!\overset{\overset{\displaystyle Cl}{|}}{C}H_2$$

Ethylene   Chloroethane

Regioselective reaction A reaction in which one direction of bond forming or breaking occurs in preference to all other directions.

Regiospecific reaction A reaction in which one direction of bond forming or breaking occurs to the exclusion of all other directions of bond forming or breaking.

Addition of HCl to propene gives 2-chloropropane (isopropyl chloride); hydrogen adds to carbon 1 of propene, and chlorine adds to carbon 2. If the orientation of addition were reversed, 1-chloropropane (propyl chloride) would be formed. The observed result is that 2-chloropropane is formed to the virtual exclusion of 1-chloropropane. We say that addition of HCl to propene is highly regioselective. A **regioselective reaction** is a reaction in which one direction of bond forming or breaking occurs in preference to all other directions of bond forming or breaking. A **regiospecific reaction** is one in which that direction is the only one that occurs.

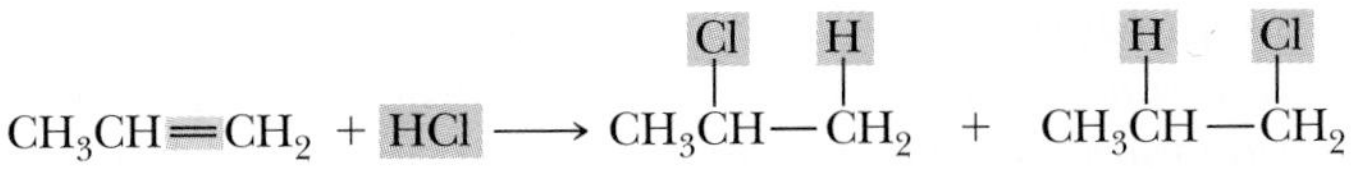

Propene 2-Chloropropane 1-Chloropropane (not observed)

This regioselectivity was noted by Vladimir Markovnikov who made the generalization known as **Markovnikov's rule:** in addition of H—X to alkenes, hydrogen adds to the double-bonded carbon that has the greater number of hydrogens already bonded to it. Although Markovnikov's rule provides a way to predict the products of many alkene addition reactions, it does not explain why one product predominates over other possible products.

Markovnikov's rule In the addition of HX, H_2O, or ROH to an alkene, hydrogen adds to the carbon of the double bond having the greater number of hydrogens.

Example 6.3

Name and draw a structural formula for the product of each alkene addition reaction.

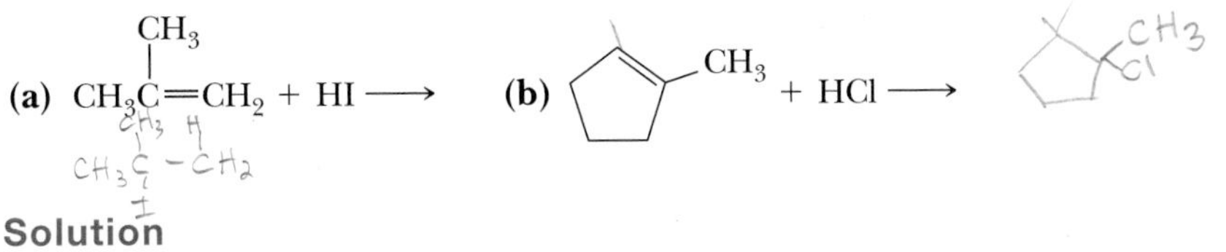

Solution

Using Markovnikov's rule, predict that 2-iodo-2-methylpropane is the product in (a) and 1-chloro-1-methylcyclopentane is the product in (b).

(a) 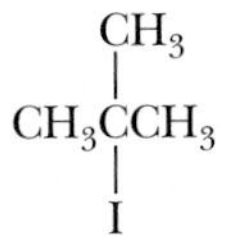 **(b)** 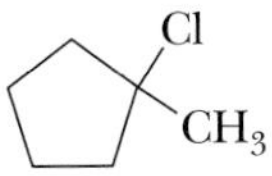

2-Iodo-2-methylpropane 1-Chloro-1-methylcyclopentane

Problem 6.3

Name and draw the structural formula for the product of each alkene addition reaction.

(a) $CH_3CH{=}CH_2 + HI \longrightarrow$ **(b)** 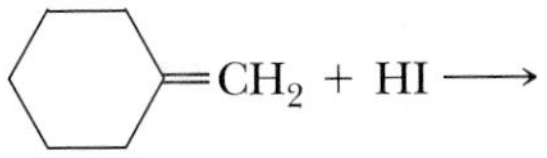

Chemists account for the addition of HX to an alkene by a two-step mechanism, which we illustrate by the reaction of 2-butene with hydrogen chloride to give 2-chlorobutane. Let us first look at this two-step mechanism in overview and then go back and study each step in detail.

In overview, addition begins with the transfer of a proton from HCl to 2-butene, as shown by the two curved arrows on the left side of Step 1. The first curved arrow shows breaking of the pi bond of the alkene and its electron pair being used to form a new covalent bond with the hydrogen atom of HCl. The second curved arrow shows breaking of the polar covalent bond in HCl and its electron pair being given entirely to chlorine, forming chloride ion. The first step in this mechanism results in formation of an organic cation and chloride ion. The second step is reaction of the organic cation with chloride ion to form a 2-chlorobutane.

Mechanism   Electrophilic Addition of HCl to 2-Butene

Step 1:   Proton transfer from HCl to 2-butene (a Lewis base) gives the *sec*-butyl cation.

$$CH_3CH\!=\!CHCH_3 + \overset{\delta+}{H}\!-\!\overset{\delta-}{\overset{..}{\underset{..}{Cl}}}: \xrightleftharpoons{\overset{\text{slow, rate}}{\text{determining}}} CH_3\overset{+}{CH}\!-\!\overset{\overset{H}{|}}{CHCH_3} + :\overset{..}{\underset{..}{Cl}}:^-$$

sec-Butyl cation
(a 2° carbocation
intermediate)

Step 2:   Reaction of the *sec*-butyl cation (a Lewis acid) with chloride ion (a Lewis base) completes the valence shell of carbon and gives 2-chlorobutane.

$$:\overset{..}{\underset{..}{Cl}}:^- \quad + \quad CH_3\overset{+}{CH}CH_2CH_3 \xrightarrow{\text{fast}} CH_3\overset{\overset{\overset{..}{\underset{..}{Cl}}:}{|}}{CH}CH_2CH_3$$

Chloride ion   *sec*-Butyl cation
(a Lewis base) (a Lewis acid)

Now that we have looked at this two-step mechanism in overview, let us go back and look at the individual steps in more detail. There is a great deal of important organic chemistry embedded in these two steps, and it is important that you understand it now.

Step 1 results in formation of an organic cation. One carbon atom in this cation has only six electrons in its valence shell and carries a charge of +1. A species containing a positively charged carbon atom is called a **carbocation** (*carb*on + *cat*ion). (Such carbon-containing cations have also been called carbonium ions and carbenium ions. Even though these terms are obsolete, you may still encounter them.) Carbocations are classified as primary (1°), secondary (2°), or tertiary (3°) depending on the number of carbon atoms bonded to the carbon bearing the positive charge. All carbocations are Lewis acids (Section 4.5). They are also **electrophiles.**

In a carbocation, the carbon bearing the positive charge is bonded to three other atoms and, as predicted by the VSEPR model, the three bonds about it are coplanar and form bond angles of approximately 120°. According to the molecular orbital model, the electron-deficient carbon of a carbocation uses sp^2 hybrid orbitals to form sigma bonds to the three attached groups. The unhybridized $2p$ orbital lies perpendicular to the sigma bond framework and contains no electrons. Figure 6.3 shows a Lewis structure and an orbital overlap diagram for the *tert*-butyl cation.

Figure 6.4 shows an energy diagram for the two-step reaction of 2-butene with HCl. The slower, rate-determining step (the one that crosses the higher energy barrier) is Step 1, which leads to formation of the 2° carbocation intermediate. This carbocation intermediate lies in an energy minimum between the transition states for Steps 1 and 2. As soon as the carbocation intermediate (a Lewis acid) is formed, it re-

Carbocation A species in which a carbon atom has only six electrons in its valence shell and bears a positive charge.

Electrophile From the Greek meaning electron loving. Any species that can accept a pair of electrons to form a new covalent bond; a Lewis acid.

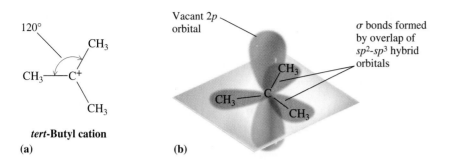

Figure 6.3
The structure of the *tert*-butyl cation.
(a) Lewis structure and (b) an orbital picture.

acts with chloride ion (a Lewis base) in a Lewis acid-base reaction to give 2-chlorobutane. Note that the energy level for 2-chlorobutane (the product) is lower than the energy level for 2-butene and HCl (the reactants). Thus, in this alkene addition reaction, heat is released; the reaction is exothermic.

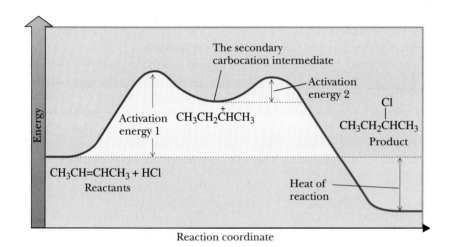

Figure 6.4
An energy diagram for the two-step addition of HCl to 2-butene. The reaction is exothermic.

Relative Stabilities of Carbocations — Regioselectivity and Markovnikov's Rule

Reaction of HX and an alkene can, at least in principle, give two different carbocation intermediates depending on which of the doubly bonded carbon atoms the proton is transferred to, as illustrated by the reaction of HCl with propene.

The observed product is 2-chloropropane, indicating that the 2° carbocation intermediate is formed in preference to the 1° carbocation intermediate.

Similarly, in the reaction of HCl with 2-methylpropene, proton transfer to the carbon-carbon double bond might form either the isobutyl cation (a 1° carbocation) or the *tert*-butyl cation (a 3° carbocation).

CH$_3$
|
CH$_3$C=CH$_2$ + H—Cl: ⟶ CH$_3$CHCH$_2$:Cl:⁻ ⟶ CH$_3$CHCH$_2$Cl:
+

2-Methylpropene   Isobutyl cation   1-Chloro-2-methylpropane
(a 1° carbocation)   (not formed)

CH$_3$
|
CH$_3$C=CH$_2$ + H—Cl: ⟶ CH$_3$CCH$_3$:Cl:⁻ ⟶ CH$_3$CCH$_3$
+ |
                              :Cl:

2-Methylpropene   *tert*-Butyl cation   2-Chloro-2-methylpropane
(a 3° carbocation)   (product formed)

In this reaction, the observed product is 2-chloro-2-methylpropane, indicating that the 3° carbocation is formed in preference to the 1° carbocation.

From such experiments and a great amount of other experimental evidence, we know that a 3° carbocation is both more stable and requires a lower activation energy for its formation than a 2° carbocation, which is, in turn, more stable and requires a lower activation energy for its formation than a 1° carbocation. Methyl and 1° carbocations are so unstable they are never formed in solution. It follows, then, that a more stable carbocation intermediate forms faster than a less stable carbocation intermediate. Following is the order of stability of four types of alkyl carbocations.

Supporting Concepts; Carbocations

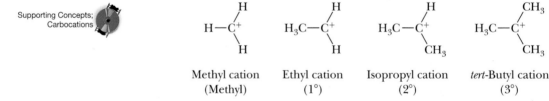

H
|
H—C⁺
|
H

H
|
H$_3$C—C⁺
|
H

H
|
H$_3$C—C⁺
|
CH$_3$

CH$_3$
|
H$_3$C—C⁺
|
CH$_3$

Methyl cation   Ethyl cation   Isopropyl cation   *tert*-Butyl cation
(Methyl)   (1°)   (2°)   (3°)

Order of increasing carbocation stability ⟶

Now that we know the order of stability of carbocations, how do we account for that order? The principles of physics teach us that a system bearing a charge (either positive or negative) is more stable if the charge is delocalized. Using this principle, we can explain the order of stability of carbocations if we assume that alkyl groups bonded to a positively charged carbon are electron releasing and thereby help delocalize the charge on the cation. The electron-releasing ability of alkyl groups bonded to a cationic carbon is accounted for by two effects: the inductive effect and hyperconjugation.

The inductive effect operates in the following way. The electron deficiency of the carbon atom bearing a positive charge exerts an electron-withdrawing **inductive effect** that polarizes electrons from adjacent sigma bonds toward it. Thus, the positive charge of the cation is not localized on the trivalent carbon, but rather is delocalized over nearby atoms. The larger the volume over which the positive charge is delocalized, the greater the stability of the cation. Thus, as the number of alkyl groups

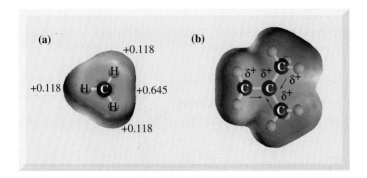

(a)   **(b)**

Figure 6.5
Electrostatic potential plots showing the distribution of the positive charge in (a) the methyl cation and (b) the *tert*-butyl cation. Electron donation by the methyl groups decreases the positive charge (blue area) on the central carbon of the *tert*-butyl cation.

bonded to the cationic carbon increases, the stability of the cation increases. Figure 6.5 illustrates the electron-withdrawing inductive effect of the positively charged carbon and the resulting delocalization of charge. According to quantum mechanical calculations, the charge on carbon in the methyl cation is approximately +0.645, and the charge on each of the hydrogen atoms is +0.118. Thus, even in the methyl cation, the positive charge is not localized on carbon. Rather, it is delocalized over the volume of space occupied by the entire ion. Polarization of electron density and delocalization of charge is even more extensive in the *tert*-butyl cation.

The second effect operating to stabilize carbocations is hyperconjugation. **Hyperconjugation** involves partial overlapping of the sigma bonding orbital of an adjacent C—H or C—C bond with the vacant $2p$ orbital of the cationic carbon (Figure 6.6). In this way, electrons of the adjacent sigma bond are partially delocalized, which also partially delocalizes the positive charge of the cation. Replacing any C—H bond to the cationic carbon by an alkyl group increases the possibility for electron delocalization by hyperconjugation and increases the stability of the carbocation. Note that, because of the geometry of the cationic carbon and the C—H bonds on an adjacent carbon, only electrons in an adjacent C—H bond can participate in hyperconjugation.

Hyperconjugation Interaction of electrons in a sigma bonding orbital with the vacant $2p$ orbital of an adjacent positively charged carbon.

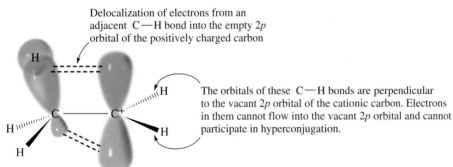

Delocalization of electrons from an adjacent C—H bond into the empty $2p$ orbital of the positively charged carbon

The orbitals of these C—H bonds are perpendicular to the vacant $2p$ orbital of the cationic carbon. Electrons in them cannot flow into the vacant $2p$ orbital and cannot participate in hyperconjugation.

Figure 6.6
Hyperconjugation.

Example 6.4

Arrange these carbocations in order of increasing stability.

(a)   **(b)**   **(c)**

Solution

Carbocation (a) is secondary, (b) is tertiary, and (c) is primary. In order of increasing stability they are c < a < b.

Problem 6.4

Arrange these carbocations in order of increasing stability.

(a) (image of cyclohexyl cation with +CH₃) **(b)** (image of cyclohexyl cation with CH₃) **(c)** (image of cyclohexyl cation with +CH₂)

Example 6.5

Propose a mechanism for the addition of HI to methylidenecyclohexane to give 1-iodo-1-methylcyclohexane. Which step in your mechanism is rate determining?

Solution

Propose a two-step mechanism similar to that proposed for the addition of HCl to propene.

Step 1: A rate-determining proton transfer from HI to the carbon-carbon double bond gives a 3° carbocation intermediate.

$$\text{(Methylidenecyclohexane)} =CH_2 + H-\ddot{I}: \xrightarrow[\text{determining}]{\text{slow, rate}} \text{(cyclohexyl)}^+-CH_3 + :\ddot{I}:^-$$

Methylidenecyclohexane   A 3° carbocation
(Methylenecyclohexane)   intermediate

Step 2: Reaction of the carbocation intermediate (a Lewis acid) with iodide ion (a Lewis base) completes the valence shell of carbon and gives the product.

$$^+-CH_3 + :\ddot{I}:^- \xrightarrow{\text{fast}} \overset{:\ddot{I}:}{\underset{CH_3}{}}$$

1-Iodo-1-methylcyclohexane

Problem 6.5

Propose a mechanism for addition of HI to 1-methylcyclohexene to give 1-iodo-1-methylcyclohexane. Which step in your mechanism is rate determining?

Mechanisms: Alkenes;
Propene hydration

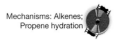

Hydration Addition of water.

B. Addition of Water—Acid-Catalyzed Hydration

In the presence of an acid catalyst, most commonly concentrated sulfuric acid, water adds to an alkene to give an alcohol. Addition of water is called **hydration.** In the case of simple alkenes, H adds to the carbon of the double bond with the greater number of hydrogens, and OH adds to the carbon with the fewer hydrogens. Thus, H—OH adds to alkenes in accordance with Markovnikov's rule.

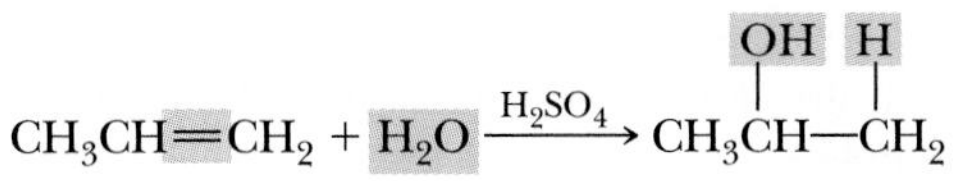

$$CH_3CH{=}CH_2 + \boxed{H_2O} \xrightarrow{H_2SO_4} CH_3\underset{\substack{| \\ }}{\overset{\boxed{OH}}{C}H}{-}\overset{\boxed{H}}{C}H_2$$

Propene   2-Propanol

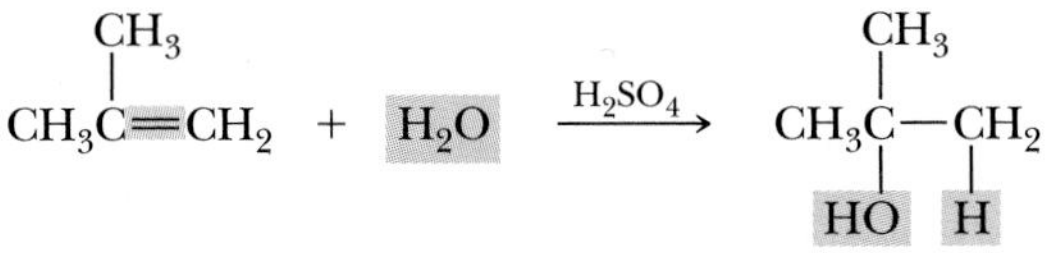

$$\overset{\overset{\displaystyle CH_3}{|}}{CH_3C}{=}CH_2 \;+\; \boxed{H_2O} \xrightarrow{H_2SO_4} \overset{\overset{\displaystyle CH_3}{|}}{CH_3C}\underset{\underset{\displaystyle \boxed{HO}\;\boxed{H}}{|\;\;\;|}}{-CH_2}$$

2-Methylpropene   2-Methyl-2-propanol

Example 6.6

Draw a structural formula for the product of acid-catalyzed hydration of 1-methyl-cyclohexene.

Solution

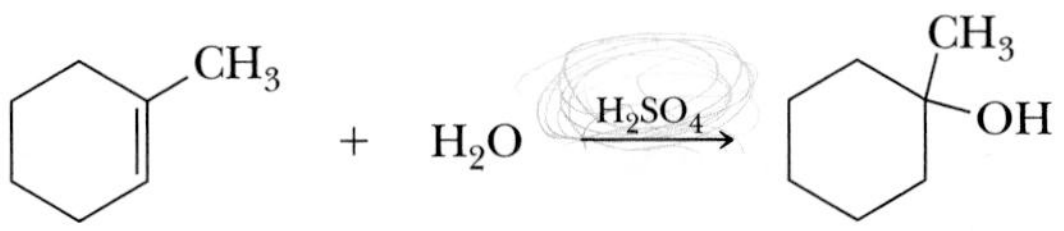

1-Methylcyclohexene   1-Methylcyclohexanol

Problem 6.6

Draw a structural formula for the product of each alkene hydration reaction.

(a) 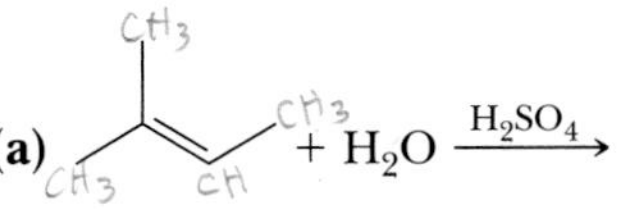 [structure] $+ H_2O \xrightarrow{H_2SO_4}$ (b) 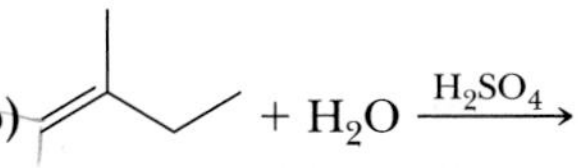 [structure] $+ H_2O \xrightarrow{H_2SO_4}$

The mechanism for acid-catalyzed hydration of alkenes is quite similar to what we already proposed for addition of HCl, HBr, and HI to alkenes and is illustrated by conversion of propene to 2-propanol. Formation of the carbocation intermediate in Step 1 is the rate-determining step. This mechanism is consistent with the fact that acid is a catalyst. An H_3O^+ is consumed in Step 1 but another is generated in Step 3.

Mechanism   Acid-Catalyzed Hydration of Propene

Step 1:   Proton transfer from H_3O^+ to propene gives a 2° carbocation intermediate.

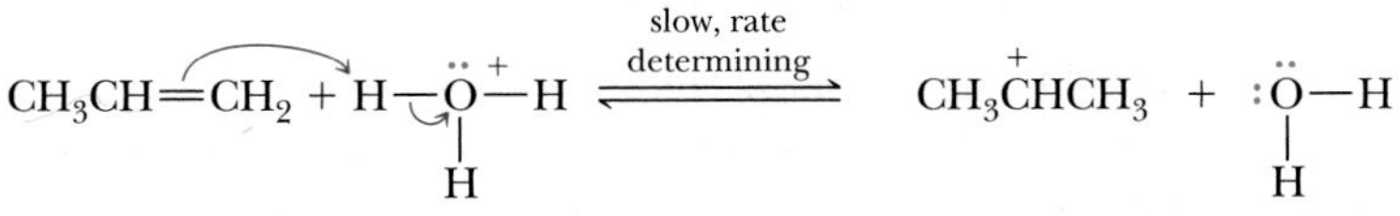

$$CH_3CH{=}CH_2 + H{-}\overset{\overset{\displaystyle}{..}}{\underset{\underset{\displaystyle H}{|}}{O}}{}^{+}{-}H \underset{\text{determining}}{\overset{\text{slow, rate}}{\rightleftharpoons}} CH_3\overset{+}{C}HCH_3 + \overset{..}{:O}{-}H$$
$$\underset{\underset{\displaystyle H}{|}}{}$$

A 2° carbocation
intermediate

Step 2:   The carbocation intermediate (a Lewis acid) completes its valence shell by forming a new covalent bond with an unshared pair of electrons of the oxygen atom of water (a Lewis base) and gives an **oxonium ion.**

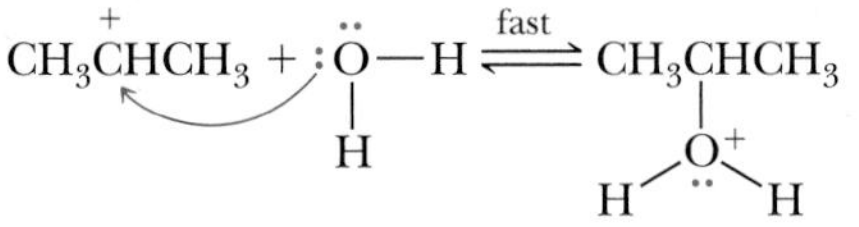

Oxonium ion   An ion in which oxygen bears a positive charge.

An oxonium ion

Step 3:   Proton transfer from the oxonium ion to water gives the alcohol and generates a new acid catalyst.

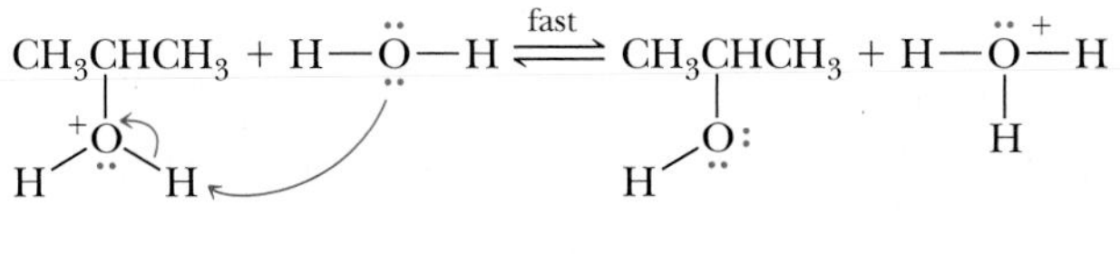

Example 6.7

Propose a mechanism for the acid-catalyzed hydration of methylidenecyclohexane to give 1-methylcyclohexanol. Which step in your mechanism is rate determining?

Solution

Propose a three-step mechanism similar to that for the acid-catalyzed hydration of propene. Formation of the 3° carbocation intermediate in Step 1 is rate determining.

Step 1:   Proton transfer from the acid catalyst to the alkene gives a 3° carbocation intermediate.

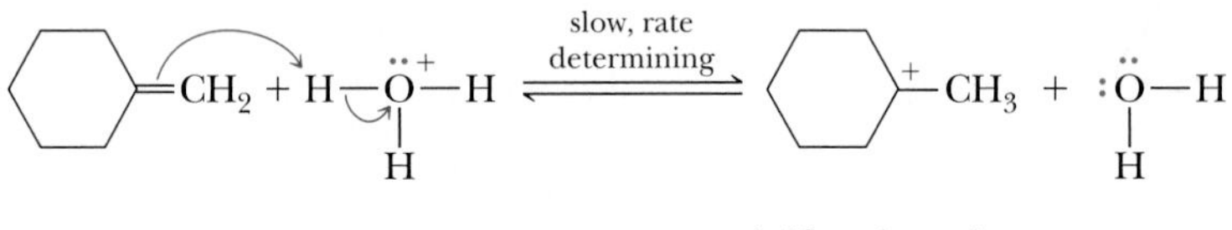

A 3° carbocation
intermediate

Step 2:   Reaction of this intermediate (a Lewis acid) with water (a Lewis base) completes the valence shell of carbon and gives an oxonium ion.

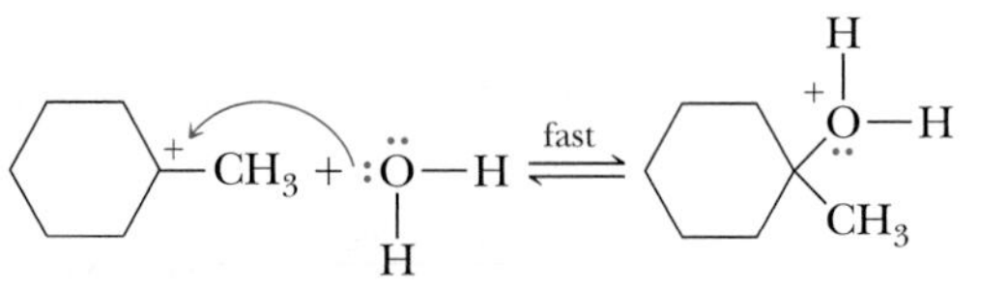

An oxonium ion

Step 3:   Proton transfer from the oxonium ion to water gives the alcohol and generates a new acid catalyst.

[Chemical structure diagram showing the protonation/deprotonation equilibrium with "fast" label]

Problem 6.7

Propose a mechanism for the acid-catalyzed hydration of 1-methylcyclohexene to give 1-methylcyclohexanol. Which step in your mechanism is rate determining?

C.   Carbocation Rearrangements

As we have seen in the preceding discussions, the expected product of electrophilic addition to a carbon-carbon double bond involves rupture of the pi bond and formation of two new sigma bonds in its place. In addition of HCl to 3-methyl-1-butene, however, only 40% of the expected product is formed. The major product is 2-chloro-2-methylbutane, a compound with a different connectivity (a different order of attachment of its atoms) compared with the connectivity of atoms in the starting alkene. We say that formation of 2-chloro-2-methylbutane has involved a **rearrangement.** Typically, either a hydrogen or an alkyl group migrates with its bonding electrons to an adjacent electron-deficient atom. In the rearrangements we examine in this chapter, migration is to an electron-deficient carbon atom bearing a positive charge.

Rearrangement A change in connectivity of the atoms in a product compared with the connectivity of the same atoms in the starting material.

$$CH_3CHCH=CH_2 + HCl \longrightarrow CH_3CHCHCH_3 \quad + \quad CH_3CCH_2CH_3$$

| 3-Methyl-1-butene | 2-Chloro-3-methylbutane (the expected product) (40%) | 2-Chloro-2-methylbutane (a product of rearrangement) (60%) |

Formation of the rearranged product in this reaction can be accounted for by the following mechanism, the key step of which is a type of rearrangement called a **1,2-shift.** In the rearrangement shown in Step 2, the migrating group is a hydride ion (a hydrogen nucleus with two valence electrons).

1,2-Shift A type of rearrangement in which an atom or group of atoms with its bonding electrons moves from one atom to an adjacent electron-deficient atom.

Mechanism   Carbocation Rearrangement in the Addition of HCl to an Alkene

Step 1:   Proton transfer to the alkene gives a 2° carbocation intermediate.

[Chemical mechanism diagram showing]

$$CH_3CCH=CH_2 + H-\ddot{C}l: \xrightarrow[\text{determining}]{\text{slow, rate}} CH_3C-CHCH_3 + :\ddot{C}l:^-$$

3-Methyl-1-butene

A 2° carbocation intermediate

Step 2:   Migration of a hydride ion from an adjacent carbon to the positively charged carbon of the 2° carbocation gives a more stable 3° carbocation. In this rearrangement, the major movement is that of the bonding electron pair.

$$\underset{\substack{\\ \text{H}}}{\overset{\substack{\text{CH}_3 \\ |}}{\text{CH}_3\text{C}-\text{CHCH}_3}} \xrightarrow{\text{fast}} \underset{\substack{\\ \text{H}}}{\overset{\substack{\text{CH}_3 \\ |}}{\text{CH}_3\text{C}-\text{CHCH}_3}}$$

A 3° carbocation
intermediate

Step 3:   Reaction of the 3° carbocation intermediate (a Lewis acid) with chloride ion (a Lewis base) gives the rearranged product.

$$\overset{\substack{\text{CH}_3 \\ |}}{\text{CH}_3\text{C}-\text{CH}_2\text{CH}_3} + :\ddot{\text{Cl}}:^{-} \xrightarrow{\text{fast}} \underset{\substack{| \\ :\ddot{\text{Cl}}:}}{\overset{\substack{\text{CH}_3 \\ |}}{\text{CH}_3\text{C}-\text{CH}_2\text{CH}_3}}$$

2-Chloro-2-methylbutane

The driving force for this rearrangement is the fact that the less stable 2° carbocation originally formed is converted to a more stable 3° carbocation. From the study of this and other carbocation rearrangements, we find that 2° carbocations rearrange to more stable 2° or 3° carbocations. They rarely rearrange in the opposite direction. 1° Carbocations are never formed in reactions taking place in solution.

Rearrangements also occur in acid-catalyzed hydration of alkenes, especially when the carbocation formed in the first step can rearrange to a more stable carbocation. For example, acid-catalyzed hydration of 3-methyl-1-butene gives 2-methyl-2-butanol.

$$\overset{\substack{\text{CH}_3 \\ |}}{\text{CH}_3\text{CHCH}=\text{CH}_2} + \text{H}_2\text{O} \xrightarrow{\text{H}_2\text{SO}_4} \underset{\substack{| \\ \text{OH}}}{\overset{\substack{\text{CH}_3 \\ |}}{\text{CH}_3\text{CCH}_2\text{CH}_3}}$$

3-Methyl-1-butene   2-Methyl-2-butanol

Example 6.8

In addition of HCl to 3,3-dimethyl-1-butene, only 17% of the unrearranged product is formed. Formation of 2-chloro-2,3-dimethylbutane, the major product, involves a rearrangement. Propose a mechanism for the formation of this rearranged product.

$$\underset{\substack{| \\ \text{CH}_3}}{\overset{\substack{\text{CH}_3 \\ |}}{\text{CH}_3\text{CCH}=\text{CH}_2}} + \text{HCl} \longrightarrow \underset{\substack{| \\ \text{CH}_3\text{Cl}}}{\overset{\substack{\text{CH}_3 \\ |}}{\text{CH}_3\text{C}-\text{CHCH}_3}} + \underset{\substack{| \quad | \\ \text{Cl} \quad \text{CH}_3}}{\overset{\substack{\text{CH}_3 \\ |}}{\text{CH}_3\text{C}-\text{CHCH}_3}}$$

3,3-Dimethyl-1-butene   2-Chloro-3,3-   2-Chloro-2,3-
                              dimethylbutane   dimethylbutane
                                 (17%)   (83%)

Solution

Following is a three-step mechanism for the formation of 2-chloro-2,3-dimethylbutane.

Step 1: Proton transfer from HCl to the double bond of the alkene gives a 2° carbocation intermediate.

$$CH_3C—CH{=}CH_2 + H—Cl: \xrightarrow[\text{determining}]{\text{slow, rate}} CH_3—C—CH—CH_3 + :Cl:^-$$

A 2° carbocation
intermediate

Step 2: The less stable 2° carbocation rearranges to a more stable 3° carbocation by migration of an adjacent methyl group with its pair of bonding electrons (a 1,2-methyl shift).

$$CH_3C—CH—CH_3 \xrightarrow{\text{fast}} CH_3C—CH—CH_3$$

Step 3: Reaction of the 3° carbocation (a Lewis acid) with chloride ion (a Lewis base) completes the valence shell of carbon and gives the rearranged product.

$$:Cl:^- + CH_3C—CH—CH_3 \xrightarrow{\text{fast}} CH_3C—CH—CH_3$$

Problem 6.8

Acid-catalyzed hydration of 3-methyl-1-butene gives 2-methyl-2-butanol as the major product. Propose a mechanism for the formation of this alcohol.

D. Addition of Bromine and Chlorine

Supporting Concepts;
Electrophilic addition
reactions

Chlorine, Cl_2, and bromine, Br_2, react with alkenes at room temperature by addition of halogen atoms to the two carbon atoms of the double bond with formation of two new carbon-halogen bonds. Fluorine, F_2, also adds to alkenes, but because its reactions are very fast and difficult to control, this reaction is not a useful laboratory procedure. Iodine, I_2, also adds but the reaction is not preparatively useful.

Halogenation with bromine or chlorine is generally carried out either with the pure reagents or by mixing them in an inert solvent such as CCl_4 or CH_2Cl_2.

$$CH_3CH{=}CHCH_3 + Br_2 \xrightarrow{CCl_4} CH_3CH—CHCH_3$$

2-Butene   2,3-Dibromobutane

Addition of bromine or chlorine to a cycloalkene gives a *trans*-dihalocycloalkane. Addition of bromine to cyclohexene, for example, gives *trans*-1,2-dibromocyclohexane; the cis isomer is not formed. Thus, addition of a halogen to an alkene is stereospecific. A **stereospecific reaction** is a reaction in which one stereoisomer is formed or

Stereospecific reaction A reaction in which one stereoisomer is formed or destroyed to the exclusion of all others.

Stereoselective reaction A reaction in which one stereoisomer is formed or destroyed in preference to all others.

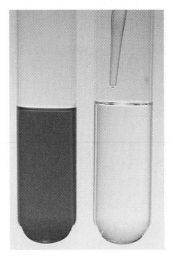

A solution of Br_2 in carbon tetrachloride is red. Add a few drops of an alkene and the color disappears. *(Charles D. Winters)*

destroyed to the exclusion of all others that might be formed or destroyed; a **stereoselective reaction** is one in which one stereoisomer is formed or destroyed in preference to all others.

Cyclohexene + Br_2 $\xrightarrow{CCl_4}$ *trans*-1,2-Dibromocyclohexane

Reaction of bromine with an alkene is a particularly useful qualitative test for the presence of a carbon-carbon double bond. If we dissolve bromine in carbon tetrachloride, the solution is red. Both alkenes and dibromoalkanes are colorless. If we now mix a few drops of the bromine solution with an alkene, a dibromoalkane is formed, and the red solution becomes colorless.

Example 6.9

Complete these reactions, showing the stereochemistry of the product.

(a) [cyclopentene] + Br_2 $\xrightarrow{CH_2Cl_2}$ **(b)** [1-methylcyclohexene with CH_3] + Cl_2 $\xrightarrow{CH_2Cl_2}$

Solution

Addition of both Br_2 and Cl_2 is stereospecific. The halogen atoms are trans to each other in each product.

(a) [cyclopentene] + Br_2 $\xrightarrow{CH_2Cl_2}$ [trans-1,2-dibromocyclopentane]   **(b)** [methylcyclohexene] + Cl_2 $\xrightarrow{CH_2Cl_2}$ [product with H_3C, Cl, Cl]

Problem 6.9

Complete these reactions.

(a) $CH_3\underset{\underset{CH_3}{|}}{\overset{\overset{CH_3}{|}}{C}}CH=CH_2$ + Br_2 $\xrightarrow{CH_2Cl_2}$ **(b)** [methylenecyclohexane with CH_2] + Cl_2 $\xrightarrow{CH_2Cl_2}$

Stereospecificity and Bridged Halonium Ion Intermediates

We explain the addition of bromine and chlorine to alkenes and their stereospecificity by the following two-step mechanism. Reaction is initiated in Step 1 by interaction of the pi electrons of the alkene with bromine to form an intermediate in which bromine bears a positive charge. A bromine atom bearing a positive charge is called a **bromonium ion,** and the cyclic structure of which it is a part is called a **bridged bromonium ion.** A bridged bromonium ion may look odd to you because it has two bonds to bromine, but it is an acceptable Lewis structure. Calculation of formal charge places a positive charge on bromine. Then, in Step 2, a bromide ion reacts with this bridged intermediate from the side opposite to that occupied by the bromine atom to give the dibromoalkane. Thus, bromine atoms are added from opposite faces of the carbon-carbon double bond. We say that this addition occurs with **anti stereospecificity.**

Anti stereospecificity The addition of atoms or groups of atoms to opposite faces of a carbon-carbon double bond.

Mechanism   Addition of Bromine with Anti Stereospecificity

Step 1:   Reaction of the pi electrons of the carbon-carbon double bond with bromine gives a bridged bromonium ion intermediate. This intermediate is shown as a hybrid of three contributing structures, of which the bridged bromonium ion is the most important. The open carbocation structures are only minor contributors.

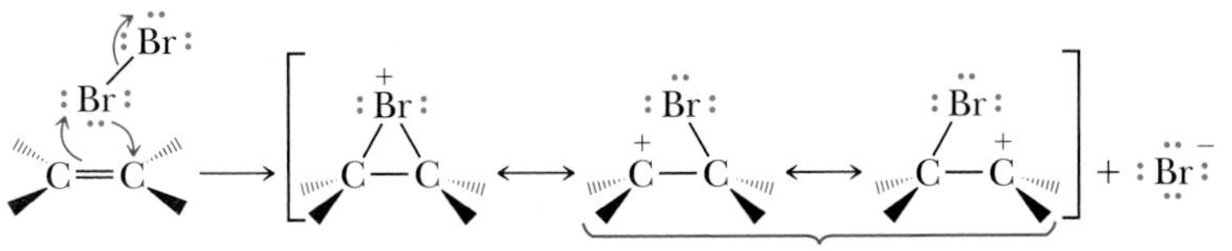

The bridged bromonium ion is the most important contributing structure.

These carbocations are only minor contributing structures.

Step 2:   Attack of bromide ion (a Lewis base) on carbon (a Lewis acid) from the side opposite the bromonium ion opens the three-membered ring to give the anti product. Both carbon atoms of the bromonium ion are attacked, but the products are identical. The bromines are not only anti but also in the same plane (coplanar). Thus, we call this an anti-coplanar attack.

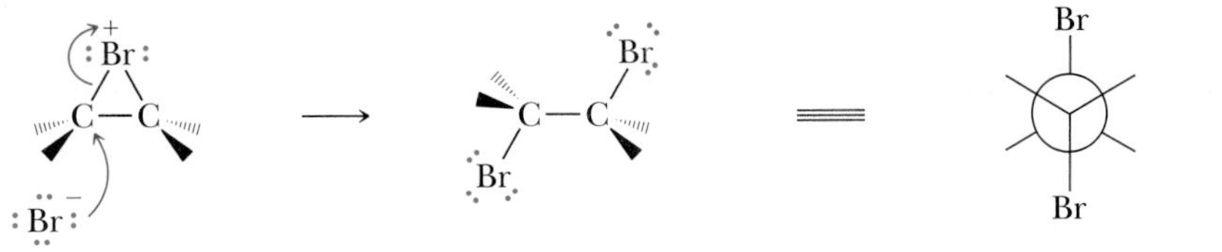

Anti (trans-coplanar) orientation of added bromine atoms

Newman projection of the product

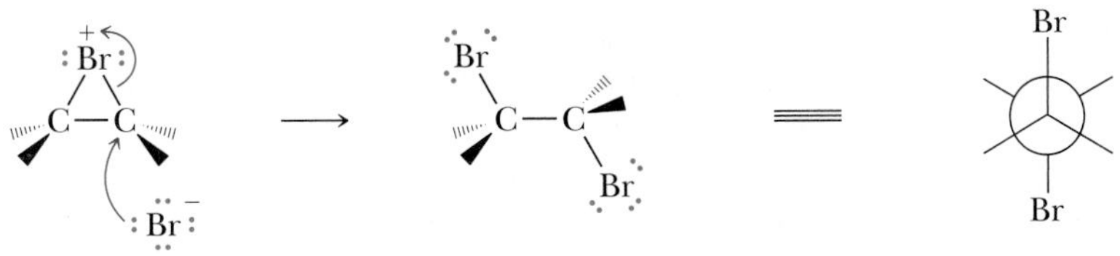

Newman projection of the product

Addition of chlorine or bromine to cyclohexene and its derivatives gives a trans diaxial product because only axial positions on adjacent atoms of a cyclohexane ring are anti and coplanar. The initial trans diaxial conformation of the product is in equilibrium with the trans diequatorial conformation, and, in simple derivatives of cyclohexane, the latter is more stable and predominates.

trans Diaxial   trans Diequatorial
(more stable)

In derivatives of cyclohexane in which interconversion between one chair conformation and the other is not possible or is severely restricted, the trans diaxial product is isolated. If a cyclohexane ring contains a bulky alkyl group, such as *tert*-butyl (Section 2.6), then the molecule exists overwhelmingly in a conformation in which the *tert*-butyl group is equatorial. Bromination of 4-*tert*-butylcyclohexene gives 1,2-dibromo-4-*tert*-butylcyclohexane. Attack of bromine occurs at both faces of the six-membered ring, but each bromonium ion intermediate reacts with bromide ion to give the same product. In the favored chair conformation of this product, *tert*-butyl is equatorial, and the bromine atoms remain axial.

$(CH_3)_3C$ —⬡ + Br_2 ⟶ $(CH_3)_3C$

4-*tert*-Butylcyclohexene   1,2-Dibromo-
4-*tert*-butylcyclohexane

E.   Addition of HOCl and HOBr

Treatment of an alkene with Br_2 or Cl_2 in the presence of water results in addition of OH and Br, or OH and Cl, to the carbon-carbon double bond to give a **halohydrin.**

Halohydrin A compound containing a halogen atom and a hydroxyl group on adjacent carbons; those containing Br and OH are bromohydrins, and those containing Cl and OH are chlorohydrins.

$$CH_3CH{=}CH_2 + Cl_2 + H_2O \longrightarrow \underset{\text{1-Chloro-2-propanol}}{CH_3\overset{\text{HO}}{\underset{}{CH}}{-}\overset{\text{Cl}}{\underset{}{CH_2}}} + HCl$$

Propene   1-Chloro-2-propanol
(a chlorohydrin)

Addition of HOCl and HOBr is regiospecific (halogen adds to the less substituted carbon atom) and anti stereospecific. Both the regiospecificity and anti stereospecificity are illustrated by the addition of HOBr to 1-methylcyclopentene. Bromine and the hydroxyl group add anti to each other with Br bonding to the less substituted carbon and OH bonding to the more substituted carbon.

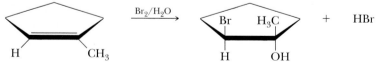

1-Methylcyclopentene   2-Bromo-1-methylcyclopentanol
(anti addition of —OH and —Br)

To account for the regiospecificity and anti stereospecificity of these reactions, chemists propose a three-step mechanism. Step 1 involves reaction of halogen with the pi bond of the alkene to form the same bridged halonium ion intermediate as in the halogenation of an alkene. This intermediate has some of the character of a carbocation (to account for the regiospecificity) and some of the character of a halonium ion (to account for the stereospecificity). Reaction of this halonium ion intermediate with H_2O in Step 2 followed by proton transfer in Step 3 completes the reaction.

Mechanism   Halohydrin Formation and Its Anti Stereospecificity

Step 1: Reaction of the pi electrons of the carbon-carbon double bond with bromine gives a bridged bromonium ion intermediate. The secondary carbocation makes only a minor contribution. The primary carbocation, not shown, is even higher in energy and makes no contribution.

The bridged bromonium ion is the most important contributing structure.

This carbocation is only a minor contributing structure.

Step 2: Attack of H_2O (a Lewis base) on the more substituted carbon of the bridged bromonium ion (a Lewis acid) opens the three-membered ring. The regiospecificity is determined by the minor contributing structure, which places the positive charge on the more substituted carbon.

Anti-stereospecific addition of Br and H_2O

Step 3: Proton transfer to water completes the reaction.

Example 6.10

Draw the structure of the bromohydrin formed when 2-methylpropene is treated with Br_2/H_2O.

Solution

Addition is regiospecific, with —OH adding to the more substituted carbon and —Br adding to the less substituted carbon.

$$\underset{\text{2-Methylpropene}}{\underset{\displaystyle |}{CH_3-\overset{\displaystyle \overset{CH_3}{|}}{C}=CH_2}} + Br_2 + H_2O \longrightarrow \underset{\underset{\text{1-Bromo-2-methyl-}}{\text{2-propanol}}}{\underset{\displaystyle \overset{|}{HO}\ \overset{|}{Br}}{CH_3-\overset{\overset{CH_3}{|}}{C}-CH_2}} + HBr$$

Problem 6.10

Draw the structure of the chlorohydrin formed when 1-methylcyclohexene is treated with Cl_2/H_2O.

F.   Oxymercuration-Reduction

Hydration of alkenes can be accomplished by treating an alkene with mercury(II) acetate (mercuric acetate) in water followed by reduction of the resulting product with sodium borohydride, $NaBH_4$. In the following structural formulas for mercury(II) acetate, the acetate group is written in full in the first formula. In the second formula, it is abbreviated as AcO.

$$\underset{\substack{\text{Mercury(II) acetate}\\\text{(Mercuric acetate)}}}{\overset{\displaystyle O \qquad\qquad O}{\overset{\displaystyle \|\qquad\qquad\ \|}{CH_3CO-Hg-OCCH_3 \qquad AcO-Hg-OAc}}}$$

Oxymercuration, the addition of mercury(II) to one carbon of the double bond and oxygen to the other, is illustrated by the first step in the two-step conversion of 2-butene to 2-butanol.

Oxymercuration:

$$\underset{\text{2-Butene}}{CH_3CH=CHCH_3} + \underset{\substack{\text{Mercury(II)}\\\text{acetate}}}{Hg(OAc)_2} + H_2O \longrightarrow \underset{\substack{\text{An organomercury}\\\text{compound}}}{\underset{\displaystyle \overset{|}{HgOAc}}{CH_3\overset{\overset{OH}{|}}{C}HCHCH_3}} + \underset{\text{Acetic acid}}{\overset{\displaystyle O}{\overset{\displaystyle \|}{CH_3COH}}}$$

The initial organomercury compound is reduced by sodium borohydride, $NaBH_4$, to replace Hg by H.

Reduction:

$$\underset{\substack{\text{An organomercury}\\\text{compound}}}{\underset{\displaystyle \overset{|}{HgOAc}}{CH_3\overset{\overset{OH}{|}}{C}HCHCH_3}} \xrightarrow{NaBH_4} \underset{\text{2-Butanol}}{\underset{\displaystyle \overset{|}{H}}{CH_3\overset{\overset{OH}{|}}{C}HCHCH_3}} + \underset{\text{Acetic acid}}{\overset{\displaystyle O}{\overset{\displaystyle \|}{CH_3COH}}} + Hg$$

Oxymercuration-reduction   A method for converting an alkene to an alcohol. The alkene is treated with mercury(II) acetate followed by reduction with sodium borohydride.

Oxymercuration is regiospecific: HgOAc becomes bonded to the less substituted carbon of the alkene and OH of water becomes bonded to the more substituted carbon. The result of oxymercuration followed by sodium borohydride reduction is Markovnikov addition of H—OH to an alkene.

$$CH_3CH_2CH_2CH_2CH{=}CH_2 \xrightarrow[\text{2. NaBH}_4]{\text{1. Hg(OAc)}_2,\ H_2O} CH_3CH_2CH_2CH_2\overset{\overset{\displaystyle OH}{|}}{C}HCH_3$$

1-Hexene   2-Hexanol

$$CH_3\overset{\overset{\displaystyle CH_3}{|}}{C}HCH{=}CH_2 \xrightarrow[\text{2. NaBH}_4]{\text{1. Hg(OAc)}_2,\ H_2O} CH_3\overset{\overset{\displaystyle CH_3}{|}}{C}H\underset{\underset{\displaystyle OH}{|}}{C}HCH_3$$

3-Methyl-1-butene   3-Methyl-2-butanol

Oxymercuration of 3-methyl-1-butene followed by $NaBH_4$ reduction gives 3-methyl-2-butanol exclusively and illustrates a very important feature of this reaction sequence: it occurs without rearrangement. You might compare the product of oxymercuration-reduction of 3-methyl-1-butene with the product formed by acid-catalyzed hydration of the same alkene (Section 6.3C). In the former, no rearrangement occurs. In the latter, the major product is 2-methyl-2-butanol, a compound formed by rearrangement. The fact that no rearrangement occurs during oxymercuration-reduction indicates that no free carbocation intermediate is formed.

The stereospecificity of oxymercuration is illustrated by the reaction of mercury(II) acetate in the presence of water with cyclopentene.

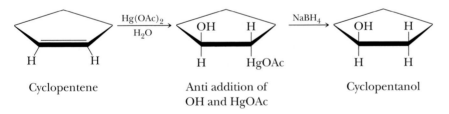

Cyclopentene   Anti addition of   Cyclopentanol
                      OH and HgOAc

The fact that oxymercuration is both regiospecific and stereospecific has led chemists to propose the following mechanism for this reaction, which is closely analogous to that for the addition of Br_2 and Cl_2 to an alkene (Section 6.3D). Dissociation of mercury(II) acetate in Step 1 gives acetate ion (a Lewis base) and $AcOHg^+$ (a Lewis acid). Reaction then proceeds by interaction of $AcOHg^+$ with the electron pair of the pi bond to give a **bridged mercurinium ion** intermediate. This bridged intermediate closely resembles a bridged bromonium ion intermediate (Section 6.3D) with the difference that, unlike bromine, mercury has no electron pair to donate to form fully covalent bonds in the intermediate. Rather, in the bridged mercurinium ion intermediate, the two pi electrons of the carbon-carbon double bond now form a ring containing three atoms bonded by two electrons. The open cation structure, which places the positive charge on the carbon giving the more stable carbocation, is a minor contributing structure to the mercurinium ion intermediate. The carbocation on the primary carbon is a negligible contributor.

Mechanism   Oxymercuration-Reduction of an Alkene

Step 1: Dissociation of mercury(II) acetate gives $AcOHg^+$ (a Lewis acid) and acetate ion (a Lewis base).

$$AcO-Hg-OAc \longrightarrow AcO-Hg^+ + AcO^-$$

$$\text{(a Lewis acid)}$$

Step 2: Attack of $AcOHg^+$ on the carbon-carbon double bond of the alkene forms a bridged mercurinium ion intermediate. In this intermediate, the two electrons of the pi bond now form a two-electron three-center bond, here indicated by dashed lines, in which each atom participating in the three-center bond bears a fraction of the positive charge.

An open carbocation intermediate (a minor contributor)   A bridged mercurinium ion intermediate (the major contributor)   An open carbocation intermediate (a negligible contributor)

Step 3: Anti stereospecific attack of water (a Lewis acid) on the bridged mercurinium ion (a Lewis base) occurs at the more substituted carbon to open the three-membered ring. Proton transfer from this product to water completes oxymercuration of the alkene.

Anti stereospecific addition of HgOAc and HOH

Step 4: Reduction of the C—HgOAc bond gives the final product and metallic mercury.

The fact that oxymercuration occurs without rearrangement indicates that the intermediate formed in Step 2 is not a true carbocation but rather a resonance hybrid with largely the character of a bridged intermediate. Furthermore, the bridged structure allows us to account for the fact that the stereochemistry of this electrophilic addition is predominantly anti; the nucleophile attacks the bridged intermediate from the face opposite that occupied by mercury, as shown in Step 3.

Yet the fact that the electrophile, $AcOHg^+$, adds to the less substituted carbon and that the nucleophile, HOH, adds to the more substituted carbon indicates that the intermediate must have some carbocation character. It is probable that, in the actual intermediate, mercury is bonded partially to each carbon, thereby preventing rearrangement. Of the two carbon atoms of the mercurinium ion, the more substituted one has a greater degree of partial positive charge and is the one attacked by the nucleophile, H_2O. This is well accounted for by resonance theory; only the more stable carbocation structure participates appreciably in the resonance-stabilized mercurinium ion intermediate.

6.4 Hydroboration-Oxidation

Mechanisms: (Electrophilic) Addition; Hydroboration of an alkene

Hydroboration-oxidation of an alkene is an extremely valuable laboratory method for the regioselective and stereospecific hydration of an alkene. Furthermore, this sequence of reactions occurs without rearrangement.

Hydroboration-oxidation A method for converting an alkene to an alcohol. The alkene is treated with borane (BH_3) to give a trialkylborane, which is then oxidized with alkaline hydrogen peroxide to give the alcohol.

Hydroboration is the addition of borane, BH_3, to an alkene to form a trialkylborane. The overall reaction occurs in three successive steps. BH_3 reacts with one molecule of alkene, then a second, and finally a third until all three hydrogens of borane have been replaced by alkyl groups.

$$H-B\begin{matrix}H\\\\H\end{matrix} + 3CH_2{=}CH_2 \longrightarrow CH_3CH_2-B\begin{matrix}CH_2CH_3\\\\CH_2CH_3\end{matrix}$$

Borane Triethylborane
(A trialkylborane)

Borane cannot be prepared as a pure compound because it dimerizes to diborane, B_2H_6, a toxic gas that ignites spontaneously in air.

$$2BH_3 \rightleftharpoons B_2H_6$$

Borane Diborane

However, BH_3 exists as a stable complex with an ether, such as tetrahydrofuran (THF), by formation of a stable Lewis acid-base complex. Borane is most often used as a commercially available solution of BH_3 in THF.

$$2\ \overset{}{\bigcirc}\!\!:\!O\!: \ + \ B_2H_6 \rightleftharpoons 2\ \overset{}{\bigcirc}\!\!:\!\overset{+}{O}-\overset{-}{B}H_3$$

Tetrahydrofuran
(THF)

$BH_3 \cdot THF$

Boron, atomic number 5, has three electrons in its valence shell. To bond with three other atoms, boron uses sp^2 hybrid orbitals. The unoccupied $2p$ orbital of boron is perpendicular to the plane created by boron and the three other atoms to which it is bonded. An example of a stable compound in which boron is bonded to three other atoms is boron trifluoride, BF_3, a planar molecule with F—B—F bond angles of 120° (Section 1.2F). Because of the vacant $2p$ orbital in the valence shell of boron, BH_3, BF_3, and all other tricovalent compounds of boron are Lewis acids and act as electron-pair acceptors. These compounds of boron closely resemble carbocations, except that they are electrically neutral.

Addition of borane to alkenes is regioselective and stereospecific.

> Regioselectivity: In addition of borane to an unsymmetrical alkene, boron becomes bonded predominantly to the less substituted carbon of the double bond.

> Stereospecificity: Hydrogen and boron add from the same face of the double bond; the reaction is **syn** (from the same side) **stereospecific.**

Syn stereospecific The addition of atoms or groups of atoms to the same face of a carbon-carbon double bond.

Both the regioselectivity and stereospecificity are illustrated by hydroboration of 1-methylcyclopentene.

1-Methylcyclopentene

Syn addition of BH_3
(R = 2-methylcyclopentyl)

Addition of borane to an alkene is initiated by coordination of the vacant $2p$ orbital of boron (a Lewis acid) with the electron pair of the pi bond (a Lewis base). We account for the stereospecificity of hydroboration by proposing formation of a cyclic, four-center transition state. Boron and hydrogen are added simultaneously and from the same face of the double bond.

Mechanism   Hydroboration — A Concerted Regioselective and Stereospecific Addition

Reaction occurs in one step. Bond breaking and bond forming occur simultaneously, and boron adds regioselectively to the less substituted carbon atom of the double bond.

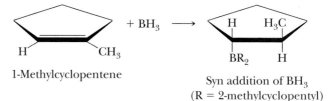

An acceptable mechanism for hydroboration must account for both the regioselectivity and the stereospecificity. We account for the regioselectivity by a combination of steric and electronic factors. In terms of steric effects, boron, the larger part of the reagent, adds selectively to the less hindered carbon of the double bond, and hydrogen, the smaller part of the reagent, adds to the more hindered carbon. It is believed that the observed regioselectivity is due largely to steric effects.

Electronic effects probably also influence the regioselectivity. The electronegativity of hydrogen (2.1) is slightly greater than that of boron (2.0); hence, there is a small degree of polarity (approximately 5%) to each B—H bond, with boron bearing a partial positive charge and hydrogen a partial negative charge. It is proposed that there is some degree of carbocation character in the transition state and that the partial positive charge is on the carbon better able to accommodate it, the more substituted one.

$$\underset{\text{Starting reagents}}{CH_3CH_2CH_2CH=CH_2} \quad \overset{\overset{\delta-}{H}-\overset{\delta+}{B}}{\longrightarrow} \quad \underset{\substack{\text{Transition state} \\ \text{(some carbocation character} \\ \text{exists in the transition state)}}}{CH_3CH_2CH_2\overset{\delta+}{CH}\text{---}CH_2}$$

Trialkylboranes are rarely isolated. Rather, they are converted directly to other products formed by substitution of another atom (H, O, N, C, or halogen) for boron. One of the most important reactions of trialkylboranes is with hydrogen peroxide in aqueous sodium hydroxide. Hydrogen peroxide is an oxidizing agent and, under these conditions, oxidizes trialkylboranes to form an alcohol and sodium borate, Na_3BO_3.

Mechanism   Oxidation of a Trialkylborane by Alkaline Hydrogen Peroxide

Step 1:   Donation of a pair of electrons from hydroperoxide ion (a Lewis base) to the boron atom of the trialkylborane (a Lewis acid) gives an intermediate in which boron has a filled valence shell and bears a negative charge.

$$\underset{\substack{\text{Hydrogen} \\ \text{peroxide}}}{HOOH} + OH^- \rightleftharpoons \underset{\substack{\text{Hydroperoxide} \\ \text{ion}}}{HOO^-} + HOH$$

$$\underset{\substack{\text{A trialkylborane} \\ \text{(a Lewis acid)}}}{R-\overset{\overset{\displaystyle R}{|}}{\underset{\underset{\displaystyle R}{|}}{B}}} + \underset{\substack{\text{Hydroperoxide ion} \\ \text{(a Lewis base)}}}{:\ddot{O}-\ddot{O}-H} \longrightarrow R-\overset{\overset{\displaystyle R}{|}}{\underset{\underset{\displaystyle R}{|}}{\overset{-}{B}}}-\ddot{O}-\ddot{O}-H$$

Step 2:   Rearrangement of an R group with its pair of bonding electrons to an adjacent oxygen (a 1,2-shift) results in ejection of hydroxide ion.

$$R-\overset{\overset{\displaystyle R}{|}}{\underset{\underset{\displaystyle R}{|}}{\overset{-}{B}}}-\overset{..}{\underset{..}{O}}\overset{\frown}{-}\overset{..}{\underset{..}{O}}-H \longrightarrow R-\overset{\overset{\displaystyle R}{|}}{\underset{\underset{\displaystyle R}{|}}{B}}-\overset{..}{\underset{..}{O}}: + \ \overset{..}{\underset{..}{:O}}-H$$

Two more reactions with hydroperoxide ion followed by rearrangements give a trialkylborate.

$$R-\overset{\overset{\displaystyle R}{|}}{\underset{\underset{\displaystyle R}{|}}{B}}-\overset{..}{\underset{..}{O}}: \ \xrightarrow[\text{NaOH}]{H_2O_2} \ \xrightarrow[\text{NaOH}]{H_2O_2} \ \ \overset{..}{\underset{..}{:O}}-\overset{\overset{\displaystyle :\overset{..}{O}-R}{|}}{\underset{\underset{\displaystyle R}{|}}{B}}-\overset{..}{\underset{\underset{\displaystyle R}{|}}{O}}:$$

A trialkylborate
(a triester of boric acid)

Step 3:   Reaction of the trialkylborate with aqueous NaOH gives the alcohol and sodium borate.

$$(RO)_3B + 3NaOH \longrightarrow 3ROH + Na_3BO_3$$

A trialkylborate   Sodium borate

The net reaction from hydroboration and subsequent oxidation is hydration of a carbon-carbon double bond. Because hydrogen is added to the more substituted carbon of the double bond, we refer to the results of hydroboration and subsequent oxidation as anti-Markovnikov hydration. (The key step in this reaction, the addition, is actually a Markovnikov addition.)

$$(CH_3CH_2CH_2CH_2CH_2CH_2)_3B + 3H_2O_2 + 3NaOH \longrightarrow$$

Trihexylborane

$$3CH_3CH_2CH_2CH_2CH_2CH_2OH + Na_3BO_3 + 3H_2O$$

1-Hexanol   Sodium
borate

Hydrogen peroxide oxidation of a trialkylborane is stereospecific in that the configuration of the alkyl group is retained; whatever the position of boron in relation to other groups in the trialkylborane, the OH group by which it is replaced occupies the same position. Thus, the net result of hydroboration-oxidation of an alkene is syn addition of H and OH to a carbon-carbon double bond.

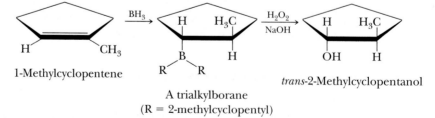

1-Methylcyclopentene

A trialkylborane
(R = 2-methylcyclopentyl)

trans-2-Methylcyclopentanol

Example 6.11

Draw structural formulas for the trialkylborane and alcohol formed in the following reaction sequences.

(a) $\overset{\overset{\displaystyle CH_3}{|}}{CH_3C}{=}CHCH_3$ $\xrightarrow{BH_3}$ a trialkylborane $\xrightarrow[\text{NaOH}]{H_2O_2}$ an alcohol

(b) [structure: 1-methylcyclohexene] $\xrightarrow{BH_3}$ a trialkylborane $\xrightarrow[\text{NaOH}]{H_2O_2}$ an alcohol

Solution

(a) $\overset{\overset{\displaystyle CH_3}{|}}{CH_3C}{=}CHCH_3$ $\xrightarrow{BH_3}$ $\left(\overset{\overset{\displaystyle CH_3}{|}}{\underset{\underset{\displaystyle CH_3}{|}}{CH_3CHCH}} {-} B \right)_3$ $\xrightarrow[\text{NaOH}]{H_2O_2}$ $\overset{\overset{\displaystyle CH_3}{|}}{\underset{\underset{\displaystyle OH}{|}}{CH_3CHCHCH_3}}$

(b) [structure: 1-methylcyclohexene] $\xrightarrow{BH_3}$ [structure with CH₃ and BR₂] $\xrightarrow[\text{NaOH}]{H_2O_2}$ [structure with CH₃ and OH]

(R = 2-Methylcyclohexyl)   *trans*-2-Methylcyclohexanol

Problem 6.11

Draw structural formulas for the trialkylborane and alkene that give the following alcohols under the reaction conditions shown.

(a) An alkene $\xrightarrow{BH_3}$ a trialkylborane $\xrightarrow[\text{NaOH}]{H_2O_2}$ $\overset{\overset{\displaystyle CH_3}{|}}{CH_3CHCH_2CH_2OH}$

(b) An alkene $\xrightarrow{BH_3}$ a trialkylborane $\xrightarrow[\text{NaOH}]{H_2O_2}$ [cyclopentane]$-CH_2OH$

6.5 Oxidation

Reactivity Explorer; Alkenes

In this section and Section 6.6, we study oxidation and reduction of alkenes. We begin with a general method by which you can recognize oxidation-reduction reactions and then consider three oxidation-reduction reactions of alkenes.

A. How to Recognize Oxidation-Reduction

In the following reactions, propene is transformed into three different compounds by reactions we study in this chapter. The first reaction involves reduction, the third involves oxidation, and the second involves neither oxidation nor reduction. These equations are not complete because they do not specify any reactant other than propene, and they do not specify what reagents are necessary to bring about the particular transformation. Each does specify, however, that the carbon atoms of the products are derived from those of propene.

$$
CH_3CH=CH_2
\begin{cases}
\xrightarrow{\text{reduction}} & \overset{\underset{|}{H}}{C}H_3\overset{\underset{|}{H}}{C}H-CH_2 & \text{Propane} \\
\xrightarrow{\text{neither}} & \overset{\underset{|}{OH}}{C}H_3\overset{\underset{|}{H}}{C}H-CH_2 & \text{2-Propanol} \\
\xrightarrow{\text{oxidation}} & \overset{\underset{|}{OH}}{C}H_3\overset{\underset{|}{OH}}{C}H-CH_2 & \text{1,2-Propanediol}
\end{cases}
$$

It is possible to decide if transformations such as these involve oxidation, reduction, or neither by the use of **balanced half-reactions**. To write a balanced half-reaction:

1. Write a half-reaction showing the organic reactant(s) and product(s).
2. Complete a material balance; that is, balance the number of atoms on each side of the half-reaction. To balance the number of oxygens and hydrogens for a reaction taking place in acid solution, use H_2O for oxygens and then H^+ for hydrogens. For a reaction taking place in basic solution, use OH^- and H_2O.
3. Complete a charge balance; that is, balance the charge on both sides of the half-reaction. To balance the charge, add electrons, e^-, to one side or the other. The equation completed in this step is a balanced half-reaction.

Oxidation The loss of electrons. Alternatively, either the loss of hydrogens, the gain of oxygens, or both.

Reduction The gain of electrons. Alternatively, either the gain of hydrogen, loss of oxygens, or both.

Oxidation is the loss of electrons. If electrons appear on the right side of a balanced half-reaction, the reactant has given up electrons and has been oxidized. **Reduction** is the gain of electrons. If electrons appear on the left side of a balanced half-reaction, the reactant has gained electrons and has been reduced. If no electrons appear in the balanced half-reaction, then the transformation involves neither oxidation nor reduction. Let us apply these steps to the transformation of propene to propane.

Step 1: Half-reaction $\quad CH_3CH=CH_2 \longrightarrow CH_3CH_2CH_3$

Step 2: Material balance $\quad CH_3CH=CH_2 + 2H^+ \longrightarrow CH_3CH_2CH_3$

Step 3: Balanced half-reaction $\quad CH_3CH=CH_2 + 2H^+ + 2e^- \longrightarrow CH_3CH_2CH_3$

Because two electrons appear on the left side of the balanced half-reaction (Step 3), conversion of propene to propane is a two-electron reduction. To bring it about requires use of a reducing agent.

Following is a balanced half-reaction for the transformation of propene to 2-propanol:

$$
\text{Balanced half-reaction:} \quad CH_3CH=CH_2 + H_2O \longrightarrow CH_3\overset{\underset{|}{OH}}{C}HCH_3
$$
$$
\text{Propene} \qquad\qquad\qquad \text{2-Propanol}
$$

Because no electrons are required to achieve an electrical balance, conversion of propene to 2-propanol is neither oxidation nor reduction.

A balanced half-reaction for the transformation of propene to 1,2-propanediol requires two electrons on the right side of the equation for a charge balance; this transformation is a two-electron oxidation.

$$
\text{Balanced half-reaction:} \quad CH_3CH=CH_2 + 2H_2O \longrightarrow CH_3\overset{\underset{|}{OH}}{\overset{HO}{C}}HCH_2 + 2H^+ + 2e^-
$$
$$
\text{Propene} \qquad\qquad\qquad\qquad \text{1,2-Propanediol}
$$

It is important to realize that this strategy for recognizing oxidation and reduction is only that, a strategy. In no way does it give any indication of how a particular oxidation or reduction might be carried out in the laboratory. For example, the balanced half-reaction for the transformation of propene to propane requires $2H^+$ and $2e^-$. Yet by far the most common laboratory procedure for reducing propene to propane does not involve H^+ at all; rather it involves molecular hydrogen, H_2, and a transition metal catalyst.

It is also possible to recognize oxidation and reduction in most organic compounds by the following guidelines:

> Oxidation: The loss of hydrogen, the gain of oxygen, or both
>
> Reduction: The gain of hydrogen, the loss of oxygen, or both

These guidelines, which focus on either the loss or gain of hydrogens or oxygens, will suffice to recognize most oxidations or reductions of organic compounds. Where this method fails, then you must use balanced half-reactions.

Example 6.12

Use a balanced half-reaction to show that each transformation involves an oxidation.

(a) $CH_3CH_2CH_2OH \longrightarrow CH_3CH_2\overset{O}{\overset{\|}{C}}H$ (b) $CH_3CH=CH_2 \longrightarrow CH_3\overset{O}{\overset{\|}{C}}H + H\overset{O}{\overset{\|}{C}}H$

Solution

First complete a material balance and then a charge balance. The first transformation is a two-electron oxidation; the second is a four-electron oxidation. To bring each about requires an oxidizing agent.

(a) $CH_3CH_2CH_2OH \longrightarrow CH_3CH_2\overset{O}{\overset{\|}{C}}H + 2H^+ + 2e^-$

(b) $CH_3CH=CH_2 + 2H_2O \longrightarrow CH_3\overset{O}{\overset{\|}{C}}H + H\overset{O}{\overset{\|}{C}}H + 4H^+ + 4e^-$

Problem 6.12

Use a balanced half-reaction to show that each transformation involves a reduction.

(a) [cyclohexanone] $\longrightarrow$ [cyclohexanol] (b) $CH_3CH_2\overset{O}{\overset{\|}{C}}OH \longrightarrow CH_3CH_2CH_2OH$

B. OsO₄ — Oxidation of an Alkene to a Glycol

Osmium tetroxide, OsO_4, and certain other transition metal oxides are effective oxidizing agents for the conversion of an alkene to a **glycol,** a compound with two hydroxyl groups on adjacent carbons. Oxidation of an alkene by OsO_4 is stereospecific in that it involves syn addition of OH groups to the carbons of the double bond. For example,

Glycol A compound with two hydroxyl (—OH) groups on adjacent carbons.

oxidation of cyclopentene gives *cis*-1,2-cyclopentanediol, a cis glycol. Note that both cis and trans isomers are possible for this glycol but that only the cis glycol is formed.

A cyclic osmic ester

cis-1,2-Cyclopentanediol
(a cis glycol)

The stereospecificity of osmium tetroxide oxidation of alkenes is accounted for by the formation of a cyclic osmic ester in which oxygen atoms of OsO_4 form new covalent bonds with each carbon of the double bond in such a way that the five-membered osmium-containing ring is bonded in a cis configuration to the original alkene. Osmic esters can be isolated and characterized. Usually, however, the cyclic osmic ester is treated directly with a reducing agent, such as $NaHSO_3$, which cleaves osmium-oxygen bonds to give the cis glycol and reduced forms of osmium.

The drawbacks of OsO_4 are that it is both expensive and highly toxic. One strategy to circumvent the high cost is to use it in catalytic amounts along with stoichiometric amounts of another oxidizing agent that reoxidizes the reduced forms of osmium and, thus, recycles the osmium reagent. Oxidizing agents commonly used for this purpose are hydrogen peroxide and *tert*-butyl hydroperoxide. When this procedure is used, there is no need for a reducing step using $NaHSO_3$.

$$HOOH$$

$$\begin{array}{c} CH_3 \\ | \\ CH_3COOH \\ | \\ CH_3 \end{array}$$

Hydrogen
peroxide

tert-Butyl hydroperoxide
(*t*-BuOOH)

C.   Ozone — Cleavage of a Carbon-Carbon Double Bond (Ozonolysis)

Treatment of an alkene with ozone, O_3, followed by a suitable work-up cleaves the carbon-carbon double bond and forms two carbonyl (C=O) groups in its place. The alkene is dissolved in an inert solvent, such as CH_2Cl_2, and a stream of ozone is bubbled through the solution. The products isolated from ozonolysis depend on the reaction conditions. Hydrolysis of the reaction mixture with water yields hydrogen peroxide, an oxidizing agent that can bring about further oxidations. To prevent side reactions caused by reactive peroxide intermediates, a weak reducing agent is added during the work-up to reduce peroxides to water. The reducing agent most commonly used for this purpose is dimethyl sulfide, $(CH_3)_2S$, as illustrated in the following example.

$$\underset{\text{2-Methyl-2-pentene}}{\overset{\displaystyle CH_3}{\underset{|}{CH_3C}}{=}CHCH_2CH_3} \xrightarrow[\text{2. } (CH_3)_2S]{\text{1. } O_3} \underset{\substack{\text{Propanone} \\ \text{(a ketone)}}}{\overset{O}{\overset{\|}{CH_3CCH_3}}} + \underset{\substack{\text{Propanal} \\ \text{(an aldehyde)}}}{\overset{O}{\overset{\|}{HCCH_2CH_3}}}$$

The initial product of reaction of an alkene with ozone is an adduct called a molozonide, which rearranges under the conditions of the reaction to an isomeric ozonide. Low-molecular-weight ozonides are explosive and are rarely isolated. They are treated directly with a weak reducing agent to give the carbonyl-containing products.

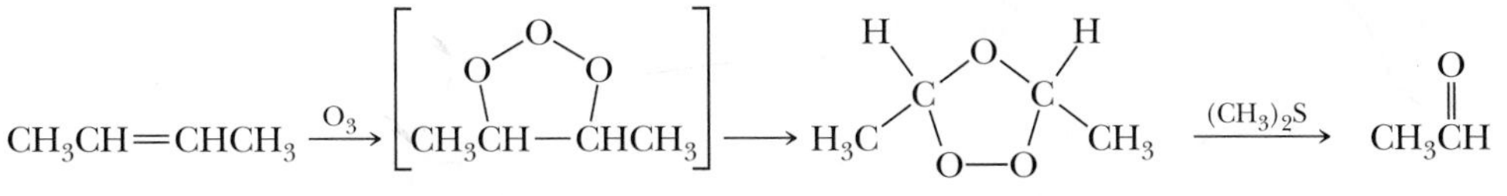

2-Butene   A molozonide   An ozonide   Acetaldehyde

To understand how an ozonide is formed, we must first examine the structure of ozone. This molecule can be written as a hybrid of four contributing structures, all of which show separation of unlike charge; it is not possible to write a Lewis structure for ozone without separation of charge. Because of the positive formal charge on the terminal oxygens, this molecule is very electrophilic.

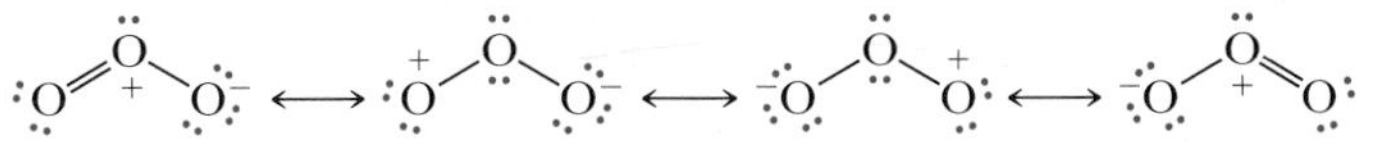

Mechanism   Formation of an Oxide

Step 1:   Ozone reacts with the alkene first as an electrophile and then as a nucleophile to give a molozonide. Even though this interaction is shown here occurring in two steps, both steps are probably simultaneous or "concerted."

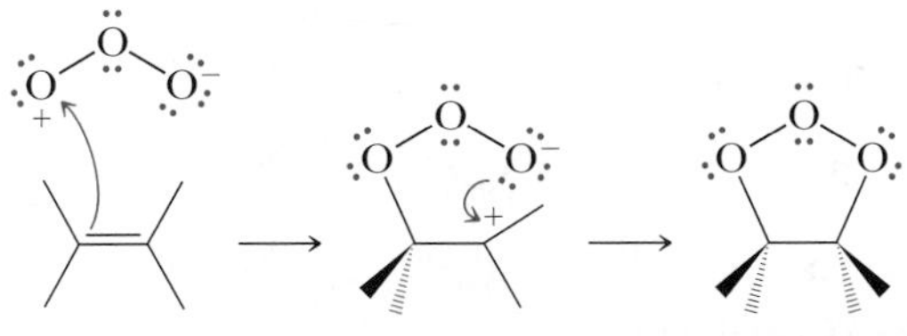

A molozonide

Step 2:   Rearrangement of valence electrons in the molozonide results in cleavage of one carbon-carbon and one oxygen-oxygen bond. The resulting fragments then recombine to form a secondary ozonide.

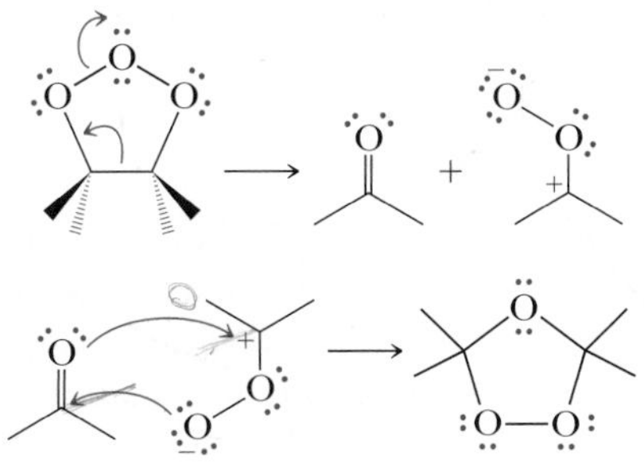

Example 6.13

Draw structural formulas for the products of the following ozonolysis reactions. Name the new functional groups formed in each oxidation.

(a)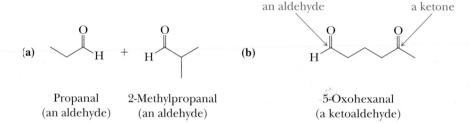

$\frac{1.\ O_3}{2.\ (CH_3)_2S}$

(b) $\frac{1.\ O_3}{2.\ (CH_3)_2S}$

Solution

an aldehyde   a ketone

(a)   +

(b)

Propanal   2-Methylpropanal   5-Oxohexanal
(an aldehyde)   (an aldehyde)   (a ketoaldehyde)

Problem 6.13

What alkene of molecular formula C_6H_{12}, when treated with ozone and then dimethyl sulfide, gives the following product(s)?

(a)   (b)   (c)

(only product)   (equal moles of each)   (only product)

6.6 Reduction

Most alkenes react quantitatively with molecular hydrogen, H_2, in the presence of a transition metal catalyst to give alkanes. Commonly used transition metals include platinum, palladium, ruthenium, and nickel. Yields are usually quantitative or nearly so. Because conversion of an alkene to an alkane involves reduction by hydrogen in the presence of a catalyst, the process is called **catalytic reduction** or, alternatively, **catalytic hydrogenation.**

$$+ \quad H_2 \quad \xrightarrow[25°C,\ 3\ atm]{Pd} \quad$$

Cyclohexene   Cyclohexane

Monosubstituted and disubstituted carbon-carbon double bonds react readily at room temperature under a few atmospheres pressure of hydrogen. Trisubstituted carbon-carbon double bonds require slightly elevated temperatures and pressures of up to 100 psi (pounds per square inch). Tetrasubstituted carbon-carbon double bonds are difficult to reduce and may require temperatures up to 275°C and hydrogen pressures of 1000 psi.

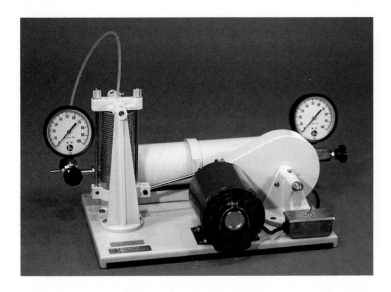

Paar shaker-type hydrogenation apparatus. *(Paar Instrument Co., Moline, IL)*

Although addition of hydrogen to an alkene is exothermic [the heat of hydrogenation of ethylene is -137.1 kJ $(-32.8$ kcal)/mol and that of cyclohexene is -119.5 kJ $(-28.6$ kcal)/mol], reduction is immeasurably slow in the absence of a catalyst. The metal catalyst is used as a finely powdered solid or may be supported on some inert material, such as finely powdered charcoal or alumina. The reaction is usually carried out by dissolving the alkene in ethanol or another nonreacting organic solvent, adding the solid catalyst, and then shaking the mixture under hydrogen gas at pressures of from 1 to 50 atm. Alternatively, the metal may be complexed with certain organic molecules and used in the form of a soluble complex.

Catalytic reduction is stereoselective, with the most common pattern being syn addition of hydrogens to the carbon-carbon double bond. Catalytic reduction of 1,2-dimethylcyclohexene, for example, yields predominantly *cis*-1,2-dimethylcyclohexane. Along with the cis isomer are formed lesser amounts of *trans*-1,2-dimethylcyclohexane.

$$\text{1,2-Dimethylcyclohexene} + H_2 \xrightarrow{\text{Pt}} \text{cis-1,2-Dimethylcyclohexane} + \text{trans-1,2-Dimethylcyclohexane}$$

1,2-Dimethyl-cyclohexene		70% to 85% *cis*-1,2-Dimethyl-cyclohexane		30% to 15% *trans*-1,2-Dimethyl-cyclohexane

A. Mechanism of Catalytic Reduction

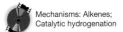
Mechanisms: Alkenes; Catalytic hydrogenation

Certain transition metals are able to adsorb large quantities of hydrogen onto their surfaces, probably by forming metal-hydrogen sigma bonds. Similarly, alkenes are also adsorbed on metal surfaces with formation of carbon-metal bonds. Addition of hydrogen atoms to the alkene occurs in two steps (Figure 6.7).

If addition of hydrogens is syn stereospecific, how then is the formation of a trans product accounted for? It is proposed that before a second hydrogen can be

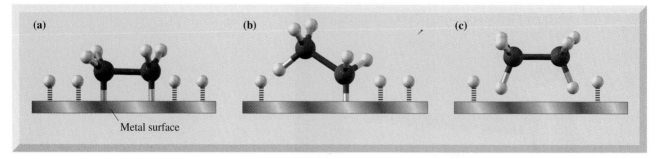

Metal surface

Figure 6.7

Addition of hydrogen to an alkene involving a transition metal catalyst. (a) Hydrogen and the alkene are adsorbed on the metal surface and (b) one hydrogen atom is transferred to the alkene forming a new C—H bond. The other carbon remains adsorbed on the metal surface. (c) A second C—H bond is formed, and the alkane is desorbed.

delivered from the metal surface to complete the reduction, there is transfer of a hydrogen from a carbon atom adjacent to the original double bond to the metal surface. This hydrogen transfer in effect reverses the first step and forms a new alkene that is isomeric with the original alkene. As shown in the following equation, 1,2-dimethylcyclohexene undergoes isomerization on the metal surface to form 1,6-dimethylcyclohexene. This alkene then leaves the metal surface. When it is later readsorbed and reduced, hydrogens are still added to it with syn stereospecificity, but not necessarily from the same side as the original hydrogen. This is mainly a problem with tetrasubstituted alkenes; less substituted alkenes react more stereoselectively.

$$\text{1,2-Dimethyl-} \quad \xrightarrow{\text{H}_2/\text{Pt}} \quad \text{} \quad \longrightarrow \quad \text{1,6-Dimethyl-}$$

1,2-Dimethyl-
cyclohexene

1,6-Dimethyl-
cyclohexene

B.   Heats of Hydrogenation and the Relative Stabilities of Alkenes

The **heat of hydrogenation** of an alkene is defined as its heat of reaction, ΔH^0, with hydrogen to form an alkane. Table 6.2 lists heats of reaction for the catalytic hydrogenation of several alkenes. Three important points are derived from the information given in Table 6.2.

1. Reduction of an alkene to an alkane is an exothermic process. This observation is consistent with the fact that, during hydrogenation, there is net conversion of a weaker pi bond to a stronger sigma bond; that is, one sigma bond (H—H) and one pi bond (C=C) are broken, and two new sigma bonds (C—H) are formed. For a comparison of the relative strengths of sigma and pi bonds, refer to Section 1.8F.

2. Heats of hydrogenation depend on the degree of substitution of the carbon-carbon double bond; the greater the substitution, the lower the heat of hydrogenation. Compare, for example, heats of hydrogenation of ethylene (no substituents), propene (one substituent), 1-butene (one substituent), and the cis and trans isomers of 2-butene (two substituents).

3. The heat of hydrogenation of a trans alkene is lower than that of the isomeric cis

Table 6.2 Heats of Hydrogenation of Several Alkenes

Name	Structural Formula	ΔH^0 [kJ(kcal)/mol]
Ethylene	$CH_2=CH_2$	$-137\ (-32.8)$
Propene	$CH_3CH=CH_2$	$-126\ (-30.1)$
1-Butene	$CH_3CH_2CH=CH_2$	$-127\ (-30.3)$
cis-2-Butene	H₃C, CH₃ C=C, H H	$-119.7\ (-28.6)$
trans-2-Butene	H₃C, H C=C, H CH₃	$-115.5\ (-27.6)$
2-Methyl-2-butene	H₃C, CH₃ C=C, H₃C H	$-113\ (-26.9)$
2,3-Dimethyl-2-butene	H₃C, CH₃ C=C, H₃C CH₃	$-111\ (-26.6)$

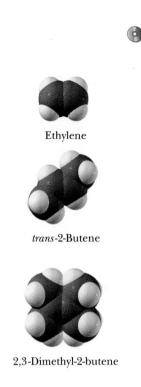

Ethylene

trans-2-Butene

2,3-Dimethyl-2-butene

alkene. Compare, for example, the heats of hydrogenation of *cis*-2-butene and *trans*-2-butene. Because reduction of each alkene gives butane, any difference in heats of hydrogenation must be due to a difference in relative energy between the two alkenes (Figure 6.8). The alkene with the lower (less negative) value of ΔH^0 is the more stable.

We explain the greater stability of trans alkenes relative to cis alkenes in terms of nonbonded interaction strain, which can be visualized using space-filling models (Figure 6.9). In *cis*-2-butene, the two —CH₃ groups are sufficiently close to each

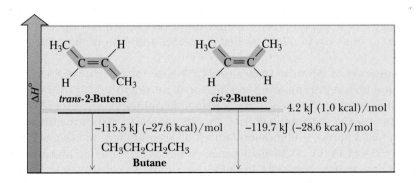

Figure 6.8

Heats of hydrogenation of *cis*-2-butene and *trans*-2-butene. *Trans*-2-butene is more stable than *cis*-2-butene by 4.2 kJ (1.0 kcal)/mol.

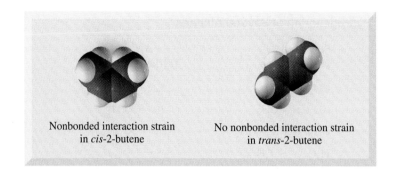

Figure 6.9
Space-filling models of (a) *cis*-2-butene and (b) *trans*-2-butene.

Nonbonded interaction strain
in *cis*-2-butene

No nonbonded interaction strain
in *trans*-2-butene

other that there is strain caused by repulsion between the electron clouds of each. This repulsion is reflected in the larger heat of hydrogenation (decreased stability) of *cis*-2-butene compared with *trans*-2-butene [approximately 4.2 kJ (1.0 kcal)/mol]. Thus hydrogenation allows measurement of the strain energy of *cis*-2-butene directly.

6.7   Molecules Containing Stereocenters as Reactants or Products

As the structure of an organic compound is altered in the course of a reaction, one or more stereocenters, usually at carbon, may be created, inverted, or destroyed. In part A of this section, we consider two alkene addition reactions in which a chiral molecule is created in an achiral environment. In doing so, we will illustrate the point that an enantiomerically pure compound can never be produced from achiral starting materials reacting in an achiral environment. Then, in part B, we consider reaction of achiral starting materials reacting in a chiral environment, in this case in the presence of a chiral catalyst. We shall see that under the proper experimental conditions, an enantiomerically pure product may be produced from achiral reagents.

A.   Reaction of Achiral Starting Materials in an Achiral Environment

Addition of bromine to 2-butene (Section 6.3D) gives 2,3-dibromobutane, a molecule with two stereocenters. Three stereoisomers are possible for this compound: a meso compound and a pair of enantiomers (Section 3.4). We now ask, "Is the product one enantiomer, a pair of enantiomers, the meso compound, or a mixture of all three stereoisomers?" A partial answer is that the product formed depends on the configuration of the alkene. Let us first examine addition of bromine to *cis*-2-butene.

Attack of bromine on *cis*-2-butene from either face of the planar part of the molecule gives the same bridged bromonium ion intermediate (Figure 6.10). Although this intermediate has two stereocenters, it has a plane of symmetry and is, therefore, meso. Attack of Br$^-$ on this intermediate from the side opposite that of the bromonium ion bridge gives a pair of enantiomers. Attack of bromide ion on carbon 2 of this meso intermediate gives the 2*S*,3*S* enantiomer. Attack of bromide ion on carbon 3 gives the 2*R*,3*R* enantiomer. Attack of bromide ion occurs at equal rates at each

Step 1: Reaction of *cis*-2-butene with bromine forms bridged bromonium ions which are meso and identical.

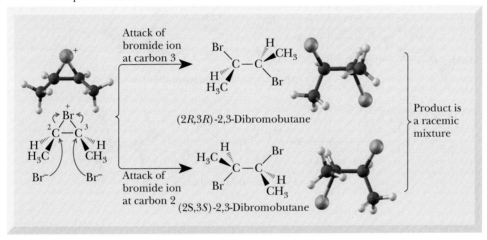

cis-2-Butene
(achiral)

(meso bridged bromonium ion intermediate)

Step 2: Attack of bromide at carbons 2 and 3 occurs with equal probability to give enantiomeric products in a racemic mixture.

Attack of bromide ion at carbon 3

(2*R*,3*R*)-2,3-Dibromobutane

Attack of bromide ion at carbon 2

(2*S*,3*S*)-2,3-Dibromobutane

Product is a racemic mixture

Figure 6.10

Anti addition of bromine to *cis*-2-butene gives 2,3-dibromobutane as a racemic mixture.

carbon; therefore, the enantiomers are formed in equal amounts, and 2,3-dibromobutane is obtained as a racemic mixture (Figure 6.10). We have shown attack of Br—Br from one side of the carbon-carbon double bond. Attack from the opposite side followed by opening of the resulting bromonium ion intermediate also produces these same two stereoisomers.

Addition of Br_2 to *trans*-2-butene leads to two enantiomeric bridged bromonium ion intermediates. Attack by Br^- at either carbon atom of either bromonium ion intermediate gives the meso product, which is optically inactive (Figure 6.11).

In Section 6.5B we studied oxidation of alkenes by osmium tetroxide in the presence of hydrogen peroxide. This oxidation results in syn stereospecific hydroxylation of the alkene to form a glycol. In the case of cycloalkenes, the product is a cis glycol. The first step in each oxidation involves formation of a cyclic osmic ester and is followed immediately by reaction with water to give a glycol. As shown in the following

Step 1: Reaction of *trans*-2-butene with bromine forms bridged bromonium ions which are enantiomers

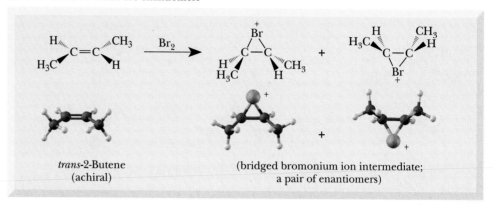

trans-2-Butene
(achiral)

(bridged bromonium ion intermediate;
a pair of enantiomers)

Step 2: Attack of bromide on either carbon of either enantiomer gives meso-2,3-dibromobutane

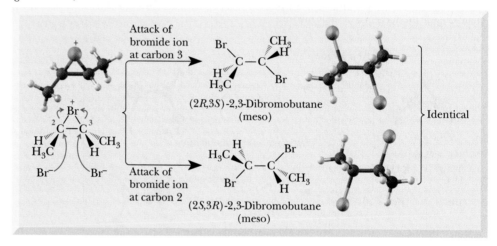

Attack of bromide ion at carbon 3

(2*R*,3*S*)-2,3-Dibromobutane
(meso)

Attack of bromide ion at carbon 2

(2*S*,3*R*)-2,3-Dibromobutane
(meso)

Identical

Figure 6.11
Anti addition of bromine to
trans-2-butene gives
meso-2,3-dibromobutane.

sequences, syn hydroxylation of *cis*-2-butene gives meso-2,3-butanediol. Syn hydroxylation of *trans*-2-butene gives racemic 2,3-butanediol.

cis-2-Butene
(achiral)

OsO_4
ROOH

(2*S*,3*R*)-2,3-Butanediol

(2*R*,3*S*)-2,3-Butanediol

identical;
a meso
compound

(2R,3R)-2,3-Butanediol

Note that the stereochemistry of the product of osmium tetroxide oxidation of *trans*-2-butene is opposite to that formed on addition of bromine to *trans*-2-butene. Osmium tetroxide oxidation gives the glycol as a pair of enantiomers forming a racemic mixture. Addition of bromine to *trans*-2-butene gives the dibromoalkane as a meso compound. A similar difference is observed between the stereochemical outcome of these reactions with *cis*-2-butene. The difference in outcomes is because bromination of an alkene involves anti addition, whereas oxidation by osmium tetroxide involves syn addition.

In this section, we have seen three examples of reactions in which achiral starting materials give chiral products. In each case, the chiral product is formed as a racemic mixture, or else an achiral meso compound is formed. This illustrates a very important point about the creation of chiral molecules: an enantiomerically pure compound can never be produced from achiral starting materials and reagents. Although the molecules of the product may be chiral, their enantiomers are produced in equal amounts as a racemic mixture.

We will encounter many reactions throughout the remainder of this course where achiral starting materials are converted into chiral products under achiral reaction conditions. For convenience, we often draw just one of the enantiomeric products, but we must always keep in mind that, under these experimental conditions, both are formed in equal amounts if only achiral reagents are used.

B. Reaction of Achiral Starting Materials in a Chiral Environment

Reduction of a carbon-carbon double bond can be carried out using hydrogen in the presence of a transition metal catalyst. Because hydrogen atoms are delivered to either side of the double bond with equal probability, if a new stereocenter is created, equal amounts for both the R and S configurations will be produced.

Within the last three decades, chemists have discovered ways to embed transition metal hydrogenation catalysts in chiral molecules with the result that hydrogen can be delivered to only one face of the alkene. In catalytic reductions where a new stereocenter is formed, a large enantiomeric excess of one enantiomer may be formed. The first of these chiral hydrogenation catalysts to be developed was 2,2-*bis*-(diphenylphosphanyl)-1,1'-binaphthyl, more commonly known as BINAP. BINAP has been resolved into its R and S enantiomers. The fact that BINAP can be resolved depends on restricted rotation about the single bond joining the two naphthalene rings.

BINAP

(S)-(−)-BINAP
$[\alpha]_D^{25}$ −223°

(R)-(+)-BINAP
$[\alpha]_D^{25}$ +223°

Treatment of either enantiomer of BINAP with ruthenium(II) chloride forms a complex in which ruthenium(II) is bound as a complex ion in the chiral environment of the larger organic molecule. This complex is soluble in dichloromethane, CH_2Cl_2, and can be used as a homogeneous hydrogenation catalyst.

$$(R)\text{-BINAP} + RuCl_2 \longrightarrow (R)\text{-BINAP-Ru(II)}$$

This catalyst is used to produce (S)-naproxen in greater than 98% enantiomeric excess. (S)-Naproxen is the anti-inflammatory and pain reliever in Aleve and several other over-the-counter medications. Note that, in this reduction, neither the benzene rings of the naphthyl group nor the carboxyl group are reduced. We will have more to say about the reduction of these groups in later chapters.

(S)-Naproxen

Ruthenium(II) BINAP complexes are somewhat specific for the types of carbon-carbon double bonds they reduce. To be reduced, the double bond must have some kind of neighboring functional group that serves as a directing group during the reduction. The most common of these directing groups are the carboxyl group of carboxylic acids and esters and the hydroxyl group of alcohols. As seen in the following example, only the carbon-carbon double nearer the —OH group is reduced.

(E)-4,8-Dimethyl-3,7-nonadien-1-ol

(R)-4,8-Dimethyl-7-nonen-1-ol

Summary

A **reaction mechanism** (Section 6.2) is a detailed description of how and why a chemical reaction occurs as it does, which bonds are broken and which new ones are formed, the order in which the various bond-breaking and bond-forming steps take place and their relative rates, the role of the solvent if the reaction takes place in solution, and the role of the catalyst if the reaction involves a catalyst. Transition state theory (Section 6.2A) provides a model for understanding the relationships among molecular structure, reaction rates, and energetics. An **energy diagram** shows the changes in energy that occur in going from reactants to products. Energy is measured on the vertical axis, and the change in position of the atoms during reaction is measured on the horizontal axis (the reaction coordinate). A **transition state** is a point on the reaction coordinate at which the energy is a maximum. The difference in energy between reactants and a transition state is called the **activation energy,** $\Delta G^{\ddagger}$. A **reaction intermediate** corresponds to an energy minimum between two transition states. In a multistep reaction, the step that crosses the highest energy barrier is called the **rate-determining step.** The relationship between activation energy and rate constant for a reaction is given by the equation, $k = Ce^{-\Delta G^{\ddagger}/RT}$ (Section 6.2B).

A characteristic reaction of alkenes is addition, during which a pi bond is broken and sigma bonds to two new atoms are formed. A **regioselective reaction** is a reaction in which one direction of bond forming or bond breaking occurs in preference to all other directions (Section 6.3A); in a **regiospecific reaction,** only one direction occurs. According to **Markovnikov's rule** (Section 6.3A), in addition of HX or H_2O to an alkene, hydrogen adds to the carbon of the double bond having the greater number of hydrogens. An **electrophile** (Section 6.3A) is any atom, molecule, or ion that can accept a pair of electrons to form a new covalent bond. By definition, an electrophile is also a Lewis acid (Section 4.5). The rate-determining step in electrophilic addition to an alkene is formation of a **carbocation** intermediate. A carbocation is a positively charged ion that contains a carbon atom with only six electrons in its valence shell. Carbocations are planar with bond

angles of approximately 120° about the positive carbon. The order of stability of carbocations is 3° > 2° > 1° > methyl. Carbocations are stabilized by the electron-releasing **inductive effect** of alkyl groups bonded to the cationic carbon and by **hyperconjugation** (Section 6.3A).

A **stereoselective reaction** is a reaction in which one stereoisomer is formed or destroyed in preference to all others that might be formed or destroyed (Section 6.3D); in a **stereospecific reaction,** only one stereoisomer is formed or destroyed. Addition of new atoms or groups of atoms from opposite sides or faces of a double bond is called anti addition. In cyclic systems, anti addition is equivalent to trans-coplanar addition. Syn addition is the addition of atoms or groups of atoms to the same side or face of a double bond.

The driving force for **carbocation rearrangement** (Section 6.3C) is conversion of a 2° carbocation to a more stable 2° or 3° carbocation. Rearrangement is by a 1,2-shift in which an atom or group of atoms moves with its bonding electrons from one atom to an adjacent atom within the same molecule.

To determine if a transformation is an oxidation, a reduction, or neither, use a **balanced half-reaction** (Section 6.5A). **Oxidation** is the loss of electrons; **reduction** is the gain of electrons.

From **heats of hydrogenation** of a series of alkenes (Section 6.6B), we conclude that in general (1) the greater the degree of substitution of the carbon-carbon double bond, the more stable the alkene, and (2) a trans alkene is more stable than a cis alkene.

Key Reactions

1. Addition of HX (Section 6.3A)

Addition is regiospecific and follows Markovnikov's rule. Reaction involves a carbocation intermediate (a Lewis acid), which may rearrange before it combines with a halide ion (a Lewis base) to complete the addition.

$$\text{(cyclopentene with } CH_3) + HCl \longrightarrow \text{(cyclopentane with } Cl \text{ and } CH_3)$$

2. Acid-Catalyzed Hydration (Section 6.3B)

Addition is regiospecific and follows Markovnikov's rule. Reaction involves a carbocation intermediate (a Lewis acid), which may rearrange before it combines with water (a Lewis base) to complete the hydration.

$$\underset{\displaystyle \overset{CH_3}{|}}{CH_3C}=CH_2 + H_2O \xrightarrow{H_2SO_4} CH_3\underset{\displaystyle \underset{OH}{|}}{\overset{\displaystyle \overset{CH_3}{|}}{C}}CH_3$$

3. Addition of Bromine and Chlorine (Section 6.3D)

Addition is anti stereospecific; it involves anti addition of halogen atoms by way of a bridged halonium ion intermediate with no rearrangement.

4. Addition of HOCl and HOBr (Section 6.3E)

Addition is regiospecific (—X adds to the less substituted carbon via a bridged halonium ion intermediate and —OH adds to the more substituted carbon), anti stereospecific, and occurs without rearrangement.

5. Oxymercuration-Reduction (Section 6.3F)

Oxymercuration is regiospecific (HgOAc adds to the less substituted carbon, and OH adds to the more substituted carbon) and anti stereospecific. Anti addition occurs via a bridged mercurinium ion intermediate with no rearrangement. The result of oxymercuration-reduction is Markovnikov hydration of an alkene.

6. Hydroboration-Oxidation (Section 6.4)

Addition of BH_3 is syn stereospecific and regioselective (boron adds to the less substituted carbon and hydrogen to the more substituted carbon). Hydroboration-oxidation results in anti-Markovnikov hydration of the alkene without rearrangement.

7. Oxidation to a Glycol by OsO₄ (Section 6.5B)

Oxidation gives a glycol resulting from syn addition of —OH groups to the double bond via a cyclic osmic ester with no rearrangement. Oxidation is stereospecific for syn hydroxylation.

8. Oxidation by Ozone (Section 6.5C)

Treatment with ozone followed by dimethyl sulfide cleaves a carbon-carbon double bond and gives two carbonyl groups in its place.

9. Addition of H₂; Catalytic Reduction (Section 6.6)

Catalytic reduction is stereoselective and involves predominantly syn addition of hydrogens.

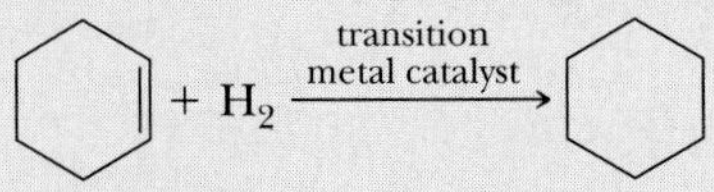

10. Enantioselective Reduction (Section 6.7)

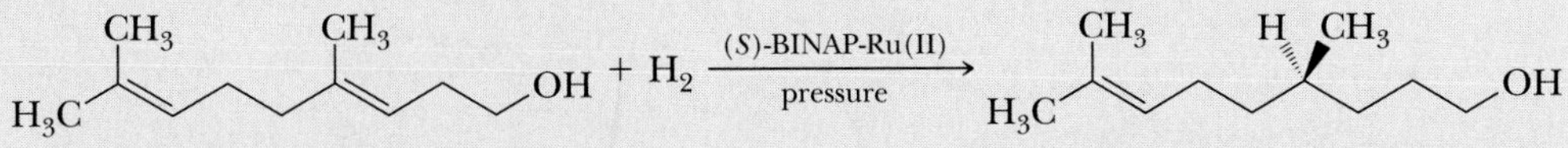

(E)-4,8-Dimethyl-3,7-nonadien-1-ol   (R)-4,8-Dimethyl-7-nonadien-1-ol

Problems

Energetics of Chemical Reactions

$\left(\Sigma \text{BDE}_R - \Sigma \text{BDE}_P \right)$

6.14 Most chemical reactions can occur as written if they are exothermic, that is, if the bonds that are formed in the products are stronger than the ones broken in the starting materials. To determine if a reaction is exothermic as written, add the bond dissociation energies of all bonds broken in the starting materials (it costs energy to break bonds). Subtract from this the total of bond dissociation energies of all bonds formed in the products (formation of bonds liberates energy). If the sum of these numbers is negative, the reaction is exothermic (energy is liberated), and the reaction proceeds to the right as written. If the sum of these numbers is positive, the reaction is endothermic (it requires energy), and it does not proceed to the right as written. Using the table of average bond dissociation energies at 25°C, determine which of the following reactions are energetically favorable at room temperature, that is, if a suitable catalyst could be found, which would proceed to the right as written?

Bond	Bond Dissociation Energy kJ (kcal)/mol	Bond	Bond Dissociation Energy kJ (kcal)/mol
H—H	435 (104)	C—Si	301 (72)
O—H	463 (111)	C=C	611 (146)
C—H	413 (98.7)	C=O (aldehyde)	728 (174)
N—H	391 (93.4)	C=O (CO₂)	803 (192)
Si—H	318 (76)	C≡O (CO)	1075 (257)
C—C	346 (82.6)	N≡N	950 (227)
C—N	305 (73)	C≡C	837 (200)
C—O	358 (85.5)	O=O	498 (119)
C—I	213 (51)		

(a) $CH_2{=}CH_2 + 2H_2 + N_2 \longrightarrow H_2NCH_2CH_2NH_2$
(b) $CH_2{=}CH_2 + CH_4 \longrightarrow CH_3CH_2CH_3$
(c) $CH_2{=}CH_2 + (CH_3)_3SiH \longrightarrow CH_3CH_2Si(CH_3)_3$
(d) $CH_2{=}CH_2 + CHI_3 \longrightarrow CH_3CH_2CI_3$

(e) $CH_2{=}CH_2 + CO + H_2 \longrightarrow CH_3CH_2\overset{\overset{\displaystyle O}{\|}}{C}H$

(f) [structure] $+ CH_2{=}CH_2 \longrightarrow$ [structure]

(g) [structure] $+ CO_2 \longrightarrow$ [structure]

(h) $HC{\equiv}CH + O_2 \longrightarrow HC{-}CH$ (with two O's double-bonded)

(i) $2CH_4 + O_2 \longrightarrow 2CH_3OH$

Electrophilic Additions

6.15 Draw structural formulas for the isomeric carbocation intermediates formed on treatment of each alkene with HCl. Label each carbocation primary, secondary, or tertiary, and state which of the isomeric carbocations is formed more readily.

(a) $CH_3CH_2\overset{\overset{\displaystyle CH_3}{|}}{C}{=}CHCH_3$ (b) $CH_3CH_2CH{=}CHCH_3$ (c) [cyclopentene with CH_3]   (d) [cyclohexane]${=}CH_2$

6.16 Arrange the alkenes in each set in order of increasing rate of reaction with HI, and explain the basis for your ranking. Draw the structural formula of the major product formed in each case.

(a) $CH_3CH{=}CHCH_3$ and $CH_3\overset{\overset{\displaystyle CH_3}{|}}{C}{=}CHCH_3$ (b) [cyclohexene with methyl]   and   [cyclohexene]

6.17 Predict the organic product(s) of the reaction of 2-butene with each reagent.

(a) H_2O (H_2SO_4)   (b) Br_2 (c) Cl_2
(d) Br_2 in H_2O (e) HI   (f) Cl_2 in H_2O
(g) $Hg(OAc)_2$, H_2O (h) product (g) + $NaBH_4$

6.18 Draw a structural formula of an alkene that undergoes acid-catalyzed hydration to give each alcohol as the major product. More than one alkene may give each alcohol as the major product.

(a) 3-Hexanol   (b) 1-Methylcyclobutanol
(c) 2-Methyl-2-butanol (d) 2-Propanol

6.19 Reaction of 2-methyl-2-pentene with each reagent is regiospecific. Draw a structural formula for the product of each reaction, and account for the observed regiospecificity.

(a) HI   (b) HBr
(c) H_2O in the presence of H_2SO_4 (d) Br_2 in H_2O
(e) $Hg(OAc)_2$ in H_2O

6.20 Reaction of 1-methylcyclopentene with each reagent is regiospecific and stereospecific. Account for the observed regiospecificity and stereospecificity.

(a) BH_3 (b) Br_2 in H_2O (c) $Hg(OAc)_2$ in H_2O

6.21 Draw a structural formula for an alkene with the indicated molecular formula that gives the compound shown as the major product. Note that more than one alkene may give the same compound as the major product.

(a) $C_5H_{10} + H_2O$ $\xrightarrow{H_2SO_4}$ [structure with OH]   (b) $C_5H_{10} + Br_2 \longrightarrow$ [structure with Br, Br]

(c) $C_7H_{12} + HCl \longrightarrow$ [cyclohexane with CH_3 and Cl]

6.22 Account for the fact that addition of HCl to 1-bromopropene gives exclusively 1-bromo-1-chloropropane.

$$CH_3CH{=}CHBr + HCl \longrightarrow CH_3CH_2CHBrCl$$

1-Bromopropene   1-Bromo-1-chloropropane

6.23 Propenoic acid (acrylic acid) reacts with HCl to give 3-chloropropanoic acid. It does not produce 2-chloropropanoic acid. Account for this result.

$$CH_2{=}CHCOH + HCl \longrightarrow ClCH_2CH_2COH \qquad \left[\begin{array}{c} Cl\ O \\ |\ \ || \\ CH_3CHCOH \end{array}\right]$$

Propenoic acid   3-Chloropropanoic acid   2-Chloropropanoic acid
(Acrylic acid)   (this product is not formed)

6.24 Draw a structural formula for the alkene of molecular formula C_5H_{10} that reacts with Br_2 to give each product.

 (a) Br   (b) Br   (c) Br

6.25 Draw alternative chair conformations for the product formed by addition of bromine to 4-*tert*-butylcyclohexene. The Gibbs free energy differences between equatorial and axial substituents on a cyclohexane ring are 21 kJ (4.9 kcal)/mol for *tert*-butyl and 2.0–2.6 kJ (0.48–0.62 kcal)/mol for bromine. Estimate the relative percentages of the alternative chair conformations you drew in the first part of this problem.

6.26 Draw a structural formula for the cycloalkene of molecular formula C_6H_{10} that reacts with Cl_2 to give each compound.

(a) [cyclohexane with Cl, Cl]   (b) [cyclopentane with Cl, Cl, CH₃]   (c) [cyclopentane with H₃C, Cl, Cl]   (d) [cyclopentane with Cl, CH₂Cl]

6.27 Reaction of this bicycloalkene with bromine in carbon tetrachloride gives a trans dibromide. In both (a) and (b), the bromine atoms are trans to each other. However, only one of these products is formed. Which trans dibromide is formed? How do you account for the fact that it is formed to the exclusion of the other trans dibromide?

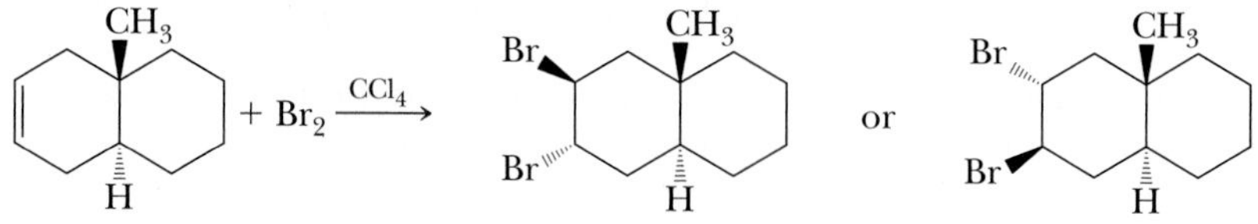

(a) (b)

6.28 Terpin, prepared commercially by the acid-catalyzed hydration of limonene (Figure 5.5), is used medicinally as an expectorant for coughs (see *The Merck Index,* 12th ed., #9314).

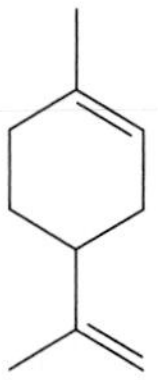

$$+\ 2H_2O \xrightarrow{H_2SO_4} C_{10}H_{20}O_2$$

Terpin

Limonene

(a) Propose a structural formula for terpin and a mechanism for its formation.
(b) How many cis,trans isomers are possible for the structural formula you proposed?

6.29 Propose a mechanism for this reaction. In so doing, account for its regioselectivity.

$$CH_3-\underset{\displaystyle |}{\overset{\displaystyle CH_3}{C}}=CH_2 + ICl \longrightarrow CH_3\underset{\displaystyle Cl}{\overset{\displaystyle CH_3}{C}}-CH_2I$$

6.30 Treatment of 2-methylpropene with methanol in the presence of sulfuric acid gives *tert*-butyl methyl ether. Propose a mechanism for the formation of this product.

$$CH_3\overset{\displaystyle CH_3}{C}=CH_2 + CH_3OH \xrightarrow{H_2SO_4} CH_3\underset{\displaystyle CH_3}{\overset{\displaystyle CH_3}{C}}-OCH_3$$

6.31 When 2-pentene is treated with Cl_2 in methanol, three products are formed. Account for the formation of each product. (You need not be concerned, however, with explaining their relative percentages.)

$$CH_3CH=CHCH_2CH_3 \xrightarrow[CH_3OH]{Cl_2}$$

$$CH_3\overset{\displaystyle Cl}{C}H\overset{\displaystyle OCH_3}{C}HCH_2CH_3 + CH_3\overset{\displaystyle H_3CO}{C}H\overset{\displaystyle Cl}{C}HCH_2CH_3 + CH_3\overset{\displaystyle Cl}{C}H\overset{\displaystyle Cl}{C}HCH_2CH_3$$

50% 35% 15%

6.32 Treatment of cyclohexene with HBr in the presence of acetic acid gives bromocyclohexane (85%) and cyclohexyl acetate (15%). Propose a mechanism for the formation of the latter product.

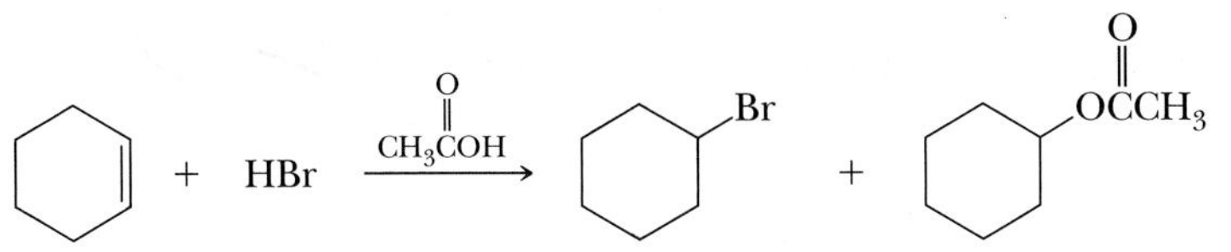

Cyclohexene   Bromocyclohexane   Cyclohexyl acetate
(85%)   (15%)

6.33 Propose a mechanism for this reaction.

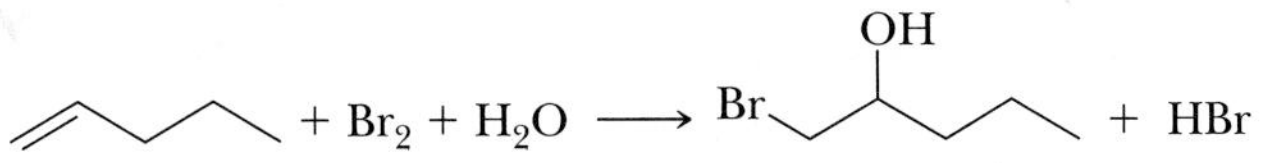

1-Pentene   1-Bromo-2-pentanol

6.34 Treatment of 4-penten-1-ol with bromine in water forms a cyclic bromoether. Account for the formation of this product rather than a bromohydrin as was formed in Problem 6.33.

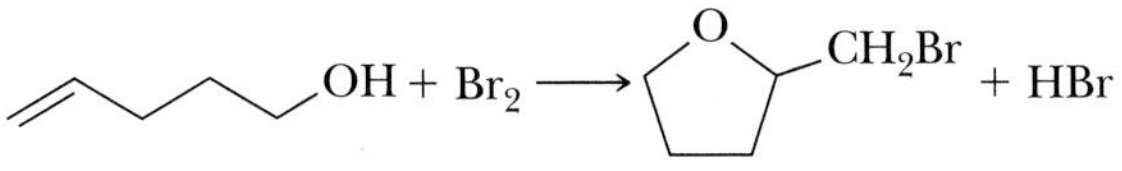

4-Penten-1-ol

6.35 Provide a mechanism for each reaction.

(a)

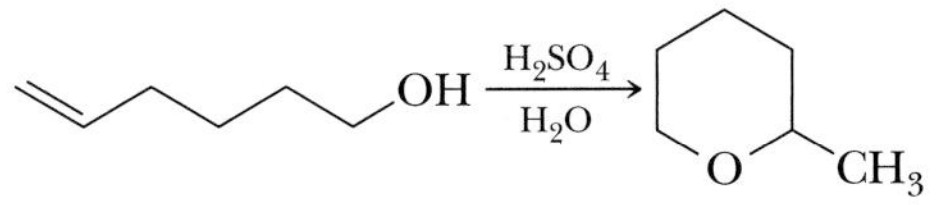

(b)

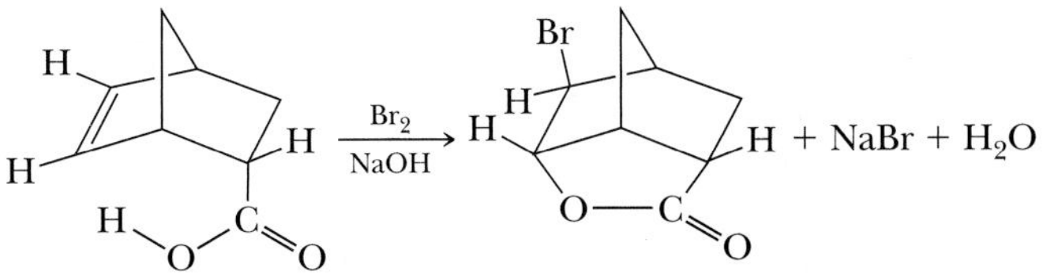

6.36 Treatment of 1-methyl-1-vinylcyclopentane with HCl gives mainly 1-chloro-1,2-dimethylcyclohexane. Propose a mechanism for the formation of this product.

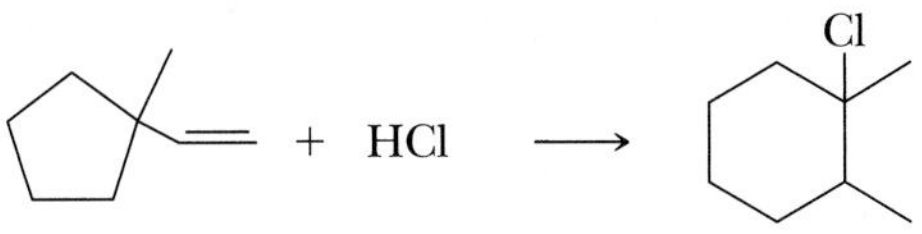

1-Methyl-1-vinyl-   1-Chloro-1,2-dimethyl-
cyclopentane   cyclohexane

Hydroboration

6.37 Each alkene is treated with diborane in tetrahydrofuran (THF) to form a trialkylborane that is then oxidized with hydrogen peroxide in aqueous sodium hydroxide. Draw a structural formula for the alcohol formed in each case. Specify stereochemistry where appropriate.

(a) [cyclohexane with =CH$_2$]   (b) [cyclohexene with CH$_3$]   (c) CH$_3$C=CHCH$_2$CH$_3$ with CH$_3$ substituent

(d) CH$_2$=CH(CH$_2$)$_5$CH$_3$ (e) (CH$_3$)$_3$CCH=CH$_2$

6.38 Reaction of α-pinene with diborane followed by treatment of the resulting trialkylborane with alkaline hydrogen peroxide gives an alcohol with the following structural formula. Of the four possible cis,trans isomers, one is formed in over 85% yield.

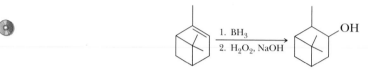

α-Pinene

(a) Draw structural formulas for the four possible cis,trans isomers of the bicyclic alcohol.
(b) Which is the structure of the isomer formed in 85% yield? How do you account for its formation? *Hint:* Examine the Chem3D model of α-pinene, and determine from which face of the double bond borane is more likely to approach.

Oxidation

6.39 Write structural formulas for the major organic product(s) formed by reaction of 1-methylcyclohexene with each oxidizing agent.

(a) OsO$_4$/H$_2$O$_2$ (b) O$_3$ followed by (CH$_3$)$_2$S

6.40 Each alkene is treated with ozone and then with dimethyl sulfide. Draw the structural formula of the organic product(s) formed from each.

(a) [structure]   (b) [structure]

(c) α-Pinene (Figure 5.5)   (d) Limonene (Figure 5.5)
(e) Caryophyllene (Figure 5.6)

6.41 Draw the structural formula of the alkene that reacts with ozone followed by dimethyl sulfide to give each product or set of products.

(a) C$_7$H$_{12}$ $\xrightarrow[\text{2. (CH}_3)_2\text{S}]{\text{1. O}_3}$ [diketone structure]

(b) C$_{10}$H$_{18}$ $\xrightarrow[\text{2. (CH}_3)_2\text{S}]{\text{1. O}_3}$ [ketone] + [ketone] + [dialdehyde]

(c) C$_{10}$H$_{18}$ $\xrightarrow[\text{2. (CH}_3)_2\text{S}]{\text{1. O}_3}$ [keto-aldehyde structure]

6.42 (a) Draw a structural formula for the bicycloalkene of molecular formula C$_8$H$_{12}$ that, on treatment with ozone followed by dimethyl sulfide, gives cyclohexane-1,4-dicarbaldehyde.
(b) Do you predict the product to be the cis isomer, the trans isomer, or a mixture of cis and trans isomers? Explain.

(c) Draw a suitable stereorepresentation for the more stable chair conformation of the di-carbaldehyde formed in this oxidation.

$$C_8H_{12} \xrightarrow[\text{2. } (CH_3)_2S]{\text{1. } O_3}$$

Cyclohexane-1,4-dicarbaldehyde

Reduction

6.43 Predict the major organic product(s) of the following reactions. Show stereochemistry where appropriate.

(a)

Geraniol

$+ \, 2H_2 \xrightarrow{Pt}$

(b)

$+ \, H_2 \xrightarrow{Pt}$

α-Pinene

6.44 The heat of hydrogenation of *cis*-2,2,5,5-tetramethyl-3-hexene is -154 kJ $(-36.7$ kcal)/mol, whereas that of the trans isomer is only -113 kJ $(-26.9$ kcal)/mol.

(a) Why is the heat of hydrogenation of the cis isomer so much larger (more negative) than that of the trans isomer?

(b) If a catalyst could be found that allowed equilibration of the cis and trans isomers at room temperature (such catalysts do exist), what would be the ratio of trans to cis isomers?

Synthesis

6.45 Show how to convert ethylene to these compounds.

(a) Ethane	**(b)** Ethanol	**(c)** Bromoethane
(d) 2-Chloroethanol	**(e)** 1,2-Dibromoethane	**(f)** 1,2-Ethanediol
(g) Chloroethane		

6.46 Show how to convert cyclopentene into these compounds.

(a) *trans*-1,2-Dibromocyclopentane	**(b)** *cis*-1,2-Cyclopentanediol
(c) Cyclopentanol	**(d)** Iodocyclopentane
(e) Cyclopentane	**(f)** Pentanedial

Reactions That Produce Chiral Compounds

6.47 State the number and kind of stereoisomers formed when (R)-3-methyl-1-pentene is treated with these reagents.

(R)-3-Methyl-1-pentene

(a) $Hg(OAc)_2$, H_2O followed by $NaBH_4$ **(b)** H_2/Pt
(c) BH_3 followed by H_2O_2 in $NaOH$ **(d)** Br_2 in CCl_4

6.48 Describe the stereochemistry of the bromohydrin formed in each reaction.
   (a) *cis*-3-Hexene + Br_2/H_2O **(b)** *trans*-3-Hexene + Br_2/H_2O

6.49 In each of these reactions, the organic starting material is achiral. The structural formula of the product is given. For each reaction, determine the following.
   (1) How many stereoisomers are possible for the product
   (2) Which of the possible stereoisomers is/are formed in the reaction shown
   (3) Whether the product is optically active or optically inactive

(a)

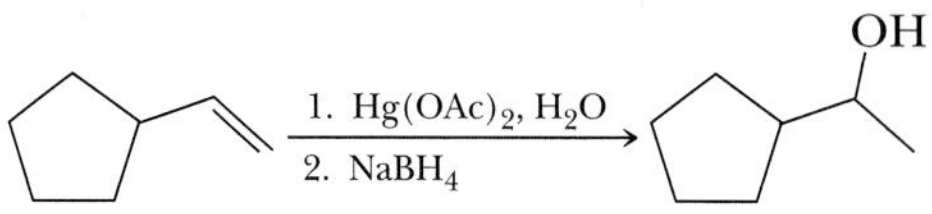

(b)

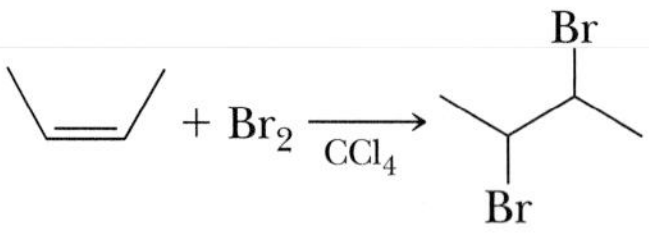

(c)

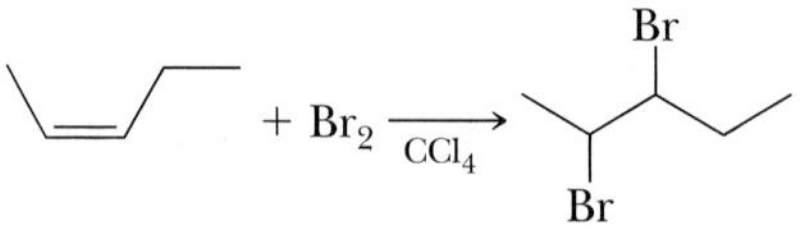

(d)

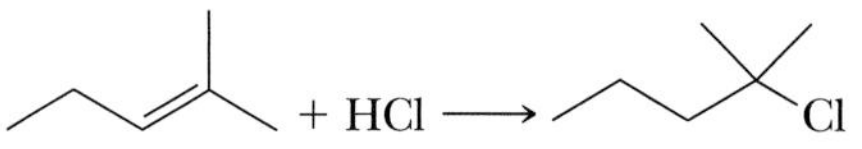

(e)

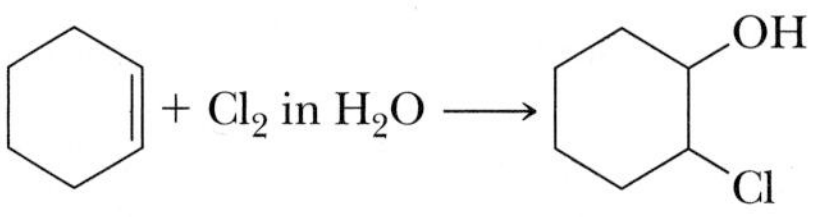

(f)

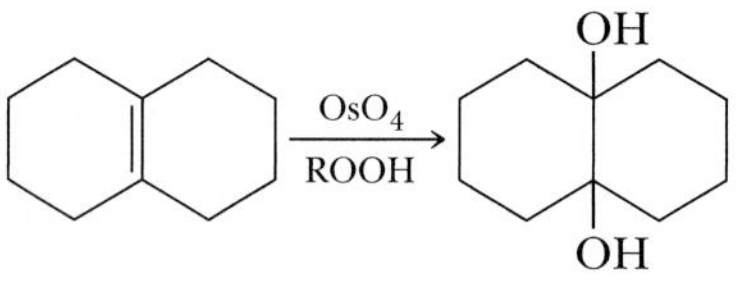

(g)

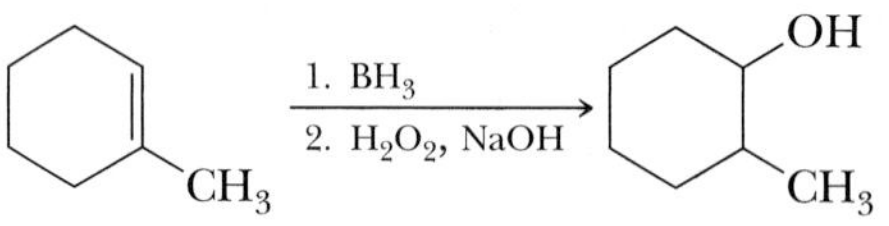

(h)

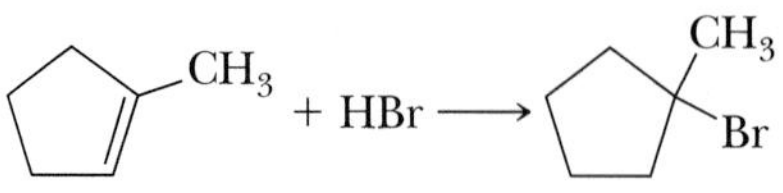

A Conversation with . . .
Carl Djerassi

C arl Djerassi, professor of chemistry at Stanford University, is one of a few modern-day Renaissance men. His prolific career includes many notable contributions as an industrial research chemist, a college professor, an author of poetry, drama, and fiction, and a patron and collector of fine art. Djerassi was born in Vienna in 1923.

His first job was with Ciba Pharmaceutical Corporation in New Jersey. He quickly established himself as a research chemist by synthesizing Pyribenzamine, one of the first antihistamines used to control allergies.

At Syntex, a small pharmaceutical company then located in Mexico City, Djerassi and George Rosenkranz directed the research group that in 1951 first synthesized the important hormone cortisone, which is used as an anti-inflammatory drug. A short time later, he led the research team that synthesized the first oral contraceptive, norethindrone, one that is still widely used.

Djerassi has published more than 1000 research papers and seven books on steroids, alkaloids, antibiotics, lipids, and terpenoids. His many research awards include the National Medal of Science (1973) for the first synthesis of an oral contraceptive, the

National Medal of Technology (1991) for promoting novel approaches to insect control, and the Priestley Medal (1992), the highest award given by the American Chemical Society. He is a member of the National Academy of Science and the National Inventors' Hall of Fame.

In addition to his career as a chemist, Djerassi has published a collection of short stories, five novels, a collection of poems, a collection of essays *(From the Lab into the World: A Pill for People, Pets, and Bugs)*, and two autobiographies outlining his career: *The Pill, Pygmy Chimps, and Degas' Horse* and *Steroids Made It Possible* (http://www.djerassi.com).

Djerassi is in the process of expanding his literary endeavors to science-in-theater, where he plans a trilogy of plays that presents realistic science through an attention-gathering plot and the dialog of the actors. The first play, *An Immaculate Misconception*, premiered at the 1998 Edinburgh Fringe Festival.

As a collector and patron of the arts, Djerassi began to acquire in the 1960s the works of Swiss artist Paul Klee (1879–1940). The collection, one of the largest private holdings of Klee's works, is on permanent display in the San Francisco Museum of Modern Art. In 1979, Djerassi established a resident artists' colony near Woodside, California. This program, which has served 1000 artists thus far, provides residence and studio space for visual, literary, choreographic, and musical artists.

Early Scientific Endeavors
"I was not a kid who had any interest in chemistry. I didn't have a chemistry set at home. I didn't blow up any basements. When I lived in Vienna, my education was based much more in the humanities than in the sciences. But both my parents were physicians and it was always assumed that I would become a physician too. But then Hitler came, and I left Vienna after the *Anschluss.*

"While I was waiting for my U.S. visa, I went to an American school for a year in Sofia, Bulgaria, where my father lived. There I learned English. Since I already had a very good education and could speak English very well, when in America, I went straight to Newark Junior College rather than completing the last two years of high school. I began a pre-med curriculum, which included chemistry. I had a superb chemistry teacher, Nathan Washton, and there is no question he got me very interested in chemistry."

> I have always been very much an intellectual polygamist. In this age of tremendous specialization, I think it is very important that people are interested in more than just one intellectual area.

Educational and Career Motivations

"After a year at Newark Junior College, I transferred for one semester to Tarkio College [in Missouri] and then went on to Kenyon College [in Ohio]. My father had stayed behind in Bulgaria and my mother, who didn't have an American license to practice medicine, took a job as a physician's assistant in upstate New York. So I was on my own.

"At Kenyon, there were only two chemistry professors, but I got a first-class education and began to do some research. I also applied to pharmaceutical companies for jobs, and I ended up as a junior chemist at Ciba. It was near the end of my days at Kenyon and certainly when I was at Ciba and did research in medicinal chemistry on antihistamines that I really got hooked on chemistry."

Rewards of Industrial Research

"Doing research for Ciba was really a seminal event. Consider that I had graduated from college in 1942 and wasn't quite 19 at the time. I was assigned to work with a senior chemist, Charles Huttrer, who himself was an immigrant from Vienna. He treated me much more like a colleague than an assistant. We started on a brand-new project to synthesize antihistamines because there were none in the United States at that time—the basic research had started in France. We were lucky. The research went very fast and within a year we had synthesized Pyribenzamine, one of the first two antihistamines produced in this country.

"When the patent was filed I was one of the inventors and was one of the authors on a research paper in the *Journal of the American Chemical Society*. Thus in just one year I became a participant in the development of a drug that was very soon taken by hundreds of thousands of people to treat allergies. That was very heady stuff. It made me a real optimist. Success, of course, is the greatest incentive there is. So when I went to graduate school at Wisconsin I was already a person who had made an important discovery."

Interest in the Chemistry of Steroids

"I assumed all along that, after graduate work, I would go back to Ciba and didn't even consider doing anything else at that time—although in the back of my mind I knew I wanted to go into academia. Ciba's Swiss headquarters was a powerhouse in steroids at the time, and since steroid hormones were such a familiar concept in the organization—even though I was working on antihistamines—I started reading about steroids in Louis Fieser's famous text, *Natural Products Related to Phenanthrene*.

"I decided I wanted to do my PhD thesis on steroids. There were two young assistant professors working in this area, William S. Johnson and Alfred L. Wilds, and I ended up working with the latter. I worked on the conversion of androgens to estrogens, a very tough chemical problem at that time. With the advantage of a full-time research fellowship and the year of experience at Ciba, I finished the program in slightly over two years."

Joining Syntex and Working in Mexico City

"I returned to Ciba as a senior chemist and continued working on antihistamines and antispasmodics. That was in the late 1940s when the therapeutic properties of cortisone were discovered and that became one of the hottest and most competitive topics in the field of organic chemistry. And I wanted to work on the synthesis of cortisone, but was not permitted to do so at Ciba since that was a project reserved primarily for the Swiss laboratories.

"Then one day I got a letter in the mail asking if I would consider a job at a Mexican firm called Syntex. They invited me for an interview and I was really charmed by Mexico City and by George Rosenkranz, the firm's technical director. He said that they wanted to develop a synthesis of cortisone from a Mexican yam and that I would be in charge of that project but could work on other steroid projects as well. The time I spent there from late 1949 until early 1952 was probably the most productive period of my scientific life."

Synthesizing an Oral Contraceptive

"Our initial goal was to develop an orally active progestational hormone—in other words, a compound that would mimic the biological properties of progesterone. At that time, progesterone was clinically used for menstrual disorders and infertility, but there were ideas about using it as a contraceptive because it is progesterone that naturally stops further ovulation after an ovum is fertilized. However, progesterone itself is not active by mouth, and daily injections would be needed. By combining ideas discovered by previous investigators in Europe and the States, we set out to synthesize a steroid that would not only be active by mouth but would also have enhanced progestational activity.

"This compound was 19-nor-17α-ethynyltestosterone (norethindrone), whose synthesis we completed on October 15, 1951. It was first tested for menstrual disorders and fertility problems and then as an oral contraceptive. Nearly 50 years after its synthesis, it is still the active ingredient in about a third of all the oral contraceptives used throughout the world."

From Contraceptives to Insect Control

"Conceptually, there was a relationship between our work on oral contraceptives and insect control. In a way, you could say that steroid oral contraceptives were true biorational methods of human birth control since progesterone—our conceptional lead compound—is really nature's contraceptive. That was a model on which insect control could be based on their counterpart, the insect juvenile hormones. At this time (the late 1960s), I was in charge of research at Syntex in addition to being a professor at Stanford University. Governments and the public realized that conventional methods of insect control—largely spraying with chlorinated hydrocarbons such as DDT—were damaging the environment.

"We formed a new company in the Stanford Industrial Park called Zoecon to synthesize insect-controlling hormones. In the 1960s, a juvenile hormone based on a sesquiterpene skeleton had been discovered, and we decided to focus on it. Insects pass through a juvenile stage controlled by the juvenile hormone, whose production is later shut off by another hormone so that the insect can then mature. Our biorational approach was to synthesize an artificial juvenile hormone that would continue to be applied to immature insects, so that the insect would never reach the stage at which it could reproduce. This new approach to controlling mosquitoes, fleas, cockroaches, and other insects that do their damage as adults was approved by the Environmental Protection Agency for public use."

Exploring Scientific Issues with Fiction

"We as scientists pay very little attention to our behavior and practices. We do not teach such topics in our undergraduate or even in our graduate courses. We don't have courses on how to become a scientist or how to behave as a scientist. We learn how through osmosis from our mentors and colleagues. As chemists, we are very analytical about the work around ourselves, but we do very little introspective self-analysis.

"Writing a special type of fiction, which I call science-in-fiction, really did that for me. It was a very good idea for me to have started writing fiction later in life, because I think that fiction-writing is one area where it helps to have experience, a basis for comparison, and a historical perspective. In many respects writing fiction has enabled me to talk about aspects of my science that I really had not thought about that much. Then I became convinced of its importance pedagogically.

"Instead of telling readers, 'Let me tell you about my science,' because many nonscientists who pay no attention to it get scared the moment you say 'science,' I now say, 'Let me tell you a story.' And as I tell them a story, I teach them a lot about science and about scientists. And I am exceptionally accurate about this, which is why I call it science-in-fiction rather than science fiction. A recent graduate course I taught, 'Ethical Discourse Through Science-in-Fiction,' showed that it is much easier to demonstrate that point in an anonymous way [*Nature*, **393:** 511 (1998)]. By discussing these aspects of scientific behavior in the cloak of fiction we can illustrate ethical dilemmas that people would otherwise never discuss openly because of discretion, embarrassment, or fear of retribution."

Balancing a Three-Pronged Career

"I have always been very much an intellectual polygamist. In this age of tremendous specialization, I think it is very important that people are interested in more than just one intellectual area. And I don't mean just in different areas of chemistry and not just in different areas of science. I have always been interested in doing different things at the same time, which is why I use the word 'polygamy,' because that refers to a man who has several wives at the same time. I'm not using it of course in a sexual context; I am using it in an intellectual one. In true polygamy, each wife is more or less equal. I would say that I don't have a favorite wife, even though I have several wives at this stage: academic, industrial, literary, artistic, and so on. They are all important components of my life."

7

HALOALKANES, ALKENES, AND ARENES

Compounds containing a halogen atom covalently bonded to an sp^3 hybridized carbon atom are named haloalkanes or, in the common system of nomenclature, alkyl halides. Several haloalkanes are important laboratory and industrial solvents. In addition, haloalkanes are invaluable building blocks for organic synthesis. In this chapter, we begin with structure and the physical properties of haloalkanes. We then study one method for the preparation of haloalkanes, radical halogenation of alkanes, as a vehicle to introduce an important type of reaction mechanism, namely the mechanism of radical chain reactions.

■ Compact disks are made of poly(vinyl chloride). *(Charles D. Winters)* Inset: A model of vinyl chloride.

7.1 Structure

Haloalkane (alkyl halide) A compound containing a halogen atom covalently bonded to an sp^3 hybridized carbon atom. Given the symbol R—X.

Haloalkene (vinylic halide) A compound containing a halogen bonded to one of the carbons of a carbon-carbon double bond.

Haloarene (aryl halide) A compound containing a halogen atom bonded to a benzene ring. Given the symbol Ar—X.

The general symbol for a **haloalkane** is R—X, where —X may be —F, —Cl, —Br, or —I. If halogen is bonded to a doubly bonded carbon of an alkene, the compound belongs to a class called **haloalkenes.** If it is bonded to a benzene ring, the compound belongs to a class called a **haloarene,** often given the generic symbol Ar—X.

| A haloalkane (an alkyl halide) | A haloalkene (an alkenyl or vinylic halide) | A haloarene (an aryl halide) Ar—X |

7.2 Nomenclature

A. IUPAC System

IUPAC names for haloalkanes are derived by naming the parent alkane according to the rules given in Section 2.3A.

- The parent chain is numbered from the direction that gives the first substituent encountered the lowest number, whether it is a halogen or an alkyl group.
- Halogen substituents are indicated by the prefixes fluoro-, chloro-, bromo-, and iodo- and are listed in alphabetical order with other substituents.
- The location of each halogen atom on the parent chain is given by a number preceding the name of the halogen.
- In haloalkenes, numbering the parent hydrocarbon is determined by the location of the carbon-carbon double bond. Numbering is done in the direction to give the carbon atoms of the double bond and substituents the lowest set of numbers.

| 3-Bromo-2-methylpentane | *trans*-2-Chlorocyclohexanol | 4-Bromocyclohexene |

B. Common Names

Common names of haloalkanes and haloalkenes consist of the common name of the alkyl group followed by the name of the halide as a separate word. Hence, the name alkyl halide is a common name for this class of compounds. In the following examples, the IUPAC name of the compound is given first, and then its common name is given in parentheses.

| 2-Bromobutane (*sec*-Butyl bromide) | Chloroethene (Vinyl chloride) | 3-Chloropropene (Allyl chloride) |

Several polyhaloalkanes are important solvents and are generally referred to by their common names. Dichloromethane (methylene chloride) is the most widely used haloalkane solvent. Compounds of the type CHX_3 are called **haloforms.** The common name for $CHCl_3$, for example, is chloroform. It is from the name chloroform that the common name methyl chloroform is derived for the compound CH_3CCl_3. Methyl chloroform and trichloroethylene (trichlor) are common solvents for commercial dry cleaning.

Haloform A compound of the type CHX_3 where X is a halogen.

CH_2Cl_2	$CHCl_3$	CH_3CCl_3	$CCl_2{=}CHCl$
Dichloromethane (Methylene chloride)	Trichloromethane (Chloroform)	1,1,1-Trichloroethane (Methyl chloroform)	Trichloroethylene (Trichlor)

Hydrocarbons in which all hydrogens are replaced by halogens are commonly named as perhaloalkanes or perhaloalkenes. Perchloroethylene, commonly known as perc, is the most common dry cleaning solvent in use today.

Perchloroethane   Perfluoropropane   Perchloroethylene

Example 7.1

Write the IUPAC name and, where possible, the common name of each compound. Show stereochemistry where relevant.

(a) CH_3CCH_2Br with CH_3 groups   (b)   (c) $CH_3(CH_2)_6$, $C{-}Br$, H, CH_3

Solution

(a) 1-Bromo-2,2-dimethylpropane. Its common name is neopentyl bromide.
(b) (*E*)-4-bromo-3-methyl-2-pentene.
(c) (*S*)-2-Bromononane.

Problem 7.1

Write the IUPAC name, and where possible, the common name of each compound. Show stereochemistry where relevant.

(a)   (b) 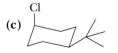   (c)   (d)

7.3 Physical Properties of Haloalkanes

A. Polarity

Fluorine, chlorine, and bromine are all more electronegative than carbon (Table 1.5), and, as a result, C—X bonds with these atoms are polarized with a partial negative charge on halogen and a partial positive charge on carbon. Table 7.1 shows that each of the halomethanes has a substantial dipole moment. The magnitude of a dipole moment depends on the size of the partial charges, the distance between them, and the polarizability of the three pairs of unshared electrons on each halogen. For the halomethanes, dipole moment increases as the electronegativity of halogen and the bond length increase. These two trends run counter to each other, with the net effect that chloromethane has the largest dipole moment of the series.

B. Boiling Point

Haloalkanes are associated in the liquid state by a combination of dipole-dipole, dipole–induced dipole, and induced dipole–induced dipole (dispersion) forces. These forces are grouped together under the term **van der Waals forces,** in honor of J. D. van der Waals, the 19th century Dutch physicist who pioneered our modern understanding of the effects of these forces on the physical properties of compounds. Van der Waals attractive forces pull molecules together. As atoms or molecules are brought closer and closer, van der Waals attractive forces are overcome by repulsive forces between electron clouds of adjacent atoms. The energy minimum is where the net attractive forces are the strongest. Nonbonded interatomic and intermolecular distances at these minima can be measured by x-ray crystallography of solid compounds, and each atom and group of atoms can be assigned an atomic or molecular radius called a **van der Waals radius.** Nonbonded atoms in a molecule cannot approach each other closer than the van der Waals radius without causing steric strain. Van der Waals radii for selected atoms and groups of atoms are given in Table 7.2. Notice from Table 7.2 that the van der Waals radius of fluorine is only slightly greater than that of hydrogen and that, among the halogens, only iodine has a larger van der Waals radius than methyl.

Boiling points of several low-molecular-weight haloalkanes and the alkanes from which they are derived are given in Table 7.3. There are several trends to be noticed from these data. First, as with hydrocarbons, constitutional isomers with branched chains have lower boiling points than their unbranched-chain isomers (Section 2.8). Compare, for example, the boiling points of unbranched-chain 1-bromobutane

Van der Waals forces A group of intermolecular attractive forces including dipole-dipole, dipole–induced dipole, and induced dipole–induced dipole (dispersion) forces.

Table 7.1 **Dipole Moments (Gas Phase) of Halomethanes**

Halomethane	Electronegativity of Halogen	Carbon-Halogen Bond Length (pm)	Dipole Moment (debyes, D)
CH_3F	4.0	139	1.85
CH_3Cl	3.0	178	1.87
CH_3Br	2.8	193	1.81
CH_3I	2.5	214	1.62

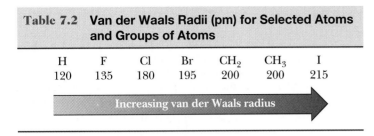

Table 7.2   Van der Waals Radii (pm) for Selected Atoms and Groups of Atoms

H	F	Cl	Br	CH$_2$	CH$_3$	I
120	135	180	195	200	200	215

Increasing van der Waals radius

(butyl bromide, bp 100°C) with the more branched and compact 2-bromo-2-methyl-propane (*tert*-butyl bromide, bp 72°C). Branched-chain constitutional isomers have lower boiling points because they have a more spherical shape and, therefore, decreased surface area, leading to smaller van der Waals forces between their molecules.

Second, for an alkane and haloalkane of comparable size and shape, the haloalkane has a higher boiling point. Compare, for example, the boiling points of ethane (bp −89°C) and bromomethane (bp 4°C). Although both molecules are roughly the same size (the van der Waals radii of —CH$_3$ and —Br are almost identical) and have roughly the same effective contact area, the boiling point of bromomethane is considerably higher than that of ethane. This difference is due almost entirely to the greater **polarizability** of the three unshared pairs of electrons on the halogen compared with the shared electron pairs in the hydrocarbon of comparable size and shape. Recall from Section 2.8 that the strength of dispersion forces, the weakest of all intermolecular forces, depends on the polarizability of electrons, which, in turn, depends on how tightly they are held to the nucleus. The farther electrons are from the nucleus, the greater their polarizability. In addition, unshared electron pairs, being less tightly bound, have a higher polarizability than electrons shared in a covalent bond.

Polarizability   A measure of the ease of distortion of the distribution of electron density about an atom or group in response to interaction with other molecules or ions. Fluorine, which has a high electronegativity, holds its electrons tightly and has a very low polarizability. Iodine, which has a lower electronegativity and holds its electrons less tightly, has a very high polarizability.

Table 7.3   Boiling Points of Some Low-Molecular-Weight Alkanes and Alkyl Halides

Alkyl Group	Name	Boiling Point (°C)				
		H	F	Cl	Br	I
CH$_3$—	Methyl	−161	−78	−24	4	43
CH$_3$CH$_2$—	Ethyl	−89	−37	13	38	72
CH$_3$(CH$_2$)$_2$—	Propyl	−45	3	46	71	102
(CH$_3$)$_2$CH—	Isopropyl	−45	−11	35	60	89
CH$_3$(CH$_2$)$_3$—	Butyl	0	32	77	100	130
CH$_3$CH$_2$(CH$_3$)CH—	*sec*-Butyl	0	25	67	90	119
(CH$_3$)$_2$CHCH$_2$—	Isobutyl	−1	16	68	91	120
(CH$_3$)$_3$C—	*tert*-Butyl	−1	12	51	72	98
CH$_3$(CH$_2$)$_4$—	Pentyl	36	63	108	129	157
CH$_3$(CH$_2$)$_5$—	Hexyl	69	92	134	155	181

Table 7.4   Densities of Some Low-Molecular-Weight Haloalkanes

| Alkyl Group | Name | Density of Liquid (g/mL) at 25°C | | |
		Cl	Br	I
CH_3—	Methyl	—	—	2.279
CH_3CH_2—	Ethyl	—	1.460	1.936
$CH_3(CH_2)_2$—	Propyl	0.891	1.354	1.749
$(CH_3)_2CH$—	Isopropyl	0.862	1.314	1.703
$CH_3(CH_2)_3$—	Butyl	0.886	1.276	1.615
$(CH_3)_3C$—	*tert*-Butyl	0.842	1.221	1.545
$CH_3(CH_2)_5$—	Hexyl	0.879	1.174	1.440

Third, the boiling points of fluoroalkanes are even lower than those of hydrocarbons of comparable molecular weight. Compare, for example, the boiling points of hexane (MW 86.2, bp 69°C) and 1-fluoropentane (MW 90.1, bp 63°C) and the boiling points of 2-methylpropane (MW 58.1, bp −1°C) and 2-fluoropropane (MW 62.1, bp −11°C). This difference is due to the small size of fluorine, the tightness with which its electrons are held, and their particularly low polarizability. The distinctive properties of fluorocarbons, for example, the nonstick properties of polytetrafluoroethylene (PTFE), one consumer end product use of which is Teflon, are also a consequence of the uniquely low polarizability of the three unshared electron pairs on fluorine.

C.   Density

The densities of liquid haloalkanes are greater than those of hydrocarbons of comparable molecular weight because of the halogens' large mass-to-volume ratio. A bromine atom and a methyl group have almost identical van der Waals radii, but bromine has a mass of 79.9 atomic mass units (amu) compared with 15 amu for methyl. Table 7.4 gives densities for some low-molecular-weight haloalkanes that are liquid at 25°C. The densities of all liquid bromoalkanes and iodoalkanes are greater than that of water. Although the densities of liquid chloroalkanes are less than that of water, further substitution of chlorine for hydrogen increases the density to the point where di- and polychloroalkanes have a greater density than water (Table 7.5). These compounds sink in water and form the lower layer when mixed with water because they are insoluble.

Table 7.5   Density of Polyhalomethanes

| Haloalkane | X = | Density of Liquid (g/mL) at 25°C | | |
		Cl	Br	I
CH_2X_2		1.327	2.497	3.325
CHX_3		1.483	2.890	4.008
CX_4		1.594	3.273	4.23

D. Bond Lengths and Bond Strengths

With the exception of C—F bonds, C—X bonds are weaker than C—H bonds as measured by bond dissociation energies (BDE), which are one of the best measures of bond strength. A table of BDE values for many bonds is given in Appendix 3. **Bond dissociation energy** is defined as the amount of heat required to break a bond homolytically into two radicals in the gas phase at 25°C.

$$A \overset{\frown\frown}{} B \longrightarrow A \cdot + \cdot B$$

This reaction is a "virtual" one because it can't actually be carried out in most cases. Instead, the extensive tables of BDEs are collected from thermochemical data on heats of combustion, hydrogenation, and other reactions. The useful thing about these data is that by adding and subtracting them, heats of reaction for reactions that have never been measured can be calculated with remarkable accuracy.

C—X BDEs are tabulated in Table 7.6. As the size of the halogen atom increases, the C—X bond length increases, and its strength decreases. These relationships between bond strength and bond length help us to understand the difference in the ease with which haloalkanes undergo reactions that involve carbon-halogen bond breaking. Fluoroalkanes, for example, with the strongest and shortest C—X bonds, are highly resistant to bond breaking under most conditions. This characteristic inertness is one of the factors that make perfluoroalkanes such as Teflon such useful materials.

Of all the fluoroalkanes, **chlorofluorocarbons (CFCs)** manufactured under the trade name Freons are the most widely known. CFCs are nontoxic, nonflammable, odorless, and noncorrosive and seemed to be ideal replacements for the hazardous compounds such as ammonia and sulfur dioxide formerly used as heat-transfer agents in refrigeration systems. Among the CFCs most widely used for this purpose were trichlorofluoromethane (CCl_3F, Freon-11) and dichlorodifluoromethane (CCl_2F_2 Freon-12).

The CFCs found wide use as industrial cleaning solvents to prepare surfaces for coatings, to remove cutting oils and waxes from millings, and to remove protective coatings. CFCs were also used as propellants for aerosol sprays.

Concern about the environmental impact of CFCs arose in the 1970s when it was shown that more than 4.5×10^5 kg/yr of these compounds were being emitted into the atmosphere. Then, in 1974 Sherwood Rowland of the University of California, Irvine, and Mario Molina, now at the Massachusetts Institute of Technology, announced their theory, which has since been amply confirmed, of ozone destruction by these compounds. When released into the air, CFCs escape to the lower atmosphere, but because of their inertness, they do not decompose there. Slowly, they find their way to

Table 7.6	**Average Bond Dissociation Energies for C—H and C—X Bonds**	
Bond	**Bond Length (pm)**	**Bond Dissociation Energy [kJ (kcal)/mol]**
C—H	109	377–418 (90–100)
C—F	142	439 (105)
C—Cl	178	345 (80)
C—Br	193	272 (65)
C—I	214	209 (50)

the stratosphere where they absorb ultraviolet radiation from the sun and then decompose. As they do so, they set up a chemical reaction that leads to the destruction of the stratospheric ozone layer, which acts as a shield for the earth against short-wavelength ultraviolet radiation from the sun. An increase in short-wavelength ultraviolet radiation reaching the earth is believed to lead to destruction of certain crops and agricultural species, and even to increased incidence of skin cancer in light-skinned individuals.

The results of this concern were two conventions, one in Vienna in 1985 and the other in Montreal in 1987, held by the United Nations Environmental Program. The 1987 meeting produced the Montreal Protocol, which set limits on the production and use of ozone-depleting CFCs and urged complete phaseout of their production by the year 1996. This phaseout resulted in enormous costs and is not yet complete in developing countries. The fact that an international agreement on the environment that set limits on the production of any substance could be reached is indeed amazing and bodes well for the health of the planet.

Rowland, Molina, and Paul Crutzen, a Dutch chemist at the Max Planck Institute for Chemistry in Germany, were awarded the 1995 Nobel Prize for chemistry. As noted in the award citation by the Royal Swedish Academy of Sciences, "By explaining the chemical mechanisms that affect the thickness of the ozone layer, these three researchers have contributed to our salvation from a global environmental problem that could have catastrophic consequences."

The chemical industry responded by developing non-ozone-depleting alternatives to CFCs, among which are the hydrofluorocarbons (HFCs) and hydrochlorofluorocarbons (HCFCs). These compounds are much more chemically reactive in the atmosphere than the Freons and are destroyed before reaching the stratosphere. However, they tend to act as "greenhouse gases" and may contribute to global warming.

$$
\begin{array}{cc}
\quad\ \ \overset{\displaystyle F}{\underset{\displaystyle F}{|}}\ \ \overset{\displaystyle F}{\underset{\displaystyle F}{|}} & \quad\ \ \overset{\displaystyle Cl}{\underset{\displaystyle Cl}{|}}\ \ \overset{\displaystyle F}{\underset{\displaystyle F}{|}} \\
H-C-C-F & H-C-C-F
\end{array}
$$

HFC-134a   HCFC-123

We must not assume, however, that haloalkanes are introduced into the environment only by human action. It is estimated, for example, that annual production of bromomethane from natural sources is 2.7×10^8 kg, largely from marine algae, giant kelp, and volcanoes. Furthermore, global emission of chloromethane is estimated to be 4.5×10^9 kg/yr, most of it from terrestrial and marine biomass. These haloalkanes, however, have only short atmospheric lifetimes, and only a tiny fraction of them reach the stratosphere. The CFCs are the problem; they have longer atmospheric lifetimes, reach the stratosphere, and do their damage there.

Reactivity Explorer;
Alkanes

7.4   Preparation of Haloalkanes by Halogenation of Alkanes

As we saw in Sections 6.3A and 6.3D, haloalkanes can be prepared by addition of HX and X_2 to alkenes. They are also prepared by replacement of the —OH group of alcohols by halogen (Section 9.5). Many of the simpler, low-molecular-weight haloalkanes are prepared by the halogenation of alkanes, illustrated here by the treatment of 2-methylpropane with bromine at an elevated temperature.

$$
\underset{\substack{\text{2-Methylpropane}\\(\text{Isobutane})}}{\overset{\displaystyle CH_3}{\underset{\displaystyle CH_3}{\underset{\displaystyle |}{\overset{\displaystyle |}{CH_3CH}}}}} + Br_2 \xrightarrow{\ heat\ } \underset{\substack{\text{2-Bromo-2-methylpropane}\\(\textit{tert}\text{-Butyl bromide})}}{\overset{\displaystyle CH_3}{\underset{\displaystyle CH_3}{\underset{\displaystyle |}{\overset{\displaystyle |}{CH_3CBr}}}}} + HBr
$$

Halogenation of alkanes is common with Br_2 and Cl_2. Fluorine, F_2, is seldom used because its reactions with alkanes are so exothermic that they are difficult to control and can actually cause C—C bond cleavage and even explosions. Iodine, I_2, is seldom used because the reaction is endothermic and the position of equilibrium favors alkane and I_2 rather than iodoalkane and HI.

If a mixture of methane and chlorine gas is kept in the dark at room temperature, no detectable change occurs. If, however, the mixture is heated or exposed to light, a reaction begins almost at once with evolution of heat. The products are chloromethane and hydrogen chloride. What occurs is a **substitution** reaction, in this case substitution of a chlorine atom for a hydrogen atom in methane and the production of an equivalent amount of hydrogen chloride.

Substitution A reaction in which an atom or group of atoms in a compound is replaced by another atom or group of atoms.

$$
\underset{\text{Methane}}{CH_4} + \underset{}{Cl_2} \xrightarrow{\ heat\ } \underset{\substack{\text{Chloromethane}\\(\text{Methyl chloride})}}{CH_3Cl} + HCl \qquad \Delta H^0 = -100 \text{ kJ } (-23.9 \text{ kcal})/\text{mol}
$$

If chloromethane is allowed to react with more chlorine, further chlorination produces a mixture of dichloromethane (methylene chloride), trichloromethane (chloroform), and tetrachloromethane (carbon tetrachloride).

$$
\underset{\substack{\text{Chloromethane}\\(\text{Methyl chloride})}}{CH_3Cl} + Cl_2 \xrightarrow{\ heat\ } \underset{\substack{\text{Dichloromethane}\\(\text{Methylene chloride})}}{CH_2Cl_2} + HCl
$$

$$
\underset{\substack{\text{Dichloromethane}\\(\text{Methylene chloride})}}{CH_2Cl_2} \xrightarrow[\ heat\ ]{\ Cl_2\ } \underset{\substack{\text{Trichloromethane}\\(\text{Chloroform})}}{CHCl_3} \xrightarrow[\ heat\ ]{\ Cl_2\ } \underset{\substack{\text{Tetrachloromethane}\\(\text{Carbon tetrachloride})}}{CCl_4}
$$

Notice that in the last equation, the reagent Cl_2 is placed over the reaction arrow and the equivalent amount of HCl formed is not shown. Placing reagents over reaction arrows and omitting byproducts is commonly done to save space.

Treatment of ethane with chlorine gives chloroethane (ethyl chloride); treatment with bromine gives bromoethane (ethyl bromide).

$$
\underset{\text{Ethane}}{CH_3CH_3} + Br_2 \xrightarrow{\ heat\ } \underset{\substack{\text{Bromoethane}\\(\text{Ethyl bromide})}}{CH_3CH_2Br} + HBr
$$

A. Regioselectivity

Supporting Concepts;
Radicals: Stability

Treatment of propane with bromine gives a mixture consisting of approximately 8% of 1-bromopropane and 92% of 2-bromopropane.

$$\text{CH}_3\text{CH}_2\text{CH}_3 + \text{Br}_2 \xrightarrow[\text{or light}]{\text{heat}} \underset{\substack{\text{2-Bromopropane}\\(92\%)}}{\text{CH}_3\overset{\overset{\text{Br}}{|}}{\text{C}}\text{HCH}_3} + \underset{\substack{\text{1-Bromopropane}\\(8\%)}}{\text{CH}_3\text{CH}_2\text{CH}_2\text{Br}} + \text{HBr}$$

Propane contains eight hydrogens: one set of six equivalent primary hydrogens and one set of two equivalent secondary hydrogens (Section 2.3B). Substitution of bromine for a primary hydrogen gives 1-bromopropane; substitution of bromine for a secondary hydrogen gives 2-bromopropane. On the basis of random substitution of any one of the eight hydrogens in propane, we would predict the isomeric bromopropanes to be formed in the ratio of 6:2 or 75% 1-bromopropane and 25% 2-bromopropane. In fact, in the bromination of propane, substitution of a secondary hydrogen is strongly favored over a primary hydrogen, and 2-bromopropane is the major product.

Product Distribution	$\text{CH}_3\text{CH}_2\text{CH}_2\text{Br}$	$\text{CH}_3\overset{\overset{\text{Br}}{|}}{\text{C}}\text{HCH}_3$
Prediction based on ratio of six 1° H to two 2° H	75%	25%
Experimental observation	8%	92%

Other experiments have shown that substitution at a tertiary hydrogen is favored over both secondary and primary hydrogens. For example, monobromination of 2-methylpentane is nearly regiospecific and gives almost exclusively 2-bromo-2-methylpentane.

$$\underset{\text{2-Methylpentane}}{\text{CH}_3\overset{\overset{\text{CH}_3}{|}}{\text{C}}\text{HCH}_2\text{CH}_2\text{CH}_3} + \text{Br}_2 \xrightarrow[\text{or light}]{\text{heat}} \underset{\text{2-Bromo-2-methylpentane}}{\text{CH}_3\overset{\overset{\text{CH}_3}{|}}{\underset{\underset{\text{Br}}{|}}{\text{C}}}\text{CH}_2\text{CH}_2\text{CH}_3} + \text{HBr}$$

The reaction of bromine with an alkane occurs regioselectively in the order 3° hydrogen > 2° hydrogen > 1° hydrogen. Chlorination of alkanes is also regioselective, but much less so than bromination. For example, treatment of propane with chlorine gives a mixture consisting of approximately 57% 2-chloropropane and 43% 1-chloropropane.

$$\text{CH}_3\text{CH}_2\text{CH}_3 + \text{Cl}_2 \xrightarrow[\text{or light}]{\text{heat}} \underset{\substack{\text{2-Chloropropane}\\(57\%)}}{\text{CH}_3\overset{\overset{\text{Cl}}{|}}{\text{C}}\text{HCH}_3} + \underset{\substack{\text{1-Chloropropane}\\(43\%)}}{\text{CH}_3\text{CH}_2\text{CH}_2\text{Cl}} + \text{HCl}$$

Thus, we can conclude that, although both bromine and chlorine are regioselective in hydrogen replacement in the order 3° > 2° > 1°, regioselectivity is far greater for bromination than for chlorination. From data on product distribution, it has been determined that regioselectivity per hydrogen for bromination is approximately

1600:80:1, whereas for chlorination, it is approximately 5:4:1. We will discuss reasons for this regioselectivity in Section 7.5.

Example 7.2

Name and draw structural formulas for all monobromination products formed by treatment of 2-methylpropane with Br_2. Predict the major product based on the regioselectivity of the reaction of Br_2 with alkanes.

$$
\begin{array}{c}
CH_3 \\
| \\
CH_3CHCH_3 + Br_2 \xrightarrow[\text{heat}]{\text{light or}} \text{monobromoalkanes} + HBr
\end{array}
$$

Solution

In 2-methylpropane, there are nine equivalent primary hydrogens and one tertiary hydrogen. Substitution of bromine for a primary hydrogen gives 1-bromo-2-methylpropane; substitution for the tertiary hydrogen gives 2-bromo-2-methylpropane. Given that the regioselectivity per hydrogen of bromination for $3° > 2° > 1°$ hydrogens is approximately 1600:80:1, it is necessary to correct for the number of hydrogens: nine methyl and one tertiary. The result is that 99.4% of the product is 2-bromo-2-methylpropane and 0.6% is 1-bromo-2-methylpropane.

$$
\begin{array}{ccc}
CH_3 & & CH_3 & & CH_3 \\
| & & | & & | \\
CH_3CHCH_3 & + Br_2 \xrightarrow[\text{or light}]{\text{heat}} & CH_3CCH_3 & + & CH_3CHCH_2Br + HBr \\
& & | & & \\
& & Br & &
\end{array}
$$

| 2-Methylpropane | 2-Bromo-2-methylpropane (major product) | 1-Bromo-2-methylpropane |

$$
\text{Predicted \% 2-bromo-2-methylpropane} = \frac{1 \times 1600}{(1 \times 1600) + (9 \times 1)} \times 100 = 99.4\%
$$

Problem 7.2

Name and draw structural formulas for all monochlorination products formed by treatment of 2-methylpropane with Cl_2. Predict the major product based on the regioselectivity of the reaction of Cl_2 with alkanes.

B. Energetics

Supporting Concepts; Radicals: Reactions

We can learn a lot about these reactions by careful consideration of the energetics of each step. A selection of C—H BDE values is given in Table 7.7. Note that the BDE values for saturated hydrocarbons depend on the type of hydrogen being abstracted and are in the order methane $> 1° > 2° > 3°$. This order follows because the resulting radicals are electron deficient (they have only seven electrons in the valence shell of the carbon bearing the radical); like carbocations, the more alkyl groups they have, the more stable they are. Note also that an sp^2 C—H bond is particularly strong, as mentioned in Section 1.8F.

Using data from Table 7.7 and Appendix 3, we can calculate the heat of reaction, ΔH^0, for the halogenation of an alkane. In this calculation for the chlorination of

Table 7.7   Bond Dissociation Energies for Selected C—H Bonds

Hydrocarbon	Radical	Name of Radical	Type of Radical	ΔH^0 [kJ (kcal)/mol]
$CH_2=CHCH_2-H$	$CH_2=CHCH_2\cdot$	Allyl	Allylic	360 (86)
$C_6H_5CH_2-H$	$C_6H_5CH_2\cdot$	Benzyl	Benzylic	368 (88)
$(CH_3)_3C-H$	$(CH_3)_3C\cdot$	*tert*-Butyl	3°	389 (93)
$(CH_3)_2CH-H$	$(CH_3)_2CH\cdot$	Isopropyl	2°	402 (96)
CH_3CH_2-H	$CH_3CH_2\cdot$	Ethyl	1°	418 (100)
CH_3-H	$CH_3\cdot$	Methyl	Methyl	439 (105)
$CH_2=CH-H$	$CH_2=CH\cdot$	Vinyl	Vinylic	444 (106)

Radical stability →

methane, energy is required to break CH_3—H and Cl—Cl bonds [439 and 247 kJ (105 and 59 kcal)/mol, respectively]. Energy is released in making the CH_3—Cl and H—Cl bonds [−355 and −431 kJ (−85 and −103 kcal)/mol, respectively]. Summing these energies, we calculate that chlorination of methane to form chloromethane and hydrogen chloride liberates 100 kJ (24 kcal)/mol.

$$CH_4 + Cl_2 \longrightarrow CH_3Cl + HCl \qquad \Delta H^0 = -100 \text{ kJ/mol}$$

					(-24 kcal/mol)
BDE, kJ/mol	+439	+247	−355	−431	
(kcal/mol)	(105)	(59)	(−85)	(−103)	

Example 7.3

Using the table of bond dissociation energies in Appendix 3, calculate ΔH^0 for bromination of propane to give 2-bromopropane and hydrogen bromide.

Solution

Under each molecule is given the energy for breaking or forming each corresponding bond. The calculated heat of reaction is −59 kJ (−14 kcal)/mol.

$$\underset{H}{\overset{}{CH_3CHCH_3}} + Br-Br \longrightarrow \underset{Br}{\overset{}{CH_3CHCH_3}} + H-Br \qquad \Delta H^0 = -59 \text{ kJ/mol}$$

					(-14 kcal/mol)
BDE, kJ/mol	+402	+192	−285	−368	
(kcal/mol)	(+96)	(+46)	(−68)	(−88)	

Problem 7.3

Using the table of bond dissociation energies in Appendix 3, calculate ΔH^0 for bromination of propane to give 1-bromopropane and hydrogen bromide.

Mechanisms: Alkanes Radical Halogenation of an Alkane

7.5   Mechanism of Halogenation of Alkanes

Radical Any chemical species that contains one or more unpaired electrons.

From detailed studies of the conditions and products for halogenation of alkanes, chemists have concluded that these reactions occur by a type of mechanism called a radical chain mechanism. A **radical,** sometimes called a free radical, is any chemical species that contains one or more unpaired electrons.

A. Formation of Radicals

Radicals are produced from a molecule by cleavage of a bond in such a way that each atom or fragment participating in the bond retains one electron. In the following equations, we use single-headed or **fishhook arrows** to show the change in position of single electrons. BDEs of these reactions are shown on the right.

Fishhook arrow A barbed curved arrow used to show the change in position of a single electron.

$$CH_3CH_2\overset{\frown\frown}{O-O}CH_2CH_3 \xrightarrow{80°} CH_3CH_2O\cdot + \cdot OCH_2CH_3 \qquad \Delta H^0 = +150 \text{ kJ/mol}$$
$$(+36 \text{ kcal/mol})$$

Diethyl peroxide   Ethoxy radicals

$$Cl\overset{\frown\frown}{-}Cl \xrightarrow{\text{light}} Cl\cdot + \cdot Cl \qquad \Delta H^0 = +247 \text{ kJ/mol}$$
$$(+59 \text{ kcal/mol})$$

Chlorine   Chlorine atoms

$$CH_3\overset{\frown\frown}{-}CH_3 \xrightarrow{\text{heat}} CH_3\cdot + \cdot CH_3 \qquad \Delta H^0 = +377 \text{ kJ/mol}$$
$$(+90 \text{ kcal/mol})$$

Ethane   Methyl radicals

Energy to cause bond cleavage and generation of radicals can be supplied either by light or heat. The energy of visible and ultraviolet radiation (wavelength from 200 to 700 nm) falls in the range of 585 to 167 kJ (140 to 40 kcal)/mol and is of the same order of magnitude as the bond dissociation energies of halogen-halogen covalent bonds. The bond dissociation energy of Br_2 is 192 kJ (46 kcal)/mol; that for Cl_2 is 247 kJ (59 kcal)/mol. Dissociation of these halogens can also be brought about by heating to temperatures above 350°C.

Oxygen-oxygen single bonds in peroxides (ROOR) and hydroperoxides (ROOH) have dissociation energies in the range of 125 to 167 kJ (30 to 40 kcal)/mol, and compounds containing these bonds are cleaved to radicals at considerably lower temperatures than those required for rupture of carbon-carbon bonds. Diethyl peroxide, for example, begins to dissociate to ethoxy radicals at 80°C.

B. A Radical Chain Mechanism

To account for the products formed from halogenation of alkanes, chemists propose a radical chain mechanism involving three types of steps: (1) chain initiation, (2) chain propagation, and (3) chain termination. We illustrate radical halogenation of alkanes by the reaction of chlorine with ethane.

Mechanism   **Radical Halogenation of Ethane**

Chain initiation: formation of radicals from nonradical species.

Step 1: $$Cl\overset{\frown\frown}{-}Cl \xrightarrow[\text{or light}]{\text{heat}} Cl\cdot + \cdot Cl$$

Chain propagation: reaction of a radical and a molecule to form a new radical.

Step 2: $$CH_3\overset{\frown}{CH_2-}H + \cdot Cl \longrightarrow CH_3CH_2\cdot + H-Cl$$
Step 3: $$CH_3CH_2\cdot + Cl\overset{\frown}{-}Cl \longrightarrow CH_3CH_2-Cl + \cdot Cl$$

Chain termination: destruction of radicals. The first three possible chain termination steps involve coupling of radicals to form a new covalent bond. The fourth chain termination step, called disproportionation, involves transfer of a hydrogen atom from the beta position of one radical to another radical and formation of an alkane and an alkene.

Step 4: $CH_3CH_2 \cdot + \cdot CH_2CH_3 \longrightarrow CH_3CH_2—CH_2CH_3$

Step 5: $CH_3CH_2 \cdot + \cdot Cl \longrightarrow CH_3CH_2—Cl$

Step 6: $Cl \cdot + \cdot Cl \longrightarrow Cl—Cl$

Step 7: $CH_3CH_2 \cdot + \overset{H}{\underset{|}{CH_2}}—CH_2 \cdot \longrightarrow CH_3CH_3 + CH_2{=}CH_2$

Chain initiation   A step in a chain reaction characterized by the formation of reactive intermediates (radicals, anions, or cations) from nonradical or noncharged molecules.

Chain propagation   A step in a chain reaction characterized by the reaction of a reactive intermediate and a molecule to give a new reactive intermediate and a new molecule.

The characteristic feature of a **chain initiation** step is formation of radicals from nonradical compounds. In the case of chlorination of ethane, chain initiation is by thermal or light-induced cleavage of the Cl—Cl bond to give two chlorine atoms (radicals).

The characteristic feature of a **chain propagation** step is reaction of a radical and a molecule to give a new radical. A chlorine atom is consumed in Step 2, but an ethyl radical is produced. Similarly, an ethyl radical is consumed in Step 3, but a chlorine radical is produced. Steps 2 and 3 can repeat thousands of times as long as neither radical is removed by a different reaction.

A second characteristic feature of chain propagation steps is that, when added together, they give the observed stoichiometry of the reaction. Adding Steps 2 and 3 and canceling structures that appear on both sides of the equation gives the balanced equation for the radical chlorination of ethane.

Steps 2 + 3 $CH_3CH_3 + Cl_2 \longrightarrow CH_3CH_2Cl + HCl$

Chain length   The number of times the cycle of chain propagation steps repeats in a chain reaction.

Chain termination   A step in a chain reaction that involves destruction of reactive intermediates.

The number of times a cycle of chain propagation steps repeats is called **chain length.**

A characteristic feature of a **chain termination** step is destruction of radicals. Among the most important chain termination reactions during halogenation of alkanes are radical couplings, illustrated by Steps 4, 5, 6, and 7 in the mechanism for the halogenation of ethane.

The structures, geometries, and relative stabilities of simple alkyl radicals are similar to those of alkyl carbocations. They are planar, or almost so, with bond angles of 120° about the carbon with the unpaired electron. This geometry indicates that carbon is sp^2 hybridized and that the unpaired electron occupies the unhybridized $2p$ orbital. The order of stability of alkyl radicals, like alkyl carbocations, is 3° > 2° > 1° > methyl.

C.   Energetics of Chain Propagation Steps

After the radical chain is initiated, the heat of reaction is derived entirely from the heat of reaction of the individual chain propagation steps. In Step 2 of radical chlorination of ethane, for example, energy is required to break the CH_3CH_2—H bond [418 kJ (100 kcal)/mol], but energy is released on formation of the H—Cl bond [−431 kJ (−103 kcal)/mol]. Similarly, energy is required in Step 3 for breaking the Cl—Cl bond [247 kJ (59 kcal)/mol], but energy is released on formation of the CH_3CH_2—Cl bond [−335 kJ (−80 kcal)/mol]. We see that, just as the sum of the chain propagation steps for radical halogenation gives the observed stoichiometry, the sum of the heats of reaction for each propagation step is equal to the observed heat of reaction.

Reaction Step			ΔH^0, kJ/mol (kcal/mol)
Step 2: CH_3CH_2—H + ·Cl ⟶ CH_3CH_2· + H—Cl			−13
+418		−431	(−3)
(+100)		(−103)	
Step 3: CH_3CH_2· + Cl—Cl ⟶ CH_3CH_2—Cl + ·Cl			−88
	+247	−335	(−21)
	(+59)	(−80)	
Sum: CH_3CH_2—H + Cl—Cl ⟶ CH_3CH_2—Cl + H—Cl			−101
			(−24)

Example 7.4

Using the table of bond dissociation energies in Appendix 3, calculate ΔH^0 for each propagation step in the radical bromination of propane to give 2-bromopropane and HBr.

Solution

Here are the two chain propagation steps along with bond dissociation energies for the bonds broken and the bonds formed. The first chain propagation step is endothermic, the second is exothermic, and the overall reaction is exothermic by 58 kJ (14 kcal)/mol.

ΔH^0, kJ/mol (kcal/mol)

Step 2:   CH₃ĊHCH₃ (H) + Br· ⟶ CH₃ĊHCH₃ + H—Br
+402   −368
(+96)   (−88)

+34 kJ/mol
(+8 kcal/mol)

Step 3:   CH₃ĊHCH₃ + Br—Br ⟶ CH₃CHCH₃ (Br) + Br·
+192   −285
(+46)   (−68)

−93 kJ/mol
(−22 kcal/mol)

Sum: CH₃CHCH₃ (H) + Br—Br ⟶ CH₃CHCH₃ (Br) + H—Br

−59 kJ/mol
(−14 kcal/mol)

Problem 7.4

Write a pair of chain propagation steps for the radical bromination of propane to give 1-bromopropane, and calculate ΔH^0 for each propagation step, and for the overall reaction.

D. Regioselectivity of Bromination Versus Chlorination—Hammond's Postulate

The regioselectivity in halogenation of alkanes can be accounted for in terms of the relative stabilities of radicals (3° > 2° > 1° > methyl). But how do we account for the greater regioselectivity in bromination of alkanes compared with chlorination of alkanes? To do so, we need to consider **Hammond's postulate,** a refinement of transition state theory proposed in 1955 by George Hammond, then at Iowa State University. According to this postulate, the structure of the transition state for an exothermic reaction step looks more like the reactants of that step than like the products. Conversely, the structure of the transition state for an endothermic reaction step looks more like the products of that step than like the reactants. It is important to realize that we cannot observe a transition state directly; we can only infer its existence, structure, and stability. What Hammond's postulate does is give us a reasonable way of deducing something about the structure of a transition state by examining things we can observe: the structure of reactants and products and heats of reaction. Hammond's postulate applies equally well to multistep reactions. The transition state of any exothermic step in a multistep reaction looks like the starting material(s) of that step; the transition state of any endothermic step in the sequence looks like the product(s) of that step. Shown in Figure 7.1 are energy diagrams for a highly exothermic reaction and a highly endothermic reaction, each occurring in one step.

Now let us apply Hammond's postulate to explain the relative regioselectivities of chlorination versus bromination of alkanes. In applying Hammond's postulate, we deal with the rate-determining step, which, in radical halogenation of alkanes, is the abstraction of a hydrogen atom by a halogen radical. Given in Table 7.8 are heats of reaction, ΔH^0, for the hydrogen abstraction step in chlorination and bromination of the different hydrogens of 2-methylpropane (isobutane). Also given under the formulas of isobutane, HCl, and HBr are bond dissociation energies for the bonds broken (1° and 3° C—H) and formed (H—Cl and H—Br) in each step. Because the 3° radical is more stable than the 1° radical, the BDE for the 3° H is lower than that of a 1° H by about 29 kJ (7 kcal)/mol.

Abstraction of hydrogen by chlorine is exothermic, which, according to Hammond's postulate, means that the transition state for H abstraction by Cl· is reached early in the course of the reaction [Figure 7.2(a)]. Therefore, the structure of the transition state for this step resembles the reactants, namely the alkane and a chlorine atom. Because there is relatively little radical character on carbon in this

Figure 7.1

Hammond's postulate. Energy diagrams for two one-step reactions. In the exothermic reaction, the structure of the transition state resembles that of the reactants. In the endothermic reaction, the structure of the transition state resembles that of the products.

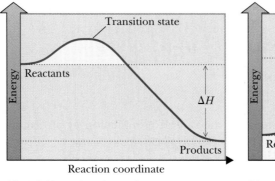

(a) Highly exothermic reaction

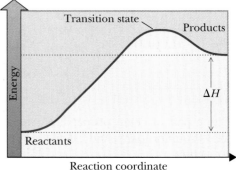

(b) Highly endothermic reaction

Table 7.8   Heats of Reaction for Hydrogen Abstraction in 2-Methylpropane

Reaction Step	ΔH^0 [kJ (kcal)/mol]
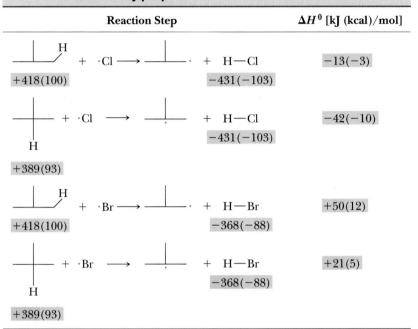	−13(−3)
	−42(−10)
	+50(12)
	+21(5)

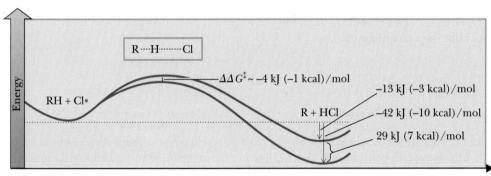

(a)   Chlorination

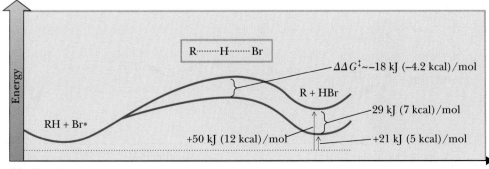

(b)   Bromination

Figure 7.2

Transition states and energetics for hydrogen abstraction in the radical chlorination and bromination of 2-methylpropane (isobutane).

transition state, regioselectivity in radical chlorination is only slightly influenced by the relative stabilities of possible radical intermediates. Products are determined more by whether a chlorine atom happens to collide with a 1°, 2°, or 3° hydrogen.

In Section 7.4A, we were given the fact that the selectivity for abstraction of a 3° H compared to a 1° H in chlorination is 5:1. Using this ratio of reaction rates and the relationship between $\Delta G^{\ddagger}$ and rate constants (Section 6.2B), we can calculate that the difference in activation energies, $\Delta\Delta G^{\ddagger}$, for the abstraction of a 3° H versus a 1° H is about 4 kJ (1 kcal)/mol. However, we can calculate from the bond energies [Figure 7.2(a)] that $\Delta\Delta H^0$ for the two reactions is about 29 kJ (7 kcal)/mol. Thus the difference in product stabilities is only slightly reflected in the transition states. Contrast this reaction with bromination [Figure 7.2(b)]. For bromination, the selectivity of 3° H to 1° H is 1600:1, which corresponds to $\Delta\Delta G^{\ddagger}$ of approximately 18 kJ (4.2 kcal)/mol. The $\Delta\Delta H^0$ for the two reactions is the same as for chlorination (it is just the difference in BDE of the two radicals). Thus bromination, because it is endothermic, has a transition state much more like the product radical, and the transition state has much more of the energy difference of the primary and tertiary radicals. The larger $\Delta\Delta G^{\ddagger}$ is the reason for the much larger regioselectivity in radical bromination than in radical chlorination.

BDE values are ΔH^0 and not ΔG^0. However, because we are dealing with similar reactions, we can assume that entropy differences between them will be nearly zero and, therefore, $\Delta\Delta G^0 \sim \Delta\Delta H^0$, which allows us to make these comparisons.

E. Stereochemistry of Radical Halogenation

When radical halogenation produces a stereocenter or takes place at a hydrogen on an existing stereocenter, the product is an equal mixture of R and S enantiomers. Consider, for example, radical bromination of butane, which produces 2-bromobutane. In this example, both of the starting materials are achiral and, as is true for any reaction of achiral starting materials that gives a chiral product (Section 6.7A), the product is a racemic mixture.

$$\underset{\text{Butane}}{CH_3CH_2CH_2CH_3} + Br_2 \xrightarrow[\text{or light}]{\text{heat}} \underset{(R,S)\text{-2-Bromobutane}}{CH_3CH_2\overset{\overset{\displaystyle Br}{|}}{C}HCH_3} + HBr$$

In the case of the *sec*-butyl radical, the carbon bearing the unpaired electron is sp^2 hybridized, and the unpaired electron lies in the unhybridized $2p$ orbital. Reaction of the alkyl radical intermediate with halogen in the second chain propagation step occurs with equal probability from either face to give an equal mixture of the R and S configurations at the newly created stereocenter.

The unhybridized $2p$ orbital

Radical intermediate   Br—Br   (R)-2-Bromobutane   (S)-2-Bromobutane

7.6 Allylic Halogenation

We saw in Section 6.3D that propene and other alkenes react with Br_2 and Cl_2 at room temperature by addition to the carbon-carbon double bond. If, however, propene and one of these halogens are allowed to react at a high temperature, an entirely different reaction takes place, namely substitution of a halogen occurs at the **allylic carbon.** We illustrate **allylic substitution** by the reaction of propene with chlorine at high temperature.

Allylic carbon A carbon adjacent to a carbon-carbon double bond.
Allylic substitution Any reaction in which an atom or group of atoms is substituted for another atom or group of atoms at an allylic carbon.

$$CH_2=CHCH_3 + Cl_2 \xrightarrow{350°C} CH_2=CHCH_2Cl + HCl$$

Propene   3-Chloropropene
                        (Allyl chloride)

A comparable reaction takes place when propene is treated with bromine at an elevated temperature.

To predict which of the various C—H bonds in propene is most likely to break when a mixture of propene and bromine or chlorine is heated, we need to look at bond dissociation energies. We find that the bond dissociation energy of an allylic C—H bond in propene (Table 7.7) is approximately 84 kJ (20 kcal)/mol less than that of a vinylic C—H bond and 58 kJ (14 kcal)/mol less than a C—H bond of ethane. The reason that the allylic C—H bond is weaker is discussed in Section 7.6B. Note from Table 7.7 that the benzylic radical $C_6H_5CH_2\cdot$ is stabilized in exactly the same way as the allylic radical and for the same reason; benzylic compounds undergo many of the same reactions as allylic compounds (Section 20.6B).

$+360$ kJ (86 kcal)/mol

$+444$ kJ (106 kcal)/mol

Treatment of propene with bromine or chlorine at elevated temperatures illustrates a very important point about organic reactions: it is often possible to change the product(s) of a reaction by changing the reaction conditions and thereby changing the mechanism of the reaction.

$$CH_2=CHCH_3 + Br_2 \xrightarrow{\text{high temp.}} CH_2=CHCH_2Br + HBr \qquad \text{(allylic substitution)}$$

Propene

$$CH_2=CHCH_3 + Br_2 \xrightarrow[CCl_4]{\text{room temp.}} \underset{\underset{Br \quad Br}{|\quad\;|}}{CH_2CHCH_3} \qquad \text{(electrophilic addition)}$$

A very useful way to carry out allylic bromination in the laboratory at or slightly above room temperature is to use the reagent *N*-bromosuccinimide (NBS) in carbon tetrachloride. Reaction between an alkene and NBS is most commonly initiated by light.

In the NBS reaction, a double substitution occurs: bromine for a hydrogen in the alkene and hydrogen for bromine in NBS.

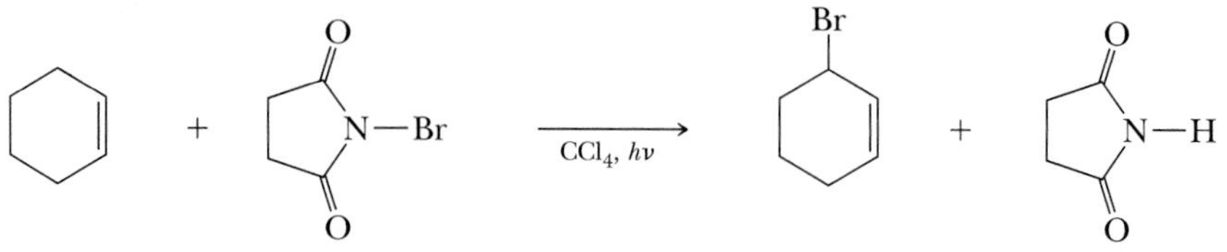

| Cyclohexene | N-Bromosuccinimide (NBS) | | 3-Bromocyclohexene | Succinimide |

A. Mechanism of Allylic Halogenation

Allylic bromination and chlorination proceed by a radical chain reaction involving the same type of chain initiation, chain propagation, and chain termination steps involved in the radical halogenation of alkanes.

Following is a mechanism for allylic bromination of propene.

Mechanism Allylic Bromination of Propene

Chain initiation: the formation of radicals from NBS.

Step 1:

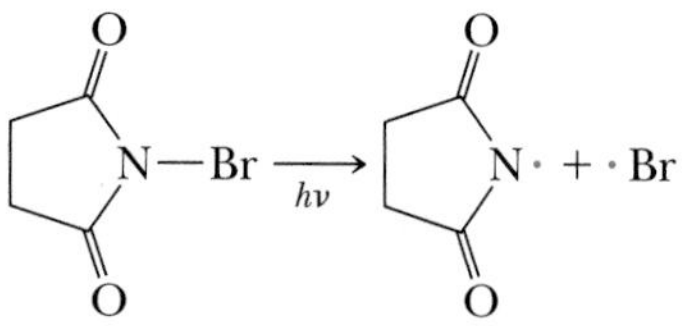

Chain propagation: formation of products. Reaction of a radical and a nonradical gives a new radical. (Both radicals can abstract hydrogen atoms. We show only the Br· reaction.) In the first propagation step, a bromine atom abstracts an allylic hydrogen (the weakest C—H bond in propene) to produce an allyl radical. The allyl radical, in turn, reacts with a bromine molecule to form allyl bromide and a new bromine atom. Note that, as must be the case, this combination of chain propagation steps adds up to the observed stoichiometry. This reaction is exactly like halogenation of alkanes but is strongly regioselective for an allylic hydrogen because of its weaker bond.

Step 2: $CH_2{=}CHCH_2{-}H + \cdot Br \longrightarrow CH_2{=}CHCH_2\cdot + H{-}Br$

Step 3: $CH_2{=}CHCH_2\cdot + Br{-}Br \longrightarrow CH_2{=}CHCH_2{-}Br + \cdot Br$

Chain termination: the destruction of radicals. Propagation of the chain reaction continues until termination steps produce nonradical products and thus stop further reaction.

Step 4: $Br \cdot + \cdot Br \longrightarrow Br-Br$

Step 5: $CH_2=CHCH_2 \cdot + \cdot Br \longrightarrow CH_2=CHCH_2-Br$

Step 6: $CH_2=CHCH_2 \cdot + \cdot CH_2CH=CH_2 \longrightarrow CH_2=CHCH_2-CH_2CH=CH_2$

The Br_2 necessary for allylic bromination is formed by reaction of HBr with NBS.

Step 7:

$$\text{N-Br} + HBr \longrightarrow \text{N-H} + Br_2$$

Bromine formed in this step then reacts with an allyl radical to continue the chain propagation reactions. Thus, in effect, NBS reacts with the HBr formed in the first chain propagation step to yield Br_2, which then continues the chain reaction.

The mechanism we described for allylic bromination by NBS poses the following problem. NBS is the indirect source of Br_2, which then takes part in chain propagation. But if Br_2 is present in the reaction mixture, why does it not react instead with the carbon-carbon double bond by electrophilic addition? In other words, why is the observed reaction allylic substitution rather than addition to the double bond? The answer is that when radicals are present, the rates of the chain propagation steps are much faster than the rate of electrophilic addition of bromine to the alkene. Furthermore, the concentration of Br_2 is very low throughout the time of the reaction, which slows the rate of electrophilic addition.

B.   Structure of the Allyl Radical

The allyl radical can be represented as a hybrid of two contributing structures. Here fishhook arrows show the redistribution of single electrons and how the contributing structure is converted to the other.

$$\overset{\cdot}{C}H_2=CH-CH_2 \longleftrightarrow CH_2-CH=CH_2$$

Equivalent contributing structures

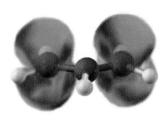

Figure 7.3
Computed electron spin density for the allyl radical. Unpaired electron density appears only on carbons 1 and 3.

The eight atoms of the allyl radical lie in a plane, and all bond angles are approximately 120°. Each carbon atom is sp^2 hybridized, and the three $2p$ orbitals participating in resonance stabilization of the radical are parallel to one another as shown in Figure 7.3. Because of delocalization of electrons in the allyl radical (the two electrons of the pi bond and the single electron of the radical), the allyl radical is considerably more stable than might be expected by looking at any one of its contributing structures.

Based on bond dissociation energies, we conclude that an allyl radical is even more stable than a 3° alkyl radical. Note that because of the larger amount of s-character in its carbon sp^2 hybrid orbital, a vinylic hydrogen is stronger (has a larger bond dissociation energy) than any sp^3 C—H bond and is never abstracted in homolytic reactions.

According to the molecular orbital description, the conjugated system of the allyl radical involves the formation of three molecular orbitals by overlap of three $2p$ atomic orbitals (Figure 7.4). The lowest energy molecular orbital (MO) has zero nodes, the next MO has one node, and the highest energy MO has two nodes. The molecular orbital of intermediate energy in this case leads to neither net stabilization

Figure 7.4
Molecular orbital model of covalent bonding in the allyl radical. Combination of three $2p$ atomic orbitals gives three pi molecular MOs. The lowest, a pi-bonding MO, has zero nodes. The next in energy, a pi-nonbonding MO, has one node, and the highest in energy, a pi-antibonding MO, has two nodes.

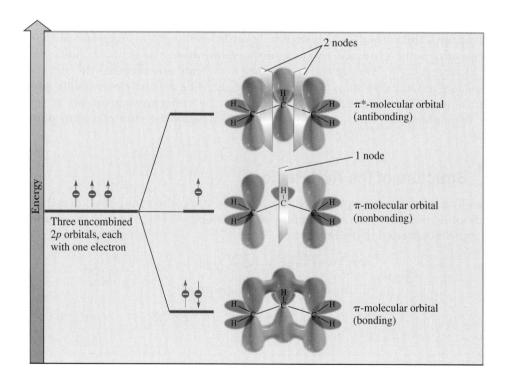

2 nodes

π*-molecular orbital (antibonding)

1 node

π-molecular orbital (nonbonding)

π-molecular orbital (bonding)

Three uncombined $2p$ orbitals, each with one electron

Energy

nor destabilization and is, therefore, called a nonbonding MO. The lowest pi MO is at a lower energy than the isolated $2p$ orbitals.

In the lowest energy (ground) state of the allyl radical, two electrons of the pi system lie in the pi-bonding MO, and the third lies in the pi-nonbonding MO; the pi-antibonding MO is unoccupied. Because the lowest pi MO is at a lower energy than the isolated $2p$ atomic orbitals, putting two electrons in this MO releases considerable energy, which accounts for the stability of the allyl radical. The lone electron of the allyl radical is associated with the pi nonbonding MO, which places electron density on carbons 1 and 3 only. Thus both the resonance model and molecular orbital theory are equivalent in predicting radical character on carbons 1 and 3 of the allyl radical but no radical character on carbon 2.

Example 7.5

Account for the fact that allylic bromination of 1-octene by NBS gives these products.

$$CH_3(CH_2)_4CH_2CH{=}CH_2 \xrightarrow[\text{CCl}_4]{\text{NBS}} CH_3(CH_2)_4\overset{\overset{\text{Br}}{|}}{C}HCH{=}CH_2 + CH_3(CH_2)_4CH{=}CHCH_2Br$$

1-Octene	3-Bromo-1-octene	1-Bromo-2-octene
	(17%)	(83%)

Solution

The rate-determining step of this radical chain mechanism is hydrogen abstraction from the allylic position on 1-octene to give a 2° allylic radical. This radical is stabilized by delocalization of the two pi electrons and the unpaired electron as shown in these contributing structures. Reaction of the radical at carbon 1 gives the major product. Reaction of the radical at carbon 3 gives the minor product. The more substituted (and more stable) alkene isomer predominates.

$$CH_3(CH_2)_4\overset{\overset{\text{H}}{|}}{C}HCH{=}CH_2 \xrightarrow{\cdot Br} CH_3(CH_2)_4\overset{\cdot}{C}H{-}CH{=}CH_2 \longleftrightarrow CH_3(CH_2)_4CH{=}CH{-}\overset{\cdot}{C}H_2 + HBr$$

1-Octene

$$\downarrow Br_2 \qquad\qquad \downarrow Br_2$$

$$CH_3(CH_2)_4\overset{\overset{\text{Br}}{|}}{C}H{-}CH{=}CH_2 \quad + \quad CH_3(CH_2)_4CH{=}CH{-}\overset{\overset{\text{Br}}{|}}{C}H_2 + \cdot Br$$

3-Bromo-1-octene	1-Bromo-2-octene
	(major product)

Problem 7.5

Given the solution to Example 7.5, predict the structure of the product(s) formed when 3-hexene is treated with NBS.

C H E M I S T R Y   I N   A C T I O N

Radical Autoxidation

One of the most important reactions for materials, foods, and also for living systems is called **autoxidation,** that is, oxidation requiring oxygen and no other reactant. This reaction takes place by a radical chain mechanism very similar to that for allylic bromination. If you open a bottle of cooking oil that has stood for a long time, you will notice the hiss of air entering the bottle. This is because there is a negative pressure caused by the consumption of oxygen by autoxidation of the oil.

Cooking oil contains polyunsaturated fatty acid esters (Section 26.1). The most common of these compounds have 16 or 18 carbon chains containing 1,4-diene functional groups, as shown in the following structure. (Both double bonds are cis; the nature of R_1 and R_2 need not concern us at this stage.) The hydrogens on the CH_2 group between the double bonds are doubly allylic; that is, they are allylic with respect to two different double bonds. As you might expect, the radical formed by abstraction of one of these hydrogens is unusually stable because it is even more delocalized (stabilized by resonance) than an allylic radical. An allylic C—H bond is much weaker than a corresponding alkane C—H bond; the doubly allylic C—H is even weaker.

Autoxidation begins when a radical initiator, $X \cdot$, formed either by light activation of an impurity in the oil or by thermal decomposition of peroxide impurities, abstracts a doubly allylic hydrogen to form a radical. This radical is stabilized by resonance with both double bonds (1 and 2 in the following structure).

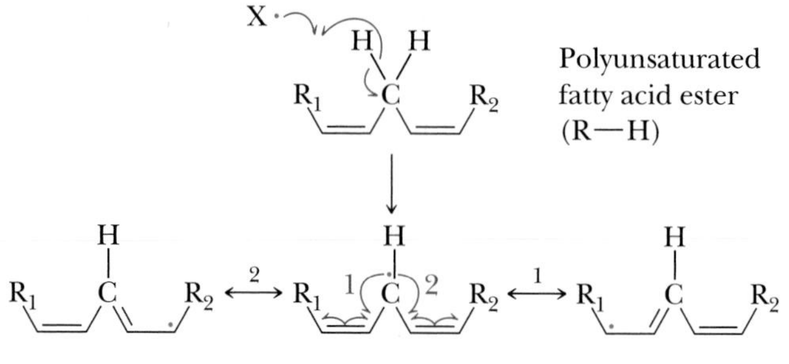

This radical reacts with oxygen, itself a diradical, to form a peroxy radical, which then reacts with another 1,4-diene fatty acid ester (R—H) to give a new radical (R·) and a hydroperoxide. Hydroperoxides are formed on both sides by reactions with the resonance hybrid; only one is shown. The new radical reacts again with oxygen, causing a radical chain reaction in which hundreds of molecules of fatty acid ester are oxidized for each initiator radical.

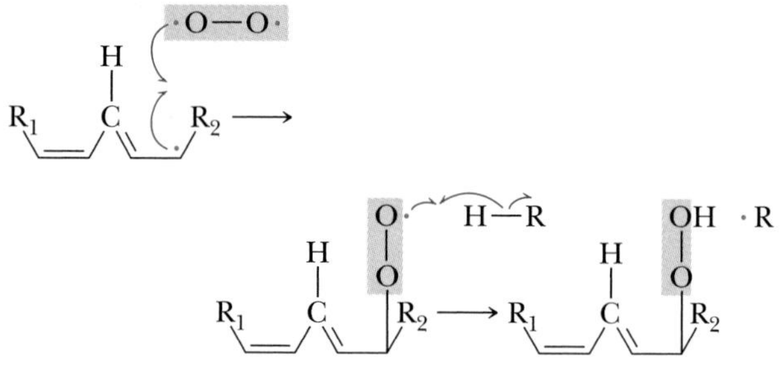

Peroxy radical

The ultimate fate of the peroxide, and some of the peroxy radical as well, is complex. Some autoxidation products degrade to short-chain aldehydes and carboxylic acids with unpleasant "rancid" smells familiar to anyone who has smelled old cooking oil or aged foods that contain polyunsaturated oils. It has been suggested that some products of autoxidation of oils are toxic and/or carcinogenic. Oils lacking the 1,4-diene structure are much less easily oxidized. Similar oxidative degradation of materials in low-density lipoproteins (LDL, Section 26.4A) deposited on the walls of arteries leads to cardiovascular disease in humans. Many effects of aging in humans and damage to materials such as rubber and plastic occur by similar mechanisms.

Many plants contain polyunsaturated fatty acid esters in their leaves or seeds. In the natural state, they are protected against autoxidation by a variety of agents. One of the most important of these agents is α-tocopherol (vitamin E, Section 26.6C). This compound is a phenol (Section 20.5). The characteristic of phenols that makes them effective as protective agents against autoxidation is their O—H bond, which is even weaker than the doubly allylic C—H bond. Vitamin E reacts preferentially with the initial peroxy radical to give a resonance-stabilized phenoxy radical, which is very unreactive and survives to scavenge another peroxy radical, ROO·. The resulting peroxide is stable under the conditions. The antioxidant thus removes two molecules of peroxy radical and stops the radical chain oxidation dead in its tracks.

Because vitamin E is removed in the processing of many food products, similar phenols such as BHT (Problem 21.23) are often added to foods to prevent spoilage by autoxidation. These compounds are all

Vitamin E

A phenoxy radical

a peroxide group

A peroxide derived from vitamin E

called **radical inhibitors** because of their ability to terminate radical chains. Similar compounds are also added to many materials, such as plastics and rubber, to protect them against autoxidation.

C. Walling, "Autoxidation," *Active Oxygen in Chemistry*, C. S. Foote, J. S. Valentine, A. Greenberg, and J. F. Liebman, Eds.; Chapman and Hall, London, 1995; pp. 24–65.

Summary

Haloalkanes contain a halogen covalently bonded to an sp^3 hybridized carbon (Section 7.1). In the IUPAC system, halogen atoms are named fluoro-, chloro-, bromo-, and iodo- substituents and are listed in alphabetical order with other substituents (Section 7.2A). In the common system, they are named alkyl halides. **Haloalkenes** contain a halogen covalently bonded to an sp^2 hybridized carbon of an alkene. In the common system, they are named alkenyl or vinylic halides. **Haloarenes** contain a halogen atom bonded to a benzene ring.

The **van der Waals radius** of fluorine is only slightly greater than that of hydrogen, and, among the other halogens, only iodine has a larger van der Waals radius than methyl.

Among alkanes and chloro-, bromo-, and iodoalkanes of comparable size and shape, the haloalkanes have the higher boiling points because of the greater polarizability of the unshared electrons of the halogen atom (Section 7.3B). Boiling points of fluoroalkanes are generally lower than those of alkanes of comparable size and shape because of the uniquely low polarizability of the valence electrons of fluorine. The density of liquid haloalkanes is greater than that of hydrocarbons of comparable molecular weight because of the halogen's larger mass-to-volume ratio (Section 7.3C).

A **radical chain reaction** consists of three types of steps: chain initiation, chain propagation, and chain termination (Section 7.5B). In **chain initiation,** radicals are formed from nonradical compounds. In a **chain propagation** step, a radical and a molecule react to give a new radical. When summed, chain propagation steps give the observed stoichiometry of the reaction. **Chain length** is the number of times a cycle of chain propagation steps repeats. In a **chain termination** step, radicals are destroyed. Simple alkyl radicals are planar or almost so with bond angles of 120° about the carbon with the unpaired electron. Heats of reaction for a radical reaction and for individual chain initiation, propagation, and termination steps can be calculated from bond dissociation energies.

According to **Hammond's postulate** (Section 7.5D), the structure of the transition state of an exothermic reaction step looks more like the reactants of that step than like the products. Conversely, the structure of the transition state of an endothermic reaction step looks more like the products of that step than like the reactants. Hammond's postulate accounts for the fact that bromination of an alkane shows greater regioselectivity than chlorination. For both bromination and chlorination of alkanes, the rate-determining step is hydrogen abstraction to form an alkyl radical. Hydrogen abstraction is endothermic for bromination and exothermic for chlorination.

Allylic substitution is any reaction in which an atom or group of atoms is substituted for another atom or group of atoms at a carbon adjacent to a carbon-carbon double bond (Section 7.6). **Allylic halogenation** proceeds by a radical chain mechanism. Because of delocalization of electrons, the allyl radical is more stable than the *tert*-butyl radical.

Key Reactions

1. Chlorination and Bromination of Alkanes (Section 7.4)

Chlorination and bromination of alkanes are regioselective in the order 3° H > 2° H > 1° H. Bromination has a higher regioselectivity than chlorination.

$$CH_3CH_2CH_3 + Br_2 \xrightarrow[\text{or light}]{\text{heat}} CH_3\overset{\overset{\displaystyle Br}{|}}{C}HCH_3 + CH_3CH_2CH_2{-}Br$$

$$92\% \qquad\qquad 8\%$$

2. Allylic Bromination and Chlorination (Section 7.6)

These reactions occur at high temperatures (heat is the radical initiator) using the halogens themselves. Bromination occurs using *N*-bromosuccinimide (NBS) at slightly above room temperature.

Problems

Nomenclature

7.6 Give IUPAC names for the following compounds. Where stereochemistry is shown, include a designation of configuration in your answer.

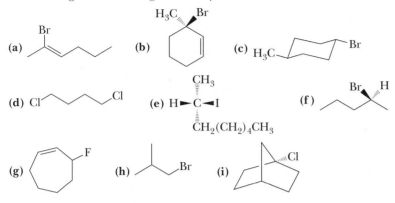

7.7 Draw structural formulas for the following compounds.

(a) 3-Iodo-1-propene   (b) (*R*)-2-Chlorobutane

(c) *meso*-2,3-Dibromobutane   (d) *trans*-1-Bromo-3-isopropylcyclohexane

(e) 1-Iodo-2,2-dimethylpropane   (f) Bromocyclobutane

Physical Properties

7.8 Water and methylene chloride are insoluble in each other. When each is added to a test tube, two layers form. Which layer is water and which is methylene chloride?

7.9 The boiling point of methylcyclohexane (C_7H_{14}, MW 98.2) is 101°C. The boiling point of perfluoromethylcyclohexane (C_7F_{14}, MW 350) is 76°C. Account for the fact that although the molecular weight of perfluoromethylcyclohexane is over three times that of methylcyclohexane, its boiling point is lower than that of methylcyclohexane.

7.10 Account for the fact that, among the chlorinated derivatives of methane, chloromethane has the largest dipole moment and tetrachloromethane has the smallest dipole moment.

Name	Molecular Formula	Dipole Moment (debyes, D)
Chloromethane	CH_3Cl	1.87
Dichloromethane	CH_2Cl_2	1.60
Trichloromethane	$CHCl_3$	1.01
Tetrachloromethane	CCl_4	0

Halogenation of Alkanes

7.11 Name and draw structural formulas for all possible monohalogenation products that might be formed in the following reactions.

(a) ⬠ + Cl_2 $\xrightarrow{\text{light}}$ (b) ⟍⟋⟍⟋ + Cl_2 $\xrightarrow{\text{light}}$ (c) ⟍⟋⟍ + Br_2 $\xrightarrow{\text{light}}$

7.12 Which compounds can be prepared in high yield by regioselective halogenation of an alkane?

(a) 2-Chloropentane   (b) Chlorocyclopentane
(c) 2-Bromo-2-methylheptane   (d) 2-Bromo-3-methylbutane
(e) 2-Bromo-2,4,4-trimethylpentane   (f) Iodoethane

7.13 There are three constitutional isomers of molecular formula C_5H_{12}. When treated with chlorine at 300°C, isomer A gives a mixture of four monochlorination products. Under the same conditions, isomer B gives a mixture of three monochlorination products, and isomer C gives only one monochlorination product. From this information, assign structural formulas to isomers A, B, and C.

7.14 Following is a balanced equation for bromination of propane.

$$CH_3CH_2CH_3 + Br_2 \longrightarrow CH_3\overset{\overset{\displaystyle Br}{|}}{C}HCH_3 + HBr$$

(a) Using the values for bond dissociation energies given in Appendix 3, calculate ΔH^0 for this reaction.
(b) Propose a pair of chain propagation steps, and show that they add up to the observed reaction.
(c) Calculate ΔH^0 for each chain propagation step.
(d) Which propagation step is rate determining?

7.15 Write a balanced equation and calculate ΔH^0 for reaction of CH_4 and I_2 to give CH_3I and HI. Explain why this reaction cannot be used as a method of preparation of iodomethane.

7.16 Following are balanced equations for fluorination of propane to produce a mixture of 1-fluoropropane and 2-fluoropropane.

$$CH_3CH_2CH_3 + F_2 \longrightarrow CH_3CH_2CH_2F + HF$$

Propane   1-Fluoropropane

$$CH_3CH_2CH_3 + F_2 \longrightarrow CH_3\overset{\overset{\displaystyle F}{|}}{C}HCH_3 + HF$$

Propane   2-Fluoropropane

Assume that each product is formed by a radical chain mechanism.

(a) Calculate ΔH^0 for each reaction.
(b) Propose a pair of chain propagation steps for each reaction, and calculate ΔH^0 for each step.
(c) Reasoning from the Hammond postulate, predict the regioselectivity of radical fluorination relative to that of radical chlorination and bromination.

7.17 As you demonstrated in Problem 7.16, radical fluorination of alkanes is highly exothermic. As per Hammond's postulate, assume that the transition state for radical fluorination is almost identical to the starting material. With this assumption, estimate the fraction of each monofluoro product formed in the fluorination of 2-methylbutane.

7.18 Cyclobutane reacts with bromine to give bromocyclobutane, but bicyclo[1.1.0]butane reacts with bromine to give 1,3-dibromocyclobutane. Account for the differences between the reactions of these two compounds.

Cyclobutane   Bromocyclobutane

Bicyclo[1.1.0]butane   1,3-Dibromocyclobutane

7.19 The first chain propagation step of all radical halogenation reactions we considered in Section 7.5B is abstraction of hydrogen by the halogen atom to give an alkyl radical and HX, as for example

$$CH_3CH_3 + \cdot Br \longrightarrow CH_3CH_2 \cdot + HBr$$

Suppose, instead, that radical halogenation occurs by an alternative pair of chain propagation steps, beginning with this step:

$$CH_3CH_3 + \cdot Br \longrightarrow CH_3CH_2Br + H \cdot$$

(a) Propose a second chain propagation step. Remember that a characteristic of chain propagation steps is that they add to the observed reaction.
(b) Calculate the heat of reaction, ΔH^0, for each propagation step.
(c) Compare the energetics and relative rates of the set of chain propagation steps in Section 7.5B with the set proposed here.

Allylic Halogenation

7.20 Following is a balanced equation for the allylic bromination of propene.

$$CH_2=CHCH_3 + Br_2 \longrightarrow CH_2=CHCH_2Br + HBr$$

(a) Calculate the heat of reaction, ΔH^0, for this conversion.
(b) Propose a pair of chain propagation steps, and show that they add up to the observed stoichiometry.
(c) Calculate the ΔH^0 for each chain propagation step, and show that they add up to the observed ΔH^0 for the overall reaction.

7.21 Using the table of bond dissociation energies (Appendix 3), estimate the bond dissociation energy of each indicated bond in cyclohexene.

7.22 Propose a series of chain initiation, propagation, and termination steps for this reaction, and estimate its heat of reaction.

7.23 The major product formed when methylenecyclohexane is treated with NBS in carbon tetrachloride is 1-bromomethylcyclohexene. Account for the formation of this product.

7.24 Draw the structural formula of the products formed when each alkene is treated with one equivalent of NBS in CCl_4 in the presence of benzoyl peroxide.

(a) $CH_3CH=CHCH_2CH_3$

(b)

(c)

7.25 The activation energy for hydrogen abstraction from ethane by a chlorine atom is 4.2 kJ/mol; that for hydrogen abstraction by a bromine atom is 55 kJ/mol. Calculate the ratio of rate constants, k_{Cl}/k_{Br}, for these two reactions.

7.26 Calculate the ΔH^0 for the following reaction step. What can you say regarding the possibility of bromination at a vinylic hydrogen?

$$CH_2=CH_2 + Br\cdot \longrightarrow CH_2=CH\cdot + HBr$$

Synthesis

7.27 Show reagents and conditions to bring about these conversions.

(a)

(b) $CH_3CH=CHCH_3 \longrightarrow CH_3CH=CHCH_2Br$

(c) $CH_3CH=CHCH_3 \longrightarrow CH_3CH-CHCH_3$
$\underset{Br}{|}\underset{Br}{|}$

(d)

(e)

NUCLEOPHILIC SUBSTITUTION AND β-ELIMINATION

Nucleophilic substitution is any reaction in which one **nucleophile** is substituted for another as illustrated by the following general equation in which Nu:⁻ is the nucleophile and X is the leaving group. Nucleophilic substitution is an acid-base reaction in which the nucleophile acts as a Lewis base and the substrate, the electrophile, acts as a Lewis acid. By far the most carefully studied chemical behavior of

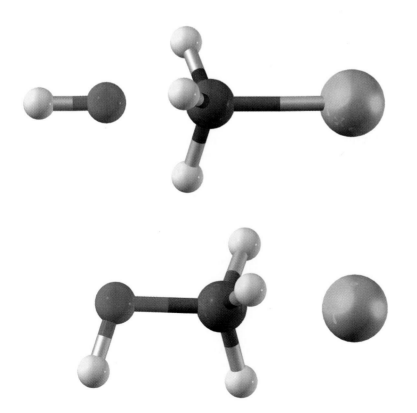

■ Hydroxide ion reacts with bromomethane (upper models) to give methanol and bromide ion (lower models) by an S_N2 mechanism (Section 8.3).

269

haloalkanes is their reaction with nucleophilic reagents, such as HO^-, $CH_3CH_2O^-$, and NH_3.

$$Nu\!:^- \ + \ -\overset{|}{\underset{|}{C}}-X \xrightarrow{\text{nucleophilic substitution}} -\overset{|}{\underset{|}{C}}-Nu \ + \ :X^-$$

Nucleophile   leaving group

β-Elimination A reaction in which a small molecule, such as HCl, HBr, HI, or HOH, is split out or eliminated from adjacent carbons.

Because all nucleophiles are also bases, nucleophilic substitution and base-promoted **β-elimination** are competing reactions. The ethoxide ion, for example, is both a nucleophile and a base. With bromocyclohexane, it reacts as a nucleophile to give ethoxycyclohexane (cyclohexyl ethyl ether) and as a base to give cyclohexene and ethanol.

As a nucleophile, ethoxide ion attacks this carbon.

As a base, ethoxide ion attacks this hydrogen.

In this chapter, we study these two types of organic reactions. Using them, haloalkanes can be converted to compounds with other functional groups including alcohols, ethers, thiols, sulfides, amines, nitriles, alkenes, and alkynes. Thus, an understanding of nucleophilic substitution and β-elimination opens entirely new areas of organic chemistry.

8.1 Nucleophilic Aliphatic Substitution

Nucleophile From the Greek, meaning nucleus loving. A molecule or ion that donates a pair of electrons to another atom or ion to form a new covalent bond; a Lewis base.

Nucleophilic substitution is one of the most important reactions of haloalkanes and can lead to a wide variety of new functional groups, many of which are illustrated in Table 8.1. As you study the entries in this table, note these points.

1. If the nucleophile is negatively charged, as for example OH^- and $HC \equiv C^-$, then the atom donating the pair of electrons in a substitution reaction becomes neutral in the product.
2. If the nucleophile is uncharged, as for example NH_3 and CH_3OH, then the atom donating the pair of electrons in the substitution reaction becomes positively charged in the product.

Example 8.1

Complete these nucleophilic substitution reactions. In each reaction, show the electron pair on the nucleophile that forms the new bond to carbon.

(a) $\diagdown\diagup\diagdown Br + CH_3O^-Na^+ \xrightarrow[\text{methanol}]{}$ (b) $\diagdown\diagup\diagdown Cl + NH_3 \xrightarrow[\text{ethanol}]{}$

Table 8.1 Some Nucleophilic Substitution Reactions*

Reaction: $Nu:^- + CH_3Br \longrightarrow CH_3Nu + :Br^-$

Nucleophile	Product	Class of Compound Formed
$HO:^- \longrightarrow$	CH_3OH	An alcohol
$RO:^- \longrightarrow$	CH_3OR	An ether
$HS:^- \longrightarrow$	CH_3SH	A thiol (a mercaptan)
$RS:^- \longrightarrow$	CH_3SR	A sulfide (a thioether)
$HC\equiv C:^- \longrightarrow$	$CH_3C\equiv CH$	An alkyne
$N\equiv C:^- \longrightarrow$	$CH_3C\equiv N$	A nitrile
$I:^- \longrightarrow$	CH_3I	An alkyl iodide
$:\ddot{N}=\overset{+}{N}=\ddot{N}:^- \longrightarrow$	$CH_3-\ddot{N}=\overset{+}{N}=\ddot{N}:^-$	An alkyl azide
$:NH_3 \longrightarrow$	$CH_3NH_3{}^+$	An alkylammonium ion
$:O-H$ with $H \longrightarrow$	$CH_3\overset{+}{O}-H$ with H	An alcohol (after proton transfer)
$:O-CH_3$ with $H \longrightarrow$	$CH_3\overset{+}{O}-CH_3$ with H	An ether (after proton transfer)

*Except for azide ion, the only electron pair shown in each nucleophile is the one that forms the new bond to carbon in the substitution reaction.

Solution

(a) Methoxide ion is the nucleophile, and bromine is the leaving group.

$$\text{(1-Bromobutane)}\;\diagup\!\!\diagup\!\!\diagdown\!\!Br + CH_3O:^-Na^+ \xrightarrow[\text{methanol}]{} \diagup\!\!\diagup\!\!\diagdown\!\!OCH_3 + Na^+Br^-$$

1-Bromobutane   Sodium methoxide   1-Methoxybutane (Butyl methyl ether)   Sodium bromide

(b) Ammonia is the nucleophile, and chlorine is the leaving group.

$$\diagup\!\!\diagup\!\!\diagdown\!\!Cl + :NH_3 \xrightarrow[\text{ethanol}]{} \diagup\!\!\diagup\!\!\diagdown\!\!NH_3{}^+ Cl^-$$

1-Chlorobutane   Ammonia   Butylammonium chloride

Problem 8.1

Complete the following nucleophilic substitution reactions. In each, show the electron pair on the nucleophile that forms the new bond to carbon.

(a) [cyclohexyl-Cl] $+ CH_3\overset{O}{\overset{\|}{C}}O^-Na^+ \xrightarrow[\text{ethanol}]{}$

(b) [isopropyl with I] $+$ [S^-Na^+] $\xrightarrow[\text{acetone}]{}$

(c) [branched chain with Br] $+ H_3C-C\equiv C^-Li^+ \xrightarrow[\text{dimethyl sulfoxide}]{}$

Table 8.2 **Common Protic Solvents**

Solvent	Structure	Dielectric Constant (25°C)
Water	H_2O	79
Formic acid	HCOOH	59
Methanol	CH_3OH	33
Ethanol	CH_3CH_2OH	24
Acetic acid	CH_3COOH	6

8.2 Solvents for Nucleophilic Substitution Reactions

Solvents provide a medium in which the reactants are dissolved and in which nucleophilic substitution takes place. Common solvents for these reactions can be divided into two groups: **protic** and **aprotic.**

Furthermore, solvents are classified as polar and nonpolar based on their **dielectric constant.** The greater the value of the dielectric constant of a solvent, the smaller the interaction between ions of opposite charge dissolved in it. As an arbitrary guideline, we say that a solvent is **polar** if it has a dielectric constant of 15 or greater. A solvent is **nonpolar** if it has a dielectric constant of less than 15.

The common protic solvents for nucleophilic substitution reactions are water, low-molecular-weight alcohols, and low-molecular-weight carboxylic acids (Table 8.2). Each is able to solvate ionic substances by electrostatic interactions between its partially negatively charged oxygen(s) and cations and between its partially positively

Protic solvent A solvent that is a hydrogen-bond donor; the most common protic solvents contain —OH groups. Common protic solvents are water and low-molecular-weight alcohols such as ethanol.

Aprotic solvent A solvent that cannot serve as a hydrogen-bond donor; nowhere in the molecule is a hydrogen bonded to an atom of high electronegativity. Common aprotic solvents are dichloromethane, diethyl ether, and dimethyl sulfoxide.

Dielectric constant A measure of a solvent's ability to insulate opposite charges from one another.

Table 8.3 **Common Aprotic Solvents**

Solvent	Structure	Dielectric Constant
Polar		
Dimethyl sulfoxide (DMSO)	$CH_3\overset{\overset{\displaystyle O}{\|}}{S}CH_3$	48.9
Acetonitrile	$CH_3C{\equiv}N$	37.5
N,N-Dimethylformamide (DMF)	$H\overset{\overset{\displaystyle O}{\|}}{C}N(CH_3)_2$	36.7
Acetone	$CH_3\overset{\overset{\displaystyle O}{\|}}{C}CH_3$	20.7
Nonpolar		
Dichloromethane	CH_2Cl_2	9.1
Diethyl ether	$CH_3CH_2OCH_2CH_3$	4.3
Toluene	⬡—CH_3	2.3
Hexane	$CH_3(CH_2)_4CH_3$	1.9

Increasing solvent polarity →

charged hydrogen and anions. By our arbitrary guideline, water, formic acid, methanol, and ethanol are classified as polar protic solvents. Because of its smaller dielectric constant, acetic acid is classified as a nonpolar protic solvent.

The aprotic solvents most commonly used for nucleophilic substitution reactions are given in Table 8.3. Of these, dimethyl sulfoxide (DMSO), acetonitrile, *N,N*-dimethylformamide (DMF), and acetone are classified as polar aprotic solvents. Dichloromethane, diethyl ether, toluene, and hexane are classified as nonpolar aprotic solvents.

8.3 Mechanisms of Nucleophilic Aliphatic Substitution

Supporting Concepts;
Nucleophilic Substitution
Reactions

On the basis of a wealth of experimental observations developed over a 70-year period, two limiting mechanisms for nucleophilic substitutions have been proposed. A fundamental difference between them is the timing of bond breaking between carbon and the leaving group and of bond forming between carbon and the nucleophile. At one extreme, the two processes are concerted, meaning that bond breaking and bond forming occur simultaneously. Thus departure of the leaving group is assisted by the incoming nucleophile. This mechanism is designated **S_N2.** Here S stands for *Substitution,* N for *Nucleophilic,* and 2 for ***bi*molecular reaction.** This type of substitution reaction is classified as bimolecular because both the haloalkane and the nucleophile are involved in the rate-determining step.

Following is an S_N2 mechanism for the reaction of hydroxide ion and bromomethane to form methanol and bromide ion. We will present evidence in the Section 8.4C to show that backside attack is a necessary characteristic of all S_N2 reactions.

S_N2 reaction A bimolecular nucleophilic substitution reaction.
Bimolecular reaction A reaction in which two species are involved in the rate-determining step.

Mechanism An S_N2 Reaction

Supporting Concepts;
Nucleophilic Substitution
Reactions: S_N2 Reaction

The nucleophile attacks the reactive center from the side opposite the leaving group; that is, an S_N2 reaction involves backside attack of the nucleophile.

$$HO:^- + \quad \begin{array}{c} H \\ | \\ C-Br \\ H \end{array} \longrightarrow \left[\begin{array}{c} H \\ \delta- \; | \; \delta- \\ HO\text{---}C\text{---}Br \\ H \; H \end{array} \right] \longrightarrow \begin{array}{c} H \\ | \\ HO-C \\ H \; H \end{array} + :Br^-$$

Transition state with simultaneous
bond breaking and bond forming

Figure 8.1 shows an energy diagram for an S_N2 reaction.

In the other limiting mechanism, called S_N1, bond breaking between carbon and the leaving group is entirely completed before bond forming with the nucleophile begins. In the designation **S_N1,** S stands for *Substitution,* N for *Nucleophilic,* and 1 for ***uni*molecular reaction.** This type of substitution is classified as unimolecular because only the haloalkane is involved in the rate-determining step.

S_N1 reaction A unimolecular nucleophilic substitution reaction.
Unimolecular reaction A reaction in which only one species is involved in the rate-determining step.

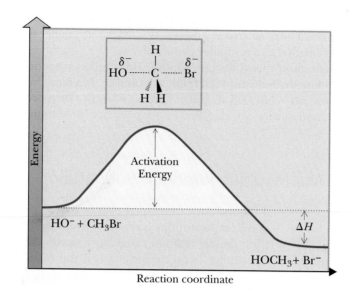

Figure 8.1
An energy diagram for an S_N2 reaction. There is one transition state and no reactive intermediate.

Solvolysis A nucleophilic substitution in which the solvent is also the nucleophile.

An S_N1 mechanism is illustrated by the **solvolysis** reaction between 2-bromo-2-methylpropane (*tert*-butyl bromide) and methanol to form 2-methoxy-2-methylpropane (*tert*-butyl methyl ether) and HBr. The last step in this three-step mechanism is an acid-base reaction following the S_N1 reaction.

Supporting Concepts;
Nucleophilic Substitution
Reactions: S_N1 Reaction

Mechanism   An S_N1 Reaction

Step 1:   Ionization of the C—X bond forms a carbocation intermediate.

$$\underset{\text{A carbocation intermediate;}}{\underset{\text{its shape is trigonal planar}}{}}$$

Step 2:   Reaction of the carbocation intermediate (a Lewis acid and electrophile) with methanol (a Lewis base and nucleophile) gives an oxonium ion. Attack of the nucleophile occurs with equal probability from either face of the planar carbocation intermediate.

Nucleophile   Lewis acid
(Lewis base)

Step 3: Proton transfer from the oxonium ion to methanol completes the reaction and gives *tert*-butyl methyl ether.

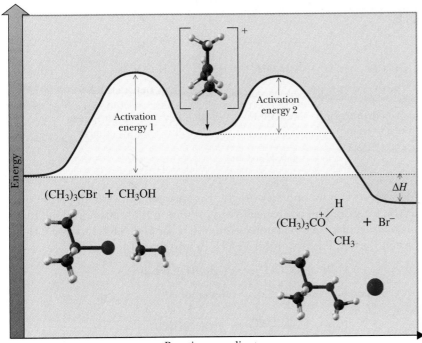

Figure 8.2 shows an energy diagram for the S$_N$1 reaction of 2-bromo-2-methylpropane with methanol.

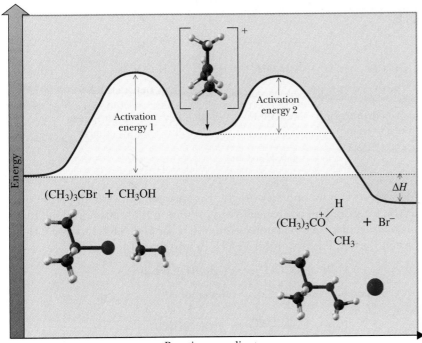

Figure 8.2
An energy diagram for an S$_N$1 reaction. There is one transition state leading to formation of the carbocation intermediate in Step 1 and a second transition state for the reaction of the carbocation intermediate with methanol in Step 2 to give the oxonium ion. Step 1 crosses the higher energy barrier and, therefore, is rate determining.

8.4 Experimental Evidence for S$_N$1 and S$_N$2 Mechanisms

Let us now examine the experimental evidence on which these two contrasting mechanisms are based. As we do, we will consider the following questions.

1. What effect does the structure of the nucleophile have on the rate of reaction?
2. What is the stereochemical outcome of nucleophilic substitution when the leaving group is displaced from a stereocenter?

3. What effect does the structure of the haloalkane have on the rate of reaction?
4. What effect does the structure of the leaving group have on the rate of reaction?
5. What is the role of the solvent?
6. Under what conditions are skeletal rearrangements observed?

A.   Kinetics

The kinetic order of nucleophilic substitutions can be studied by measuring the effect on rate of varying the concentrations of haloalkane and nucleophile. Those reactions whose rate is dependent only on the concentration of haloalkane are classified as S_N1; those reactions whose rate is dependent on the concentration of both haloalkane and nucleophile are classified as S_N2.

Because the transition state for formation of the carbocation intermediate in an S_N1 mechanism involves only the haloalkane and not the nucleophile, it is a unimolecular process. The result is a first-order reaction. In this instance, the rate of reaction is expressed as the rate of disappearance of the starting material, 2-bromo-2-methylpropane.

$$
\begin{array}{ccccccc}
& \text{CH}_3 & & & & \text{CH}_3 & \\
& | & & & & | & \\
\text{CH}_3\text{CBr} & & + & \text{CH}_3\text{OH} & \longrightarrow & \text{CH}_3\text{COCH}_3 & + & \text{HBr} \\
& | & & & & | & \\
& \text{CH}_3 & & & & \text{CH}_3 &
\end{array}
$$

2-Bromo-2-methylpropane	Methanol	2-Methoxy-2-methylpropane
(*tert*-Butyl bromide)		(*tert*-Butyl methyl ether)

$$
\text{Rate} = -\frac{d[(\text{CH}_3)_3\text{CBr}]}{dt} = k[(\text{CH}_3)_3\text{CBr}]
$$

By contrast, there is only one step in the S_N2 mechanism. For the reaction of OH^- and CH_3Br, for example, both species are present in the transition state; that is, the reaction is bimolecular. The reaction between CH_3Br and $NaOH$ to give CH_3OH and $NaBr$ is second order; it is first order in CH_3Br and first order in OH^-.

$$
\text{CH}_3\text{Br} + \text{Na}^+\text{OH}^- \longrightarrow \text{CH}_3\text{OH} + \text{Na}^+\text{Br}^-
$$

Bromomethane   Methanol

$$
\text{Rate} = -\frac{d[\text{CH}_3\text{Br}]}{dt} = k[\text{CH}_3\text{Br}][\text{OH}^-]
$$

B.   Structure of the Nucleophile

Nucleophilicity is a kinetic property and is measured by relative rates of reaction. Relative nucleophilicities for a series of nucleophiles are established by measuring the rate at which each displaces a leaving group from an haloalkane, for example, the rate at which each displaces bromide ion from ethyl bromide in ethanol at 25°C.

$$
\text{CH}_3\text{CH}_2\text{Br} + \text{NH}_3 \longrightarrow \text{CH}_3\text{CH}_2\overset{+}{\text{N}}\text{H}_3 + \text{Br}^-
$$

From these studies, we can then make correlations between the structure of a nucleophile and its relative nucleophilicity. Listed in Table 8.4 are the types of nucleophiles we deal with most commonly in this text.

Nucleophilicity A kinetic property measured by the rate at which a nucleophile causes nucleophilic substitution on a reference compound under a standardized set of experimental conditions.

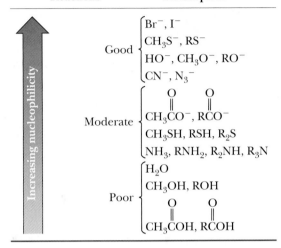

Table 8.4 **Common Nucleophiles and Their Relative Nucleophilicities**

Because all nucleophiles are bases as well (see Chapter 4), we also study correlations between nucleophilicity and basicity. **Basicity** also has a strong correlation with leaving group ability, which we shall see in Section 8.4F.

The solvent in which nucleophilic substitution is carried out has a marked effect on relative nucleophilicities. For a fuller understanding of the role of the solvent, let us consider nucleophilic substitution reactions carried out in polar aprotic solvents and in polar protic solvents. An organizing principle for substitution reactions is the following: all other factors being equal, the freer the nucleophile, the greater its nucleophilicity. Conversely, the greater the interaction of the nucleophile with the solvent, the lower its nucleophilicity.

Basicity An equilibrium property measured by the position of equilibrium in an acid-base reaction, as for example the acid-base reaction between ammonia and water.

Nucleophilicity in Polar Aprotic Solvents

The most commonly used polar aprotic solvents (DMSO, acetone, acetonitrile, and DMF) are very effective in solvating cations (in addition to the attraction to the negative end of the dipole, the lone pairs on oxygen and nitrogen act as Lewis bases) but are not nearly as effective in solvating anions. Consider, for example, acetone. Because the negative end of its dipole and the lone pairs on oxygen can come close to the center of positive charge in a cation, acetone is effective in solvating cations. The positive end of its dipole, however, is shielded by surrounding groups (two methyls) and is less effective in solvating anions. The sodium ion of sodium iodide, for example, is effectively solvated by acetone and DMSO, but the iodide ion is only poorly solvated. Because anions are only poorly solvated in polar aprotic solvents, they participate readily in nucleophilic substitution reactions. Their relative nucleophilicities parallel their relative basicities. The relative nucleophilicities of halide ions in polar aprotic solvents, for example, are $F^- > Cl^- > Br^- > I^-$.

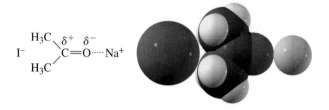

Solvation of NaI in acetone

Nucleophilicity in Polar Protic Solvents

The relative nucleophilicities of halide ions in polar protic solvents are quite different from those in polar aprotic solvents (Table 8.5). In polar protic solvents, iodide ion, the least basic of the halide ions, has the greatest nucleophilicity. Conversely, fluoride ion, the most basic of the halide ions, has the smallest nucleophilicity. The reason for this reversal of correlation between nucleophilicity and basicity lies in the degree of solvation of anions in protic solvents compared with aprotic solvents.

- In polar aprotic solvents, anions are only weakly solvated and, therefore, relatively free to participate in nucleophilic substitution reactions.
- In polar protic solvents, anions are highly solvated by hydrogen bonding with solvent molecules and, therefore, are less able to participate in nucleophilic substitution reactions.

The negative charge on the fluoride ion, the smallest of the halide ions, is concentrated in a small volume, and the very tightly held solvent shell constitutes a barrier between fluoride ion and substrate. The fluoride ion must be at least partially removed from its tightly held solvation shell before it can participate in nucleophilic substitution. The following structure shows a fluoride ion hydrogen bonded by methanol (several solvent molecules are probably involved). The negative charge on the iodide ion, the largest of the halide ions, is far less concentrated, the solvent shell is less tightly held, and iodide is considerably freer to participate in nucleophilic substitution reactions.

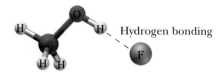

Hydrogen bonding

Table 8.5	Relative Nucleophilicities of Halide Ions in Aprotic and Protic Solvents	
Solvent	Increasing nucleophilicity →	
Polar aprotic	$I^- < Br^- < Cl^- < F^-$	
Polar protic	$F^- < Cl^- < Br^- < I^-$	

Table 8.6	**Relative Nucleophilicities of Atoms Within a Period**
Period	**Increasing nucleophilicity** →
Period 2	$F^- < OH^- < NH_2^- < CH_3^-$
Period 3	$Cl^- < SH^- < PH_2^-$

We can make the following additional generalizations about nucleophilicity:

1. Within a period of the Periodic Table, nucleophilicity increases from right to left (Table 8.6); that is, it increases with basicity.
2. In a series of reagents with the same nucleophilic atom, anionic reagents are stronger nucleophiles than neutral reagents (Table 8.7). This trend also parallels the basicity of the nucleophile.
3. When comparing groups of reagents in which the nucleophilic atom is the same, the stronger the base, the greater the nucleophilicity. The oxygen nucleophiles in Table 8.8 are listed in order of increasing nucleophilicity. Below each, for comparison, is given the formula and pK_a of its conjugate acid. In this series, the carboxylic acid is the strongest acid; consequently, its anion is the weakest base and the poorest nucleophile.

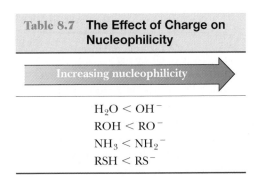

Table 8.7	**The Effect of Charge on Nucleophilicity**
	Increasing nucleophilicity →
	$H_2O < OH^-$
	$ROH < RO^-$
	$NH_3 < NH_2^-$
	$RSH < RS^-$

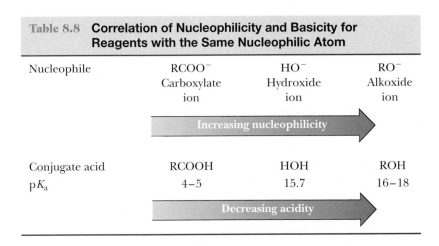

Table 8.8	**Correlation of Nucleophilicity and Basicity for Reagents with the Same Nucleophilic Atom**		
Nucleophile	$RCOO^-$ Carboxylate ion	HO^- Hydroxide ion	RO^- Alkoxide ion
	Increasing nucleophilicity →		
Conjugate acid	RCOOH	HOH	ROH
pK_a	4–5	15.7	16–18
	Decreasing acidity →		

Supporting Concepts;
Nucleophilic Substitution
Reactions: S$_N$1 Reaction,
S$_N$2 Reaction

C.   Stereochemistry

S$_N$1 Reactions

Experiments in which nucleophilic substitution takes place at a stereocenter provide us with information about the stereochemical outcome of the reaction. One of the compounds studied to determine the stereochemistry of an S$_N$1 reaction was the following chloroalkane. When either enantiomer of this molecule undergoes nucleophilic substitution by an S$_N$1 pathway, the product is almost completely racemic. Ionization of this secondary halide forms an achiral carbocation. Attack of the nucleophile can occur on it from either side. Attack from the right gives the *R* enantiomer; attack from the left gives the *S* enantiomer. The *R* and *S* enantiomers are formed in equal amounts, and the product is a racemic mixture.

R enantiomer	Planar carbocation (achiral)	*S* enantiomer	*R* enantiomer

A racemic mixture

The S$_N$1 mechanism as initially described requires complete racemization of any product where the carbon at which substitution takes place is a stereocenter. Although examples of complete racemization have been observed, it is common to find only partial racemization, with the predominant product being the one with inversion of configuration at the stereocenter. Although bond breaking between carbon and the leaving group is complete, the leaving group (chloride ion in this example) remains associated for a short time with the carbocation in an ion pair.

Approach of the nucleophile from this side is less hindered. → C^+ Cl^- ← Approach of the nucleophile from this side is partially blocked by chloride ion, which remains associated with the carbocation as an ion pair.

To the extent that the leaving group remains associated with the carbocation as an ion pair, it hinders approach of the nucleophile (methanol in this example) from that side of the carbocation. The result is that somewhat more than 50% of the product is formed by attack of the nucleophile from the side of the carbocation opposite that of the leaving group.

S$_N$2 Reactions

That every S$_N$2 reaction proceeds with inversion of configuration was shown in an ingenious experiment designed by the English chemists E. D. Hughes and C. K. Ingold. They studied the exchange reaction between enantiomerically pure 2-iodooctane and iodine-131, a radioactive isotope of iodine. Iodine-127, the naturally occurring isotope of iodine, is stable and does not undergo radioactive decay.

Hughes and Ingold first demonstrated that the reaction is second order: first order in 2-iodooctane and first order in iodide ion. Therefore, the reaction proceeds by an S_N2 mechanism. They observed further that the rate of racemization of enantiomerically pure 2-iodooctane is exactly twice the rate of incorporation of iodine-131. This observation must mean, they reasoned, that each displacement of iodine-127 by iodine-131 proceeds with inversion of configuration, as illustrated in the following equation.

(S)-2-Iodooctane   (R)-2-Iodooctane

Substitution with inversion of configuration in one molecule cancels the rotation of one molecule that has not reacted so that, for each molecule undergoing inversion, one racemic pair is formed. Inversion of configuration in 50% of the molecules results in 100% racemization.

Example 8.2

Complete these S_N2 reactions, showing the configuration of each product.

Solution

S_N2 reactions occur with inversion of configuration at the stereocenter. In (a), the starting material is the trans isomer; the product is the cis isomer. In (b), the starting material is the R enantiomer; the product is the S enantiomer.

Problem 8.2

Complete these S_N2 reactions, showing the configuration of each product.

D.   Structure of the Haloalkane

S_N1 reactions are governed mainly by electronic factors, namely the relative stabilities of carbocation intermediates. S_N2 reactions, on the other hand, are governed mainly by steric factors and their transition states are particularly sensitive to **steric hindrance** about the site of reaction.

Relative Stabilities of Carbocations

Supporting Concepts;
Carbocations

As we learned in Section 6.3A, 3° carbocations are the most stable (lowest activation energy for their formation), whereas 1° are the least stable (highest activation energy for their formation). In fact, 1° carbocations are so unstable that they rarely if ever are formed in solution. Therefore, 3° haloalkanes are most likely to react by carbocation formation; 2° haloalkanes are less likely to react by carbocation formation, and methyl and 1° haloalkanes never react in this manner.

Steric Hindrance

To complete a substitution reaction, the nucleophile must approach the substitution center and begin to form a new covalent bond to it. If we compare the ease of approach to the substitution center of a 1° haloalkane with that of a 3° haloalkane, we see that the approach is considerably easier in the case of a primary haloalkane than in that of the tertiary haloalkane. Two hydrogen atoms and one alkyl group screen the backside of the substitution center of a primary haloalkane. On the other hand, three alkyl groups screen the backside of the substitution center of a tertiary haloalkane.

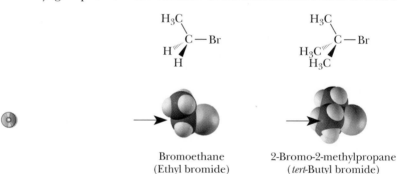

Bromoethane
(Ethyl bromide)

2-Bromo-2-methylpropane
(*tert*-Butyl bromide)

Given the competition between electronic and steric factors, we find that tertiary haloalkanes react by an S_N1 mechanism because 3° carbocation intermediates are particularly stable and 3° alkyl halides are very hindered toward backside attack; tertiary haloalkanes never react by an S_N2 mechanism. Halomethanes and primary haloalkanes have little crowding around the reaction site and react by an S_N2 mechanism; they never react by an S_N1 mechanism because methyl and primary carbocations are so unstable. Secondary haloalkanes may react by either S_N1 or S_N2 mechanisms, depending on the nucleophile and solvent. The competition between electronic and steric factors and their effects on relative rates of nucleophilic substitution reactions of haloalkanes are summarized in Figure 8.3.

We see a similar effect of steric hindrance on S_N2 reactions in molecules with branching at the β-carbon (the carbon bearing the halogen in a haloalkane is called the α-carbon, and the next carbon is the β-carbon). Table 8.9 shows relative rates of S_N2 reactions on a series of primary bromoalkanes. In these data, the rate of nucleophilic substitution of bromoethane is taken as a reference and is given the value 1.0.

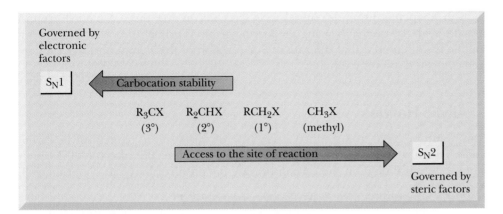

Figure 8.3
Effect of steric and electronic factors in competition between S_N1 and S_N2 reactions of haloalkanes. Methyl and primary haloalkanes react only by the S_N2 mechanism; they do not react by S_N1. Tertiary haloalkanes do not react by S_N2; they only react by S_N1. Secondary haloalkanes may be made to react by either S_N1 or S_N2 mechanisms depending on the solvent and the choice of nucleophile.

Table 8.9 Effect of β-Branches on the Rate of S_N2 Reactions

Alkyl bromide	⟍—Br	⟍⟋—Br	⟍⟋—Br	⟍⟋—Br
β-Branches	0	1	2	3
Relative rate	1.0	4.1×10^{-1}	1.2×10^{-3}	1.2×10^{-5}

As CH_3 branches are added to the β-carbon, the relative rate of reaction decreases. Compare the relative rates of ethyl bromide (no β-branch) with that of 1-bromo-2,2-dimethylpropane (neopentyl bromide), a compound with three β-branches. The rate of S_N2 substitution of this compound is only 10^{-5} that of bromoethane. For all practical purposes, primary halides with three β-branches do not undergo S_N2 reactions.

As shown in Figure 8.4, the carbon of the C—Br bond in bromoethane is unhindered and open to attack by a nucleophile in an S_N2 reaction. The corresponding

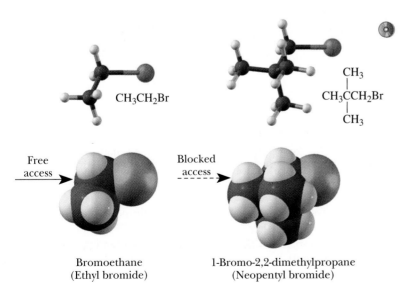

Free
access →

Blocked
access ⇢

Bromoethane
(Ethyl bromide)

1-Bromo-2,2-dimethylpropane
(Neopentyl bromide)

CH_3CH_2Br

CH_3CH_2Br with CH₃ ... CH₃ branches:
$$CH_3\overset{\overset{\displaystyle CH_3}{|}}{\underset{\underset{\displaystyle CH_3}{|}}{C}}CH_2Br$$

Figure 8.4
The effect of β-branching in S_N2 reactions on a primary haloalkane. With bromoethane, attack of the nucleophile is unhindered. With 1-bromo-2,2-dimethylpropane, the three β-branches block approach of the nucleophile to the backside of the C—Br bond, thus drastically reducing the rate of S_N2 reaction of this compound.

carbon in neopentyl bromide, however, is screened by the three β-methyl groups. Thus, although the carbon bearing the leaving group is primary, approach of the nucleophile is so hindered that the rate of S_N2 reaction of this compound is greatly reduced compared to bromoethane.

E. Allylic Halides

Allylic   Next to a carbon-carbon double bond.

Allylic carbocation   Any carbocation in which an allylic carbon bears the positive charge.

At this point, we need to introduce a new type of carbocation, namely **allylic carbocations.** The allyl cation is the simplest allylic carbocation. Because the allyl cation has only one substituent on the carbon bearing the positive charge, it is a primary allylic carbocation.

Allylic carbocations are considerably more stable than comparably substituted alkyl carbocations because of the resonance interaction between the positively charged carbon and the vinyl group. The allyl cation, for example, can be represented as a hybrid of two equivalent contributing structures.

$$CH_2{=}CH{-}\overset{+}{C}H_2 \longleftrightarrow \overset{+}{C}H_2{-}CH{=}CH_2$$

Allyl cation
(a hybrid of two equivalent contributing structures)

Electrostatic potential map for the allyl cation. The positive charge is on carbons 1 and 3.

Because of this delocalization of both the pi electrons of the double bond and the positive charge in the hybrid, the allyl cation is considerably more stable than a primary carbocation. It has been determined experimentally that the presence of one vinyl group provides approximately as much stabilization as two alkyl groups. Thus, the allyl cation and isopropyl cation are of comparable stability.

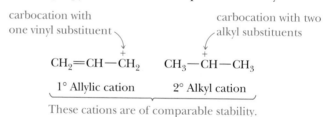

carbocation with
one vinyl substituent

carbocation with two
alkyl substituents

$$CH_2{=}CH{-}\overset{+}{C}H_2 \qquad CH_3{-}\overset{+}{C}H{-}CH_3$$

1° Allylic cation   2° Alkyl cation

These cations are of comparable stability.

The classification of allylic cations as 1°, 2°, and 3° is determined by the location of the positive charge in the more important contributing structure. Following are examples of 2° and 3° allylic carbocations.

$$CH_2{=}CH{-}\overset{+}{C}H{-}CH_3 \qquad CH_2{=}CH{-}\overset{+}{\underset{\underset{\displaystyle CH_3}{|}}{C}}{-}CH_3$$

A 2° allylic carbocation   A 3° allylic carbocation

Benzylic carbocations show approximately the same stability as allylic carbocations. Both are stabilized by resonance delocalization of the positive charge.

Benzyl cation
(a benzylic carbocation)

$$C_6H_5{-}CH_2{}^+$$

The benzyl cation is also written
in this abbreviated form

In Section 6.3, we presented the order of stability of methyl, 1°, 2°, and 3° carbocations. We can now expand this order to include 1°, 2°, and 3° allylic and benzylic carbocations.

$$\text{methyl} < 1° \text{ alkyl} < \left\{\begin{array}{l} 2° \text{ alkyl} \\ 1° \text{ allylic} \\ 1° \text{ benzylic} \end{array}\right\} < \left\{\begin{array}{l} 3° \text{ alkyl} \\ 2° \text{ allylic} \\ 2° \text{ benzylic} \end{array}\right\} < \left\{\begin{array}{l} 3° \text{ allylic} \\ 3° \text{ benzylic} \end{array}\right\}$$

Increasing stability of carbocations →

Example 8.3

Write an additional contributing structure for each carbocation, and state which of the two makes the greater contribution to the resonance hybrid. Classify each as a 1°, 2°, or 3° allylic cation.

(a)   (b)

Solution

The additional contributing structure in each case is a 3° allylic cation. The contributing structure having the greater degree of substitution on the positively charged carbon makes the greater contribution to the hybrid.

(a) ⟷   (b) ⟷

   Greater contribution   Greater contribution

Problem 8.3

Write an additional contributing structure for each carbocation, and state which of the two makes the greater contribution to the resonance hybrid.

(a) $=CH_2$ (b)

Primary allylic and benzylic halides can be made to react by either S_N1 or S_N2 mechanisms depending on the solvent and the nucleophile. They are primary (the steric factor favoring S_N2), and, at the same time, they can lose a halide ion to form a stable allylic or benzylic carbocation (the electronic factor favoring S_N1). In polar protic solvents, they undergo solvolysis by an S_N1 mechanism. They can be made to undergo S_N2 reactions in aprotic solvents by treatment with good nucleophiles. Secondary and tertiary allylic and benzylic halides react almost exclusively by an S_N1 mechanism.

F.   The Leaving Group

In the transition state for nucleophilic substitution on a haloalkane, the leaving group develops a partial negative charge in both S_N1 and S_N2 reactions; therefore, the ability of a group to function as a leaving group is related to how stable it is as an anion. The most stable anions and the best leaving groups are the conjugate bases of strong acids. Thus we can use the information on the relative strengths of organic and inorganic acids in Table 4.1 to determine which anions are the best leaving groups. This order is shown here.

Reactivity as a leaving group ←

$$I^- > Br^- > Cl^- \gg F^- > CH_3\overset{\displaystyle O}{\overset{\|}{C}}O^- > HO^- > CH_3O^- > NH_2^-$$

← **Stability of anion; strength of conjugate acid**

The best leaving groups in this series are the halogens, I^-, Br^-, and Cl^-. Hydroxide ion, OH^-, methoxide ion, CH_3O^-, and amide ion, NH_2^-, are such poor leaving groups that they rarely, if ever, are displaced in nucleophilic aliphatic substitution. However, we shall see in Chapters 9 and 10 that protonation of —OH and —OR converts them to better leaving groups.

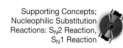

G.   The Solvent

To appreciate the important role of the solvent in nucleophilic substitution reactions, we need to be specific about whether the substitution is S_N2 or S_N1. Let us first take up the effect of solvent on S_N2 reactions.

The Effect of Solvent on S_N2 Reactions

The most common type of S_N2 reaction involves a negative nucleophile and a negative leaving group.

negatively charged nucleophile   negative charge dispersed in the transition state   negatively charged leaving group

$$Nu\!:^- + \overset{}{C}-X \longrightarrow \left[\overset{\delta-}{Nu}\text{---}\overset{|}{C}\text{---}\overset{\delta-}{X} \right] \longrightarrow Nu-C + :X^-$$

Transition state

The stronger the solvation of the nucleophile, the greater the energy required to remove the nucleophile from its solvation shell to reach the transition state, and hence the lower the rate of the S_N2 reaction.

Because of their good ability to solvate cations but only poor ability to solvate anions (nucleophiles), polar aprotic solvents have a particularly dramatic effect on the rate of S_N2 reactions. Reactions of the same substrate and nucleophile are accelerated (often by several orders of magnitude) when carried out in aprotic solvents compared with the rate obtained in protic solvents. Table 8.10 shows ratios of specific rate

Table 8.10   Rates of an S_N2 Reaction as a Function of Solvent

$$\text{Br} + \text{N}_3^- \xrightarrow[\text{solvent}]{S_N2} \text{N}_3 + \text{Br}^-$$

Solvent Type	Solvent	$\dfrac{k_{(\text{solvent})}}{k_{(\text{methanol})}}$
Polar aprotic	$CH_3C\equiv N$	5000
	$(CH_3)_2NCHO$	2800
	$(CH_3)_2S{=}O$	1300
Polar protic	H_2O	7
	CH_3OH	1

constants for the S_N2 reaction of 1-bromobutane with sodium azide as a function of solvent. The rate of reaction in methanol is taken as a reference.

The Effect of Solvent on S_N1 Reactions

Nucleophilic substitution by an S_N1 pathway involves creation and separation of opposite charges in the transition state of the rate-determining step. For this reason, the rate of S_N1 reactions depends on the ability of the solvent to keep opposite charges separated as well as its ability to stabilize both positive and negative sites by solvation. The solvents that meet these requirements best are polar protic solvents such as H_2O, ROH, and, to a lesser degree, RCOOH. As seen in Table 8.11, the rate of solvolysis of

Table 8.11   Rates of an S_N1 Reaction as a Function of Solvent

$$\underset{\substack{| \\ CH_3}}{\overset{\substack{CH_3 \\ |}}{CH_3CCl}} + \text{ROH} \xrightarrow{\text{solvolysis}} \underset{\substack{| \\ CH_3}}{\overset{\substack{CH_3 \\ |}}{CH_3COR}} + \text{HCl}$$

Solvent	$\dfrac{k_{(\text{solvent})}}{k_{(\text{ethanol})}}$
Water	100,000
80% water: 20% ethanol	14,000
40% water: 60% ethanol	100
Ethanol	1

Increasing polarity of solvent →

Increasing rate of solvolysis →

2-chloro-2-methylpropane (*tert*-butyl chloride) increases by a factor of 10^5 when the solvent is changed from ethanol to water.

H. Skeletal Rearrangement

As we saw in Section 6.3C, skeletal rearrangement is typical of reactions involving a carbocation intermediate that can rearrange to a more stable one. Because there is little or no carbocation character at the substitution center, S_N2 reactions are free of rearrangement. In contrast, S_N1 reactions often proceed with rearrangement. An example of an S_N1 reaction involving rearrangement is solvolysis of 2-chloro-3-phenylbutane in methanol, a polar protic solvent and a weak nucleophile. The major substitution product is the ether with rearranged structure. The chlorine atom in the starting material is on a 2° carbon, but the methoxyl group in the product is on the adjacent 3° carbon.

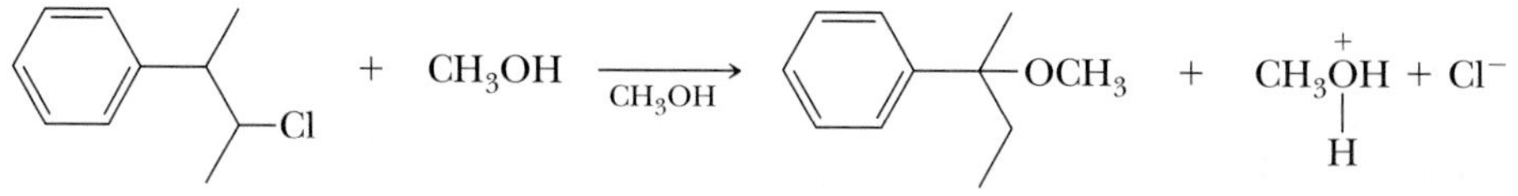

2-Chloro-3-phenylbutane   2-Methoxy-2-phenylbutane

As seen in the following mechanism, reaction is initiated by ionization of the carbon-chlorine bond to form a 2° carbocation, which rearranges to a considerably more stable 3° carbocation by shift of a hydrogen with its pair of electrons (a hydride ion) from the adjacent benzylic carbon. Note that not only is the rearranged carbocation tertiary, but it is also benzylic, which adds resonance stabilization.

Mechanism   **Rearrangement During Solvolysis of 2-Chloro-3-phenylbutane**

Step 1: Ionization of the C—Cl bond gives a 2° carbocation intermediate.

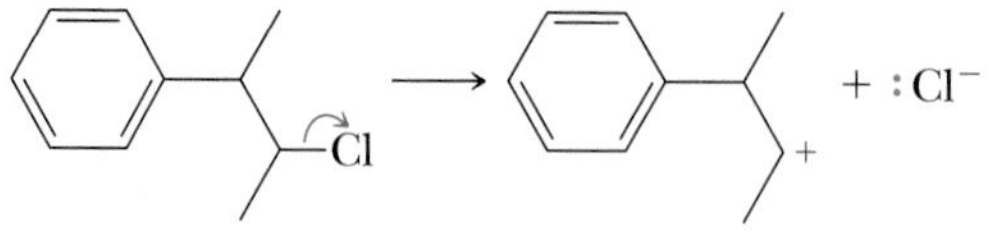

A 2° carbocation

Step 2: Migration of a hydrogen atom with its bonding electrons (a hydride ion) gives a more stable 3° benzylic carbocation intermediate.

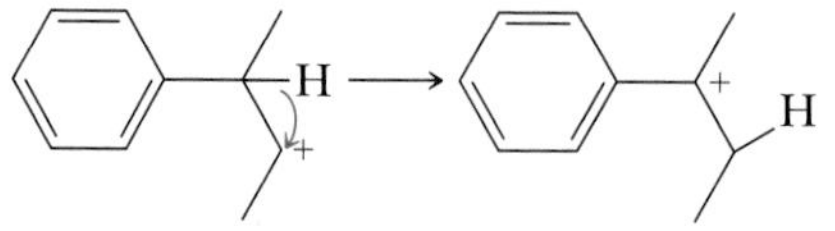

A 3° benzylic
carbocation

Step 3: Reaction of the 3° benzylic carbocation intermediate (a Lewis acid) with methanol (a Lewis base) forms an oxonium ion.

An oxonium ion

Step 4: Proton transfer to solvent (in this case, methanol) gives the final product.

In general, migration of a hydrogen atom or alkyl group with its bonding electrons occurs when a more stable carbocation can be formed.

The factors favoring S$_N$1 or S$_N$2 reactions are summarized in Table 8.12. Also shown is the expected configurational result when substitution takes place at a stereocenter.

Table 8.12 Summary of S$_N$1 Versus S$_N$2 Reactions of Haloalkanes

Type of Alkyl Halide	S$_N$2	S$_N$1
Methyl CH$_3$X	S$_N$2 is favored.	S$_N$1 does not occur. The methyl cation is so unstable that it is never observed in solution.
Primary RCH$_2$X	S$_N$2 is favored.	S$_N$1 rarely occurs. Primary cations are so unstable that they are rarely observed in solution.
Secondary R$_2$CHX	S$_N$2 is favored in aprotic solvents with good nucleophiles.	S$_N$1 is favored in protic solvents with poor nucleophiles. Carbocation rearrangements may occur.
Tertiary R$_3$CX	S$_N$2 does not occur because of steric hindrance around the reaction center.	S$_N$1 is favored because of the ease of formation of tertiary carbocations.
Substitution at a stereocenter	Inversion of configuration. The nucleophile attacks the stereocenter from the side opposite the leaving group.	Racemization is favored. The carbocation intermediate is planar, and attack of the nucleophile occurs with equal probability from either side. There is often some net inversion of configuration.

8.5   Neighboring Group Participation

Thus far we have considered two limiting mechanisms for nucleophilic substitutions that focus on the degree of covalent bonding between the nucleophile and the substitution center during departure of the leaving group. In an S_N2 mechanism, the leaving group is assisted in its departure by the nucleophile. In an S_N1 mechanism, the leaving group is not assisted in this way. An essential criterion for distinguishing between these two pathways is the order of reaction. Nucleophile-assisted substitutions are second order: first order in RX and first order in nucleophile. Nucleophile-unassisted substitutions are first order: first order in RX and zero order in nucleophile.

Chemists recognize that certain nucleophilic substitutions, which have the kinetic characteristics of first-order (S_N1) substitution, in fact involve two successive displacement reactions. A characteristic feature of a great many of these reactions is the presence of an internal nucleophile, most commonly sulfur, nitrogen, or oxygen, on the carbon atom beta to the leaving group. This neighboring nucleophile participates in the departure of the leaving group to give an intermediate, which then reacts with an external nucleophile to complete the reaction.

The mustard gases are one group of compounds that react by participation of a neighboring group. The characteristic structural feature of a mustard gas is a two-carbon chain, with a halogen on one carbon and a divalent sulfur or trivalent nitrogen on the other carbon (S—C—C—X or N—C—C—X). An example of a mustard gas is bis(2-chloroethyl)sulfide, a poison gas used extensively in World War I. This compound is a deadly vesicant (blistering agent) and quickly causes conjunctivitis and blindness.

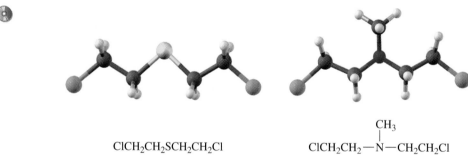

$ClCH_2CH_2SCH_2CH_2Cl$

Bis(2-chloroethyl)sulfide
(a sulfur mustard gas)

$$\overset{\displaystyle CH_3}{\underset{\displaystyle |}{ClCH_2CH_2-N-CH_2CH_2Cl}}$$

Bis(2-chloroethyl)methylamine
(a nitrogen mustard gas)

Bis(2-chloroethyl)sulfide and bis(2-chloroethyl)methylamine are not gases at all. They are oily liquids. They do, however, have a high vapor pressure, hence the designation "gas." Nitrogen and sulfur mustards react very rapidly with moisture in the air and in the mucous membranes of the eye, nose, and throat to produce HCl, which then burns and blisters these sensitive tissues. What is unusual about the reactivity of the mustard gases is that they react very rapidly with water, a very poor nucleophile.

$$Cl\diagup\!\!\!\diagdown\!\!\!S\!\diagdown\!\!\!\diagup\!\!\!Cl + 2H_2O \longrightarrow HO\!\diagup\!\!\!\diagdown\!\!\!S\!\diagdown\!\!\!\diagup\!\!\!OH + 2HCl$$

Mustard gases also react rapidly with other nucleophiles, such as those in biological molecules, which makes them particularly dangerous chemicals. For insight into how mustard gases were recognized as potential starting points for the synthesis of effective drugs for the treatment of certain kinds of cancer, see the Chemistry in Action box "Mustard Gases and the Treatment of Neoplastic Diseases" in Chapter 28.

The reason for the extremely rapid hydrolysis of the sulfur mustards is neighboring group participation by sulfur in the ionization of the carbon-chlorine bond to form a cyclic sulfonium ion. At this point, you should review halogenation and oxymercuration of alkenes (Sections 6.3D and 6.3F) and compare the cyclic halonium and mercurinium ions formed there with the cyclic sulfonium ion formed here.

The cyclic sulfonium ion contains a highly strained three-membered ring and reacts rapidly with an external nucleophile to open the ring followed by proton transfer to H_2O to give H_3O^+. The net effect of these reactions is nucleophilic substitution of OH for Cl.

Of the two steps in the mechanism of the hydrolysis of a sulfur mustard, the first is the slower and rate determining. As a result, the rate of reaction is proportional to the concentration of the mustard but independent of the concentration of the external nucleophile. Thus, although this reaction has the kinetic characteristics of an S_N1 reaction, it actually involves two successive displacement reactions.

Mechanism **Hydrolysis of a Sulfur Mustard—Participation by a Neighboring Group**

Step 1:   An internal S_N2 reaction in which ionization of the C—Cl bond, assisted by the neighboring sulfur atom, gives a highly strained three-membered ring.

A cyclic sulfonium ion

Step 2:   Reaction of the cyclic sulfonium ion (a Lewis acid) with water (a Lewis base) opens the three-membered ring. In this S_N2 reaction, H_2O is the nucleophile, and sulfur is the leaving group.

Step 3:   Proton transfer to water completes the reaction.

We continue to use the terms S_N2 and S_N1 to describe nucleophilic substitution reactions. You should realize, however, that these designations do not adequately describe all nucleophilic substitution reactions.

Example 8.4

Write a mechanism for the hydrolysis of the nitrogen mustard bis(2-chloro-ethyl)methylamine.

Solution

Following is a three-step mechanism.

Step 1: An internal S_N2 reaction in which ionization of the C—Cl bond is assisted by the neighboring nitrogen atom to form a highly strained three-membered ring.

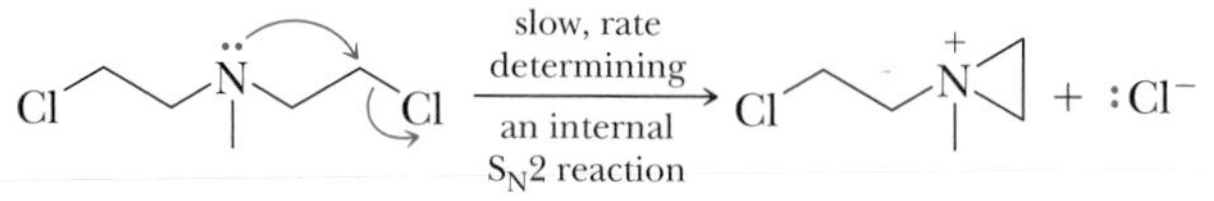

A cyclic
ammonium ion

Step 2: Reaction of the cyclic ammonium ion (a Lewis acid) with water (a Lewis base) opens the three-membered ring. In this S_N2 reaction, H_2O is the nucleophile, and nitrogen is the leaving group.

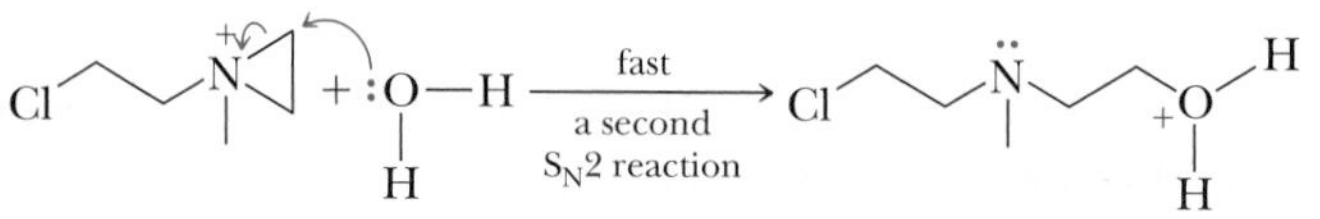

Step 3: Proton transfer to water completes the reaction.

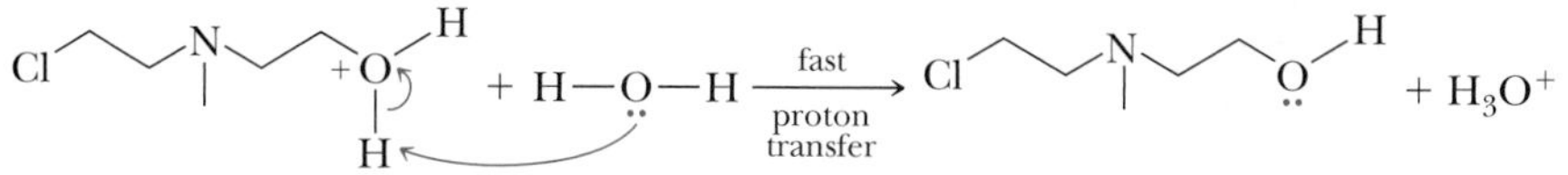

Problem 8.4

Knowing what you do about the regioselectivity of S_N2 reactions, predict the product of hydrolysis of this compound.

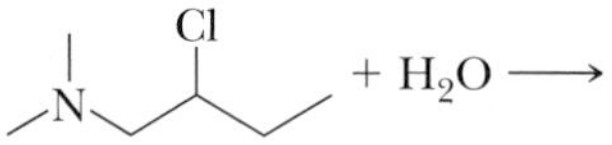

8.6 Analysis of Several Nucleophilic Substitution Reactions

Predictions about the mechanism for a particular nucleophilic substitution reaction must be based on considerations of the structure of the haloalkane, the nucleophile, the leaving group, and the solvent. Following are five nucleophilic substitution reactions and an analysis of the factors that favor an S_N1 or S_N2 mechanism for each.

Nucleophilic Substitution 1

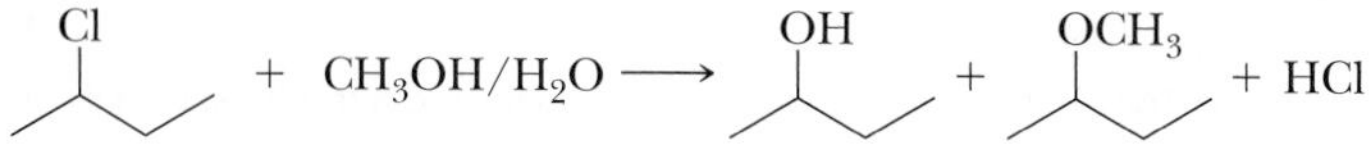

R enantiomer

The mixture of methanol and water is a polar protic solvent and a good ionizing solvent in which to form carbocations. 2-Chlorobutane ionizes in this solvent to form a fairly stable 2° carbocation intermediate. Both water and methanol are poor nucleophiles. From this analysis, we predict that reaction is by an S_N1 mechanism, a nucleophile-unassisted ionization of the 2° chloroalkane to give a carbocation intermediate, which then reacts with either water or methanol as the nucleophile to give the observed products. Each product is formed as an approximately 50:50 mixture of R and S enantiomers.

Nucleophilic Substitution 2

This is a primary bromoalkane with two beta branches in the presence of cyanide ion, a good nucleophile. Dimethyl sulfoxide, a polar aprotic solvent, is a particularly good solvent in which to carry out nucleophile-assisted substitution reactions because its ability to solvate cations (in this case, Na^+) is good, but its ability to solvate anions (in this case, CN^-) is poor. From this analysis, we predict that this reaction goes by an S_N2 mechanism.

Nucleophilic Substitution 3

Bromine is a good leaving group, and it is on a 2° carbon. The methylsulfide ion is a good nucleophile. Acetone, a polar aprotic solvent, is a good medium in which to carry out S_N2 reactions but a poor medium in which to carry out S_N1 reactions. From this analysis, we predict that this reaction goes by an S_N2 mechanism and that the product is the S enantiomer.

Nucleophilic Substitution 4

Ionization of the carbon-bromine bond forms a resonance-stabilized 2° allylic carbocation. Acetic acid is a poor nucleophile, which reduces the likelihood of an S_N2 reaction. Further, acetic acid is a protic (hydroxylic) solvent that favors S_N1 reaction. From this analysis, we predict that this reaction goes by an S_N1 mechanism.

Nucleophilic Substitution 5

The bromoalkane is primary, and bromine is a good leaving group. Trivalent compounds of phosphorus, a third row element, are good nucleophiles. Toluene is a nonpolar aprotic solvent. Given the combination of a primary halide, a good leaving group, a good nucleophile, and a nonpolar aprotic solvent, we predict reaction by an S_N2 pathway.

Example 8.5

Write the expected substitution product(s) for each reaction, and predict the mechanism by which each product is formed.

(a) 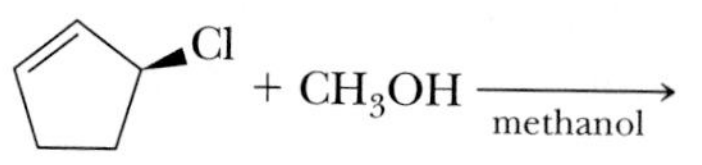 + CH$_3$OH $\xrightarrow[\text{methanol}]{}$

S enantiomer

(b) 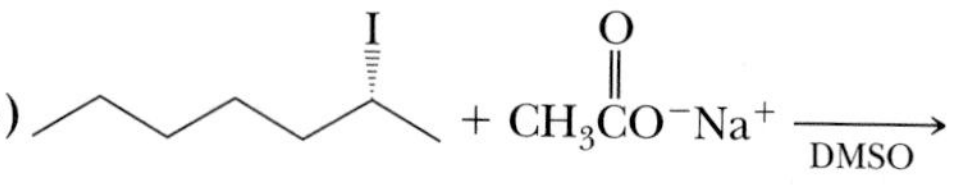 + CH$_3$CO$^-$Na$^+$ $\xrightarrow[\text{DMSO}]{}$

R enantiomer

Solution

(a) This 2° allylic chloride is treated with methanol, a poor nucleophile and a polar protic solvent. Ionization of the carbon-chlorine bond forms a resonance-stabilized secondary allylic cation. Therefore, we predict reaction by an S$_N$1 mechanism and formation of the product as a racemic mixture.

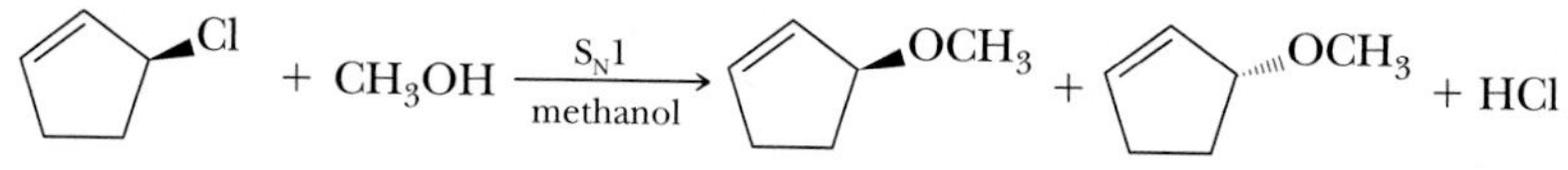

+ CH$_3$OH $\xrightarrow[\text{methanol}]{S_N1}$ + + HCl

S enantiomer   *S* enantiomer   *R* enantiomer
                              (50%)   (50%)

(b) Iodide is a good leaving group on a moderately accessible secondary carbon. Acetate ion dissolved in a polar aprotic solvent is a moderate nucleophile. We predict substitution by an S$_N$2 pathway with inversion of configuration at the stereo-center.

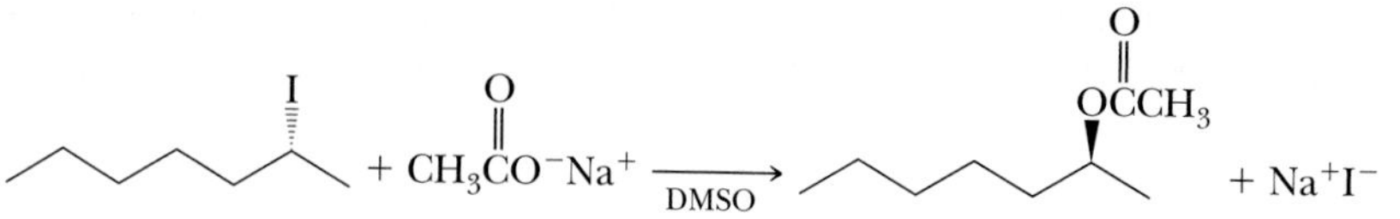

+ CH$_3$CO$^-$Na$^+$ $\xrightarrow[\text{DMSO}]{}$ + Na$^+$I$^-$

R enantiomer   *S* enantiomer

Problem 8.5

Write the expected substitution product(s) for each reaction and predict the mechanism by which each product is formed.

(a) 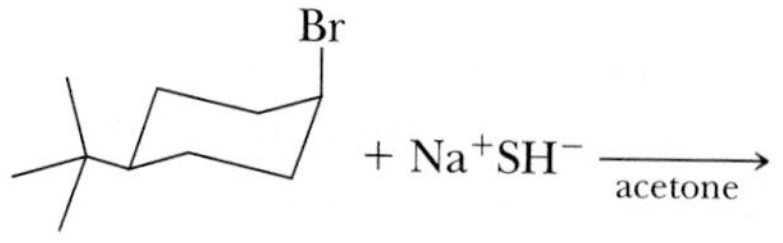 + Na$^+$SH$^-$ $\xrightarrow[\text{acetone}]{}$

(b) 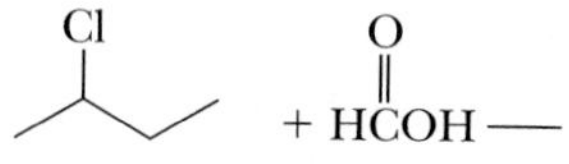 + HCOH $\longrightarrow$

R enantiomer

8.7 Phase-Transfer Catalysts

Very often, nucleophilic substitution involves reactions between a covalent organic compound and an ionic compound, as for example between 1-bromooctane and sodium cyanide.

$$CH_3(CH_2)_6CH_2Br + Na^+CN^- \longrightarrow CH_3(CH_2)_6CH_2CN + Na^+Br^-$$

1-Bromooctane   Nonanenitrile

This reaction is simple to write, but for it to occur, both compounds must be brought together so that they can react. The solubility characteristics of these reactants are quite different. Sodium cyanide is an ionic solid, soluble in water and a few other polar solvents but insoluble in nonpolar organic solvents such as dichloromethane. 1-Bromooctane, on the other hand, is insoluble in water but quite soluble in dichloromethane and other nonpolar organic solvents. One way to bring these reactants together is to dissolve them in DMSO, DMF, or other polar aprotic solvents. The advantages of DMSO and DMF are that each dissolves both organic and ionic compounds. When 1-bromooctane and sodium cyanide are dissolved in DMSO, reaction between them occurs very readily.

Although DMSO and DMF are excellent solvents in which to carry out organic reactions, they have certain disadvantages. Both are several times more expensive than solvents such as dichloromethane and ethanol, and, on an industrial scale, solvent cost can be an important consideration. Furthermore, because DMSO and DMF are so soluble in water, it is often difficult to recover them from mixtures with water. Finally, because they have higher boiling points than other common solvents (189°C for DMSO and 153°C for DMF compared with 78°C for ethanol and 40°C for dichloromethane), it is often difficult to remove them entirely from organic reaction products by distillation.

Another way to bring about reaction between 1-bromooctane and cyanide ion is by using a **phase-transfer catalyst.** The characteristics necessary for an effective phase-transfer catalyst for anions are a balanced combination of (1) **hydrophilic** character to dissolve in water and form an ion pair with the anion to be transported and (2) **hydrophobic** character to dissolve in the organic phase and transport the anion into it. One commonly used phase-transfer catalyst is the water-soluble salt tetrabutylammonium chloride.

$$(CH_3CH_2CH_2CH_2)_4N^+Cl^-$$

Tetrabutylammonium chloride
($Bu_4N^+Cl^-$)

The tetrabutylammonium ion, Bu_4N^+, has both hydrophilic and hydrophobic regions. The positively charged nitrogen atom of this ion is a hydrophilic site that interacts with water, with other polar molecules, and with anions. Its four butyl groups are hydrophobic sites and do not interact with water. Thus, because of its hydrophilic positively charged nitrogen atom, the Bu_4N^+ ion is soluble in water, and because of its four hydrophobic butyl groups, it is also soluble in nonpolar organic solvents.

The way a phase-transfer catalyst works is illustrated using tetrabutylammonium chloride. Suppose sodium cyanide is dissolved in water, 1-bromooctane is dissolved in dichloromethane, and the solutions are mixed. Because water and dichloromethane are immiscible, a two-phase system results [Figure 8.5(a)]. No reaction takes place between 1-bromooctane and cyanide ion because they are in different phases.

When added to this two-phase system, $Bu_4N^+Cl^-$ dissolves in the aqueous phase [Figure 8.5(a)]. Bu_4N^+ and CN^- then form an ion pair that is transferred into the organic phase. CN^- displaces the bromine atom, and Br^- is transferred as an ion pair with Bu_4N^+ to the aqueous phase. This process is repeated until all cyanide ions have been transferred to the organic phase and have reacted with 1-bromooctane.

Phase-transfer catalyst   A substance that transfers ions from an aqueous phase into an organic phase and vice versa.
Hydrophilic   From the Greek, meaning water loving.
Hydrophobic   From the Greek, meaning water fearing.

Figure 8.5

Phase-transfer catalysis. (a) NaCN is dissolved in water (the upper phase), 1-bromo-octane is dissolved in dichloromethane (the lower phase), the phase-transfer cata-lyst $Bu_4N^+Cl^-$ is added, and the mixture is shaken. (b) CN^- is transferred from the aqueous phase to the organic phase as an ion pair with Bu_4N^+ and (c) displaces Br^- from 1-bromooctane. (d) Br^- forms an ion pair with Bu_4N^+ and is transferred from the or-ganic phase to the aqueous phase.

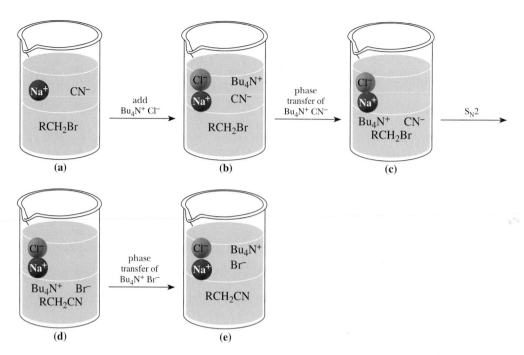

Supporting Concepts;
Elimination Reactions

8.8   β-Elimination

Dehydrohalogenation Removal of —H and —X from adjacent carbons; a type of β-elimination.

Here, we study a type of β-elimination called **dehydrohalogenation.** In the presence of base, halogen is removed from one carbon of a haloalkane, and hydrogen is re-moved from an adjacent carbon to form an alkene.

$$
\underset{\text{A haloalkane}}{\overset{\overset{\displaystyle H}{|}}{-\underset{|}{\overset{\beta}{C}}-\underset{\underset{\displaystyle X}{|}}{\overset{\alpha}{C}}-}} + \underset{\text{Base}}{CH_3CH_2O^-Na^+} \xrightarrow[CH_3CH_2OH]{} \underset{\text{An alkene}}{\overset{\diagdown}{\underset{\diagup}{C}}=\overset{\diagup}{\underset{\diagdown}{C}}} + CH_3CH_2OH + Na^+X^-
$$

It is important to keep in mind that β-elimination and nucleophilic substitution are competing reactions. In this section as well as in Sections 8.9 and 8.10, we concen-trate on β-elimination. In Section 8.11, we examine the results of competition be-tween these two types of reactions.

Strong bases that serve effectively in β-eliminations of haloalkanes are OH^-, OR^-, and NH_2^-. Following are three examples of base-promoted β-elimination reac-tions. In the first example, the base is shown as a reactant. In the second and third examples, the base is a reactant but is shown over the reaction arrow.

 $+ \ t\text{-BuO}^-K^+ \longrightarrow$  $+ \ t\text{-BuOH} + K^+Br^-$

1-Bromodecane   Potassium   1-Decene
               tert-butoxide

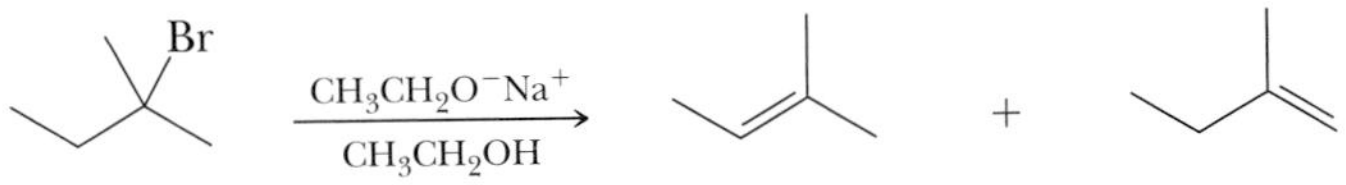

2-Bromo-2-methylbutane → 2-Methyl-2-butene (major product) + 2-Methyl-1-butene

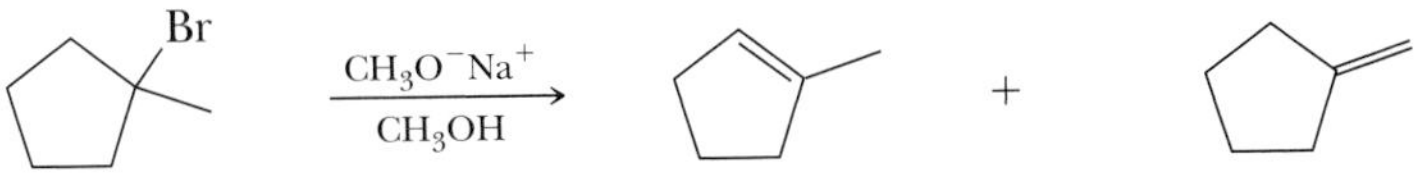

1-Bromo-1-methylcyclopentane → 1-Methylcyclopentene (major product) + Methylenecyclopentane

In the second and third illustrations, there are two nonequivalent β-carbons, each bearing a hydrogen; therefore, two alkenes are possible. In each case, the major product of this and most other β-elimination reactions is the more substituted (and therefore more stable) alkene (Section 6.6B). Each reaction is said to follow **Zaitsev's rule,** or to undergo Zaitsev elimination. Not all alkenes, however, undergo β-elimination to give the more stable alkene, and these exceptions give us important clues to the mechanism of β-elimination, as we shall see when we discuss the regio- and stereochemistry of β-elimination reactions.

Zaitsev's rule A rule stating that the major product of a β-elimination reaction is the most stable alkene; that is, it is the alkene with the greatest number of substituents on the carbon-carbon double bond.

Example 8.6

Predict the β-elimination product(s) formed when each bromoalkane is treated with sodium ethoxide in ethanol. If two or more products might be formed, predict which is the major product.

(a) 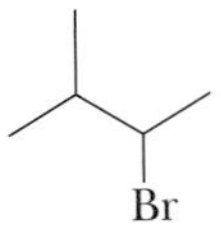   (b)

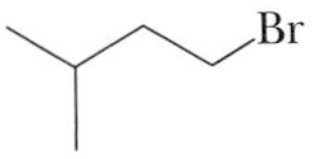

Solution

(a) There are two nonequivalent β-carbons in this bromoalkane, and two alkenes are possible. 2-Methyl-2-butene, the more substituted alkene, is the major product.

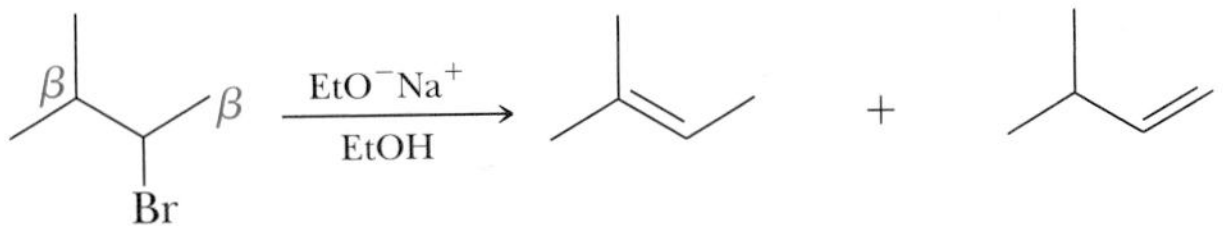

2-Methyl-2-butene (major product)   3-Methyl-1-butene

(b) There is only one β-carbon in this bromoalkane, and only one alkene is possible.

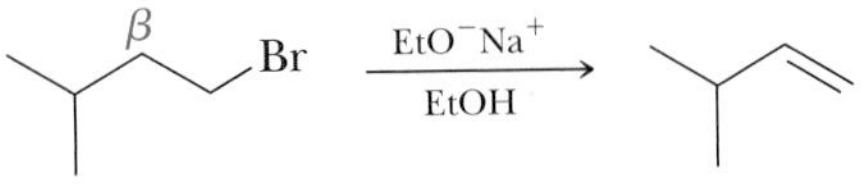

3-Methyl-1-butene

Problem 8.6

Predict the β-elimination product(s) formed when each chloroalkane is treated with sodium ethoxide in ethanol. If two or more products might be formed, predict which is the major product.

(a)   **(b)**   **(c)**

8.9   Mechanisms of β-Elimination

There are two limiting mechanisms for β-eliminations. A fundamental difference between them is the timing of the bond-breaking and bond-forming steps. Recall that we made the same statement about the two limiting mechanisms for nucleophilic substitution reactions (Section 8.3).

A.   E1 Mechanism

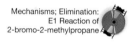

Mechanisms; Elimination: E1 Reaction of 2-bromo-2-methylpropane

At one extreme, breaking of the C—X bond is complete before any reaction occurs with the base to lose a hydrogen and form the carbon-carbon double bond. This mechanism is designated **E1,** where E stands for *e*limination and 1 stands for *uni*molecular; one species, in this case the haloalkane, is involved in the rate-determining step. The mechanism of an E1 reaction is illustrated here by the reaction of 2-bromo-2-methylpropane to form 2-methylpropene. In this two-step mechanism, the rate-determining step is ionization of the carbon-halogen bond to form a carbocation intermediate.

Mechanism   **E1 Reaction of 2-Bromo-2-methylpropane**

Step 1:   Rate-determining ionization of the C—Br bond gives a carbocation intermediate.

$$CH_3-\underset{\underset{Br}{|}}{\overset{\overset{CH_3}{|}}{C}}-CH_3 \xrightarrow[\text{determining}]{\text{slow, rate}} CH_3-\underset{+}{\overset{\overset{CH_3}{|}}{C}}-CH_3 \ + \ :Br^-$$

A carbocation intermediate

Step 2:   Proton transfer from the carbocation intermediate to solvent (in this case methanol) gives the alkene.

$$\underset{H_3C}{\overset{H}{\diagdown}}O: + H-CH_2-\underset{+}{\overset{\overset{CH_3}{|}}{C}}-CH_3 \xrightarrow{\text{fast}} \underset{H_3C}{\overset{H}{\diagdown}}\overset{+}{O}-H \ + \ CH_2=\underset{}{\overset{\overset{CH_3}{|}}{C}}-CH_3$$

In an E1 mechanism, one transition state exists for the formation of the carbocation in Step 1 and a second exists for the loss of a hydrogen in Step 2 (Figure 8.6). Forma-

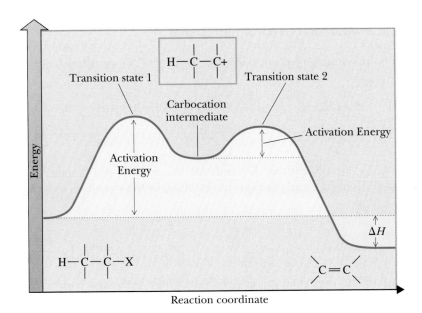

Figure 8.6
An energy diagram for an E1 reaction showing two transition states and one carbocation intermediate.

tion of the carbocation intermediate in Step 1 crosses the higher energy barrier and is the rate-determining step.

B.   E2 Mechanism

At the other extreme is a concerted process. In an E2 reaction, here illustrated by the reaction of 1-bromopropane with sodium ethoxide, proton transfer to the base, formation of the carbon-carbon double bond, and ejection of bromide ion occur simultaneously; all bond-breaking and bond-forming steps are concerted. Because base removes a β-hydrogen at the same time the C—Br bond is broken to form a halide ion, the transition state has considerable double bond character (Figure 8.7).

Mechanisms: Elimination
Alkyl halide
dehydrohalogenation

Figure 8.7
An energy diagram for an E2 reaction. There is considerable double bond character in the transition state.

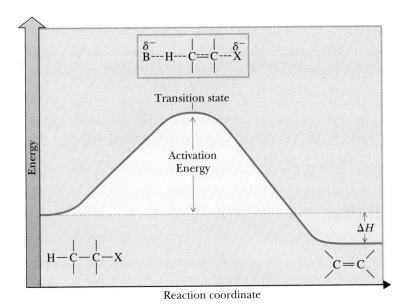

Mechanism   E2 Reaction of 1-Bromopropane

Bond breaking and bond forming are concerted; that is, they occur simultaneously.

$$CH_3CH_2O{:}^- + H{-}\underset{\underset{CH_3}{|}}{CH}{-}CH_2{-}Br \longrightarrow CH_3CH_2O{-}H + CH_3CH{=}CH_2 + {:}\,Br^-$$

This mechanism is designated **E2,** where E stands for *e*limination and 2 stands for *bi*molecular; both the haloalkane and the base are involved in the transition state for the rate-determining step.

Supporting Concepts;
Elimination Reactions

8.10   Experimental Evidence for E1 and E2 Mechanisms

As we examine some of the experimental evidence on which these two contrasting mechanisms are based, we consider the following questions:

1. What are the kinetics of base-promoted β-eliminations?
2. Where two or more alkenes are possible, what factors determine the ratio of the possible products?
3. What is the stereochemistry of β-elimination?

A.   Kinetics

E1 Reactions

The rate-determining step in an E1 reaction is ionization of the halide to form a carbocation. Because this step involves only the haloalkane, the reaction is said to be unimolecular and follows first-order kinetics.

$$\text{Rate} = -\frac{d[RX]}{dt} = k[RX]$$

Recall that the first step in an S_N1 reaction is also formation of a carbocation. Thus, for both S_N1 and E1 reactions, formation of the carbocation is the rate-determining step.

E2 Reactions

Only one step occurs in an E2 mechanism, and the transition state is bimolecular. The reaction is second order: first order in haloalkane and first order in base.

$$\text{Rate} = -\frac{d[RX]}{dt} = k[RX][\text{Base}]$$

B.   Regioselectivity

E1 Reactions

The major product in E1 reactions is the more stable alkene; that is, the alkene with the more highly substituted carbon-carbon double bond (Zaitsev's rule). After the carbocation is formed in the rate-determining step of an E1 reaction, it may lose a hydrogen to complete β-elimination, or it may rearrange and then lose a hydrogen.

E2 Reactions

For E2 reactions using strong bases and in which the leaving group is a halide ion, the major product is that formed following Zaitsev's rule. Double bond character is so highly developed in the transition state that the relative stability of possible alkenes determines which alkene is the major product. Thus the transition state of lowest energy is that leading to the most stable, most highly substituted alkene.

| 2-Bromohexane | 2-Hexene (74%) | 1-Hexene (26%) |

With larger, sterically hindered bases such as *tert*-butoxide, however, where isomeric alkenes are possible, the major product is often the less substituted alkene.

C.   Stereoselectivity

An E2 reaction is most favorable when —X and —H are oriented anti (at a dihedral angle of 180°) to one another.

—H and —X are anti and coplanar
(dihedral angle 180°)

This requirement can be demonstrated clearly in chlorocyclohexanes. In these molecules, anti and coplanar correspond to trans, diaxial. Consider the E2 reaction of the cis and trans isomers of 1-chloro-2-isopropylcyclohexane. From the cis isomer, the major product is 1-isopropylcyclohexene, the more substituted cycloalkene. From the trans isomer, only 3-isopropylcyclohexene, the less substituted alkene, is formed.

cis-1-Chloro-2- isopropylcyclohexane   1-Isopropylcyclohexene   3-Isopropylcyclohexene
(major product)

trans-1-Chloro-2-isopropylcyclohexane   3-Isopropylcyclohexene

In the more stable chair conformation of the cis isomer, the considerably larger isopropyl group is equatorial, and the smaller chlorine is axial. In this chair conformation, —H on carbon 2 and —Cl on carbon 1 are anti and coplanar. Concerted E2 elimination gives 1-isopropylcyclohexene, a trisubstituted alkene, as the major product. Note that —H on carbon 6 and —Cl are also anti and coplanar. Dehydrohalogenation of this combination of —H and —Cl gives 3-isopropylcyclohexene, a disubstituted and, therefore, less stable alkene.

Mechanism   **E2 Reaction of *cis*-1-Chloro-2-isopropylcyclohexane**

In the more stable chair conformation of the cis isomer, —H and —Cl are anti and coplanar. All bond-breaking and bond-forming steps are concerted.

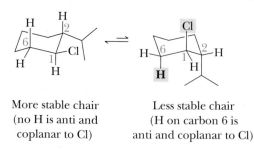

1-Isopropylcyclohexene

In the more stable chair conformation of the trans isomer, both isopropyl and chlorine are equatorial. In this conformation, the hydrogen atom on carbon 2 is cis to the chlorine atom. One of the hydrogen atoms on carbon 6 is trans to —Cl, but it is not anti and coplanar. In the alternative, less stable chair conformation of the trans isomer, both isopropyl and chlorine are axial. In this conformation, the axial hydrogen in carbon 6 is anti and coplanar to chlorine, and E2 β-elimination can occur to give 3-isopropylcyclohexene. Thus even though the diaxial conformation is less stable, the reaction goes through this conformation because it is the only one with an anti-coplanar arrangement of the Cl and a β—H.

Mechanism   **E2 Reaction of *trans*-1-Chloro-2-isopropylcyclohexane**

Step 1:   The more stable chair conformation is converted to the less stable chair conformation.

More stable chair
(no H is anti and
coplanar to Cl)

Less stable chair
(H on carbon 6 is
anti and coplanar to Cl)

Step 2: E2 reaction can take place now that an —H and —Cl are anti and coplanar.

3-Isopropylcyclohexene

The rate at which the cis isomer undergoes E2 reaction is considerably greater than the rate for the trans isomer. We can account for this observation in the following manner. The more stable chair conformation of the cis isomer has —H and —Cl anti and coplanar, and the activation energy for the reaction is that required to reach the E2 transition state. The more stable chair conformation of the trans isomer, however, cannot undergo anti elimination. To react, it must first be converted to the less stable chair. Thus the activation energy for the trans isomer includes (1) the energy necessary to convert the more stable chair to the less stable chair, and (2) the energy to reach the E2 transition state from this conformation.

Example 8.7

Treatment of 1,2-dibromo-1,2-diphenylethane with sodium methoxide in methanol gives 1-bromo-1,2-diphenylethylene. The meso isomer of the starting material gives (E)-1-bromo-1,2-diphenylethylene, whereas the racemic mixture of the starting material gives (Z)-1-bromo-1,2-diphenylethylene. How do you account for the stereospecificity of these β-eliminations?

meso-1,2-Dibromo-
1,2-diphenylethane

(E)-1-Bromo-
1,2-diphenylethylene

racemic-1,2-Dibromo-
1,2-diphenylethane

(Z)-1-Bromo-
1,2-diphenylethylene

Solution

It is a requirement for an E2 reaction that —H and —X be anti and coplanar. Following is a stereorepresentation of the meso isomer, drawn to show the plane of symmetry. Rotation of the left carbon by 60° brings —H and —Br into an anti and coplanar relationship. E2 reaction on this conformation gives the (E)-alkene. E2 reaction of the proper conformation of either enantiomer of the racemic mixture gives the (Z)-alkene.

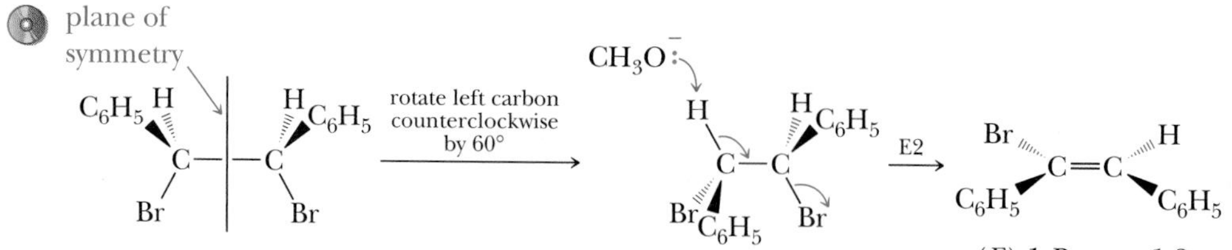

Meso isomer

(E)-1-Bromo-1,2-diphenylethylene

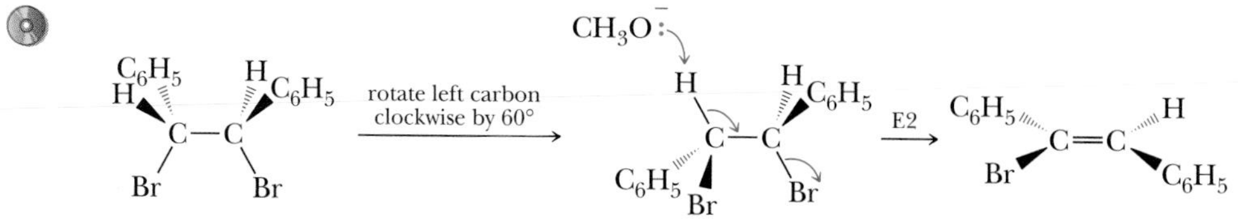

One enantiomer of the racemic mixture

(Z)-1-Bromo-1,2-diphenylethylene

Problem 8.7

1-Chloro-4-isopropylcyclohexane exists as two stereoisomers: one cis and one trans. Treatment of either isomer with sodium ethoxide in ethanol gives 4-isopropylcyclohexene by an E2 reaction.

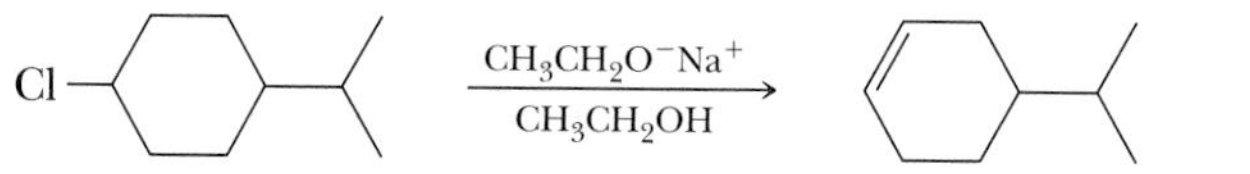

1-Chloro-4-isopropylcyclohexane

4-Isopropylcyclohexene

The cis isomer undergoes E2 reaction several orders of magnitude faster than the trans isomer. How do you account for this experimental observation?

The factors favoring E1 or E2 are summarized in Table 8.13.

Table 8.13   Summary of E1 Versus E2 Reactions for Haloalkanes

Alkyl Halide	E1	E2
Primary RCH_2X	E1 does not occur. Primary carbocations are so unstable that they are never observed in solution.	E2 is favored.
Secondary R_2CHX	Main reaction with weak bases such as H_2O, ROH.	Main reaction with strong bases such as OH^- and OR^-.
Tertiary R_3CX	Main reaction with weak bases such as H_2O, ROH.	Main reaction with strong bases such as OH^- and OR^-.

8.11 Substitution Versus Elimination

Supporting Concepts;
S_N1/S_N2 vs. E1/E2

Thus far, we have considered two types of reactions of haloalkanes, namely nucleophilic substitution and β-elimination. Many of the nucleophiles we have considered, for example hydroxide ion and alkoxide ions, are also strong bases. Thus nucleophilic substitution and β-elimination often compete with each other, and the ratio of products formed by these reactions depends on the relative rates of the two reactions.

A. S_N1 Versus E1 Reactions

Reactions of secondary and tertiary haloalkanes in polar protic solvents give mixtures of substitution and elimination products. In both reactions, Step 1 is the formation of a carbocation intermediate. This step is then followed by one or more characteristic carbocation reactions: (1) loss of a hydrogen (E1) to give an alkene, (2) reaction with solvent (S_N1) to give a substitution product, or (3) rearrangement followed by reaction (1) or (2). In polar protic solvents, the products formed depend only on the structure of the particular carbocation. For example, *tert*-butyl chloride and *tert*-butyl iodide in 80% aqueous ethanol both react with solvent giving the same mixture of substitution and elimination products. Because iodide ion is a better leaving group than chloride ion, *tert*-butyl iodide reacts over 100 times faster than *tert*-butyl chloride. Yet the ratio of products is the same.

It is difficult to predict the ratio of substitution to elimination products for first-order reactions of haloalkanes. For the majority of cases, however, S_N1 predominates over E1.

B. S_N2 Versus E2 Reactions

It is considerably easier to predict the ratio of substitution to elimination products for second-order reactions of haloalkanes with reagents that act both as nucleophiles and bases. The guiding principles are:

1. Branching at the α-carbon or β-carbon(s) increases steric hindrance about the α-carbon and significantly retards S_N2 reactions. Conversely, branching at the α-carbon or β-carbon(s) increases the rate of E2 reaction because of the increased stability of the alkene product.

2. The greater the nucleophilicity of the attacking reagent, the greater the S_N2-to-E2 ratio. Conversely, the greater the basicity of the attacking reagent, the greater the E2-to-S_N2 ratio.

Attack of base on a β-hydrogen by E2 is only slightly affected by branching at the α-carbon.

S_N2 attack of a nucleophile is impeded by branching at the α- and β-carbons.

Primary halides react with bases/nucleophiles to give predominantly substitution products. With strong bases, such as hydroxide ion and ethoxide ion, a percentage of the product is formed by an E2 reaction, but it is generally small compared with that formed by an S_N2 reaction. With strong bulky bases, such as *tert*-butoxide ion, the

Table 8.14 Summary of Substitution Versus Elimination Reactions of Haloalkanes

Halide	Reaction	Comments
Methyl CH$_3$X	S_N2	S_N1 reactions of methyl halides are never observed. The methyl cation is so unstable that it is never formed in solution.
Primary RCH$_2$X	S_N2	The main reaction with good nucleophiles/weak bases such as I$^-$ and CH$_3$COO$^-$.
	E2	The main reaction with strong, bulky bases such as potassium *tert*-butoxide.
	S_N1/E1	Primary cations are never formed in solution, and, therefore, S_N1 and E1 reactions of primary halides are never observed.
Secondary R$_2$CHX	S_N2	The main reaction with bases/nucleophiles where pK_a of the conjugate acid is 11 or less, as for example I$^-$ and CH$_3$COO$^-$.
	E2	The main reaction with bases/nucleophiles where the pK_a of the conjugate acid is 11 or greater, as for example OH$^-$ and CH$_3$CH$_2$O$^-$.
	S_N1/E1	Common in reactions with weak nucleophiles in polar protic solvents, such as water, methanol, and ethanol.
Tertiary R$_3$CHX	E2	Main reaction with strong bases such as HO$^-$ and RO$^-$.
	S_N1/E1	Main reactions with poor nucleophiles/weak bases.
	S_N2	S_N2 reactions of tertiary halides are rarely observed because of the extreme crowding around the 3° carbon.

E2 product becomes the major product. Tertiary halides react with all strong bases/good nucleophiles to give only elimination products.

Secondary halides are borderline, and substitution or elimination may be favored depending on the particular base/nucleophile, solvent, and temperature at which the reaction is carried out. Elimination is favored with strong bases/nucleophiles (the pK_a of the conjugate acid is 11 or above), as for example hydroxide ion and ethoxide ion. Substitution is favored with weak bases/nucleophiles (the pK_a of the conjugate acid is 11 or below, as for example acetate ion). The reason for this change in the ratio of $S_N2/E2$ is that as the pK_a of the conjugate acid increases, basicity increases faster than nucleophilicity. These generalizations about substitution versus elimination reactions of methyl, primary, secondary, and tertiary haloalkanes are summarized in Table 8.14.

Example 8.8

Predict whether each reaction proceeds predominantly by substitution (S_N1 or S_N2) or elimination (E1 or E2) or whether the two compete. Write structural formulas for the major organic product(s).

(a) [structure: 2-chloro-2-methylbutane] + NaOH $\xrightarrow[H_2O]{80°C}$

(b) [structure: 1-bromo-3-methylbutane] + $(C_2H_5)_3N$ $\xrightarrow[CH_2Cl_2]{30°C}$

Solution

(a) A 3° halide is heated with a strong base/good nucleophile. Elimination by an E2 reaction predominates to give 2-methyl-2-butene as the major product.

[structure] + NaOH $\xrightarrow[H_2O]{80°C}$ [structure 2-methyl-2-butene] + NaCl + H_2O

(b) Reaction of a 1° halide with this moderate nucleophile/weak base gives substitution by an S_N2 reaction.

[structure] + $(C_2H_5)_3N$ $\xrightarrow[CH_2Cl_2]{30°C}$ [structure] $\overset{+}{N}(C_2H_5)_3$ Br^-

Problem 8.8

Predict whether each reaction proceeds predominantly by substitution (S_N1 or S_N2) or elimination (E1 or E2) or whether the two compete. Write structural formulas for the major organic product(s).

(a) [structure with Br] + $CH_3O^-Na^+$ $\xrightarrow[methanol]{}$

(b) [cyclohexane structure with Cl] + Na^+I^- $\xrightarrow[acetone]{}$

(c) C_6H_5 [structure with Br] + Na^+CN^- $\xrightarrow[methanol]{}$

Summary

A **nucleophile** is a molecule or ion that donates a pair of electrons to another atom or ion to form a new covalent bond; that is, a nucleophile is a Lewis base. **Nucleophilic substitution** is any reaction in which one nucleophile is substituted for another. There are two limiting mechanisms for nucleophilic substitution, namely S_N2 and S_N1.

Protic solvents are hydrogen-bond donors (Section 8.2). The most common protic solvents are those containing —OH groups. **Aprotic solvents** cannot serve as hydrogen-bond donors. Common aprotic solvents are acetone, diethyl ether, dimethyl sulfoxide, and N,N-dimethylformamide. **Polar solvents** interact strongly with ions and polar molecules. **Nonpolar solvents** do not interact strongly with ions and polar molecules. The **dielectric constant** is the most commonly used measure of solvent polarity.

Solvolysis is a nucleophilic substitution reaction in which the solvent is the nucleophile (Section 8.3).

The **nucleophilicity** of a reagent is measured by the rate of its reaction in a reference nucleophilic substitution (Section 8.4B). Relative nucleophilicities depend on whether a reaction is carried out in a polar protic solvent or a polar aprotic solvent. A general principle is that the freer a nucleophile is from a surrounding solvation shell, the greater its nucleophilicity.

The ability of a group to function as a leaving group is related to its stability as an anion (Section 8.4F). The most stable anions and the best leaving groups are the conjugate bases of strong acids.

Certain nucleophilic displacements that have the kinetic characteristic of S_N1 reactions (first order in haloalkane and zero order in nucleophile) involve two successive S_N2 reactions. Many such reactions involve participation of a neighboring nucleophile. The mustard gases are one group of compounds whose nucleophilic substitution reactions involve neighboring group participation (Section 8.5).

A **phase-transfer catalyst** is a substance that transports ions from an aqueous phase into an organic phase and vice versa (Section 8.7). Effective phase-transfer catalysts for anions have a balanced combination of (1) **hydrophilic** character to dissolve in water and form an ion pair with the anion to be transferred and (2) **hydrophobic** character to dissolve in the organic phase and transfer the anion into it.

A **β-elimination reaction** (Section 8.8) involves removal of atoms or groups of atoms from adjacent carbon atoms. β-Elimination to give the more highly substituted alkene is called **Zaitsev elimination.**

Key Reactions

1. Nucleophilic Aliphatic Substitution: S_N2 (Section 8.3)

S_N2 reactions occur in one step; departure of the leaving group is assisted by the incoming nucleophile, and both nucleophile and leaving group are involved in the transition state. The nucleophile may be negatively charged as in the first example or neutral as in the second example. S_N2 reactions result in inversion of configuration at the reaction center. They are accelerated in polar aprotic solvents compared with polar protic solvents. S_N2 reactions are governed by steric factors, namely the degree of crowding around the site of reaction.

2. Nucleophilic Aliphatic Substitution: S_N1 (Section 8.3)

An S_N1 reaction occurs in two steps. Step 1 is a slow, rate-determining ionization of the C—X bond to form a carbocation intermediate followed in Step 2 by rapid reaction of the carbocation intermediate with a nucleophile to complete the substitution. Reaction at a stereocenter gives largely racemization, often accompanied with a slight excess of inversion of configuration. Reactions often involve carbocation rearrangements and are accelerated by polar protic solvents. S_N1 reactions are governed by electronic factors, namely the relative stabilities of carbocation intermediates.

$$C_6H_5 \overset{\displaystyle |}{\underset{\displaystyle Cl}{\big\langle}} + CH_3OH \longrightarrow C_6H_5 \overset{\displaystyle |}{\underset{\displaystyle OCH_3}{\big\langle}} + HCl$$

3. Neighboring Group Participation (Section 8.5)

Neighboring group participation is characterized by first-order kinetics and participation of an internal nucleophile in the departure of the leaving group, as in hydrolysis of a sulfur or nitrogen mustard gas. The mechanism for their solvolysis involves two successive nucleophilic displacements.

$$Cl \diagdown\diagup^N\diagdown\diagup Cl + 2H_2O \longrightarrow HO \diagdown\diagup^N\diagdown\diagup OH + 2HCl$$

4. β-Elimination: E1 (Section 8.9A)

An E1 reaction occurs in two steps: slow, rate-determining breaking of the C—X bond to form a carbocation intermediate followed by rapid proton transfer to solvent to form an alkene. An E1 reaction is first order in haloalkane and zero order in base. Skeletal rearrangements are common.

$$\overset{\displaystyle |}{\underset{\displaystyle Cl}{\big\langle}} \xrightarrow[\text{CH}_3\text{COOH}]{\text{E1}} \diagup\diagdown + HCl$$

5. β-Elimination: E2 (Section 8.9B)

An E2 reaction occurs in one step: simultaneous reaction with base to remove a hydrogen, formation of the alkene, and departure of the leaving group. Elimination is stereospecific, requiring an anti and coplanar arrangement of the groups being eliminated.

$$\xrightarrow[\text{CH}_3\text{CH}_2\text{OH}]{\text{CH}_3\text{CH}_2\text{O}^-\text{Na}^+}$$

Problems

Nucleophilic Aliphatic Substitution

8.9   Draw a structural formula for the most stable carbocation of each molecular formula.

(a) $C_4H_9^+$ (b) $C_3H_7^+$ (c) $C_8H_{15}^+$ (d) $C_3H_7O^+$

8.10   The reaction of 1-bromopropane and sodium hydroxide in ethanol follows an S_N2 mechanism. What happens to the rate of this reaction under the following conditions?

(a) The concentration of NaOH is doubled.

(b) The concentrations of both NaOH and 1-bromopropane are doubled.

(c) The volume of the solution in which the reaction is carried out is doubled.

8.11 From each pair, select the stronger nucleophile.

(a) H_2O or OH^- (b) CH_3COO^- or OH^-
(c) CH_3SH or CH_3S^- (d) Cl^- or I^- in DMSO
(e) Cl^- or I^- in methanol   (f) CH_3OCH_3 or CH_3SCH_3

8.12 Draw a structural formula for the product of each S_N2 reaction. Where configuration of the starting material is given, show the configuration of the product.

(a) $CH_3CH_2CH_2Cl + CH_3CH_2ONa \xrightarrow{\text{ethanol}}$ (b) $(CH_3)_3N + CH_3I \xrightarrow{\text{acetone}}$

(c) $-CH_2Br + NaCN \xrightarrow{\text{acetone}}$ (d) $Cl + CH_3SNa \xrightarrow{\text{ethanol}}$

(e) $CH_3CH_2CH_2Cl + CH_3C{\equiv}C^-Li^+ \xrightarrow{\text{ether}}$ (f) $-CH_2Cl + NH_3 \xrightarrow{\text{ethanol}}$

(g) $O\underset{}{\diagdown}NH + CH_3(CH_2)_6CH_2Cl \xrightarrow{\text{ethanol}}$ (h) $CH_3CH_2CH_2Br + NaCN \xrightarrow{\text{acetone}}$

8.13 You were told that each reaction in Problem 8.12 proceeds by an S_N2 mechanism. Suppose that you were not told the mechanism. Describe how you could conclude from the structure of the haloalkane, the nucleophile, and the solvent that each reaction is in fact an S_N2 reaction.

8.14 Treatment of 1,3-dichloropropane with potassium cyanide results in the formation of propanedinitrile. The rate of this reaction is about 1000 times greater in DMSO than it is in ethanol. Account for this difference in rate.

1,3-Dichloropropane   Propanedinitrile

8.15 Treatment of 1-aminoadamantane, $C_{10}H_{17}N$, with methyl 2,4-dibromobutanoate involves two successive S_N2 reactions and gives compound A. Propose a structural formula for compound A.

1-Aminoadamantane   Methyl   A
                      2,4-dibromobutanoate

8.16 Select the member of each pair that shows the greater rate of S_N2 reaction with KI in acetone.

8.17 Select the member of each pair that shows the greater rate of S_N2 reaction with KN_3 in acetone.

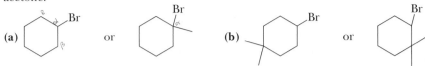

(a)   or   (b)   or

8.18 What hybridization best describes the reacting carbon in the S_N2 transition state?

8.19 Each carbocation is capable of rearranging to a more stable carbocation. Limiting yourself to a single 1,2-shift, suggest a structure for the rearranged carbocation.

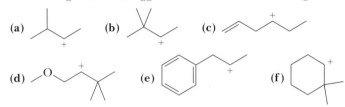

(a)   (b)   (c)

(d)   (e)   (f)

8.20 Attempts to prepare optically active iodides by nucleophilic displacement on optically active compounds with I^- normally produce racemic iodoalkanes. Why are the product iodoalkanes racemic?

8.21 Draw a structural formula for the product of each S_N1 reaction. Where configuration of the starting material is given, show the configuration of the product.

(a)   Cl $+ \ CH_3CH_2OH \xrightarrow{\text{ethanol}}$ (b) $+ \ CH_3OH \xrightarrow{\text{methanol}}$ Cl

       S enantiomer

(c) $CH_3CCl + CH_3COH \xrightarrow{\text{acetic acid}}$ (d) $-Br + CH_3OH \xrightarrow{\text{methanol}}$

     CH_3 O

     CH_3

8.22 Suppose that you were told that each reaction in Problem 8.21 is a substitution reaction but were not told the mechanism. Describe how you could conclude from the structure of the haloalkane or cycloalkene, the nucleophile, and the solvent that each reaction is in fact an S_N1 reaction.

8.23 Alkenyl halides such as vinyl bromide, $CH_2{=}CHBr$, undergo neither S_N1 nor S_N2 reactions. What factors account for this lack of reactivity?

8.24 Select the member of each pair that undergoes S_N1 solvolysis in aqueous ethanol more rapidly.

(a)   Cl   or $-$Cl   (b) $-$Cl   or $-$Br

(c)   Cl   or   Cl   (d)   Cl   or   Cl

(e)   Cl   or   Cl   (f)   Br   or   Br

8.25 Account for the following relative rates of solvolysis under experimental conditions favoring S_N1 reaction.

Relative rate of   0.2   1 10^9
solvolysis (S_N1)

8.26 Not all tertiary halides undergo S_N1 reactions readily. For example, 1-iodobicyclo[2.2.2]octane is very unreactive under S_N1 conditions. What feature of this molecule is responsible for such lack of reactivity? You will find it helpful to examine the CD model of this compound and the cation it would form.

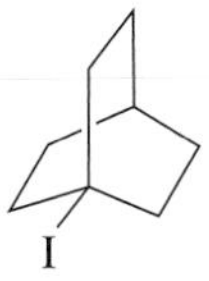

1-Iodobicyclo[2.2.2]octane

8.27 Show how you might synthesize the following compounds from a haloalkane and a nucleophile.

(a) 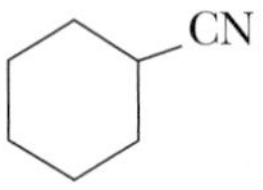   (b) 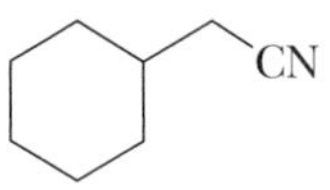   (c) 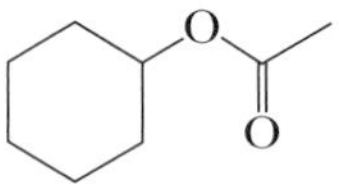   (d)

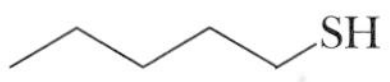

(e) 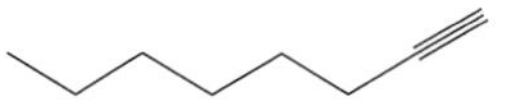   (f) 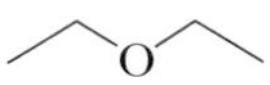   (g)

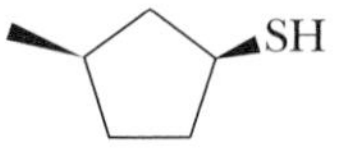

8.28 3-Chloro-1-butene reacts with sodium ethoxide in ethanol to produce 3-ethoxy-1-butene. The reaction is second order; first order in 3-chloro-1-butene and first order in sodium ethoxide. In the absence of sodium ethoxide, 3-chloro-1-butene reacts with ethanol to produce both 3-ethoxy-1-butene and 1-ethoxy-2-butene. Explain these results.

8.29 1-Chloro-2-butene undergoes hydrolysis in warm water to give a mixture of these allylic alcohols. Propose a mechanism for their formation.

$$CH_3CH{=}CHCH_2Cl \xrightarrow{H_2O} CH_3CH{=}CHCH_2OH + CH_3\overset{\displaystyle OH}{\underset{\displaystyle |}{C}}HCH{=}CH_2$$

1-Chloro-2-butene   2-Buten-1-ol   3-Buten-2-ol

8.30 The following nucleophilic substitution occurs with rearrangement. Suggest a mechanism for formation of the observed product. If the starting material has the S configuration, what is the configuration of the stereocenter in the product?

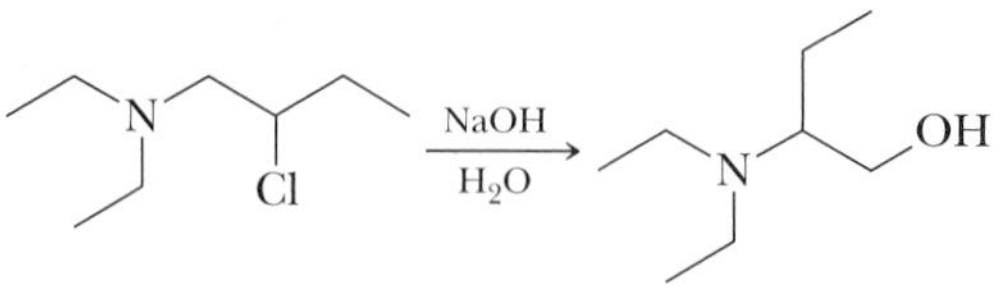

8.31 Propose a mechanism for the formation of these products in the solvolysis of this bromobicycloalkane.

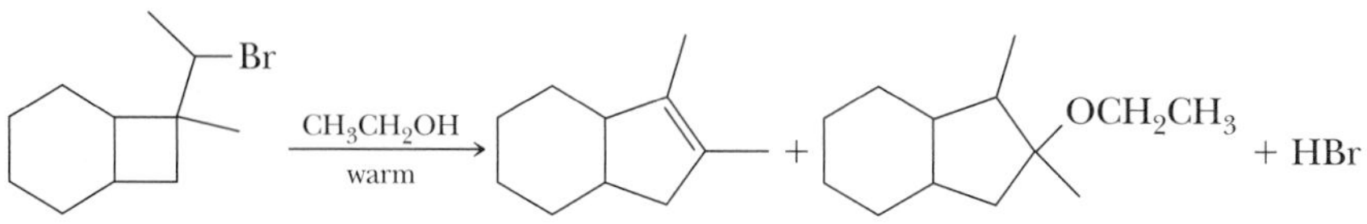

8.32 Solvolysis of the following bicyclic compound in acetic acid gives a mixture of products, two of which are shown. The leaving group is the anion of a sulfonic acid, $ArSO_3H$. A sulfonic acid is a strong acid, and its anion, $ArSO_3^-$, is a weak base and a good leaving group. Propose a mechanism for this reaction.

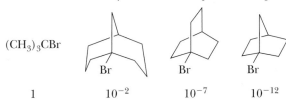

8.33 Which compound in each set undergoes more rapid solvolysis when refluxed in ethanol? Show the major product formed from the more reactive compound.

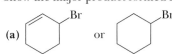

8.34 Account for the relative rates of solvolysis of these compounds in aqueous acetic acid.

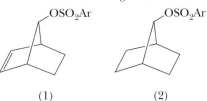

8.35 A comparison of the rates of S_N1 solvolysis of the bicyclic compounds (1) and (2) in acetic acid shows that compound (1) reacts 10^{11} times faster than compound (2). Furthermore, solvolysis of (1) occurs with complete retention of configuration: the nucleophile occupies the same position on the one-carbon bridge as did the leaving OSO_2Ar group.

(1)   (2)

(a) Draw structural formulas for the products of solvolysis of each compound.
(b) Account for the difference in rate of solvolysis of (1) and (2).
(c) Account for complete retention of configuration in the solvolysis of (1).

β-Eliminations

8.36 Draw structural formulas for the alkene(s) formed by treatment of each haloalkane or halocycloalkane with sodium ethoxide in ethanol. Assume that elimination occurs by an E2 mechanism.

    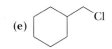 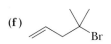

8.37 Draw structural formulas of all chloroalkanes that undergo dehydrohalogenation when treated with KOH to give each alkene as the major product. For some parts, only one chloroalkane gives the desired alkene as the major product. For other parts, two chloroalkanes may work.

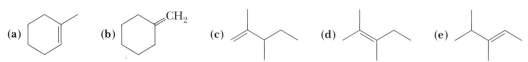

8.38 Following are diastereomers (A) and (B) of 3-bromo-3,4-dimethylhexane. On treatment with sodium ethoxide in ethanol, each gives 3,4-dimethyl-3-hexene as the major product. One diastereomer gives the (*E*)-alkene, and the other gives the (*Z*)-alkene. Which diastereomer gives which alkene? Account for the stereospecificity of each β-elimination.

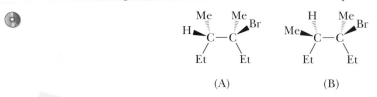

8.39 Treatment of the following stereoisomer of 1-bromo-2,3-diphenylpropane with sodium ethoxide in ethanol gives a single stereoisomer of 1,2-diphenylpropene. Predict whether the product has the *E* configuration or the *Z* configuration.

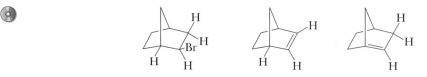

1-Bromo-2,3-diphenylpropane   1,2-Diphenylpropene

8.40 Elimination of HBr from 2-bromonorbornane gives only 2-norbornene and no 1-norbornene. How do you account for the regiospecificity of this dehydrohalogenation? In answering this question, you will find it helpful to look at molecular models of both 1-norbornene and 2-norbornene on the CD and to analyze the strain in each.

2-Bromonorbornane   2-Norbornene   1-Norbornene

8.41 Which isomer reacts faster when refluxed with potassium *tert*-butoxide in *tert*-butyl alcohol, *cis*-1-bromo-3-isopropylcyclohexane, or *trans*-1-bromo-3-isopropylcyclohexane? Draw the structure of the expected product from the faster reacting compound.

Substitution Versus Elimination

8.42 Consider the following statements in reference to S$_N$1, S$_N$2, E1, and E2 reactions of haloalkanes. To which mechanism(s), if any, does each statement apply?

(a) Involves a carbocation intermediate.
(b) Is first order in haloalkane and first order in nucleophile.
(c) Involves inversion of configuration at the site of substitution.
(d) Involves retention of configuration at the site of substitution.
(e) Substitution at a stereocenter gives predominantly a racemic product.

(f) Is first order in haloalkane and zero order in base.
(g) Is first order in haloalkane and first order in base.
(h) Is greatly accelerated in protic solvents of increasing polarity.
(i) Rearrangements are common.
(j) Order of reactivity of haloalkanes is $3° > 2° > 1°$.
(k) Order of reactivity of haloalkanes is methyl $> 1° > 2° > 3°$.

8.43 Arrange these haloalkanes in order of increasing ratio of E2 to S_N2 products observed on reaction of each with sodium ethoxide in ethanol.

                      CH_3 CH_3 CH_3
                      | | |
(a) CH_3CH_2Br **(b)** CH_3CHCH_2Br **(c)** $CH_3CCH_2CH_3$ **(d)** $CH_3CHCH_2CH_2Br$
                                           |
                                         Cl

8.44 Draw a structural formula for the major organic product of each reaction and specify the most likely mechanism by which each is formed.

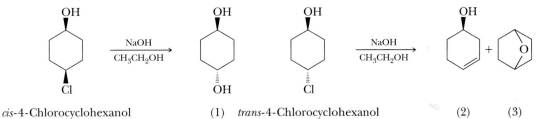

8.45 When *cis*-4-chlorocyclohexanol is treated with sodium hydroxide in ethanol, it gives only the substitution product *trans*-1,4-cyclohexanediol (1). Under the same reaction conditions, *trans*-4-chlorocyclohexanol gives 3-cyclohexenol (2) and the bicyclic ether (3).

cis-4-Chlorocyclohexanol   (1)   *trans*-4-Chlorocyclohexanol   (2)   (3)

(a) Propose a mechanism for formation of product (1), and account for its configuration.
(b) Propose a mechanism for formation of product (2).
(c) Account for the fact that the bicyclic ether (3) is formed from the trans isomer but not from the cis isomer.

Synthesis

8.46 Show how to convert the given starting material into the desired product. Note that some syntheses require only one step, whereas others require two or more.

8.47 The Williamson ether synthesis involves treatment of a haloalkane with a metal alkoxide. Following are two reactions intended to give *tert*-butyl ethyl ether. One reaction gives the ether in good yield, the other reaction does not. Which reaction gives the ether? What is the product of the other reaction, and how do you account for its formation?

8.48 The following ethers can, in principle, be synthesized by two different combinations of haloalkane or halocycloalkane and metal alkoxide. Show one combination that forms ether bond (1) and another that forms ether bond (2). Which combination gives the higher yield of ether?

8.49 Propose a mechanism for this reaction.

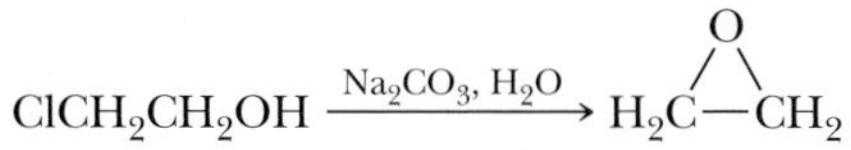

$$ClCH_2CH_2OH \xrightarrow{Na_2CO_3, H_2O} H_2C\overset{\displaystyle O}{\underset{\diagdown\quad\diagup}{\text{—}}}CH_2$$

8.50 Each of these compounds can be synthesized by an S_N2 reaction. Suggest a combination of haloalkane and nucleophile that will give each product.

(a) CH_3OCH_3

(b) CH_3SH

(c) $CH_3CH_2CH_2PH_2$

(d) CH_3CH_2CN

(e) $CH_3SCH_2C(CH_3)_3$

(f) $(CH_3)_3NH^+Cl^-$

(g) $C_6H_5\overset{\displaystyle O}{\overset{\|}{C}}OCH_2C_6H_5$

(h) $(R)\text{-}CH_3\overset{\displaystyle N_3}{\overset{|}{C}}HCH_2CH_2CH_3$

(i) $CH_2{=}CHCH_2OCH(CH_3)_2$

(j) $CH_2{=}CHCH_2OCH_2CH{=}CH_2$

(k)

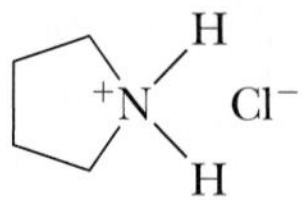

(l)

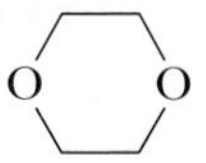

9

ALCOHOLS AND THIOLS

In this chapter, we study the physical and chemical properties of alcohols, a class of oxygen-containing compounds. We also study thiols, a class of sulfur-containing compounds. A **thiol** is like an alcohol in structure, except that it contains an —SH group rather than an —OH group.

$$CH_3CH_2OH \qquad CH_3CH_2SH$$

Ethanol   Ethanethiol

(an alcohol)   (a thiol)

■ Fermentation vats of wine grapes at the Beaulieu Vineyards, California. *(Earl Roberge/Photo Researchers, Inc.)* Inset: A model of ethanol.

Ethanol is the fuel additive in gasohol, the alcohol in alcoholic beverages, and an important industrial solvent. Ethanethiol, like all other low-molecular-weight thiols, has a stench; such smells from skunks, rotten eggs, and sewage are caused by thiols or H_2S.

Alcohols are particularly important in the laboratory and biochemical transformations of organic compounds. They can be converted into many other types of compounds, such as alkenes, alkyl halides, aldehydes, ketones, carboxylic acids, and esters. Not only can alcohols be converted to these compounds, but they can also be prepared from them. Thus alcohols play a central role in the interconversion of organic functional groups.

Because sulfur and oxygen are both Group 6 elements, thiols and alcohols undergo many of the same types of reactions. Sulfur, a third row element, however, can expand its valence shell to include more than eight electrons; therefore, thiols undergo some reactions that are not possible for alcohols. In addition, sulfur's electronegativity and basicity are less than those of oxygen.

9.1 Structure and Nomenclature of Alcohols

A. Structure

The functional group of an alcohol is an —OH (hydroxyl) group bonded to an sp^3 hybridized carbon (Section 1.3A). The oxygen atom of an alcohol is also sp^3 hybridized. Two sp^3 hybrid orbitals of oxygen form sigma bonds to atoms of carbon and hydrogen, and the remaining two sp^3 hybrid orbitals each contain an unshared pair of electrons. Figure 9.1 shows a Lewis structure and a ball-and-stick model of methanol, CH_3OH, the simplest alcohol. The measured H—C—O bond angle in methanol is 108.9°, very close to the tetrahedral angle of 109.5°.

B. Nomenclature

Nomenclature
Alcohols

In the IUPAC system, the longest chain of carbon atoms containing the —OH group is selected as the parent alkane and numbered from the end closer to —OH. To show that the compound is an alcohol, change the suffix -e of the parent alkane to -ol (Section 2.5), and use a number to show the location of the —OH group. The location of the —OH group takes precedence over alkyl and halogen groups in numbering the parent chain. For cyclic alcohols, numbering begins with the carbon bearing the —OH group. Because the —OH group is understood to be on carbon 1 of the ring, there is no need give its location a number. In complex alcohols, the number for the hydroxyl group is often placed between the infix and the suffix. Thus, for example, 2-methyl-1-propanol and 2-methylpropan-1-ol are both acceptable

(a)   (b)

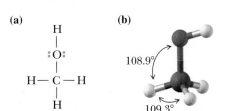

Figure 9.1
Methanol, CH_3OH. (a) Lewis structure and (b) ball-and-stick model.

names. For bicyclic alcohols, the numbering of the bicyclic ring takes precedence over the location of the —OH group, and a number must be given to locate the —OH group.

Common names for alcohols are derived by naming the alkyl group attached to —OH and then adding the word "alcohol." Here are IUPAC names and, in parentheses, common names for several low-molecular-weight alcohols.

$$CH_3CH_2CH_2\text{\underline{OH}}$$

1-Propanol
(Propyl alcohol)

$$\overset{\text{OH}}{\underset{\mid}{CH_3CHCH_3}}$$

2-Propanol
(Isopropyl alcohol)

$$CH_3CH_2CH_2CH_2\text{\underline{OH}}$$

1-Butanol
(Butyl alcohol)

$$\overset{\text{OH}}{\underset{\mid}{CH_3CH_2CHCH_3}}$$

2-Butanol
(*sec*-butyl alcohol)

$$\overset{\text{CH}_3}{\underset{\mid}{CH_3CHCH_2\text{\underline{OH}}}}$$

2-Methyl-1-propanol
(Isobutyl alcohol)

$$\overset{\text{CH}_3}{\underset{\underset{\text{CH}_3}{\mid}}{\overset{\mid}{CH_3\text{C}\text{\underline{OH}}}}}$$

2-Methyl-2-propanol
(*tert*-butyl alcohol)

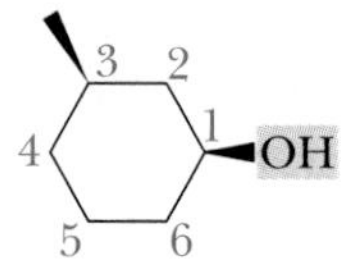

cis-3-methylcyclohexanol

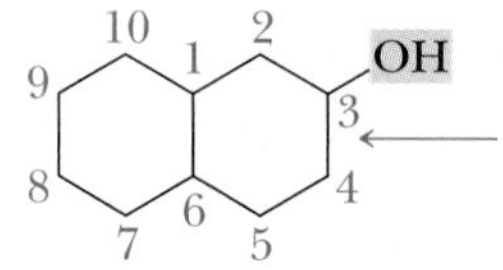

Bicyclo[4.4.0]decan-3-ol

Numbering of the bicyclic ring takes precedence over the location of —OH.

Example 9.1

Write IUPAC names for these alcohols.

(a) 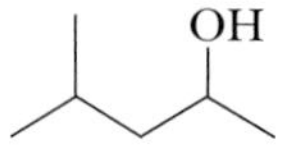   (b) 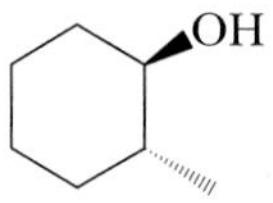   (c)

Solution

(a) 4-Methyl-2-pentanol   (b) *trans*-2-Methylcyclohexanol
(c) Bicyclo[2.2.1]heptan-1-ol

Problem 9.1

Write IUPAC names for these alcohols, and include the configuration for (a).

(a) 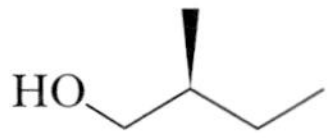   (b) 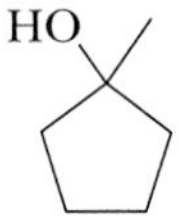   (c)

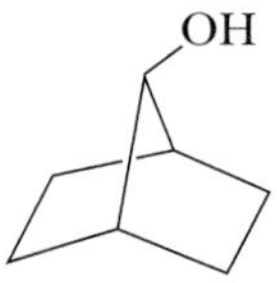

We classify alcohols as primary (1°), secondary (2°), or tertiary (3°) depending on whether the —OH group is on a primary, secondary, or tertiary carbon.

$$
\begin{array}{ccc}
H & R' & R' \\
| & | & | \\
R-C-OH & R-C-OH & R-C-OH \\
| & | & | \\
H & H & R''
\end{array}
$$

Primary (1°) Secondary (2°) Tertiary (3°)

Example 9.2

Classify each alcohol as primary, secondary, or tertiary.

(a) [cyclohexyl—C(OH)(CH₃)(H) structure with OH] (b) $CH_3\overset{\displaystyle CH_3}{\underset{\displaystyle CH_3}{C}OH}$ (c) [cyclopentyl—CH₂OH]

Solution

(a) Secondary (2°) (b) Tertiary (3°) (c) Primary (1°)

Problem 9.2

Classify each alcohol as primary, secondary, or tertiary.

(a) $CH_3\overset{\displaystyle CH_3}{\underset{\displaystyle CH_3}{C}CH_2OH}$ (b) [cyclopropyl]—OH (c) $CH_2{=}CHCH_2OH$ (d) [cyclopentane with CH₃ and OH]

Commercial antifreeze contains ethylene glycol. (*Charles D. Winters*)

In the IUPAC system, a compound containing two hydroxyl groups is named as a **diol,** one containing three hydroxyl groups is named as a **triol,** and so on. In IUPAC names for diols, triols, and so on, the final -e (the suffix) of the parent alkane name is retained, as for example in the name 1,2-ethanediol. As with many organic compounds, common names for certain diols and triols have persisted. Compounds containing two hydroxyl groups on adjacent carbons are often referred to as **glycols** (Section 6.5B). Ethylene glycol and propylene glycol are synthesized from ethylene and propylene, respectively, hence their common names.

Diol A compound containing two hydroxyl groups.
Triol A compound containing three hydroxyl groups.
Glycol A compound containing two hydroxyl groups on adjacent carbons.

$$
\underset{\substack{| \quad | \\ OH \quad OH}}{CH_2CH_2}
$$

1,2-Ethanediol
(Ethylene glycol)

$$
\underset{\substack{| \quad | \\ HO \quad OH}}{CH_3CHCH_2}
$$

1,2-Propanediol
(Propylene glycol)

$$
\underset{\substack{| \quad | \quad | \\ HO \quad HO \quad OH}}{CH_2CHCH_2}
$$

1,2,3-Propanetriol
(Glycerol, Glycerine)

Compounds containing —OH and C=C groups are often referred to as **unsaturated alcohols** because of the presence of the carbon-carbon double bond. In the IUPAC system, the double bond is shown by changing the infix of the parent alkane from -an- to -en- (Section 2.5), and the hydroxyl group is shown by changing the suffix of the parent alkane from -e to -ol. Numbers must be used to show the location of both the carbon-carbon double bond and the hydroxyl group. The parent alkane is numbered to give the —OH group the lowest possible number; that is, the group shown by a suffix (in this case, -ol) takes precedence over the group shown by an infix (in this case, -en-).

Example 9.3

Write IUPAC names for these unsaturated alcohols.

(a) CH_2=$CHCH_2OH$ (b) HO⌒⌒⌒

Solution

(a) 2-Propen-1-ol. Its common name is allyl alcohol.
(b) (*E*)-2-Hexen-1-ol (*trans*-2-hexen-1-ol).

Problem 9.3

Write IUPAC names for these unsaturated alcohols.

(a) ⌒⌒OH   (b) ⌒⌒OH
with OH

9.2   Physical Properties of Alcohols

Because of the presence of the polar —OH group, alcohols are polar compounds, with partial positive charges on carbon and hydrogen and a partial negative charge on oxygen (Figure 9.2).

The attraction between the positive end of one dipole and the negative end of another is called **dipole-dipole interaction.** When the positive end of one of the dipoles is a hydrogen atom bonded to F, O, or N (atoms of high electronegativity) and the negative end of the other dipole is an F, O, or N atom, the attractive interaction between dipoles is particularly strong and is given the special name of **hydrogen bonding.** The length of a hydrogen bond in water is 177 pm, about 80% longer than an O—H covalent bond. The strength of a hydrogen bond in water is approximately 21 kJ (5 kcal)/mol. For comparison, the strength of the O—H covalent bond in water is approximately 498 kJ (119 kcal)/mol. As can be seen by comparing these numbers, an O----H hydrogen bond is considerably weaker than an O—H covalent bond. The presence of a large number of hydrogen bonds in liquid water, however, has an important effect on the physical properties of water. Because of hydrogen bonding, extra energy is required to separate each water molecule from its neighbors, hence the relatively high boiling point of water.

Similarly, there is extensive hydrogen bonding between alcohol molecules in the pure liquid. Figure 9.3 shows the association of ethanol molecules by hydrogen bond-

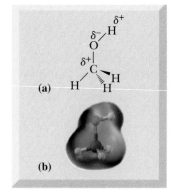

Figure 9.2
Methanol. (a) Polarity and (b) an electrostatic potential map.

Dipole-dipole interaction   The attraction between the positive end of one dipole and the negative end of another.

Hydrogen bonding   The attractive interaction between a hydrogen atom bonded to an atom of high electronegativity (most commonly F, O, or N) and a lone pair of electrons on another atom of high electronegativity (again, most commonly F, O, or N).

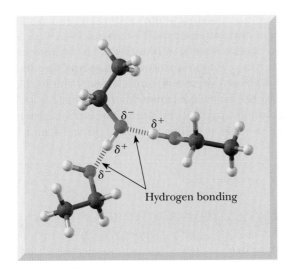

Figure 9.3
The association of ethanol molecules in the liquid state. Each O—H can participate in up to three hydrogen bonds (one through hydrogen and two through oxygen). Only two of these three possible hydrogen bonds per molecule are shown here.

ing between the partially negative oxygen atom of one ethanol molecule and the partially positive hydrogen atom of another ethanol molecule.

Table 9.1 lists the boiling points and solubilities in water for several groups of alcohols and hydrocarbons of similar molecular weight. Of the compounds compared in each group, the alcohols have the higher boiling points because more energy is needed to overcome the attractive forces of hydrogen bonding between their polar —OH groups. The presence of additional hydroxyl groups in a molecule further increases the extent of hydrogen bonding, as can be seen by comparing the boiling points of hexane (bp 69°C), 1-pentanol (bp 138°C), and 1,4-butanediol (bp 230°C), all of which have approximately the same molecular weight. Because of increased dispersion forces between larger molecules, boiling points of all types of compounds,

Table 9.1   Boiling Points and Solubilities in Water of Five Groups of Alcohols and Hydrocarbons of Similar Molecular Weight

Structural Formula	Name	Molecular Weight	bp (°C)	Solubility in Water
CH_3OH	Methanol	32	65	Infinite
CH_3CH_3	Ethane	30	− 89	Insoluble
CH_3CH_2OH	Ethanol	46	78	Infinite
$CH_3CH_2CH_3$	Propane	44	− 42	Insoluble
$CH_3CH_2CH_2OH$	1-Propanol	60	97	Infinite
$CH_3CH_2CH_2CH_3$	Butane	58	0	Insoluble
$CH_3CH_2CH_2CH_2OH$	1-Butanol	74	117	8 g/100 g
$CH_3CH_2CH_2CH_2CH_3$	Pentane	72	36	Insoluble
$HOCH_2CH_2CH_2CH_2OH$	1,4-Butanediol	90	230	Infinite
$CH_3CH_2CH_2CH_2CH_2OH$	1-Pentanol	88	138	2.3 g/100 g
$CH_3CH_2CH_2CH_2CH_2CH_3$	Hexane	86	69	Insoluble

including alcohols, increase with increasing molecular weight. Compare, for example, the boiling points of ethanol, 1-propanol, 1-butanol, and 1-pentanol.

The effect of hydrogen bonding in alcohols is illustrated dramatically by comparing the boiling points of ethanol (bp 78°C) and its constitutional isomer dimethyl ether (bp − 24°C). The difference in boiling point between these two compounds is due to the presence of a polar O—H group in the alcohol, which is capable of forming intermolecular hydrogen bonds. This hydrogen bonding increases the attractive forces between molecules of ethanol, and, thus, ethanol has a higher boiling point than dimethyl ether.

$$CH_3CH_2OH \qquad CH_3OCH_3$$

Ethanol   Dimethyl ether
bp 78°C   bp − 24°C

Because alcohols can interact by hydrogen bonding with water, they are more soluble in water than alkanes and alkenes of comparable molecular weight. Methanol, ethanol, and 1-propanol are soluble in water in all proportions. As molecular weight increases, the physical properties of alcohols become more like those of hydrocarbons of comparable molecular weight. Alcohols of higher molecular weight are much less soluble in water because of the increase in size of the hydrocarbon portion of the molecule.

Example 9.4

Following are three alcohols of molecular formula $C_4H_{10}O$. Their boiling points, listed from lowest to highest, are 82.3°C, 99.5°C, and 117°C. Which alcohol has which boiling point?

$$CH_3CH_2CH_2CH_2OH \qquad CH_3CH_2\overset{\displaystyle OH}{\underset{\displaystyle |}{C}}HCH_3 \qquad CH_3\overset{\displaystyle CH_3}{\underset{\displaystyle CH_3}{\overset{|}{\underset{|}{C}}}}OH$$

1-Butanol   2-Butanol   2-Methyl-2-propanol

Solution

Boiling points of these constitutional isomers depend on the strength of intermolecular hydrogen bonding. The primary —OH group of 1-butanol is the most accessible for intermolecular hydrogen bonding; this alcohol has the highest boiling point, 117°C. The tertiary —OH group of 2-methyl-2-propanol is the least accessible for intermolecular hydrogen bonding; this alcohol has the lowest boiling point, 82.3°C.

Problem 9.4

Arrange these compounds in order of increasing boiling point.

Example 9.5

Arrange these compounds in order of increasing solubility in water.

Solution

Hexane, C_6H_{14}, a nonpolar hydrocarbon, has the lowest solubility in water. Both 1-pentanol and 1,4-butanediol are polar compounds due to the presence of —OH groups, and each interacts with water molecules by hydrogen bonding. Because 1,4-butanediol has more sites within its molecules for hydrogen bonding than 1-pentanol, it is more soluble in water than 1-pentanol. The water solubilities of these compounds are given in Table 9.1.

Insoluble	2.3 g/100 mL	Infinitely soluble

Problem 9.5

Arrange these compounds in order of increasing solubility in water.

9.3   Acidity and Basicity of Alcohols

Supporting Concepts;
Acids and Bases

In dilute aqueous solution, alcohols are very weakly acidic as shown by the ionization of methanol.

$$CH_3\overset{\frown}{O}-H + :\overset{..}{O}-H \rightleftharpoons CH_3\overset{..}{O}:^- + H-\overset{+}{O}-H \qquad K_a = \frac{[CH_3O^-][H_3O^+]}{[CH_3OH]} = 10^{-15.5}$$

$$\underset{H}{} \qquad \underset{H}{} \qquad pK_a = 15.5$$

Shown in Table 9.2 are acid ionization constants for several low-molecular-weight alcohols. Methanol and ethanol are about as acidic as water. Higher molecular-weight, water-soluble alcohols are slightly weaker acids than water. Thus, although alcohols have some acidity, they are not strong enough acids to react with weak bases such as sodium bicarbonate or sodium carbonate. (At this point, it would be wise to review Section 4.4 and the discussion of the position of equilibrium in acid-base reactions.)

For simple alcohols, such as methanol and ethanol, acidity depends primarily on the degree of solvation and stabilization of the alkoxide ion by water molecules. The negatively charged oxygen atoms of the methoxide and ethoxide ions are almost

Table 9.2	pK$_a$ Values for Selected Alcohols in Dilute Aqueous Solution*		
Compound	**Structural Formula**	**pK$_a$**	
Acetic acid	CH_3COOH	4.8	Stronger acid
Methanol	CH_3OH	15.5	
Water	**H_2O**	**15.7**	
Ethanol	CH_3CH_2OH	15.9	
2-Propanol	$(CH_3)_2CHOH$	17	
2-Methyl-2-propanol	$(CH_3)_3COH$	18	Weaker acid

*Also given for comparison are pK$_a$ values for water and acetic acid.

equally accessible for solvation as is the hydroxide ion; therefore, these alcohols are about as acidic as water. As the bulk of the alkyl group bonded to oxygen increases, the ability of water molecules to solvate the alkoxide ion decreases. 2-Methyl-2-propanol (*tert*-butyl alcohol) is a weaker acid than either methanol or ethanol, primarily because of the bulk of the *tert*-butyl group, which reduces solvation of the *tert*-butoxide anion by surrounding water molecules.

In the presence of strong acids, the oxygen atom of an alcohol is a base and reacts with an acid by proton transfer to form an oxonium ion.

$$CH_3CH_2-O-H \;+\; H-\overset{+}{\underset{\underset{H}{|}}{O}}-H \;\overset{H_2SO_4}{\rightleftharpoons}\; CH_3CH_2-\overset{+}{\underset{\underset{H}{|}}{O}}-H \;+\; :O-H$$

Ethanol	Hydronium ion	Ethyloxonium ion
	(pK_a -1.7)	(pK_a -2.4)

Thus, we see that alcohols can function as both weak acids and weak bases.

9.4 Reaction of Alcohols with Active Metals

Alcohols react with Li, Na, K, and other active metals to liberate hydrogen and form metal alkoxides. In this oxidation/reduction reaction, Na is oxidized to Na^+, and H^+ is reduced to H_2.

$$2CH_3OH + 2Na \longrightarrow 2CH_3O^-Na^+ + H_2$$

Sodium methoxide
(MeO$^-$Na$^+$)

To name a metal alkoxide, name the cation first, followed by the name of the anion. The name of the anion is derived from the prefix showing the number of carbon atoms and their arrangement (meth-, eth-, isoprop-, *tert*-but-, and so on) followed by the suffix -oxide.

Alkoxide ions range from nearly the same to somewhat more basic than the hydroxide ion. In addition to sodium methoxide, the following metal salts of alcohols are commonly used in organic reactions requiring a strong base in a nonaqueous solvent, as for example sodium ethoxide in ethanol and potassium *tert*-butoxide in 2-methyl-2-propanol (*tert*-butyl alcohol).

$$CH_3CH_2O^-Na^+ \qquad\qquad CH_3\overset{\overset{\textstyle CH_3}{|}}{\underset{\underset{\textstyle CH_3}{|}}{C}}O^-K^+$$

Sodium ethoxide	Potassium *tert*-butoxide
(EtO$^-$Na$^+$)	(*t*-BuO$^-$K$^+$)

Alcohols can also be converted to salts by reaction with bases stronger than alkoxide ions. One such base is sodium hydride, NaH. Hydride ion, H:$^-$, the conjugate base of H_2, is an extremely strong base.

$$CH_3CH_2OH \;+\; Na^+H^- \longrightarrow CH_3CH_2O^-Na^+ \;+\; H_2$$

Ethanol	Sodium hydride	Sodium ethoxide

Sodium metal reacts with methanol with the evolution of hydrogen gas. *(Charles D. Winters)*

Reactivity Explorer;
Alcohols

Reactions of sodium hydride with compounds containing acidic hydrogens are irreversible and driven to completion by the formation of H_2, which is given off as a gas.

Example 9.6

Write a balanced equation for the reaction of cyclohexanol with sodium metal.

Solution

$$2 \langle \text{cyclohexyl} \rangle -OH + 2Na \longrightarrow 2 \langle \text{cyclohexyl} \rangle -O^-Na^+ + H_2$$

Problem 9.6

Predict the position of equilibrium for this acid-base reaction.

$$CH_3CH_2O^-Na^+ + CH_3\overset{\overset{\displaystyle O}{\|}}{C}OH \rightleftharpoons CH_3CH_2OH + CH_3\overset{\overset{\displaystyle O}{\|}}{C}O^-Na^+$$

9.5 Conversion of Alcohols to Haloalkanes

Conversion of an alcohol to a haloalkane involves substitution of halogen for —OH at a saturated carbon. The most common reagents for this conversion are the halogen acids (HCl, HBr, and HI) and the inorganic halides PBr_3 and $SOCl_2$.

A. Reaction with HCl, HBr, and HI

Mechanisms: Alcohols
Reaction of an alcohol
with HCl

Tertiary alcohols react very rapidly with HCl, HBr, and HI. Mixing a tertiary alcohol with concentrated hydrochloric acid for a few minutes at room temperature results in conversion of the alcohol to a chloroalkane. Reaction is evident by formation of a water-insoluble chloroalkane that separates from the aqueous layer. Low-molecular-weight, water-soluble primary and secondary alcohols are unreactive under these conditions.

$$\underset{\text{2-Methyl-2-propanol}}{CH_3\overset{\overset{\displaystyle CH_3}{|}}{\underset{\underset{\displaystyle CH_3}{|}}{C}}OH} + HCl \xrightarrow{25°C} \underset{\text{2-Chloro-2-methylpropane}}{CH_3\overset{\overset{\displaystyle CH_3}{|}}{\underset{\underset{\displaystyle CH_3}{|}}{C}}Cl} + H_2O$$

Water-insoluble tertiary alcohols are converted to tertiary chlorides by bubbling gaseous HX through a solution of the alcohol dissolved in diethyl ether or tetrahydrofuran (THF).

$$\underset{\text{1-Methylcyclohexanol}}{\langle \text{OH} \rangle} + HCl \xrightarrow[\text{ether}]{0°C} \underset{\substack{\text{1-Chloro-1-methyl-}\\\text{cyclohexane}}}{\langle \text{Cl} \rangle} + H_2O$$

Water-insoluble primary and secondary alcohols react only slowly under these conditions.

Primary and secondary alcohols are converted to bromoalkanes and iodoalkanes by treatment with hydrobromic and hydroiodic acids. For example, when heated to reflux with concentrated HBr, 1-butanol is converted smoothly to 1-bromobutane.

$$\text{\raisebox{0pt}{OH}} + \text{HBr} \xrightarrow[\text{reflux}]{\text{H}_2\text{O}} \text{\raisebox{0pt}{Br}} + \text{H}_2\text{O}$$

| 1-Butanol | 1-Bromobutane |

Many secondary alcohols give at least some rearranged product, evidence for the formation of carbocation intermediates during their reaction. For example, reaction of 3-pentanol with HBr gives 3-bromopentane as the major product, along with some 2-bromopentane.

a product of rearrangement

$$\text{3-Pentanol} + \text{HBr} \xrightarrow{\text{heat}} \text{3-Bromopentane} + \text{2-Bromopentane} + \text{H}_2\text{O}$$

| 3-Pentanol | 3-Bromopentane (major product) | 2-Bromopentane |

Primary alcohols with extensive β-branching give large amounts of a product derived from rearrangement. For example, treatment of 2,2-dimethyl-1-propanol (neopentyl alcohol) with HBr gives a rearranged product almost exclusively.

$$\text{2,2-Dimethyl-1-propanol} + \text{HBr} \longrightarrow \text{2-Bromo-2-methylbutane} + \text{H}_2\text{O}$$

| 2,2-Dimethyl-1-propanol | 2-Bromo-2-methylbutane (a product of rearrangement) |

Based on observations of the relative ease of reaction of alcohols with HX (3° > 2° > 1°) and the occurrence of rearrangements, chemists have concluded that conversion of tertiary and secondary alcohols to haloalkanes by concentrated HX occurs by an S_N1 mechanism and involves formation of a carbocation intermediate.

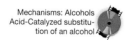

Mechanisms: Alcohols Acid-Catalyzed substitution of an alcohol

Mechanism   Reaction of a 3° Alcohol with HCl—An S_N1 Reaction

Step 1:   Rapid and reversible proton transfer from the acid to —OH of the alcohol gives an oxonium ion. A result of this proton transfer is to convert OH⁻, a poor leaving group, into H_2O, a better leaving group.

$$\text{CH}_3-\underset{\underset{\text{CH}_3}{|}}{\overset{\overset{\text{CH}_3}{|}}{\text{C}}}-\ddot{\text{O}}-\text{H} + \text{H}-\overset{+}{\underset{\text{H}}{\text{O}}}-\text{H} \underset{\text{reversible}}{\overset{\text{rapid and}}{\rightleftharpoons}} \text{CH}_3-\underset{\underset{\text{CH}_3}{|}}{\overset{\overset{\text{CH}_3}{|}}{\text{C}}}-\overset{+}{\underset{\text{H}}{\text{O}}}\overset{\text{H}}{\diagup} + \text{:O}-\text{H}$$

| 2-Methyl-2-propanol (*tert*-Butyl alcohol) | An oxonium ion |

Step 2: Loss of water gives a 3° carbocation intermediate.

$$CH_3-\underset{\underset{CH_3}{|}}{\overset{\overset{CH_3}{|}}{C}}-\overset{+}{\overset{\curvearrowleft}{O}}\underset{H}{\overset{H}{\diagup}} \xrightarrow[\text{S}_\text{N}1]{\substack{\text{slow, rate} \\ \text{determining}}} CH_3-\underset{\underset{CH_3}{|}}{\overset{\overset{CH_3}{|}}{C^+}} \quad + \quad :\overset{H}{\underset{H}{\diagdown\diagup}}O$$

A 3° carbocation
intermediate

Step 3: Reaction of the 3° carbocation (a Lewis acid) with chloride ion (a Lewis base) gives the product.

$$CH_3-\underset{\underset{CH_3}{|}}{\overset{\overset{CH_3}{|}}{C^+}} \overset{\curvearrowleft}{+} :Cl^- \xrightarrow{\text{fast}} CH_3-\underset{\underset{CH_3}{|}}{\overset{\overset{CH_3}{|}}{C}}-Cl$$

2-Chloro-2-methylpropane
(*tert*-Butyl chloride)

Primary alcohols react with HX by an S_N2 mechanism. In the rate-determining step, halide ion reacts at the carbon bearing the oxonium ion to displace H_2O and form the C—X bond.

Mechanism   Reaction of a 1° Alcohol with HBr—An S_N2 Reaction

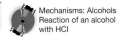
Mechanisms: Alcohols
Reaction of an alcohol
with HCl

Step 1: Rapid and reversible proton transfer gives an oxonium ion. A result of this step is to transform OH^-, a poor leaving group, into H_2O, a better leaving group.

$$CH_3CH_2CH_2CH_2-\overset{..}{\underset{..}{O}}-H + H-\overset{+}{\underset{\underset{H}{|}}{O}}-H \xrightleftharpoons[\text{reversible}]{\text{rapid and}} CH_3CH_2CH_2CH_2-\overset{+}{\underset{\underset{H}{|}}{O}}\overset{H}{\diagup} + \overset{..}{\underset{H}{\diagdown}}O-H$$

An oxonium ion

Step 2: Nucleophilic displacement of H_2O by Br^- gives the bromoalkane.

$$Br\overset{-}{:} + CH_3CH_2CH_2CH_2-\overset{+}{\underset{\underset{H}{\diagdown}}{O}}\overset{H}{\diagup} \xrightarrow[\text{S}_\text{N}2]{\substack{\text{slow, rate} \\ \text{determining}}} CH_3CH_2CH_2CH_2-Br + :\overset{H}{\underset{H}{\diagdown\diagup}}O$$

It is difficult, if not impossible, for primary alcohols with extensive β-branching, such as 2,2-dimethyl-1-propanol, to react by direct displacement of H_2O from the primary carbon. Furthermore, formation of a primary carbocation is also difficult, if not impossible. Primary alcohols with extensive β-branching react by a mechanism involving formation of an intermediate 3° carbocation by simultaneous loss of H_2O and migration of an alkyl group, as illustrated by the conversion of 2,2-dimethyl-1-propanol to 2-chloro-2-methylbutane. Note that two changes take place simultaneously in Step 2; the C—O bond breaks and a methyl group with its pair of electrons migrates to the site occupied by the departing H_2O group. Because the rate-determining step of this transformation involves only one reactant, namely the protonated alcohol, it is classified as an S_N1 reaction.

Mechanism   Rearrangement upon Treatment of Neopentyl Alcohol with HX

Step 1: Rapid and reversible proton transfer gives an oxonium ion. This step converts OH^-, a poor leaving group, into H_2O, a better leaving group.

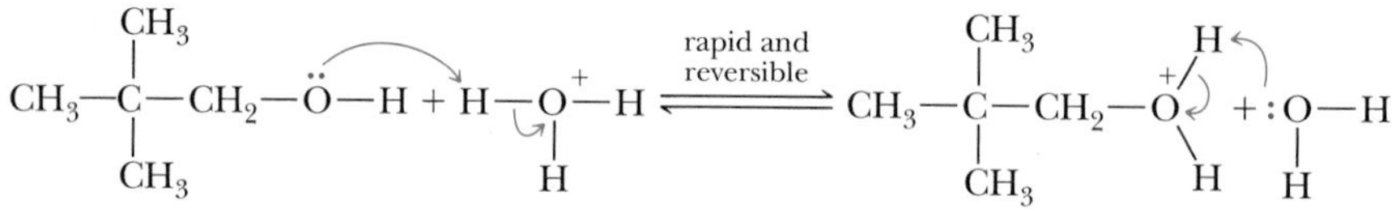

2,2-Dimethyl-1-propanol   An oxonium ion

Step 2: Simultaneous loss of H_2O and migration of an adjacent methyl group with its bonding electrons gives a 3° carbocation.

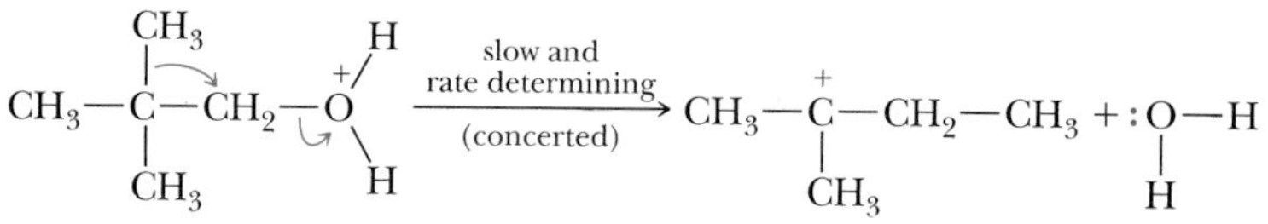

A 3° carbocation
intermediate

Step 3: Reaction of the 3° carbocation (a Lewis acid) with chloride ion (a Lewis base) gives the 3° alkyl halide.

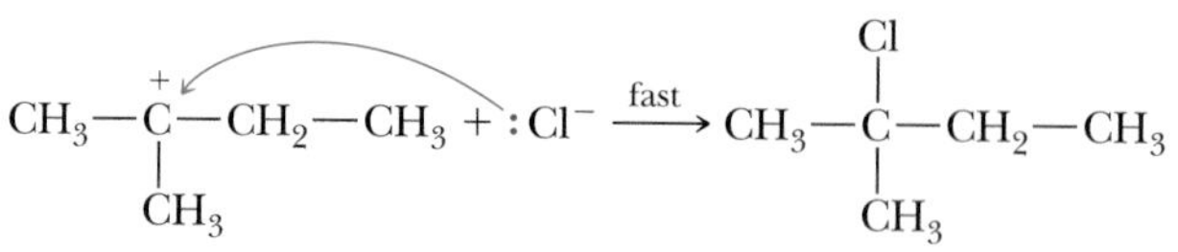

2-Chloro-2-methylbutane

In summary, preparation of haloalkanes by treatment of ROH with HX is most useful for primary and tertiary alcohols. Because of the possibility of rearrangement, this process is less useful for secondary alcohols (except for simple cycloalkanols) and for primary alcohols with branching on the β-carbon.

B.   Reaction with Phosphorus Tribromide

An alternative method for the synthesis of bromoalkanes from primary and secondary alcohols is through the use of phosphorus tribromide, PBr_3. This method of preparation of bromoalkanes takes place under milder conditions than treatment with HBr. Although rearrangement sometimes occurs with PBr_3, the extent is considerably less than that with HBr, especially when the reaction mixture is kept at or below 0°C.

2-Methyl-1-propanol   Phosphorous   1-Bromo-2-methylpropane   Phosphorous
(Isobutyl alcohol)   tribromide   (Isobutyl bromide)   acid

Conversion of an alcohol to a bromoalkane takes place in two steps. The essential feature of Step 1 is to convert OH, a very poor leaving group, to a dibromophosphite, $HOPBr_2$, a good leaving group. Displacement by bromide ion in Step 2 occurs by an S_N2 mechanism to give the alkyl bromide. The other two bromine atoms on phosphorus are replaced in similar reactions, giving three moles of RBr and one of phosphorous acid, H_3PO_3.

Mechanism   **Reaction of a Primary Alcohol with PBr₃**

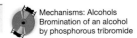

Mechanisms: Alcohols
Bromination of an alcohol
by phosphorous tribromide

Step 1:   Nucleophilic displacement on phosphorus by the oxygen atom of the alcohol gives a protonated dibromophosphite group, which converts OH^-, a poor leaving group, into a good leaving group.

OH^-, a poor leaving group, is converted into this good leaving group.

$$R{-}CH_2{-}\overset{..}{O}{-}H + Br{-}\underset{\underset{Br}{|}}{P}{-}Br \longrightarrow R{-}CH_2{-}\overset{+}{\underset{\underset{H}{|}}{O}}{-}PBr_2 + :Br^-$$

Step 2:   Nucleophilic displacement of the protonated dibromophosphite group by bromide ion gives the bromoalkane.

$$Br:^- + R{-}CH_2{-}\overset{+}{\underset{\underset{H}{|}}{O}}{-}PBr_2 \xrightarrow{S_N2} R{-}CH_2{-}Br + H\overset{..}{O}{-}PBr_2$$

C.   Reaction with Thionyl Chloride

The most widely used reagent for the conversion of primary and secondary alcohols to chloroalkanes is thionyl chloride, $SOCl_2$. Yields are high, and rearrangements are seldom observed. The byproducts of this conversion are HCl and SO_2.

OH + $SOCl_2$ $\xrightarrow{\text{pyridine}}$ Cl + SO_2 + HCl

1-Heptanol   Thionyl   1-Chloroheptane
                    chloride

Reactions are most commonly carried out in the presence of pyridine (Section 22.6C) or a tertiary amine such as triethylamine $(C_2H_5)_3N$. The function of the amine (a weak base) is twofold. First, it catalyzes the reaction by forming a small amount of the alkoxide in equilibrium. The alkoxide is more reactive than the alcohol as a nucleophile. In addition, the amine neutralizes the HCl generated during the reaction and, in this way, prevents unwanted side reactions.

$$RO-H + \text{(Pyridine)} \rightleftharpoons RO^- + \text{(pyridinium)}$$

Pyridine   H

$$\text{(Pyridine)} + H-Cl \longrightarrow \text{(pyridinium)} Cl^-$$

Pyridinium chloride

A particular value of the reaction of alcohols with thionyl chloride is that it is stereospecific; it occurs with inversion of configuration. Reaction of thionyl chloride with (S)-2-octanol, for example, in the presence of a tertiary amine occurs with inversion of configuration and gives (R)-2-chlorooctane.

$$\underset{\substack{H_3C \\ H}}{CH_3(CH_2)_5}C-OH + SOCl_2 \xrightarrow{\text{tertiary amine}} \underset{\substack{CH_3 \\ H}}{(CH_2)_5CH_3}Cl-C + SO_2 + HCl$$

(S)-2-Octanol   Thionyl   (R)-2-Chlorooctane
                        chloride

A key feature in the mechanism of this reaction is conversion of OH^-, a poor leaving group, into a chlorosulfite ester, a good leaving group. This reaction is analogous to the first step of the reaction of an alcohol with PBr_3. If reaction between the alcohol and thionyl chloride is carried out at 0°C or below, the alkyl chlorosulfite ester can be isolated. Nucleophilic displacement of this leaving group by chloride ion in Step 3 gives the product.

Mechanisms: Alcohols
Reaction of an alcohol
with thionyl chloride
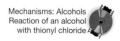

Mechanism   Reaction of an Alcohol with Thionyl Chloride

Step 1:   Nucleophilic displacement of chlorine by oxygen gives a protonated alkyl chlorosulfite ester.

$$\underset{\substack{H \\ R_2}}{R_1}C-O-H + Cl-\overset{O}{\underset{}{S}}-Cl \longrightarrow \underset{\substack{H \\ R_2}}{R_1}C-\overset{+}{O}\overset{H}{\underset{\substack{S=O \\ Cl}}{}} + :Cl^-$$

Thionyl
chloride

Step 2:   Proton transfer to the tertiary amine gives an alkyl chlorosulfite. The result of Steps 1 and 2 is to transform OH^-, a poor leaving group, into a chlorosulfite ester, a good leaving group.

A tertiary   An alkyl chlorosulfite
amine

Step 3:   Nucleophilic displacement of the chlorosulfite group by chloride ion occurs with inversion of configuration and gives the alkyl chloride. Both SO_2 and Cl^- are good leaving groups.

D.   Formation of Alkyl Sulfonates

As we have just seen, alcohols react with thionyl chloride with displacement of chloride ion to form alkyl chlorosulfites. They also react with compounds called sulfonyl chlorides to form alkyl sulfonates. Sulfonyl chlorides are derived from sulfonic acids, compounds that are very strong acids, comparable in strength to sulfuric acid. What is important for us at this point is that a sulfonate anion is a very weak base and stable anion and, therefore, a very good leaving group in nucleophilic substitution reactions.

A sulfonyl chloride   A sulfonic acid   A sulfonate anion
                        (a very strong acid)   (a very weak base and stable anion;
                                                 a very good leaving group)

Two of the most commonly used sulfonyl chlorides are *p*-toluenesulfonyl chloride (abbreviated tosyl chloride, Ts-Cl) and methanesulfonyl chloride (abbreviated mesyl chloride, Ms-Cl). Treatment of ethanol with *p*-toluenesulfonyl chloride in the presence of pyridine gives ethyl *p*-toluenesulfonate (ethyl tosylate). Pyridine is added to catalyze the reaction and to neutralize the HCl formed as a byproduct. Cyclohexanol is converted to cyclohexyl methanesulfonate (cyclohexyl mesylate) by a similar reaction of cyclohexanol with methanesulfonyl chloride. In each case, the reaction involves breaking the O—H bond of the alcohol; it does not affect the C—O bond in any way. If the carbon bearing the —OH group is a stereocenter, sulfonate ester formation takes place with retention of configuration.

$$CH_3CH_2OH + Cl-\overset{\overset{O}{\|}}{\underset{\underset{O}{\|}}{S}}-\text{⟨ ⟩}-CH_3 \xrightarrow{\text{pyridine}} CH_3CH_2O-\overset{\overset{O}{\|}}{\underset{\underset{O}{\|}}{S}}-\text{⟨ ⟩}-CH_3 + HCl$$

Ethanol	p-Toluenesulfonyl chloride	Ethyl p-toluenesulfonate (Ethyl tosylate)

$$\text{⟨ ⟩}-OH + Cl-\overset{\overset{O}{\|}}{\underset{\underset{O}{\|}}{S}}-CH_3 \xrightarrow{\text{pyridine}} \text{⟨ ⟩}-O-\overset{\overset{O}{\|}}{\underset{\underset{O}{\|}}{S}}-CH_3 + HCl$$

Cyclohexanol	Methanesulfonyl chloride	Cyclohexyl methanesulfonate (Cyclohexyl mesylate)

A particular advantage of sulfonate esters is that, through their use, a hydroxyl group, a very poor leaving group, can be converted to a tosylate or mesylate group, often shown as OTs and OMs, respectively, both very good leaving groups that are readily displaced by nucleophilic substitution.

$$I^{-} + CH_3CH_2-O-\overset{\overset{O}{\|}}{\underset{\underset{O}{\|}}{S}}-\text{⟨ ⟩}-CH_3 \xrightarrow[\text{acetone}]{S_N2} CH_3CH_2I + {:}O^{-}-\overset{\overset{O}{\|}}{\underset{\underset{O}{\|}}{S}}-\text{⟨ ⟩}-CH_3$$

Ethyl p-toluenesulfonate	p-Toluenesulfonate anion (a good leaving group)

Following is a two-step sequence for conversion of (S)-2-octanol to (R)-2-octyl acetate via a tosylate. The first step involves cleavage of the O—H bond and proceeds with retention of configuration at the stereocenter. The second step involves nucleophilic displacement of tosylate by acetate ion and proceeds with inversion of configuration at the stereocenter.

$$\overset{CH_3(CH_2)_5}{\underset{CH_3}{\overset{|}{\underset{H^{\text{''''}}}{C}}}}-OH + Cl-Ts \xrightarrow{\text{pyridine}} \overset{CH_3(CH_2)_5}{\underset{CH_3}{\overset{|}{\underset{H^{\text{''''}}}{C}}}}-OTs + HCl$$

(S)-2-Octanol	(S)-2-Octyl tosylate

$$CH_3\overset{\overset{O}{\|}}{C}O^{-}Na^{+} + \overset{CH_3(CH_2)_5}{\underset{CH_3}{\overset{|}{\underset{H^{\text{''''}}}{C}}}}-OTs \xrightarrow[\text{ethanol}]{S_N2} CH_3\overset{\overset{O}{\|}}{C}O-\overset{(CH_2)_5CH_3}{\underset{CH_3}{\overset{|}{\underset{H}{C}}}} + Na^{+}OTs^{-}$$

(S)-2-Octyl tosylate	(R)-2-Octyl acetate

Example 9.7

Show how to convert *trans*-4-methylcyclohexanol to *cis*-1-iodo-4-methylcyclohexane via a tosylate.

Solution

Treat the alcohol with *p*-toluenesulfonyl chloride in pyridine to form a tosylate with retention of configuration. Then treat the tosylate with sodium iodide in acetone. The S_N2 reaction with inversion of configuration gives the product.

Problem 9.7

Show how to convert (*R*)-2-butanol to (*S*)-2-butanethiol via a tosylate.

$$(R)\text{-2-Pentanol} \qquad (S)\text{-2-Pentanethiol}$$

9.6   Acid-Catalyzed Dehydration of Alcohols

An alcohol can be converted to an alkene by **dehydration;** that is, elimination of a molecule of water from adjacent carbon atoms. In the laboratory, dehydration of an alcohol is most often brought about by heating it with either 85% phosphoric acid or concentrated sulfuric acid. Primary alcohols are the most difficult to dehydrate and generally require heating in concentrated sulfuric acid at temperatures as high as 180°C. Secondary alcohols undergo acid-catalyzed dehydration at somewhat lower temperatures. Acid-catalyzed dehydration of tertiary alcohols often requires temperatures only slightly above room temperature.

Dehydration   Elimination of water.

H_3PO_4 or H_2SO_4

$$CH_3CH_2OH \xrightarrow[180°C]{H_2SO_4} CH_2{=}CH_2 + H_2O$$

$$\text{Cyclohexanol} \qquad\qquad \text{Cyclohexene}$$

$$\underset{\substack{\text{2-Methyl-2-propanol}\\(\textit{tert}\text{-butyl alcohol})}}{} \qquad \underset{\substack{\text{2-Methylpropene}\\(\text{Isobutylene})}}{}$$

Thus the ease of acid-catalyzed dehydration of alcohols is in this order:

$$1° \text{ alcohol } < \text{ } 2° \text{ alcohol } < \text{ } 3° \text{ alcohol}$$

Ease of dehydration of alcohols

When isomeric alkenes are obtained in acid-catalyzed dehydration of an alcohol, the alkene having the greater number of substituents on the double bond (the more stable alkene) generally predominates (Zaitsev's rule, Section 8.8).

$$CH_3CH_2\overset{\overset{\displaystyle OH}{|}}{C}HCH_3 \xrightarrow[\text{heat}]{85\% \text{ } H_3PO_4} CH_3CH=CHCH_3 + CH_3CH_2CH=CH_2 + H_2O$$

2-Butanol   2-Butene   1-Butene
                        (80%)   (20%)

Example 9.8

Draw structural formulas for the alkenes formed on acid-catalyzed dehydration of these alcohols. Where isomeric alkenes are possible, predict which alkene is the major product.

(a) $\xrightarrow{H_2SO_4}$ (b) $\xrightarrow{H_2SO_4}$

Solution

(a) Elimination of H_2O from carbons 2−3 gives 2-methyl-2-butene; elimination of H_2O from carbons 1−2 gives 3-methyl-1-butene. 2-Methyl-2-butene, with three alkyl groups (three methyl groups) on the double bond, is the major product. 3-Methyl-1-butene, with only one alkyl group (an isopropyl group) on the double bond, is the minor product. A small amount of 2-methyl-1-butene is formed by rearrangement.

3-Methyl-2-butanol   2-Methyl-2-butene   3-Methyl-1-butene
                        (major product)

(b) The major product, 1-methylcyclopentene, has three alkyl substituents on the carbon-carbon double bond. 3-Methylcyclopentene has only two substituents on the double bond.

2-Methylcyclopentanol   1-Methylcyclopentene   3-Methylcyclopentene
                           (major product)

Problem 9.8

Draw structural formulas for the alkenes formed by acid-catalyzed dehydration of these alcohols. Where isomeric alkenes are possible, predict which is the major product.

(a) $\xrightarrow[\text{dehydration}]{\text{acid-catalyzed}}$ (b) $\xrightarrow[\text{dehydration}]{\text{acid-catalyzed}}$

Dehydration of primary and secondary alcohols is often accompanied by re-arrangement. Acid-catalyzed dehydration of 3,3-dimethyl-2-butanol, for example, gives a mixture of two alkenes, each of which is the result of a rearrangement.

$\xrightarrow[140°-170°C]{H_2SO_4}$

3,3-Dimethyl-2-butanol   2,3-Dimethyl-2-butene   2,3-Dimethyl-1-butene
                                       (80%)   (20%)

Based on the relative rates of dehydration of alcohols (3° > 2° > 1°) and the prevalence of rearrangement, particularly among primary and secondary alcohols, chemists propose a three-step mechanism for acid-catalyzed dehydration of sec-ondary and tertiary alcohols. This mechanism involves formation of a carbocation in the rate-determining step and, therefore, is classified as an E1 mechanism.

Mechanism   Acid-Catalyzed Dehydration of 2-Butanol— An E1 Reaction

Mechanisms: Alcohols
Dehydration of an alcohol

Step 1:   Proton transfer from H_3O^+ to the OH group of the alcohol gives an oxonium ion. A result of this step is to convert OH^-, a poor leaving group, into H_2O, a better leaving group.

$HO: $ $H \overset{+}{O} H$

$CH_3CHCH_2CH_3 + H - \overset{+}{O} - H \underset{\text{rapid and reversible}}{\rightleftharpoons} CH_3CHCH_2CH_3 + :O-H$

An oxonium ion

Step 2:   Breaking of the C—O bond gives a 2° carbocation intermediate and H_2O.

$H \overset{+}{O} H$

$CH_3CHCH_2CH_3 \underset{\text{slow, rate determining}}{\rightleftharpoons} CH_3\overset{+}{C}HCH_2CH_3 + H_2O$

A 2° carbocation
intermediate

Step 3:   Proton transfer from a carbon adjacent to the positively charged carbon to H_2O gives the alkene. The sigma electrons of the C—H bond become the pi electrons of the carbon-carbon double bond.

$CH_3 - \overset{+}{C}H - CH - CH_3 + :O-H \underset{\text{rapid and reversible}}{\rightleftharpoons} CH_3 - CH = CH - CH_3 + H - \overset{+}{O} - H$

We account for acid-catalyzed dehydration accompanied by rearrangement in the same manner we did for the rearrangements that accompany S_N1 and E1 reactions

(Section 8.4H). Rearrangement occurs through formation of a carbocation intermediate followed by migration of an atom or group, with its pair of electrons, from the β-carbon to the carbon bearing the positive charge. The driving force for rearrangements of this type is conversion of a less stable carbocation to a more stable one. Proton transfer to H_2O gives the alkenes. As in other cases of acid-catalyzed dehydration of alkenes, the Zaitsev rule applies, and the more substituted alkene predominates.

3,3-Dimethyl-2-butanol → A 2° carbocation intermediate → A 3° carbocation intermediate → 2,3-Dimethyl-2-butene + H_3O^+ / 2,3-Dimethyl-1-butene + H_3O^+

Primary alcohols with little or no β-branching undergo acid-catalyzed dehydration to give a terminal alkene and rearranged alkenes. Acid-catalyzed dehydration of 1-butanol, for example, gives only 12% of 1-butene. The major product is a mixture of the trans and cis isomers of 2-butene.

1-Butanol $\xrightarrow[140°-170°C]{H_2SO_4}$ *trans*-2-Butene (56%) + *cis*-2-Butene (32%) + 1-Butene (12%)

We account for the formation of these products by a combination of E1 and E2 mechanisms.

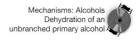

Mechanisms: Alcohols Dehydration of an unbranched primary alcohol

Mechanism   Acid-Catalyzed Dehydration of an Unbranched Primary Alcohol

Step 1: Proton transfer from H_3O^+ to the OH group gives an oxonium ion.

1-Butanol + H—O—H (rapid and reversible) ⇌ ... O—H + :O—H

Step 2: Simultaneous proton transfer to solvent and loss of H_2O gives the carbon-carbon double bond of the terminal alkene.

H—O: + ... O—H $\xrightarrow{E2}$ H—O—H + 1-Butene

Step 3: Simultaneous shift of a hydride ion from the β-carbon to the α-carbon and loss of H_2O gives a carbocation intermediate.

A 2° carbocation

Step 4: Transfer of a proton from a carbon adjacent to the carbocation to solvent gives the re-arranged alkenes.

trans-2-butene   *cis*-2-butene

Example 9.9

Propose a mechanism to account for this acid-catalyzed dehydration.

Solution

Proton transfer to the OH group to form an oxonium ion followed by loss of H_2O gives a 2° carbocation intermediate. Migration of a methyl group with its pair of electrons from the adjacent carbon to the positively charged carbon gives a more stable 3° carbocation. Loss of H^+ from this intermediate gives the observed product.

A 2° carbocation intermediate   A 3° carbocation intermediate

Problem 9.9

Propose a mechanism to account for this acid-catalyzed dehydration.

In Section 6.3B we discussed the acid-catalyzed hydration of alkenes to give alcohols. In the present section, we discussed the acid-catalyzed dehydration of alcohols to give alkenes. In fact, hydration-dehydration reactions are reversible. Alkene hydration and alcohol dehydration are competing processes, and the following equilibrium exists.

$$\underset{\text{An alkene}}{\diagup\!\!\!\overset{\diagdown}{C}=\overset{\diagup}{C}\!\!\!\diagdown} + \boxed{H_2O} \; \underset{\longleftarrow}{\overset{\text{acid catalyst}}{\longrightarrow}} \; \underset{\text{An alcohol}}{-\overset{|}{\underset{|}{C}}-\overset{|}{\underset{|}{C}}-}$$

Large amounts of water (use of dilute aqueous acid) favor alcohol formation, whereas scarcity of water (use of concentrated acid) or experimental conditions where water is removed (heating the reaction mixture above 100°C) favor alkene formation. Thus, depending on experimental conditions, it is possible to use the hydration-dehydration equilibrium to prepare either alcohols or alkenes in high yields.

This hydration-dehydration equilibrium illustrates a very important principle in the study of reaction mechanisms—the **principle of microscopic reversibility.** According to this principle, the sequence of transition states and reactive intermediates (that is, the mechanism) for any reversible reaction must be the same, but in reverse order, for the backward reaction as for the forward reaction.

To apply the principle of microscopic reversibility to acid-catalyzed hydration-dehydration equilibria, the mechanism we presented in this section for the acid-catalyzed dehydration of 2-butanol to give 2-butene is exactly the reverse of that presented in Section 6.3D for the acid-catalyzed hydration of 2-propene to give 2-propanol.

Principle of microscopic reversibility The sequence of transition states and reactive intermediates in the mechanism of any reversible reaction must be the same, but in reverse order, for the backward reaction as for the forward reaction.

9.7 The Pinacol Rearrangement

Compounds containing two hydroxyl groups on adjacent carbon atoms are called 1,2-diols, or alternatively, glycols. Such compounds can be synthesized by a variety of methods, including oxidation of alkenes by OsO_4 (Section 6.5B). The products of acid-catalyzed dehydration of glycols are quite different from those of acid-catalyzed dehydration of alcohols. For example, treatment of 2,3-dimethyl-2,3-butanediol (commonly called pinacol) with concentrated sulfuric acid gives 3,3-dimethyl-2-butanone (commonly called pinacolone).

$$\underset{\substack{\text{2,3-Dimethyl-2,3-butanediol}\\ \text{(Pinacol)}}}{CH_3-\overset{\overset{\displaystyle HO}{|}}{\underset{\underset{\displaystyle H_3C}{|}}{C}}-\overset{\overset{\displaystyle OH}{|}}{\underset{\underset{\displaystyle CH_3}{|}}{C}}-CH_3} \;\; \overset{H_2SO_4}{\longrightarrow} \;\; \underset{\substack{\text{3,3-Dimethyl-2-butanone}\\ \text{(Pinacolone)}}}{CH_3-\overset{\overset{\displaystyle O}{\|}}{C}-\overset{\overset{\displaystyle CH_3}{|}}{\underset{\underset{\displaystyle CH_3}{|}}{C}}-CH_3} \; + \; H_2O$$

Note two features of this reaction. It involves (1) dehydration of a glycol to form a ketone and (2) migration of a methyl group from one carbon to an adjacent carbon. Acid-catalyzed conversion of pinacol to pinacolone is an example of a type of reaction called the **pinacol rearrangement.** We account for the conversion of pinacol to pinacolone in a three-step mechanism.

Mechanism   The Pinacol Rearrangement of 2,3-Dimethyl-2,3-Butanediol (Pinacol)

Step 1:   Proton transfer from the acid catalyst to one of the —OH groups gives an oxonium ion.

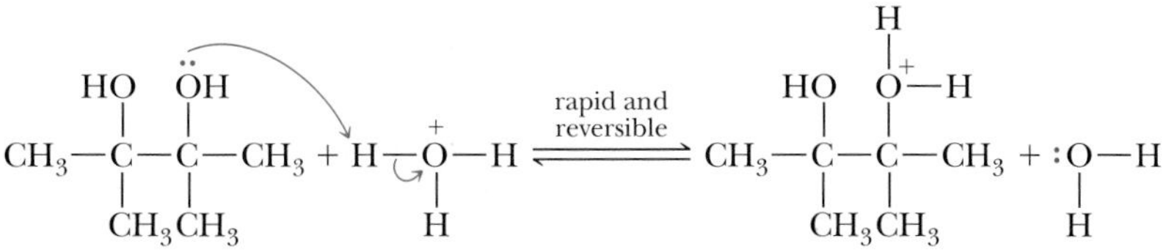

An oxonium ion

Step 2:   Loss of H_2O from the oxonium ion gives a 3° carbocation intermediate.

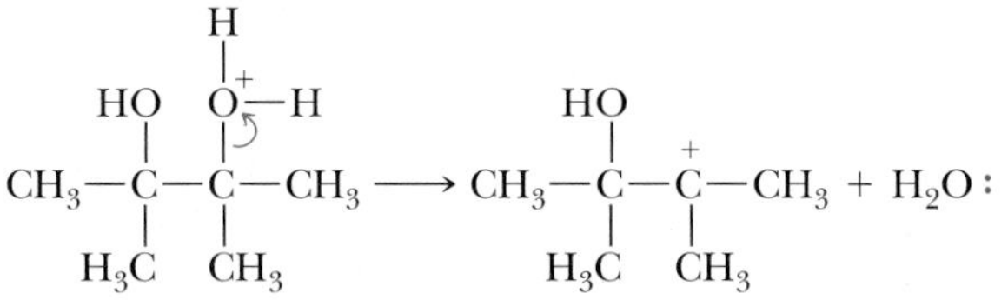

A 3° carbocation
intermediate

Step 3:   Migration of a methyl group from the adjacent carbon with its bonding electrons gives a new, more stable cation intermediate. Of the two contributing structures drawn for it, the one on the right makes the greater contribution because, in it, both carbon and oxygen have complete octets of valence electrons (Section 1.6D).

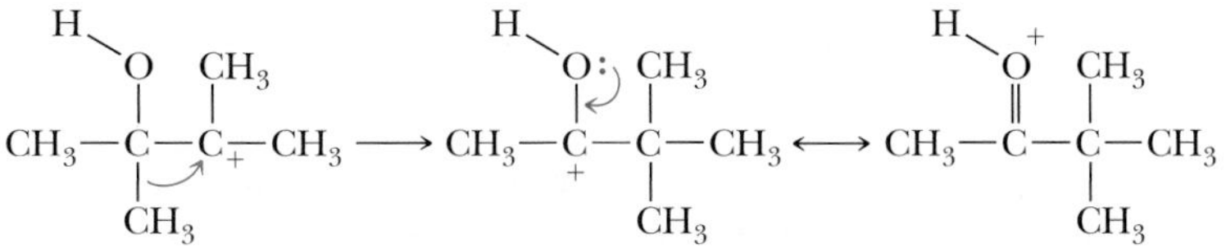

A resonance-stabilized cation intermediate

Step 4:   Proton transfer to solvent gives pinacolone.

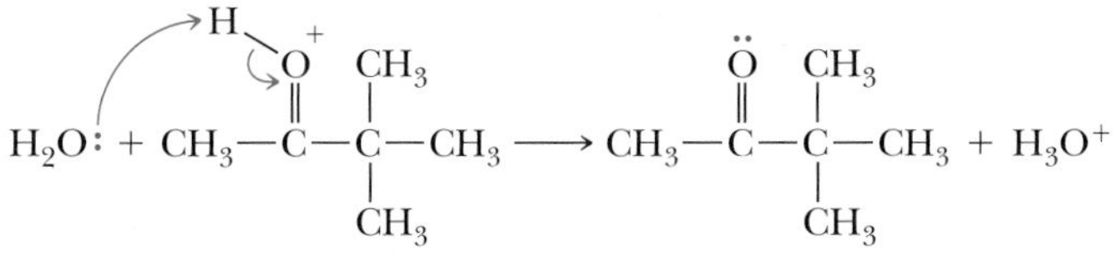

The pinacol rearrangement is general for all 1,2-diols. In the rearrangement of pinacol, a symmetrical diol, equivalent carbocations are formed no matter which —OH is protonated and leaves. Studies of unsymmetrical 1,2-diols have revealed that the —OH group that is protonated and leaves is the one that gives rise to the more stable carbocation. For example, treatment of 2-methyl-1,2-propanediol with cold concentrated sulfuric acid gives a tertiary carbocation. Subsequent migration of hydride ion (H:⁻) followed by transfer of a proton from the new cation to solvent gives 2-methylpropanal.

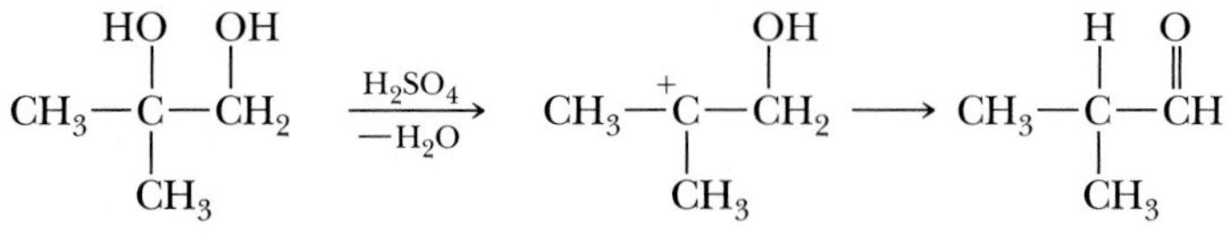

2-Methyl-1,2-propanediol   A 3° carbocation   2-Methylpropanal
                                   intermediate

Example 9.10

Predict the product of treatment of each glycol with H_2SO_4.

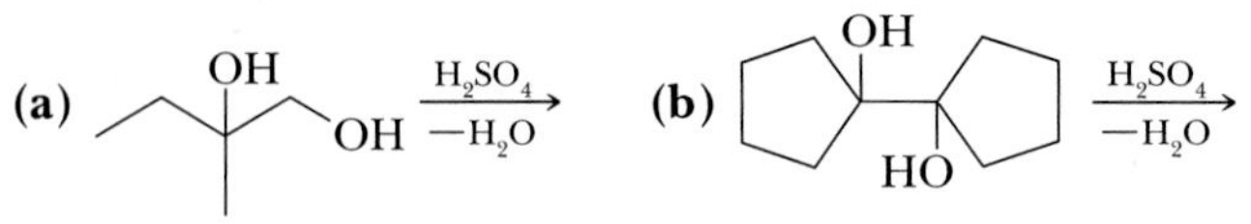

Solution

(a) *Step 1:*   Protonation of the tertiary hydroxyl group followed by loss of a molecule of water gives a tertiary carbocation intermediate.

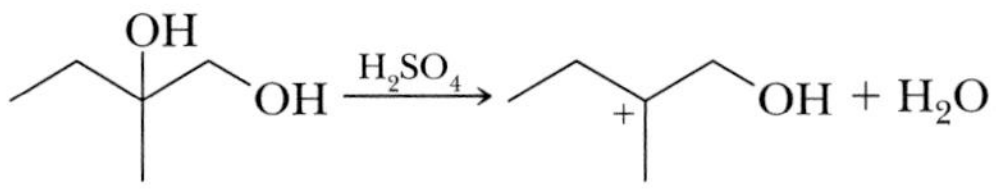

A 3° carbocation
intermediate

Step 2:   Migration of a hydride ion from the adjacent carbon gives a resonance-stabilized cation intermediate. The contributing structure on the right has filled valence shells on both carbon and oxygen and, therefore, makes the greater contribution to the hybrid.

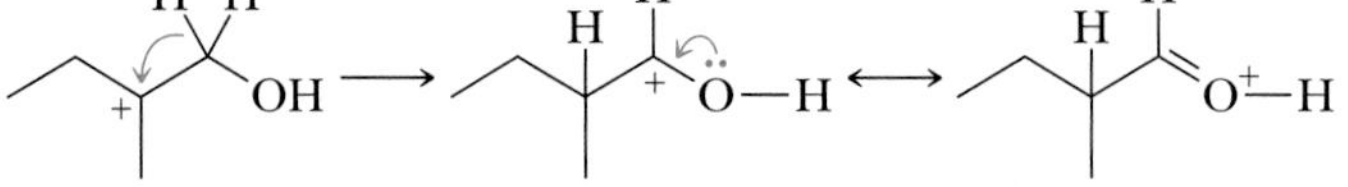

A resonance-stabilized cation intermediate

Step 3:   Proton transfer from the resonance-stabilized cation intermediate to water completes the reaction to give 2-methylbutanal.

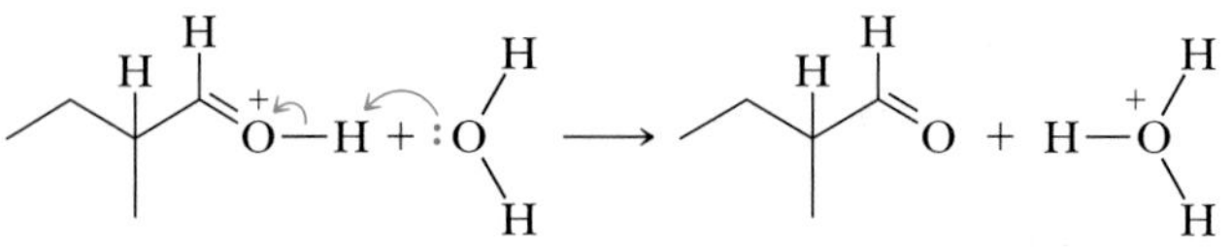

2-Methylbutanal

(b) Protonation of either hydroxyl group followed by loss of water gives a tertiary carbocation. The group that then migrates is a CH_2 group of the five-membered ring,
and the product is a bicyclic ketone. This compound belongs to the class of compounds called spiro compounds, in which two rings share only one carbon atom.

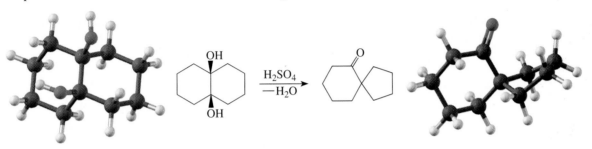

Compound (b) $\xrightarrow[-H_2O]{H_2SO_4}$

A 3° carbocation

redraw

a spiro carbon

Spiro[4.5]decan-6-one

Problem 9.10

Propose a mechanism to account for the following transformation:

$\xrightarrow[-H_2O]{H_2SO_4}$

9.8   Oxidation of Alcohols

Reactivity Explorer;
Alcohols

Oxidation of a primary alcohol gives an aldehyde or a carboxylic acid, depending on experimental conditions. Secondary alcohols are oxidized to ketones. Tertiary alcohols are
not oxidized. Following is a series of transformations in which a primary alcohol is oxidized first to an aldehyde and then to a carboxylic acid. The fact that each transformation involves oxidation is indicated by the symbol O in brackets over the reaction arrow.

$$CH_3-\overset{\overset{\displaystyle OH}{|}}{\underset{\underset{\displaystyle H}{|}}{C}}-H \xrightarrow{[O]} CH_3-\overset{\overset{\displaystyle O}{\|}}{C}-H \xrightarrow{[O]} CH_3-\overset{\overset{\displaystyle O}{\|}}{C}-OH$$

A primary   An aldehyde   A carboxylic
alcohol   acid

Inspection of balanced half-reactions (Section 6.5A) shows that each transformation
in this series is a two-electron oxidation.

$$CH_3-\overset{\overset{\displaystyle OH}{|}}{\underset{\underset{\displaystyle H}{|}}{C}}-H \longrightarrow CH_3-\overset{\overset{\displaystyle O}{\|}}{C}-H + 2H^+ + \boxed{2e^-}$$

$$CH_3-\overset{\overset{\displaystyle O}{\|}}{C}-H + H_2O \longrightarrow CH_3-\overset{\overset{\displaystyle O}{\|}}{C}-OH + 2H^+ + \boxed{2e^-}$$

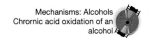

A. Chromic Acid

The reagent most commonly used in the laboratory for the oxidation of a primary alcohol to a carboxylic acid is chromic acid, H_2CrO_4. Chromic acid is prepared by dissolving either chromium(VI) oxide or potassium dichromate in aqueous sulfuric acid.

$$CrO_3 + H_2O \xrightarrow{H_2SO_4} H_2CrO_4$$

Chromium(VI)   Chromic acid
oxide

$$K_2Cr_2O_7 \xrightarrow{H_2SO_4} H_2Cr_2O_7 \xrightarrow{H_2O} 2H_2CrO_4$$

Potassium   Chromic acid
dichromate

Oxidation of 1-octanol using chromic acid in aqueous sulfuric acid gives octanoic acid in high yield. These experimental conditions are more than sufficient to oxidize the intermediate aldehyde to a carboxylic acid.

$$CH_3(CH_2)_6CH_2OH \xrightarrow[H_2SO_4,\ H_2O]{CrO_3} \left[CH_3(CH_2)_6\overset{\overset{\displaystyle O}{\|}}{C}H \right] \longrightarrow CH_3(CH_2)_6\overset{\overset{\displaystyle O}{\|}}{C}OH$$

1-Octanol   Octanal   Octanoic acid
                               (not isolated)

B.   Pyridinium Chlorochromate

The form of Cr(VI) commonly used for oxidation of a primary alcohol to an aldehyde is prepared by dissolving CrO_3 in aqueous HCl and adding pyridine to precipitate pyridinium chlorochromate (PCC) as a solid. PCC oxidations are carried out in aprotic solvents, most commonly dichloromethane, CH_2Cl_2.

$$CrO_3 + HCl + \underset{\text{Pyridine}}{\bigcirc\!\!\!\!\!\text{N}} \longrightarrow \underset{\substack{\text{Pyridinium chlorochromate} \\ \text{(PCC)}}}{\bigcirc\!\!\!\!\!\overset{+}{\text{N}}\text{H}\ ClCrO_3^-}$$

pyridinium ion

chlorochromate ion

This reagent not only is very selective for the oxidation of primary alcohols to aldehydes but also has little effect on carbon-carbon double bonds or other easily oxidized functional groups. In the following example, geraniol, a primary terpene alcohol, is oxidized to geranial without affecting either carbon-carbon double bond.

$$\xrightarrow{PCC}$$

Geraniol   Geranial

Secondary alcohols are oxidized to ketones by both chromic acid and PCC.

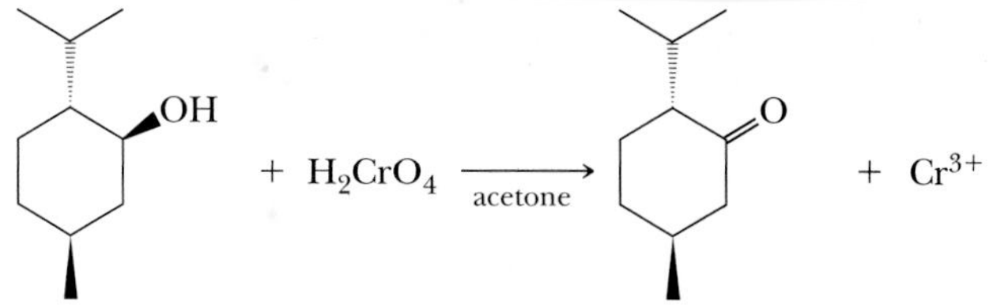

2-Isopropyl-5-methyl-
cyclohexanol
(Menthol)

2-Isopropyl-5-methyl-
cyclohexanone
(Menthone)

Tertiary alcohols are resistant to oxidation because the carbon bearing the —OH is already bonded to three carbon atoms and, therefore, cannot form an additional carbon-oxygen bond.

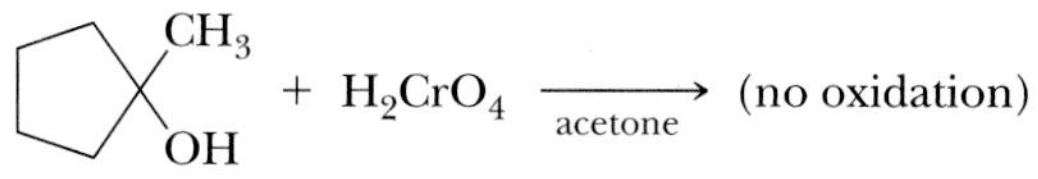

1-Methylcyclopentanol

Note that the lack of an alpha-hydrogen in tertiary alcohols prevents such a reaction from taking place. The essential feature of this oxidation is the presence of at least one H on the carbon bearing the OH group.

Mechanism   Chromic Acid Oxidation of an Alcohol

Step 1:   Reaction of the alcohol and chromic acid gives an alkyl chromate ester by a mechanism similar to that for the formation of a carboxylic ester (Section 17.8). There is no change in oxidation state of either carbon or chromium as a result of this reaction.

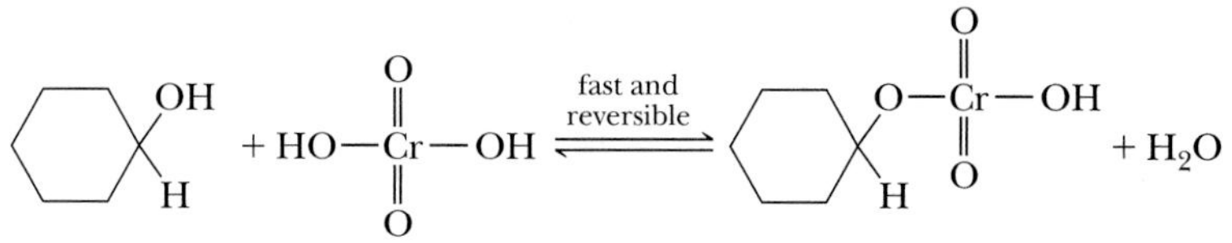

Cyclohexanol

An alkyl chromate ester

Step 2:   Reaction of the alkyl chromate ester with a base, here shown as a water molecule, results in cleavage of a C—H bond, formation of the carbonyl group, and reduction of chromium(VI) to chromium(IV). This is the oxidation-reduction step; carbon undergoes a two-electron oxidation, and chromium undergoes a two-electron reduction. Chromium(IV) then participates in further oxidations by a similar mechanism and eventually is transformed to Cr(III).

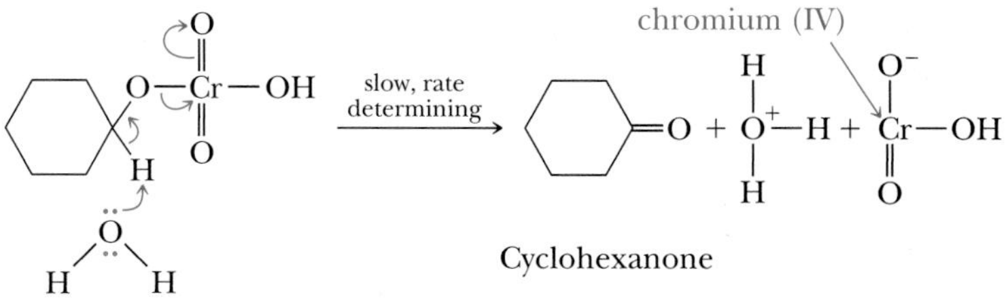

Cyclohexanone

We have shown that in aqueous chromic acid, a primary alcohol is oxidized first to an aldehyde and then to a carboxylic acid. In the second step, it is not the aldehyde that is oxidized but rather the aldehyde hydrate formed by addition of a molecule of water to the aldehyde carbonyl group (hydration). We will study the hydration of aldehydes and ketones in more detail in Section 16.8. It is an —OH of the aldehyde hydrate that now forms an ester with chromic acid to complete the oxidation of the aldehyde to a carboxylic acid.

$$R-\overset{\overset{\displaystyle O}{\|}}{C}-H + H_2O \underset{\text{reversible}}{\overset{\text{fast and}}{\rightleftharpoons}} R-\overset{\overset{\displaystyle OH}{|}}{\underset{\underset{\displaystyle H}{|}}{C}}-OH$$

An aldehyde   An aldehyde
                                  hydrate

$$R-\overset{\overset{\displaystyle OH}{|}}{\underset{\underset{\displaystyle H}{|}}{C}}-OH \overset{H_2CrO_4}{\longrightarrow} R-\overset{\overset{\displaystyle O-CrO_3H}{|}}{\underset{\underset{\displaystyle H}{|}}{C}}-OH \longrightarrow R-\overset{\overset{\displaystyle O}{\|}}{C}-OH$$

                        $H_2O\quad H$

                      An alkyl   A carboxylic
                   chromate ester   acid

This is the reason why PCC is specific for the oxidation of primary alcohols to aldehydes; it does not oxidize aldehydes further because the PCC reagent is not used in water but rather in an organic solvent, usually CH_2Cl_2. Only the hydrate is susceptible to further oxidation by Cr(VI).

Example 9.11

Draw the product of treatment of each alcohol with PCC.

(a) 1-Hexanol   **(b)** 2-Hexanol   **(c)** Cyclohexanol

Solution

1-Hexanol, a primary alcohol, is oxidized to hexanal. 2-Hexanol, a secondary alcohol, is oxidized to 2-hexanone. Cyclohexanol, a secondary alcohol, is oxidized to cyclohexanone.

(a)   **(b)**   **(c)**

Hexanal   2-Hexanone   Cyclohexanone

Problem 9.11

Draw the product of treatment of each alcohol in Example 9.11 with chromic acid.

CHEMISTRY IN ACTION

Blood Alcohol Screening

Potassium dichromate oxidation of ethanol to acetic acid is the basis for the original breath alcohol screening test used by law enforcement agencies to determine a person's blood alcohol content (BAC). The test is based on the difference in color between the dichromate ion (reddish orange) in the reagent and the chromium(III) ion (green) in the product. Thus, color change can be used as a measure of the quantity of ethanol present in a breath sample.

$$CH_3CH_2OH \quad + \quad Cr_2O_7^{2-} \xrightarrow[H_2O]{H_2SO_4}$$

Ethanol   Dichromate ion
            (reddish orange)

$$\underset{\text{Acetic acid}}{CH_3\overset{\overset{\displaystyle O}{\|}}{C}OH} \quad + \quad \underset{\substack{\text{Chromium(III) ion}\\ \text{(green)}}}{Cr^{3+}}$$

In its simplest form, a breath alcohol screening test consists of a sealed glass tube containing a potassium dichromate-sulfuric acid reagent impregnated on silica gel. To administer the test, the ends of the tube are broken off, a mouthpiece is fitted to one end, and the other end is inserted into the neck of a plastic bag. The person being tested then blows into the mouthpiece until the plastic bag is inflated.

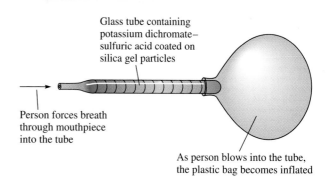

Glass tube containing potassium dichromate–sulfuric acid coated on silica gel particles

Person forces breath through mouthpiece into the tube

As person blows into the tube, the plastic bag becomes inflated

As breath containing ethanol vapor passes through the tube, reddish orange dichromate is reduced to green chromium(III). The concentration of ethanol in the breath is then estimated by measuring how far the green color extends along the length of the tube. When the green color extends beyond the halfway point, the person is judged to have a sufficiently high blood alcohol content to warrant further, more precise testing.

The Breathalyzer, a more accurate testing device, operates on the same principle as the simplified screening test. In a Breathalyzer test, a measured volume of breath is bubbled through a solution of potassium dichromate in aqueous sulfuric acid, and the color change is measured spectrophotometrically.

These tests measure alcohol in the breath. The legal definition of being under the influence of alcohol, however, is based on blood alcohol content, not breath alcohol content. The chemical correlation between these two measurements is that air deep within the lungs is in equilibrium with blood passing through the pulmonary arteries, and an equilibrium is established between blood alcohol and breath alcohol. It has been determined by tests in a person drinking alcohol that 2100 mL of breath contains the same amount of ethanol as 1.00 mL of blood.

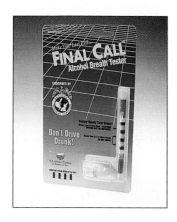

A device for testing the presence of ethanol on the breath. The device works because ethanol is oxidized by potassium dichromate. The yellow-orange color of dichromate turns to blue-green as it is reduced to Cr(III) ion. (Charles D. Winters)

C. Periodic Acid

Periodic acid, H_5IO_6 (or alternatively $HIO_4 \cdot 2H_2O$), is a white crystalline solid, mp 122°C. Its major use in organic chemistry is for the cleavage of glycols to carbonyl-containing compounds (a two-electron oxidation). In the process, periodic acid is reduced to iodic acid (a two-electron reduction).

$$\begin{array}{ll} -\overset{|}{\underset{|}{C}}-OH \\ -\overset{|}{\underset{|}{C}}-OH \end{array} \longrightarrow \begin{array}{l} -C=O \\ -C=O \end{array} + 2H^+ + 2e^- \quad \text{(a two-electron oxidation)}$$

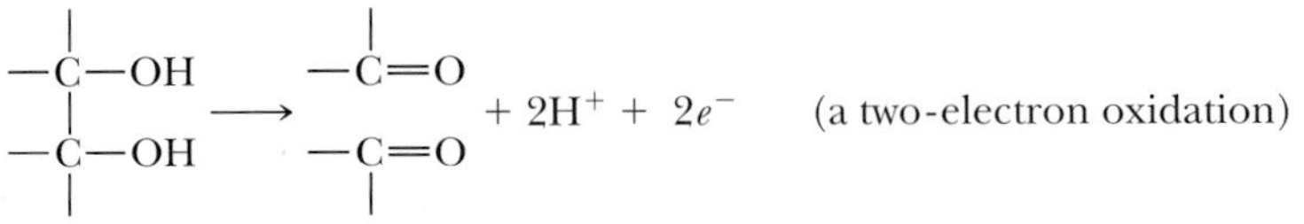

$$HIO_4 \;+\; 2H^+ + 2e^- \longrightarrow HIO_3 \;+\; H_2O \quad \text{(a two-electron reduction)}$$

Periodic acid   Iodic acid

For example, when treated with periodic acid, 1,2-cyclohexanediol is oxidized to hexanedial.

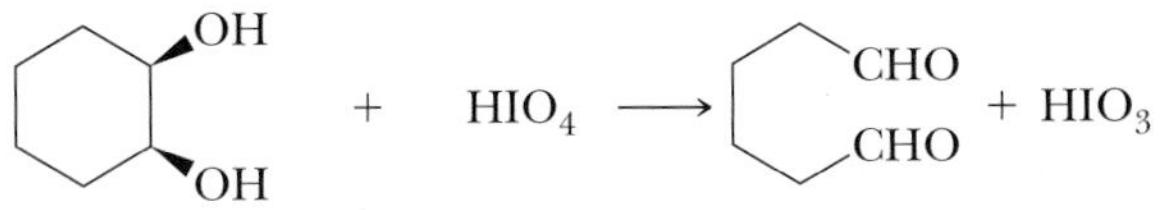

cis-1,2-cyclohexanediol   Periodic   Hexanedial   Iodic
                     acid   acid

Step 1 of the mechanism of periodic acid oxidation of a glycol involves formation of a five-membered cyclic periodic diester between the diol and HIO_4. Redistribution of electrons within this cyclic ester in Step 2 generates the two carbonyl groups and HIO_3.

Mechanism   Oxidation of a Glycol by Periodic Acid

Step 1: Reaction of the glycol with periodic acid gives a five-membered cyclic periodic ester. There is no change in the oxidation state of either iodine or the glycol as a result of ester formation.

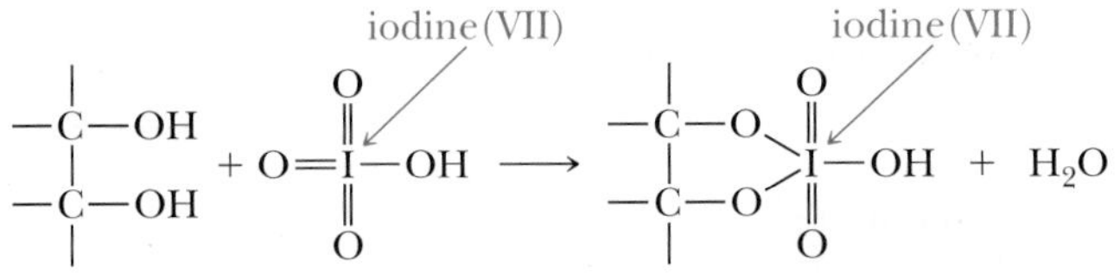

A cyclic periodic ester

Step 2: Redistribution of valence electrons within the cyclic periodic ester gives HIO_3 and two carbonyl groups. A result of this electron redistribution is a two-electron oxidation of the organic component and a two-electron reduction of the iodine-containing component.

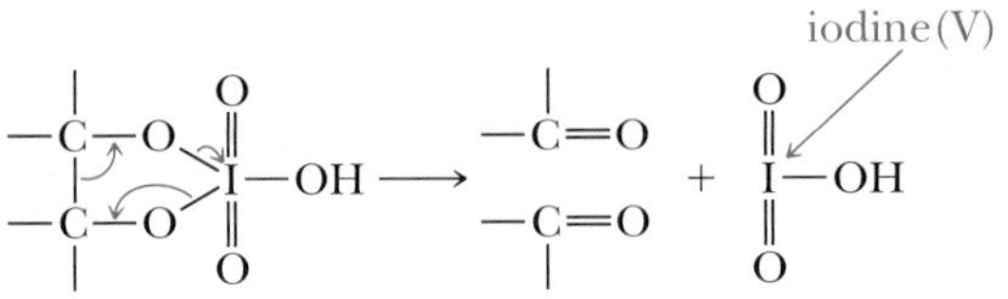

HIO_3
(Iodic acid)

This mechanism is consistent with the fact that HIO_4 oxidations are restricted to glycols that can form a five-membered cyclic periodic diester. Any glycol that cannot form such a cyclic diester is not oxidized by periodic acid. Following are structural formulas for two isomeric decalindiols. Only the cis glycol can form a cyclic ester with periodic acid, and only it is oxidized by this reagent.

The trans isomer is unreactive toward periodic acid.

The cis isomer forms a cyclic ester with periodic acid and is cleaved.

Example 9.12

What products are formed when each glycol is treated with HIO_4?

(a) (structure with OH and CH₂OH)

(b) $(CH_3)_2C-CHCH_3$ with HO and OH

Solution

The bond between the carbons bearing the —OH groups is cleaved and each —OH group is converted to a carbonyl group.

(a) =O + H—C—H

(b) $\begin{matrix}H_3C\\H_3C\end{matrix}C=O$ + H—C—CH₃

Problem 9.12

α-Hydroxyketones and α-hydroxyaldehydes are also oxidized by treatment with periodic acid. It is not the α-hydroxyketone or aldehyde, however, that undergoes reaction with periodic acid but the hydrate formed by addition of water to the carbonyl group of the α-hydroxyketone or aldehyde. Write a mechanism for the oxidation of this α-hydroxyaldehyde by HIO_4.

$$CH_3CH_2CH_2CH-CH \xrightarrow[2. HIO_4]{1. H_2O}$$

9.9 Thiols

A. Structure

The functional group of a **thiol** is an —SH (sulfhydryl) group bonded to an sp^3 hybridized carbon. Figure 9.4 shows a Lewis structure and a ball-and-stick model of methanethiol, CH_3SH, the simplest thiol. The C—S—H bond angle in methanethiol

Thiol A compound containing an —SH (sulfhydryl) group bonded to an sp^3 hybridized carbon.

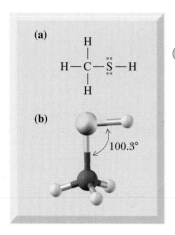

Figure 9.4
Methanethiol, CH_3SH. (a) Lewis structure and (b) ball-and-stick model.

Mercaptan A common name for a thiol; that is, any compound containing an —SH (sulfhydryl) group.

Mushrooms, onions, garlic, and coffee all contain sulfur compounds. One of these present in the aroma of coffee is

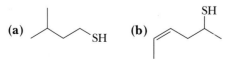

(Charles D. Winters)

is 100.3°. By way of comparison, the H—S—H bond angle in H_2S is 93.3°. If a sulfur atom were bonded to two other atoms by sp^3 hybrid orbitals, bond angles about sulfur would be approximately 109.5°. If, instead, a sulfur atom were bonded to two other atoms by $3p$ orbitals, bond angles would be approximately 90°. The fact that the C—S—H bond angle in methanethiol is 100.3° and the H—S—H bond angle in H_2S is 93.3° indicates that there is considerably more p-character in the bonding orbitals of divalent sulfur than there is in those of divalent oxygen.

B. Nomenclature

In the older literature, thiols are often referred to as **mercaptans,** which literally means mercury capturing. Thiols react with Hg^{2+} in aqueous solution to give sulfide salts as insoluble precipitates. Thiophenol, C_6H_5SH, for example, gives $(C_6H_5S)_2Hg$.

According to the IUPAC system, thiols are named by selecting as the parent alkane the longest chain of carbon atoms that contains the —SH group. To show that the compound is a thiol, the final -e in the name of the parent alkane is retained and the suffix -thiol is added. The location of the —SH group takes precedence over alkyl groups and halogens in numbering the parent chain. In compounds containing other functional groups of higher precedence, the presence of an —SH group is indicated by the IUPAC prefix sulfanyl-. Alternatively, it may be indicated by the common-name prefix mercapto-. According to the IUPAC system, —OH takes precedence over —SH in both numbering and naming.

Common names for simple thiols are derived by naming the alkyl group attached to —SH and adding the word "mercaptan."

$$CH_3CH_2SH$$

Ethanethiol
(Ethyl mercaptan)

$$CH_3CHCH_2SH$$
$$CH_3$$

2-Methyl-1-propanethiol
(Isobutyl mercaptan)

$$HSCH_2CH_2OH$$

2-Sulfanylethanol
(Mercaptoethanol)

Example 9.13

Write names for these thiols (including stereochemistry if relevant).

(a) ⌇⌇⌇SH   (b) ⌇⌇SH

Solution

(a) 1-Pentanethiol (pentyl mercaptan)
(b) (*E*)-2-Butene-1-thiol (*trans*-2-butene-1-thiol)

Problem 9.13

Write IUPAC names for these thiols (including stereochemistry if relevant).

(a) (structure with SH)   (b) (structure with SH)

C. Physical Properties

The most outstanding physical characteristic of low-molecular-weight thiols is their stench. Traces of ethanethiol are added to natural gas as an odorant so that gas leaks can be detected. The scent of skunks is due primarily to these two thiols:

$$CH_3$$
$$|$$
$$CH_3CHCH_2CH_2SH$$
3-Methyl-1-butanethiol

$$CH_3CH{=}CHCH_2SH$$
2-Butene-1-thiol

Because of the very low polarity of the S—H bond, thiols show little association by hydrogen bonding. Consequently, they have lower boiling points and are less soluble in water and other polar solvents than alcohols of comparable molecular weights. Table 9.3 gives names and boiling points for three low-molecular-weight thiols. Shown for comparison are boiling points for alcohols of the same number of carbon atoms. In Section 9.2 we illustrated the importance of hydrogen bonding in alcohols by comparing the boiling points of ethanol (bp 78°C) and its constitutional isomer dimethyl ether (bp − 24°C). By comparison, the boiling point of ethanethiol is 35°C, and that of its constitutional isomer dimethyl sulfide is 37°C. The fact that the boiling points of these constitutional isomers are almost identical indicates that little or no association by hydrogen bonding occurs between thiol molecules.

$$CH_3CH_2SH \qquad CH_3SCH_3$$

Ethanethiol   Dimethyl sulfide
bp 35°C   bp 37°C

D. Preparation

The most common preparation of thiols, RSH, depends on the high nucleophilicity of the hydrosulfide ion, HS⁻ (Section 8.4B). Sodium hydrosulfide is prepared by bubbling H_2S through a solution of NaOH in water or aqueous ethanol. Reaction of HS⁻ with a haloalkane gives a thiol.

$$CH_3(CH_2)_8CH_2I + Na^+SH^- \xrightarrow[\text{ethanol}]{S_N2} CH_3(CH_2)_8CH_2SH + Na^+I^-$$

1-Iododecane   Sodium   1-Decanethiol
              hydrosulfide

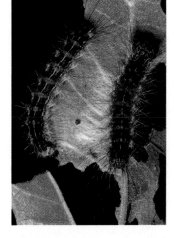

Disparlure is the sex attractant of the gypsy moth, *porthetria dispar*. *(Animals, Animals/William D. Griffin)*

Table 9.3	Boiling Points of Thiols and Alcohols of the Same Number of Carbon Atoms		
Thiol	**bp (°C)**	**Alcohol**	**bp (°C)**
Methanethiol	6	Methanol	65
Ethanethiol	35	Ethanol	78
1-Butanethiol	98	1-Butanol	117

In practice, the scope and limitations of this reaction are governed by the limitations of the S_N2 reaction and by competition between substitution and β-elimination (Section 8.11). The reaction is most useful for preparation of thiols from primary halides. Yields are lower from secondary halides because of the competing β-elimination (E2) reaction. With tertiary halides, β-elimination (E2) predominates, and the alkene formed by dehydrohalogenation is the major product.

In a commercial application of thiol formation by this nucleophilic substitution, the sodium salt of thioglycolic acid is prepared by the reaction of sodium hydrosulfide and sodium iodoacetate.

$$
\text{Na}^+\,\text{SH}^- \;+\; \underset{\substack{\text{Sodium iodoacetate}}}{\text{ICH}_2\overset{\overset{\displaystyle O}{\|}}{\text{C}}\text{O}^-\,\text{Na}^+} \;\xrightarrow{S_N2}\; \underset{\substack{\text{Sodium mercaptoacetate}\\\text{(Sodium thioglycolate)}}}{\text{HSCH}_2\overset{\overset{\displaystyle O}{\|}}{\text{C}}\text{O}^-\,\text{Na}^+} \;+\; \text{Na}^+\,\text{I}^-
$$

Sodium
hydrosulfide

The sodium and ammonium salts of thioglycolic acid are used in cold waving of hair. These compounds work by breaking peptide disulfide bonds, which maintain protein structure (Section 27.6C). These disulfide bonds can subsequently be reformed by oxidation (Section 9.9F). The calcium salt of thioglycolic acid is used as a depilatory, that is, to remove body hair.

$$(\text{HSCH}_2\text{COO}^-)_2\text{Ca}^{2+}$$

Calcium mercaptoacetate
(Calcium thioglycolate)

E. Acidity

Hydrogen sulfide is a stronger acid than water.

$$\text{H}_2\text{O} + \text{H}_2\text{O} \rightleftharpoons \text{HO}^- + \text{H}_3\text{O}^+ \qquad pK_a = 15.7$$

$$\text{H}_2\text{S} + \text{H}_2\text{O} \rightleftharpoons \text{HS}^- + \text{H}_3\text{O}^+ \qquad pK_a = 7.0$$

Similarly, thiols are stronger acids than alcohols. Compare, for example, the pK_a values of ethanol and ethanethiol in dilute aqueous solution.

$$\text{CH}_3\text{CH}_2\text{OH} + \text{H}_2\text{O} \rightleftharpoons \text{CH}_3\text{CH}_2\text{O}^- + \text{H}_3\text{O}^+ \qquad pK_a = 15.9$$

$$\text{CH}_3\text{CH}_2\text{SH} + \text{H}_2\text{O} \rightleftharpoons \text{CH}_3\text{CH}_2\text{S}^- + \text{H}_3\text{O}^+ \qquad pK_a = 8.5$$

Thiols are sufficiently strong acids that when dissolved in aqueous sodium hydroxide, they are converted completely to alkylsulfide salts.

$$\text{CH}_3\text{CH}_2\text{SH} \;+\; \text{Na}^+\text{OH}^- \;\longrightarrow\; \text{CH}_3\text{CH}_2\text{S}^-\text{Na}^+ \;+\; \text{H}_2\text{O}$$

$pK_a = 8.5$		$pK_a = 15.7$	
Stronger acid	Stronger base	Weaker base	Weaker acid

To name salts of thiols, give the name of the cation first, followed by the name of the alkyl group to which is attached the suffix -sulfide. For example, the sodium salt derived from ethanethiol is named sodium ethylsulfide.

F. Oxidation

Many of the chemical properties of thiols stem from the fact that the sulfur atom of a thiol is oxidized easily to several higher oxidation states, the most important of which are summarized in Table 9.4. Each reaction requires a specific oxidizing agent to avoid overoxidation. The relationships between the relative oxidation states of a thiol, a disulfide, a sulfinic acid, and a sulfonic acid are shown in the form of a sequence of balanced half-reactions. There are other oxidation states too, but they are not very stable. Note that the valence shell of sulfur contains 8 electrons in a thiol and a disulfide, 10 electrons in a sulfinic acid, and 12 electrons in a sulfonic acid (Section 1.2F).

The most common oxidation-reduction reaction of sulfur compounds in biological systems is interconversion between a thiol and a disulfide. The functional group of a disulfide is an —S—S— group.

$$2RSH + I_2 \longrightarrow RSSR + 2HI$$

A thiol   A disulfide

Thiols are also oxidized to disulfides by molecular oxygen. In fact, thiols are so susceptible to oxidation that they must be protected from contact with air during storage.

$$2RSH + \tfrac{1}{2}O_2 \longrightarrow RSSR + H_2O$$

A thiol   A disulfide

The disulfide bond is an important structural feature stabilizing the tertiary structure of many proteins (Section 27.6C).

Table 9.4 Functional Groups Formed by Oxidation of a Thiol

A thiol can be converted to the following oxidation products:

$$2R{-}S{-}H \longrightarrow R{-}S{-}S{-}R + 2H^+ + 2e^-$$

A thiol   A disulfide

$$R{-}S{-}H + 2H_2O \longrightarrow R{-}\overset{\overset{\displaystyle O}{\|}}{S}{-}O{-}H + 4H^+ + 4e^-$$

A thiol   A sulfinic acid

$$R{-}S{-}H + 3H_2O \longrightarrow R{-}\overset{\overset{\displaystyle O}{\|}}{\underset{\underset{\displaystyle O}{\|}}{S}}{-}O{-}H + 6H^+ + 6e^-$$

A thiol   A sulfonic acid

Summary

The functional group of an alcohol (Section 9.1A) is an —OH (hydroxyl) group bonded to an sp^3 hybridized carbon. IUPAC names of alcohols (Section 9.1B) are derived by changing the suffix of the parent alkane from -e to -ol. The chain is numbered from the direction that gives the carbon bearing —OH the lower number. In compounds containing other functional groups of higher precedence, the presence of —OH is indicated by the prefix hydroxy-. Common names for alcohols are derived by naming the alkyl group bonded to —OH and adding the word "alcohol." Alcohols are classified as 1°, 2°, or 3° (Section 9.1B) depending on whether the —OH group is bonded to a primary, secondary, or tertiary carbon. Alcohols are polar compounds (Section 9.2) with oxygen bearing a partial negative charge and both the α-carbon and the hydrogen bearing partial positive charges. Because of intermolecular association by **hydrogen bonding,** the boiling points of alcohols are higher than those of hydrocarbons of comparable molecular weight. Because of increased dispersion forces, the boiling points of alcohols increase with increasing molecular weight.

Alcohols interact with water by hydrogen bonding and, therefore, are more soluble in water than hydrocarbons of comparable molecular weight.

According to the **principle of microscopic reversibility** (Section 9.6), the sequence of transition states and reactive intermediates (that is, the mechanism) for any reversible reaction must be the same, but in reverse order, for the backward reaction as for the forward reaction.

A thiol (Section 9.9A) is a sulfur analog of an alcohol; it contains an —SH (sulfhydryl) group in place of an —OH group. Thiols (Section 9.9B) are named in the same manner as alcohols, but the suffix -e of the parent alkane is retained and -thiol is added. Common names for thiols are derived by naming the alkyl group bonded to —SH and adding the word "mercaptan." In compounds containing other functional groups of higher precedence, the presence of —SH is indicated by the prefix mercapto-. Because the S—H bond is almost nonpolar, the physical properties of thiols are more like those of hydrocarbons of comparable molecular weight.

Key Reactions

1. Acidity of Alcohols (Section 9.3)

In dilute aqueous solution, methanol and ethanol are comparable in acidity to water. Secondary and tertiary alcohols are weaker acids.

$$CH_3OH + H_2O \rightleftharpoons CH_3O^- + H_3O^+ \qquad pK_a = 15.5$$

2. Reaction with Active Metals (Section 9.4)

Alcohols react with <u>Li, Na, K,</u> and other active metals to form metal alkoxides, which are nearly the same or somewhat stronger bases than the alkali metal hydroxides such as NaOH and KOH.

$$CH_3CH_2OH + Na \longrightarrow CH_3CH_2O^-Na^+ + H_2$$

⭐3. Reaction with HCl, HBr, and HI (Section 9.5A)

Primary alcohols react by an S_N2 mechanism.

$$CH_3CH_2CH_2CH_2OH + HBr \xrightarrow{\text{reflux}} CH_3CH_2CH_2CH_2Br + H_2O$$

Tertiary alcohols react by an S_N1 mechanism with formation of a <u>carbocation intermediate</u>.

$$\underset{\underset{CH_3}{|}}{\overset{\overset{CH_3}{|}}{CH_3COH}} + HCl \xrightarrow{25°C} \underset{\underset{CH_3}{|}}{\overset{\overset{CH_3}{|}}{CH_3CCl}} + H_2O$$

Secondary alcohols may react by an S_N2 or an S_N1 mechanism, depending on the alcohol and experimental conditions. Primary alcohols with extensive β-branching react by an S_N1 mechanism involving formation of a rearranged carbocation.

$$\underset{\underset{\underset{CH_3}{|}}{\overset{\overset{CH_3}{|}}{CH_3CCH_2OH}}}{} + HBr \longrightarrow \underset{\underset{Br}{|}}{\overset{\overset{CH_3}{|}}{CH_3CCH_2CH_3}} + H_2O$$

4. Reaction with PBr$_3$ (Section 9.5B)

Although some rearrangement may occur with this reagent, it is less likely than in the reaction of an alcohol with HBr.

$$3\underset{\overset{|}{OH}}{CH_3CHCH_3} + PBr_3 \longrightarrow 3\underset{\overset{|}{Br}}{CH_3CHCH_3} + H_3PO_3$$

5. Reaction with SOCl$_2$ (Section 9.5C)

This is often the method of choice for converting a primary or secondary alcohol to an alkyl chloride.

$$\text{~~~~~}OH + SOCl_2 \xrightarrow{\text{pyridine}} \text{~~~~~}Cl + SO_2 + HCl$$

6. Acid-Catalyzed Dehydration (Section 9.6)

When isomeric alkenes are possible, the major product is generally the more substituted alkene (Zaitsev's rule). Rearrangements are common with secondary alcohols and also with primary alcohols with extensive β-branching.

$$\underset{\overset{|}{OH}}{CH_3CH_2CHCH_3} \xrightarrow[\text{heat}]{H_3PO_4} \underset{\text{Major product}}{CH_3CH=CHCH_3} + CH_3CH_2CH=CH_2 + H_2O$$

7. Pinacol Rearrangement (Section 9.7)

Dehydration of a glycol involves formation of a carbocation intermediate, rearrangement, and loss of H^+ to give an aldehyde or ketone.

$$\underset{\underset{CH_3}{\overset{|}{\underset{|}{CH_3}}}}{\overset{\overset{HO~~~OH}{|~~~~|}}{CH_3-C-C-CH_3}} \xrightarrow{H_2SO_4} \underset{\underset{\overset{|}{CH_3}}{}}{\overset{\overset{O~~~~CH_3}{\|~~~~|}}{CH_3-C-C-CH_3}} + H_2O$$

8. Oxidation of a Primary Alcohol to an Aldehyde (Section 9.8)

This oxidation is most conveniently carried out using pyridinium chlorochromate (PCC).

$$\text{⬠}-CH_2OH \xrightarrow{PCC} \text{⬠}-\overset{\overset{O}{\|}}{CH}$$

⭐ **9. Oxidation of a Primary Alcohol to a Carboxylic Acid (Section 9.8)**

A primary alcohol is oxidized to a carboxylic acid by chromic acid.

$$\text{OH} + H_2CrO_4 \xrightarrow[\text{acetone}]{H_2O} \text{OH} + Cr^{3+}$$

⭐ **10. Oxidation of a Secondary Alcohol to a Ketone (Section 9.8)**

A secondary alcohol is oxidized to a ketone by chromic acid or by PCC.

$$\text{OH} + H_2CrO_4 \xrightarrow[\text{acetone}]{H_2O} + Cr^{3+}$$

11. Oxidative Cleavage of a Glycol (Section 9.8C)

HIO_4 reacts with a glycol to form a five-membered cyclic intermediate that undergoes carbon-carbon bond cleavage to form two carbonyl groups.

$$\text{(OH, OH)} + HIO_4 \longrightarrow \text{(CHO, CHO)} + HIO_3$$

12. Acidity of Thiols (Section 9.9E)

Thiols are weak acids, pK_a 8–9, but considerably stronger than alcohols, pK_a 16–18.

$$CH_3CH_2SH + H_2O \rightleftharpoons CH_3CH_2S^- + H_3O^+ \qquad pK_a = 8.5$$

13. Oxidation of Thiols (Section 9.9F)

Oxidation by weak oxidizing agents such as O_2 and I_2 gives disulfides.

$$2RSH + \tfrac{1}{2}O_2 \longrightarrow RSSR + H_2O$$

Problems

Structure and Nomenclature

9.14 Which are secondary alcohols?

(a) [cyclohexane with CH₃ and OH] 3°   (b) $(CH_3)_3COH$ 1°   (c) [structure with OH] 2°   (d) [cyclopentane with OH] 2°

9.15 Name these compounds.

(a) [chain with OH]   (b) HO—[chain]—OH   (c) [chain with OH]

(d) [chain with OH and Cl]   (e) HO—[chain]—OH   (f) [chain with OH] (R)

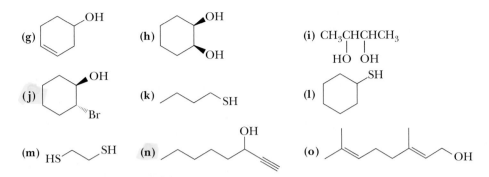

9.16 Write structural formulas for these compounds.

(a) Isopropyl alcohol
(b) Propylene glycol
(c) 5-Methyl-2-hexanol
(d) 2-Methyl-2-propyl-1,3-propanediol
(e) 1-Chloro-2-hexanol
(f) *cis*-3-Isobutylcyclohexanol
(g) 2,2-Dimethyl-1-propanol
(h) 2-Mercaptoethanol
(i) Allyl alcohol
(j) *trans*-2-Vinylcyclohexanol
(k) (Z)-5-Methyl-2-hexen-1-ol
(l) 2-Propyn-1-ol
(m) 3-Chloro-1,2-propanediol
(n) *cis*-3-Pentene-1-ol
(o) Bicyclo[2.2.1]heptan-7-ol

9.17 Name and draw structural formulas for the eight isomeric alcohols of molecular formula $C_5H_{12}O$. Classify each as primary, secondary, or tertiary. Which are chiral?

Physical Properties of Alcohols

9.18 Arrange these compounds in order of increasing boiling point. (Values in °C are − 42, 78, 117, and 198.)

(a) $CH_3CH_2CH_2CH_2OH$ (b) CH_3CH_2OH (c) $HOCH_2CH_2OH$
(d) $CH_3CH_2CH_3$

9.19 Arrange these compounds in order of increasing boiling point. (Values in °C are − 42, − 24, 78, and 118.)

(a) CH_3CH_2OH (b) CH_3OCH_3 (c) $CH_3CH_2CH_3$ (d) CH_3COOH

9.20 Compounds that contain an N—H group associate by hydrogen bonding.

(a) Do you expect this association to be stronger or weaker than that of compounds containing an O—H group?
(b) Based on your answer to part (a), which would you predict to have the higher boiling point, 1-butanol or 1-butanamine?

9.21 Which compounds can participate in hydrogen bonding with water?

(a) $CH_3CH_2CH_2OCH_2CH_2OH$ (b) $(CH_3CH_2)_2NH$ (c) $CH_3CH=CHCH_3$

(d) $CH_3\overset{\overset{\text{O}}{\|}}{C}CH_3$ (e) CH_3SCH_3 (f) $CH_3CH_2\overset{\overset{\text{O}}{\|}}{C}OH$

9.22 From each pair of compounds, select the one that is more soluble in water.

(a) CH_2Cl_2 or CH_3OH

(b) $CH_3\overset{\overset{\text{O}}{\|}}{C}CH_3$ or $CH_3\overset{\overset{\text{CH}_2}{\|}}{C}CH_3$

(c) CH_3CH_2Cl or NaCl

(d) $CH_3CH_2CH_2SH$ or $CH_3CH_2CH_2OH$

(e) $CH_3CH_2\overset{\overset{\text{OH}}{|}}{C}HCH_2CH_3$ or $CH_3CH_2\overset{\overset{\text{O}}{\|}}{C}CH_2CH_3$

9.23 Arrange the compounds in each set in order of decreasing solubility in water.

(a) Ethanol, butane, diethyl ether   (b) 1-Hexanol, 1,2-hexanediol, hexane

9.24 Each compound given in this problem is a common organic solvent. From each pair of compounds, select the solvent with the greater solubility in water.

(a) CH_2Cl_2 or CH_3CH_2OH (b) $CH_3CH_2OCH_2CH_3$ or CH_3CH_2OH

(c)  or $CH_3CH_2OCH_2CH_3$

(d) $CH_3CH_2OCH_2CH_3$ or $CH_3(CH_2)_3CH_3$

9.25 The decalinols A and B can be equilibrated using aluminum isopropoxide in 2-propanol (isopropyl alcohol) containing a small amount of acetone. Assume a value of ΔG^0 (equatorial to axial) for cyclohexanol is 4.0 kJ (0.95 kcal)/mol; calculate the percent of each isomer in the equilibrium mixture at 25°C.

A   B

Acid-Base Reactions of Alcohols

9.26 Complete the following acid-base reactions. In addition, show all valence electrons on the interacting atoms and show by the use of curved arrows the flow of electrons in each reaction.

(a) $CH_3CH_2OH + H\overset{+}{\underset{H}{O}}H \longrightarrow$ (b) $CH_3CH_2OCH_2CH_3 + HO\overset{O}{\underset{O}{\overset{\|}{S}}}OH \longrightarrow$

(c) $CH_3CH_2CH_2CH_2CH_2OH + HI \longrightarrow$

(d) $CH_3CH_2CH_2\overset{O}{\overset{\|}{C}}OH + HO\overset{O}{\underset{O}{\overset{\|}{S}}}OH \longrightarrow$ (e) ⬡—$OH + BF_3 \longrightarrow$

(f) $CH_3CH=\overset{+}{CH}CHCH_3 + HOH \longrightarrow$

9.27 Select the stronger acid from each pair, and explain your reasoning. For each stronger acid, write a structural formula for its conjugate base.

(a) H_2O or H_2CO_3 (b) CH_3OH or CH_3COOH
(c) CH_3CH_2OH or $CH_3C\equiv CH$ (d) CH_3CH_2OH or CH_3CH_2SH

9.28 From each pair, select the stronger base. For each stronger base, write the structural formula of its conjugate acid.

(a) OH^- or CH_3O^- (each in H_2O)   (b) $CH_3CH_2O^-$ or $CH_3C\equiv C^-$
(c) $CH_3CH_2S^-$ or $CH_3CH_2O^-$ (d) $CH_3CH_2O^-$ or NH_2^-

9.29 In each equilibrium, label the stronger acid, the stronger base, the weaker acid, and the weaker base. Also estimate the position of each equilibrium.

(a) $CH_3CH_2O^- + CH_3C\equiv CH \rightleftharpoons CH_3CH_2OH + CH_3C\equiv C^-$
(b) $CH_3CH_2O^- + HCl \rightleftharpoons CH_3CH_2OH + Cl^-$
(c) $CH_3COOH + CH_3CH_2O^- \rightleftharpoons CH_3COO^- + CH_3CH_2OH$

Reactions of Alcohols

9.30 Write equations for the reaction of 1-butanol with each reagent. Where you predict no reaction, write NR.

(a) Na metal   (b) HBr, heat   (c) HI, heat

(d) PBr$_3$ (e) SOCl$_2$, pyridine   (f) K$_2$Cr$_2$O$_7$, H$_2$SO$_4$, H$_2$O, heat

(g) HIO$_4$ (h) PCC   (i) CH$_3$SO$_2$Cl, pyridine

9.31 Write equations for the reaction of 2-butanol with each reagent listed in Problem 9.30. Where you predict no reaction, write NR.

9.32 Complete these equations. Show structural formulas for the major products, but do not balance.

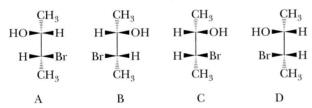

9.33 When (R)-2-butanol is left standing in aqueous acid, it slowly loses its optical activity. Account for this observation.

9.34 Two diastereomeric sets of enantiomers, A/B and C/D, exist for 3-bromo-2-butanol. When enantiomer A or B is treated with HBr, only racemic 2,3-dibromobutane is formed; no meso isomer is formed. When enantiomer C or D is treated with HBr, only meso 2,3-dibromobutane is formed; no racemic 2,3-dibromobutane is formed. Account for these observations.

9.35 Acid-catalyzed dehydration of 3-methyl-2-butanol gives three alkenes: 2-methyl-2-butene, 3-methyl-1-butene, and 2-methyl-1-butene. Propose a mechanism to account for the formation of each product.

9.36 Show how you might bring about the following conversions. For any conversion involving more than one step, show each intermediate compound formed.

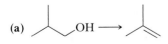

(c)

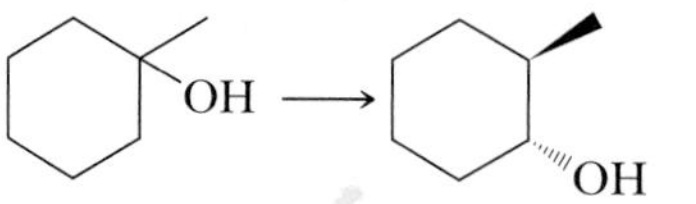

(d)

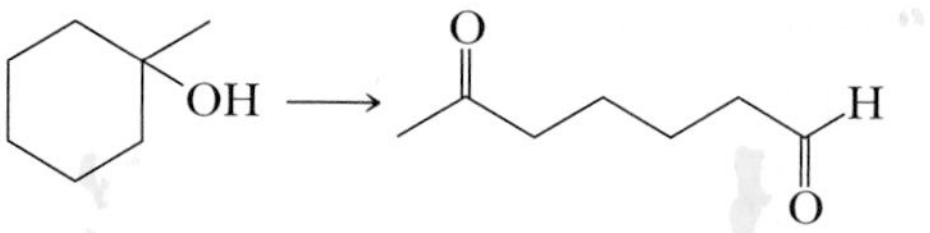

(e)

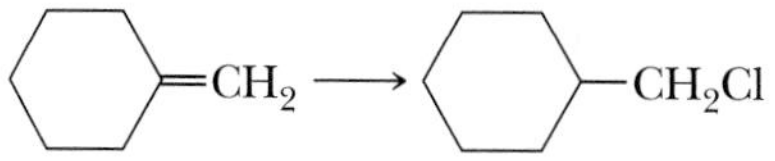

(f)

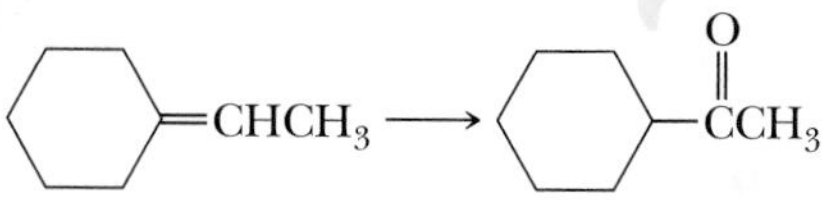

(g) $CH_3(CH_2)_6CH_2OH \longrightarrow CH_3(CH_2)_6\overset{\text{O}}{\overset{\|}{C}}H$

(h)

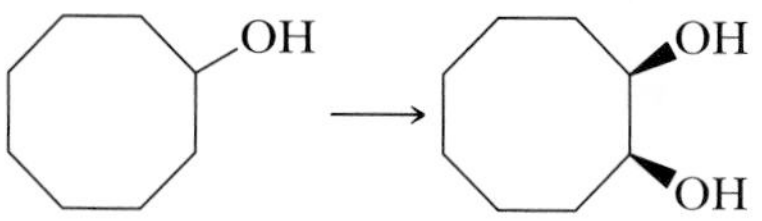

(i)

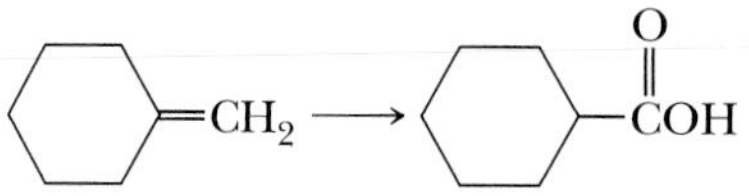

Pinacol Rearrangement

9.37 Propose a mechanism for the following pinacol rearrangement catalyzed by boron trifluoride etherate.

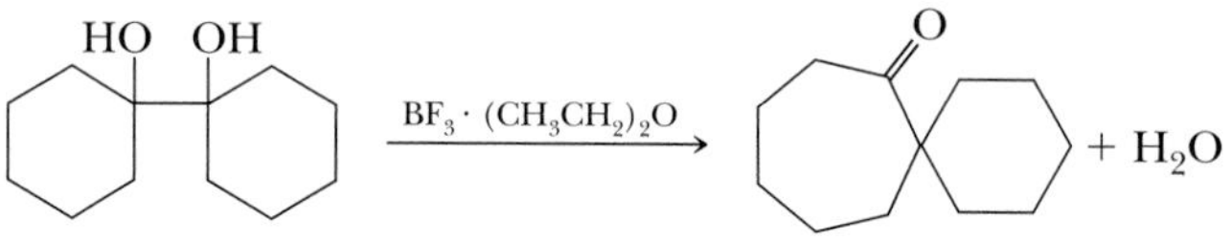

Spiro[5.6]dodecan-7-one

Synthesis

9.38 Alkenes can be hydrated to form alcohols by (1) hydroboration followed by oxidation and (2) oxymercuration followed by reduction. Compare the products formed from these alkenes by sequence (1) to those formed from sequence (2).

(a) Propene   (b) *cis*-2-Butene   (c) *trans*-2-Butene   (d) Cyclopentene
(e) 1-Methylcyclohexene

9.39 Give reactions for the synthesis of each alcohol or diol from a suitable alkene.

(a) 2-Pentanol   (b) 1-Pentanol
(c) 2-Methyl-2-pentanol (d) 2-Methyl-2-butanol
(e) 3-Pentanol   (f) 3-Ethyl-3-pentanol
(g) 1,2-Hexanediol

9.40 Dihydropyran is synthesized by treating tetrahydrofurfuryl alcohol with acid. Propose a mechanism for this conversion.

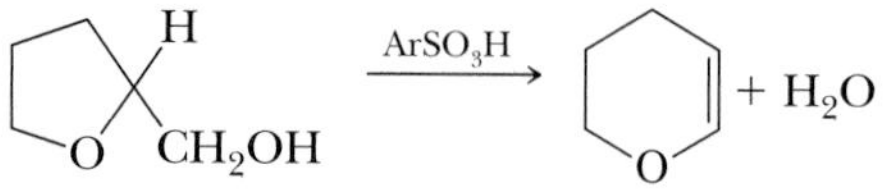

Tetrahydrofurfuryl   Dihydropyran
alcohol

9.41 Show how to convert propene to each of these compounds. Use any inorganic reagents as necessary.

(a) Propane
(b) 1,2-Propanediol
(c) 1-Propanol
(d) 2-Propanol
(e) Propanal
(f) Propanone
(g) Propanoic acid
(h) 1-Bromo-2-propanol
(i) 3-Chloropropene
(j) 1,2,3-Trichloropropane
(k) 1-Chloropropane
(l) 2-Chloropropane
(m) 2-Propen-1-ol
(n) Propenal

9.42 (a) How many stereoisomers are possible for 4-methyl-1,2-cyclohexanediol?
(b) Which of the possible stereoisomers are formed by oxidation of 4-methylcyclohexene with osmium tetroxide?
(c) Is the product formed in part (b) optically active or optically inactive?

9.43 Show how to bring about this conversion in good yield.

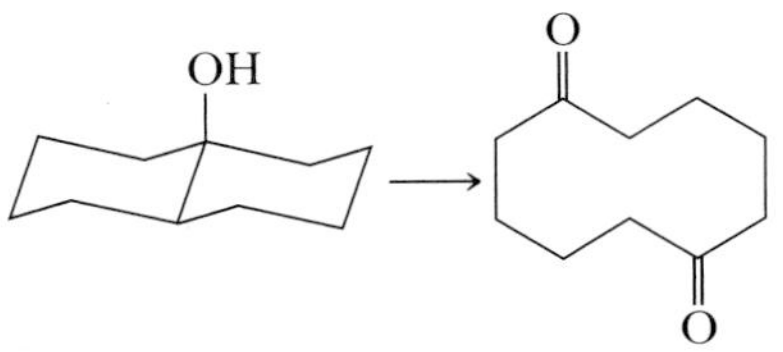

9.44 The tosylate of a primary alcohol normally undergoes an S_N2 reaction with hydroxide ion to give a primary alcohol. Reaction of this tosylate, however, gives a compound of molecular formula $C_7H_{12}O$. Propose a structural formula for this compound and a mechanism for its formation.

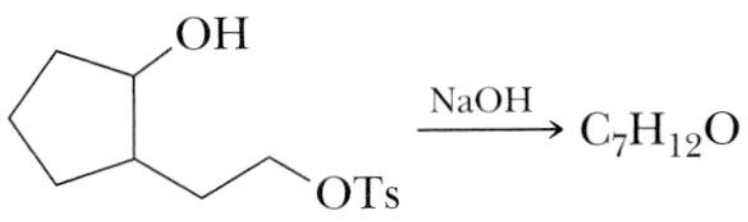

9.45 Chrysanthemic acid occurs as a mixture of esters in flowers of the chrysanthemum (pyrethrum) family. For a model of one of these esters, see the front cover. Reduction of chrysanthemic acid to its alcohol (Section 17.6A) followed by conversion of the alcohol to its tosylate gives chrysanthemyl tosylate. Solvolysis (Section 8.3) of the tosylate gives a mixture of artemesia and yomogi alcohols. Propose a mechanism for the formation of these alcohols from chrysanthemyl tosylate.

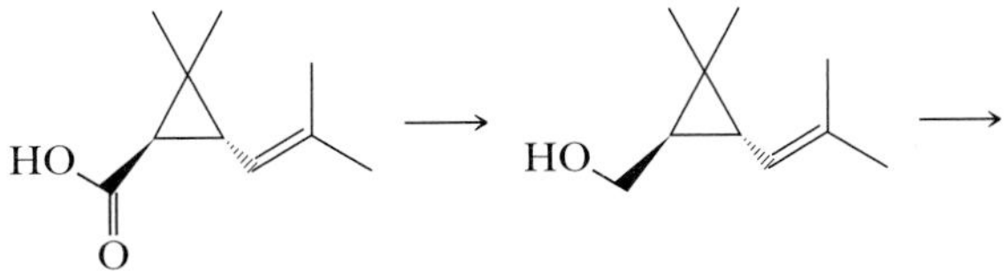

Chrysanthemic acid   Chrysanthemyl alcohol

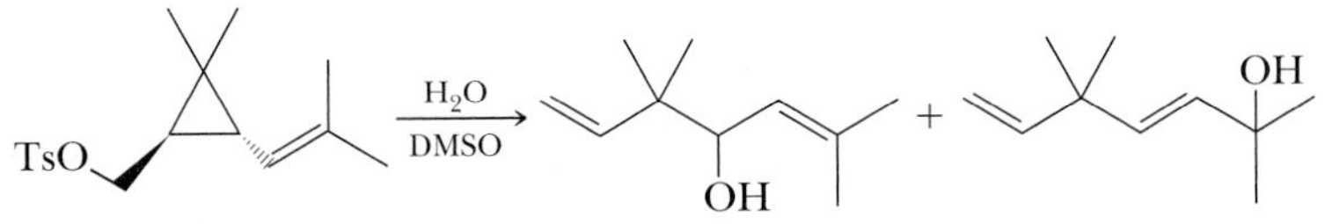

Chrysanthemyl tosylate   Artemesia alcohol   Yomogi alcohol

9.46 Show how to convert cyclohexene to each compound in good yield.

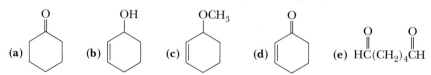

9.47 Oxymercuration of the following bicycloalkene followed by reduction is stereospecific in addition to being regiospecific. The product is a single alcohol in better than 95% yield. Propose a structural formula for this alcohol and account for the stereospecificity of its formation. *Hint:* Examine the Chem3D model of this molecule and see if you can determine which face of the double bond is more accessible to the oxymercuration reagent.

ALKYNES

I n this chapter, we continue our discussion of the chemistry of carbon-carbon pi bonds as we now consider the chemistry of **alkynes.** Just as alkenes and alkynes are similar in that the multiple bond in each is a combination of sigma and pi bonds, they also show similarities in the types of chemical reactions they undergo. Alkynes undergo electrophilic additions of X_2, HX, and H_2O. They undergo hydroboration/oxidation, and the carbon-carbon triple bond can be reduced first to a double bond and then to a single bond. Another important reaction of alkynes is the

■ Cutting with an oxyacetylene torch. (*T. J. Florian, Rainbow*) Inset: A model of acetylene.

363

conversion of terminal alkynes to their alkali metal salts, which are good nucleophiles and, therefore, valuable building blocks for the construction of larger molecules.

10.1 Structure

Alkyne An unsaturated hydrocarbon that contains one or more carbon-carbon triple bonds.

The functional group of an **alkyne** is a carbon-carbon triple bond. The simplest alkyne is ethyne, C_2H_2, more commonly named acetylene (Figure 10.1). Acetylene is a linear molecule; all bond angles are 180°. The carbon-carbon bond length in acetylene is 121 pm (Table 5.1). By comparison, the length of the carbon-carbon double bond in ethylene is 134 pm and that of the carbon-carbon single bond in ethane is 154 pm. Thus, triple bonds are shorter than double bonds, which, in turn, are shorter than single bonds. The bond dissociation energy of the carbon-carbon triple bond in acetylene [837 kJ (200 kcal)/mol] is also considerably larger than that for the carbon-carbon double bond in ethylene [720 kJ (172 kcal)/mol] and the carbon-carbon single bond in ethane [377 kJ (90 kcal)/mol].

A triple bond is described in terms of overlap of sp hybrid orbitals of adjacent carbon atoms to form a sigma bond, overlap of parallel $2p_y$ orbitals to form one pi bond, and overlap of parallel $2p_z$ orbitals to form a second pi bond (Figure 1.21). In acetylene, each carbon-hydrogen bond is formed by overlap of a $1s$ orbital of hydrogen with an sp orbital of carbon. Because of the 50% s-character of the acetylenic C—H bond, it is unusually strong (see Table 1.11 and related text).

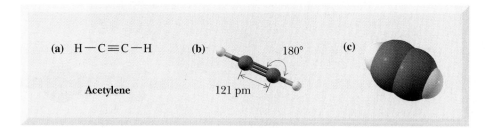

(a) H—C≡C—H   **(b)**   180°   **(c)**

Acetylene   121 pm

Figure 10.1
Acetylene. (a) Lewis structure, (b) ball-and-stick model, and (c) space-filling model.

Nomenclature
Alkynes

10.2 Nomenclature

A. IUPAC Names

According to the rules of the IUPAC system, the infix -yn- is used to show the presence of a carbon-carbon triple bond (Section 2.5). Thus, HC≡CH is named ethyne and $CH_3C≡CH$ is named propyne. The IUPAC system retains the name acetylene and, therefore, there are two acceptable names for HC≡CH, ethyne and acetylene. Of these two names, acetylene is used much more frequently. There is no need to use a number to locate the position of the triple bond in ethyne and propyne; there is only one possible location for it in each compound.

For larger molecules, number the longest carbon chain that contains the triple bond from the end that gives the triply bonded carbons the lower numbers. Show the location of the triple bond by the number of the first carbon of the triple bond. If a hydrocarbon chain contains more than one triple bond, the infixes -adiyn-, -atriyn-, and so forth, are used.

3-Methyl-1-butyne   6,6-Dimethyl-3-heptyne   1,6-Heptadiyne

Example 10.1

Write the IUPAC name of each compound.

(a)   (b)   (c)

Solution

(a) 2-Pentyne   (b) 3-Chloropropyne
(c) 5-Hexyn-1-ol. The hydroxyl group is indicated by the suffix -ol, and its location determines the numbering of the carbon chain.

Problem 10.1

Write the IUPAC name of each compound.

(a)   (b)   (c) OH

B.   Common Names

Common names for alkynes are derived by prefixing the names of substituents on the carbon-carbon triple bond to the word "acetylene." Note in the third example that when a carbon-carbon double bond (indicated by -en-) and a carbon-carbon triple bond (indicated by -yn-) are both present in the same molecule, the IUPAC rules specify that the location of the double bond takes precedence in numbering the compound.

	$CH_3C{\equiv}CH$	$CH_3C{\equiv}CCH_3$	$CH_2{=}CHC{\equiv}CH$
IUPAC name:	Propyne	2-Butyne	1-Buten-3-yne
Common name:	Methylacetylene	Dimethylacetylene	Vinylacetylene

Alkynes in which the triple bond is between carbons 1 and 2 are commonly referred to as **terminal alkynes.** Examples of terminal alkynes are propyne and 1-butyne.

Example 10.2

Write the common name of each alkyne.

(a) 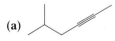   (b)    (c)

Solution

(a) Isobutylmethylacetylene   (b) *sec*-Butylacetylene   (c) *tert*-Butylacetylene

Problem 10.2

Write the common name of each alkyne.

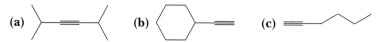

(a)   (b)   (c)

The smallest cycloalkyne that has been isolated is cyclooctyne. This molecule is quite unstable and polymerizes rapidly at room temperature. The $C—C\equiv C$ bond angle in cyclooctyne is approximately 155°, indicating a high degree of angle strain. Cyclononyne has also been prepared and is stable at room temperature. The $C—C\equiv C$ bond angles in this cycloalkyne are approximately 160°, which still represents a considerable distortion from the optimal $C—C\equiv C$ bond angle of 180°. You can see the distortion of the $C—C\equiv C$ bond angles in the accompanying molecular model. You can also see the degree to which $C—C$ and $C—H$ bonds on adjacent carbons are staggered, thus minimizing torsional strain.

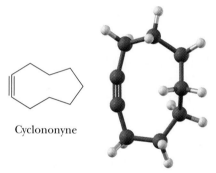

Cyclononyne

10.3 Physical Properties

The physical properties of alkynes are quite similar to those of alkanes and alkenes with similar carbon skeletons. The lower molecular-weight alkynes are gases at room temperature. Those that are liquids at room temperature have densities less than

Table 10.1 Physical Properties of Some Low-Molecular-Weight Alkynes

Name	Formula	Melting Point (°C)	Boiling Point (°C)	Density at 20°C (g/mL)
Ethyne	$HC\equiv CH$	−81	−84	(a gas)
Propyne	$CH_3C\equiv CH$	−102	−23	(a gas)
1-Butyne	$CH_3CH_2C\equiv CH$	−126	8	(a gas)
2-Butyne	$CH_3C\equiv CCH_3$	−32	27	0.691
1-Pentyne	$CH_3(CH_2)_2C\equiv CH$	−90	40	0.690
1-Hexyne	$CH_3(CH_2)_3C\equiv CH$	−132	71	0.716
1-Octyne	$CH_3(CH_2)_5C\equiv CH$	−79	125	0.746
1-Decyne	$CH_3(CH_2)_7C\equiv CH$	−36	174	0.766

1.0 g/mL (they are less dense than water). Listed in Table 10.1 are melting points, boiling points, and densities of several low-molecular-weight alkynes. Because, like alkanes and alkenes, alkynes are nonpolar compounds, they are insoluble in water and other polar solvents. They are soluble in each other and in other nonpolar organic solvents.

10.4   Acidity of 1-Alkynes

One of the major differences between the chemistry of alkynes and that of alkenes or alkanes is that a hydrogen attached to a triply bonded carbon atom of a terminal alkyne is sufficiently acidic that it can be removed by a strong base such as sodium amide (Section 4.4E). The reaction can be very slow, however, and a little dimethyl sulfoxide, an acid intermediate in acidity between ammonia and acetylene, is often used to catalyze it.

$$HC{\equiv}CH + {^-}NH_2 \rightleftharpoons HC{\equiv}C^- + NH_3 \qquad K_{eq} = 10^{13}$$

pK$_a$ 25			pK$_a$ 38
Stronger	Stronger	Weaker	Weaker
acid	base	base	acid

Other strong bases commonly used to form acetylide anions are sodium hydride and lithium diisopropylamide (LDA).

$$Na^+H^- \qquad\qquad [(CH_3)_2CH]_2N^-Li^+$$

Sodium hydride   Lithium diisopropylamide
(LDA)

Water is a stronger acid than acetylene; therefore, the hydroxide ion is not a strong enough base to convert acetylene to the acetylide anion.

$$HC{\equiv}CH + OH^- \rightleftharpoons HC{\equiv}C^- + H_2O \qquad K_{eq} = 10^{-9.3}$$

pK$_a$ 25			pK$_a$ 15.7
Weaker	Weaker	Stronger	Stronger
acid	base	base	acid

The pK$_a$ values for alkene and alkane hydrogens are so large (they are so weakly acidic) that neither the commonly used alkali metal hydroxides nor sodium hydride, sodium amide, or lithium diisopropylamide are strong enough bases to remove a proton from them.

10.5   Alkylation of Acetylide Anions

Acetylide anions are both strong bases and good nucleophiles. As nucleophiles, they undergo S$_N$2 reactions with alkyl halides, alkyl tosylates, and alkyl mesylates to form new carbon-carbon bonds to alkyl groups; that is, they undergo **alkylation** reactions. For example, treatment of sodium acetylide with 1-bromobutane gives 1-hexyne.

Alkylation   Any reaction in which a new carbon-carbon bond to an alkyl group is formed.

$$HC{\equiv}C^-Na^+ + CH_3(CH_2)_2CH_2{-}Br \xrightarrow{S_N2} HC{\equiv}CCH_2(CH_2)_2CH_3 + Na^+Br^-$$

Sodium   1-Bromobutane   1-Hexyne
acetylide

Because of the ready availability of acetylene and the ease with which it is converted to a good nucleophilic anion, alkylation of the acetylide anion is the most convenient laboratory method for the synthesis of terminal alkynes. The process of alkylation can be repeated, and a terminal alkyne can, in turn, be converted to an internal alkyne.

$$CH_3CH_2C{\equiv}C^- \, Na^+ + CH_3CH_2{-}Br \xrightarrow{S_N2} CH_3CH_2C{\equiv}CCH_2CH_3 + Na^+Br^-$$

Sodium butynide　　　Bromoethane　　　　　　3-Hexyne

Because acetylide anions are strong bases as well as good nucleophiles, alkylation of acetylide anions is practical only with methyl and primary halides, tosylates, and mesylates. With secondary and tertiary halides, E2 elimination becomes the main reaction (Section 8.11B).

Sodium　　　　Bromocyclohexane　　　Acetylene　　Cyclohexene
acetylide

10.6　Preparation

In addition to alkylation of salts of terminal alkynes, there are several other important types of reactions by which alkynes can be made. Starting materials for these methods include alkyl halides and alkenes.

Alkynes can be synthesized from alkenes by the combination of an addition reaction followed by two successive β-elimination reactions. In the addition reaction, the alkene is treated with bromine or chlorine to form a dibromoalkane or dichloroalkane (Section 6.3D).

$$RCH{=}CHR + Br_2 \xrightarrow{CCl_4} \overset{\displaystyle Br \quad Br}{\underset{\displaystyle}{RCH{-}CHR}}$$

An alkene　　　　　　　　A vicinal dibromide

Treatment of a dibromide of this type with two moles of strong base, most commonly sodium amide, results in the elimination of two moles of HBr by successive E2 reactions and formation of an alkyne. The two halogens may be on adjacent carbons (a **vicinal dihalide,** from the Latin: *vicinalis,* neighbor), or they may be on the same carbon (a **geminal dihalide,** from the Latin: *geminatus,* twin).

Vicinal (vic) dihalide　Vicinal from the Latin: *vicinalis,* neighbor. A compound containing two halogen atoms on adjacent carbon atoms.

Geminal (gem) dihalide　Geminal from the Latin: *geminatus,* twin. A compound containing two halogens on the same carbon atom.

$$R{-}\overset{\displaystyle Br}{\underset{\displaystyle}{CH}}{-}\overset{\displaystyle Br}{\underset{\displaystyle}{CH}}{-}R + 2NaNH_2 \xrightarrow[-33°C]{NH_3(l)} R{-}C{\equiv}C{-}R + 2NaBr + 2NH_3$$

A vicinal　　　　Sodium　　　　　　An alkyne
dibromide　　　　amide

Double dehydrohalogenation to form an alkyne may also be carried out using potassium *tert*-butoxide in dimethyl sulfoxide (DMSO).

$$R-\underset{\underset{H}{|}}{\overset{\overset{H}{|}}{C}}-\underset{\underset{Br}{|}}{\overset{\overset{Br}{|}}{C}}-R + 2(CH_3)_3CO^-K^+ \xrightarrow[DMSO]{} R-C\equiv C-R + 2(CH_3)_3COH + 2K^+Br^-$$

A geminal dibromide	Potassium *tert*-butoxide	An alkyne	2-Methyl-2-propanol

With a strong base such as sodium amide, both dehydrohalogenations occur readily. However, with weaker bases, such as sodium hydroxide or potassium hydroxide in ethanol, it is often possible to stop the reaction after the first dehydrohalogenation and isolate the haloalkene. In practice, it is much more common to use a stronger base and go directly to the alkyne.

Given the ease of converting alkenes to dihaloalkanes followed by base-promoted double dehydrohalogenation, alkenes are valuable starting materials for the preparation of alkynes. The following equations show the conversion of 1-hexene to 1-hexyne. Note that three moles of sodium amide are used in this sequence. Two moles are required for the double dehydrohalogenation reaction, which gives 1-hexyne. As soon as any 1-hexyne (a weak acid, pK_a 25) is formed, it reacts with sodium amide (a strong base) to form an alkyne salt. Thus, a third mole of sodium amide is required to complete the dehydrohalogenation of the remaining bromoalkene. Addition of water or aqueous acid completes the sequence and gives 1-hexyne.

$$CH_3(CH_2)_3CH{=}CH_2 \xrightarrow{Br_2} CH_3(CH_2)_3\underset{}{\overset{\overset{Br}{|}}{C}H}{-}\underset{}{\overset{\overset{Br}{|}}{C}H_2} \xrightarrow[-2HBr]{2NaNH_2} CH_3(CH_2)_3C{\equiv}CH \xrightarrow{NaNH_2}$$

1-Hexene	1,2-Dibromohexane	1-Hexyne

$$CH_3(CH_2)_3C{\equiv}C^-Na^+ \xrightarrow{H_2O} CH_3(CH_2)_3C{\equiv}CH$$

Sodium salt of 1-hexyne	1-Hexyne

In dehydrohalogenation of a haloalkene with at least one hydrogen on each adjacent carbon, a side reaction occurs, namely, the formation of an allene.

$$\underset{\underset{\underset{R}{|}}{RC}}{\overset{\overset{H}{|}}{}}-\underset{}{\overset{\overset{X}{|}}{C}}{=}\underset{}{\overset{\overset{H}{|}}{CR}} \xrightarrow[-HBr]{NaNH_2}$$

A haloalkene (a vinylic halide)

$$\underset{R}{\overset{R}{}}C{=}C{=}\underset{R}{\overset{H}{}}C$$

An allene

$$\underset{\underset{R}{|}}{\overset{\overset{H}{|}}{RCC}}{\equiv}CR$$

An alkyne

An **allene** has two adjacent carbon-carbon double bonds; that is, C=C=C. The simplest allene is 1,2-propadiene, commonly named allene. In it, each end carbon is sp^2 hybridized, and the middle carbon is sp hybridized. Each carbon-carbon sigma bond is formed by the overlap of sp and sp^2 hybrid orbitals. One pi bond is formed by the

Allene The compound $CH_2{=}C{=}CH_2$. Any compound that contains adjacent carbon-carbon double bonds; that is, the arrangement C=C=C.

overlap of parallel $2p_y$ orbitals; the other, by the overlap of parallel $2p_z$ orbitals. The two pi-bonding molecular orbitals are in planes perpendicular to each other, as are the two CH_2 groups.

CH_2=C=CH_2

1,2-Propadiene
(Allene)

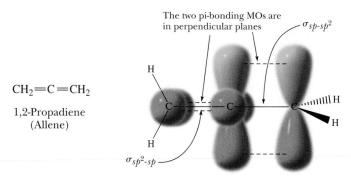

Most allenes are less stable than their isomeric alkynes. For example, allene itself is less stable by 6.7 kJ (1.6 kcal)/mol than its isomer propyne, and 1,2-butadiene is less stable by 16.7 kJ (4.0 kcal)/mol than its isomer 2-butyne.

$$CH_2=C=CH_2 \longrightarrow CH_3C\equiv CH \qquad \Delta H^0 = -6.7 \text{ kJ } (-1.6 \text{ kcal})/\text{mol}$$

$$CH_2=C=CHCH_3 \longrightarrow CH_3C\equiv CCH_3 \qquad \Delta H^0 = -16.7 \text{ kJ } (-4.0 \text{ kcal})/\text{mol}$$

Because of their lower stability relative to isomeric alkynes, allenes are generally only minor products of alkyne-forming dehydrohalogenation reactions.

Example 10.3

Show how you might prepare 2-butyne from each starting material.

(a) (b)

Solution

(a) This synthesis can be done in two steps: treatment of 2-butene with one mole of bromine followed by dehydrohalogenation with two moles of sodium amide.

(b) This synthesis requires three steps. First, dehydrohalogenation of 2-chlorobutane with sodium hydroxide (Section 8.8), followed by addition of chlorine (or bromine), and then a double dehydrohalogenation as in part (a) of this problem.

Problem 10.3

Draw a structural formula for an alkene and dichloroalkane of the given molecular formula that yields the indicated alkyne by each reaction sequence.

(a) $C_6H_{12} \xrightarrow{Cl_2} C_6H_{12}Cl_2 \xrightarrow{2NaNH_2}$

(b) $C_7H_{14} \xrightarrow{Cl_2} C_7H_{14}Cl_2 \xrightarrow{2NaNH_2}$

10.7 Reduction

Reactivity Explorer
Alkynes

Two types of reactions are used to convert alkynes to alkenes and alkanes: catalytic reduction and chemical reduction.

A. Catalytic Reduction

Treatment of an alkyne with hydrogen in the presence of a transition metal catalyst, most commonly palladium, platinum, or nickel, results in the addition of two moles of hydrogen to the alkyne and its conversion to an alkane. Catalytic reduction of an alkyne can be brought about at or slightly above room temperature and with moderate pressures of hydrogen gas.

$$CH_3C{\equiv}CCH_3 + 2H_2 \xrightarrow[\text{3 atm}]{\text{Pd, Pt, or Ni}} CH_3CH_2CH_2CH_3$$

2-Butyne   Butane

Reduction of an alkyne occurs in two stages: first addition of one mole of hydrogen to form an alkene and then addition of the second mole to the alkene to form the alkane. In most cases, it is not possible to stop the reaction at the alkene stage.

However, by careful choice of catalyst, it is possible to stop the reduction after the addition of one mole of hydrogen. The catalyst most commonly used for this purpose consists of finely powdered palladium metal deposited on solid calcium carbonate that has been specially modified with lead salts. This combination is known as the **Lindlar catalyst.** Reduction (hydrogenation) of alkynes over a Lindlar catalyst is stereoselective; syn addition of two hydrogen atoms to the carbon-carbon triple bond gives a cis alkene.

Lindlar catalyst Finely powdered palladium metal deposited on solid calcium carbonate that has been specially modified with lead salts. Its particular use is as a catalyst for the reduction of an alkyne to a cis alkene.

$$CH_3C{\equiv}CCH_3 + H_2 \xrightarrow[\text{catalyst}]{\text{Lindlar}} \begin{array}{c} H_3C \qquad CH_3 \\ \diagdown \quad \diagup \\ C{=}C \\ \diagup \quad \diagdown \\ H \qquad H \end{array}$$

2-Butyne   *cis*-2-Butene

Because addition of hydrogen in the presence of the Lindlar catalyst is stereoselective for syn addition, it has been proposed that reduction proceeds by simultaneous or nearly simultaneous transfer of two hydrogen atoms from the surface of the metal catalyst to the alkyne. We presented a similar mechanism in Section 6.6A for the catalytic reduction of an alkene to an alkane.

B.   Chemical Reduction

Alkynes can also be reduced to alkenes by using either sodium or lithium metal in liquid ammonia or in low-molecular-weight primary or secondary amines. The alkali metal is the reducing agent and in the process is oxidized to M^+. Reduction of an alkyne to an alkene by lithium or sodium in liquid ammonia, $NH_3(l)$, is stereoselective; it involves mainly anti addition of two hydrogen atoms to the triple bond.

$$R\!-\!\!\!\equiv\!\!\!-R' + 2Na \xrightarrow[NH_3(l)]{} \underset{R}{\overset{H}{\diagdown}}C\!=\!C\underset{H}{\overset{R'}{\diagup}} + 2NaNH_2$$

$$\xrightarrow[NH_3(l)]{2Na}$$

4-Octyne   *trans*-4-Octene

Catalytic and alkali metal reduction of alkynes are complementary reactions; by the proper choice of reagents and reaction conditions, it is possible to reduce an alkyne to either a cis alkene or a trans alkene.

The stereoselectivity of alkali metal reduction of alkynes to alkenes is accounted for by the following mechanism. As you study it, note that it involves two one-electron reductions and two proton-transfer reactions. The stereochemistry of the alkene is determined in Step 3. Adding the four steps and canceling species that appear on both sides of the equation give the overall equation for the reaction.

Mechanism   Reduction of an Alkyne by Sodium in Liquid Ammonia

Step 1:   A one-electron reduction of the alkyne gives an alkenyl radical anion (an ion containing an unpaired electron on one carbon and a negative charge on an adjacent carbon).

$$R\!-\!C\!\equiv\!C\!-\!R + \!{\cdot}Na \longrightarrow R\!-\!\overset{\cdot}{C}\!=\!\overset{\cdot\cdot -}{C}\!-\!R \longleftrightarrow R\!-\!\overset{\cdot\cdot -}{C}\!=\!\overset{\cdot}{C}\!-\!R + Na^+$$

A resonance-stabilized alkenyl radical anion

Step 2:   The alkenyl radical anion (a very strong base) abstracts a proton from a molecule of ammonia (under these conditions, a weak acid) to give an alkenyl radical.

$$R\!-\!\overset{\cdot}{C}\!=\!\overset{\cdot\cdot -}{C}\!-\!R + H\!-\!\underset{H}{\overset{|}{N}}\!-\!H \longrightarrow \underset{R}{\overset{\cdot}{\diagup}}C\!=\!C\underset{H}{\overset{R}{\diagup}} + :N\!-\!H$$

An alkenyl   Amide
radical   ion

Step 3: A one-electron reduction of the alkenyl radical gives an alkenyl anion. The trans alkenyl anion is more stable than its cis isomer, and it is at this step that the stereochemistry of the final product is determined.

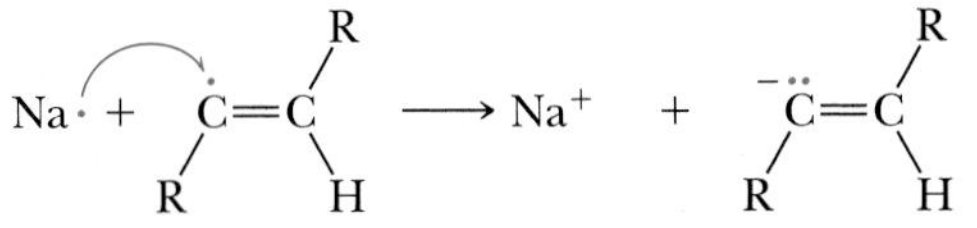

An alkenyl anion

Step 4: A second proton-transfer reaction completes the reduction and gives the trans alkene.

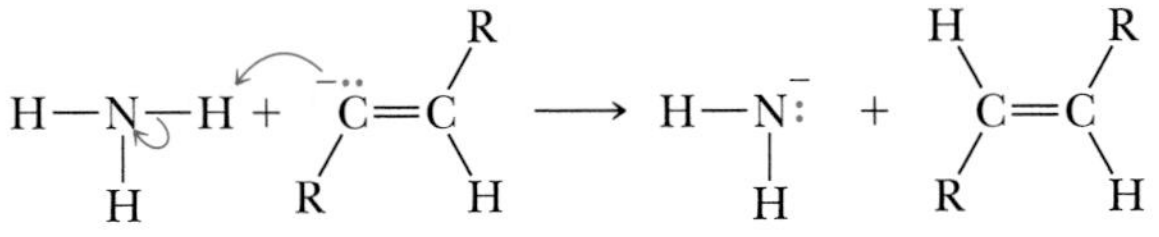

Amide ion   A trans alkene

10.8 Hydroboration

Borane adds readily to an internal alkyne as illustrated by its reaction with 3-hexyne. The product of hydroboration of an alkyne is a **trialkenylborane** (the infix -enyl- shows the presence of a carbon-carbon double bond on the carbon attached to boron).

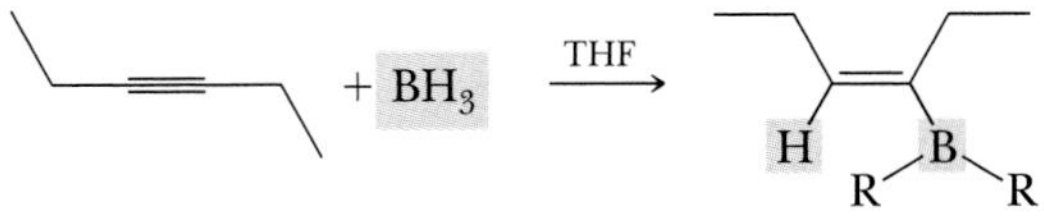

3-Hexyne   A trialkenylborane
                    (R = *cis*-3-hexenyl group)

As with hydroboration of alkenes (Section 6.4), hydroboration of alkynes is also stereo-selective; it involves syn addition of hydrogen and boron.

Treatment of a trialkenylborane with a carboxylic acid, such as acetic acid, results in stereospecific replacement of boron by hydrogen: a cis alkenyl group attached to a boron is converted to a cis alkene.

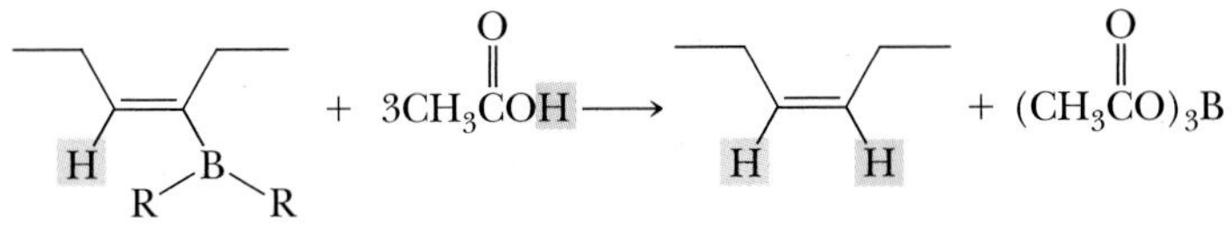

A trialkenylborane   *cis*-3-Hexene

The net effect of hydroboration of an internal alkyne followed by treatment with acetic acid is reduction of the alkyne to a cis alkene. Thus, hydroboration-protonoly-sis and catalytic reduction over a Lindlar catalyst provide alternative schemes for con-version of an alkyne to a cis alkene.

Terminal alkynes react regioselectively with borane to form trialkenylboranes. In practice, however, the reaction is difficult to stop at this stage because the alkenyl group reacts further with borane to undergo a second hydroboration.

It is possible to prevent the second hydroboration step and, in effect, stop the reaction at the alkenylborane stage by using a sterically hindered disubstituted borane. One of the most widely used of these is disiamylborane, $(sia)_2BH$, prepared by treatment of borane with 2 equivalents of 2-methyl-2-butene (amyl is an older common name for pentyl).

Di-*sec*-**iso**amylborane
[$(\mathbf{sia})_2$BH]

Reaction of this sterically hindered dialkylborane with a terminal alkyne results in a single addition and formation of an alkenylborane.

1-Octyne   An alkenylborane

Note that, as for unsymmetrical alkenes, the addition of $(sia)_2BH$ to a carbon-carbon triple bond of a terminal alkene is regioselective; boron adds to the less substituted carbon.

Treatment of an alkenylborane with hydrogen peroxide in aqueous sodium hydroxide gives a product that corresponds to hydration of an alkyne; that is, it corresponds to addition of H to one carbon of the triple bond and OH to the other as illustrated by the hydroboration-oxidation of 2-butyne. To this point, hydroboration/oxidation of alkynes is identical to that of alkenes (Section 6.4).

| 2-Butyne | 2-Buten-2-ol (an enol) | 2-Butanone (a ketone) | (for keto-enol tautomerism) |

Enol A compound containing a hydroxyl group bonded to a doubly bonded carbon atom.

The initial product of hydroboration-oxidation of an alkyne is an **enol,** a compound containing a hydroxyl group bonded to a carbon of a carbon-carbon double bond. The name enol is derived from the fact that it is both an alkene (-en-) and an alcohol (-ol). Enols are in equilibrium with an isomer formed by migration of a hydrogen atom from oxygen to carbon and rearrangement of the carbon-carbon double bond to form a carbon-oxygen double bond.

The keto and enol forms of 2-butanone are said to be tautomers. **Tautomers** are constitutional isomers that are in equilibrium with each other but that differ only in the location of a hydrogen or other atom and a double bond relative to a heteroatom, most commonly O, N, and S. This type of isomerism is called tautomerism. Because the type of tautomerism we are dealing with in this section involves keto (from ketone) and enol forms, it is commonly called **keto-enol tautomerism.** As can be seen from the value of K_{eq}, 2-butanone (the keto form) is much more stable than its enol. We discuss keto-enol tautomerism in more detail in Section 16.11B.

Tautomers Constitutional isomers in equilibrium with each other that differ in the location of a hydrogen atom and a double bond relative to a heteroatom, most commonly O, N, or S.

Keto-enol tautomerism A type of isomerism involving keto (from ketone) and enol tautomers.

Thus, the product isolated after hydroboration-oxidation of 2-butyne is 2-butanone and that from hydroboration-oxidation of 3-hexyne is 3-hexanone.

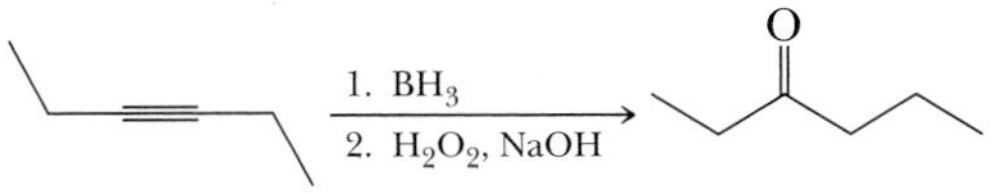

3-Hexyne 3-Hexanone

Hydroboration of a terminal alkyne using disiamylborane followed by oxidation in alkaline hydrogen peroxide gives an enol that is in equilibrium with the more stable aldehyde. Thus, hydroboration-oxidation of a terminal alkyne under these conditions gives an aldehyde.

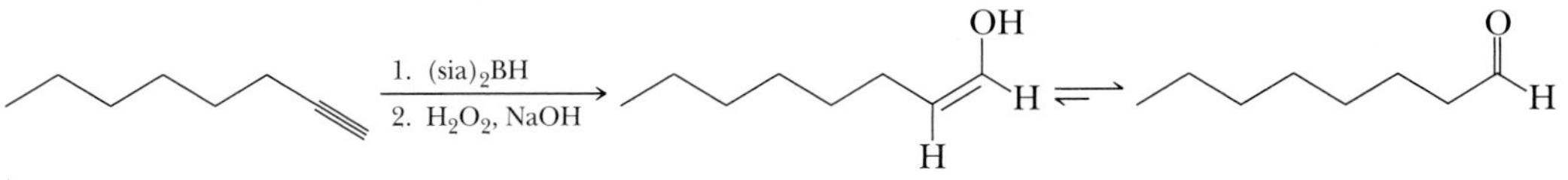

1-Octyne An enol Octanal

Example 10.4

Hydroboration-oxidation of 2-pentyne gives a mixture of two ketones, each of molecular formula $C_5H_{10}O$. Propose structural formulas for these two ketones and for the enol from which each is derived.

Solution

Because each carbon of the triple bond in 2-pentyne has the same degree of substitution, very little regioselectivity occurs during hydroboration. Two enols are formed, and from them, isomeric ketones.

$$CH_3CH_2C{\equiv}CCH_3 \xrightarrow[\text{2. } H_2O_2,\ NaOH]{\text{1. } BH_3} CH_3CH_2\overset{OH}{\overset{|}{C}}{=}CHCH_3 + CH_3CH_2CH{=}\overset{OH}{\overset{|}{C}}CH_3$$

2-Pentyne — 2-Pentene-3-ol (an enol) — 2-Pentene-2-ol (an enol)

$$\downarrow \qquad\qquad \downarrow$$

$$CH_3CH_2\overset{O}{\overset{\|}{C}}CH_2CH_3 + CH_3CH_2CH_2\overset{O}{\overset{\|}{C}}CH_3$$

3-Pentanone 2-Pentanone

Problem 10.4

Draw the structural formula for a hydrocarbon of the given molecular formula that undergoes hydroboration-oxidation to give the indicated product.

(a) C_7H_{10} $\xrightarrow[\text{2. H}_2\text{O}_2,\text{ NaOH}]{\text{1. (sia)}_2\text{BH}}$ [cyclopentane ring]—$CH_2\overset{\displaystyle O}{\overset{\displaystyle \|}{C}}H$

(b) C_7H_{12} $\xrightarrow[\text{2. H}_2\text{O}_2,\text{ NaOH}]{\text{1. BH}_3}$ [cyclopentane ring]—CH_2CH_2OH

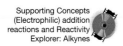

Supporting Concepts
(Electrophilic) addition
reactions and Reactivity
Explorer: Alkynes

10.9 Electrophilic Additions

Alkynes undergo many of the same electrophilic additions as alkenes. In this section, we study addition of bromine and chlorine, the hydrogen halides, and water.

A. Addition of Bromine and Chlorine

Addition of one mole of Br_2 to an alkyne gives a dibromoalkene. As illustrated by the reaction of 2-butyne with one mole of Br_2, addition of bromine to a triple bond is stereoselective; the major product corresponds to anti addition of the two bromine atoms. Doing the bromination in acetic acid with added bromide ion, for example LiBr, increases the preference for anti addition significantly.

$$CH_3C{\equiv}CCH_3 + Br_2 \xrightarrow[\text{anti addition}]{\text{CH}_3\text{COOH, LiBr}}$$

[structure: H_3C and Br on one carbon, Br and CH_3 on the other carbon, C=C double bond]

2-Butyne   (E)-2,3-Dibromo-2-butene

Addition of bromine to alkynes follows much the same type of mechanism as it does for addition to alkenes (Section 6.3D), namely, formation of a bridged bromonium ion intermediate, which is then attacked by bromide ion from the face opposite that occupied by the positively charged bromine atom. Alkynes similarly undergo addition of Cl_2, although less stereoselectively than with Br_2.

[mechanism: $H_3C-C{\equiv}C-CH_3$ with $Br-Br$ $\longrightarrow$ bromonium ion intermediate with Br^- $\longrightarrow$ H_3C, Br / Br, CH_3 alkene product]

Addition of a second mole of Br_2 gives a tetrabromoalkane.

[structure: H_3C, Br / Br, CH_3 alkene] $+ Br_2 \longrightarrow$ $CH_3\underset{Br}{\overset{Br}{C}}-\underset{Br}{\overset{Br}{C}}CH_3$

2,2,3,3-Tetrabromobutane

B.   Addition of Hydrogen Halides

Alkynes add either one or two moles of HBr and HCl, depending on the ratios in which the alkyne and halogen acid are mixed.

$$CH_3C{\equiv}CH \xrightarrow{\ HBr\ } CH_3\overset{\overset{\displaystyle Br}{|}}{C}{=}CH_2 \xrightarrow{\ HBr\ } CH_3\overset{\overset{\displaystyle Br}{|}}{\underset{\underset{\displaystyle Br}{|}}{C}}CH_3$$

Propyne   2-Bromopropene   2,2-Dibromopropane

As shown in this equation, additions of both the first and second moles of HBr are regioselective. They follow Markovnikov's rule (Section 6.3A); hydrogen adds to the carbon that bears the greater number of hydrogens.

We can account for this regioselectivity of addition of HX by proposing a two-step mechanism for each. In Step 1, reaction of the pair of electrons of a pi bond of the alkyne with HBr forms a **vinylic carbocation.** The more stable 2° vinylic carbocation is formed in preference to a less stable 1° vinylic carbocation. The vinylic carbocation then reacts with bromide ion to give the observed product.

Vinylic carbocation A carbocation in which the positive charge is on one of the carbons of a carbon-carbon double bond.

Mechanism   Addition of HBr to an Alkyne

Step 1:   Proton transfer from HBr to the alkyne gives a vinylic carbocation; the more stable 2° vinylic carbocation is formed in preference to the less stable 1° vinylic carbocation.

$$CH_3C{\equiv}CH + H{-}Br \longrightarrow CH_3\overset{+}{C}{=}CH_2 + {:}Br^-$$

A 2° vinylic
carbocation

Step 2:   Reaction of the vinylic carbocation (a Lewis acid) with bromide ion (a Lewis base) gives the vinylic bromoalkene.

$$CH_3\overset{+}{C}{=}CH_2 + {:}Br^- \longrightarrow CH_3\overset{\overset{\displaystyle Br}{|}}{C}{=}CH_2$$

Alkynes are considerably less reactive toward most electrophilic additions than are alkenes. The major reason for this difference is the instability of the *sp*-hybridized vinylic carbocation intermediate formed from an alkyne compared with the *sp*²-hybridized alkyl carbocation formed from an alkene.

In the case of addition of the second mole of HX, Step 1 is again the reaction of the electron pair of the remaining pi bond with HBr to form a carbocation. Of the two possible carbocations, the one with the positive charge on the carbon bearing the halogen is a 2° vinylic carbocation and, therefore, is favored over the 1° vinylic carbocation. The 2° vinylic carbocation is also favored because of the possibility for resonance stabilization by the adjacent halogen atom.

Resonance-stabilized 2° carbocation

Addition of one mole of HCl to acetylene gives chloroethene (vinyl chloride), a compound of considerable industrial importance.

$$HC \equiv CH + HCl \longrightarrow CH_2 = CHCl$$

Ethyne   Chloroethene
(Acetylene)   (Vinyl chloride)

In 1995, the United States produced 15 billion pounds of vinyl chloride for use as a monomer in the production of the polymer poly(vinyl chloride), abbreviated PVC. PVC dominates much of the plumbing and construction market for plastics. Approximately 67% of all pipe, fittings, and conduit, along with 42% of all plastics used in construction at the present time, are fabricated from PVC. We will describe the synthesis of this polymer and its properties in Chapter 24. Our purpose here is to describe the synthesis of vinyl chloride.

Vinyl chloride   Poly(vinyl chloride)
                                     (PVC)

At one time, hydrochlorination of acetylene was the major source of vinyl chloride. As the cost of production of acetylene increased, however, manufacturers of vinyl chloride sought other routes to this material. The starting material chosen was ethylene. What was sought was a way to convert ethylene to vinyl chloride.

$$CH_2 = CH_2 \xrightarrow{\text{???}} CH_2 = CHCl$$

Ethylene   Vinyl chloride

One answer to this problem was to treat ethylene with chlorine gas to form 1,2-dichloroethane. When heated in the presence of charcoal or another catalyst, 1,2-dichloroethane loses a molecule of HCl to form vinyl chloride.

Ethylene   1,2-Dichloroethane   Vinyl chloride

However, the byproduct HCl is corrosive and presented problems in handling and disposal. Although the problem of how to process this HCl was new, the solution had been discovered almost a century earlier. In 1868 Henry Deacon discovered that HCl can be oxidized to Cl_2 and H_2O by the oxygen in air when the gaseous mixture is passed over a copper(II) catalyst.

$$4HCl \ + \ O_2 \xrightarrow{\text{CuCl}_2} 2H_2O \ + \ 2Cl_2 \qquad \text{(Deacon process)}$$

This process has been improved by the use of new technology, and today vinyl chloride is made by passing ethylene, hydrogen chloride, and air over a copper(II) chloride–potassium chloride catalyst to give 1,2-dichloroethane which is then thermally cracked to vinyl chloride and HCl. HCl is then recycled. The three reactions in this scheme and the net result are as follows.

$$4HCl + O_2 \longrightarrow 2H_2O + 2Cl_2$$
$$2CH_2{=}CH_2 + 2Cl_2 \longrightarrow 2ClCH_2CH_2Cl$$
$$\underline{2ClCH_2CH_2Cl \longrightarrow 2CH_2{=}CHCl + 2HCl}$$

Net reaction: $2CH_2{=}CH_2 + 2HCl + O_2 \longrightarrow 2CH_2{=}CHCl + 2H_2O$

We described the production of vinyl chloride first from acetylene and then from ethylene to illustrate an important point about industrial organic chemistry. The aim is to produce a desired chemical from readily available and inexpensive starting materials by reactions in which byproducts can be recycled in some useful way. The objective, increasingly sought after by all chemical companies, is to minimize both costs and production of materials that require disposal or can harm the environment.

C.   Addition of Water — Hydration

In the presence of concentrated sulfuric acid and Hg(II) salts as catalysts, alkynes undergo the addition of water. The Hg(II) salts most often used for this purpose are the sulfates or acetates. For terminal alkynes, addition of water follows Markovnikov's rule; hydrogen adds to the carbon atom of the triple bond bearing the hydrogen. The resulting enol is in equilibrium with a keto form (Section 10.8), and the product isolated is a ketone (an aldehyde in the case of acetylene itself).

$$CH_3C{\equiv}CH + H_2O \xrightarrow[\text{HgSO}_4]{\text{H}_2\text{SO}_4} \overset{\overset{\displaystyle OH}{\displaystyle |}}{CH_3C}{=}CH_2 \rightleftharpoons \overset{\overset{\displaystyle O}{\displaystyle \|}}{CH_3C}CH_3$$

Propyne   1-Propen-2-ol   Propanone
                           (An enol)   (Acetone)

The mechanism of this reaction is illustrated by the hydration of propyne to give propanone (acetone). The first step in this Hg^{2+}-catalyzed hydration is formation of a bridged mercurinium ion intermediate in a reaction similar to that between mercury(II) acetate and an alkene (Section 6.3F). Reaction of this mercurinium ion intermediate with water followed by proton transfer to solvent gives an organomercury enol. Note that this addition follows Markovnikov's rule: the electrophile, Hg^{2+}, adds to the less substituted carbon, and the nucleophile, H_2O, adds to the more substituted carbon.

Mechanism   HgSO₄/H₂SO₄ Catalyzed Hydration of an Alkyne

Step 1:   Attack of Hg^{2+} (a Lewis acid) on the carbon-carbon triple bond (a Lewis base) gives a bridged mercurinium ion intermediate, which contains a three-center/two-electron bond.

A bridged mercurinium
ion intermediate

Step 2:   Attack of water (a Lewis base) on the bridged mercurinium ion intermediate (a Lewis acid) from the side opposite the bridge opens the three-membered ring. The mercurinium ion intermediate is a hybrid of contributing structures. Because the 2° vinylic cation structure makes a greater contribution to the hybrid than the 1° vinylic cation structure, attack of water occurs preferentially at the more substituted carbon, which accounts for the observed regioselectivity of the reaction.

Step 3:   Proton transfer to solvent gives an organomercury enol.

Step 4:   Tautomerism of the enol gives the keto form.

Enol form   Keto form

Step 5:   Proton transfer to the carbonyl group of the ketone gives an oxonium ion.

Steps 5 and 6: Loss of Hg^{2+} from the oxonium ion gives the enol form of the final product. Tautomerism of the enol gives the ketone.

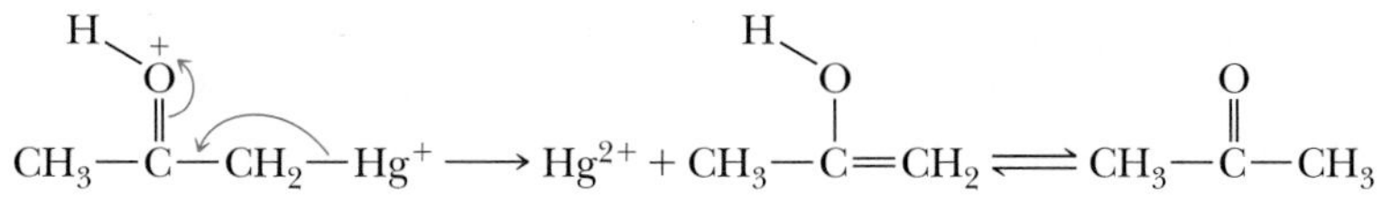

Example 10.5

Show reagents to bring about the following conversions:

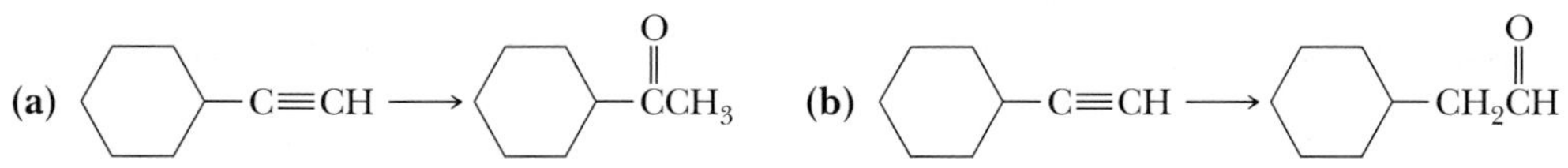

Solution

(a) Hydration of this monosubstituted alkyne using a mercuric catalyst gives an enol that is in equilibrium with the more stable keto form.

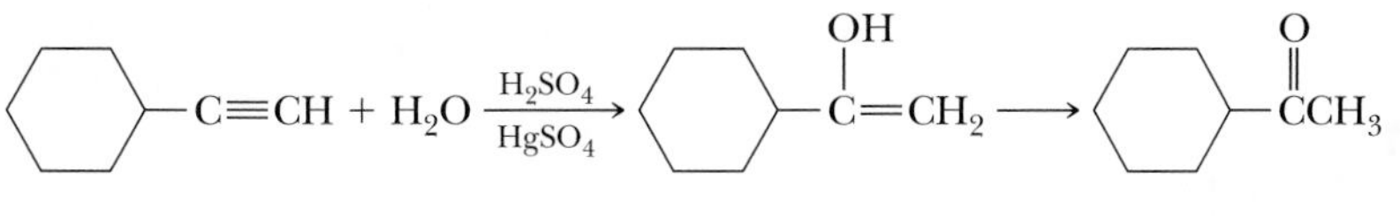

(b) Hydroboration using disiamylborane followed by treatment with alkaline hydrogen peroxide gives an enol that is in equilibrium with the more stable aldehyde.

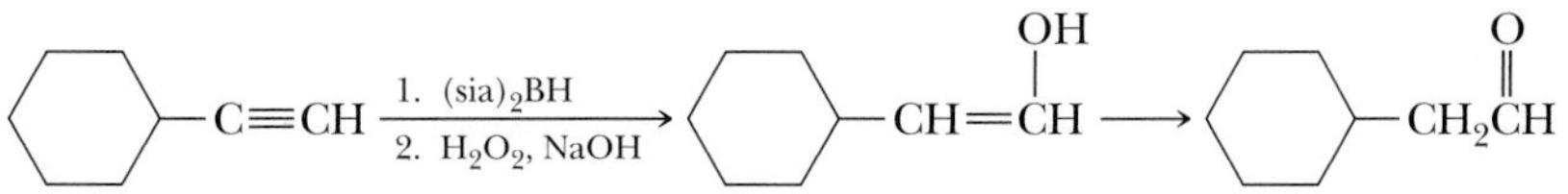

Problem 10.5

Hydration of 2-pentyne gives a mixture of two ketones, each of molecular formula $C_5H_{10}O$. Propose structural formulas for these two ketones and for the enol from which each is derived.

D. Addition of Acetic Acid — Formation of Vinylic Esters

In the presence of sulfuric acid and Hg(II) salts as catalysts, carboxylic acids add to alkynes to form enol esters. Perhaps the most important example of this reaction occurs between acetylene and acetic acid to form the ester with the common name vinyl acetate.

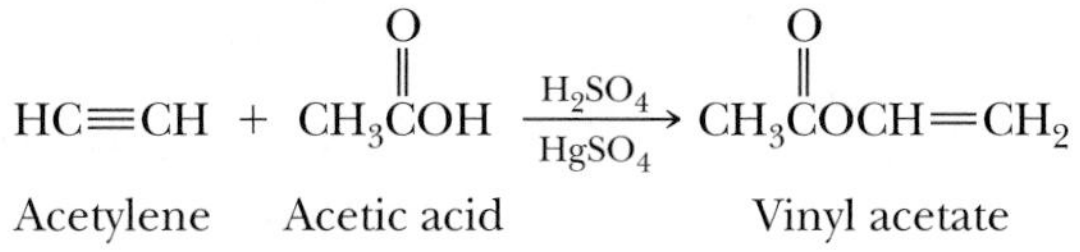

Acetylene   Acetic acid   Vinyl acetate

Annual production of vinyl acetate, the monomer for the production of poly(vinyl acetate), in the United States in 1995 was 2.9 billion pounds. The major use of poly(vinyl acetate) is as an adhesive in the construction and packaging industries and in the paint and coatings industry.

Until 1967, the bulk of vinyl acetate manufactured in the United States was made by the sulfuric acid–mercury(II) sulfate catalyzed addition of acetic acid to acetylene. As the price of acetylene rose, manufacturers turned to ethylene as a starting material for the production of vinyl acetate. Using a modification of technology originally developed by Wacker-Chemie in 1959 and known as the **Wacker process,** ethylene, acetic acid, and molecular oxygen react at elevated temperatures and in the presence of a palladium(II) chloride–copper(II) chloride catalyst to produce vinyl acetate.

$$2CH_2{=}CH_2 + 2CH_3\overset{\displaystyle O}{\overset{\|}{C}}OH + O_2 \xrightarrow[\text{(Wacker process)}]{PdCl_2 \cdot CuCl_2} 2CH_3\overset{\displaystyle O}{\overset{\|}{C}}OCH{=}CH_2 + 2H_2O$$

Ethylene   Acetic acid   Vinyl acetate

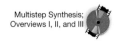

Multistep Synthesis;
Overviews I, II, and III

10.10   Organic Synthesis

We have now seen how to prepare both terminal and internal alkynes from acetylene and substituted acetylenes, and we have seen several common reactions of alkynes, including additions (HX, X_2, and H_2O), hydroboration/oxidation, and reduction. Now let us move a step farther to consider what might be called the art of **organic synthesis.** Synthesis is an important objective of organic chemists for the preparation of compounds for use as pharmaceuticals, agrochemicals, plastics, elastomers, or textile fibers. A successful synthesis must provide the desired product in maximum yield and with maximum control of stereochemistry at all stages of the synthesis. Furthermore, there is increasing desire to develop "green" syntheses, that is syntheses that do not produce or release byproducts harmful to the environment.

> **Organic synthesis**   A series of reactions by which a set of organic starting materials is converted to a more complicated structure.

Our goal in this section is to develop an ability to plan a successful synthesis. The best strategy is to work backward from the desired product. First, we analyze the desired product in the following way.

1. The carbon skeleton. How to build the carbon skeleton is often the most challenging problem. Here you need to consider what carbon-carbon bond-forming reactions are available to you. At this stage in the course, you have only one reaction, namely alkylation of alkynide anions (Section 10.5).
2. The functional groups. What are the functional groups and how can they be changed to facilitate formation of the carbon skeleton? How can they then be changed to give the final set of functional groups in the desired product?

Target Molecule: *cis*-3-Hexene.   As readily available starting materials, we use acetylene and haloalkanes.

Target molecule:

cis-3-Hexene

Analysis: The functional group in the product is a cis carbon-carbon double bond, which can be prepared by catalytic reduction of a carbon-carbon triple bond using the Lindlar catalyst. We then disconnect the carbon skeleton into possible starting materials, which we can later reconnect by known reactions during the synthesis. In the example here, we disconnect at the two carbon-carbon single bonds adjacent to the triple bond. These bonds can be formed during the synthesis by alkylation of the acetylide dianion using a haloalkane (Section 10.5). This type of scheme, in which we work from the desired product back to a set of starting materials, is called a **retrosynthesis.** We use an open arrow to symbolize a step in a retrosynthesis.

Retrosynthesis   A process of reasoning backwards from a target molecule to a suitable set of starting materials.

$$\text{cis-3-Hexene} \implies \overset{\text{disconnect here}}{\text{3-Hexyne}} \xrightarrow{S_N2} \;\; :C\equiv C:^- \;\; + \;\; 2CH_3CH_2Br$$

| *cis*-3-Hexene | 3-Hexyne | Acetylide dianion | Bromoethane |

Synthesis: Thus, our starting materials for this synthesis of *cis*-3-hexene are acetylene and bromoethane, both readily available compounds. This synthesis is carried out in five steps as follows.

$$HC\equiv CH \xrightarrow[\text{2. } CH_3CH_2Br]{\text{1. } NaNH_2} \;\;\xrightarrow[\text{4. } CH_3CH_2Br]{\text{3. } NaNH_2} \;\;\xrightarrow[\substack{\text{Lindlar} \\ \text{catalyst}}]{\text{5. } H_2} $$

| Acetylene | 1-Butyne | 3-Hexyne | *cis*-3-Hexene |

Target Molecule: 2-Heptanone.   Let us use the same approach to devise a synthesis for 2-heptanone, a compound with a penetrating odor responsible for the "peppery" odor of cheeses of the Roquefort type. As readily available starting materials, we again use acetylene and alkyl halides.

$$\text{Target molecule:} \quad \overset{\displaystyle O}{\overset{\|}{CH_3CCH_2CH_2CH_2CH_2CH_3}}$$

2-Heptanone

Analysis: The functional group in the target molecule is a ketone, which can be prepared by hydration of a carbon-carbon triple bond. Hydration of 1-heptyne gives only 2-heptanone. Hydration of 2-heptyne gives a mixture of 2-heptanone and 3-heptanone. Therefore, we choose a functional group interconversion via 1-heptyne.

An acid-catalyzed hydration gives a mixture of 2-heptanone and 3-heptanone.

2-Heptanone

2-Heptyne

1-Heptyne $\implies HC\equiv C:^- + Br$

Acetylide anion   1-Bromopentane

Synthesis: Our starting materials are acetylene and 1-bromopentane, and the synthesis is carried out in three steps as follows.

1-Heptyne   2-Heptanone

Example 10.6

How might the scheme for the synthesis of 2-heptanone be modified so that the product is heptanal?

Solution

Steps 1 and 2 are the same and give 1-heptyne. Instead of acid-catalyzed hydration of 1-heptyne, treat the alkyne with $(sia)_2BH$ followed by alkaline hydrogen peroxide.

1-Heptyne   Heptanal

Problem 10.6

Show how the synthetic scheme in Example 10.6 might be modified to give the following.
(a) 1-Heptanol   **(b)** 2-Heptanol

Summary

Alkynes contain one or more carbon-carbon triple bonds. The triple bond is a combination of one sigma bond formed by overlap of sp hybrid orbitals and two pi bonds formed by overlap of two sets of parallel $2p$ orbitals (Section 10.1). In the IUPAC system (Section 10.2A), the infix -yn- is used to show the presence of the carbon-carbon triple bond. Common names of alkynes are derived by prefixing the names of substituents on the carbon-carbon triple bond to the word "acetylene" (Section 10.2B). The physical properties of alkynes (Section 10.3) are similar to those of alkanes and alkenes of comparable carbon skeleton.

The pK_a values of terminal alkynes are approximately 25 (Section 10.4); they are less acidic than water but more acidic than alkanes, alkenes, and ammonia. Hydrogen attached to a carbon-carbon triple bond is sufficiently acidic that it can be removed by a strong base, most commonly $NaNH_2$, NaH, or LDA.

Tautomers (Section 10.8) are constitutional isomers that are in equilibrium with each other but that differ in the location of a hydrogen and a double bond relative to a heteroatom, most commonly O, N, and S. **Keto-enol tautomerism** is the most common type of tautomerism we encounter in this course. The functional group of an enol is an —OH group attached to a double-bonded carbon atom. The enol form is in equilibrium with the keto form, and equilibrium most commonly lies far on the side of the keto form.

Key Reactions

1. Acidity of Alkynes (Section 10.4)

Terminal alkynes react with strong bases, most commonly $NaNH_2$, NaH, or LDA, to give salts.

$$CH_3C{\equiv}CH + Na^+NH_2^- \longrightarrow CH_3C{\equiv}C^-\ Na^+ + NH_3$$

2. Alkylation of Acetylide Anions (Section 10.5)

S_N2 reactions using acetylide anions are effective using only methyl and 1° alkyl halides. With 2° and 3° alkyl halides, E2 elimination is the major reaction.

$$HC{\equiv}C^-Na^+ + CH_3(CH_2)_2CH_2{-}Br \xrightarrow{S_N2} HC{\equiv}CCH_2(CH_2)_2CH_3 + Na^+Br^-$$

3. Synthesis of an Alkyne from an Alkene (Section 10.6)

Addition of Br_2 or Cl_2 to an alkene followed by a double dehydrohalogenation using $NaNH_2$ or other strong base gives an alkyne.

$$RCH{=}CHR \xrightarrow[CCl_4]{Br_2} \overset{\displaystyle Br\quad Br}{\underset{}{RCH{-}CHR}} \xrightarrow[NH_3(l)]{2NaNH_2} RC{\equiv}CR$$

4. Catalytic Reduction (Section 10.7A)

Reaction of an alkyne with two moles of H_2 under moderate pressure in the presence of a transition metal catalyst at room temperature gives an alkane.

Reaction in the presence of the Lindlar catalyst is stereoselective; syn addition of one mole of H_2 to an internal alkyne gives a cis alkene.

5. Reduction Using Na or Li Metal in NH₃(l) (Section 10.7B)

Alkali metal reduction is stereoselective: anti addition of hydrogens to an internal alkyne gives a trans alkene.

6. Hydroboration-Oxidation (Section 10.8)

Hydroboration of an internal alkyne is stereoselective; syn addition of BH_3 occurs. Oxidation using $H_2O_2/NaOH$ gives an enol that is in equilibrium by keto-enol tautomerism with a ketone.

Hydroboration of a terminal alkyne using a hindered dialkylborane followed by oxidation using H_2O_2/NaOH and then keto-enol tautomerism gives an aldehyde.

$$\xrightarrow[\text{2. } H_2O_2, \text{ NaOH}]{\text{1. } (sia)_2BH}$$

7. Keto-Enol Tautomerism (Section 10.8)

In an equilibrium between a keto form and an enol form, the keto form generally predominates.

$$\underset{\text{Enol form}}{CH_3CH=CCH_3} \rightleftharpoons \underset{\text{Keto form}}{CH_3CH_2CCH_3}$$

(OH on enol form, O double bond on keto form)

8. Addition of Br_2 and Cl_2 (Section 10.9A)

Addition of one mole of Br_2 or Cl_2 is stereoselective; anti addition gives an (E)-dihaloalkene. Addition of a second mole of halogen gives a tetrahaloalkane.

$$CH_3C\equiv CCH_3 \xrightarrow{Br_2} \underset{Br}{\overset{H_3C}{C=C}}\underset{CH_3}{\overset{Br}{}} \xrightarrow{Br_2} CH_3-\underset{Br}{\overset{Br}{C}}-\underset{Br}{\overset{Br}{C}}-CH_3$$

9. Addition of HX (Section 10.9B)

Addition of HX is regioselective. Reaction by way of a vinylic carbocation intermediate follows Markovnikov's rule. Addition of 2HX gives a geminal dihaloalkane.

$$CH_3C\equiv CH \xrightarrow{HBr} \underset{}{\overset{Br}{CH_3C=CH_2}} \xrightarrow{HBr} \underset{Br}{\overset{Br}{CH_3CCH_3}}$$

10. Acid-Catalyzed Hydration (Section 10.9C)

Acid-catalyzed addition of water in the presence of Hg(II) salts is regioselective. Ketoenol tautomerism of the resulting enol gives a ketone.

$$CH_3C\equiv CH + H_2O \xrightarrow[HgSO_4]{H_2SO_4} \underset{}{\overset{OH}{CH_3C=CH_2}} \rightleftharpoons CH_3CCH_3$$

Problems

Structure and Nomenclature

10.7   Write the IUPAC name for each compound.

(a) $CH_3C\equiv CCCH_3$ (with CH_3 above and CH_3 below the third carbon)   **(b)** $HC\equiv CCH_2Br$ **(c)** cyclopentyl—$C\equiv CH$

(d) $HC\equiv CCH_2CH_2CH_2C\equiv CH$ (e) $CH_3(CH_2)_5C\equiv CCH_2OH$
(f) $CH_3(CH_2)_6C\equiv CH$ (g) $CH_3C\equiv CCH_2OH$
(h) $CH_3(CH_2)_7C\equiv C(CH_2)_7COOH$

10.8 Draw a structural formula for each compound.

(a) 3-Hexyne (b) But-1-en-3-yne (vinylacetylene)
(c) 3-Chloro-1-butyne (d) 5-Isopropyl-3-octyne
(e) 3-Pentyn-2-ol (f) 2-Butyne-1,4-diol
(g) 2,2,5-Trimethyl-3-pentyne
(h) 4,4-Dimethyl-2-pentyne (*tert*-butylmethylacetylene)
(i) Cyclodecyne

10.9 Predict all bond angles about each highlighted atom.

(a) $CH_3C\equiv \!{\small\bigcirc}\!CCH_3$ (b) $CH_2\!=\!{\small\bigcirc}\!CHC\equiv CH$

(c) $CH_2\!=\!{\small\bigcirc}\!C\!=\!CHCH_3$ (d) ${\small\bigcirc}\!CH_2\!=\!CHCH\!=\!CH_2$

10.10 State the orbital hybridization of each highlighted atom.

(a) $CH_3C\equiv \!{\small\bigcirc}\!CCH_3$ (b) $CH_2\!=\!{\small\bigcirc}\!CHC\equiv CH$

(c) $CH_2\!=\!{\small\bigcirc}\!C\!=\!CHCH_3$ (d) $O\!=\!{\small\bigcirc}\!C\!=\!O$

10.11 Describe each highlighted carbon-carbon bond in terms of the overlap of atomic orbitals.

(a) $CH_3C\!\equiv\!{\small\bigcirc}\!CCH_3$ (b) $CH_2\!=\!CH\!-\!C\!\equiv\!CH$

(c) $CH_2\!=\!{\small\bigcirc}\!C\!=\!CHCH_3$ (d) $CH_2\!=\!CH\!-\!CH\!=\!CH_2$

10.12 Enanthotoxin (see *The Merck Index,* 12th ed., #3608) is an extremely poisonous organic compound found in hemlock water dropwart, which is reputed to be the most poisonous plant in England. It is believed that no British plant has been responsible for more fatal accidents. The most poisonous part of the plant is the roots, which resemble small white carrots, giving the plant the name "five finger death." Also poisonous are its leaves, which look like parsley. Enanthotoxin is thought to interfere with the Na^+ current in nerve cells, which leads to convulsions and death.

How many stereoisomers are possible for enanthotoxin?

Preparation of Alkynes

10.13 Show how to prepare each alkyne from the given starting material. In part (c), D indicates deuterium. Deuterium-containing reagents such as BD_3, D_2O, and CH_3COOD are available commercially.

(a) $CH_3CH_2CH_2CH\!=\!CH_2 \longrightarrow CH_3CH_2CH_2C\equiv CH$
(b) $CH_3(CH_2)_5\underset{\underset{\displaystyle Cl}{|}}{C}HCH_3 \longrightarrow CH_3(CH_2)_4C\equiv CCH_3$

(c) $CH_3CH_2CH_2C\equiv CH \longrightarrow CH_3CH_2CH_2C\equiv CD$

10.14 If a catalyst could be found that would establish an equilibrium between 1,2-butadiene and 2-butyne, what would be the ratio of the more stable isomer to the less stable isomer at 25°C?

$$CH_2\!=\!C\!=\!CHCH_3 \rightleftharpoons CH_3C\equiv CCH_3 \qquad \Delta G^0 = -16.7 \text{ kJ } (-4.0 \text{ kcal})/\text{mol}$$

Reactions of Alkynes

10.15 Complete each acid-base reaction and predict whether the position of equilibrium lies toward the left or the right.

(a) $CH_3C{\equiv}CH + (CH_3)_3CO^-K^+ \underset{(CH_3)_3COH}{\overset{(CH_3)_3COH}{\rightleftharpoons}}$ (b) $CH_2{=}CH_2 + Na^+NH_2^- \underset{NH_3(l)}{\overset{NH_3(l)}{\rightleftharpoons}}$

(c) $CH_3C{\equiv}CCH_2OH + Na^+NH_2^- \underset{NH_3(l)}{\rightleftharpoons}$

10.16 Draw structural formulas for the major product(s) formed by reaction of 3-hexyne with each of these reagents. Where you predict no reaction, write NR.

(a) H_2(excess)/Pt
(b) H_2/Lindlar catalyst
(c) Na in $NH_3(l)$
(d) BH_3 followed by H_2O_2/NaOH
(e) BH_3 followed by CH_3COOH
(f) BH_3 followed by CH_3COOD
(g) Cl_2 (one mole)
(h) $NaNH_2$ in $NH_3(l)$
(i) HBr (one mole)
(j) HBr (two moles)
(k) H_2O in H_2SO_4/$HgSO_4$
(l) CH_3COOH in H_2SO_4/$HgSO_4$

10.17 Draw the structural formula of the enol formed in each alkyne hydration reaction, and then draw the structural formula of the carbonyl compound with which each enol is in equilibrium.

(a) $CH_3(CH_2)_5C{\equiv}CH + H_2O \xrightarrow[H_2SO_4]{HgSO_4}$ (an enol) $\longrightarrow$

(b) $CH_3(CH_2)_5C{\equiv}CH \xrightarrow[\text{2. NaOH/H}_2O_2]{\text{1. (sia)}_2BH}$ (an enol) $\longrightarrow$

10.18 Propose a mechanism for this reaction.

$$\underset{\text{Acetylene}}{HC{\equiv}CH} + \underset{\text{Acetic acid}}{CH_3\overset{\displaystyle O}{\overset{\|}{C}}OH} \xrightarrow[HgSO_4]{H_2SO_4} \underset{\text{Vinyl acetate}}{CH_3\overset{\displaystyle O}{\overset{\|}{C}}OCH{=}CH_2}$$

Syntheses

10.19 Show how to convert 9-octadecynoic acid to the following:

$$CH_3(CH_2)_7C{\equiv}C(CH_2)_7COOH$$

9-Octadecynoic acid

(a) (*E*)-9-Octadecenoic acid (eliadic acid)
(b) (*Z*)-9-Octadecenoic acid (oleic acid)
(c) 9,10-Dihydroxyoctadecanoic acid
(d) Octadecanoic acid (stearic acid)

10.20 For small-scale and consumer welding applications, many hardware stores sell cylinders of MAAP gas, which is a mixture of propyne (methylacetylene) and 1,2-propadiene (allene), with other hydrocarbons. How would you prepare the methylacetylene/allene mixture from propene in the laboratory?

10.21 Show reagents and experimental conditions you might use to convert propyne into each product. Some of these syntheses can be done in one step. Others require two or more steps.

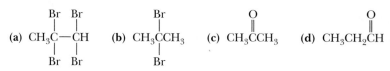

10.22 Show reagents and experimental conditions you might use to convert each starting material into the desired product. Some of these syntheses can be done in one step. Others require two or more steps.

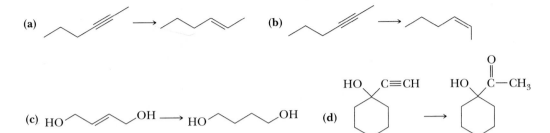

(c) HO⟶⟶⟶OH ⟶ HO⟶⟶⟶OH

10.23 Show how to convert 1-butyne to each of these compounds.

(a) $CH_3CH_2C{\equiv}C^-$ Na^+ **(b)** $CH_3CH_2C{\equiv}CD$

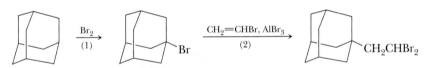

10.24 Rimantadine (see *The Merck Index*, 12th ed., #8390) was among the first antiviral drugs to be licensed in the United States for use against the influenza A virus and in treating established illnesses. It is thought to exert its antiviral effect by blocking a late stage in the assembly of the virus. Rimantadine is synthesized from adamantane by the following sequence. We discuss the chemistry of Step 5 in Section 16.10A.

(a) Describe experimental conditions to bring about Step 1. By what type of mechanism does this reaction occur? Account for the regioselectivity of bromination in Step 1.

(b) Propose a mechanism for Step 2. *Hint:* As we shall see in Section 21.1A, reaction of a bromoalkane such as 1-bromoadamantane with aluminum bromide (a Lewis acid, Section 4.5) results in the formation of a carbocation and $AlBr_4^-$. Assume that adamantyl cation is formed in Step 2, and proceed from there to describe a mechanism.

(c) Account for the regioselectivity of carbon-carbon bond formation in Step 2.

(d) Describe experimental conditions to bring about Step 3.

(e) Describe experimental conditions to bring about Step 4.

10.25 Show reagents and experimental conditions to bring about the following transformations.

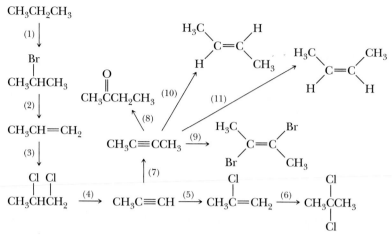

10.26 Show reagents to bring about each conversion.

$$CH_3(CH_2)_5C \equiv C^- Na^+ \xrightarrow{\text{(b)}} CH_3(CH_2)_5C \equiv CCH_2CH_3$$

$$CH_3(CH_2)_5C \equiv CH \xrightarrow[\text{(e)}]{\substack{\text{(a)}\\\text{(d)}}}$$

with products including:

$$CH_3(CH_2)_6\overset{O}{\overset{\|}{C}}H$$

$$CH_3(CH_2)_5\overset{O}{\overset{\|}{C}}CH_3$$

$$\underset{\text{(c)}}{CH_3(CH_2)_5} \underset{H}{\overset{CH_2CH_3}{C=C}} H$$

10.27 Which of these alkynes can be prepared in good yield by monoalkylation or dialkylation of acetylene? For each that cannot, explain why not.

(a) 3-Methyl-1-butyne   (b) 4,4-Dimethyl-1-pentyne   (c) 2-Octyne

10.28 Propose a synthesis for (*Z*)-9-tricosene (muscalure, see *The Merck Index*, 12th ed., #6388), the sex pheromone for the common house fly (*Musca domestica*) starting with acetylene and haloalkanes as sources of carbon atoms.

$$\underset{H}{\overset{CH_3(CH_2)_7}{C=C}}\underset{H}{\overset{(CH_2)_{12}CH_3}{}}$$

Muscalure

10.29 Propose a synthesis of each compound starting from acetylene and any necessary organic and inorganic reagents.

(a) 4-Octyne   (b) 4-Octanone   (c) *cis*-4-Octene
(d) *trans*-4-Octene   (e) 4-Octanol   (f) meso-4,5-Octanediol

10.30 Show how to prepare each compound from ethylene.

(a) 1,2-Dichloroethane   (b) Chloroethene (vinyl chloride)   (c) 1,1-Dichloroethane

10.31 Show how to bring about this conversion.

11

ETHERS, SULFIDES, AND EPOXIDES

I n this chapter, we first discuss the structure, nomenclature, and physical properties of ethers and then compare their physical properties with those of isomeric alcohols. Next, we study the preparation and reactions of ethers. As we shall see, their most important reactions involve nucleophilic substitution. This chapter continues the discussion of S_N1 and S_N2 reaction mechanisms begun in Chapter 8 and continued in Chapters 9 and 10.

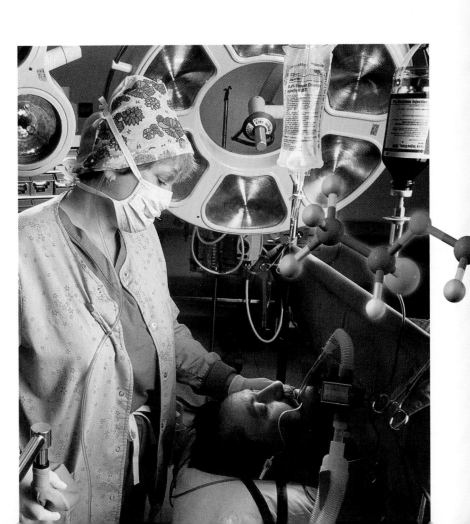

■ The discovery that inhaling ethers could make a patient insensitive to pain revolutionized the practice of medicine. *(Allan Levenson/Stone)* Inset: A model of isoflurane, $CF_3CHClOCHF_2$, a halogenated ether widely used as an inhalation anesthetic in both human and veterinary medicine.

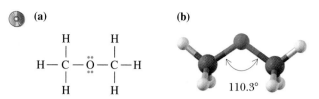

Figure 11.1
Dimethyl ether, CH_3OCH_3.
(a) Lewis structure and
(b) ball-and-stick model.

11.1 Structure of Ethers

Ether A compound containing an oxygen atom bonded to two carbon atoms.

The functional group of an **ether** is an atom of oxygen bonded to two carbon atoms. Figure 11.1 shows a Lewis structure and a ball-and-stick model of dimethyl ether, CH_3OCH_3, the simplest ether. In dimethyl ether, two sp^3 hybrid orbitals of oxygen form sigma bonds to the two carbon atoms. The other two sp^3 hybrid orbitals each contain an unshared pair of electrons. The C—O—C bond angle in dimethyl ether is 110.3°, a value close to the tetrahedral angle of 109.5°.

In ethyl vinyl ether, the ether oxygen is bonded to one sp^3 and one sp^2-hybridized carbon.

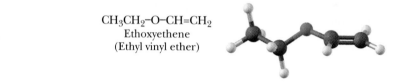

$$CH_3CH_2-O-CH=CH_2$$
Ethoxyethene
(Ethyl vinyl ether)

11.2 Nomenclature of Ethers

Nomenclature
Ethers

Alkoxy group An —OR group, where R is an alkyl group.

In the IUPAC system, ethers are named by selecting the longest carbon chain as the parent alkane and naming the —OR group bonded to it as an **alkoxy** substituent. Common names are derived by listing the alkyl groups bonded to oxygen in alphabetical order and adding the word "ether." Following are the IUPAC names and, in parentheses, the common names for three low-molecular-weight ethers.

$CH_3CH_2OCH_2CH_3$	*trans*-2-Ethoxycyclohexanol	2-Methoxy-2-methylpropane
Ethoxyethane		(*tert*-Butyl methyl ether)
(Diethyl ether)		

Chemists almost invariably use common names for low-molecular-weight ethers. For example, although ethoxyethane is the IUPAC name for $CH_3CH_2OCH_2CH_3$, it is rarely called that; rather, it is called diethyl ether, ethyl ether, or even more commonly, simply ether. The abbreviation for *tert*-butyl methyl ether is MTBE, after the common name of methyl *tert*-butyl ether.

Three other ethers deserve special mention. 2-Methoxyethanol and 2-ethoxyethanol, more commonly known as Methyl Cellosolve and Cellosolve, are good polar protic solvents in which to carry out organic reactions and are also used commercially in some paint strippers. *Diethylene glycol dimethyl ether*, more commonly known by its acronym, diglyme, is a common solvent for hydroboration and $NaBH_4$ reductions.

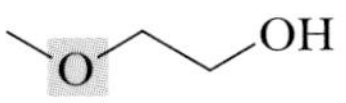

2-Methoxyethanol
(Methyl Cellosolve)

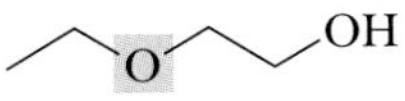

2-Ethoxyethanol
(Cellosolve)

Diethylene glycol dimethyl ether
(Diglyme)

Cyclic ethers are given special names. The presence of oxygen in a saturated ring is indicated by the prefix ox- and ring sizes from three to six are indicated by the endings -irane, -etane, -olane, and -ane, respectively. Several of these smaller ring cyclic ethers are more often referred to by their common names, here shown in parentheses. Numbering of the atoms of the ring begins with the oxygen atom. These compounds and others in which a heteroatom (non-carbon atom) are in the ring are called **heterocycles.**

Heterocycle A cyclic compound whose ring contains more than one kind of atom. Ethylene oxide, for example, is a heterocycle whose ring contains two carbon atoms and one oxygen atom.

Ethylene oxide

Tetrahydrofuran

Tetrahydropyran

1,4-Dioxane

Example 11.1

Write IUPAC and common names for these ethers.

(a) $CH_3\overset{\overset{\displaystyle CH_3}{|}}{\underset{\underset{\displaystyle CH_3}{|}}{C}}OCH_2CH_3$ (b) 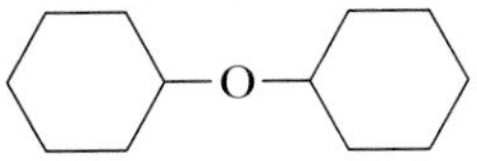 ⬡—O—⬡   (c) $CH_2{=}CHOCH_3$

Solution

(a) 2-Ethoxy-2-methylpropane. Its common name is *tert*-butyl ethyl ether.
(b) Cyclohexoxycyclohexane. Its common name is dicyclohexyl ether.
(c) Methoxyethene. Its common name is methyl vinyl ether.

Problem 11.1

Write IUPAC and common names for these ethers.

(a) $CH_3\overset{\overset{\displaystyle CH_3}{|}}{CH}CH_2OCH_2CH_3$ (b) $CH_3OCH_2CH_2OCH_3$ (c) 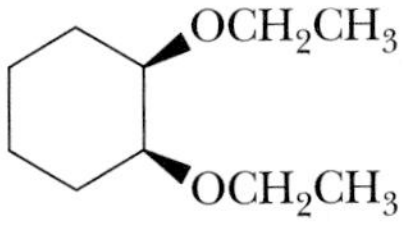
⬡ with OCH_2CH_3 and OCH_2CH_3

11.3 Physical Properties of Ethers

Ethers are polar molecules in which oxygen bears a partial negative charge and each attached carbon bears a partial positive charge (Figure 11.2). However, only weak dipole-dipole interactions exist between their molecules in the liquid state (Figure

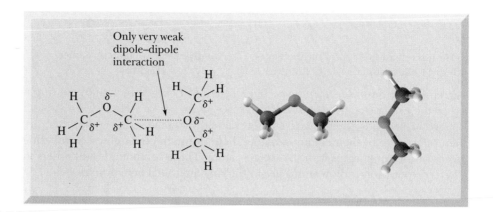

Figure 11.2
Ethers are polar compounds, but there are only weak dipole-dipole interactions between their molecules in the liquid state.

11.2). Consequently, boiling points of ethers are much lower than those of alcohols of comparable molecular weight (Table 11.1) and are close to those of hydrocarbons of comparable molecular weight (compare Tables 2.5 and 11.1). Because ethers cannot act as hydrogen bond donors, they are much less soluble in water than alcohols (Table 11.1). However, they can act as hydrogen bond acceptors (Figure 11.3), which makes them more water soluble than hydrocarbons of comparable molecular weight and shape (compare data in Tables 2.5 and 11.1).

Example 11.2

Arrange these compounds in order of increasing solubility in water.

Ethylene glycol   Diethyl ether   Hexane
dimethyl ether

Table 11.1 **Boiling Points and Solubilities in Water of Some Ethers and Alcohols of Comparable Molecular Weight**

Structural Formula	Name	Molecular Weight	bp (°C)	Solubility in Water
CH_3CH_2OH	Ethanol	46	78	Infinite
CH_3OCH_3	Dimethyl ether	46	-24	7.8 g/100 g
$CH_3CH_2CH_2CH_2OH$	1-Butanol	74	117	7.4 g/100 g
$CH_3CH_2OCH_2CH_3$	Diethyl ether	74	35	8.0 g/100 g
$HOCH_2CH_2CH_2CH_2OH$	1,4-Butanediol	90	230	Infinite
$CH_3CH_2CH_2CH_2CH_2OH$	1-Pentanol	88	138	2.3 g/100 g
$CH_3OCH_2CH_2OCH_3$	Ethylene glycol dimethyl ether	90	84	Infinite
$CH_3CH_2CH_2CH_2OCH_3$	Butyl methyl ether	88	71	Slight

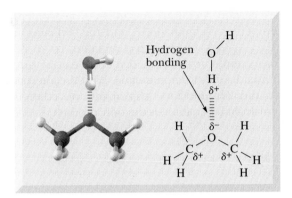

Figure 11.3
Ethers are hydrogen bond acceptors only. They are not hydrogen bond donors.

Solution

Water is a polar solvent. Hexane, a nonpolar hydrocarbon, has the lowest solubility in water. Both diethyl ether and ethylene glycol dimethyl ether are polar compounds due to the presence of the polar C—O—C bond, and each interacts with water as a hydrogen bond acceptor. Because ethylene glycol dimethyl ether has more sites within its molecules for hydrogen bonding than diethyl ether, it is more soluble in water than diethyl ether.

Insoluble   8g/100g water   Soluble in all proportions

Problem 11.2

Arrange these compounds in order of increasing boiling point.

11.4 Preparation of Ethers

A. Williamson Ether Synthesis

The most common general method for the synthesis of ethers, the **Williamson ether synthesis,** involves second-order nucleophilic displacement of a halide ion or other good leaving group by an alkoxide ion.

$$
\underset{\substack{\text{Sodium} \\ \text{isopropoxide}}}{\underset{\text{CH}_3\text{CHO}^-\text{Na}^+}{\overset{\text{CH}_3}{|}}} + \underset{\substack{\text{Iodomethane} \\ \text{(Methyl iodide)}}}{\text{CH}_3\text{I}} \xrightarrow{\text{S}_\text{N}2} \underset{\substack{\text{2-Methoxypropane} \\ \text{(Isopropyl methyl ether)}}}{\underset{\text{CH}_3\text{CHOCH}_3}{\overset{\text{CH}_3}{|}}} + \text{Na}^+\text{I}^-
$$

> Williamson ether synthesis   A general method for the synthesis of ethers by an S_N2 reaction between an alkyl halide and an alkoxide ion.

In planning a Williamson ether synthesis, it is essential to use a combination of reactants that maximizes nucleophilic substitution and minimizes the competing β-elimination (E2) reaction (Section 8.11B). Yields of ether are highest when the halide to be displaced is on a methyl or a primary carbon. Yields are low in the displacement from secondary halides (because of competing β-elimination), and the Williamson ether synthesis fails altogether with tertiary halides, in which β-elimination by an E2 mechanism is the exclusive reaction. For example, *tert*-butyl methyl ether can be prepared by the reaction of potassium *tert*-butoxide and bromomethane. With the alternative combination of sodium methoxide and 2-bromo-2-methylpropane, no ether is formed. 2-Methylpropene, formed by dehydrohalogenation, is the only product.

$$
\underset{\substack{\text{Potassium} \\ \textit{tert}\text{-butoxide}}}{\overset{\displaystyle CH_3}{\underset{\displaystyle CH_3}{CH_3CO^-K^+}}} \;+\; \underset{\substack{\text{Bromomethane} \\ \text{(Methyl bromide)}}}{CH_3Br} \;\xrightarrow{\;S_N2\;}\; \underset{\substack{\text{2-Methoxy-2-methylpropane} \\ (\textit{tert}\text{-Butyl methyl ether})}}{\overset{\displaystyle CH_3}{\underset{\displaystyle CH_3}{CH_3COCH_3}}} \;+\; K^+Br^-
$$

$$
\underset{\substack{\text{2-Bromo-2-} \\ \text{methylpropane}}}{\overset{\displaystyle CH_3}{\underset{\displaystyle CH_3}{CH_3CBr}}} \;+\; \underset{\substack{\text{Sodium} \\ \text{methoxide}}}{CH_3O^-Na^+} \;\xrightarrow{\;E2\;}\; \underset{\text{2-Methylpropene}}{\overset{\displaystyle CH_3}{CH_3C{=}CH_2}} \;+\; CH_3OH + Na^+Br^-
$$

Example 11.3

Show the combination of alcohol and haloalkane that can best be used to prepare these ethers by the Williamson ether synthesis.

(a)   (b)

Solution

(a) Treat 2-propanol with sodium metal to form sodium isopropoxide. Then treat this metal alkoxide with 1-bromobutane. The alternative combination of sodium butoxide and 2-bromopropane gives considerably more elimination product.

1-Bromobutane   Sodium isopropoxide

(b) Treat (*S*)-2-butanol with sodium metal to form the sodium alkoxide. This reaction involves only the O—H bond and does not affect the stereocenter. Then, treat this sodium alkoxide with a haloethane to give the desired product.

(*S*)-2-Butanol   (*S*)-2-Ethoxyethane

An alternative synthesis is to convert the (R)-2-butanol to its tosylate followed by treatment with sodium ethoxide. This synthesis, however, gives only a low yield of the desired product. Recall from Section 8.11B that when a 2° halide or tosylate is treated with a strong base/good nucleophile, E2 is the major reaction.

$$\underset{\text{(R)-2-Butanol}}{\overset{\text{OH}}{\vert}} \xrightarrow[\text{2. CH}_3\text{CH}_2\text{O}^-\text{Na}^+]{\text{1. TsCl, pyridine}} \underset{\substack{\text{Product of E2} \\ \text{(major product)}}}{\text{CH}_3\text{CH}{=}\text{CHCH}_3} + \underset{\substack{\text{(S)-2-Ethoxyethane} \\ \text{(product of S}_\text{N}\text{2 reaction;} \\ \text{a minor product)}}}{}$$

Problem 11.3

Show how you might use the Williamson ether synthesis to prepare these ethers.

(a) **(b)**

B. Acid-Catalyzed Dehydration of Alcohols

Diethyl ether and several other commercially available ethers are synthesized on an industrial scale by acid-catalyzed dehydration of primary alcohols. Intermolecular dehydration of ethanol to give ethylene is a competing reaction, but it requires a higher temperature. In practice, experimental conditions can be adjusted to favor formation of either product.

$$2\text{CH}_3\text{CH}_2\text{OH} \xrightarrow[140°\text{C}]{\text{H}_2\text{SO}_4} \underset{\text{Diethyl ether}}{\text{CH}_3\text{CH}_2\text{OCH}_2\text{CH}_3} + \text{H}_2\text{O}$$

Ethanol

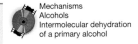

Mechanisms
Alcohols
Intermolecular dehydration
of a primary alcohol

Mechanism   Acid-Catalyzed Intermolecular Dehydration of a Primary Alcohol

Step 1: Proton transfer from the acid catalyst to the hydroxyl group gives an oxonium ion. This step converts OH⁻, a poor leaving group, into H₂O, a better leaving group.

$$\text{CH}_3\text{CH}_2{-}\overset{..}{\underset{..}{\text{O}}}{-}\text{H} + \text{H}{-}\overset{\overset{\text{O}}{\|}}{\underset{\overset{\|}{\text{O}}}{\text{S}}}{-}\text{O}{-}\text{H} \overset{\text{proton}}{\underset{\text{transfer}}{\rightleftharpoons}} \text{CH}_3\text{CH}_2{-}\overset{+}{\underset{\underset{\text{H}}{\vert}}{\text{O}}}{-}\text{H} + \overset{-}{:}\text{O}{-}\overset{\overset{\text{O}}{\|}}{\underset{\overset{\|}{\text{O}}}{\text{S}}}{-}\text{O}{-}\text{H}$$

An oxonium ion

Step 2: Nucleophilic displacement of H₂O by the OH group of a second alcohol molecule gives a new oxonium ion.

$$\text{CH}_3\text{CH}_2{-}\overset{..}{\text{O}}{-}\text{H} + \text{CH}_3\text{CH}_2{-}\overset{+}{\underset{\underset{\text{H}}{\vert}}{\text{O}}}{-}\text{H} \xrightarrow{\text{S}_\text{N}2} \text{CH}_3\text{CH}_2{-}\overset{+}{\underset{\underset{\text{H}}{\vert}}{\text{O}}}{-}\text{CH}_2\text{CH}_3 + :\overset{}{\underset{\underset{\text{H}}{\vert}}{\text{O}}}{-}\text{H}$$

An new oxonium ion

Step 3:   Proton transfer from the oxonium ion to H_2O completes the reaction.

$$CH_3CH_2\overset{+}{\underset{\underset{H}{|}}{O}}-CH_2CH_3 + :\underset{\underset{H}{|}}{O}-CH_2CH_3 \overset{\text{proton}}{\underset{\text{transfer}}{\rightleftharpoons}} CH_3CH_2-\overset{..}{\underset{..}{O}}-CH_2CH_3 + H-\overset{+}{\underset{\underset{H}{|}}{O}}-H$$

Yields of ethers from acid-catalyzed intermolecular dehydration of alcohols are highest for symmetrical ethers formed from unbranched primary alcohols. Examples of symmetrical ethers formed in good yield by this method are dimethyl ether, diethyl ether, and dibutyl ether. From secondary alcohols, yields of ether are lower because of competition from acid-catalyzed dehydration (Section 9.6). In the case of tertiary alcohols, dehydration is the only reaction.

Example 11.4

Explain why this reaction does not give a good yield of ethyl hexyl ether.

 $\text{OH} + \text{HO}\diagdown\diagup \xrightarrow[\text{heat}]{\text{H}_2\text{SO}_4} \diagup\diagdown\diagup\diagdown\diagup\text{O}\diagup + \text{H}_2\text{O}$

Solution

From this reaction, we expect a mixture of three ethers: diethyl ether, ethyl hexyl ether, and dihexyl ether.

Problem 11.4

Show how ethyl hexyl ether might be prepared by a Williamson ether synthesis.

Mechanisms
Alcohols
Acid-catalyzed addition of
an alcohol to an alkene

C.   Acid-Catalyzed Addition of Alcohols to Alkenes

Under suitable conditions, alcohols can be added to the carbon-carbon double bond of an alkene to give an ether. The usefulness of this method of ether synthesis is limited to the interaction of alkenes that can form stable carbocations and methanol or primary alcohols. An example is the commercial synthesis of *tert*-butyl methyl ether. 2-Methylpropene and methanol are passed over an acid catalyst to give the ether.

$$\underset{\underset{CH_3C=CH_2}{|}}{\overset{\overset{CH_3}{|}}{}} + CH_3OH \xrightarrow[\text{catalyst}]{\text{acid}} \underset{\underset{CH_3}{|}}{\overset{\overset{CH_3}{|}}{CH_3COCH_3}}$$

2-Methoxy-2-methylpropane
(*tert*-Butyl methyl ether)

The mechanism for this ether synthesis involves proton transfer from the acid catalyst to the alkene to give a carbocation intermediate followed by addition of the nucleophile to give an oxonium ion. Proton transfer from the oxonium ion regenerates the acid catalyst and gives the final product.

Mechanism   Acid-Catalyzed Addition of an Alcohol to an Alkene

Step 1:   Proton transfer from the acid catalyst to the alkene gives a carbocation intermediate.

$$CH_3C{=}CH_2 + H{-}\overset{+}{\underset{H}{O}}{-}CH_3 \rightleftharpoons CH_3\overset{+}{C}CH_3 + :\underset{H}{O}{-}CH_3$$

(with CH_3 substituents on the carbons)

Step 2:   Reaction between the carbocation intermediate (a Lewis acid) and the alcohol (a Lewis base) gives an oxonium ion.

$$CH_3\overset{+}{C}CH_3 + HO\overset{\cdot\cdot}{}CH_3 \rightleftharpoons CH_3CCH_3$$

(with CH_3 substituents; product bearing $\overset{+}{O}$ with H and CH_3)

Step 3:   Proton transfer to solvent (in this case methanol) completes the reaction.

$$CH_3{-}\overset{\cdot\cdot}{O}{-}H + CH_3CCH_3 \rightleftharpoons CH_3{-}\overset{\cdot\cdot\,+}{\underset{H}{O}}{-}H + CH_3CCH_3$$

(with CH_3 substituents; oxonium $\overset{+}{O}$ bearing H and CH_3 on left, and :O bearing CH_3 on right)

As an octane-improving additive, MTBE is superior to ethanol (the additive in gasohol). A blend of 15% MTBE with gasoline improves octane rating by approximately 5 units. The U.S. production of MTBE in 1995 was 3.8×10^{10} kg. This ether is of current interest because it was added to gasoline under an EPA mandate to add "oxygenates," which make gasoline burn more smoothly (it raises the octane number) and lower exhaust emissions. Unfortunately, because it is much more soluble in water than gasoline, it has gotten into the water table in many places, in some cases because of leaky gas station storage tanks. It has been detected in lakes, reservoirs, and water supplies, in some cases at concentrations that exceed limits for both "taste and odor" and human health. Consequently, its use as a gasoline additive is being phased out.

11.5 Reactions of Ethers

Ethers resemble hydrocarbons in their resistance to chemical reaction. They do not react with oxidizing agents such as potassium dichromate or potassium permanganate. They are stable toward even very strong bases, and, except for tertiary alkyl ethers, they are not affected by most weak acids at moderate temperatures. Because of their good solvent properties and general inertness to chemical reaction, ethers are excellent solvents in which to carry out many organic reactions.

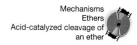

A. Acid-Catalyzed Cleavage by Concentrated HX

Cleavage of dialkyl ethers requires both a strong acid and a good nucleophile, hence the use of concentrated aqueous HI (57%) or HBr (48%). Dibutyl ether, for example, reacts with hot concentrated HBr to give two molecules of 1-bromobutane.

$$(CH_3CH_2CH_2CH_2)_2O + 2HBr \xrightarrow[\text{heat}]{} 2CH_3CH_2CH_2CH_2Br + H_2O$$

Dibutyl ether   1-Bromobutane

Concentrated HCl (38%) is far less effective in cleaving dialkyl ethers, primarily because Cl^- is a weaker nucleophile in water than either I^- or Br^-.

The mechanism of acid-catalyzed cleavage of dialkyl ethers depends on the nature of the carbons bonded to oxygen. If both carbons are primary, cleavage involves an S_N2 reaction in which a halide ion is the nucleophile. Otherwise, it is an S_N1 reaction.

Mechanism **Acid-Catalyzed Cleavage of a Dialkyl Ether**

Step 1: Proton transfer from the acid catalyst to the oxygen atom of the ether gives an oxonium ion.

$$CH_3CH_2{-}\overset{..}{\underset{..}{O}}{-}CH_2CH_3 + H{-}\overset{+}{\underset{\underset{H}{|}}{O}}{-}H \xrightleftharpoons[]{\text{proton transfer}} CH_3CH_2{-}\overset{+}{\underset{\underset{H}{|}}{O}}{-}CH_2CH_3 + \overset{..}{:}\underset{\underset{H}{|}}{O}{-}H$$

An oxonium ion

Step 2: Nucleophilic displacement by halide ion on the primary carbon cleaves the C—O bond. The leaving group is CH_3CH_2OH, a weak base, and a poor nucleophile.

$$Br\overset{..}{:}{}^{-} + CH_3CH_2{-}\overset{+}{\underset{\underset{H}{|}}{O}}{-}CH_2CH_3 \xrightarrow{S_N2} CH_3CH_2{-}Br + \overset{..}{:}\underset{\underset{H}{|}}{O}{-}CH_2CH_3$$

This cleavage produces one molecule of bromoalkane and one molecule of alcohol. In the presence of excess concentrated HBr, the alcohol is converted to a second molecule of bromoalkane by another S_N2 process (Section 8.3).

Tertiary, allylic, and benzylic ethers are particularly susceptible to cleavage by acid, often under quite mild conditions. Tertiary butyl ethers, for example, are cleaved by aqueous HCl at room temperature.

$$\langle\text{ring}\rangle{-}O{-}\overset{\overset{CH_3}{|}}{\underset{\underset{CH_3}{|}}{C}}CH_3 + HCl \xrightarrow{S_N1} \langle\text{ring}\rangle{-}OH + Cl{-}\overset{\overset{CH_3}{|}}{\underset{\underset{CH_3}{|}}{C}}{-}CH_3$$

Proton transfer from the acid to the oxygen atom of the ether produces an oxonium ion, which then cleaves to produce a particularly stable 3°, allylic, or benzylic carbocation.

Example 11.5

Account for the fact that reaction of most methyl ethers with concentrated HI gives CH_3I and ROH as the initial major products rather than CH_3OH and RI. For example,

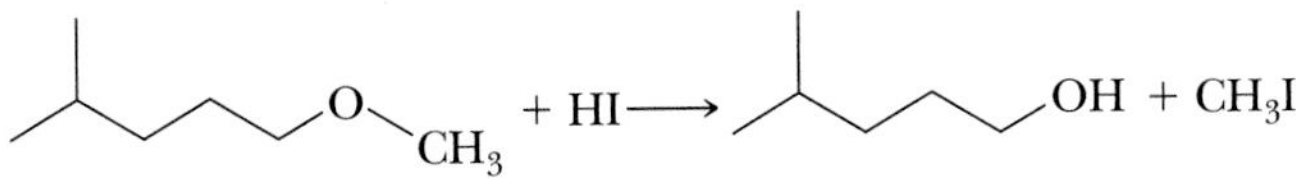

+ HI $\longrightarrow$ OH + CH_3I

Solution

The first step is protonation of the ether oxygen to give an oxonium ion. Cleavage is by an S_N2 pathway on the less hindered methyl carbon.

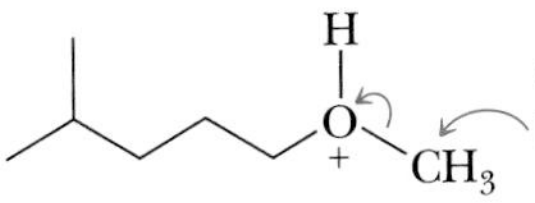

site of attack by nucleophile

Problem 11.5

Account for the fact that treatment of *tert*-butyl methyl ether with a limited amount of concentrated HI gives methanol and *tert*-butyl iodide rather than methyl iodide and *tert*-butyl alcohol.

Example 11.6

Draw structural formulas for the major products of the following reactions.

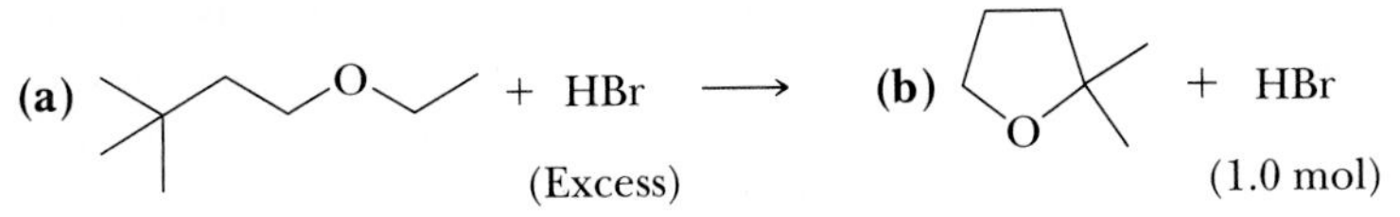

(a) + HBr $\longrightarrow$
(Excess)

(b) + HBr
(1.0 mol)

Solution

(a) Cleavage on either side of the ether oxygen by an S_N2 mechanism gives an alcohol and a bromoalkane. Reaction of the alcohol then gives a second molecule of alkyl bromide.

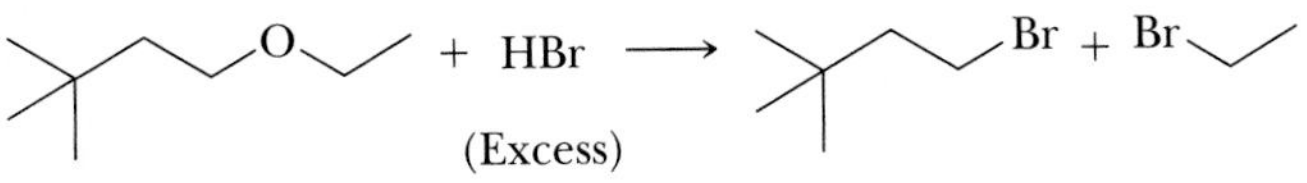

+ HBr $\longrightarrow$ Br + Br
(Excess)

(b) Proton transfer to the ether oxygen followed by cleavage gives a 3° carbocation intermediate, which may then (1) react with bromide ion to give a bromoalcohol or (2) lose a proton to give an unsaturated alcohol.

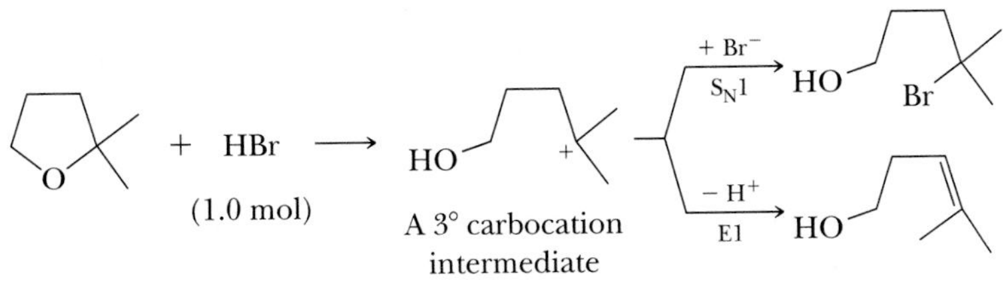

+ HBr $\longrightarrow$ HO

+ Br^- / S_N1 → HO Br

− H^+ / E1 → HO

(1.0 mol) A 3° carbocation intermediate

Problem 11.6

Draw structural formulas for the major products of these reactions.

(a) CH₃COCH₃ + HBr ⟶   (b) [hexane ring with O] + HBr ⟶
     (with CH₃ groups)   (Excess)   (Excess)

B.   Oxidation of Ethers—Formation of Hydroperoxides

Two hazards must be avoided when working with diethyl ether and other low-molecular-weight ethers. First, they are highly flammable. Consequently, open flames and electric appliances with sparking contacts must be avoided where ethers are being used (lab refrigerators and ovens are frequent causes of ignition). Because diethyl ether is so volatile (its boiling point is 35°C), it should be used in a fume hood to prevent the build-up of vapors and possible explosion. Second, anhydrous ethers react with molecular oxygen at a C—H bond adjacent to the ether oxygen to form explosive **hydroperoxides.**

Hydroperoxide A compound containing an —OOH group.

$$CH_3CH_2OCH_2CH_3 + \boxed{O_2} \longrightarrow CH_3CH_2OCHCH_3$$

            Diethyl ether   A hydroperoxide (OOH)

Hydroperoxidation proceeds by a radical chain mechanism (see the Chemistry in Action box "Radical Autoxidation" in Chapter 7). Rates of hydroperoxide formation increase dramatically if the C—H bond adjacent to oxygen is secondary, as for example in diisopropyl ether, because of favored generation of a relatively stable radical intermediate. This hydroperoxide precipitates from solution as a waxy solid.

$$CH_3CHOCHCH_3 + O_2 \longrightarrow CH_3CHOCCH_3$$

with CH₃, CH₃ substituents (left) and CH₃, CH₃ and OOH (right)

         Diisopropyl ether   A hydroperoxide

Hydroperoxides are dangerous because they are explosive. Peroxides can be detected in ethers by shaking a small amount of the ether with an acidified 10% aqueous solution of potassium iodide, KI, or by using starch iodine paper with a drop of acetic acid. Peroxides oxidize iodide ion to iodine, I_2, which gives a yellow color to the solution. This is converted to a deep blue-purple color by the addition of starch, which forms a complex with iodine (a simple test is to use acidified starch-iodide paper strips, which are sold for this purpose). Hydroperoxides can be removed by treatment with a reducing agent. One effective procedure is to shake the hydroperoxide-contaminated ether with a solution of iron(II) sulfate in dilute aqueous sulfuric acid. Bottles containing diisopropyl ether with precipitated hydroperoxide should be removed and disposed of by a bomb squad!

11.6   Ethers as Protecting Groups

When dealing with organic compounds containing two or more functional groups, it is often necessary to protect one functional group (to prevent its reaction) while carrying out a reaction at another functional group. Suppose, for example, that you wish to convert 4-pentyn-1-ol to 4-heptyn-1-ol.

$$HC{\equiv}CCH_2CH_2CH_2OH \xrightarrow{\ ?\ } CH_3CH_2C{\equiv}CCH_2CH_2CH_2OH$$

4-Pentyn-1-ol   4-Heptyn-1-ol

The new carbon-carbon bond in the product can be formed by alkylation of the acetylide anion from 4-pentyn-1-ol (Section 10.5) with bromoethane. 4-Pentyn-1-ol, however, contains two acidic hydrogens, one on the hydroxyl group (pK_a 16–18) and the other on the carbon-carbon triple bond (pK_a 25). Treatment of this compound with one equivalent of $NaNH_2$ forms the alkoxide anion (the —OH group is the stronger acid) rather than the acetylide anion.

pK_a 25 pK_a 16–18

$$HC{\equiv}CCH_2CH_2CH_2OH + Na^+NH_2^- \longrightarrow HC{\equiv}CCH_2CH_2CH_2O^-Na^+ + NH_3$$

4-Pentyn-1-ol

It is necessary, therefore, to protect the —OH group to prevent its reaction with sodium amide. A good protecting group is

- Easily added to the sensitive functional group.
- Resistant to the reagents used to transform the unprotected functional group or groups.
- Easily removed to regenerate the original functional group.

Chemists have devised protecting groups for many functional groups, and we will encounter several of them in this text. In this section, we concentrate on two hydroxyl protecting groups.

The first of these is the *tert*-butyl group, formed by treatment of an alcohol with 2-methylpropene (isobutylene) in the presence of an acid catalyst. This treatment of 4-pentyn-1-ol with 2-methylpropene in the presence of sulfuric acid as a catalyst gives the *tert*-butyl ether. Treatment of this ether with sodium amide followed by bromoethane forms the new carbon-carbon bond. The protecting group is then removed by treatment with aqueous acid. Thus, the *tert*-butyl protecting group is added in the presence of an acid catalyst, is stable under neutral and basic conditions, and is removed by treatment with aqueous acid.

$$HC{\equiv}CCH_2CH_2CH_2OH \xrightarrow[\text{H}_2\text{SO}_4]{\text{1. CH}_2{=}\text{C(CH}_3)_2} HC{\equiv}CCH_2CH_2CH_2OCCH_3 \overset{CH_3}{\underset{CH_3}{|}} \xrightarrow[\text{3. CH}_3\text{CH}_2\text{Br}]{\text{2. Na}^+\text{NH}_2^-}$$

4-Pentyn-1-ol

$$CH_3CH_2C{\equiv}CCH_2CH_2CH_2OCCH_3 \overset{CH_3}{\underset{CH_3}{|}} \xrightarrow{\text{4. H}_3\text{O}^+/\text{H}_2\text{O}}$$

$$CH_3CH_2C{\equiv}CCH_2CH_2CH_2OH + H_2C{=}C(CH_3)_2$$

4-Heptyn-1-ol

Example 11.7

When the *tert*-butyl protecting group is removed by treatment with aqueous acid, its products are 2-methyl-2-propanol and 2-methylpropene. Propose a mechanism for the acid-catalyzed removal of the *tert*-butyl protecting group and the formation of these two products.

Solution

Proton transfer from the acid catalyst to the oxygen of the ether gives an oxonium ion. Cleavage on the side of the *tert*-butyl group gives the more stable *tert*-butyl cation, which then either transfers a proton to water to give 2-methylpropene (E1) or adds water to give 2-methyl-2-propanol (S_N1).

$$RCH_2-O-C(CH_3)_3 \xrightarrow{H_3O^+} RCH_2-\overset{H}{\overset{+}{O}}-\underset{CH_3}{\overset{CH_3}{C}}CH_3 \longrightarrow RCH_2-OH + CH_3\overset{CH_3}{\underset{CH_3}{C^+}}$$

An oxonium ion

$$(CH_3)_2C=CH_2 + H_3O^+ \qquad CH_3\underset{CH_3}{\overset{CH_3}{C}}OH + H_3O^+$$

Problem 11.7

Why is the use of the *tert*-butyl protecting group limited to the protection of primary alcohols?

An —OH group can also be protected by converting it to a trimethylsilyl ether, —OSi(CH_3)$_3$, by treating the alcohol with chlorotrimethylsilane in the presence of a tertiary amine, such as triethylamine or pyridine. The function of the tertiary amine is to catalyze the reaction by forming some of the more nucleophilic alkoxide ion and to neutralize the HCl formed during the reaction of the alcohol with chlorotrimethylsilane.

$$RCH_2OH + Cl-\underset{CH_3}{\overset{CH_3}{Si}}-CH_3 \xrightarrow{(CH_3CH_2)_3N} RCH_2O-\underset{CH_3}{\overset{CH_3}{Si}}-CH_3$$

Chlorotrimethylsilane A trimethylsilyl ether

Like the *tert*-butyl ether protecting group, the trimethylsilyl ether group is removed by treatment with aqueous acid or, alternatively, by using fluoride ion, F⁻, in the form of tetrabutylammonium fluoride, ($CH_3CH_2CH_2CH_2)_4N^+F^-$.

$$RCH_2O-\underset{CH_3}{\overset{CH_3}{Si}}-CH_3 + H_2O \xrightarrow{H^+} RCH_2OH + HO-\underset{CH_3}{\overset{CH_3}{Si}}-CH_3$$

A trimethylsilyl ether

Many other silyl protecting groups are in common use and are chosen by synthetic chemists to give just the degree of reactivity versus stability needed for a given set of reaction conditions. Among them is the *tert*-butyldimethylsilyl (TBDMS) group.

11.7 Epoxides—Structure and Nomenclature

Although **epoxides** are technically classed as ethers, we discuss them separately because of their exceptional chemical reactivity compared with other ethers. Simple epoxides are named as derivatives of oxirane, the parent epoxide. Where the epoxide is a part of another ring system, it is named using the prefix epoxy-.

Epoxide A cyclic ether in which oxygen is one atom of a three-membered ring.

| Oxirane (Ethylene oxide) | *cis*-2,3-Dimethyloxirane (*cis*-2-Butene oxide) | 1,2-Epoxycyclohexane (Cyclohexene oxide) |

Common names of epoxides are derived by giving the name of the alkene from which the epoxide is formally derived followed by the word "oxide"; an example is *cis*-2-butene oxide.

11.8 Synthesis of Epoxides

A. Ethylene Oxide

Ethylene oxide, one of the few epoxides manufactured on an industrial scale, is prepared by passing a mixture of ethylene and air (or oxygen) over a silver catalyst. In the United States, the 1995 production of ethylene oxide by this method was 17×10^9 kg.

Oxirane
(Ethylene oxide)

This method fails when applied to other low-molecular-weight alkenes.

B. Oxidation of Alkenes with Peroxycarboxylic Acids

The most common laboratory method for the synthesis of epoxides from alkenes is oxidation with a peroxycarboxylic acid (a peracid). Three of the most widely used peroxyacids are *meta*-chloroperoxybenzoic acid (MCPBA), the magnesium salt of monoperoxyphthalic acid (MMPP), and peroxyacetic acid.

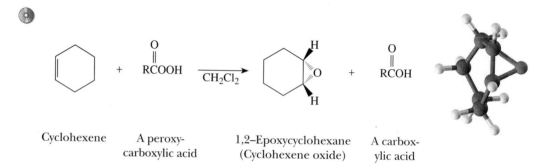

meta-Chloroperoxybenzoic acid
(MCPBA)

Magnesium monoperoxyphthalate
(MMPP)

Peroxyacetic acid
(Peracetic acid)

Following is a balanced equation for the epoxidation of cyclohexene by a peroxycarboxylic acid, RCO_3H. In the process, the peroxycarboxylic acid is reduced to a carboxylic acid.

Cyclohexene   A peroxy-   1,2–Epoxycyclohexane   A carbox-
carboxylic acid   (Cyclohexene oxide)   ylic acid

Epoxidation of an alkene is stereospecific. Epoxidation of *cis*-2-butene, for example, yields only *cis*-2,3-dimethyloxirane, and epoxidation of *trans*-2-butene yields only *trans*-2,3-dimethyloxirane.

cis-2-Butene   *cis*-2,3-Dimethyloxirane

trans-2-Butene   *trans*-2,3-Dimethyloxirane

A mechanism for epoxidation by a peroxyacid must take into account the following facts. (1) The reaction takes place in nonpolar solvents, which means that the reaction cannot involve the formation of ions or any species with significant separation of unlike charges. (2) The reaction is stereospecific, with retention of the alkene configuration, which means that even though the pi bond of the carbon-carbon double bond is broken, at no time is there free rotation about the remaining sigma bond. Following is a mechanism consistent with these observations.

Mechanism Epoxidation of an Alkene by RCO$_3$H

Arrow 1: Interaction of the pi electrons of the carbon-carbon double bond with the end oxygen atom of the peroxyacid and formation of a new C—O bond.

Arrows 2 and 3: Shifts of electron pairs within the peroxyacid.

Arrow 4: Formation of the second carbon-oxygen bond.

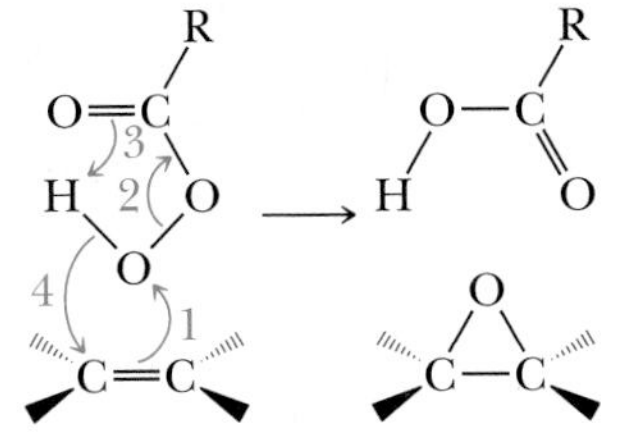

The numbering of these arrows does not imply an order in which covalent bonds are broken and made. Rather, they are meant as a guide to help you understand the mechanism. It is thought that the entire combination of bond-making and bond-breaking steps is concerted, or nearly so.

C. Internal Nucleophilic Substitution in Halohydrins

A second general method for the preparation of epoxides from alkenes involves (1) treatment of the alkene with chlorine or bromine in water to form a chlorohydrin or bromohydrin followed by (2) treatment of the halohydrin with base to bring about intramolecular displacement of X$^-$. By these steps, propene is first converted to 1-chloro-2-propanol and then to methyloxirane (propylene oxide).

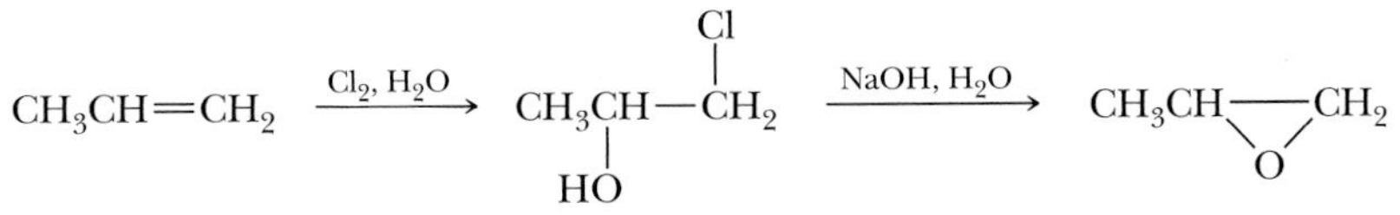

| Propene | 1-Chloro-2-propanol (a chlorohydrin) | Methyloxirane (Propylene oxide) |

We studied the reaction of alkenes with chlorine or bromine in water to form halohydrins (Section 6.3E) and saw that it is both regiospecific and stereospecific. Conversion of a halohydrin to an epoxide with base is stereospecific as well and can be viewed as an internal S$_N$2 reaction. Hydroxide ion or other base abstracts a proton from the halohydrin hydroxyl group to form an alkoxide ion, a nucleophile, which then displaces halogen on the adjacent carbon. As with all S$_N$2 reactions, attack of the nucleophile is from the backside of the C—X bond and causes inversion of configuration at the site of substitution.

An epoxide

Note that this displacement of halide by the alkoxide ion can also be viewed as an intramolecular variation of the Williamson ether synthesis we studied in Section 11.4A. In this case, the displacing alkoxide and leaving halide ions are on adjacent carbon atoms.

Example 11.8

Conversion of an alkene to a halohydrin and internal displacement of a halide ion by an alkoxide ion are both stereospecific. Use this information to demonstrate that the configuration of the alkene is preserved in the epoxide. As an illustration, show that reaction of *cis*-2-butene by this two-step sequence gives *cis*-2,3-dimethyloxirane (*cis*-2-butene oxide).

Solution

Addition of HOCl to an alkene occurs by anti addition of —OH and —Cl to the double bond (Section 6.3E). The conformation of this product is also the conformation necessary for backside displacement of the halide ion by alkoxide ion. Thus, a cis alkene gives a cis-disubstituted oxirane and a trans alkene gives a trans-disubstituted oxirane.

cis-2-Butene   A chlorohydrin   *cis*-2,3-Dimethyloxirane

Problem 11.8

Consider the possibilities for stereoisomerism in the halohydrin and epoxide formed in Example 11.8.

(a) How many stereoisomers are possible for the chlorohydrin? Which of the possible chlorohydrins are formed in this step?

(b) How many stereoisomers are possible for the epoxide? Which of the possible stereoisomers is/are formed in this three-step sequence?

D.   Sharpless Asymmetric Epoxidation

One of the most useful organic reactions discovered in the last 25 years is the titanium-catalyzed asymmetric epoxidation of primary allylic alcohols developed by Professor Barry Sharpless, then at the Massachusetts Institute of Technology. The reagent consists of *tert*-butyl hydroperoxide, titanium tetraisopropoxide, and diethyl tartrate. Recall from Section 3.5 that tartaric acid has two stereocenters and exists as three stereoisomers: a pair of enantiomers and a meso compound (Figure 3.5). The form of tartaric acid used in the Sharpless epoxidation is either pure (+)-diethyl tartrate or its enantiomer, (−)-diethyl tartrate. The *tert*-butyl hydroperoxide is the oxidizing agent and must be present in molar amounts. Titanium isopropoxide and diethyl tartrate are catalysts and present in lesser amounts, generally 5 to 10 mole percent.

What is remarkable about the Sharpless epoxidation is that it is stereospecific; either enantiomer of the epoxide can be produced depending on which enantiomer of diethyl tartrate is used. If (+)-diethyl tartrate is used as in the catalyst, the product is enantiomer A. If (−)-diethyl tartrate is used, the product is enantiomer B.

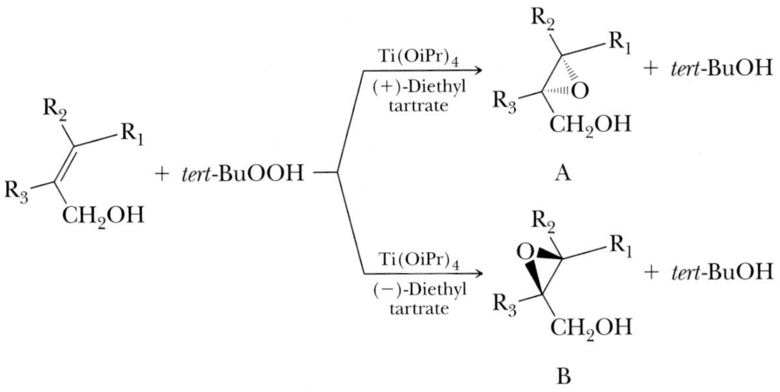

Example 11.9

Draw the expected products of Sharpless epoxidation of each allylic alcohol using (+)-diethyl tartrate as the chiral catalyst.

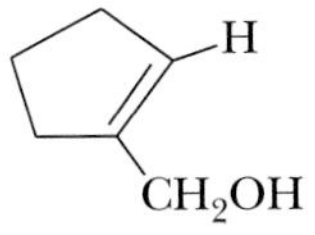

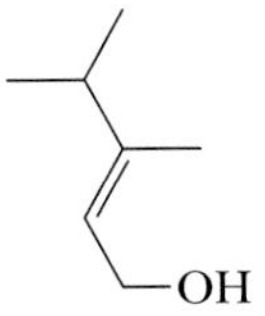

Solution

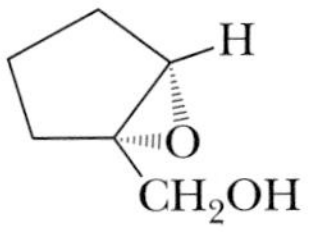

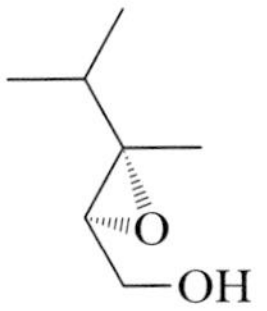

Problem 11.9

Draw the expected products of Sharpless epoxidation of each allylic alcohol using (+)-diethyl tartrate as the chiral catalyst.

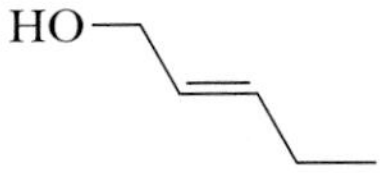

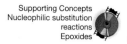

11.9   Reactions of Epoxides

Because of the strain associated with the three-membered ring, epoxides undergo a variety of ring-opening reactions, the characteristic feature of which is nucleophilic substitution at one of the carbons of the epoxide ring with the oxygen atom as the leaving group.

Characteristic reaction of epoxides:

$$C-C \overset{O}{\diagdown} + HNu: \longrightarrow \underset{HO}{C}-C\overset{Nu}{\diagup}$$

A.   Acid-Catalyzed Ring Opening

In the presence of an acid catalyst, such as sulfuric acid, epoxides are hydrolyzed to glycols. As an example, acid-catalyzed hydrolysis of oxirane gives 1,2-ethanediol (ethylene glycol). Production of ethylene glycol in the United States in 1995 was 11.4×10^9 kg.

$$H_2C\overset{O}{\diagup\diagdown}CH_2 + H_2O \xrightarrow{H^+} HOCH_2CH_2OH$$

Oxirane
(Ethylene oxide)

1,2-Ethanediol
(Ethylene glycol)

Mechanism   Acid-Catalyzed Hydrolysis of an Epoxide

Step 1:   Proton transfer from the acid catalyst to oxygen of the epoxide gives a bridged oxonium ion intermediate.

Step 2:   Backside attack of H_2O on the protonated epoxide opens the three-membered ring.

Step 3:   Proton transfer to solvent completes formation of the glycol.

Attack of a nucleophile on a protonated epoxide shows a stereospecificity typical of S_N2 reactions; the nucleophile attacks anti to the leaving hydroxyl group, and the —OH groups in the glycol thus formed are anti. As a result, hydrolysis of an epoxycycloalkane yields a *trans*-1,2-cycloalkanediol.

1,2-Epoxycyclopentane
(Cyclopentene oxide)

trans-1,2-Cyclopentanediol

Note the similarity in ring opening of this bridged oxonium ion intermediate, the bridged halonium ion intermediate in electrophilic addition of halogens or X_2/H_2O to an alkene (Sections 6.3D and 6.3E), and the bridged mercurinium ion intermediate in oxymercuration (Section 6.3F). In each case, the intermediate is a three-membered ring with a heteroatom bearing a positive charge, and attack of the nucleophile is anti to the leaving group.

Because there is some carbocation character developed in the transition state for an acid-catalyzed epoxide ring opening, attack of the nucleophile on unsymmetrical epoxides occurs preferentially at the carbon better able to bear a partial positive charge.

1-Methyl-1,2-epoxycyclohexane

2-Methoxy-2-methylcyclohexanol

Thus, the stereochemistry of acid-catalyzed ring openings is S_N2-like in that attack of the nucleophile is from the side opposite the bridged oxonium ion intermediate. The regiochemistry, however, is S_N1-like. Because of the partial carbocation character of the transition state, attack of the nucleophile on the oxonium ion intermediate occurs preferentially at the more substituted carbon; that is, at the one better able to bear the partial positive charge that develops on carbon in the transition state.

At this point, let us compare the stereochemistry of the glycol formed by acid-catalyzed hydrolysis of an epoxide with that formed by oxidation of an alkene with osmium tetroxide (Section 6.5B). Each reaction sequence is stereospecific but gives a different stereoisomer. Acid-catalyzed hydrolysis of cyclopentene oxide gives *trans*-1,2-cyclopentanediol; osmium tetroxide oxidation of cyclopentene gives *cis*-1,2-cyclopentanediol. Thus, a cycloalkene can be converted to either a cis glycol or a trans glycol by the proper choice of reagents.

trans-1,2-Cyclopentanediol

cis-1,2-Cyclopentanediol

B. Nucleophilic Ring Opening

Ethers are not normally susceptible to reaction with nucleophiles. Because of the strain associated with a three-membered ring, however, epoxides undergo ring-opening reactions with a variety of nucleophiles. Good nucleophiles attack an epoxide ring by an S_N2 mechanism and show a regioselectivity for attack of the nucleophile at the less hindered carbon. Following is an equation for the reaction of methyloxirane (propylene oxide) with sodium methoxide in methanol; attack of the nucleophile occurs on the primary carbon.

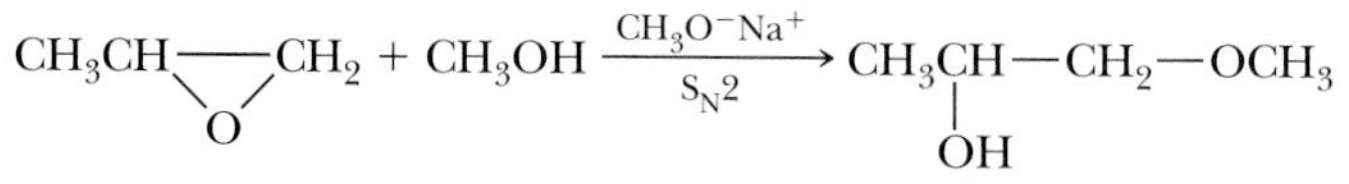

$$CH_3CH \overset{\textstyle \diagup\diagdown}{\underset{O}{\quad}} CH_2 + CH_3OH \xrightarrow[S_N2]{CH_3O^-Na^+} CH_3CH-CH_2-OCH_3$$
$$\underset{OH}{|}$$

Mechanism   Nucleophilic Opening of an Epoxide Ring

Step 1: Backside attack of the nucleophile on the less hindered carbon of the highly strained epoxide opens the ring and displaces O^-.

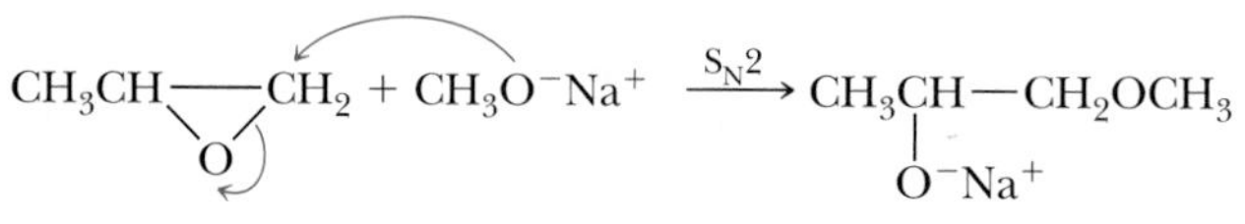

$$CH_3CH \overset{\textstyle \diagup\diagdown}{\underset{O}{\quad}} CH_2 + CH_3O^-Na^+ \xrightarrow{S_N2} CH_3CH-CH_2OCH_3$$
$$\underset{O^-Na^+}{|}$$

Step 2: Proton transfer completes the reaction.

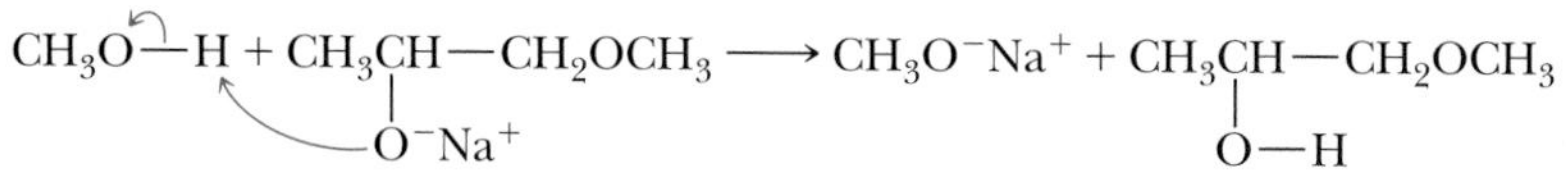

$$CH_3O-H + CH_3CH-CH_2OCH_3 \longrightarrow CH_3O^-Na^+ + CH_3CH-CH_2OCH_3$$
$$\underset{O^-Na^+}{|} \qquad\qquad\qquad\qquad \underset{O-H}{|}$$

The nucleophilic ring opening of epoxides is also stereospecific; as expected of an S_N2 reaction, attack of the nucleophile is anti to the leaving group. An illustration is the reaction of cyclohexene oxide with sodium methoxide in methanol to give *trans*-2-methoxycyclohexanol.

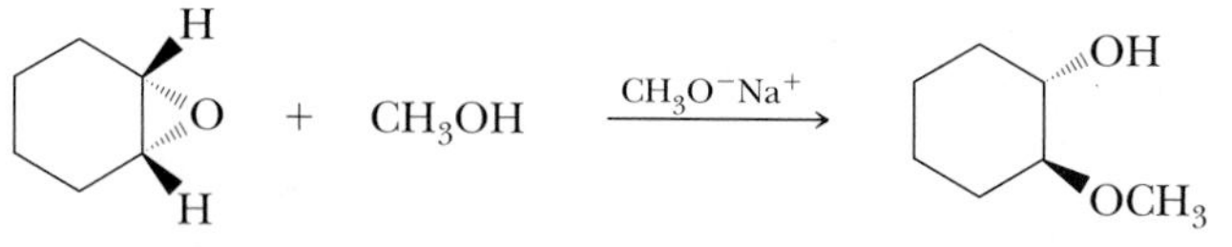

Cyclohexene oxide   *trans*-2-Methoxycyclohexanol

The value of epoxides lies in the number of nucleophiles that bring about ring opening and the combinations of functional groups that can be prepared from them. The most important of these ring-opening reactions are summarized in the following chart.

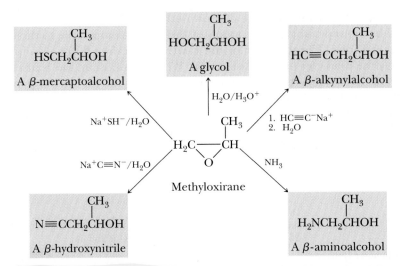

$$HSCH_2CHOH$$ (with CH₃ above)

A β-mercaptoalcohol

$$HOCH_2CHOH$$ (with CH₃ above)

A glycol

$$HC{\equiv}CCH_2CHOH$$ (with CH₃ above)

A β-alkynylalcohol

$$N{\equiv}CCH_2CHOH$$ (with CH₃ above)

A β-hydroxynitrile

$$H_2NCH_2CHOH$$ (with CH₃ above)

A β-aminoalcohol

Methyloxirane

Finally, treatment with LiAlH₄ reduces an epoxide to an alcohol. Lithium aluminum hydride is similar to sodium borohydride, NaBH₄, in that it is a donor of hydride ion, H:⁻, which is both a strong base and a good nucleophile. In the reduction of a substituted epoxide by LiAlH₄, preferential attack of the hydride-reducing agent occurs at the less hindered carbon of the epoxide, an observation consistent with S$_N$2 reactivity.

$$\text{Ph—CH—CH}_2 \xrightarrow[\text{2. }H_2O]{\text{1. LiAlH}_4} \text{Ph—CHCH}_2\text{—H}$$

Phenyloxirane
(Styrene oxide)

1-Phenylethanol

11.10 Crown Ethers

In the early 1960s, Charles Pedersen of Du Pont discovered a family of cyclic polyethers derived from ethylene glycol and substituted ethylene glycols. Compounds of this structure were given the name **crown ethers** because one of their most stable conformations resembles the shape of a crown. These ethers are named by the system devised by Pedersen. The parent name "crown" is preceded by a number describing the size of the ring and followed by a number describing the number of oxygen atoms in the ring, as for example, 18-crown-6. For his work, Pedersen shared the 1987 Nobel Prize for chemistry with Donald J. Cram of the United States and Jean-Marie Lehn of France.

Crown ether A cyclic polyether derived from ethylene glycol and substituted ethylene glycols. Crown ethers are excellent phase-transfer catalysts.

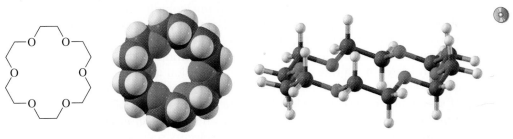

18-Crown-6
(a cyclic hexamer)

Space-filling model,
viewed from above

Ball-and-stick model,
viewed through an edge

A remarkable structural feature of crown ethers is that the diameter of the cavity created by the repeating oxygen atoms of the ring is comparable to the diameter of alkali metal ions. The diameter of the cavity in 18-crown-6, for example, is approximately the diameter of a potassium ion. When a potassium ion is inserted into the cavity of 18-crown-6, the unshared electron pairs on the six oxygens of the crown ether are close enough to the potassium ion to provide very effective solvation for K^+. 18-Crown-6 forms somewhat weaker complexes with rubidium ion (a somewhat larger ion) and with sodium ion (a somewhat smaller ion). It does not coordinate to any appreciable degree with lithium ion (a considerably smaller ion). 12-Crown-4, however, with its smaller cavity, does form a strong complex with lithium ion.

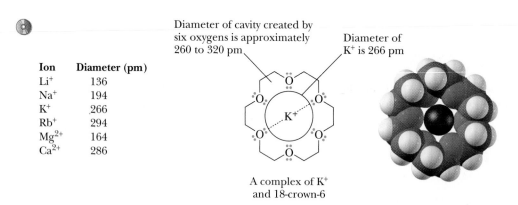

Ion	Diameter (pm)
Li^+	136
Na^+	194
K^+	266
Rb^+	294
Mg^{2+}	164
Ca^{2+}	286

Diameter of cavity created by six oxygens is approximately 260 to 320 pm

Diameter of K^+ is 266 pm

A complex of K^+ and 18-crown-6

The cavity of a crown ether is a polar region, and the unshared pairs of electrons on the oxygen atoms lining the cavity provide effective solvation for alkali metal ions. The outer surface of the crown is nonpolar and hydrocarbon-like, and, thus, crown ethers and their alkali metal ion complexes dissolve readily in nonpolar organic solvents.

Crown ethers have proven to be particularly valuable for the same reasons as phase-transfer catalysts (Section 8.7), namely, their ability to cause inorganic salts to dissolve in nonpolar aprotic organic solvents such as methylene chloride, hexane, and benzene. Potassium permanganate, for example, does not dissolve in benzene. If 18-crown-6 is added to benzene, the solution takes on the purple color characteristic of permanganate ion. The crown-potassium ion complex is soluble in benzene and brings permanganate ion into solution with it. The resulting "purple benzene" is a valuable reagent for the oxidation of water-insoluble organic compounds.

Crown ethers have also proven valuable in nucleophilic displacement reactions. The cations of potassium salts, such as KF, KCN, or KN_3, are very tightly bound within the solvation cavity of 18-crown-6 molecules. The anions, however, are only weakly solvated, and because of the geometry of cation binding within the cavity of the crown, only loose ion pairing occurs between the anion and cation. Thus, in nonpolar aprotic solvents, these anions are without any appreciable solvent shell and are, therefore, highly reactive as nucleophiles. The nucleophilicity of F^-, CN^-, N_3^-, and other anions in nonpolar aprotic solvents containing an 18-crown-6 equals and often exceeds their nucleophilicity in polar aprotic solvents such as DMSO and acetonitrile.

11.11 Thioethers

A. Nomenclature

To derive the IUPAC name of a thioether, select the longest carbon chain as the parent alkane and name the sulfur-containing substituent as an alkylsulfanyl group. To derive a common name, list the groups bonded to sulfur and add the word "**sulfide**" to show the presence of the —S— group.

Sulfide The sulfur analog of an ether; a molecule containing a sulfur atom bonded to two carbon atoms.

$$CH_3CH_2SCH_2CH_3$$

Ethylsulfanylethane
(Diethyl sulfide)

$$CH_3CH_2SCHCH_3 \atop {\overset{\displaystyle CH_3}{|}}$$

2-Ethylsulfanylpropane
(Ethyl isopropyl sulfide)

The functional group of a **disulfide** is an —S—S— group. IUPAC names of disulfides are derived by selecting the longest carbon chain as the parent alkane and indicating the disulfide-containing substituent as an alkyldisulfanyl group. Common names of disulfides are derived by listing the names of the groups attached to sulfur and adding the word "disulfide."

Disulfide A molecule containing an —S—S— group.

$$H_3C \diagdown ^S \diagdown _S \diagup ^{CH_3}$$

Methyldisulfanylmethane
(Dimethyl disulfide)

B. Preparation of Sulfides

Symmetrical sulfides, RSR (also called symmetrical thioethers), are prepared by the treatment of one mole of Na_2S (where S^{2-} is the nucleophile) with two moles of haloalkane.

$$2RX + Na_2S \longrightarrow RSR + 2NaX$$

A sulfide

This same reaction can also be used to prepare five- and six-membered cyclic sulfides. Treatment of a 1,4-dihaloalkane with Na_2S gives a five-membered cyclic sulfide; treatment of a 1,5-dihaloalkane with Na_2S gives a six-membered one.

1,4-Dichlorobutane   Thiolane
(Tetrahydrothiophene)

1,5-Dichloropentane   Thiane
(Tetrahydrothiopyran)

Unsymmetrical sulfides, RSR', are prepared by converting a thiol to a sodium salt with either sodium hydroxide or sodium ethoxide and then allowing the salt to react with a haloalkane. This method of thioether formation is the sulfur analog of the Williamson ether synthesis (Section 11.4A).

$$CH_3(CH_2)_8CH_2S^-Na^+ + CH_3I \xrightarrow{S_N2} CH_3(CH_2)_8CH_2SCH_3 + Na^+I^-$$

Sodium 1-decanethiolate   1-Methylsulfanyldecane
(Decyl methyl sulfide)

Note that all these reactions leading to sulfides (thioethers) are direct applications of nucleophilic substitution reactions (Chapter 8).

C. Oxidation of Sulfides

Many of the properties of sulfides stem from the fact that divalent sulfur is a reducing agent; it is easily oxidized to two higher oxidation states. Treatment of a sulfide with one mole of 30% aqueous hydrogen peroxide at room temperature gives a sulfoxide, as illustrated by oxidation of methyl phenyl sulfide to methyl phenyl sulfoxide. Several other oxidizing agents, including sodium periodate, $NaIO_4$, also bring about the same conversion. Treatment of a sulfoxide with $NaIO_4$ brings about its oxidation to a sulfone.

Methyl phenyl sulfide   Methyl phenyl sulfoxide   Methyl phenyl sulfone

Dimethyl sulfoxide (DMSO) is manufactured on an industrial scale by air oxidation of dimethyl sulfide in the presence of oxides of nitrogen.

$$2CH_3-S-CH_3 + O_2 \xrightarrow[\text{nitrogen}]{\text{oxides of}} 2CH_3-\overset{\displaystyle O}{\overset{\|}{S}}-CH_3$$

Dimethyl sulfide   Dimethyl sulfoxide

Summary

An **ether** (Section 11.1) contains an atom of oxygen bonded to two carbon atoms. In the IUPAC name, the parent chain is named and the —OR group is named as an **alkoxy** substituent (Section 11.2). Common names are derived by naming the two groups attached to oxygen followed by the word "ether." Heterocyclic ethers have an oxygen atom as one of the members of a ring. Ethers are weakly polar compounds (Section 11.3) and associate by weak dipole-dipole interactions and dispersion forces. Boiling points of ethers are close to those of hydrocarbons of comparable molecular weight but much lower than those of the corresponding alcohols. Because ethers are hydrogen bond acceptors, they are more soluble in water than are hydrocarbons of comparable molecular weight.

Crown ethers (Section 11.10) are cyclic polyethers having 12 or more atoms in a ring. The cavity of a crown ether is a polar region, and the unshared pairs of electrons on the ether oxygens can solvate alkali metal ions. The cavity of 18-crown-6, for example, has approximately the same diameter as potassium ion. The outer surface of a crown ether is nonpolar and hydrocarbon-like. Crown ethers are valuable for their ability to cause ionic compounds to dissolve in nonpolar organic solvents.

Thioethers (Section 11.11) are named as alkylsulfanyl alkanes. Common names for thioethers are derived by naming the two groups bonded to sulfur followed by the word "sulfide."

Key Reactions

1. Williamson Ether Synthesis (Section 11.4A)

The Williamson ether synthesis is a general method for the synthesis of dialkyl ethers by an S_N2 reaction between an alkyl halide and an alkoxide ion. Yields are highest with 1° alkyl halides and considerably lower with 2° halides because of competition from E2 elimination. The reaction fails altogether with 3° halides.

$$CH_3CO^-K^+ + CH_3Br \xrightarrow{S_N2} CH_3COCH_3 + K^+Br^-$$

(with CH_3 substituents shown above and below each central carbon)

2. Acid-Catalyzed Dehydration of Alcohols (Section 11.4B)

Yields are highest for symmetrical ethers formed from unbranched primary alcohols.

$$2CH_3CH_2OH \xrightarrow[140°C]{H_2SO_4} CH_3CH_2OCH_2CH_3 + H_2O$$

Ethanol   Diethyl ether

3. Acid-Catalyzed Addition of Alcohols to Alkenes (Section 11.4C)

Proton transfer to the alkene generates a carbocation. Nucleophilic addition of an alcohol to the carbocation followed by proton transfer to the solvent gives the ether. The *tert*-butyl group can be used to protect primary alcohols. It is removed by treatment with acid.

$$CH_3C{=}CH_2 + CH_3OH \xrightarrow[catalyst]{acid} CH_3COCH_3$$

(with CH_3 substituents shown above and below each central carbon)

4. Acid-Catalyzed Cleavage of Dialkyl Ethers (Section 11.5A)

Cleavage of ethers requires both a strong acid and a good nucleophile, hence the use of concentrated HBr and HI. Cleavage of primary and secondary alkyl ethers is by an S_N2 pathway. Cleavage of tertiary alkyl ethers is by an S_N1 pathway.

$$\text{⬡}{-}OCH_3 + HI \text{ (excess)} \longrightarrow \text{⬡}{-}I + CH_3I + H_2O$$

5. Reaction of Alcohols with Chlorotrimethylsilane (Section 11.6)

The trimethylsilyl group is used as a protecting group for primary and secondary alcohols. It is removed by treatment with fluoride ion to regenerate the original alcohol.

$$RCH_2OH + ClSi(CH_3)_3 \xrightarrow{(CH_3CH_2)_3N} RCH_2OSi(CH_3)_3$$

Chlorotrimethylsilane   A trimethylsilyl ether

6. Oxidation of Alkenes by Peroxycarboxylic Acids (Section 11.8B)

Three commonly used peroxycarboxylic acid oxidizing agents are *meta*-chloroperoxybenzoic acid (MCPBA), the magnesium salt of monoperoxyphthalic acid (MMPP), and peroxyacetic acid. Each oxidizes an alkene to an epoxide.

7. Synthesis of Epoxides from Halohydrins (Section 11.8C)

Both formation of the halohydrin and the following intramolecular S_N2 reaction are stereospecific.

8. Sharpless Asymmetric Epoxidation (Section 11.8D)

Oxidation of an alkene by *tert*-butyl hydroperoxide in the presence of a chiral catalyst consisting of (+)- or (−)-diethyl tartrate and titanium isopropoxide gives an enantiomerically pure epoxide. The enantiomer formed depends on which enantiomer of diethyl tartrate is used in the catalyst.

9. Acid-Catalyzed Hydrolysis of Epoxides (Section 11.9A)

Hydrolysis of an epoxide derived from a cycloalkene gives a trans glycol.

10. Nucleophilic Ring Opening of Epoxides (Section 11.9B)

Attack on the epoxide is regiospecific with the nucleophile attacking the less substituted carbon of the epoxide.

11. Reduction of an Epoxide to an Alcohol (Section 11.9B)

Regioselective hydride ion transfer from lithium aluminum hydride to the less hindered carbon of the epoxide gives an alcohol.

12. Oxidation of Sulfides (Section 11.11C)

Oxidation of a sulfide gives either a sulfoxide or a sulfone, depending on the oxidizing agent and experimental conditions. Air oxidation of dimethyl sulfide is a commercial route to dimethyl sulfoxide, a polar aprotic solvent.

$$CH_3-S-CH_3 + O_2 \xrightarrow[\text{nitrogen}]{\text{oxides of}} CH_3-\overset{\overset{\displaystyle O}{\|}}{S}-CH_3$$

Problems

Structure and Nomenclature

11.10 Write names for these compounds. Where possible, write both IUPAC names and common names.

(a) 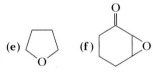   (b) 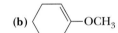   (c) 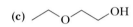 (d)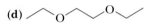

(e) 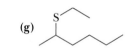   (f)   (g)   (h)

11.11 Draw structural formulas for these compounds.

(a) 2-(1-Methylethoxy)propane   (b) *trans*-2,3-Diethyloxirane
(c) *trans*-2-Ethoxycyclopentanol   (d) Etheneoxyethene
(e) Cyclohexene oxide   (f) 3-Cyclopropyloxy-1-propene
(g) (*R*)-2-Methyloxirane   (h) 1,1-Dimethoxycyclohexane

Physical Properties

11.12 Each compound given in this problem is a common organic solvent. From each pair of compounds, select the solvent with the greater solubility in water.

(a) CH_2Cl_2 and CH_3CH_2OH

(b) $CH_3CH_2OCH_2CH_3$ and CH_3CH_2OH

(c) $CH_3\overset{\overset{\displaystyle O}{\|}}{C}CH_3$ and $CH_3CH_2OCH_2CH_3$

(d) $CH_3CH_2OCH_2CH_3$ and $CH_3(CH_2)_3CH_3$

11.13 Following are structural formulas and solubilities in water for diethyl ether and tetrahydrofuran (THF). Account for the fact that tetrahydrofuran is much more soluble in water than diethyl ether.

Diethyl ether   Tetrahydrofuran
(8 g/100 mL water)   (very soluble in water)

11.14 Because of the Lewis base properties of ether oxygen atoms, crown ethers are excellent complexing agents for Na^+, K^+, and NH_4^+. What kind of molecule might serve as a complexing agent for Cl^- or Br^-?

Preparation of Ethers

11.15 Write equations to show a combination of reactants to prepare each ether. Which ethers can be prepared in good yield by a Williamson ether synthesis? If there are any that cannot be prepared by the Williamson method, explain why not.

(a) 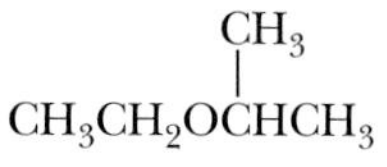 $CH_3CH_2OCHCH_3$ with CH_3 substituent

(b) 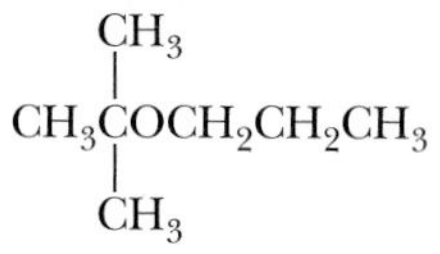 $CH_3COCH_2CH_2CH_3$ with CH_3 substituents

(c) 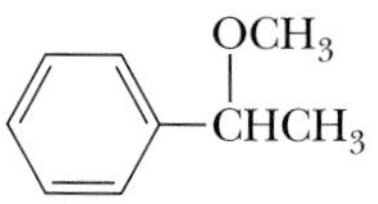 phenyl–$CHCH_3$ with OCH_3

(d) 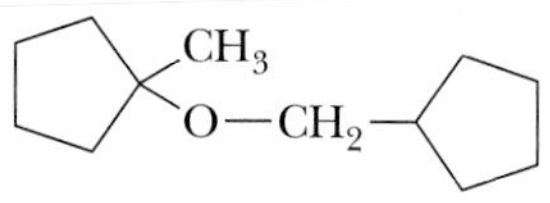 cyclopentane with CH_3 and $O-CH_2-$ cyclopentane

(e) 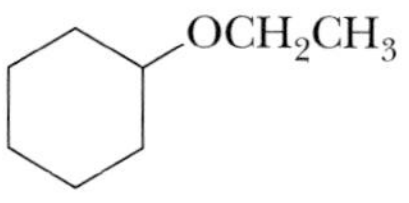 cyclohexane with OCH_2CH_3

(f) 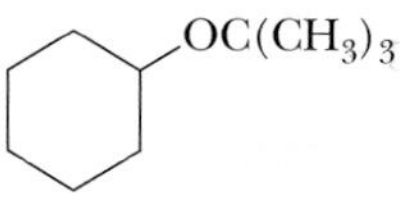 cyclohexane with $OC(CH_3)_3$

11.16 Propose a mechanism for this reaction.

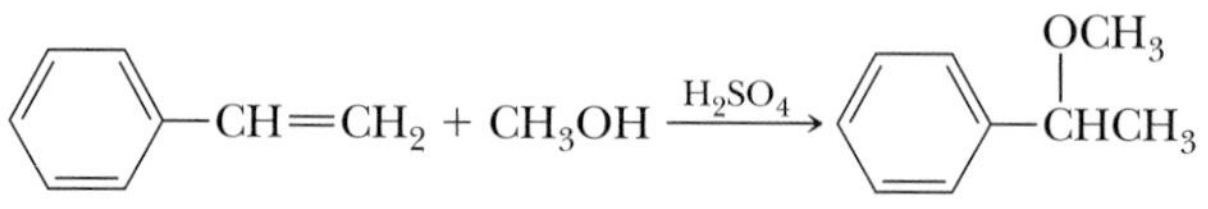

$$phenyl-CH=CH_2 + CH_3OH \xrightarrow{H_2SO_4} phenyl-CHCH_3 \text{ with } OCH_3$$

Reactions of Ethers

11.17 Draw structural formulas for the products formed when each compound is refluxed in concentrated HI.

(a) 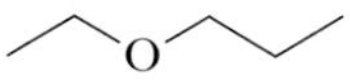 propyl–O–ethyl ether   (b) 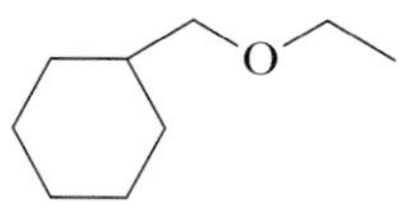 cyclohexyl–CH_2–O–ethyl   (c) 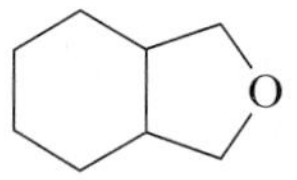 bicyclic ether with O   (d) 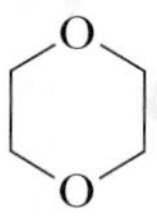 1,4-dioxane

11.18 Following is an equation for the reaction of diisopropyl ether and oxygen to form a hydroperoxide.

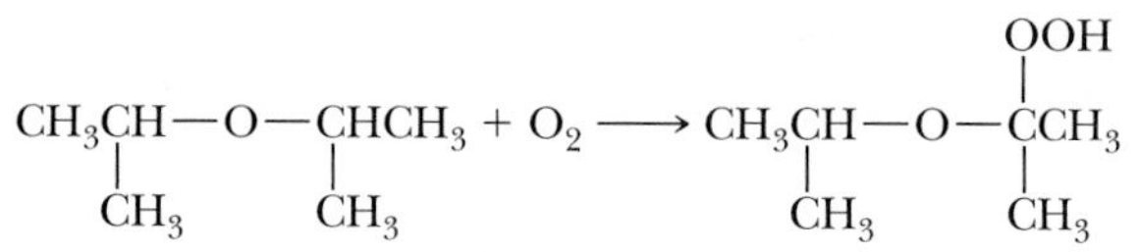

$$CH_3CH-O-CHCH_3 + O_2 \longrightarrow CH_3CH-O-\overset{OOH}{\underset{CH_3}{C}}CH_3$$

with CH_3 substituents on left; CH_3 on right

Diisopropyl ether   A hydroperoxide

Formation of an ether hydroperoxide is a radical chain reaction.

(a) Write a pair of chain propagation steps that accounts for the formation of this ether hydroperoxide. Assume that initiation is by a radical, $R \cdot$.

(b) Account for the fact that hydroperoxidation of ethers is regioselective; that is, it occurs preferentially at a carbon adjacent to the ether oxygen.

Synthesis and Reactions of Epoxides

11.19 Triethanolamine (TEA) is a widely used biological buffer, with maximum buffering capacity at pH 7.8. Propose a synthesis of this compound from ethylene oxide and ammonia. The structural formula of triethanolamine is $(HOCH_2CH_2)_3N$.

11.20 Ethylene oxide is the starting material for the synthesis of both Methyl Cellosolve and Cellosolve, two important industrial solvents. Propose a mechanism for these reactions.

$$H_2C\text{—}CH_2 + CH_3OH \xrightarrow{H_2SO_4} CH_3OCH_2CH_2OH$$
$$\underset{O}{\diagdown\diagup}$$

Oxirane   2-Methoxyethanol
(Ethylene oxide)   (Methyl Cellosolve)

$$H_2C\text{—}CH_2 + CH_3CH_2OH \xrightarrow{H_2SO_4} CH_3CH_2OCH_2CH_2OH$$
$$\underset{O}{\diagdown\diagup}$$

Oxirane   2-Ethoxyethanol
(Ethylene oxide)   (Cellosolve)

11.21 Ethylene oxide is the starting material for the synthesis of 1,4-dioxane. Propose a mechanism for each step in this synthesis.

1,4-Dioxane

11.22 Propose a synthesis for each ether starting with ethylene oxide and any readily available alcohols.

(a)   **(b)**

11.23 Propose a synthesis for 18-crown-6. If a base is used in your synthesis, does it make a difference whether it is lithium hydroxide or potassium hydroxide?

11.24 Predict the structural formula of the major product of the reaction of 2,2,3-trimethyloxirane with each set of reagents.

(a) $CH_3OH/CH_3O^-Na^+$ **(b)** CH_3OH/H^+

11.25 The following equation shows the reaction of *trans*-2,3-diphenyloxirane with hydrogen chloride in benzene to form 2-chloro-1,2-diphenylethanol.

$$C_6H_5\text{—}\underset{O}{\overset{H}{\underset{C\text{—}C}{}}}\text{—}\overset{C_6H_5}{\underset{H}{}} + HCl \longrightarrow C_6H_5CH\text{—}CHC_6H_5$$
$$\underset{HO}{|}\quad\underset{Cl}{|}$$

trans-2,3-Diphenyloxirane   2-Chloro-1,2-diphenylethanol

(a) How many stereoisomers are possible for 2-chloro-1,2-diphenylethanol?
(b) Given that opening of the epoxide ring in this reaction is stereoselective, predict which of the possible stereoisomers of 2-chloro-1,2-diphenylethanol is/are formed in the reaction.

11.26 Propose a mechanism to account for this rearrangement.

Tetramethyloxirane   3,3-Dimethyl-2-butanone

11.27 Following is the structural formula for an epoxide derived from 9-methyldecalin. Acid-catalyzed hydrolysis of this epoxide gives a trans diol. Of the two possible trans diols, only one is formed. How do you account for this stereospecificity?

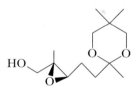

Only this glycol   This glycol is
is formed.   not formed.

11.28 Following are two reaction sequences for converting 1,2-diphenylethylene into 2,3-diphenyloxirane.

$$PhCH{=}CHPh \xrightarrow{RCO_3H} PhCH{-}CHPh$$

1,2-Diphenylethylene   2,3-Diphenyloxirane

$$PhCH{=}CHPh \xrightarrow[\text{2. } CH_3O^-Na^+]{\text{1. } Cl_2, H_2O} PhCH{-}CHPh$$

1,2-Diphenylethylene   2,3-Diphenyloxirane

Suppose that the starting alkene is *trans*-1,2-diphenylethylene.

(a) What is the configuration of the oxirane formed in each sequence?
(b) Does the oxirane formed in either sequence rotate the plane of polarized light? Explain.

11.29 The following enantiomer of a chiral epoxide is an intermediate in the synthesis of the insect pheromone frontalin. Show how this enantiomer can be prepared from an allylic alcohol precursor, using the Sharpless epoxidation.

11.30 Human white cells produce an enzyme called myeloperoxidase. This enzyme catalyzes the reaction between hydrogen peroxide and chloride ion to produce hypochlorous acid, HOCl, which reacts as if it were Cl^+OH^-. When attacked by white cells, cholesterol gives a chlorohydrin as the major product.

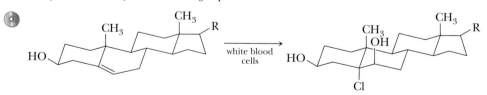

Cholesterol

(a) Propose a mechanism for this reaction. Account for both the regioselectivity and the stereoselectivity.

(b) On standing or (much more rapidly) on treatment with base, the chlorohydrin is converted to an epoxide. Show the structure of the epoxide and a mechanism for its formation.

11.31 Propose a mechanism for the following acid-catalyzed rearrangement.

Synthesis

11.32 Show reagents and experimental conditions to synthesize the following compounds from 1-propanol. Any derivative of 1-propanol prepared in an earlier part of this problem may then be used for a later synthesis.

(a) Propanal
(b) Propanoic acid
(c) Propene
(d) 2-Propanol
(e) 2-Bromopropane
(f) 1-Chloropropane
(g) 1,2-Dibromopropane
(h) Propyne
(i) 2-Propanone
(j) 1-Chloro-2-propanol
(k) Methyloxirane
(l) Dipropyl ether
(m) Isopropyl propyl ether
(n) 1-Mercapto-2-propanol
(o) 1-Amino-2-propanol
(p) 1,2-Propanediol

11.33 Starting with *cis*-3-hexene, show how to prepare the following.

(a) Meso 3,4-hexanediol   **(b)** Racemic 3,4-hexanediol

11.34 Show reagents and experimental conditions to convert cycloheptene to the following. Any compound made in an earlier part of this problem may be used as an intermediate in any following conversion.

(a)   **(b)** 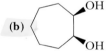   **(c)**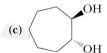

(d) 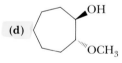   **(e)** 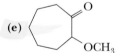   **(f)**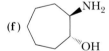

(g) 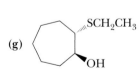   **(h)**   **(i)**

(j) 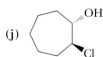   **(k)** 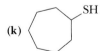   **(l)**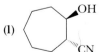

(m)   **(n)** HC(CH$_2$)$_5$CH with two C=O groups

interchapter "Problems from Medicinal Chemistry." Epichlorohydrin is synthesized industrially in the following series of three reactions.

Step 1:

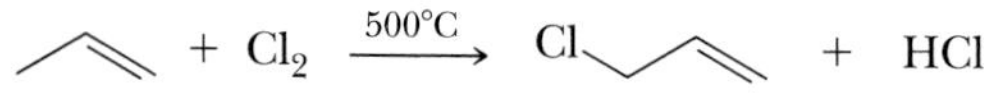

Propene   3-Chloropropene
(Allyl chloride)

Step 2:

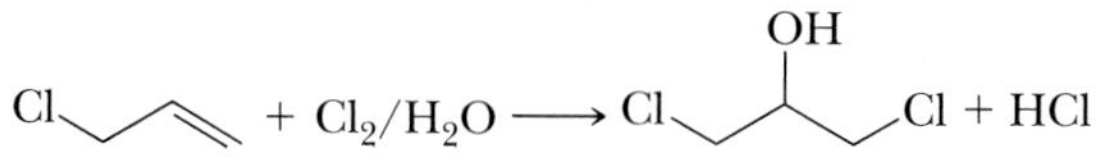

Step 3:

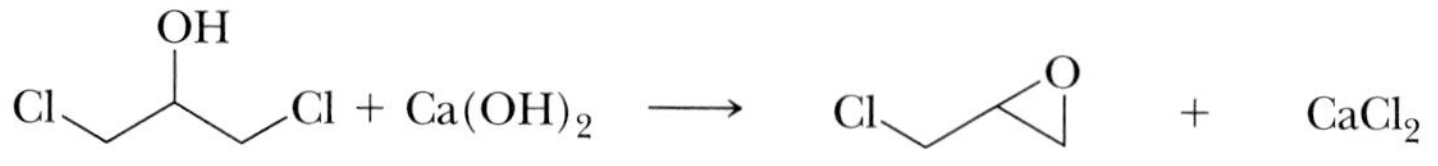

3-Chloro-1,2-epoxypropane
(Epichlorohydrin)

(a) Describe the type of mechanism by which each step occurs.
(b) Account for the regiospecificity of Steps 1 and 2.

INFRARED SPECTROSCOPY

Determination of molecular structure is one of the central themes of organic chemistry. For this purpose, chemists today rely almost exclusively on instrumental methods, four of which we discuss in this text. We begin in this chapter with **infrared spectroscopy.** Then, in Chapters 13 and 14, we introduce nuclear magnetic resonance (NMR) spectroscopy and mass spectrometry (MS). A brief introduction to ultra-violet (UV)-visible spectroscopy is contained in Chapter 23, which deals with conjugated systems.

■ A scientist working with a Fourier-Transform infrared spectroscopy instrument. *(Chris Taylor/CSIRO/Science Photo Library/Photo Researchers, Inc.)* Inset: A model of 3-methyl-2-butanone. For an IR spectrum of this compound, see Figure 12.2.

In this chapter, we first develop a basic understanding of the theory behind molecular spectroscopy, and then we concentrate on the interpretation of spectra and the information they can provide about details of molecular structure.

12.1 Electromagnetic Radiation

Electromagnetic radiation Light and other forms of radiant energy.

Gamma rays, x-rays, ultraviolet light, visible light, infrared radiation, microwaves, and radio waves are all types of **electromagnetic radiation.** Because electromagnetic radiation behaves as a wave traveling at the speed of light, it can be described in terms of its wavelength and its frequency. Wavelengths, frequencies, and energies of various regions of the electromagnetic spectrum are summarized in Table 12.1.

Wavelength (λ) The distance between consecutive peaks on a wave.

Wavelength is given the symbol λ (Greek lambda) and is usually expressed in the SI base unit of meters. Other derived units commonly used to express wavelength are given in Table 12.2. The wavelengths of visible light fall only in the range 400–700 nm. Infrared radiation (which can be felt as heat but not seen) falls in the range 2–15 μm.

Frequency The number of full cycles of a wave that pass a fixed point in a second; given the symbol ν (Greek nu) and reported in hertz (Hz).

Frequency, the number of full cycles of a wave that pass a given point in a second, is given the symbol ν (Greek nu) and is reported in **hertz (Hz),** which has the units s^{-1}. Wavelength and frequency are inversely proportional, and one can be calculated from the other using the following relationship:

$$\lambda\nu = c$$

Hertz (Hz) The unit in which frequency is measured; s^{-1} (read "per second").

where c is the velocity of light, 3.00×10^8 m/s. For example, consider the infrared radiation of wavelength 1.5×10^{-5} m (15 μm). The frequency of this radiation is 2.0×10^{13} Hz.

$$\nu = \frac{3 \times 10^8 \text{ m/s}}{1.5 \times 10^{-5} \text{ m}} = 2.0 \times 10^{13} \text{ Hz}$$

Table 12.1 **Wavelengths, Frequencies, and Energies of Some Regions of the Electromagnetic Spectrum**

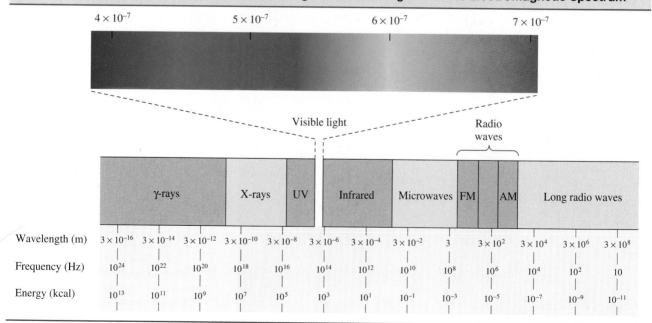

	γ-rays	X-rays	UV	Infrared	Microwaves	FM	AM	Long radio waves

Wavelength (m)	3×10^{-16}	3×10^{-14}	3×10^{-12}	3×10^{-10}	3×10^{-8}	3×10^{-6}	3×10^{-4}	3×10^{-2}	3	3×10^{2}	3×10^{4}	3×10^{6}	3×10^{8}
Frequency (Hz)	10^{24}	10^{22}	10^{20}	10^{18}	10^{16}	10^{14}	10^{12}	10^{10}	10^{8}	10^{6}	10^{4}	10^{2}	10
Energy (kcal)	10^{13}	10^{11}	10^{9}	10^{7}	10^{5}	10^{3}	10^{1}	10^{-1}	10^{-3}	10^{-5}	10^{-7}	10^{-9}	10^{-11}

Table 12.2 Common Units Used to Express Wavelength (λ)	
Unit	Relation to Meter
Meter (m)	—
Millimeter (mm)	$1 \text{ mm} = 10^{-3} \text{ m}$
Micrometer (μm)	$1 \, \mu\text{m} = 10^{-6} \text{ m}$
Nanometer (nm)	$1 \text{ nm} = 10^{-9} \text{ m}$
Angstrom (Å)	$1 \text{ Å} = 10^{-10} \text{ m}$

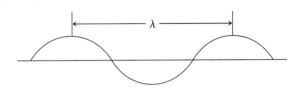

An alternative way to describe electromagnetic radiation is in terms of its properties as a stream of particles called **photons.** The energy in a mole of photons is related to the frequency of the radiation by the equations

$$E = h\nu = h\frac{c}{\lambda}$$

where E is the energy in kilojoules (kilocalories) per mole and h is Planck's constant, 3.99×10^{-13} kJ (9.537×10^{-14} kcal)s·mol^{-1}.

Example 12.1

Calculate the energy in kilojoules per mole of radiation of wavelength 2.50 μm. What type of radiant energy is this? (Refer to Table 12.1.)

Solution

Make certain that dimensions for distance are consistent; if the dimension of wavelength is meters, then express the velocity of light in meters per second. First, convert 2.50 μm to meters.

$$2.50 \, \mu\text{m} \times \frac{10^{-6} \text{ m}}{1 \, \mu\text{m}} = 2.50 \times 10^{-6} \text{ m}$$

Now substitute this value in the equation $E = hc/\lambda$.

$$E = h\frac{c}{\lambda} = 3.99 \times 10^{-13} \frac{\text{kJ} \cdot \text{s}}{\text{mol}} \times 3.00 \times 10^{8} \frac{\text{m}}{\text{s}} \times \frac{1}{2.50 \times 10^{-6} \text{ m}}$$

$$= 47.9 \text{ kJ } (11.45 \text{ kcal})/\text{mol}$$

Electromagnetic radiation with energy of 47.9 kJ/mol corresponds to radiation in the infrared region.

Problem 12.1

Calculate the energy of red light (680 nm) in kilojoules per mole. Which form of radiation carries more energy, infrared radiation of wavelength 2.50 μm or red light of wavelength 680 nm?

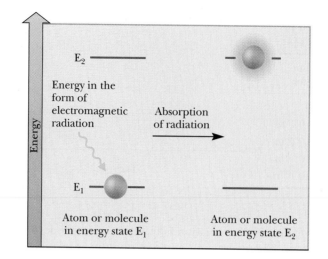

Figure 12.1
Absorption of energy in the form of electromagnetic radiation causes an atom or molecule in energy state E_1 to change to a higher energy state E_2.

12.2 Molecular Spectroscopy

Organic molecules are flexible. As we discussed in Chapter 2, atoms and groups of atoms can rotate about single covalent bonds. In addition, covalent bonds can stretch and bend just as if their atoms were joined by flexible springs. Furthermore, electrons within molecules can move from one electronic energy level to another, for example by promotion of an electron from a pi-bonding molecular orbital to a pi-antibonding molecular orbital. Finally, certain nuclei behave as if they are spinning magnetic particles and can change from one nuclear spin energy level to another.

An atom or molecule can be made to undergo a transition from energy state E_1 to a higher energy state E_2 by irradiating it with electromagnetic radiation corresponding to the energy difference between states E_1 and E_2 as illustrated schematically in Figure 12.1. When the atom or molecule returns from E_2 to the ground state E_1, an equivalent amount of energy is emitted.

Molecular spectroscopy is the experimental process of measuring which frequencies of radiation are absorbed or emitted by a particular substance and then attempt-

Molecular spectroscopy The study of which frequencies of electromagnetic radiation are absorbed or emitted by substances and the correlation between these frequencies and specific types of molecular structure.

Table 12.3 **Types of Energy Transitions Resulting from Absorption of Energy from Three Regions of the Electromagnetic Spectrum**

Region of Electromagnetic Spectrum	Frequency (hertz)	Type of Spectroscopy	Absorption of Electromagnetic Radiation Results in Transitions Between
Radio frequency	$3 \times 10^7 - 9 \times 10^8$	Nuclear magnetic resonance	Nuclear spin levels
Infrared	$1 \times 10^{13} - 1 \times 10^{14}$	Infrared	Vibrational energy levels
Ultraviolet-visible	$2.5 \times 10^{14} - 1.5 \times 10^{15}$	Ultraviolet-visible	Electronic energy levels

ing to correlate patterns of energy absorption or emission with details of molecular structure. The regions of the electromagnetic spectrum of most interest to us and the relationships of each to changes in atomic and molecular energy levels are summarized in Table 12.3.

12.3   Infrared Spectroscopy

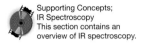

Supporting Concepts;
IR Spectroscopy
This section contains an overview of IR spectroscopy.

A.   The Vibrational Infrared Spectrum

The **vibrational infrared region,** which extends from 2.5×10^{-6} to 2.5×10^{-5} m, is used for **infrared spectroscopy.** Radiation in this region is most commonly referred to by its **wavenumber,** the number of waves per centimeter, with units cm^{-1} (read "reciprocal centimeters"). The frequency in wavenumbers is the reciprocal of the wavelength in centimeters, or the frequency (ν) in hertz divided by c.

$$\bar{\nu} = \frac{1}{\lambda \text{ (cm)}} = \frac{10^{-2} \text{ (m} \cdot \text{cm}^{-1})}{\lambda \text{ (m)}} = \frac{\nu}{c}$$

When expressed in frequencies, the vibrational region of the infrared spectrum extends from 4000 to 400 cm^{-1}.

$$\bar{\nu} = \frac{10^{-2} \text{ m} \cdot \text{cm}^{-1}}{2.5 \times 10^{-6} \text{ m}} = 4000 \text{ cm}^{-1} \qquad \bar{\nu} = \frac{10^{-2} \text{ m} \cdot \text{cm}^{-1}}{2.5 \times 10^{-5} \text{ m}} = 400 \text{ cm}^{-1}$$

An advantage of using frequencies is that they are directly proportional to energy; the higher the frequency, the higher the energy of the radiation.

Shown in Figure 12.2 is an infrared spectrum of 3-methyl-2-butanone. The horizontal axis at the bottom of the chart paper is calibrated in frequencies (cm^{-1}); that at the top is calibrated in wavelength (micrometers, μm). The frequency scale is often divided into two or more regions. For all spectra reproduced in this text, it is divided into three linear regions: 4000–2200 cm^{-1}, 2200–1000 cm^{-1}, and 1000–400 cm^{-1}. The vertical axis measures transmittance (the fraction of light transmitted), with 100% at the top and 0% at the bottom. Thus, the baseline for an infrared

Vibrational infrared region The portion of the infrared region that extends from 4000 to 400 cm^{-1}.

Infrared spectroscopy A spectroscopic technique in which a compound is irradiated with infrared radiation, absorption of which causes covalent bonds to change from a lower vibrational energy level to a higher one. Infrared (IR) spectroscopy is particularly valuable for determining the kinds of functional groups present in an organic molecule.

Wavenumber ($\bar{\nu}$) The frequency of electromagnetic radiation expressed as the number of waves per centimeter.

Figure 12.2
Infrared spectrum of 3-methyl-2-butanone.

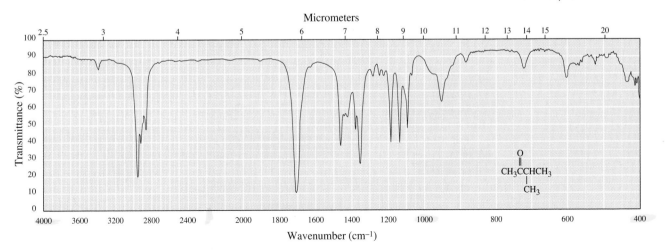

spectrum (100% transmittance of radiation through the sample, 0% absorption) is at the top of the chart paper, and absorption of radiation corresponds to a trough or valley. Strange as it may seem, we commonly refer to infrared absorptions as peaks, even though they are conventionally displayed pointing downward.

The spectrum in Figure 12.2 was recorded on a **neat** sample, which means the pure liquid. A few drops of the liquid are placed between two sodium chloride discs and spread to give a thin film through which infrared radiation is then passed. Liquid and solid samples may also be dissolved in carbon tetrachloride or another solvent with minimal infrared absorption and analyzed in a liquid sampling cell. Still another way to obtain the infrared spectrum of a solid is to mix it with potassium bromide and then compress the mixture under high pressure to give a thin transparent wafer, which is placed in the beam of the spectrophotometer. Both NaCl and KBr are transparent to infrared radiation and can be used as windows and optics for infrared spectroscopy. These materials are very easily damaged by moisture, however! Infrared spectra of gas samples are determined using specially constructed gas-handling cells.

Example 12.2

Some infrared spectrophotometers are calibrated to record spectra on an ordinate that is linear in wavelength (μm), whereas others record them on an ordinate that is linear in frequencies (cm^{-1}). Carry out the following conversions (note the convenient formula for converting between μm and cm^{-1}: $\bar{\nu}\lambda = 10^4$).

(a) 7.05 μm to cm^{-1} **(b)** 3.35 μm to cm^{-1} **(c)** 3280 cm^{-1} to μm

Solution

(a) 1418 cm^{-1} **(b)** 2985 cm^{-1} **(c)** 3.05 μm

Problem 12.2

Which is higher in energy?

(a) Infrared radiation of 1715 cm^{-1} or of 2800 cm^{-1}?
(b) Radio-frequency radiation of 300 MHz or of 60 MHz?

B. Molecular Vibrations

Atoms joined by covalent bonds are not permanently fixed in one position but rather undergo continual vibrations relative to each other. The energies associated with these vibrations are quantized, which means that, within a molecule, only specific vibrational energy levels are allowed. The energies associated with transitions between vibrational energy levels in most covalent molecules correspond to frequencies in the infrared region, 4000–400 cm^{-1}.

For a molecule to absorb this radiation, the bond undergoing vibration must be polar (Section 1.2C), and its vibration must cause a periodic change in the bond dipole moment. If two charges are connected by a spring, a change in distance between the charges corresponds to a change in dipole moment. Any vibration that meets these criteria is said to be **infrared active.** The greater the polarity of the bond, the more intense the absorption. Covalent bonds that do not meet these criteria are said

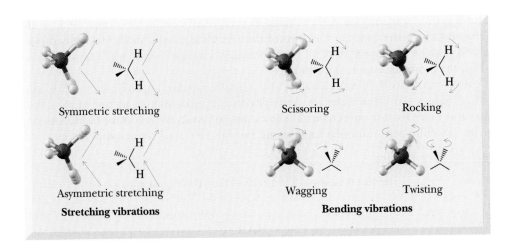

Symmetric stretching

Scissoring

Rocking

Asymmetric stretching

Wagging

Twisting

Stretching vibrations

Bending vibrations

Figure 12.3

Fundamental modes of vibration for a methylene group.

to be **infrared inactive.** The carbon-carbon double and triple bonds in symmetrically substituted alkenes and alkynes, for example, do not absorb infrared radiation because they are not polar bonds.

2,3-Dimethyl-2-butene

2-Butyne

For a nonlinear molecule containing n atoms, $3n - 6$ allowed fundamental vibrations exist. For a molecule as simple as ethanol, CH_3CH_2OH, there are 21 fundamental vibrations, and for hexanoic acid, $CH_3(CH_2)_4COOH$, there are 54. Thus, for even relatively simple molecules, a large number of vibrational energy levels exist, and the patterns of energy absorption for these and larger molecules are very complex.

The simplest vibrational motions in molecules giving rise to absorption of infrared radiation are stretching and bending motions. Illustrated in Figure 12.3 are the fundamental stretching and bending vibrations for a methylene group.

C. Characteristic Absorption Patterns

Analysis of the modes of vibration for a molecule is very complex because all the atoms contribute to the vibrational modes. However, we can make useful generalizations about where absorptions due to particular vibrational modes will appear in the infrared spectrum by considering each individual bond and ignoring other bonds in the molecule. As a simplifying assumption, let us consider two covalently bonded atoms as two vibrating masses connected by a spring. The total energy is proportional to the frequency of vibration. The frequency of a stretching vibration is given by the following equation, which is derived from Hooke's law for a vibrating spring.

$$\bar{\nu} = 4.12 \sqrt{\frac{K}{\mu}}$$

where $\bar{\nu}$ is the frequency of the vibration in reciprocal centimeters (cm^{-1}); K is the force constant of the bond, a measure of the bond's strength, in dynes per centimeter; and μ is the "reduced mass" of the two atoms, $(m_1 m_2)/(m_1 + m_2)$, where m is the mass of the atoms in atomic mass units.

Force constants for single, double, and triple bonds are approximately 5, 10, and 15×10^5 dynes per centimeter, respectively, thus approximately in the ratio $1:2:3$. Using the value for the force constant for a single bond, we calculate the frequency for the stretching vibration of a single bond between ^{12}C and ^1H as follows:

For ^{12}C—^1H stretching:

$$\text{Reduced mass} = 12 \times 1/(12 + 1) = 0.923 \text{ g/atom}$$

And

$$\bar{\nu} = 4.12 \sqrt{\frac{K}{\mu}} = 4.12 \sqrt{\frac{5 \times 10^5}{0.923}} = 3032 \text{ cm}^{-1}$$

The experimentally determined value for the frequency of an alkyl C—H stretching vibration is approximately 2900 cm^{-1}. Given the simplifying assumptions made in this calculation and the fact that the value of the force constant for a single bond is an average value, the agreement between the calculated value and the experimental value is remarkably good. Although frequencies calculated in this manner can be remarkably accurate, they are generally not accurate enough for precise determination of molecular structure.

Hooke's law predicts that the *position* of absorption of a stretching vibration in an IR spectrum depends both on the strength of the vibrating bond and on the masses of the atoms connected by the bond. The stronger the bond and the lighter the atoms, the higher the frequency of the stretching vibration. As we saw earlier, the *intensity* of the absorption depends primarily on the polarity of the vibrating bond.

Example 12.3

Calculate the stretching frequency in reciprocal centimeters for a carbon-carbon double bond. Assume that each carbon is the most abundant isotope, namely ^{12}C.

Solution

Assume a force constant of 10×10^5 dynes per centimeter for C=C. The calculated frequency is 1682 cm^{-1}, a value close to the experimental value of 1630 cm^{-1}.

$$\bar{\nu} = \sqrt{\frac{10 \times 10^5}{12 \times 12/(12 + 12)}} = 1682 \text{ cm}^{-1}$$

Problem 12.3

Without doing the calculation, which member of each pair do you expect to occur at the higher frequency?

(a) C=O or C=C stretching   **(b)** C=O or C—O stretching
(c) C≡C or C=O stretching   **(d)** C—H or C—Cl stretching

Detailed interpretation of most infrared spectra is difficult because of the complexity of vibrational modes. In addition to the fundamental vibrational modes we have described, other types of absorptions occur, resulting in so-called overtone and coupling peaks that are usually quite weak. Only a fraction of the peaks in an infrared spectrum are useful for interpretation.

To one skilled in the interpretation of infrared spectra, the absorption patterns can yield an enormous amount of information about chemical structure. However, we have neither the time nor the need to develop this level of competence. The value of infrared spectra for us is that they can be used to determine the presence or absence of certain functional groups. A carbonyl group, for example, typically shows strong absorption at approximately $1630-1820$ cm^{-1}. The position of absorption for a particular carbonyl group depends on whether it is an aldehyde, a ketone, a carboxylic acid, or an ester and, if in a ring, on the size of the ring. In this chapter, we discuss how structural variations, such as ring size or other factors, affect this value.

D. Correlation Tables

Data on absorption patterns of functional groups are collected in tables called **correlation tables.** Table 12.4 is a listing of infrared absorptions for the types of bonds and functional groups we deal with most often. With each new functional group introduced in the following chapters, we also present more detailed correlation tables for that functional group. A cumulative correlation table can be found in Appendix 6. In these tables, the intensity of a particular absorption is often referred to as strong (s), medium (m), or weak (w). In general, bonds between C and O where the electronegativity is largest have the largest dipole moments and tend to give the strongest infrared absorptions.

In general, we pay most attention to the region from 3500 to 1000 cm^{-1} because the stretching and bending vibrations for most functional groups are found in this region. Vibrations in the region 1000 to 400 cm^{-1} are much more complex and far more difficult to analyze. Because even slight variations in molecular structure and absorption patterns are most obvious in this region, it is often called the **fingerprint region.** If two compounds have even slightly different structures, the differences in their infrared spectra are most clearly discernible in this region.

Fingerprint region The portion of the vibrational infrared region that extends from 1000 to 400 cm^{-1}.

Table 12.4	Infrared Stretching Absorptions of Selected Functional Groups	
Bond	**Frequency (cm^{-1})**	**Intensity**
O—H	3200–3650	Weak to strong (strongest when H-bonded)
N—H	3100–3550	Medium
C—H	2700–3300	Weak to medium
C=C	1600–1680	Weak to medium
C=O	1630–1820	Strong
C—O	1000–1250	Strong

Example 12.4

What functional group is most likely present if a compound shows IR absorption at these frequencies?

(a) 1705 cm^{-1} **(b)** 2950 cm^{-1}

Solution

(a) A C=O group   **(b)** An aliphatic C—H group

Problem 12.4

A compound shows strong, very broad IR absorption in the region 3300–3600 cm^{-1} and strong, sharp absorption at 1715 cm^{-1}. What functional group accounts for both of these absorptions?

Example 12.5

Propanone and 2-propen-1-ol are constitutional isomers. Show how to distinguish between them by IR spectroscopy.

$$
\underset{\substack{\text{Propanone}\\\text{(Acetone)}}}{CH_3-\overset{\displaystyle O}{\overset{\|}{C}}-CH_3}
\qquad
\underset{\substack{\text{2-Propen-1-ol}\\\text{(Allyl alcohol)}}}{CH_2=CH-CH_2-OH}
$$

Solution

Only propanone shows strong absorption in the C=O stretching region, 1630–1820 cm^{-1}. Alternatively, only 2-propen-1-ol shows strong absorption in the O—H stretching region, 3200–3650 cm^{-1}.

Problem 12.5

Propanoic acid and methyl ethanoate are constitutional isomers. Show how to distinguish between them by IR spectroscopy.

$$
\underset{\substack{\text{Propanoic acid}}}{CH_3CH_2\overset{\displaystyle O}{\overset{\|}{C}}OH}
\qquad
\underset{\substack{\text{Methyl ethanoate}\\\text{(Methyl acetate)}}}{CH_3\overset{\displaystyle O}{\overset{\|}{C}}OCH_3}
$$

12.4   Interpreting Infrared Spectra

A.   Alkanes

Infrared spectra of alkanes are usually simple with few peaks, the most common of which are given in Table 12.5.

Shown in Figure 12.4 is an infrared spectrum of decane. The strong peak with multiple splitting between 2850 and 3000 cm^{-1} is characteristic of alkane C—H

Table 12.5   Infrared Absorptions of Alkanes, Alkenes, and Alkynes

Hydrocarbon	Vibration	Frequency (cm^{-1})	Intensity
Alkane			
C—H	Stretching	2850–3000	Medium
CH_2	Bending	1450	Medium
CH_3	Bending	1375 and 1450	Weak to medium
C—C	(Not useful for interpretation—too many bands)		
Alkene			
C—H	Stretching	3000–3100	Weak to medium
C=C	Stretching	1600–1680	Weak to medium
Alkyne			
C—H	Stretching	3300	Medium to strong
C≡C	Stretching	2100–2250	Weak

stretching; it is strong in this spectrum because there are so many C—H bonds and no other functional groups. The other prominent peaks are methylene bending absorption at 1465 cm^{-1} and methyl bending absorption at 1380 cm^{-1}.

B.   Alkenes

An easily recognized alkene absorption is the vinylic C—H stretching slightly to the left of 3000 cm^{-1} (i.e., higher frequency). Also characteristic of alkenes is C=C stretching at 1600–1680 cm^{-1}. This vibration, however, is often weak and difficult to observe: the more symmetrical the alkene, the weaker the absorption.

Both vinylic C—H stretching and C=C stretching can be seen in the infrared spectrum of cyclohexene (Figure 12.5). Also visible are the aliphatic C—H stretching near 2900 cm^{-1} and methylene bending near 1440 cm^{-1}.

Figure 12.4
Infrared spectrum of decane (neat, salt plates).

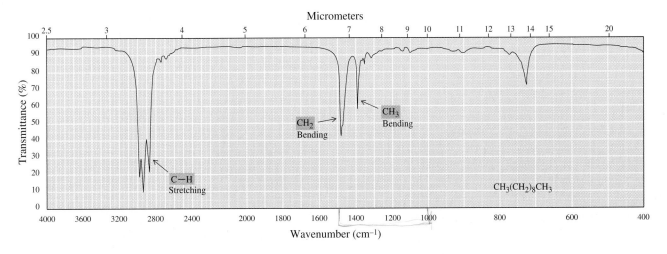

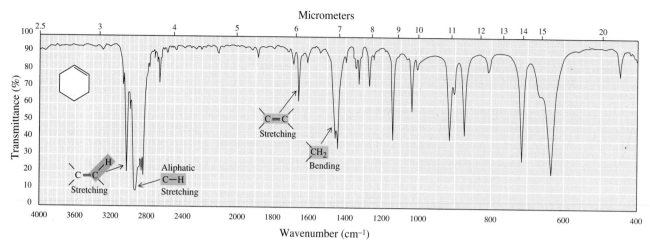

Figure 12.5

Infrared spectrum of cyclo-hexene (neat, salt plates).

C. Alkynes

Alkyne ≡C—H stretching occurs near 3300 cm^{-1}, at higher frequency than for either alkyl C—H or vinylic C—H. This peak is usually sharp and strong. The strong *sp* CH bond has a higher force constant than for alkenes, which in turn absorb at higher frequency than the even weaker alkane (*sp^3*) C—H bonds. Also observed in terminal alkynes is absorption near 2150 cm^{-1} due to C≡C stretching. Both of these peaks can be seen in the infrared spectrum of 1-octyne (Figure 12.6). For internal alkynes, the C≡C stretching absorption is often very weak or completely absent because stretching of this bond results in little or no change in the bond dipole moment (Section 12.3B).

D. Aromatics (Benzene and Its Derivatives)

Aromatic rings show a medium to weak peak in the C—H stretching region at approximately 3030 cm^{-1} characteristic of *sp^2* C—H bonds. In addition, aromatic rings

Figure 12.6

Infrared spectrum of 1-octyne.

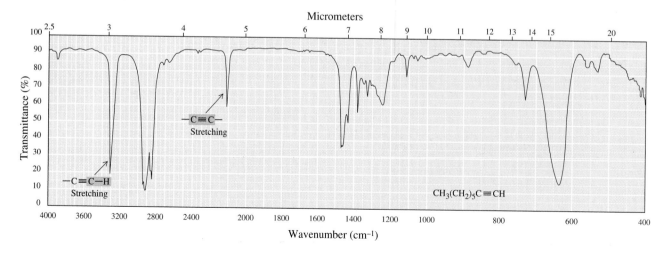

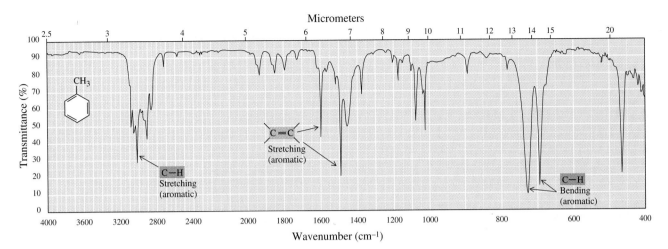

Figure 12.7
Infrared spectrum of toluene.

show strong absorption in the region 690–900 cm^{-1} due to out-of-plane C—H bending. Finally, these compounds show several absorptions due to C=C stretching between 1450 and 1600 cm^{-1}. Actually, these are complex vibrational modes of the entire ring; some involve all atoms moving in and out (breathing), whereas others involve some atoms moving in and others moving out. The intensities of these peaks can vary depending on the symmetry of ring substitution patterns. Each of these characteristic absorption patterns can be seen in the infrared spectrum of toluene (Figure 12.7).

E. Alcohols

Both the position of the O—H stretching absorption and its intensity depend on the extent of hydrogen bonding. Under normal conditions (in concentrated solutions or without solvent), where there is extensive hydrogen bonding between alcohol molecules, O—H stretching absorption occurs as a broad peak at 3200–3500 cm^{-1}. The variety of hydrogen-bonded states is responsible for this broadening. The "free" O—H stretch near 3650 cm^{-1} is seen only in very dilute solution in non-hydrogen-bonding solvents. The C—O stretching absorption appears in the range 1000–1250 cm^{-1} (Table 12.6).

Shown in Figure 12.8 is an infrared spectrum of neat 1-hexanol. The hydrogen-bonded O—H stretching appears as a broad band of medium intensity centered at 3340 cm^{-1}. The C—O stretching appears at 1058 cm^{-1}, a value characteristic of primary alcohols.

Table 12.6	**Infrared Absorptions of Alcohols**	
Bond	**Frequency (cm^{-1})**	**Intensity**
O—H (free)	3600–3650	Weak
O—H (hydrogen bonded)	3200–3500	Medium, broad
C—O	1000–1250	Medium

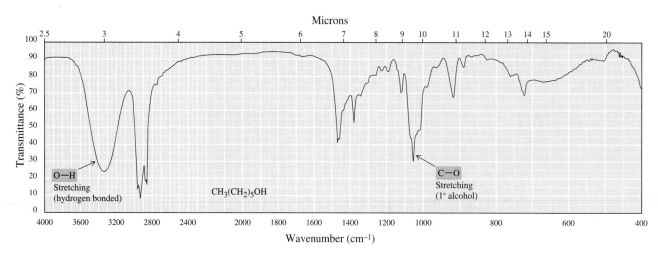

Figure 12.8
Infrared spectrum of 1-hexanol (neat, salt plates).

F. Ethers

Ethers have an oxygen atom bonded to two carbon atoms. Either or both of the carbon atoms may be sp^3 hybridized, sp^2 hybridized, or sp hybridized. In the simplest ether, dimethyl ether, both carbons are sp^3 hybridized. In diphenyl ether, both carbons are sp^2 hybridized, and in ethyl vinyl ether, one carbon is sp^3 hybridized and the other is sp^2 hybridized.

$$CH_3CH_2-O-CH_2CH_3 \qquad CH_3CH_2-O-CH=CH_2$$

Ethoxyethane   Ethoxyethene   Diphenyl ether
(Diethyl ether)   (Ethyl vinyl ether)

The C—O stretching absorptions of ethers are similar to those observed in alcohols. Dialkyl ethers typically show a single absorption in the region between 1000 and 1100 cm^{-1} as can be seen in the infrared spectrum of dibutyl ether (Figure 12.9).

The presence or absence of O—H stretching at 3200 to 3500 cm^{-1} for hydrogen-bonded O—H can be used to distinguish between an ether and an isomeric alcohol.

Figure 12.9
Infrared spectrum of dibutyl ether.

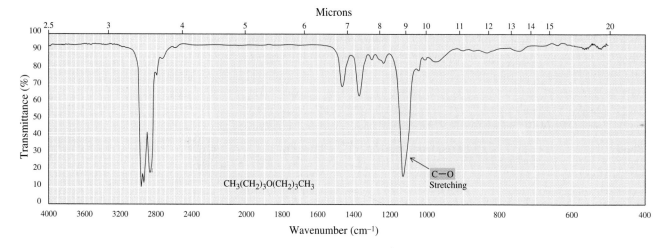

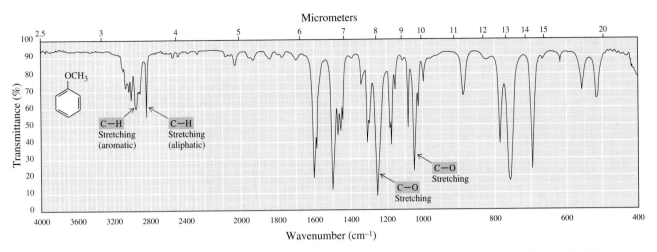

Figure 12.10

Infrared spectrum of anisole.

A C—O stretching absorption is also present in esters (see Section 12.4J). In this case, the presence or absence of C=O stretching can be used to distinguish between an ether and an ester.

Aromatic and vinyl ethers typically show two C—O stretching vibrations, one at either end of the range for C—O stretching. Anisole (Figure 12.10), for example, shows C—O stretching vibrations at 1050 cm^{-1} (sp^3 C—O) and 1250 cm^{-1} (sp^2 C—O). Ethers in which one of the bonds is attached to an sp^2 hybridized carbon typically also have a band in the region between 1200–1250 cm^{-1}.

G. Amines

The most important and readily observed infrared absorptions of primary and secondary amines are due to N—H stretching vibrations and appear in the region 3300–3500 cm^{-1}. Like O—H bonds, N—H bonds become broader and shift to longer wavelength on hydrogen bonding. Primary amines have two bands in this region: one due to symmetric stretching and the other due to asymmetric stretching. The two N—H stretching absorptions characteristic of a primary amine can be seen in the IR spectrum of 1-butanamine (Figure 12.11). Secondary amines give only one

Figure 12.11

Infrared spectrum of 1-butanamine, a primary amine.

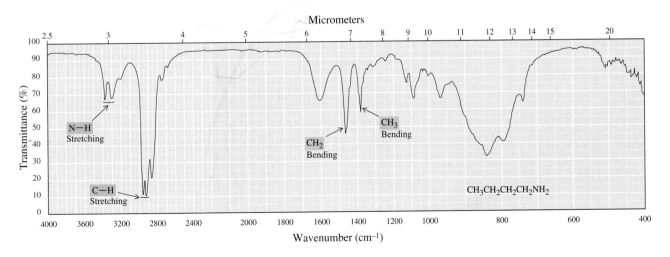

absorption in this region. Tertiary amines have no N—H and, therefore, are transparent in this region of the infrared spectrum.

H.   Aldehydes and Ketones

Aldehydes and ketones show characteristic strong infrared absorption between 1630 and 1820 cm^{-1} associated with the stretching vibration of the carbon-oxygen double bond. The stretching vibration for the carbonyl group of menthone occurs at 1705 cm^{-1} (Figure 12.12).

Because few other bond vibrations absorb energy between 1630 and 1820 cm^{-1}, absorption in this region of the spectrum is a reliable means for confirming the presence of a carbonyl group. Because several different functional groups contain a carbonyl group, it is often not possible to tell from absorption in this region alone whether the carbonyl-containing compound is an aldehyde, a ketone, a carboxylic acid (Section 12.4I), an ester (Section 12.4J), an amide, or an anhydride. However, other absorptions such as the C—O stretch in esters can help distinguish these groups. Aldehydes frequently have a weak but very distinctive absorption at 2720 cm^{-1} due to stretching of the C—H of the CHO group.

The position of the C=O stretching vibration is quite sensitive to the molecular environment of the carbonyl group, as illustrated by comparison of these cycloalkanones. Cyclohexanone and larger cyclic ketones in which there is little or no angle strain show absorption near 1715 cm^{-1}. As ring size decreases and angle strain increases, the C=O absorption shifts to a higher frequency as shown in the following series.

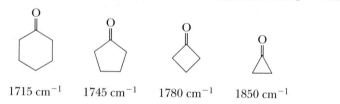

$$1715 \text{ cm}^{-1} \qquad 1745 \text{ cm}^{-1} \qquad 1780 \text{ cm}^{-1} \qquad 1850 \text{ cm}^{-1}$$

Conjugation   A situation in which two multiple bonds are separated by a single bond. Alternatively, a series of overlapping p orbitals. 1,3-Butadiene, for example, is a conjugated diene, and 3-buten-2-one is a conjugated enone.

The presence of a carbon-carbon double bond or benzene ring in **conjugation** with the carbonyl group results in a shift of the C=O absorption to a lower frequency, as seen by comparing the carbonyl stretching frequencies of the following molecules.

Figure 12.12

Infrared spectrum of menthone.

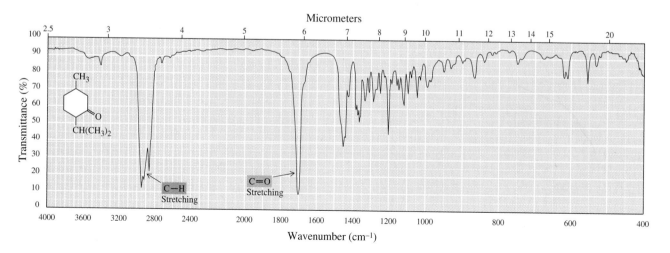

$$1717 \text{ cm}^{-1} \qquad 1690 \text{ cm}^{-1} \qquad 1700 \text{ cm}^{-1}$$

I. Carboxylic Acids

A carboxyl group gives rise to two characteristic absorptions in the infrared spectrum. One of these occurs in the region $1700-1725 \text{ cm}^{-1}$ and is associated with the stretching vibration of the carbonyl group. This is essentially the same range of absorption as for the carbonyl group of aldehydes and ketones, but it is usually broader in the case of the carboxyl carbonyl because of intermolecular hydrogen bonding. The other infrared absorption characteristic of a carboxyl group is a peak between 2500 and 3300 cm^{-1} due to the stretching vibration of the O—H group, which often overlaps the C—H stretching absorptions. This absorption is generally very broad due to hydrogen bonding between molecules of the carboxylic acid. Both C=O and O—H stretches can be seen in the infrared spectrum of pentanoic acid in Figure 12.13.

J. Carboxylic Esters

The functional group of a carboxylic ester, most commonly referred to as simply an ester, is a carbonyl group bonded to an —OR group.

$$\text{CH}_3\overset{\displaystyle O}{\overset{\|}{\text{C}}}-\text{OCH}_2\text{CH}_3 \qquad \text{CH}_3(\text{CH}_2)_3\overset{\displaystyle O}{\overset{\|}{\text{C}}}-\text{OCH(CH}_3)_2$$

Ethyl ethanoate   Isopropyl pentanoate
(Ethyl acetate)

We discussed the structure of esters in Section 1.3E and will discuss their nomenclature in detail in Chapter 18. At this point in the course, all you need be concerned with is the fact that a carboxylic ester contains both a C=O group and a C—O—C

Figure 12.13
Infrared spectrum of pentanoic acid.

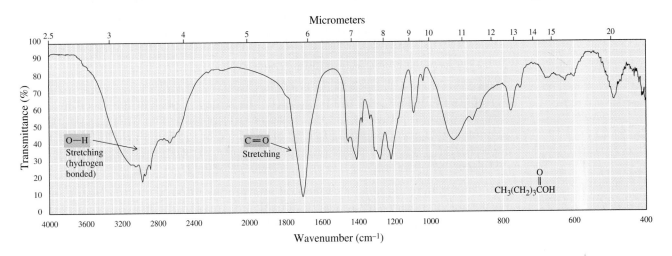

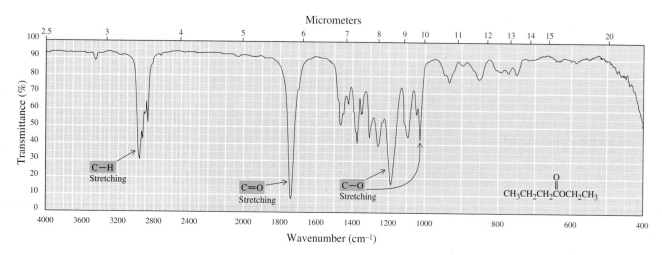

Figure 12.14
Infrared spectrum of ethyl butanoate.

group. Note that one of the carbons of the C—O—C group is sp^2 hybridized and the other is sp^3 hybridized. Esters display strong C=O stretching absorption in the region between 1735 and 1800 cm^{-1}. As in ketones, this band is shifted to higher frequency in smaller rings, and to lower frequency by conjugation. In addition, esters also display strong C—O stretching absorptions in the region 1000–1100 cm^{-1} for the sp^3 C—O stretch and 1200–1250 cm^{-1} for the sp^2 C—O stretch (Figure 12.14). Ethers in which one of the carbons attached to oxygen is sp^2 hybridized also show this band.

12.5 Solving Infrared Spectral Problems

The following steps may prove helpful as a systematic approach to solving IR problems.

Step 1: Check the region around 3000 cm^{-1}; absorption is due to C—H stretching. Absorption is generally to the right of 3000 cm^{-1} for sp^3 C—H stretching of alkanes and to the left for the sp^2 C—H stretching of alkenes and aromatic rings.

Step 2: Is there a strong, broad band in the region of 3500 cm^{-1}? If yes, then the molecule contains an —OH group either of an alcohol or a carboxylic acid. If there is no absorption around 1700 cm^{-1} (due to a carbonyl group), then the functional group is a hydroxyl group. If there is a peak around 1700 cm^{-1}, then the functional group may be a carboxyl group. One or two peaks in the 3500 cm^{-1} region may indicate the presence of a 2° or 1° amine, respectively.

Step 3: Is there a sharp peak in the region 1630–1820 cm^{-1}? If yes, there is a C=O group present. This peak will probably be the strongest peak in the spectrum. If no peak is present in this region, there is no C=O present. The type of carbonyl-containing functional group can often be determined by looking for the presence or absence of an aldehyde C—H stretch at 2720 cm^{-1}, a carboxyl O—H stretch around 3500 cm^{-1}, and a C—O stretch around 1000–1250 cm^{-1}.

Finally, here is a note of caution on interpreting infrared spectra. Even though it is possible to obtain a great deal of valuable information about a compound from its infrared spectrum, it is often very difficult to determine its structure based solely on this information.

Summary

The **vibrational infrared** spectrum (Section 12.3A) extends from 4000 to 400 cm^{-1}. To be **infrared active** (Section 12.3B), a bond must be polar, and its vibration must change the dipole moment of the bond. There are $3n - 6$ allowed fundamental vibrations for a nonlinear molecule containing n atoms. The simplest vibrations that give rise to absorption of infrared radiation are stretching and bending vibrations. Stretching may be symmetrical or asymmetrical. Scissoring, rocking, wagging, and twisting are names given to types of bending vibrations.

The frequency of vibration for an infrared-active bond can be derived from Hooke's law for the vibration of a simple

harmonic oscillator (Section 12.3C). Hooke's law predicts that the frequency of vibration increases when (1) the bond strength increases and (2) the reduced mass of the vibrating system decreases.

A **correlation table** (Section 12.3D) is a list of the absorption patterns of functional groups. The intensity of a peak is referred to as strong (s), medium (m), or weak (w). Bending and stretching vibrations for most functional groups appear in the region 3650–1000 cm^{-1}. The region 1000–400 cm^{-1} is called the **fingerprint region.**

Problems

Infrared Spectra

12.6 Following are infrared spectra of methylenecyclopentane and 2,3-dimethyl-2-butene. Assign each compound its correct spectrum.

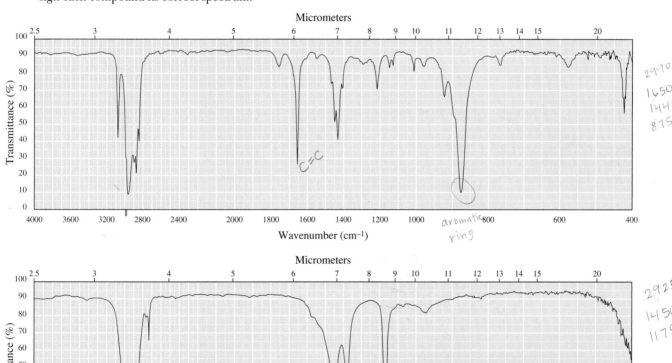

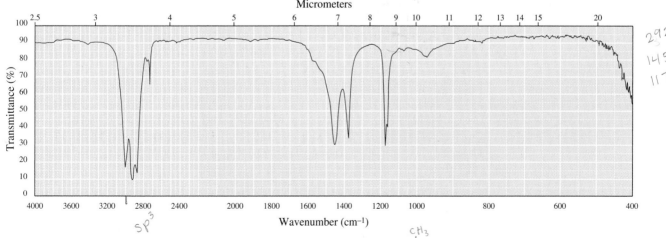

12.7 Following are infrared spectra of nonane and 1-hexanol. Assign each compound its correct spectrum.

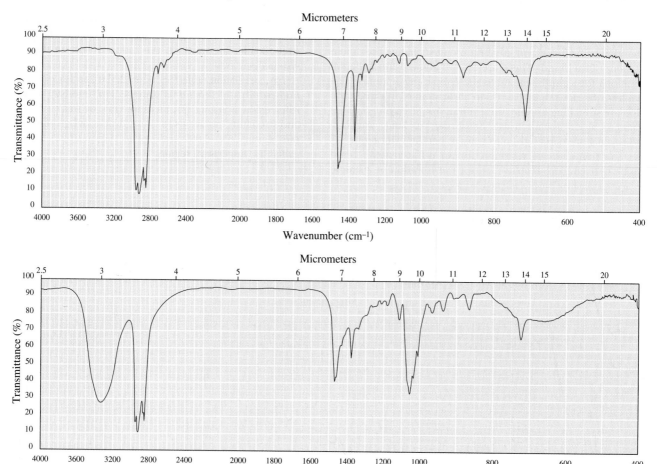

12.8 Following are infrared spectra of 2-methyl-1-butanol and *tert*-butyl methyl ether. Assign each compound its correct spectrum.

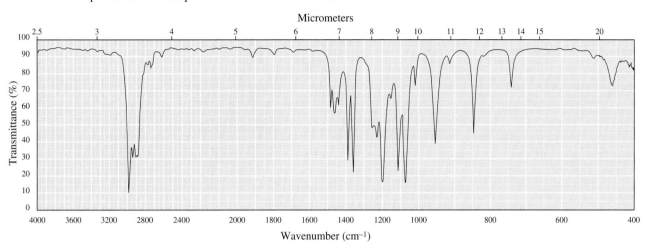

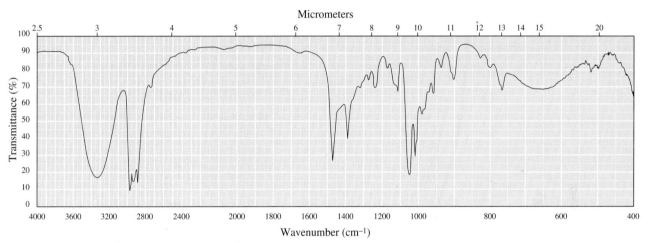

12.9 The IR C≡C stretching absorption in symmetrical alkynes are usually absent. Why is this so?

12.10 Explain the fact that the C—O stretch in ethers and esters occurs at 1000–1100 cm^{-1} when the C is sp^3 hybridized, but at 1250 cm^{-1} when it is sp^2 hybridized.

12.11 A compound has strong infrared absorption bands at the following frequencies. Suggest likely functional groups that may be present.

(a) 1735, 1250, and 1100 cm^{-1}

(b) 1745 cm^{-1} but not 1000–1250 cm^{-1}

(c) 1710 and 2500–3400 (broad) cm^{-1}

(d) A single band at about 3300 cm^{-1}

(e) 3600 and 1050 cm^{-1}

(f) 1100 cm^{-1} but not 3300–3650 cm^{-1}

12.12 Show how IR spectroscopy can be used to distinguish between the compounds in each set.

(a) [structure: CH₃CH₂CH₂OH] and [structure: ethyl ether]

(b) [structure: aldehyde] and [structure: alkene]

(c) [structure: cyclopentanol] and [structure: cyclopentene]

(d) [structure: cyclopentanone] and [structure: methoxycyclopentane]

(e) [structure: ethyl benzoate] and [structure: ethoxy benzoic acid]

(f) [structure: hexene] and [structure: tetrahydropyran]

(g) [structure: cinnamaldehyde] and [structure: cinnamic acid]

(h) [structure: alkene] and [structure: alkyne]

(i) [structure: cyclohexene oxide] and [structure: cyclohexanol]

(j) [structure: cyclohexylmethanol] and [structure: cyclohexanecarboxylic acid]

A Conversation with . . .

Jacquelyn Gervay-Hague

Jacquelyn Gervay-Hague is a young faculty member moving up fast in her field and at the university where she teaches. An organic chemist and an associate professor of chemistry at the University of Arizona in Tucson, she was recently honored with the American Chemical Society's (ACS) Horace Isbel Award. The award, given by the ACS Division of Carbohydrate Chemistry, recognizes each year one young scientist who shows the most promise for having a fruitful career in that field.

However, when reached for this conversation, she was neither working at the university nor doing research on carbohydrate chemistry. She was enjoying one of the special privileges of academia, the sabbatical. This year-long leave of absence from the university, which a faculty member can apply for every seven years, is allowing Gervay-Hague to try something different.

A Sabbatical Year

"I've never had the pleasure of working in industry," she said. "And, most of my students want to work in industry. For my sabbatical, I wanted to go to a company and learn what it's like so that I would be able to tell my students and, perhaps, be

MODERN CARBOHYDRATE CHEMISTRY

better able to train them for industry. I've been really fortunate to be given the opportunity to come to Roche Bioscience in Palo Alto, California, to work."

She added, "Here, I'm actually doing research in an area completely unrelated to the one in which I work at the university. I'm still leading my research program at the university on therapeutics that target HIV infection and cancer but I'm also working with Roche scientists on osteoporosis research."

Gervay-Hague manages to do research in two places at the same time by keeping in touch with her students via e-mail and returning to the university once a month. She explains, "My graduate students and postdoctoral research staff are very

independent. We communicate daily by phone or e-mail and have everything set up so that I can follow their work. The Web has really made this possible."

When her sabbatical year is over, Gervay-Hague will go back to teaching several courses, managing her well-established and award-winning research program, writing grant proposals, serving on academic committees, and working with undergraduate as well as graduate students and postdoctoral associates in the laboratory. It's teaching that she misses most now, she says.

Meanwhile, Gervay-Hague is enjoying learning about how industry uses combinatorial chemistry and parallel synthesis techniques, subjects on which she will develop a course when she returns to the university. She is also learning about factors that bridge a medicinal compound's passage from the laboratory bench to the drug store.

She says, "There are many more factors to consider in industry in developing a drug than I was aware of. Although the chemical challenges are the same, there are other factors in industry to take into account. For example, there are marketability factors, factors having to do with how much of a compound to synthesize in order to conduct clinical trials, and ethical questions about

449

clinical trials. So many aspects of drug development have little to do with the chemistry of making the compound."

Gervay-Hague characterizes her research program at the university as a "bridging" activity, too. She bridges chemistry, biology, and virology to understand biological processes at the molecular level. Her current projects include the synthesis of carbohydrates and novel helical materials; the development of targeted drug delivery systems for treatment of cancer and HIV infections; and the development of new nuclear magnetic resonance techniques for understanding reaction mechanisms and molecular recognition.

> Suddenly organic chemistry made sense, and I became captivated by the subject matter.

She says, "It's very important for us to understand how the compounds we've developed in the laboratory work in a biological system. In order to do so, we introduce them into live cells. We start out with an idea of how the compound should work and we test that hypothesis in cells to see if we get the desired effect. Do we get antiviral activity, for example. If we change the functional groups on the compound, do we get the same activity, or is it enhanced or diminished?

"Right now, we have compounds—carbohydrate-based materials—that inhibit HIV viral entry into host cells. We're still at the level of trying to find out how to get optimal inhibition—a goal based on

our molecular biology perspective. However, the inhibitory effects we're seeing from these compounds are very positive. It's very exciting."

Gervay-Hague's research accomplishments have brought her several honors besides the 1999 Horace Isbel Award. She was named an Eli Lilly Academic awardee in 1997 and was appointed a Fellow of the Alfred P. Sloan Foundation in 1998. In 1997, she also won the University of Arizona, College of Science Innovation in Teaching Award in recognition of her teaching accomplishments in organic chemistry.

Gervay-Hague received BS and PhD degrees in chemistry from the University of California at Los Angeles and held a postdoctoral position in the Yale University Laboratory of synthetic organic chemist Samuel J. Danishefsky.

A Late Start
"My interest in chemistry got started late. I never took chemistry in high school; in fact, when I entered UCLA, I planned to major in psychobiology. The degree required chemistry, so I enrolled in the introductory courses my first year at UCLA. I did fine, although I considered it nothing more than work I had to complete to meet other goals. Then in my second year, I took organic chemistry. Even though I worked very hard, I just couldn't get it. So I understand students who have difficulty with organic chemistry in spite of their efforts. I waited until my senior year to finish the organic sequence. The instructor, Mike [Michael E.] Jung, was a terrific teacher. Suddenly organic chemistry made sense, and I became captivated by the subject matter. What turned me on was hearing about and discussing the re-

search Mike was conducting in the laboratory. Research made chemistry a living science for me. It was during that course that I decided to become a chemistry professor. But since it was so late in my undergraduate career, I had to take another year and a half of chemistry courses to graduate as a chemistry major. That was a great time for me because I knew exactly what I wanted to do."

A Taste of Graduate School
"After I decided to become a chemistry major, I wanted to get some hands-on lab experience. I approached Professor Christopher Foote (of UCLA) and asked if I could work in his lab over the summer. He had an opening and let me join his group. There, I worked with singlet oxygen. I looked at the singlet oxygen 'ene' reaction and other kinds of photochemical experiments; I found the research fun and challenging. Following that experience, I joined Mike Jung's research group to get experience in the synthesis of organic compounds. When it came time to apply to graduate school, I applied only to UCLA because I wanted to continue my work in Mike Jung's group."

Organic Synthesis
"In graduate school, I spent a good two years trying to develop a route to the synthesis of reserpine [a natural product with tranquilizing properties] via an intramolecular Diels-Alder reaction. Unfortunately, we could not overcome the unreactive nature of the Diels-Alder dienophile. Then Mike [Jung] came back from a conference with an interesting idea about doing an intramolecular Diels-Alder cyclization, and we decided to look at the effects

of substituents on the rate of the reaction. I synthesized a number of test compounds with different alkyl groups near the reaction site and then studied their cyclization. We did all of the classic experiments to prove the reactions were essentially irreversible, and then we published a paper reporting this somewhat surprising result. Later I started doing the reactions in a different solvent and saw some amazing and unexpected differences in reaction rates. We were really intrigued, and in the end, we discovered that during the reaction some unfavorable dipole interactions took place, interactions accentuated in polar solvents that accelerated the reaction."

A New Interest Sends Gervay-Hague East

"For postdoctoral study, I went to Yale University to work with Samuel J. Danishefsky, a noted synthetic organic chemist. There I wanted to work on a molecule called sialic acid, which is a 3-deoxy sugar with a hydroxyl and a carboxyl group at the anomeric carbon.

Sialic acid

This compound is found in cell glycolipids, and I was captivated by sialic acid, both for its biological role and as a synthetic project.

"My years as a postdoctoral student were very formative. Almost everything I now study in my lab is based on the work I did at Yale. There I learned carbohydrate chemistry and helped to complete the total synthesis of Sialyl-Lewis X glycal [a tetrasaccharide that can be converted to compounds known to be important in tumor metastasis and in attracting white blood cells to damaged tissue]. Also exciting for me was interacting with graduate students and other postdocs, who had come there from all over the world."

On Being a Professor

"In the fall of 1992, I joined the faculty of the University of Arizona. The most rewarding part of being a professor is working with students. I always learn something; they always learn something. Using grant money from the National Science Foundation a few years ago, I developed an undergraduate course to help students realize that, given a basic understanding of chemistry, they can come up with original ideas for research and that they can learn to think critically about important areas of organic chemistry. I gave the students original research articles and asked them to propose some ideas for further research. Many of the ideas they came up with had been done and were in the literature and that was exciting to them. They couldn't believe that they—just sophomores in organic chemistry—could think of ideas that scientists studied and wrote about in journal articles.

"That course was extremely successful and I taught it for three years before going back to graduate courses. Next fall, when I return from my sabbatical, I'll be developing another new course that will be part of a new chemistry sequence. Organic chemistry is typically taught in the sophomore year but we're going to try offering it in the freshman year. What we're trying to do with this approach is speed up the students' introduction to organic chemistry because many of them find it to be a really interesting course. I'll be bringing current research ideas in organic chemistry to the forefront in the course. That will be the basis on which we'll explore the basic principles of organic chemistry.

"I've noticed changes in my students in the eight years since I started teaching. Students today are increasingly more practical and more concerned about what's going on in the world. They want to know about things that they see on the Web and hear about on the news and in the newspaper. Students are very interested in being able to talk about the latest scientific discoveries, such as the human genome project, which have a basis in organic chemistry. I think the popular press does a lot to help people—not only my students but people like my husband who is in transportation—to understand science. If you can bring that into the classroom, it can be very exciting."

On Being a Woman Scientist

"I know that some women face barriers in science, but that hasn't been my personal experience. Some of my women students have told me it makes a difference to them that I'm a woman chemist. They say that they are much more comfortable talking to me than to a male professor, especially when they don't understand something. I have the sense that they don't feel intimidated when they approach me. If talking to women students helps them reach their potential, then I am happy to help. I have male students, too, who come to talk with me. It is rewarding to be able to help both female and male students with many of the challenges they face in college."

Advice on Science for Students

"A lot of students who read this book send me e-mail when they are going through tough times in their organic chemistry courses. They read about me and how I didn't do well at first in organic chemistry, and it helps them to keep trying.

"I'm happy that my story encourages them to go on with their studies but I don't want to make everyone a chemist. I don't think that's prudent. My goal is to make everybody in my classes appreciate chemistry. At the same time, I tell my sophomore students that chemistry, and certainly organic chemistry, is not strictly intuitive. If you have difficulty understanding the concepts, it's not because you're not smart; organic chemistry is challenging, much like a foreign language. So, I think a large part of being successful is just sticking to it.

"For students who are contemplating a career in chemistry, one of the most important things to do is to get laboratory experience in a research setting, not just lab classes. In fact, it's a good idea to have a few different research experiences in case one doesn't work out. That is how you'll know if a career in chemistry is for you."

NUCLEAR MAGNETIC RESONANCE SPECTROSCOPY

N uclear magnetic resonance (NMR) spectroscopy was developed in the late 1950s and, within a decade, became the single most important technique available to chemists for the determination of molecular structure. In this chapter, we first develop a basic understanding of the theory behind this type of spectroscopy, and then we concentrate on the interpretation of spectra and the information they provide us about molecular structure.

We discussed the electromagnetic spectrum in Chapter 12. In this chapter, we concentrate on absorption of radio-frequency radiation by

■ Magnetic resonance imaging is a useful medical diagnostic tool. *(Paul Shambroom/Science Source/Photo Researchers, Inc.)* Inset: A model of methyl acetate. For a ^{1}H-NMR spectrum of methyl acetate, see Figure 13.5.

453

nuclei and the resulting transitions between energy levels, called **nuclear magnetic resonance spectroscopy.** Felix Bloch and Edward Purcell, both of the United States, first detected the phenomenon of nuclear magnetic resonance in 1946. They shared the 1952 Nobel Prize for physics. Nuclear magnetic resonance spectroscopy gives us information about the number and types of atoms in a molecule: for example, about hydrogens using ^{1}H-NMR spectroscopy and about carbons using ^{13}C-NMR spectroscopy. It can also give us substantial information about the connectivity of the atoms and, in many cases, can allow determination of the structure of a molecule with no additional information.

13.1 Nuclear Spin States

You are already familiar from general chemistry with the concept that an electron has a spin quantum number of $\frac{1}{2}$, with allowed values of $+\frac{1}{2}$ and $-\frac{1}{2}$, and that a spinning charge creates an associated magnetic field. In effect, an electron behaves as if it is a tiny bar magnet and has a magnetic moment. The same effect holds for certain atomic nuclei.

Any atomic nucleus that has an odd mass number, an odd atomic number, or both also has a spin and a resulting nuclear magnetic moment. The allowed nuclear spin states are determined by the spin quantum number, I, of the nucleus. A nucleus with spin quantum number I has $2I + 1$ spin states. Our focus in this chapter is on nuclei of ^{1}H and ^{13}C, isotopes of the two elements most common to organic compounds. Each has a nuclear spin quantum number of $\frac{1}{2}$ and therefore has $2(\frac{1}{2}) + 1 = 2$ allowed spin states. Quantum numbers and allowed nuclear spin states for these nuclei and those of other elements common to organic compounds are shown in Table 13.1. Note that ^{12}C, ^{16}O, and ^{32}S each have a spin quantum number of zero and only one allowed nuclear spin state; these nuclei are inactive in NMR spectroscopy.

Table 13.1   **Spin Quantum Numbers and Allowed Nuclear Spin States for Selected Isotopes of Elements Common to Organic Compounds**

Element	^{1}H	^{2}H	^{12}C	^{13}C	^{14}N	^{16}O	^{31}P	^{32}S
Nuclear spin quantum number (I)	$\frac{1}{2}$	1	0	$\frac{1}{2}$	1	0	$\frac{1}{2}$	0
Number of spin states	2	3	1	2	3	1	2	1

Supporting Concepts
NMR Spectroscopy
Introduction
The CD-ROM contains an extensive tutorial explaining the theory of NMR spectroscopy and the manner in which NMR spectra are interpreted.

13.2 Orientation of Nuclear Spins in an Applied Magnetic Field

Within a collection of ^{1}H and ^{13}C atoms, the spins of their tiny nuclear bar magnets are completely random in orientation. When placed between the poles of a powerful magnet of field strength B_0, however, interactions between the nuclear spins and the applied magnetic field are quantized, with the result that only certain orientations of

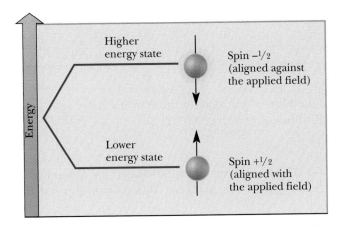

Figure 13.1

^1H and ^{13}C nuclei with spin $+\frac{1}{2}$ are aligned with the applied magnetic field, B_0, and are in the lower spin energy state; those with spin $-\frac{1}{2}$ are aligned against the applied magnetic field and are in the higher spin energy state.

nuclear magnetic moments are allowed. For ^1H and ^{13}C nuclei, only two orientations are allowed as illustrated in Figure 13.1. By convention, nuclei with spin $+\frac{1}{2}$ are aligned with the applied field and are in the lower energy state; nuclei with spin $-\frac{1}{2}$ are aligned against the applied field and are in the higher energy state.

The difference in energy between nuclear spin states increases with the strength of the applied field (Figure 13.2). At an applied field strength of 7.05 T, which is readily available with present-day superconducting electromagnets, the difference in energy between nuclear spin states for ^1H is approximately 0.120 J (0.0286 cal)/mol, which corresponds to electromagnetic radiation of 300 MHz. At 7.05 T, the energy difference in nuclear spin states for ^{13}C nuclei is approximately 0.030 J (0.00715 cal)/mol, which corresponds to radiation of 75 MHz. Advanced commercial instruments can now be purchased with fields nearly three times this; the operating frequencies are proportional to the field. Sensitivities are more than proportionally higher!

To put these values for nuclear spin energy levels in perspective, energies for transitions between vibrational energy levels observed in infrared spectroscopy are 8 to 63 kJ (2 to 15 kcal)/mol. Those between electronic energy levels in ultraviolet-

Note: The SI unit for magnetic field strength is the tesla (T). A unit still in common use, however, is the gauss (G). Values of T and G are related by the equation 1 T = 10^4 G.

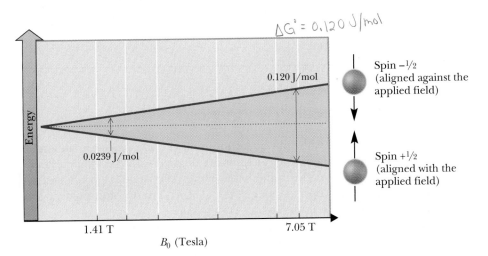

Figure 13.2

The energy difference between the allowed nuclear spin states increases linearly with applied field strength. Values shown here are for ^1H nuclei.

visible spectroscopy are 167 to 585 kJ (40 to 140 kcal)/mol. Nuclear transitions involve only tiny energies!

Example 13.1

Calculate the ratio of nuclei in the higher spin state to those in the lower spin state, N_h/N_l, for ^1H at 25°C in an applied field strength of 7.05 T.

Solution

Use the equation given in Section 2.6A for the relationship between the difference in energy states and equilibrium constant. In this problem, this relationship has the form

$$\Delta G^0 = -2.303 \, RT \log \frac{N_h}{N_l}$$

The difference in energy between the higher and lower nuclear spin states in an applied field of 7.05 T is approximately 0.120 J/mol, and the temperature is 25 + 273 = 298 K. Substituting these values in this equation gives

$$\log \frac{N_h}{N_l} = \frac{-\Delta G^0}{2.303 \, RT} = \frac{-0.120 \, \text{J} \cdot \text{mol}^{-1}}{2.303 \times 8.314 \, \text{J} \cdot \text{K}^{-1} \cdot \text{mol}^{-1} \times 298 \, \text{K}} = -2.097 \times 10^{-5}$$

$$\frac{N_h}{N_l} = 0.9999517 = \frac{1.000000}{1.000048}$$

From this calculation, we determine that, for every 1,000,000 hydrogen atoms in the higher energy state in this applied field, there are 1,000,048 in the lower energy state. The excess population of the lower energy state under these conditions is only 48 per million! What is important about this number is that the strength of an NMR signal is proportional to the population difference. The greater the difference, the stronger the signal.

Problem 13.1

Calculate the ratio of nuclei in the higher spin state to those in the lower spin state, N_h/N_l, for ^{13}C at 25°C in an applied field strength of 7.05 T. The difference in energy between the higher and lower nuclear spin states in this applied field is approximately 0.030 J (0.00715 cal)/mol.

13.3 Nuclear Magnetic "Resonance"

As we have seen, when nuclei with spin quantum number $\frac{1}{2}$ are placed in an applied magnetic field, a small majority of nuclear spins are aligned with the applied field in the lower energy state. When nuclei in the lower energy spin state are irradiated with a radio frequency of the appropriate energy, they absorb energy, and nuclear spins flip from the lower energy state to the higher energy state.

To understand the mechanism by which a spinning nucleus absorbs energy and the meaning of resonance in this context, imagine a spinning nucleus. When an ap-

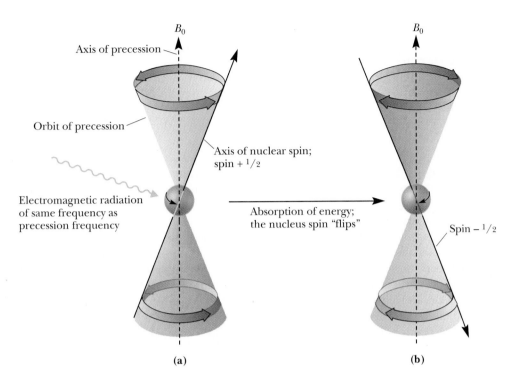

(a)   (b)

Figure 13.3
The origin of nuclear magnetic "resonance." (a) Precession of a spinning nucleus in an applied magnetic field. (b) Absorption of electromagnetic radiation occurs when the frequency of radiation is equal to the frequency of precession.

plied field of strength B_0 is turned on, the nucleus becomes aligned with the applied field in an allowed spin energy state. The nucleus then begins to precess as shown in Figure 13.3(a) and traces out a cone-shaped surface in much the same manner as a spinning top or gyroscope traces out a cone-shaped surface as it precesses in the earth's gravitational field. We can express the rate of precession as a frequency in hertz.

If the precessing nucleus is irradiated with electromagnetic radiation at the precession frequency, then the two frequencies couple, energy is absorbed, and the nuclear spin "flips" from spin state $+\frac{1}{2}$ (with the applied field) to spin state $-\frac{1}{2}$ (against the applied field) as illustrated in Figure 13.3(b). For ^{1}H in an applied magnetic field of 7.05 T, the frequency of precession is approximately 300 MHz. For ^{13}C in the same field, it is approximately 75 MHz. **Resonance** in this context is the absorption of electromagnetic radiation by a precessing nucleus and the resulting flip of its nuclear spin from the lower energy state to the higher energy state. The spectrometer detects this coupling of precession frequency and electromagnetic radiation and records it as a **signal.**

If we were dealing with ^{1}H nuclei isolated from all other atoms and electrons, any combination of applied field and electromagnetic radiation that produces a signal for one hydrogen nucleus would produce a signal for all hydrogen nuclei. In other words, hydrogens would be indistinguishable. Hydrogens in an organic molecule, however, are not isolated; they are surrounded by electrons, which are caused to circulate by an applied magnetic field. You may think of this electron circulation as you would the flow of electrons in a wire. We know from the laws of physics that the flow of electric current through a wire generates a magnetic field. In the case of electrons about a hydrogen nucleus, their circulation generates a magnetic field opposed to

Resonance, in NMR spectroscopy The absorption of electromagnetic radiation by a precessing nucleus and the resulting flip of its nuclear spin from a lower energy state to a higher energy state.
Signal A recording in an NMR spectrum of a nuclear magnetic resonance.

the applied field and thereby partially shields the hydrogen nucleus from the applied field. The circulation of electrons around a nucleus in an applied field is called a **diamagnetic current,** and the nuclear shielding resulting from it is called **diamagnetic shielding.** Although the diamagnetic shielding created by circulating electrons is orders of magnitude weaker than the applied fields used in NMR spectroscopy, it is nonetheless significant at the molecular level.

The degree of **shielding** depends on several factors, which we will take up soon. For the moment, however, it is sufficient to realize that the greater the shielding of a particular nucleus by local magnetic fields, the greater the strength of the applied field required to bring it into resonance. Conversely, the less the shielding of a nucleus, or as it is more commonly expressed, the greater its **deshielding,** the lower the strength of the applied field required to bring it into resonance. This difference is called the **chemical shift.**

The differences in resonance frequencies among the various hydrogen nuclei within a molecule due to shielding/deshielding are generally very small. The difference between the resonance frequencies of hydrogens in chloromethane compared with those in fluoromethane, for example, under an applied field of 7.05 T is only 360 Hz. Considering that the radio-frequency radiation used at this applied field is approximately 300 MHz, the difference in resonance frequencies between these two sets of hydrogens is only slightly greater than 1 ppm compared with the irradiating frequency.

$$\frac{360 \text{ Hz}}{300 \times 10^6 \text{ Hz}} = \frac{1.2}{10^6} = 1.2 \text{ ppm}$$

It is customary to measure the resonance frequencies of individual nuclei relative to the resonance frequency of nuclei in a reference compound. The reference compound now universally accepted for ^{1}H-NMR and ^{13}C-NMR spectroscopy is tetramethylsilane (TMS), which is arbitrarily assigned a chemical shift of 0 ppm.

$$
\begin{array}{c}
\text{CH}_3 \\
| \\
\text{H}_3\text{C}-\text{Si}-\text{CH}_3 \\
| \\
\text{CH}_3
\end{array}
$$

Tetramethylsilane (TMS)

To standardize reporting of NMR data for both ^{1}H and ^{13}C spectra, the chemical shift (δ), in parts per million, is defined as the frequency shift from either the hydrogens or the carbons in TMS divided by the operating frequency of the spectrometer. Thus, by definition, chemical shift is independent of the operating frequency of the spectrometer. On the chart paper used to record NMR spectra, chemical shift values are shown in increasing order to the left of the TMS signal.

13.4 An NMR Spectrometer

The essential elements of an NMR spectrometer are a powerful magnet, a radio-frequency generator, a radio-frequency detector, and a sample tube (Figure 13.4).

The sample is dissolved in a solvent, most commonly carbon tetrachloride (CCl_4), deuterochloroform ($CDCl_3$), or deuterium oxide (D_2O), which have no pro-

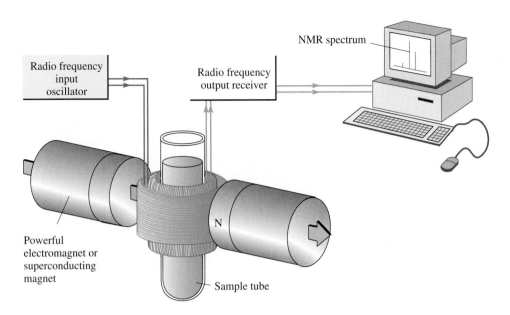

NMR spectrum

Radio frequency
input
oscillator

Radio frequency
output receiver

N

Powerful
electromagnet or
superconducting
magnet

Sample tube

Figure 13.4
Schematic diagram of a nuclear
magnetic resonance spectrometer.

tons and do not interfere in ^{1}H spectra. The sample cell is a small glass tube suspended in the magnetic field and set spinning on its long axis to ensure that all parts of the sample experience a homogeneous applied field.

Modern Fourier Transform NMR (FT-NMR) spectrometers operate in the following way. The magnetic field is held constant, and the sample is irradiated with a short pulse (approximately 10^{-5} s) of radio-frequency energy that flips the spins of all susceptible nuclei simultaneously. The process by which each nucleus returns to its equilibrium state emits a sine wave at the frequency of its resonance. The intensity of the sine wave decays with time and falls to zero as nuclei return to their equilibrium state. A computer records this intensity-versus-time information and then uses a mathematical algorithm called a Fourier transform (FT) to convert it to intensity-versus-frequency information. An FT-NMR spectrum can be recorded in less than two seconds. A particular advantage of FT-NMR spectroscopy is that a large number of spectra (as many as several thousand per sample) can be recorded and digitally summed to give a time-averaged spectrum. Instrumental electronic noise is random and partially cancels out when spectra are time-averaged, but sample signals accumulate and become much stronger relative to the noise than those from a single spectrum.

All NMR spectra shown in this text were recorded and displayed using FT techniques. All ^{1}H-NMR spectra were recorded at an applied magnetic field strength of 7.05 T using a radio frequency of 300 MHz, except as noted. All ^{13}C-NMR spectra were recorded at 7.05 T and 75 MHz, except as noted.

Figure 13.5 shows a 300 MHz ^{1}H-NMR spectrum of methyl acetate. The lower axis is δ, in parts per million. The small signal at δ 0 is due to the hydrogens of the TMS reference. The remainder of the spectrum consists of two signals: one for the three hydrogens on the methyl adjacent to oxygen and one for the three hydrogens on the methyl adjacent to the carbonyl group. It is not our purpose at the moment to determine which hydrogens give rise to which signal, but only to recognize the

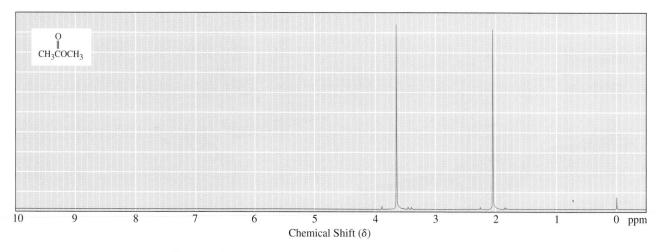

Figure 13.5
^{1}H-NMR spectrum of methyl acetate.

Downfield The shift of an NMR signal to the left on the chart paper.
Upfield The shift of an NMR signal to the right on the chart paper.

form in which an NMR spectrum is recorded and the origin of the calibration marks.

Here is a note on terminology. If a signal is shifted toward the left on the chart paper, we say that it is shifted **downfield.** Conversely, if a signal is shifted toward the right on the chart paper, we say that it is shifted **upfield.**

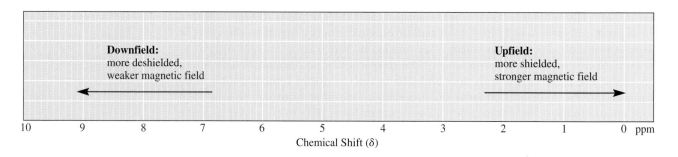

Supporting Concepts
NMR Spectroscopy
Equivalent Hydrogens

13.5 Equivalent Hydrogens

Equivalent hydrogens Hydrogens that have the same chemical environment.

Given the structural formula of a compound, how do we know how many signals to expect in its ^{1}H-NMR spectrum? The answer is that equivalent hydrogens give the same ^{1}H-NMR signal; nonequivalent hydrogens give different ^{1}H-NMR signals. **Equivalent hydrogens** have the same chemical environment.

Example 13.2

State the number of sets of equivalent hydrogens in each compound and the number of hydrogens in each set.

(a) 2-Methylpropane   **(b)** 2-Methylbutane

Solution

(a) 2-Methylpropane contains two sets of equivalent hydrogens: a set of nine equivalent primary hydrogens and one tertiary hydrogen.

(b) 2-Methylbutane contains four sets of equivalent hydrogens. Nine primary hydrogens are in this molecule: one set of three and one set of six. To see that there are two sets, note that replacement by chlorine of any hydrogen in the set of three gives 1-chloro-3-methylbutane. Replacement by chlorine of any hydrogen in the set of six gives 1-chloro-2-methylbutane. In addition, the molecule contains a set of two equivalent secondary hydrogens and one tertiary hydrogen.

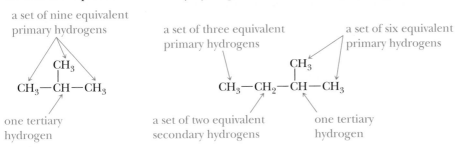

Problem 13.2

State the number of sets of equivalent hydrogens in each compound and the number of hydrogens in each set.

(a) 3-Methylpentane   (b) 2,2,4-Trimethylpentane

Here are four organic compounds, each of which has one set of equivalent hydrogens and gives one signal in its ^{1}H-NMR spectrum.

$$CH_3\overset{\overset{\displaystyle O}{\|}}{C}CH_3 \qquad ClCH_2CH_2Cl$$

Propanone   1,2-Dichloroethane   Cyclopentane   2,3-Dimethyl-2-butene
(Acetone)

Molecules with two or more sets of equivalent hydrogens give rise to a different resonance signal for each set.

$$CH_3\overset{\overset{\displaystyle Cl}{|}}{C}HCl$$

1,1-Dichloroethane   Cyclopentanone   (Z)-1-Chloropropene   Cyclohexene
(2 signals)   (2 signals)   (3 signals)   (3 signals)

You should see immediately that valuable information about molecular structure can be obtained simply by counting the number of signals in the ^{1}H-NMR spectrum of a compound. Consider, for example, the two constitutional isomers of molecular formula $C_2H_4Cl_2$. The compound 1,2-dichloroethane has one set of equivalent hydrogens and one signal in its ^{1}H-NMR spectrum. Its constitutional isomer

1,1-dichloroethane has two sets of equivalent hydrogens and two signals in its ^{1}H-NMR spectrum. Thus, simply counting signals allows you to distinguish between these two compounds.

Example 13.3

Each compound gives only one signal in its ^{1}H-NMR spectrum. Propose a structural formula for each compound.

(a) C_2H_6O (b) $C_3H_6Cl_2$ (c) C_6H_{12} (d) C_4H_6

Solution

(a) CH_3OCH_3 (b) $\begin{array}{c} Cl \\ | \\ CH_3CCH_3 \\ | \\ Cl \end{array}$ (c) ⬡ or $\begin{array}{c} H_3C \\ \;\;\;C=C\;\;\; \\ H_3C \end{array} \begin{array}{c} CH_3 \\ \\ CH_3 \end{array}$ (d) $CH_3C{\equiv}CCH_3$

Problem 13.3

Each compound gives only one signal in its ^{1}H-NMR spectrum. Propose a structural formula for each compound.

(a) C_3H_6O (b) C_5H_{10} (c) C_5H_{12} (d) $C_4H_6Cl_4$

13.6 Signal Areas

Figure 13.6

^{1}H-NMR spectrum of *tert*-butyl acetate showing the integration. The total vertical rise of 90 chart divisions corresponds to 12 hydrogens, 9 in one set and 3 in the other.

We have just seen that the number of signals in a ^{1}H-NMR spectrum gives us information about the number of sets of equivalent hydrogens. The relative areas of these signals provide additional information. As a spectrum is being run, the instrument's computer numerically measures the area under each signal. In the spectra shown in this text, this information is displayed in the form of a line of integration superposed on the original spectrum. The vertical rise of the line of integration over

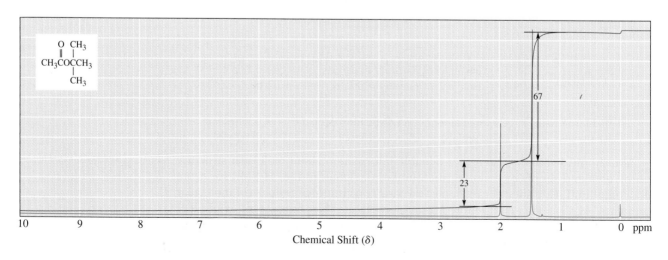

each signal is proportional to the area under that signal, which, in turn, is proportional to the number of hydrogens giving rise to that signal. Figure 13.6 shows an integrated ^{1}H-NMR spectrum of *tert*-butyl acetate, $C_6H_{12}O_2$. The spectrum shows signals at δ 1.44 and 1.95. The integrated signal heights are $23 + 67$, or 90 chart divisions, which correspond to 12 hydrogens. From these numbers we calculate that $(23/90) \times 12$ or 3 hydrogens are in one set, and $(67/90) \times 12$ or 9 hydrogens are in the second set.

Example 13.4

Following is a ^{1}H-NMR spectrum for a compound of molecular formula $C_9H_{10}O_2$. From the integration, calculate the number of hydrogens giving rise to each signal.

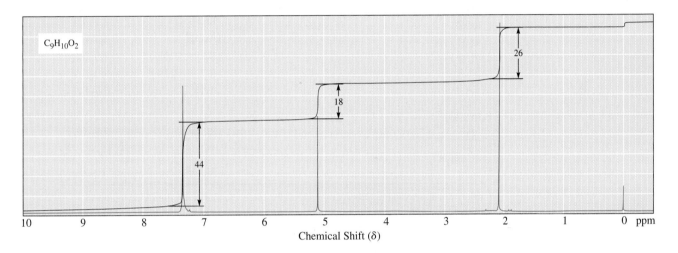

Solution

The total vertical rise in the line of integration is 88 chart divisions and corresponds to 10 hydrogens. From these numbers, we calculate that $44/88 \times 10$, or 5, of the hydrogens give rise to the signal at δ 7.34. By similar calculations, the signals at δ 5.08 and 2.06 correspond to two hydrogens and three hydrogens, respectively.

Problem 13.4

The line of integration of the two signals in the ^{1}H-NMR spectrum of a ketone of molecular formula $C_7H_{14}O$ rises 62 and 10 chart divisions, respectively. Calculate the number of hydrogens giving rise to each signal, and propose a structural formula for this ketone.

13.7 Chemical Shift

Supporting Concepts
NMR Spectroscopy:
Chemical shift

The chemical shift for a signal in a ^{1}H-NMR spectrum can give valuable information about the type of hydrogens giving rise to that absorption. Hydrogens on methyl groups bonded to sp^3-hybridized carbons, for example, give signals near δ 0.8–1.0 (compare Figure 13.6). Hydrogens on methyl groups bonded to a carbonyl carbon

Table 13.2 Average Values of Chemical Shifts of Representative Types of Hydrogens

Type of Hydrogen (R = alkyl, Ar = aryl)	Chemical Shift (δ)*	Type of Hydrogen (R = alkyl, Ar = aryl)	Chemical Shift (δ)*
$(CH_3)_4Si$	0 (by definition)	$RCOCH_3$ (C=O)	3.7–3.9
RCH_3	0.8–1.0		
RCH_2R	1.2–1.4	$RCOCH_2R$ (C=O)	4.1–4.7
R_3CH	1.4–1.7	RCH_2I	3.1–3.3
$R_2C=CRCHR_2$	1.6–2.6	RCH_2Br	3.4–3.6
$RC\equiv CH$	2.0–3.0	RCH_2Cl	3.6–3.8
$ArCH_3$	2.2–2.5	RCH_2F	4.4–4.5
$ArCH_2R$	2.3–2.8	$ArOH$	4.5–4.7
ROH	0.5–6.0	$R_2C=CH_2$	4.6–5.0
RCH_2OH	3.4–4.0	$R_2C=CHR$	5.0–5.7
RCH_2OR	3.3–4.0	ArH	6.5–8.5
R_2NH	0.5–5.0		
$RCCH_3$ (C=O)	2.1–2.3	RCH (C=O)	9.5–10.1
$RCCH_2R$ (C=O)	2.2–2.6	$RCOH$ (C=O)	10–13

* Values are approximate. Other atoms within the molecule may cause the signal to appear outside the ranges.

give signals near δ 2.1–2.3 (compare Figures 13.5 and 13.6), and hydrogens on a methyl group bonded to oxygen give signals near δ 3.7–3.9 (Figure 13.5). Tabulated in Table 13.2 are average chemical shifts for most of the types of hydrogens we deal with in this course. Notice that most of these values fall within a rather narrow range of 0–10 δ units (ppm).

Example 13.5

Following are structural formulas for two constitutional isomers of molecular formula $C_6H_{12}O_2$.

$$\begin{array}{cc} \underset{\displaystyle\overset{\displaystyle |}{CH_3}}{CH_3COCCH_3} & \underset{\displaystyle\overset{\displaystyle |}{CH_3}}{CH_3OC-CCH_3} \\ (1) & (2) \end{array}$$

(a) Predict the number of signals in the ^{1}H-NMR spectrum of each isomer.

(b) Predict the ratio of areas of the signals in each spectrum.

(c) Show how you can distinguish between these isomers on the basis of chemical shift.

Solution

(a, b) The ^{1}H-NMR spectrum of each consists of two signals in the ratio 9:3, or 3:1.

(c) Distinguish between these constitutional isomers by the chemical shift of the single —CH_3 group. The hydrogens of CH_3O are deshielded (appear farther downfield) than the hydrogens of CH_3C=O. See Table 13.2 for approximate values for each chemical shift. Experimentally determined values are

$$\delta\ 1.95 \qquad \underset{CH_3}{\underset{|}{\underset{\underset{CH_3}{|}}{CH_3CO\overset{O}{\overset{\|}{C}}CH_3}}} \qquad \delta\ 1.44 \qquad\qquad \delta\ 3.67 \qquad \underset{CH_3}{\underset{|}{\underset{\underset{CH_3}{|}}{CH_3O\overset{O}{\overset{\|}{C}}-CCH_3}}} \qquad \delta\ 1.20$$

(1) (2)

Problem 13.5

Following are two constitutional isomers of molecular formula $C_4H_8O_2$.

$$\underset{(1)}{CH_3CH_2O\overset{O}{\overset{\|}{C}}CH_3} \qquad \underset{(2)}{CH_3CH_2\overset{O}{\overset{\|}{C}}OCH_3}$$

(a) Predict the number of signals in the ^{1}H-NMR spectrum of each isomer.

(b) Predict the ratio of areas of the signals in each spectrum.

(c) Show how you can distinguish between these isomers on the basis of chemical shift.

The chemical shift of a particular type of hydrogen depends primarily on three factors: (A) the electronegativity of nearby atoms, (B) the hybridization of the adjacent atoms, and (C) magnetic induction within an adjacent pi bond. Let us consider these factors one at a time.

A. Electronegativity of Nearby Atoms

As illustrated in Table 13.3 for the chemical shift of methyl hydrogens in the series CH_3—X, the greater the electronegativity of X, the greater the chemical shift. The

Table 13.3 Dependence of Chemical Shift of CH_3X on the Electronegativity of X

CH_3—X	Electronegativity of X	Chemical Shift (δ) of Methyl Hydrogens
CH_3F	4.0	4.26
CH_3OH	3.5	3.47
CH_3Cl	3.1	3.05
CH_3Br	2.8	2.68
CH_3I	2.5	2.16
$(CH_3)_4C$	2.1	0.86
$(CH_3)_4Si$	1.8	0.00 (by definition)

effect of an electronegative substituent falls off quickly with distance. The effect of an electronegative substituent two atoms away is only about 10% of that when it is on the adjacent atom. The effect of an electronegative substituent three atoms away is almost negligible.

Electronegativity and chemical shift are related in the following way. The presence of an electronegative atom reduces electron density on atoms bonded to it and therefore their shielding. This effect deshields nearby atoms and causes them to resonate farther downfield, that is, with a larger chemical shift.

B. Hybridization of Adjacent Atoms

Hydrogens bonded to an sp^3-hybridized carbon typically absorb at δ 0.8 to 1.7. Vinylic hydrogens (those on a carbon of a carbon-carbon double bond) are considerably deshielded and resonate at δ 4.6 to 5.7 (Table 13.4). Part of the explanation for the greater deshielding of vinylic hydrogens compared with alkyl hydrogens lies in the hybridization of carbon. Because a sigma-bonding orbital of an sp^2-hybridized carbon has more s-character than a sigma-bonding orbital of an sp^3-hybridized carbon (33% compared with 25%), an sp^2-hybridized carbon atom is more electronegative. Vinylic hydrogens are deshielded by this electronegativity effect and resonate farther downfield relative to alkyl hydrogens. Similarly, acetylene and aldehyde hydrogens also appear farther downfield compared with alkyl hydrogens.

However, differences in chemical shifts of vinylic and acetylenic hydrogens cannot be accounted for on the basis of the hybridization of carbon alone. If the chemical shift of vinylic hydrogens (δ 4.6–5.7) were due entirely to the hybridization of carbon, then the chemical shift of acetylenic hydrogens should be even greater than that of vinylic hydrogens. Yet the chemical shift of acetylenic hydrogens is only δ 2.0 to 3.0. It seems that either the chemical shift of acetylenic hydrogens is abnormally small or the chemical shift of vinylic hydrogens is abnormally large. In either case, another factor must be contributing to the magnitude of the chemical shift. Theoretical and experimental evidence (neither of which we take the time to develop here) suggest that the chemical shifts of hydrogens attached to pi-bonded carbons are influenced not only by the relative electronegativities of the sp^2- and sp-hybridized carbon atoms but also by magnetic induction from pi bonds.

Table 13.4	The Effect of Hybridization on Chemical Shift	
Type of Hydrogen (R = alkyl)	**Name of Hydrogen**	**Chemical Shift (δ)**
RCH_3, R_2CH_2, R_3CH	Alkyl	0.8–1.7
$R_2C=C(R)CHR_2$	Allylic	1.6–2.6
$RC\equiv CH$	Acetylenic	2.0–3.0
$R_2C=CHR$, $R_2C=CH_2$	Vinylic	4.6–5.7
$RCHO$	Aldehydic	9.5–10.1

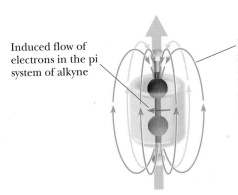

Induced flow of electrons in the pi system of alkyne

Induced local magnetic field of the pi electrons is against the applied field; it requires a greater applied field to bring an acetylenic hydrogen into resonance.

Applied field, B_0

Figure 13.7
A magnetic field induced in the pi bonds of a carbon-carbon triple bond shields an acetylenic hydrogen and shifts its signal upfield.

C. Diamagnetic Effects from Pi Bonds

To understand the influence of pi bonds on the chemical shift of an acetylenic hydrogen, imagine that the carbon-carbon triple bond is oriented as shown in Figure 13.7 with respect to the applied field. In accord with the laws of magnetic induction, the applied field induces a circulation of the pi electrons, which in turn produces an induced magnetic field. Given the geometry of an alkyne and the cylindrical nature of its pi electron cloud, the induced magnetic field is shielding in the vicinity of the acetylenic hydrogen. Therefore, a stronger applied field is required to make an acetylenic hydrogen resonate; the local magnetic field induced in the pi bonds shifts the signal of an acetylenic hydrogen upfield to a smaller δ value.

The effect of the induced circulation of pi electrons on a vinylic hydrogen (Figure 13.8) is opposite to that on an acetylenic hydrogen. The direction of the induced magnetic field in the pi bond of a carbon-carbon double bond is parallel to the applied field in the region of the vinylic hydrogens. The induced magnetic field deshields vinylic hydrogens and, thus, shifts their signal downfield to a larger δ value. The presence of the pi electrons in the carbonyl group has a similar effect on the chemical shift of the hydrogen of an aldehyde group.

The effects of the pi electrons in benzene are even more dramatic than in alkenes. All six hydrogens of benzene are equivalent, and its ^{1}H-NMR spectrum is a

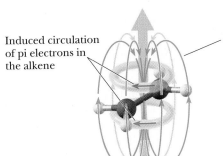

Induced circulation of pi electrons in the alkene

Induced local magnetic field of the pi electrons reinforces the applied field and provides part of the field necessary to bring a vinyl hydrogen into resonance.

Applied field, B_0

Figure 13.8
A magnetic field induced in the pi bond of a carbon-carbon double bond deshields vinylic hydrogens and shifts their signals downfield.

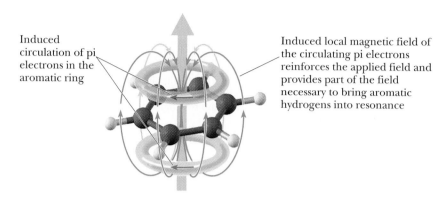

Induced circulation of pi electrons in the aromatic ring

Induced local magnetic field of the circulating pi electrons reinforces the applied field and provides part of the field necessary to bring aromatic hydrogens into resonance

Applied field

Figure 13.9

The magnetic field induced by circulation of pi electrons in an aromatic ring reinforces the applied field near aromatic hydrogens and shifts their signals downfield.

sharp singlet at δ 7.27. Hydrogens attached to a substituted benzene ring appear in the region δ 6.5–8.5. Few other hydrogens absorb in this region, and, thus, aryl hydrogens are quite easily identifiable by their distinctive chemical shifts, as much as 2 ppm lower than comparably substituted alkenes.

That aryl hydrogens absorb even farther downfield than vinylic hydrogens is accounted for by the existence of a **ring current,** a special property of aromatic rings (Figure 13.9). When the plane of an aromatic ring tumbles in an applied magnetic field, the applied field causes the pi electrons to circulate around the ring, which constitutes the so-called ring current. This induced ring current has associated with it a magnetic field that opposes the applied field in the middle of the ring but reinforces the applied field on the outside of the ring. Thus, given the position of aromatic hydrogens relative to the induced ring current, they come into resonance at a larger chemical shift.

Supporting Concepts
NMR Spectroscopy
Peak Splitting

13.8   Signal Splitting and the (*n* + 1) Rule

We have now seen three kinds of information that can be derived from examination of a ^{1}H-NMR spectrum:

1. From the number of signals, we can determine the number of sets of equivalent hydrogens.
2. From integration of signal areas, we can determine the relative numbers of hydrogens giving rise to each signal.
3. From the chemical shift of each signal, we derive information about the types of hydrogens in each set.

A fourth kind of information can be derived from the splitting pattern of each signal. Consider, for example, the ^{1}H-NMR spectrum of 1,1-dichloroethane shown in Figure 13.10. This molecule contains two sets of hydrogens. According to what we have learned so far, we predict two signals with relative areas 3:1 corresponding to the three hydrogens of the —CH$_3$ group and the one hydrogen of the —CHCl$_2$ group. You see from the spectrum, however, that there are, in fact, six **peaks.** These peaks are named by how the signal is split: two peaks are a doublet, three peaks are a triplet, and so on. The grouping of two peaks at δ 2.1 is the signal for the

Peak   The units into which an NMR signal is split; two in a doublet, three in a triplet, four in a quartet, and so on.

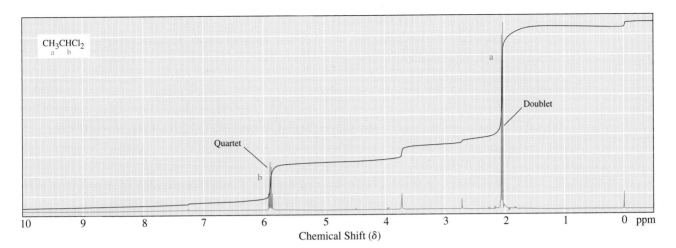

CH$_3$CHCl$_2$
a   b

Quartet

b

Doublet

a

10   9   8   7   6   5   4   3   2   1   0 ppm

Chemical Shift (δ)

three hydrogens of the —CH$_3$ group, and the grouping of four peaks at δ 5.9 is the signal for the single hydrogen of the —CHCl$_2$ group. We say that the CH$_3$ resonance at δ 2.1 is split into a doublet and that the CH resonance at δ 5.9 is split into a quartet.

The degree of **signal splitting** can be predicted on the basis of the **($n + 1$) rule.** According to this rule, if a hydrogen has n hydrogens nonequivalent to it but equivalent among themselves on the same or adjacent atom(s), its ^{1}H-NMR signal is split into ($n + 1$) peaks.

Let us apply the ($n + 1$) rule to the analysis of the spectrum of 1,1-dichloroethane. The three hydrogens of the —CH$_3$ group have one nonequivalent neighbor hydrogen ($n = 1$), and, therefore, their signal is split into a doublet. The single hydrogen of the —CHCl$_2$ group has a set of three nonequivalent neighbor hydrogens ($n = 3$), and its signal is split into a quartet.

Signal splitting Splitting of an NMR signal into a set of peaks by the influence of nonequivalent nuclei on the same or adjacent atom(s).

($n + 1$) Rule The ^{1}H-NMR signal of a hydrogen or set of equivalent hydrogens is split into ($n + 1$) peaks by a nonequivalent set of n equivalent neighboring hydrogens.

For these hydrogens, $n = 1$; their signal is split into (1 + 1) or two peaks—a doublet.

For this hydrogen, $n = 3$; its signal is split into (3 + 1) or four peaks—a quartet.

$$CH_3 - CH - Cl$$
$$|$$
$$Cl$$

Example 13.6

Predict the number of signals and the splitting pattern of each signal in the ^{1}H-NMR spectrum of each molecule.

$$\text{(a) } CH_3\overset{\displaystyle O}{\overset{\|}{C}}CH_2CH_3 \qquad \text{(b) } CH_3CH_2\overset{\displaystyle O}{\overset{\|}{C}}CH_2CH_3 \qquad \text{(c) } CH_3\overset{\displaystyle O}{\overset{\|}{C}}CH(CH_3)_2$$

Solution

The sets of equivalent hydrogens in each molecule are labeled *a, b,* and *c.* Molecule (a) has three sets of equivalent hydrogens; its ^{1}H-NMR spectrum shows a singlet, a quartet, and a triplet in the ratio 3:2:3. Molecule (b) has two sets of equivalent hydrogens; its ^{1}H-NMR spectrum shows a triplet and a quartet in the ratio 3:2. Molecule (c) has three sets of equivalent hydrogens; its ^{1}H-NMR spectrum show a singlet, a septet, and a doublet in the ratio 3:1:6.

(a) singlet — quartet — triplet

$$\underset{a}{CH_3}-\overset{\overset{\displaystyle O}{\|}}{C}-\underset{b}{CH_2}-\underset{c}{CH_3}$$

(b) triplet — quartet — triplet

$$\underset{a}{CH_3}-\underset{b}{CH_2}-\overset{\overset{\displaystyle O}{\|}}{C}-\underset{b}{CH_2}-\underset{a}{CH_3}$$

(c) singlet — septet — doublet

$$\underset{a}{CH_3}-\overset{\overset{\displaystyle O}{\|}}{C}-\underset{b}{CH}\underset{c}{(CH_3)_2}$$

Problem 13.6

Following are pairs of constitutional isomers. Predict the number of signals in the ^{1}H-NMR spectrum of each isomer and the splitting pattern of each signal.

(a) $CH_3OCH_2\overset{\overset{\displaystyle O}{\|}}{C}CH_3$ and $CH_3CH_2\overset{\overset{\displaystyle O}{\|}}{C}OCH_3$

(b) $CH_3\underset{\underset{\displaystyle Cl}{|}}{\overset{\overset{\displaystyle Cl}{|}}{C}}CH_3$ and $ClCH_2CH_2CH_2Cl$

13.9 The Origins of Signal Splitting

Coupling The magnetic interaction of the nuclear spins of nearby atoms.

When the chemical shift of one nucleus is influenced by the spin of another, the two are said to be **coupled.** Consider, for example, the situation in which nonequivalent hydrogens, H_a and H_b, exist on adjacent carbons.

$$\overset{\overset{\displaystyle H_a \quad H_b}{|\quad\ |}}{-\overset{|}{C}-\overset{|}{C}-}$$

The chemical shift of H_a is influenced by whether the spin of H_b is aligned with the applied field or aligned against it. If the field caused by H_b adds to the applied field, then H_a absorbs at a lower applied field. If, on the other hand, the field caused by H_b subtracts from it, then H_a absorbs at a higher applied field.

As a result of spin-spin coupling between nonequivalent hydrogens H_a and H_b, the signal of hydrogen H_a is split into two peaks. Because there is an almost equal probability of H_a experiencing the $+\frac{1}{2}$ and $-\frac{1}{2}$ spin states of H_b, each peak of the doublet is of equal area and each is half the area of what the H_a signal would be if hydrogen H_b were not present (Figure 13.11). Note that spin-spin coupling and the resulting signal splitting is reciprocal; if H_a is split by H_b, then H_b is equally split by H_a. If neither of these hydrogens has any other neighbors, then their ^{1}H-NMR signals are two doublets of equal area.

Splitting patterns for a hydrogen with zero, one, two, and three equivalent neighbors are summarized in Table 13.5. The ratios of the areas of the peaks in these and any other splitting patterns can be derived from an analysis of possible spin combinations for adjacent hydrogens. With three adjacent hydrogens, for example, areas within the resulting quartet are 1:3:3:1. Alternatively, the ratio of peak areas in any multiplet can be derived from a mathematical mnemonic device called **Pascal's triangle** (Figure 13.12). Here is a note of caution in counting the number of

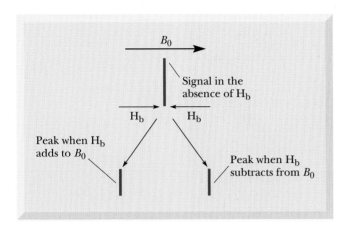

Figure 13.11
The signal of H_a is split into two peaks of equal area (a doublet) by its nonequivalent neighbor H_b.

peaks in a multiplet. If the signal of a particular hydrogen is of low intensity compared with others in the spectrum, it may not be possible to distinguish some of the smaller side peaks because of electronic noise in the baseline. It should also be noted here that all the nuclei of adjacent hydrogens couple. It is only when coupling is between nonequivalent hydrogens that it results in signal splitting; coupling between equivalent hydrogens, whether they are on the same or adjacent carbons, does not result in signal splitting.

Table 13.5	Observed Signal Splitting Patterns for a Hydrogen with Zero, One, Two, and Three Equivalent Neighboring Hydrogens	
Structure	**Spin States of H_b**	**Signal of H_a***
H_a | | —C—C— | |		(singlet)
H_a H_b | | —C—C— | |	↑ ↓	1 1
H_a H_b | | —C—C—H_b | |	↑↓ ↑↑ ↑↓ ↓↓	1 2 1
H_a H_b | | —C—C—H_b | | H_b	↓↑↑ ↓↓↑ ↑↑↑ ↑↓↑ ↓↓↓	1 3 3 1 ⌐ δ H_a

* The total area of integration is the same for all four signals.

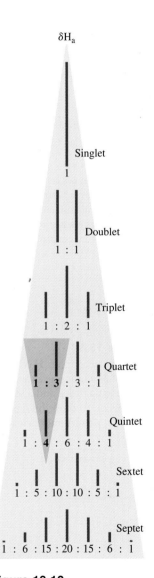

δH_a

Singlet
1

Doublet
1 : 1

Triplet
1 : 2 : 1

Quartet
1 : 3 : 3 : 1

Quintet
1 : 4 : 6 : 4 : 1

Sextet
1 : 5 : 10 : 10 : 5 : 1

Septet
1 : 6 : 15 : 20 : 15 : 6 : 1

Figure 13.12
Pascal's triangle. As illustrated by the highlighted entries, each entry is the sum of the values immediately above it to the left and the right.

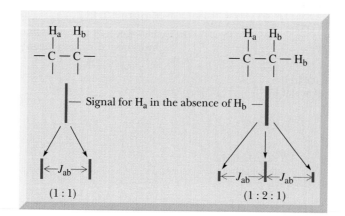

Figure 13.13

Measurement of the coupling constant, J_{ab}, for a doublet (ratio 1:1) and a triplet (ratio 1:2:1) derived from splitting the H_a signal by nonequivalent hydrogen(s) H_b.

13.10   Coupling Constants

Coupling constant (_J_)   The distance between peaks in a split signal, expressed in hertz. The value of _J_ is a quantitative measure of the magnetic interaction of nuclei whose spins are coupled.

A **coupling constant** (_J_) is the separation between adjacent peaks in a multiplet and is a quantitative measure of the shielding/deshielding influence of the magnetic moments of adjacent hydrogens. The magnitude of a coupling constant is expressed in hertz and, for protons in ^{1}H-NMR spectroscopy, is generally in the range 0–18 Hz. The value of _J_ depends only on fields caused by magnetic atoms within a molecule and is independent of the external field strength. Measurements of coupling constants for hydrogens with one and two equivalent neighboring hydrogens are illustrated in Figure 13.13. The coupling constant for two hydrogens on adjacent sp^3-hybridized carbon atoms is approximately 7 Hz. For a spectrometer operating at 300 MHz, a coupling constant of 10 Hz corresponds to only 0.023 ppm. Because peaks with this and comparable values of _J_ are so narrowly spaced, splitting patterns from spectra taken at 300 MHz and higher are often very difficult to determine. It is, therefore, common practice to retrace certain signals in expanded form so that splitting patterns are easier to observe.

In cases where signal-splitting patterns might be difficult to determine, we show expansion of appropriate signals. Shown in Figure 13.14 is a ^{1}H-NMR spectrum of 3-pentanone showing the quartet-triplet pattern of the ethyl group. Shown above the

Figure 13.14

The quartet-triplet ^{1}H-NMR signals of 3-pentanone showing the original trace and a scale expansion to show the signal-splitting pattern more clearly.

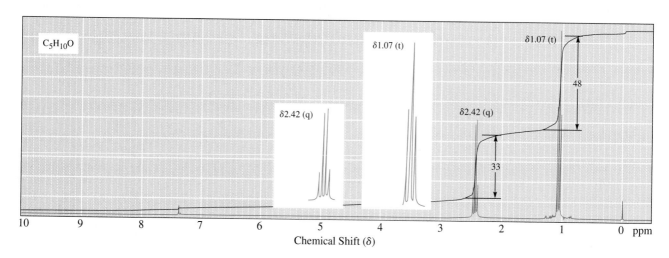

Table 13.6 Approximate Values of *J* for Compounds Containing Alkyl and Alkenyl Groups

| 6–8 Hz | 8–14 Hz | 0–5 Hz | 0–5 Hz |

| 11–18 Hz | 5–10 Hz | 0–5 Hz | 8–11 Hz |

original trace of each signal is a scale expansion in which you can see the triplet-quartet pattern more easily. You should note also that the types of splitting patterns we have described hold only when the separation between coupled signals is much greater than the coupling constant. When this is not the case, spectra can become much more complex.

Given in Table 13.6 are approximate values for coupling constants for hydrogens on singly and doubly bonded carbon atoms.

13.11 Stereochemistry and Topicity

The discussion of the number of equivalent hydrogens given in the beginning of this chapter is slightly oversimplified because stereochemistry can affect chemical shift. Depending on the symmetry of the molecule, otherwise equivalent atoms may be **homotopic, enantiotopic,** or **diastereotopic.** The simplest way to visualize the topicity of a molecule (i.e., which of these classes it falls into) is by mentally substituting one of the atoms or groups of atoms by an isotope and then deciding whether the resulting compound would be (a) the same or (b) different from its mirror image or whether (c) diastereomers are possible. Depending on the outcome of the test, the atoms or groups are homotopic, enantiotopic, or diastereotopic, respectively.

Consider the following molecules:

Dichloromethane (achiral) → Substitute one H by D → Achiral. Substitution does not create a stereocenter; therefore hydrogens are homotopic.

Chlorofluoromethane (achiral) → Substitute one H by D → Chiral. Substitution creates a stereocenter; therefore, hydrogens are enantiotopic. Both hydrogens are prochiral; one is pro-R-chiral, the other is pro-S-chiral.

Homotopic groups Atoms or groups on an atom that give an achiral molecule when one of the groups is replaced by another group. The hydrogens of the CH_2 group of propane, for example, are homotopic. Homotopic atoms or groups have identical chemical shifts under all conditions.

Enantiotopic groups Atoms or groups on an atom that give a stereocenter when one of the atoms or groups is replaced by another group. A pair of enantiomers results. The hydrogens of a CH_2 group of ethanol, for example, are enantiotopic. Enantiotopic groups have identical chemical shifts in achiral environments. In chiral environments, they have different chemical shifts.

Diastereotopic groups Atoms or groups on an atom that is bonded to two nonidentical groups, one of which contains a stereocenter. When one of the atoms or groups is replaced by another group, a new stereocenter is created and a set of diastereomers results. The hydrogens of the CH_2 group of 2-butanol, for example, are diastereotopic. Diastereotopic groups have different chemical shifts under all conditions.

Prochiral hydrogens Refers to two hydrogens bonded to a carbon atom. When one or the other of them is replaced by a different atom, the carbon atom becomes a stereocenter. The hydrogens of the CH_2 group of ethanol, for example, are prochiral. Replacing one of them by deuterium gives (R)-1-deuteroethanol; replacing the other by deuterium gives (S)-1-deuteroethanol.

Pro-R-hydrogen Replacing this hydrogen by deuterium gives a stereocenter with an R configuration.

Pro-S-hydrogen Replacing this hydrogen by deuterium gives a stereocenter with an S configuration.

If one hydrogen in dichloromethane is substituted with a deuterium, an achiral compound results. This compound is identical to its mirror image, and the two hydrogens in dichloromethane are homotopic and are identical. Homotopic groups have identical chemical shifts in all environments.

If one hydrogen in chlorofluoromethane is substituted with a deuterium, the resulting compound is chiral, and not identical to its mirror image. The two hydrogens in this compound are, therefore, enantiotopic. Enantiotopic hydrogens have identical chemical shifts except in chiral environments. In a chiral solvent, for example, the two hydrogens would have different chemical shifts. Even though the distinction between homotopic and enantiotopic compounds is of little practical consequence in NMR spectroscopy, the two hydrogens in chlorofluoromethane can be distinguished by enzymes, which also provide a chiral environment. The CH_2 hydrogens in this molecule are said to be **prochiral.**

The compound 2-butanol presents a more complex situation.

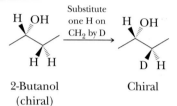

Substitute one H on CH_2 by D

2-Butanol (chiral)   Chiral

Substitution creates a stereocenter which is diastereomeric; therefore, hydrogens are diastereotopic. Both hydrogens are prochiral; one is pro-R-chiral, the other is pro-S-chiral.

If a hydrogen on one of the methyl groups on carbon-3 of 3-methyl-2-butanol is substituted with a deuterium, a new stereocenter is created. Because there is already one stereocenter, diastereomers are now possible. Thus the methyl groups on carbon-3 of 3-methyl-2-butanol are diastereotopic. Diastereotopic hydrogens have different chemical shifts under all conditions, which can lead to unexpected complexity in spectra of simple compounds. The ^1H-NMR spectrum of 3-methyl-2-butanol is shown in Figure 13.15. The methyl groups on carbon-3 are nonequivalent and give two doublets rather than one doublet of twice the intensity, which would be expected if they were equivalent.

Any molecule with a chiral center near two otherwise identical groups on a carbon with a third substituent has the potential for diastereotopicity. Of course, like any other nonequivalent groups, diastereotopic groups may have very similar or accidentally identical chemical shifts. Generally, the shift differences fall off rapidly with increasing distance from the chiral center.

Figure 13.15

^1H-NMR spectrum of 3-methyl-2-butanol (500 MHz, Ping Kang, UCLA). The methyl groups on carbon-3 are diastereotopic and therefore nonequivalent. They appear as two doublets.

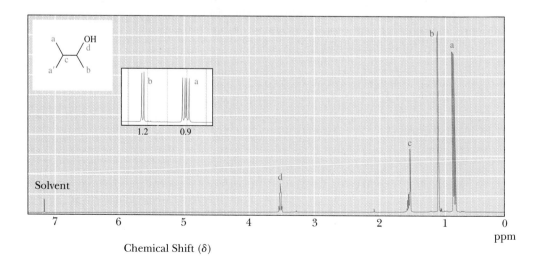

Chemical Shift (δ)

C H E M I S T R Y   I N   A C T I O N

Magnetic Resonance Imaging

The NMR phenomenon was discovered and explained by physicists in the 1950s, and by the 1960s it had become an invaluable analytical tool for chemists. It was realized by the early 1970s that imaging of parts of the body using NMR could be a valuable addition to diagnostic medicine. Because the term "nuclear magnetic resonance" sounds to many people as if the technique might involve radioactive material, health care personnel call the technique magnetic resonance imaging (MRI).

The body contains several nuclei that, in principle, could be used for MRI. Of these, hydrogens, most of which come from the hydrogens of water, triglycerides (Section 26.1), and membrane phospholipids (Section 26.5), give the most useful signals. Phosphorus MRI is also used in diagnostic medicine.

Recall that in NMR spectroscopy, energy in the form of radio-frequency radiation is absorbed by nuclei in the sample. The relaxation time is the characteristic time at which excited nuclei give up this energy and relax to their ground state.

In 1971, it was discovered that relaxation of water in certain cancerous tumors takes much longer than the relaxation of water in normal cells. Thus, if a relaxation image of the body could be obtained, it might be possible to identify tumors at an early stage. Subsequent work demonstrated that many tumors can be identified in this way. Another important application of MRI is in the examination of the brain and spinal cord. White and gray matter are easily distinguished by MRI, which is useful in the study of such diseases as multiple sclerosis. Magnetic resonance imaging and x-ray imaging are, in many cases, complementary. The hard, outer layer of bone is essentially invisible to MRI but shows up extremely well in x-ray images, whereas soft tissue is nearly transparent to x-rays but shows up in MRI.

The key to any medical imaging technique is to know which part of the body gives rise to which signal. In MRI, spatial information is encoded using magnetic field gradients. Recall that a linear relationship exists between the frequency at which a nucleus resonates and the intensity of the magnetic field. In ^{1}H-NMR spectroscopy, we use a homogeneous magnetic field in which all equivalent hydrogens absorb at the same radio frequency and have the same chemical shift. In MRI, the patient is placed in a magnetic field gradient that can be varied from place to place. Nuclei in the weaker magnetic field gradient absorb at a lower frequency. Nuclei elsewhere in the stronger magnetic field absorb at a higher frequency. In a magnetic field gradient, a correlation exists between the absorption frequency of a nucleus and its position in space. A gradient along a single axis images a plane. Two mutually perpendicular gradients image a line segment, and three mutually perpendicular gradients image a point. In practice, more complicated procedures are used to obtain magnetic resonance images, but they are all based on the idea of magnetic field gradients.

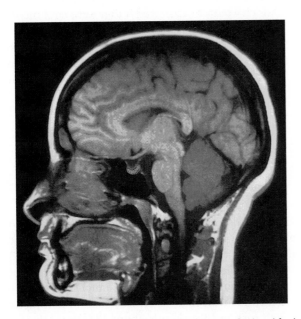

Computer-enhanced MRI scan of a normal human brain with pituitary gland highlighted. (Scott Camazine/Photo Researchers)

Example 13.7

Indicate whether the highlighted hydrogens in the following compounds are homotopic, enantiotopic, or diastereotopic.

(a) 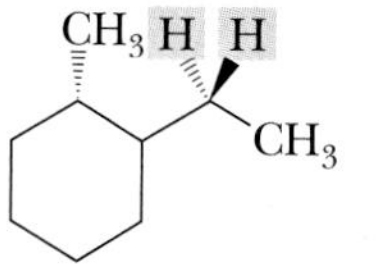

(b) CH_3CH_2Br

(c) 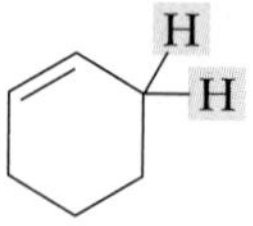

Solution

(a) Diastereotopic (near a chiral center). These hydrogens will have different chemical shifts.

(b) Enantiotopic. These hydrogens will have the same chemical shift except in chiral environments.

(c) Enantiotopic [see part (b)].

Problem 13.7

Following is a ^{1}H-NMR spectrum of 2-butanol. Explain why the CH_2 protons appear as a complex multiplet rather than as a simple quintet.

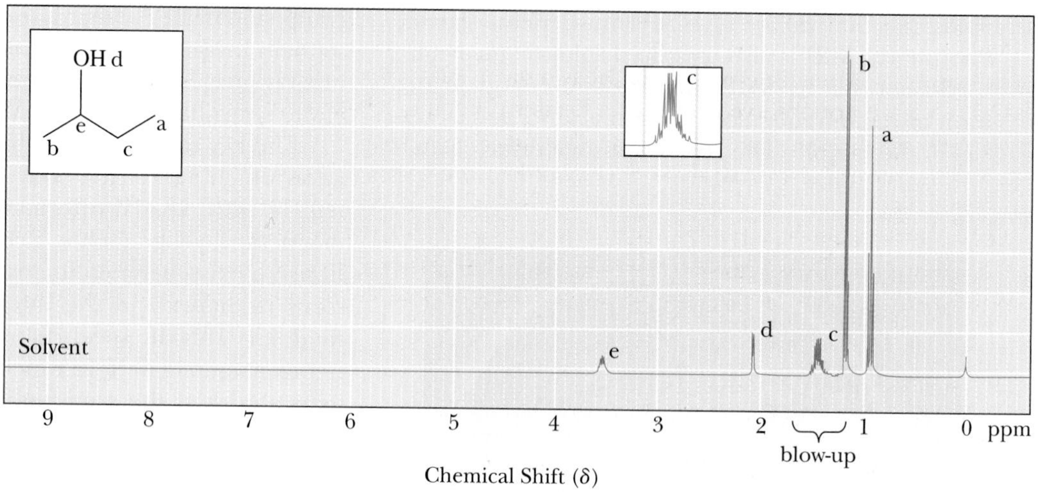

Supporting Concepts
NMR Spectroscopy
^{13}C-NMR

13.12 ^{13}C-NMR

The development of ^{13}C-NMR spectroscopy lagged behind ^{1}H-NMR spectroscopy primarily because of two problems. One is the particularly low natural abundance of ^{13}C (only 1.1%) and the resulting weak signal. The second problem is that the magnetic moment of ^{13}C is considerably smaller than that of ^{1}H, which causes the population of the higher and lower nuclear spin states to differ by much less than for ^{1}H. Taken in combination, these two factors mean that ^{13}C-NMR signals in natural samples (those not artificially enriched with carbon-13) are only about 10^{-4} times the strength of ^{1}H-NMR signals. Whereas ^{1}H-NMR spectroscopy became a routine analytical tool in the mid-1960s, it was not until 20 years later with the development of FT-NMR techniques that ^{13}C-NMR spectroscopy became widely available as a routine analytical tool.

As with ^{1}H-NMR spectra, splitting patterns in ^{13}C-NMR spectra are also explained according to the $(n + 1)$ rule. Because, in natural abundance, only 1.1% of carbon atoms are ^{13}C, almost all ^{13}C atoms in a molecule have only magnetically inactive ^{12}C next to them; therefore, ^{13}C—^{13}C signal splitting is not normally observed. However, the signal from a ^{13}C nucleus is split by the hydrogens bonded to it. The signal for a carbon-13 atom with three attached hydrogens is split to a quartet, that for an atom of carbon-13 with two attached hydrogens is split to a triplet, and so on. The ^{13}C—H signal splitting provides important information about the number of hydrogen atoms bonded to carbon. The disadvantage of ^{13}C—H signal splitting is that coupling constants of 100–250 Hz are common. Coupling constants of this magnitude correspond to 1.33–3.33 ppm at 75 MHz, which means that there can be significant overlap between signals and that splitting patterns are very often difficult to determine. In addition, there are smaller but significant couplings from hydrogens that are not directly bonded to the carbon, but are separated by two or three bonds. This extensive splitting causes the already weak signals of the ^{13}C to split into many smaller peaks that are easily lost in the noise. For this reason, the most common mode of operation of a ^{13}C-NMR spectrometer is a hydrogen-decoupled mode. (See Problem 13.30 for an interesting problem on the use of coupling constants to determine orbital hybridization.)

In the hydrogen-decoupled mode, the sample is irradiated with two different radio frequencies. The first radio frequency is used to excite ^{13}C nuclei. The second is a broad spectrum of frequencies that causes all hydrogens in the molecule to undergo rapid transitions among their nuclear spin states. On the time scale of a ^{13}C-NMR spectrum, each hydrogen is in a time-average of the two states, with the result that ^{1}H-^{13}C spin-spin interactions are not observed. The term for this process is spin-spin decoupling. In a hydrogen-decoupled spectrum, all ^{13}C signals appear as singlets. The hydrogen-decoupled ^{13}C-NMR spectrum of 1-bromobutane (Figure 13.16) consists of four singlets.

Table 13.7 shows approximate chemical shifts for carbon-13 spectroscopy. Notice how much wider the range of chemical shifts is for ^{13}C-NMR spectroscopy than for ^{1}H-NMR spectroscopy. Whereas most chemical shifts for ^{1}H-NMR spectroscopy fall within a rather narrow range of 0–10 ppm, those for ^{13}C-NMR spectroscopy cover 0–220 ppm.

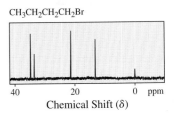

$CH_3CH_2CH_2CH_2Br$

Chemical Shift (δ)

Figure 13.16
Hydrogen-decoupled ^{13}C-NMR spectrum of 1-bromobutane.

Table 13.7 ^{13}C-NMR Chemical Shifts

Type of Carbon	Chemical Shift (δ)*	Type of Carbon	Chemical Shift (δ)*
RCH_3	10–40	⬡C—R	110–160
RCH_2R	15–55		
R_3CH	20–60		
RCH_2I	0–40	$RCOR$ (O)	160–180
RCH_2Br	25–65		
RCH_2Cl	35–80	$RCNR_2$ (O)	165–180
R_3COH	40–80		
R_3COR	40–80	$RCOH$ (O)	175–185
$RC\equiv CR$	65–85		
$R_2C=CR_2$	100–150	RCH, RCR (O)(O)	180–215

Because of this expanded scale, it is very unusual to find any two nonequivalent carbons in the same molecule with identical chemical shifts. Most commonly, each different type of carbon within a molecule has a distinct signal clearly resolved from all other signals.

Notice further that the chemical shift of carbonyl carbons is quite distinct from those of sp^3-hybridized carbons and of other types of sp^2-hybridized carbons. The presence or absence of a carbonyl carbon is quite easy to recognize in a ^{13}C-NMR spectrum. Note that signals from sp^2-hybridized carbons fall in a distinctive range of 100–160 ppm.

A great advantage of ^{13}C-NMR spectroscopy is that it generally makes it possible to count the number of types of carbon in a molecule. There is one caution here, however. Because of certain complications, including the long relaxation times of ^{13}C nuclei, it is generally not possible to determine the number of carbons of each type by integration of signal areas.

Example 13.8

Predict the number of signals in the proton-decoupled ^{13}C-NMR spectrum of each compound.

(a) $CH_3\overset{\overset{\displaystyle O}{\|}}{C}OCH_3$ (b) $CH_3CH_2CH_2\overset{\overset{\displaystyle O}{\|}}{C}CH_3$ (c) $CH_3CH_2\overset{\overset{\displaystyle O}{\|}}{C}CH_2CH_3$

Solution

Following are the number of signals in a proton-decoupled spectrum of each compound, along with the chemical shifts of each signal. The chemical shifts of the carbonyl carbons are quite distinctive (Table 13.7) and in these examples occur at δ 171.37, 208.85, and 211.97.

(a) Methyl acetate: three signals (δ 171.37, 51.53, and 20.63)
(b) 2-Pentanone: five signals (δ 208.85, 45.68, 29.79, 17.35, and 13.68)
(c) 3-Pentanone: three signals (δ 211.97, 35.45, and 7.92)

Problem 13.8

Explain how to distinguish between the members of each pair of constitutional isomers based on the number of signals in the proton-decoupled ^{13}C-NMR spectrum of each member.

13.13 The DEPT Method

We saw in Section 13.12 that there is spin-spin coupling between a ^{13}C atom and its attached hydrogens, but, because coupling constants are large and overlap of peaks is considerable, proton-coupled ^{13}C-NMR spectra are often very difficult to interpret. For these reasons, ^{13}C-NMR spectra are commonly run in the proton-decoupled mode, in which case information on C/H ratios is lost. **Distortionless Enhancement by Polarization Transfer,** or **DEPT** as it is more commonly known, provides a way to reacquire this information with good signal strength. DEPT uses complex sequences of pulses in both the ^1H and ^{13}C resonance ranges with the result that the ^{13}C signals for CH_3, CH_2, and CH exhibit different "phases." Signals for CH_3 and CH carbons are recorded as positive signals, and those for CH_2 carbons are recorded as negative signals with one pulse sequence. Using a slightly different pulse sequence, CH_3 and CH signals can be distinguished. In the DEPT technique, a carbon with no attached hydrogens, such as a carbonyl carbon or a quaternary carbon, gives no signal.

DEPT spectra may be run in several ways. In one variation, a first trace records only CH_3 signals, a second trace records only CH_2 signals, and a third trace records only CH signals. In the most common variation, CH_3, CH_2, and CH signals are recorded on one spectrum. The first trace shows CH_3 and CH as positive signals, and the second trace shows CH_2 as negative signals. Usually, the CH carbons absorb at lower field than the CH_3, so there is no ambiguity. Shown in Figure 13.17(a) is a proton-decoupled ^{13}C-NMR spectrum of isopentyl acetate showing six signals. Figure 13.17(b) is a DEPT spectrum showing color-coded signals corresponding to CH_3, CH_2, and CH groups. Note that the carbonyl carbon appears in the

DEPT-NMR Distortionless Enhancement by Polarization Transfer. A spectroscopic technique for distinguishing among ^{13}C signals for CH_3, CH_2, CH, and quaternary carbons in ^{13}C-NMR.

Figure 13.17
^{13}C-NMR spectra of isopentyl acetate. (a) Proton-decoupled spectrum and (b) DEPT spectrum.

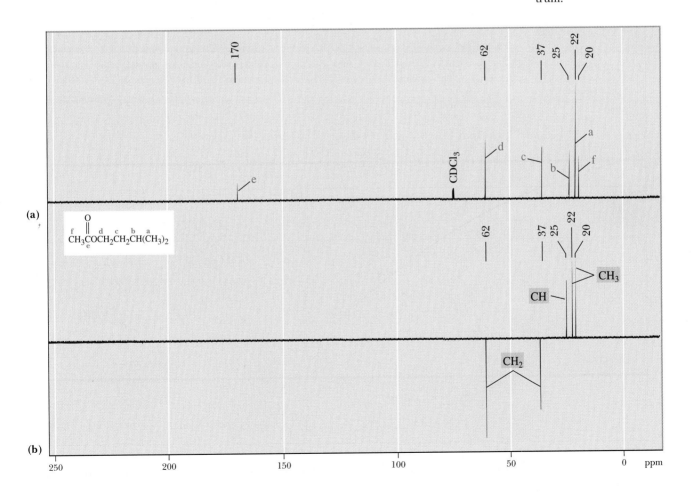

proton-decoupled spectrum but does not appear in the DEPT spectrum because it has no attached hydrogens.

Example 13.9

Assign all signals in the ^{13}C-NMR spectrum of isopentyl acetate.

Solution

The positive DEPT signals at δ 20, 22, and 24 represent CH_3 and CH groups. The taller methyl signal at δ 22 represents the two equivalent methyl groups (a), and the shorter methyl signal at δ 20 represents the single methyl group (f). The signal at δ 24, the only other positive DEPT signal, represents the CH group (b). The signal at δ 62 represents (d), the CH_2 group nearer to and more deshielded by the adjacent oxygen atom. The signal at δ 37 represents (c), the other CH_2 group. The signal at δ 170, which is not present in the DEPT spectrum, represents the carbon (e) of the carbonyl group.

Problem 13.9

Assign all signals in the ^{13}C-NMR spectrum of 4-methyl-2-pentanone.

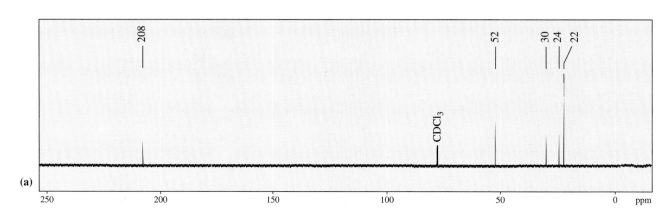

(a)

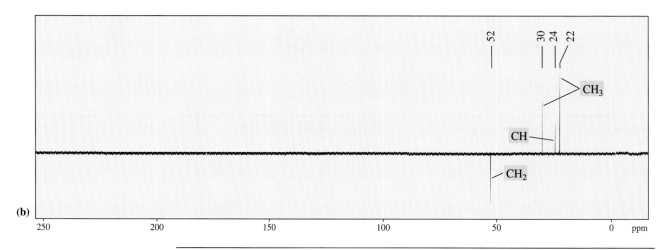

(b)

13.14 Interpreting NMR Spectra

Supporting Concepts
NMR Spectroscopy
Interpretation

A. Alkanes

All hydrogens in alkanes are in very similar chemical environments; therefore, ^{1}H-NMR chemical shifts of alkane hydrogens fall within a narrow range of δ 0.8–1.7. Chemical shifts for alkane carbons in ^{13}C-NMR spectroscopy fall within the considerably wider range of δ 0–60.

B. Alkenes

The chemical shifts of vinylic hydrogens are larger than those of alkane hydrogens and typically fall in the range δ 4.6–5.7. Vinylic hydrogens are deshielded by the sp^2-hybridized carbons of the double bond (Section 13.7B) and the local magnetic field induced in the pi bond of alkenes (Section 13.7C). The splitting pattern observed in the ^{1}H-NMR spectrum of vinyl acetate (Figure 13.18) is typical of monosubstituted alkenes. The singlet at δ 2.12 represents the three hydrogens of the methyl group. The terminal vinylic hydrogens appear at δ 4.58 and δ 4.90. The internal vinylic hydrogen, which normally appears in the range δ 5.0–5.7, is shifted farther downfield to δ 7.30 due to deshielding by the adjacent electronegative oxygen atom of the ester.

As shown in Table 13.6, coupling constants are generally larger for trans vinylic hydrogens (11–18 Hz) than for cis vinylic hydrogens (5–10 Hz), and it is often possible to distinguish between cis and trans alkenes. It is also possible to distinguish between vinylic hydrogens and geminal hydrogens ($=CH_2$), coupling constants for which are generally 0–5 Hz.

Shown in Figure 13.19 is a graphical analysis of the spectrum of vinyl acetate. The signal of each vicinal hydrogen is a doublet of doublets. The signal for H_c, for example, is split to a doublet by coupling with H_a and further split to a doublet of doublets by coupling with H_b. The sp^2-hybridized carbons of alkenes give ^{13}C-NMR signals in the range δ 100–160 ppm (Table 13.7), which is considerably downfield from sp^3-hybridized carbons.

Figure 13.18
^{1}H-NMR spectrum of vinyl acetate.

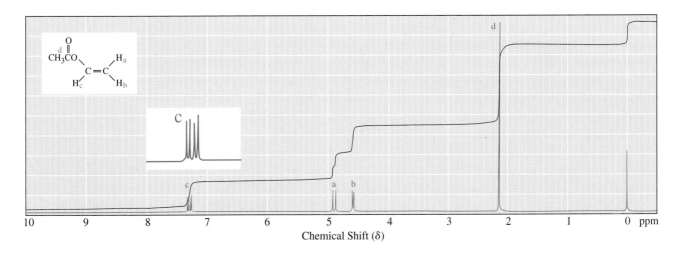

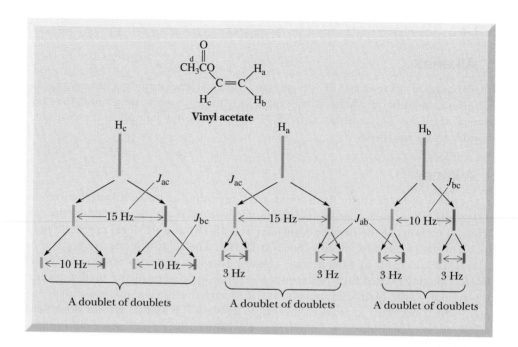

Figure 13.19
Graphical analysis of the signal-splitting patterns of the three vinylic hydrogens in vinyl acetate. The largest coupling constant, J_{ac} (15 Hz), is that for the two trans vinylic hydrogens. The smallest coupling constant, J_{ab} (3 Hz), is that for the two terminal vinylic hydrogens.

C. Alcohols

The chemical shift of a hydroxyl hydrogen in a ^{1}H-NMR spectrum is variable and depends on the purity of the sample, the solvent, the concentration, and the temperature. It often appears in the range δ 3.0–4.0, but depending on experimental conditions, it may appear as high as δ 0.5. Hydrogens on the carbon bearing the —OH group are deshielded by the electron-withdrawing inductive effect of the oxygen atom, and their absorptions also typically appear in the range δ 3.4–4.0. Shown in Figure 13.20 is the ^{1}H-NMR spectrum of 1-propanol. This spectrum consists of four signals. The hydroxyl hydrogen appears at δ 3.18 as a narrowly spaced triplet. The sig-

Figure 13.20
^{1}H-NMR spectrum of 1-propanol.

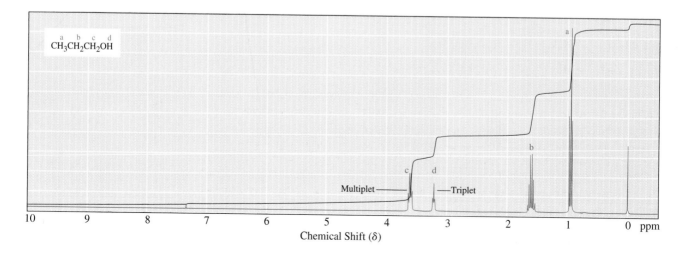

nal of hydrogens on carbon 1 in 1-propanol appears as a quartet (split by the two CH_2 hydrogens and the one OH hydrogen) at δ 3.56.

Signal splitting between the hydrogen on O—H and its neighbors on the adjacent —CH_2— group is seen in the ¹H-NMR spectrum of 1-propanol. However, this splitting is rarely seen. The reason is that most samples of alcohol contain traces of acid, base, or other impurities that catalyze the transfer of the hydroxyl proton from the oxygen of one alcohol molecule to the oxygen of another alcohol molecule. This transfer, which is very fast compared to the time scale required to make an NMR measurement, decouples the hydroxyl proton from all other protons in the molecule. For this same reason, the hydroxyl proton does not usually split the signal of any α-hydrogens.

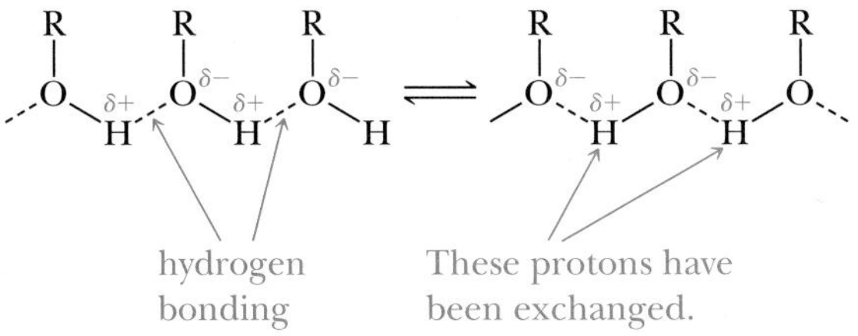

hydrogen bonding

These protons have been exchanged.

D. Ethers

The most distinctive feature of the ¹H-NMR spectrum of ethers is the chemical shift of hydrogens on the carbon is attached to the ether oxygen. Resonance signals for this type of hydrogen fall in the range δ 3.3–4.0, which corresponds to a downfield shift of approximately 2.4 units compared with their normal position in alkanes. The chemical shifts of **H—C—O—** hydrogens in ethers are similar to those seen for comparable **H—C—OH** hydrogens of alcohols.

E. Aldehydes and Ketones

Aldehyde hydrogens typically appear at δ 9.5–10.1 in the ¹H-NMR. Because almost nothing else absorbs in this region, it is very useful for identification. Hydrogens on an α-carbon of an aldehyde or ketone appear around δ 2.2–2.6. The carbonyl carbons of aldehydes and ketones have characteristic positions in the ¹³C-NMR between δ 180 and δ 215 (and can be distinguished from carboxylic acid derivatives, which absorb at higher field). The NMR spectra of aldehydes and ketones are discussed in more depth in Section 16.4.

F. Carboxylic Acids and Esters

Hydrogens on the α-carbon to a carboxyl group in acids and esters appear in the ¹H-NMR spectrum in the range δ 2.0–2.6. The hydrogen of the carboxyl group gives a very distinctive signal in the range δ 10–13, at lower field than most other types of hydrogens, even lower than an aldehyde hydrogen (δ 9.5–10.1), and serves to distinguish carboxyl hydrogens from most other types of hydrogens. The ¹H-NMR signal for the carboxyl hydrogen of 2-methylpropanoic acid, for example, appears at δ 12.2 and is shown at the left in Figure 13.21. The ¹³C absorption of the carboxyl carbon in acids and esters appears in the range δ 160–180 and is distinctly lower than that in ketones. Hydrogens α to an ester oxygen are strongly deshielded and resonate at δ 3.6–4.7, lower than in alcohols and ethers.

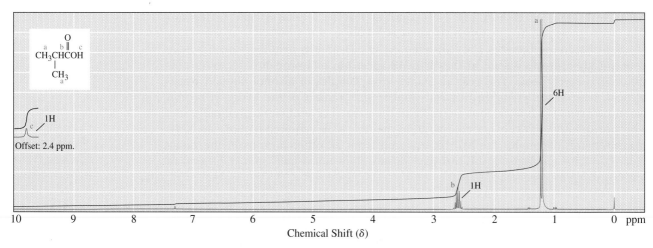

Figure 13.21
^{1}H-NMR spectrum of 2-methylpropanoic acid (isobutyric acid).

G. Amines

The chemical shifts of amine hydrogens, like those of hydroxyl hydrogens (Section 13.14C), vary between δ 0.5 and δ 5.0, depending on conditions. As in alcohols, exchange is fast enough that spin-spin splitting between amine hydrogens and hydrogens on adjacent α-carbons is averaged. Thus, amine hydrogens generally appear as broad singlets. Coupling to ^{14}N (beyond the scope of this text) causes these signals to broaden. Hydrogens α to the amine nitrogen appear around δ 2.5 ppm, about 1 ppm higher than for hydrogens α to oxygen in ethers and alcohols.

Carbons bonded to nitrogen appear in the ^{13}C-NMR spectrum approximately 20 ppm lower than in alkanes of comparable structure, but about 20 ppm above carbons attached to oxygen in ethers or alcohols. The NMR spectra of amines is discussed in more detail in Section 22.5B.

13.15 Solving NMR Problems

One of the first steps in determining molecular structure is establishing the molecular formula. In the past, this was most commonly done by elemental analysis, combustion analysis to determine percent composition, molecular weight determination, and so forth. More commonly today, molecular weight and molecular formula are determined by mass spectrometry (Chapter 14). In the examples that follow, we assume that the molecular formula of any unknown compound has already been determined, and we proceed from that point using spectral analysis to determine a structural formula.

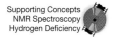

A. Index of Hydrogen Deficiency

Valuable information about the structural formula of an unknown compound can be obtained from inspection of its molecular formula. In addition to learning the number of atoms of carbon, hydrogen, oxygen, nitrogen, and so forth in a molecule of the compound, we can also determine what is called its index of hydrogen deficiency. For each ring and pi bond, the molecular formula has two fewer hydrogens. The **index of hydrogen deficiency** is the sum of the number of rings and pi bonds in a molecule. It is determined by comparing the number of hydrogens in the molecular formula of a compound whose structure is to be determined with the number of hydrogens in a reference alkane of the same number of carbon atoms. The molecular formula of a reference alkane is C_nH_{2n+2} (Section 2.1).

$$\text{Index of hydrogen deficiency} = \frac{(H_{reference} - H_{molecule})}{2}$$

Example 13.10

Calculate the index of hydrogen deficiency for 1-hexene, C_6H_{12}, and account for this deficiency by reference to its structural formula.

Solution

The molecular formula of the reference hydrocarbon of six carbon atoms is C_6H_{14}. The index of hydrogen deficiency of 1-hexene is $(14 - 12)/2 = 1$ and is accounted for by the one pi bond in 1-hexene.

Problem 13.10

Calculate the index of hydrogen deficiency of cyclohexene, C_6H_{10}, and account for this deficiency by reference to its structural formula.

To compare the molecular formula for a compound containing elements besides carbon and hydrogen, write the formula of the reference hydrocarbon with the same number of carbon atoms, and make the following adjustments to the number of hydrogen atoms in the unknown.

1. For each atom of a Group 7 element (F, Cl, Br, I), replace it by one hydrogen; halogen substitutes for hydrogen and reduces the number of hydrogens by one per halogen. The general formula of an acyclic monochloroalkane, for example, is $C_nH_{2n+1}Cl$; the general formula of the corresponding acyclic alkane is C_nH_{2n+2}.
2. No correction is necessary for the addition of atoms of Group 6 elements (O, S, Se). Insertion of a divalent Group 6 element into a hydrocarbon does not change the number of hydrogens.
3. For each atom of a Group 5 element (N, P, As) added, subtract one hydrogen, because insertion of a trivalent Group 5 element adds one hydrogen to the molecular formula. The general molecular formula for an acyclic alkylamine, for example, is $C_nH_{2n+3}N$.

Example 13.11

Isopentyl acetate, a compound with a banana-like odor, is a component of the alarm pheromone of honeybees. The molecular formula of isopentyl acetate is $C_7H_{14}O_2$. Calculate the index of hydrogen deficiency of this compound.

Solution

The molecular formula of the reference hydrocarbon is C_7H_{16}. Adding oxygens does not require any correction in the number of hydrogens. The index of hydrogen deficiency is $(16 - 14)/2 = 1$, indicating either one ring or one pi bond. Following is the structural formula of isopentyl acetate. It contains one pi bond, in this case, a carbon-oxygen pi bond.

Isopentyl acetate

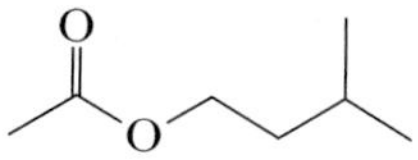

Problem 13.11

The index of hydrogen deficiency of niacin is 5. Account for this index of hydrogen deficiency by reference to the structural formula of niacin.

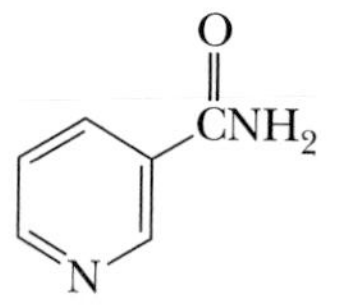

Nicotinamide
(Niacin)

Spectroscopy
The CD-ROM contains a database of nearly 200 compounds for which NMR spectra can be viewed and analyzed.

B.   From a ^{1}H-NMR Spectrum to a Structural Formula

The following steps may prove helpful as a systematic approach to solving spectral problems.

Step 1:   *Molecular formula and index of hydrogen deficiency.* Examine the molecular formula, calculate the index of hydrogen deficiency, and deduce what information you can about the presence or absence of rings or pi bonds.

Step 2:   *Number of signals.* Count the number of signals to determine the number of sets of equivalent hydrogens present in the compound.

Step 3:   *Integration.* Use the integration and the molecular formula to determine the numbers of hydrogens present in each set.

Types of Hydrogens	Descriptive Name	Typically Absorb in the Range (δ)
RCH_3 RCH_2R R_3CH	Alkyl hydrogens	0.8–1.7
$R_2C{=}CRCHR_2$	Allylic hydrogens	1.6–2.6
RCH_2OH RCH_2OR	Hydrogens on a carbon adjacent to an sp^3-hybridized oxygen	3.3–4.0
$R_2C{=}CH_2$ $R_2C{=}CHR$	Vinylic hydrogens	4.6–5.7
ArH	Aromatic hydrogens	6.8–8.5
$\overset{\displaystyle O}{\overset{\displaystyle \|}{RCH}}$	Aldehyde hydrogens	9.5–10.1
$\overset{\displaystyle O}{\overset{\displaystyle \|}{RCOH}}$	Carboxyl hydrogens	10–13

Step 4: *Pattern of chemical shifts.* Examine the NMR spectrum for signals characteristic of the types of equivalent hydrogens shown in the table on the previous page. Keep in mind that these are broad ranges and that hydrogens of each type may be shifted either farther upfield or farther downfield, depending on the details of the molecular structure.

Step 5: *Signal-splitting patterns.* Examine splitting patterns for information about the number of nearest nonequivalent hydrogen neighbors.

Step 6: *Structural formula.* Write a structural formula consistent with the previous information.

Spectral Problem 1

Molecular formula $C_5H_{10}O$.

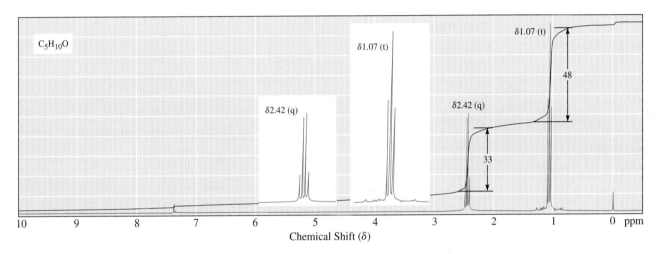

Analysis of Spectral Problem 1

Step 1: *Molecular formula and index of hydrogen deficiency.* The reference compound is C_5H_{12}; therefore, the index of hydrogen deficiency is 1, and the molecule contains either one ring or one pi bond.

Step 2: *Number of signals.* There are two signals (a triplet and a quartet) and, therefore, two sets of equivalent hydrogens.

Step 3: *Integration.* From the integration, the hydrogens in each set are in the ratio 3:2. Because there are 10 hydrogens, 6H must give rise to the signal at δ 1.07, and 4H must give rise to the signal at δ 2.42.

Step 4: *Pattern of chemical shifts.* The signal at δ 1.07 is in the alkyl region and, based on its chemical shift, most probably indicates a methyl group. No signal occurs between δ 4.6 and δ 5.7; there are no vinylic hydrogens. If a carbon–carbon double bond is in the molecule, no hydrogens are on it (that is, it is tetrasubstituted). The chemical shift of the four protons at δ 2.42 is consistent with two CH_2 groups next to a carbonyl group.

Step 5: *Signal-splitting patterns.* The methyl signal at δ 1.07 is split into a triplet (t); it must have two neighbors, indicating $-CH_2CH_3$. The signal at δ 2.42 is split into a quartet (q); it must have three neighbors. An ethyl group accounts for these two signals. No other signals occur in the spectrum; therefore, there are no other types of hydrogens in the molecule.

Step 6: *Structural formula.* Put this information together to arrive at the following structural formula. The chemical shift of the methylene group (—CH$_2$—) at δ 2.42 is consistent with an alkyl group adjacent to a carbonyl group.

$$CH_3-CH_2-\overset{\overset{\textstyle O}{\|}}{C}-CH_2-CH_3$$

δ 2.42 (q)
δ 1.07 (t)

3-Pentanone

Spectral Problem 2

Molecular formula C$_7$H$_{14}$O.

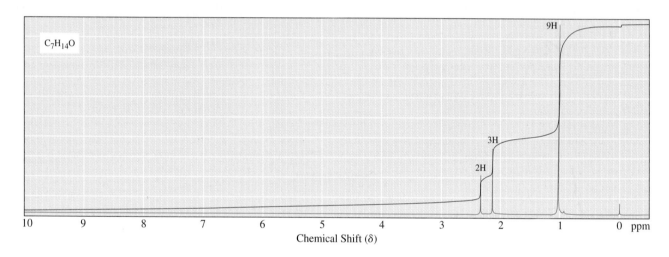

Analysis of Spectral Problem 2

Step 1: *Molecular formula and index of hydrogen deficiency.* The index of hydrogen deficiency is 1; the compound contains one ring or one pi bond.

Step 2: *Number of signals.* There are three signals and, therefore, three sets of equivalent hydrogens.

Step 3: *Integration.* Reading from right to left, the number of hydrogens in these signals are 9, 3, and 2.

Step 4: *Pattern of chemical shifts.* The signal at δ 1.01 is characteristic of a methyl group adjacent to an sp^3-hybridized carbon. The signals at δ 2.11 and δ 2.32 are characteristic of alkyl groups adjacent to a carbonyl group.

Step 5: *Signal-splitting pattern.* All signals are singlets (s). Therefore, none of the groups has hydrogens on neighboring carbons.

Step 6: *Structural formula.* The compound is 4,4-dimethyl-2-pentanone.

$$\delta\ 1.01\ (s) \searrow \quad \overset{CH_3}{\underset{CH_3}{\underset{|}{\overset{|}{CH_3-C-CH_2-C-CH_3}}}} \quad \overset{\delta\ 2.32\ (s)}{\overset{O}{\overset{\|}{}}} \nearrow \delta\ 2.11\ (s)$$

4,4-Dimethyl-2-pentanone

Summary

Nuclei of ^1H and ^{13}C have a nuclear spin quantum number of $\frac{1}{2}$ and allowed nuclear spin states of $+\frac{1}{2}$ and $-\frac{1}{2}$ (Section 13.1). In the presence of an applied magnetic field, B_0, nuclei with spin $+\frac{1}{2}$ are aligned with the applied field and are in the lower energy state; nuclei with spin $-\frac{1}{2}$ are aligned against the applied field and are in the higher energy state (Section 13.2).

When placed in a powerful magnetic field (Section 13.3), ^1H and ^{13}C nuclei become aligned in an allowed spin state and precess about the applied field. **Resonance** is the absorption of electromagnetic radiation by a precessing nucleus and the resulting flip of its nuclear spin from the lower energy spin state to the higher energy spin state. An NMR spectrometer records such resonance as a **signal.**

The experimental conditions required to cause nuclei to resonate are affected by the local chemical and magnetic environment. Electrons around a hydrogen or carbon create local magnetic fields that shield the nuclei of these atoms from the applied field (Section 13.3). Any factor that increases the exposure of nuclei to an applied field is said to **deshield** them and shift their signal downfield to a larger δ value. Conversely, any factor that decreases the exposure of nuclei to an applied field is said to **shield** them and shift their signal upfield to a smaller δ value.

Equivalent hydrogens within a molecule have identical chemical shifts (Section 13.5). The chemical shift of a particular set of equivalent hydrogens depends primarily on three factors: (1) nearby electronegative atoms have a deshielding effect; (2) the greater the percent of s-character in a hybrid orbital, the greater the deshielding effect of the atom to which the orbital belongs; and (3) induced local magnetic fields in pi bonds either add to or subtract from the applied field. The area of a ^1H-NMR signal is proportional to the number of equivalent hydrogens giving rise to that signal (Section 13.6).

The resonance signals in ^1H-NMR spectra are reported by how far they are shifted from the resonance signal of the 12 equivalent hydrogens in tetramethylsilane (TMS). The resonance signals in ^{13}C-NMR spectra are reported by how far they are shifted from the resonance signal of the four equivalent carbons in TMS (Section 13.3). Chemical shift, δ (Section 13.7), is defined as the frequency shift from TMS divided by the operating frequency of the spectrometer.

According to the **(n + 1) rule,** if a hydrogen has n hydrogens nonequivalent to it but equivalent among themselves on the same or adjacent atom(s), its ^1H-NMR signal is split into (n + 1) peaks (Section 13.8). Splitting patterns are commonly referred to as singlets (s), doublets (d), triplets (t), quartets (q), quintets, and multiplets (m). The relative intensities of peaks in a multiplet can be predicted from an analysis of spin combinations for adjacent hydrogens (Table 13.5) or from the mnemonic device called **Pascal's triangle** (Figure 13.12).

A **coupling constant (J)** is the distance between adjacent peaks in a multiplet and is reported in hertz (Section 13.10). The value of J depends only on internal fields within a molecule and is independent of spectrometer field.

Groups of atoms in which substitution of one atom by an isotope creates an achiral molecule are called **homotopic** (Section 13.11). Those in which such substitution produces a chiral molecule are **enantiotopic.** Those molecules that can give rise to diastereomers on such substitution are called **diastereotopic.** Homotopic groups always have identical chemical shifts. Enantiotopic groups also do, except in a chiral environment. Diastereotopic groups, however, are nonequivalent in all environments.

There are four important types of structural information obtained from a ^1H-NMR spectrum.

- From the number of signals, we can determine the number of sets of equivalent hydrogens.
- From the integration of signal areas, we can determine the relative numbers of hydrogens in each set.
- From the chemical shift of each signal, we can derive information about the chemical environment of the hydrogens in each set.
- From the splitting pattern of each signal, we can derive information about the number and chemical equivalency of hydrogens on the same and adjacent carbon atoms.

^{13}C-NMR spectra (Section 13.12) are commonly recorded in a hydrogen-decoupled instrumental mode. In this mode, all ^{13}C signals appear as singlets. The **DEPT** method can be used to identify CH_3, CH_2, and CH signals separately (Section 13.13).

Problems

Index of Hydrogen Deficiency

13.12 Complete the following table.

Class of Compound	General Molecular Formula	Index of Hydrogen Deficiency	Reason for Hydrogen Deficiency
Alkane	C_nH_{2n+2}	0	(Reference hydrocarbon)
Alkene	C_nH_{2n}	1	One pi bond
Alkyne	_____	___	_____
Alkadiene	_____	___	_____
Cycloalkane	_____	___	_____
Cycloalkene	_____	___	_____
Bicycloalkane	_____	___	_____

13.13 Calculate the index of hydrogen deficiency of these compounds.

(a) Aspirin, $C_9H_8O_4$ (b) Ascorbic acid (vitamin C), $C_6H_8O_6$
(c) Pyridine, C_5H_5N (d) Urea, CH_4N_2O
(e) Cholesterol, $C_{27}H_{46}O$ (f) Dopamine, $C_8H_{11}NO_2$

Interpretation of ^{1}H-NMR and ^{13}C-NMR Spectra

13.14 Complete the following table. Which nucleus requires the least energy to flip its spin at this applied field? Which nucleus requires the most energy?

Nucleus	Applied Field (tesla, T)	Radio Frequency (MHz)	Energy (J/mol)
^{1}H	7.05	300	_____
^{13}C	7.05	75.5	_____
^{19}F	7.05	282	_____

13.15 The natural abundance of ^{13}C is only 1.1%. Furthermore, its sensitivity in NMR spectroscopy (a measure of the energy difference between a spin aligned with or against an external magnetic field) is only 1.6% that of ^{1}H. What are the relative signal intensities expected for the ^{1}H-NMR and ^{13}C-NMR spectra of the same sample of $Si(CH_3)_4$?

13.16 Following are structural formulas for three constitutional isomers of molecular formula $C_7H_{16}O$ and three sets of ^{13}C-NMR spectral data. Assign each constitutional isomer its correct spectral data.

(a) $CH_3CH_2CH_2CH_2CH_2CH_2CH_2OH$

(b) $CH_3\overset{\text{OH}}{\underset{\text{CH}_3}{C}}CH_2CH_2CH_2CH_3$

(c) $CH_3CH_2\overset{\text{OH}}{\underset{\text{CH}_2CH_3}{C}}CH_2CH_3$

Spectrum 1	Spectrum 2	Spectrum 3
74.66	70.97	62.93
30.54	43.74	32.79
7.73	29.21	31.86
	26.60	29.14
	23.27	25.75
	14.09	22.63
		14.08

13.17 Following are structural formulas for the cis isomers of 1,2-, 1,3-, and 1,4-dimethylcyclo-hexane and three sets of ^{13}C-NMR spectral data. Assign each constitutional isomer its correct spectral data.

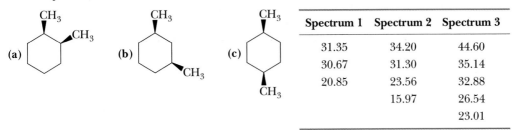

	Spectrum 1	Spectrum 2	Spectrum 3
	31.35	34.20	44.60
	30.67	31.30	35.14
	20.85	23.56	32.88
		15.97	26.54
			23.01

13.18 Following are structural formulas, dipole moments, and ^{1}H-NMR chemical shifts for acetonitrile, fluoromethane, and chloromethane.

$$CH_3C\equiv N \qquad CH_3F \qquad CH_3Cl$$

Acetonitrile	Fluoromethane	Chloromethane
3.92 D	1.85 D	1.87 D
δ 1.97	δ 4.26	δ 3.05

(a) How do you account for the fact that the dipole moments of fluoromethane and chloromethane are almost identical even though fluorine is considerably more electronegative than chlorine?

(b) How do you account for the fact that the dipole moment of acetonitrile is considerably greater than that of either fluoromethane or chloromethane?

(c) How do you account for the fact that the chemical shift of the methyl hydrogens in acetonitrile is considerably less than that for either fluoromethane or chloromethane?

13.19 Following are three compounds of molecular formula $C_4H_8O_2$, and three ^{1}H-NMR spectra. Assign each compound its correct spectrum and assign all signals to their corresponding hydrogens.

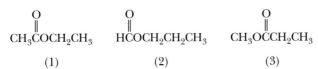

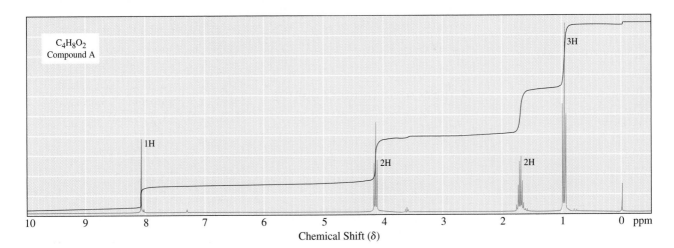

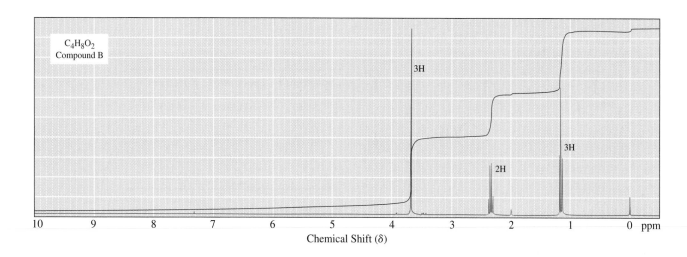

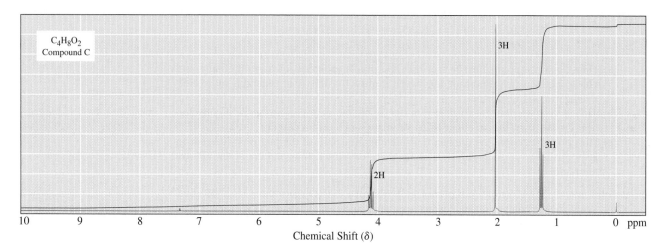

13.20 Following are ^{1}H-NMR spectra for compounds D, E, and F, each of molecular formula C_6H_{12}. Each readily decolorizes a solution of Br_2 in CCl_4. Propose structural formulas for compounds D, E, and F, and account for the observed patterns of signal splitting.

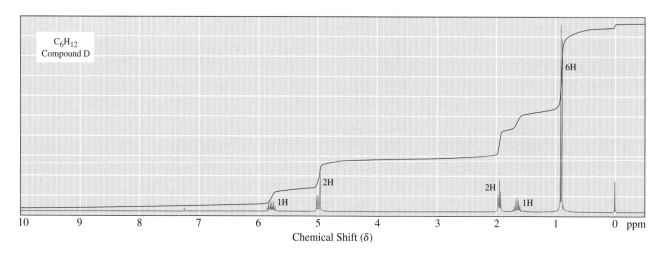

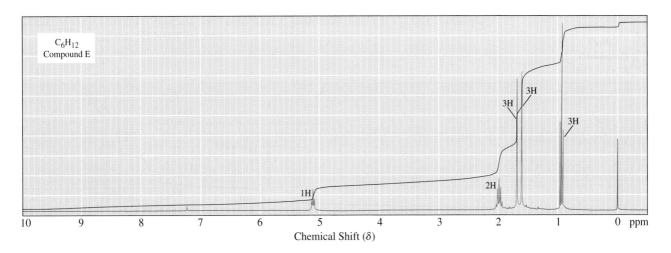

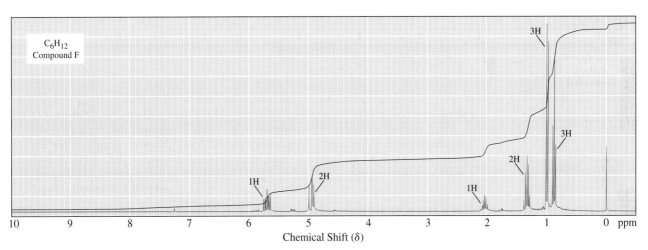

13.21 Following are ^{1}H-NMR spectra for compounds G, H, and I, each of molecular formula $C_5H_{12}O$. Each is a liquid at room temperature, is slightly soluble in water, and reacts with sodium metal with the evolution of a gas.

   (a) Propose structural formulas of compounds G, H, and I.

   (b) Explain why there are four lines between δ 0.86 and 0.90 in Spectrum G.

   (c) Explain why the 2H multiplets at 1.5 and 3.5 ppm for compound H are so complex.

13.25 Following is the ^{1}H-NMR spectrum of compound O, molecular formula C_7H_{12}. Compound O reacts with bromine in carbon tetrachloride to give a compound of molecular formula $C_7H_{12}Br_2$. The ^{13}C-NMR spectrum of compound O shows signals at 150.12, 106.43, 35.44, 28.36, and 26.36. Deduce the structural formula of compound O.

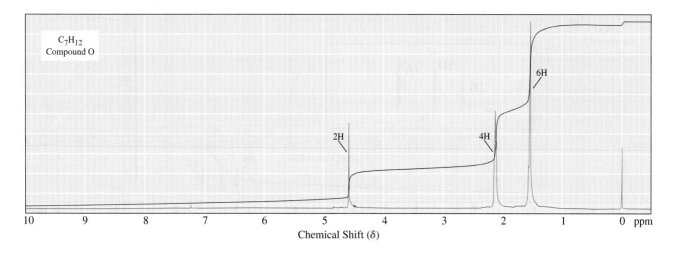

13.26 Treatment of compound P with BH_3 followed by $H_2O_2/NaOH$ gives compound Q. Following are ^{1}H-NMR spectra for compounds P and Q along with ^{13}C-NMR spectral data. From this information, deduce structural formulas for compounds P and Q.

$$C_7H_{12} \xrightarrow[\text{2. H}_2\text{O}_2,\ \text{NaOH}]{\text{1. BH}_3} C_7H_{14}O$$

(P)   (Q)

^{13}C-NMR	
(P)	**(Q)**
132.38	72.71
32.12	37.59
29.14	28.13
27.45	22.68

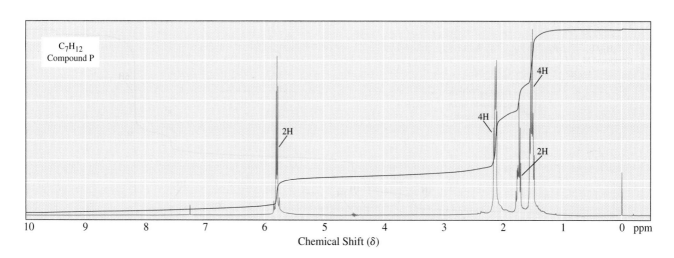

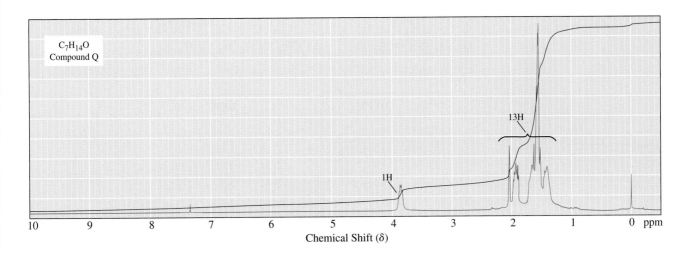

C$_7$H$_{14}$O
Compound Q

13H

1H

10   9   8   7   6   5   4   3   2   1   0 ppm

Chemical Shift (δ)

13.27 The ^{1}H-NMR of compound R, C$_6$H$_{14}$O, consists of two signals: δ 1.1 (doublet) and δ 3.6 (septet) in the ratio 6:1. Propose a structural formula for compound R consistent with this information.

13.28 Following are eight structural formulas along with the ^{13}C-NMR and DEPT-NMR spectral information. Given this information, assign each carbon in each compound its correct ^{13}C chemical shift.

(a) CH$_3$CH$_2$CH$_2$CHCH$_3$ (with Br on the CH)

(b) CH$_3$CH$_2$C=CH$_2$ (with CH$_3$ on the C)

(c) CH$_2$=CHCH$_2$CHCH$_3$ (with CH$_3$ on the CH)

^{13}C	DEPT		^{13}C	DEPT		^{13}C	DEPT
51.55	CH		147.70	—		137.81	CH
43.22	CH$_2$		108.33	CH$_2$		115.26	CH$_2$
26.46	CH$_2$		30.56	CH$_2$		43.35	CH$_2$
21.00	CH$_3$		22.47	CH$_3$		28.12	CH$_3$
13.40	CH$_3$		12.23	CH$_3$		22.26	CH$_3$

(d) CH$_3$CCH$_2$Br (with CH$_3$ groups above and below the central C)

(e) CH$_3$CH$_2$CCH$_2$CH$_3$ (with O double bonded to central C)

(f) CH$_3$CH$_2$CCH$_3$ (with O double bonded to C)

^{13}C	DEPT		^{13}C	DEPT		^{13}C	DEPT
49.02	CH$_2$		207.8	—		208.7	—
33.15	—		35.1	CH$_2$		37.6	CH$_2$
28.72	CH$_3$		7.5	CH$_2$		30.1	CH$_3$
						9.2	CH$_3$

$$\text{(g)} \quad CH_3CHCOCH_3$$
with O double bond on C, CH₃ substituent

$$\text{(h)} \quad CH_3COCH_2CH_2CHCH_3$$
with O double bond, CH₃ substituent

^{13}C	DEPT
177.48	—
51.50	CH_3
33.94	CH
19.01	CH_3

^{13}C	DEPT
171.17	—
63.12	CH_2
37.21	CH_2
25.05	CH
24.45	CH_3
21.02	CH_3

13.29 Write structural formulas for the following compounds.

(a) $C_2H_4Br_2$: δ 2.5 (d, 3H) and 5.9 (q, 1H)
(b) $C_4H_8Cl_2$: δ 1.60 (d, 3H), 2.15 (m, 2H), 3.72 (t, 2H), and 4.27 (m, 1H)
(c) $C_5H_8Br_4$: δ 3.6 (s, 8H)
(d) C_4H_8O: δ 1.0 (t, 3H), 2.1 (s, 3H), and 2.4 (quartet, 2H)
(e) $C_4H_8O_2$: δ 1.2 (t, 3H), 2.1 (s, 3H), and 4.1 (quartet, 2H); contains an ester
(f) $C_4H_8O_2$: δ 1.2 (t, 3H), 2.3 (quartet, 2H), and 3.6 (s, 3H); contains an ester
(g) C_4H_9Br: δ 1.1 (d, 6H), 1.9 (m, 1H), and 3.4 (d, 2H)
(h) $C_6H_{12}O_2$: δ 1.5 (s, 9H) and 2.0 (s, 3H)
(i) $C_7H_{14}O$: δ 0.9 (t, 6H), 1.6 (sextet, 4H), and 2.4 (t, 4H)
(j) $C_5H_{10}O_2$: δ 1.2 (d, 6H), 2.0 (s, 3H), and 5.0 (septet, 1H)
(k) $C_5H_{11}Br$: δ 1.1 (s, 9H) and 3.2 (s, 2H)
(l) $C_7H_{15}Cl$: δ 1.1 (s, 9H) and 1.6 (s, 6H)

13.30 The percent s-character of carbon participating in a C—H bond can be established by measuring the ^{13}C—^1H coupling constant and using the relationship

$$\text{Percent } s\text{-character} = 0.2\, J(^{13}C-^1H)$$

The ^{13}C—^1H coupling constant observed for methane, for example, is 125 Hz, which gives 25% s-character, the value expected for an sp^3-hybridized carbon atom.

(a) Calculate the expected ^{13}C—^1H coupling constant in ethylene and acetylene.
(b) In cyclopropane, the ^{13}C—^1H coupling constant is 160 Hz. What is the hybridization of carbon in cyclopropane?

13.31 Ascaridole is a natural product that has been used to treat intestinal worms. Explain why the two methyls on the isopropyl group in ascaridole appear in its ^1H-NMR spectrum as four lines of equal intensity, with two sets of two each separated by 7 Hz.

13.32 The ^{13}C-NMR spectrum of 3-methyl-2-butanol shows signals at δ 17.88 (CH_3), 18.16 (CH_3), 20.01 (CH_3), 35.04 (carbon-3), and 72.75 (carbon-2). Account for the fact that each methyl group in this molecule gives a different signal.

Ascaridole

MASS SPECTROMETRY

Mass spectrometry is an analytical technique for measuring the mass-to-charge ratio (m/z) of ions, most commonly positive ions. The principles of mass spectrometry were first recognized in 1898. In 1911, J. J. Thomson recorded the first mass spectrum, that of neon, and discovered that this element can be separated into a more abundant isotope, ^{20}Ne, and a less abundant isotope, ^{22}Ne. Using improved instrumentation, F. W. Aston showed that most of the naturally occurring elements are mixtures of isotopes. It was found, for example, that approximately 75% of chlorine atoms in nature are ^{35}Cl, and 25% are ^{37}Cl. Mass spectrometry did not come into common use, however, until the 1950s at which time commercial instruments that offered high resolution, reliability, and relatively inexpensive maintenance became available. Today, mass spectrometry is our most valuable analytical tool for the determination of accurate molecular weights. Furthermore, extensive

■ Crystals of dopamine viewed under polarized light. For a partial mass spectrum of dopamine, see Figure 14.2. *(Herb Charles Ohlmeyer/Fran Heyl Associates)* Inset: A model of dopamine.

information about the molecular formula and structure of a compound can be obtained from analysis of its mass spectrum. Mass spectroscopy is becoming increasingly important in biochemistry as well; sequencing of proteins using this technique alone allows protein structures to be determined on a virtually single-cell scale.

14.1 A Mass Spectrometer

A mass spectrometer (Figure 14.1) is designed to do three things:

1. Convert neutral atoms or molecules into a beam of positive or negative ions.
2. Separate the ions on the basis of their mass-to-charge (m/z) ratio.
3. Measure the relative abundance of each type of ion.

From this information, we can determine both the molecular weight and the molecular formula of an unknown compound. In addition, we can obtain valuable clues about the molecular structure of the compound.

There are many types of mass spectrometers; we have space in this text to describe only the simplest. In the most common type of spectrometer, a vaporized sample in an evacuated ionization chamber is bombarded with high-energy electrons that cause electrons to be stripped from molecules of the sample. [Increasingly, radical anions (in which an extra electron has been added to a molecule) are studied; these are beyond the scope of this text.] The resulting positive ions are accelerated by a series of negatively charged accelerator plates into an analyzing chamber about which is placed a magnetic (electric in some spectrometers) field perpendicular to the direction of the ion beam. The magnetic field causes the ion beam to curve. The radius of curvature of each ion depends on the charge on the ion (z), its mass (m), the accelerating voltage, and the strength of the magnetic field. A mass spectrum is a plot of relative ion abundance versus m/z ratio.

Samples of gases and volatile liquids can be introduced directly into the ionization chamber. Because the interior of a mass spectrometer is kept at a high vacuum, volatile liquids and even some solids are vaporized instantly. For less volatile liquids and solids, the sample may be placed on the tip of a heated probe that is then in-

Figure 14.1

Schematic diagram of an electron ionization mass spectrometer.

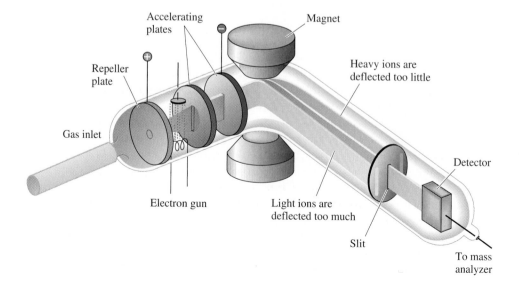

serted directly into the ionization chamber. Another extremely useful method for introducing a sample into the ionization chamber is to link a gas chromatograph (GC) directly to the mass spectrometer. Each fraction eluted from the GC is monitored and passed directly into the ionization chamber of the mass spectrometer.

Once in the ionization chamber, molecules of the sample are bombarded with a stream of high-energy electrons that are emitted from a hot filament and then accelerated by an electric field to energies of approximately 70 eV [1 eV = 96.49 kJ (23.05 kcal)/mol]. Collisions between molecules of the sample and these high-energy electrons result in loss of electrons from sample molecules to form positive ions. A **molecular ion, M‡**, is the species formed by removal of a single electron from a molecule. A molecular ion belongs to a class of ions called **radical cations.** When methane, for example, is bombarded with high-energy electrons, an electron is dislodged from a molecule to give a molecular ion of m/z 16.

$$
\begin{array}{c}
\text{H} \\
| \\
\text{H} \!-\! \overset{\displaystyle |}{\underset{\displaystyle |}{\text{C}}} \!-\! \text{H} + e^- \\
\text{H}
\end{array}
\longrightarrow
\left[
\begin{array}{c}
\text{H} \\
| \\
\text{H} \!-\! \overset{\displaystyle |}{\underset{\displaystyle |}{\text{C}}} \!-\! \text{H} \\
\text{H}
\end{array}
\right]^{\ddagger}
+ \; 2e^-
$$

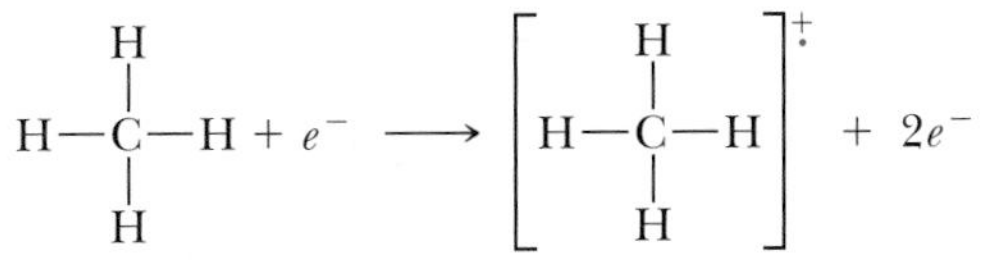

Molecular ion
(a radical cation)

> **Molecular ion (M‡)**   The cation formed by removal of a single electron from a parent molecule in a mass spectrometer.
> **Radical cation**   A species formed when a neutral molecule loses one electron; it contains both an odd number of electrons and a positive charge.

Which electron is lost in forming the molecular ion is determined by the **ionization potential** of the atom or molecule. Ionization potentials for most organic molecules are between 8 and 15 eV. They are at the lower end of this range for nonbonding electrons of oxygen and nitrogen, and for pi electrons in unsaturated compounds such as alkenes, alkynes, and aromatic hydrocarbons. Ionization potentials for sigma electrons, such as those of C—C, C—H, and C—O sigma bonds, are at the higher end of the range.

> **Ionization potential (IP)**   The minimum energy required to remove an electron from an atom or molecule to a distance where there is no electrostatic interaction between the resulting ion and electron.

For our purposes, it doesn't matter which electron is lost because, in general, the radical cation is delocalized throughout the molecule. Therefore, we write the molecular formula of the parent molecule in brackets with a plus sign to show that it is a cation and with a dot to show that it has an odd number of electrons. See, for example, the molecular ion for ethyl isopropyl ether, shown on the left. At times, however, we will find it useful to depict the radical cation in a certain position to better understand its reactions as in the formula on the right.

$[\text{CH}_3\text{CH}_2\text{OCH(CH}_3)_2]^{\ddagger}$ $\text{CH}_3\text{CH}_2\overset{\cdot+}{\underset{\cdot\cdot}{\text{O}}}\text{CH(CH}_3)_2$

As we shall see in Section 14.2D, a molecular ion can undergo fragmentation to form a variety of smaller cations (which themselves may undergo further fragmentation), as well as radicals and smaller molecules. Of these smaller fragments, the most common types of mass spectrometer detect only cations.

Once molecular ions and their fragmentation cations have been formed, a positively charged repeller plate directs them toward a series of negatively charged accelerator plates, producing a rapidly traveling ion beam. The ion beam is then focused by one or more slits and passed into a mass analyzer where it is placed in a magnetic field perpendicular to the direction of the ion beam. The magnetic field causes the ion beam to curve. Cations with larger values of m/z are deflected less than those with smaller m/z values. By varying either the accelerating voltage or the strength of the magnetic field, cations of the same m/z ratio can be focused on a detector, where the cation current is recorded. Modern detectors are capable of detecting single cations and of scanning a desired mass-to-charge region in a few tenths of a second or less.

Mass spectrum A plot of the relative abundance of ions versus their mass-to-charge ratio.

Base peak The peak due to the most abundant ion in a mass spectrum; the most intense peak. It is assigned an arbitrary intensity of 100.

A **mass spectrum** is a plot of the relative abundance of each cation versus mass-to-charge ratio. The peak due to the most abundant cation is called the **base peak** and is assigned an arbitrary intensity of 100. The relative abundances of all other cations in a mass spectrum are reported as percentages of the base peak. Figure 14.2 shows a partial mass spectrum of dopamine, a neurotransmitter in the brain's caudate nucleus, a center involved with coordination and integration of fine muscle movement. A deficiency of dopamine is an underlying biochemical defect in Parkinson's disease.

As can be seen in the following table, the number of peaks recorded depends on the sensitivity of the detector. If we record all peaks with intensity equal to or greater than 0.5% of the base peak, as in Figure 14.2, we find 45 peaks for dopamine. If we record all peaks with intensity equal to or greater than 0.05% of the base peak, we find 120 peaks.

**Number of Peaks
Recorded in a Mass
Spectrum of Dopamine**

Peak Intensity Relative to Base Peak (%)	Number of Peaks Recorded
>5	8
>1	31
>0.5	45
>0.05	120

The technique we have described is called **electron ionization mass spectrometry (EI MS).** This technique was the first developed and for a time the one most widely used. It is limited, however, to relatively low-molecular-weight compounds that are vaporized easily in the evacuated ionization chamber. In recent years, a revolution in ionization techniques has extended the use of mass spectrometry to very-high-molecular-weight compounds and others that cannot be vaporized directly. Among the new techniques is fast-atom bombardment (FAB), which uses high-energy particles, such as xenon atoms, to bombard a dispersion of a compound in a nonvolatile matrix, producing ions of the compound and expelling them into the gas phase. A second technique is matrix-assisted laser desorption ionization (MALDI), which uses photons from an energetic laser for the same purpose. A third technique is chemical ionization (CI), which uses gas-phase acid-base reactions to produce ions. CI is particularly useful for identifying the molecular mass of a base (Brønsted-Lowry or Lewis) as its conjugate acid, MH^+. In addition, electrospray mass spectroscopy, in which a solution of the analyte is intro-

Figure 14.2

A partial mass spectrum of dopamine showing all peaks with intensity equal to or greater than 0.5% of the base peak.

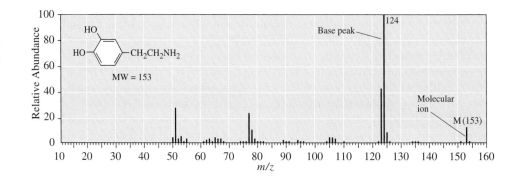

duced directly (for example, from a liquid chromatograph) through a charged capillary into a high vacuum, is particularly effective for biological macromolecules.

14.2 Features of a Mass Spectrum

To understand the complexity of a mass spectrum, we need to understand some of the relationships between mass spectra and resolution, the presence of isotopes, and the fragmentation of molecules and molecular ions in both the ionization chamber and the analyzing chamber.

A. Resolution

An important operating characteristic of a mass spectrometer is its **resolution,** that is, how well it separates ions of different mass. **Low-resolution mass spectrometry** refers to instruments capable of distinguishing among ions of different nominal mass; that is, ions that differ by one or more atomic mass units (amu). **High-resolution mass spectrometry** refers to instruments capable of distinguishing among ions that differ in precise mass by as little as 0.0001 amu.

To illustrate, compounds of molecular formulas C_3H_6O and C_3H_8O have nominal masses of 58 and 60 and can be resolved by low-resolution mass spectrometry. The compounds C_3H_8O and $C_2H_4O_2$, however, have the same nominal mass of 60 and cannot be distinguished by low-resolution mass spectrometry. If we calculate the precise mass of each compound using the data in Table 14.1, we see that they differ

Resolution In mass spectrometry, a measure of how well a mass spectrometer separates ions of different mass.

Low-resolution mass spectrometry Use of instrumentation that gives data capable of distinguishing only between ions that differ by 1 amu or more.

High-resolution mass spectrometry Use of instrumentation that gives data capable of distinguishing between ions that differ in mass by as little as 0.0001 amu.

Table 14.1 Precise Masses and Natural Abundances of Isotopes Relative to 100 Atoms of the Most Abundant Isotope

Element	Atomic Weight	Isotope	Precise Mass (amu)	Relative Abundance	
Hydrogen	1.0079	^1H	1.00783	100	
		^2H	2.01410	0.016	
Carbon	12.011	^{12}C	12.0000	100	
		^{13}C	13.0034	1.11	
Nitrogen	14.007	^{14}N	14.0031	100	
		^{15}N	15.0001	0.38	
Oxygen	15.999	^{16}O	15.9949	100	
		^{17}O	16.9991	0.04	
		^{18}O	17.9992	0.20	
Sulfur	32.066	^{32}S	31.9721	100	95% abund.
		^{33}S	32.9715	0.78	
		^{34}S	33.9679	4.40	4.2% abund.
Chlorine	35.453	^{35}Cl	34.9689	100	
		^{37}Cl	36.9659	32.5	
Bromine	79.904	^{79}Br	78.9183	100	
		^{81}Br	80.9163	98.0	

by 0.03642 amu and can be distinguished by high-resolution mass spectrometry. Observation of a molecular ion with a mass of 60.058 or 60.021 would establish the molecular formula of the unknown compound.

Molecular Formula	Nominal Mass	Precise Mass
C_3H_8O	60	60.05754
$C_2H_4O_2$	60	60.02112

B.   The Presence of Isotopes

In the mass spectrum of dopamine (Figure 14.2), the molecular ion appears at m/z 153. If you look more closely at this mass spectrum, you see a small peak at m/z 154, due to an ion 1 amu heavier than the molecular ion of dopamine. This peak is actually the sum of four separate peaks, each of amu 154 and each corresponding to the presence in the ion of a single heavier isotope of H, C, N, or O in dopamine. Because this peak corresponds to an ion 1 amu heavier than the molecular ion, it is called an M + 1 peak. We are concerned in this section primarily with M + 1 and M + 2 peaks.

Virtually all the elements common to organic compounds, including H, C, N, O, S, Cl, and Br, are mixtures of isotopes. Exceptions are fluorine, phosphorus, and iodine, which occur in nature exclusively as ^{19}F, ^{31}P, and ^{127}I. Table 14.1 shows average atomic weights for the elements most common to organic compounds along with the masses and relative abundances in nature of the stable isotopes of each. In this table, the relative abundances are tabulated according to the number of atoms of heavier isotope per 100 atoms of the most abundant isotope. Carbon in nature, for example, is 98.90% ^{12}C and 1.10% ^{13}C. Thus, there are 1.11 atoms of carbon-13 in nature for every 100 atoms of carbon-12.

$$1.10 \times \frac{100}{98.90} = 1.11 \text{ atoms } ^{13}C \text{ per 100 atoms } ^{12}C$$

Example 14.1

Calculate the precise mass of each ion to five significant figures. Unless otherwise indicated, use the mass of the most abundant isotope of each element.

(a) $[CH_2Cl_2]^{+\cdot}$ **(b)** $[^{13}CH_2Cl_2]^{+\cdot}$ **(c)** $[CH_2Cl^{37}Cl]^{+\cdot}$

Solution

(a) $12.0000 + 2(1.00783) + 2(34.9689) = 83.953$ **(b)** 84.957   **(c)** 85.951

Problem 14.1

Calculate the nominal mass of each ion. Unless otherwise indicated, use the mass of the most abundant isotope of each element.

(a) $[CH_3Br]^{+\cdot}$ **(b)** $[CH_3{}^{81}Br]^{+\cdot}$ **(c)** $[^{13}CH_3Br]^{+\cdot}$

C. Relative Abundance of M, M + 2, and M + 1 Peaks

The most common elements giving rise to significant M + 2 peaks are chlorine, bromine, and oxygen. Chlorine in nature is 75.77% ^{35}Cl and 24.23% ^{37}Cl. Thus, a ratio of M to M + 2 peaks of approximately 3:1 indicates the presence of a single chlorine atom in the compound. Similarly, bromine in nature is 50.5% ^{79}Br and 49.5% ^{81}Br; a ratio of M to M + 2 of approximately 1:1 indicates the presence of a single bromine atom in the compound. The contribution ^{18}O is only 0.2% but makes the major contribution to the M + 2 peak in compounds containing only C, H, N, and O. Sulfur is the only other element common to organic compounds that gives a significant M + 2 peak.

Let us use pentane, C_5H_{12}, to illustrate the relationship between M and M + 1 peaks. Pentane has a nominal mass of 72, and its molecular ion appears at m/z 72. In any sample of pentane, there is a probability that there will be a molecule in which one of the atoms of carbon is ^{13}C, the heavier isotope of carbon present in nature. This molecule has a nominal mass of 73, and its molecular ion will appear at m/z 73. Similarly, there is a probability that there will be a molecule in which one of the atoms of hydrogen is the heavier isotope of hydrogen, namely deuterium, ^2H. The probability of each of these isotope substitutions occurring is related to the natural abundance of each isotope in the following way:

$$\%(M + 1) = \Sigma \text{ (Abundance of heavier isotope} \times \text{Number of atoms in the formula)}$$

Using this formula, we calculate that the relative intensity of the M + 1 peak for pentane is

$$(M + 1) = \Sigma (1.11 \times 5C + 0.016 \times 12H) = 5.55 + 0.19 = 5.74\% \text{ of molecular ion peak}$$

Notice that the M + 1 peak for pentane is due almost entirely to ^{13}C. The same is true for other common compounds containing only C and H. Because M + 1 peaks are relatively low in intensity compared to the molecular ion peak and often difficult to measure with any precision, they are generally not useful for accurate determinations of molecular formulas. M + 1 and M + 2 peaks, however, can be useful for getting a rough idea of the number of carbons, oxygens, sulfurs, and halogens.

D. Fragmentation of Molecular Ions

To attain high efficiency of molecular ion formation and to give reproducible mass spectra, it is common to use electrons with energies of 70 eV [approximately 6750 kJ (1600 kcal)/mol]. This energy is sufficient not only to dislodge one or more electrons from a molecule but also to cause extensive fragmentation because it is well in excess of bond energies in organic molecules. These fragments may be unstable as well and, in turn, break apart into even smaller fragments.

The molecular ions for some compounds have a sufficiently long lifetime in the analyzing chamber that they are observed in the mass spectrum, sometimes as the base (most intense) peak. Molecular ions of other compounds have a shorter lifetime and are observed either in only low abundance or not at all. As a result, the mass spectrum of a compound typically (but not always) consists of a peak for the molecular ion and series of peaks for fragment ions. The fragmentation pattern and relative abundances of ions are unique for each compound under a given set of ionizing conditions and are characteristic of that compound. Fragmentation patterns give us valuable information about molecular structure.

A great deal of the chemistry of ion fragmentation can be understood in terms of the formation and relative stabilities of carbocations in solution. Where fragmentation occurs forming new carbocations, the mode of fragmentation that gives the most stable carbocation is favored. Thus, the probability of fragmentation to form a new carbocation increases in the following order:

$$CH_3^+ < 1° < 2° \cong 1° \text{ allylic/benzylic} < 3° \cong 2° \text{ allylic/benzylic} < 3° \text{ allylic/benzylic}$$

Increasing carbocation stability →

Molecular rearrangements are also characteristic of certain types of functional groups. We concentrate in this chapter on the fragmentation patterns common to the classes of compounds we have already encountered and describe fragmentation patterns of new classes of compounds as we encounter them in later chapters.

14.3 Interpreting Mass Spectra

Spectroscopy Questions
MS: The CD-ROM contains a set of interactive MS interpretation exercises.

Chemists often use mass spectra primarily for the determination of molecular weight and molecular formula. Very rarely do they attempt a full interpretation of a mass spectrum, which can be very time consuming and difficult. The mass spectrum of dopamine (Figure 14.2), for example, contains at least 45 peaks with intensity equal to or greater than 0.5% of the intensity of the base peak. We have neither the need nor the time to attempt to interpret this level of complexity. Rather, we concentrate in this section on the fragmentation mechanisms giving rise to major peaks.

As we now look at typical mass spectra of the classes of organic compounds we have seen so far, keep the following two points in mind. They provide you with valuable information about the molecular composition of an unknown compound.

1. The only elements giving rise to significant M + 2 peaks are ^{18}O (0.2%), ^{34}S (4.40%), ^{37}Cl (32.5%), and ^{81}Br (98%). If no large M + 2 peak is present, then these elements are absent.

Nitrogen rule A compound with an odd number of nitrogen atoms has an odd m/z ratio; if zero or an even number of nitrogen atoms, the molecular ion has an even m/z ratio.

2. Is the mass of the molecular ion odd or even? According to the **nitrogen rule,** if a compound has an odd number of nitrogen atoms, its molecular ion will appear at an odd m/z value. Conversely, if a compound has an even number of nitrogen atoms, its molecular ion will appear at an even m/z value. This rule is most helpful when there is an odd number of nitrogens. Where there is an even number of nitrogens, you may need additional experimental information to establish their presence.

A. Alkanes

Two rules will help you interpret the mass spectra of alkanes. (1) Fragmentation tends to occur toward the middle of unbranched chains rather than at the ends. (2) The differences in energy among allylic, benzylic, tertiary, secondary, primary, and methyl carbocations in the gas phase are much greater than the differences among comparable radicals. Therefore, where alternative modes of fragmentation are possible, the more stable carbocation tends to form in preference to the more stable radical.

Unbranched alkanes fragment to form a series of cations differing by 14 amu (a CH_2 group), with each fragment formed by a one-bond cleavage having an odd mass number. The mass spectrum of octane (Figure 14.3), for example, shows a peak for the molecular ion (m/z 114), as well as peaks for $C_6H_{13}^+$ (m/z 85), $C_5H_{11}^+$ (m/z 71), $C_4H_9^+$ (m/z 57), $C_3H_7^+$ (m/z 43), and $C_2H_5^+$ (m/z 29). Fragmentation of the CH_2—CH_3 bond is not observed; there is no peak corresponding to a methyl cation (m/z 15), nor is there one corresponding to a heptyl cation (m/z 99). In mass spectrometry, fragmentations are shown by lines through the bond that is cleaved with an angled part toward the fragment that bears the charge.

Fragmentation of branched-chain alkanes leads preferentially to the formation of secondary and tertiary carbocations, and, because these cations are more easily formed than methyl and primary carbocations, extensive fragmentation is likely. For this reason, the molecular ion of branched-chain hydrocarbons is often very weak or absent entirely from the spectrum. The molecular ion corresponding to m/z 114 is not observed, for example, in the mass spectrum of the highly branched 2,2,4-trimethylpentane (Figure 14.4). The base peak for this hydrocarbon is at m/z 57, due to the *tert*-butyl cation ($C_4H_9^+$). Other prominent peaks are at m/z 43, due to the isopropyl cation, and m/z 41, due to the allyl cation (CH_2=$CHCH_2^+$).

Sometimes peaks occur in a mass spectrum, the origin of which seems to defy any of the rules of chemical logic we have encountered thus far. For example, the prominent peak at m/z 29 in the mass spectrum of 2,2,4-trimethylpentane (Figure 14.4) is due to the ethyl cation, $CH_3CH_2^+$. There is, however, no ethyl group in the parent molecule! We can only conclude that this cation must be formed by some combination of fragmentation and rearrangement quite beyond anything that we have seen up to this point.

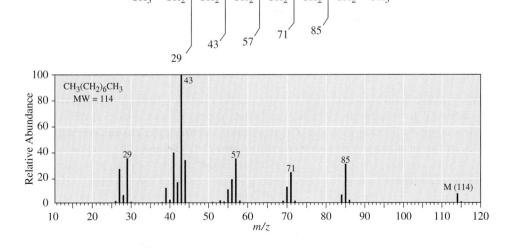

Figure 14.3

Mass spectrum of octane.

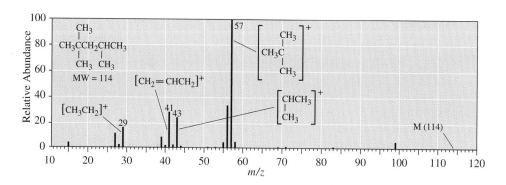

Figure 14.4

Mass spectrum of 2,2,4-trimethylpentane. The peak due to the molecular ion is of such low intensity that it does not appear in this spectrum.

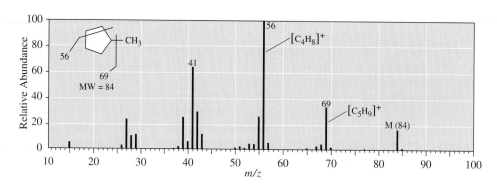

Figure 14.5

Mass spectrum of methylcyclo-pentane.

The most common fragmentation patterns of cycloalkanes are loss of side chains and loss of ethylene, $CH_2=CH_2$. The peak at m/z 69 in the mass spectrum of methyl-cyclopentane (Figure 14.5) is due to loss of the one-carbon side chain to give the cy-clopentyl cation, $C_5H_9^+$. The base peak at m/z 56 is due to loss of ethylene and corre-sponds to a cation of molecular formula $C_4H_8^+$. Note here that one-carbon cleavages of cycloalkanes (and alkanes as well) give fragments with odd mass numbers; two-bond cleavages give fragments with even mass numbers.

Example 14.2

The base peak at m/z 56 in the mass spectrum of methylcyclopentane corresponds to loss of ethylene to give a radical cation of molecular formula $C_4H_8^{+\cdot}$. Propose a struc-tural formula for this radical cation and show how it might be formed.

Solution

Following is a structural formula for a molecular ion that might be formed in the ion-izing chamber. In it, a single electron has been dislodged from a carbon-carbon sin-gle bond to give a 1° radical and a 2° carbocation. Rearrangement of bonding elec-trons in this radical cation gives ethylene and a new radical cation.

Molecular ion
(a radical cation, m/z 84)   Ethylene   A new radical cation
(a 2° carbocation, m/z 56)

Problem 14.2

Propose a structural formula for the cation of m/z 41 observed in the mass spectrum of methylcyclopentane.

B. Alkenes

Alkenes characteristically show a strong molecular ion peak, most probably formed by removal of one pi electron from the double bond. Furthermore, they cleave readily to form resonance-stabilized allylic cations, such as the allyl cation seen at m/z 41 in the mass spectrum of 1-butene (Figure 14.6).

Cyclohexenes undergo fragmentation to give a 1,3-diene and an alkene in a process that is the reverse of a Diels-Alder reaction (Section 23.3). The terpene, limonene, a disubstituted cyclohexene, for example, fragments by a reverse Diels-Alder reaction to give two molecules of 2-methyl-1,3-butadiene (isoprene): one formed as a neutral diene and the other formed as a diene radical cation. Note here that the two-bond cleavage of this hydrocarbon gives fragments with even mass numbers.

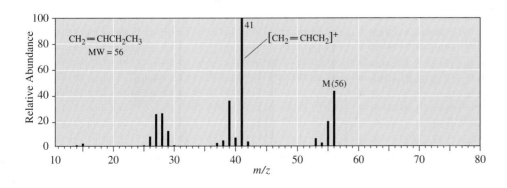

Limonene	A neutral diene	A radical cation
(m/z 136)	(m/z 68)	(m/z 68)

C. Alkynes

As with alkenes, alkynes show a strong peak due to the molecular ion. Their fragmentation patterns are also similar to those of alkenes. One of the most prominent peaks in the mass spectrum of most alkynes is due to the resonance-stabilized 3-propynyl (propargyl) cation (m/z 39) or a substituted propargyl cation.

$$HC\equiv C\overset{\frown}{-}CH_2{}^+ \longleftrightarrow H\overset{+}{C}=C=CH_2$$

3-Propynyl cation
(Propargyl cation)

Figure 14.6
Mass spectrum of 1-butene.

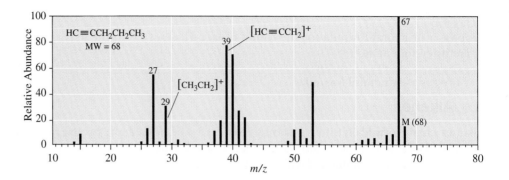

Figure 14.7
Mass spectrum of 1-pentyne.

Both the molecular ion, m/z 68, and the propargyl cation, m/z 39, are seen in the mass spectrum of 1-pentyne (Figure 14.7). Also seen is the ethyl cation, m/z 29.

D. Alcohols

The intensity of the molecular ion from primary and secondary alcohols is normally quite low, and there usually is no molecular ion detectable for tertiary alcohols. One of the most common fragmentation patterns for alcohols is loss of a molecule of water to give a peak corresponding to the molecular ion minus 18 (M − 18). Another common pattern is loss of an alkyl group from the carbon bearing the —OH group to form a resonance-stabilized oxonium ion and an alkyl radical. The oxonium ion is particularly stable because of delocalization of charge.

$$
\underset{\substack{\text{Molecular ion} \\ \text{(a radical cation)}}}{R'-\overset{\overset{\displaystyle R}{|}}{\underset{\displaystyle R''}{C}}-\overset{+}{\underset{\cdot\cdot}{O}}-H} \longrightarrow \underset{\substack{\text{A radical}}}{R\cdot} + \underset{\substack{\text{A resonance-stabilized cation}}}{R'-\overset{+}{\underset{\displaystyle R''}{C}}=\overset{}{\underset{\cdot\cdot}{O}}-H \longleftrightarrow R'-\overset{+}{\underset{\displaystyle R''}{C}}-\overset{\cdot\cdot}{\underset{\cdot\cdot}{O}}-H}
$$

Each of these patterns is found in the mass spectrum of 1-butanol (Figure 14.8). The molecular ion appears at m/z 74. The prominent peak at m/z 56 corresponds to loss of a molecule of water from the molecular ion (M − 18). The base peak at m/z 31 corresponds to cleavage of a propyl group (M − 43) from the carbon bearing the —OH group. The propyl cation is visible at m/z 43.

Figure 14.8
Mass spectrum of 1-butanol.

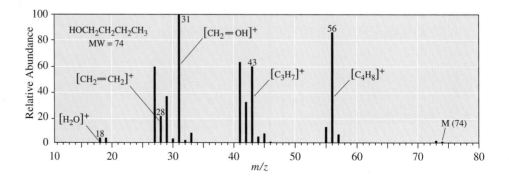

Example 14.3

A low-resolution mass spectrum of 2-methyl-2-butanol (MW 88) shows 16 peaks. The molecular ion is absent. Account for the formation of peaks at m/z 73, 70, 59, and 55, and propose a structural formula for each cation.

Solution

The peak at m/z 73 (M − 15) corresponds to loss of a methyl radical from the molecular ion. The peak at m/z 59 (M − 29) corresponds to loss of an ethyl radical. Loss of water as a neutral molecule from the molecular ion gives an alkene of m/z 70 (M − 18) as a radical cation. Loss of methyl from this radical cation gives an allylic carbocation of m/z 55.

Problem 14.3

The low-resolution mass spectrum of 2-pentanol shows 15 peaks. Account for the formation of the peaks at m/z 73, 70, 55, 45, 43, and 41.

E. Aldehydes and Ketones

A characteristic fragmentation pattern of aliphatic aldehydes and ketones is cleavage of one of the bonds to the carbonyl group (α-cleavage). α-Cleavage of 2-octanone, for example, gives carbonyl-containing ions of m/z 43 and 113. α-Cleavage of the aldehyde proton gives an M − 1 peak, which is often quite distinct and provides a useful way to distinguish between an aldehyde and a ketone.

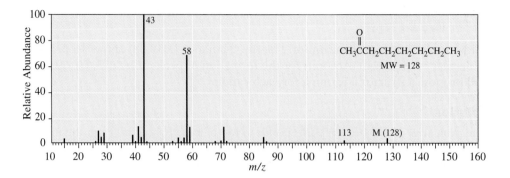

Figure 14.9

Mass spectrum of 2-octanone. Ions of m/z 43 and 113 result from α-cleavage. The ion at m/z 58 results from McLafferty rearrangement.

Aldehydes and ketones with a sufficiently long carbon chain show a fragmentation called a McLafferty rearrangement. In a **McLafferty rearrangement** of an aldehyde or ketone, the carbonyl oxygen abstracts a hydrogen five atoms away to give an alkene and a new radical cation. Reaction occurs through a six-membered ring transition state. McLafferty rearrangement of 2-heptanone, for example, gives 1-butene and a radical cation of m/z 58, which is the enol of acetone. Because McLafferty rearrangements involve cleavage of two bonds, molecules of even m/z give fragments of even m/z.

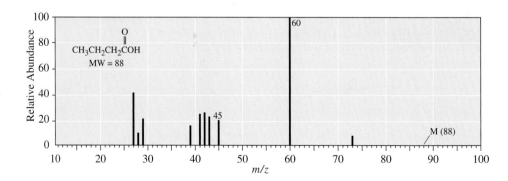

The results of both α-cleavage and McLafferty rearrangement can be seen in the mass spectrum of 2-octanone (Figure 14.9).

F.   Carboxylic Acids and Esters

The molecular ion peak from a carboxylic acid is generally observed, although it is often very weak. The most common fragmentation patterns are α-cleavage of the carboxyl group to give the ion $[\text{COOH}]^+$ of m/z 45 and McLafferty re-

Figure 14.10

Mass spectrum of butanoic acid. Common fragmentation patterns of carboxylic acids are α-cleavage to give the ion $[\text{COOH}]^+$ of m/z 45 and McLafferty rearrangement.

arrangement. The base peak is very often that due to the McLafferty rearrangement product.

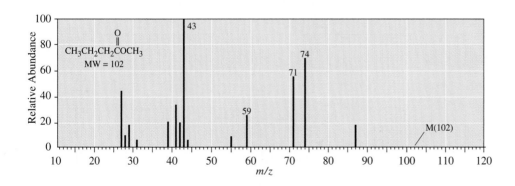

Molecular ion
m/z 88

α-cleavage

+ [O=C—O—H]⁺

m/z 45

Mechanism   **McLafferty Rearrangement of a Carboxylic Acid**

Molecular ion
m/z 88

McLafferty rearrangement

m/z 60

Each of these patterns is seen in the mass spectrum of butanoic acid (Figure 14.10). Esters and amides also generally show discernible molecular ion peaks. Like carboxylic acids, their most characteristic fragmentation patterns are α-cleavage and McLafferty rearrangement, both of which can be seen in the mass spectrum of methyl butanoate (Figure 14.11). Peaks at *m/z* 71 and 59 are the result of α-cleavage.

Molecular ion
m/z 102

α-cleavage

+ + ·OCH₃

m/z 71

+ + ·OCH₃

m/z 59

$CH_3CH_2CH_2COCH_3$
MW = 102

Figure 14.11

Mass spectrum of methyl butanoate. Characteristic fragmentation patterns of esters are α-cleavage and McLafferty rearrangement.

The peak at m/z 74 is the result of McLafferty rearrangement.

Molecular ion
m/z 102

m/z 74

G. Aromatic Hydrocarbons

The mass spectra of most aromatic hydrocarbons show an intense molecular ion peak. The mass spectrum of toluene (Figure 14.12), for example, shows an intense molecular ion peak at m/z 92.

The mass spectra of toluene and most other alkylbenzenes show a fragment ion of m/z 91. Although it might seem that the most likely structure for this ion is that of the benzyl cation, experimental evidence suggests a molecular rearrangement to form the more stable tropylium ion (Section 20.2E). In the tropylium ion, an aromatic cation, the positive charge is delocalized equally over all seven carbon atoms of the cycloheptatrienyl ring.

Toluene radical
cation

Tropylium cation
(m/z 91)

H. Amines

Of the compounds containing C, H, N, O, and the halogens, only those containing an odd number of nitrogen atoms have a molecular ion of odd m/z ratio (the nitrogen rule. Thus, mass spectrometry can be a particularly valuable tool for identifying amines. The molecular ion for aliphatic amines, however, is often very weak. The most characteristic fragmentation of amines, and the one often giving rise to the base peak, is β-cleavage. Where alternative possibilities for β-cleavage exist, it is generally the largest R group that is lost. In contrast to nitrogen-free molecules, single-bond fragments from compounds that have one nitrogen (or an odd number) give compounds with even mass. The most prominent peak by far in the mass spectrum of

Figure 14.12

Mass spectrum of toluene. Prominent are the intense molecular ion peak at m/z 92 and the tropylium cation at m/z 91.

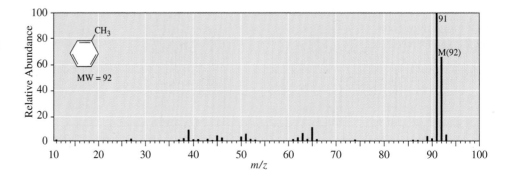

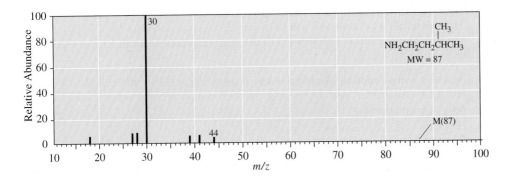

Figure 14.13

Mass spectrum of 3-methyl-1-butanamine (isopentylamine). The most characteristic fragmentation pattern of aliphatic amines is β-cleavage.

3-methyl-1-butanamine (Figure 14.13) is due to $[CH_2{=}NH_2]^+$, m/z 30, resulting from β-cleavage. β-Cleavage is also characteristic of secondary and tertiary amines. Complex rearrangement and fragmentation processes give the m/z 30 peak as a major fragment even from secondary and tertiary amines.

$$CH_3{-}\overset{\underset{|}{CH_3}}{CH}{-}CH_2{-}CH_2{-}\overset{\cdot}{\underset{}{N}}H_2 \xrightarrow{\text{β-cleavage}} CH_3{-}\overset{\underset{|}{CH_3}}{CH}{-}CH_2\cdot + CH_2{=}\overset{+}{N}H_2$$

$$(m/z\ 30)$$

Summary

A **mass spectrum** (Section 14.1) is a plot of relative ion abundance versus mass-to-charge (m/z) ratio. The **base peak** is the most intense peak in a mass spectrum. A **molecular ion, M^+,** is a radical cation derived from the parent molecule by loss of one electron.

Low-resolution mass spectrometry distinguishes between ions that differ in nominal mass, that is, ions that differ by 1 amu (Section 14.2A). **High-resolution mass spectrometry** distinguishes between ions that differ by as little as 0.0001 amu.

M + 1 and higher peaks in a mass spectrum are due to the presence of heavier isotopes (Section 14.2B). The abundance of these higher mass-to-charge peaks relative to the molecular ion peak provides information about the elemental composition of the molecular ion (Section 14.2C). The presence of a single chlorine atom, for example, is indicated by M to M + 2 peaks in a ratio of 3:1. According to the **nitrogen rule,** if a compound has an odd number of nitrogen atoms, its molecular ion will have an odd m/z value (Section 14.3).

The mass spectrum of a compound typically consists of a peak for the molecular ion and a series of peaks for fragment ions (Section 14.2D). The fragmentation pattern and relative abundances of ions are unique for each compound and are characteristic of that compound. Fragments formed by cleavage of one bond have odd mass if they contain no nitrogen; those from cleavage of two bonds have even mass. Many of the observed fragmentation patterns can be understood in terms of the relative stability of carbocations (Section 14.2D). Where alternative modes of fragmentation are possible, the more stable carbocation tends to be formed in preference to the more stable radical.

Problems

14.4 Draw acceptable Lewis structures for the molecular ion (radical cation) formed from the following molecules when each is bombarded by high-energy electrons in a mass spectrometer.

14.5   The molecular ion for compounds containing only C, H, and O always has an even mass-to-charge value. Why is this so? What can you say about mass-to-charge ratio of ions that arise from fragmentation of one bond in the molecular ion? From fragmentation of two bonds in the molecular ion?

14.6   For which compounds containing a heteroatom (an atom other than carbon or hydrogen) does a molecular ion have an even-numbered mass and for which does it have an odd-numbered mass?

(a) A chloroalkane of molecular formula $C_nH_{2n+1}Cl$
(b) A bromoalkane of molecular formula $C_nH_{2n+1}Br$
(c) An alcohol of molecular formula $C_nH_{2n+1}OH$
(d) A primary amine of molecular formula $C_nH_{2n+1}NH_2$
(e) A thiol of molecular formula $C_nH_{2n+1}SH$

14.7   The so-called nitrogen rule states that if a compound has an odd number of nitrogen atoms, the value of m/z for its molecular ion will be an odd number. Why is this so?

14.8   Both $C_6H_{10}O$ and C_7H_{14} have the same nominal mass, namely 98. Show how these compounds can be distinguished by the m/z ratio of their molecular ions in high-resolution mass spectrometry.

14.9   Show how the compounds of molecular formula C_6H_9N and C_5H_5NO can be distinguished by the m/z ratio of their molecular ions in high-resolution mass spectrometry.

14.10   What rule would you expect for the m/z values of fragment ions resulting from the cleavage of one bond in a compound with an odd number of nitrogen atoms?

14.11   Determine the probability of the following in a natural sample of ethane.

(a) One carbon in an ethane molecule is ^{13}C.
(b) Both carbons in an ethane molecule are ^{13}C.
(c) Two hydrogens in an ethane molecule are replaced by deuterium atoms.

14.12   The molecular ions of both $C_5H_{10}S$ and $C_6H_{14}O$ appear at m/z 102 in low-resolution mass spectrometry. Show how determination of the correct molecular formula can be made from the appearance and relative intensity of the M + 2 peak of each compound.

14.13   In Section 14.3, we saw several examples of fragmentation of molecular ions to give resonance-stabilized cations. Make a list of these resonance-stabilized cations, and write important contributing structures of each. Estimate the relative importance of the contributing structures in each set.

14.14   Carboxylic acids often give a strong fragment ion at m/z (M − 17). What is the likely structure of this cation? How might it be formed? Show by drawing contributing structures that it is stabilized by resonance.

14.15   For primary amines with no branching on the carbon bearing the nitrogen, the base peak occurs at m/z 30. What cation does this peak represent? How is it formed? Show by drawing contributing structures that this cation is stabilized by resonance.

14.16   The base peak in the mass spectrum of propanone (acetone) occurs at m/z 43. What cation does this peak represent?

14.17   A characteristic peak in the mass spectrum of most aldehydes occurs at m/z 29. What cation does this peak represent? (No, it is not an ethyl cation, $CH_3CH_2^+$.)

14.18   Predict the relative intensities of the M and M + 2 peaks for the following.

(a) CH_3CH_2Cl (b) CH_3CH_2Br (c) $BrCH_2CH_2Br$ (d) CH_3CH_2SH

14.19 The mass spectrum of compound A shows the molecular ion at m/z 85, an M + 1 peak at m/z 86 of approximately 6% abundance relative to M, and an M + 2 peak at m/z 87 of less than 0.1% abundance relative to M.

(a) Propose a molecular formula for compound A.
(b) Draw at least ten possible structural formulas for this molecular formula.

14.20 The mass spectrum of compound B, a colorless liquid, shows these peaks in its mass spectrum. Determine the molecular formula of compound B, and propose a structural formula for it.

m/z	Relative Abundance
43	100 (base)
M 78	23.6 (M)
79	1.00
M+2 80	7.55
81	0.25

14.21 Write molecular formulas for the five possible molecular ions of m/z 88 containing the elements C, H, N, and O. (*Note:* Because the value of the mass of this set of molecular ions is an even number, members of the set must have either no nitrogen atoms or an even number of nitrogen atoms.)

14.22 Write molecular formulas for the five possible molecular ions of m/z 100 containing only the elements C, H, N, and O.

14.23 The molecular ion in the mass spectrum of 2-methyl-1-pentene appears at m/z 84. Propose structural formulas for the prominent peaks at m/z 69, 55, 41, and 29.

14.24 Following is the mass spectrum of 1,2-dichloroethane.

(a) Account for the appearance of an M + 2 peak with approximately two thirds the intensity of the molecular ion peak.
(b) Predict the intensity of the M + 4 peak.
(c) Propose structural formulas for the cations of m/z 64, 63, 62, 51, 49, 27, and 26.

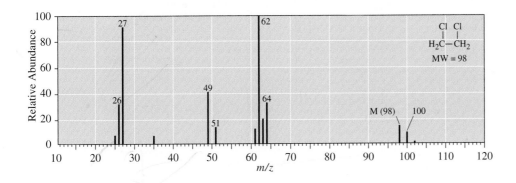

14.25 Following is the mass spectrum of 1-bromobutane.

    (a) Account for the appearance of the M + 2 peak of approximately 95% of the intensity of the molecular ion peak.

    (b) Propose structural formulas for the cations of m/z 57, 41, and 29.

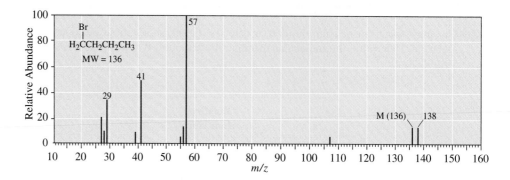

14.26 Following is the mass spectrum of bromocyclopentane. The molecular ion m/z 148 is of such low intensity that it does not appear in this spectrum. Assign structural formulas for the cations of m/z 69 and 41.

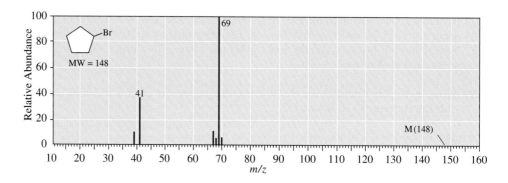

14.27 Following is the mass spectrum of 3-methyl-2-butanol. The molecular ion m/z 88 does not appear in this spectrum. Propose structural formulas for the cations of m/z 45, 43, and 41.

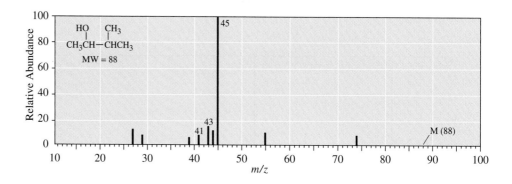

14.28 The following is the mass spectrum of compound C, C_3H_8O. Compound C is infinitely soluble in water, undergoes reaction with sodium metal with the evolution of a gas, and undergoes reaction with thionyl chloride to give a water-insoluble chloroalkane. Propose a structural formula for compound C, and write equations for each of its reactions.

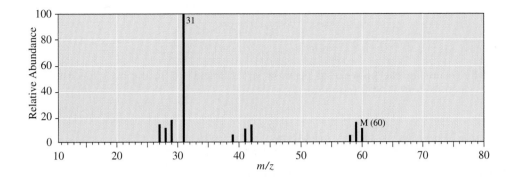

14.29 Following are mass spectra for the constitutional isomers 2-pentanol and 2-methyl-2-butanol. Assign each isomer its correct spectrum.

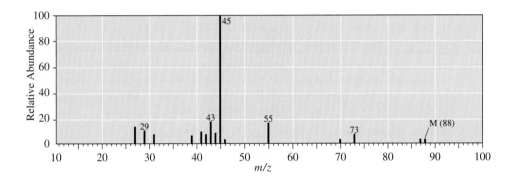

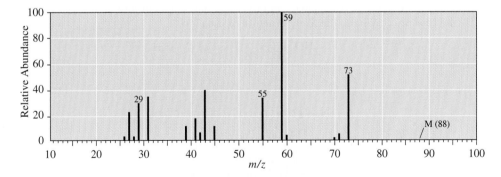

14.30 2-Methylpentanal and 4-methyl-2-pentanone are constitutional isomers of molecular formula $C_6H_{12}O$. Each shows a molecular ion peak in its mass spectrum at m/z 100. Spectrum A shows significant peaks at m/z 85, 58, 57, 43, and 42. Spectrum B shows significant peaks at m/z 71, 58, 57, 43, and 29. Assign each compound its correct spectrum.

14.31 Account for the presence of peaks at m/z 135 and 107 in the mass spectrum of 4-methoxybenzoic acid (*p*-anisic acid).

14.32 Account for the presence of the following peaks in the mass spectrum of hexanoic acid.

(a) m/z 60

(b) A series of peaks differing by 14 amu at m/z 45, 59, 73, and 87

(c) A series of peaks differing by 14 amu at m/z 29, 43, 57, and 71

14.33 All methyl esters of long-chain aliphatic acids (for example, methyl tetradecanoate, $C_{13}H_{27}COOCH_3$) show significant fragment ions at m/z 74, 59, and 31. What are the structures of these ions? How are they formed?

14.34 Propylbenzene, $C_6H_5CH_2CH_2CH_3$, and isopropyl benzene, $C_6H_5CH(CH_3)_2$, are constitutional isomers of molecular formula C_9H_{12}. One of these compounds shows prominent peaks in its mass spectrum at m/z 120 and 105. The other shows prominent peaks at m/z 120 and 91. Which compound has which spectrum?

14.35 Account for the formation of the base peaks in these mass spectra.

(a) Isobutylamine, m/z 44   (b) Diethylamine, m/z 58

14.36 Because of the sensitivity of mass spectrometry, it is often used to detect the presence of drugs in blood, urine, or other biological fluids. Tetrahydrocannabinol (nominal mass 314), a component of marijuana, exhibits two strong fragment ions at m/z 246 and 231 (the base peak). What is the likely structure of each ion?

14.37 Electrospray mass spectrometry is a recently developed technique for looking at large molecules with a mass spectrometer. In this technique, molecular ions, each associated with one or more H^+ ions, are prepared under mild conditions in the mass spectrometer. As an example, a protein (P) with a molecular weight of 11,812 gives clusters of the type $(P + 8H)^{8+}$, $(P + 7H)^{7+}$, and $(P + 6H)^{6+}$. At what mass-to-charge values do these three clusters appear in the mass spectrum?

15

ORGANOMETALLIC COMPOUNDS

I n this chapter, we undertake our first discussion of a broad class of organic compounds called **organometallic compounds,** compounds that contain a carbon-metal bond. In recent years, there has been an enormous explosion in our understanding of their chemistry, particularly as stereospecific (and often enantioselective) reagents for synthetic chemistry. We have already seen one example in the Sharpless enantioselective epoxidation of alkenes (Section 11.8D).

Two extremely important reactions of metals and metal compounds are **oxidative addition** and its complement, **reductive elimination.** In oxidative addition, a reagent adds to a metal, causing its coordination to

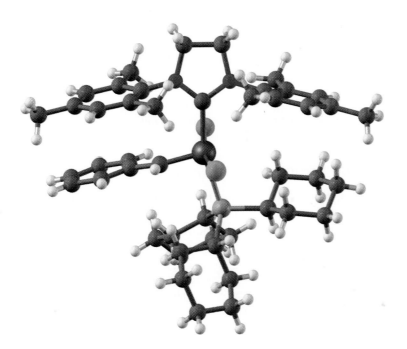

■ A ruthenium-containing organometallic catalyst for alkene metathesis reactions. See Section 15.4E for the structure of this catalyst and a discussion of this reaction. See also the CD accompanying this text for a Chem3D model of this catalyst which you can manipulate to see its structure more clearly.

521

increase by two; reductive elimination is the reverse. These reactions are called oxidative and reductive because the formal charge of the metal changes by two during the reaction. Oxidative addition can occur with a metal coordinated with one or more **ligands (L);** it can also occur with a free metal, M(0). Alkyl halides, hydrogen, halogens, and many other types of compounds can take part in these reactions. The reactivity of different substrates depends greatly on the metal.

> **Ligand (L)** A Lewis base bonded to a metal atom in a coordination compound. It may bond strongly or weakly. Refer to your general chemistry text for a more detailed description of types of ligand and coordination compounds.

$$ML_n + X_2 \underset{\substack{\text{reductive} \\ \text{elimination}}}{\overset{\substack{\text{oxidative} \\ \text{addition}}}{\rightleftharpoons}} \begin{matrix} X \\ \diagdown \\ ML_n \\ \diagup \\ X \end{matrix}$$

This chapter cannot possibly cover the wealth of organometallic reagents and catalysts that have been developed for synthetic organic chemistry, particularly during the last few years. We focus, therefore, on transformations that are fundamental to synthetic chemistry or illustrate the power of modern organometallic chemistry. Organomagnesium, lithium, and copper reagents have been selected because of their historical importance and their continued use in modern organic synthesis. The Simmons-Smith reaction is one of the few reactions for the formation of substituted cyclopropanes. The Heck and alkene metathesis reactions illustrate the power of organometallic catalysts to bring about transformations that cannot be accomplished in any other way. The interface between organic and inorganic chemistry and the development of new organometallic reagents and catalysts is one of the most exciting areas of chemical research and development today.

Reactivity Explorer
Alkyl Halides

15.1 Organomagnesium and Organolithium Compounds

We begin with organomagnesium and organolithium compounds and concentrate on their formation and basicity. We discuss their use in organic synthesis in more detail in later chapters, particularly in Chapters 16 and 18.

RMgX   RLi

An organomagnesium   An organolithium
compound   compound
(a Grignard reagent)

A. Formation and Structure

Organomagnesium compounds are among the most readily available, easily prepared, and easily handled organometallics. They are commonly named **Grignard reagents** after the French chemist Victor Grignard (1871–1935), who was awarded the 1912 Nobel Prize for chemistry for their discovery and application to organic synthesis.

Grignard reagents are typically prepared by the slow addition of an alkyl, aryl, or alkenyl (vinylic) halide to a stirred suspension of a slight excess of magnesium metal in an ether solvent, most commonly diethyl ether or tetrahydrofuran (THF). Organic iodides and bromides generally react very rapidly under these conditions, whereas most organic chlorides react more slowly. Bromides are the most common starting materials for preparation of Grignard reagents. It is common to use the higher boiling THF (bp 67°C) to prepare Grignard reagents from the less reactive organic halides. Generally there is an induction period at the beginning of the reaction due to the presence of traces of moisture and a thin oxide film coated on the surface of the magnesium. When reaction starts, it is exothermic, and the remaining organic halide is added at a rate sufficient to maintain a gentle reflux of the ether.

Butylmagnesium bromide, for example, is prepared by treating 1-bromobutane in diethyl ether with magnesium metal. Aryl Grignard reagents, such as phenylmagnesium bromide, are prepared in the same manner. These reactions are oxidative additions because they result in an increase in the formal oxidation state of magnesium by two, that is, from Mg(0) to Mg(II).

$$\text{Br} + \text{Mg} \xrightarrow{\text{ether}} \text{MgBr}$$

1-Bromobutane   Butylmagnesium bromide
                       (an alkyl Grignard reagent)

$$\text{Br} + \text{Mg} \xrightarrow{\text{ether}} \text{MgBr}$$

Bromobenzene   Phenylmagnesium
                      bromide
             (an aryl Grignard reagent)

Even though the equation for formation of Grignard reagents looks like a simple oxidative addition, the mechanism is considerably more complicated and involves radicals. We have no need in this course to discuss the mechanism for their formation.

Grignard reagents form on the surface of the metal and dissolve as coordination complexes solvated by ether. In this ether-soluble complex, magnesium acts as a Lewis acid and the ether acts as a Lewis base.

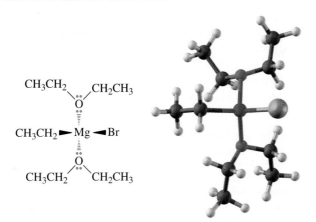

CH₃CH₂   CH₂CH₃

CH₃CH₂—Mg—Br

CH₃CH₂   CH₂CH₃

Ethylmagnesium bromide dietherate

Organolithium reagents are prepared by the reaction of an alkyl, aryl, or alkenyl halide with two equivalents of lithium metal, as illustrated by the preparation of butyllithium. In this reaction, a solution of 1-chlorobutane in pentane is added to lithium wire at −10°C.

$$\text{Cl} + 2\text{Li} \xrightarrow{\text{pentane}} \text{Li} + \text{LiCl}$$

1-Chlorobutane   Butyllithium

Organolithium compounds are very reactive as nucleophiles in carbonyl addition reactions even at very low temperatures. They are also powerful and effective bases. For these reasons, they are now widely used in modern synthetic chemistry. However, they react rapidly with atmospheric oxygen and moisture and, therefore, must be used under an inert atmosphere of N_2 or Ar, which decreases their convenience.

The carbon-metal bonds in Grignard and organolithium reagents are best described as polar covalent, with carbon bearing a partial negative charge and the metal bearing a partial positive charge. Shown in Table 15.1 are electronegativity differences (Pauling scale, Table 1.5) between carbon and various metals. From this difference, we can estimate the percent ionic character of each carbon-metal bond. Organolithium and organomagnesium bonds have the highest partial ionic character, whereas that of organocopper and organomercury compounds is lower. In spite of the polarity of the carbon-metal bond, these compounds do not behave as salts. Organolithium reagents, for example, which have the highest percent partial ionic character, dissolve in nonpolar hydrocarbon solvents such as pentane because they self-assemble into well-ordered aggregates, $(RLi)_x$, that present a nonpolar surface to the surrounding solvent.

The feature that makes Grignard and organolithium reagents so valuable in synthetic organic chemistry is the fact that the carbon atom bearing the halogen is transformed from an electrophilic center in the haloalkane, alkene, or arene to a nucleophilic center in the organometallic compound. In the structural formula of butylmagnesium bromide on the right, the carbon-magnesium bond is shown as ionic to emphasize its nucleophilic character.

As nucleophiles, these compounds react with the electrophilic carbon atom of the carbonyl groups of aldehydes and ketones (Chapter 16) and of carboxylic esters and acid chlorides (Chapter 18). Herein lies the value of organomagnesium and organolithium reagents in synthetic organic chemistry—the formation of new carbon-carbon bonds.

Table 15.1	Percent Ionic Character of Some C—M Bonds	
C—M Bond	**Difference in Electronegativity**	**Percent Ionic Character***
C—Li	2.5 − 1.0 = 1.5	60
C—Mg	2.5 − 1.2 = 1.3	52
C—Al	2.5 − 1.5 = 1.0	40
C—Zn	2.5 − 1.6 = 0.9	36
C—Sn	2.5 − 1.8 = 0.7	28
C—Cu	2.5 − 1.9 = 0.6	24
C—Hg	2.5 − 1.9 = 0.6	24

*Percent ionic character $= \dfrac{E_C - E_M}{E_C} \times 100$.

B.   Reaction with Proton Acids

Both Grignard and organolithium compounds are very strong bases and react readily with any acid (proton donor) stronger than the alkane from which they are derived. Ethylmagnesium bromide, for example, reacts instantly with water, which donates a proton to give ethane and magnesium salts. This reaction is an example of a stronger acid and a stronger base reacting to give a weaker acid and a weaker base (Section 4.4). Ethane is evolved from the reaction mixture as a gas.

$$\overset{\delta-}{CH_3CH_2}\!-\!\overset{\delta+}{MgBr} + \overset{\delta+}{H}\!-\!\overset{\delta-}{OH} \longrightarrow CH_3CH_2\!-\!H + Mg^{2+} + OH^- + Br^- \qquad pK_{eq} = -35$$
$$K_{eq} = 10^{35}$$

Stronger base	Stronger acid	Weaker acid	Weaker base
	pK_a 15.7	pK_a 51	

Following are several classes of proton donors that react readily with Grignard and organolithium reagents. Because they react so readily with these types of compounds, Grignard and organolithium compounds cannot be prepared from any organohalogen that also contains one of these functional groups. Nor can they be prepared from any organohalogen compound that also contains a nitro or carbonyl group because they also react with these groups.

R_2NH	$RC{\equiv}CH$	ROH	HOH	$ArOH$	RSH	$RCOOH$
pK_a 38–40	pK_a 25	pK_a 16–18	pK_a 15.7	pK_a 9–10	pK_a 8–9	pK_a 4–5
1° and 2° amines	Terminal alkynes	Alcohols	Water	Phenols	Thiols	Carboxylic acids

---

Example 15.1

Write an equation for the acid-base reaction between ethylmagnesium iodide and an alcohol. Use curved arrows to show the flow of electrons in this reaction. In addition, show by using appropriate pK_a values that this reaction is an example of a stronger acid and stronger base reacting to form a weaker acid and weaker base.

Solution

The alcohol is the stronger acid, and the partially negatively charged ethyl group is the stronger base.

$$CH_3CH_2\!-\!MgI \;+\; H\!-\!OR \longrightarrow CH_3CH_2\!-\!H \;+\; RO^-MgI^+$$

A Grignard reagent (stronger base)	An alcohol pK_a 16–18 (stronger acid)	An alkane pK_a 51 (weaker acid)	A magnesium alkoxide (weaker base)

Problem 15.1

Explain how these Grignard reagents would react with molecules of their own kind to "self-destruct."

(a) $HOCH_2CH_2CH_2MgBr$ **(b)** $HC{\equiv}C(CH_2)_4CH_2MgBr$

C.   Reaction with Oxiranes

As we saw in Section 11.9, the oxirane ring is so strained that it undergoes ring-opening reactions with a variety of nucleophiles. We can now add to the list of reactive nucleophiles Grignard and organolithium reagents. Butylmagnesium bromide, for example, reacts with oxirane (ethylene oxide) to give a magnesium alkoxide, which, on treatment with aqueous acid, gives 1-hexanol.

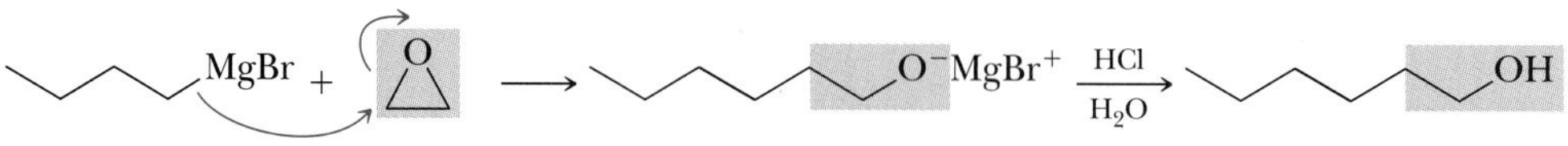

| Butylmagnesium bromide | Ethylene oxide | A magnesium alkoxide | 1-Hexanol |

As illustrated in this example, the product of treatment of a Grignard reagent with oxirane followed by protonation of the alkoxide is a primary alcohol with a carbon chain two carbons longer than the original chain. In reaction of a substituted oxirane, the major product corresponds to attack of the Grignard reagent on the less hindered carbon of the three-membered ring in an S_N2-like reaction. Treatment of methyloxirane (propylene oxide) with phenylmagnesium bromide, for example, followed by workup in aqueous acid gives 1-phenyl-2-propanol.

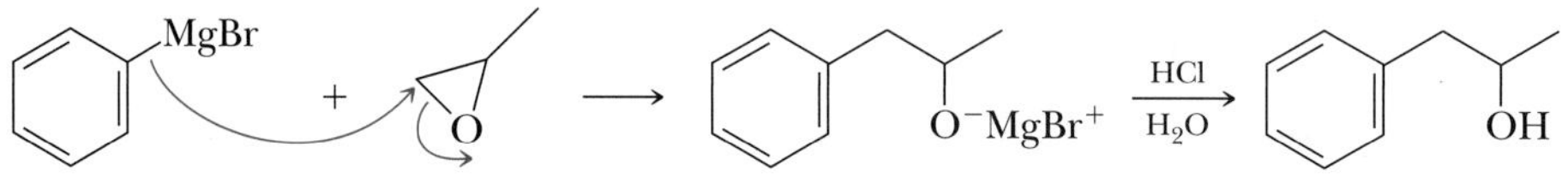

| Phenylmagnesium bromide | Methyloxirane (Propylene oxide) | A magnesium alkoxide | 1-Phenyl-2-propanol |

Example 15.2

Show how to prepare each alcohol from an organohalogen compound and an oxirane.

(a) 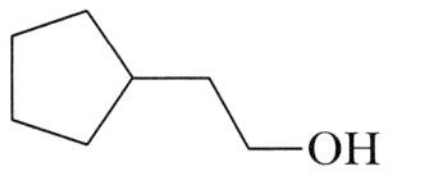   **(b)**

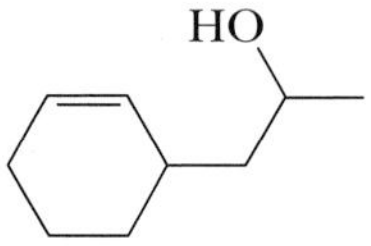

Solution

Shown is a retrosynthetic analysis for each compound followed by a synthesis.

(a)

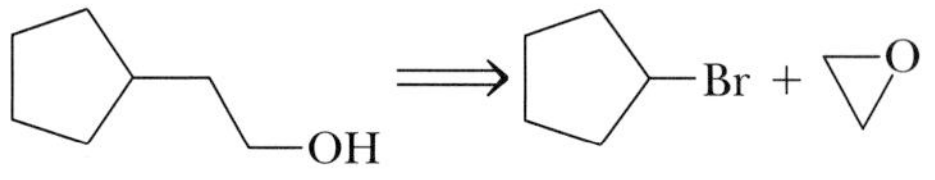

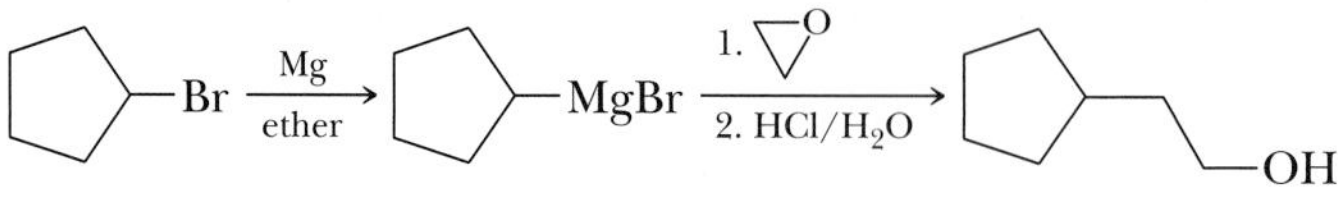

(b)

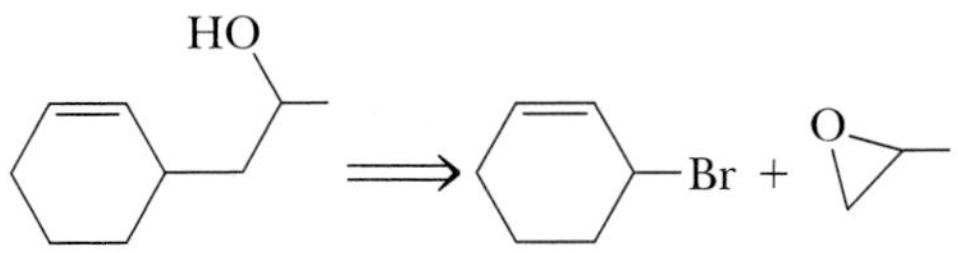

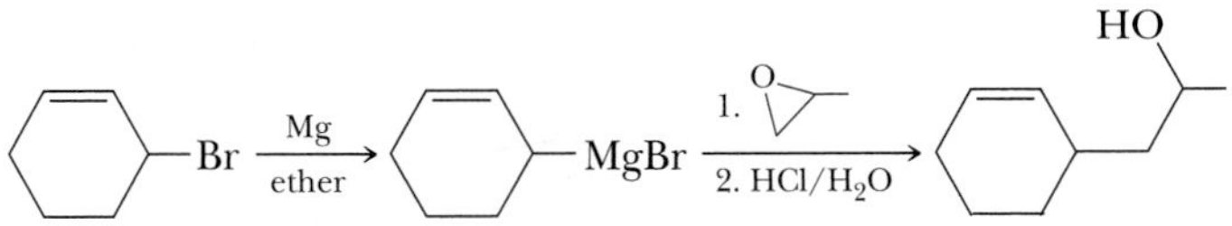

Problem 15.2

Recalling the reactions of alcohols from Chapter 9, show how to synthesize each compound from an organohalogen compound and an oxirane, followed by a transformation of the resulting hydroxyl group to the desired oxygen-containing functional group.

(a)

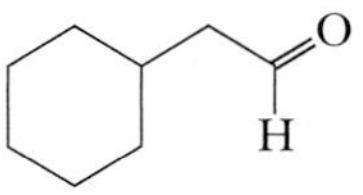

(b)

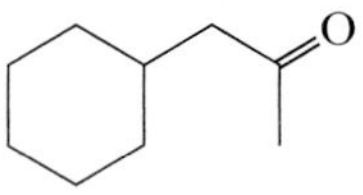

15.2 Lithium Diorganocopper (Gilman) Reagents

A. Formation and Structure

An important use of organolithium reagents (Section 15.1) is in the preparation of diorganocopper reagents, often called Gilman reagents after Henry Gilman (1893–1986) of Iowa State University who was the first to develop their chemistry. They are easily prepared by treatment of an alkyl, aryl, or alkenyl lithium compound with copper(I) iodide, as illustrated by the preparation of lithium dibutylcopper from butyllithium.

$$2CH_3CH_2CH_2CH_2Li + CuI \xrightarrow[\text{or THF}]{\text{diethyl ether}} (CH_3CH_2CH_2CH_2)_2Cu^- Li^+ + LiI$$

Butyllithium	Copper(I) iodide	Lithium dibutylcopper (a Gilman reagent)

Gilman reagents consist of two organic groups associated with a copper(I) ion giving a negatively charged species, which is the source of the carbon nucleophile. Lithium ion is associated with this negatively charged species as the counter ion.

B. Coupling with Organohalogen Compounds

Gilman reagents are especially valuable for the formation of new carbon-carbon bonds by a coupling reaction with an alkyl chloride, bromide, or iodides (fluorides are unreactive under these conditions) as illustrated by the following preparation of 2-methyl-1-dodecene. Notice that only one of the Gilman-reagent alkyl groups is transferred in the reaction. Because Gilman reagents are ultimately prepared from halides, this leads to effective coupling of two halides.

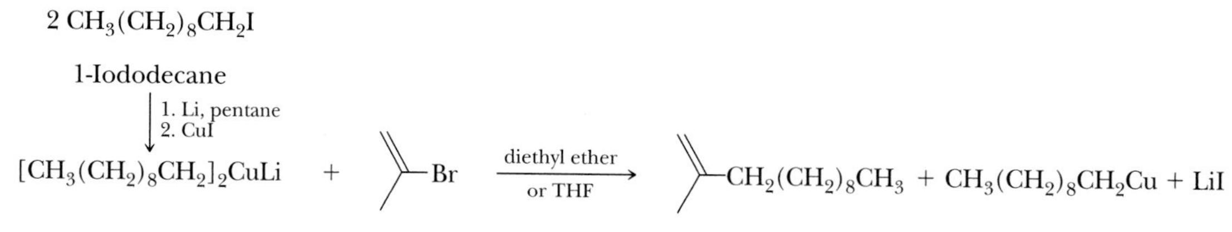

$2\ CH_3(CH_2)_8CH_2I$

1-Iododecane

| 1. Li, pentane
| 2. CuI

$[CH_3(CH_2)_8CH_2]_2CuLi$ + (2-bromopropene) $\xrightarrow[\text{or THF}]{\text{diethyl ether}}$ (2-methyl-1-dodecene) + $CH_3(CH_2)_8CH_2Cu$ + LiI

Lithium
didecylcopper

2-Bromopropene

2-Methyl-1-dodecene

This example illustrates the coupling of an alkyl halide, a nucleophile, with a vinylic halide, an electrophile. Vinylic halides are normally quite unreactive toward nucleophilic displacement. Thus, the lithium diorganocopper reaction shown here is unique.

Gilman reagents giving the best yields of coupling products are those prepared from methyl, primary alkyl, allylic, vinylic, and aryl halides via the corresponding organolithium compounds. Yields are lower with secondary and tertiary alkyl halides.

Coupling with a vinylic halide is stereospecific; the configuration of the carbon-carbon double bond is preserved, as illustrated by the synthesis of *trans*-5-tridecene.

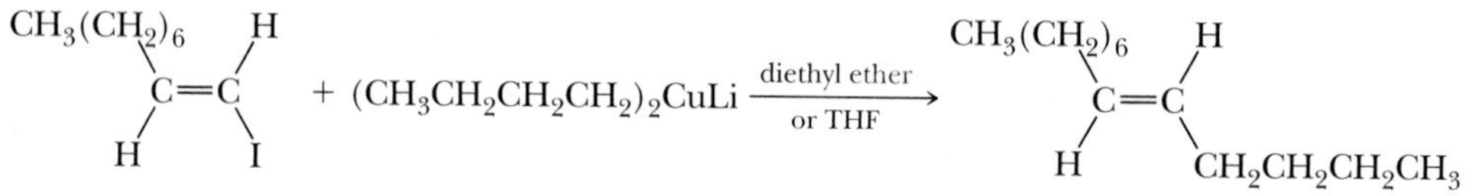

$CH_3(CH_2)_6$ H
     C=C
   H   I
+ $(CH_3CH_2CH_2CH_2)_2CuLi$ $\xrightarrow[\text{or THF}]{\text{diethyl ether}}$
$CH_3(CH_2)_6$ H
     C=C
   H $CH_2CH_2CH_2CH_3$

trans-Iodo-1-nonene

trans-5-Tridecene

A variation on the preparation of Gilman reagents is to use a Grignard reagent in the presence of a catalytic amount of Cu(I). Zoecon Corporation has developed a synthesis of 150-kg batches of the housefly sex attractant muscalure by treating (Z)-1-bromo-9-octadecene with pentylmagnesium bromide in the presence of catalytic amounts of Cu(I). The starting bromoalkene is easily prepared from the readily available (Z)-9-octadecenoic acid (oleic acid, Section 26.1). Yields of muscalure are reported to be nearly quantitative.

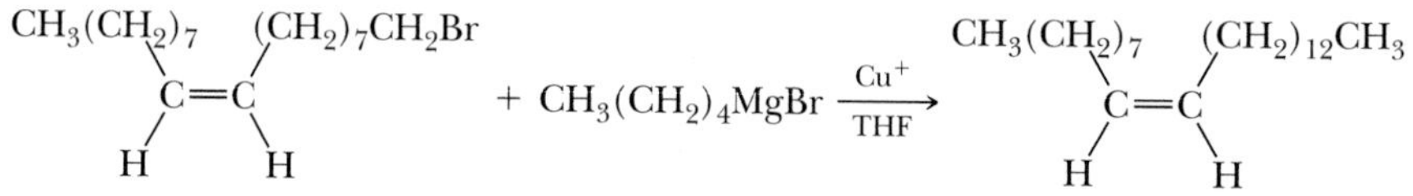

$CH_3(CH_2)_7$ $(CH_2)_7CH_2Br$
     C=C
   H   H
+ $CH_3(CH_2)_4MgBr$ $\xrightarrow[\text{THF}]{Cu^+}$
$CH_3(CH_2)_7$ $(CH_2)_{12}CH_3$
     C=C
   H   H

(Z)-1-Bromo-9-octadecene

(Z)-9-Tricosene
(Muscalure)

The mechanism of these coupling reactions is not fully understood and is the subject of active investigation.

Example 15.3

Show how to bring about each conversion using a lithium diorganocopper reagent.

(a) 1-Bromocyclohexene to 1-methylcyclohexene

(b) 1-Bromo-2-methylpropane to 2,5-dimethylhexane using the bromide as the only source of carbon

Solution

(a) Treat 1-bromocyclohexene with lithium dimethylcopper.

| 1-Bromocyclohexene | Lithium dimethylcopper | 1-Methylcyclohexene |

$$\text{Br} + (CH_3)_2CuLi \longrightarrow \text{—}CH_3 + CH_3Cu + LiBr$$

1-Bromocyclohexene   Lithium dimethylcopper   1-Methylcyclohexene

(b) Treat 1-bromo-2-methylpropane with lithium diisobutylcopper, itself prepared from 1-bromo-2-methylpropane.

$$2(CH_3)_2CHCH_2Br$$
$$\downarrow \quad \begin{array}{l}1.\ Li,\ hexane \\ 2.\ CuI\end{array}$$

$$[(CH_3)_2CHCH_2]_2CuLi \ + \ (CH_3)_2CHCH_2Br \longrightarrow CH_3CHCH_2\text{—}CH_2CHCH_3$$

Lithium diisobutyl copper   1-Bromo-2-methylpropane (Isobutyl bromide)   2,5-Dimethylhexane

Problem 15.3

Show how to bring about each conversion using a lithium diorganocopper reagent.

(a) **(b)**

C. Reaction with Oxiranes

The reaction of epoxides with Gilman reagents is an important method for the formation of new carbon-carbon bonds. Like organolithium compounds and Grignard reagents, these compounds bring about regioselective ring opening of substituted epoxides at the less substituted carbon to give alcohols. Treatment of styrene oxide with lithium divinylcopper, for example, followed by workup in aqueous acid gives 1-phenyl-3-buten-1-ol.

Styrene oxide   1. $(CH_2{=}CH)_2CuLi$ 2. $H_2O,\ HCl$ 1-Phenyl-3-buten-1-ol

Example 15.4

Show two combinations of epoxide and Gilman reagent that can be used to prepare 1-phenyl-3-hexanol.

Solution

The carbon bearing the hydroxyl group must have been one of the carbon atoms of the epoxide ring. The second carbon of the epoxide was either the one to the right of the carbon now bearing the —OH or the one to the left of it. In these solutions, the phenyl group is written C_6H_5—. Either route would be satisfactory.

Problem 15.4

Show how to prepare each Gilman reagent in Example 15.4 from an appropriate alkyl halide and each epoxide from an appropriate alkene.

15.3   Organopalladium Reagents — The Heck Reaction

A.   The Nature of the Reaction

In the early 1970s, Richard Heck at the Hercules Company and later at the University of Delaware discovered a palladium-catalyzed reaction in which the carbon group of a haloalkene or haloarene is substituted for a hydrogen on the carbon-carbon double bond (a vinylic hydrogen) of an alkene. This reaction, now known as the **Heck reaction,** is particularly valuable in synthetic organic chemistry because it is the only general method yet discovered for this type of substitution.

Substitution for a vinylic hydrogen by the Heck reaction is highly regioselective; formation of the new carbon-carbon bond most commonly occurs at the less substituted carbon of the double bond. In addition, where a cis or trans configuration is possible at the new carbon-carbon double bond of the product, the Heck reaction is highly stereoselective, often giving almost exclusively the E configuration of the resulting alkene.

In addition, the Heck reaction is stereospecific with regard to the haloalkene; the configuration of the double bond in the haloalkene is preserved.

(*Z*)-3-Iodo-3-hexene   Phenylethene   (1*E*,3*Z*)-1-Phenyl-3-ethyl-1,3-hexadiene
                         (Styrene)

(*E*)-3-Iodo-3-hexene   Phenylethene   (1*E*,3*E*)-1-Phenyl-3-ethyl-1,3-hexadiene
                         (Styrene)

Preparation of the Catalyst

The form of the palladium catalyst most commonly added to the reaction medium is palladium(II) acetate, $Pd(OAc)_2$. This and other Pd(II) compounds are better termed precatalysts because the catalytically active form of the metal is a complex of Pd(0) formed in situ by reduction of Pd(II) to Pd(0). Reaction of Pd(0) with good ligands, L, gives the actual catalyst, PdL_2.

The Organic Halogen Compound

The most common halides used in Heck reactions are aryl, heterocyclic, benzylic, and vinylic iodides and bromides, with iodides being generally more reactive. Contrast this behavior with nucleophilic substitution reactions, where substrates with leaving groups on sp^2 carbons are the least reactive. Alkyl halides in which there is an easily eliminated beta-hydrogen are rarely used because of the ease with which they undergo β-elimination under conditions of the Heck reaction to form alkenes. Triflates (trifluoromethanesulfonates), which are easily prepared by treating an alcohol with trifluoromethanesulfonyl chloride, are also excellent substrates.

Trifluoromethanesulfonyl chloride   Alcohol   A trifluoromethanesulfonate
                                                (a triflate)

A particular advantage of the Heck reaction is the wide range of functional groups, including alcohols, ethers, aldehydes, ketones, and esters, that may be present elsewhere in the organic halogen compound or alkene without reacting themselves or affecting the Heck reaction.

The Alkene

The reactivity of the alkene is a function of steric crowding about the carbon-carbon double bond. Ethylene and monosubstituted alkenes are the most reactive; the greater the degree of substitution on the double bond, the slower the reaction and the lower the yield of product.

The Base

Commonly used bases are tertiary amines such as triethylamine, $(CH_3CH_2)_3N$, sodium or potassium acetate, and sodium hydrogen carbonate.

The Solvent

Polar aprotic solvents such as N,N-dimethylformamide [$HCON(CH_3)_2$, DMF], acetonitrile (CH_3CN), and dimethyl sulfoxide (CH_3SOCH_3, DMSO) are commonly used. It is also possible to carry out some Heck reactions in aqueous methanol. The polar solvents are needed to dissolve the $Pd(OAc)_2$ at the beginning of the reaction.

The Ligands Coordinating with Pd(0)

Among the most common ligands, L, used for coordination of the Pd(0) is triphenylphosphine, $(C_6H_5)_3P$. Many other ligands such as BINAP (Section 6.7) can be used as well, including chiral ones that can lead to chiral products.

B. Mechanism of the Reaction

The mechanism of the Heck reaction is divided into two stages: formation of the Heck catalyst and the catalytic cycle. As you study the catalytic cycle, note in particular that both Steps 2 and 4 are syn stereospecific; the reaction will not proceed if these syn relationships cannot be obtained. Step 2 involves syn addition of R and PdL_2X to the double bond. Step 4 involves syn elimination of H and the Pd(II) species to regenerate a double bond. These additions and eliminations contrast with most of the addition and elimination reactions we have seen, which prefer the anti geometry. Reaction of boron hydrides and osmium tetroxide are examples of syn additions.

Mechanism   **The Heck Reaction**

Stage 1: Formation of the Heck Catalyst, PdL_2
A two-electron reduction of Pd(II) to Pd(0) accompanied by its complex formation with two molecules of a ligand, L, gives the Heck catalyst, PdL_2. A common reducing agent is triethylamine or, as in the following example, the alkene itself. Because the catalyst is present only in small amounts, an insignificant amount of the alkene is lost to this reaction.

Pd (OAc)$_2$	An alkene	Triphenylphosphine	A Pd(0) complex,
Palladium (II) acetate			abbreviated PdL_2

Stage 2: The Catalytic Cycle

Oxidative addition of the haloalkene or haloarene, RX, to PdL_2 gives a tetracoordinated Pd(0) complex containing both R and X groups bonded to Pd. Syn addition of R and PdL_2X to the alkene in Step 2 gives an intermediate in which Pd is bonded to the more substituted carbon. This intermediate must undergo internal rotation about the central carbon-carbon single bond in Step 3 to place H and PdL_2X syn to each other.

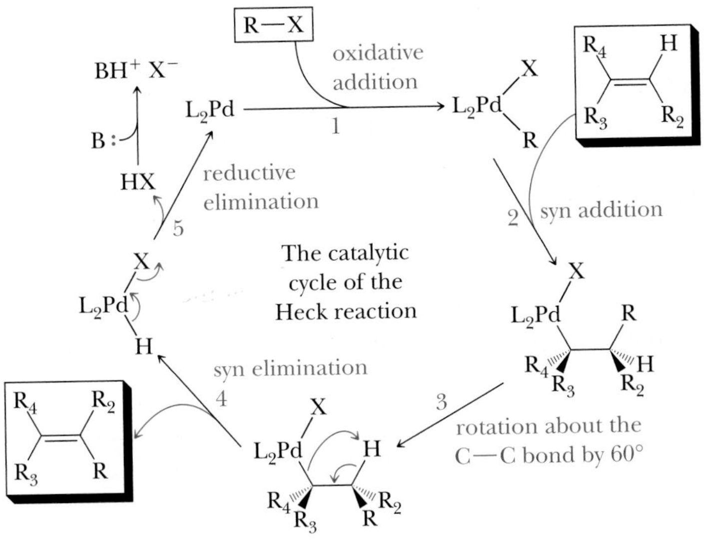

As is seen in this cycle, the alkene, organohalogen compound, and base are required in equimolar amounts; the Pd(0) species is required in only a catalytic amount. Note also the inversion of configuration (R_2 and R_3 are originally cis to each other but in the product are trans). This inversion is a consequence of the consecutive syn addition and elimination. The complete mechanism for this reaction has additional intermediates (involving pi complexes of the alkene with the palladium), but those shown here are the important ones for understanding the reaction and its stereochemistry.

Example 15.5

Complete these Heck reactions.

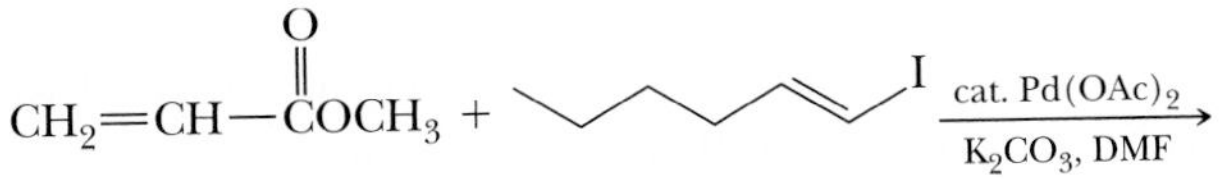

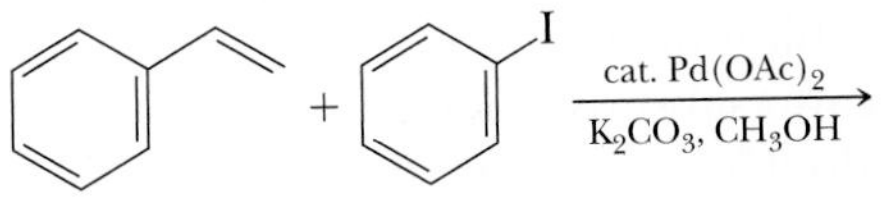

Solution

In (a), 1-iodohexene has the E configuration, and this double bond retains its E configuration in the product. Furthermore, the carbon-carbon double bond adjacent to the ester in the product now has the possibility for cis,trans isomerism. The Heck reaction is highly stereoselective, and this double bond has the more stable E configuration as well. In (b), the major product is (E)-1,2-diphenylethene.

(a)

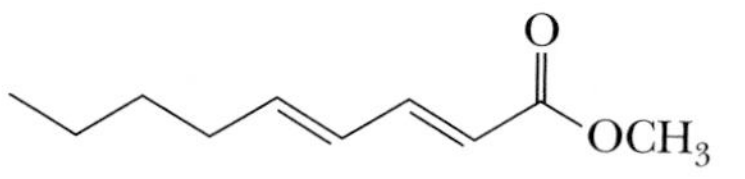

Methyl (2E,4E)-2,4-nonadienoate

(b)

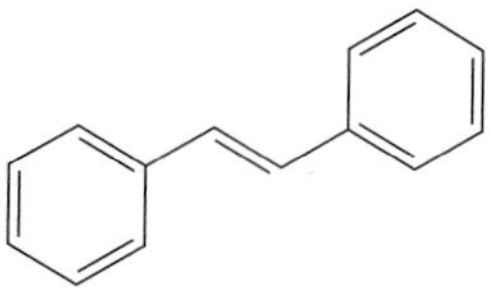

(E)-1,2-Diphenylethene
(*trans*-Stilbene)

Problem 15.5

Show how you might prepare each compound by a Heck reaction using methyl 2-propenoate as the starting alkene.

$$CH_2=CH-\overset{\overset{\displaystyle O}{\|}}{C}OCH_3$$

Methyl 2-propenoate
(Methyl acrylate)

(a)

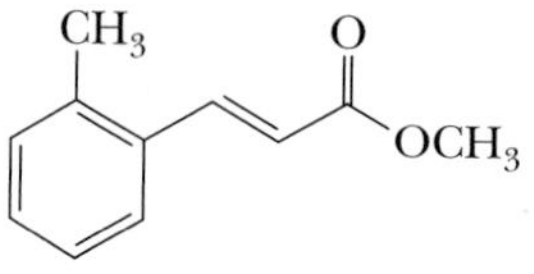

(the *E* isomer)

(b)

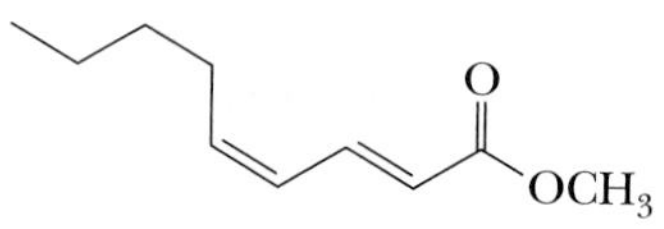

(the 2E,4Z isomer)

The usual pattern in a Heck reaction of acyclic alkenes is replacement of one of the hydrogens on the double bond by an organo group. If the organopalladium group attacks the double bond so that the R group is bonded to a carbon that lacks a hydrogen, or if the only syn hydrogen is on a neighboring carbon, the double bond shifts away from the original position. Note that the product contains a stereocenter, but, because it is formed from achiral reagents and in an achiral environment, it is formed as a racemic mixture.

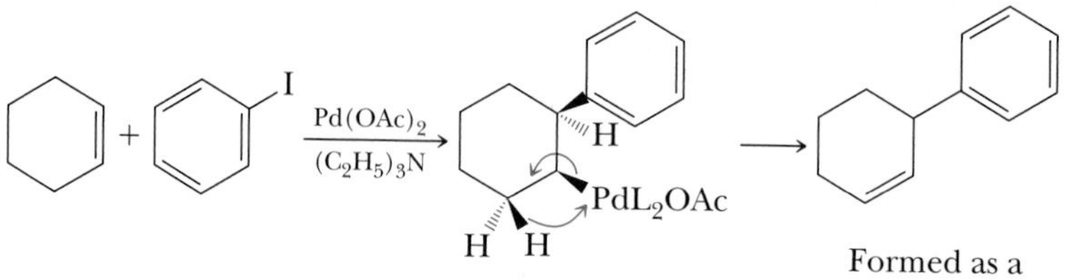

Formed as a
racemic mixture

A particularly valuable feature of the Heck reaction is that, when used with a chiral ligand, it can give chiral products in significant enantiomeric excess (ee). In the following example, the chirality is provided by the chiral ligand (R)-BINAP (Section 6.7B). For this reaction to yield a chiral product, the hydrogen eliminated cannot be on the carbon on which the organo substituent ends up, because, if this were the case, it would be attached to a double bond and the product would be achiral.

This enantiomer
is formed in 71% ee.

Because of the chiral ligand, the activation energy for the transition state in the syn addition to the alkene (Step 2 of the catalytic cycle) is different depending on which side of the alkene the metal complex approaches (the two transition states are diastereomers). This difference in activation energy means that approach to one side of the alkene is favored and results in an excess of one enantiomer of the product.

15.4 Carbenes, Carbenoids, and Alkene Metathesis

A **carbene**, R_2C:, is a neutral molecule in which a carbon atom is surrounded by only six valence electrons. Because they are electron deficient, carbenes are highly reactive and behave as electrophiles. As we will see, one of their most important types of reactions is with alkenes (nucleophiles) to give cyclopropanes.

Carbene A neutral molecule that contains a carbon atom surrounded by only six valence electrons (R_2C:).

A. Methylene

The simplest carbene is methylene, CH_2, prepared by **photolysis** (cleavage by light) or **thermolysis** (cleavage on heating) of diazomethane, CH_2N_2, an explosive, toxic gas.

$$H_2\overset{..}{C}=\overset{+}{N}=\overset{..}{\underset{..}{N}}:^- \longleftrightarrow H_2\overset{..}{C}-\overset{+}{N}\equiv N: \xrightarrow{h\nu} H_2C: + :N\equiv N:$$

Methylene
(the simplest
carbene)

In the lowest electronic state of most carbenes, carbon is sp^2 hybridized with the unshared pair of electrons occupying the third sp^2 orbital. The unhybridized $2p$ orbital lies perpendicular to the plane created by the three sp^2 orbitals. Note that this orbital description of methylene is very much like that of a carbocation (Section 6.3A). In both species, carbon is sp^2 hybridized with a vacant $2p$ orbital. Methylene in this electronic state resembles a hybrid of a carbocation and a carbanion in that it has both a vacant p orbital and a lone pair.

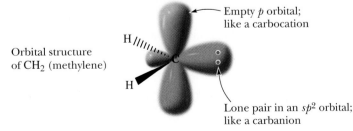

Orbital structure
of CH_2 (methylene)

Empty p orbital;
like a carbocation

Lone pair in an sp^2 orbital;
like a carbanion

Methylene generated in this manner reacts with all C—H and C=C bonds and is so nonselective that it is of little synthetic use.

B.   Dichlorocarbene

Although we often think of chlorine atoms as electron-withdrawing substituents, dichlorocarbene is much more stable and chemoselective than free methylene because resonance with the lone pairs on chlorine partially satisfies the electron deficiency on carbon.

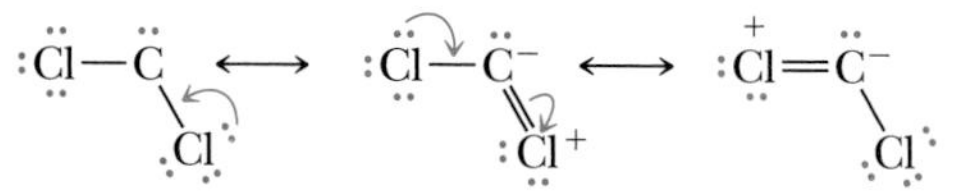

Dichlorocarbene can be prepared by treatment of chloroform with potassium *tert*-butoxide. The resulting carbene reacts cleanly with alkenes to give dichlorocyclopropanes. Addition of a dihalocarbene to an alkene shows syn stereospecificity.

$$CHCl_3 \quad + \quad (CH_3)_3CO^-K^+ \quad \longrightarrow \quad Cl_2C: \quad + \quad (CH_3)_3COH + K^+Cl^-$$

Trichloromethane (Chloroform)	Potassium *tert*-butoxide	Dichlorocarbene	*tert*-Butyl alcohol

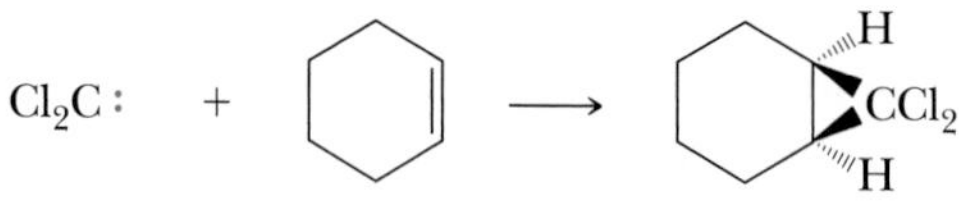

Reaction of a cis alkene with a dihalocarbene gives only a *cis*-dihalocyclopropane as illustrated by the reaction of *cis*-3-hexene with dichlorocarbene. Similarly, reaction of a trans alkene gives only a *trans*-dihalocyclopropane.

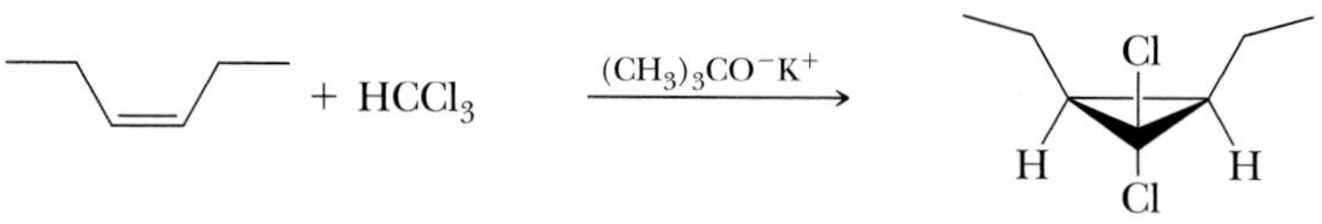

cis-3-Hexene   *cis*-1,1-Dichloro-2,3-diethylcyclopropane

Mechanism   **Formation of Dichlorocarbene and Its Reaction with Cyclohexene**

Taken together, Steps 1 and 2 result in α-elimination of H and Cl; that is, both atoms are eliminated from the same carbon. We have seen many examples of β-elimination, where hydrogen and a leaving group are eliminated from neighboring carbons. There are very few examples of α-elimination, and they are possible only where no β-hydrogen exists.

Step 1:   Treatment of chloroform, which is somewhat acidic because of its three electron-withdrawing chlorine atoms, with potassium *tert*-butoxide gives the trichloromethide anion.

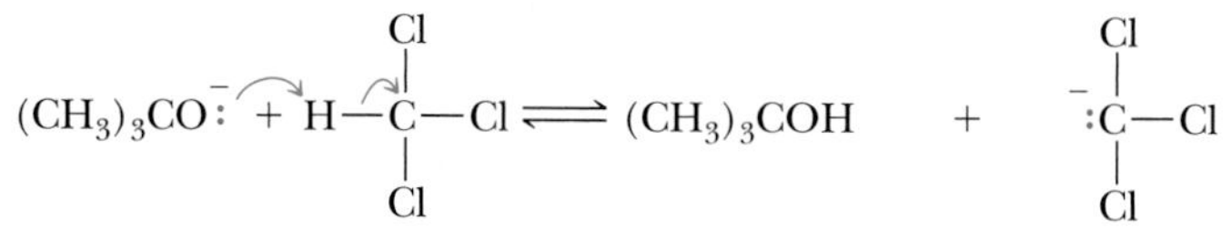

Trichloromethide anion

Step 2:   Elimination of Cl^- from CCl_3^- gives dichlorocarbene.

$$:\overset{\displaystyle Cl}{\underset{\displaystyle Cl}{C}}-Cl \longrightarrow :C-Cl \ + \ :Cl^-$$

Dichlorocarbene

Step 3:   Syn addition of dichlorocarbene to cyclohexene gives a dichlorocyclopropane.

$$Cl_2C: \quad + \qquad \longrightarrow$$

Dichlorocarbene   Cyclohexene   A dichlorocyclopropane
(an electrophile)   (a nucleophile)

Example 15.6

Predict the product from the following reaction.

$$CHBr_3 + (CH_3)_3CO^-K^+ \quad + \qquad \longrightarrow$$

(Z)-3-Methyl-2-pentene

Solution

Bromoform gives dibromocarbene, which reacts stereospecifically with the alkene to give a dibromocyclopropane.

$$CHBr_3 + (CH_3)_3CO^-K^+ + \qquad \longrightarrow$$

$$Br_2C: + (CH_3)_3COH + K^+Br^-$$

Problem 15.6

Predict the product of the following reaction.

$$=CH_2 + CHBr_3 + (CH_3)_3CO^-K^+ \longrightarrow$$

C.   The Simmons-Smith Reaction

Reactivity Explorer
Alkenes

Although methylene prepared from diazomethane itself is not synthetically useful, addition of methylene to an alkene can be accomplished using a reaction first reported by the American chemists Howard Simmons and Ronald Smith. The

Simmons-Smith reaction uses diiodomethane and zinc dust activated by a small amount of copper to produce iodomethylzinc iodide, in a reaction reminiscent of a Grignard reaction. Even though we show the Simmons-Smith reagent here as ICH_2ZnI, its structure is considerably more complex and not fully understood.

$$CH_2I_2 \quad + \quad Zn(Cu) \quad \xrightarrow{\text{diethyl ether}} \quad ICH_2ZnI$$

Diiodomethane   Zinc-copper   Iodomethylzinc iodide
                       couple   (Simmons-Smith reagent)

This organozinc compound reacts with a wide variety of alkenes to give cyclopropanes.

Methylenecyclopentane   Spiro[4.2]heptane

2-Cyclohexenone   Bicyclo[4.1.0]heptan-2-one

Mechanism   The Simmons-Smith Reaction with an Alkene

Carbenoid A compound that delivers the elements of a carbene without actually producing a free carbene.

Although an α-elimination from the Simmons-Smith reagent to give methylene would in principle be possible, the reagent is much more selective than free methylene. Instead, the organozinc compound reacts directly with the alkene by a concerted mechanism to give the cyclopropane-containing product. The Simmons-Smith reagent is an example of a **carbenoid,** a compound that delivers the elements of a carbene without actually producing a free carbene.

Example 15.7

Draw a structural formula for the product of treating each alkene with the Simmons-Smith reagent.

Solution

Reaction at each carbon-carbon double bond forms a cyclopropane ring.

(a)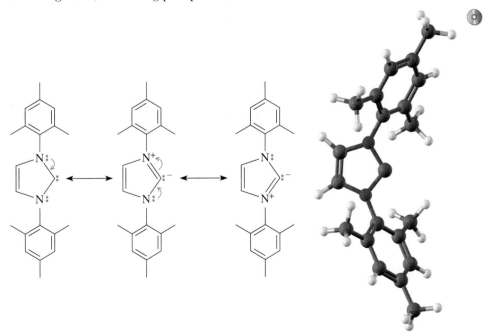

(b)

Problem 15.7

Show how the following compound could be prepared from any compound containing ten or fewer carbons.

D. Stable Nucleophilic Carbenes

Certain carbenes with strongly electron-donating substituents are particularly stable. Their stability can be enhanced further by adding sterically bulky substituents that hinder self-reactions. For example, the following cyclic carbene is stable enough to isolate. In this case, the large 2,4,6-trimethylbenzene (mesitylene) substituents protect the carbene from attack by nucleophiles or oxygen. Rather than being electron-deficient like most carbenes, these compounds are nucleophiles because of the strong electron donation by the nitrogens. Because of their nucleophilicity, they are excellent ligands (resembling phosphines) for certain transition metals.

Key Reactions

1. Formation of Organomagnesium (Grignard) and Organolithium Compounds (Section 15.1A)

Organomagnesium compounds are prepared by treating an alkyl, aryl, or alkenyl (vinylic) halide with magnesium in diethyl ether or THF. Organolithium compounds are prepared by treating an alkyl, aryl, or alkenyl halide with lithium in pentane or other hydrocarbon solvent.

$$\text{Br} + \text{Mg} \xrightarrow{\text{ether}} \text{MgBr}$$

$$\text{Cl} + 2\text{Li} \xrightarrow{\text{pentane}} \text{Li} + \text{LiCl}$$

2. Reaction of RMgX and RLi with Proton Donors (Section 15.1B)

Both organomagnesium and organolithium compounds are strong bases and react with any proton donor stronger than the alkane from which the organolithium or magnesium compound is derived. Water must be completely excluded during their preparation and use.

$$CH_3CH_2MgBr + H_2O \longrightarrow CH_3CH_3 + Mg(OH)Br$$

3. Reaction of a Grignard Reagent with an Epoxide (Section 15.1C)

Treatment of a Grignard reagent with an epoxide followed by hydrolysis of the magnesium alkoxide salt in aqueous acid gives an alcohol with its carbon chain extended by two carbon atoms.

$$\text{MgBr} \xrightarrow[\text{2. HCl, } H_2O]{\text{1. } \triangle O} \text{OH}$$

4. Formation of Gilman Reagents (Section 15.2A)

Lithium diorganocopper (Gilman) reagents are prepared by treating an organolithium compound with copper(I) iodide.

$$2CH_3CH_2CH_2CH_2Li + CuI \xrightarrow[\text{or THF}]{\text{ether}} (CH_3CH_2CH_2CH_2)_2Cu^- Li^+ + LiI$$

5. Treatment of a Gilman Reagent with an Alkyl, Aryl, or Alkenyl Halide (Section 15.2B)

Coupling of a Gilman reagent with an alkyl, alkenyl, or aryl halide results in formation of a new carbon-carbon bond.

$$\text{Br} + (CH_3)_2CuLi \longrightarrow \text{CH}_3 + CH_3Cu + LiBr$$

6. The Heck Reaction (Section 15.3)

In a palladium(0)-catalyzed reaction, the carbon group of a haloalkene or haloarene is substituted for a hydrogen on the carbon-carbon double bond (a vinylic hydrogen) of an alkene. Reaction generally proceeds with a high degree of both stereoselectivity and regioselectivity.

$$\text{Br} + CH_2{=}CHCOCH_3 \xrightarrow{\text{Pd catalyst}} \text{COCH}_3$$

⭐ **7. Reaction of Dichloro- or Dibromocarbene with an Alkene (Section 15.4B)**

The dihalocarbene is generated by treatment of $CHCl_3$ or $CHBr_3$ with a strong base such as potassium *tert*-butoxide. Addition of the dihaloalkene to an alkene shows syn stereospecificity.

$$CHBr_3 + \quad \xrightarrow{(CH_3)_3CO^-K^+}$$

⭐**8. The Simmons-Smith Reaction (Section 15.4C)**

Treatment of CH_2I_2 with a zinc-copper couple generates an organozinc compound, known as the Simmons-Smith reagent, that reacts with alkenes to give cyclopropanes.

$$+ CH_2I_2 \xrightarrow[\text{diethyl ether}]{Zn(Cu)} \quad CH_2 + ZnI_2$$

9. Alkene Metathesis Reaction (Section 15.4E)

In this organometallic-catalyzed reaction, two alkenes exchange carbons of their double bonds. In a ring-closing alkene methathesis reaction, both alkenes are in the same mole-cule, and the product is a cycloalkene. Catalysts with Ru and Mo are often used; a nucle-ophilic carbene complex of Ru is particularly useful.

EtOOC COOEt $\xrightarrow{\text{catalyst}}$ EtOOC COOEt $+ CH_2{=}CH_2$

Problems

15.9 Complete these reactions involving lithium diorganocopper (Gilman) reagents.

(a) $\bigcirc\!\!-\!Br + \left(\!\!\diagup\!\!\right)_2 CuLi \xrightarrow{\text{ether}}$

(b) (cyclohexenyl bromide) $+ \left(\diagdown\!\!\diagup\!\!\diagdown\right)_2 CuLi \xrightarrow{\text{ether}}$

(c) $\diagdown\!\!\diagup\!\!\diagdown\!\!-I + \left(\!\!\diagup\!\!\right)_2 CuLi \xrightarrow{\text{ether}}$

(d) $\bigcirc\!\!=\!C\!\!\stackrel{H}{\underset{Cl}{}} + \left(\!\!\stackrel{H_3C}{\underset{H}{}}C{=}C\stackrel{H}{\underset{CH_2}{}}\right)_2 CuLi \xrightarrow{\text{ether}}$

15.10 Show how to convert 1-bromopentane to each of these compounds using a lithium diorganocopper (Gilman) reagent. Write an equation, showing structural formulas, for each synthesis.

(a) Nonane   (b) 3-Methyloctane   (c) 2,2-Dimethylheptane
(d) 1-Heptene   (e) 1-Octene

15.11 In Problem 15.10, you used a series of lithium diorganocopper (Gilman) reagents. Show how to prepare each Gilman reagent from an appropriate alkyl or vinylic halide.

15.12 Show how to prepare each compound from the given starting compound through the use of a lithium diorganocopper (Gilman) reagent.

(a) 4-Methylcyclopentene from 4-bromocyclopentene
(b) (Z)-2-Undecene from (Z)-1-bromopropene
(c) 1-Butylcyclohexene from 1-iodocyclohexene
(d) 1-Decene from 1-iodooctane
(e) 1,8-Nonadiene from 1,5-dibromopentane

15.13 The following is a retrosynthetic scheme for the preparation of *trans*-2-allylcyclohexanol. Show reagents to bring about the synthesis of this compound from cyclohexane.

15.14 Complete these equations.

(a) $CH_3CH_2CH_2C{\equiv}CH + CH_3CH_2MgBr \xrightarrow{\text{diethyl ether}}$

(b) $+ CH_2I_2 \xrightarrow[\text{diethyl ether}]{\text{Zn(Cu)}}$

(c) $+ CHBr_3 + (CH_3)_3CO^-K^+ \longrightarrow$

(d) $+ CH_2I_2 \xrightarrow[\text{diethyl ether}]{\text{Zn(Cu)}}$

(e) $-CH{=}CH_2 + CH_2I_2 \xrightarrow[\text{diethyl ether}]{\text{Zn(Cu)}}$

15.15 Reaction of following cycloalkene with the Simmons-Smith reagent is stereospecific and gives only the isomer shown. Suggest a reason for this stereospecificity.

15.16 Show how the following compound can be prepared in good yield.

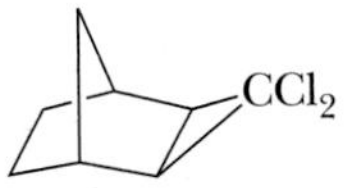

15.17 Show the product of the following reaction (do not concern yourself with stereochemistry).

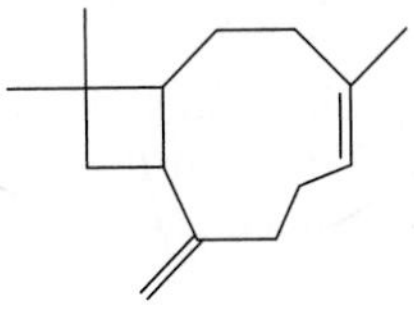

$$+ \quad CH_2I_2 \quad \xrightarrow{Zn(Cu)}$$

(excess)

Caryophyllene

15.18 As has been demonstrated in the text, when the starting alkene has CH_2 as its terminal group, the Heck reaction is highly stereoselective for formation of the *E* isomer, as illustrated in this example. Here, the benzene ring is abbreviated C_6H_5—. Show how the mechanism proposed in the text allows you to account for this stereoselectivity.

$$CH_2=CH-\overset{\overset{\displaystyle O}{\|}}{C}OCH_3 + C_6H_5Br \xrightarrow[\text{(CH}_3\text{CH}_2)_3\text{N}]{\text{Pd(OAc)}_2, \text{2Ph}_3\text{P}} C_6H_5 \text{~~~~} OCH_3$$

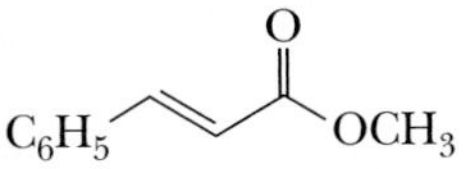

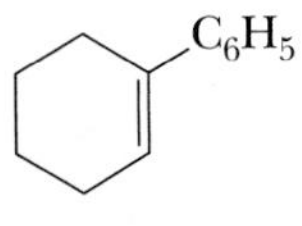

15.19 Heck reaction of bromobenzene and (*E*)-3-hexene gives a mixture of 3-phenyl-3-hexene and 4-phenyl-2-hexene in roughly equal amounts. Account for the formation of these two products.

$$+ C_6H_5Br \xrightarrow[\text{(CH}_3\text{CH}_2)_3\text{N}]{\text{Pd(OAc)}_2, \text{2Ph}_3\text{P}}$$

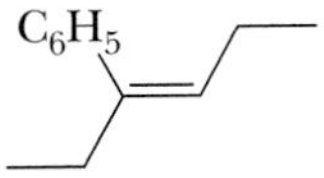

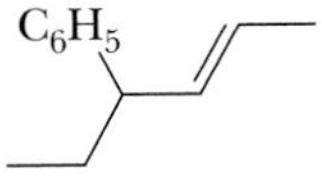

(*E*)-3-Hexene

(*Z*)-3-Phenyl-3-hexene   (*E*)-4-Phenyl-2-hexene

15.20 Complete these Heck reactions.

(a) $2C_6H_5CH=CH_2 + I-$$-I \xrightarrow[\text{(CH}_3\text{CH}_2)_3\text{N}]{\text{Pd(OAc)}_2, \text{2Ph}_3\text{P}}$

(b) $CH_2=CH\overset{\overset{\displaystyle O}{\|}}{C}OCH_3 +$ $\xrightarrow[\text{(CH}_3\text{CH}_2)_3\text{N}]{\text{Pd(OAc)}_2, \text{2Ph}_3\text{P}}$

15.21 Treatment of cyclohexene with iodobenzene under the conditions of the Heck reaction might be expected to give 1-phenylcyclohexene. The exclusive product, however, is 3-phenylcyclohexene. Account for the formation of this product.

$$+ C_6H_5I \xrightarrow[\text{(CH}_3\text{CH}_2)_3\text{N}]{\text{Pd(OAc)}_2, \text{2Ph}_3\text{P}}$$

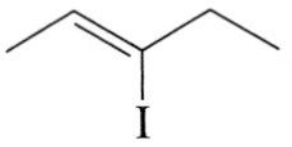

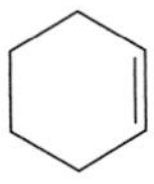

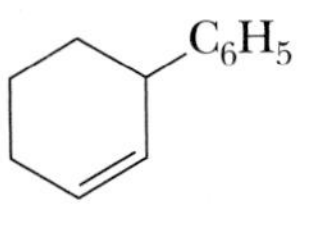

3-Phenylcyclohexene   1-Phenylcyclohexene
(the only product)   (not formed)

15.22 Account for the formation of the product and for the cis stereochemistry of its ring junction. (The function of silver carbonate is to enhance the rate of reaction.)

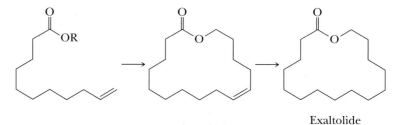

$$\xrightarrow[\text{Ag}_2\text{CO}_3,\ \text{CH}_3\text{CN}]{\text{Pd(OAc)}_2,\ \text{PPh}_3}$$

86%

15.23 Account for the formation of the following product, including the cis stereochemistry at the ring junction. (Tf is an abbreviation for the trifluoromethanesulfonate group, CF_3SO_2-, which can be used in the place of a halogen atom in a Heck reaction.)

$$\xrightarrow[\text{K}_2\text{CO}_3]{\text{Pd(OAc)}_2,\ (R)\text{-BINAP}}$$

15.24 The cyclic ester (lactone) Exaltolide (see *The Merck Index*, 12th ed., #3956) has a musk-like fragrance and is used as a fixative in perfumery. Show how this compound could be synthesized from the indicated starting material. Give the structure of R.

Exaltolide

15.25 Show how the spiro[2.2]pentane can be prepared in one step from organic compounds containing three carbons or less and any necessary inorganic reagents or solvents.

15.26 Predict the product of each alkene metathesis reaction using a Ru-nucleophilic carbene catalyst.

(a)

$$\xrightarrow[\text{CH}_2\text{Cl}_2,\ 40°\text{C},\ 30\ \text{min}]{5\ \text{mole}\ \%\ \text{Ru catalyst}}$$

(b)

$$\xrightarrow[\text{CH}_2\text{Cl}_2,\ 40°\text{C},\ 30\ \text{min}]{5\ \text{mole}\ \%\ \text{Ru catalyst}}$$

ALDEHYDES AND KETONES

In this and several of the following chapters, we study the physical and chemical properties of compounds containing the carbonyl group, C=O. Because the carbonyl group is the functional group of aldehydes, ketones, and carboxylic acids and their derivatives, it is one of the most important functional groups in organic chemistry. The chemical properties of this functional group are straightforward, and an understanding of its few characteristic reaction themes leads very quickly to an understanding of a wide variety of organic reactions.

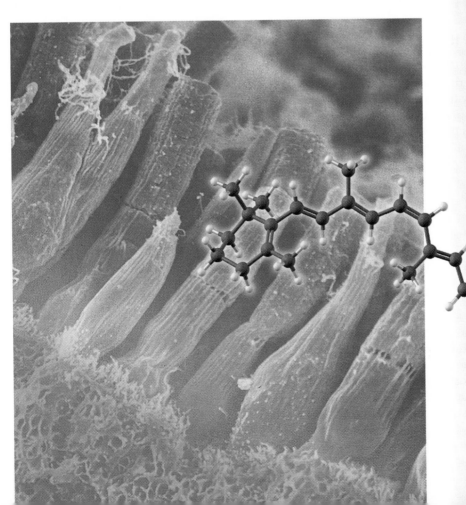

■ Rod cells in the human eye. *(Omikron/Photo Researchers, Inc.)* Inset: A model of 11-*cis*-retinal, an oxidized form of vitamin A. For the reaction of 11-*cis*-retinal with opsin to form visual purple, see Section 16.10A.

16.1   Structure and Bonding

The functional group of an **aldehyde** is a carbonyl group bonded to a hydrogen atom and a carbon atom (Section 1.3C). In methanal (always called formaldehyde), the simplest aldehyde, the carbonyl group is bonded to two hydrogen atoms. In other aldehydes, it is bonded to one hydrogen atom and one carbon atom. Following are Lewis structures for formaldehyde and ethanal (always called acetaldehyde). The functional group of a **ketone** is a carbonyl group bonded to two carbon atoms (Section 1.3C). Following is a Lewis structure for propanone (always called acetone), the simplest ketone.

O   O   O

‖   ‖   ‖

HCH CH_3CH CH_3CCH_3

Methanal Ethanal Propanone

(Formaldehyde) (Acetaldehyde) (Acetone)

A carbon-oxygen double bond consists of one sigma bond formed by overlap of sp^2 hybrid orbitals of carbon and oxygen and one pi bond formed by the overlap of parallel $2p$ orbitals. The two nonbonding pairs of electrons on oxygen lie in the remaining sp^2 hybrid orbitals (Figure 1.19).

Nomenclature
Aldehydes and Ketones

16.2   Nomenclature

A.   IUPAC Nomenclature

The IUPAC system of nomenclature for aldehydes and ketones follows the familiar pattern of selecting as the parent alkane the longest chain of carbon atoms that contains the functional group. The aldehyde group is shown by changing the suffix -e of the parent alkane to -al (Section 2.5). Because the carbonyl group of an aldehyde can only appear at the end of a parent chain and because numbering must start with it as carbon-1, its position is unambiguous; there is no need to use a number to locate it.

For unsaturated aldehydes, the presence of a carbon-carbon double or triple bond is indicated by the infix -en- or -yn-. As with other molecules with both an infix and a suffix, the location of the group corresponding to the suffix determines the numbering pattern.

3-Methylbutanal 2-Propenal (2*E*)-3,7-Dimethyl-2,6-octadienal

(Acrolein)   (Geranial)

For cyclic molecules in which —CHO is bonded directly to the ring, the molecule is named by adding the suffix -carbaldehyde to the name of the ring. The atom of the ring to which the aldehyde group is attached is numbered 1 unless the ring (as, for example, a bicyclic ring) has some other fixed numbering pattern, in which case the —CHO group is given a number as low as possible.

Cyclopentane-
carbaldehyde

trans-4-Hydroxycyclo-
hexanecarbaldehyde

Among the aldehydes for which the IUPAC system retains common names are benzaldehyde and cinnamaldehyde, as well as formaldehyde and acetaldehyde. Note here the alternative ways of writing the phenyl group. In benzaldehyde, it is written as a line-angle drawing; in cinnamaldehyde, it is written C_6H_5-.

Benzaldehyde

trans-3-Phenyl-2-propenal
(Cinnamaldehyde)

In the IUPAC system, ketones are named by selecting as the parent alkane the longest chain that contains the carbonyl group and then indicating its presence by changing the suffix from -e to -one (Section 2.5). The parent chain is numbered from the direction that gives the carbonyl carbon the smaller number. The IUPAC system retains the common names acetone, acetophenone, and benzophenone.

CH_3CCH_3

Propanone
(Acetone)

Acetophenone

Benzophenone

1-Phenyl-1-pentanone

Benzaldehyde is found in the kernels of bitter almonds. Cinnamaldehyde is found in Ceylon and Chinese cinnamon oils. (*Charles D. Winters*)

Example 16.1

Write IUPAC names for each compound. Specify configuration of all stereocenters in (a) and (c).

(a)　(b)　(c)

Solution

(a) The longest chain has six carbons, but the longest chain that contains the aldehyde group has five carbons. Therefore, the parent chain is pentane. The name is (*S*)-2-ethyl-3-methylpentanal.

(b) Number the six-membered ring beginning with the carbonyl carbon. The IUPAC name is 3-methyl-2-cyclohexenone.

(c) The name (2S,5R)-2-isopropyl-5-methylcyclohexanone provides a complete description of both the configuration of each stereocenter and also the trans relationship between the isopropyl and methyl groups. The common name of this compound is menthone.

Problem 16.1

Write the IUPAC name for each compound. Specify configuration in (c).

(a) ![structure with CHO] **(b)** ![cyclohexane-1,3-dione structure] **(c)** C_6H_5—C (CHO, CH$_3$, H)

Example 16.2

Write structural formulas for all ketones of molecular formula $C_6H_{12}O$, and give each its IUPAC name. Which of these ketones is/are chiral?

Solution

Following are line-angle drawings and IUPAC names for the six ketones of this molecular formula. Only 3-methyl-2-pentanone is chiral; the R enantiomer is drawn here.

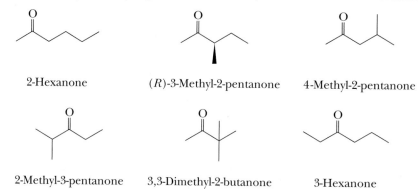

2-Hexanone   (R)-3-Methyl-2-pentanone   4-Methyl-2-pentanone

2-Methyl-3-pentanone   3,3-Dimethyl-2-butanone   3-Hexanone

Problem 16.2

Write structural formulas for all aldehydes of molecular formula $C_6H_{12}O$, and give each its IUPAC name. Which of these aldehydes is/are chiral?

B.   IUPAC Names for More Complex Aldehydes and Ketones

Order of precedence of functions
A ranking of functional groups in order of priority for the purposes of IUPAC nomenclature.

In naming compounds that contain more than one functional group that might be indicated by a suffix, the IUPAC system has established an **order of precedence of functions.** Table 16.1 gives the order of precedence for the functional groups we have studied so far.

Table 16.1   **Increasing Order of Precedence of Six Functional Groups**

Functional Group	Suffix if Higher in Precedence	Prefix if Lower in Precedence
—COOH	-oic acid	—
—CHO	-al	oxo-
C=O	-one	oxo-
—OH	-ol	hydroxy-
—SH	-thiol	-sulfanyl
—NH$_2$	-amine	-amino

increasing precedence ↑

Example 16.3

Write the IUPAC name for each compound. Specify configuration for (b) and (c).

(a) 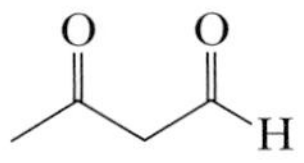   (b) 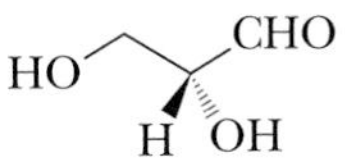 (c)

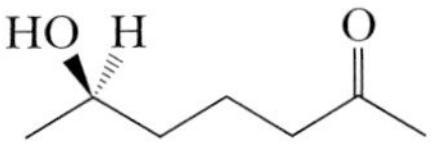

Solution

(a) 3-Oxobutanal. An aldehyde has higher precedence than a ketone. The presence of the carbonyl group of the ketone is indicated by the prefix oxo- (Table 16.1).
(b) (R)-2,3-Dihydroxypropanal. Its common name is glyceraldehyde. Glyceraldehyde is the simplest carbohydrate (Section 25.1).
(c) (R)-6-Hydroxy-2-heptanone.

Problem 16.3

Write the IUPAC name for each compound.

(a) 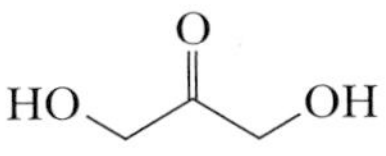   (b) 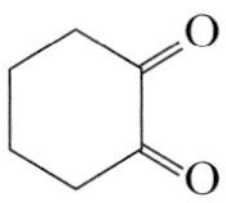   (c)

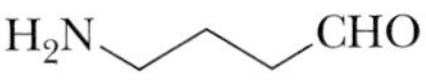

C.   Common Names

The common name for an aldehyde is derived from the common name of the corresponding carboxylic acid by dropping the word "acid" and changing the suffix -ic or -oic to -aldehyde. Because we have not yet studied common names for carboxylic acids, we are not in a position to discuss common names for aldehydes. We can illustrate how they are derived, however, by reference to a few common names with which you are familiar. The name formaldehyde is derived from formic acid; the name acetaldehyde is derived from acetic acid.

$$\underset{\text{Formaldehyde}}{\overset{\displaystyle O}{\underset{\displaystyle \|}{HCH}}} \qquad \underset{\text{Formic acid}}{\overset{\displaystyle O}{\underset{\displaystyle \|}{HCOH}}} \qquad \underset{\text{Acetaldehyde}}{\overset{\displaystyle O}{\underset{\displaystyle \|}{CH_3CH}}} \qquad \underset{\text{Acetic acid}}{\overset{\displaystyle O}{\underset{\displaystyle \|}{CH_3COH}}}$$

Common names for ketones are derived by naming the two alkyl or aryl groups bonded to the carbonyl group as separate words, followed by the word "ketone."

Ethyl isopropyl ketone   Diethyl ketone   Dicyclohexyl ketone

16.3   Physical Properties

Oxygen is more electronegative than carbon (3.5 compared with 2.5); therefore, a carbon-oxygen double bond is polar, with oxygen bearing a partial negative charge and carbon bearing a partial positive charge. In addition, the resonance structure shown on the right emphasizes the reactivity of the carbonyl oxygen as a Lewis base and the carbonyl carbon as a Lewis acid. The bond moment of a carbonyl group is 2.3 D (Table 1.7).

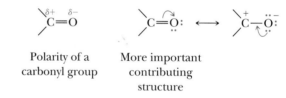

Polarity of a
carbonyl group

More important
contributing
structure

Because of the polarity of the carbonyl group, aldehydes and ketones are polar compounds and interact in the pure liquid by dipole-dipole interactions; they have higher boiling points than nonpolar compounds of comparable molecular weight (Table 16.2).

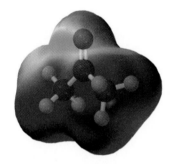

An electrostatic potential map for acetone. Note the large negative charge (red) on the carbonyl oxygen and the positive charges (blue) on the three carbons.

Table 16.2	Boiling Points of Six Compounds of Comparable Molecular Weight		
Name	**Structural Formula**	**Molecular Weight**	**bp (°C)**
Diethyl ether	$CH_3CH_2OCH_2CH_3$	74	34
Pentane	$CH_3CH_2CH_2CH_2CH_3$	72	36
Butanal	$CH_3CH_2CH_2CHO$	72	76
2-Butanone	$CH_3CH_2COCH_3$	72	80
1-Butanol	$CH_3CH_2CH_2CH_2OH$	74	117
Propanoic acid	CH_3CH_2COOH	74	141

Table 16.3   Physical Properties of Selected Aldehydes and Ketones

IUPAC Name	Common Name	Structural Formula	bp (°C)	Solubility (g/100 g water)
Methanal	Formaldehyde	HCHO	−21	Infinite
Ethanal	Acetaldehyde	CH_3CHO	20	Infinite
Propanal	Propionaldehyde	CH_3CH_2CHO	49	16
Butanal	Butyraldehyde	$CH_3CH_2CH_2CHO$	76	7
Hexanal	Caproaldehyde	$CH_3(CH_2)_4CHO$	129	Slight
Propanone	Acetone	CH_3COCH_3	56	Infinite
2-Butanone	Ethyl methyl ketone	$CH_3COCH_2CH_3$	80	26
3-Pentanone	Diethyl ketone	$CH_3CH_2COCH_2CH_3$	101	5

Pentane and diethyl ether have the lowest boiling points. Diethyl ether is a polar molecule, but because of steric hindrance, only weak dipole-dipole interactions exist between its molecules (Section 11.3). Both butanal and 2-butanone are polar compounds, and, because of the intermolecular attraction between their carbonyl groups, their boiling points are higher than those of pentane and diethyl ether. Alcohols (Section 9.2) and carboxylic acids (Section 17.3) are polar compounds, and their molecules associate by hydrogen bonding; the boiling points of 1-butanol and propanoic acid are higher than those of butanal and 2-butanone, compounds whose molecules cannot associate by hydrogen bonding.

The carbonyl groups of aldehydes and ketones interact with water molecules by hydrogen bonding; therefore, low-molecular-weight aldehydes and ketones are more soluble in water than are nonpolar compounds of comparable molecular weight. Listed in Table 16.3 are boiling points and solubilities in water for several low-molecular-weight aldehydes and ketones.

16.4   Spectroscopic Properties

Spectroscopy
Aldehydes, Ketones
The CD-ROM contains NMR, IR, and mass spectra for 20 aldehydes and ketones.

^{1}H-NMR spectroscopy is an important means for identifying aldehydes and for distinguishing between aldehydes and other carbonyl-containing compounds. Just as the pi system of a carbon-carbon double bond causes a downfield shift in the signal of a vinylic hydrogen (Figure 13.8), the pi system of a carbon-oxygen double bond causes a large downfield shift in the signal of an aldehyde hydrogen, typically to δ 9.5–10.1. Coupling constants between this hydrogen and those on the adjacent α-carbon are small (approximately 1–3 Hz); consequently, the aldehyde hydrogen signal often appears as a singlet rather than a doublet or triplet as the case may be. The ^{1}H-NMR spectrum of butanal, for example, shows a singlet at δ 9.78 for the aldehyde hydrogen if the spectrum is not expanded (Figure 16.1).

Hydrogens on an α-carbon of an aldehyde or ketone typically appear around δ 2.1–2.6. Because coupling constants between α-hydrogens and an aldehyde hydrogen are small (approximately 1–3 Hz), the ^{1}H-NMR signals of α-hydrogens may appear unsplit by aldehyde hydrogen. The carbonyl carbons of aldehydes and ketones

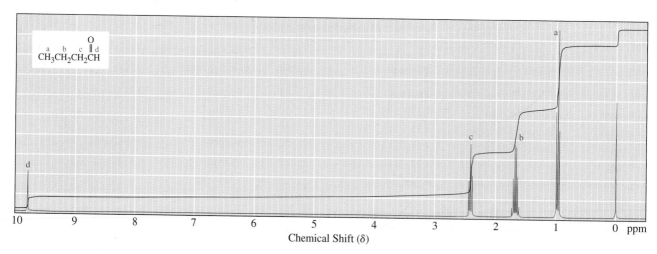

Figure 16.1
^{1}H-NMR spectrum of butanal.

are readily identifiable in ^{13}C-NMR spectroscopy by the position of their signal between δ 180 and 215.

The infrared spectroscopy of aldehydes and ketones was discussed in Section 12.4H. They show characteristic strong infrared absorption between 1630 and 1810 cm^{-1} associated with the stretching vibration of the carbon-oxygen double bond.

16.5 Reactions

Reactivity Explorer
Ketones and Aldehydes
Supporting Concepts
Carbonyl Chemistry
This is a comprehensive overview
of reactivity of compounds containing the carbonyl group.

One of the most common reaction themes of a carbonyl group is addition of a nucleophile to form a tetrahedral carbonyl addition compound. In the following general reaction, the nucleophilic reagent is written as Nu:$^-$ to emphasize the presence of its unshared pair of electrons.

Tetrahedral carbonyl
addition compound

A second common reaction theme of a carbonyl group is reaction with a proton or Lewis acid to form a resonance-stabilized cation. Protonation increases the electron deficiency of the carbonyl carbon and makes it more reactive toward nucleophiles. This cation then reacts with nucleophiles to give a tetrahedral carbonyl addition compound.

$$B:^- + H-Nu: \quad \underset{R}{\overset{R}{\diagdown}}C=\overset{+}{\overset{\cdot\cdot}{O}}-H \quad \xrightarrow{\text{slow}} \quad Nu-\underset{\underset{R}{\overset{|}{R}}}{\overset{\overset{\cdot\cdot}{O}-H}{C}} \quad + \quad H-B$$

Tetrahedral carbonyl
addition compound

16.6   Addition of Carbon Nucleophiles

In this section, we examine reactions of aldehydes and ketones with the following
types of carbon nucleophiles:

RMgX   RLi $RC{\equiv}C:^-$ $:C{\equiv}N:^-$

A Grignard   An organolithium   An anion of a   Cyanide ion
reagent   reagent   terminal alkyne

From the perspective of the organic chemist, addition of a carbon nucleophile is the
most important type of carbonyl addition reaction because a new carbon-carbon
bond is formed in the process.

1,3-Dihydroxypropanone,
more commonly known as dihydroxyacetone, is the active
ingredient in artificial tanning
agents such as Man-Tan and
Magic Tan. *(George Semple)*

A.   Addition of Grignard Reagents

The special value of Grignard reagents (Section 15.1) is that they provide excellent
ways to form new carbon-carbon bonds. Given the difference in electronegativity between carbon and magnesium $(2.5 - 1.2 = 1.3)$, the carbon-magnesium bond of a
Grignard reagent is polar covalent with carbon bearing a partial negative charge and
magnesium bearing a partial positive charge. In its reactions, a Grignard reagent behaves as a carbanion. A **carbanion** is a good nucleophile and adds to the carbonyl
group of an aldehyde or ketone to form a tetrahedral carbonyl addition compound.
The driving force for these reactions is the attraction of the partial negative charge
on the carbon of the organometallic compound for the partial positive charge on the
carbonyl carbon. In the following examples, the magnesium-oxygen bond is written
$-O^-[MgBr]^+$ to emphasize its ionic character. The alkoxide ions formed in these reactions are strong bases (Section 9.5) and, when treated with an aqueous acid such as
HCl or aqueous NH_4Cl during workup, form alcohols.

Carbanion   An anion in which
carbon has an unshared pair of
electrons and bears a negative
charge.

Addition to Formaldehyde Gives a Primary Alcohol

Treatment of a Grignard reagent with formaldehyde followed by hydrolysis in aqueous acid gives a primary alcohol.

$$CH_3CH_2-MgBr \; + \; H-\overset{\overset{\displaystyle O}{\|}}{C}-H \; \xrightarrow{\text{ether}} \; CH_3CH_2-\overset{\overset{\displaystyle O^-[MgBr]^+}{|}}{CH_2} \; \xrightarrow[\text{H}_2\text{O}]{\text{HCl}} \; CH_3CH_2-\overset{\overset{\displaystyle OH}{|}}{CH_2} \; + \; Mg^{2+}$$

Formaldehyde   A magnesium   1-Propanol
                              alkoxide   (a primary alcohol)

Addition to an Aldehyde (Except Formaldehyde) Gives a Secondary Alcohol

Treatment of a Grignard reagent with any other aldehyde followed by hydrolysis in aqueous acid gives a secondary alcohol.

Acetaldehyde
(an aldehyde)

A magnesium
alkoxide

1-Cyclohexylethanol
(a secondary alcohol)

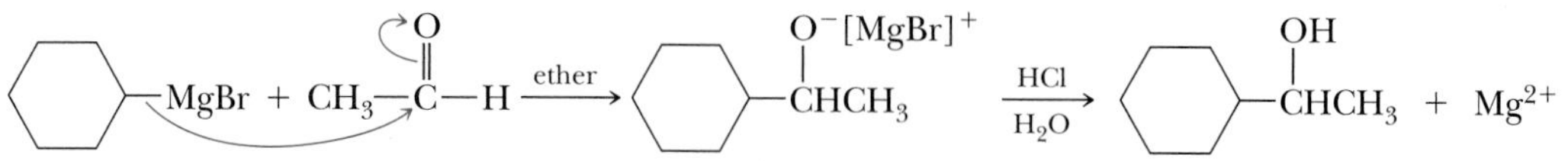

Addition to a Ketone Gives a Tertiary Alcohol

Treatment of a Grignard reagent with a ketone followed by hydrolysis in aqueous acid gives a tertiary alcohol.

Acetone
(a ketone)

A magnesium alkoxide

2-Phenyl-2-propanol
(a tertiary alcohol)

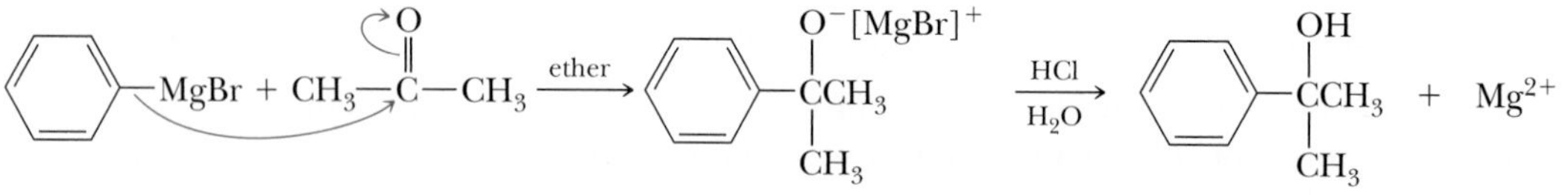

Example 16.4

2-Phenyl-2-butanol can be synthesized by three different combinations of a Grignard reagent and a ketone. Show each combination.

Solution

In each solution, curved arrows show formation of the new carbon-carbon bond and the alkoxide ion. The new carbon-carbon bond formed by each set of reagents is labeled in the final product.

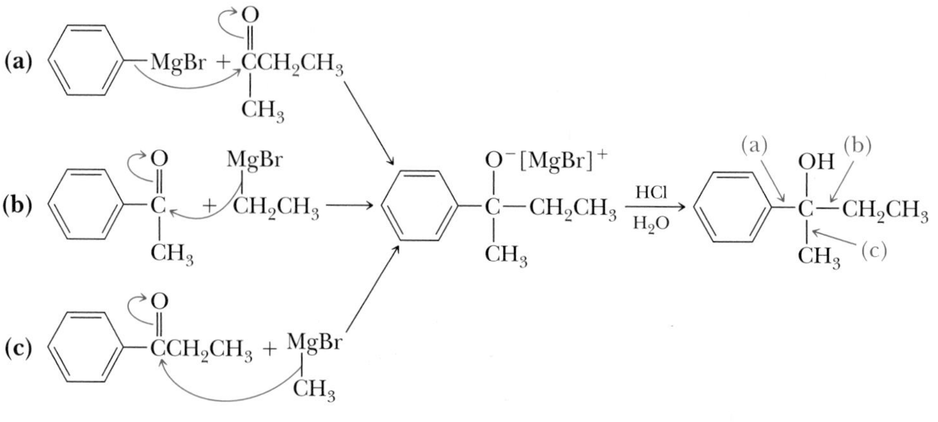

Problem 16.4

Show how these four products can be synthesized from the same Grignard reagent.

(a)   (b)   (c)   (d)

B. Addition of Organolithium Compounds

Organolithium compounds are generally more reactive in carbonyl addition reactions than organomagnesium compounds and typically give higher yields of products. They are more troublesome to use, however, because they must be prepared and used under an atmosphere of nitrogen or other inert gas. Following is a synthesis illustrating the use of an organolithium compound to form a sterically hindered tertiary alcohol.

Phenyllithium   3,3-Dimethyl-2-butanone   A lithium alkoxide   3,3-Dimethyl-2-phenyl-2-butanol

C. Addition of Salts of Terminal Alkynes

The anion of a terminal alkyne is a nucleophile (Section 10.5) and adds to the carbonyl group of an aldehyde or ketone to form a tetrahedral carbonyl addition compound. These addition compounds contain both a hydroxyl group and a carbon-carbon triple bond, each of which can be further modified. In the following example, addition of sodium acetylide to cyclohexanone followed by hydrolysis in aqueous acid gives 1-ethynylcyclohexanol.

Sodium acetylide   Cyclohexanone   A sodium alkoxide   1-Ethynylcyclohexanol

Acid-catalyzed hydration of this hydroxyalkyne (Section 10.9C) gives an α-hydroxyketone. Alternatively, hydroboration followed by oxidation with alkaline hydrogen peroxide (Section 10.8) gives a β-hydroxyaldehyde.

An α-hydroxyketone

A β-hydroxyaldehyde

This example illustrates two of the most valuable reactions of alkynes: (1) addition of the anion of a terminal alkyne to the carbonyl group of an aldehyde or ketone gives an alkynyl alcohol and (2) hydration of a terminal alkyne gives either an aldehyde or ketone depending on the method of hydration.

D.   Addition of Hydrogen Cyanide

Cyanohydrin   A molecule containing an —OH group and a —CN group bonded to the same carbon.

Hydrogen cyanide, HCN, adds to the carbonyl group of an aldehyde or ketone to form a tetrahedral carbonyl addition compound called a **cyanohydrin.** For example, HCN adds to acetaldehyde to form acetaldehyde cyanohydrin in 75% yield. We study the naming of compounds containing the nitrile group in Section 18.1E.

$$CH_3\overset{\displaystyle O}{\overset{\|}{C}}H + HC\equiv N \rightleftharpoons CH_3\overset{\displaystyle OH}{\underset{\displaystyle H}{\overset{|}{\underset{|}{C}}}}-C\equiv N$$

2-Hydroxypropanenitrile
(Acetaldehyde cyanohydrin)

Addition of hydrogen cyanide proceeds by way of cyanide ion. Because HCN is a weak acid, pK_a 9.31, the concentration of cyanide ion in aqueous HCN is too low for cyanohydrin formation to proceed at a reasonable rate. For this reason, cyanohydrin formation is generally carried out by dissolving NaCN or KCN in water and adjusting the pH of the solution to approximately 10.0, giving a solution in which HCN and CN^- are present in comparable concentrations. Reaction is initiated by nucleophilic addition of cyanide ion to the carbonyl group to form an alkoxide ion that in turn reacts with HCN to form the cyanohydrin and CN^-.

Mechanisms
Ketones
Formation of a cyanohydrin

Mechanism   Formation of a Cyanohydrin

Step 1:   Nucleophilic addition of cyanide ion to the carbonyl carbon gives a tetrahedral carbonyl addition compound.

Step 2: Proton transfer from HCN gives the cyanohydrin.

For aldehydes and most aliphatic ketones, the position of equilibrium favors cyanohydrin formation. For many aryl ketones and sterically hindered aliphatic ketones, however, the position of equilibrium favors starting materials, and cyanohydrin formation is not a useful reaction for these types of compounds. The following synthesis of ibuprofen, for example, failed because the cyanohydrin was formed only in low yield.

4-Isobutylacetophenone   A cyanohydrin   Ibuprofen

Benzaldehyde cyanohydrin (mandelonitrile) provides an interesting example of a chemical defense mechanism in the biological world. This substance is synthesized by millipedes (*Apheloria corrigata*) and stored in special glands. When a millipede is threatened, the cyanohydrin is released from the storage gland and undergoes enzyme-catalyzed reversal of cyanohydrin formation to produce HCN, which is then released to ward off predators. The quantity of HCN emitted by a single millipede is sufficient to kill a small mouse. Mandelonitrile is also found in bitter almond and peach pits. Its function there is unknown, as is how millipedes survive exposure to hydrogen cyanide.

Benzaldehyde cyanohydrin   Benzaldehyde

The main value of cyanohydrins lies in the new functional groups into which they can be converted. First, the secondary or tertiary alcohol group of a cyanohydrin may undergo acid-catalyzed dehydration to form an unsaturated nitrile. For example, acid-catalyzed dehydration of acetaldehyde cyanohydrin gives acrylonitrile, the monomer from which polyacrylonitrile (Orlon, Table 24.1) is made.

2-Hydroxypropanenitrile   Propenenitrile
(Acetaldehyde cyanohydrin)   (Acrylonitrile)

Second, a nitrile is reduced to a primary amine by hydrogen in the presence of nickel or other transition metal catalyst (Section 18.11C). Catalytic reduction of benzaldehyde cyanohydrin, for example, gives 2-amino-1-phenylethanol.

Benzaldehyde cyanohydrin   2-Amino-1-phenylethanol

As we shall see in Section 18.5E, hydrolysis of a nitrile in the presence of an acid catalyst gives a carboxylic acid. Thus, even though nitriles are little used themselves, they are valuable intermediates for the synthesis of other functional groups.

16.7 The Wittig Reaction

Ylide A neutral molecule with positive and negative charges on adjacent atoms.

In 1954, Georg Wittig reported a method for the synthesis of alkenes from aldehydes and ketones using compounds called phosphonium **ylides** (see Chemistry in Action box "The Octet Rule," Chapter 1). For his pioneering study and development of this reaction into a major synthetic tool, Professor Wittig shared the 1979 Nobel Prize for chemistry. (The other recipient was Herbert C. Brown for his studies of hydroboration and the chemistry of organoboron compounds.) A Wittig synthesis is illustrated by the conversion of cyclohexanone to methylenecyclohexane. In this reaction, a C=O double bond is converted to a C=C double bond.

Cyclohexanone   A phosphonium ylide   Methylenecyclohexane   Triphenylphosphine oxide

We study the Wittig reaction in two stages: first, the formation and structure of phosphonium ylides and, second, the reaction of a phosphonium ylide with the carbonyl group of an aldehyde or ketone to give an alkene.

Phosphorus is the second element in Group 5A of the Periodic Table and, like nitrogen, has five electrons in its valence shell. Examples of trivalent phosphorus compounds are phosphine, PH_3, and triphenylphosphine, Ph_3P. Phosphine is a highly toxic, flammable gas. Triphenylphosphine is a colorless, odorless solid.

Because phosphorus is below nitrogen in the Periodic Table, phosphines are weaker bases than amines and also good nucleophiles (Section 8.4B). Treatment of a phosphine with a primary or secondary alkyl halide gives a phosphonium salt by an S_N2 pathway.

Triphenylphosphine   Methyltriphenylphosphonium iodide (an alkyltriphenylphosphonium salt)

Because phosphines are also weak bases, treatment of a tertiary halide with a phosphine gives largely an alkene by an E2 pathway.

α-Hydrogen atoms on the alkyl group of an alkyltriphenylphosphonium ion are weakly acidic and can be removed by reaction with a very strong base, typically butyllithium, sodium hydride (NaH), or sodium amide ($NaNH_2$).

$$CH_3CH_2CH_2CH_2 : Li^+ + H-CH_2-\overset{+}{P}Ph_3 \; I^- \longrightarrow CH_3CH_2CH_2CH_3 + \; \overset{-}{:}CH_2-\overset{+}{P}Ph_3 + LiI$$

Butyllithium   An alkyltriphenyl-phosphonium iodide   Butane   A phosphonium ylide

The product of removal of hydrogen from an alkyltriphenylphosphonium ion is a phosphonium ylide.

Phosphonium ylides react with the carbonyl groups of aldehydes and ketones by a cycloaddition reaction to form a four-membered ring called an **oxaphosphetane.** The name for this ring system is derived by combination of the following units: oxa-to show that it contains oxygen, -phosph- to show that it contains phosphorus, -et- to show that it is a four-membered ring, and -ane to show only carbon-carbon single bonds in the ring. Oxaphosphetanes can be isolated at low temperature. On warming to room temperature, they fragment to give triphenylphosphine oxide and an alkene. The driving force for this reaction is formation of the very strong phosphorus-oxygen bond in triphenylphosphine oxide.

Mechanism The Wittig Reaction

Mechanisms
Ketones
The Wittig reaction

Step 1:   Reaction of a phosphonium ylide with the carbonyl group of an aldehyde or ketone gives a dipolar intermediate called a **betaine** that then collapses to a four-membered oxaphosphetane ring.

Betaine   A neutral molecule with nonadjacent positive and negative charges. An example of a betaine is the intermediate formed by addition of a Wittig reagent to an aldehyde or ketone.

$$
\begin{array}{ccc}
O{=}CR_2 & & \\
\overset{+}{Ph_3P}{-}\overset{..\,-}{CH_2} & \longrightarrow & \left[\begin{array}{c} :\overset{-}{O}{-}CR_2 \\ \overset{+}{Ph_3P}{-}CH_2 \end{array}\right] & \longrightarrow & \begin{array}{c} O{-}CR_2 \\ | \quad\; | \\ Ph_3P{-}CH_2 \end{array}
\end{array}
$$

A betaine   An oxaphosphetane

Step 2:   Decomposition of the oxaphosphetane gives triphenylphosphine oxide and an alkene.

$$
\begin{array}{c}
O{-}CR_2 \\
|\quad | \\
Ph_3P{-}CH_2
\end{array}
\longrightarrow \quad Ph_3P{=}O \quad + \quad R_2C{=}CH_2
$$

Triphenylphosphine oxide   An alkene

The Wittig reaction is effective with a wide variety of aldehydes and ketones and with ylides derived from a wide variety of primary, secondary, and allylic halides as shown by the following examples.

$$CH_3\overset{\overset{O}{\|}}{C}CH_3 + Ph_3\overset{+}{P}-\overset{-}{C}H(CH_2)_3CH_3 \longrightarrow CH_3\overset{\overset{CH_3}{|}}{C}=CH(CH_2)_3CH_3 + Ph_3P=O$$

2-Methyl-2-heptene

$$PhCH_2\overset{\overset{O}{\|}}{C}H + Ph_3\overset{+}{P}-\overset{-}{C}HCH_3 \longrightarrow PhCH_2CH=CHCH_3 + Ph_3P=O$$

1-Phenyl-2-butene
(87% *Z* isomer, 13% *E* isomer)

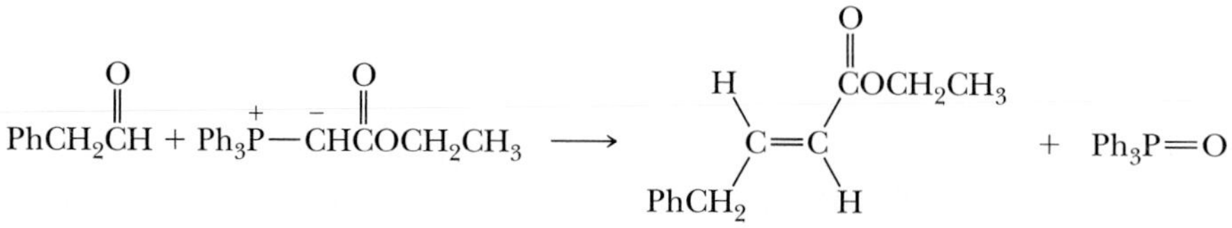

$$PhCH_2\overset{\overset{O}{\|}}{C}H + Ph_3\overset{+}{P}-\overset{-}{C}H\overset{\overset{O}{\|}}{C}OCH_2CH_3 \longrightarrow$$

Ethyl (*E*)-4-phenyl-2-butenoate
(only the *E* isomer is formed)

Example 16.5

Show how this alkene can be synthesized by a Wittig reaction.

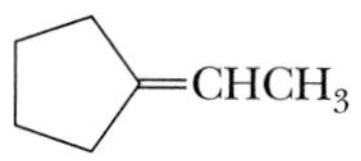

Solution

Starting materials are either cyclopentanone and the triphenylphosphonium ylide derived from bromoethane, or acetaldehyde and the triphenylphosphonium ylide derived from chlorocyclopentane. Either route is satisfactory.

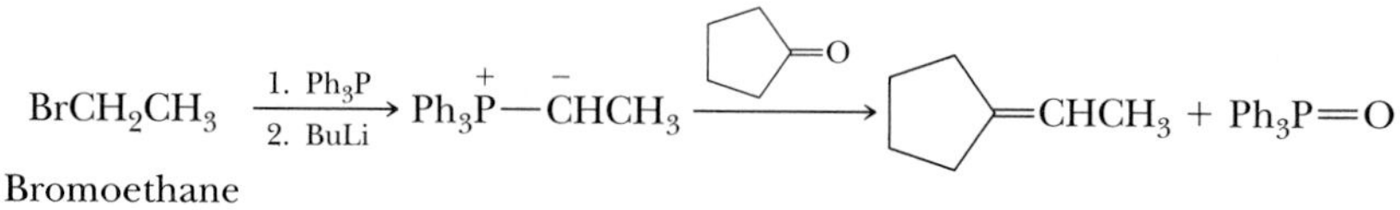

$$BrCH_2CH_3 \xrightarrow[\text{2. BuLi}]{\text{1. Ph}_3\text{P}} Ph_3\overset{+}{P}-\overset{-}{C}HCH_3 \longrightarrow \bigcirc\!\!=CHCH_3 + Ph_3P=O$$

Bromoethane

$$\bigcirc\!\!-Cl \xrightarrow[\text{2. NaH}]{\text{1. Ph}_3\text{P}} \bigcirc\!\!-\overset{-}{\underset{+}{P}}Ph_3 \xrightarrow{CH_3\overset{\overset{O}{\|}}{C}H} \bigcirc\!\!=CHCH_3 + Ph_3P=O$$

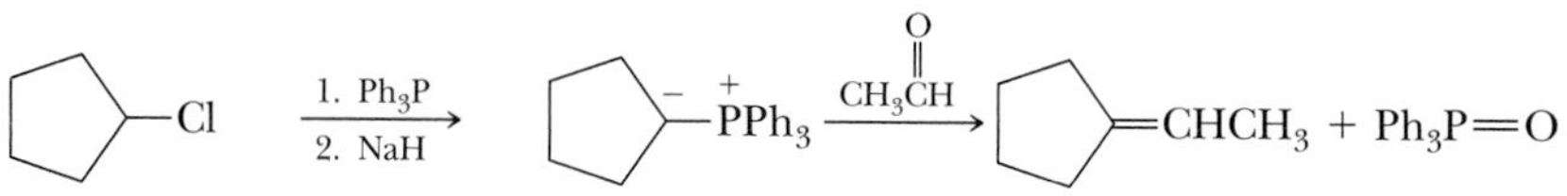

Chlorocyclopentane

Problem 16.5

Show how each alkene can be synthesized by a Wittig reaction. There are two routes to each.

(a) 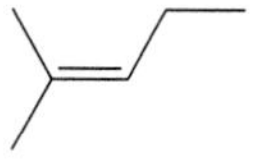   **(b)**

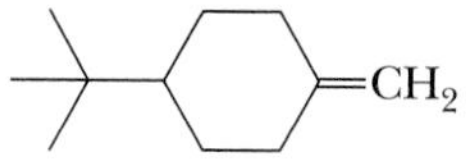

16.8 Addition of Oxygen Nucleophiles

A. Addition of Water

Addition of water (hydration) to a carbonyl group of an aldehyde or ketone forms a **geminal** or gem-diol. A gem-diol is commonly referred to as the hydrate of the corresponding aldehyde or ketone. These compounds are unstable and are rarely isolated. This reaction is catalyzed by acids and bases. The mechanism is identical to that for the addition of alcohols, which is discussed next.

Geminal From the Latin: *geminatus*, twin. Refers to two identical groups on the same carbon.

$$\begin{array}{c}\diagdown\\ \diagup\end{array} C = O \ + \ H_2O \xrightleftharpoons[\text{base}]{\text{acid or}} \ C \begin{array}{c}OH\\ \diagup\\ \diagdown\\ OH\end{array}$$

Carbonyl group of an aldehyde or ketone   A hydrate

These reactions are readily reversible and the diol can eliminate water to regenerate the aldehyde or ketone. Although equilibrium strongly favors the carbonyl group, for a few simple aldehydes the gem-diol is favored. For example, when formaldehyde is dissolved in water at $20°C$, the position of equilibrium is such that it is more than 99% hydrated. A 37% solution of formaldehyde in water is called formalin. Formalin is commonly used to preserve biological specimens.

$$\begin{array}{c}O\\ \parallel\\ HCH\end{array} \ + \ H_2O \ \rightleftharpoons \ \begin{array}{c}OH\\ \mid\\ HCOH\\ \mid\\ H\end{array}$$

Formaldehyde   Formaldehyde hydrate (>99%)

In contrast, an aqueous solution of acetone consists of less than 0.1% of the gem-diol at equilibrium.

$$\begin{array}{c}H_3C\\ \diagdown\\ \diagup\\ H_3C\end{array} C = O \ + \ H_2O \ \rightleftharpoons \ \begin{array}{c}H_3C\\ \diagdown \quad \diagup OH\\ C\\ \diagup \quad \diagdown\\ H_3C \qquad OH\end{array}$$

Acetone (99.9%)   2,2-Propanediol (0.1%)

B. Addition of Alcohols — Formation of Acetals

Alcohols add to aldehydes and ketones in the same manner as described for water. Addition of one molecule of alcohol to the carbonyl group of an aldehyde or ketone forms a **hemiacetal** (a half-acetal). This reaction is also catalyzed by either acid or base.

Hemiacetal A molecule containing an —OH and an —OR or —OAr group bonded to the same carbon.

$$\begin{array}{c}O \qquad\quad H\\ \parallel \qquad\quad \mid\\ CH_3CCH_3 \ + \ OCH_2CH_3\end{array} \xrightleftharpoons[\text{base}]{\text{acid or}} \begin{array}{c}OH\\ \mid\\ CH_3COCH_2CH_3\\ \mid\\ CH_3\end{array}$$

A hemiacetal

The functional group of a hemiacetal is a carbon bonded to an —OH group and an —OR group.

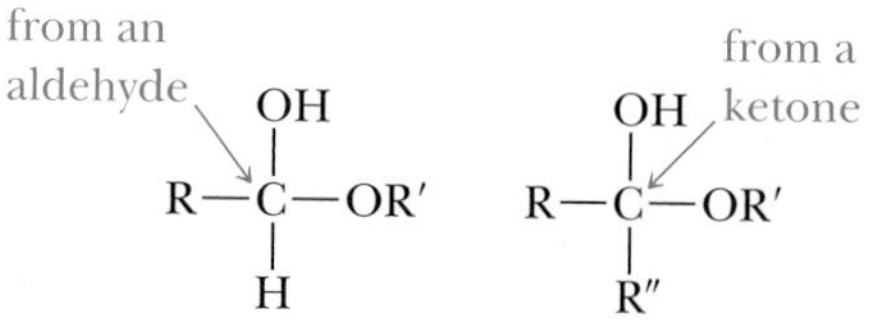

from an
aldehyde   OH   OH   from a
            | | ketone
       R—C—OR'   R—C—OR'
            | |
            H   R''

Hemiacetals

Hemiacetals are generally unstable and are only minor components of an equilibrium mixture, except in one very important type of compound. When a hydroxyl group is part of the same molecule that contains the carbonyl group, and a five- or six-membered ring can form, the compound exists almost entirely in the cyclic hemiacetal form. We have much more to say about cyclic hemiacetals when we consider the chemistry of carbohydrates in Chapter 25.

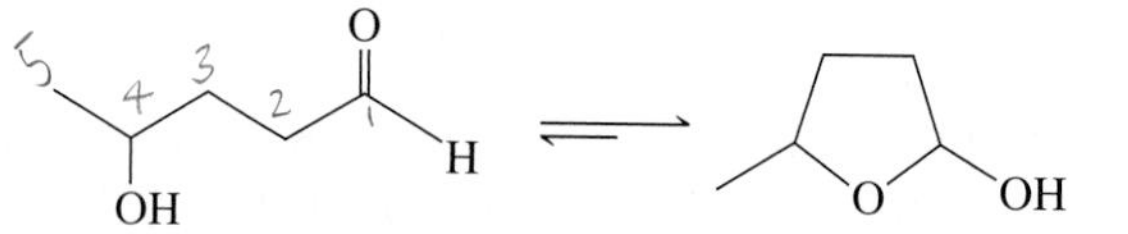

4-Hydroxypentanal   A cyclic hemiacetal
                                     (major form present
                                     at equilibrium)

Formation of hemiacetals is catalyzed by bases such as hydroxide or alkoxide. The function of the catalyst is to remove a proton from the alcohol, making it a better nucleophile.

Mechanism   Base-Catalyzed Formation of a Hemiacetal

Step 1:   Proton transfer from HOR to the base gives the alkoxide.

$$B: + H—OR \rightleftharpoons B—H + :OR$$

Step 2:   Attack of RO⁻ on the carbonyl carbon gives a tetrahedral carbonyl addition compound.

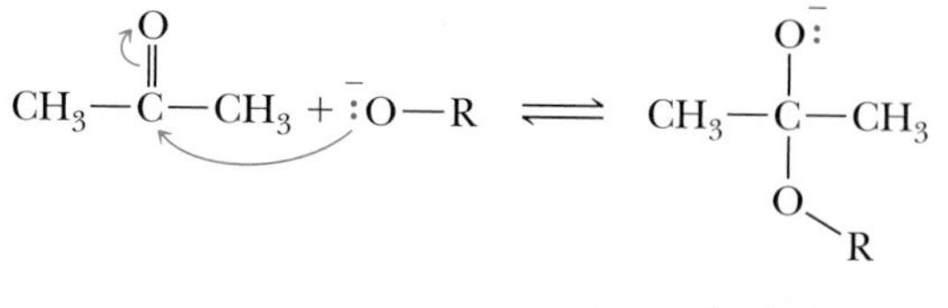

$$CH_3—\overset{O}{\underset{}{C}}—CH_3 + :O—R \rightleftharpoons CH_3—\underset{\underset{R}{O}}{\overset{\overset{-}{O}}{C}}—CH_3$$

A tetrahedral carbonyl
addition compound

Step 3:   Proton transfer from the alcohol to the O⁻ gives the hemiacetal and regenerates the base catalyst.

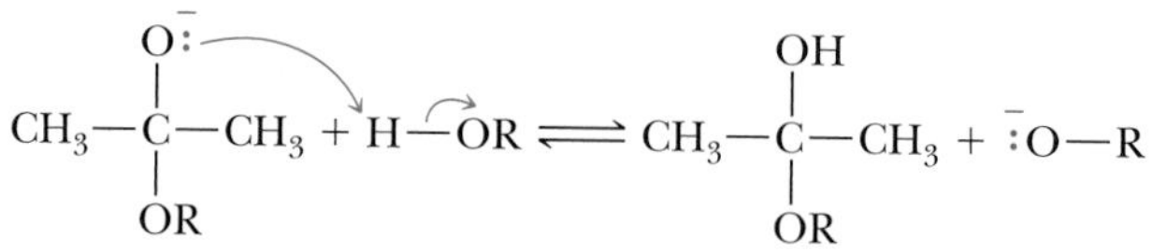

$$CH_3—\underset{OR}{\overset{\overset{-}{O}}{C}}—CH_3 + H—OR \rightleftharpoons CH_3—\underset{OR}{\overset{OH}{C}}—CH_3 + :O—R$$

Formation of hemiacetals can also be catalyzed by acid, most commonly sulfuric acid, *p*-toluenesulfonic acid, or hydrogen chloride. The function of the acid catalyst, here represented by H—A, is to protonate the carbonyl oxygen, thus rendering the carbonyl carbon more electrophilic and more susceptible to attack by the nucleophilic oxygen atom of the alcohol.

Mechanism   Acid-Catalyzed Formation of a Hemiacetal

Mechanisms
Ketones
Acid-catalyzed formation of a hemiacetal

Step 1:   Proton transfer from HA, the acid catalyst, to the carbonyl oxygen gives the conjugate acid of the aldehyde or ketone.

$$CH_3-\overset{\overset{\displaystyle O:}{\|}}{C}-CH_3 + H-A \rightleftharpoons CH_3-\overset{\overset{\displaystyle \overset{+}{O}-H}{\|}}{C}-CH_3 + :A^-$$

Step 2:   Attack of ROH on the carbonyl carbon gives a tetrahedral carbonyl addition compound.

$$CH_3-\overset{\overset{\displaystyle \overset{+}{O}-H}{\|}}{C}-CH_3 + H-O-R \rightleftharpoons CH_3-\overset{\overset{\displaystyle :OH}{|}}{\underset{\underset{\displaystyle H \quad R}{\overset{+}{O}}}{C}}-CH_3$$

A tetrahedral carbonyl
addition compound

Step 3:   Proton transfer from the oxonium ion to A⁻ gives the hemiacetal and regenerates the acid catalyst.

$$CH_3-\overset{\overset{\displaystyle OH}{|}}{\underset{\underset{\displaystyle H \quad R}{\overset{+}{O}}}{C}}-CH_3 \rightleftharpoons CH_3-\overset{\overset{\displaystyle OH}{|}}{\underset{\underset{\displaystyle :OR}{|}}{C}}-CH_3 + H-A$$

$$A:\curvearrowright$$

Hemiacetals react further with alcohols to form **acetals** plus a molecule of water. This reaction and its reverse can only be catalyzed by acids, not by bases, because the OH group cannot be displaced by nucleophiles.

Acetal   A molecule containing two —OR or —OAr groups bonded to the same carbon.

$$CH_3\overset{\overset{\displaystyle OH}{|}}{\underset{\underset{\displaystyle CH_3}{|}}{C}}OCH_2CH_3 + CH_3CH_2OH \underset{}{\overset{H^+}{\rightleftharpoons}} CH_3\overset{\overset{\displaystyle OCH_2CH_3}{|}}{\underset{\underset{\displaystyle CH_3}{|}}{C}}OCH_2CH_3 + H_2O$$

A hemiacetal   A diethyl acetal

The functional group of an acetal is a carbon bonded to two —OR groups.

from an
aldehyde   OR′   OR′   from a
                  | | ketone
$$R-\overset{\overset{\displaystyle OR'}{|}}{\underset{\underset{\displaystyle H}{|}}{C}}-OR' \qquad R-\overset{\overset{\displaystyle OR'}{|}}{\underset{\underset{\displaystyle R''}{|}}{C}}-OR'$$

Acetals

The mechanism for the acid-catalyzed conversion of a hemiacetal to an acetal is divided into four steps. Note that acid H—A is a true catalyst in this reaction. It is used in Step 1, but another H—A is generated in Step 4. The latter steps of this reaction are very similar to those for hemiacetal formation.

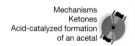

Mechanism   Acid-Catalyzed Formation of an Acetal

Step 1:   Proton transfer from the acid, H—A, to the hemiacetal OH group gives an oxonium ion.

$$R-\overset{\overset{\displaystyle HO:}{|}}{\underset{\underset{\displaystyle H}{|}}{C}}-OCH_3 + H-A \rightleftharpoons R-\overset{\overset{\displaystyle \overset{H\;\;+\;\;H}{\diagdown O \diagup}}{|}}{\underset{\underset{\displaystyle H}{|}}{C}}-OCH_3 + A:^-$$

An oxonium ion

Step 2:   Loss of water gives a new, resonance-stabilized cation.

$$R-\overset{\overset{\displaystyle \overset{H\;\;+\;\;H}{\diagdown O \diagup}}{|}}{\underset{\underset{\displaystyle H}{|}}{C}}-\ddot{O}CH_3 \rightleftharpoons R-\overset{}{\underset{\underset{\displaystyle H}{|}}{C}}\overset{+}{=}OCH_3 \longleftrightarrow R-\overset{\overset{\displaystyle +}{}}{\underset{\underset{\displaystyle H}{|}}{C}}-\ddot{O}CH_3 + H_2O$$

A resonance-stabilized cation

Step 3:   Reaction of the resonance-stabilized cation (a Lewis acid) with methanol (a Lewis base) gives the conjugate acid of the acetal.

$$CH_3-\overset{}{\underset{}{\ddot{O}}:} + R-\overset{}{\underset{\underset{\displaystyle H}{|}}{C}}\overset{+}{=}OCH_3 \rightleftharpoons R-\overset{\overset{\displaystyle \overset{H\;\;+\;\;CH_3}{\diagdown O \diagup}}{|}}{\underset{\underset{\displaystyle H}{|}}{C}}-\ddot{O}CH_3$$

A protonated acetal

Step 4:   Proton transfer from the protonated acetal to A⁻ gives the acetal and generates a new molecule of the acid catalyst.

$$A: + R-\underset{\underset{H}{|}}{\overset{\overset{H}{\underset{}{\overset{+}{O}}}\diagup^{CH_3}}{C}}-OCH_3 \xrightarrow{\text{(4)}} R-\underset{\underset{H}{|}}{\overset{\overset{:O-CH_3}{|}}{C}}-OCH_3 + H-A$$

An acetal

Formation of acetals is often carried out using the alcohol as the solvent and dissolving either dry HCl (hydrogen chloride gas) or *p*-toluenesulfonic acid in the alcohol. Because the alcohol is both a reactant and solvent, it is present in large molar excess, which forces the equilibrium to the right and favors acetal formation. Note that this reaction is completely reversible. Addition of excess water to an acetal causes hydrolysis to the ketone.

An excess of alcohol pushes the equilibrium toward formation of the acetal.

Removal of water favors formation of the acetal.

$$R-\overset{\overset{O}{\|}}{C}-R + 2CH_3CH_2OH \underset{}{\overset{H^+}{\rightleftharpoons}} R-\underset{\underset{R}{|}}{\overset{\overset{OCH_2CH_3}{|}}{C}}-OCH_2CH_3 + H_2O$$

A diethyl acetal

In another experimental technique to force the equilibrium to the right, water is removed from the reaction vessel as an **azeotrope** by distillation using a Dean-Stark trap (Figure 16.2). In this method for preparing an acetal, the aldehyde or ketone, alcohol, acid catalyst, and benzene are brought to reflux. The component in this mixture with the lowest boiling point is an azeotrope, bp 69°C, consisting of 91% benzene and 9% water. This vapor is condensed and collected in a side trap where it separates into two layers. At room temperature, the composition of the upper, less dense layer is 99.94% benzene and 0.06% water. The composition of the lower, more dense layer is almost the reverse, 0.07% benzene and 99.93% water. As reflux continues, benzene from the top layer is returned to the refluxing mixture, and water is drawn off at the bottom through a stopcock. A Dean-Stark trap "pumps" water out of the reaction mixture, thus forcing the equilibrium to the right. This same apparatus is used in many other reactions where water needs to be removed, as for example in formation of enamines (Section 16.10A).

Azeotrope   A liquid mixture of constant composition with a boiling point that is different from that of any of its components.

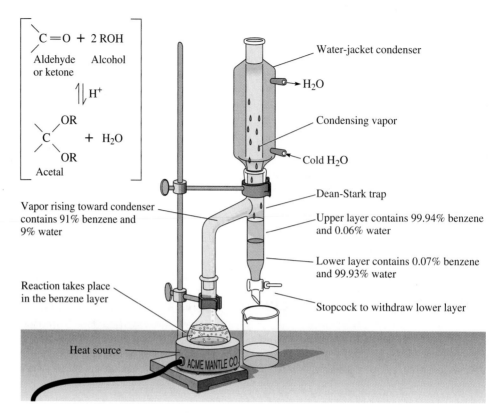

Figure 16.2
A Dean-Stark trap for removing water by azeotropic distillation with benzene. Toluene or xylene can be used if a higher reaction temperature is desired.

Example 16.6

Show the reaction of the carbonyl group of each aldehyde or ketone with one molecule of alcohol to give a hemiacetal and then with a second molecule of alcohol to give an acetal. Note that in part (b), ethylene glycol is a diol, and one molecule of it provides both —OH groups.

(a) [structure] + 2C$_2$H$_5$OH $\xrightleftharpoons{\text{H}^+}$ **(b)** [structure] =O + HO[structure]OH $\xrightleftharpoons{\text{H}^+}$

Solution

Given are structural formulas of the hemiacetal and then the acetal.

(a) [structure: HO OC$_2$H$_5$] → [structure: C$_2$H$_5$O OC$_2$H$_5$]   **(b)** [structure with OH] → [cyclic acetal structure]

Hemiacetal   Acetal   Hemiacetal   Cyclic acetal

Problem 16.6

Hydrolysis of an acetal in aqueous acid gives an aldehyde or ketone and two molecules of alcohol. Following are structural formulas for three acetals. Draw the structural formulas for the products of hydrolysis of each in aqueous acid.

Like ethers (Section 11.5), acetals are also unreactive to bases, hydride-reducing agents such as $LiAlH_4$ and $NaBH_4$, Grignard and other organometallic reagents, oxidizing agents (except, of course, for those involving the use of aqueous acid), and catalytic reduction. Because of their lack of reactivity toward these reagents, acetals are often used to protect the carbonyl groups of aldehydes and ketones while reactions are carried out on functional groups in other parts of the molecule.

C. Acetals as Carbonyl-Protecting Groups

The use of acetals as carbonyl-protecting groups is illustrated by the synthesis of 5-hydroxy-5-phenylpentanal from benzaldehyde and 4-bromobutanal.

One obvious way to form a new carbon-carbon bond between these two molecules is to treat benzaldehyde with the Grignard reagent from 4-bromobutanal. A Grignard reagent formed from this compound, however, reacts immediately with the carbonyl group of another molecule of 4-bromobutanal, causing it to self-destruct during preparation. A way to avoid this problem is to protect the carbonyl group of 4-bromobutanal by conversion to an acetal. Cyclic acetals are often used because they are particularly easy to prepare.

Treatment of the protected bromoaldehyde with magnesium in diethyl ether followed by addition of benzaldehyde gives a magnesium alkoxide.

Treatment of the magnesium alkoxide with aqueous acid accomplishes two things. First, protonation of the alkoxide anion gives the desired hydroxyl group, and, second, hydrolysis of the cyclic acetal regenerates the aldehyde group.

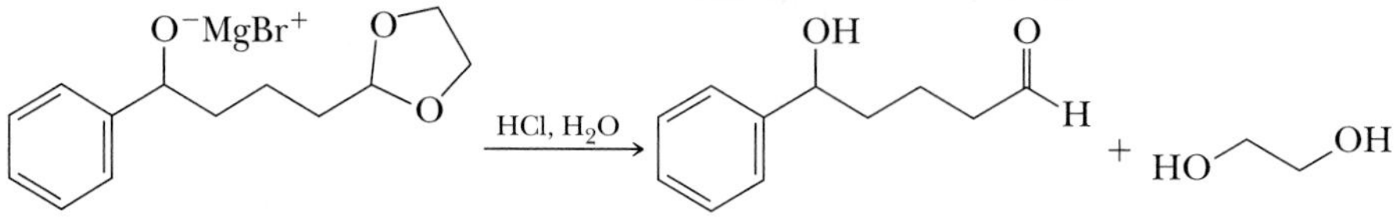

D. Tetrahydropyranyl Ether — Protecting an Alcohol as an Acetal

We have just seen in Section 16.8C that an aldehyde or ketone can be protected by conversion to an acetal. A similar strategy can be used to protect a primary or secondary alcohol. Treatment of the alcohol with dihydropyran in the presence of an acid catalyst, commonly anhydrous HCl or a sulfonic acid, RSO_3H, converts the alcohol into a tetrahydropyranyl (THP) ether.

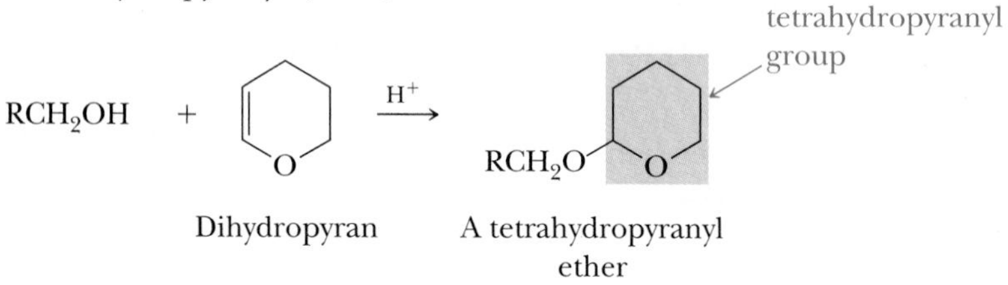

Dihydropyran   A tetrahydropyranyl ether

Because the THP group is an acetal, it is stable in neutral and basic solutions and to most oxidizing and reducing agents. It is removed easily by treatment with dilute aqueous acid to regenerate the original primary or secondary alcohol.

Example 16.7

Write a mechanism for the formation of a THP ether from a primary alcohol RCH_2OH catalyzed by a sulfonic acid $ArSO_3H$.

Solution

Step 1:   Dihydropyran (a vinyl or enol ether) is weakly basic and is protonated to give a resonance-stabilized cation.

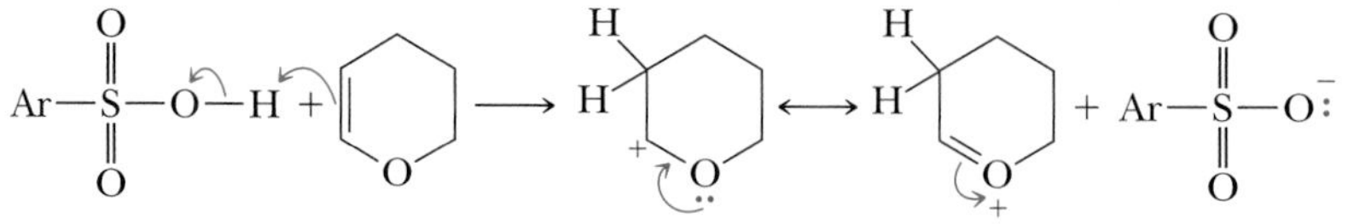

A resonance-stabilized cation

Step 2:   Reaction of the resonance-stabilized cation with the alcohol gives an oxonium ion.

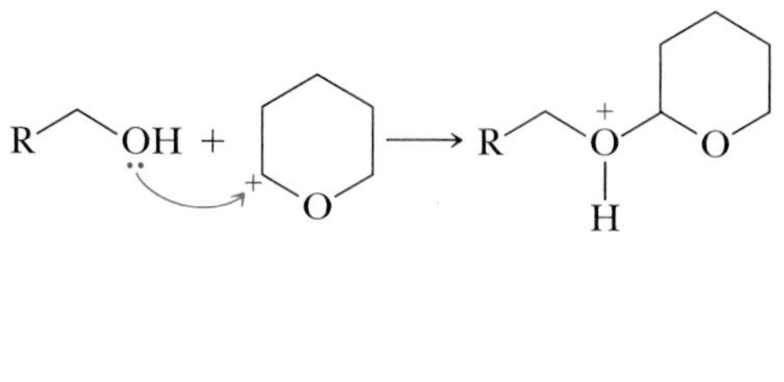

Step 3: Proton transfer from the oxonium ion to $ArSO_3^-$ completes the reaction.

Problem 16.7

Write a mechanism for the acid-catalyzed hydrolysis of a THP ether to regenerate the original alcohol. Into what compound is the THP group converted?

16.9 Addition of Sulfur Nucleophiles

The sulfur atom of a thiol is a better nucleophile than the oxygen atom of an alcohol (Section 8.4B). Thiols, like alcohols, add to the carbonyl group of aldehydes and ketones to form tetrahedral carbonyl addition compounds. A common sulfur nucleophile used for this purpose is 1,3-propanedithiol. The carbonyl groups of both aldehydes and ketones react with this compound in the presence of an acid catalyst to form cyclic thioacetals. Products of this reaction are also called 1,3-dithianes; "1,3-dithi-" indicates that atoms of sulfur occur at positions 1 and 3 of the ring, and "-ane" indicates that the ring is six-membered (Section 11.2). Following is an equation for formation of a 1,3-dithiane from an aldehyde.

| An aldehyde | 1,3-Propane-dithiol | A 1,3-dithiane (a cyclic thioacetal) | |

The special value of 1,3-dithianes derived from an aldehyde is that the hydrogen atom on carbon 2 of the ring is weakly acidic; it has a pK_a of approximately 31. In the presence of a very strong base, such as butyllithium, a 1,3-dithiane derived from an aldehyde can be converted into a lithium salt.

| A 1,3-dithiane (stronger acid) pK_a 31 | Butyllithium (stronger base) | A lithio-1,3-dithiane (weaker base) | Butane (weaker acid) pK_a 50 |

A 1,3-dithiane anion is an excellent nucleophile and reacts by an S_N2 mechanism with a primary alkyl, allylic, or benzylic halide to give a disubstituted dithiane. Treatment of this product with mercuric chloride, $HgCl_2$, in aqueous acetonitrile brings about hydrolysis of the dithiane to give a ketone and a mercury salt of the dithiol.

Lithium salt of
a 1,3-dithiane

The result of this series of reactions is conversion of an aldehyde to a ketone.

Anions derived from 1,3-dithianes also add to the carbonyl groups of aldehydes and ketones to form tetrahedral carbonyl addition compounds. Hydrolysis of the lithium salt in aqueous acetonitrile containing mercury(II) chloride gives an α-hydroxyketone.

Lithium salt of
a 1,3-dithiane

An α-hydroxyketone

The result of this series of reactions is replacement of the hydrogen atom of the original aldehyde by either a secondary alcohol group or, if carbonyl addition takes place on a ketone, a tertiary alcohol group.

These two important reactions of 1,3-dithianes provide examples of what is called by the German name **Umpolung,** which means pole reversal. Under normal circumstances, the carbon atom of a carbonyl group bears a partial positive charge and is, therefore, an electrophile. When converted to a 1,3-dithiane and then treated with butyllithium, the same carbon becomes an anion and, therefore, a nucleophile.

Example 16.8

Give structural formulas for the lettered compounds in each reaction sequence.

Solution

(a)

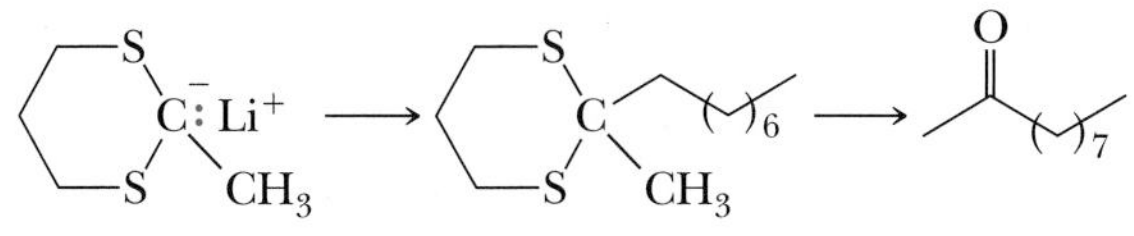

A   B   C

(b)

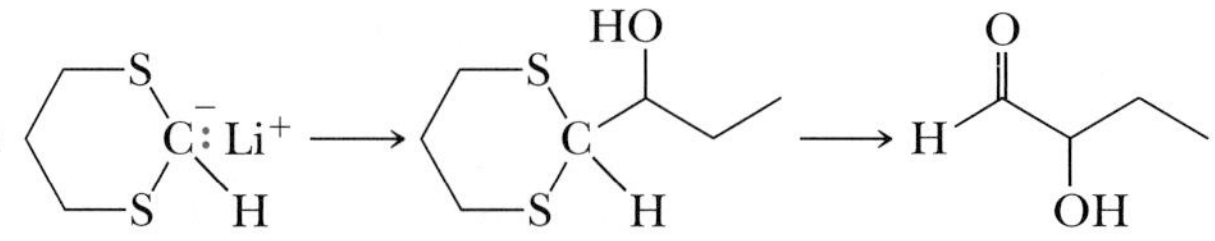

D   E   F

Problem 16.8

Show how to convert the given starting material into the indicated product using a 1,3-dithiane as an intermediate.

(a) $HCH \longrightarrow HCCH_2(CH_2)_6CH_3$ (b) 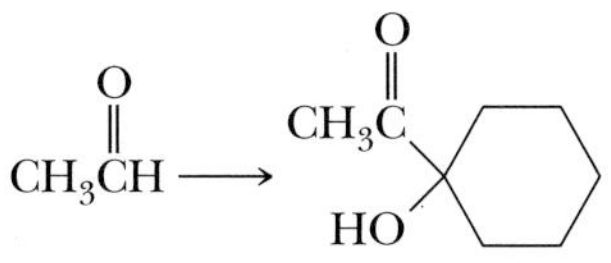 $CH_3CH \longrightarrow$

16.10   Addition of Nitrogen Nucleophiles

A.   Ammonia and Its Derivatives

Ammonia, primary aliphatic amines (RNH_2), and primary aromatic amines ($ArNH_2$) react with the carbonyl group of aldehydes and ketones in the presence of an acid catalyst to give an **imine** or, alternatively, a **Schiff base.**

> **Imine**   A compound containing a carbon-nitrogen double bond, $R_2C=NR'$; also called a Schiff base.
> **Schiff base**   An alternative name for an imine.

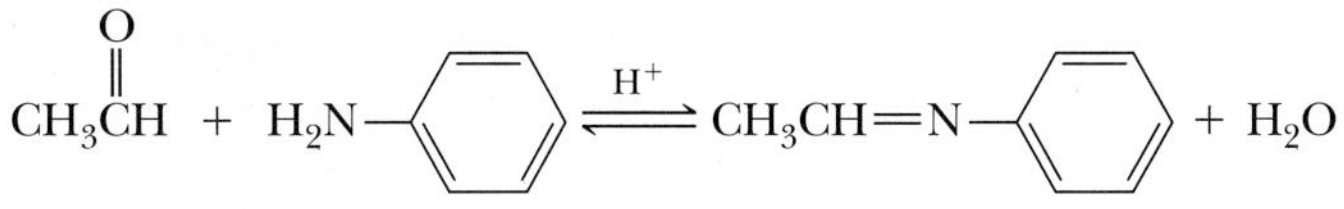

Acetaldehyde   Aniline   An imine
                                         (a Schiff base)

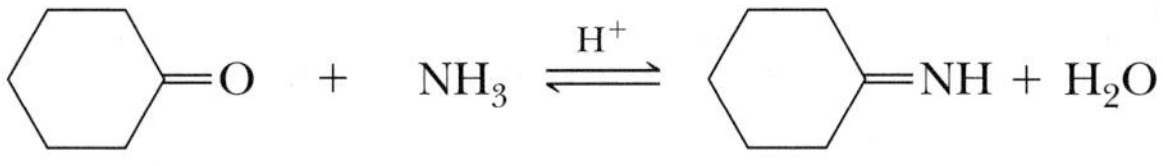

Cyclohexanone   Ammonia   An imine
                                     (a Schiff base)

The mechanism of imine formation can be divided into two steps. In Step 1, the nitrogen atom of ammonia or a primary amine, both good nucleophiles, adds to the carbonyl carbon to give a tetrahedral carbonyl addition compound. This reaction is the same as with water and alcohols. Acid-catalyzed dehydration of this addition compound in Step 2 is the slow, rate-determining step.

Mechanisms
Ketones
Formation of an imine from
a ketone

Mechanism   Formation of an Imine from an Aldehyde or Ketone

Step 1: Addition of the nucleophilic nitrogen to the carbonyl carbon followed by proton transfer gives a tetrahedral carbonyl addition compound.

A tetrahedral carbonyl
addition compound

Step 2: Protonation of the OH group followed by loss of water and proton transfer to solvent gives the imine.

An imine

One of the chief values of imines is that the carbon-nitrogen double bond can be reduced by hydrogen in the presence of a nickel or other transition metal catalyst to a carbon-nitrogen single bond. By this two-step reaction, called **reductive amination,** a primary amine is converted to a secondary amine by way of an imine as illustrated by the conversion of cyclohexylamine to dicyclohexylamine. Imines are usually unstable unless the C=N group is part of an extended system of conjugation (for example, rhodopsin) and are generally not isolated.

Reductive amination A method for preparing amines by treating an aldehyde or ketone with an amine in the presence of a reducing agent.

Cyclohexanone   Cyclohexylamine   (An imine)   Dicyclohexylamine

To give but one example of the importance of imines in biological systems, the active form of vitamin A aldehyde (retinal) is bound to the protein opsin in the human retina in the form of an imine. The primary amino group for this reaction is provided by the side chain of the amino acid lysine (Section 27.1). The imine is called rhodopsin or visual purple.

11-*cis*-Retinal

Rhodopsin
(Visual purple)

Example 16.9

Write structural formulas for the imines formed in these reactions.

(a)

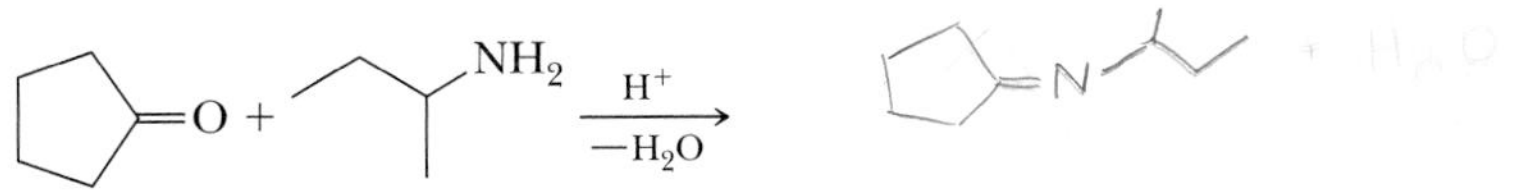

(b)

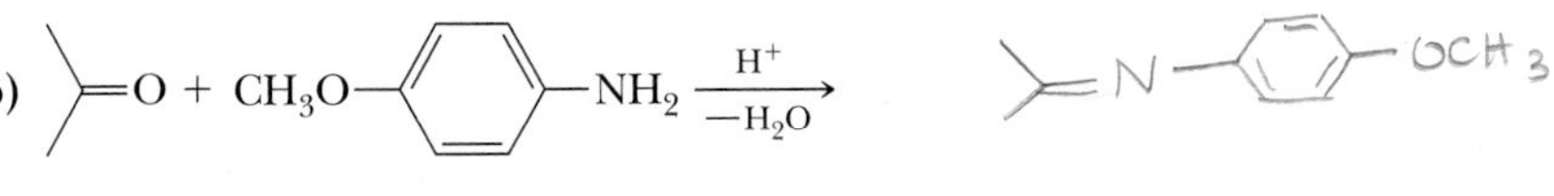

Solution

Given is a structural formula for each imine.

(a)

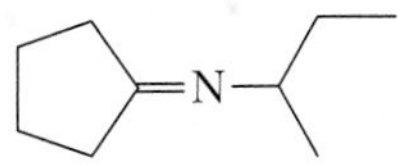

(b)

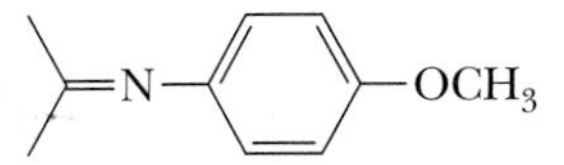

Problem 16.9

Acid-catalyzed hydrolysis of an imine gives an amine and an aldehyde or ketone. When one equivalent of acid is used, the amine is converted to an ammonium salt. Write structural formulas for the products of hydrolysis of the following imines using one equivalent of HCl.

(a) 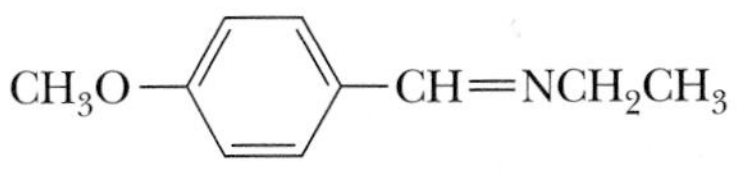 CH_3O—⟨benzene ring⟩—CH=NCH₂CH₃ (b)

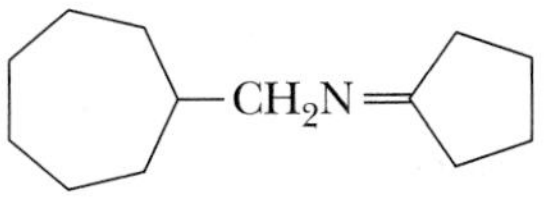

Secondary amines react with aldehydes and ketones to form enamines. The name **enamine** is derived from -en- to indicate the presence of a carbon-carbon double bond and -amine to indicate the presence of an amino group. An example is enamine formation between cyclohexanone and piperidine, a cyclic secondary amine. Water is removed by a Dean-Stark trap (Section 16.8B), which forces the equilibrium to the right.

Enamine An unsaturated compound derived by the reaction of an aldehyde or ketone and a secondary amine followed by loss of H_2O; $R_2C=CR—NR_2$.

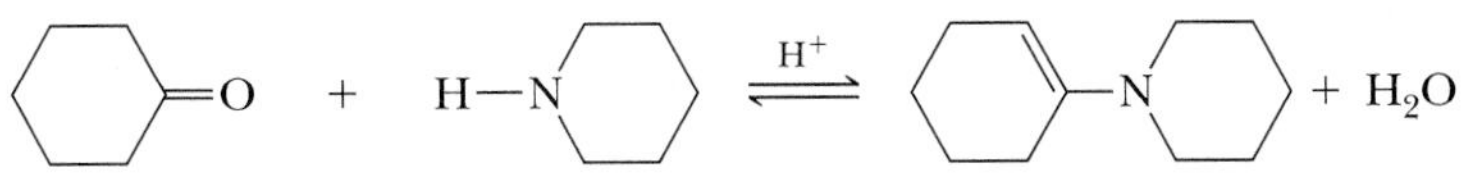

Cyclohexanone   Piperidine   An enamine
(a secondary amine)

The mechanism for formation of an enamine is very similar to that for formation of an imine. In Step 1, nucleophilic addition of the secondary amine to the carbonyl carbon of the aldehyde or ketone followed by proton transfer from nitrogen to oxygen gives a tetrahedral carbonyl addition compound. Acid-catalyzed dehydration in Step 2 gives the enamine. It is at this stage that enamine formation differs from imine formation. The nitrogen has no proton to lose. Instead, a proton is lost from the α-carbon of the ketone or aldehyde portion of the molecule in an elimination reaction.

| Phenylacetaldehyde | Piperidine | A tetrahedral carbonyl addition compound | An enamine |

We will return to the chemistry of enamines and their use in synthesis in Section 19.5.

B. Hydrazine and Related Compounds

Aldehydes and ketones react with hydrazine to form compounds called hydrazones in a reaction similar to imine formation, as illustrated by the treatment of cyclopentanone with hydrazine. A common use of hydrazones is as intermediates in the Wolff-Kishner reduction of carbonyl groups to methylene groups (Section 16.14C).

Hydrazine   A hydrazone

Other derivatives of ammonia and hydrazine that react with aldehydes and ketones to give imines are listed in Table 16.4. The chief value of the nitrogen nucleophiles listed in Table 16.4 is that most aldehydes and ketones react with them to give crystalline solids with sharp melting points. Historically, these derivatives often provided a convenient way to identify liquid aldehydes or ketones by comparison with literature values. Now, these compounds are more readily identified by IR and NMR spectroscopy.

Table 16.4   Derivatives of Ammonia and Hydrazine Used for Forming Imines

Reagent, H_2N—R	Name of Reagent	Name of Derivative Formed
H_2N—OH	Hydroxylamine	An oxime
H_2N—NH— (phenyl)	Phenylhydrazine	A phenylhydrazone
H_2N—NH— (2,4-dinitrophenyl)	2,4-Dinitrophenylhydrazine	A 2,4-dinitrophenylhydrazone
H_2N—NHCNH$_2$ (O)	Semicarbazide	A semicarbazone

16.11 Keto-Enol Tautomerism

A. Acidity of α-Hydrogens

A carbon atom adjacent to a carbonyl group is called an **α-carbon,** and hydrogen atoms bonded to it are called **α-hydrogens.**

α-Carbon   A carbon atom adjacent to a carbonyl group.

α-Hydrogen   A hydrogen on a carbon alpha to a carbonyl group.

$$\alpha\text{-hydrogens} \qquad \underset{\text{α-carbons}}{CH_3-\overset{\overset{\displaystyle O}{\|}}{C}-CH_2-CH_3}$$

Because carbon and hydrogen have comparable electronegativities, a C—H bond normally has little polarity, and a hydrogen bonded to carbon shows very low acidity. The situation is different, however, for hydrogens alpha to a carbonyl group. α-Hydrogens are more acidic than acetylenic, vinylic, and alkane hydrogens but less acidic than —OH hydrogens of alcohols.

Type of Bond	pK_a
CH_3CH_2O-H	16
$CH_3\overset{\overset{\displaystyle O}{\|}}{C}CH_2-H$	20
$CH_3C\equiv C-H$	25
$CH_2=CH-H$	44
CH_3CH_2-H	51

Hydrogens alpha to the carbonyl group of an aldehyde or ketone are more acidic than hydrogens of alkanes, alkenes, and alkynes. This greater acidity arises because the negative charge on the resulting **enolate anion** is delocalized by resonance, thus stabilizing it relative to an alkane, alkene, or alkyne anion. It is also stabilized by the electron-withdrawing inductive effect of the electronegative oxygen. Recall that we used these same factors in Section 4.4C to explain the greater acidity of carboxylic acids compared with alcohols.

Enolate anion   An anion derived by loss of a hydrogen from a carbon alpha to a carbonyl group; the anion of an enol.

$$CH_3-\overset{\overset{\displaystyle O}{\|}}{C}-CH_2-H + :A^- \rightleftharpoons CH_3-\overset{\overset{\displaystyle O}{\|}}{C}-\overset{..}{C}H_2^- \longleftrightarrow CH_3-\overset{\overset{\displaystyle O^-}{\|}}{C}=CH_2 + H-A$$

Resonance-stabilized enolate anion

Example 16.10

Predict the position of the following equilibrium.

$$\underset{\text{Acetophenone}}{C_6H_5-\overset{\overset{\displaystyle O}{\|}}{C}-CH_3} + CH_3CH_2O^- \underset{\text{ethanol}}{\rightleftharpoons} C_6H_5-\overset{\overset{\displaystyle O^-}{\|}}{C}=CH_2 + CH_3CH_2OH$$

Solution

The pK_a of ethanol is 16 (Table 4.1, rounded to two significant figures). Assume that the pK_a of acetophenone is approximately equal to that of acetone, namely approximately 20. Ethanol is the stronger acid; therefore, the equilibrium lies to the left.

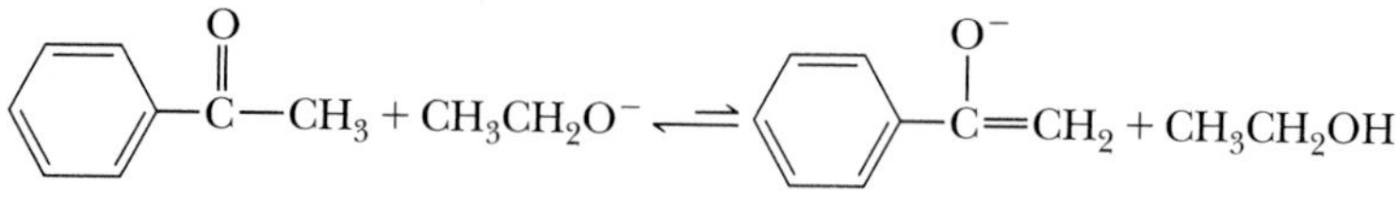

Weaker acid
pK_a 20

Stronger acid
pK_a 16

Problem 16.10

Predict the position of the following equilibrium.

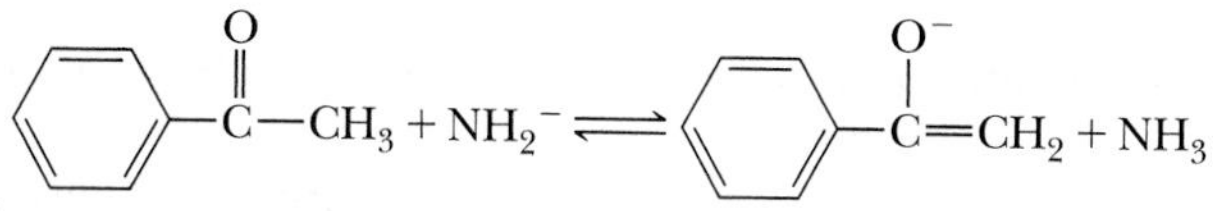

Acetophenone

When an enolate anion reacts with a proton donor, it may do so either on oxygen or on the α-carbon. Protonation of the enolate anion on the α-carbon gives the original molecule in what is called the **keto form.** Protonation on oxygen gives an **enol form.** Because the anion can be derived by loss of a proton from the enol form, it is called an enolate anion.

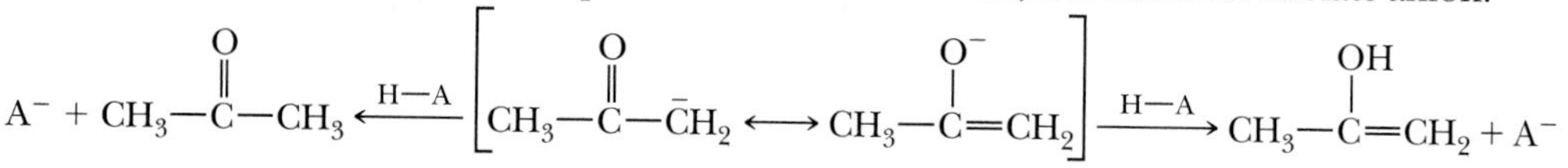

Keto form

Resonance-stabilized enolate anion

Enol form

Enol formation can also be catalyzed by acid. Step 1 is rapid and reversible proton transfer from the acid, H—A, to the carbonyl oxygen. This is then followed by a second proton transfer from the α-carbon to A^- to give the enol. The only difference between the base and acid-catalyzed reactions is the order of proton addition and elimination. In acid-catalyzed reactions, a proton is added first; in base-catalyzed reactions, a proton is removed first.

Mechanism   Acid-Catalyzed Equilibration of Keto and Enol Tautomers

Step 1:   Proton transfer from the acid catalyst, H—A, to the carbonyl oxygen gives the conjugate acid of the ketone as a resonance-stabilized oxonium ion.

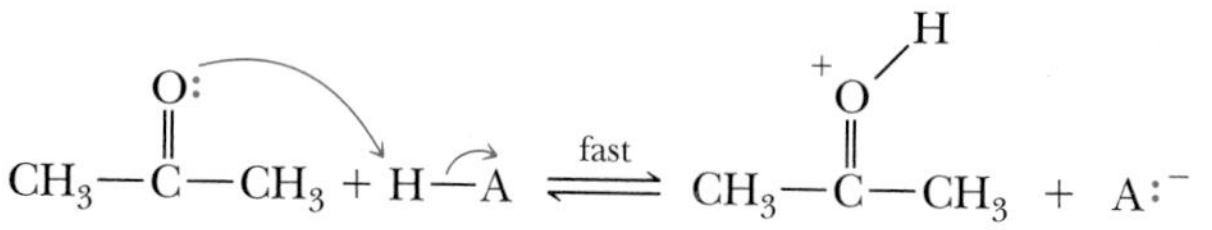

Keto form

The conjugate acid
of the ketone

Step 2: Proton transfer from the α-carbon to the base, A$^-$, gives the enol and generates a new molecule of the acid catalyst.

$$\underset{\text{Enol form}}{CH_3-\overset{\overset{\displaystyle \overset{+}{O}-H}{\|}}{C}-CH_2-H + :A^- \overset{\text{slow}}{\rightleftharpoons} CH_3-\overset{\overset{\displaystyle :OH}{|}}{C}=CH_2 + H-A}$$

B. The Position of Equilibrium in Keto-Enol Tautomerism

Aldehydes and ketones with at least one α-hydrogen are in equilibrium with their enol forms. We first encountered this type of equilibrium in Section 10.8 in our study of the hydroboration/oxidation and acid-catalyzed hydration of alkynes. As seen in Table 16.5, the position of keto-enol equilibrium for simple aldehydes and ketones lies far on the side of the keto form because a carbon-oxygen double bond is stronger than a carbon-carbon double bond.

Table 16.5 The Position of Keto-Enol Equilibrium for Some Simple Aldehydes and Ketones*

Keto Form	Enol Form	% Enol at Equilibrium
$CH_3\overset{\overset{\displaystyle O}{\|}}{C}H \rightleftharpoons$	$CH_2=\overset{\overset{\displaystyle OH}{\|}}{C}H$	6×10^{-5}
$CH_3\overset{\overset{\displaystyle O}{\|}}{C}CH_3 \rightleftharpoons$	$CH_3\overset{\overset{\displaystyle OH}{\|}}{C}=CH_2$	6×10^{-7}
cyclopentanone $\rightleftharpoons$	cyclopentenol	1×10^{-6}
cyclohexanone $\rightleftharpoons$	cyclohexenol	4×10^{-5}

*Data from J. March, *Advanced Organic Chemistry*, 4th ed., Wiley Interscience, New York, 1992, p. 70.

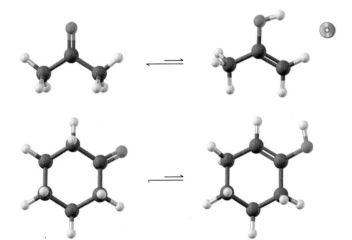

For certain types of molecules, the enol form may be the major form and, in some cases, the only form present at equilibrium. For β-diketones, such as 1,3-cyclohexanedione and 2,4-pentanedione, where an α-carbon is between two carbonyl groups, the position of equilibrium shifts in favor of the enol form.

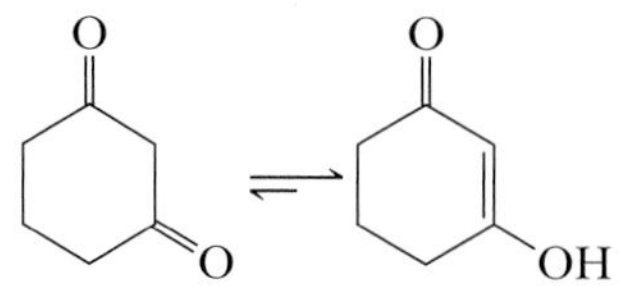

1,3-Cyclohexanedione

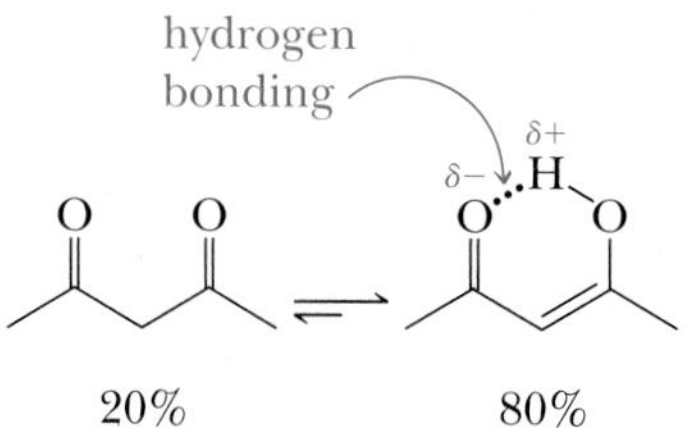

20%   80%

2,4-Pentanedione
(Acetylacetone)

These enols are stabilized by conjugation of the pi systems of the carbon-carbon double bond and the carbonyl group. The enol of 2,4-pentanedione, an open-chain β-diketone, is further stabilized by intramolecular hydrogen bonding.

Example 16.11

Write two enol forms for each compound. Which enol of each predominates at equilibrium?

(a) 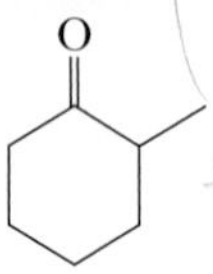   (b)

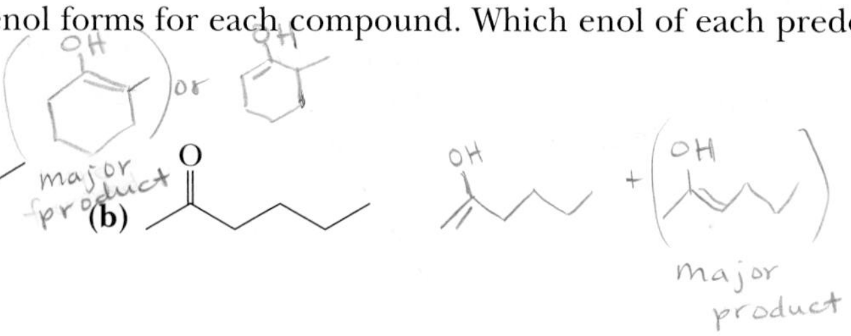

Solution

In each case, the major enol form has the more substituted double bond.

(a) 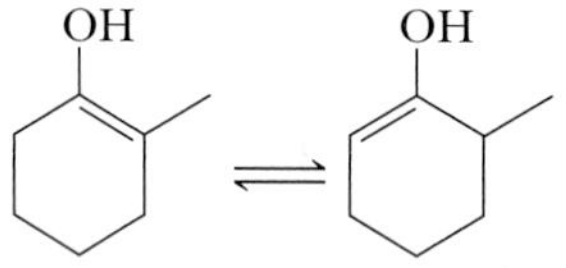   (b)

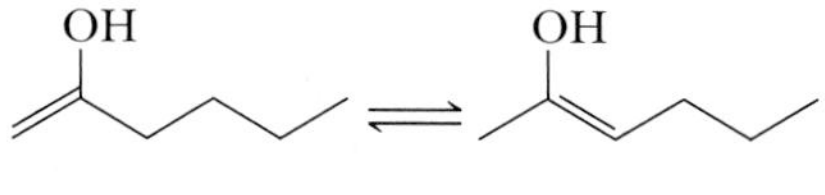

Major enol   Major enol

Problem 16.11

Draw the structural formula for the keto form of each enol.

(a) 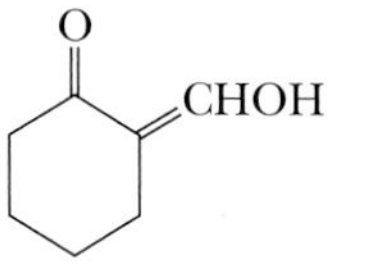   (b) 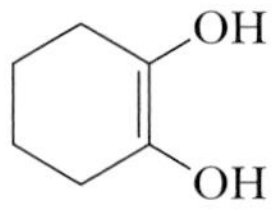   (c)

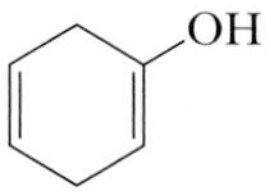

16.12 Reactions at an α-Carbon

A. Racemization

When enantiomerically pure (either R or S) 3-phenyl-2-butanone is dissolved in ethanol, no change occurs in the optical activity of the solution over time. If, however, a trace of either acid (for example, aqueous or gaseous HCl) or base (for example, sodium ethoxide) is added, the optical activity of the solution begins to decrease and gradually drops to zero. When 3-phenyl-2-butanone is isolated from this solution, it is found to be a racemic mixture (Section 3.7C). Furthermore, the rate of race-mization is proportional to the concentration of acid or base. These observations can be explained by a rate-determining acid- or base-catalyzed formation of an achiral enol intermediate. Tautomerism of the achiral enol to the chiral keto form generates the *R* and *S* enantiomers with equal probability.

| (*R*)-3-Phenyl-2-butanone | An achiral enol | (*S*)-3-Phenyl-2-butanone |

Racemization by this mechanism occurs only at α-carbon stereocenters with at least one α-hydrogen.

B.   Deuterium Exchange

When an aldehyde or ketone with one or more α-hydrogens is dissolved in an aque-ous solution enriched with D_2O and also containing catalytic amounts of either D^+ or OD^-, exchange of α-hydrogens occurs at a rate proportional to the concentration of the acid or base catalyst. We account for incorporation of deuterium by proposing a rate-determining acid- or base-catalyzed enolization followed by incorporation of deuterium as the enol form converts to the keto form.

$$\underset{\text{Acetone}}{CH_3\overset{O}{\overset{\|}{C}}CH_3} + 6D_2O \underset{}{\overset{D^+ \text{ or } OD^-}{\rightleftharpoons}} \underset{\text{Acetone-}d_6}{CD_3\overset{O}{\overset{\|}{C}}CD_3} + 6HOD$$

In naming compounds, the presence of deuterium is shown by the symbol "d," and the number of deuterium atoms is shown by a subscript following it.

Deuterium exchange has two values. First, by observing changes in hydrogen ra-tios before and after deuterium exchange, it is possible to determine the number of exchangeable α-hydrogens in a molecule. Second, exchange of α-hydrogens is a con-venient way to introduce an isotopic label into molecules. In addition to acetone-d_6, more than 225 deuterium-labeled compounds are available commercially, in isotopic enrichments of up to 99.8 atom % D. Among these are

$CDCl_3$	$CD_3\overset{O}{\overset{\|}{C}}OD$	$CH_3\overset{O}{\overset{\|}{C}}OD$	$NaBD_4$	CH_3CH_2OD
Chloroform-d	Acetic-d_3 acid-d	Acetic acid-d	Sodium borodeuteride	Ethanol-d

Deuterated solvents, such as $CDCl_3$, acetone-d_6, and benzene-d_6, are used as solvents in ^1H-NMR spectroscopy because they lack protons that might otherwise obscure the spectrum of the compound of interest.

C.   α-Halogenation

Aldehydes and ketones with at least one α-hydrogen react at the α-carbon with bromine and chlorine to form α-haloaldehydes and α-haloketones as illustrated by bromination of acetophenone.

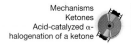

Acetophenone

Bromination or chlorination at an α-carbon is catalyzed by both acid and base. For acid-catalyzed halogenation, acid generated by the reaction catalyzes further reaction. The slow step of acid-catalyzed halogenation is formation of an enol. This is followed by rapid reaction of the double bond with halogen to give the α-haloketone.

Mechanism   Acid-Catalyzed α-Halogenation of a Ketone

Step 1: Acid-catalyzed keto-enol tautomerism gives the enol.

Step 2: Nucleophilic attack of the enol on the halogen molecule, X_2, gives an α-haloketone.

The slow step in base-promoted α-halogenation is removal of an α-hydrogen by base to form an enolate anion, which then reacts with halogen by nucleophilic displacement to form the final product. This procedure for α-halogenation produces HX as a byproduct, and, in order to keep the solution basic, it is necessary to add slightly more than one mole of base per mole of aldehyde or ketone. Because base is a reactant required in equimolar amounts, we say that this reaction is base promoted rather than base catalyzed.

Mechanism   Base-Promoted α-Halogenation of a Ketone

Step 1: Proton transfer from the α-carbon to the base gives a resonance-stabilized enolate anion.

Resonance-stabilized enolate anion

Step 2:   Nucleophilic attack of the enolate anion on halogen gives an α-bromoketone.

$$\overset{\overset{\textstyle O:^-}{}}{\underset{R'}{C}}{=}\underset{R}{\overset{R}{C}} \quad + \overset{\frown}{Br{-}Br} \quad \xrightarrow{\text{fast}} \quad \overset{O}{\underset{R'}{\overset{\|}{C}}}{-}\underset{\underset{R}{|}}{\overset{\overset{Br}{|}}{C}}{-}R \; + \; :Br^-$$

A major difference exists between acid-catalyzed and base-promoted α-halogenation. In principle, both can lead to polyhalogenation. In practice, the rate of acid-catalyzed introduction of a second halogen is considerably less than the rate of the first halogenation because the electronegative α-halogen decreases the basicity of the carbonyl oxygen toward protonation. Thus, it is generally possible to stop acid-catalyzed halogenation at a single substitution. For base-promoted halogenation, each successive halogenation is more rapid than the previous one because introduction of an electronegative halogen atom on an α-carbon further increases the acidity of remaining α-hydrogens, and, thus, each successive α-hydrogen is removed more rapidly than the previous one. For this reason, base-promoted halogenation is generally not a useful synthetic reaction.

However, one circumstance in which base-promoted halogenation is useful is the halogenation of methyl ketones. In the presence of base, a methyl ketone reacts with three equivalents of halogen to form a 1,1,1-trihaloketone, which then reacts with an additional mole of hydroxide ion to form a carboxylic salt and a trihalomethane. Reaction of the carboxylic salt with aqueous HCl or other strong acid gives the carboxylic acid.

$$\overset{O}{\overset{\|}{R C}}{CH_3} + 3Br_2 + 3NaOH \longrightarrow \overset{O}{\overset{\|}{R C}}{CBr_3} + 3NaBr + 3H_2O$$

$$\overset{O}{\overset{\|}{R C}}{CBr_3} + NaOH \longrightarrow \overset{O}{\overset{\|}{R C}}{O^-Na^+} \quad + \quad CHBr_3$$

$$\text{Tribromomethane}$$
$$\text{(Bromoform)}$$

Common names for the trihalomethanes are chloroform, bromoform, and iodoform. For this reason, reaction of a methyl ketone with a halogen in base is called the **haloform reaction.**

Mechanism   The Final Step of the Haloform Reaction

Step 1:   Hydroxide ion adds to the carbonyl carbon to form a tetrahedral carbonyl addition intermediate, which collapses to give a haloform anion and a carboxyl group. A CX_3^- carbanion is stabilized by three electron-withdrawing halogen atoms, making it a good leaving group (Section 8.4F). The haloform reaction is one of the rare cases where a carbanion is a leaving group.

$$RC(=O)-CBr_3 + {}^-{:}OH \rightleftharpoons RC(-O^-)(CBr_3)(OH) \longrightarrow RC(=O)OH + {:}CBr_3$$

Conjugate base
of bromoform

Step 2: Proton transfer from the carboxyl group to the bromoform anion completes the reaction.

$$RC(=O)-O-H + {:}CBr_3 \longrightarrow RC(=O)-O^{-}{:} + H-CBr_3$$

Bromoform

The haloform reaction is an indirect way to oxidize a methyl ketone to a carboxylic acid as illustrated by the oxidation of the following unsaturated methyl ketone to an unsaturated carboxylic acid.

$$\xrightarrow[\text{2. HCl/H}_2\text{O}]{\text{1. Cl}_2/\text{NaOH}}$$

5-Methyl-3-hexen-2-one   4-Methyl-2-pentenoic acid   + $CHCl_3$
                             Trichloromethane
                             (Chloroform)

Reactivity Explorer
Ketones and Aldehydes

16.13 Oxidation

A. Oxidation of Aldehydes

Aldehydes are oxidized to carboxylic acids by a variety of common oxidizing agents, including chromic acid and molecular oxygen. In fact, aldehydes have one of the most easily oxidized of all functional groups. Oxidation by chromic acid is illustrated by the conversion of hexanal to hexanoic acid. For the mechanism of this oxidation, review Section 9.8.

$$\text{CHO} \xrightarrow{\text{H}_2\text{CrO}_4} \text{COOH}$$

Hexanal   Hexanoic acid

Aldehydes are also oxidized to carboxylic acids by Ag(I) ion. One laboratory procedure is to shake a solution of the aldehyde in aqueous ethanol or tetrahydrofuran with a slurry of Ag_2O.

$$\text{CH}_3\text{O} \dots \text{CHO} + Ag_2O \xrightarrow[\text{NaOH}]{\text{THF, H}_2\text{O}} \xrightarrow[\text{H}_2\text{O}]{\text{HCl}} \text{CH}_3\text{O} \dots \text{COOH} + Ag$$

Vanillin   Vanillic acid

Tollens' reagent, another form of Ag(I), is prepared by dissolving silver nitrate in water, adding sodium hydroxide to precipitate Ag(I) as Ag_2O, and then adding aqueous ammonia to redissolve silver(I) as the silver-ammonia complex ion.

$$Ag^+NO_3^- + 2NH_3 \underset{}{\overset{NH_3, H_2O}{\rightleftharpoons}} Ag(NH_3)_2^+NO_3^-$$

When Tollens' reagent is added to an aldehyde, the aldehyde is oxidized to a carboxylate anion, and Ag(I) is reduced to metallic silver. If this reaction is carried out properly, silver precipitates as a smooth, mirror-like deposit, hence the name silver-mirror test. Ag(I) is rarely used at the present time for the oxidation of aldehydes because of the cost of silver and because other, more convenient methods exist for this oxidation. This reaction, however, is still used for silvering glassware, including mirrors. In this process, formaldehyde or glucose (Section 25.1) is generally used as the aldehyde to reduce Ag(I).

Aldehydes are also oxidized to carboxylic acids by molecular oxygen and by hydrogen peroxide. Reaction with oxygen is a radical chain reaction (see Chemistry in Action box "Radical Autoxidation," Chapter 7) made possible because the aldehyde C—H bond is unusually weak.

$$2 \, \text{C}_6\text{H}_5\text{—CH} \, (=O) + O_2 \longrightarrow 2 \, \text{C}_6\text{H}_5\text{—COH} \, (=O)$$

Benzaldehyde → Benzoic acid

Molecular oxygen is the least expensive and most readily available of all oxidizing agents. On an industrial scale, air oxidation of organic compounds, including aldehydes, is very common. Air oxidation of aldehydes can also be a problem. Aldehydes that are liquid at room temperature are so sensitive to oxidation by molecular oxygen that they must be protected from contact with air during storage. Often this is done by sealing the aldehyde in a container under an atmosphere of nitrogen.

Tollens' reagent A solution prepared by dissolving Ag_2O in aqueous ammonia; used for selective oxidation of an aldehyde to a carboxylic acid.

A silver mirror has been deposited on the inside of this flask by the reaction of an aldehyde with Tollens' reagent. *(Charles D. Winters)*

Example 16.12

Draw a structural formula for the product formed by treating each compound with Tollens' reagent followed by acidification with aqueous HCl.

(a) Pentanal **(b)** Cyclopentanecarbaldehyde

Solution

The aldehyde group in each compound is oxidized to a carboxyl group.

(a) $CH_3(CH_2)_3COH$ (=O) **(b)** cyclopentane-COH (=O)

Pentanoic acid Cyclopentanecarboxylic acid

Problem 16.12

Complete these oxidations.

(a) Hexanal + $H_2O_2 \longrightarrow$ **(b)** 3-Phenylpropanal + Tollens' reagent $\longrightarrow$

B. Oxidation of Ketones

Ketones, in contrast to aldehydes, are oxidized only under rather special conditions. For example, they are not normally oxidized by chromic acid or potassium permanganate. In fact, chromic acid is used routinely to oxidize secondary alcohols to ketones in good yield (Section 9.8A).

Ketones undergo oxidative cleavage, via their enol form, by potassium dichromate and potassium permanganate and other oxidants at higher temperatures and higher concentrations of acid or base. The carbon-carbon double bond of the enol is cleaved to form two carboxyl or ketone groups, depending on the substitution pattern of the original ketone. An important industrial application of this reaction is oxidation by nitric acid of cyclohexanone to hexanedioic acid (adipic acid), one of the two monomers required for the synthesis of the polymer nylon 66 (Section 24.5B).

Cyclohexanone (keto form)   Cyclohexanone (enol form)   Hexanedioic acid (Adipic acid)

As this example shows, this reaction is most useful for oxidation of symmetrical cycloalkanones, which yield a single product. Most other ketones give mixtures of products.

16.14 Reduction

Aldehydes are reduced to primary alcohols, and ketones are reduced to secondary alcohols. In addition, both aldehyde and ketone carbonyl groups can be reduced to —CH$_2$— groups.

A. Catalytic Reduction

The carbonyl group of an aldehyde or ketone is reduced to a hydroxyl group by hydrogen in the presence of a transition metal catalyst, most commonly finely divided platinum or nickel. Reductions are generally carried out at temperatures from 25 to 100°C and at pressures of hydrogen from 1 to 5 atm. Under such conditions, cyclohexanone is reduced to cyclohexanol.

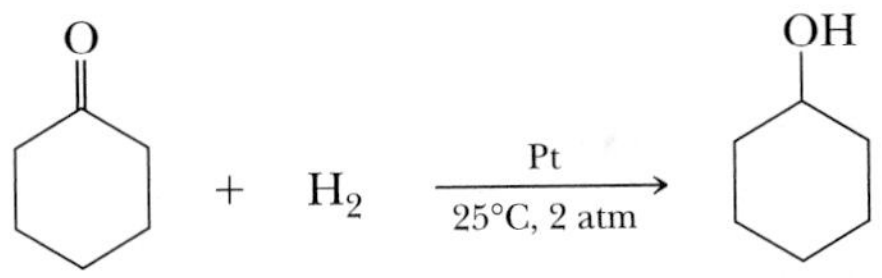

Cyclohexanone Cyclohexanol

Catalytic reduction of aldehydes and ketones is simple to carry out, yields are generally very high, and isolation of the final product is very easy. A disadvantage is that some other functional groups are also reduced under these conditions, for example, carbon-carbon double and triple bonds.

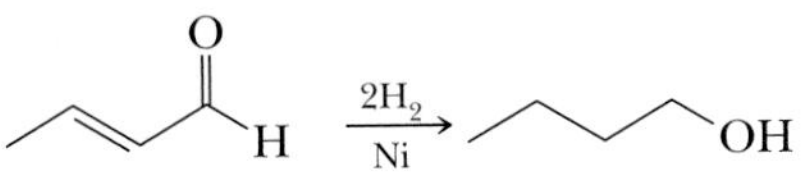

trans-2-Butenal (Crotonaldehyde) 1-Butanol

It is generally easier, however, to hydrogenate a carbon-carbon double bond than the carbon-oxygen double bond of an aldehyde or ketone. For this reason, it is often possible, by proper choice of metal catalyst and reaction conditions, to bring about selective catalytic reduction of a C=C bond in the presence of an aldehyde or ketone carbonyl group.

B. Metal Hydride Reductions

By far the most common laboratory reagents for reduction of the carbonyl group of an aldehyde or ketone to a hydroxyl group are sodium borohydride and lithium aluminum hydride (LAH). These compounds behave as sources of **hydride ion,** a very strong nucleophile.

Hydride ion A hydrogen atom with two electrons in its valence shell; $H:^-$.

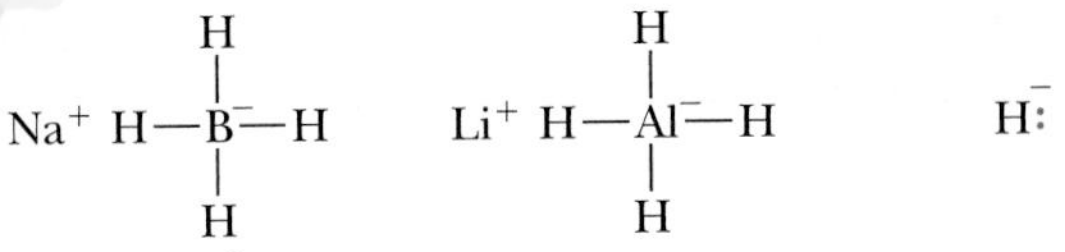

Sodium borohydride Lithium aluminum hydride Hydride ion

Lithium aluminum hydride is a very powerful reducing agent; it reduces not only the carbonyl groups of aldehydes and ketones rapidly but also those of carboxylic acids (Section 17.6A) and their functional derivatives (Section 18.11). Sodium borohydride is a much more selective reagent, reducing only aldehydes and ketones rapidly.

Reductions using sodium borohydride are most commonly carried out in aqueous methanol, in pure methanol, or in ethanol. The initial product of reduction is a tetraalkyl borate, which, on warming with water, is converted to an alcohol and sodium borate salts. One mole of sodium borohydride reduces four moles of aldehyde or ketone.

$$4R\overset{O}{\overset{\|}{C}}H + NaBH_4 \xrightarrow{\text{methanol}} (RCH_2O)_4B^-\ Na^+ \xrightarrow{H_2O} 4RCH_2OH + \text{borate salts}$$

A tetraalkyl borate

The key step in the metal hydride reduction of an aldehyde or ketone is transfer of a hydride ion from the reducing agent to the carbonyl carbon to form a tetrahedral carbonyl addition compound. In the reduction of an aldehyde or ketone to an alcohol, only the hydrogen atom bonded to carbon comes from the hydride reducing agent; the hydrogen atom bonded to oxygen comes from water during hydrolysis of the metal alkoxide salt.

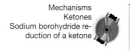

Mechanisms
Ketones
Sodium borohydride reduction of a ketone

Mechanism   Sodium Borohydride Reduction of an Aldehyde or Ketone

$$Na^{+}H-\underset{\underset{H}{|}}{\overset{\overset{H}{|}}{B}}\bar{}H + R-\overset{\overset{O}{\|}}{C}-R' \longrightarrow R-\overset{\overset{O-\bar{B}H_3\ Na^+}{|}}{\underset{\underset{H}{|}}{C}}-R' \xrightarrow{H_2O} R-\overset{\overset{O-H}{|}}{\underset{\underset{H}{|}}{C}}-R'$$

This H comes from water during hydrolysis.

This H comes from the hydride reducing agent.

Unlike sodium borohydride, LAH reacts violently with water, methanol, and other protic solvents to liberate hydrogen gas and form metal hydroxides. Therefore, reductions of aldehydes and ketones using this reagent must be carried out in aprotic solvents, most commonly diethyl ether or tetrahydrofuran. The stoichiometry for LAH reductions is the same as that for sodium borohydride reductions: one mole of LAH per four moles of aldehyde or ketone. Because of the formation of gelatinous aluminum salts, aqueous acid or base workup is usually used to dissolve these salts.

$$4R\overset{\overset{O}{\|}}{C}R + LiAlH_4 \xrightarrow{ether} (R_2CHO)_4Al^-\ Li^+ \xrightarrow[H^+\ or\ OH^-]{H_2O} 4R\overset{\overset{OH}{|}}{C}HR + aluminum\ salts$$

A tetraalkyl aluminate

The following equations illustrate selective reduction of a carbonyl group in the presence of a carbon-carbon double bond and, alternatively, selective reduction of a carbon-carbon double bond in the presence of a carbonyl group using rhodium on powdered charcoal as a catalyst.

Selective reduction of a carbonyl group:

$$RCH{=}CH\overset{\overset{O}{\|}}{C}R' \xrightarrow[\text{2. H}_2\text{O}]{\text{1. NaBH}_4} RCH{=}CH\overset{\overset{OH}{|}}{C}HR'$$

Selective reduction of carbon-carbon double bond:

$$RCH{=}CH\overset{\overset{O}{\|}}{C}R' + H_2 \xrightarrow{Rh} RCH_2CH_2\overset{\overset{O}{\|}}{C}R'$$

Example 16.13

Complete these reductions.

Solution

The carbonyl group of the aldehyde in (a) is reduced to a primary alcohol, and the carbonyl group of the ketone in (b) is reduced to a secondary alcohol.

Problem 16.13

What aldehyde or ketone gives these alcohols on reduction with $NaBH_4$?

C.   Reduction of a Carbonyl Group to a Methylene Group

Several methods are available for reducing the carbonyl group of an aldehyde or ketone to a methylene group (—CH_2—). One of the first discovered involves refluxing the aldehyde or ketone with amalgamated zinc (zinc with a surface layer of mercury) in concentrated HCl. This reaction is known as the **Clemmensen reduction** after the German chemist, E. Clemmensen, who developed it in 1912. The mechanism of Clemmensen reduction is not well understood but involves electrons from the Zn.

Clemmensen reduction   Reduction of the C=O group of an aldehyde or ketone to a CH_2 group using Zn(Hg) and HCl.

Because the Clemmensen reduction requires the use of concentrated HCl, it cannot be used to reduce a carbonyl group in a molecule that also contains acid-sensitive groups, as for example, a tertiary alcohol that might undergo dehydration or an acetal that is hydrolyzed and the resulting carbonyl group also reduced.

The **Wolff-Kishner reduction,** discovered independently by N. Kishner in 1911 and L. Wolff in 1912 and reported within months of Clemmensen's discovery, is an alternative method for reduction of a carbonyl group to a methylene group. In this reduction, a mixture of the aldehyde or ketone, hydrazine, and concentrated

Wolff-Kishner reduction   Reduction of the C=O group of an aldehyde or ketone to a CH_2 group using hydrazine and base.

potassium hydroxide is refluxed in a high-boiling solvent such as diethylene glycol (bp 245°C).

$$\underset{\text{Hydrazine}}{\text{Ph}-\overset{\overset{\displaystyle O}{\|}}{C}CH_3 + H_2NNH_2} \xrightarrow[\substack{\text{diethylene glycol}\\ \text{(reflux)}}]{\text{KOH}} \text{Ph}-CH_2CH_3 + N_2 + H_2O$$

More recently it has been found possible to bring about the same reaction in dimethyl sulfoxide (DMSO) with potassium *tert*-butoxide and hydrazine at room temperature.

Mechanisms
Ketones
Wolff-Kishner reduction

Mechanism   Wolff-Kishner Reduction

Step 1: Reaction of the carbonyl group of the aldehyde or ketone with hydrazine gives a hydrazone (Section 16.10B).

$$\underset{R}{\overset{R}{>}}C=O + H_2N-NH_2 \longrightarrow \underset{R}{\overset{R}{>}}C=N-NH_2 + H_2O$$

Hydrazine   A hydrazone

Step 2: Base-catalyzed tautomerism gives an isomer with an N=N double bond. Compare keto-enol tautomerism (Section 16.11B).

$$\underset{R}{\overset{R}{>}}C=N-\overset{H}{\underset{\;}{N}}-H + \ddot{:}O-H \rightleftharpoons \underset{R}{\overset{R}{>}}C=N-\overset{..}{\underset{H}{N}}^- \longleftrightarrow \underset{R}{\overset{R}{>}}\overset{..}{C}-N=N-H + H-O-H$$

$$\Updownarrow$$

$$\underset{R}{\overset{H}{>}}C-N=N-H + {}^-\!\ddot{:}O-H$$

Step 3: Proton transfer to hydroxide ion followed by loss of N_2 gives a carbanion; proton transfer from water to the carbanion gives the hydrocarbon and hydroxide ion.

$$\underset{R}{\overset{H}{>}}C-N=N-H + {}^-\ddot{:}O-H \rightleftharpoons \underset{R}{\overset{H}{>}}C-N=\overset{..}{N}{}^- + H_2O$$

$$\downarrow \text{loss of } N\equiv N$$

$$\underset{R}{\overset{H}{>}}C\overset{..}{:}{}^- + H-O-H \longrightarrow \underset{R}{\overset{H}{>}}C-H + {}^-\ddot{:}O-H$$

A carbanion

Each of the reductions has its special conditions, advantages, and disadvantages. The Clemmensen reduction cannot be used in the presence of groups sensitive to concentrated acid; the Wolff-Kishner reduction cannot be used in the presence of groups sensitive to concentrated base. However, the carbonyl group of almost any aldehyde or ketone can be reduced to a methylene group by one of these methods.

West Indian vanilla. *Vanilla pompona. (Jane Grushow from Grant Heilman)*

Example 16.14

Complete the following reactions.

(a)

CHO

OCH$_3$

OH

Vanillin
(from vanilla beans)

$\xrightarrow[\text{heat}]{\text{Zn/Hg, HCl}}$

CH$_3$

OCH$_3$

OH

(b)

O

Camphor

$\xrightarrow[\substack{\text{diethylene glycol,}\\\text{heat}}]{\text{N}_2\text{H}_4,\ \text{KOH}}$

Solution

Reaction (a) is a Clemmensen reduction and reaction (b) is a Wolff-Kishner reduction.

(a)

CH$_3$

OCH$_3$

OH

(b)

O

Problem 16.14

Complete the following reactions.

(a)

Civetone
(from the civet cat; used in perfumery)

O

$\xrightarrow[\text{heat}]{\text{Zn/Hg, HCl}}$

(b)

O

H

Citronellal
(from citronella and lemon grass oils)

$\xrightarrow[\substack{\text{diethylene glycol,}\\\text{heat}}]{\text{N}_2\text{H}_4,\ \text{KOH}}$

CH$_3$

Summary

An **aldehyde** (Section 16.1) contains a carbonyl group bonded to a hydrogen atom and a carbon atom. A **ketone** contains a carbonyl group bonded to two carbons. An aldehyde is named by changing -e of the parent alkane to -al (Section 16.2). A ketone is named by changing -e of the parent alkane to -one and using a number to locate the carbonyl group. In naming compounds that contain more than one functional group, the IUPAC system has established an **order of precedence of functions** (Section 16.2B). If the carbonyl group of an aldehyde or ketone is lower in precedence than other functional groups in the molecule, it is indicated by the infix -oxo-.

Aldehydes and ketones are polar compounds (Section 16.3) and interact in the pure state by dipole-dipole interactions; they have higher boiling points and are more soluble in water than nonpolar compounds of comparable molecular weight.

One of the most common reaction themes of aldehydes and ketones is addition of a nucleophile to the carbonyl carbon to form a tetrahedral carbonyl addition compound (Section 16.5).

The carbon atom adjacent to a carbonyl group is called an **α-carbon** (Section 16.11A), and a hydrogen bonded to it is called an **α-hydrogen.** The pK_a of an α-hydrogen of an aldehyde or ketone is approximately 20, which makes it less acidic than alcohols but more acidic than terminal alkynes.

Key Reactions

1. Reaction with Grignard Reagents (Section 16.6A)

Treatment of formaldehyde with a Grignard reagent followed by hydrolysis gives a primary alcohol. Similar treatment of any other aldehyde gives a secondary alcohol. Treatment of a ketone gives a tertiary alcohol.

2. Reaction with Organolithium Reagents (Section 16.6B)

Reactions of aldehydes and ketones with organolithium reagents are similar to those with Grignard reagents.

3. Reaction with Anions of Terminal Alkynes (Section 16.6C)

Treatment of an aldehyde or ketone with the alkali metal salt of a terminal alkyne followed by hydrolysis gives an α-alkynylalcohol.

4. Reaction with HCN to Form Cyanohydrins (Section 16.6D)

For aldehydes and most sterically unhindered aliphatic ketones, equilibrium favors formation of the cyanohydrin. For aryl ketones, equilibrium favors starting materials, and little cyanohydrin is obtained.

$$\underset{\text{O}}{\overset{\text{O}}{\underset{||}{\text{C}_6\text{H}_5\text{CH}}}} + \text{HC}\equiv\text{N} \xrightarrow{\text{NaCN}} \underset{\overset{|}{\text{OH}}}{\text{C}_6\text{H}_5\text{CHC}}\equiv\text{N}$$

5. The Wittig Reaction (Section 16.7)

Treatment of an aldehyde or ketone with a triphenylphosphonium ylide gives an oxaphosphetane intermediate, which fragments to give triphenylphosphine oxide and an alkene.

6. Hydration (Section 16.8A)

The degree of hydration is greater for aldehydes than for ketones.

$$\underset{\text{HCH}}{\overset{\text{O}}{\overset{||}{\text{HCH}}}} + \text{H}_2\text{O} \rightleftharpoons \underset{\overset{|}{\text{H}}}{\overset{\overset{\text{OH}}{|}}{\text{HCOH}}}$$

(>99%)

7. Addition of Alcohols to Form Hemiacetals (Section 16.8B)

Hemiacetals are only minor components of an equilibrium mixture of aldehyde or ketone and alcohol, except where the —OH and the C=O are parts of the same molecule and a five- or six-membered ring can form. The reaction is catalyzed by acid or base.

8. Addition of Alcohols to Form Acetals (Section 16.8B)

Formation of acetals is catalyzed by acid. Acetals are stable to water and aqueous base but are hydrolyzed in aqueous acid. Acetals are valuable as carbonyl-protecting groups.

9. Addition of Sulfur Nucleophiles: Formation of 1,3-Dithianes (Section 16.9)

The most commonly used thiol for preparation of thioacetals is 1,3-propanedithiol. The product is a 1,3-dithiane.

10. Alkylation of Anions Derived from Aldehyde 1,3-Dithianes (Section 16.9)

Treatment of an aldehyde 1,3-dithiane (pK_a 31) with butyllithium gives an anion. This anion can enter into substitution reactions with primary alkyl, allylic, and benzylic halides and addition reactions with the carbonyl group of aldehydes and ketones.

$$\text{dithiane-C:}^-\text{Li}^+ \xrightarrow[\text{2. H}_2\text{O, HgCl}_2]{\text{1. C}_6\text{H}_5\text{CH}_2\text{Cl}} \text{CH}_3\overset{\overset{\textstyle O}{\|}}{\text{C}}\text{CH}_2\text{C}_6\text{H}_5$$

11. Addition of Ammonia and Its Derivatives: Formation of Imines (Section 16.10A)

Addition of ammonia or a primary amine to the carbonyl group of an aldehyde or ketone forms a tetrahedral carbonyl addition compound. Loss of water from this intermediate gives an imine.

$$\text{cyclopentanone}=\text{O} + \text{H}_2\text{NCH}_3 \rightleftharpoons[\text{H}^+] \text{cyclopentane}=\text{NCH}_3 + \text{H}_2\text{O}$$

12. Addition of Secondary Amines: Formation of Enamines (Section 16.10A)

Addition of a secondary amine to the carbonyl group of an aldehyde or ketone forms a tetrahedral carbonyl addition intermediate. Acid-catalyzed dehydration of this intermediate gives an enamine.

$$\text{cyclohexanone}=\text{O} + \text{H}-\text{N(piperidine)} \rightleftharpoons[\text{H}^+] \text{cyclohexene}-\text{N(piperidine)} + \text{H}_2\text{O}$$

13. Addition of Hydrazine and Its Derivatives (Section 16.10B)

Treatment of an aldehyde or ketone with hydrazine gives a hydrazone. Derivatives of hydrazine react similarly.

$$\text{cyclopentanone}=\text{O} + \text{H}_2\text{NNH}_2 \longrightarrow \text{cyclopentane}=\text{NNH}_2 + \text{H}_2\text{O}$$

14. Keto-Enol Tautomerism (Section 16.11B)

The keto form predominates at equilibrium, except for those aldehydes and ketones in which the enol is stabilized by resonance or hydrogen bonding.

$$\text{CH}_3\overset{\overset{\textstyle O}{\|}}{\text{C}}\text{CH}_3 \rightleftharpoons \text{CH}_3\overset{\overset{\textstyle OH}{|}}{\text{C}}=\text{CH}_2$$

Keto form   Enol form
(approx 99.9%)

15. Deuterium Exchange at an α-Carbon (Section 16.12B)

Acid- or base-catalyzed deuterium exchange at an α-carbon involves formation of an enol or enolate anion intermediate.

$$\text{CH}_3\overset{\overset{\textstyle O}{\|}}{\text{C}}\text{CH}_3 + 6\text{D}_2\text{O} \underset{\text{DCl}}{\rightleftharpoons} \text{CD}_3\overset{\overset{\textstyle O}{\|}}{\text{C}}\text{CD}_3 + 6\text{HOD}$$

16. Halogenation at an α-Carbon (Section 16.12C)

The rate-determining step in acid-catalyzed α-halogenation is the formation of an enol. In base-promoted α-halogenation, it is formation of an enolate anion.

17. The Haloform Reaction (Section 16.12C)

The haloform reaction oxidizes a methyl ketone to a carboxylic acid.

18. Oxidation of an Aldehyde to a Carboxylic Acid (Section 16.13A)

The aldehyde group is among the most easily oxidized functional groups. Oxidizing agents include $KMnO_4$, $K_2Cr_2O_7$, Tollens' reagent, and O_2.

19. Catalytic Reduction (Section 16.14A)

Catalytic reduction of the carbonyl group of an aldehyde or ketone to a hydroxyl group is simple to carry out and yields of alcohols are high. A disadvantage of this method is that some other functional groups, including carbon-carbon double and triple bonds, may also be reduced.

20. Metal Hydride Reduction (Section 16.14B)

Both $LiAlH_4$ and $NaBH_4$ are selective in that neither reduces isolated carbon-carbon double or triple bonds.

21. Clemmensen Reduction (Section 16.14C)

Reduction of the carbonyl group of an aldehyde or ketone using amalgamated zinc in the presence of concentrated hydrochloric acid gives a methylene group.

22. Wolff-Kishner Reduction (Section 16.14C)

Formation of a hydrazone followed by treatment with base, commonly KOH in diethylene glycol or potassium *tert*-butoxide in dimethyl sulfoxide, reduces the carbonyl group of an aldehyde or ketone to a methylene group.

Problems

Structure and Nomenclature

16.15 Name these compounds. Show stereochemistry where relevant.

16.16 Draw structural formulas for these compounds.

(a) 1-Chloro-2-propanone
(b) 3-Hydroxybutanal
(c) 4-Hydroxy-4-methyl-2-pentanone
(d) 3-Methyl-3-phenylbutanal
(e) 1,3-Cyclohexanedione
(f) 3-Methyl-3-buten-2-one
(g) 5-Oxohexanal
(h) 2,2-Dimethylcyclohexanecarbaldehyde
(i) 3-Oxobutanoic acid

Spectroscopy

16.17 The infrared spectrum of compound A, $C_6H_{12}O$, shows a strong, sharp peak at 1724 cm^{-1}. From this information and its ^{1}H-NMR spectrum, deduce the structure of compound A.

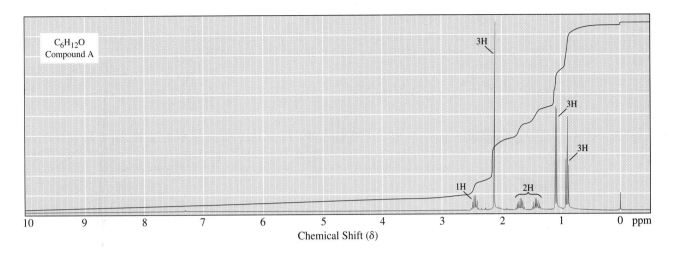

C$_6$H$_{12}$O
Compound A

3H

3H

3H

1H

2H

16.18 Following are ^{1}H-NMR spectra for compounds B, C$_6$H$_{12}$O$_2$, and C, C$_6$H$_{10}$O. On warming
in dilute acid, compound B is converted to compound C. Deduce the structural formu-
las for compounds B and C.

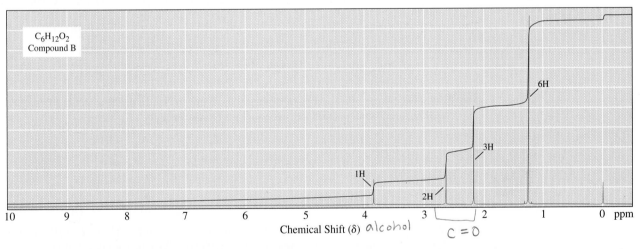

C$_6$H$_{12}$O$_2$
Compound B

6H

3H

1H

2H

Chemical Shift (δ) alcohol   C = O

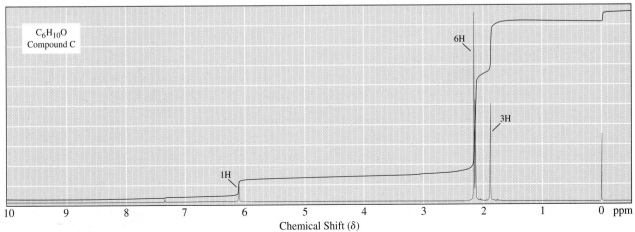

C$_6$H$_{10}$O
Compound C

6H

3H

1H

Chemical Shift (δ)

Addition of Carbon Nucleophiles

16.19 Draw structural formulas for the product formed by treating each compound with propylmagnesium bromide followed by aqueous HCl.

(a) CH_2O (b) [epoxide structure]   (c) [ketone structure]   (d) [cyclopentenone structure with =O]

16.20 Suggest a synthesis for the following alcohols starting from an aldehyde or ketone and an appropriate Grignard reagent. Below each target molecule is the number of combinations of Grignard reagent and aldehyde or ketone that might be used.

(a) [structure with OH]   (b) [structure with OH]   (c) [structure with OH, OCH_3]

3 Combinations   2 Combinations   2 Combinations

16.21 Show how to synthesize the following alcohol using 1-bromopropane, propanal, and ethylene oxide as the only sources of carbon atoms. It can be done using each compound only once.

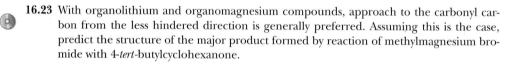

16.22 1-Phenyl-2-butanol is used in perfumery. Show how to synthesize this alcohol from bromobenzene, 1-butene, and any necessary inorganic reagents.

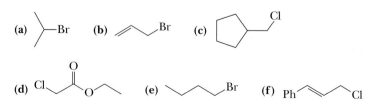

Bromobenzene   1-Butene   1-Phenyl-2-butanol

16.23 With organolithium and organomagnesium compounds, approach to the carbonyl carbon from the less hindered direction is generally preferred. Assuming this is the case, predict the structure of the major product formed by reaction of methylmagnesium bromide with 4-*tert*-butylcyclohexanone.

Wittig Reaction

16.24 Draw structural formulas for (1) the alkyltriphenylphosphonium salt formed by treatment of each haloalkane with triphenylphosphine, (2) the phosphonium ylide formed by treatment of each phosphonium salt with butyllithium, and (3) the alkene formed by treatment of each phosphonium ylide with acetone.

(a) [structure]—Br   (b) [structure]Br   (c) [cyclopentane with CH_2Cl]

(d) Cl[ester structure with O]   (e) [structure]Br   (f) Ph[structure]Cl

16.25 Show how to bring about the following conversions using a Wittig reaction.

(a) →

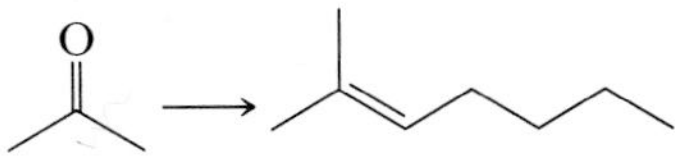

(b) →

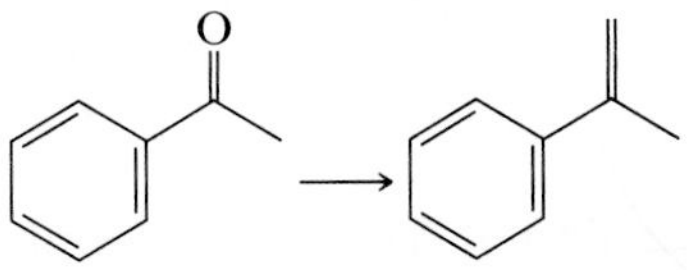

(c)

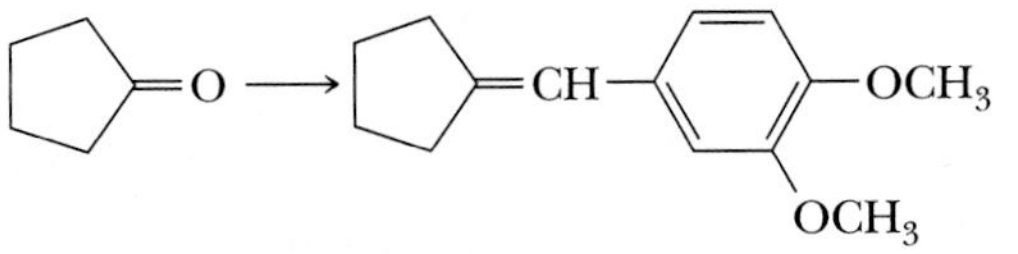

16.26 The Wittig reaction can be used for the synthesis of conjugated dienes, as for example 1-phenyl-1,3-pentadiene. Propose two sets of reagents that might be combined in a Wittig reaction to give this conjugated diene.

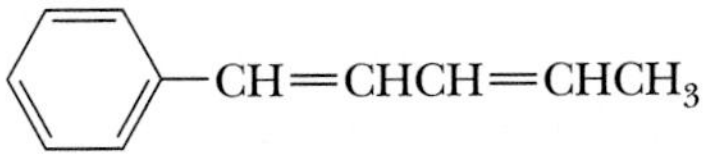

1-Phenyl-1,3-pentadiene

16.27 Wittig reactions with the following α-chloroethers can be used for the synthesis of aldehydes and ketones.

$$ClCH_2OCH_3 \qquad \overset{\displaystyle CH_3}{\underset{}{Cl\overset{|}{C}HOCH_3}}$$

(A) (B)

(a) Draw the structure of the triphenylphosphonium salt and Wittig reagent formed from each chloroether.
(b) Draw the structural formula of the product formed by treatment of each Wittig reagent with cyclopentanone. Note that the functional group is an enol ether or, alternatively, a vinyl ether.
(c) Draw the structural formula of the product formed on acid-catalyzed hydrolysis of each enol ether from part (b).

16.28 It is possible to generate sulfur ylides in a manner similar to that used to produce phosphonium ylides. For example, treating a sulfonium salt with a strong base gives the sulfur ylide.

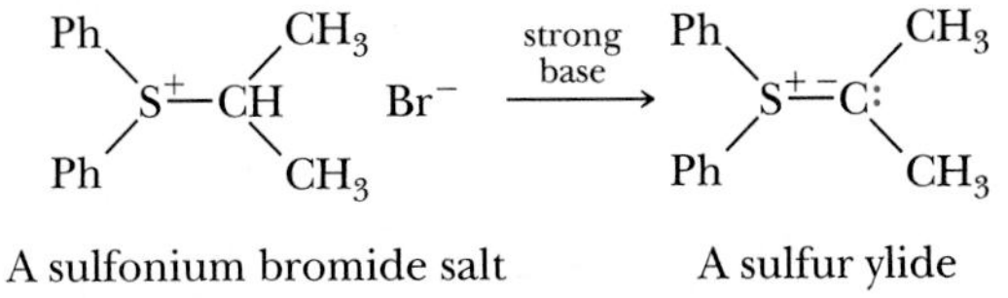

A sulfonium bromide salt   A sulfur ylide

Sulfur ylides react with ketones to give epoxides. Suggest a mechanism for this reaction.

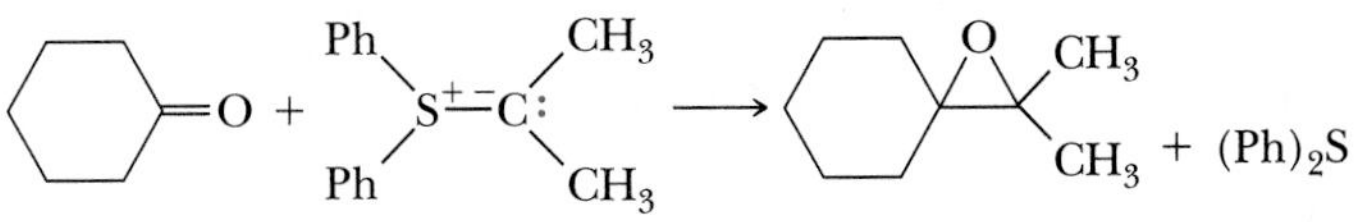

16.29 Propose a structural formula for compound D and for the product, $C_9H_{14}O$, formed in this reaction sequence.

Addition of Oxygen Nucleophiles

16.30 5-Hydroxyhexanal forms a six-membered cyclic hemiacetal, which predominates at equilibrium in aqueous solution.

5-Hydroxyhexanal

(a) Draw a structural formula for this cyclic hemiacetal.
(b) How many stereoisomers are possible for 5-hydroxyhexanal?
(c) How many stereoisomers are possible for this cyclic hemiacetal?
(d) Draw alternative chair conformations for each stereoisomer and label groups axial or equatorial. Also predict which of the alternative chair conformations for each stereoisomer is the more stable.

16.31 Draw structural formulas for the hemiacetal and then the acetal formed from each pair of reactants in the presence of an acid catalyst.

16.32 Draw structural formulas for the products of hydrolysis of the following acetals in aqueous HCl.

16.33 Propose a mechanism to account for the formation of a cyclic acetal from 4-hydroxypentanal and one equivalent of methanol. If the carbonyl oxygen of 4-hydroxypentanal is enriched with oxygen-18, do you predict that the oxygen label appears in the cyclic acetal or in the water?

4-Hydroxypentanal   A cyclic acetal

16.34 Propose a mechanism for this acid-catalyzed hydrolysis.

16.35 All rearrangements we have discussed so far involve generation of an electron-deficient atom followed by a 1,2-shift of an atom or group of atoms to it. A mechanism that can be written for the following reaction also involves generation of an electron-deficient oxygen followed by a 1,2-shift from an adjacent carbon to it.

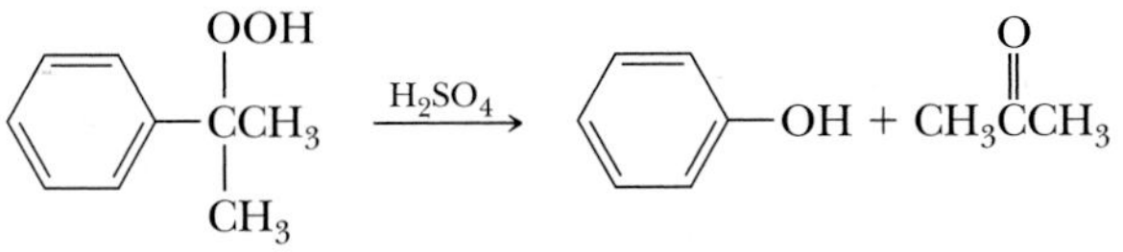

Cumene hydroperoxide   Phenol   Acetone

Propose a mechanism for the acid-catalyzed rearrangement of cumene hydroperoxide to phenol and acetone based on the previous rationale. In completing a mechanism, you will want to review the characteristic structural features of a hemiacetal (Section 16.8B) and its equilibration with a ketone by loss of an alcohol.

16.36 In Section 11.6, we saw that ethers, such as diethyl ether and tetrahydrofuran, are quite resistant to the action of dilute acids and require hot concentrated HI or HBr for cleavage. However, acetals, in which two ether groups are linked to the same carbon, undergo hydrolysis readily even in dilute aqueous acid. How do you account for this marked difference in chemical reactivity toward dilute aqueous acid between ethers and acetals?

16.37 Show how to bring about the following conversion.

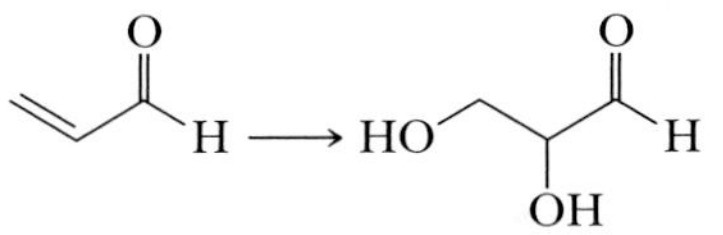

16.38 A primary or secondary alcohol can be protected by conversion to its tetrahydropyranyl ether. Why is formation of THP ethers by this reaction limited to primary and secondary alcohols?

16.39 Which of these molecules will cyclize to give the insect pheromone frontalin?

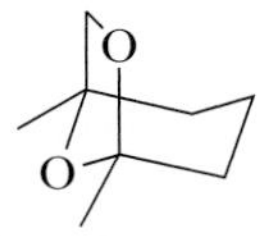

Frontalin 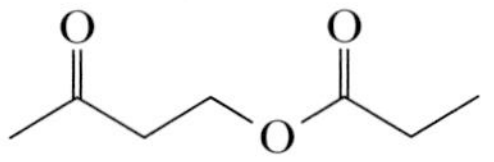   (A) 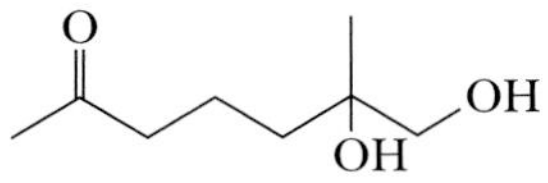   (B) 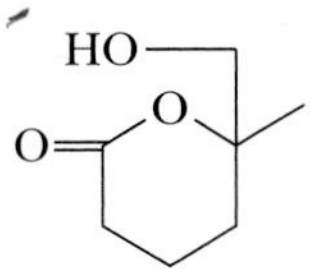   (C)

Addition of Sulfur Nucleophiles
16.40 Draw a structural formula for the product of reaction of benzaldehyde with the following dithiols in the presence of an acid catalyst.

    (a) 1,2-Ethanedithiol   **(b)** 1,3-Propanedithiol

16.41 Draw a structural formula for the product formed by treating each of these compounds with (1) the lithium salt of the 1,3-dithiane derived from acetaldehyde and then (2) H_2O, $HgCl_2$.

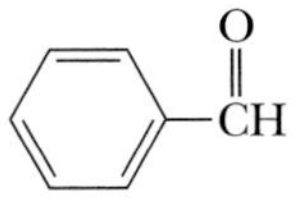

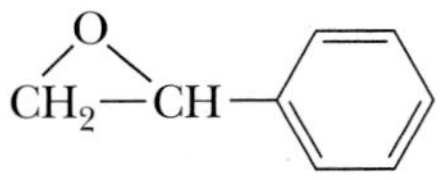

16.42 Show how to bring about the following conversions using a 1,3-dithiane.

(a) (b) (c)

Addition of Nitrogen Nucleophiles

16.43 Draw structural formulas for the products of the following acid-catalyzed reactions.

(a) Phenylacetaldehyde + hydrazine $\longrightarrow$
(b) Cyclopentanone + semicarbazide $\longrightarrow$
(c) Acetophenone + 2,4-dinitrophenylhydrazine $\longrightarrow$
(d) Benzaldehyde + hydroxylamine $\longrightarrow$

16.44 Following are structural formulas for amphetamine and methamphetamine (see *The Merck Index*, 12th ed., #623 and #6015). The major central nervous system effects of amphetamine and amphetamine-like drugs are locomotor stimulation, euphoria and excitement, stereotyped behavior, and anorexia. Show how each of these drugs can be synthesized by reductive amination of an appropriate aldehyde or ketone and amine. For structural formulas of several more anorexics, see MC.42 and MC.43.

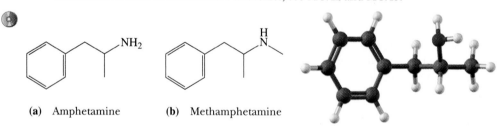

(a) Amphetamine   (b) Methamphetamine

16.45 Following is the final step in the synthesis of the antiviral drug Rimantadine. Describe experimental conditions to bring this conversion.

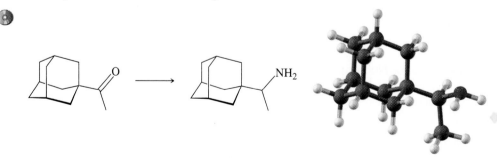

Keto-Enol Tautomerism

16.46 The following molecule belongs to a class of compounds called enediols; each carbon of the double bond carries an —OH group. Draw structural formulas for the α-hydroxyketone and the α-hydroxyaldehyde with which this enediol is in equilibrium.

$$\alpha\text{-Hydroxyaldehyde} \rightleftharpoons \underset{\underset{\displaystyle CH_3}{|}}{\overset{\overset{\displaystyle CH-OH}{\|}}{C-OH}} \rightleftharpoons \alpha\text{-Hydroxyketone}$$

An enediol

16.47 In dilute aqueous base, (R)-glyceraldehyde is converted into an equilibrium mixture of (R,S)-glyceraldehyde and dihydroxyacetone. Propose a mechanism for this isomerization.

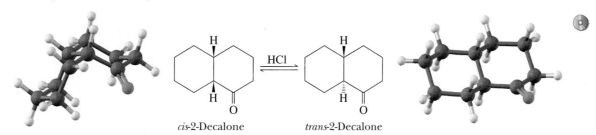

(R)-Glyceraldehyde   (R,S)-Glyceraldehyde   Dihydroxyacetone

16.48 When *cis*-2-decalone is dissolved in ether containing a trace of HCl, an equilibrium is established with *trans*-2-decalone. The latter ketone predominates in the equilibrium mixture. Propose a mechanism for this isomerization and account for the fact that the trans isomer predominates at equilibrium.

cis-2-Decalone   *trans*-2-Decalone

Reactions at an α-Carbon

16.49 The following bicyclic ketone has two α-carbons and three α-hydrogens. When this molecule is treated with D_2O in the presence of an acid catalyst, only two α-hydrogens exchange with deuterium. The α-hydrogen at the bridgehead does not exchange. How do you account for the fact that two α-hydrogens do exchange but the third does not? You will find it helpful to look at the models on the CD of this molecule and of the enols by which exchange of α-hydrogens occurs.

This α-hydrogen does not exchange.   These two α-hydrogens exchange.

16.50 Propose a mechanism for this reaction.

16.51 The base-promoted rearrangement of an α-haloketone to a carboxylic acid, known as the Favorskii rearrangement, is illustrated by the conversion of 2-chlorocyclohexanone

to cyclopentanecarboxylic acid. It is proposed that NaOH first converts the α-haloketone to the substituted cyclopropanone shown in brackets, and then to the sodium salt
of cyclopentanecarboxylic acid.

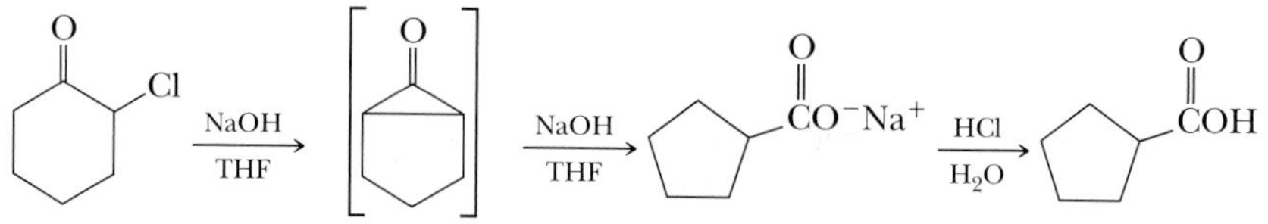

A proposed
intermediate

(a) Propose a mechanism for base-promoted conversion of 2-chlorocyclohexanone to
the proposed intermediate.
(b) Propose a mechanism for base-promoted conversion of the proposed intermediate
to sodium cyclopentanecarboxylate.

16.52 If the Favorskii rearrangement 2-chlorocyclohexanone is carried out using sodium
ethoxide in ethanol, the product is ethyl cyclopentanecarboxylate. Propose a mechanism for this reaction.

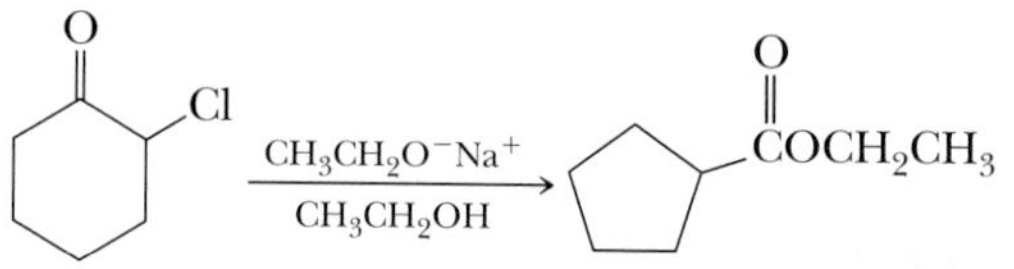

16.53 (R)-(+)-Pulegone is converted to (R)-citronellic acid by addition of HCl followed by
treatment with NaOH. Propose a mechanism for each step in this transformation, including the regioselectivity of HCl addition.

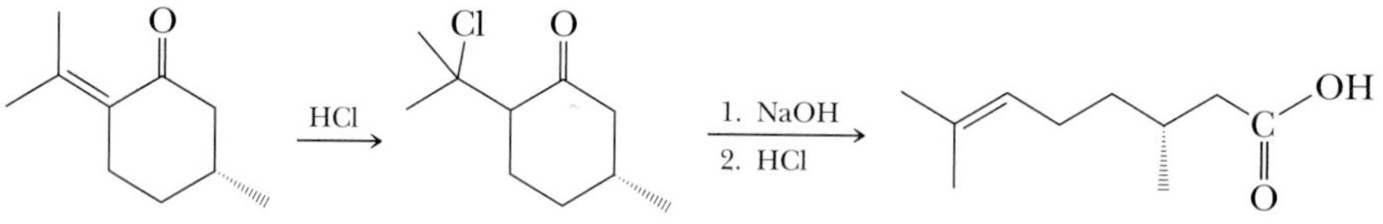

(R)-(+)-Pulegone

(R)-3,7-Dimethyl-6-octenoic acid
(R)-Citronellic acid

Oxidation/Reduction of Aldehydes and Ketones

16.54 Draw structural formulas for the products formed by treatment of butanal with the following reagents.

(a) $LiAlH_4$ followed by H_2O (b) $NaBH_4$ in CH_3OH/H_2O
(c) H_2/Pt (d) $Ag(NH_3)_2^+$ in NH_3/H_2O
(e) H_2CrO_4, heat   (f) $HOCH_2CH_2OH$, HCl
(g) $Zn(Hg)/HCl$ (h) N_2H_4, KOH at 250°C
(i) $C_6H_5NH_2$ (j) $C_6H_5NHNH_2$

16.55 Draw structural formulas for the products of the reaction of acetophenone with the
reagents given in Problem 16.54.

Synthesis

16.56 Starting with cyclohexanone, show how to prepare these compounds. In addition to the
given starting material, use any other organic or inorganic reagents as necessary.

(a) Cyclohexanol   (b) Cyclohexene
(c) *cis*-1,2-Cyclohexanediol   (d) 1-Methylcyclohexanol
(e) 1-Methylcyclohexene   (f) 1-Phenylcyclohexanol
(g) 1-Phenylcyclohexene   (h) Cyclohexene oxide
(i) *trans*-1,2-Cyclohexanediol

16.57 Show how to convert cyclopentanone to these compounds. In addition to cyclopentanone, use any other organic or inorganic reagents as necessary.

16.58 Disparlure is a sex attractant of the gypsy moth (*Porthetria dispar*). It has been synthesized in the laboratory from the following *Z*-alkene.

(*Z*)-2-Methyl-7-octadecene   Disparlure

(a) Propose two sets of reagents that might be combined in a Wittig reaction to give the indicated *Z*-alkene. Note that, at least for simple phosphonium ylides, the product of a Wittig reaction is predominantly the *Z*-alkene.
(b) How might the *Z*-alkene be converted to disparlure?
(c) How many stereoisomers are possible for disparlure? How many are formed in the sequence you chose?

16.59 Propose structural formulas for compounds A, B, and C in the following conversion. Show also how to prepare compound C by a Wittig reaction.

16.60 Following is a retrosynthetic scheme for the synthesis of *cis*-3-penten-2-ol. Write a synthesis for this compound from the indicated starting materials.

16.61 Following is the structural formula of the tranquilizer Oblivon (meparfynol, *The Merck Index*, 12th ed., #5890). Propose a synthesis for this compound starting with acetylene and a ketone.

Oblivon

16.62 Following is the structural formula of surfynol, a defoaming surfactant. Describe the synthesis of this compound from acetylene and a ketone.

Surfynol

16.63 Propose a mechanism for this isomerization.

$$\xrightarrow{H^+}$$

16.64 Propose a mechanism for this isomerization.

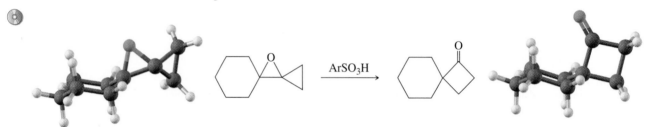

$$\xrightarrow{ArSO_3H}$$

16.65 Starting with acetylene and 1-bromobutane as the only sources of carbon atoms, show how to synthesize the following.

(a) meso-5,6-Decanediol   (b) racemic 5,6-Decanediol
(c) 5-Decanone   (d) 5,6-Epoxydecane
(e) 5-Decanol   (f) Decane
(g) 6-Methyl-5-decanol   (h) 6-Methyl-5-decanone

16.66 Following are the final steps in one industrial synthesis of vitamin A acetate.

Pseudoionone $\xrightarrow[(1)]{H_2SO_4}$ β-Ionone $\xrightarrow{(2)}$ ...OH $\xrightarrow{(3)}$

...OH $\xrightarrow[(4)]{Ph_3P, HBr}$ ...$\overset{+}{P}Ph_3$ Br^- $\xrightarrow{(5)}$

Vitamin A acetate

(a) Propose a mechanism for the acid-catalyzed cyclization in Step 1.

(b) Propose reagents to bring about Step 2.

(c) Propose reagents to bring about Step 3.

(d) Propose a mechanism for formation of the phosphonium salt in Step 4.

(e) Show how Step 5 can be completed by a Wittig reaction.

16.67 Following is the structural formula of the principal sex pheromone of the Douglas fir tussock moth (*Orgyia pseudotsugata*), a severe defoliant of the fir trees of western North America.

$$CH_3(CH_2)_9\overset{\overset{\displaystyle O}{\|}}{C}(CH_2)_3 \quad (CH_2)_4CH_3$$
$$\underset{\underset{\displaystyle H}{|}}{C}=\underset{\underset{\displaystyle H}{|}}{C}$$

(*Z*)-6-Heneicosene-11-one

Several syntheses of this compound have been reported in the literature, starting materials for three of which are given here. Show a series of steps by which each set of starting materials could be converted into the preceding target molecule.

(a) $CH_3(CH_2)_9\overset{\overset{\displaystyle O}{\|}}{CH} + CH_3(CH_2)_4C\equiv C(CH_2)_3OH$

(b) $CH_3(CH_2)_9\overset{\overset{\displaystyle O}{\|}}{C}(CH_2)_3C\equiv CH + CH_3(CH_2)_4Br$

(c) [structure] $+ CH_3(CH_2)_5Br + CH_3(CH_2)_9Br$

16.68 Both (*S*)-citronellal and isopulegol are naturally occurring terpenes (Section 5.4). When (*S*)-citronellal is treated with tin(IV) chloride (a Lewis acid) followed by neutralization with aqueous ammonium chloride, isopulegol is obtained in 85% yield.

(*S*)-Citronellal   Isopulegol
($C_{10}H_{18}O$)   ($C_{10}H_{18}O$)

1. SnCl$_4$, CH$_2$Cl$_2$
2. NH$_4$Cl

(a) Show that both compounds are terpenes.

(b) Propose a mechanism for the conversion of (*S*)-citronellal to isopulegol.

(c) How many stereocenters are present in isopulegol? How many stereoisomers are possible for a molecule with this number of stereocenters?

(d) Isopulegol is formed as a single, enantiomerically pure stereoisomer. Account for the fact that only a single stereoisomer is formed.

17

CARBOXYLIC ACIDS

The most important chemical property of carboxylic acids, another class of organic compounds containing the carbonyl group, is their acidity. Furthermore, carboxylic acids form numerous important derivatives, including esters, amides, anhydrides, and acid halides. In this chapter we study carboxylic acids themselves; in Chapter 18, we study their derivatives.

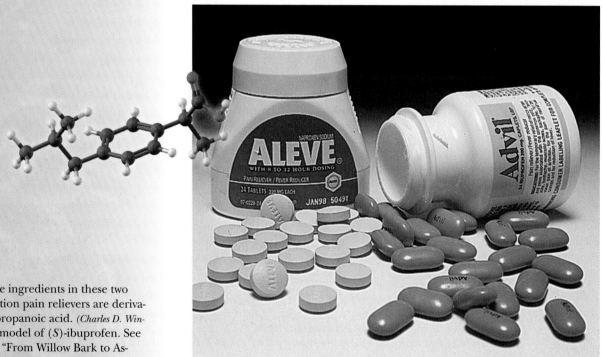

■ The active ingredients in these two nonprescription pain relievers are derivatives of arylpropanoic acid. *(Charles D. Winters)* Inset: A model of (*S*)-ibuprofen. See also the box "From Willow Bark to Aspirin and Beyond" in the Medicinal Chemistry interchapter following Chapter 23.

17.1 Structure

The functional group of a carboxylic acid is the **carboxyl group** (Section 1.3D), so named because it is made up of a carbonyl group and a hydroxyl group. Following is a Lewis structure of the carboxyl group as well as two alternative representations for it:

$$-\overset{\overset{\ddot{O}:}{\|}}{\underset{\underset{:\ddot{O}-H}{}}{C}} \qquad -COOH \qquad -CO_2H$$

Alternative representations of a carboxyl group

The general formula for an aliphatic carboxylic acid is RCOOH; the general formula for an aromatic carboxylic acid is ArCOOH.

17.2 Nomenclature

 Nomenclature
Carboxylic Acids

A. IUPAC System

The IUPAC name of a carboxylic acid is derived from that of the longest carbon chain that contains the carboxyl group by dropping the final -e from the name of the parent alkane and adding the suffix -oic followed by the word "acid" (Section 2.5). The chain is numbered beginning with the carbon of the carboxyl group. Because the carboxyl carbon is understood to be carbon 1, there is no need to give it a number. The IUPAC system retains the common names formic acid and acetic acid, which are always used to refer to these acids.

HCOOH   CH₃COOH

Methanoic acid   Ethanoic acid   3-Methylbutanoic acid
(Formic acid)   (Acetic acid)   (Isovaleric acid)

If the carboxylic acid contains a carbon-carbon double or triple bond, change the infix from -an- to -en- or -yn- to indicate the presence of the multiple bond, and show the location of the multiple bond by a number.

Propenoic acid   *trans*-2-Butenoic acid   *trans*-3-Phenylpropenoic acid
(Acrylic acid)   (Crotonic acid)   (Cinnamic acid)

In the IUPAC system, a carboxyl group takes precedence over most other functional groups (Table 16.1), including hydroxyl groups, amino groups, and the carbonyl groups of aldehydes and ketones. As illustrated in the following examples, an

Formic acid was first obtained in 1670 from the destructive distillation of ants, whose Latin genus is *Formica*. It is one of the components of the venom injected by stinging ants. *(Ted Nelson/Dembinsky Photo Associates)*

—OH group is indicated by the prefix hydroxy-; an —NH$_2$ group, by amino-; and the C=O group of an aldehyde or ketone, by oxo-.

5-Hydroxyhexanoic acid 5-Oxohexanoic acid 4-Aminobutanoic acid

Dicarboxylic acids are named by adding the suffix -dioic acid to the name of the carbon chain that contains both carboxyl groups. The numbers of the carboxyl carbons are not indicated because they can be only at the ends of the parent chain. Following are IUPAC and common names for several important aliphatic dicarboxylic acids.

Ethanedioic acid
(Oxalic acid)

Propanedioic acid
(Malonic acid)

Butanedioic acid
(Succinic acid)

Pentanedioic acid
(Glutaric acid)

Hexanedioic acid
(Adipic acid)

The name "oxalic acid" is derived from one of its sources in the biological world, namely, plants of the genus *Oxalis*, one of which is rhubarb. Adipic acid is one of the two monomers required for the synthesis of the polymer nylon 66 (Section 24.5A).

A carboxylic acid containing a carboxyl group bonded to a cycloalkane ring is named by giving the name of the ring and adding the suffix -carboxylic acid. The atoms of the ring are numbered beginning with the carbon bearing the —COOH group.

2-Cyclohexenecarboxylic
acid

trans-1,3-Cyclopentanedicarboxylic
acid

The simplest aromatic carboxylic acid is benzoic acid. Derivatives are named by using numbers to show the location of substituents relative to the carboxyl group. Certain aromatic carboxylic acids have common names by which they are more usually known. For example, 2-hydroxybenzoic acid is more often called salicylic acid, a name derived from the fact that this aromatic carboxylic acid was first isolated from the bark of the willow, a tree of the genus *Salix*.

| Benzoic acid | 2-Hydroxybenzoic acid (Salicylic acid) | 1,2-Benzenedicarboxylic acid (Phthalic acid) | 1,4-Benzenedicarboxylic acid (Terephthalic acid) |

Aromatic dicarboxylic acids are named by adding the words "dicarboxylic acid" to "benzene," for example, 1,2-benzenedicarboxylic acid and 1,4-benzenedicarboxylic acid. Each is more usually known by its common name: phthalic acid and terephthalic acid respectively. Terephthalic acid is one of the two organic components required for the synthesis of the textile fiber known as Dacron polyester, or Dacron (Section 24.5B).

B. Common Names

Aliphatic carboxylic acids, many of which were known long before the development of structural theory and IUPAC nomenclature, are named according to their source or for some characteristic property. Table 17.1 lists several of the unbranched aliphatic carboxylic acids found in the biological world along with the common name and Latin or Greek derivation of each. Those of 16, 18, and 20 carbon atoms are particularly abundant in fats and oils (Section 26.1) and in the phospholipid components of biological membranes (Section 26.5).

Table 17.1 Several Aliphatic Carboxylic Acids and Their Common Names and Derivations

Structure	IUPAC Name	Common Name	Derivation
$HCOOH$	Methanoic acid	Formic acid	Latin: *formica*, ant
CH_3COOH	Ethanoic acid	Acetic acid	Latin: *acetum*, vinegar
CH_3CH_2COOH	Propanoic acid	Propionic acid	Greek: *propion*, first fat
$CH_3(CH_2)_2COOH$	Butanoic acid	Butyric acid	Latin: *butyrum*, butter
$CH_3(CH_2)_3COOH$	Pentanoic acid	Valeric acid	Latin: *valeriana*, a flowering plant
$CH_3(CH_2)_4COOH$	Hexanoic acid	Caproic acid	Latin: *caper*, goat
$CH_3(CH_2)_6COOH$	Octanoic acid	Caprylic acid	Latin: *caper*, goat
$CH_3(CH_2)_8COOH$	Decanoic acid	Capric acid	Latin: *caper*, goat
$CH_3(CH_2)_{10}COOH$	Dodecanoic acid	Lauric acid	Latin: *laurus*, laurel
$CH_3(CH_2)_{12}COOH$	Tetradecanoic acid	Myristic acid	Greek: *myristikos*, fragrant
$CH_3(CH_2)_{14}COOH$	Hexadecanoic acid	Palmitic acid	Latin: *palma*, palm tree
$CH_3(CH_2)_{16}COOH$	Octadecanoic acid	Stearic acid	Greek: *stear*, solid fat
$CH_3(CH_2)_{18}COOH$	Eicosanoic acid	Arachidic acid	Greek: *arachis*, peanut

When common names are used, the Greek letters α, β, γ, δ, and so forth, are often added as a prefix to locate substituents. The α-position in a carboxylic acid is the one next to the carboxyl group; an α-substituent in a common name is equivalent to a 2-substituent in an IUPAC name. GABA is an inhibitory neurotransmitter in the central nervous system of humans. Alanine is one of the 20 protein-derived amino acids.

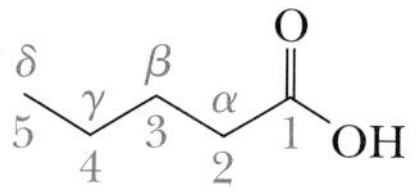

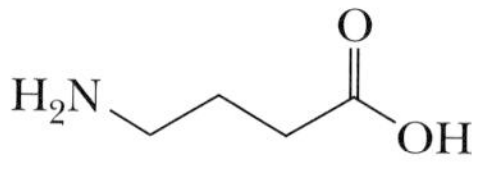

4-Aminobutanoic acid
(γ-Aminobutyric acid, GABA)

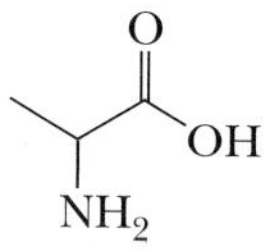

2-Aminopropanoic acid
(α-Aminopropionic acid;
alanine)

In common names, the presence of a ketone carbonyl in a substituted carboxylic acid is indicated by the prefix keto-, illustrated by the common name β-ketobutyric acid. This substituted carboxylic acid is also named acetoacetic acid. In deriving this name, 3-oxobutanoic acid is regarded as a substituted acetic acid. In the common nomenclature, the substituent is named an **aceto group,** $CH_3CO—$.

Aceto group   A $CH_3CO—$ group; also called an acetyl group.

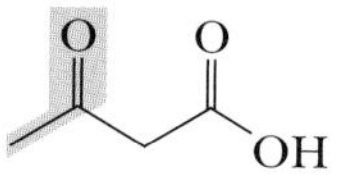

3-Oxobutanoic acid
(β-Ketobutyric acid;
acetoacetic acid)

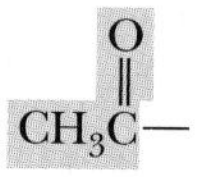

Acetyl group
(aceto group)

Example 17.1

Write IUPAC names for the following carboxylic acids:

(a) 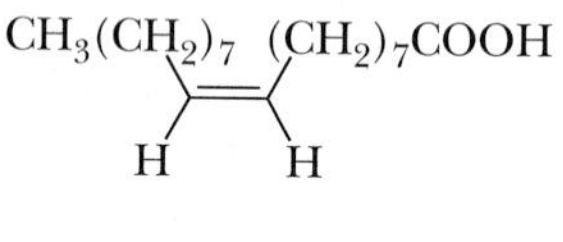 $CH_3(CH_2)_7$ $(CH_2)_7COOH$

(b)

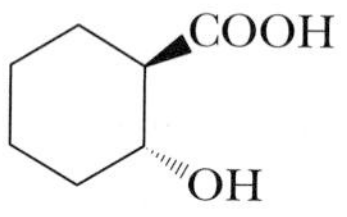

(c)

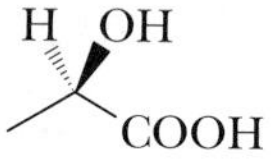

(d)

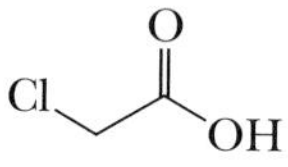

Solution

(a) (Z)-9-Octadecenoic acid (oleic acid)
(b) *trans*-2-Hydroxycyclohexanecarboxylic acid
(c) (R)-2-Hydroxypropanoic acid [(R)-lactic acid]
(d) Chloroacetic acid

Problem 17.1

Each of these carboxylic acids has a well-recognized common name. A derivative of glyceric acid is an intermediate in glycolysis. Maleic acid is an intermediate in the tricarboxylic acid (TCA) cycle. Mevalonic acid is an intermediate in the biosynthesis of steroids. Write the IUPAC name for each compound. Be certain to specify configuration.

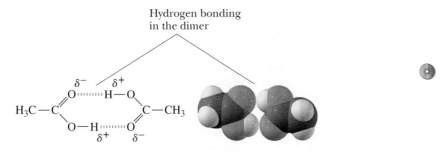

(a) Glyceric acid   (b) Maleic acid   (c) Mevalonic acid

17.3 Physical Properties

In the liquid and solid states, carboxylic acids are associated by hydrogen bonding into dimeric structures as shown for acetic acid in the liquid state.

Hydrogen bonding in the dimer

Carboxylic acids have significantly higher boiling points than other types of organic compounds of comparable molecular weight, such as alcohols, aldehydes, and ketones. For example, butanoic acid (Table 17.2) has a higher boiling point than either 1-pentanol or pentanal. The higher boiling points of carboxylic acids result from their polarity and from the fact that they form very strong hydrogen-bonded dimers.

Carboxylic acids also interact with water molecules by hydrogen bonding through both the carbonyl and hydroxyl groups. Because of greater hydrogen-bonding

Table 17.2 Boiling Points and Solubilities in Water of Selected Carboxylic Acids, Alcohols, and Aldehydes of Comparable Molecular Weight

Structure	Name	Molecular Weight	Boiling Point (°C)	Solubility (g/100 mL H_2O)
CH_3COOH	Acetic acid	60.1	118	Infinite
$CH_3CH_2CH_2OH$	1-Propanol	60.1	97	Infinite
CH_3CH_2CHO	Propanal	58.1	48	16.0
$CH_3(CH_2)_2COOH$	Butanoic acid	88.1	163	Infinite
$CH_3(CH_2)_3CH_2OH$	1-Pentanol	88.1	137	2.3
$CH_3(CH_2)_3CHO$	Pentanal	86.1	103	Slight
$CH_3(CH_2)_4COOH$	Hexanoic acid	116.2	205	1.0
$CH_3(CH_2)_5CH_2OH$	1-Heptanol	116.2	176	0.2
$CH_3(CH_2)_5CHO$	Heptanal	114.1	153	0.1

interactions, carboxylic acids are more soluble in water than alcohols, ethers, aldehydes, and ketones of comparable molecular weight. The solubility of a carboxylic acid in water decreases as its molecular weight increases. We account for this trend in the following way. A carboxylic acid consists of two regions of distinctly different polarity: a polar hydrophilic carboxyl group and, except for formic acid, a nonpolar hydrophobic hydrocarbon chain. The **hydrophilic** carboxyl group increases water solubility; the **hydrophobic** hydrocarbon chain decreases water solubility.

Hydrophilic Water loving.

Hydrophobic Water hating.

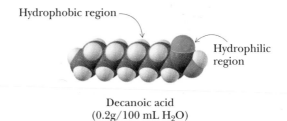

Hydrophobic region

Hydrophilic region

Decanoic acid
(0.2g/100 mL H₂O)

The first four aliphatic carboxylic acids (formic, acetic, propanoic, and butanoic acids) are infinitely soluble in water because the hydrophobic character of the hydrocarbon chain is more than counterbalanced by the hydrophilic character of the carboxyl group. As the size of the hydrocarbon chain increases relative to the size of the hydrophilic group, water solubility decreases. The solubility of hexanoic acid is 1.0 g/100 mL of water. That of decanoic acid is only 0.2 g/100 mL of water.

One other physical property of carboxylic acids must be mentioned. The liquid carboxylic acids from propanoic acid to decanoic acid have extremely foul odors, about as bad as those of thiols, though different. Butanoic acid is found in stale perspiration and is a major component of "locker room odor." Pentanoic acid smells even worse, and goats, which secrete C_6, C_8, and C_{10} acids, are not famous for their pleasant odors.

Supporting Concepts
Acids and Bases

17.4 Acidity

A. Acid Ionization Constants

Carboxylic acids are weak acids. Values of K_a for most unsubstituted aliphatic and aromatic carboxylic acids fall within the range 10^{-4} to 10^{-5}. The value of K_a for acetic acid, for example, is 1.74×10^{-5}. Its pK_a is 4.76.

$$CH_3COOH + H_2O \rightleftharpoons CH_3COO^- + H_3O^+ \qquad K_a = \frac{[CH_3COO^-][H_3O^+]}{[CH_3COOH]} = 1.74 \times 10^{-5}$$

$$pK_a = 4.76$$

As we discussed in Section 4.4C, the greater acidity of carboxylic acids (pK_a 4–5) compared with alcohols (pK_a 16–18) is because resonance stabilizes the carboxylate anion by delocalizing its negative charge. There is no comparable resonance stabilization of alkoxide ions.

We also saw in Section 4.4D that substitution at the α-carbon of an atom or group of atoms of higher electronegativity than carbon further increases the acidity of carboxylic acids by the inductive effect. Compare, for example, the acidity of acetic acid,

pK_a 4.76, and chloroacetic acid, pK_a 2.86. To see the effects of multiple halogen substitution, compare the values of pK_a for acetic acid with its mono-, di-, and trichloroderivatives. A single chlorine substituent increases acid strength by nearly 100.
Trichloroacetic acid, the strongest of the three acids, is a stronger acid than H_3PO_4.

Formula:	CH_3COOH	$ClCH_2COOH$	$Cl_2CHCOOH$	Cl_3CCOOH
Name:	Acetic acid	Chloroacetic acid	Dichloroacetic acid	Trichloroacetic acid
pK_a:	4.76	2.86	1.48	0.70

Increasing acid strength →

We also see an example of the inductive effect in a comparison of the relative
acidities of benzoic acid and acetic acid. Because of the stronger electron-withdrawing inductive effect of the sp^2-hybridized carbon of the benzene ring compared with
the sp^3-hybridized carbon of the methyl group, benzoic acid is a stronger acid than
acetic acid; its K_a is approximately four times that of acetic acid.

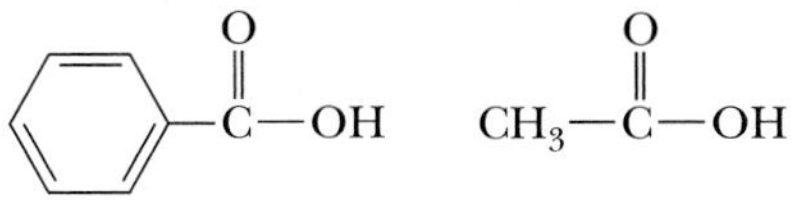

Benzoic acid
pK_a 4.19

Acetic acid
pK_a 4.76

Example 17.2

Which is the stronger acid in each pair?

(a) Benzoic acid   or   4-Nitrobenzoic acid

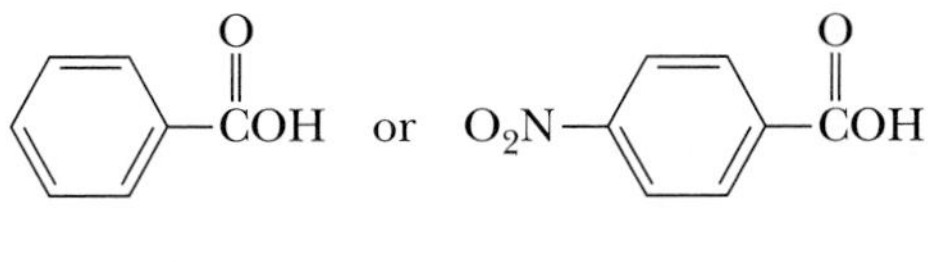

(b) 2-Hydroxypropanoic acid (Lactic acid)   or   Propanoic acid

Solution

(a) 4-Nitrobenzoic acid (pK_a 3.42) is a considerably stronger acid than benzoic acid
   (pK_a 4.19) because of the electron-withdrawing inductive effect of the nitro
   group.
(b) 2-Hydroxypropanoic acid (pK_a 3.08) is a stronger acid than propanoic acid (pK_a
   4.87) because of the electron-withdrawing inductive effect of the adjacent hydroxyl oxygen.

Problem 17.2

Which is the stronger acid in each pair?

(a) CH_3COOH or CH_3SO_3H

Acetic acid   Methanesulfonic
acid

(b) 2-Oxopropanoic acid
(Pyruvic acid)   or   Propanoic acid

B. Reaction with Bases

All carboxylic acids, whether soluble or insoluble in water, react with NaOH, KOH, and other strong bases to form water-soluble salts.

Benzoic acid
(slightly soluble in water)   Sodium benzoate
(60 g/100 mL water)

Sodium benzoate and calcium propanoate are used as preservatives in baked goods. *(Charles D. Winters)*

Sodium benzoate, a fungal growth inhibitor, is often added to baked goods "to retard spoilage." Calcium propanoate is also used for the same purpose. Carboxylic acids also form water-soluble salts with ammonia and amines.

Benzoic acid
(slightly soluble in water)   Ammonium benzoate
(20 g/100 mL water)

As described in Section 4.3, carboxylic acids react with sodium bicarbonate and sodium carbonate to form water-soluble sodium salts and carbonic acid (a weaker acid). Carbonic acid, in turn, decomposes to give water and carbon dioxide, which evolves as a gas.

$$CH_3COOH + NaHCO_3 \longrightarrow CH_3COO^- Na^+ + CO_2 + H_2O$$

Salts of carboxylic acids are named in the same manner as the salts of inorganic acids; the cation is named first and then the anion. The name of the anion is derived from the name of the carboxylic acid by dropping the suffix -ic acid and adding the suffix -ate.

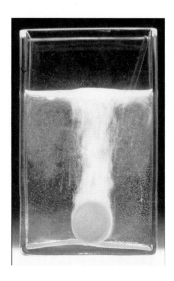

A commercial remedy for excess stomach acid. The bubbles are carbon dioxide, formed from the reaction between citric acid and sodium bicarbonate. *(Charles D. Winters)*

Example 17.3

Complete each acid-base reaction and name the carboxylic salt formed.

(a) COOH + NaOH $\longrightarrow$ **(b)** COOH + $NaHCO_3$ $\longrightarrow$

Solution

Each carboxylic acid is converted to its sodium salt. In (b), carbonic acid is formed; it decomposes to carbon dioxide and water.

(a) ⌒⌒COOH + NaOH ⟶ ⌒⌒COO⁻Na⁺ + H_2O

     Butanoic acid   Sodium butanoate

(b)

        OH   OH

     ⋀COOH   + $NaHCO_3$ ⟶ ⋀COO⁻Na⁺   + H_2O + CO_2

   2-Hydroxypropanoic acid   Sodium 2-hydroxypropanoate
        (Lactic acid)   (Sodium lactate)

Problem 17.3

Write equations for the reaction of each acid in Example 17.3 with ammonia, and name the carboxylic salt formed.

A consequence of the water solubility of carboxylic acid salts is that water-insoluble carboxylic acids can be converted to water-soluble ammonium or alkali metal salts and extracted into aqueous solution. The salt, in turn, can be transformed back to the free carboxylic acid by addition of HCl, H_2SO_4, or other strong acid. These reactions allow an easy separation of carboxylic acids from water-insoluble nonacidic compounds.

Shown in Figure 17.1 is a flow chart for the separation of benzoic acid, a water-insoluble carboxylic acid, from benzyl alcohol, a nonacidic compound. First, the mixture of benzoic acid and benzyl alcohol is dissolved in diethyl ether. When the ether solution is shaken with aqueous NaOH or other strong base, benzoic acid is converted to its water-soluble salt. Then the ether and aqueous phases are separated. The ether solution is distilled, yielding first diethyl ether (bp 35°C) and then benzyl alcohol (bp 205°C). The aqueous solution is acidified with HCl, and benzoic acid precipitates as a crystalline solid (mp 122°C) and is recovered by filtration.

17.5 Preparation of Carboxylic Acids

We have already seen how carboxylic acids are prepared by oxidation of primary alcohols (Section 9.8) and aldehydes (Section 16.13A). We mention here two additional methods, a general one and an important industrial one.

A. Addition of Grignard Reagents to Carbon Dioxide

Treatment of a **Grignard reagent** with carbon dioxide gives the magnesium salt of a carboxylic acid, which, on treatment with aqueous acid, gives a carboxylic acid. Thus, carbonation of a Grignard reagent is a convenient way to convert an alkyl or aryl halide to a carboxylic acid.

            Carbon   Cyclopentane-
            dioxide   carboxylic acid

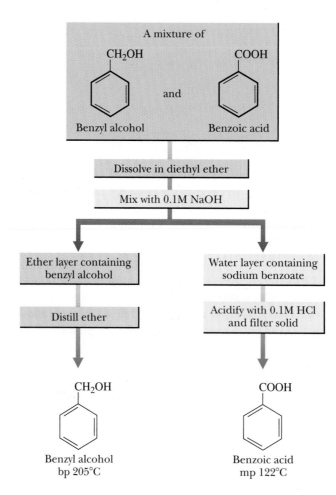

Figure 17.1

Flow chart for separation of benzoic acid from benzyl alcohol.

B. Industrial Synthesis of Acetic Acid — Transition Metal Catalysis

In 1995, acetic acid production in the United States totaled approximately 10^7 kg, a volume that ranked it 19th in the list of all organics manufactured by the U.S. chemical industry. The first industrial synthesis of acetic acid was commercialized in 1916 in Canada and Germany, using acetylene as a feedstock. The process involved two stages: (1) hydration of acetylene to acetaldehyde, followed by (2) oxidation of acetaldehyde to acetic acid by molecular oxygen, catalyzed by cobalt(III) acetate.

$$HC\equiv CH + H_2O \xrightarrow[\text{HgSO}_4]{\text{H}_2\text{SO}_4} [CH_2=CHOH] \longrightarrow CH_3\overset{\overset{\displaystyle O}{\|}}{C}H \xrightarrow[\text{Co}^{3+}]{\text{O}_2} CH_3\overset{\overset{\displaystyle O}{\|}}{C}OH$$

Acetylene   Enol of   Acetaldehyde   Acetic acid
                     acetaldehyde

The technology of producing acetic acid from acetylene is simple, and yields are high; these factors made this the major route to acetic acid for over 50 years. Acetylene was prepared by the reaction of calcium carbide with water. Calcium carbide, in turn, was prepared by heating calcium oxide (from limestone, $CaCO_3$) with coke (from coal) to between 2000 and 2500°C in an electric furnace.

$$CaO \;+\; 2C \xrightarrow{2500°C} CaC_2 \xrightarrow{2H_2O} HC{\equiv}CH \;+\; Ca(OH)_2$$

| Calcium oxide | Calcium carbide | Acetylene |

The preparation of calcium carbide required enormous amounts of energy and, as the cost of energy rose, acetylene ceased to be an economical feedstock from which to manufacture acetic acid.

As an alternative feedstock, chemists and chemical engineers turned to ethylene, already available in huge quantities from the refining of natural gas and petroleum. The process of producing acetic acid from ethylene depends on the fact, known since 1894, that in the presence of catalytic amounts of Pd^{2+} and Cu^{2+} salts, ethylene is oxidized by molecular oxygen to acetaldehyde.

$$2CH_2{=}CH_2 + O_2 \xrightarrow[\text{(Wacker process)}]{Pd^{2+},\,Cu^{2+}} 2CH_3\overset{\displaystyle O}{\overset{\|}{C}}H$$

The first chemical plant to use ethylene oxidation for the manufacture of acetaldehyde was built in Germany by Wacker-Chemie in 1959, and the process itself became known as the **Wacker process.**

In another approach to the synthesis of acetic acid, chemists turned to a route based on carbon monoxide, a readily available raw material. The carbonylation of methanol is exothermic. The challenge was to find a catalytic system that would bring about this reaction.

$$CH_3OH + \boxed{CO} \longrightarrow CH_3\overset{\displaystyle O}{\overset{\|}{C}}OH \qquad \Delta H^0 = -138 \text{ kJ}(33\text{kcal})/\text{mol}$$

In 1973, the Monsanto Company in the United States developed a process for the carbonylation of methanol in the presence of small amounts of soluble rhodium(III) salts, HI, and H_2O. In this process, methanol and CO react continuously in the liquid phase at 150–160°C under a pressure of 30 atm. In the following mechanism, Steps 1 and 2 represent preparation of species for the catalytic cycle. Steps 3 and 4 are the catalytic cycle; in them methanol and carbon monoxide are converted to acetic acid.

Mechanism   Rhodium-Catalyzed Carbonylation of Methanol

Step 1: Reaction of methanol with HI gives methyl iodide.

$$CH_3OH + HI \longrightarrow CH_3I$$

Step 2: Methyl iodide and the rhodium-carbonyl complex undergo oxidative addition to form a complex containing a methyl-rhodium bond.

$$CH_3I + Rh(CO)I_2 \longrightarrow [CH_3{-}Rh(CO)I_3]^{-}$$

A methyl-rhodium carbonyl complex

Step 3:   Carbon monoxide is inserted into the methyl-rhodium bond to give an acyl-rhodium bond.

$$[CH_3-Rh(CO)I_3]^- + CO \longrightarrow [CH_3\overset{\overset{\displaystyle O}{\|}}{C}-Rh(CO)I_3]^-$$

An acyl-rhodium
carbonyl complex

Step 4:   Methanolysis of the acyl-rhodium bond gives acetic acid and regenerates the methyl-rhodium carbonyl complex.

$$[CH_3\overset{\overset{\displaystyle O}{\|}}{C}-Rh(CO)I_3]^- + CH_3OH \longrightarrow CH_3\overset{\overset{\displaystyle O}{\|}}{C}OH + [CH_3-Rh(CO)I_3]^-$$

The **Monsanto process** is based on methanol, which is readily available by catalytic reduction of carbon monoxide. Carbon monoxide and hydrogen, a mixture called **synthesis gas** (Section 2.10C), are in turn available from the reaction of water with methane, coal, and various petroleum products. Synthesis gas is now the major source of both methanol and acetic acid, and it is likely that it will be a major feedstock for the production of other organics in the decades ahead.

Reactivity Explorer
Carboxylic Acids

17.6   Reduction

The carboxyl group is one of the organic functional groups most resistant to reduction. It is not affected by catalytic hydrogenation under conditions that easily reduce aldehydes and ketones to alcohols and that reduce alkenes and alkynes to alkanes. The most common reagent for the reduction of carboxylic acids to primary alcohols is the very powerful reducing agent lithium aluminum hydride (Section 16.14B).

A.   Lithium Aluminum Hydride

Lithium aluminum hydride, $LiAlH_4$ (LAH), reduces a carboxylic acid to a primary alcohol in excellent yield. Reduction is most commonly carried out in diethyl ether or tetrahydrofuran (THF). The initial product is an aluminum alkoxide, which is then treated with water to give the primary alcohol and lithium and aluminum hydroxides. These hydroxides are insoluble in diethyl ether or THF and are removed by filtration. Evaporation of the solvent then yields the primary alcohol.

$$\xrightarrow[\text{2. } H_2O]{\text{1. } LiAlH_4,\ \text{ether}}$$

$$+ \quad LiOH + Al(OH)_3$$

3-Cyclopentene-
carboxylic acid   4-Hydroxymethylcyclopentene

Alkenes are generally not affected by metal hydride reducing reagents. These reagents function as hydride ion donors, that is, as nucleophiles, and alkenes are not attacked by nucleophiles. The less reactive sodium borohydride (Section 16.14B) does not reduce carboxylic acids. In the reduction of a carboxyl group by lithium aluminum hydride, the first hydride ion reacts with the carboxyl hydrogen to give H_2. The three remaining hydride ions are then used for carboxyl reduction.

Following are balanced equations for treatment of a carboxylic acid with LAH to form a tetraalkyl aluminate ion, followed by its hydrolysis in water.

$$4RCOH + 3LiAlH_4 \longrightarrow [(RCH_2O)_4Al]Li + 2LiAlO_2 + 4H_2$$

$$[(RCH_2O)_4Al]Li + 4H_2O \longrightarrow 4RCH_2OH + LiOH + Al(OH)_3$$

In the reduction of a carboxyl group, two hydrogens from $LiAlH_4$ are delivered to the carbonyl group. The hydrogen on the hydroxyl group of the product is provided by water or by aqueous acid during workup. The mechanism of lithium aluminum hydride reduction of carboxyl derivatives in presented in Section 18.11.

B. Selective Reduction of Other Functional Groups

Because carboxyl groups are not affected by the conditions of catalytic hydrogenation, which normally reduce aldehydes, ketones, alkenes, and alkynes, it is possible to selectively reduce these functional groups to alcohols or alkanes in the presence of carboxyl groups.

5-Oxohexanoic acid $+ H_2 \xrightarrow[\text{25°C, 2 atm}]{\text{Pt}}$ 5-Hydroxyhexanoic acid

We saw in Section 16.14B that aldehydes and ketones are reduced to alcohols by both $LiAlH_4$ and $NaBH_4$. Only $LiAlH_4$, however, reduces carboxyl groups. Thus, it is possible to reduce an aldehyde or ketone carbonyl group selectively in the presence of a carboxyl group by using the less reactive $NaBH_4$ as the reducing agent. An example is the selective reduction of the following ketoacid to a hydroxyacid.

5-Oxo-5-phenylpentanoic acid $\xrightarrow[\text{2. } H_2O]{\text{1. } NaBH_4}$ 5-Hydroxy-5-phenylpentanoic acid

17.7 Esterification

$$R\overset{O}{\underset{}{C}}OH + COH \longrightarrow ester$$

A. Fischer Esterification

Esters can be prepared by treating a carboxylic acid with an alcohol in the presence of an acid catalyst, most commonly concentrated sulfuric acid or gaseous HCl. Conversion of a carboxylic acid and alcohol to an ester is given the special name **Fischer esterification** after the German chemist, Emil Fischer (1852–1919), whose name is also firmly established in the chemistry of carbohydrates. As an example of Fischer

Fischer esterification The process of forming an ester by refluxing a carboxylic acid and an alcohol in the presence of an acid catalyst, commonly H_2SO_4 or HCl.

esterification, treatment of acetic acid with ethanol in the presence of concentrated sulfuric acid gives ethyl acetate and water.

These products contain ethyl acetate. *(Charles D. Winters)*

Ethanoic acid	Ethanol	Ethyl ethanoate
(Acetic acid)	(Ethyl alcohol)	(Ethyl acetate)

We study the structure, nomenclature, and reactions of esters in detail in Chapter 18. In this chapter, we discuss only their preparation from carboxylic acids.

Acid-catalyzed esterification is reversible, and generally the quantities of both carboxylic acid and alcohol remaining at equilibrium are appreciable. If, for example, 60.1 g (1.00 mol) of acetic acid and 60.1 g (1.00 mol) of 1-propanol are heated under reflux in the presence of a few drops of concentrated sulfuric acid until equilibrium is reached, the reaction mixture contains approximately 0.67 mol each of propyl acetate and water, and 0.33 mol each of acetic acid and 1-propanol. Thus, at equilibrium, about 67% of the carboxylic acid and alcohol are converted to the desired ester.

	Acetic acid	1-Propanol	Propyl acetate	
Initial:	1.00 mol	1.00 mol	0.00 mol	0.00 mol
Equilibrium:	0.33 mol	0.33 mol	0.67 mol	0.67 mol

By control of reaction conditions, it is possible to use Fischer esterification to prepare esters in high yields. If the alcohol is inexpensive compared with the carboxylic acid, a large excess of it can be used to drive the equilibrium to the right and achieve a high conversion of carboxylic acid to its ester. Alternatively, water can be removed by azeotropic distillation and a Dean-Stark trap (Figure 16.2).

Example 17.4

Complete these Fischer esterifications.

(a) C_6H_5—COH + CH_3OH $\xrightarrow{H^+}$

(b) HO—...—OH + 2 —OH $\xrightarrow{H^+}$

Solution

Following is the structural formula for the ester produced in each reaction.

(a) C_6H_5—COCH$_3$ (b)

Problem 17.4

Complete these Fischer esterifications:

(a) 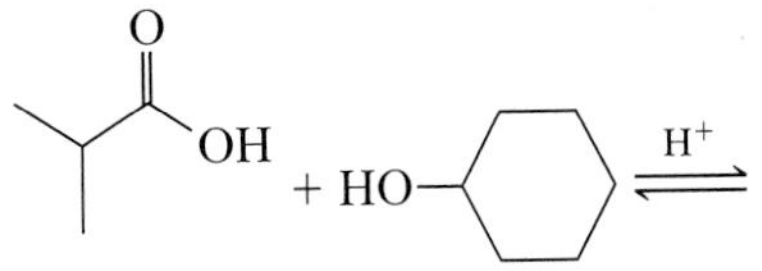
+ HO— ⇌ (with H⁺)

(b) 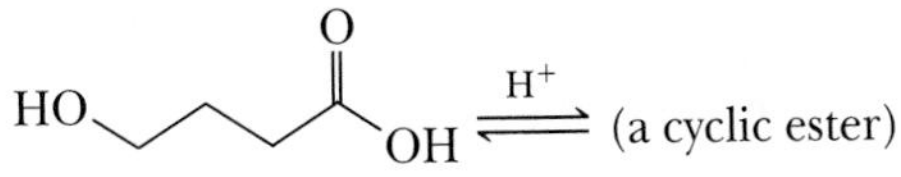 HO⟋⟍⟋ —OH ⇌ (a cyclic ester) (with H⁺)

CHEMISTRY IN ACTION

The Pyrethrins — Natural Insecticides of Plant Origin

Pyrethrum is a natural insecticide obtained from the powdered flower heads of several species of *Chrysanthemum*, particularly *C. cinerariaefolium*. The active substances in pyrethrum, principally pyrethrins I and II, are contact poisons for insects and cold-blooded vertebrates. Because their concentrations in the pyrethrum powder used in chrysanthemum-based insecticides are nontoxic to plants and higher animals, pyrethrum powder has found wide use in household and livestock sprays as well as in dusts for edible plants.

Pyrethrins I and II are esters of chrysanthemic acid. A model of Pyrethrin I is shown as an inset on the cover of this text.

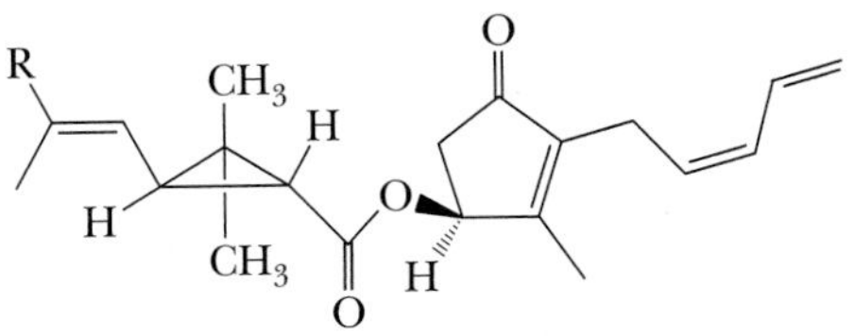

in the environment. In an effort to develop synthetic compounds as effective as these natural insecticides but with a greater biostability, chemists prepared a series of esters related in structure to chrysanthemic acid. Among the synthetic pyrethrenoids now in common use in household and agricultural products are the following.

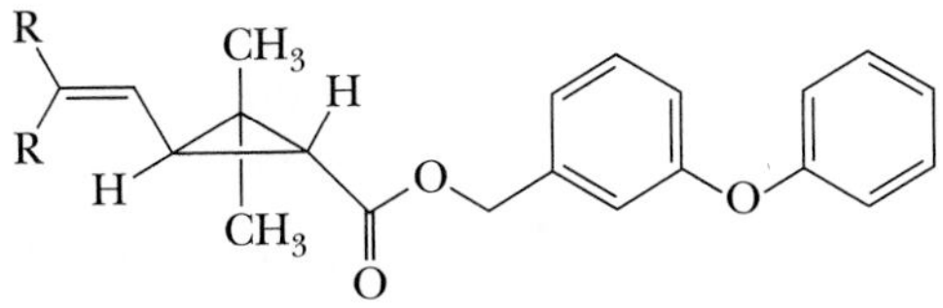

Permethrin R = Cl
Phenothrin R = CH₃

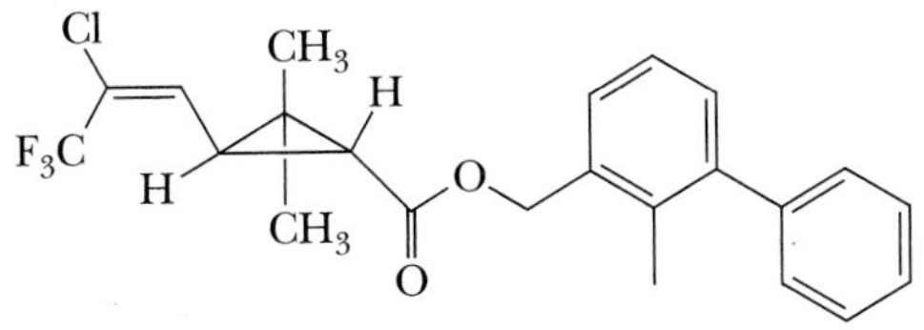

Pyrethrin I R = CH₃
Pyrethrin II R = COOCH₃

Bifenthrin

While pyrethrum powders are effective insecticides, the active substances in them are destroyed rapidly

B. Mechanism of Fischer Esterification

It is important that you understand the mechanism of Fischer esterification thoroughly because it is a model for many other reactions of carboxylic acids presented in this chapter as well as for the reactions of functional derivatives of carboxylic acids

Mechanism   Fischer Esterification

1 Proton transfer from the acid catalyst to the carbonyl oxygen increases the electrophilicity of the carbonyl carbon . . .

2 which is then attacked by the nucleophilic oxygen atom of the alcohol . . .

3 to form an oxonium ion.

4 Proton transfer from the oxonium ion to a second molecule of alcohol . . .

5 gives a tetrahedral carbonyl addition intermediate (TCAI).

6 Proton transfer to one of the —OH groups of the TCAI . . .

7 gives a new oxonium ion.

8 Loss of water from this oxonium ion . . .

9 gives the ester and water.

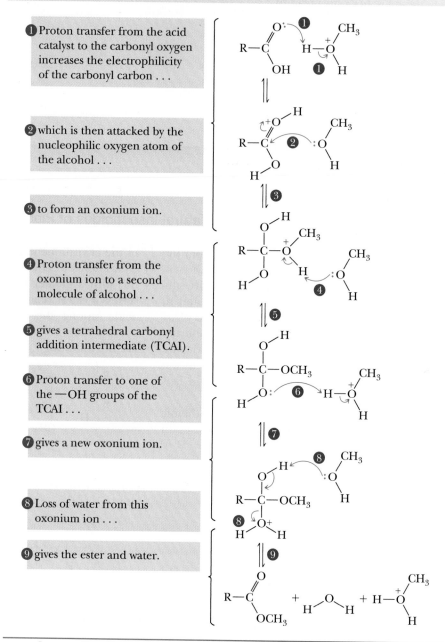

presented in Chapter 18. Steps 1–5 of Fischer esterification are closely analogous to the acid-catalyzed reaction of aldehydes and ketones with alcohols to form hemiacetals (Section 16.8B). Steps 6–9 are closely analogous to acid-catalyzed dehydration of alcohols, except that in Fischer esterification, H^+ is lost from oxygen; in acid-catalyzed dehydration of an alcohol, H^+ is lost from carbon. Loss of H^+ from oxygen

CHEMISTRY IN ACTION

Esters as Flavoring Agents

Flavoring agents are the largest class of food additives. At the present time, over a thousand synthetic and natural flavors are available. The majority of these are concentrates or extracts from the material whose flavor is desired. These flavoring agents are often complex mixtures of tens to hundreds of compounds. A number of flavoring agents, many of them esters, however, are synthesized industrially. Many of these synthetic flavoring agents are major components of the natural flavors and adding only one or a few of them is sufficient to make ice cream, soft drinks, or candy taste naturally flavored.

Structure	Name	Flavor
	Ethyl formate	Rum
	(3-Methyl)butyl acetate (Isopentyl acetate)	Banana
	Octyl acetate	Orange
	Methyl butanoate	Apple
	Ethyl butanoate	Pineapple
	Methyl 2-aminobenzoate (Methyl anthranilate)	Grape

is much easier than from carbon. Note that the overall result of Fischer esterification is nucleophilic substitution, but the mechanism is an addition-elimination sequence and is quite different from either the S_N2 or S_N1 substitution mechanisms we studied in Chapter 8.

C. Formation of Methyl Esters Using Diazomethane

Treatment of a carboxylic acid with diazomethane, usually in ether solution, converts the carboxylic acid under mild conditions and in very high yield to its methyl ester.

$$
\underset{\text{RCOH}}{\overset{\overset{\displaystyle O}{\|}}{}} \; + \; \underset{\text{Diazomethane}}{\text{CH}_2\text{N}_2} \; \xrightarrow{\text{ether}} \; \underset{\text{A methyl ester}}{\overset{\overset{\displaystyle O}{\|}}{\text{RCOCH}_3}} \; + \; \text{N}_2
$$

Diazomethane, a potentially explosive, toxic yellow gas, is best represented as a resonance hybrid of two contributing structures.

$$
\text{H}-\overset{\overset{\displaystyle \text{H}}{|}}{\underset{}{\text{C}}}-\overset{+}{\text{N}}\equiv\text{N}\colon \longleftrightarrow \text{H}-\overset{\overset{\displaystyle \text{H}}{|}}{\underset{}{\text{C}}}=\overset{+}{\text{N}}=\overset{-}{\ddot{\text{N}}}\colon
$$

Diazomethane
(a resonance hybrid of two
important contributing structures)

The reaction of a carboxylic acid with diazomethane occurs in two steps.

Mechanisms
Carboxylic Acids
Formation of a methyl ester
with diazomethane

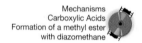

Mechanism   Formation of a Methyl Ester Using Diazomethane

Step 1: Proton transfer from the carboxyl group to diazomethane gives a carboxylate anion and methyldiazonium cation.

$$
\underset{}{\overset{\overset{\displaystyle O}{\|}}{\text{R}-\text{C}-\text{O}-\text{H}}} + \text{:}\bar{\text{C}}\text{H}_2-\overset{+}{\text{N}}\equiv\text{N} \longrightarrow \underset{\substack{\text{A carboxylate} \\ \text{anion}}}{\overset{\overset{\displaystyle O}{\|}}{\text{R}-\text{C}-\text{O}\colon^{-}}} + \underset{\substack{\text{methyldiazonium} \\ \text{cation}}}{\text{CH}_3-\overset{+}{\text{N}}\equiv\text{N}}
$$

Step 2: Nucleophilic displacement of N_2, an extraordinarily good leaving group, gives the methyl ester.

$$
\underset{}{\overset{\overset{\displaystyle O}{\|}}{\text{R}-\text{C}-\text{O}\colon^{-}}} + {}^{+}\text{CH}_3-\overset{+}{\text{N}}\equiv\text{N} \xrightarrow{\text{S}_\text{N}2} \underset{}{\overset{\overset{\displaystyle O}{\|}}{\text{R}-\text{C}-\text{O}-\text{CH}_3}} + \text{N}\equiv\text{N}
$$

Because of the hazards associated with the use of diazomethane, it is used only where other means of preparation of methyl esters are too harsh, and, even then, only in small quantities.

17.8   Conversion to Acid Chlorides

The functional group of an acid halide is a carbonyl group bonded to a halogen atom. Among the acid halides, acid chlorides are the most frequently used in the laboratory and in industrial organic chemistry.

$$\underset{\begin{array}{c}\text{Functional group}\\\text{of an acid halide}\end{array}}{\overset{\overset{\displaystyle O}{\|}}{-C-X}}\qquad\underset{\text{Acetyl chloride}}{\overset{\overset{\displaystyle O}{\|}}{CH_3CCl}}\qquad\underset{\text{Benzoyl chloride}}{\overset{\overset{\displaystyle O}{\|}}{CCl}}$$

We study the nomenclature, structure, and characteristic reactions of acid halides in Chapter 18. Here, we are concerned only with their synthesis from carboxylic acids.

Acid chlorides are most often prepared by the reaction of a carboxylic acid with thionyl chloride, the same reagent used to convert an alcohol to a chloroalkane (Section 9.5C).

$$\underset{\text{Butanoic acid}}{\overset{\overset{\displaystyle O}{\|}}{\diagdown\diagup\diagdown\diagup^{C}\diagdown_{OH}}}\;+\;\underset{\text{Thionyl chloride}}{SOCl_2}\;\longrightarrow\;\underset{\text{Butanoyl chloride}}{\overset{\overset{\displaystyle O}{\|}}{\diagdown\diagup\diagdown\diagup^{C}\diagdown_{Cl}}}\;+\;SO_2\;+\;HCl$$

The mechanism for the reaction of thionyl chloride with a carboxylic acid to form an acid chloride is very similar to that presented in Section 9.5C for the conversion of an alcohol to a haloalkane.

Mechanism   Reaction of a Carboxylic Acid with Thionyl Chloride

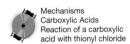

Mechanisms
Carboxylic Acids
Reaction of a carboxylic
acid with thionyl chloride

Step 1: An acyl chlorosulfite ester is formed. The effect of this step is to convert OH^-, a poor leaving group, into a chlorosulfite group, a good leaving group.

$$\overset{\overset{\displaystyle O}{\|}}{R-C}-O-H + \overset{\overset{\displaystyle O}{\|}}{Cl-S}-Cl \longrightarrow \underset{\begin{array}{c}\text{A chlorosulfite}\\\text{group}\end{array}}{R-\overset{\overset{\displaystyle O}{\|}}{C}-\underline{O-\overset{\overset{\displaystyle O}{\|}}{S}-Cl}} + H-Cl$$

Step 2: Attack of chloride ion on the carbonyl carbon of the chlorosulfite ester gives a tetrahedral carbonyl addition intermediate. Its collapse gives the acid chloride, sulfur dioxide, and chloride ion.

$$R-\overset{\overset{\displaystyle O}{\|}}{C}-O-\overset{\overset{\displaystyle O}{\|}}{S}-Cl + :Cl^- \longrightarrow \underset{\begin{array}{c}\text{A tetrahedral carbonyl}\\\text{addition intermediate}\end{array}}{\left[R-\overset{\overset{\displaystyle \overset{-}{O}:}{|}}{\underset{\underset{\displaystyle Cl}{|}}{C}}-O-\overset{\overset{\displaystyle O}{\|}}{S}-Cl \right]} \longrightarrow R-\overset{\overset{\displaystyle O}{\|}}{C}-Cl + SO_2 + :Cl^-$$

Note that, in the second step of this mechanism, chloride ion adds to the carbonyl carbon to form a tetrahedral carbonyl addition intermediate, which then collapses by loss of the chlorosulfite group to give the acid chloride. Addition to a carbonyl carbon to form a tetrahedral carbonyl addition intermediate followed by its collapse is a theme common to a great many reactions of carboxylic acids and their derivatives, and we shall see much more of this theme in Chapters 18 and 19.

Example 17.5

Complete the following equations.

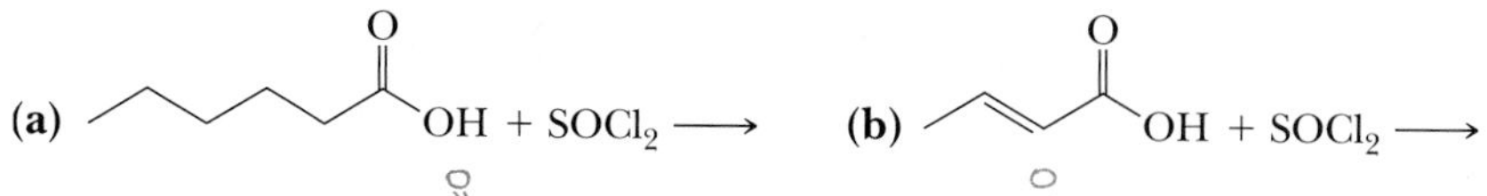

(a)   [structure: pentanoic acid] OH + $SOCl_2$ ⟶   (b)   [structure: unsaturated acid] OH + $SOCl_2$ ⟶

Solution   [structure] Cl + SO_2 + HCl   [structure] Cl + SO_2 + HCl

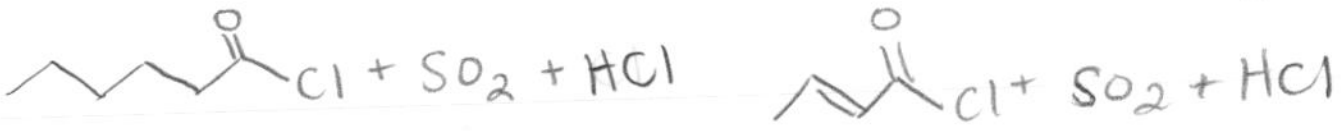

Following are the products of each reaction.

(a)   [structure] Cl + SO_2 + HCl   (b)   [structure] Cl + SO_2 + HCl

Problem 17.5

Complete the following equations.

(a)   [structure with COOH and OCH_3] + $SOCl_2$ ⟶   (b)   [cyclohexanol structure] OH + $SOCl_2$ ⟶

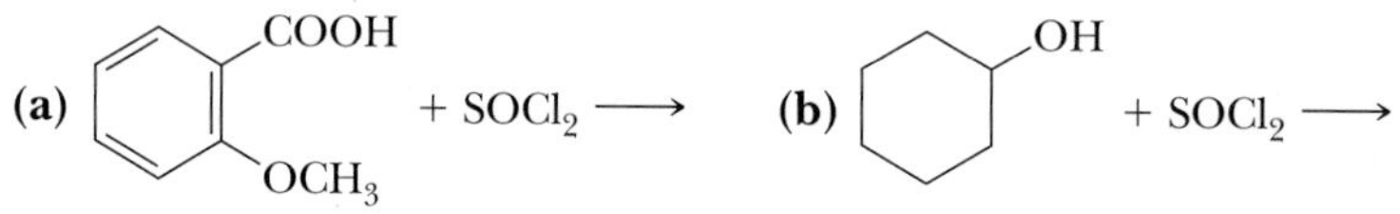

17.9   Decarboxylation

A.   β-Ketoacids

Decarboxylation Loss of CO_2 from a carboxyl group.

Decarboxylation is the loss of CO_2 from the carboxyl group of a molecule. Almost any carboxylic acid, heated to a very high temperature, undergoes thermal decarboxylation.

$$RCOH \xrightarrow[\text{heat}]{\text{decarboxylation}} RH + CO_2$$

Most carboxylic acids, however, are quite resistant to moderate heat and melt or even boil without decarboxylation. Exceptions are carboxylic acids that have a carbonyl group β to the carboxyl group. This type of carboxylic acid undergoes decarboxylation quite readily on mild heating. For example, when 3-oxobutanoic acid is warmed, it undergoes decarboxylation to give acetone and carbon dioxide.

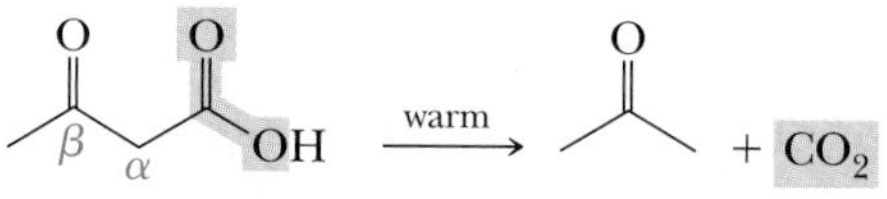

[structure: 3-oxobutanoic acid with β and α labels] OH $\xrightarrow{\text{warm}}$ [acetone structure] + CO_2

3-Oxobutanoic acid   Acetone
(Acetoacetic acid)

Decarboxylation on moderate heating is a unique property of 3-oxocarboxylic acids (β-ketoacids) and is not observed with other classes of ketoacids.

Mechanisms
Carboxylic Acids
Decarboxylation of a
β-Ketocarboxylic acid

Mechanism   Decarboxylation of a β-Ketocarboxylic Acid

Step 1: Redistribution of six electrons in a cyclic six-membered transition state gives carbon dioxide and an enol.

Step 2: Keto-enol tautomerism (Section 16.11B) of the enol gives the more stable keto form of the product.

(A cyclic six-membered transition state)   Enol of a ketone   + CO_2

An important example of decarboxylation of a β-ketoacid in the biological world occurs during the oxidation of foodstuffs in the tricarboxylic acid (TCA) cycle. One of the intermediates in this cycle is oxalosuccinic acid, which undergoes spontaneous decarboxylation to produce α-ketoglutaric acid. Only one of the three carboxyl groups of oxalosuccinic acid has a carbonyl group in the β-position to it, and it is this carboxyl group that is lost as CO_2.

Only this carboxyl has a C=O beta to it.

Oxalosuccinic acid   α-Ketoglutaric acid

B. Malonic Acid and Substituted Malonic Acids

The presence of a ketone or aldehyde carbonyl group β to the carboxyl group is sufficient to facilitate decarboxylation. In the more general reaction, decarboxylation is facilitated by the presence of any carbonyl group at the β-position, including that of a carboxyl group or ester. Malonic acid and substituted malonic acids, for example, undergo thermal decarboxylation, as illustrated by the decarboxylation of malonic acid when it is heated slightly above its melting point of 135–137°C.

$$HOCCH_2COH \xrightarrow{140°-150°C} CH_3COH + CO_2$$

Propanedioic acid
(Malonic acid)

CHEMISTRY IN ACTION

Ketone Bodies and Diabetes Mellitus

3-Oxobutanoic acid (acetoacetic acid) and its reduction product, 3-hydroxybutanoic acid, are synthesized in the liver from acetyl-CoA, a product of the metabolism of fatty acids and certain amino acids. 3-Hydroxybutanoic acid and 3-oxobutanoic acid are known collectively as ketone bodies.

The concentration of ketone bodies in the blood of healthy, well-fed humans is approximately 0.01 mM/L. However, in persons suffering from starvation or diabetes mellitus, the concentration of ketone bodies may increase to as much as 500 times normal. Under these conditions, the concentration of acetoacetic acid increases to the point where it undergoes spontaneous decarboxylation to form acetone and carbon dioxide. Acetone is not metabolized by humans and is excreted through the kidneys

and the lungs. The odor of acetone is responsible for the characteristic "sweet smell" of the breath of severely diabetic patients.

$$CH_3CCH_2COH$$

3-Oxobutanoic acid
(Acetoacetic acid)

$$CH_3CHCH_2COH$$

3-Hydroxybutanoic acid
(β-Hydroxybutyric acid)

The mechanism of decarboxylation of malonic acids is very similar to what we have just seen for the decarboxylation of β-ketoacids. Formation of a cyclic, six-membered transition state involving rearrangement of three electron pairs gives the enol form of a carboxylic acid, which is in turn isomerized to the carboxylic acid.

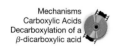

Mechanisms
Carboxylic Acids
Decarboxylation of a
β-dicarboxylic acid

Mechanism   Decarboxylation of a β-Dicarboxylic Acid

Step 1:   Redistribution of six electrons in a cyclic six-membered transition state gives carbon dioxide and the enol form of a carboxyl group.

Step 2:   Keto-enol tautomerism (Section 16.11B) of the enol gives the more stable keto form of the carboxyl group.

A cyclic six-membered
transition state

Enol of a
carboxylic acid

$+ CO_2$

Example 17.6

Each of these carboxylic acids undergoes thermal decarboxylation. Draw a structural formula for the enol intermediate and final product formed in each reaction.

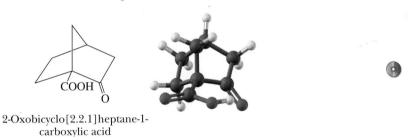

Solution

Following is a structural formula for the enol intermediate and the final product of each decarboxylation.

(a) [structures: OH-cyclohexenol → cyclohexanone]

(b) [structures: cyclobutylidene-C(OH)₂ → cyclobutyl-COOH]

(c) [structures: enol → ketone]

Problem 17.6

Account for the observation that the following β-ketoacid does not undergo thermal decarboxylation. It can be heated for extended periods at temperatures above its melting point without noticeable decomposition.

2-Oxobicyclo[2.2.1]heptane-1-
carboxylic acid

Summary

The functional group of a carboxylic acid (Section 17.1) is the **carboxyl group,** —COOH. IUPAC names of carboxylic acids (Section 17.2A) are derived from the parent alkane by dropping the suffix -e and adding -oic acid. Dicarboxylic acids are named as -dioic acids.

Carboxylic acids are polar compounds (Section 17.3) and, in the liquid and solid states, are associated by hydrogen bonding into dimers. Carboxylic acids have higher boiling points and are more soluble in water than alcohols, aldehydes, ketones, and ethers of comparable molecular weight. A carboxylic acid consists of two regions of distinctly different polarity; a polar **hydrophilic** carboxyl group, which increases solubility in water, and a nonpolar **hydrophobic** hydrocarbon chain, which decreases solubility in water.

The first four aliphatic carboxylic acids are infinitely soluble in water because the hydrophilic carboxyl group more than counterbalances the hydrophobic hydrocarbon chain. As the size of the carbon chain increases, however, the hydrophobic group becomes dominant, and solubility in water decreases.

Values of pK_a for aliphatic carboxylic acids are in the range 4.0–5.0 (Section 17.4A). The greater acidity of carboxylic acids compared with alcohols is explained by greater resonance stabilization of a carboxylate anion relative to an alkoxide anion. Electron-withdrawing substituents near the carboxyl group increase its acidity.

Key Reactions

1. Acidity of Carboxylic Acids (Section 17.4A)

Values of pK_a for most unsubstituted aliphatic and aromatic carboxylic acids are within the range pK_a 4–5. Substitution by electron-withdrawing groups near the carboxyl group decreases its pK_a (increases its acidity).

$$CH_3COH + H_2O \rightleftharpoons CH_3CO^- + H_3O^+ \qquad K_a = 1.74 \times 10^{-5}$$

2. Reaction of Carboxylic Acids with Bases (Section 17.4B)

Carboxylic acids form water-soluble salts with alkali metal hydroxides, carbonates, and bicarbonates, as well as with ammonia and aliphatic and aromatic amines.

$$\text{C}_6\text{H}_5-\text{COOH} + \text{NaOH} \xrightarrow{\text{H}_2\text{O}} \text{C}_6\text{H}_5-\text{COO}^-\text{Na}^+ + \text{H}_2\text{O}$$

$$\text{C}_6\text{H}_5-\text{COOH} + \text{NH}_3 \xrightarrow{\text{H}_2\text{O}} \text{C}_6\text{H}_5-\text{COO}^-\text{NH}_4^+$$

3. Carbonation of a Grignard Reagent (Section 17.5A)

Adding CO_2 to a Grignard reagent followed by acidification provides a useful route to carboxylic acids.

$$\text{C}_5\text{H}_9-\text{Br} \xrightarrow{\text{Mg/ether}} \text{C}_5\text{H}_9-\text{MgBr} \xrightarrow[\text{2. HCl/H}_2\text{O}]{\text{1. CO}_2} \text{C}_5\text{H}_9-\overset{\text{O}}{\underset{}{\text{C}}}-\text{OH} + \text{Mg}^{2+}$$

Cyclopentane-
carboxylic acid

4. Industrial Preparation of Acetic Acid from Carbon Monoxide (Section 17.5B)

$$\text{CO} \xrightarrow[\text{reduction}]{\text{Catalytic}} \text{CH}_3\text{OH} \xrightarrow[\text{HI, H}_2\text{O}]{\text{rhodium(III)}} \text{CH}_3\text{COH}$$

5. Reduction by Lithium Aluminum Hydride (Section 17.6A)

Lithium aluminum hydride reduces a carboxyl group to a primary alcohol.

6. Fischer Esterification (Section 17.7A)

Fischer esterification is reversible. To achieve high yields of ester, it is necessary to force the equilibrium to the right. One way to accomplish this is to use an excess of alcohol; another is to remove water by azeotropic distillation using a Dean-Stark trap.

7. Reaction with Diazomethane (Section 17.7C)

Diazomethane is used to form methyl esters from carboxylic acids. Because diazomethane is explosive and poisonous, it is used only when other means of preparation of methyl esters are not suitable.

8. Conversion to Acid Halides (Section 17.8)

Acid chlorides, the most common and widely used of the acid halides, are prepared by treatment of a carboxylic acid with thionyl chloride.

9. Decarboxylation of β-Ketoacids (Section 17.9A)

The mechanism of decarboxylation involves redistribution of bonding electrons in a cyclic, six-membered transition state.

10. Decarboxylation of β-Dicarboxylic Acids (Section 17.9B)

The mechanism of decarboxylation of a β-dicarboxylic acid is similar to that for decarboxylation of a β-ketoacid.

$$HOCCH_2COH \xrightarrow{\text{heat}} CH_3COH + CO_2$$

Problems

Structure and Nomenclature

17.7 Write the IUPAC name of each compound, showing stereochemistry where relevant.

(a) [cyclohexene ring]—COOH (b) [structure with H, OH] (c) [structure with COOH]

(d) [cyclopentane structure with COOH] (e) [structure] COO⁻NH₄⁺ (f) HO [structure with H, COOH and COOH]

17.8 Draw structural formulas for these carboxylic acids.

(a) Phenylacetic acid
(b) 4-Aminobutanoic acid
(c) 3-Chloro-4-phenylbutanoic acid
(d) Propenoic acid (acrylic acid)
(e) (Z)-3-Hexenedioic acid
(f) 2-Pentynoic acid
(g) 2-Oxocyclohexanecarboxylic acid
(h) 2,2-Dimethylpropanoic acid

17.9 Megatomoic acid, the sex attractant of the female black carpet beetle, has the following structure.

[structure] COOH

(a) What is its IUPAC name?
(b) State the number of stereoisomers possible for this compound.

17.10 Draw structural formulas for these salts.

(a) Sodium benzoate
(b) Lithium acetate
(c) Ammonium acetate
(d) Disodium adipate
(e) Sodium salicylate
(f) Calcium butanoate

17.11 The monopotassium salt of oxalic acid is present in certain leafy vegetables, including rhubarb. Both oxalic acid and its salts are poisonous in high concentrations. Draw the structural formula of monopotassium oxalate.

17.12 Potassium sorbate is added as a preservative to certain foods to prevent bacteria and molds from causing food spoilage and to extend the foods' shelf life. The IUPAC name of potassium sorbate is potassium (2E,4E)-2,4-hexadienoate. Draw the structural formula of potassium sorbate.

17.13 Zinc 10-undecenoate, the zinc salt of 10-undecenoic acid, is used to treat certain fungal infections, particularly *Tinea pedis* (athlete's foot). Draw a structural formula of this zinc salt.

17.14 On a cyclohexane ring, an axial carboxyl group has a conformational energy of 5.9 kJ (1.4 kcal)/mol relative to an equatorial carboxyl group. Consider the equilibrium for the alternative chair conformations of *trans*-1,4-cyclohexanedicarboxylic acid. Draw the less stable chair conformation on the left of the equilibrium arrows and the more stable chair on the right. Calculate ΔG^0 for the equilibrium as written, and calculate the ratio of the more stable chair to the less stable chair at 25°C.

Physical Properties

17.15 Arrange the compounds in each set in order of increasing boiling point.

(a) $CH_3(CH_2)_5COOH$ $CH_3(CH_2)_6CHO$ $CH_3(CH_2)_6CH_2OH$

(b)

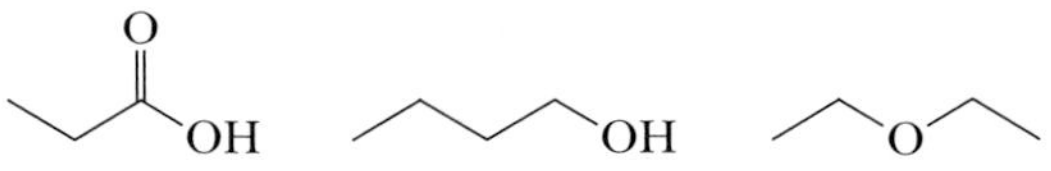

17.16 Acetic acid has a boiling point of 118°C, whereas its methyl ester has a boiling point of 57°C. Account for the fact that the boiling point of acetic acid is higher than that of its methyl ester, even though acetic acid has a lower molecular weight.

Spectroscopy

17.17 Given here are ^{1}H-NMR and ^{13}C-NMR spectral data for nine compounds. Each compound shows strong absorption between 1720 and 1700 cm^{-1}, and strong, broad absorption over the region 2500–3500 cm^{-1}. Propose a structural formula for each compound. Refer to Appendices 3, 4, and 5 for spectral correlation tables.

(a) $C_5H_{10}O_2$

^{1}H-NMR	^{13}C-NMR
0.94 (t, 3H)	180.71
1.39 (m, 2H)	33.89
1.62 (m, 2H)	26.76
2.35 (t, 2H)	22.21
12.0 (s, 1H)	13.69

(b) $C_6H_{12}O_2$

^{1}H-NMR	^{13}C-NMR
1.08 (s, 9H)	179.29
2.23 (s, 2H)	47.82
12.1 (s, 1H)	30.62
	29.57

(c) $C_5H_8O_4$

^{1}H-NMR	^{13}C-NMR
0.93 (t, 3H)	170.94
1.80 (m, 2H)	53.28
3.10 (t, 1H)	21.90
12.7 (s, 2H)	11.81

(d) $C_5H_8O_4$

^{1}H-NMR	^{13}C-NMR
1.29 (s, 6H)	174.01
12.8 (s, 2H)	48.77
	22.56

(e) $C_4H_6O_2$

^{1}H-NMR	^{13}C-NMR
1.91 (d, 3H)	172.26
5.86 (d, 1H)	147.53
7.10 (m, 1H)	122.24
12.4 (s, 1H)	18.11

(f) $C_3H_4Cl_2O_2$

^{1}H-NMR	^{13}C-NMR
2.34 (s, 3H)	171.82
11.3 (s, 1H)	79.36
	34.02

(g) $C_5H_8Cl_2O_2$

^{1}H-NMR	^{13}C-NMR
1.42 (s, 6H)	180.15
6.10 (s, 1H)	77.78
12.4 (s, 1H)	51.88
	20.71

(h) $C_5H_9BrO_2$

^{1}H-NMR	^{13}C-NMR
0.97 (t, 3H)	176.36
1.50 (m, 2H)	45.08
2.05 (m, 2H)	36.49
4.25 (t, 1H)	20.48
12.1 (s, 1H)	13.24

(i) $C_4H_8O_3$

^{1}H-NMR	^{13}C-NMR
2.62 (t, 2H)	177.33
3.38 (s, 3H)	67.55
3.68 (s, 2H)	58.72
11.5 (s, 1H)	34.75

Preparation of Carboxylic Acids

17.18 Complete these reactions.

(a) 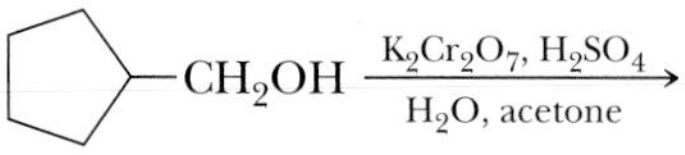 cyclopentyl—CH_2OH $\xrightarrow[\text{H}_2\text{O, acetone}]{\text{K}_2\text{Cr}_2\text{O}_7,\ \text{H}_2\text{SO}_4}$

(b) 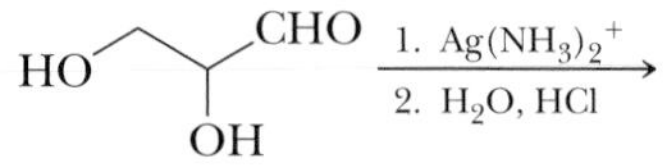 HO—CH(OH)—CHO $\xrightarrow[\text{2. H}_2\text{O, HCl}]{\text{1. Ag(NH}_3)_2{}^+}$

(c) 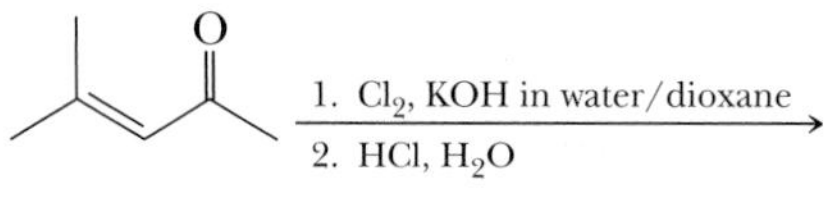 $\xrightarrow[\text{2. HCl, H}_2\text{O}]{\text{1. Cl}_2,\ \text{KOH in water/dioxane}}$

(d) 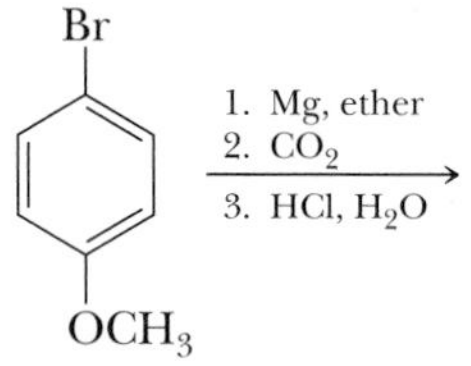 $\xrightarrow[\text{3. HCl, H}_2\text{O}]{\begin{array}{l}\text{1. Mg, ether}\\\text{2. CO}_2\end{array}}$

17.19 Show how to bring about each conversion in good yield.

Grignard

(a) 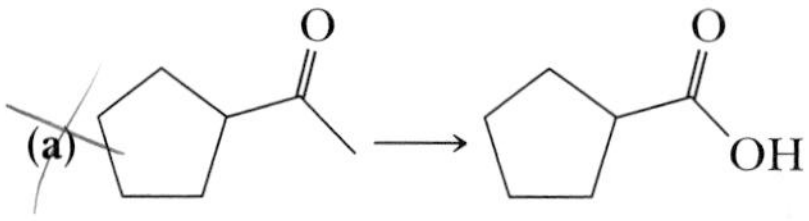 **(b)** 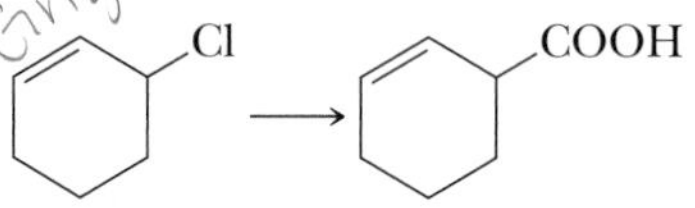   —COOH

(c) 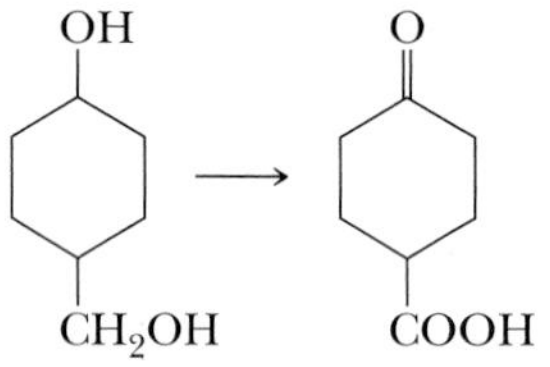 **(d)**  C_6H_5—CH$_2$CH$_2$OH $\longrightarrow$ C_6H_5—CH$_2$COOH

17.20 Show how to prepare pentanoic acid from these compounds.

(a) 1-Pentanol **(b)** Pentanal **(c)** 1-Pentene **(d)** 1-Butanol
(e) 1-Bromopropane **(f)** 2-Hexanone **(g)** 1-Hexene

17.21 Draw the structural formula of a compound of the given molecular formula that, on oxidation by potassium dichromate in aqueous sulfuric acid, gives the carboxylic acid or dicarboxylic acid shown.

(a) 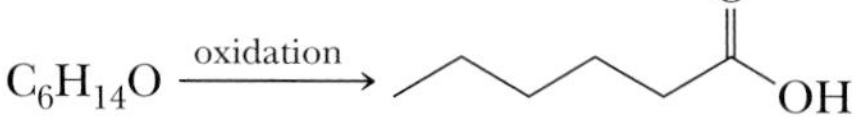 $C_6H_{14}O$ $\xrightarrow{\text{oxidation}}$

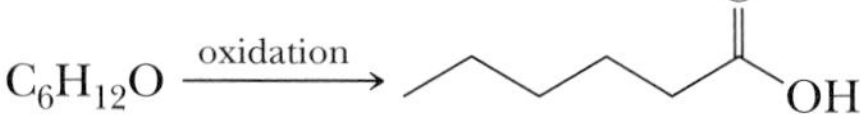

(b) $C_6H_{12}O$ $\xrightarrow{\text{oxidation}}$

(c) $C_6H_{14}O_2$ $\xrightarrow{\text{oxidation}}$

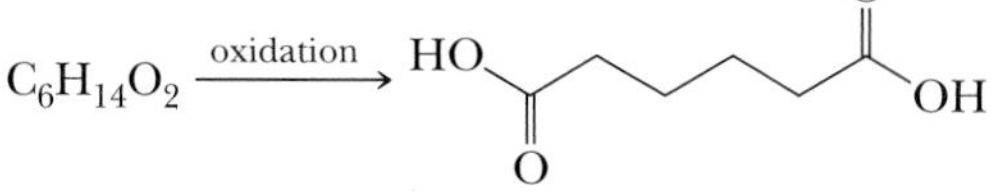

17.22 Show the reagents and experimental conditions necessary to bring about each conversion in good yield.

(a) [cyclopentane-OH] $\longrightarrow$ [cyclopentane-COOH]

(b) $CH_3\overset{\overset{\displaystyle CH_3}{|}}{\underset{\underset{\displaystyle CH_3}{|}}{C}}OH \longrightarrow CH_3\overset{\overset{\displaystyle CH_3}{|}}{\underset{\underset{\displaystyle CH_3}{|}}{C}}COOH$

(c) $CH_3\overset{\overset{\displaystyle CH_3}{|}}{\underset{\underset{\displaystyle CH_3}{|}}{C}}OH \longrightarrow CH_3\overset{\overset{\displaystyle CH_3}{|}}{\underset{}{C}}HCOOH$

(d) $CH_3\overset{\overset{\displaystyle CH_3}{|}}{\underset{\underset{\displaystyle CH_3}{|}}{C}}OH \longrightarrow CH_3\overset{\overset{\displaystyle CH_3}{|}}{\underset{}{C}}HCH_2COOH$

(e) $CH_3CH{=}CHCH_3 \longrightarrow CH_3CH{=}CHCH_2COOH$

17.23 Succinic acid can be synthesized by the following series of reactions from acetylene. Show the reagents and experiential conditions necessary to carry out this synthesis.

[reaction scheme: acetylene $\longrightarrow$ 2-butyne-1,4-diol $\longrightarrow$ 1,4-butanediol $\longrightarrow$ butanedioic acid]

Acetylene   2-Butyne-1,4-diol   1,4-Butanediol   Butanedioic acid
(Succinic acid)

17.24 The reaction of an α-diketone with concentrated sodium or potassium hydroxide to give the salt of an α-hydroxyacid is given the general name benzil-benzilic acid rearrangement. It is illustrated by the conversion of benzil to sodium benzilate and then to benzilic acid.

$$Ph{-}\overset{\overset{\displaystyle O}{||}}{C}{-}\overset{\overset{\displaystyle O}{||}}{C}{-}Ph + NaOH \xrightarrow{\ H_2O\ } Ph{-}\overset{\overset{\displaystyle HO}{|}}{\underset{\underset{\displaystyle Ph}{|}}{C}}{-}\overset{\overset{\displaystyle O}{||}}{C}{-}O^-Na^+ \xrightarrow[H_2O]{HCl} Ph{-}\overset{\overset{\displaystyle HO}{|}}{\underset{\underset{\displaystyle Ph}{|}}{C}}{-}\overset{\overset{\displaystyle O}{||}}{C}{-}OH$$

Benzil   Sodium benzilate   Benzilic acid
(an α-diketone)

Propose a mechanism for the rearrangement of benzil to sodium benzilate.

Acidity of Carboxylic Acids

17.25 Select the stronger acid in each set.

(a) Phenol (pK_a 9.95) and benzoic acid (pK_a 4.17)
(b) Lactic acid (K_a 8.4 × 10^{-4}) and ascorbic acid (K_a 7.9 × 10^{-5})

17.26 Assign the acid in each set its appropriate pK_a.

(a) [benzene-COOH]   and   [benzene-SO$_3$H]   (pK_a 4.19 and 0.70)

(b) $CH_3\overset{\overset{\displaystyle O}{||}}{C}CH_2COOH$ and $CH_3\overset{\overset{\displaystyle O}{||}}{C}COOH$ (pK_a 3.58 and 2.49)

(c) CH_3CH_2COOH and $N{\equiv}CCH_2COOH$ (pK_a 4.78 and 2.45)

17.27 Low-molecular-weight dicarboxylic acids normally exhibit two different pK_a values. Ionization of the first carboxyl group is easier than the second. This effect diminishes with molecular size, and, for adipic acid and longer chain dicarboxylic acids, the two acid ionization constants differ by about one pK unit.

Dicarboxylic Acid	Structural Formula	pK_{a1}	pK_{a2}
Oxalic	HOOCCOOH	1.23	4.19
Malonic	$HOOCCH_2COOH$	2.83	5.69
Succinic	$HOOC(CH_2)_2COOH$	4.16	5.61
Glutaric	$HOOC(CH_2)_3COOH$	4.31	5.41
Adipic	$HOOC(CH_2)_4COOH$	4.43	5.41

Why do the two pK_a values differ more for the shorter chain dicarboxylic acids than for the longer chain dicarboxylic acids?

17.28 Complete the following acid-base reactions.

(a) —CH_2COOH + NaOH $\longrightarrow$ (b) CH_3CH=$CHCH_2COOH$ + $NaHCO_3$ $\longrightarrow$

(c) + $NaHCO_3$ $\longrightarrow$ (d) $CH_3\overset{OH}{\underset{|}{C}}HCOOH$ + $H_2NCH_2CH_2OH$ $\longrightarrow$

(e) CH_3CH=$CHCH_2COO^-$ Na^+ + HCl $\longrightarrow$
(f) $CH_3CH_2CH_2CH_2Li$ + CH_3COOH $\longrightarrow$
(g) $CH_3CH_2CH_2CH_2MgBr$ + CH_3CH_2OH $\longrightarrow$

17.29 The normal pH range for blood plasma is 7.35–7.45. Under these conditions, would you expect the carboxyl group of lactic acid (pK_a 3.07) to exist primarily as a carboxyl group or as a carboxylic anion? Explain.

17.30 The K_{a1} of ascorbic acid is 7.94×10^{-5}. Would you expect ascorbic acid dissolved in blood plasma to exist primarily as ascorbic acid or as ascorbate anion? Explain.

17.31 Excess ascorbic acid is excreted in the urine, the pH of which is normally in the range 4.8–8.4. What form of ascorbic acid would you expect to be present in urine of pH 8.4, free ascorbic acid or ascorbate anion? Explain.

Reactions of Carboxylic Acids

17.32 Give the expected organic product when phenylacetic acid, $PhCH_2COOH$, is treated with each reagent.

(a) $SOCl_2$ (b) $NaHCO_3$, H_2O
(c) NaOH, H_2O (d) CH_3MgBr (1 equivalent)
(e) $LiAlH_4$ followed by H_2O (f) CH_2N_2
(g) CH_3OH + H_2SO_4 (catalyst)

17.33 Show how to convert *trans*-3-phenyl-2-propenoic acid (cinnamic acid) to these compounds.

(a) C_6H_5 ⌣⌣ OH   (b) C_6H_5 ⌣⌣ OH   (c) C_6H_5 ⌣⌣ OH

17.34 Show how to convert 3-oxobutanoic acid (acetoacetic acid) to these compounds.

(a)   (b)   (c)

17.35 Complete these examples of Fischer esterification. Assume that the alcohol is present in excess.

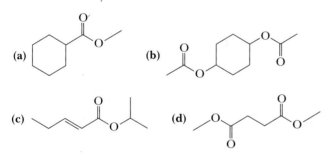

(a)   OH + HO $\xrightarrow{H^+}$

(b)   COOH / COOH + CH_3OH $\xrightarrow{H^+}$

(c) HO   OH +   OH $\xrightarrow{H^+}$

17.36 Benzocaine, a topical anesthetic, is prepared by treatment of 4-aminobenzoic acid with ethanol in the presence of an acid catalyst followed by neutralization. Draw the structural formula of benzocaine.

17.37 Name the carboxylic acid and alcohol from which each ester is derived.

(a)   (b)   (c)   (d)

17.38 When 4-hydroxybutanoic acid is treated with an acid catalyst, it forms a lactone (a cyclic ester). Draw the structural formula of this lactone, and propose a mechanism for its formation.

17.39 Fischer esterification cannot be used to prepare *tert*-butyl esters. Instead, carboxylic acids are treated with 2-methylpropene and an acidic catalyst to generate them.

(a) Why does the Fischer esterification fail for the synthesis of *tert*-butyl esters?

(b) Propose a mechanism for the 2-methylpropene method.

$$\underset{\text{RCOH}}{\overset{O}{\|}} + \underset{\underset{\text{2-Methylpropene}}{\text{(Isobutylene)}}}{CH_2{=}\overset{CH_3}{\underset{}{C}}CH_3} \xrightarrow{H^+} \underset{\underset{\underset{\text{A }\textit{tert}\text{-butyl ester}}{}}{CH_3}}{\overset{O}{\underset{\|}{RCO}}\overset{CH_3}{\underset{}{C}}CH_3}$$

2-Methylpropene (Isobutylene)   A *tert*-butyl ester

17.40 Draw the product formed on thermal decarboxylation of each compound.

(a) $C_6H_5\overset{\overset{\text{O}}{\|}}{C}CH_2COOH$ (b) $C_6H_5CH_2\overset{\overset{\text{COOH}}{|}}{C}HCOOH$ (c) [cyclopentane ring bearing $\overset{\overset{\text{O}}{\|}}{C}CH_3$ and $COOH$ on the same carbon]

17.41 When heated, carboxylic salts in which there is a good leaving group on the carbon beta to the carboxylate group undergo decarboxylation/elimination to give an alkene. Propose a mechanism for this type of decarboxylation/elimination. Compare the mechanism of these decarboxylations with the mechanism for decarboxylation of β-ketoacids; in what way(s) are the mechanisms similar?

(a) $Br-CH_2-\overset{\overset{\text{CH}_3}{|}}{\underset{\underset{\text{CH}_3}{|}}{C}}-COO^-Na^+ \xrightarrow{\text{heat}} CH_2{=}C(CH_3)_2 + CO_2 + Na^+Br^-$

(b) [benzene ring]$-\overset{\underset{\underset{\text{Br}}{|}}{}}{C}H-\overset{\underset{\underset{\text{Br}}{|}}{}}{C}H-COO^-Na^+ \xrightarrow{\text{heat}}$ [benzene ring]$-CH{=}CHBr + CO_2 + Na^+Br^-$

17.42 A mnemonic phrase for remembering the common names for the dicarboxylic acids oxalic through adipic is "Oh my, such good apples." Explain.

17.43 Show how cyclohexanecarboxylic acid could be synthesized from cyclohexane in good yield.

[cyclohexane ring]$-COOH \Longrightarrow$ [cyclohexane ring]

FUNCTIONAL DERIVATIVES OF CARBOXYLIC ACIDS

I n this chapter, we study five classes of organic compounds, each related to the carboxyl group: acid halides, acid anhydrides, esters, amides, and nitriles. Under the general formula of each functional group is a drawing to help you see how it is formally related to a carboxylic acid. Loss of —OH from a carboxyl group and H— from H—Cl, for example, gives an acid chloride. Similarly, loss of —OH from a carboxyl group and H— from ammonia gives an amide.

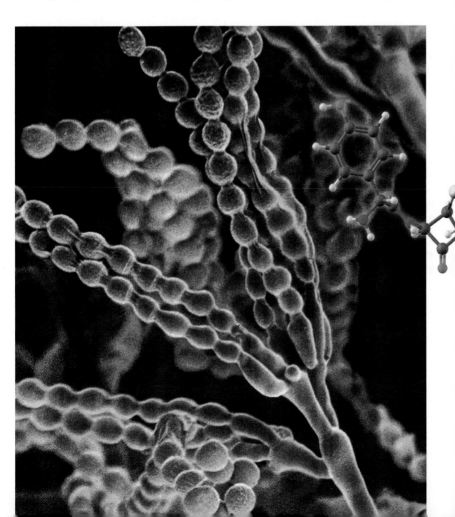

■ Colored scanning electron micrograph of *Penicillium s.* fungus. The stalk-like objects are condiophores to which are attached numerous round condia. The condia are the fruiting bodies of the fungus. *(SCIMAT/Science Source/Photo Researchers, Inc.)* Inset: A model of amoxicillin. See the Chemistry in Action box "The Penicillins and Cephalosporins— β-Lactam Antibiotics."

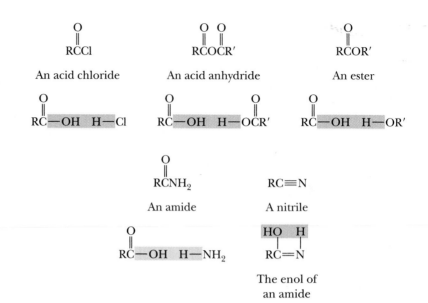

18.1 Structure and Nomenclature

A. Acid Halides

Acyl group An RCO— or
ArCO— group.

The functional group of an **acid halide** (acyl halide) is an **acyl group** (RCO—)
bonded to a halogen atom. Acid chlorides are the most common acid halides. Acid
halides are named by changing the suffix -ic acid in the name of the parent carboxylic acid to -yl halide.

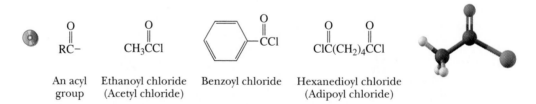

| An acyl group | Ethanoyl chloride (Acetyl chloride) | Benzoyl chloride | Hexanedioyl chloride (Adipoyl chloride) |

Replacement of —OH in a sulfonic acid by chlorine gives a derivative called a
sulfonyl chloride. Following are structural formulas for two sulfonic acids and the
acid chloride derived from each.

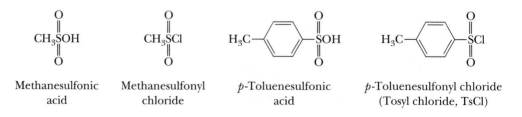

| Methanesulfonic acid | Methanesulfonyl chloride | p-Toluenesulfonic acid | p-Toluenesulfonyl chloride (Tosyl chloride, TsCl) |

B. Acid Anhydrides

Carboxylic Anhydrides

The functional group of a **carboxylic anhydride** is two acyl groups bonded to an oxygen atom. These compounds are called acid anhydrides because they are formally derived from two carboxylic acid molecules by loss of water. The anhydride may be symmetrical (two identical acyl groups), or it may be mixed (two different acyl groups). Anhydrides are named by replacing the word "acid" in the name of the parent carboxylic acid with the word "anhydride."

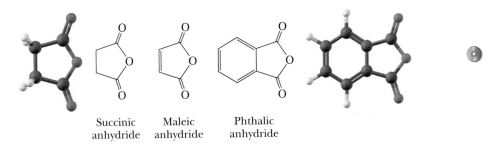

Acetic anhydride   Benzoic anhydride

Cyclic anhydrides are named from the dicarboxylic acids from which they are derived. Here are the cyclic anhydrides derived from succinic acid, maleic acid, and phthalic acid.

Succinic anhydride   Maleic anhydride   Phthalic anhydride

Phosphoric Anhydrides

Because of the special importance of anhydrides of phosphoric acid in biochemical systems (Chapters 28 and 29), we include them here to show their similarity with the anhydrides of carboxylic acids. The functional group of a **phosphoric anhydride** is two phosphoryl groups bonded to an oxygen atom. Here are structural formulas for two anhydrides of phosphoric acid and the ions derived by ionization of each acidic hydrogen.

Diphosphoric acid
(Pyrophosphoric acid)

Diphosphate ion
(Pyrophosphate ion)

Triphosphoric acid

Triphosphate ion

C. Esters

Esters of Carboxylic Acids

The functional group of a **carboxylic ester** is an acyl group bonded to —OR or —OAr. Both IUPAC and common names of esters are derived from the names of the parent carboxylic acids. The alkyl or aryl group bonded to oxygen is named first, followed by the name of the acid in which the suffix -ic acid is replaced by the suffix -ate.

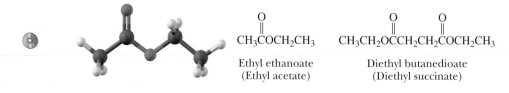

$$CH_3COCH_2CH_3$$

Ethyl ethanoate
(Ethyl acetate)

$$CH_3CH_2OCCH_2CH_2COCH_2CH_3$$

Diethyl butanedioate
(Diethyl succinate)

Lactones — Cyclic Esters

Lactone A cyclic ester.

Cyclic esters are called **lactones.** The IUPAC system has developed a set of rules for naming these compounds. Nonetheless, the simplest lactones are still named by dropping the suffix -ic or -oic acid from the name of the parent carboxylic acid and adding the suffix -olactone. The location of the oxygen atom in the ring is indicated by a number, if the IUPAC name of the acid is used, or by a Greek letter α, β, γ, δ, ε, and so forth, if the common name of the acid is used.

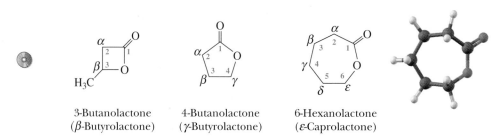

3-Butanolactone
(β-Butyrolactone)

4-Butanolactone
(γ-Butyrolactone)

6-Hexanolactone
(ε-Caprolactone)

Esters of Phosphoric Acid

Phosphoric acid has three —OH groups and forms mono-, di-, and triesters, which are named by giving the name(s) of the alkyl or aryl group(s) attached to oxygen followed by the word "phosphate," as for example dimethyl phosphate. In more complex phosphoric esters, it is common to name the organic molecule and then indicate the presence of the phosphoric ester using either the word "phosphate" or the prefix phospho-. On the right are two phosphoric esters, each of special importance in the biological world. The first is an intermediate in glycolysis, the metabolic pathway by which glucose is converted to pyruvate (Section 29.6). Pyridoxal phosphate is one of the metabolically active forms of vitamin B_6. Each of these esters is shown as it is ionized at pH 7.4, the pH of blood plasma; the two hydroxyl groups of these phosphoryl groups are ionized giving each a charge of -2.

From Moldy Clover to a Blood Thinner

In 1933, a disgruntled farmer delivered a bale of moldy clover, a pail of unclotted blood, and a dead cow to the laboratory of Dr. Carl Link at the University of Wisconsin. Six years and many bales of moldy clover later, Link and his collaborators isolated the anticoagulant dicoumarol, a substance that delays or prevents blood clotting. When cows are fed moldy clover, they ingest dicoumarol, their blood clotting is inhibited, and they bleed to death from minor cuts and scratches. Dicoumarol exerts its anticoagulation effect by interfering with vitamin K activity (Section 26.6D). Within a few years after its discovery, dicoumarol became widely used to treat victims of heart attack and others at risk for developing blood clots.

The powerful anticoagulant dicoumarol was first isolated from moldy clover. *(Grant Heilman/Grant Heilman Photography, Inc.)*

Coumarin
(from sweet clover)

as sweet clover
becomes moldy

Dicoumarol
(an anticoagulant)

Dicoumarol is a derivative of coumarin, a lactone that gives sweet clover its pleasant smell. Coumarin, which does not interfere with blood clotting, is converted to dicoumarol as sweet clover becomes moldy.

In a search for even more potent anticoagulants, Link developed warfarin (named for the Wisconsin Alumni Research Foundation), now used primarily as a rat poison. When rats consume it, their blood fails to clot, and they bleed to death. Warfarin is also used as a blood anticoagulant in humans. The *S* enantiomer shown here is more active than the *R* enantiomer. The commercial product is sold as a racemic mixture. The synthesis of racemic warfarin is described in Problem 19.60. (See *The Merck Index*, 12th ed., #10174.)

Warfarin
(A synthetic anticoagulant)

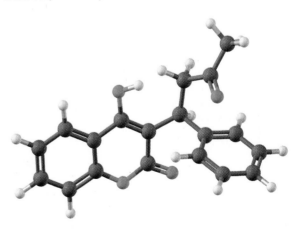

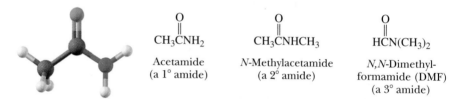

Dimethyl
phosphate

Glyceraldehyde
3-phosphate

Pyridoxal phosphate

D. Amides and Imides of Carboxylic Acids

The functional group of an **amide** is an acyl group bonded to a nitrogen atom. Amides are named by dropping the suffix -oic acid from the IUPAC name of the parent acid, or -ic acid from its common name, and adding -amide. If the nitrogen atom of an amide is bonded to an alkyl or aryl group, the group is named, and its location on nitrogen is indicated by N-. Two alkyl or aryl groups on nitrogen are indicated by N,N-di-. N,N-Dimethylformamide (DMF) is a widely used polar aprotic solvent (Section 8.2).

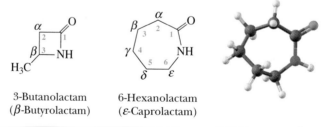

$$CH_3\overset{\overset{\displaystyle O}{\|}}{C}NH_2$$

Acetamide
(a 1° amide)

$$CH_3\overset{\overset{\displaystyle O}{\|}}{C}NHCH_3$$

N-Methylacetamide
(a 2° amide)

$$H\overset{\overset{\displaystyle O}{\|}}{C}N(CH_3)_2$$

N,N-Dimethyl-
formamide (DMF)
(a 3° amide)

Lactam A cyclic amide.

Cyclic amides are given the special name **lactam.** Their names are derived in a manner similar to those of lactones, with the difference that the suffix -lactone is replaced by -lactam.

3-Butanolactam
(β-Butyrolactam)

6-Hexanolactam
(ε-Caprolactam)

Imide A functional group in which two acyl groups, RCO— or ArCO—, are bonded to a nitrogen atom.

The functional group of an **imide** is two acyl groups bonded to nitrogen. Both succinimide and phthalimide are cyclic imides.

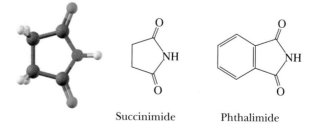

Succinimide

Phthalimide

Example 18.1

Write IUPAC names for these compounds.

(a) [structure with OCH₃] *methyl-3methylbutanoate*

(b) [structure with OC₂H₅] *ethyl-3-oxobutanoate / ethyl-β-ketobutyrate*

(c) H₂N [structure] NH₂ *hexanediamide*

(d) Ph [anhydride structure] Ph

Solution

Given first are IUPAC names and then, in parentheses, common names.

(a) Methyl 3-methylbutanoate (methyl isovalerate, from isovaleric acid).
(b) Ethyl 3-oxobutanoate (ethyl β-ketobutyrate, from β-ketobutyric acid). It is also named ethyl acetoacetate (Section 19.6).
(c) Hexanediamide (adipamide, from adipic acid).
(d) Phenylethanoic anhydride (phenylacetic anhydride, from phenylacetic acid).

Problem 18.1

Draw structural formulas for these compounds.

(a) *N*-Cyclohexylacetamide   (b) 1-Methylpropyl methanoate
(c) Cyclobutyl butanoate   (d) *N*-(1-Methylheptyl)succinate
(e) Diethyl adipate   (f) 2-Aminopropanamide

E.   Nitriles

The functional group of a **nitrile** is a cyano (C≡N) group bonded to a carbon atom. IUPAC names follow the pattern alkanenitrile: for example, ethanenitrile. Common names are derived by dropping the suffix -ic or -oic acid from the name of the parent carboxylic acid and adding the suffix -onitrile.

Nitrile   A compound containing a —C≡N (cyano) group bonded to a carbon atom.

CH₃C≡N

[benzene ring]—C≡N

[benzene ring]—CH₂C≡N

Ethanenitrile
(Acetonitrile)

Benzonitrile

Phenylethanenitrile
(Phenylacetonitrile)

18.2   Acidity of Amides, Imides, and Sulfonamides

Following are structural formulas of a primary amide, a sulfonamide, and two cyclic imides, along with pK_a values for each.

C H E M I S T R Y   I N   A C T I O N

The Penicillins and Cephalosporins — β-Lactam Antibiotics

The penicillins were discovered in 1928 by the Scottish bacteriologist Sir Alexander Fleming. As a result of the brilliant experimental work of Sir Howard Florey, an Australian pathologist, and Ernst Chain, a German chemist who fled Nazi Germany, penicillin G was introduced into the practice of medicine in 1943. For their pioneering work in developing one of the most effective antibiotics of all time, Fleming, Florey, and Chain were awarded the 1945 Nobel Prize for medicine or physiology. The mold from which Fleming discovered penicillin was *Penicillium notatum*, a strain that gives a relatively low yield of penicillin. It was replaced in commercial production of the antibiotic by *P. chrysogenum*, a strain cultured from a mold found growing on a grapefruit in a market in Peoria, Illinois.

The structural feature common to all penicillins is a β-lactam ring fused to a five-membered thiazolidine ring. The penicillins owe their antibacterial activity to a common mechanism that inhibits the biosynthesis of a vital part of bacterial cell walls.

Soon after the penicillins were introduced into medical practice, penicillin-resistant strains of bacteria began to appear and have since proliferated. One approach to combating resistant strains is to synthesize newer, more effective penicillins. Among those developed are ampicillin, methicillin, and amoxicillin. Another approach is to search for newer, more effective β-lactam antibiotics. At the present time, the most effective of these are the cephalosporins, the first of which was isolated from the fungus *Cephalosporium acremonium*. This class of β-lactam antibiotics has an even broader spectrum of antibacterial activity than the penicillins and is effective against many penicillin-resistant bacterial strains. However, resistance to the cephalosporins is now also widespread. The most common mechanism of resistance in bacteria involves their production of a specific enzyme that cleaves the β-lactam. Several compounds have been found that inhibit this enzyme, and it is hoped that drugs based on these can be taken in combination with penicillins and cephalosporins to restore their effectiveness.

The penicillins differ in the group bonded to the acyl carbon

β-lactam

Amoxicillin
(a β-lactam antibiotic)

The cephalosporins differ in the group attached to the acyl carbon and the side chain attached to the thiazine ring.

β-lactam

A cephalosporin, a newer generation β-lactam antibiotic.

Acetamide	Benzenesulfonamide	Succinimide	Phthalimide
pK_a 15–17	pK_a 10	pK_a 9.7	pK_a 8.3

Values of pK_a for amides of carboxylic acids are in the range of 15–17, which means that they are comparable in acidity to alcohols. Amides show no evidence of acidity in aqueous solution; that is, water-insoluble amides do not react with aqueous solutions of NaOH or other alkali metal hydroxides to form water-soluble salts.

Imides (pK_a 8–10) are considerably more acidic than amides and readily dissolve in 5% aqueous NaOH by forming water-soluble salts. We account for the acidity of imides in the same manner as for the acidity of carboxylic acids (Section 17.4A), namely the imide anion is stabilized by delocalization of the negative charge. The more important contributing structures for the anion formed by ionization of an imide delocalize the negative charge on nitrogen and the two carbonyl oxygens.

Phthalimide   A resonance-stabilized anion

Sulfonamides derived from ammonia and primary amines are also sufficiently acidic to dissolve in aqueous solutions of NaOH or other alkali metal hydroxides by forming water-soluble salts. The pK_a of benzenesulfonamide is approximately 10. We account for the acidity of sulfonamides in the same manner as for imides, namely the resonance stabilization of the resulting anion.

Benzenesulfonamide   A resonance-stabilized anion

Example 18.2

Phthalimide is insoluble in water. Will phthalimide dissolve in aqueous NaOH?

Solution

Phthalimide is the stronger acid and NaOH is the stronger base. The position of equilibrium, therefore, lies to the right. $K_{eq} = 2.5 \times 10^7$. Phthalimide dissolves in aqueous NaOH by forming a water-soluble sodium salt.

$$\text{(phthalimide)} + \text{NaOH} \rightleftharpoons \text{(phthalimide } N^-\text{Na}^+) + \text{H}_2\text{O} \qquad pK_{eq} = -7.4$$
$$K_{eq} = 2.5 \times 10^7$$

pK$_a$ 8.3   pK$_a$ 15.7
(stronger acid)   (stronger base)   (weaker base)   (weaker acid)

Problem 18.2

Will phthalimide dissolve in aqueous sodium bicarbonate?

The noncaloric artificial sweetener, saccharin, is an imide. The imide hydrogen of saccharin is sufficiently acidic that it reacts with sodium hydroxide and aqueous ammonia to form water-soluble salts. The ammonium salt is used to make liquid sweeteners. Saccharin is used in solid form as the Ca^{2+} salt. (See *The Merck Index,* 12th ed., #8463.)

Saccharin

Saccharin is approximately 500 times sweeter than sugar, and at one time, was the most important noncaloric sweetener used in foods. At the present time, the most widely used noncaloric artificial sweetener is aspartame (Nutrasweet, see Problem 27.49).

Spectroscopy
Acid Derivatives

18.3   Spectroscopic Properties

Esters and amides are the most common functional derivatives of carboxylic acids and also the most commonly analyzed. Therefore, we concentrate on the spectroscopic properties of these two functional derivatives.

A.   Infrared Spectroscopy

The most important infrared absorption of carboxylic acids and their derivatives is due to the C=O stretching vibration. Infrared spectroscopic data for these compounds are summarized in Table 18.1.

The carbonyl stretching of amides occurs at 1630–1680 cm^{-1}, at a lower frequency than for other carbonyl compounds. Primary and secondary amides show N—H stretching in the region 3200–3400 cm^{-1}; primary amides (RCONH$_2$) usually show two N—H absorptions, whereas secondary amides (RCONHR) show only a single N—H absorption (Figure 18.1).

Table 18.1 Infrared Absorptions for Carboxylic Acids and Their Functional Derivatives

Compound	Stretching Absorption (cm^{-1})	Additional Absorptions (cm^{-1})
$R\overset{O}{\overset{\|}{C}}Cl$	1790–1800	
$R\overset{O}{\overset{\|}{C}}O\overset{O}{\overset{\|}{C}}R$	1740–1760 and 1800–1850	C—O stretching at 900–1300
$R\overset{O}{\overset{\|}{C}}OR$	1735–1800	C—O stretching at 1000–1100 and 1200–1250
$R\overset{O}{\overset{\|}{C}}OH$	1700–1725	O—H stretching at 2400–3400 C—O stretching at 1210–1320
$R\overset{O}{\overset{\|}{C}}NH_2$	1630–1680	N—H stretching at 3200 and 3400 (1° amides have two N—H peaks) (2° amides have one N—H peak)
RC≡N	2200–2250	

Esters display strong C═O stretching absorption in the region between 1735 and 1800 cm^{-1} (Section 12.4J). Like cyclic ketones, they show strong ring size effects; six-membered lactones absorb near where acyclic esters do (around 1735 cm^{-1}); five- and four-membered ring lactones absorb at progressively higher frequencies. In addition, esters and lactones also display strong C—O stretching absorptions in the region 1000–1100 cm^{-1} for the sp^3 C—O stretch and 1200–1250 cm^{-1} for the sp^2 C—O stretch.

Figure 18.1
Infrared spectrum of *N*-methylpropanamide (a secondary amide).

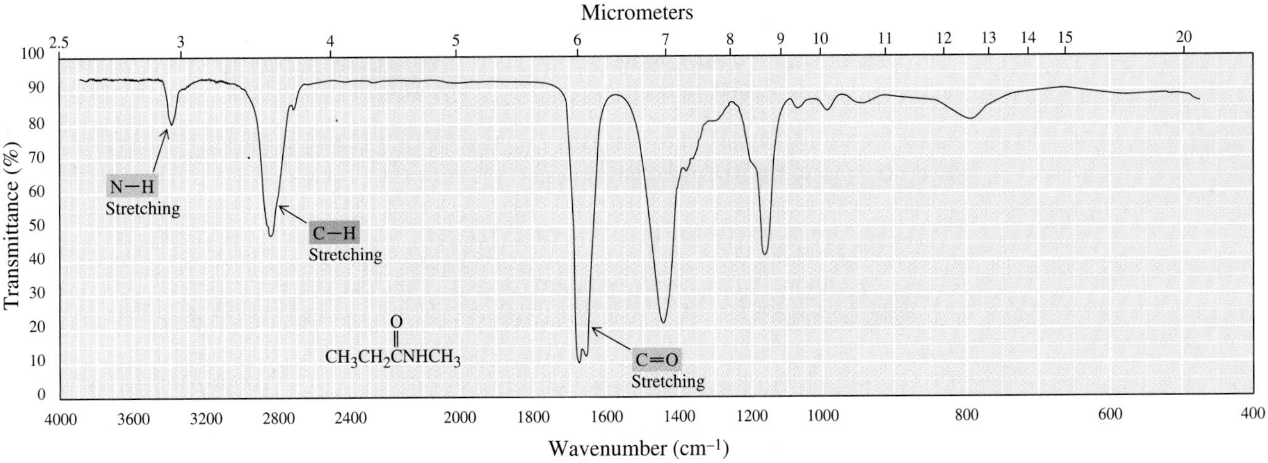

Anhydrides have two carbonyl stretching absorptions, one near 1760 cm^{-1} and the other near 1810 cm^{-1}. In addition, anhydrides display strong C—O stretching absorption in the region $900–1300 \text{ cm}^{-1}$. Nitriles can be distinguished by strong C≡N stretching absorption at $2200–2250 \text{ cm}^{-1}$.

B. Nuclear Magnetic Resonance Spectroscopy

Hydrogens on an α-carbon to a carbonyl group are slightly deshielded and come into resonance at $\delta\ 2.1–2.6$. Hydrogens on the carbon attached to the ester oxygen are more strongly deshielded and come into resonance at $\delta\ 3.7–4.7$. It is possible to distinguish between ethyl acetate and its constitutional isomer, methyl propanoate, by the chemical shifts of either the —CH$_3$ singlet absorption (compare $\delta\ 2.04$ and 3.68), or the —CH$_2$— quartet absorption (compare $\delta\ 4.11$ and 2.33). Carbonyl carbons of esters and other carboxylic acid derivatives show characteristic ^{13}C-NMR absorption at $\delta\ 160–180$.

$$\delta\ 2.04(s) \qquad \delta\ 4.11(q) \qquad \delta\ 2.33(q) \qquad \delta\ 3.68(s)$$

CH$_3$—C(=O)—O—CH$_2$—CH$_3$ \qquad CH$_3$—CH$_2$—C(=O)—O—CH$_3$

Ethyl acetate \qquad\qquad Methyl propanoate

Supporting Concepts
Carbonyl Chemistry
The CD-ROM contains a comprehensive overview of the reactivity of compounds containing the carbonyl group.

18.4 Characteristic Reactions

In this and subsequent sections, we examine the interconversions of various carboxylic acid derivatives. All these reactions begin with formation of a tetrahedral carbonyl addition intermediate. The first step of this reaction is exactly analogous to the addition of alcohols to aldehydes and ketones (Section 16.8B). This reaction can be carried out under basic conditions, in which a negative or reactive neutral nucleophile adds directly to the carbonyl carbon. The tetrahedral carbonyl addition intermediate formed then adds H$^+$. The result of this reaction is nucleophilic addition.

Nucleophilic addition (basic conditions):

A carboxylic acid derivative \qquad Tetrahedral carbonyl addition intermediate \qquad Addition product

As with aldehydes and ketones, this reaction can also be catalyzed by acid, in which case protonation of the carbonyl oxygen precedes the attack of the nucleophile.

Nucleophilic addition (acid conditions):

A carboxylic acid derivative \qquad\qquad Tetrahedral carbonyl addition intermediate

For functional derivatives of carboxylic acids, the fate of the tetrahedral carbonyl addition intermediate is quite different from that of aldehydes and ketones; the intermediate collapses to expel the leaving group Y and regenerate the carbonyl group. The result of this addition-elimination sequence is **nucleophilic acyl substitution.**

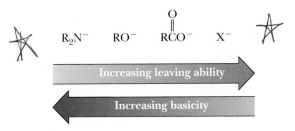

Nucleophilic acyl substitution:

Tetrahedral carbonyl addition intermediate

Substitution product

Nucleophilic acyl substitution A reaction in which a nucleophile bonded to the carbon of an acyl group is replaced by another nucleophile.

The major difference between these two types of carbonyl addition reactions is that aldehydes and ketones do not have a group that can leave as a relatively stable anion. They undergo only nucleophilic acyl addition. The four carboxylic acid derivatives we study in this chapter do have a group Y that can leave as a relatively stable anion or as a neutral species; they undergo nucleophilic acyl substitution.

In this general reaction, we show the nucleophile and the leaving group as anions. This need not be the case. Neutral molecules, such as water, alcohols, ammonia, and amines, may also serve as nucleophiles and leaving groups in this reaction, mainly when it is carried out under acid-catalyzed conditions. We show the leaving groups here as anions, however, to illustrate an important point about leaving groups: the weaker the base (the more stable the anion), the better the leaving group (Section 8.4F).

$$R_2N^- \quad RO^- \quad \overset{\overset{O}{\parallel}}{RCO^-} \quad X^-$$

Increasing leaving ability →

← **Increasing basicity**

The weakest base in the series, and the best leaving group, is halide ion; acid halides are the most reactive toward nucleophilic acyl substitution. The strongest base, and the poorest leaving group, is amide ion; amides are the least reactive toward nucleophilic acyl substitution. Acid halides and acid anhydrides are so reactive that they are not found in nature. Esters and amides, however, are universally present.

$$\underset{\text{Amide}}{\overset{\overset{O}{\parallel}}{RCNH_2}} \quad \underset{\text{Ester}}{\overset{\overset{O}{\parallel}}{RCOR'}} \quad \underset{\text{Anhydride}}{\overset{\overset{O}{\parallel}\;\overset{O}{\parallel}}{RCOCR'}} \quad \underset{\text{Acid halide}}{\overset{\overset{O}{\parallel}}{RCX}}$$

Increasing reactivity toward nucleophilic acyl substitution →

In addition, many reactions of carboxyl derivatives occur by acid catalysis. In these reactions, the carbonyl group is protonated in the first step, which increases its electrophilicity. Then the leaving group is protonated in a later step to decrease its basicity and make it a better leaving group. We will see detailed mechanisms for many examples of both acid- and base-catalyzed reactions in this chapter. We have already seen the acid-catalyzed reversible reaction of acids with alcohols in Section 17.7A.

18.5   Reaction with Water — Hydrolysis

A.   Acid Chlorides

Low-molecular-weight acid chlorides react very rapidly with water to form carboxylic acids and HCl. Higher molecular-weight acid halides are less soluble and, consequently, react less rapidly with water. Because the mechanisms for hydrolysis of acid chlorides and anhydrides are identical, we show the mechanism only for acid anhydrides (Section 18.5B).

$$
\underset{\substack{\text{O} \\ \parallel}}{CH_3CCl} + H_2O \longrightarrow \underset{\substack{\text{O} \\ \parallel}}{CH_3COH} + HCl
$$

B.   Acid Anhydrides

Anhydrides are generally less reactive than acid chlorides. However the lower molecular-weight anhydrides also react readily with water to form two molecules of carboxylic acid.

$$
\underset{\substack{\text{O O} \\ \parallel\parallel}}{CH_3COCCH_3} + H_2O \longrightarrow \underset{\substack{\text{O} \\ \parallel}}{CH_3COH} + \underset{\substack{\text{O} \\ \parallel}}{HOCCH_3}
$$

The following mechanism is divided into two stages: first, formation of a tetrahedral carbonyl addition intermediate, and second, collapse of this intermediate by elimination of acetate ion, a moderate base, and a good leaving group.

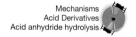
Mechanisms
Acid Derivatives
Acid anhydride hydrolysis

Mechanism   Hydrolysis of an Acid Anhydride

Step 1:   Addition of H_2O to one of the carbonyl groups gives a tetrahedral carbonyl addition intermediate. This reaction is acid catalyzed but will occur without added acid; water is strong enough acid to protonate a trace of the anhydride. Once some anhydride has reacted, more acid is produced.

Tetrahedral carbonyl
addition intermediate

Step 2:   Protonation of the leaving group followed by collapse of the tetrahedral carbonyl addition intermediate gives two molecules of carboxylic acid.

C.   Esters

Esters are hydrolyzed only very slowly, even in boiling water. Hydrolysis becomes considerably more rapid, however, when they are refluxed in aqueous acid or base. We already discussed acid-catalyzed (Fischer) esterification in Section 17.7A and pointed out that it is an equilibrium reaction. Hydrolysis of esters in aqueous acid is also an equilibrium reaction and proceeds by the same mechanism as esterification, except in reverse. The role of the acid catalyst is to protonate the carbonyl oxygen. In doing so, it increases the electrophilic character of the carbonyl carbon toward attack by water to form a tetrahedral carbonyl addition intermediate. Collapse of this intermediate gives the carboxylic acid and an alcohol. In this reaction, acid is a catalyst; it is consumed in the first step but another is generated at the end of the reaction.

Tetrahedral carbonyl
addition intermediate

Although formation of a tetrahedral carbonyl addition intermediate is the most common mechanism for the hydrolysis of esters in aqueous acid, alternative mechanistic pathways are followed in special cases. Such a case occurs when the alkyl group attached to oxygen can form an especially stable carbocation. Then protonation of the carbonyl oxygen is followed by cleavage of the O—C bond to give a carboxylic acid and a carbocation. Benzyl and *tert*-butyl esters readily undergo this type of ester hydrolysis.

Mechanisms
Acid Derivatives
Acid hydrolysis of a *tert*-butyl ester

Mechanism   **Hydrolysis of a *tert*-Butyl Ester in Aqueous Acid**

Step 1:   Proton transfer to the carbonyl oxygen gives a cation, which is the conjugate acid of the ester.

Step 2: Rearrangement of electron pairs in the cation intermediate gives the carboxylic acid and *tert*-butyl cation. This cation then either reacts with water to give *tert*-butyl alcohol (S$_N$1) or transfers a proton to water to give 2-methylpropene (E1).

$$CH_3-\underset{\underset{}{\overset{\overset{\displaystyle :O}{\|}}{}}}{C}-O-\underset{\underset{CH_3}{|}}{\overset{\overset{CH_3}{|}}{C}}-CH_3 \xrightarrow{(1)} CH_3-\overset{\overset{\displaystyle \overset{+}{O}-H}{|}}{C}-O-\underset{\underset{CH_3}{|}}{\overset{\overset{CH_3}{|}}{C}}-CH_3 \xrightarrow{(2)} CH_3-\overset{\overset{\displaystyle :O-H}{\|}}{C}=O \ + \ \overset{\overset{CH_3}{|}}{\underset{\underset{CH_3}{|}}{\overset{+}{C}}}-CH_3$$

tert-Butyl cation

Hydrolysis of esters may also be carried out using hot aqueous base, such as aqueous NaOH. Hydrolysis of esters in aqueous base is often called **saponification,** a reference to the use of this reaction in the manufacture of soaps (Section 26.2A). Although the carbonyl carbon of an ester is not strongly electrophilic, hydroxide ion is a good nucleophile and adds to the carbonyl carbon to form a tetrahedral carbonyl addition intermediate, which in turn collapses to give a carboxylic acid and an alkoxide ion. The carboxylic acid reacts with the alkoxide ion or other base present to form a carboxylic acid anion. Thus, each mole of ester hydrolyzed requires one mole of base as shown in the following balanced equation.

$$R\overset{\overset{\displaystyle O}{\|}}{C}OCH_3 + NaOH \xrightarrow{H_2O} R\overset{\overset{\displaystyle O}{\|}}{C}O^-Na^+ + CH_3OH$$

Mechanism **Hydrolysis of an Ester in Aqueous Base**

Mechanisms
Acid Derivatives
Ester Hydrolysis in
Aqueous Base

Saponification

Step 1: Addition of hydroxide ion to the carbonyl carbon of the ester gives a tetrahedral carbonyl addition intermediate.

Step 2: Collapse of this intermediate gives a carboxylic acid and an alkoxide ion.

Step 3: Proton transfer between the carboxyl group and the alkoxide ion gives the carboxylic anion. This strongly exothermic acid-base reaction drives the whole reaction to completion.

$$R-\overset{\overset{\displaystyle O}{\|}}{C}-O-CH_3 + \ ^-OH \underset{}{\overset{(1)}{\rightleftharpoons}} R-\overset{\overset{\displaystyle :O}{|}}{\underset{\underset{OH}{|}}{C}}-O-CH_3 \overset{(2)}{\rightleftharpoons} R-\overset{\overset{\displaystyle O}{\|}}{\underset{\underset{O-H}{|}}{C}} + \ ^-\!:O-CH_3 \xrightarrow{(3)} R-\overset{\overset{\displaystyle O}{\|}}{\underset{\underset{O:}{|}}{C}} + HO-CH_3$$

There are two major differences between hydrolysis of esters in aqueous acid and aqueous base.

1. For hydrolysis of an ester in aqueous acid, acid is required in only catalytic amounts. For hydrolysis in aqueous base, base is required in stoichiometric amounts because it is a reactant, not just a catalyst.
2. Hydrolysis of an ester in aqueous acid is reversible, but hydrolysis in aqueous base is irreversible because a carboxylic acid anion (weakly electrophilic, if at all) is not attacked by ROH (a weak nucleophile).

Other acid derivatives react with base in an identical manner to esters.

Example 18.3

Complete and balance equations for the hydrolysis of each ester in aqueous sodium hydroxide. Show all products as they are ionized under these conditions.

(a) 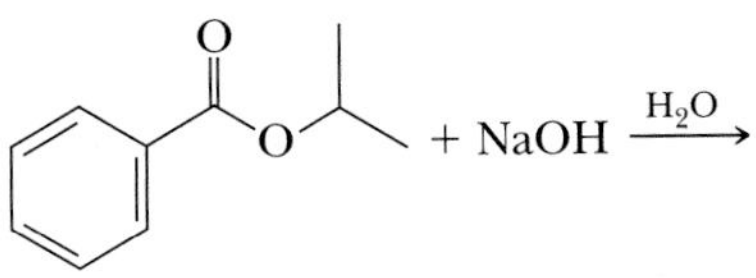

(b) 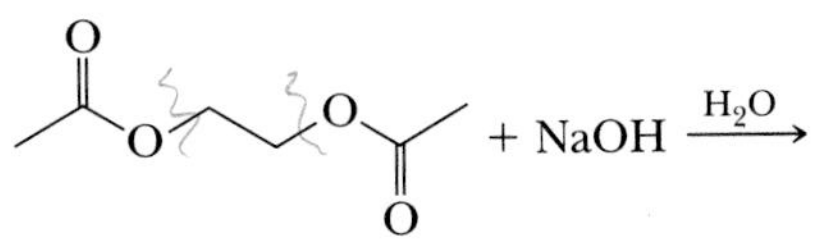

Solution

The products of hydrolysis of (a) are benzoic acid and 2-propanol. In aqueous NaOH, benzoic acid is converted to its sodium salt. Therefore, one mole of NaOH is required for hydrolysis of one mole of this ester. Compound (b) is a diester of ethylene glycol. Two moles of NaOH are required for its hydrolysis.

(a) 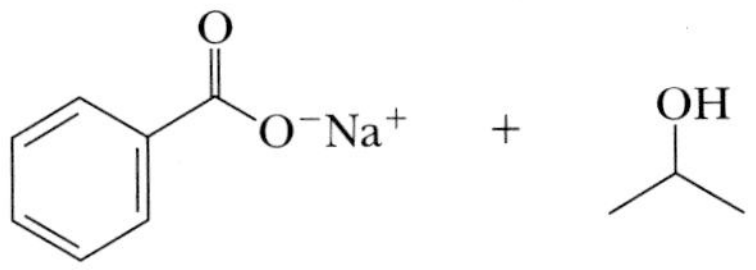

Sodium benzoate

2-Propanol (Isopropyl alcohol)

(b) 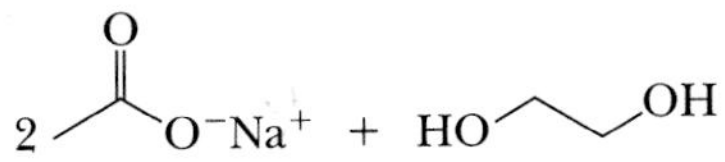

Sodium acetate

1,2-Ethanediol (Ethylene glycol)

Problem 18.3

Complete and balance equations for the hydrolysis of each ester in aqueous solution. Show each product as it is ionized under the indicated experimental conditions.

(a) 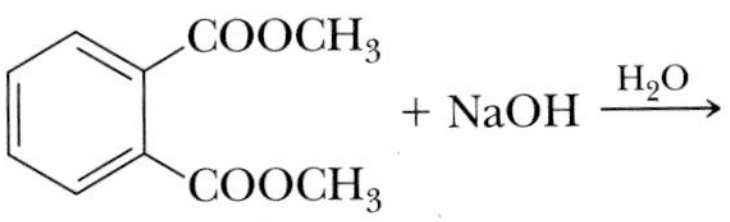

(b) 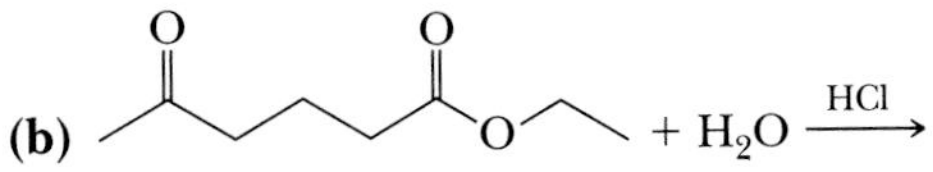

D. Amides

Amides require considerably more vigorous conditions for hydrolysis in both acid and base than esters. Amides undergo hydrolysis in hot aqueous acid to give a carboxylic acid and an ammonium ion. Hydrolysis is driven to completion by the acid-base reaction between ammonia or the amine and acid to form an ammonium ion. One mole of acid is required per mole of amide.

2-Phenylbutanamide

2-Phenylbutanoic acid

In aqueous base, the products of amide hydrolysis are a carboxylic acid salt and ammonia or an amine. Hydrolysis in aqueous base is driven to completion by the acid-base reaction between the resulting carboxylic acid and base to form a salt. One mole of base is required per mole of amide.

N-Phenylethanamide
(N-Phenylacetamide,
acetanilide)

Sodium acetate   Aniline

The steps in the mechanism for the hydrolysis of amides in aqueous acid are similar to those for the hydrolysis of esters in aqueous acid. The role of hydrogen ion in Step 1 is to protonate the carbonyl oxygen to increase the electrophilic character of the carbonyl carbon. Following protonation, the polarized carbonyl group reacts with a molecule of water in Step 2 to give a tetrahedral carbonyl addition intermediate. Collapse of this intermediate in Step 3 completes the reaction. Note that the leaving group in this case is a neutral amine (a weaker base), a far better leaving group than amide ion (a much stronger base).

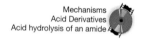

Mechanisms
Acid Derivatives
Acid hydrolysis of an amide

Mechanism   Hydrolysis of an Amide in Aqueous Acid

Step 1:   Protonation of the carbonyl oxygen gives a resonance-stabilized cation intermediate.

Resonance-stabilized cation

Step 2:   Addition of water to the carbonyl carbon of the cation intermediate followed by proton transfer gives a tetrahedral carbonyl addition intermediate.

proton transfer
from oxygen
to nitrogen

Tetrahedral carbonyl
addition intermediate

Step 3:   Collapse of the tetrahedral carbonyl addition intermediate coupled with proton transfer gives a carboxylic acid and ammonium ion.

$$R—\overset{\overset{\displaystyle O—H}{|}}{\underset{\underset{\displaystyle OH}{|}}{C}}—NH_3{}^+ \longrightarrow R—\overset{\overset{\displaystyle {}^+O—H}{||}}{\underset{\underset{\displaystyle OH}{|}}{C}} \ + \ \overset{..}{N}H_3 \longrightarrow R—\overset{\overset{\displaystyle ..O..}{||}}{C}—OH \ + \ NH_4{}^+$$

The mechanism for the hydrolysis of amides in aqueous base is similar to that for the hydrolysis of esters in aqueous base. Loss of nitrogen and proton transfer from water to nitrogen are concerted so that the leaving group is not amide ion, $NH_2{}^-$, a stronger base and poorer leaving group, but rather ammonia, NH_3, a weaker base and better leaving group. (The NH_2 group is basic enough to be protonated in part, so it may actually be neutral ammonia that leaves.)

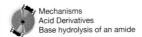

Mechanisms
Acid Derivatives
Base hydrolysis of an amide

Mechanism   Hydrolysis of an Amide in Aqueous Base

Step 1:   Addition of hydroxide ion to the carbonyl carbon gives a tetrahedral carbonyl addition intermediate as with esters.

Step 2:   Collapse of the tetrahedral carbonyl addition intermediate gives a carboxylic acid and ammonia.

$$R—\overset{\overset{\displaystyle :O^-}{|}}{\underset{\underset{\displaystyle OH}{|}}{C}}—NH_2 \ + \ H—\overset{\frown}{O}—H \longrightarrow R—\overset{\overset{\displaystyle O}{||}}{\underset{\underset{\displaystyle OH}{|}}{C}} \ + \ NH_3 \ + \ :OH^-$$

Tetrahedral carbonyl
addition intermediate

Step 3:   Proton transfer forms the carboxylate anion and water. Hydrolysis is driven to completion by this acid-base reaction.

$$R—\overset{\overset{\displaystyle O}{||}}{C}—O—H + :O—H \longrightarrow R—\overset{\overset{\displaystyle O}{||}}{C}—O:^- + H—O—H$$

Example 18.4

Write equations for hydrolysis of these amides in concentrated aqueous HCl. Show all products as they exist in aqueous HCl, and the number of moles of HCl required for hydrolysis of each amide.

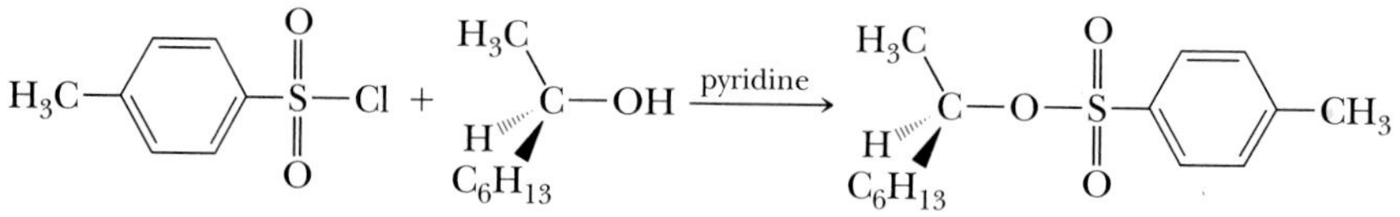

p-Toluenesulfonyl chloride
(Tosyl chloride; TsCl)

(R)-2-Octanol

(R)-2-Octyl p-toluenesulfonate
[(R)-2-Octyl tosylate]

As discussed in Section 9.6D, a special value of p-toluenesulfonic (tosylate) and methanesulfonic (mesylate) esters is that, in forming them, an —OH is converted from a poor leaving group (hydroxide ion) in nucleophilic displacement to an excellent leaving group, the p-toluenesulfonate (tosylate) or methanesulfonate (mesylate) anions.

B. Acid Anhydrides

Acid anhydrides react with alcohols to give one mole of ester and one mole of a carboxylic acid. Thus, the reaction of an alcohol with an anhydride is a useful method for the synthesis of esters. This reaction is also catalyzed by acids and tertiary amines.

$$\underset{\text{Acetic anhydride}}{CH_3COCCH_3} + \underset{\text{Ethanol}}{HOCH_2CH_3} \longrightarrow \underset{\text{Ethyl acetate}}{CH_3COCH_2CH_3} + \underset{\text{Acetic acid}}{CH_3COH}$$

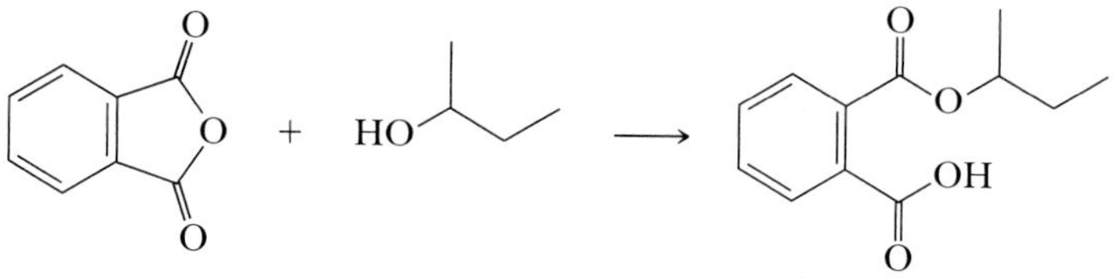

Phthalic anhydride

2-Butanol
(sec-Butyl alcohol)

1-Methylpropyl hydrogen phthalate
(sec-Butyl hydrogen phthalate)

Aspirin is synthesized on an industrial scale by the reaction of acetic anhydride and salicylic acid.

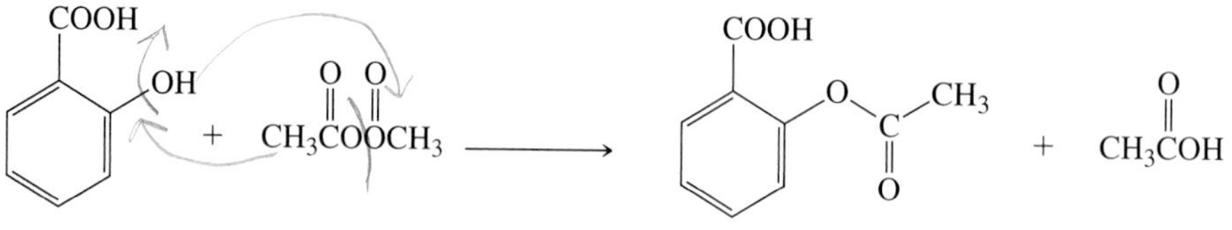

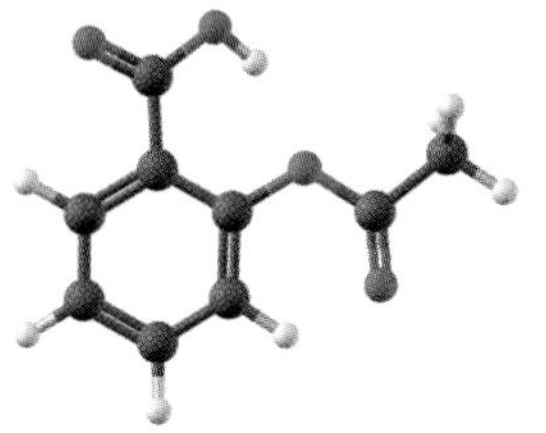

2-Hydroxybenzoic acid
(Salicylic acid)

Acetic anhydride

Acetylsalicylic acid
(Aspirin)

Acetic acid

Transesterification Exchange of the —OR or —OAr group of an ester for another —OR or —OAr group.

C. Esters

Esters react with alcohols in an acid-catalyzed reaction called **transesterification**. For example, it is possible to convert methyl acrylate to butyl acrylate by heating the methyl ester with 1-butanol in the presence of an acid catalyst. The acids most commonly used are HCl as a gas bubbled into the reaction medium and *p*-toluenesulfonic acid.

| Methyl propenoate (Methyl acrylate) (bp 81°C) | 1-Butanol (bp 117°C) | Butyl propenoate (Butyl acrylate) (bp 147°C) | Methanol (bp 65°C) |

Transesterification is an equilibrium reaction and can be driven in either direction by control of experimental conditions. For example, in the reaction of methyl acrylate with 1-butanol, transesterification is carried out at a temperature slightly above the boiling point of methanol (the lowest boiling component in the mixture). Methanol distills from the reaction mixture, thus shifting the position of equilibrium in favor of butyl acrylate. Conversely, reaction of butyl acrylate with a large excess of methanol shifts the equilibrium to favor formation of methyl acrylate.

Example 18.6

Complete the following transesterification reactions. The stoichiometry of each is given in the problem.

Solution

Problem 18.6

Complete the following transesterification reaction. The stoichiometry is given in the equation.

D. Amides   Not used to make ester

Amides, the least reactive of the functional derivatives of carboxylic acids, do not react with alcohols. Thus, the reaction of an amide with an alcohol cannot be used to prepare an ester.

18.7 Reactions with Ammonia and Amines

A. Acid Halides

Acid halides react readily with ammonia and primary or secondary amines to form amides. For complete conversion of an acid halide to an amide, two equivalents of ammonia or amine are used: one to form the amide and one to neutralize the hydrogen halide formed.

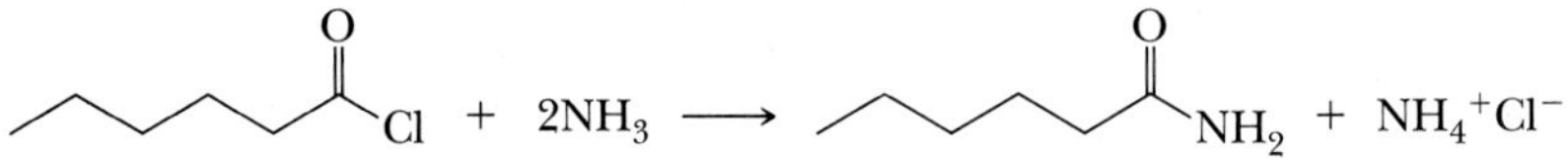

Hexanoyl chloride — Ammonia — Hexanamide — Ammonium chloride

The mechanism for the reaction between an acid chloride and ammonia or a primary or secondary amine begins in Step 1 by addition of ammonia or the amine to the carbonyl carbon to form a tetrahedral carbonyl addition intermediate. This intermediate collapses in Step 2 to eject chloride, a good leaving group, and give the amide.

Mechanism Reaction of Acetyl Chloride and Ammonia

Step 1: Nucleophilic addition of ammonia to the carbonyl carbon followed by a proton transfer gives a tetrahedral carbonyl addition intermediate.

Step 2: Collapse of this intermediate gives chloride ion and an amide.

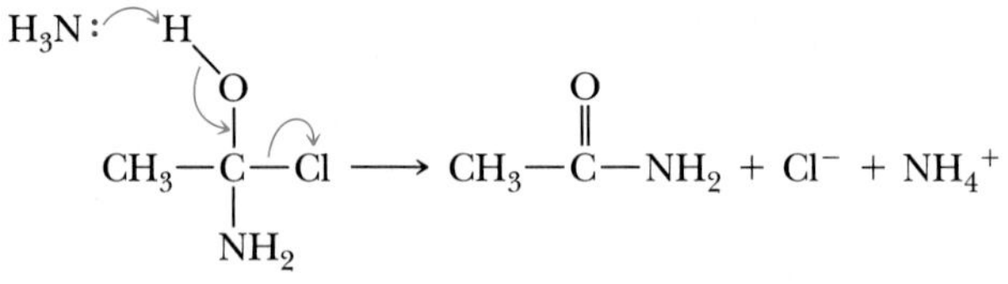

Tetrahedral carbonyl addition intermediate — An amide

B. Acid Anhydrides

Acid anhydrides react with ammonia and primary and secondary amines to form amides. As with acid halides, two moles of amine are required: one mole to form the amide and one mole to neutralize the carboxylic acid byproduct.

$$CH_3\overset{O}{\overset{\|}{C}}O\overset{O}{\overset{\|}{C}}CH_3 + 2NH_3 \longrightarrow CH_3\overset{O}{\overset{\|}{C}}NH_2 + CH_3\overset{O}{\overset{\|}{C}}O^-\,NH_4^+$$

Acetic anhydride — Ammonia — Ethanamide (Acetamide) — Ammonium acetate

Alternatively, if the amine used to make the amide is expensive, a tertiary amine such as triethylamine may be used to neutralize the carboxylic acid.

C. Esters

Esters react with ammonia and primary or secondary amines to form amides. Because an alkoxide anion is a poor leaving group compared with either a halide or carboxylate ion, esters are less reactive toward ammonia, primary amines, and secondary amines than are acid halides or acid anhydrides.

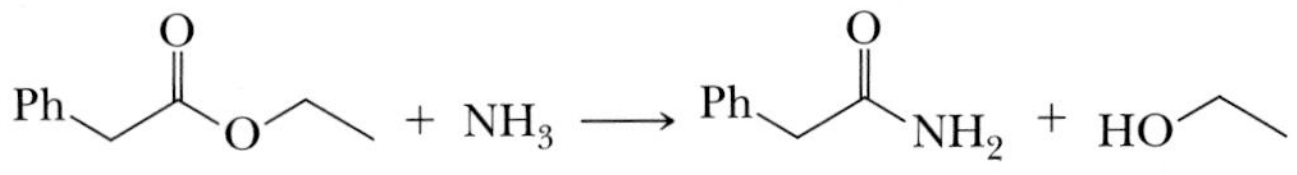

Ethyl phenylacetate   Phenylacetamide   Ethanol

D. Amides

Amides do not react with ammonia or primary or secondary amines.

Example 18.7

Complete the following reactions. The stoichiometry of each reaction is given in the equation.

(a) 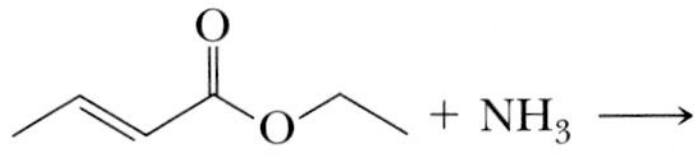 + NH$_3$ ⟶   (b) + 2NH$_3$ ⟶

   Ethyl 2-butenoate   Diethyl carbonate

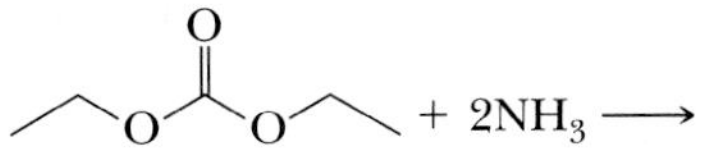

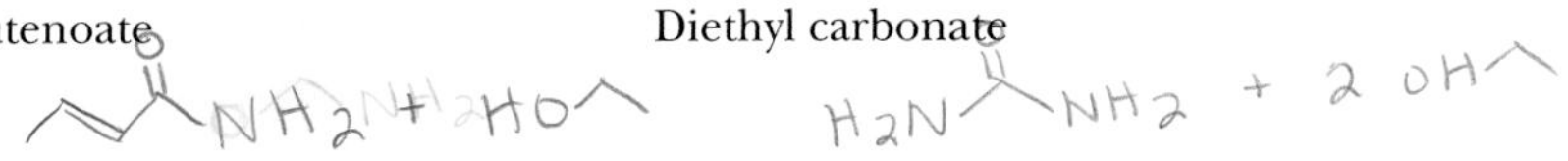

Solution

(a) 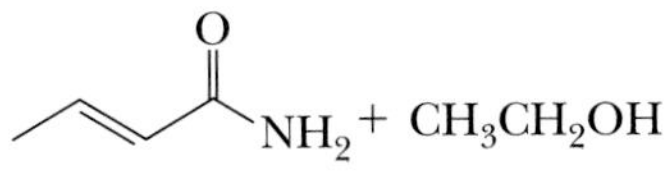 + CH$_3$CH$_2$OH   (b) H$_2$N⌿C⌿NH$_2$ + 2CH$_3$CH$_2$OH

   2-Butenamide   Urea

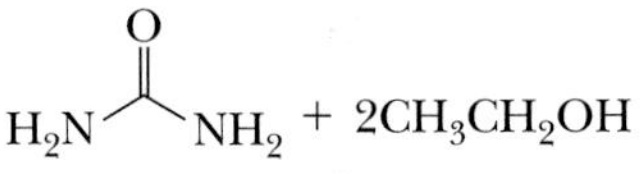

Problem 18.7

Complete and balance equations for the following reactions. The stoichiometry of each reaction is given in the equation.

(a) 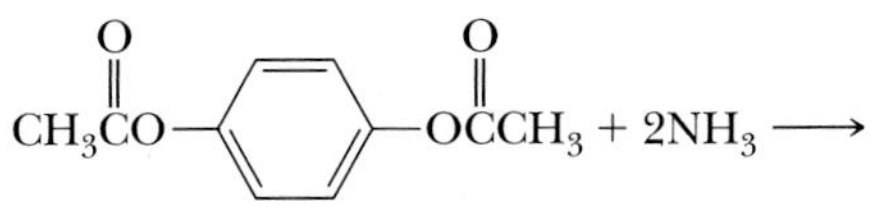 + 2NH$_3$ ⟶   (b) 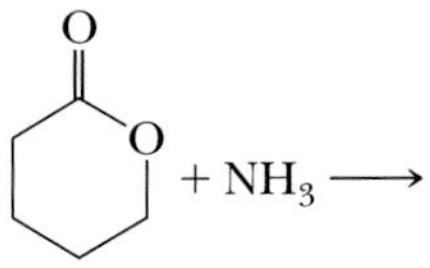 + NH$_3$ ⟶

18.8   Reaction of Acid Chlorides with Salts of Carboxylic Acids

Acid chlorides react with salts of carboxylic acids to give anhydrides. Most commonly used are the sodium or potassium salts.

$$CH_3\overset{\overset{\displaystyle O}{\|}}{C}Cl \;+\; Na^+\;{}^-O\overset{\overset{\displaystyle O}{\|}}{C}\!-\!\!\bigcirc \;\longrightarrow\; CH_3\overset{\overset{\displaystyle O}{\|}}{C}O\overset{\overset{\displaystyle O}{\|}}{C}\!-\!\!\bigcirc \;+\; Na^+Cl^-$$

Acetyl chloride   Sodium benzoate   Acetic benzoic anhydride

Reaction of an acid halide with the anion of a carboxylic acid is a particularly useful method for synthesis of mixed anhydrides.

18.9 Reactions with Organometallic Compounds

A. Grignard Reagents

Treatment of a formic ester with two moles of a Grignard reagent followed by hydrolysis of the magnesium alkoxide salt in aqueous acid gives a secondary alcohol. Treatment of an ester other than a formate with a Grignard reagent gives a tertiary alcohol in which two of the groups bonded to the carbon bearing the —OH group are the same.

$$HC\overset{\overset{\displaystyle O}{\|}}{O}CH_3 + 2RMgX \longrightarrow \begin{matrix} \text{magnesium} \\ \text{alkoxide} \\ \text{salt} \end{matrix} \xrightarrow{H_2O,\,HCl} \underset{\underset{\displaystyle R}{|}}{\overset{\overset{\displaystyle OH}{|}}{HC}}\!\!-\!R \;+\; CH_3OH$$

An ester of   A secondary
formic acid   alcohol

$$CH_3\overset{\overset{\displaystyle O}{\|}}{C}OCH_3 \;+\; 2RMgX \longrightarrow \begin{matrix} \text{magnesium} \\ \text{alkoxide} \\ \text{salt} \end{matrix} \xrightarrow{H_2O,\,HCl} \underset{\underset{\displaystyle R}{|}}{\overset{\overset{\displaystyle OH}{|}}{CH_3C}}\!\!-\!R \;+\; CH_3OH$$

An ester of any acid   A tertiary
other than formic acid   alcohol

Reaction of an ester with a Grignard reagent involves formation of two successive tetrahedral carbonyl addition intermediates. The first collapses to give a new carbonyl compound: an aldehyde from a formic ester and a ketone from all other esters. The second intermediate is stable and, when protonated, gives the final alcohol. It is important to realize that it is not possible to use RMgX and an ester to prepare a ketone; the intermediate ketone is more reactive than the ester and reacts immediately with the Grignard reagent to give a tertiary alcohol.

Mechanisms
Acid Derivatives
Reaction of an ester
with a Grignard
reagent

Mechanism   Reaction of an Ester with a Grignard Reagent

Steps 1 & 2:   Reaction begins in Step 1 with addition of one mole of Grignard reagent to the carbonyl carbon to form a tetrahedral carbonyl addition intermediate. Because alkoxide ion is a moderately good leaving group from a tetrahedral intermediate, this intermediate collapses in Step 2 to give a new carbonyl-containing compound and a magnesium alkoxide salt.

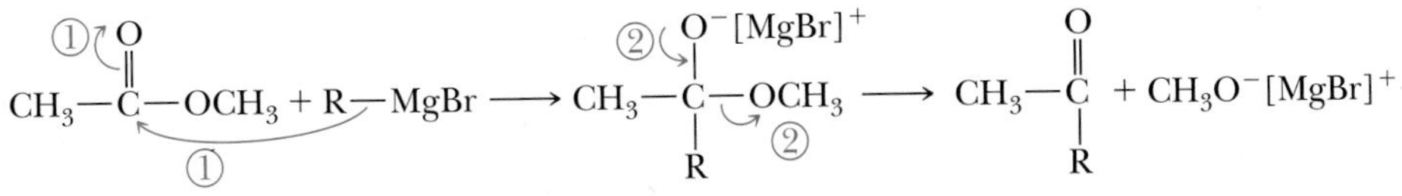

A magnesium salt
(a tetrahedral carbonyl
addition intermediate)

A ketone

Steps 3 & 4:   This new carbonyl-containing compound then reacts in Step 3 with a second mole of Grignard reagent to form a second tetrahedral carbonyl addition compound, which, after hydrolysis in aqueous acid (Step 4), gives a tertiary alcohol (or a secondary alcohol if the starting ester was a formate).

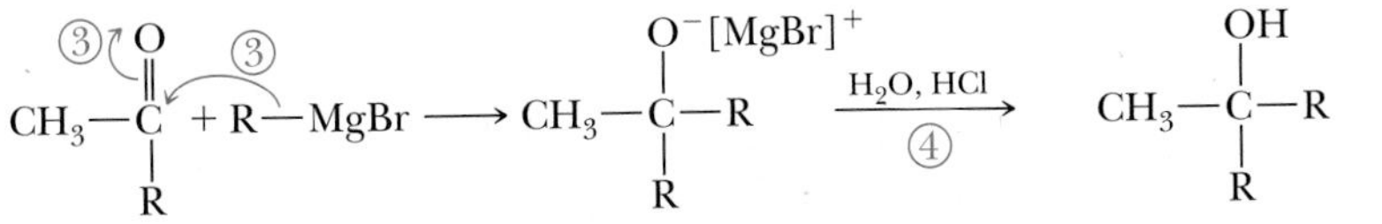

A ketone   Magnesium salt   A tertiary alcohol

Example 18.8

Complete these Grignard reactions.

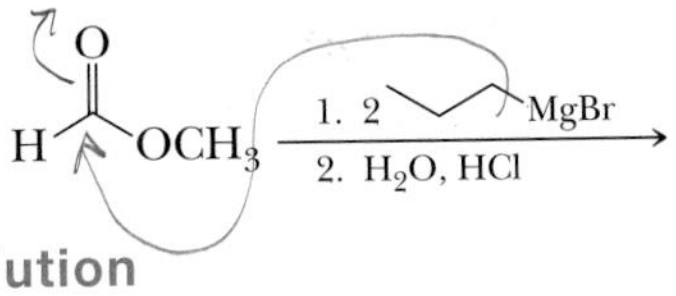

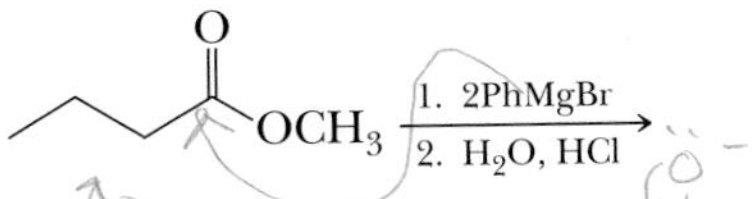

Solution

Sequence (a) gives a secondary alcohol, and sequence (b) gives a tertiary alcohol.

(a) 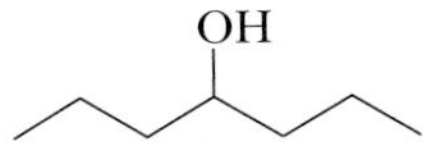   **(b)**

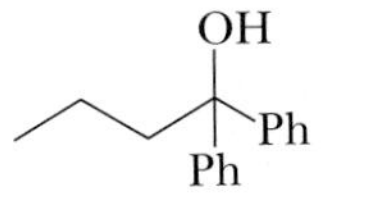

Problem 18.8

Show how to prepare these alcohols by treatment of an ester with a Grignard reagent.

(a) 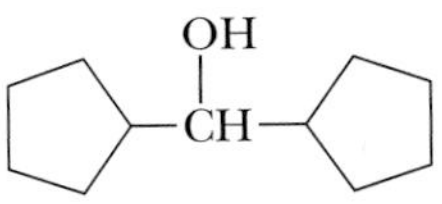   **(b)**

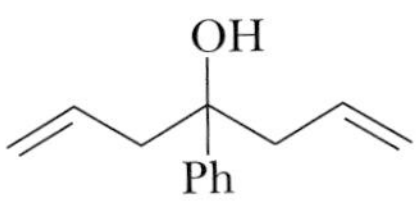

B.   Organolithium Compounds

Organolithium compounds are even more powerful nucleophiles than Grignard reagents, and they react with esters to give the same types of secondary and tertiary alcohols as shown for Grignard reagents, often in higher yields.

$$RCOCH_3 \xrightarrow[\text{2. } H_2O, HCl]{\text{1. } 2R'Li} R-\overset{\overset{\displaystyle OH}{|}}{\underset{\underset{\displaystyle R'}{|}}{C}}-R'$$

C. Lithium Diorganocuprates

※ Only react w/ acid chlorides

Acid chlorides react readily with lithium diorganocopper (Gilman) reagents to give ketones, as illustrated by the conversion of pentanoyl chloride to 2-hexanone. The reaction is carried out at − 78°C in either diethyl ether or tetrahydrofuran. Following hydrolysis in aqueous acid, the ketone is isolated in good yield.

Pentanoyl chloride   2-Hexanone

Notice that, under these conditions, the ketone does not react further. This contrasts with the reaction of an ester with a Grignard reagent or organolithium compound, where the intermediate ketone reacts with a second mole of the organometallic compound to give an alcohol. The reason for this difference in reactivity is that the tetrahedral carbonyl addition intermediate is stable at − 78°C; it survives until the workup causes it to decompose to the ketone.

R$_2$CuLi reagents react readily only with the very reactive acid chlorides; they do not react with aldehydes, ketones, esters, amides, acid anhydrides, or nitriles. The following compound contains both an acid chloride and an ester group. When treated with lithium dimethylcopper, only the acid chloride reacts.

Example 18.9

Show how to bring about these conversions in good yield.

Solution

(a) Treat the acid chloride with lithium dimethylcopper followed by H$_2$O.

(b) Treat the carboxylic acid with thionyl chloride to form the acid chloride, followed by treatment with lithium diallylcopper and then aqueous acid.

$$\text{Ph} \diagdown \diagup \diagdown \!\!\! \underset{\text{OH}}{\overset{\text{O}}{\diagdown}} \xrightarrow[\text{3. H}_2\text{O}]{\substack{\text{1. SOCl}_2 \\ \text{2. (CH}_2\!=\!\text{CHCH}_2)_2\text{CuLi}}} \text{Ph} \diagdown \diagup \diagdown \!\!\! \underset{}{\overset{\text{O}}{\diagdown}} \diagup \diagdown$$

Problem 18.9

Show how to bring about these conversions in good yield.

(a) PhCOOH $\longrightarrow$ $\text{Ph} \diagdown \!\!\! \underset{}{\overset{\text{O}}{\diagdown}} \diagdown \diagup \diagdown \diagup$

(b) CH$_2$=CHCl $\longrightarrow$ $\diagdown \diagup \!\!\! \underset{}{\overset{\text{O}}{\diagdown}} \diagdown \diagup \diagdown$

18.10 Interconversion of Functional Derivatives

We have seen throughout the past several sections that acid chlorides are the most reactive toward nucleophilic acyl substitution and that carboxylic salts are the least reactive. Another useful way to think about the reactions of functional derivatives of carboxylic acids is summarized in Figure 18.2. The reactivities also parallel the overall direction of the reaction, that is, the equilibrium constant. Any functional group lower in this figure can be prepared from any functional group above it by treatment with an appropriate oxygen or nitrogen nucleophile. An acid chloride, for example, can be converted to an acid anhydride, an ester, an amide, or a carboxylic acid. Acid anhydrides, esters, and amides, however, do not react with chloride ion to give acid chlorides.

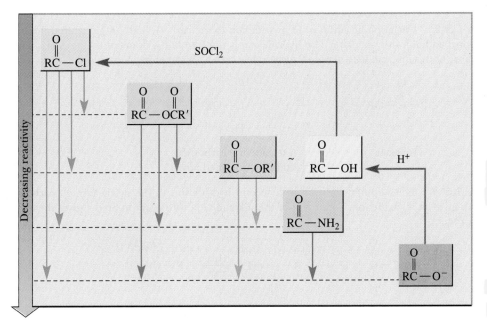

Figure 18.2
Reactivities of carboxyl derivatives toward nucleophilic acyl substitution. A more reactive derivative may be converted to a less reactive derivative by treatment with an appropriate reagent. Treatment of a carboxylic acid with thionyl chloride (the acid chloride of sulfurous acid) converts it to the more reactive acid chloride. Carboxylic acids are about as reactive as esters under acidic conditions, but they are converted to the unreactive carboxylates under basic conditions.

18.11   Reduction

Most reductions of carbonyl compounds, including aldehydes and ketones, are now accomplished by transfer of hydride ions from boron or aluminum hydrides. We have already seen the use of sodium borohydride to reduce the carbonyl group of aldehydes and ketones to hydroxyl groups (Section 16.14B) and the use of lithium aluminum hydride to reduce not only aldehyde and ketone carbonyl groups but also carboxyl groups to hydroxyl groups (Section 17.6A).

A.   Esters

An ester is reduced by lithium aluminum hydride to two alcohols; the alcohol derived from the acyl group is primary and is usually the objective of the reduction.

(handwritten note: LiAlH4 gets rid of O=C ; just OH left)

$$\text{Ph}\overset{\text{O}}{\underset{}{\text{—C—}}}\text{OCH}_3 \xrightarrow[\text{2. H}_2\text{O, HCl}]{\text{1. LiAlH}_4,\ \text{ether}} \text{Ph—CH—CH}_2\text{OH} + \text{CH}_3\text{OH}$$

Methyl 2-phenylpropanoate   2-Phenyl-1-propanol   Methanol

Reduction of an ester by lithium aluminum hydride involves hydride ion transfer to the carbonyl carbon to form a tetrahedral carbonyl addition intermediate. This intermediate then collapses by loss of alkoxide ion to an aldehyde. A second hydride ion transfer to the new carbonyl group completes the reduction. Treatment of the resulting alkoxide ion with water or aqueous acid gives the product of the reduction. This reaction is closely analogous to the reaction of Grignard reagents, with the exception that a hydride ion rather than an alkyl anion is being added to the carbonyl carbon.

Mechanisms
Acid Derivatives
Reduction of an ester by LAH

Mechanism   Reduction of an Ester by Lithium Aluminum Hydride

Steps 1 & 2:   Nucleophilic addition of hydride ion to the carbonyl carbon gives a tetrahedral carbonyl addition intermediate. The hydride ion is not free but is donated by the AlH_4^- ion. Collapse of this intermediate in Step 2 gives a new carbonyl-containing compound.

$$\text{R—C(=O)—OR'} + :\text{H}^- \xrightarrow{(1)} \text{R—}\overset{:\text{O}^-}{\underset{\text{H}}{\text{C}}}\text{—OR'} \xrightarrow{(2)} \text{R—}\overset{\text{O}}{\underset{\text{H}}{\text{C}}} + :\text{OR'}$$

An ester   A tetrahedral carbonyl
                                    addition intermediate

Steps 3 & 4:   Nucleophilic addition of a second hydride ion to the newly formed carbonyl group gives an alkoxide ion. Treatment of this alkoxide ion with water in Step 4 gives the alcohol product.

$$\text{R—}\overset{\text{O}}{\underset{\text{H}}{\text{C}}} + :\text{H}^- \xrightarrow{(3)} \text{R—}\overset{:\text{O}^-}{\underset{\text{H}}{\text{C}}}\text{—H} \xrightarrow[(4)]{\text{H}_2\text{O}} \text{R—}\overset{\text{O—H}}{\underset{\text{H}}{\text{C}}}\text{—H}$$

A primary alcohol

Sodium borohydride is not normally used to reduce esters because the reaction is very slow. Because of this lower reactivity of sodium borohydride toward esters, it is possible to reduce the carbonyl group of an aldehyde or ketone to an alcohol group with this reagent without reducing an ester or carboxyl group in the same molecule.

Reduction of an ester to a primary alcohol can be viewed as two successive hydride ion transfers, as shown in the mechanism we have just presented. Chemists wondered if it might be possible to modify the structure of the reducing agent so as to reduce an ester to an aldehyde and no further. The most useful modified hydride reducing agent developed for this purpose is diisobutylaluminum hydride (**DIBALH**).

$$[(CH_3)_2CHCH_2]_2AlH$$

Diisobutylaluminum hydride

Reductions are typically carried out in toluene or hexane at $-78°C$ (dry ice/acetone temperature) followed by warming to room temperature and addition of aqueous acid to hydrolyze the aluminum salts and liberate the aldehyde. Reduction of esters using DIBALH has become a valuable method for the synthesis of aldehydes, as illustrated by the synthesis of hexanal.

Methyl hexanoate   Hexanal

If reduction of an ester using **DIBALH** is carried out at room temperature, the ester is reduced to a primary alcohol. At low temperature, the tetrahedral carbonyl addition intermediate does not eliminate alkoxide ion, and the more reactive aldehyde is not formed until after workup, when the hydride ion has been destroyed. Thus, temperature control is critical for the selective reduction of an ester to an aldehyde.

B. Amides

Lithium aluminum hydride reduction of amides can be used to prepare primary, secondary, or tertiary amines, depending on the degree of substitution of the amide.

Octanamide   1-Octanamine

N,N-Dimethylbenzamide   N,N-Dimethylbenzylamine

The mechanism for the reduction of an amide to an amine is shown here divided into four steps.

Mechanism   Reduction of an Amide by Lithium Aluminum Hydride

In Step 1, a hydride ion is transferred from AlH_4^- to the carbonyl carbon. The electrons of the pi bond flow to oxygen and in Step 2 form an oxygen-aluminum bond. Rearrangement of electron pairs in Step 3 generates an iminium ion, which then adds a second hydride ion in Step 4 to complete the reduction. H_3AlO^- is a reasonably good leaving group; recall that $Al(OH)_3$ is somewhat acidic.

An iminium ion

Example 18.10

Show how to bring about the following conversions in good yield.

(a) $C_6H_5\overset{O}{\overset{\|}{C}}OH \longrightarrow C_6H_5CH_2-N\bigcirc$ (b) [cyclohexane]$-\overset{O}{\overset{\|}{C}}OH \longrightarrow$ [cyclohexane]$-CH_2NHCH_3$

Solution

The key in each part is to convert the carboxylic acid to an amide and then to reduce the amide with $LiAlH_4$. The amide can be prepared by treating the carboxylic acid with $SOCl_2$ to form the acid chloride (Section 17.8) and then treating the acid chloride with an amine (Section 18.7A). Alternatively, the carboxylic acid can be converted to an ethyl ester by Fischer esterification (Section 17.7A), and the ester can be treated with an amine to give the amide. Solution (a) uses the acid chloride route and solution (b) uses the ester route.

(a) $C_6H_5\overset{O}{\overset{\|}{C}}OH \xrightarrow{SOCl_2} C_6H_5\overset{O}{\overset{\|}{C}}Cl \xrightarrow{HN\bigcirc} C_6H_5\overset{O}{\overset{\|}{C}}-N\bigcirc \xrightarrow[\text{2. } H_2O]{\text{1. } LiAlH_4} C_6H_5CH_2-N\bigcirc$

(b) [cyclohexane]$-\overset{O}{\overset{\|}{C}}OH \xrightarrow[\text{2. } CH_3NH_2]{\text{1. } CH_3CH_2OH,\ H^+}$ [cyclohexane]$-\overset{O}{\overset{\|}{C}}NHCH_3 \xrightarrow[\text{4. } H_2O]{\text{3. } LiAlH_4}$ [cyclohexane]$-CH_2NHCH_3$

Problem 18.10

Show how to convert hexanoic acid to each amine in good yield.

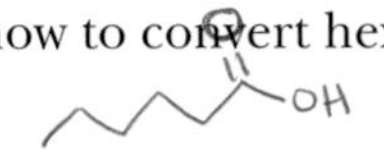

(a) 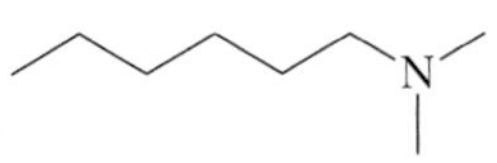 **(b)**

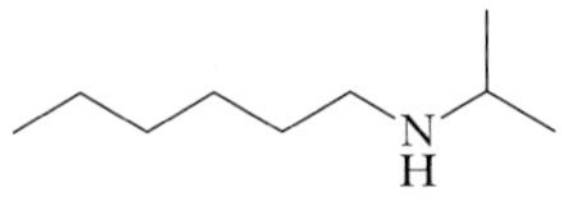

Example 18.11

Show how to convert phenylacetic acid to these compounds.

(a) 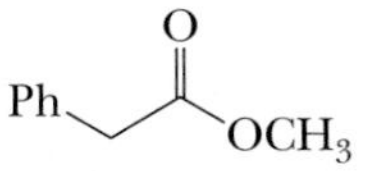 **(b)** 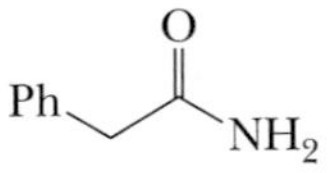 **(c)** 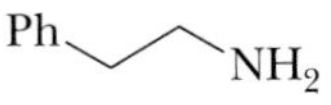 **(d)**

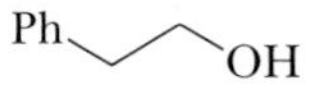

Solution

Prepare methyl ester (a) by Fischer esterification (Section 17.7A) of phenylacetic acid with methanol. Then treat this ester with ammonia to prepare amide (b). Alternatively, treat phenylacetic acid with thionyl chloride (Section 17.8) to give an acid chloride. Then treat this acid chloride with ammonia to give amide (b). Reduction of the amide (b) by LiAlH$_4$ gives the primary amine (c). Similar reduction of either phenylacetic acid or ester (a) gives alcohol (d).

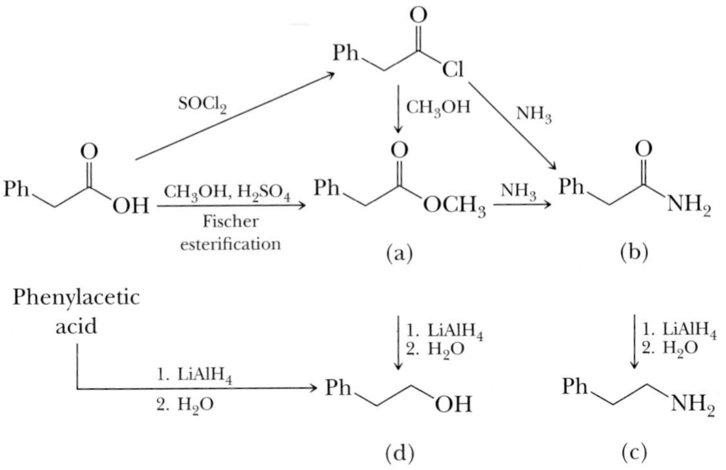

Problem 18.11

Show how to convert (R)-2-phenylpropanoic acid to these compounds.

(a)

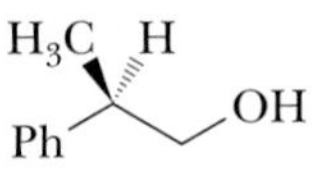

(R)-2-Phenyl-1-propanol

(b)

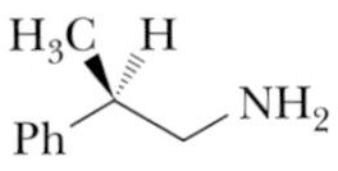

(R)-2-Phenyl-1-propanamine

C. Nitriles

The cyano group of a nitrile is reduced by lithium aluminum hydride to a primary amino group. Reduction of cyano groups is useful for the preparation of primary amines only.

$$CH_3CH=CH(CH_2)_4C\equiv N \xrightarrow[\text{2. H}_2\text{O}]{\text{1. LiAlH}_4} CH_3CH=CH(CH_2)_4CH_2NH_2$$

6-Octenenitrile   6-Octen-1-amine

Mechanisms
Acid Derivatives
Hofmann rearrangement

18.12 The Hofmann Rearrangement

When a primary amide is treated with bromine or chlorine in aqueous sodium or potassium hydroxide, the following rearrangement takes place; the amide is converted to a primary amine with one fewer carbon atom than the starting amide and the carbonyl carbon of the amide is lost as carbonate ion. Furthermore, when the migrating group is chiral, it migrates with complete retention of configuration.

$$\text{(S)-2-Phenylpropanamide} \xrightarrow[\text{H}_2\text{O}]{\text{Br}_2\text{, NaOH}} \text{(S)-1-Phenylethanamine} + Na_2CO_3$$

(S)-2-Phenylpropanamide   (S)-1-Phenylethanamine

This rearrangement of primary amides was discovered by the German chemist August Hofmann, for whom it has been named. A mechanism for the Hofmann rearrangement can be divided into four stages, each having some analogy to reactions we have already studied.

Mechanism   **The Hofmann Rearrangement of Primary Amides**

Stage 1: An acid-base reaction between the amide and hydroxide ion gives an amide anion, which is both a base and a nucleophile. Reaction of this anion with bromine by nucleophilic displacement gives an N-bromoamide.

$$R-\overset{O}{\overset{\|}{C}}-\overset{..}{N}-H \underset{}{\overset{HO^-}{\rightleftharpoons}} R-\overset{O}{\overset{\|}{C}}-\overset{..}{\overset{-}{N}}: \xrightarrow{Br-Br} R-\overset{O}{\overset{\|}{C}}-\overset{..}{N}-Br + Br^-$$

An amide anion   An N-bromoamide

Stage 2: The second amide hydrogen is abstracted by base followed by elimination of Br⁻ (an α-elimination) to give an acyl nitrene, an unstable species containing a neutral, electron-deficient nitrogen atom. Nitrenes resemble carbenes (Section 15.4) both electronically and in their reactivity. Migration of the adjacent R group of the nitrene with its bonding electrons to the electron-deficient nitrogen gives an isocyanate, a molecule in which all atoms have complete valence shells.

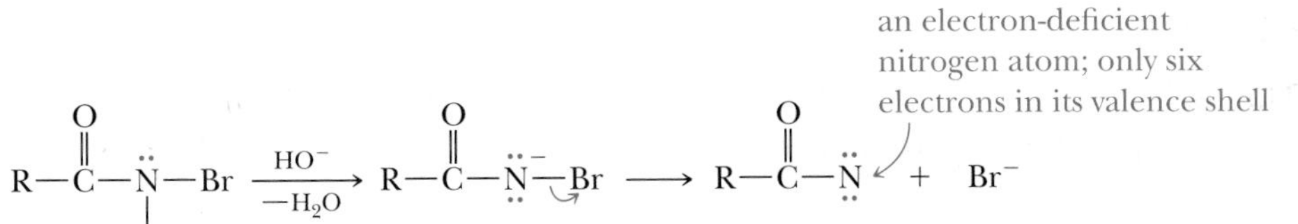

an electron-deficient
nitrogen atom; only six
electrons in its valence shell

An acyl nitrene

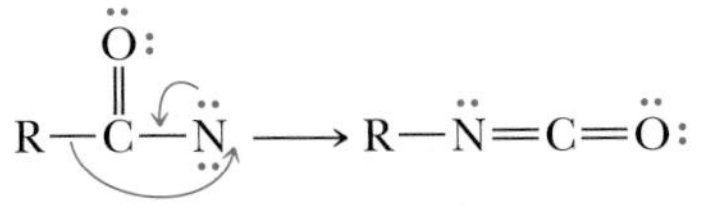

An isocyanate

Stage 3: Reaction of the isocyanate with water gives a carbamic acid, the functional group of which is a carboxyl group bonded to nitrogen.

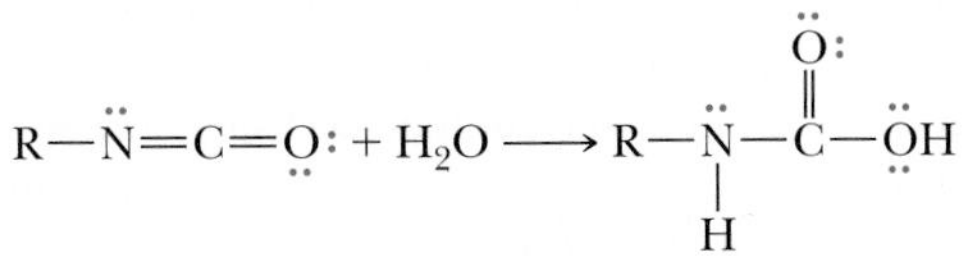

A carbamic acid

Stage 4: Carbamic acids are unstable species and undergo decarboxylation to give a primary amine and carbon dioxide.

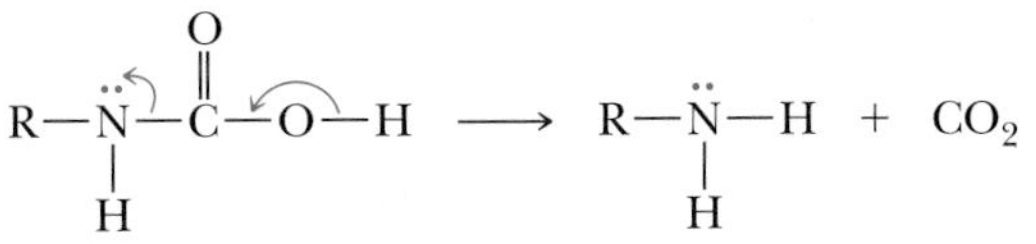

A primary amine

Example 18.12

Complete these equations.

(a)

$\xrightarrow[\text{H}_2\text{O}]{\text{Br}_2,\ \text{NaOH}}$

(S)-2-Phenylpropanamide

(b)

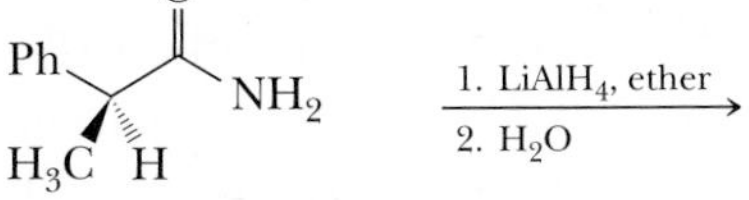

$\xrightarrow[\text{2. H}_2\text{O}]{\text{1. LiAlH}_4,\ \text{ether}}$

(S)-2-Phenylpropanamide

Solution

Each reaction proceeds with complete retention of configuration at the stereocenter.

(a)

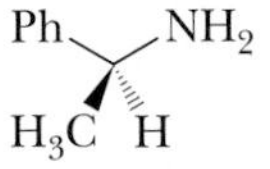

(S)-1-Phenylethanamine

(b)

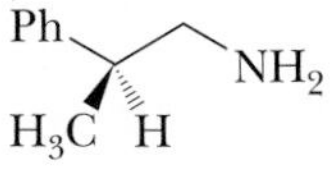

(S)-2-Phenyl-1-propanamine

Problem 18.12

Show how to convert phenylacetic acid into the following in good yield.

(a) $PhCH_2CH_2NH_2$ (b) $PhCH_2NH_2$

Summary

The functional group of an **acid halide** (Section 18.1A) is an acyl group bonded to a halogen atom. The most common and widely used of these are the acid chlorides. The functional group of a **carboxylic anhydride** (Section 18.1B) is two acyl groups bonded to an oxygen. The functional group of a **carboxylic ester** (Section 18.1C) is an acyl group bonded to —OR or —OAr. A cyclic ester is given the name **lactone.** Phosphoric acid has three —OH groups and can form mono-, di-, and tri-esters. The functional group of an **amide** (Section 18.1D) is an acyl group bonded to a nitrogen. A cyclic amide is given the name **lactam.** The functional group of an **imide** is two acyl groups bonded to a nitrogen. The functional group of a **nitrile** (Section 18.1E) is a C≡N group bonded to a carbon atom.

Values of pK_a for amides of carboxylic acids are 15–17, which means that they are comparable in acidity to alcohols (Section 9.3). Values of pK_a for imides are 8–10, which means that they dissolve in aqueous NaOH to form water-soluble salts. Sulfonamides derived from ammonia and primary amines have pK_a values of approximately 10 and dissolve in aqueous NaOH to form water-soluble salts.

A common reaction theme of functional derivatives of carboxylic acids is nucleophilic addition to the carbonyl carbon to form a tetrahedral carbonyl addition intermediate, which then collapses to regenerate the carbonyl group. The result is **nucleophilic acyl substitution** (Section 18.4). Listed in order of increasing reactivity toward nucleophilic acyl substitution, these functional derivatives are:

$$\underset{RCNH_2}{\overset{\overset{\displaystyle O}{\|}}{}} \qquad \underset{RCOR'}{\overset{\overset{\displaystyle O}{\|}}{}} \qquad \underset{RCOCR'}{\overset{\overset{\displaystyle O\ \ O}{\|\ \ \|}}{}} \qquad \underset{RCCl}{\overset{\overset{\displaystyle O}{\|}}{}}$$

Increasing chemical reactivity ⟶

Any more reactive functional derivative can be converted to any less reactive functional derivative by reaction with an appropriate oxygen or nitrogen nucleophile (Section 18.10).

Key Reactions

1. Acidity of Imides (Section 18.2)

Imides (pK_a 8–10) dissolve in aqueous NaOH by forming water-soluble salts. Imides are stronger acids than amides because the imide anion is stabilized by delocalization of the negative charge onto the two carbonyl oxygens.

Insoluble in water   A water-soluble salt

2. Acidity of Sulfonamides (Section 18.2)

Sulfonamides (pK_a 9–10), like imides, dissolve in aqueous NaOH by forming water-soluble salts. The stability of the sulfonamide anion is due to delocalization of the negative charge in the anion.

Insoluble in water   A water-soluble salt

3. Hydrolysis of an Acid Chloride (Section 18.5A)

Low-molecular-weight acid chlorides react vigorously with water. Higher molecular-weight acid chlorides react less rapidly.

$$CH_3CCl + H_2O \longrightarrow CH_3COH + HCl$$

4. Hydrolysis of an Acid Anhydride (Section 18.5B)

Low-molecular-weight acid anhydrides react readily with water. Higher molecular-weight acid anhydrides react less rapidly.

$$CH_3COCCH_3 + H_2O \longrightarrow CH_3COH + HOCCH_3$$

5. Hydrolysis of an Ester (Section 18.5C)

Esters are hydrolyzed only in the presence of acid or base. Acid is a catalyst. Base is required in an equimolar amount.

6. Hydrolysis of an Amide (Section 18.5D)

Either acid or base is required in an amount equivalent to that of the amide.

7. Hydrolysis of a Nitrile (Section 18.5E)

Either acid or base is required in an amount equivalent to that of the nitrile.

$$\text{C}_6\text{H}_5-\text{CH}_2\text{C}\equiv\text{N} + 2\text{H}_2\text{O} + \text{H}_2\text{SO}_4 \xrightarrow[\text{heat}]{\text{H}_2\text{O}} \text{C}_6\text{H}_5-\text{CH}_2\overset{\overset{\displaystyle O}{\|}}{\text{C}}\text{OH} + \text{NH}_4{}^+\text{HSO}_4{}^-$$

$$\text{CH}_3(\text{CH}_2)_9\text{C}\equiv\text{N} + \text{H}_2\text{O} + \text{NaOH} \xrightarrow[\text{heat}]{\text{H}_2\text{O}} \text{CH}_3(\text{CH}_2)_9\overset{\overset{\displaystyle O}{\|}}{\text{C}}\text{O}^-\text{Na}^+ + \text{NH}_3$$

8. Reaction of an Acid Chloride with an Alcohol (Section 18.6A)

Treatment of an acid chloride with an alcohol gives an ester plus HCl. Preparation of an acid-sensitive ester is carried out using an equimolar amount of triethylamine or pyridine to neutralize the HCl.

9. Reaction of an Acid Anhydride with an Alcohol (Section 18.6B)

Treatment of an acid anhydride with an alcohol gives one mole of ester and one mole of carboxylic acid.

$$\text{CH}_3\overset{\overset{\displaystyle O}{\|}}{\text{C}}\text{O}\overset{\overset{\displaystyle O}{\|}}{\text{C}}\text{CH}_3 + \text{HOCH}_2\text{CH}_3 \longrightarrow \text{CH}_3\overset{\overset{\displaystyle O}{\|}}{\text{C}}\text{OCH}_2\text{CH}_3 + \text{CH}_3\overset{\overset{\displaystyle O}{\|}}{\text{C}}\text{OH}$$

10. Reaction of an Ester with an Alcohol: Transesterification (Section 18.6C)

Transesterification requires an acid catalyst and an excess of alcohol to drive the reaction to completion.

11. Reaction of an Acid Chloride with Ammonia or an Amine (Section 18.7A)

Reaction requires two moles of ammonia or amine: one to form the amide and one to neutralize the HCl byproduct.

12. Reaction of an Acid Anhydride with Ammonia or an Amine (Section 18.7B)

The reaction requires two moles of ammonia or amine: one to form the amide and one to neutralize the carboxylic acid byproduct.

$$\text{CH}_3\overset{\overset{\displaystyle O}{\|}}{\text{C}}\text{O}\overset{\overset{\displaystyle O}{\|}}{\text{C}}\text{CH}_3 + 2\text{NH}_3 \longrightarrow \text{CH}_3\overset{\overset{\displaystyle O}{\|}}{\text{C}}\text{NH}_2 + \text{CH}_3\overset{\overset{\displaystyle O}{\|}}{\text{C}}\text{O}^-\text{NH}_4{}^+$$

13. Reaction of an Ester with Ammonia or an Amine (Section 18.7C)

Treatment of an ester with ammonia or a primary or secondary amine gives an amide.

14. Reaction of an Acid Chloride with a Carboxylic Acid Salt (Section 18.8)

Treatment of an acid chloride with the salt of a carboxylic acid is a valuable method for synthesizing mixed anhydrides.

$$CH_3\overset{O}{\underset{}{C}}Cl + Na^+\ ^-O\overset{O}{\underset{}{C}}-\!\!\bigcirc \longrightarrow CH_3\overset{O}{\underset{}{C}}O\overset{O}{\underset{}{C}}-\!\!\bigcirc + Na^+ Cl^-$$

15. Reaction of an Ester with a Grignard Reagent (Section 18.9A)

Treatment of a formic ester with a Grignard reagent followed by hydrolysis gives a secondary alcohol. Treatment of any other ester with a Grignard reagent gives a tertiary alcohol.

16. Reaction of an Acid Chloride with a Lithium Diorganocuprate (Section 18.9C)

Acid chlorides react readily with lithium diorganocuprates at $-78°C$ to give ketones.

17. Reduction of an Ester (Section 18.11A)

Reduction of an ester by lithium aluminum hydride gives two alcohols. Reduction by diisobutylaluminum hydride (DIBALH) at low temperature gives an aldehyde and an alcohol.

18. Reduction of an Amide (Section 18.11B)

Reduction of an amide by lithium aluminum hydride gives an amine.

19. Reduction of a Nitrile (Section 18.11C)

Reduction of a cyano group by lithium aluminum hydride gives a primary amino group.

20. The Hofmann Rearrangement of a Primary Amide (Section 18.12)

When treated with bromine and aqueous sodium hydroxide, a primary amide is converted to an amine with one fewer carbon. The carbon atom of the carbonyl group is lost as carbon dioxide. The rearrangement occurs with retention of configuration.

Problems

Structure and Nomenclature

18.13 Draw structural formulas for these compounds.

(a) Dimethyl carbonate
(b) Benzonitrile
(c) Isopropyl 3-methylhexanoate
(d) Diethyl oxalate
(e) Ethyl (Z)-2-pentenoate
(f) Butanoic anhydride
(g) Dodecanamide
(h) Ethyl 3-hydroxybutanoate
(i) Octanoyl chloride
(j) Diethyl *cis*-1,2-cyclohexanedicarboxylate
(k) Methanesulfonyl chloride
(l) *p*-Toluenesulfonyl chloride

18.14 Write the IUPAC name for each compound.

Physical Properties

18.15 Both the melting point and the boiling point of acetamide are higher than those of its *N,N*-dimethyl derivative. How do you account for these differences?

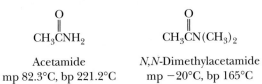

Acetamide
mp 82.3°C, bp 221.2°C

N,N-Dimethylacetamide
mp −20°C, bp 165°C

Spectroscopy

18.16 Each hydrogen of a primary amide typically has a separate ^{1}H-NMR resonance, as illustrated by the separate signals for the two amide hydrogens of propanamide, which fall at δ 6.22 and δ 6.58. Furthermore, each methyl group of *N,N*-dimethylformamide has a separate resonance (δ 3.88 and δ 3.98). How do you account for these observations?

18.17 Propose a structural formula for compound A, $C_7H_{14}O_2$, consistent with its ^{1}H-NMR and IR spectra.

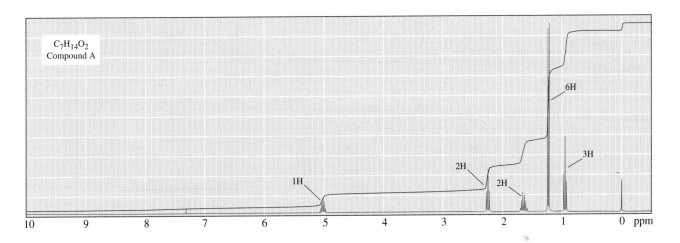

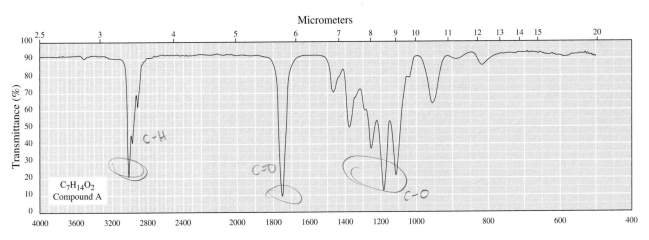

18.18 Propose a structural formula for compound B, $C_6H_{13}NO$, consistent with its ^1H-NMR and IR spectra.

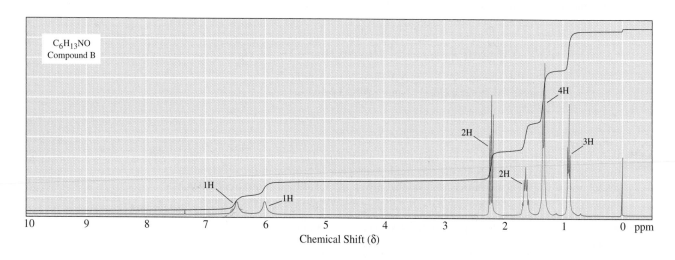

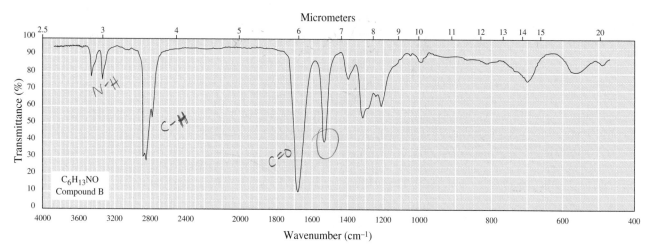

18.19 Propose a structural formula for each compound consistent with its ^1H-NMR and ^{13}C-NMR spectra.

(a) $C_5H_{10}O_2$

^1H-NMR	^{13}C-NMR
0.96 (d, 6H)	161.11
1.96 (m, 1H)	70.01
3.95 (d, 2H)	27.71
8.08 (s, 1H)	19.00

(b) $C_7H_{14}O_2$

^1H-NMR	^{13}C-NMR
0.92 (d, 6H)	171.15
1.52 (m, 2H)	63.12
1.70 (m, 1H)	37.31
2.09 (s, 3H)	25.05
4.10 (t, 2H)	22.45
	21.06

(c) $C_6H_{12}O_2$

^{1}H-NMR	^{13}C-NMR
1.18 (d, 6H)	177.16
1.26 (t, 3H)	60.17
2.51 (m, 1H)	34.04
4.13 (q, 2H)	19.01
	14.25

(d) $C_7H_{12}O_4$

^{1}H-NMR	^{13}C-NMR
1.28 (t, 6H)	166.52
3.36 (s, 2H)	61.43
4.21 (q, 4H)	41.69
	14.07

(e) $C_4H_7ClO_2$

^{1}H-NMR	^{13}C-NMR
1.68 (d, 3H)	170.51
3.80 (s, 3H)	52.92
4.42 (q, 1H)	52.32
	21.52

(f) $C_4H_6O_2$

^{1}H-NMR	^{13}C-NMR
2.29 (m, 2H)	177.81
2.59 (t, 2H)	68.58
4.36 (t, 2H)	27.79
	22.17

Reactions

18.20 Draw the structural formula of the principal product formed when benzoyl chloride is treated with the following reagents.

(a) OH

(b) OH, pyridine

(c) SH, pyridine

(d) NH$_2$ (two equivalents)

(e) O⁻Na⁺

(f) $(CH_3)_2CuLi$, then H_3O^+

(g) CH_3O——NH$_2$, pyridine

(h) C_6H_5MgBr (two equivalents), then H_3O^+

18.21 Draw the structural formula of the principal product formed when ethyl benzoate is treated with the following reagents.

(a) H_2O, NaOH, heat
(b) H_2O, H_2SO_4, heat
(c) $CH_3CH_2CH_2CH_2NH_2$
(d) DIBALH ($-78°C$), then H_2O
(e) $LiAlH_4$, then H_2O
(f) C_6H_5MgBr (two equivalents), then HCl/H_2O

18.22 The mechanism for hydrolysis of an ester in aqueous acid involves formation of a tetrahedral carbonyl addition intermediate. Evidence in support of this mechanism comes from an experiment designed by Myron Bender. He first prepared ethyl benzoate enriched with oxygen-18 in the carbonyl oxygen and then carried out acid-catalyzed hydrolysis of the ester in water containing no enrichment in oxygen-18. He discovered that if he stopped the experiment after only partial hydrolysis and isolated the remaining ester, the recovered ethyl benzoate had lost a portion of its enrichment in oxygen-18. In other words, some exchange had occurred between oxygen-18 of the ester and oxygen-16 of water. Show how this observation bears on the formation of a tetrahedral carbonyl addition intermediate during acid-catalyzed ester hydrolysis.

18.23 Predict the distribution of oxygen-18 in the products obtained from hydrolysis of ethyl benzoate labeled in the ethoxy oxygen under the following conditions.

(a) In aqueous NaOH   **(b)** In aqueous HCl
(c) What distribution would you predict if the reaction were done with the *tert*-butyl ester in HCl?

18.24 Draw the structural formula of the principal product formed when benzamide is treated with the following reagents.

(a) H_2O, HCl, heat   **(b)** NaOH, H_2O, heat
(c) $LiAlH_4$, then H_2O **(d)** Br_2, NaOH, heat

18.25 Draw the structural formula of the principal product formed when benzonitrile is treated with the following reagents.

(a) H_2O (one equivalent), H_2SO_4, heat   **(b)** H_2O (excess), H_2SO_4, heat
(c) NaOH, H_2O, heat   **(d)** $LiAlH_4$, then H_2O

18.26 Show the product expected when the following unsaturated δ-ketoester is treated with each reagent.

(a) $\dfrac{H_2 \ (1 \ mol)}{Pd, \ EtOH}$ **(b)** $\dfrac{NaBH_4}{CH_3OH}$ **(c)** $\dfrac{1. \ LiAlH_4, \ THF}{2. \ H_2O}$ **(d)** $\dfrac{1. \ DIBALH, \ -78°C}{2. \ H_2O}$

18.27 The reagent diisobutylaluminum hydride (DIBALH) reduces esters to aldehydes. When nitriles are treated with DIBALH, followed by mild acid hydrolysis, the product is also an aldehyde. Give a mechanism for this reaction.

18.28 Show the product of treatment of this anhydride with each reagent.

(a) $\dfrac{H_2O, \ HCl}{heat}$ **(b)** $\dfrac{H_2O, \ NaOH}{heat}$ **(c)** $\dfrac{1. \ LiAlH_4}{2. \ H_2O}$ **(d)** $\dfrac{CH_3OH}{}$ **(e)** $\dfrac{NH_3 \ (2 \ mol)}{}$

18.29 The analgesic acetaminophen is synthesized by treating 4-aminophenol with one equivalent of acetic anhydride. Write an equation for the formation of acetaminophen.

4-Aminophenol

18.30 Treatment of choline with acetic anhydride gives acetylcholine, a neurotransmitter. Write an equation for the formation of acetylcholine.

$$(CH_3)_3\overset{+}{N}CH_2CH_2OH$$

Choline

18.31 Nicotinic acid, more commonly named niacin, is one of the B vitamins. Show how nicotinic acid can be converted to (a) ethyl nicotinate and then to (b) nicotinamide.

Nicotinic acid   Ethyl nicotinate   Nicotinamide
(Niacin)

18.32 Complete these reactions.

(a) CH_3O—⬡—NH_2 + $CH_3\overset{O}{\overset{\|}{C}}\overset{O}{\overset{\|}{C}}CH_3$ ⟶   (b) $CH_3\overset{O}{\overset{\|}{C}}Cl$ + 2HN⬡ ⟶

(c) $CH_3\overset{O}{\overset{\|}{C}}CH_3$ + HN⬡ ⟶   (d) ⬡—NH_2 + $CH_3(CH_2)_5\overset{O}{\overset{\|}{C}}H$ ⟶

18.33 Show the product of treatment of γ-butyrolactone with each reagent.

(a) NH_3 ⟶   (b) $\dfrac{1.\ LiAlH_4}{2.\ H_2O}$ (c) $\dfrac{1.\ 2PhMgBr,\ ether}{2.\ H_2O,\ HCl}$

(d) $NaOH$ $\xrightarrow[heat]{H_2O}$ (e) $\dfrac{1.\ 2CH_3Li,\ ether}{2.\ H_2O,\ HCl}$ (f) $\dfrac{1.\ DIBALH,\ ether,\ -78°C}{2.\ H_2O,\ HCl}$

18.34 Show the product of treatment of the following γ-lactam with each reagent.

(a) $\xrightarrow[heat]{H_2O,\ HCl}$ (b) $\xrightarrow[heat]{H_2O,\ NaOH}$ (c) $\dfrac{1.\ LiAlH_4}{2.\ H_2O}$

18.35 Draw structural formulas for the products of complete hydrolysis of meprobamate, phenobarbital, and pentobarbital in hot aqueous acid. Meprobamate (see *The Merck Index*, 12th ed., #5908) is a tranquilizer prescribed under 58 different trade names, including Equanil and Miltown. Phenobarbital (see *The Merck Index*, 12th ed., #7986) is a long-acting sedative, hypnotic, and anticonvulsant. Luminal is one of over a dozen names under which it is prescribed. Pentobarbital (see *The Merck Index*, 12th ed., #7272) is a short-acting sedative, hypnotic, and anticonvulsant. Nembutal is one of several trade names under which it is prescribed.

(a) 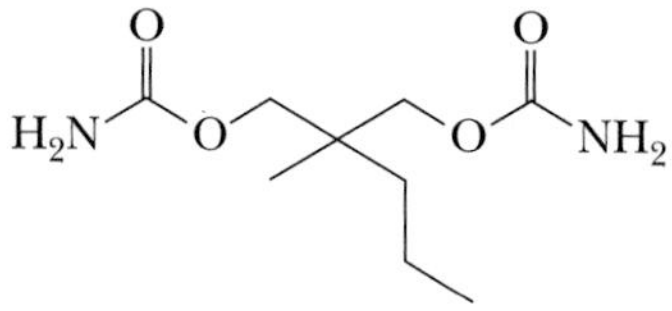 Meprobamate

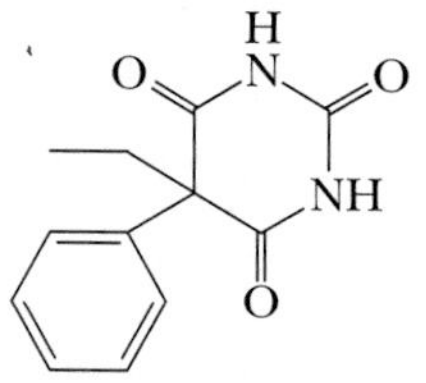

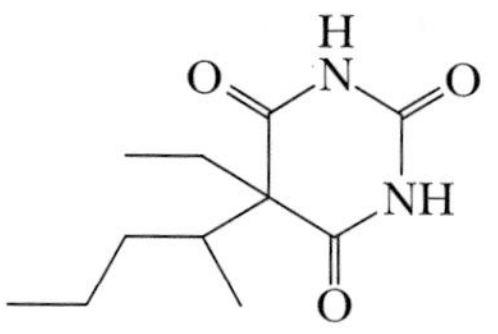

(b) Phenobarbital

(c) Pentobarbital

18.36 Following are steps in a synthesis of anthranilic acid from phthalic anhydride. Describe how you could bring about each step.

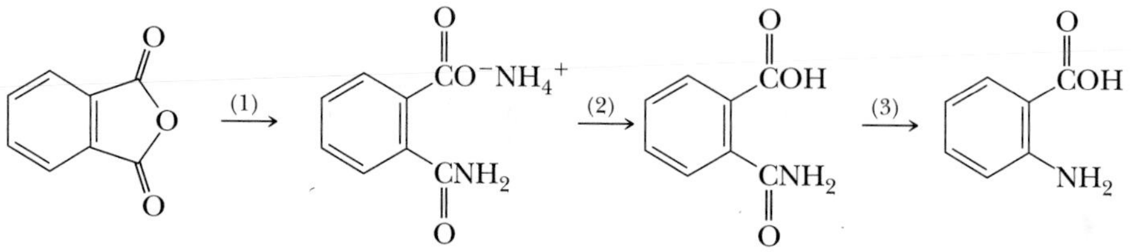

Phthalic anhydride   Anthranilic acid

18.37 Hofmann rearrangements of lower molecular-weight primary amides can be brought about using bromine in aqueous NaOH. Primary amides larger than about seven or eight carbon atoms are not sufficiently soluble in aqueous solution to react. Instead, they are dissolved in methanol or ethanol, and the corresponding sodium alkoxide is used as the base. Under these conditions, the isocyanate intermediate reacts with the alcohol to form a carbamate.

A primary amide   An isocyanate   A carbamate

Propose a mechanism for the reaction of an isocyanate with methanol to form a methyl carbamate.

18.38 (R)-(+)-Pulegone, readily available from pennyroyal oil (see *The Merck Index*, 12th ed., #8124), is an important enantiopure building block for organic syntheses. Propose a mechanism for each step in this transformation of pulegone.

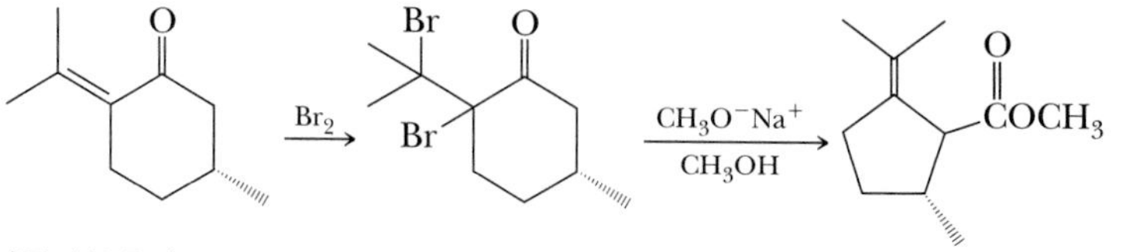

(R)-(1)-Pulegone

Synthesis

18.39 *N,N*-Diethyl *m*-toluamide (DEET) is the active ingredient in several common insect repellents. Propose a synthesis for DEET from 3-methylbenzoic acid.

18.40 Following is the structural formula of isoniazid (see *The Merck Index*, 12th ed., #5203), a drug used to treat tuberculosis. It is estimated that one third of the world's population is infected with tuberculosis, which results in approximately 3 million TB-related deaths per year. Show how to prepare isoniazid from 4-pyridine-carboxylic acid.

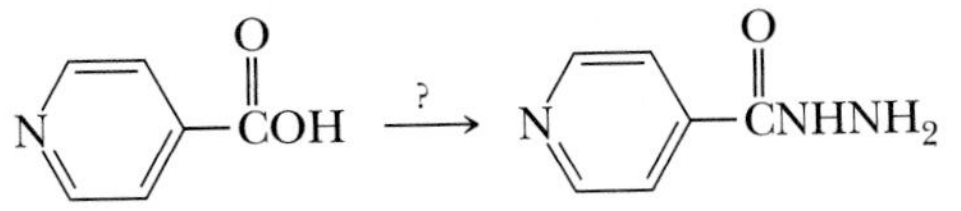

4-Pyridine-
carboxylic acid

4-Pyridine-
carboxylic acid hydrazide
(Isoniazid)

18.41 Show how to convert phenylacetylene to allyl phenylacetate.

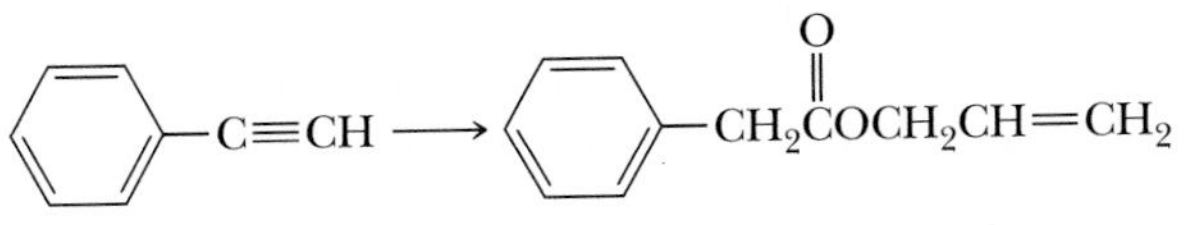

Phenylacetylene

Allyl phenylacetate

18.42 A step in a synthesis of PGE_1 (prostaglandin E_1, alprostadil; see *The Merck Index*, 12th ed., #8063) is the reaction of a trisubstituted cyclohexene with bromine to form a bromolactone. Alprostadil is used as a temporary therapy for infants born with congenital heart defects that restrict pulmonary blood flow. It brings about dilation of the ductus arteriosus, which in turn increases blood flow in the lungs and blood oxygenation. Propose a mechanism for formation of this bromolactone and account for the observed stereochemistry of each substituent on the cyclohexane ring.

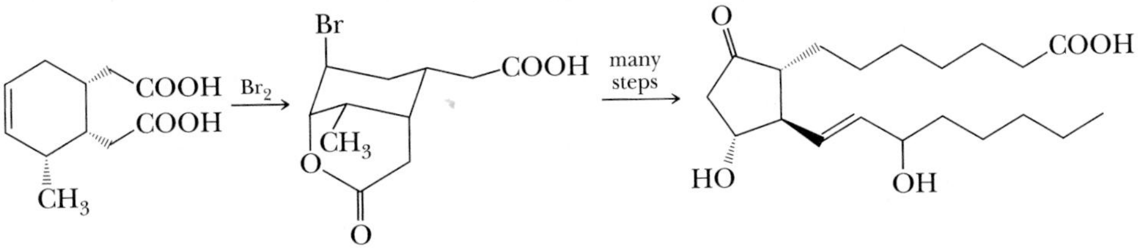

A bromolactone

PGE_1 (alprostadil)

18.43 Barbiturates are prepared by treatment of a derivative of diethyl malonate with urea in the presence of sodium ethoxide as a catalyst. Following is an equation for the preparation of barbital, a long-duration hypnotic and sedative, from diethyl diethylmalonate and urea. Barbital (see *The Merck Index*, 12th ed., #989) is prescribed under one of a dozen or more trade names.

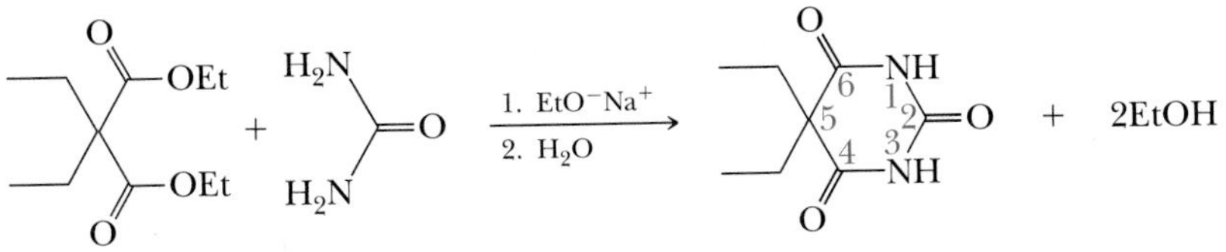

Diethyl
diethylmalonate

Urea

5,5-Diethylbarbituric acid
(Barbital)

(a) Propose a mechanism for this reaction.
(b) The pK_a of barbital is 7.4. Which is the most acidic hydrogen in this molecule? How do you account for its acidity?

18.44 The following compound is one of a group of β-chloroamines, many of which have anti-tumor activity. Describe a synthesis of this compound from anthranilic acid and ethylene oxide. This compound is a nitrogen mustard (Section 8.5).

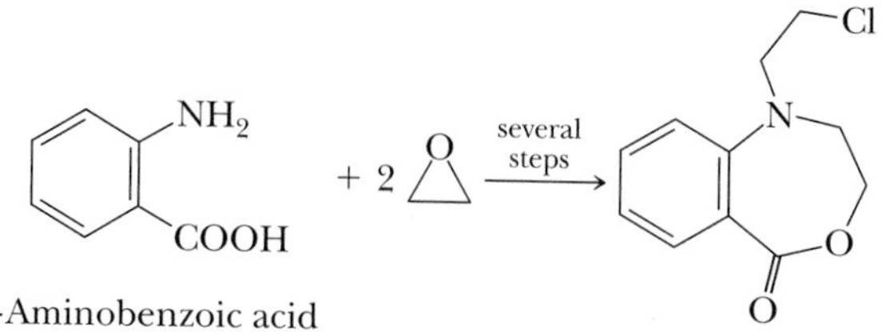

2-Aminobenzoic acid
(Anthranilic acid)

18.45 Following is a retrosynthetic scheme for the synthesis of 5-nonanone from 1-bromobutane as the only organic starting material. Show reagents and experimental conditions to bring about this synthesis.

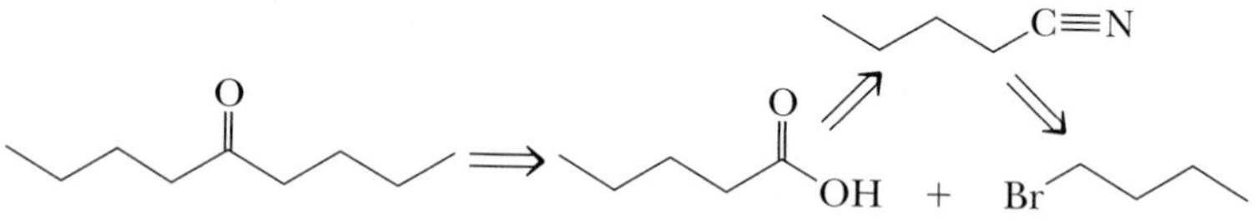

5-Nonanone

1-Bromobutane

18.46 Procaine (its hydrochloride is marketed as Novocaine) was one of the first local anesthetics for infiltration and regional anesthesia. According to this retrosynthetic scheme, procaine can be synthesized from 4-aminobenzoic acid, ethylene oxide, and diethylamine as sources of carbon atoms. Provide reagents and experimental conditions to carry out the synthesis of procaine from these three compounds. For the synthesis of several more members of the "caine" family of local anesthetics, see MC.42–MC.46. See the Chemistry in Action box "From Cocaine to Procaine and Beyond" in the Medicinal Chemistry interchapter.

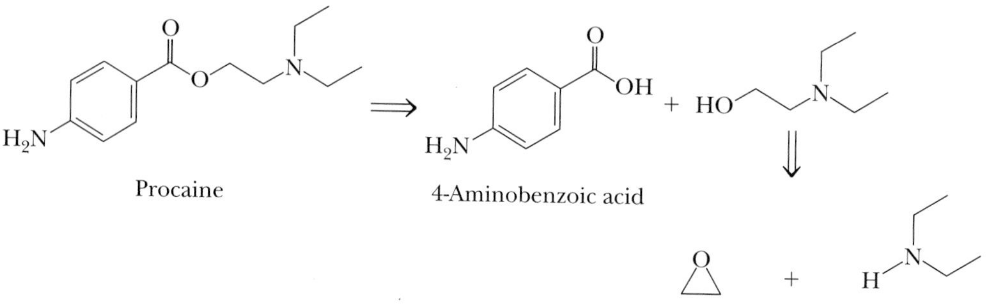

Procaine

4-Aminobenzoic acid

Ethylene oxide   Diethylamine

18.47 The following sequence of steps converts (R)-2-octanol to (S)-2-octanol. Propose structural formulas for intermediates A and B, specify the configuration of each, and account for the inversion of configuration in this sequence.

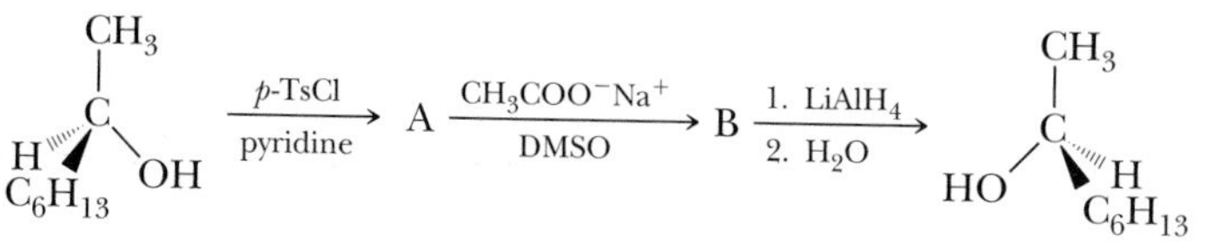

(R)-2-Octanol

(S)-2-Octanol

18.48 Reaction of a primary or secondary amine with diethyl carbonate under controlled conditions gives a carbamic ester. Propose a mechanism for this reaction.

Diethyl carbonate   Butylamine   Ethyl *N*-butylcarbamate

18.49 Several sulfonylureas, a class of compounds containing $RSO_2NHCONHR$, are useful drugs as orally active replacements for injected insulin in patients with adult-onset diabetes. It was discovered in 1942 that certain members of this class cause hypoglycemia in laboratory animals. Clinical trials of tolbutamide were begun in the early 1950s, and, since that time, more than 20 sulfonylureas have been introduced into clinical medicine. The sulfonylureas decrease blood glucose concentrations by stimulating β cells of the pancreas to release insulin and by increasing the sensitivity of insulin receptors in peripheral tissues to insulin stimulation.

Tolbutamide is synthesized by the reaction of the sodium salt of *p*-toluenesulfonamide and ethyl *N*-butylcarbamate (see Problem 18.48 for the synthesis of this carbamic ester). Propose a mechanism for the following step in the synthesis of tolbutamide.

Sodium salt of   A carbamic ester   Tolbutamide
p-toluenesulfonamide   (Oramide, Orinase)

18.50 Following are structural formulas for two more widely used sulfonylurea hypoglycemic agents. Show how each might be synthesized by converting an appropriate amine to a carbamic ester and then treating the carbamate with the sodium salt of a substituted benzenesulfonamide.

(a)   (b)

Tolazamide   Gliclazide
(Tolamide, Tolinase)   (Diamicron)

18.51 Amantadine (see *The Merck Index*, 12th ed., #389) is one of the very few available antiviral agents and is effective in preventing infections caused by the influenza A virus and in treating established illnesses. It is thought to block a late stage in the assembly of the virus. Amantadine is synthesized as follows. Treatment of 1-bromoadamantane with acetonitrile in sulfuric acid gives *N*-adamantylacetamide, which is then converted to amantadine.

1-Bromoadamantane   Amantadine

(a) Propose a mechanism for the transformation in Step 1.

(b) Describe experimental conditions to bring about Step 2.

18.52 In a series of seven steps, (*S*)-malic acid is converted to the bromoepoxide shown on the right in 50% overall yield. (*S*)-Malic acid occurs in apples and many other fruits. It is also available in enantiopure form from microbiological fermentation. This synthesis is enantioselective: of the stereoisomers possible for the bromoepoxide, only one is formed. In thinking about the chemistry of these steps, you will want to review the use of dihydropyran as an —OH protecting group (Section 16.8C) and the use the *p*-toluene-sulfonyl chloride to convert the —OH, a poor leaving group, into a tosylate, a good leaving group (Section 9.5D).

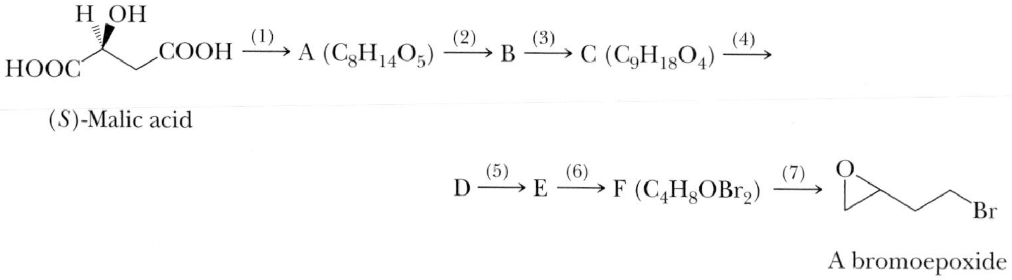

(*S*)-Malic acid

$$ D \xrightarrow{(5)} E \xrightarrow{(6)} F\ (C_4H_8OBr_2) \xrightarrow{(7)} $$

A bromoepoxide

Steps/reagents: 1. CH_3CH_2OH, H^+ 3. $LiAlH_4$, then H_2O 6. H_2O, CH_3COOH

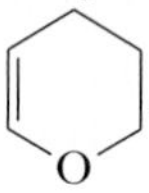

2. , H^+ 4. TsCl, pyridine   7. KOH

5. NaBr, DMSO

(a) Propose structural formulas for intermediates A through F. Also specify configuration at each stereocenter.

(b) What is the configuration of the stereocenter in the bromoepoxide? How do you account for the stereoselectivity of this seven-step conversion?

18.53 Following is a retrosynthetic analysis for the synthesis of the herbicide (*S*)-Metolachlor. According to this retrosynthetic analysis, starting materials are 2-ethyl-6-methylaniline, chloroacetic acid, acetone, and methanol. Show reagents and experimental conditions for the synthesis of Metolachlor from these four organic starting materials. Your synthesis will most likely give a racemic mixture. The chiral catalyst used by Novartis for Step 2 gives 80% enantiomeric excess of the *S* enantiomer.

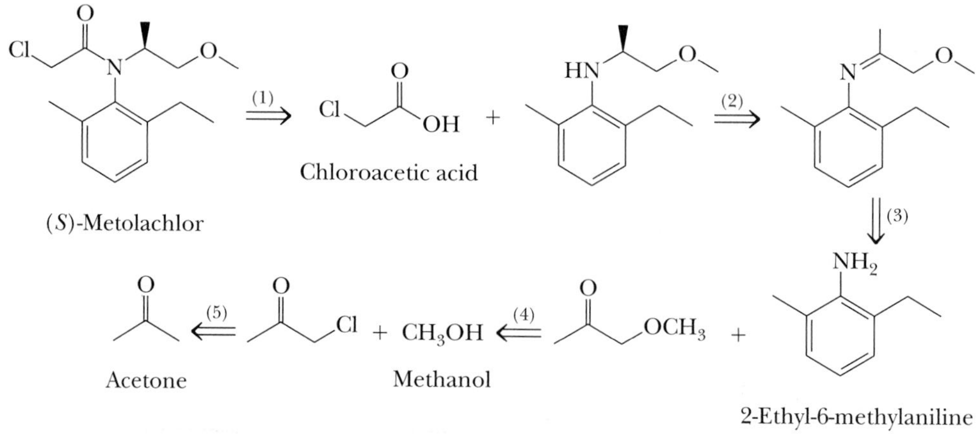

(*S*)-Metolachlor

Chloroacetic acid

Acetone   Methanol

2-Ethyl-6-methylaniline

A Conversation with . . .

Roald Hoffmann

Roald Hoffmann — Nobel Prize winning chemist, poet, playwright, science communicator, and teacher — speaks softly, gliding serenely from chatting about the differences in writing poetry versus plays to his recent research on how electron movement influences reactivity in solid state compounds. The shifts in conversation from poetry to chemistry, to theater arts, and back again are so frequent and natural that it comes as no surprise that Hoffmann was recently named the Frank H. T. Rhodes Professor of Humane Letters, in addition to his post as professor of chemistry at Cornell University. He says, "Writing and the arts are half my life."

The chemistry half of Hoffmann's life has overflowed with extraordinary achievement and honors. When he was only 44 years old, he shared the 1981 Nobel Prize in chemistry with Kenichi Fukui of Japan for work in applied theoretical chemistry. In addition, he has received awards from the American Chemical Society in both organic chemistry and inorganic chemistry, the only person to achieve this honor. And, in 1990, he was awarded the Priestley Medal, the highest award given by the American Chemical Society.

The writing half has been marked by success too. He is the author of three published collections of poetry, *The Metamict State, Gaps and Verges*, and *Memory Effects*. He recently completed a fourth collection, *Life's Currents Phased In*, and is looking for a publisher.

Hoffmann has also written three nonpoetry books: *Chemistry Imagined*, an art/science/literature collaboration with artist Vivian Torrence; *The Same and Not the Same*, thoughts on the dualities in chemistry; and *Old Wine in New Flasks: Reflections on Science and Jewish Tradition* with Shira Leibowitz Schmidt. He continues to write a column, "Marginalia," for *American Scientist* magazine. In

addition, a series of 26 half-hour TV programs, *The World of Chemistry*, which he made over 12 years ago, continues to air on PBS stations and abroad.

Most recently, fellow chemist Carl Djerassi and Hoffmann collaborated on a play via e-mail and while they were in London together. The play, called *Oxygen*, depicts what might happen if the Nobel Foundation were to decide to inaugurate a "Retro-Nobel Prize" for great discoveries of the past. To whom would they award the prize for inaugurating the modern chemical revolution? The candidates include Priestley, who today is usually called the discoverer of oxygen; Lavoisier, who first stated the modern notion of combustion, rusting, and animal respiration; and Scheele, who made oxygen gas before Priestley but published late, preferring to run his pharmacy instead. The play had ten workshop performances in San Francisco in 2000 and is slated to have its world premiere in April 2001 at the San Diego Repertory Theater while the American Chemical Society is in town for its annual meeting.

Drama from the Start

The numerous honors and applause celebrating Hoffmann's diverse achievements tell only part of the story of his life. He was born to a Polish Jewish family in Zloczow, Poland, in 1937 and was named Roald after the famous Norwegian explorer Roald Amundsen. Shortly after World War II began in 1939, the Nazis first forced him and his parents into a ghetto and then into a labor camp. However, his father smuggled Hoffmann and his mother out of the camp, and they were hidden for more than a year in the attic of a schoolhouse in the Ukraine. His father was later killed by the Nazis after trying to organize an attempt to break out of the labor camp. After the war, Hoffmann, his mother, and his stepfather made their way west to Czechoslovakia, Austria, and then Germany. They finally emigrated to the United States, arriving in New York in 1949. That Hoffmann and his mother survived these years is our good fortune. Of the 12,000 Jews living in Zloczow in 1941 when the Nazis took over, only 80 people, three of them children, survived the Holocaust. One of those three children was Roald Hoffmann.

On arriving in New York, Hoffmann learned his sixth language, English. He went to public schools in New York City and then to Stuyvesant High School, one of the city's select science schools. From there he went to Columbia University and then on to Harvard University, where he earned his PhD in 1962. Shortly thereafter, he began the work with Professor R. B. Woodward that eventually led to the Nobel Prize. He joined the Cornell faculty in 1965 and has remained there ever since.

He is the father of a son, Hillel, who is a journalist with the National Geographic Society, and a daughter, Ingrid, a physicist. Ingrid and her husband recently became the parents of Carl David, Hoffmann's first grandchild.

From Medicine to Cement to Theoretical Chemistry

"I came rather late to chemistry," he says, "I was not interested in it from childhood. I am always worried about fields in which people exhibit precocity, like music and mathematics. Precocity is some sort of evidence that you have to have talent. I don't like that. I like the idea that human beings can do anything they want to. They need to be trained sometimes. They need a teacher to awaken the intelligence within them. But to be a chemist requires no special talent, I'm glad to say. Anyone can do it, with hard work."

He took a standard chemistry course in high school. He recalls that it was a fine course, but apparently he found biology more enjoyable because, in his high school yearbook, "under the picture of me with a crew cut, it says 'medical research' under my name." Indeed, he says that "medical research was a compromise between my interest in science and the typical Jewish middle class family pressures to become a medical doctor. The same kind of pressures seem to apply to Asian-Americans today."

When he went to Columbia University, Hoffmann enrolled as a pre-med student, but says that there were several factors that shifted him away from a career in medicine. One of these was his work at the National Bureau of Standards in Washington, D.C., for two summers and then at Brookhaven National Laboratory for a third summer. He says

that these experiences gave him a feeling for the excitement of chemical research. Nonetheless, during his first summer at the Bureau of Standards, he "did some not very exciting work on the thermochemistry of cement." During his second summer there, he went over to the National Institutes of Health to find out what medical research was about. "To my amazement," he says, "most of the people had PhDs and not MDs. I just didn't know. Young people do not often know what is required for a given profession. Once I found that out, and found that I did well in chemistry, it made me feel that I didn't really have to do medicine, that I could do some research in chemistry or biology. Later, what influenced me to decide on theoretical chemistry was an excellent instructor.

"At the very same time I was being exposed to the humanities, in part because of Columbia's core curriculum—which I think is a great idea—that had so-called contemporary civilization and humanities courses. I took advantage of the liberal arts education to the hilt, and that has remained with me all my life. The humanities teachers have remained permanently fixed in my mind and have changed my ways of thinking. These were the people who really had the intellectual impact on me and helped to shape my life.

"To trace the path, I was a latecomer to chemistry and was inspired by research. I think *research* is the way in. It just gives you a different perception."

A Love for Complexity

Having discussed what brought him into chemistry, Hoffman then described the qualities he thought a student should possess to pursue a career in the field. He said, "One thing one needs to be a chemist is a love for complexity and richness. To some extent that is true of biology and natural history, too. I think one of the things that is beautiful about chemistry is that there are 10,000,000 compounds, each with different properties. What's beautiful when you make a molecule is that you can make derivatives in which you can vary substituents, the pieces of a molecule, and we know that those substituents give a molecule function, give it complexity and richness. That's why a protein or nucleic acid with all its variety is essential for life. That's why to me, intellectually, isomerism and stereochemistry in organic chemistry are at the heart of chemistry. I think we should teach that much earlier. It requires no mathematics, only a little model building; you can do this without theory. I think it is no accident that organic chemistry drew to it the intellects of its time."

Experiment and Theory

Professor Hoffmann has spent his career immersed in the theories of chemistry. However, he believes that fundamentally "chemistry is an experimental science, in spite of some of my colleagues saying otherwise. However, the educational process certainly favors theory. It's in the nature of things for teacher and student both to want to understand and then give primacy to the soluble and the understood at the expense of other things. We also have this reductionist philosophy of science, the idea that the social sciences derive from biology, that biology follows from chemistry, chemistry from physics, and so on. This notion gives an inordinate amount of impor-

tance to theoretical thinking, the more mathematical the better. Of course this is not true in reality, but it's an ideology; it is a religion of science."

There is of course a role for theory. "You can't report just the facts and nothing but the facts; by themselves they are dull. They have to be woven into a framework so that there is understanding. That's accomplished usually by a theory. It may not be mathematical, but a qualitative network of relationships." Indeed, Hoffmann believes that the incorporation of theory into chemistry "is what made American science better than that in many other countries. The emphasis in chemistry on theory and theoretical understanding is very important, but not nearly as important as the syntheses and reactions of molecules.

"Although I think chemists need to like to do experiments, that doesn't mean there is no role for people like me. It turns out that I am really an experimental chemist hiding as a theoretician. I think that is the key to my success. That is, I think I can empathize with what bothers the experimentalists. In another day I could have become an experimentalist."

Major Research Themes Today

Professor Hoffmann has worked at the forefront of surface science, inorganic chemistry, and other major chemical areas and shown how the creative application of theory yields practical results. Currently, he has turned his interests to theories about the way electrons move and how this movement influences reactivity for solid state compounds. He and his students are interested in designing new materials that are

good conductors. In a report that will soon be submitted for publication, he and graduate student Wojciech Grochala, from Poland, describe an entire new class of compounds that they predict will behave as superconductors.

Sometimes, however, the potential uses of unusual compounds that have interesting electronic and magnetic properties aren't apparent even when Hoffmann is able to decipher some of their most intimate secrets. Such is the case with compounds of the main group elements. In one of his most recent papers, he and his former graduate student Garegin Papoian, who came from Armenia to study with him, laid out a theory that extends the understanding of bonding in an important class of alloys and intermetallic compounds of the main group elements.

Antimony, tellurium, tin, and selenium (all main group elements) appear below carbon, nitrogen, and oxygen in the Periodic Table. Although compounds of these elements, such as europium and lithium antimonide, have been known for many decades, "experimentalists have said nothing about what holds them together," Hoffmann told a reporter from Cornell's news bureau recently.

Main group element compounds, in fact, blur the line between the different types of bonds that hold atoms together in a molecule or a crystal. The bonds in these compounds are actually a melange of metallic bonds, covalent bonds, and ionic bonds. The crystal structures of these compounds can be seen in a series of computer-generated drawings—not based on theory but on direct experimental work—that have an interlocking, architectural perfection. The molecular structures, ranging from simple geometries to complex lattices, reveal their bonding networks in a series of multidimensional building blocks. "Some look terribly complicated," says Hoffmann, "but take them apart, and you can see square lattices with atoms above and below, and squares forming octahedrons—fantastic structures with a certain Star Wars quality."

The drawings also help to reveal why the compounds sometimes behave in chameleon-like ways. "It's the number of electrons that determines the chemistry," says Hoffmann, "less so the identity of the specific antimony, tellurium, or tin nucleus underneath."

"What we have here is theory at its best—qualitative theory, building connections between different parts of the chemical universe, even though to outsiders these units appear not to be close to each other," Hoffmann explained. "I pride myself on seeing connections, which is what I also try to build between science and humanities. Anything I can do to connect diverse things feels worth doing."

Scientific Literacy and Democracy

Roald Hoffmann is very concerned not only about science in general and chemistry in particular, but also about our society. One of his concerns is scientific literacy because "some degree of scientific literacy is absolutely necessary today for the population at large as part of a democratic system of government. People have to make intelligent

> I pride myself on seeing connections, which is what I also try to build between science and humanities. Anything I can do to connect diverse things feels worth doing.

decisions about all kinds of technological issues." He recently offered his comments on this important issue in the *New York Times*. He wrote, "What concerns me about scientific, or humanistic, illiteracy is the barrier it poses to rational democratic governance. Democracy occasionally gives in to *technocracy*—a reliance on experts on matters such as genetic engineering, nuclear waste disposal, or the cost of medical care. That is fine, but the people must be able to vote intelligently on these issues. The less we know as a nation, the more we must rely on experts, and the more likely we are to be misled by demagogues. We must know more."

The Responsibility of Scientists

"Scientists have a great obligation to speak to the public," Hoffmann says. We have an obligation as educators to train the next generation of people. We should pay as much attention to those people who are not going to be chemists, and sometimes to make compromises about what is to be taught and what is the nature of our courses. I think scientists have an obligation to speak to the public broadly, and here I think they have been negligent. Society is paying scientists money to do research, and can demand an accounting in plain language. That's why I put in a lot of time on that television show [*The World of Chemistry*]."

A Teacher of Chemistry — And Proud of It

In the Nobel Yearbook, Professor Hoffmann wrote that the technical description of his work "does not communicate what I think is my major contribution. I am a teacher, and I am proud of it. At Cornell University I have taught primarily undergraduates. . . . I have also taught chemistry courses to nonscientists and graduate courses in bonding theory and quantum mechanics. To the chemistry community at large, and to my fellow scientists, I have tried to teach 'applied theoretical chemistry': a special blend of computations stimulated by experiment and coupled to the construction of general models—frameworks for understanding." His success in this area is unquestioned.

A perspective view of the crystal structure of neodymium distannide. The small dark spheres are neodymium and the large light spheres are tin. (Garegin Papoian © Cornell University)

ENOLATE ANIONS AND ENAMINES

This chapter is a continuation of the chemistry of carbonyl compounds. In Chapters 16–18, we concentrated on the carbonyl group itself and on nucleophilic additions to the carbonyl carbon to form tetrahedral carbonyl addition intermediates or products. In this chapter, we expand on the chemistry of carbonyl-containing compounds and consider in more detail the consequences of the acidity of α-hydrogens and the formation of enolate anions. On the following page is a resonance-stabilized **enolate anion** (Section 16.11A) formed from a ketone on treatment with base.

■ Crystals of tamoxifen (Problem 19.44 and MC.31) viewed under polarizing light. *(Herb Ohymeyer/Fran Heyl Associates)* Inset: A model of tamoxifen.

$$CH_3CCH_3 + NaOH \rightleftharpoons \left[H-\overset{O}{\underset{H}{\overset{\|}{C}}}-CCH_3 \longleftrightarrow H-\overset{O^-}{\underset{H}{\overset{|}{C}}}=CCH_3 \right] Na^+ + H_2O$$

An enolate anion

Enolate anions function as nucleophiles in S_N2 reactions as shown in this general reaction.

$$R-\overset{O}{\overset{\|}{C}}-\underset{R}{\overset{|}{C}}R + RCH_2-Br \xrightarrow[S_N2]{\text{nucleophilic substitution}} R-\overset{O}{\overset{\|}{C}}-\underset{R\ \ R}{\overset{|}{C}}CH_2R + Br^-$$

An enolate anion

They also function as nucleophiles in carbonyl addition reactions. Here, we show nucleophilic addition to the carbonyl carbon of a ketone. Enolate anions also add in this manner to the carbonyl groups of aldehydes and esters.

Supporting Concepts
Enolate
This is an overview of enol
and enolate formation and
reactivity.

$$R-\overset{O}{\overset{\|}{C}}-\underset{R}{\overset{|}{C}}R + R-\overset{O}{\overset{\|}{C}}-R \xrightarrow{\text{nucleophilic addition}} R-\overset{O}{\overset{\|}{C}}-\underset{R\ \ R}{\overset{|}{C}}-\overset{O^-}{\underset{R}{\overset{|}{C}}}R$$

An enolate anion   A ketone   A tetrahedral carbonyl
                                      addition intermediate

19.1 The Aldol Reaction

Unquestionably, the most important reaction of enolate anions derived from aldehydes and ketones is nucleophilic addition to the carbonyl group of another molecule of the same or different compound, as illustrated by the following reactions. Although such reactions may be catalyzed by either acid or base, base catalysis is more common.

$$CH_3-\overset{O}{\overset{\|}{C}}-H + \overset{H}{\underset{}{\overset{|}{C}H_2}}-\overset{O}{\overset{\|}{C}}-H \underset{\text{NaOH}}{\rightleftharpoons} CH_3-\overset{OH}{\overset{\beta|}{C}H}-\overset{\alpha}{C}H_2-\overset{O}{\overset{\|}{C}}-H$$

Ethanal   Ethanal   3-Hydroxybutanal
(Acetaldehyde)   (Acetaldehyde)   (a β-hydroxyaldehyde)

$$CH_3-\overset{O}{\overset{\|}{C}}-CH_3 + \overset{H}{\underset{}{\overset{|}{C}H_2}}-\overset{O}{\overset{\|}{C}}-CH_3 \underset{\text{Ba(OH)}_2}{\rightleftharpoons} CH_3-\overset{OH}{\underset{CH_3}{\overset{\beta|}{C}}}-\overset{\alpha}{C}H_2-\overset{O}{\overset{\|}{C}}-CH_3$$

Propanone   Propanone   4-Hydroxy-4-methyl-2-pentanone
(Acetone)   (Acetone)   (a β-hydroxyketone)

The common name of the product derived from the reaction of acetaldehyde in base is aldol, so named because it is both an *ald*ehyde and an alco*hol*. Aldol is also the generic name given to any product formed in this type of reaction. The product of an **aldol reaction** is a β-hydroxyaldehyde or a β-hydroxyketone.

The key step in a base-catalyzed aldol reaction is nucleophilic addition of the enolate anion of one carbonyl-containing molecule to the carbonyl group of another to form a tetrahedral carbonyl addition intermediate. This mechanism is illustrated by the aldol reaction between two molecules of acetaldehyde. Notice in this three-step mechanism that OH⁻ is a catalyst; an OH⁻ is used in Step 1, but another OH⁻ is generated in Step 3.

Mechanism   Base-Catalyzed Aldol Reaction

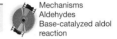
Mechanisms
Aldehydes
Base-catalyzed aldol reaction

Step 1:   Removal of an α-hydrogen by base gives a resonance-stabilized enolate anion.

$$HO^- + H - CH_2 - \overset{\overset{\textstyle O}{\|}}{C} - H \; \rightleftharpoons \; HO-H + \left[\; {}^-\overset{\alpha}{C}H_2 - \overset{\overset{\textstyle O}{\|}}{C} - H \; \longleftrightarrow \; \overset{\alpha}{C}H_2 = \overset{\overset{\textstyle -O}{|}}{C} - H \; \right]$$

Enolate anion

Step 2:   Nucleophilic addition of the enolate anion to the carbonyl carbon of another aldehyde (or ketone) gives a tetrahedral carbonyl addition intermediate.

$$CH_3 - \overset{\overset{\textstyle O}{\|}}{C} - H + {}^-\overset{\alpha}{C}H_2 - \overset{\overset{\textstyle O}{\|}}{C} - H \; \rightleftharpoons \; CH_3 - \overset{\overset{\textstyle O^-}{|}}{C}H - \overset{\alpha}{C}H_2 - \overset{\overset{\textstyle O}{\|}}{C} - H$$

A tetrahedral carbonyl
addition intermediate

Step 3:   Reaction of the tetrahedral carbonyl addition intermediate with a proton donor gives the aldol product and generates a new base catalyst.

$$CH_3 - \overset{\overset{\textstyle O^-}{|}}{C}H - CH_2 - \overset{\overset{\textstyle O}{\|}}{C} - H + H - OH \; \rightleftharpoons \; CH_3 - \overset{\overset{\textstyle OH}{|}}{C}H - CH_2 - \overset{\overset{\textstyle O}{\|}}{C} - H + OH^-$$

The mechanism of an acid-catalyzed aldol reaction involves three steps, the first two of which are preparation of the aldehyde or ketone for formation of the new carbon-carbon bond. The key step is attack of the enol of one molecule on the protonated carbonyl group of a second molecule.

Mechanism   Acid-Catalyzed Aldol Reaction

Mechanisms
Aldehydes
Acid-catalyzed aldol reaction

Step 1:   Keto and enol forms of one molecule undergo acid-catalyzed equilibration (Section 16.11B).

$$CH_3 - \overset{\overset{\textstyle O}{\|}}{C} - H \; \underset{}{\overset{HA}{\rightleftharpoons}} \; CH_2 = \overset{\overset{\textstyle OH}{|}}{C} - H$$

Step 2: Proton transfer from the acid, HA, to the carbonyl oxygen of a second molecule of aldehyde or ketone gives an oxonium ion.

$$CH_3-\overset{\overset{\displaystyle O:}{\|}}{C}-H + H-A \longrightarrow CH_3-\overset{\overset{\displaystyle \overset{+}{O}-H}{\|}}{C}-H + :A^-$$

Step 3: Attack of the enol of one molecule on the protonated carbonyl group of another molecule forms the new carbon-carbon bond. This is followed by proton transfer to A⁻ to re-generate the acid catalyst.

$$CH_3-\overset{\overset{\displaystyle \overset{+}{O}-H}{\|}}{C}-H + CH_2{=}\overset{\overset{\displaystyle O-H}{|}}{C}-H \xrightarrow{A^-} CH_3-\overset{\overset{\displaystyle O-H}{|}}{CH}-CH_2-\overset{\overset{\displaystyle O}{\|}}{C}-H + H-A$$

You might compare the mechanisms of the acid- and base-catalyzed aldol reactions. Under base catalysis, the carbon-carbon bond-forming step involves attack of a nucleophilic enolate anion on the uncharged carbonyl group of a second molecule of the aldehyde or ketone. Under acid catalysis, it involves attack of a nucleophilic enol form of one molecule on the protonated carbonyl group of the second molecule.

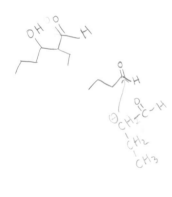

Example 19.1

Draw the product of the base-catalyzed aldol reaction of each compound.

(a) Butanal   **(b)** Cyclohexanone

Solution

The aldol product is formed by nucleophilic addition of the *α*-carbon of one molecule to the carbonyl carbon of another.

(a)

OH
⟍⟋⟍⟋⟍CHO

new
C—C bond

(b)

OH O

new
C—C bond

Problem 19.1

Draw the product of the base-catalyzed aldol reaction of each compound.

(a) Phenylacetaldehyde   **(b)** Cyclopentanone

β-Hydroxyaldehydes and *β*-hydroxyketones are very easily dehydrated, and often the conditions necessary to bring about an aldol reaction are sufficient to cause dehydration. This is particularly true in the case of acid-catalyzed aldol reactions. Alternatively, dehydration can be brought about by warming the aldol product in dilute acid. The major product from dehydration of an aldol is one in which the carbon-carbon double bond is conjugated with the carbonyl group, that is, the product is an α,β-unsaturated aldehyde or ketone. As we will see in Chapter 23, conjugated systems are more stable in general than unconjugated ones, so the product of an aldol dehydration reaction is usually the α,β-unsaturated one.

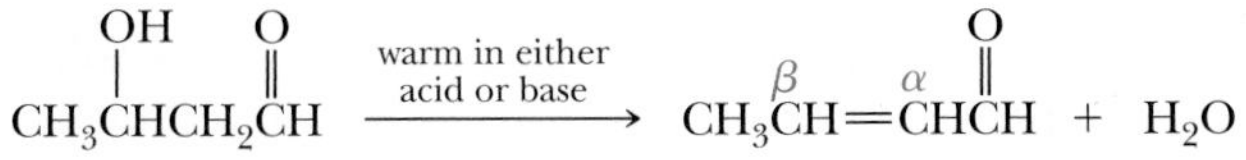

An α,β-unsaturated
aldehyde

Mechanism   Acid-Catalyzed Dehydration of an Aldol Product

Proton transfer from the acid catalyst to the enol form of the aldol product in Step 1 gives an oxonium ion which, in Step 2, undergoes concerted loss of water and formation of the conjugate acid of the final product. Proton transfer from this conjugate acid to solvent in Step 3 completes the reaction.

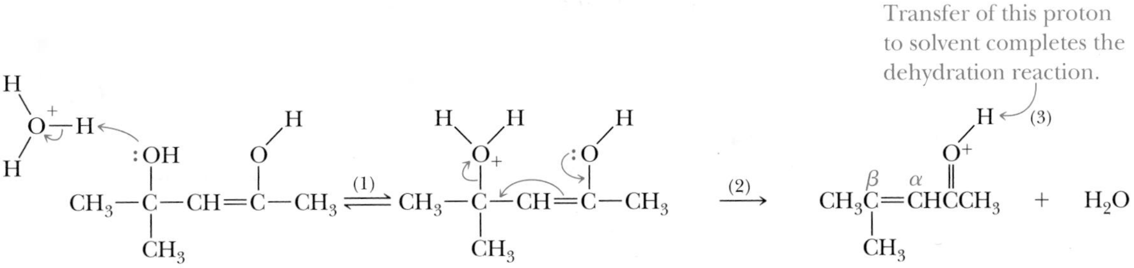

Transfer of this proton
to solvent completes the
dehydration reaction.

The enol of the ketone

Protonated α,β-unsaturated ketone

Example 19.2

Draw the product of dehydration of each aldol product in Example 19.1.

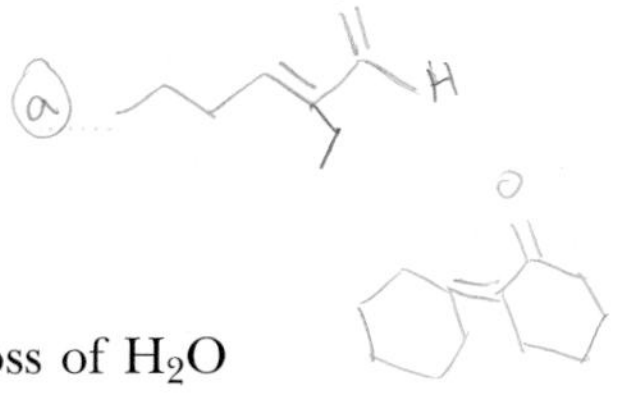

Solution

Loss of H_2O from aldol product (a) gives an α,β-unsaturated aldehyde; loss of H_2O from (b) gives an α,β-unsaturated ketone.

(a)

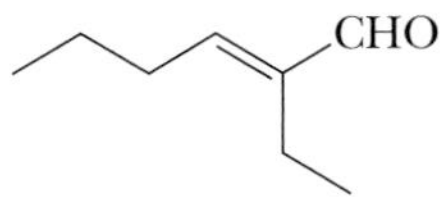

(b)

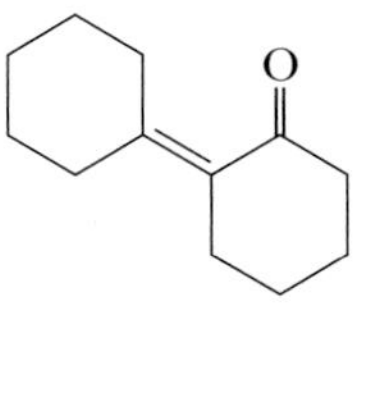

Problem 19.2

Draw the product of dehydration of each aldol product in Problem 19.1.

Aldol reactions are readily reversible, especially when catalyzed by base, and, for aldol reactions of ketones especially, there is generally little aldol product present at equilibrium. Equilibrium constants for dehydration, however, are generally large so that, if reaction conditions are sufficiently vigorous to bring about dehydration, good yields of product can be obtained.

The reactants in the key step of an aldol reaction are an enolate anion and an enolate anion acceptor. In self-reactions, both roles are played by one kind of molecule. **Crossed aldol reactions** are also possible, as for example the crossed aldol reaction between acetone and formaldehyde. Formaldehyde cannot provide an enolate anion because it has no α-hydrogen, but it can function as a particularly good anion acceptor because its carbonyl group is unhindered. Acetone forms an enolate anion but its carbonyl group, which is bonded to two alkyl groups, is less reactive than that of formaldehyde. Consequently, the crossed aldol reaction between acetone and formaldehyde gives 4-hydroxy-2-butanone.

$$CH_3CCH_3 + HCH \xrightleftharpoons[]{NaOH} CH_3CCH_2CH_2OH$$

4-Hydroxy-2-butanone

As this example illustrates, for a crossed aldol reaction to be successful, one of the two reactants should have no α-hydrogen so that an enolate anion does not form. It also helps if the compound with no α-hydrogen has the more reactive carbonyl, for example, an aldehyde. Following are examples of aldehydes that have no α-hydrogens and can be used in crossed aldol reactions. If these requirements are not met, a complex mixture of products results.

HCH   benzaldehyde   furfural $(CH_3)_3CCH$

Formaldehyde   Benzaldehyde   Furfural   2,2-Dimethylpropanal

---

Example 19.3

Draw a structural formula for the product of the base-catalyzed crossed aldol reaction between furfural and cyclohexanone and the product formed by its dehydration.

Solution

Furfural   Cyclohexanone   Aldol product

Problem 19.3

Draw the product of the base-catalyzed crossed aldol reaction between benzaldehyde and 3-pentanone and the product formed by its dehydration.

When both the enolate anion and the carbonyl group to which it adds are in the same molecule, aldol reaction results in formation of a ring. This type of intramolecular aldol reaction is particularly useful for formation of five- and six-membered rings. The intramolecular aldol reaction of 2,7-octanedione via its enolate anion at α_3 gives a five-membered ring.

Internal Aldol Rxn

2,7-Octanedione

Note that in 2,7-octanedione, two enolate anions are possible; only one of which leads to the five-membered ring just shown. Aldol reaction via enolate anion at α_1 leads to a seven-membered ring. Formation of five- and six-membered rings is favored over formation of four- and seven-membered rings.

In general, smaller rings form faster than larger rings because the reacting groups are closer together. However, the formation of three- and four-membered rings is disfavored because of the strain in these rings.

Following is another example in which either a four-membered ring (via an enolate anion at α_3) or a six-membered ring (via an enolate anion at α_1) could be formed. Because of the greater stability of six-membered rings compared to four-membered rings, the six-membered ring is formed in this intramolecular aldol reaction.

Nitro groups can be introduced into aliphatic compounds by way of an aldol reaction between the anion of a nitroalkane and an aldehyde or a ketone. The α-hydrogens of nitroalkanes are sufficiently acidic that they are removed by bases such as aqueous NaOH and KOH. The pK_a of nitromethane, for example, is 10.2. The acidity of the α-hydrogen of a nitroalkane is due to the stabilization of the resulting anion by resonance delocalization of its negative charge.

Nitromethane	Water	Resonance-stabilized anion
pK_a 10.2	pK_a 15.7	
(stronger acid)	(weaker acid)	

Following is an aldol reaction between nitromethane and cyclohexanone. Reduction of the nitro group in the aldol product thus formed is a convenient synthetic route to β-aminoalcohols.

| Cyclohexanone | Nitromethane | 1-(Nitromethyl)cyclohexanol | 1-(Aminomethyl)cyclohexanol |

19.2 Directed Aldol Reactions

A. Formation of Enolate Anions — Kinetic Versus Thermodynamic Control

Enolate anions are formed when a carbonyl compound containing an α-hydrogen is treated with base. When alkali metal hydroxides or alkoxides are used as bases, the position of equilibrium favors reactants rather than products.

With a stronger base, however, the formation of an enolate anion can be driven to completion. One of the most widely used bases for this purpose is lithium diisopropylamide (LDA).

$$[(CH_3)_2CH]_2N^-Li^+$$

Lithium diisopropylamide

Lithium diisopropylamide is prepared by dissolving diisopropylamine in tetrahydrofuran and treating this solution with butyllithium.

$$[(CH_3)_2CH]_2NH + CH_3(CH_2)_3Li \longrightarrow [(CH_3)_2CH]_2N^-Li^+ + CH_3(CH_2)_2CH_3 \quad K_{eq} = 10^{10}$$

| Diisopropylamine pK_a 40 (stronger acid) | Butyllithium (stronger base) | Lithium diisopropyl- amide (weaker base) | Butane pK_a 50 (weaker acid) |

Although LDA is a very strong base, it is a very poor nucleophile because of steric crowding around the nitrogen, and does not add to carbonyl groups. It is, therefore, ideal for generation of enolate anions from carbonyl-containing compounds, as illustrated by treatment of ethyl acetate with LDA. By using a molar equivalent of LDA, an aldehyde, ketone, or ester can be converted completely to its lithium enolate.

$$CH_3\overset{\overset{\displaystyle O}{\|}}{C}OC_2H_5 + [(CH_3)_2CH]_2N^-Li^+ \longrightarrow CH_2{=}\overset{\overset{\displaystyle O^-Li^+}{|}}{C}OC_2H_5 + [(CH_3)_2CH]_2NH \quad K_{eq} = 10^{17}$$

pK$_a$ 23	LDA	A lithium	pK$_a$ 40
(stronger acid)	(stronger base)	enolate	(weaker acid)
		(weaker base)	

For a ketone with two sets of nonequivalent α-hydrogens, the following question arises: Is formation of an enolate anion regioselective, and, if so, what are the factors that determine the degree of regioselectivity? It has been determined experimentally that a very high degree of regioselectivity often exists, and that it depends on experimental conditions. When 2-methylcyclohexanone, for example, is added to a slight excess of LDA, the ketone is converted to its lithium enolate, which consists almost entirely of the salt of the less substituted enolate anion.

2-Methyl-
cyclohexanone   99%   1%

When 2-methylcyclohexanone is treated with LDA under conditions in which the ketone is in a slight excess, then the composition of the product is quite different; it is richer in the more substituted enolate anion.

2-Methyl-
cyclohexanone   10%   90%

The most important factor determining the composition of an enolate anion mixture is whether the reaction forming it is under kinetic (rate) control or thermodynamic (equilibrium) control. In a reaction under **thermodynamic control:**

1. The reaction conditions permit the equilibration of alternative products, and
2. The composition of the product mixture is determined by the relative stabilities of the products.

Equilibrium among enolate anions is established when the ketone is in slight excess, a condition under which it is possible for proton-transfer reactions to occur. An

Thermodynamic control Experimental conditions that permit the establishment of equilibrium between two or more products of a reaction. The composition of the product mixture is determined by the relative stabilities of the products.

enolate anion can undergo proton transfer from the α-carbon of an unreacted molecule of ketone to form the alternative enolate anion, and vice versa. Thus, an equilibrium is established between alternative enolate anions.

Less stable enolate anion   More stable enolate anion

Under these conditions, the more stable enolate anion predominates. The factors that determine the relative stabilities of enolate anions are the same as those that determine the relative stabilities of alkenes (Section 6.6B); the more substituted the double bond of the enolate anion, the greater its stability. Thus, the composition of the enolate anion mixture formed under conditions of thermodynamic control reflects the relative stabilities of the individual enolate anions.

In a reaction under **kinetic control,** the composition of the product mixture is determined by the relative rates of formation of each product. In the case of formation of enolate anions, kinetic control refers to the relative rates of removal of the alternative α-hydrogens. The less hindered α-hydrogen is removed more rapidly, and thus the major product is the less substituted enolate anion. Because a slight excess of base is used, there is no ketone to serve as a proton donor, and the less stable enolate anion cannot equilibrate with the more stable one.

Kinetic control Experimental conditions under which the composition of the product mixture is determined by the relative rates of formation of each product.

B.   Using Preformed Enolate Anions in Directed Aldol Reactions

In Section 19.1, we discussed aldol reactions and pointed out a problem inherent in carrying out crossed aldol reactions. Crossed aldol reactions between an aldehyde with no α-hydrogens and a ketone generally give good yields of a single product because (1) only the ketone can form an enolate anion, and (2) the carbonyl group of an aldehyde is a better enolate anion acceptor than the more crowded carbonyl group of a ketone. A problem arises in a crossed aldol reaction when both the aldehyde and ketone have α-hydrogens. Consider, for example, the problem of how to prepare the following aldol product from phenylacetaldehyde and acetone.

Phenylacetaldehyde   Acetone   4-Hydroxy-5-phenyl-2-pentanone

Four aldol products are possible, and a mixture of the four is formed if these two compounds are mixed in the presence of $C_2H_5O^-Na^+$. However, the desired aldol reaction may be carried out successfully by treating acetone with LDA to convert it completely and irreversibly to its enolate anion. The preformed enolate anion is then treated with the aldehyde followed by workup in water to give the crossed aldol reaction product.

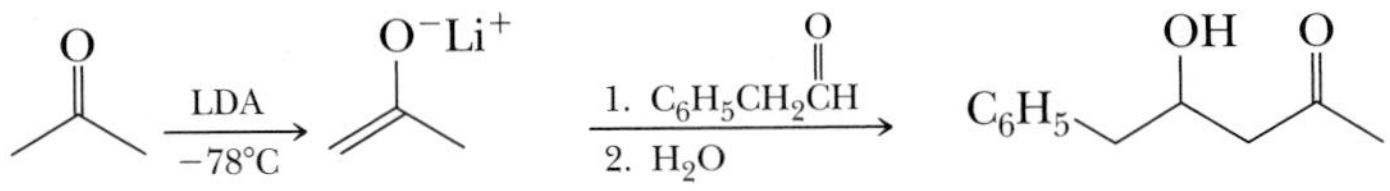

O‖ (acetone) →[LDA][−78°C] lithium enolate (O⁻Li⁺) →[1. C₆H₅CH₂CHO][2. H₂O]→ C₆H₅—CH(OH)—CH₂—CO—CH₃

A lithium enolate

4-Hydroxy-5-phenyl-2-pentanone

Example 19.4

Show how to prepare 5-hydroxy-4-methyl-3-hexanone by a directed aldol reaction.

Solution

First recognize that the two carbonyl-containing compounds to be joined in the aldol reaction are 3-pentanone and acetaldehyde. Treat the symmetrical ketone with LDA to form its lithium enolate. Treatment of this enolate anion with acetaldehyde followed by aqueous workup gives the desired aldol product.

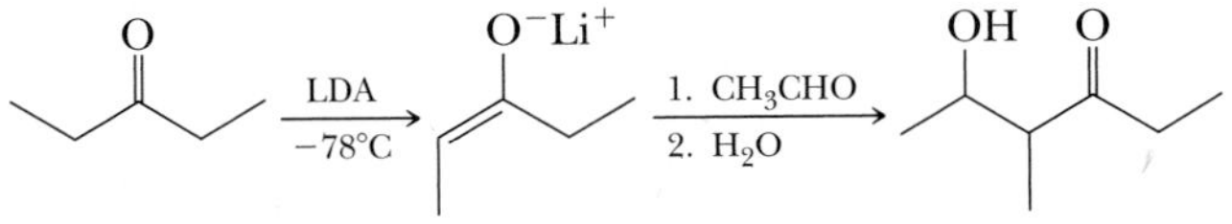

3-pentanone →[LDA][−78°C] lithium enolate (O⁻Li⁺) →[1. CH₃CHO][2. H₂O]→ aldol product (OH, O)

Problem 19.4

Show how you might prepare these compounds by directed aldol reactions.

(a) 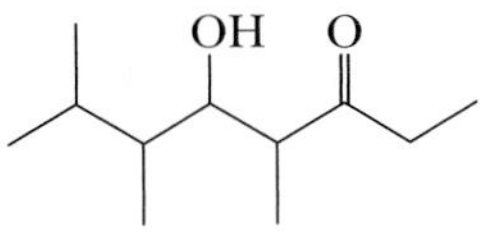 structure with OH, O

(b) 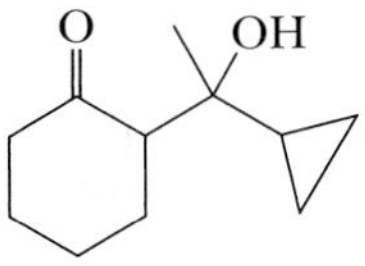 structure with O, OH, cyclopropyl

(c) 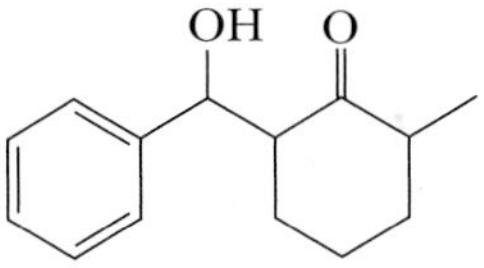 structure with OH, O, phenyl

19.3   Claisen and Dieckmann Condensations

A.   Claisen Condensation

In Chapter 18, we described reactions of esters, all of which take place at the carbonyl carbon and involve nucleophilic acyl substitution. In this section, we examine a second type of reaction characteristic of esters, namely one that involves both formation of an enolate anion and its participation in nucleophilic acyl substitution. One of the first of these reactions discovered is the **Claisen condensation,** named after the German chemist Ludwig Claisen (1851–1930). A Claisen condensation is illustrated by the reaction of two molecules of ethyl acetate in the presence of sodium ethoxide followed by acidification to give ethyl acetoacetate. Note that in this and many of the following equations, the ethyl group, C_2H_5, is written as Et.

$$2CH_3COEt \xrightarrow[\text{2. H}_2\text{O, HCl}]{\text{1. EtO}^-\text{Na}^+} CH_3CCH_2COEt + EtOH$$

Ethyl ethanoate
(Ethyl acetate)

Ethyl 3-oxobutanoate
(Ethyl acetoacetate)

Ethanol

The product of a Claisen condensation is a **β-ketoester.**

$$-\overset{|}{\underset{|}{C}}-\overset{O}{\overset{\|}{C}}\overset{\beta}{-}\overset{|}{\underset{|}{C}}-\overset{O}{\overset{\|}{C}}-OR$$

A β-ketoester

Can't use NaOH as base

Claisen condensations, like the aldol reaction, require a base. Aqueous bases, such as NaOH, however, cannot be used in Claisen condensations because they would bring about the hydrolysis of the ester. Rather, the bases most commonly used in Claisen condensations are nonaqueous bases, such as sodium ethoxide in ethanol and sodium methoxide in methanol.

Bases:
(1) EtO⁻ Na⁺ ethanol
(2) H₂O, HCl

(1) CH₃O⁻ Na⁺ methanol
(2) H₂O, HCl

Claisen condensation of two molecules of ethyl propanoate gives the following β-ketoester.

| Ethyl propanoate | Ethyl propanoate | 1. EtO⁻Na⁺
2. H₂O, HCl | Ethyl 2-methyl-3-oxopentanoate | + EtOH |

The first steps of a Claisen condensation bear a close resemblance to the first steps of an aldol reaction (Section 19.1). The carbon-carbon bond-forming step in this reaction involves nucleophilic addition of the enolate anion of one ester to the carbonyl group of another ester in Step 2 to form a tetrahedral carbonyl addition intermediate. This intermediate then collapses in Step 3 to give a β-ketoester. The position of equilibrium for Steps 1–3 lies far toward the starting ester(s). The overall reaction is driven to completion, however, by the acid-base reaction between the β-ketoester (the stronger acid) and ethoxide ion (the stronger base) to give ethanol (the weaker acid) and the anion of the β-ketoester (the weaker base).

Mechanisms
Acid Derivatives
Claisen condensation

Mechanism   Claisen Condensation

Step 1:   Removal of an α-hydrogen by base gives a resonance-stabilized enolate anion. Because the α-hydrogen of an ester is the weaker acid and ethoxide ion is the weaker base, the position of this equilibrium lies very much toward the left. The concentration of enolate anion is very low compared with that of ethoxide ion and ester.

$$C_2H_5O^- + H-CH_2-\overset{O}{\overset{\|}{C}}OEt \rightleftharpoons EtO-H + {}^-CH_2-\overset{O}{\overset{\|}{C}}OEt \longleftrightarrow CH_2=\overset{O^-}{\underset{|}{C}}OEt$$

(weaker base)   pK$_a$ 22   pK$_a$ 15.9   Resonance-stabilized enolate anion
  (weaker acid)   (stronger acid)   (stronger base)

Step 2:   Attack of the enolate anion of one ester on the carbonyl carbon of another ester gives a tetrahedral carbonyl addition intermediate. Unlike similar intermediates in aldol reactions, this intermediate (a hemiacetal anion) has an ethoxy leaving group.

$$CH_3-\overset{\overset{\displaystyle O}{\|}}{C}-OEt + {}^-CH_2-\overset{\alpha}{\overset{\overset{\displaystyle O}{\|}}{C}}OEt \rightleftharpoons \left[CH_3-\overset{\overset{\displaystyle O^-}{|}}{\underset{\underset{\displaystyle OEt}{|}}{C}}-\overset{\alpha}{CH_2}-\overset{\overset{\displaystyle O}{\|}}{C}-OEt \right]$$

A tetrahedral carbonyl
addition intermediate

Step 3: Collapse of the tetrahedral carbonyl addition intermediate and ejection of ethoxide ion gives a β-ketoester.

$$CH_3-\overset{\overset{\displaystyle O^-}{|}}{\underset{\underset{\displaystyle OEt}{|}}{C}}-\overset{\alpha}{CH_2}-\overset{\overset{\displaystyle O}{\|}}{C}-OEt \rightleftharpoons CH_3-\overset{\overset{\displaystyle O}{\|}}{C}-\overset{\alpha}{CH_2}-\overset{\overset{\displaystyle O}{\|}}{C}-OEt + EtO^-$$

Step 4: Formation of the enolate anion of the β-ketoester drives the Claisen condensation to the right.

$$CH_3-\overset{\overset{\displaystyle O}{\|}}{C}-\overset{\overset{\displaystyle H}{|}}{CH}-\overset{\overset{\displaystyle O}{\|}}{C}-OEt + EtO^- \rightleftharpoons CH_3-\overset{\overset{\displaystyle O}{\|}}{C}-\overset{-}{CH}-\overset{\overset{\displaystyle O}{\|}}{C}-OEt + EtOH$$

pK_a 10.7   (stronger base)   (weaker base) pK_a 15.9
(stronger acid)   (weaker acid)

The structural feature required for a successful Claisen condensation is an ester with two α-hydrogens: one to form the initial enolate anion and the second to form the enolate anion of the resulting β-ketoester. The β-ketoester is formed and isolated upon acidification with aqueous acid during workup.

$$CH_3-\overset{\overset{\displaystyle O}{\|}}{C}-\overset{-}{CH}-\overset{\overset{\displaystyle O}{\|}}{C}-OEt + H_3O^+ \xrightarrow{HCl, H_2O} CH_3-\overset{\overset{\displaystyle O}{\|}}{C}-CH_2-\overset{\overset{\displaystyle O}{\|}}{C}-OEt + H_2O$$

Example 19.5

Show the product of the Claisen condensation of ethyl butanoate in the presence of sodium ethoxide followed by acidification with aqueous HCl.

Solution

The new bond formed in a Claisen condensation is between the carbonyl group of one ester molecule and the α-carbon of another.

the new C—C bond

Ethyl 2-ethyl-3-oxohexanoate

Problem 19.5

Show the product of Claisen condensation of ethyl 3-methylbutanoate in the presence of sodium ethoxide followed by acidification with aqueous HCl.

B. The Dieckmann Condensation

An intramolecular Claisen condensation of a dicarboxylic ester to give a five- or six-membered ring is given the special name of **Dieckmann condensation.** In the presence of one equivalent of sodium ethoxide, for example, diethyl hexanedioate (diethyl adipate) undergoes an intramolecular condensation to form a five-membered ring.

EtO\
$\quad$ OEt $\quad \xrightarrow[\text{2. H}_2\text{O, HCl}]{\text{1. EtO}^-\text{Na}^+}$ $\quad$ OEt $\quad + \quad$ EtOH

Diethyl hexanedioate $\qquad\qquad$ Ethyl 2-oxocyclo-\
(Diethyl adipate) $\qquad\qquad\qquad$ pentanecarboxylate

Mechanisms\
Acid Derivatives\
Dieckmann condensation

The mechanism of a Dieckmann condensation is identical to the mechanism we have described for the Claisen condensation. An anion formed at the α-carbon of one ester group in Step 1 adds to the carbonyl of the other ester group in Step 2 to form a tetrahedral carbonyl addition intermediate. This intermediate ejects ethoxide ion in Step 3 to regenerate the carbonyl group. Cyclization is followed by formation of the conjugate base of the β-ketoester, just as in the Claisen condensation. The β-ketoester is isolated after acidification with aqueous acid.

C. Crossed Claisen Condensations

In a crossed Claisen condensation between two different esters, each with two α-hydrogens, a mixture of four β-ketoesters is possible; therefore, crossed Claisen condensations of this type are not synthetically useful. Like crossed aldol reactions, such condensations are useful, however, if appreciable differences in reactivity exist between the two esters, as for example when one of the esters has no α-hydrogens and can function only as an enolate anion acceptor. Following are four examples of esters without α-hydrogens.

HCOEt $\qquad$ EtOCOEt $\qquad$ EtOC—COEt $\qquad$ COEt

Ethyl formate $\quad$ Diethyl carbonate $\quad$ Diethyl ethanedioate $\quad$ Ethyl benzoate\
$\qquad\qquad\qquad\qquad\qquad\qquad$ (Diethyl oxalate)

Crossed Claisen condensations of this type are usually carried out by using the ester with no α-hydrogens in excess. In the following illustration, methyl benzoate is used in excess.

| Methyl | Methyl | Methyl 2-methyl-3-oxo- |
| benzoate | propanoate | 3-phenylpropanoate |

Example 19.6

Complete the equation for this crossed Claisen condensation.

Solution

Problem 19.6

Complete the equation for this crossed Claisen condensation.

D. Hydrolysis and Decarboxylation of β-Ketoesters

Recall from Section 18.5C that hydrolysis of an ester in aqueous sodium hydroxide (saponification) followed by acidification of the reaction mixture with aqueous HCl converts an ester to a carboxylic acid. Recall also from Section 17.9 that β-ketoacids and β-dicarboxylic acids (substituted malonic acids) readily undergo decarboxylation (lose CO_2) when heated. Both the Claisen and Dieckmann condensations yield esters of β-ketoacids. The following equations illustrate the results of a Claisen condensation followed by hydrolysis of the ester, acidification, and decarboxylation.

Claisen condensation:

Saponification followed by acidification

3. NaOH, H₂O, heat
4. H₂O, HCl

Decarboxylation:

5. heat

$+ CO_2$

The result of this five-step sequence is reaction between two molecules of ester (one furnishing a carbonyl group and the other furnishing an enolate anion) to give a ketone and carbon dioxide. In the general reaction, both ester molecules are the same, and the product is a symmetrical ketone.

from the ester furnishing
the carbonyl group

from the ester furnishing
the enolate anion

$$R-CH_2-\overset{\overset{\displaystyle O}{\|}}{C} + CH_2-\overset{\overset{\displaystyle O}{\|}}{C}-OR' \xrightarrow{\text{several steps}} R-CH_2-\overset{\overset{\displaystyle O}{\|}}{C}-CH_2-R + 2HOR' + CO_2$$
$$\quad\quad \underset{OR'}{} \quad \underset{R}{}$$

The same sequence of reactions starting with a crossed Claisen condensation gives an unsymmetrical ketone.

Example 19.7

Each set of compounds undergoes a Claisen or Dieckmann condensation followed by acidification (Steps 1, 2), saponification followed by acidification (Steps 3, 4), and thermal decarboxylation (Step 5). Draw a structural formula of the product isolated after completion of this reaction sequence.

(a) $\overset{\overset{\displaystyle O}{\|}}{PhC}OEt + CH_3\overset{\overset{\displaystyle O}{\|}}{C}OEt$ (b) $EtO\overset{\overset{\displaystyle O}{\|}}{C}(CH_2)_4\overset{\overset{\displaystyle O}{\|}}{C}OEt$

Solution

Steps 1 and 2 bring about a crossed Claisen or Dieckmann condensation to give a β-ketoester. Steps 3 and 4 bring about hydrolysis of the β-ketoester to give a β-ketoacid, and Step 5 brings about decarboxylation to give a ketone.

(a) $\xrightarrow{(1,2)} PhC\overset{\overset{\displaystyle O}{\|}}{}CH_2\overset{\overset{\displaystyle O}{\|}}{C}OEt \xrightarrow{(3,4)} PhC\overset{\overset{\displaystyle O}{\|}}{}CH_2\overset{\overset{\displaystyle O}{\|}}{C}OH \xrightarrow{(5)} Ph\overset{\overset{\displaystyle O}{\|}}{C}CH_3$

(b) $\xrightarrow{(1,2)}$ $\xrightarrow{(3,4)}$ $\xrightarrow{(5)}$

Problem 19.7

Show how to convert benzoic acid to 3-methyl-1-phenyl-1-butanone (isobutyl phenyl ketone) by the following synthetic strategies, each of which uses a different type of reaction to form the new carbon-carbon bond to the carbonyl group of benzoic acid.

Benzoic acid   3-Methyl-1-phenyl-1-butanone

(a) A lithium diorganocopper (Gilman) reagent   **(b)** A Claisen condensation

---

19.4   Claisen and Aldol Condensations in the Biological World

Carbonyl condensations are among the most widely used reactions in the biological world for the assembly of new carbon-carbon bonds in such important biomolecules as fatty acids, cholesterol, steroid hormones, and terpenes. One source of carbon atoms for the synthesis of these biomolecules is **acetyl-CoA,** a thioester of acetic acid and the thiol group of coenzyme A (Problem 25.41). In this section, we examine the series of reactions by which the carbon skeleton of acetic acid is converted to isopentenyl pyrophosphate, a key intermediate in the synthesis of terpenes, cholesterol, steroid hormones, and bile acids. Note that, in the discussion that follows, we will not be concerned with the mechanism by which each of these enzyme-catalyzed reactions occurs. Rather, our concern will be in recognizing the types of reactions that take place.

In a Claisen condensation catalyzed by the enzyme thiolase, acetyl-CoA is converted to its enolate anion, which then attacks the carbonyl group of a second molecule of acetyl-CoA to give a tetrahedral carbonyl addition intermediate. Collapse of this intermediate by loss of coenzyme A anion ($CoAS^-$) gives acetoacetyl-CoA. Subsequent proton transfer to coenzyme A anion gives coenzyme A. The mechanism of this reaction is identical to that of the Claisen condensation (Section 19.3A).

Acetyl-CoA   Acetyl-CoA   Acetoacetyl-CoA   Coenzyme A

Enzyme-catalyzed aldol reaction with a third molecule of acetyl-CoA on the ketone carbonyl of acetoacetyl-CoA gives (S)-3-hydroxy-3-methylglutaryl-CoA. Note three features of this reaction. First, reaction is stereospecific; only the S enantiomer is formed. Condensation takes place in a chiral environment created by the enzyme, 3-hydroxy-3-methylglutaryl-CoA synthetase, which induces the formation of one enantiomer of the product to the exclusion of the other. Second, hydrolysis of the thioester group of acetyl-CoA is coupled with the aldol reaction. Third, the carboxyl group is shown as it is ionized at pH 7.4, the approximate pH of blood plasma and many cellular fluids.

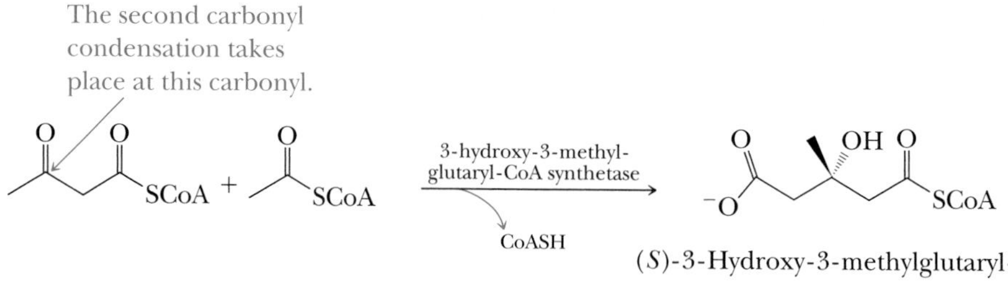

The second carbonyl condensation takes place at this carbonyl.

(S)-3-Hydroxy-3-methylglutaryl-CoA

Enzyme-catalyzed reduction by NADH of the thioester group of 3-hydroxy-3-methylglutaryl-CoA to a primary alcohol gives mevalonic acid, here shown as its anion. Note that, in this reduction, a change occurs in the designation of configuration from S to R, not because of any change in configuration at the stereocenter, but rather because there is a change in priority among the four groups bonded to the stereocenter. The priorities of the four groups bonded to the stereocenter of the S starting material and the R product are shown in the following equation.

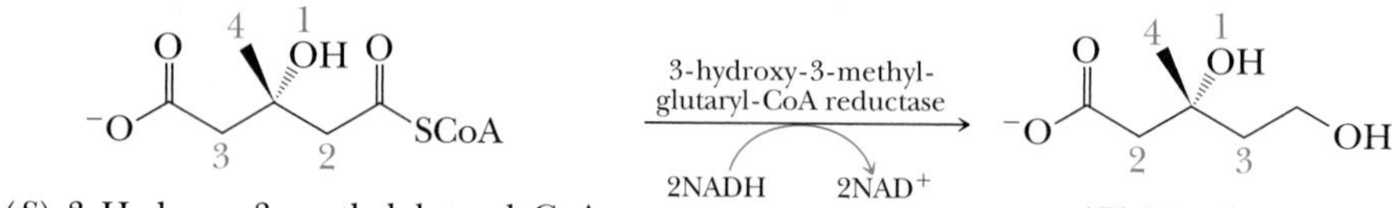

(S)-3-Hydroxy-3-methylglutaryl-CoA   (R)-Mevalonate

Enzyme-catalyzed transfer of a phosphate group from adenosine triphosphate (ATP) to the 3-hydroxyl group of mevalonate gives a phosphoric ester at carbon 3. Enzyme-catalyzed transfer of a pyrophosphate group from a second molecule of ATP gives a pyrophosphoric ester at carbon 5. Enzyme-catalyzed β-elimination from this molecule results in loss of CO_2 and PO_4^{3-}, both good leaving groups.

a phosphoric ester

a pyrophosphoric ester

(R)-3-Phospho-5-pyrophosphomevalonate   Isopentenyl pyrophosphate

Isopentenyl pyrophosphate has the carbon skeleton of isoprene, the unit into which terpenes can be divided (Section 5.4). This compound is, in fact, a key intermediate in the synthesis of terpenes, as well as of cholesterol and steroid hormones. We shall return to the chemistry of isopentenyl pyrophosphate in Section 26.4B and discuss its conversion to cholesterol and terpenes.

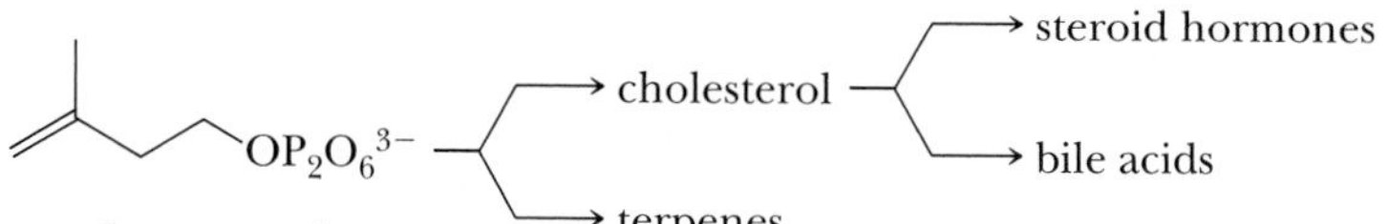

Isopentenyl pyrophosphate

19.5 Enamines

Enamines are formed by the reaction of a secondary amine with an aldehyde or ketone (Section 16.10A). The secondary amines most commonly used for this purpose are pyrrolidine and morpholine.

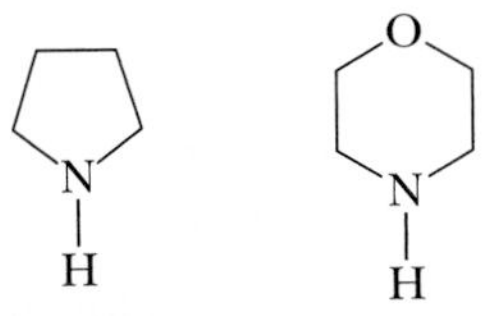

Pyrrolidine   Morpholine

Example 19.8

Draw structural formulas for the aminoalcohol and enamine formed in the following reactions.

(a) 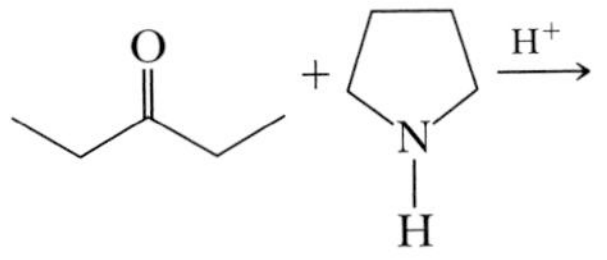   (b)

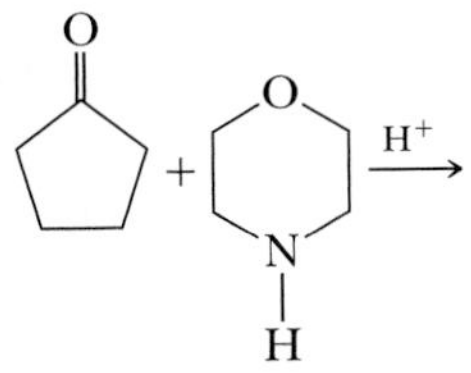

Solution

(a) 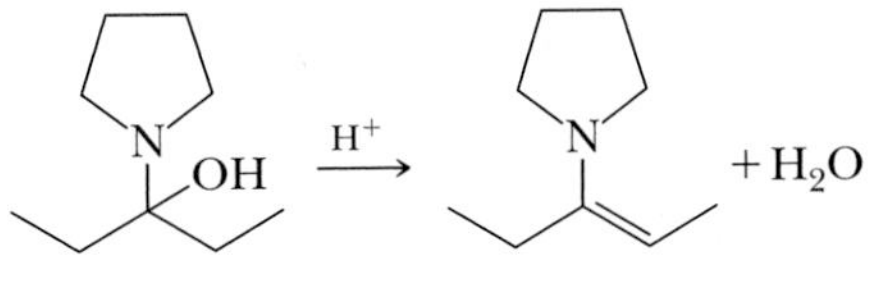   An aminoalcohol   An enamine   (b) 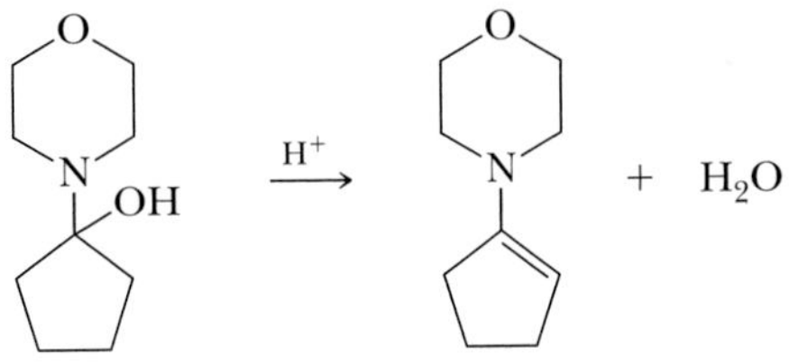   An aminoalcohol   An enamine

Problem 19.8

Following are structural formulas for two enamines. Draw structural formulas for the secondary amine and carbonyl compound from which each is derived.

(a) 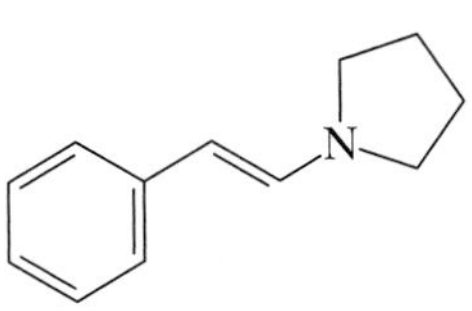   (b)

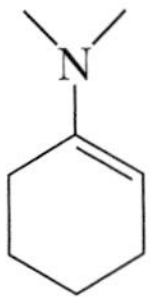

The particular value of enamines in synthetic organic chemistry is the fact that the β-carbon of an enamine is a nucleophile by virtue of the conjugation of the carbon-carbon double bond with the electron pair on nitrogen. Enamines resemble enols and enolate anions in their reactions.

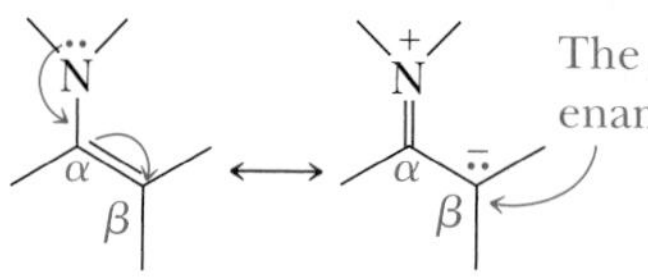

An enamine as a resonance hybrid
of two contributing structures

The β-carbon of an
enamine is nucleophilic.

The use of enamines as synthetic intermediates for the alkylation and acylation at the α-carbon of aldehydes and ketones was pioneered by Gilbert Stork of Columbia University. This use of enamines is called the Stork enamine reaction.

A. Alkylation of Enamines

Enamines readily undergo S$_N$2 reactions with methyl and primary alkyl halides, α-haloketones, and α-haloesters. Enamines are superior to enolate anions for these reactions because they are less basic and consequently give higher ratios of substitution to elimination products. In addition, they also give more alkylation on carbon than do enolate anions.

Mechanism   Alkylation of an Enamine

Step 1: Treatment of the enamine with one equivalent of an alkylating agent gives an iminium halide.

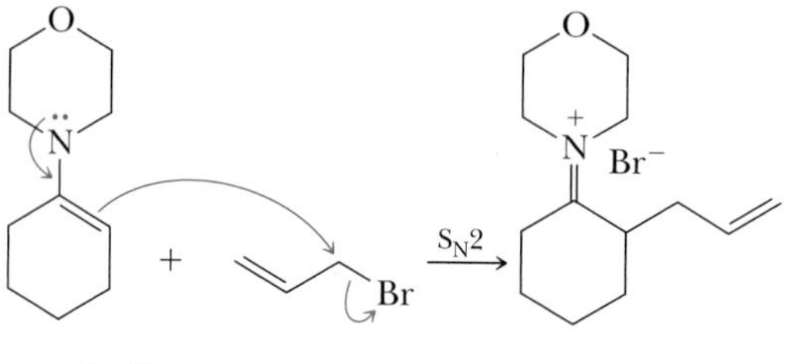

The morpholine
enamine of
cyclohexanone

Allyl
bromide

An iminium
bromide

Step 2: Hydrolysis of the iminium salt regenerates the ketone.

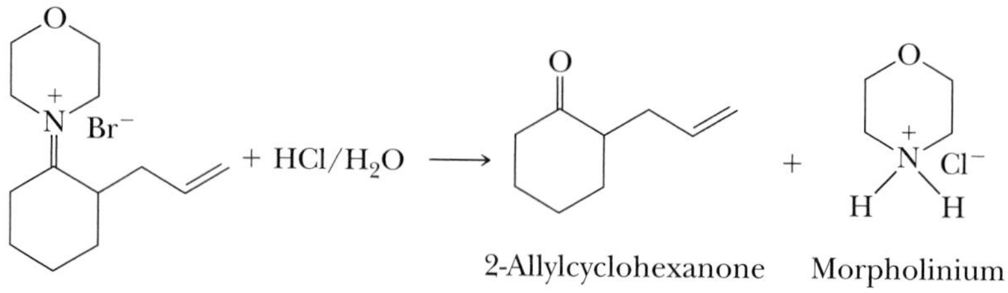

2-Allylcyclohexanone

Morpholinium
chloride

Example 19.9

Show how to use an enamine to bring about this synthesis.

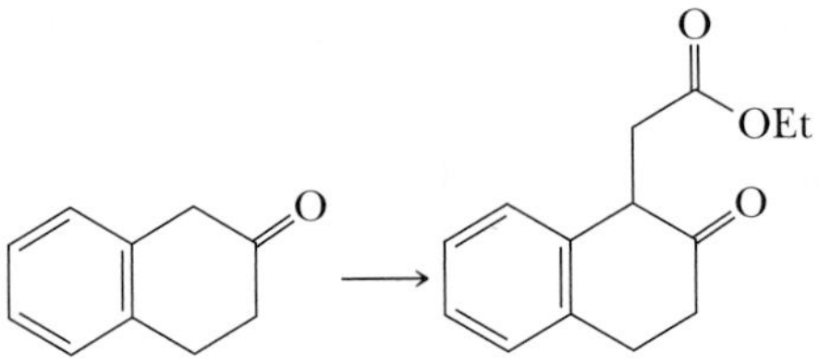

Solution

Prepare an enamine by treating the ketone with either morpholine or pyrrolidine. The intermediate aminoalcohol can undergo dehydration in two directions. The direction shown here is favored because of the stabilization gained by conjugation of the carbon-carbon double bond of the enamine with the aromatic ring. Treatment of the enamine with ethyl 2-chloroacetate followed by hydrolysis of the iminium chloride in aqueous hydrochloric acid gives the product.

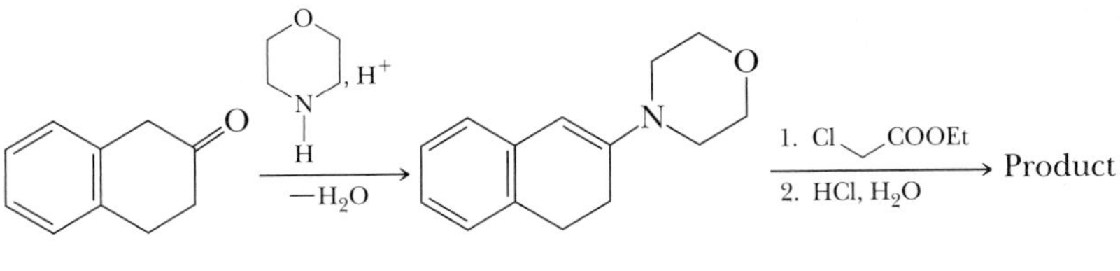

Problem 19.9

Write a mechanism for the hydrolysis of the following iminium chloride in aqueous HCl.

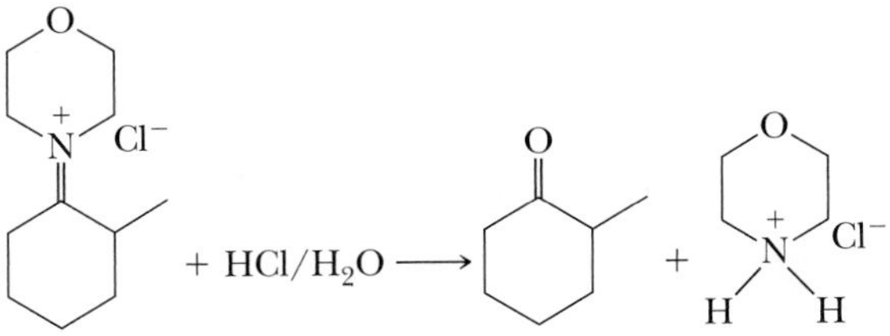

B.   Acylation of Enamines

Enamines undergo acylation when treated with acid chlorides and acid anhydrides. The reaction is a nucleophilic acyl substitution as illustrated by the conversion of cyclohexanone, via its pyrrolidine enamine, to 2-acetylcyclohexanone. Thus, we can attach an acyl group to the α-carbon of an aldehyde or ketone using its enamine as an intermediate. The process of introducing an acyl group onto an organic molecule is called **acylation.**

Acylation   The process of introducing an acyl group, RCO— or ArCO—, onto an organic molecule.

The pyrrolidine enamine of cyclohexanone + Acetyl chloride CH_3CCl → An iminium chloride $\xrightarrow{HCl/H_2O}$ 2-Acetylcyclohexanone +

Example 19.10

Show how to use an enamine to bring about this synthesis.

Solution

Treatment of cyclopentanone with pyrrolidine gives an enamine. Treatment of the enamine with hexanoyl chloride followed by hydrolysis in aqueous HCl gives the desired β-diketone.

Problem 19.10

Show how to use alkylation or acylation of an enamine to convert acetophenone to the following compounds.

(a) Ph   (b) Ph   (c) Ph ... OEt

19.6   The Acetoacetic Ester Synthesis

What makes acetoacetic ester and other β-ketoesters such versatile starting materials for formation of new carbon-carbon bonds is (1) the acidity of α-hydrogens between the two carbonyl groups (pK_a 9–13), (2) the nucleophilicity of the enolate anion resulting from loss of an α-hydrogen, and (3) the ability of the product to undergo decarboxylation after hydrolysis of the ester. The **acetoacetic ester synthesis** is useful for the preparation of monosubstituted and disubstituted acetones of the following types.

O
‖
CH₃CCH₂COEt ——→ CH₃CCH₂R   A monosubstituted acetone

Ethyl acetoacetate
(Acetoacetic ester)

O
‖
CH₃CCHR   A disubstituted acetone
|
R′

We have already seen the chemistry of the individual steps in this synthesis, but we have not put them together in this particular sequence. Let us illustrate the acetoacetic ester synthesis by choosing 5-hexen-2-one as a target molecule. The three carbons shown in color are provided by ethyl acetoacetate. The remaining three carbons represent the —R group of a substituted acetone.

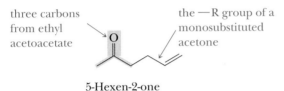

three carbons
from ethyl
acetoacetate

the —R group of a
monosubstituted
acetone

5-Hexen-2-one

1. The methylene hydrogens of ethyl acetoacetate (pK_a 10.7) are more acidic than those of ethanol (pK_a 15.9); therefore, ethyl acetoacetate is converted completely to its anion by sodium ethoxide or other alkali metal alkoxides.

+ EtO⁻Na⁺ ⟶   + EtOH

Ethyl acetoacetate   Sodium ethoxide   Sodium salt of   Ethanol
pK_a 10.7   (stronger base)   ethyl acetoacetate   pK_a 15.9
(stronger acid)   (weaker base)   (weaker acid)

2. The enolate anion of ethyl acetoacetate is a nucleophile and reacts by an S$_N$2 pathway with methyl and primary alkyl halides, α-haloketones, and α-haloesters. Secondary halides give lower yields, and tertiary halides undergo E2 elimination. In the following example, the anion of ethyl acetoacetate is alkylated with allyl bromide.

+ Br   ⟶ (S$_N$2) ⟶   + NaBr

3-Bromopropene
(Allyl bromide)

3, 4. Hydrolysis of the alkylated acetoacetic ester in aqueous NaOH followed by acidification with aqueous HCl (Section 18.5C) gives a β-ketoacid.

3. NaOH, H₂O
4. HCl, H₂O   ⟶   + EtOH

5. Heating the β-ketoacid brings about decarboxylation (Section 17.9A) to give 5-hexen-2-one.

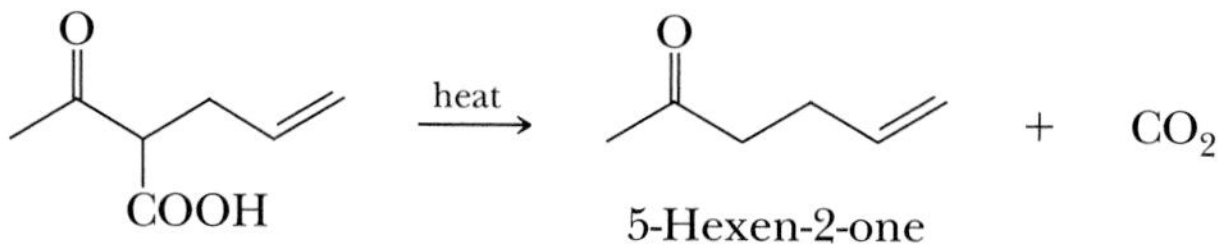

5-Hexen-2-one
(a monosubstituted acetone)

A disubstituted acetone can be prepared by interrupting this sequence after Step 2, treating the monosubstituted acetoacetic ester with a second equivalent of base, carrying out a second alkylation, and then proceeding with Steps 3–5.

1′. Treatment with a second equivalent of base gives a second enolate anion.

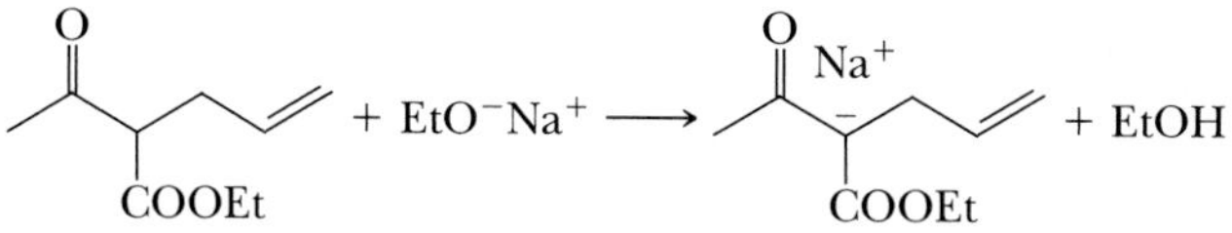

2′. Treatment of the enolate anion with an alkyl halide completes the second alkylation.

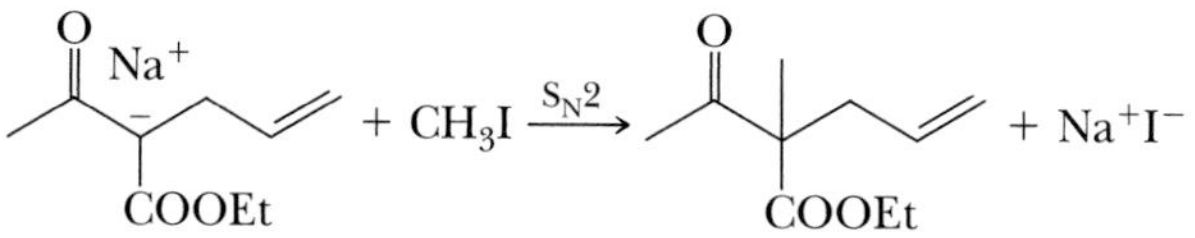

3, 4. Hydrolysis of the ester in aqueous base followed by acidification gives the β-ketoacid.

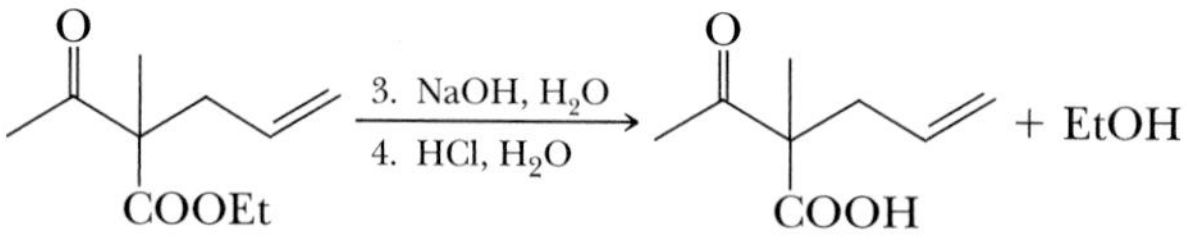

5. Decarboxylation of the β-ketoacid gives the ketone and carbon dioxide.

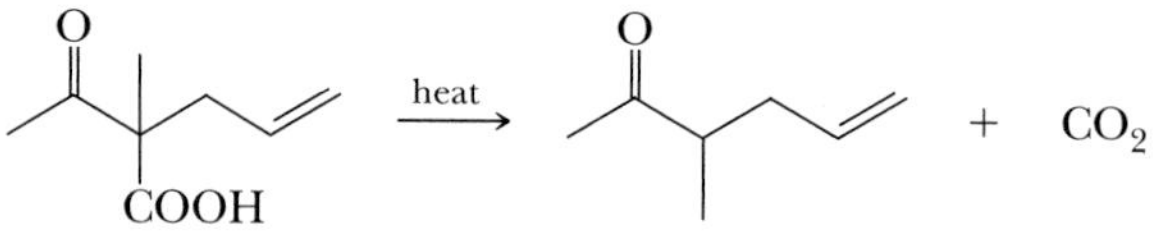

3-Methyl-5-hexen-2-one
(a disubstituted acetone)

Example 19.11

Show how the acetoacetic ester synthesis can be used to prepare this ketone.

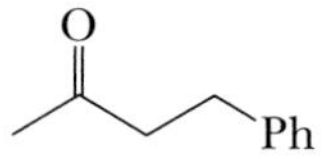

4-Phenyl-2-butanone

Solution

First determine which three carbons of the product originate from ethyl acetoacetate, then the location on the carbon chain of the —COOH lost in decarboxylation,
and finally the bond formed in the alkylation step. On the basis of this analysis, determine that the starting materials are ethyl acetoacetate and a benzyl halide.

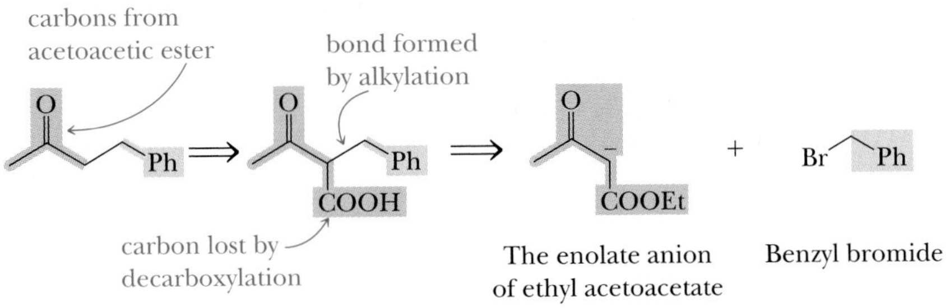

carbons from
acetoacetic ester

bond formed
by alkylation

carbon lost by
decarboxylation

The enolate anion
of ethyl acetoacetate

Benzyl bromide

Now combine these reagents in the following way to prepare the desired ketone.

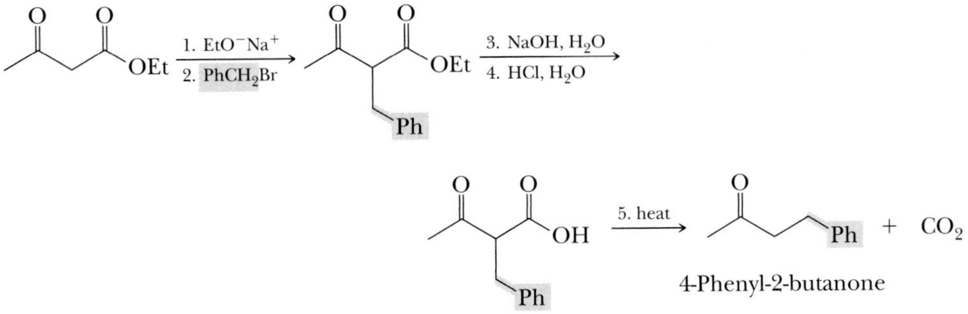

1. EtO⁻Na⁺
2. PhCH₂Br

3. NaOH, H₂O
4. HCl, H₂O

5. heat

4-Phenyl-2-butanone

Problem 19.11

Show how the acetoacetic ester synthesis can be used to prepare these compounds.

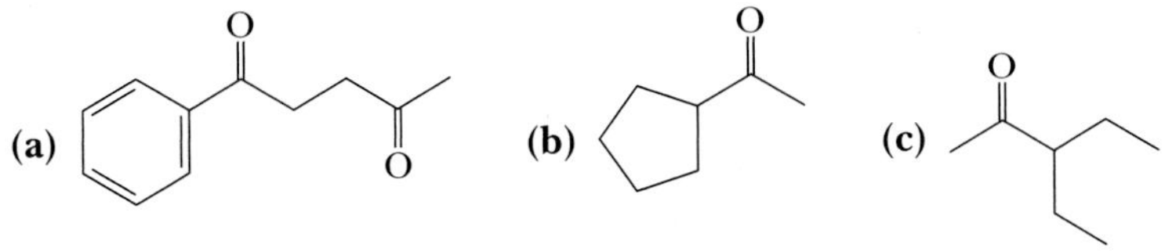

(a)   (b)   (c)

We have described what is commonly known as the acetoacetic ester synthesis and
have illustrated the use of ethyl acetoacetate as the starting reagent. This same synthetic
strategy is applicable to any β-ketoester, as for example those that are available by the
Claisen (Section 19.3A) and Dieckmann (Section 19.3B) condensations. Following are
structural formulas for two β-ketoesters that can be made to undergo (1) formation of
an enolate anion, (2) alkylation or acylation, (3) hydrolysis followed by (4) acidification, and finally (5) decarboxylation just as we have shown for ethyl acetoacetate.

Ethyl 2-oxocyclopentanecarboxylate   Ethyl 2-methyl-3-oxopentanoate

Example 19.12

Show how to convert ethyl 2-oxocyclopentanecarboxylate to 2-allylcyclopentanone.

Solution

Treat this β-ketoester with one equivalent of sodium ethoxide to form an anion followed by alkylation of the anion with one equivalent of an allyl halide. Subsequent hydrolysis of the ester in aqueous base followed by acidification and thermal decarboxylation gives the desired product.

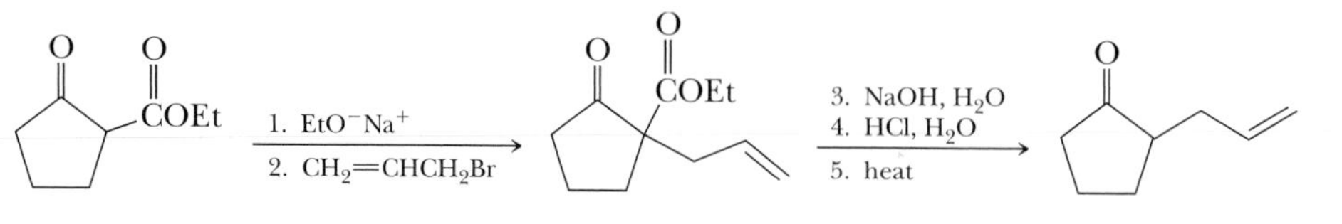

Ethyl 2-oxocyclo-
pentanecarboxylate

2-Allylcyclopentanone

Problem 19.12

Show how to convert ethyl 2-oxocyclopentanecarboxylate to this compound.

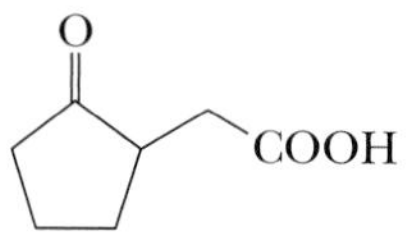

19.7 The Malonic Ester Synthesis

The factors that make malonic esters and other β-diesters such versatile starting materials for formation of new carbon-carbon bonds are the same as those we have already seen for the acetoacetic ester synthesis, namely (1) the acidity of α-hydrogens between the two carbonyl groups, (2) the nucleophilicity of the enolate anion resulting from loss of such an α-hydrogen, and (3) the ability of the product to undergo decarboxylation after hydrolysis of the ester. The **malonic ester synthesis** is useful for the preparation of monosubstituted and disubstituted acetic acids of the following types.

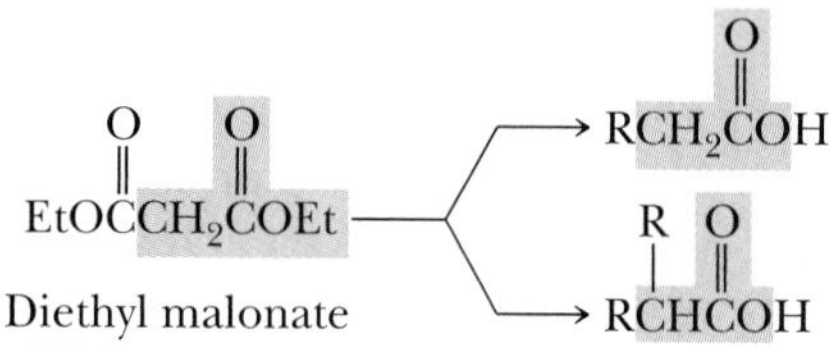

Diethyl malonate
(Malonic ester)

A monosubstituted
acetic acid

A disubstituted
acetic acid

As with the acetoacetic ester synthesis, we have already encountered all the important chemistry of the malonic ester synthesis, although not in this particular pattern. Let us illustrate this synthesis by choosing 5-methoxypentanoic acid as a target molecule. The two carbons shown in color are provided by diethyl malonate. The remaining three carbons and the methoxy group represent the —R group of a monosubstituted acetic acid.

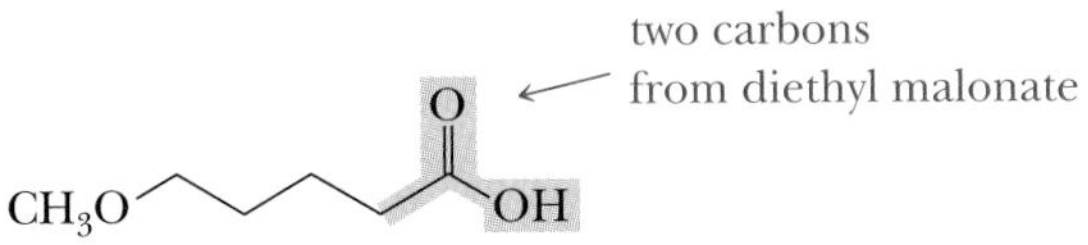

two carbons
← from diethyl malonate

$$CH_3O \diagup\diagdown\diagup \overset{O}{\underset{}{\diagdown}} OH$$

5-Methoxypentanoic acid

1. The α-hydrogens of diethyl malonate (pK_a 13.3) are more acidic than ethanol (pK_a 15.9); therefore, diethyl malonate is converted completely to its anion by sodium ethoxide or other alkali metal alkoxide.

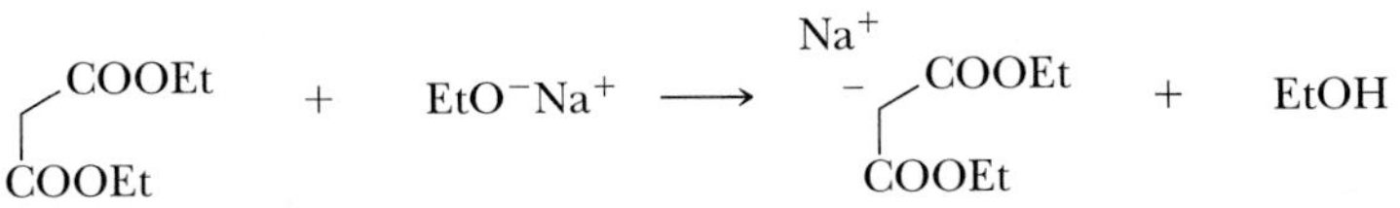

Diethyl malonate	Sodium ethoxide	Sodium salt of	Ethanol
pK_a 13.3	(stronger base)	diethyl malonate	pK_a 15.9
(stronger acid)		(weaker base)	(weaker acid)

2. The enolate anion of diethyl malonate is a nucleophile and reacts by an S_N2 pathway with methyl and primary alkyl halides, α-haloketones, and α-haloesters. In the following example, the anion of diethyl malonate is alkylated with 1-bromo-3-methoxypropane.

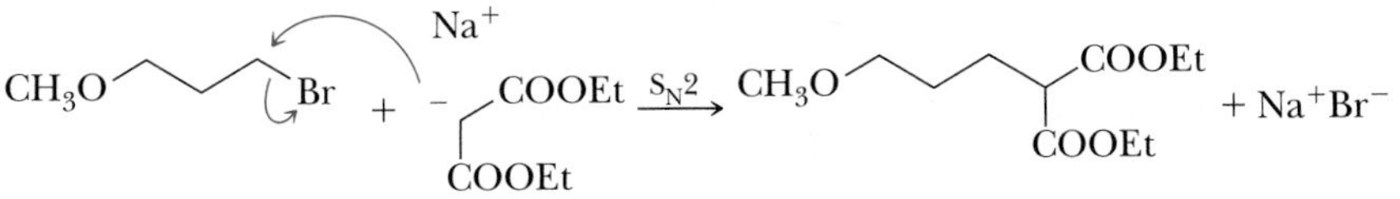

3, 4. Hydrolysis of the alkylated malonic ester in aqueous NaOH followed by acidification with aqueous HCl gives a β-dicarboxylic acid.

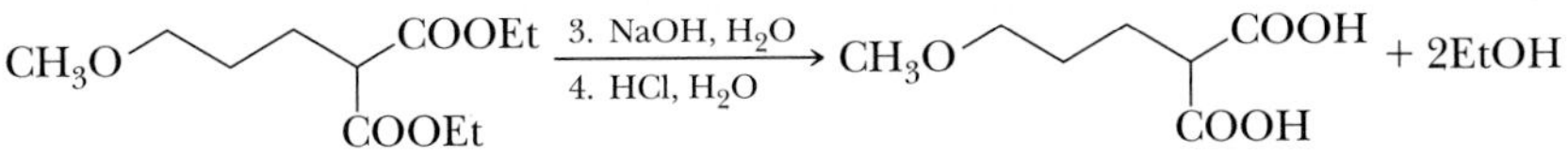

5. Heating the β-dicarboxylic acid slightly above its melting point brings about decarboxylation and gives 5-methoxypentanoic acid.

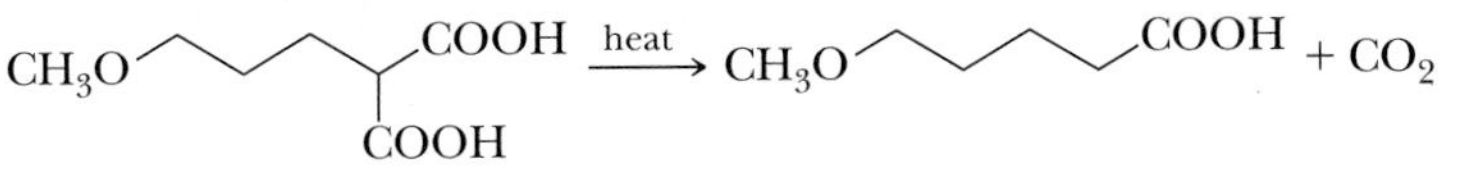

5-Methoxypentanoic acid

A disubstituted acetic acid can be prepared by interrupting the previous sequence after Step 2, treating the monosubstituted diethyl malonate with a second equivalent of base, carrying out a second alkylation, and then proceeding with Steps 3–5.

Example 19.13

Show how the malonic ester synthesis can be used to prepare 3-phenylpropanoic acid.

Solution

Determine which two carbons of the product originate from diethyl malonate, the location on the carbon chain of the —COOH lost in decarboxylation, and finally the bond formed in the alkylation step. On the basis of this analysis, determine that the starting materials are diethyl malonate and a benzyl halide.

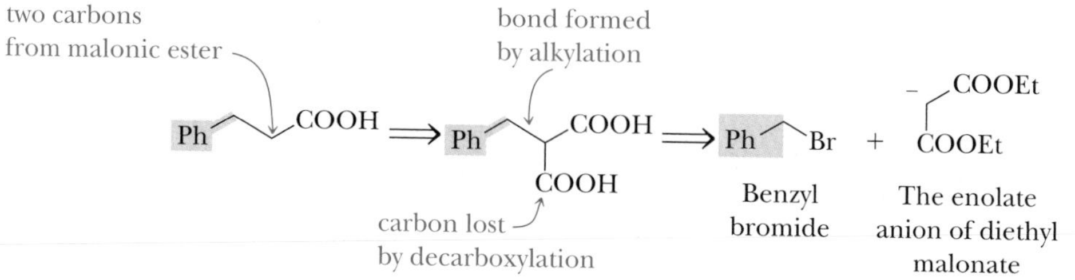

two carbons
from malonic ester

bond formed
by alkylation

Ph⁀COOH ⟹ Ph⁀COOH ⟹ Ph⁀Br + ⁻C(COOEt)(COOEt)
                            COOH

carbon lost
by decarboxylation

Benzyl
bromide

The enolate
anion of diethyl
malonate

Now combine these reagents in the following way to get the desired product.

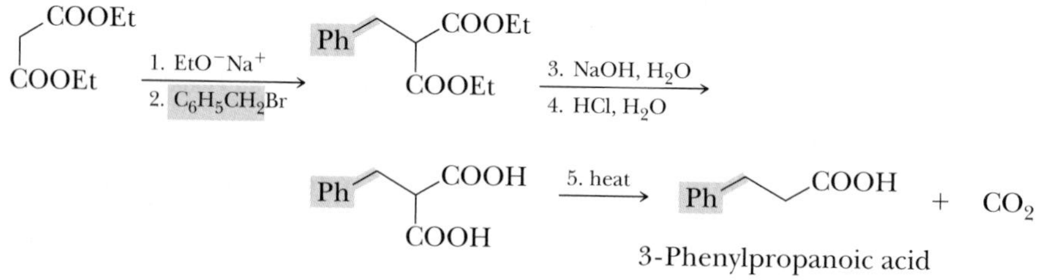

CH₂(COOEt)₂ →(1. EtO⁻Na⁺ 2. C₆H₅CH₂Br)→ Ph⁀CH(COOEt)₂ →(3. NaOH, H₂O 4. HCl, H₂O)→

Ph⁀CH(COOH)(COOH) →(5. heat)→ Ph⁀⁀COOH + CO₂

3-Phenylpropanoic acid

Problem 19.13

Show how the malonic ester synthesis can be used to prepare the following substituted acetic acids.

(a) 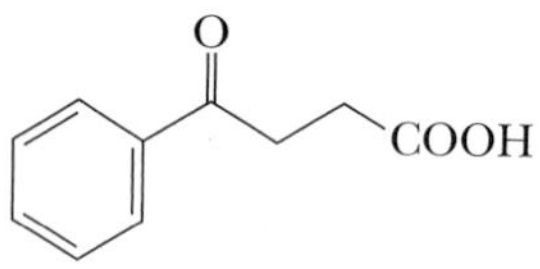 (phenyl ketone)—COOH   (b) 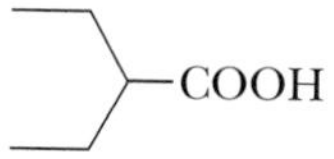 —COOH   (c) 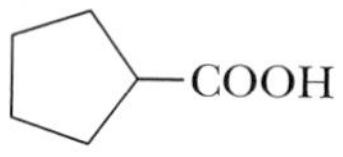 (cyclopentyl)—COOH

19.8 Conjugate Addition to α,β-Unsaturated Carbonyl Compounds

Thus far we have used a variety of carbon nucleophiles to form new carbon-carbon bonds:

1. Anions of terminal alkynes (Section 10.5)
2. Organomagnesium (Grignard) reagents, organolithium reagents, lithium diorganocopper (Gilman) reagents, and Heck reactions (Chapter 15)
3. Anions derived from 1,3-dithianes (Section 16.9)
4. Enolate anions derived from aldehydes and ketones (aldol reactions), esters (Claisen and Dieckmann condensations), β-diesters (malonic ester syntheses), and β-ketoesters (acetoacetic ester syntheses)
5. Enamines (which are synthetically equivalent to enolate anions)

These species can be used to form new carbon-carbon bonds by two synthetic strategies: (1) substitution of the carbon nucleophile in an S_N2 reaction and (2) addition of the carbon nucleophile to a carbonyl carbon. **Conjugate addition,** as it is known, presents a third synthetic strategy: addition of a carbon nucleophile to an electrophilic carbon-carbon double or triple bond conjugated with a carbonyl or other electron-withdrawing group. In this section, we study two types of conjugate additions to electrophilic double bonds: addition of enolate anions (the Michael reaction) and addition of lithium diorganocopper (Gilman) reagents.

> **Conjugate addition** Addition of a nucleophile to the β-carbon of an α,β-unsaturated carbonyl compound.

A. Michael Addition of Enolate Anions

Nucleophilic addition of enolate anions to α,β-unsaturated carbonyl compounds was first reported in 1887 by the American chemist, Arthur Michael. Following are two examples of **Michael reactions.** In the first example, the nucleophile adding to the conjugated system is the enolate anion of diethyl malonate. In the second example, the nucleophile is the enolate anion of ethyl acetoacetate.

Diethyl propanedioate (Diethyl malonate)	3-Buten-2-one (Methyl vinyl ketone)	

Ethyl 3-oxobutanoate (Ethyl acetoacetate)	2-Cyclohexenone

You will recall that nucleophiles don't add to ordinary double bonds. They are attacked not by nucleophiles but by electrophiles (Section 6.3). What activates a carbon-carbon double bond for nucleophilic attack in a Michael reaction is the presence of the adjacent carbonyl group. One important resonance structure of the α,β-unsaturated carbonyl compounds puts positive charge at the end (in this case, the β-carbon) of the double bond, making it resemble a carbonyl group in its reactivity. Thus nucleophiles can add to this type of double bond, which we call "activated" for this reason.

Note that aldol, Claisen, and Dieckmann condensations all give primary products with oxygens in a 1,3 relationship. The Michael reaction with enolate anions gives products with oxygens in a 1,5 relationship. These relationships are a consequence of the polarization of the reagents. In aldol, Claisen, and Dieckmann condensations, the carbonyl carbon is positive and the α-position is negative.

Polarization of
reactants   acceptor carbonyl

Aldol, Claisen,
Dieckmann

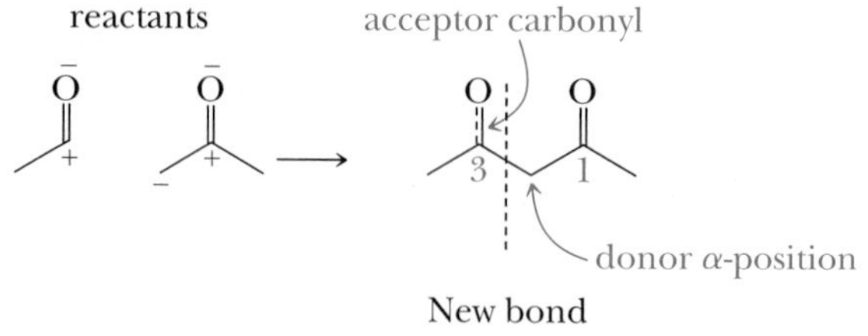

New bond

In Michael reactions, the positive polarization of the carbonyl carbon is transmitted two carbons farther by the double bond.

Michael

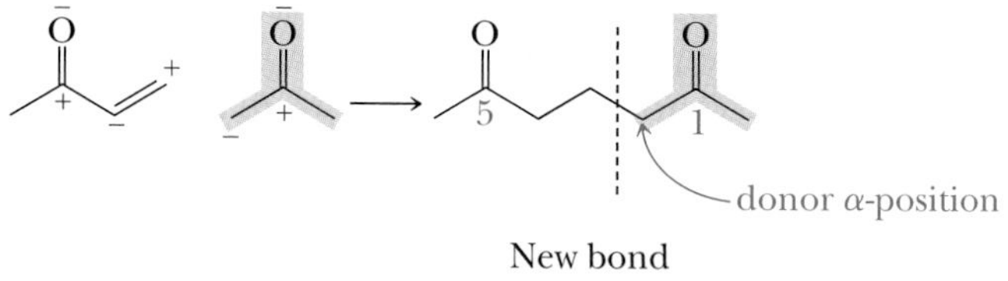

New bond

The Michael reaction takes place with a wide variety of α,β-unsaturated carbonyl compounds as well as with α,β-unsaturated nitriles and nitro compounds. The most commonly used types of nucleophiles in Michael reactions are summarized in Table 19.1. The most commonly used bases are metal alkoxides, pyridine, and piperidine.

We can write the following general mechanism for a Michael reaction. The enol formed in Step 3 corresponds to 1,4-addition to the conjugated system of the α,β-unsaturated carbonyl compound. Because this intermediate is formed, the Michael re-

Table 19.1 Combinations of Reagents for Effective Michael Reactions

These Types of α,β-Unsaturated Compounds Are Nucleophile Acceptors in Michael Reactions		These Types of Compounds Provide Effective Nucleophiles for Michael Reactions	
$CH_2{=}CHCH$ (O)	Aldehyde	$CH_3CCH_2CCH_3$ (O O)	β-Diketone
$CH_2{=}CHCCH_3$ (O)	Ketone	CH_3CCH_2COEt (O O)	β-Ketoester
$CH_2{=}CHCOEt$ (O)	Ester	CH_3CCH_2CN (O)	β-Ketonitrile
$CH_2{=}CHCNH_2$ (O)	Amide	$EtOCCH_2COEt$ (O O)	β-Diester
$CH_2{=}CHC{\equiv}N$	Nitrile	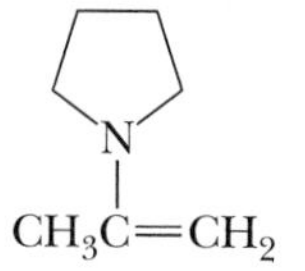 (pyrrolidine) $CH_3C{=}CH_2$	Enamine
$CH_2{=}CHNO_2$	Nitro compound		

action is classified as a 1,4- or conjugate addition. Note that in Step 3 the base, B⁻, is regenerated, in accord with the experimental observation that a Michael reaction requires only a catalytic amount of base rather than a molar equivalent.

Mechanism Michael Reaction — Conjugate Addition of Enolate Anions

Step 1: Treatment of H—Nu with base gives the nucleophile, Nu⁻.

$$Nu{-}H + B^- \rightleftharpoons Nu^- + H{-}B$$

Base

Step 2: Nucleophilic addition of Nu⁻ to the β-carbon of the conjugated system gives a resonance-stabilized enolate anion.

Resonance-stabilized enolate anion

Step 3: Proton transfer from H—B gives the enol.

An enol
(a product of 1,4-addition)

Step 4: Tautomerism (Section 16.11B) of the less stable enol form gives the more stable keto form.

Less stable enol form   More stable keto form

Example 19.14

Draw a structural formula for the product formed when each set of reactants is treated with sodium ethoxide in ethanol under conditions of the Michael reaction.

Solution

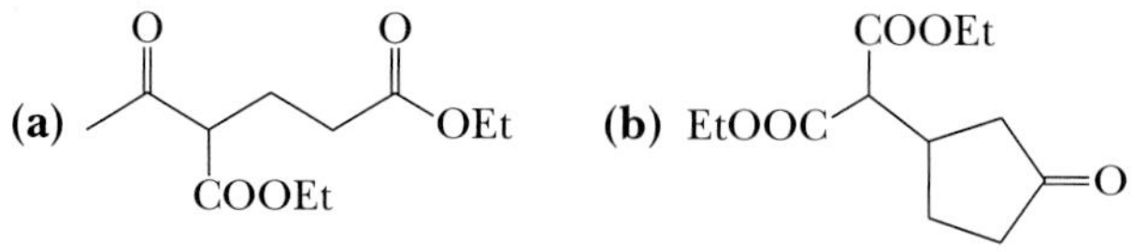

(a)   (b) EtOOC

Problem 19.14

Show the product formed from each Michael product in the solution to Example 19.14 after (1) hydrolysis in aqueous NaOH, (2) acidification, and (3) thermal decarboxylation of each β-ketoacid or β-dicarboxylic acid. These reactions illustrate the usefulness of the Michael reaction for the synthesis of 1,5-dicarbonyl compounds.

Example 19.15

Show how the series of reactions in Example 19.14 and Problem 19.14 (Michael reaction, hydrolysis, acidification, and thermal decarboxylation) can be used to prepare 2,6-heptanedione.

Solution

As shown in the following retrosynthetic analysis, this molecule can be constructed from the carbon skeletons of ethyl acetoacetate and methyl vinyl ketone.

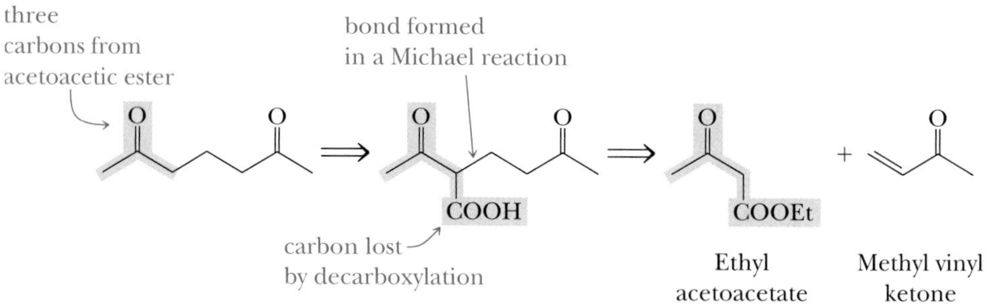

three carbons from acetoacetic ester

bond formed in a Michael reaction

carbon lost by decarboxylation

Ethyl acetoacetate

Methyl vinyl ketone

Following are the steps in their conversion to 2,6-heptanedione.

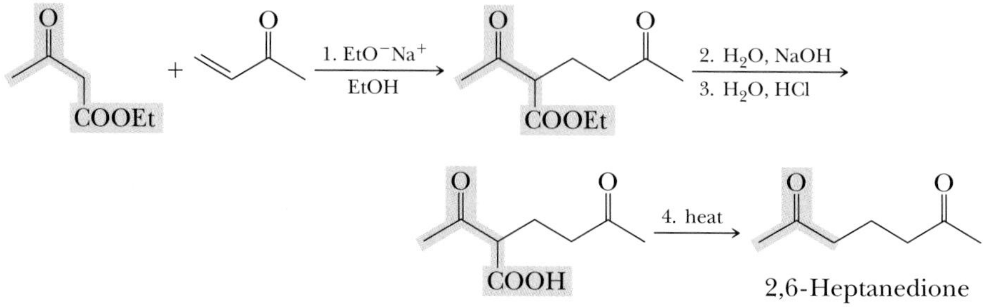

2,6-Heptanedione

Problem 19.15

Show how the sequence of Michael reaction, hydrolysis, acidification, and thermal decarboxylation can be used to prepare pentanedioic acid (glutaric acid).

As noted in Table 19.1, enamines also participate in Michael reactions as illustrated by the addition of the enamine of cyclohexanone to acrylonitrile, $CH_2=CHC\equiv N$.

Pyrrolidine
enamine of
cyclohexanone

Here is a final word about addition of nucleophiles to α,β-unsaturated carbonyl compounds. The Michael reaction is an example of 1,4-addition (conjugate addition) to an α,β-unsaturated carbonyl compound. In general, resonance-stabilized enolate anions and enamines are weak bases, react slowly, and give 1,4-addition products. Organolithium and organomagnesium compounds, on the other hand, are strong bases, react rapidly, and give primarily 1,2-addition products, that is, products formed by addition to the carbonyl carbon.

Phenyl-   4-Methyl-3-   4-Methyl-2-phenyl-
lithium   penten-2-one   3-penten-2-ol

Why do the nucleophiles listed in Table 19.1 react with conjugated carbonyl compounds by 1,4-addition rather than 1,2-addition? The answer has to do with kinetic control versus thermodynamic control of product formation. It has been shown that 1,2-addition of nucleophiles to the carbonyl carbon of α,β-unsaturated carbonyl compounds is faster than conjugate addition. If formation of the 1,2-addition product is irreversible, then that is the product observed. If, however, formation of the 1,2-addition product is reversible, then an equilibrium is established between the more rapidly formed 1,2-addition product and the more slowly formed 1,4-addition product. A carbon-oxygen double bond is stronger than a carbon-carbon double bond. Recall, for example, the relative percentages of keto and enol forms present at equilibrium for simple aldehydes and ketones (Section 16.11B). Thus, under conditions of thermodynamic (equilibrium) control, the more stable 1,4-Michael addition product is formed.

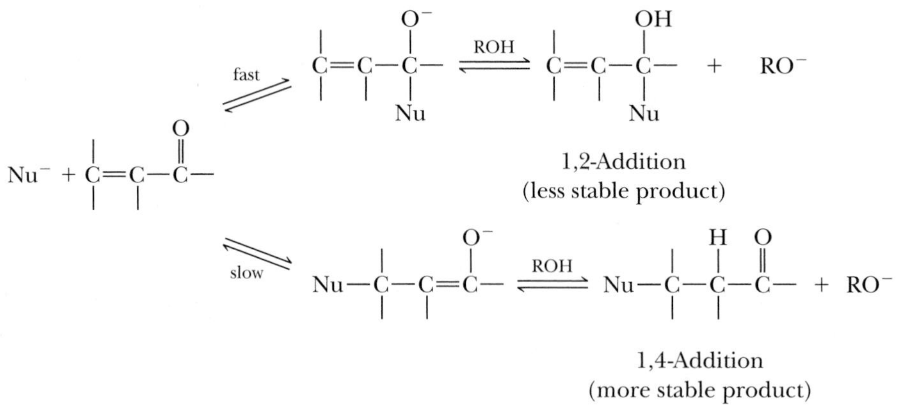

1,2-Addition
(less stable product)

1,4-Addition
(more stable product)

Michael reaction with an α,β-unsaturated ketone followed by an intramolecular aldol reaction has proven to be a valuable method for the synthesis of 2-cyclohexenones. An especially important example of a Michael-aldol sequence is the **Robinson annulation** in which treatment of a cyclic ketone, β-ketoester, or β-diketone with an α,β-unsaturated ketone in the presence of a base catalyst forms a cyclohexenone ring fused to the original ring. When the following β-ketoester, for example, is treated with methyl vinyl ketone in the presence of sodium ethoxide in ethanol, the Michael adduct is first formed and then, in the presence of sodium ethoxide, undergoes a base-catalyzed intramolecular aldol reaction followed by dehydration to give a cyclohexenone.

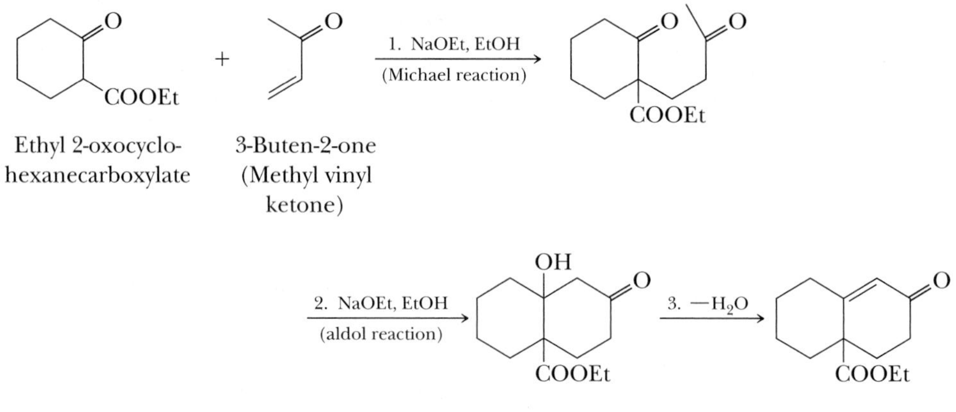

Ethyl 2-oxocyclo-
hexanecarboxylate

3-Buten-2-one
(Methyl vinyl
ketone)

Example 19.16

Draw structural formulas for the lettered compounds in the following synthetic sequence.

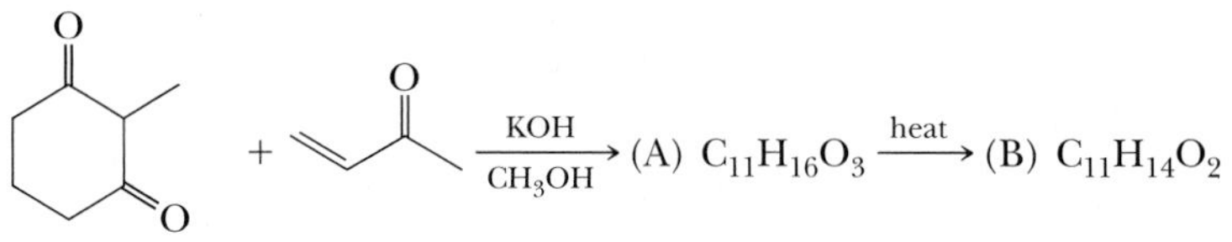

Solution

The product is the result of Michael addition to an α,β-unsaturated ketone followed by base-catalyzed aldol reaction and dehydration.

(A)   (B)

Problem 19.16

Show how to bring about the following conversion.

B.   Conjugate Addition of Lithium Diorganocopper Reagents

Lithium diorganocopper reagents undergo 1,4-addition to α,β-unsaturated aldehydes and ketones in a reaction that is closely related to the Michael reaction. Yields are highest with primary alkyl, vinyl, and aryl organocopper reagents.

3-Methyl-2-cyclohexenone   3,3-Dimethylcyclohexanone

2-Cyclohexenone   3-Phenylcyclohexanone

Lithium diorganocopper reagents are unique among organometallic compounds in that they give almost exclusively 1,4-addition, which makes them very valuable reagents in synthetic organic chemistry. The mechanism of conjugate addition of lithium diorganocopper reagents is not fully understood.

Example 19.17

Propose two syntheses of 4-octanone, each involving conjugate addition of a lithium diorganocopper reagent.

Solution

A lithium diorganocopper reagent adds to the beta carbon of an α,β-unsaturated aldehyde or ketone. Therefore, locate each beta carbon to the carbonyl group in this target molecule and disconnect at those points.

Synthesis 1:

| 4-Octanone | 1-Hexen-3-one | Bromoethane |

For this synthesis, add lithium diethylcopper to 1-hexen-3-one.

1-Hexen-3-one   4-Octanone

Synthesis 2:

| 4-Octanone | 1-Hepten-3-one | Bromomethane |

For this synthesis, add lithium dimethylcopper to 1-hepten-3-one.

1-Hepten-3-one   4-Octanone

Problem 19.17

Propose two syntheses of 4-phenyl-2-pentanone, each involving conjugate addition of a lithium diorganocopper reagent.

Summary

An **enolate anion** is an anion formed by removing an α-hydrogen from a carbonyl-containing compound (Section 19.1). Aldehydes, ketones, and esters can be converted completely to their enolate anions by treatment with a strong base such as lithium diisopropylamide.

Acetyl-CoA (Section 19.4) is the source of the carbon atoms for the synthesis of terpenes, cholesterol, steroid hormones, and fatty acids. Key intermediates in the synthesis of these biomolecules are mevalonic acid and isopentenyl pyrophosphate.

Key Reactions

1. The Aldol Reaction (Section 19.1)

The aldol reaction involves nucleophilic addition of the enolate anion of one aldehyde or ketone to the carbonyl group of another aldehyde or ketone. The product of an aldol reaction is a β-hydroxyaldehyde or a β-hydroxyketone.

2. Dehydration of the Product of an Aldol Reaction (Section 19.1)

Dehydration of the β-hydroxyaldehyde or ketone from an aldol reaction occurs very readily under acidic or basic conditions and gives an α,β-unsaturated aldehyde or ketone.

3. Formation of Enolate Anions (Section 19.2A)

Aldehydes, ketones, esters, and other compounds with acidic α-hydrogens are converted completely to their enolate anions by strong bases, such as lithium diisopropylamide (LDA).

$$CH_3COEt \ + \ (i\text{-}Pr)_2N^-Li^+ \ \longrightarrow \ CH_2{=}COEt \ + \ (i\text{-}Pr)_2NH$$

pK_a 23	LDA	A lithium enolate	pK_a 40
(stronger acid)			(weaker acid)

4. Formation of Enolate Anions Under Kinetic Control (Section 19.2A)

The composition of the enolate anion mixture is determined by the relative rates of removal of alternative α-hydrogens; the less substituted enolate anion predominates, often to the exclusion of the alternative enolate anion.

5. Formation of Enolate Anions Under Thermodynamic Control (Section 19.2A)

The composition of an enolate anion mixture is determined by the relative stabilities of the individual enolate anions; generally the more substituted enolate anion predominates.

Slight excess   10%   90%
of the ketone

6. Directed Aldol Reactions (Section 19.2B)

An enolate anion is preformed and then treated with a carbonyl compound acting as an enolate anion acceptor.

A lithium enolate

7. The Claisen Condensation (Section 19.3A)

The product of a Claisen condensation is a β-ketoester. Condensation occurs by nucleophilic acyl substitution in which the attacking nucleophile is the enolate anion of an ester.

8. The Dieckmann Condensation (Section 19.3B)

An intramolecular Claisen condensation is called a Dieckmann condensation.

9. Alkylation of an Enamine Followed by Hydrolysis (Section 19.5A)

Enamines are reactive nucleophiles with methyl and primary alkyl halides, α-haloketones, and α-haloesters.

10. Acylation of an Enamine Followed by Hydrolysis (Section 19.5B)

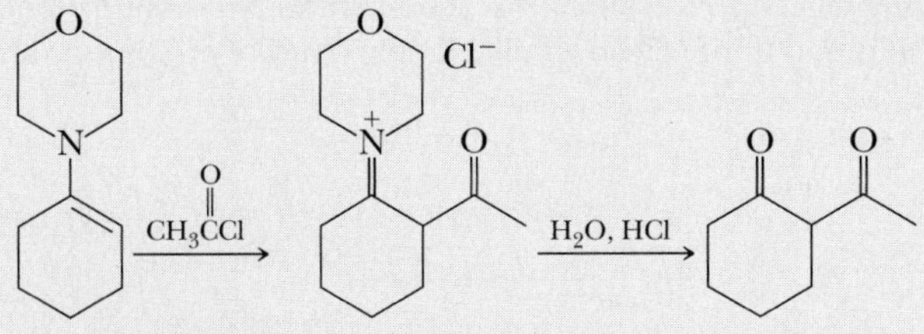

11. Acetoacetic Ester Synthesis (Section 19.6)

This sequence is useful for the synthesis of monosubstituted and disubstituted acetones.

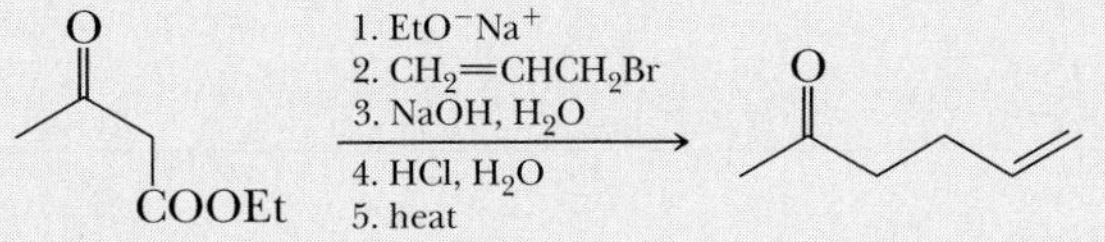

Ethyl acetoacetate

A monosubstituted
acetone

12. Malonic Ester Synthesis (Section 19.7)

This sequence is useful for the synthesis of monosubstituted and disubstituted acetic acids.

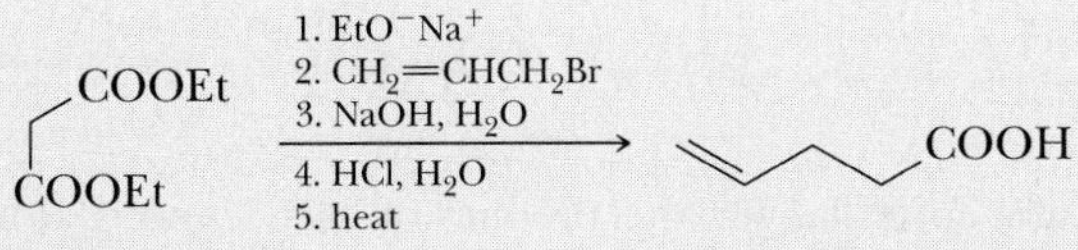

Diethyl malonate

A monosubstituted
acetic acid

13. Michael Reaction (Section 19.8A)

Addition of a relatively weakly basic nucleophile to a carbon-carbon double bond made electrophilic by conjugation with the carbonyl group of an aldehyde, ketone, or ester or with a nitro or cyano group.

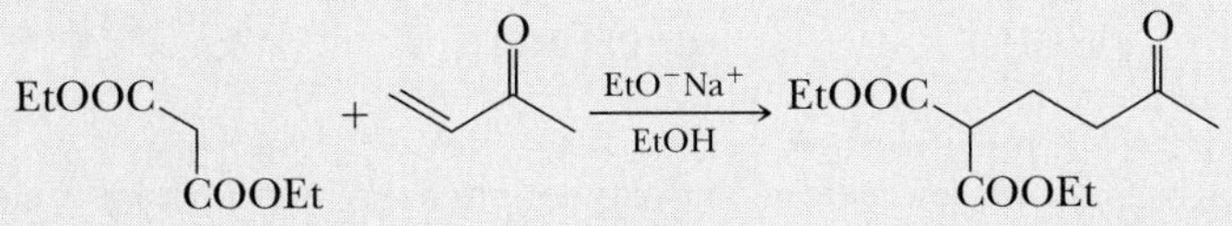

14. Robinson Annulation (Section 19.8A)

A Michael reaction followed by an intramolecular aldol reaction and dehydration forms a substituted 2-cyclohexenone.

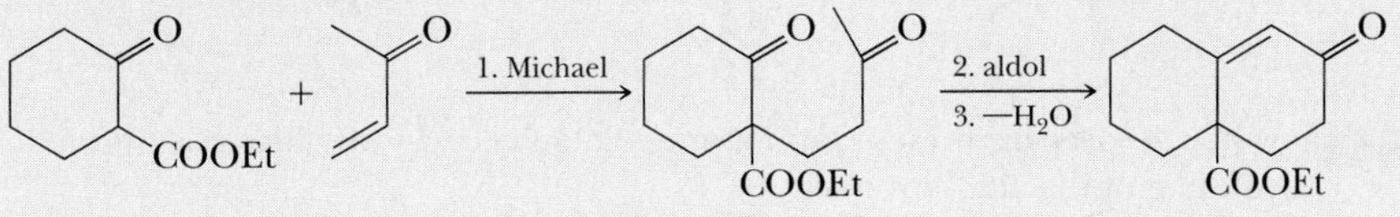

15. Conjugate Addition of Lithium Diorganocopper Reagents (Section 19.8B)

In a reaction closely related to the Michael reaction, lithium diorganocopper reagents undergo conjugate addition to the electrophilic double bond of α,β-unsaturated aldehydes and ketones.

Problems

The Aldol Reaction

19.18 Draw structural formulas for the product of the aldol reaction of each compound and for the α,β-unsaturated aldehyde or ketone formed from dehydration of each aldol product.

19.19 Draw structural formulas for the product of each crossed aldol reaction and for the compound formed by dehydration of each aldol product.

19.20 When a 1:1 mixture of acetone and 2-butanone is treated with base, six aldol products are possible. Draw a structural formula for each.

19.21 Show how to prepare these α,β-unsaturated ketones by an aldol reaction followed by dehydration of the aldol product.

19.22 Show how to prepare these α,β-unsaturated aldehydes by an aldol reaction followed by dehydration of the aldol product.

(a)

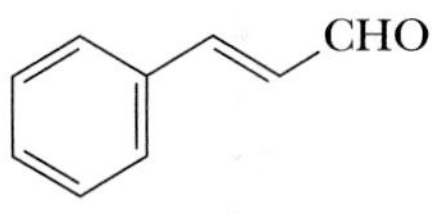

(b)

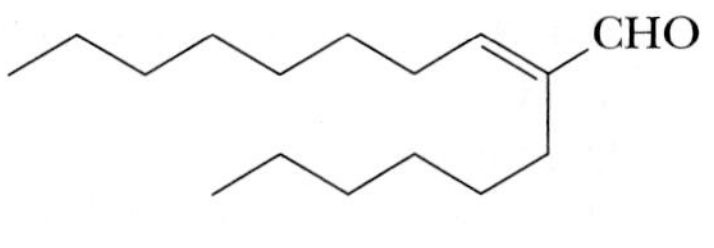

19.23 When treated with base, the following compound undergoes an intramolecular aldol reaction to give a product containing a ring (yield 78%). Propose a structural formula for this product.

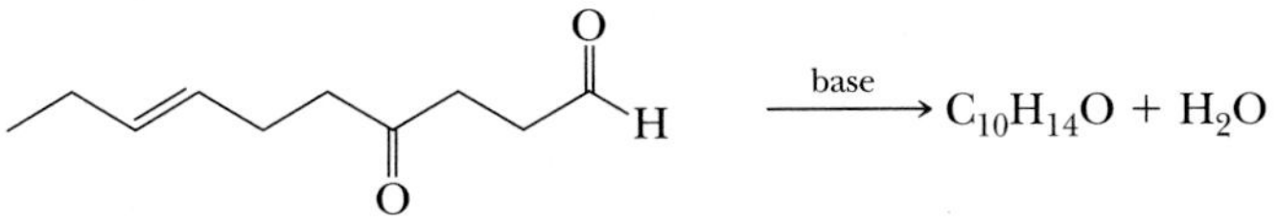

$\xrightarrow{\text{base}}$ $C_{10}H_{14}O + H_2O$

19.24 Cyclohexene can be converted to 1-cyclopentenecarbaldehyde by the following series of reactions. Propose a structural formula for each intermediate compound.

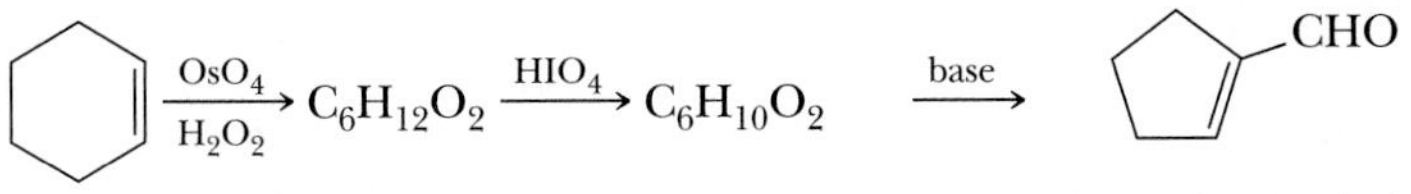

$\xrightarrow[\text{H}_2\text{O}_2]{\text{OsO}_4}$ $C_6H_{12}O_2$ $\xrightarrow{\text{HIO}_4}$ $C_6H_{10}O_2$ $\xrightarrow{\text{base}}$

1-Cyclopentenecarbaldehyde

19.25 Propose a structural formula for each lettered compound.

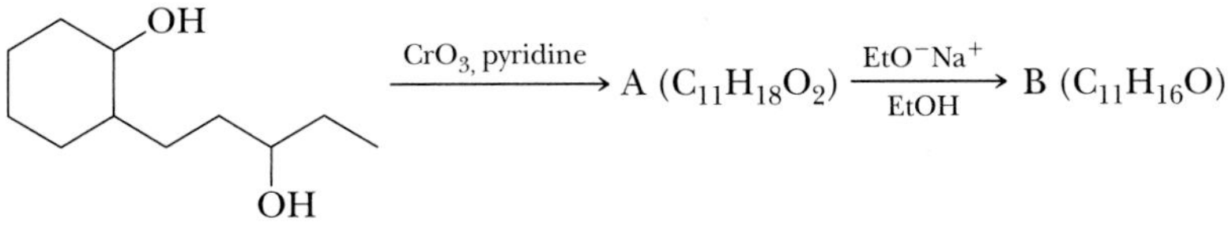

$\xrightarrow{\text{CrO}_3, \text{ pyridine}}$ A $(C_{11}H_{18}O_2)$ $\xrightarrow[\text{EtOH}]{\text{EtO}^-\text{Na}^+}$ B $(C_{11}H_{16}O)$

19.26 How might you bring about the following conversions?

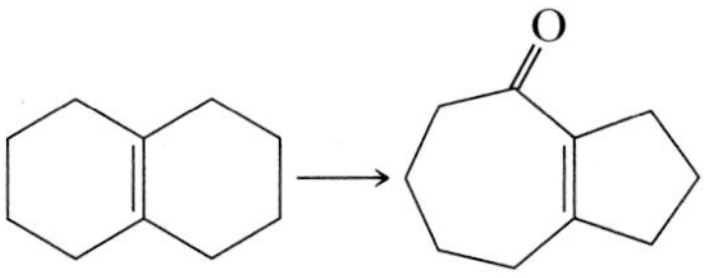

19.27 Pulegone, $C_{10}H_{16}O$, a compound from oil of pennyroyal, has a pleasant odor midway between peppermint and camphor (see *The Merck Index*, 12th ed., #8124). Treatment of pulegone with steam produces acetone and 3-methylcyclohexanone.

$+ \text{H}_2\text{O}$ $\xrightarrow{\text{steam}}$ $+$

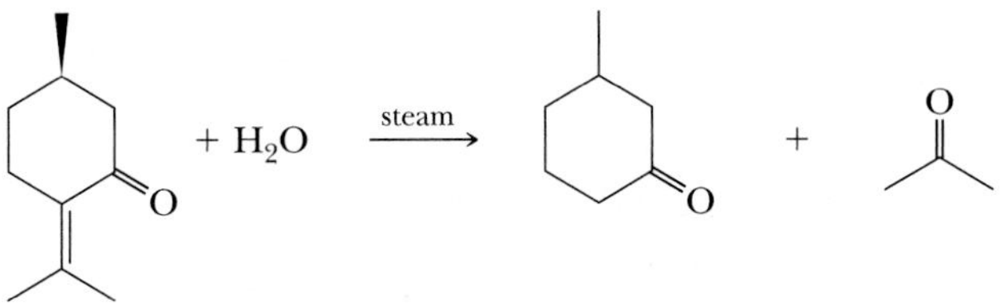

Pulegone   3-Methylcyclohexanone   Acetone

(a) Natural pulegone has the configuration shown. Assign an *R* or *S* configuration to its stereocenter.

(b) Propose a mechanism for the steam hydrolysis of pulegone to the compounds shown.

(c) In what way does this steam hydrolysis affect the configuration of the stereocenter in pulegone? Assign an *R* or *S* configuration to the 3-methylcyclohexanone formed in this reaction.

19.28 Propose a mechanism for this acid-catalyzed aldol reaction and the dehydration of the resulting aldol product.

Directed Aldol Reactions

19.29 In Section 19.2B, it was stated that four possible aldol products are formed when phenylacetaldehyde and acetone are mixed in the presence of base. Draw structural formulas for each of these aldol products.

19.30 In the synthesis of a lithium enolate from a ketone and LDA is it preferable (a) to add a solution of LDA to a solution of the ketone or (b) to add a solution of the ketone to a solution of LDA or (c) to conclude that the order in which the solutions are mixed makes no difference? Explain your answer.

The Claisen Condensation

19.31 Show the product of Claisen condensation of these esters.

(a) Ethyl phenylacetate in the presence of sodium ethoxide
(b) Methyl hexanoate in the presence of sodium methoxide

19.32 When a 1:1 mixture of ethyl propanoate and ethyl butanoate is treated with sodium ethoxide, four Claisen condensation products are possible. Draw a structural formula for each product.

19.33 Draw structural formulas for the β-ketoesters formed by Claisen condensation of ethyl propanoate with each ester.

(a) EtOC—COEt   (b) PhCOEt   (c) HCOEt

19.34 Draw a structural formula for the product of saponification, acidification, and decarboxylation of each β-ketoester formed in Problem 19.33.

19.35 The Claisen condensation can be used as one step in the synthesis of ketones, as illustrated by this reaction sequence. Propose structural formulas for compounds A, B, and the ketone formed in this sequence.

19.36 Propose a synthesis for each ketone, using as one step in the sequence a Claisen condensation and the reaction sequence illustrated in Problem 19.35.

(a) Ph $\diagdown$ $\diagup$ $\diagdown$ Ph

(b) Ph $\diagdown$ $\diagup$ Ph

(c)

19.37 Propose a mechanism for the following conversion.

$$\diagup\diagdown\diagup\diagdown\diagup\text{OEt} \xrightarrow[\text{2. } \triangle\text{O}]{\text{1. NaH}} \diagup\diagdown\diagup\diagdown \quad + \text{ EtOH}$$

19.38 Claisen condensation between diethyl phthalate and ethyl acetate followed by saponification, acidification, and decarboxylation forms a diketone, $C_9H_6O_2$. Propose structural formulas for compounds A, B, and the diketone.

$$\begin{array}{c} \text{COOEt} \\ \\ \text{COOEt} \end{array} + \text{CH}_3\text{COOEt} \xrightarrow[\text{2. HCl, H}_2\text{O}]{\text{1. EtO}^-\text{Na}^+} \text{A} \xrightarrow[\text{heat}]{\text{NaOH, H}_2\text{O}} \text{B} \xrightarrow[\text{heat}]{\text{HCl, H}_2\text{O}} C_9H_6O_2$$

Diethyl phthalate   Ethyl acetate

19.39 In 1887, the Russian chemist Sergei Reformatsky at the University of Kiev discovered that treatment of an α-haloester with zinc metal in the presence of an aldehyde or ketone followed by hydrolysis in aqueous acid results in formation of a β-hydroxyester. This reaction is similar to a Grignard reaction in that a key intermediate is an organometallic compound, in this case a zinc salt of an ester enolate anion. Grignard reagents, however, are so reactive that they undergo self-condensation with the ester.

$$\begin{array}{ccc} \text{O} & & \text{O}^-[\text{ZnBr}]^+ \\ \| & \xrightarrow[\text{benzene}]{\text{Zn}} & | \\ \text{BrCH}_2\text{COEt} & & \text{CH}_2{=}\text{COEt} \end{array} \xrightarrow[\text{2. H}_2\text{O, HCl}]{\text{1. PhCHO}} \begin{array}{c} \text{OH} \quad \text{O} \\ | \quad\quad \| \\ \text{PhCHCH}_2\text{COEt} \end{array}$$

Zinc salt of an enolate anion   A β-hydroxyester

Show how a Reformatsky reaction can be used to synthesize these compounds from an aldehyde or ketone and an α-haloester.

(a)

(b)

(c)

19.40 Many types of carbonyl condensation reactions have acquired specialized names, after the 19th century organic chemists who first studied them. Propose mechanisms for the following named condensations.

(a) Perkin condensation: Condensation of an aromatic aldehyde with a carboxylic acid anhydride.

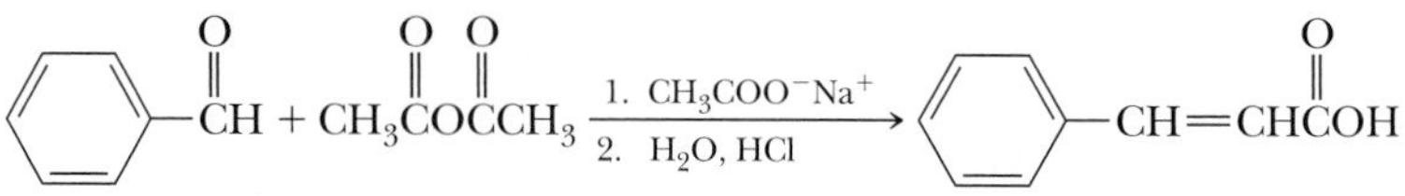

Cinnamic acid

(b) Darzens condensation: Condensation of an α-haloester with a ketone or an aromatic aldehyde.

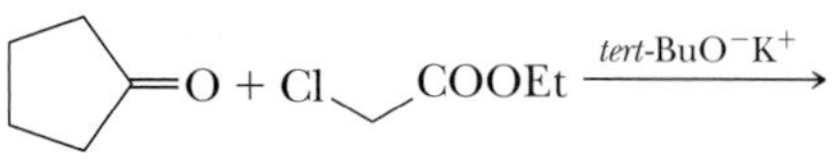

Enamines

19.41 When 2-methylcyclohexanone is treated with pyrrolidine, two isomeric enamines are formed. Why is enamine A with the less substituted double bond the thermodynamically favored product? You will find it helpful to examine models of these enamines on the CD.

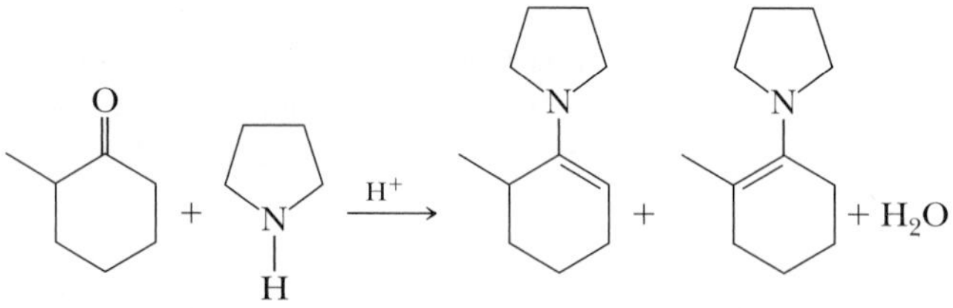

A (85%)   B (15%)

19.42 Enamines normally react with methyl iodide to give two products: one arising from alkylation at nitrogen and the second arising from alkylation at carbon. For example,

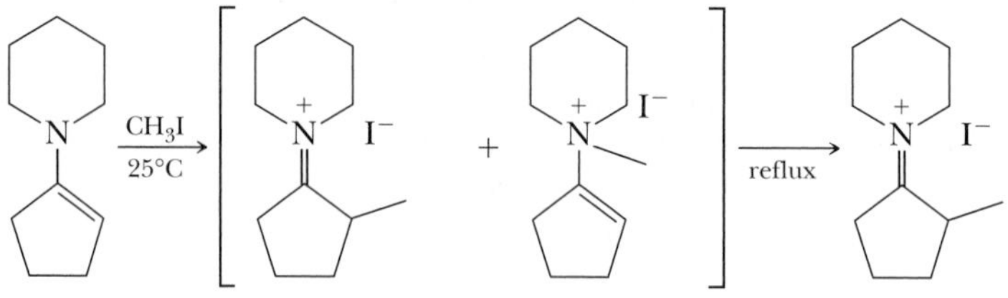

Product of   Product of
C-alkylation   N-alkylation

Heating the mixture of C-alkylation and N-alkylation products gives only the product from C-alkylation. Propose a mechanism for this isomerization.

19.43 Propose a mechanism for the following conversion.

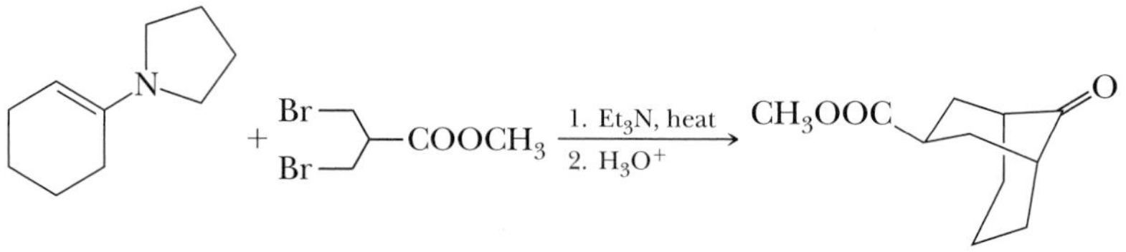

19.44 The following intermediate was needed for the synthesis of tamoxifen, a widely used antiestrogen drug. For the structural formula and synthesis of tamoxifen, see MedChem Problem MC.26. Propose a synthesis for this intermediate from compound A.

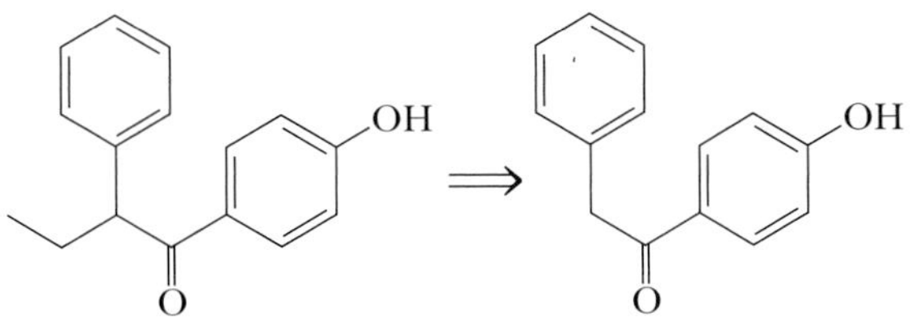

Needed for the
synthesis of tamoxifen

A

19.45 Propose a mechanism for the following reaction.

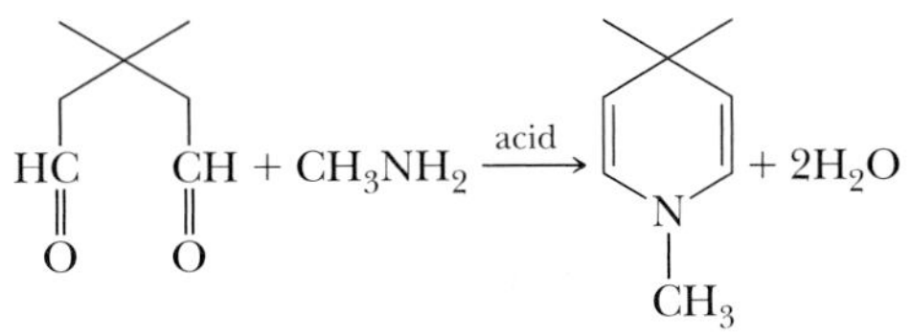

Acetoacetic Ester and Malonic Ester Syntheses

19.46 Propose syntheses of the following derivatives of diethyl malonate, each being a starting material for synthesis of a barbiturate currently available in the United States.

(a)

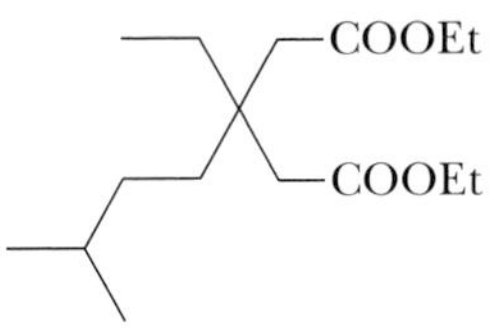

(b)

Needed for the
synthesis of amobarbital

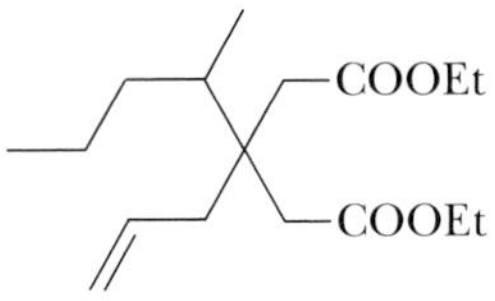

Needed for the
synthesis of secobarbital

19.47 2-Propylpentanoic acid (valproic acid; see *The Merck Index*, 12th ed., #10049) is an effective drug for treatment of several types of epilepsy, particularly absence seizures, which are generalized epileptic seizures characterized by brief and abrupt loss of consciousness. Propose a synthesis of valproic acid starting with diethyl malonate.

19.48 Show how to synthesize the following compounds using either the malonic ester synthesis or the acetoacetic ester synthesis.

(a) 4-Phenyl-2-butanone
(b) 2-Methylhexanoic acid
(c) 3-Ethyl-2-pentanone
(d) 2-Propyl-1,3-propanediol
(e) 4-Oxopentanoic acid
(f) 3-Benzyl-5-hexene-2-one
(g) Cyclopropanecarboxylic acid
(h) Cyclobutyl methyl ketone

19.49 Propose a mechanism for formation of 2-carbethoxy-4-butanolactone and then 4-butanolactone (γ-butyrolactone) in the following sequence of reactions.

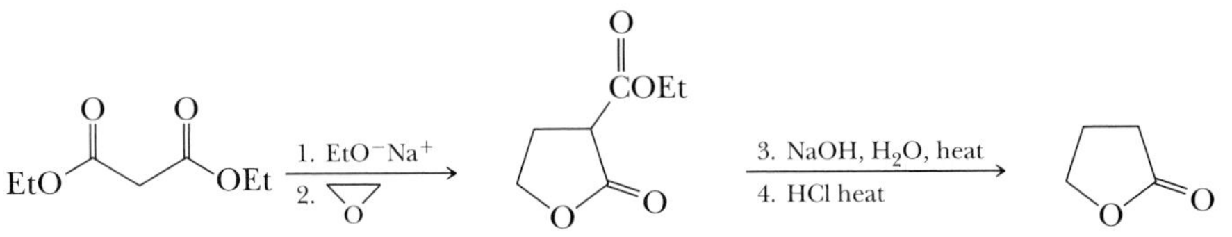

2-Carbethoxy-
4-butanolactone

4-Butanolactone
(γ-Butyrolactone)

19.57 The following β-diketone can be synthesized from cyclopentanone and an acid chloride using an enamine reaction.

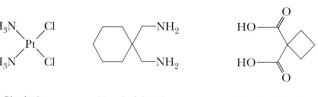

(a) Propose a synthesis of the starting acid chloride from cyclopentene.

(b) Show the steps in the synthesis of the β-diketone using a morpholine enamine.

19.58 Cisplatin (see *The Merck Index*, 12th ed., #2378) was first prepared in 1844, but it was not until 1964 that its value as an anticancer drug was realized. In that year, Barnett Rosenberg and coworkers at Michigan State University observed that when platinum electrodes are inserted into a growing bacterial culture and an electric current passed through the culture, all cell division ceased within 1 to 2 hours. The result was surprising. Equally surprising was their finding that cell division was inhibited by *cis*-diamminedichloroplatinum(II), more commonly named cisplatin, a platinum complex formed in the presence of ammonia and chloride ion. Cisplatin has a broad spectrum of anticancer activity and is particularly useful for treatment of epithelial malignancies. Evidence suggests that platinum(II) in the complex bonds to DNA and forms intrachain and interchain cross linkages. More than 1000 platinum complexes have since been prepared and tested in attempts to discover even more active cytotoxic drugs. In spiroplatin, the two NH_3 groups are replaced by primary amino groups. This drug showed excellent antileukemic activity in animal models but was disappointing in human trials. In carboplatin, the two chloride ions are replaced by carboxylate groups (see *The Merck Index*, 12 ed., #1870). In 1989, carboplatin was approved by the FDA for treatment of ovarian cancers.

Cisplatin	Needed for the synthesis of spiroplatin	Needed for the synthesis of carboplatin

(a) Devise a synthesis for the diamine needed in the synthesis of spiroplatin starting with diethyl malonate and 1,5-dibromopentane as the sources of carbon atoms.

(b) Devise a synthesis for the dicarboxylic acid needed in the synthesis of carboplatin starting with diethyl malonate and 1,3-dibromopropane as sources of carbon atoms.

19.59 Oxanamide (see *The Merck Index*, 12th ed., #7053) is a mild sedative belonging to a class of molecules called oxanamides; it contains an oxirane (epoxide) group and an amide group. As seen in this retrosynthetic scheme, the source of carbon atoms for the synthesis of oxanamide is butanal.

(a) Show reagents and experimental conditions by which oxanamide can be synthesized from butanal.

(b) How many stereocenters are there in oxanamide? How many stereoisomers are possible for this compound?

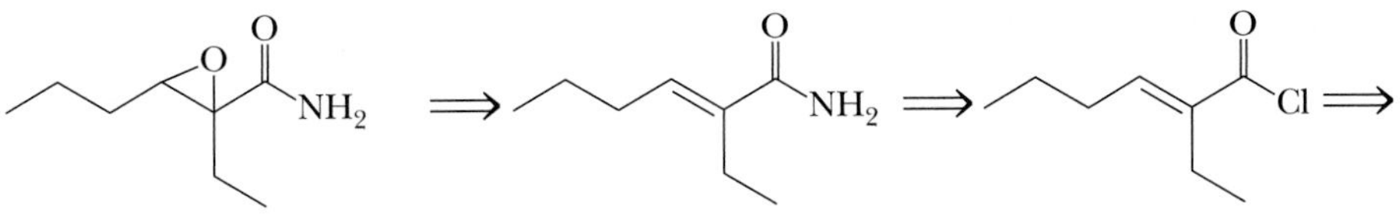

Oxanamide

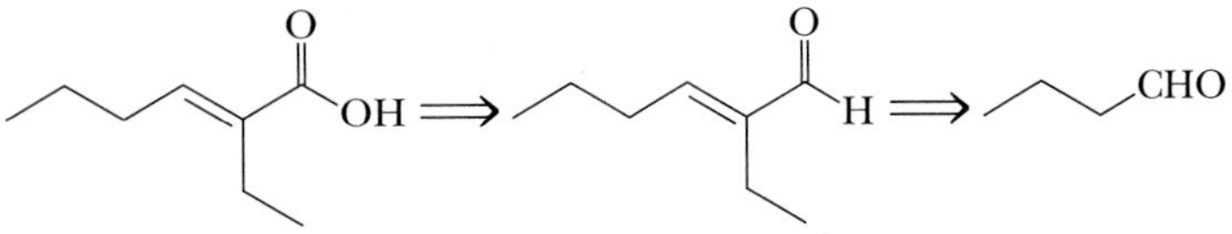

Butanal

19.60 The widely used anticoagulant warfarin (see the Chemistry in Action box "From Moldy Clover to a Blood Thinner" in Chapter 18) is synthesized from 4-hydroxycoumarin, benzaldehyde, and acetone as shown in this retrosynthesis. Show how warfarin is synthesized from these reagents.

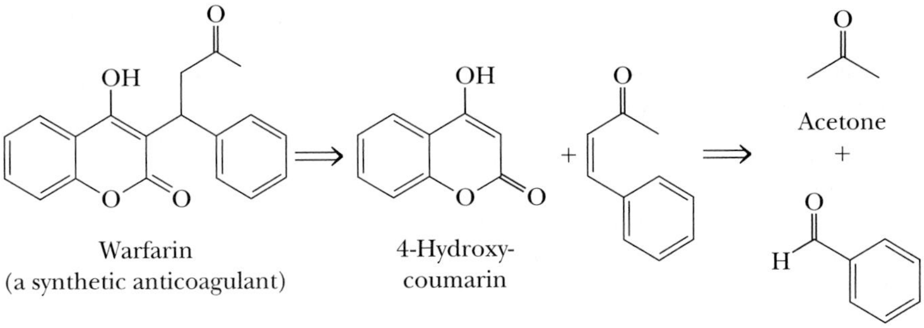

Warfarin
(a synthetic anticoagulant)

4-Hydroxy-
coumarin

Acetone
+

Benzaldehyde

19.61 Following is a retrosynthetic analysis for an intermediate in the industrial synthesis of vitamin A.

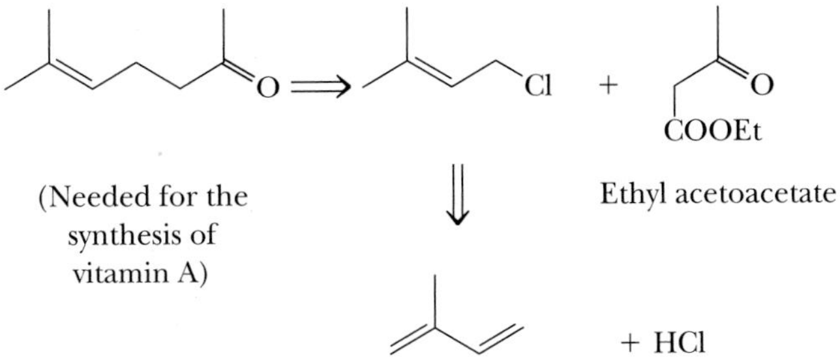

(Needed for the
synthesis of
vitamin A)

Ethyl acetoacetate

2-Methyl-1,3-butadiene
(Isoprene)

+ HCl

(a) Addition of one mole of HCl to isoprene gives 4-chloro-2-methyl-2-butene as the major product. Propose a mechanism for this addition and account for its regioselectivity.

(b) Propose a synthesis of the vitamin A precursor from this allylic chloride and ethyl acetoacetate.

19.62 Following are the steps in one of the several published syntheses of frontalin, a pheromone of the western pine beetle. See K. Mori, S. Kobayashi, and M. Matsui, *Agric. Biol. Chem.* **39:** 1889 (1975).

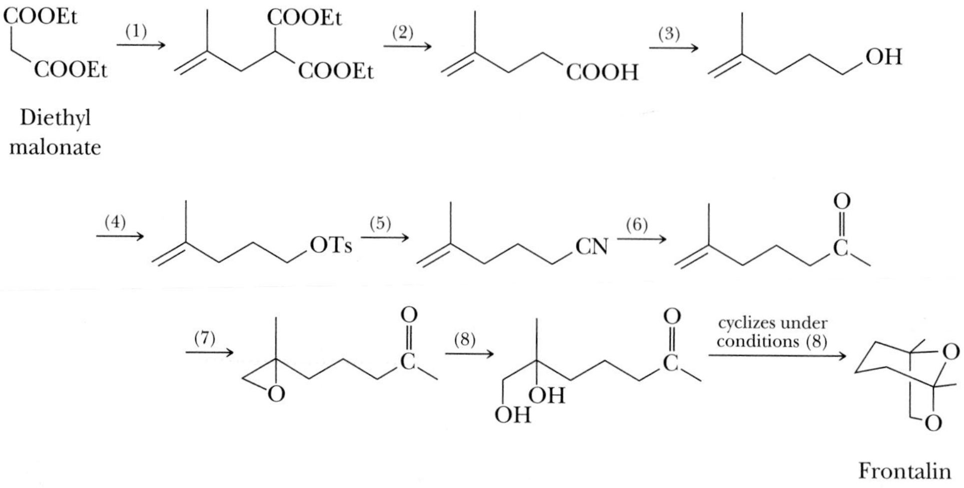

Diethyl
malonate

Frontalin

(a) Propose reagents for Steps 1–8.

(b) Propose a mechanism for the cyclization of the ketodiol from Step 8 to frontalin.

19.63 2-Ethyl-1-hexanol was needed for the synthesis of the sunscreen octyl *p*-methylcinnamate (See the Chemistry in Action box "Sunscreens and Sunblocks" in Chapter 23). Show how this alcohol could be synthesized by the following.

(a) An aldol condensation of butanal

(b) A malonic ester synthesis starting with diethyl malonate.

AROMATICS I: BENZENE AND ITS DERIVATIVES

Benzene is a colorless compound with a melting point of 6°C and a boiling point of 80°C. It was first isolated by Michael Faraday in 1825 from the oily residue that collected in the illuminating gas lines of London. Benzene's molecular formula, C_6H_6, suggests a high degree of unsaturation. Compared with an alkane of molecular formula C_6H_{14}, its index of hydrogen deficiency is four, which can be met by an appropriate combination of rings, double bonds, and triple bonds. For

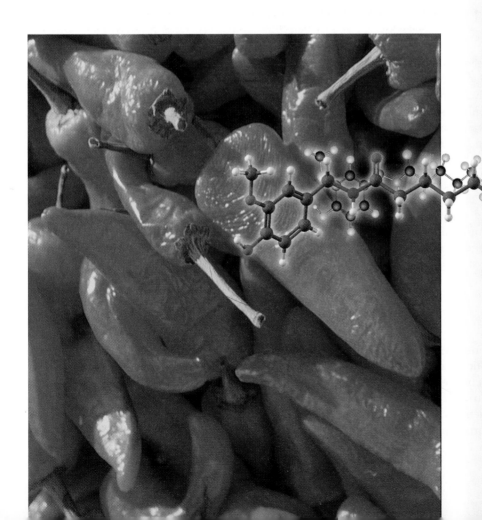

■ Peppers of the *Capsicum* family. *(Douglas Brown)* Inset: A model of capsaicin.

example, a compound of molecular formula C_6H_6 might have four double bonds, or three double bonds and one ring, or two double bonds and two rings, or one triple bond and two rings, and so on. Considering benzene's high degree of unsaturation, it might be expected to show many of the reactions characteristic of alkenes and alkynes. Yet, benzene is remarkably unreactive! It does not undergo addition, oxidation, and reduction reactions characteristic of alkenes and alkynes. For example, benzene does not react with bromine, hydrogen chloride, or other reagents that usually add to carbon-carbon double and triple bonds. It is not oxidized by potassium permanganate or chromic acid under conditions that readily oxidize alkenes and alkynes. When benzene reacts, it does so by substitution in which a hydrogen atom is replaced by another atom or group of atoms.

As noted in Chapter 5, the term "aromatic" was originally used to classify benzene and its derivatives because many of them have distinctive odors. The term "aromatic" as it is now used, refers instead to the fact that these compounds are highly unsaturated and unexpectedly stable toward reagents that attack alkenes and alkynes. The term "arene" is used to describe aromatic hydrocarbons, by analogy with alkane, alkene, and alkyne. Benzene is the parent arene. Just as a group derived by removal of an H from an alkane is called an alkyl group and given the symbol R—, a group derived by removal of an H from an arene is called an aryl group and given the symbol Ar—.

20.1   The Structure of Benzene

Let us put ourselves in the mid-19th century and examine the evidence on which chemists attempted to build a model for the structure of benzene. First, because the molecular formula of benzene is C_6H_6, it seemed clear that the molecule must be highly unsaturated. Yet, benzene does not show the chemical properties of alkenes, the only unsaturated hydrocarbons known at that time. Benzene does undergo chemical reactions, but its characteristic reaction is substitution rather than addition. When benzene is treated with bromine in the presence of ferric chloride, for example, only one compound of molecular formula C_6H_5Br is formed.

$$C_6H_6 + Br_2 \xrightarrow{\text{FeCl}_3} C_6H_5Br + HBr$$

Benzene   Bromobenzene

Chemists concluded, therefore, that all six hydrogens of benzene must be equivalent. When bromobenzene is treated with bromine in the presence of ferric chloride as a catalyst, three isomeric dibromobenzenes are formed.

$$C_6H_5Br + Br_2 \xrightarrow{\text{FeCl}_3} C_6H_4Br_2 + HBr$$

Bromobenzene   Three isomeric
dibromobenzenes

For chemists in the mid-19th century, the problem was to incorporate these observations, along with the accepted tetravalence of carbon, into a structural formula for benzene. Before we examine these proposals, we should note that the problem of the structure of benzene and other aromatic hydrocarbons has occupied the efforts of chemists for over a century. Only since the 1930s has a general understanding of this problem been realized.

A. Kekulé's Model of Benzene

The first structure for benzene was proposed by August Kekulé in 1865 and consisted of a six-membered ring with one hydrogen attached to each carbon. Although Kekulé's original structural formula provided for the equivalency of the C—H and C—C bonds, it was inadequate because all the carbon atoms were trivalent. To maintain the tetravalence of carbon, Kekulé proposed in 1872 that the ring contains three double bonds that shift back and forth so rapidly that the two forms cannot be separated. Each structure became known as a **Kekulé structure.**

Kekulé structures for benzene

Now, more than 125 years after the time of Kekulé, we are apt to misunderstand what scientists in his time knew and did not know. For example, it is a given to us that covalent bonds consist of one or more pairs of shared electrons. We must remember, however, that it was not until 1897 that J. J. Thomson, professor of physics at the Cavendish Laboratory of Cambridge University, discovered the electron. Thomson was awarded the 1906 Nobel Prize for physics. That the electron played any role in chemical bonding did not become clear for another 30 years. Thus, at the time Kekulé made his proposal for the structure of benzene, the existence of electrons and their role in chemical bonding was completely unknown.

Kekulé's proposal accounted nicely for the fact that bromination of benzene gives only one bromobenzene, and that bromination of bromobenzene gives three (and only three) isomeric dibromobenzenes.

Bromobenzene   Three isomeric dibromobenzenes

Although his proposal was consistent with many experimental observations, it did not totally solve the problem and was contested for years. The major objection was that it did not account for the unusual chemical behavior of benzene. If benzene contains three double bonds, Kekulé's critics asked, why does it not show reactions typical of alkenes? Why, for example, does benzene not add three moles of bromine to form 1,2,3,4,5,6-hexabromocyclohexane? We now understand the surprising unreactivity of benzene on the basis of two complementary descriptions, the molecular orbital model and the resonance model.

B. The Molecular Orbital Model of Benzene

The concepts of hybridization of atomic orbitals and the theory of resonance, developed by Linus Pauling in the 1930s, provided the first adequate description of the structure of benzene. The carbon skeleton of benzene forms a regular hexagon with

Figure 20.1

The molecular orbital representation of the pi bonding in benzene.

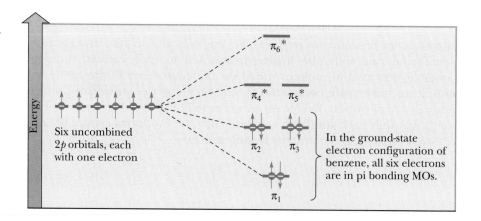

C—C—C and H—C—C bond angles of 120°. For this type of bonding, carbon uses
sp^2 hybrid orbitals. Each carbon forms sigma bonds to two adjacent carbons by overlap of sp^2-sp^2 hybrid orbitals, and one sigma bond to hydrogen by overlap of sp^2-1s orbitals. As determined experimentally, all carbon-carbon bonds are 139 pm in length,
a value almost midway between the length of a single bond between sp^3-hybridized
carbons (154 pm) and a double bond between sp^2-hybridized carbons (133 pm).

Each carbon also has a single unhybridized $2p$ orbital that is perpendicular to the
plane of the ring and contains one electron. Molecular orbital theory, which is beyond the scope of this text, shows that combination of these six parallel $2p$ atomic orbitals gives a set of six pi MOs, three pi-bonding MOs and three pi-antibonding MOs.
These six molecular orbitals and their relative energies are shown in Figure 20.1.
Note that π_2 and π_3 MOs are degenerate (they have the same energy). Similarly, π_4^*
and π_5^* are a degenerate pair of pi-antibonding MOs.

In the ground-state electron configuration of benzene, the six electrons of the pi
system occupy the three bonding MOs (Figure 20.1). The great stability of benzene
results from the fact that these three bonding MOs are of much lower energy compared with the six uncombined $2p$ atomic orbitals.

It is common to represent the pi system of benzene as one torus (a donut-shaped
region) above the plane of the ring and a second torus below the plane of the ring,
as shown in Figure 20.2. Although this picture is useful in thinking about the electron density of the pi system, you must use it with caution because it represents only
the lowest lying pi-bonding molecular orbital. Actually, π_2 and π_3 also add to a torus.

C. The Resonance Model of Benzene

One of the postulates of resonance theory is that, when a molecule or ion can be represented by two or more contributing structures, it is not adequately represented by

(a)

(b)

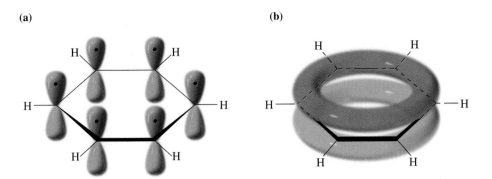

Figure 20.2
The pi system of benzene. (a)
The carbon-hydrogen framework. The six $2p$ orbitals, each
with one electron, are shown
uncombined. (b) Overlap of
parallel $2p$ orbitals forms a continuous pi cloud, shown by one
torus above the plane of the
ring and a second torus below
the plane of the ring.

any single contributing structure. We represent benzene as a hybrid of two equivalent
contributing structures, often referred to as Kekulé structures. Each Kekulé structure
makes an equal contribution to the hybrid, and thus the C—C bonds are neither single nor double bonds but something intermediate.

Benzene as a hybrid of two equivalent
contributing structures

We recognize that neither of these contributing structures exists (they are merely alternative ways to pair $2p$ orbitals with no reason to prefer one or the other) and that
the actual structure is a superposition of both. Nevertheless, chemists continue to use
a single contributing structure to represent this molecule because it is as close as we
can come to an accurate structure within the limitations of classical valence bond
structures and the tetravalence of carbon.

One way to estimate the **resonance energy** of benzene is to compare the heats of
hydrogenation of cyclohexene and benzene. Cyclohexene is readily reduced to cyclohexane by hydrogen in the presence of a transition metal catalyst (Section 6.6A).

Resonance energy The difference in energy between a resonance hybrid and the most stable
of its hypothetical contributing
structures in which electrons are
localized on particular atoms and
in particular bonds.

$$+ \ H_2 \ \xrightarrow[\text{1--2 atm}]{\text{Ni}} \qquad \Delta H^0 = -120 \text{ kJ } (-28.6 \text{ kcal})/\text{mol}$$

Cyclohexene   Cyclohexane

Benzene is reduced very slowly under these conditions to cyclohexane. It is reduced
more rapidly when heated and under a pressure of several hundred atmospheres of
hydrogen.

$$+ \ 3H_2 \ \xrightarrow[\text{200--300 atm}]{\text{Ni}} \qquad \Delta H^0 = -208 \text{ kJ } (-49.8 \text{ kcal})/\text{mol}$$

Benzene   Cyclohexane

Catalytic hydrogenation of an alkene is an exothermic reaction (Section 6.6B).
The heat of hydrogenation per double bond varies somewhat with the degree of

Figure 20.3

The resonance energy of benzene as determined by comparison of the heats of hydrogenation of cyclohexene, benzene, and the hypothetical compound 1,3,5-cyclohexatriene.

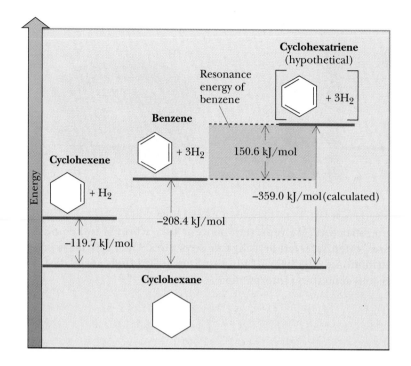

substitution of the particular alkene; for cyclohexene, $\Delta H^0 = -120$ kJ $(-28.6$ kcal$)$/mol. If we consider benzene to be 1,3,5-cyclohexatriene, a hypothetical unsaturated compound with alternating single and double bonds, we calculate that $\Delta H^0 = 3(-120$ kJ/mol$) = -359$ kJ $(-85.8$ kcal$)$/mol. The ΔH^0 for reduction of benzene to cyclohexane is only -208 kJ $(-49.8$ kcal$)$/mol, considerably less than that calculated for 1,3,5-cyclohexatriene. The difference between these values, 151 kJ (36.0 kcal)/mol, is the **resonance energy of benzene.** These experimental results are shown graphically in Figure 20.3.

There have been several other experimental determinations of the resonance energy of benzene using different model compounds, and, although these determinations differ somewhat in their results, they all agree that the resonance stabilization of benzene is large. Following are resonance energies for several other aromatic hydrocarbons.

Resonance energy [kJ (kcal)/mol]	Benzene 151 (36)	Naphthalene 255 (61)	Anthracene 347 (83)	Phenanthrene 381 (91)

20.2 The Concept of Aromaticity

Molecular orbital and resonance theories are powerful tools with which chemists can understand the unusual stability of benzene and its derivatives. According to resonance theory, benzene is best represented as a hybrid of two equivalent contributing structures. By analogy, cyclobutadiene and cyclooctatetraene can also be drawn as hybrids of two equivalent contributing structures. Is either of these compounds aromatic?

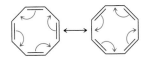

Cyclobutadiene as a
hybrid of two equivalent
contributing structures

Cyclooctatetraene as a hybrid of
two equivalent contributing structures

The answer for both compounds is no. Repeated attempts to synthesize cyclobutadiene failed. It was not synthesized until 1965, and even then it could only be detected if trapped at 4 K (− 269°C). Cyclobutadiene is a highly unstable compound and does not show any of the chemical and physical properties we associate with aromatic compounds. Cyclooctatetraene has chemical properties typical of alkenes. It reacts readily with halogens and halogen acids, as well as with mild oxidizing and reducing agents.

We are then faced with the broad question: "What are the fundamental principles underlying aromatic character?" What are the structural characteristics of unsaturated compounds that have a large resonance energy and do not undergo reactions typical of alkenes but rather undergo substitution reactions?

A.   The Hückel Criteria for Aromaticity

The underlying criteria for aromaticity were recognized in the early 1930s by Erich Hückel, a German chemical physicist. He carried out MO energy calculations for monocyclic, planar molecules in which each atom of the ring has one $2p$ orbital available for forming sets of molecular orbitals. His calculations demonstrated that monocyclic, planar molecules with a closed loop of 2, 6, 10, 14, 18, . . . pi electrons in a fully conjugated system should be aromatic. These numbers are generalized in the **($4n + 2$) pi electron rule,** where n is a positive integer (0, 1, 2, 3, 4, . . .). Conversely, monocyclic, planar molecules with $4n$ pi electrons (4, 8, 12, 16, 20, . . .) are especially unstable and are said to be antiaromatic. We will have more to say about antiaromaticity shortly. **Hückel's criteria for aromaticity** are summarized as follows. To be aromatic, a compound must:

1. Be cyclic.
2. Have one p orbital on each atom of the ring.
3. Be planar or nearly planar so that there is continuous or nearly continuous overlap of all p orbitals of the ring.
4. Have a closed loop of ($4n + 2$) pi electrons in the cyclic arrangement of p orbitals.

To appreciate the reasons for aromaticity and antiaromaticity, we must examine MO energy diagrams for the molecules and ions we will consider in this and the following section. The relative energies of the pi MOs for planar, monocyclic, fully conjugated systems can be constructed quite easily using the **Frost circle, or inscribed polygon method.** To construct such a diagram, draw a circle and then inscribe in it a polygon of the same number of sides as the ring in question. Inscribe the polygon in such a way that one of its vertices is at the bottom of the circle. The relative energies

Hückel criteria for aromaticity
To be aromatic, a monocyclic compound must have one p orbital on each atom of the ring, be planar or nearly so, and have ($4n + 2$) pi electrons in the cyclic arrangement of p orbitals.

Frost circle   A graphic method for determining the relative energies of pi MOs for planar, fully conjugated, monocyclic compounds.

Figure 20.4

Frost circles showing the number and relative energies of the pi MOs for planar, fully conjugated four-, five-, and six-membered rings.

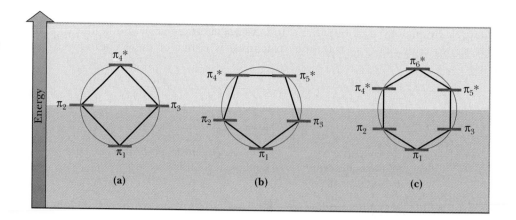

of the MOs in the ring are then given by the points where the vertices touch the circle. Those MOs below the horizontal line through the center of the circle are bonding MOs. Those on the horizontal line are nonbonding MOs, and those above the line are antibonding MOs. Shown in Figure 20.4 are Frost circles describing the MOs of monocyclic, planar, and fully conjugated four-, five-, and six-membered rings. This apparently magical method works because it reproduces geometrically the mathematical solutions to the wave equation.

Example 20.1

Construct a Frost circle for a planar seven-membered ring with one $2p$ orbital on each atom of the ring, and show the relative energies of its seven pi molecular orbitals. Which are bonding, which are antibonding, and which are nonbonding?

Solution

Of the seven pi molecular orbitals, three are bonding, and four are antibonding.

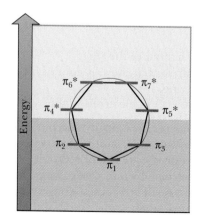

Problem 20.1

Construct a Frost circle for a planar eight-membered ring with one $2p$ orbital on each atom of the ring and show the relative energies of its eight pi molecular orbitals. Which are bonding, which are antibonding, and which are nonbonding?

B. Aromatic Hydrocarbons

Cyclobutadiene, benzene, and cyclooctatetraene are the first members of a family of molecules called annulenes. An **annulene** is a cyclic hydrocarbon with a continuous alternation of single and double bonds. The name of an annulene is derived by showing the number of atoms in the ring in brackets followed by the word "annulene." Named as annulenes, cyclobutadiene, benzene, and cyclooctatetraene are [4]annulene, [6]annulene, and [8]annulene, respectively. These compounds, however, are rarely named as annulenes.

Annulene A cyclic hydrocarbon with a continuous alternation of single and double bonds.

Beginning in the 1960s, Franz Sondheimer and his colleagues, first at the Weizmann Institute in Israel and later at the University of London, synthesized a number of larger annulenes, primarily to test the validity of Hückel's criteria for aromaticity. They found, for example, that both [14]annulene and [18]annulene are aromatic, as predicted by Hückel. [18]Annulene has a resonance energy of approximately 418 kJ (100 kcal)/mol. Notice that, for these annulenes to achieve planarity, several of the carbon-carbon double bonds in each must have the trans configuration.

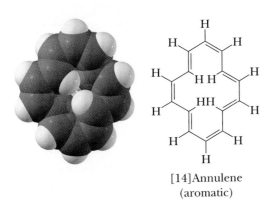

[14]Annulene
(aromatic)

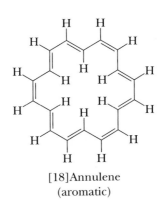

[18]Annulene
(aromatic)

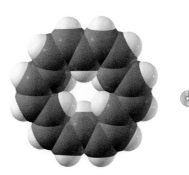

In these larger annulenes, there are two sets of equivalent hydrogens: those that point outward from the ring and those that point inward to the center of the ring. As we shall see in Section 20.4B, these two sets of equivalent hydrogens have quite different ^{1}H-NMR chemical shifts.

According to Hückel's criteria, [10]annulene should be aromatic; it is cyclic, has one $2p$ orbital on each carbon of the ring, and has $4(2) + 2 = 10$ electrons in its pi system. It has been found, however, that this molecule is not aromatic. The reason lies in the fact that the ring is too small to accommodate the two hydrogens that point inward toward the center of the ring. Nonbonded interaction between these two hydrogens forces the ring into a nonplanar conformation in which the overlap of all ten $2p$ orbitals is no longer continuous. Therefore, because [10]annulene is not planar, it is not aromatic. It shows reactions typical of alkenes and is classified as nonaromatic.

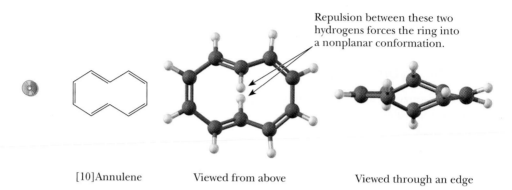

Repulsion between these two hydrogens forces the ring into a nonplanar conformation.

[10]Annulene   Viewed from above   Viewed through an edge

What is remarkable is that if the two hydrogen atoms facing inward toward the center of the ring are replaced by a —CH_2— group, the ring is now able to assume a conformation close enough to planar that it becomes aromatic.

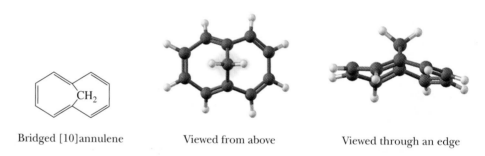

Bridged [10]annulene   Viewed from above   Viewed through an edge

C. Antiaromatic Hydrocarbons

Antiaromatic compound A monocyclic compound that is planar or nearly so, has one p orbital on each atom of the ring, and has $4n$ pi electrons in the cyclic arrangement of overlapping p orbitals, where n is an integer. Antiaromatic compounds are especially unstable.

According to the Hückel criteria, monocyclic, planar molecules with $4n$ pi electrons (4, 8, 12, 16, 20, . . .) are especially unstable and are said to be antiaromatic. By these criteria, cyclobutadiene with 4 pi electrons is **antiaromatic.** Using the Frost circle energy diagram from Figure 20.4, we can construct a molecular orbital energy diagram for cyclobutadiene (Figure 20.5). In the ground-state electron configuration of cyclobutadiene, two pi electrons fill the π_1-bonding MO. The third and fourth pi electrons are unpaired and lie in the π_2- and π_3-nonbonding MOs. The existence of these two unpaired electrons in planar cyclobutadiene makes this molecule highly unstable and reactive compared to butadiene, a noncyclic molecule containing two conjugated double bonds. It has been found that cyclobutadiene is not planar, but slightly puckered with two shorter bonds and two longer bonds, which makes the two degenerate orbitals no longer equivalent; nevertheless, it retains some apparent diradical character.

Cyclooctatetraene with eight pi electrons is not aromatic for the following reasons. X-ray studies show clearly that the most stable conformation of the molecule is a

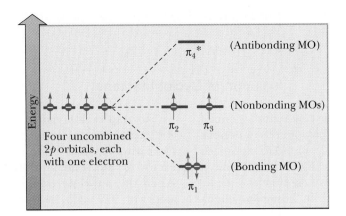

Figure 20.5
Molecular orbital energy diagram for cyclobutadiene. In the ground state, two electrons are in the low lying π_1-bonding MO. The remaining two electrons are unpaired and occupy the degenerate π_2- and π_3-nonbonding MOs.

nonplanar "tub" conformation with two distinct types of carbon-carbon bonds: four longer carbon-carbon single bonds and four shorter carbon-carbon double bonds. The four single bonds are equal in length to the single bonds between sp^2-hybridized carbons (approximately 146 pm), and the four double bonds are equal in length to double bonds in alkenes (approximately 133 pm). In the tub conformation, the overlap of $2p$ orbitals on carbons forming double bonds is excellent, but almost no overlap occurs between $2p$ orbitals at the ends of carbon-carbon single bonds because these $2p$ orbitals are not parallel. Cyclooctatetraene shows reactions typical of alkenes and is classified as nonaromatic.

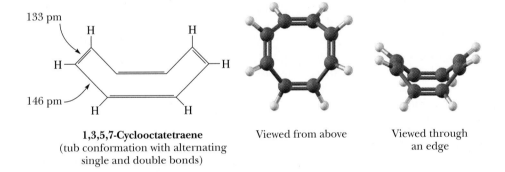

1,3,5,7-Cyclooctatetraene
(tub conformation with alternating single and double bonds)

Viewed from above

Viewed through an edge

To appreciate why planar cyclooctatetraene would be classified as antiaromatic, we need to examine the MO energy diagram for an eight-membered ring containing eight pi electrons in a cyclic, fully conjugated ring. You constructed a Frost circle for this ring in answer to Problem 20.1. Note that the most stable conformation of cyclooctatetraene is not planar, but, if it were planar, the Frost circle you constructed would be its MO energy diagram. The molecular orbital energy diagram for planar cyclooctatetraene is shown in Figure 20.6. In the ground state of planar cyclooctatetraene, six pi electrons fill the three low lying π_1-, π_2-, and π_3-bonding MOs. The remaining two pi electrons are unpaired and lie in the degenerate π_4- and π_5-nonbonding MOs. It is because of these two unpaired electrons that planar cyclooctatetraene, if it existed, would be classified as antiaromatic. Cyclooctatetraene, however, is large enough to pucker into a nonplanar conformation and become nonaromatic.

C H E M I S T R Y   I N   A C T I O N

Isolation of Cyclobutadiene

Cyclobutadiene, an antiaromatic compound, is one of the most reactive and unstable compounds in organic chemistry. By far the most unusual approach to stabilizing cyclobutadiene came from Donald J. Cram (University of California, Los Angeles), a pioneer in "host-guest chemistry." In host-guest chemistry, the host is a relatively large molecule with an internal cavity that can be occupied by a smaller, guest molecule. The tube-like molecule on the next page is an example of a host molecule called a hemicarcerand, because it is capable of incarcerating guest molecules and releasing them at higher temperatures. (In contrast, a carcerand will not release them at all.)

Notice that in this hemicarcerand there are two bands of four benzene rings, with three of the left benzene rings each linked to a different benzene ring along the right band. This molecule behaves like a clam shell. The three pairs of linked benzene rings act as hinges and the fourth benzene rings act as a mouth. At low temperatures, the mouth is closed, but as the temperature is raised to 100°C the molecule undergoes a conformational change, and the mouth opens. Molecules of the right size can enter into the cavity of the hemicarcerand, but, as the temperature is reduced to room temperature, the mouth closes, and the guest is trapped inside.

To trap cyclobutadiene inside this hemicarcerand, a precursor molecule had to be introduced into the cavity. This was done by heating it with a large excess of α-pyrone and then cooling the system. When the α-pyrone trapped in the cavity is irradiated with UV light, it loses CO_2 and forms cyclobutadiene.

α-Pyrone $\xrightarrow[\text{25°C for 30 min}]{\text{UV (200–250 nm)}}$ Cyclobutadiene $+$ CO_2

The formation of the host-guest complex was established by ¹H-NMR spectroscopy and elemental analysis. When it was irradiated with UV light at 25°C, the ¹H-NMR signals of the guest α-pyrone disappeared, and a signal for cyclobutadiene appeared at δ 2.35. Further evidence that cyclobutadiene was trapped inside the molecular cavity came when the product of irradiation was heated at 220°C. At this temperature, the mouth of the cavity opened, allowing the guest molecule to escape. Heating under these conditions gave cyclooctatetraene. Cyclooctatetraene is a major dimerization product of cyclobutadiene.

Figure 20.6

Molecular orbital energy diagram for a planar conformation of cyclooctatetraene. Three pairs of electrons fill the three low lying pi-bonding molecular orbitals. Two electrons are unpaired in degenerate pi-nonbonding molecular orbitals.

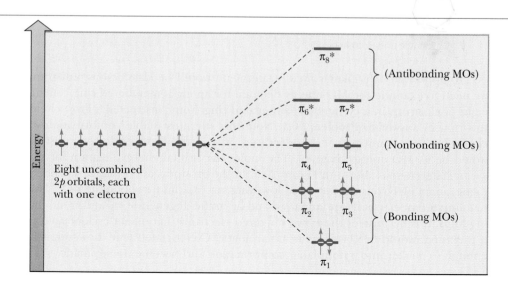

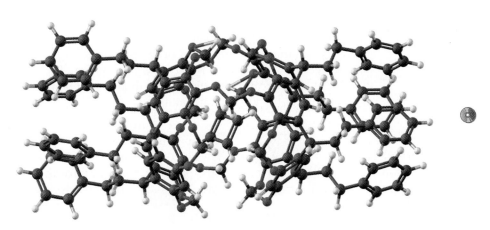

Additional spectroscopic and chemical evidence clearly established that the highly reactive cyclobutadiene had been trapped inside the cavity and was stable at room temperature in this environment (a "molecular bottle"). Because unreactive C—C and C—O bonds lined its molecular cage, cyclobutadiene could undergo no chemical reactions with its surroundings. Like some vicious beast, it could now be studied without being a danger to itself or others.

Cyclobutadiene

Cyclooctatetraene

A hemicarcerand with a molecule of cyclobutadiene trapped inside it

If [16]annulene were planar, it would be antiaromatic. The size of the ring, however, is large enough that it can pucker into a nonplanar conformation in which the double bonds are no longer fully conjugated. [16]Annulene, therefore, is nonaromatic.

[16]Annulene
(nonaromatic)

D. Heterocyclic Aromatic Compounds

Heterocyclic compound An organic compound that contains one or more atoms other than carbon in its ring.

Aromatic character is not limited to hydrocarbons; it is found in **heterocyclic compounds** as well. Pyridine and pyrimidine are heterocyclic analogs of benzene. In pyridine, one CH group of benzene is replaced by nitrogen, and, in pyrimidine, two CH groups are replaced by nitrogens.

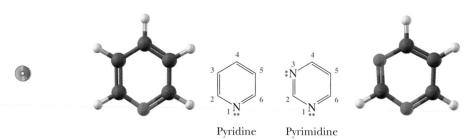

Pyridine   Pyrimidine

Each molecule meets the Hückel criteria for aromaticity. Each is monocyclic and planar, has one $2p$ orbital on each atom of the ring, and has six electrons in the pi system. In pyridine, nitrogen is sp^2 hybridized, and its unshared pair of electrons occupies an sp^2 orbital in the plane of the ring and perpendicular to the $2p$ orbitals of the pi system; thus, it is not a part of the pi system. In pyrimidine, neither unshared pair of electrons of nitrogen is part of the pi system. The resonance energy of pyridine is 134 kJ (32 kcal)/mol, slightly less than that of benzene. The resonance energy of pyrimidine is 108 kJ (26 kcal)/mol.

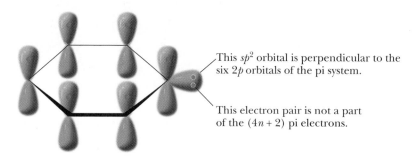

This sp^2 orbital is perpendicular to the six $2p$ orbitals of the pi system.

This electron pair is not a part of the $(4n + 2)$ pi electrons.

The five-membered ring heterocyclic compounds furan, thiophene, pyrrole, and imidazole are also aromatic.

Furan   Thiophene   Pyrrole   Imidazole

In these planar compounds, each heteroatom is sp^2 hybridized, and its unhybridized $2p$ orbital is part of a continuous cycle of five p orbitals. In furan and thiophene, one unshared pair of electrons of the heteroatom lies in the unhybridized $2p$ orbital and is a part of the pi system (Figure 20.7). The other unshared pair of electrons lies in an sp^2 hybrid orbital perpendicular to the $2p$ orbitals and is not a part of it. In pyr-

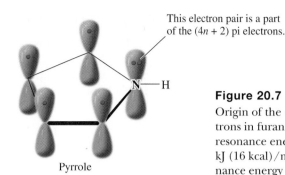

This electron pair is a part of the (4n + 2) pi electrons.

This electron pair is not a part of the (4n + 2) pi electrons.

Furan

This electron pair is a part of the (4n + 2) pi electrons.

Pyrrole

Figure 20.7
Origin of the (4n + 2) pi electrons in furan and pyrrole. The resonance energy of furan is 67 kJ (16 kcal)/mol; the resonance energy of pyrrole is 88 kJ (21 kcal)/mol.

role, the unshared pair of electrons on nitrogen is part of the aromatic sextet. In imidazole, the unshared pair on one nitrogen is part of the aromatic sextet, and the unshared pair on the other nitrogen is not.

Nature abounds with compounds having a heterocyclic ring fused to one or more other rings. Two such compounds especially important in the biological world are indole and purine.

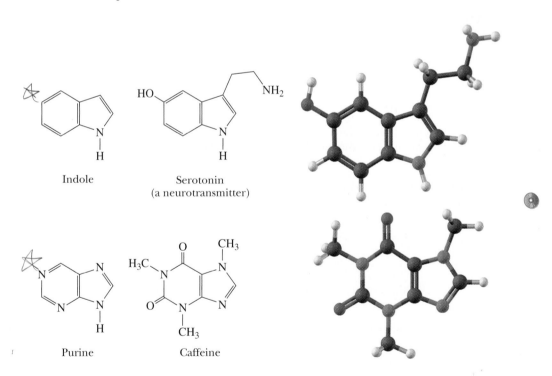

Indole

Serotonin
(a neurotransmitter)

Purine

Caffeine

Indole contains a pyrrole ring fused with a benzene ring. Compounds derived from indole include the essential amino acid L-tryptophan (Section 27.1C) and the neurotransmitter serotonin. Purine contains a six-membered pyrimidine ring fused with a five-membered imidazole ring. Caffeine is a trimethyl derivative of an oxidized purine. Compounds derived from purine and pyrimidine are building blocks of deoxyribonucleic acids (DNA) and ribonucleic acids (RNA).

E. Aromatic Hydrocarbon Ions

Any neutral monocyclic unsaturated hydrocarbon with an odd number of carbons in the ring must of necessity have at least one CH_2 group in the ring and, therefore, cannot be aromatic. Examples of such hydrocarbons are cyclopropene, cyclopentadiene, and cycloheptatriene.

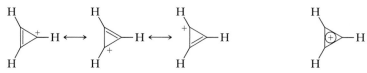

Cyclopropene   Cyclopentadiene   Cycloheptatriene

Cyclopropene has the correct number of pi electrons to be aromatic, namely $4(0) + 2 = 2$, but it does not have a continuous closed loop of $2p$ orbitals. If, however, the CH_2 group becomes a CH^+ group in which the carbon atom is sp^2 hybridized and has a vacant $2p$ orbital, thus still containing only two electrons, then the overlap of orbitals is continuous, and, according to molecular orbital theory, the **cyclopropenyl cation** should be aromatic. The cyclopropenyl cation can be drawn as a resonance hybrid of three equivalent contributing structures. That we can draw three equivalent contributing structures is not the key to the aromaticity of this cation; the key is that it meets the Hückel criteria of aromaticity. Cyclopropenyl cation may also be represented by a triangle with an inscribed circle and plus sign.

Cyclopropenyl cation represented as a hybrid of three equivalent contributing structures

Aromaticity of the cyclopropenyl cation shown by an inscribed circle and plus sign

As an example of the aromatic stabilization of this cation, 3-chlorocyclopropene reacts readily with antimony(V) chloride to form a stable salt.

3-Chloro-cyclopropene   Antimony(V) chloride (a Lewis acid)   Cyclopropenyl hexachloroantimonate

This chemical behavior is to be contrasted with that of 5-chloro-1,3-cyclopentadiene, which cannot be made to form a stable salt. In fact, a cyclic, planar, conjugated cyclopentadienyl cation has four pi electrons, and, if it were to be synthesized, it would be antiaromatic.

5-Chloro-1,3-cyclopentadiene   Cyclopentadienyl tetrafluoroborate

Note that it is possible to draw five equivalent contributing structures for the cyclopentadienyl cation. Yet this cation is not aromatic because it has only $4n$ pi electrons rather than the required $(4n + 2)$ pi electrons.

To form an aromatic ion from cyclopentadiene, it is necessary to convert the CH_2 group to a CH^- group in which the carbon becomes sp^2 hybridized and has two electrons in its unhybridized $2p$ orbital. The resulting **cyclopentadienyl anion** is aromatic. Its aromatic character may also be represented by an inscribed circle with a minus sign.

Cyclopentadienyl anion
(aromatic)

Evidence of the stability of this anion is the fact that cyclopentadiene has a pK_a of approximately 16.0 and is one of the most acidic hydrocarbons known. The acidity of cyclopentadiene is comparable to that of water (pK_a 15.7) and ethanol (pK_a 15.9). Consequently, when cyclopentadiene is treated with aqueous sodium hydroxide, an equilibrium is established in which some of the hydrocarbon is converted to its aromatic anion. K_{eq} for this equilibrium is approximately 0.5. Cyclopentadiene is converted completely to its anion by treatment with sodium amide.

$$pK_{eq} = -0.3$$
$$K_{eq} = 0.50$$

$pK_a = 16.0$ $pK_a = 15.7$

Cyclopentadiene

Cyclopentadienyl sodium

Example 20.2

Construct an MO energy diagram for the cyclopentadienyl anion and describe its ground-state electron configuration.

Solution

Refer to the Frost circle shown in Figure 20.4 for a planar, fully conjugated five-membered ring. The six pi electrons occupy the π_1, π_2, and π_3 molecular orbitals, all of which are bonding MOs.

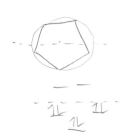

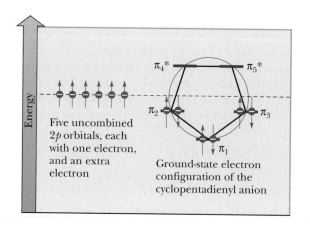

Five uncombined $2p$ orbitals, each with one electron, and an extra electron

Ground-state electron configuration of the cyclopentadienyl anion

Problem 20.2

Describe the ground-state electron configuration of the cyclopentadienyl cation and radical. Assuming each species is planar, would you expect it to be aromatic or antiaromatic?

Cycloheptatriene forms an aromatic cation by conversion of its CH_2 group to a CH^+ group with this sp^2-hybridized carbon having a vacant $2p$ atomic orbital. The **cycloheptatrienyl (tropylium) cation** is planar and has six pi electrons in seven $2p$ orbitals, one from each atom of the ring. It can be drawn as a resonance hybrid of seven equivalent contributing structures.

Cycloheptatrienyl cation
(Tropylium ion)
(aromatic)

Example 20.3

Construct an MO energy diagram for the cycloheptatrienyl cation, and describe its ground-state electron configuration.

Solution

Refer to the Frost circle you constructed in answer to Example 20.1. In the ground-state electron configuration of the cycloheptatrienyl cation, the six pi electrons occupy the π_1, π_2, and π_3 molecular orbitals, all of which are bonding.

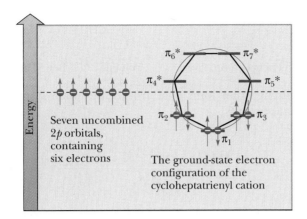

The ground-state electron configuration of the cycloheptatrienyl cation

Problem 20.3

Describe the ground-state electron configuration of the cycloheptatrienyl radical and anion. Assuming that each species is planar, would you expect it to be aromatic or antiaromatic?

20.3 Nomenclature

Nomenclature
Aromatics

A. Monosubstituted Benzenes

Monosubstituted alkylbenzenes are named as derivatives of benzene, as for example ethylbenzene. The IUPAC system retains common names for several of the simpler monosubstituted alkylbenzenes. Examples are toluene (rather than methylbenzene), cumene (rather than isopropylbenzene), and styrene (rather than vinylbenzene).

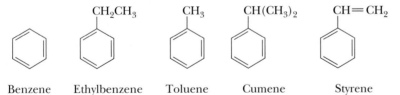

Benzene   Ethylbenzene   Toluene   Cumene   Styrene

The common names phenol, aniline, benzaldehyde, benzoic acid, and anisole are also retained by the IUPAC system.

Phenol   Aniline   Benzaldehyde   Benzoic acid   Anisole

As we noted in Section 5.1, the substituent group derived by loss of an H from benzene is a **phenyl group,** abbreviated Ph—; that derived by loss of an H from the methyl group of toluene is a **benzyl group,** abbreviated Bn—.

Benzyl group, $C_6H_5CH_2$— The group derived from toluene by removing a hydrogen from its methyl group.

Benzene   Phenyl group, Ph—   Toluene   Benzyl group, Bn—

In molecules containing other functional groups, the phenyl group and its derivatives are named as substituents.

1-Phenyl-1-pentanone   4-(3-Methoxyphenyl)-2-butanone   (Z)-2-Phenyl-2-butene

B. Disubstituted Benzenes

Ortho (*o*) Refers to groups occupying 1,2- positions on a benzene ring.

Meta (*m*) Refers to groups occupying 1,3- positions on a benzene ring.

Para (*p*) Refers to groups occupying 1,4- positions on a benzene ring.

When two substituents occur on a benzene ring, three constitutional isomers are possible. The substituents may be located by numbering the atoms of the ring or by using the locators ortho, meta, and para: 1,2- is equivalent to **ortho** (Greek, straight or correct), 1,3- is equivalent to **meta** (Greek, in the middle, between), and 1,4- is equivalent to **para** (Greek, beyond).

When one of the two substituents on the ring imparts a special name to the compound, as for example toluene, cumene, phenol, and aniline, then the compound is named as a derivative of that parent molecule. The special substituent is assumed to occupy ring position number 1. The IUPAC system retains the common name xylene for the three isomeric dimethylbenzenes.

4-Bromotoluene   3-Chloroaniline   2-Nitrobenzoic acid   *m*-Xylene
(*p*-Bromotoluene)   (*m*-Chloroaniline)   (*o*-Nitrobenzoic acid)

Where neither group imparts a special name, then the two substituents are located and listed in alphabetical order before the ending -benzene. The carbon of the benzene ring with the substituent of lower alphabetical ranking is numbered C-1.

1-Chloro-4-ethylbenzene   1-Bromo-2-nitrobenzene   1,3-Dinitrobenzene
(*p*-Chloroethylbenzene)   (*o*-Bromonitrobenzene)   (*m*-Dinitrobenzene)

C. Polysubstituted Benzenes

When three or more substituents are present on a ring, their locations are specified by numerals. If one of the substituents imparts a special name, then the compound is named as a derivative of that parent molecule. If none of the substituents imparts a special name, the substituents are numbered to give the smallest set of numbers and listed in alphabetical order before the ending -benzene. In the following examples, the first compound is a derivative of toluene, and the second is a derivative of phenol. Because there is no special name for the third compound, its three substituents are listed in alphabetical order, and the atoms of the ring are numbered using the lowest possible set of numbers.

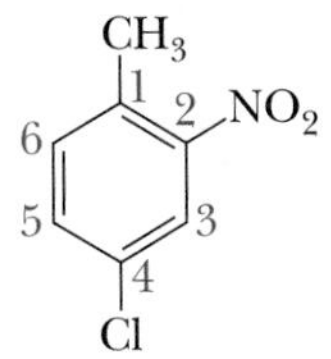

4-Chloro-2-nitrotoluene 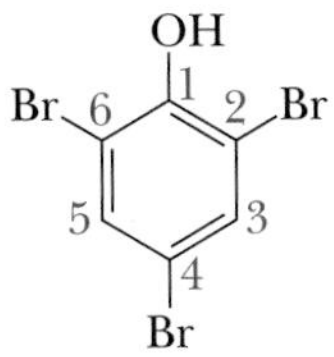   2,4,6-Tribromophenol 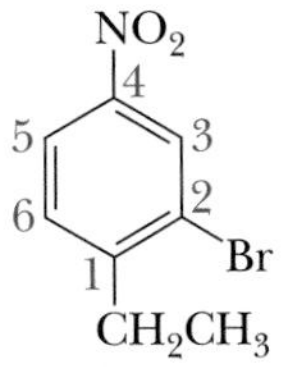   2-Bromo-1-ethyl-4-nitrobenzene

Example 20.4

Write names for these compounds.

(a)    (b) 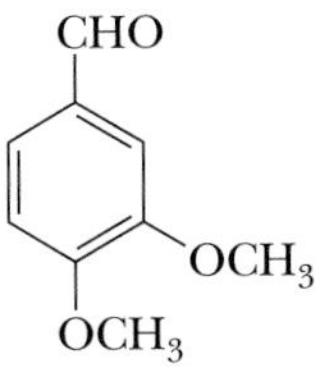   (c) 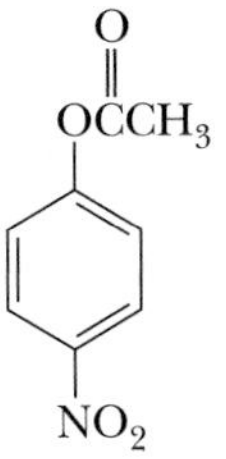   (d) 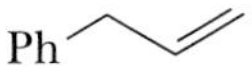 Ph

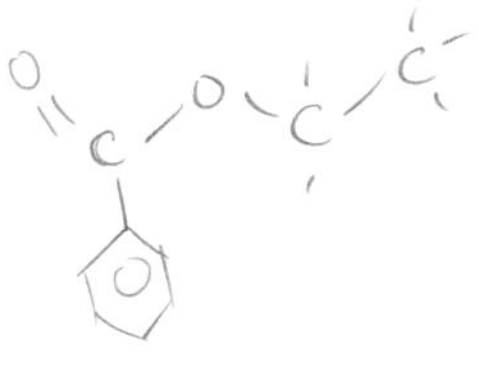

Solution

(a) Ethyl 4-aminobenzoate (ethyl *p*-aminobenzoate)
(b) 3,4-Dimethoxybenzaldehyde
(c) 4-Nitrophenyl ethanoate (*p*-nitrophenyl acetate)
(d) 3-Phenylpropene (allylbenzene)

Problem 20.4

Write names for these compounds.

(a) 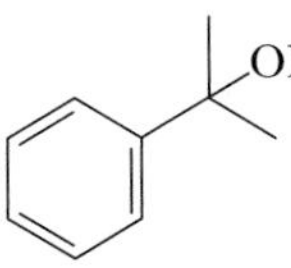   (b) 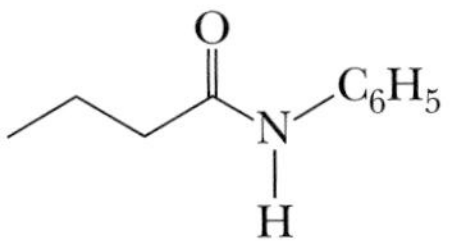   (c) 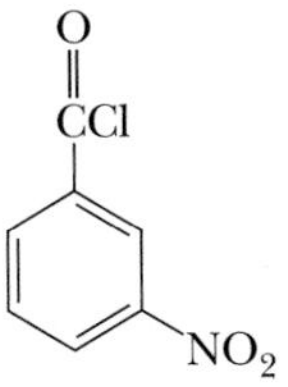   (d)

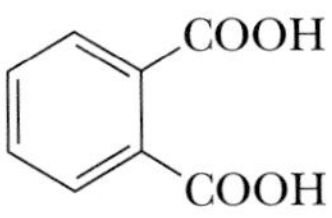

Polynuclear aromatic hydrocarbon (PAH) A hydrocarbon containing two or more fused aromatic rings.

Polynuclear aromatic hydrocarbons (PAHs) contain two or more benzene rings, each pair of which shares two ring carbon atoms. Naphthalene, anthracene, and phenanthrene, the most common PAHs, and substances derived from them are found in coal tar and high-boiling petroleum residues. At one time, naphthalene was used as a moth repellent and insecticide in preserving woolens and furs, but its use has decreased due to the introduction of chlorinated hydrocarbons, such as *p*-dichlorobenzene. In numbering PAHs, carbon atoms common to two or more rings are not numbered because they have no replaceable hydrogens.

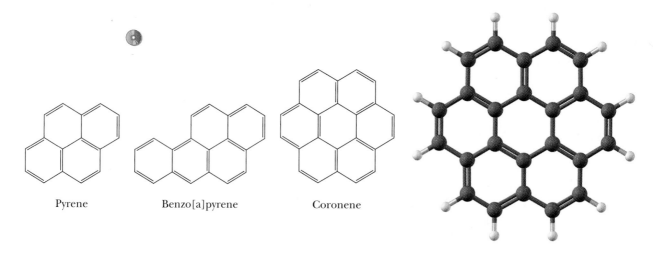

Naphthalene   Anthracene   Phenanthrene

Also found in petroleum and coal tar are lesser amounts of the following PAHs. These compounds can be found in the exhausts of gasoline-powered internal combustion engines (for example, automobile engines) and in cigarette smoke. Benzo[a]pyrene has attracted particular interest because it is a very potent carcinogen (cancer-causing substance) and mutagen.

Pyrene   Benzo[a]pyrene   Coronene

Spectroscopy
Aromatics
The CD-ROM contains
interactive NMR, IR, and
mass spectra for over 60
aromatic compounds.

Analysis of complex aromatic
NMR spectra is discussed in
Supporting Concepts
NMR Spectroscopy
Peak Splitting and Interpretation
(Example 2) sections.

20.4 Spectroscopic Properties

IR spectra of aromatic compounds are described in Chapter 12 and will not be further described here. UV-Visible spectroscopy is important for these compounds and is covered in Chapter 23.

A. NMR Spectroscopy

As mentioned in Section 13.7C, the protons on benzene and other arenes absorb at very low field because of a diamagnetic shielding from an induced ring current caused by their aromaticity. The effect of induced ring current is characteristic not only of benzene and its derivatives but also of all compounds that meet the Hückel criteria for aromaticity (Section 20.2A). Note that this concept of a circulating ring current and of an induced magnetic field correctly predicts that hydrogen atoms on the outside of the ring should come into resonance with a downfield shift. It also predicts that a hydrogen atom in the inside of the ring should come into resonance farther upfield. Of course, no hydrogens are on the inside of the benzene ring, but with larger aromatic annulenes, as for example [18]annulene, there are both "inside" hydrogens and "outside" hydrogens. The degree of the upfield chemical shift of the inside hydrogens of [18]annulene is remarkable. They come into resonance at δ − 3.00, that is at 3.00 δ units upfield (to the right) of the TMS standard.

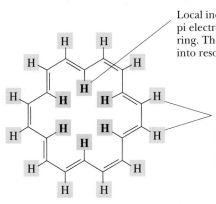

Local induced magnetic field of circulating pi electrons opposes the applied field inside the ring. The six hydrogens inside the ring come into resonance at a greater applied field; at δ –3.00.

Local induced magnetic field of circulating pi electrons reinforces the applied field outside the ring. The twelve hydrogens outside the ring come into resonance at a smaller applied field; at δ 9.3.

[18]Annulene
(aromatic)

Example 20.5

Which hydrogens have a larger chemical shift, the six hydrogens of benzene or the eight hydrogens of cyclooctatetraene? Explain.

Solution

Benzene is an aromatic compound; its six equivalent hydrogens appear as a sharp singlet at δ 7.27. Cyclooctatetraene does not meet the Hückel criteria for aromaticity because it has $4n$ pi electrons and is nonplanar. Therefore, the eight equivalent hydrogens of the cyclooctatetraene ring appear as a singlet at δ 5.8 in the region of vinylic hydrogens (δ 4.6−5.7).

Problem 20.5

Which compound gives a signal in the ^{1}H-NMR spectrum with a larger chemical shift: furan or cyclopentadiene? Explain.

In monosubstituted benzenes in which the ring substituent is neither strongly electron-withdrawing nor electron-releasing, as for example the alkylbenzenes, all

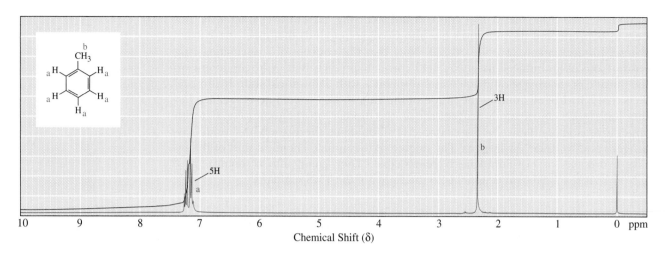

Figure 20.8

^{1}H-NMR spectrum of toluene.

Figure 20.9

The ^{1}H-NMR spectrum of 4-chloroaniline. To a first approximation, the four hydrogens of the aromatic ring appear as two doublets arising from coupling of hydrogens H$_{(b)}$ and H$_{(c)}$ with J = 8 Hz.

ring hydrogens have very similar chemical shifts. The ^{1}H-NMR spectrum of toluene (Figure 20.8) shows a 3H singlet at δ 2.3 for the three hydrogens of the methyl group and a 5H closely spaced multiplet at δ 7.3 for the five hydrogens of the aromatic ring. These hydrogens can be resolved at higher fields. The signal for the methyl hydrogens appears about 1.4 ppm farther downfield than the methyl hydrogens of an alkane. This downfield shift, like that of aryl hydrogens, is due to the effect of the induced ring current and its associated magnetic field.

When a substituent is strongly electron-releasing or electron-withdrawing, the ortho, meta, and para hydrogens have different chemical shifts, and ^{1}H-NMR spectra become significantly more complex. The ^{1}H-NMR signal for the five aromatic hydrogens of anisole, for example, consists of two complex multiplets in the ratio of 3:2. The two ortho hydrogens and the one para hydrogen are more shielded and come into resonance at the higher applied field (smaller chemical shift); the two meta hydrogens are more deshielded and come into resonance at a lower applied field (larger chemical shift). We will not attempt to analyze these complex splitting patterns here.

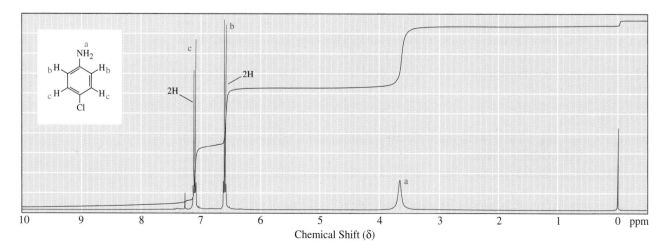

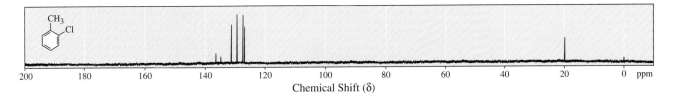

Figure 20.10
^{13}C-NMR spectrum of 2-chloro-toluene.

The splitting patterns of the aromatic protons of di-, tri-, and polysubstituted benzene rings can be very complex. One splitting pattern, however, is quite easy to recognize, namely, that of a para-disubstituted benzene ring. If the two substituents are of sufficiently different electronic effects, then to a first approximation, the spectrum appears as a pair of doublets.

singlet δ 3.65

doublet δ 6.58

doublet δ 7.10

Shown in Figure 20.9 is an ^{1}H-NMR spectrum of 4-chloroaniline.

In ^{13}C-NMR spectroscopy, carbon atoms of aromatic rings appear in the range δ 110–160. Benzene, for example, shows a single signal at δ 128. Because carbon-13 signals for alkene carbons also appear in the range δ 110–160, it is generally not possible to establish the presence of an aromatic ring by ^{13}C-NMR spectroscopy alone. ^{13}C-NMR spectroscopy is particularly useful, however, in establishing substitution patterns of aromatic rings. The ^{13}C-NMR spectrum of 2-chlorotoluene (Figure 20.10) shows six signals in the aromatic region because all its ring carbons are different; its more symmetric isomer 4-chlorotoluene (Figure 20.11) shows four signals in the aromatic region. Thus, all one needs to do is count signals to distinguish between these constitutional isomers.

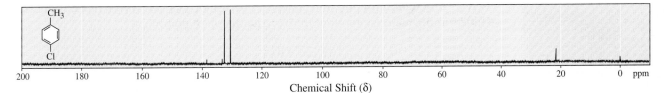

Figure 20.11
^{13}C-NMR spectrum of 4-chloro-toluene.

B. Mass Spectrometry

The mass spectrometry of aromatic compounds was discussed in Section 14.3G.

20.5 Phenols

A. Structure and Nomenclature

The functional group of a **phenol** is a hydroxyl group bonded directly to a benzene ring. Substituted phenols are named either as derivatives of phenol, as benzenols, or by common names.

Phenol A compound that contains an —OH bonded to a benzene ring; a benzenol.

Phenol   3-Methylphenol   1,2-Benzenediol   1,3-Benzenediol   1,4-Benzenediol
(*m*-Cresol)   (Catechol)   (Resorcinol)   (Hydroquinone)

Phenols are widely distributed in nature. Phenol itself and the isomeric cresols (*o*-, *m*-, and *p*-cresol) are found in coal tar and petroleum. Thymol and vanillin are important constituents of thyme and vanilla beans.

2-Isopropyl-5-methylphenol   4-Hydroxy-3-methoxybenzaldehyde
(Thymol)   (Vanillin)

Thymol is a constituent of garden thyme, *Thymus vulgaris.*
(© *Connie Toops*)

Phenol, or carbolic acid as it was once called, is a low-melting solid that is soluble in water. In sufficiently high concentrations, it is corrosive to all kinds of cells. In dilute solutions, it has some antiseptic properties and was introduced into the practice of surgery by Joseph Lister who demonstrated his technique of aseptic surgery in the surgical theater of the University of Glasgow School of Medicine in 1865. Phenol's medical use is now limited. It has been replaced by antiseptics that are both more powerful and have fewer undesirable side effects. Among these is hexylresorcinol (see *The Merck Index*, 12th ed., #4750), which is widely used in nonprescription preparations as a mild antiseptic and disinfectant. Eugenol (see *The Merck Index*, 12th ed., #3944), which can be isolated from the flower buds (cloves) of *Eugenia aromatica*, is used as a dental antiseptic and analgesic. Urushiol (see *the Merck Index*, 12th ed., #10028) is the main component of the irritating oil of poison ivy.

Poison ivy. (*Charles D. Winters*)

Hexylresorcinol   Eugenol   Urushiol

B.   Acidity of Phenols

Phenols and alcohols both contain a hydroxyl group. Phenols, however, are grouped as a separate class of compounds because their chemical properties are quite different from those of alcohols. One of the most important of these differences is that

CHEMISTRY IN ACTION

Capsaicin, for Those Who Like It Hot

Capsaicin (see *The Merck Index*, 12th ed., #1811), the pungent principle from the fruit of various species of peppers (*Capsicum* and *Solanaceae*), was isolated in 1876, and its structure was determined in 1919.

Capsaicin
(from various types of peppers)

The inflammatory properties of capsaicin are well known; as little as one drop in 5 L of water can be detected by the human tongue. We all know of the burning sensation in the mouth and sudden tearing in the eyes caused by a good dose of hot chili peppers. Capsaicin-containing extracts from these flaming foods are also used in sprays to ward off dogs or other animals that might nip at your heels while you are running or cycling.

Ironically, capsaicin is able to cause pain and relieve it as well. Currently, two capsaicin-containing creams, Mioton and Zostrix, are prescribed to treat the burning pain associated with postherpetic neuralgia, a complication of shingles. They are also prescribed for diabetics to relieve persistent foot and leg pain.

The mechanism by which capsaicin relieves these pains is not fully understood. It has been suggested, however, that after application, the nerve endings in the area responsible for the transmission of pain remain temporarily numb. Capsaicin remains bound to specific receptor sites on these pain-transmitting neurons, blocking them from further action. Eventually, capsaicin is removed from these receptor sites, but in the meantime, its presence provides needed relief from pain.

Red chili peppers being dried. (Chuck Pefley/Stone)

phenols are significantly more acidic than alcohols. The acid ionization constant of phenol is 10^6 times larger than that of ethanol.

$$\text{C}_6\text{H}_5\text{—OH} + \text{H}_2\text{O} \rightleftharpoons \text{C}_6\text{H}_5\text{—O}^- + \text{H}_3\text{O}^+ \qquad K_a = 1.1 \times 10^{-10} \qquad pK_a = 9.95$$

$$\text{CH}_3\text{CH}_2\text{OH} + \text{H}_2\text{O} \rightleftharpoons \text{CH}_3\text{CH}_2\text{O}^- + \text{H}_3\text{O}^+ \qquad K_a = 1.3 \times 10^{-16} \qquad pK_a = 15.9$$

D. Preparation of Alkyl Aryl Ethers

Alkyl aryl ethers can be prepared from a phenoxide salt and an alkyl halide (the Williamson synthesis, Section 11.4A). They cannot be prepared from an aryl halide and alkoxide salt, however, because aryl halides are quite unreactive under the conditions of Williamson synthesis; they do not undergo nucleophilic displacement by either S_N1 or S_N2 mechanisms.

Alkyl aryl ethers are often synthesized by phase-transfer catalysis (Section 8.7). Both the alkyl halide and phenol are dissolved in dichloromethane; then, the solution is mixed with an aqueous solution of sodium hydroxide, and a phase-transfer catalyst such as tetrabutylammonium bromide, $Bu_4N^+Br^-$, is added. Phenol, a poor nucleophile, reacts with sodium hydroxide in the aqueous phase to form the phenoxide ion, a good nucleophile. The phase-transfer catalyst transports phenoxide ion to the dichloromethane phase where it reacts with the alkyl halide to form an ether. The following examples illustrate the Williamson synthesis of alkyl aryl ethers. The synthesis of allyl phenyl ether involves nucleophilic displacement on a primary halide; the synthesis of anisole illustrates the use of dimethyl sulfate as a methylating agent.

| Phenol | 3-Chloropropene (Allyl chloride) | Phenyl 2-propenyl ether (Allyl phenyl ether) |

| Phenol | Dimethyl sulfate | Methyl phenyl ether (Anisole) |

An alkyl aryl ether, ArOR, is cleaved by hydrohalic acids, HX, to form an alkyl halide and a phenol. This illustrates the fact that nucleophilic substitution is not likely to occur at an aromatic carbon, and that phenols, unlike alcohols, are not converted to aryl halides by treatment with concentrated HCl, HBr, or HI.

| 2-Phenoxypropane (Isopropyl phenyl ether) | Phenol | 2-Iodopropane (Isopropyl iodide) |

E. Kolbe Carboxylation; Synthesis of Salicylic Acid

Phenoxide ions react with carbon dioxide to give a carboxylic acid salt as shown by the industrial synthesis of salicylic acid, the starting material for the production of aspirin (Section 18.6B). Phenol is dissolved in aqueous NaOH, and this solution is then saturated with CO_2 under pressure to give sodium salicylate. This process is referred to as high-pressure carboxylation of sodium phenoxide. Upon acidification of the alkaline solution, salicylic acid is isolated as a solid, mp 157–159°C.

Phenol → Sodium phenoxide → Sodium salicylate → Salicylic acid

The importance of salicylic acid in industrial organic chemistry is demonstrated by the fact that over 6×10^6 kg of aspirin are synthesized in the United States each year.

Mechanism   Kolbe Carboxylation of Phenol

The phenoxide ion reacts like an enolate ion; it is a strong nucleophile. In Step 1, nucleophilic attack of the phenoxide anion on a carbonyl group of carbon dioxide gives a substituted cyclohexadienone intermediate. Keto-enol tautomerism of this intermediate in Step 2 gives salicylate anion. Note that the enol in this case, due to its aromatic character, is the more stable of the two tautomers.

Sodium phenoxide   +   A cyclohexadienone intermediate   →(keto-enol tautomerism)→   Salicylate anion

---

F.   Oxidation to Quinones

Because of the presence of the electron-donating —OH group on the ring, phenols are susceptible to oxidation by a variety of strong oxidizing agents. For example, oxidation of phenol itself by potassium dichromate gives 1,4-benzoquinone (*p*-quinone). By definition, a quinone is a cyclohexadienedione. Those with carbonyl groups ortho to each other are called *o*-quinones; those with carbonyl groups para to each other are called *p*-quinones.

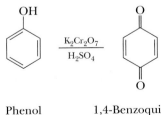

Phenol   →($K_2Cr_2O_7$ / H_2SO_4)→   1,4-Benzoquinone (*p*-Quinone)

A bombardier beetle ejects a mixture of superheated steam and *p*-quinone. (*Thomas Eisner and David Aneshansley, Cornell University*)

Mechanisms
Aromatics
Kolbe carboxylation of phenol

Quinones can also be obtained by oxidation of 1,2-benzenediol (catechol) or 1,4-benzenediol (hydroquinone).

| 1,2-Benzenediol | 1,2-Benzoquinone | 1,4-Benzenediol | 1,4-Benzoquinone |
| (Catechol) | (*o*-Quinone) | (Hydroquinone) | (*p*-Quinone) |

Perhaps the most important chemical property of quinones is that they are readily reduced to benzenediols. For example, *p*-quinone is readily reduced to hydroquinone by sodium dithionite in neutral or alkaline solution. There are other ways to carry out this reduction. The point is that it can be done very easily, as can the corresponding oxidation of a hydroquinone.

| 1,4-Benzoquinone | 1,4-Benzenediol |
| (*p*-Quinone) | (Hydroquinone) |

There are many examples in which the reversible oxidation/reduction of hydroquinones or quinones is important. One such example is coenzyme Q, alternatively known as ubiquinone. The name of this important biomolecule is derived from the Latin *ubique* (everywhere) + quinone.

| Coenzyme Q | Coenzyme Q |
| (oxidized form) | (reduced form) |

Coenzyme Q, a carrier of electrons in the respiratory chain, contains a long hydrocarbon chain of between six and ten isoprene units that serves to anchor it firmly in the nonpolar environment of the mitochondrial inner membrane. As can be seen from the balanced half-reaction, the oxidized form of coenzyme Q is a two-electron oxidizing agent. In subsequent steps of the respiratory chain, the reduced form of coenzyme Q transfers these two electrons to another link until they are eventually delivered to a molecule of oxygen, which is in turn reduced to water.

Another quinone important in biological systems is vitamin K_2. This compound was discovered in 1935 as a result of a study of newly hatched chicks with a fatal disease in which their blood was slow to clot. It was later discovered that the delayed clotting time of blood was caused by a deficiency of prothrombin, and it is now known that vitamin K$_2$ is essential to the synthesis of prothrombin in the liver. The natural form of vitamin K$_2$ has a chain of five to eight isoprene units attached to a 1,4-naphthoquinone ring. The following structure shows seven isoprene units in the side chain.

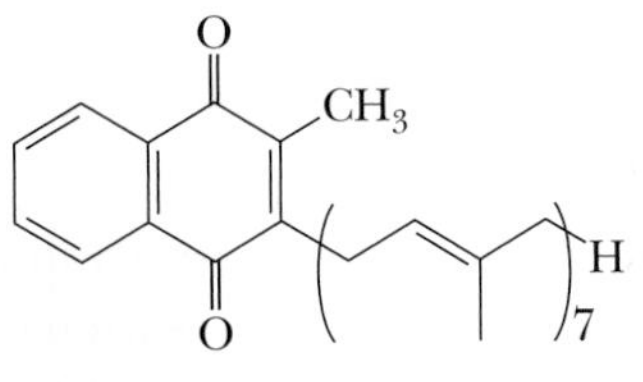

Vitamin K$_2$

The natural vitamins of the K family have for the most part been replaced by synthetic preparations in food supplements. Menadione, one such synthetic material with vitamin K activity, has only hydrogen in place of the long alkyl side chain. Menadione is prepared by chromic acid oxidation of 2-methylnaphthalene under mild conditions.

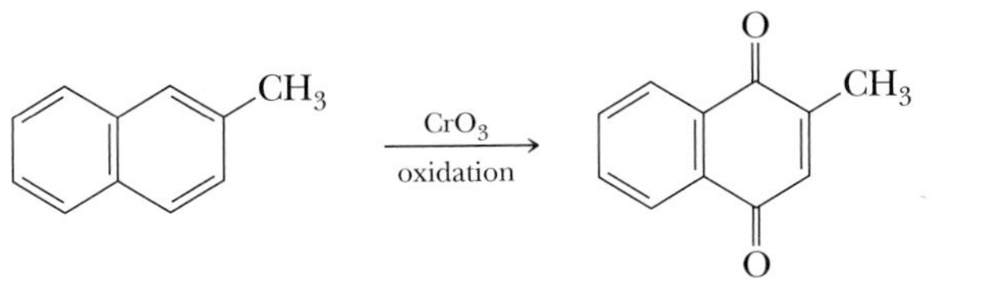

2-Methylnaphthalene   2-Methyl-1,4-naphthoquinone
                                              (Menadione)

A commercial process that uses a quinone is black-and-white photography. Black-and-white film is coated with an emulsion containing silver bromide or silver iodide crystals, which become activated by exposure to light. The activated silver ions are reduced in the developing stage to metallic silver by hydroquinone, which at the same time is oxidized to quinone. Following is an equation showing the relationship between these species.

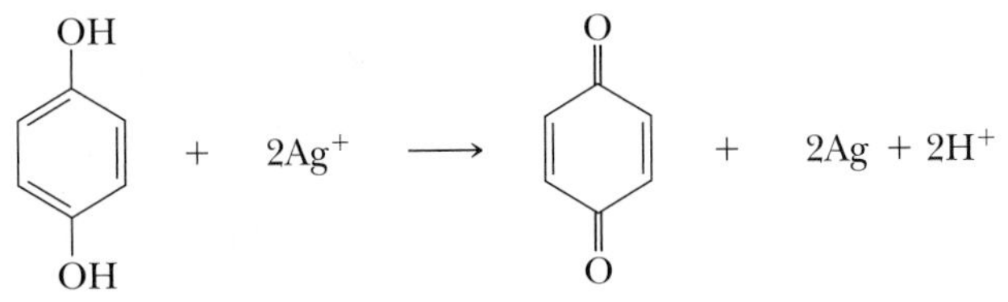

1,4-Benzenediol   1,4-Benzoquinone
(Hydroquinone)   (p-Quinone)

All silver halide not activated by light and then reduced by interaction with hydroquinone is removed in the fixing process, and the result is a black image (a negative) left by deposited metallic silver where the film has been struck by light. Other compounds are now used to reduce "light-activated" silver bromide, but the result is the same—a deposit of metallic silver in response to exposure of film to light.

20.6 Reactions at a Benzylic Position

In this section, we study two reactions of substituted aromatic hydrocarbons that occur preferentially at the **benzylic position.**

benzylic position

$$\text{—CH}_2\text{—}$$

Benzyl group

Reactions involving alkyl side chains of aromatic compounds occur preferentially at the benzylic position for two reasons. First, the benzene ring is especially resistant to reaction with many of the reagents that normally attack alkanes. Second, benzylic cations and benzylic radicals are easily formed because of resonance stabilization of these intermediates. A benzylic cation or radical is a hybrid of five contributing structures: two Kekulé structures and three that delocalize the positive charge (or the lone electron) onto carbons of the aromatic ring. Following are contributing structures for a benzylic cation. Similar contributing structures can be written for a benzylic radical and anion. Benzylic contributing structures are closely analogous to allylic structures in stabilizing cations, radicals, and anions.

The benzyl cation as a hybrid of five contributing structures

A. Oxidation

Benzene is unaffected by strong oxidizing agents, such as H_2CrO_4 and $KMnO_4$. However, when toluene is treated with these oxidizing agents under quite vigorous conditions, the side chain methyl group is oxidized to a carboxyl group to give benzoic acid.

$$\underset{\text{Toluene}}{\text{CH}_3} \xrightarrow{H_2CrO_4} \underset{\text{Benzoic acid}}{\text{COOH}}$$

Halogen and nitro substituents on an aromatic ring are unaffected by these oxidations. 2-Chloro-4-nitrotoluene, for example, is oxidized to 2-chloro-4-nitrobenzoic acid.

$$\underset{\text{2-Chloro-4-nitrotoluene}}{\text{O}_2\text{N}\quad\text{CH}_3,\ \text{Cl}} \xrightarrow{H_2CrO_4} \underset{\text{2-Chloro-4-nitrobenzoic acid}}{\text{O}_2\text{N}\quad\text{COOH},\ \text{Cl}}$$

Ethyl and isopropyl side chains are also oxidized to carboxyl groups. The side chain of *tert*-butylbenzene, however, is not oxidized.

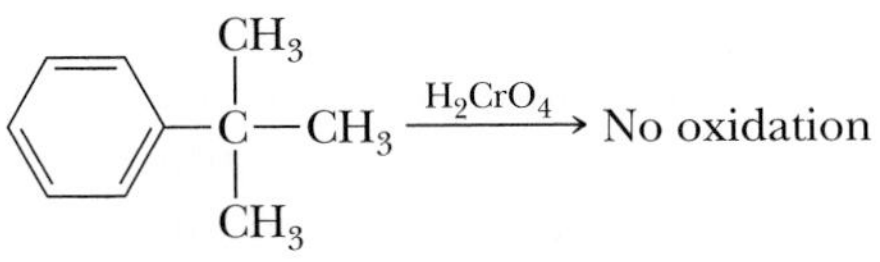

tert-Butylbenzene

From these observations, we conclude that if a benzylic hydrogen exists, then the benzylic carbon is oxidized to a carboxyl group and all other carbons of the side chain are removed. If no benzylic hydrogen exists, as in the case of *tert*-butylbenzene, no oxidation of the side chain occurs.

If more than one alkyl side chain exists, each is oxidized to —COOH. Oxidation of *m*-xylene gives 1,3-benzenedicarboxylic acid, more commonly named isophthalic acid.

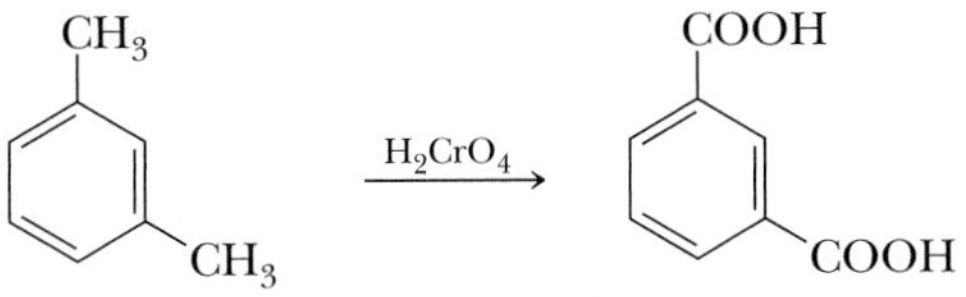

m-Xylene   1,3-Benzenedicarboxylic acid
(Isophthalic acid)

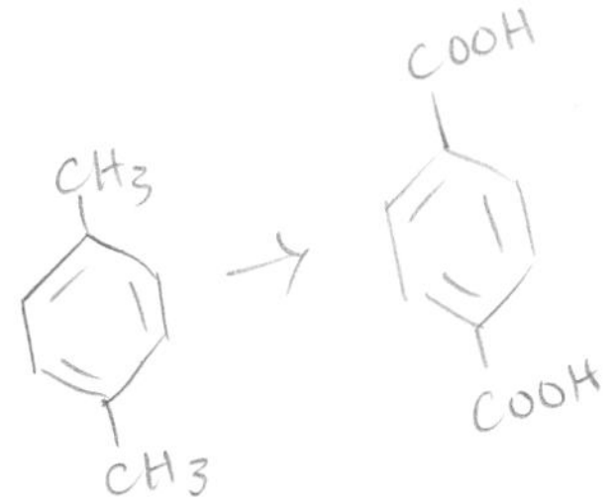

Example 20.7

Draw structural formulas for the product of vigorous oxidation 1,4-dimethylbenzene (*p*-xylene) by H_2CrO_4.

Solution

Both alkyl groups are oxidized to —COOH groups. The product is terephthalic acid, one of two monomers required for the synthesis of Dacron polyester and Mylar (Section 24.5B).

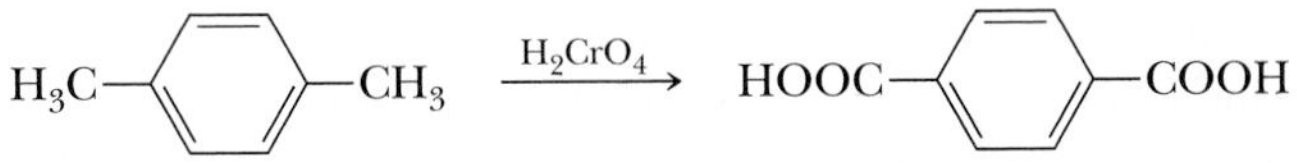

1,4-Dimethylbenzene   1,4-Benzenedicarboxylic acid
(*p*-Xylene)   (Terephthalic acid)

Problem 20.7

Predict the products resulting from vigorous oxidation of each compound by H_2CrO_4.

(a) 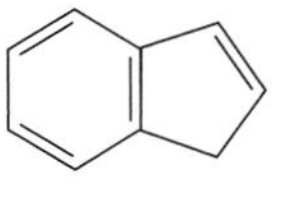   (b)

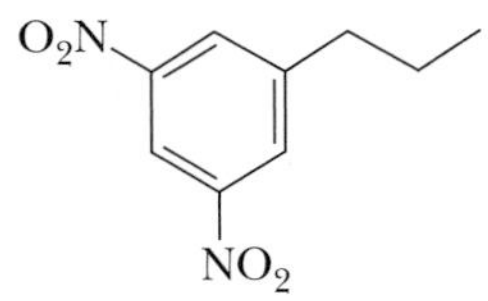

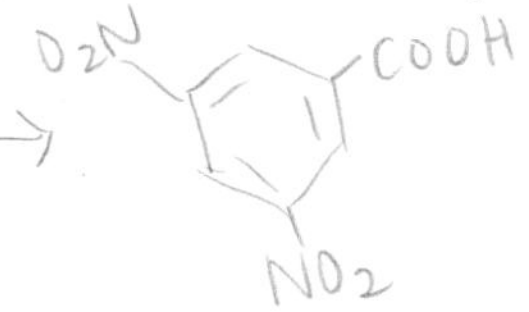

It has been difficult to study these side chain oxidations and to formulate mechanisms for them. Available evidence, however, supports the formation of unstable intermediates that are either benzylic radicals or benzylic carbocations.

Naphthalene is oxidized to phthalic acid by molecular oxygen in the presence of a vanadium(V) oxide (vanadium pentoxide) catalyst. This conversion, which is the basis for an industrial synthesis of this aromatic dicarboxylic acid, illustrates the ease of oxidation of condensed benzene rings compared with benzene itself.

Naphthalene   1,2-Benzenedicarboxylic acid
                                  (Phthalic acid)

B. Halogenation

Reaction of toluene with chlorine in the presence of heat or light results in formation of chloromethylbenzene and HCl. Bromination is easily accomplished by using *N*-bromosuccinimide (NBS) in the presence of a peroxide catalyst.

Toluene   Chloromethylbenzene
                                    (Benzyl chloride)

Toluene   *N*-Bromosuccinimide   Bromomethylbenzene   Succinimide
                                             (Benzyl bromide)

Halogenation of a larger alkyl side chain is highly regioselective, as illustrated by the halogenation of ethylbenzene. When treated with NBS, the only monobromo organic product formed is 1-bromo-1-phenylethane. This regioselectivity is dictated by the resonance stabilization of the benzylic radical intermediate. The mechanism of radical bromination at a benzylic position is identical to that for allylic bromination (Section 7.6).

Ethylbenzene   1-Bromo-1-phenylethane

When ethylbenzene is treated with chlorine under radical reaction conditions, two products are formed in the ratio of 9:1.

Ethylbenzene   1-Chloro-1-phenylethane   1-Chloro-2-phenylethane
                                   (90%)   (10%)

Thus, chlorination of alkyl side chains is also regioselective but not to the same high degree as bromination. Recall that we observed this same pattern in the regioselectivities of bromination and chlorination of alkanes (Section 7.4A).

Combining the information on product distribution for bromination and chlorination of hydrocarbons, we conclude that the order of stability of radicals is

$$\text{methyl} < 1° < 2° < 3° < \text{allylic} \cong \text{benzylic}$$

Increasing radical stability

This order reflects the C—H bond dissociation energies (BDE) for formation of these radicals (Appendix 3).

C.   Hydrogenolysis of Benzyl Ethers

Among ethers, benzylic ethers are unique in that they are cleaved under the conditions of catalytic hydrogenation as illustrated by the **hydrogenolysis** of benzyl hexyl ether. Hydrogenolysis is the cleavage of a single bond by H_2. In the hydrogenolysis of a benzylic ether, it is the single bond between the benzylic carbon and its attached oxygen that is cleaved and replaced by a carbon-hydrogen bond. In this illustration, the benzyl group is converted to toluene, and the alkyl group is converted to an alcohol.

Hydrogenolysis   Cleavage of a single bond by H_2, most commonly accomplished by treatment of a compound with H_2 in the presence of a transition metal catalyst.

$$\text{Benzyl hexyl ether} \qquad \text{Toluene} \qquad \text{1-Hexanol}$$

Benzyl ethers are formed by treatment of an alcohol or phenol with benzyl chloride in the presence of a base such as triethylamine or pyridine.

The particular value of benzylic ethers is that they can serve as protecting groups for the —OH groups of alcohols and phenols. Suppose, for example, we want to treat 2-allylphenol with diborane followed by hydrogen peroxide to bring about anti-Markovnikov hydration of the carbon-carbon double bond. This scheme will not give the desired result because the phenolic —OH group is sufficiently acidic to react with BH_3 and destroy it. The desired product can be prepared, however, by protection of the phenolic —OH group as the benzylic ether, hydroboration/oxidation of the carbon-carbon double bond, and hydrogenolysis of the benzylic ether.

2-(2-Propenyl)phenol
(2-Allylphenol)

2-(3-Hydroxypropyl)phenol

Summary

Benzene and its alkyl derivatives are classified as **aromatic hydrocarbons, or arenes.** The structure of benzene, proposed by August Kekulé in 1865, represented benzene as two rapidly interconverting Kekulé structures (Section 20.1A). The concepts of hybridization of atomic orbitals and the theory of resonance (Section 20.1B), developed by Linus Pauling in the 1930s, provided the first adequate description of the structure of benzene. According to the molecular orbital model, the six $2p$ atomic orbitals of the sp^2-hybridized ring carbon atoms combine to give three pi-bonding MOs and three pi-antibonding MOs. In the ground state, the six pi electrons of benzene lie in the three pi-bonding MOs. The **resonance energy** of benzene (Section 20.1C), as calculated from experimental values for heats of hydrogenation of benzene and cyclohexene, is approximately 151 kJ (36 kcal)/mol.

According to the **Hückel criteria for aromaticity** (Section 20.2A), a monocyclic compound is aromatic if it (1) has one p orbital on each atom of the ring, (2) is planar so that overlap of all p orbitals of the ring is continuous or nearly continuous, and (3) has $(4n + 2)$ pi electrons in the cyclic, overlapping arrangement of p orbitals. An **annulene** (Section 20.2B) is a cyclic hydrocarbon with an alternation of single and double bonds. Many have been synthesized to test the validity of Hückel's criteria for aromaticity. It has been found, for example, that [14]annulene and [18]annulene are aromatic as predicted. **Antiaromatic compounds** (Section 20.2C) have only $4n$ pi electrons in a monocyclic, planar system of continuously overlapping p orbitals.

A **heterocyclic aromatic compound** (Section 20.2D) contains one or more atoms other than carbon in an aromatic ring. Particularly abundant in the biological world are derivatives of the heterocyclic aromatic amines pyridine, pyrimidine, imidazole, and pyrrole.

The **cyclopropenyl cation,** the **cyclopentadienyl anion,** and the **cycloheptatrienyl cation** (Section 20.2E) each meet the Hückel criteria for aromaticity and are particularly stable hydrocarbon ions.

Aromatic compounds are named by the IUPAC system. The common names toluene, xylene, cumene, styrene, phenol, aniline, and benzoic acid (Section 20.3A) are retained. The C_6H_5— group is named **phenyl,** and the $C_6H_5CH_2$— group is named **benzyl.** Two substituents on a benzene ring may be located by numbering the atoms of the ring or by using the locators **ortho (o), meta (m),** and **para (p). Polynuclear aromatic hydrocarbons** (Section 20.3C) contain two or more fused benzene rings. Particularly abundant are naphthalene, anthracene, phenanthrene, and their derivatives.

The functional group of a **phenol** (Section 20.5A) is an —OH group bonded to a benzene ring. Phenol and its derivatives are weak acids, pK_a approximately 10. The greater acidity of phenols substituted with electron-withdrawing groups, for example NO_2, is accounted for by a combination of inductive and resonance effects.

Reactions of aromatic compounds containing alkyl side chains occur preferentially at the benzylic carbon (Section 20.6). Benzylic cations and radicals are especially stable because of delocalization of their positive charge or unpaired electron, respectively, onto the ortho and para positions of the aromatic ring.

Key Reactions

1. Acidity of Phenols (Section 20.5B)

Phenols are weak acids, pK_a approximately 10. Ring substituents may increase or decrease acidity by a combination of resonance and inductive effects.

$$\text{C}_6\text{H}_5\text{—OH} + \text{H}_2\text{O} \rightleftharpoons \text{C}_6\text{H}_5\text{—O}^- + \text{H}_3\text{O}^+ \qquad K_a = 1.1 \times 10^{-10} \qquad pK_a = 9.95$$

2. Reaction of Phenols with Strong Bases (Section 20.5C)

Water-insoluble phenols react quantitatively with strong bases to form water-soluble salts.

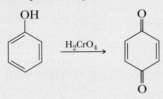

Phenol   Sodium   Sodium   Water
pK_a 9.95   hydroxide   phenoxide   pK_a 15.7
(stronger acid)   (stronger base)   (weaker base)   (weaker acid)

3. Kolbe Synthesis: Carboxylation of Phenols (Section 20.5E)

Nucleophilic addition of a phenoxide ion to carbon dioxide gives a substituted cyclohexa-dienone which then undergoes keto-enol tautomerism to regenerate the aromatic ring.

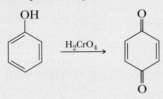

4. Oxidation of Phenols to Quinones (Section 20.5F)

Oxidation by H_2CrO_4 gives 1,2-quinones (*o*-quinones) or 1,4-quinones (*p*-quinones), de-pending on the structure of the particular phenol.

Phenol   1,4-Benzoquinone
                        (*p*-Quinone)

5. Oxidation at a Benzylic Position (Section 20.6A)

A benzylic carbon bonded to at least one hydrogen is oxidized to a carboxyl group.

6. Halogenation at a Benzylic Position (Section 20.6B)

Halogenation is regioselective for a benzylic position and occurs by a radical chain mecha-nism. Bromination shows a higher regioselectivity for a benzylic position than chlorination.

Ethylbenzene   1-Bromo-1-phenylethane

7. Hydrogenolysis of Benzylic Ethers (Section 20.6C)

Benzylic ethers are cleaved under the conditions of catalytic hydrogenation.

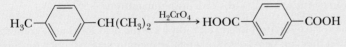

Problems

Nomenclature and Structural Formulas

20.8   Name the following molecules and ions.

(a) (4-nitro, 1-chloro benzene with NO_2 top and Cl bottom)

(b) (benzene with CH_3 and Br)

(c) (phenyl-propanol, OH)

(d) (naphthalene with NO_2 and NO_2)

(e) (phenyl group with OH and vinyl)

(f) (O_2N substituted benzene with alkyne)

(g) (phenol OH with Ph)

(h) CH_3O- (benzene) $-CH_2{}^+$

(i) $\underset{H}{\overset{C_6H_5}{}}C=C\underset{C_6H_5}{\overset{H}{}}$

(j) (cyclopropenyl cation with three C_6H_5 groups)

20.9   Draw structural formulas for these compounds.

(a) 1-Bromo-2-chloro-4-ethylbenzene   **(b)** *m*-Nitrocumene
(c) 4-Chloro-1,2-dimethylbenzene   **(d)** 3,5-Dinitrotoluene
(e) 2,4,6-Trinitrotoluene   **(f)** 4-Phenyl-2-pentanol
(g) *p*-Cresol   **(h)** Pentachlorophenol
(i) 1-Phenylcyclopropanol   **(j)** Triphenylmethane
(k) Phenylethylene (styrene)   **(l)** Benzyl bromide
(m) 1-Phenyl-1-butyne   **(n)** 3-Phenyl-2-propen-1-ol

20.10   Draw structural formulas for each of these compounds.

(a) 1-Nitronaphthalene   **(b)** 1,6-Dichloronaphthalene
(c) 9-Bromoanthracene   **(d)** 2-Methylphenanthrene

20.11   Molecules of 6,6′-dinitrobiphenyl-2,2′-dicarboxylic acid have no tetrahedral stereocenter, and yet they can be resolved to a pair of enantiomers. Account for this chirality. You will find it helpful to examine the model on the CD.

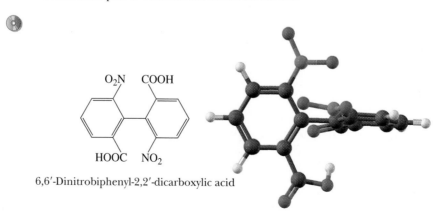

6,6′-Dinitrobiphenyl-2,2′-dicarboxylic acid

Resonance in Aromatic Compounds

20.12 Following each name is the number of Kekulé structures that can be drawn for it. Draw these Kekulé structures, and show, using curved arrows, how the first contributing structure for each molecule is converted to the second and so forth.

(a) Naphthalene (3)   (b) Phenanthrene (5)

20.13 Each molecule in this problem can be drawn as a hybrid of five contributing structures: two Kekulé structures and three that involve creation and separation of unlike charges. For (a) and (b), the creation and separation of unlike charges places a positive charge on the substituent and a negative charge on the ring. For (c), a positive charge is placed on the ring and an additional negative charge is placed on the $-NO_2$ group. Draw these five contributing structures for each molecule.

(a) Chlorobenzene   (b) Phenol   (c) Nitrobenzene

20.14 Following are structural formulas for furan and pyridine.

Furan   Pyridine

(a) Write contributing structures for the furan hybrid that place a positive charge on oxygen and a negative charge first on carbon 3 of the ring and then on each other carbon of the ring.
(b) Write contributing structures for the pyridine hybrid that place a negative charge on nitrogen and a positive charge first on carbon 2, then on carbon 4, and finally carbon 6.

The Concept of Aromaticity

20.15 State the number of p orbital electrons in each of the following.

20.16 Which of the molecules and ions given in Problem 20.15 are aromatic according to the Hückel criteria? Which, if planar, would be antiaromatic?

20.17 Construct MO energy diagrams for the cyclopropenyl cation, radical, and anion. Which of these species is aromatic according to the Hückel criteria?

20.18 Naphthalene and azulene are constitutional isomers of molecular formula $C_{10}H_8$. Naphthalene is a colorless solid with a dipole moment of zero. Azulene is a solid with an intense blue color and a dipole moment of 1.0 D. Account for the difference in dipole moments of these constitutional isomers.

Spectroscopy

20.19 Compound A, molecular formula C_9H_{12}, shows prominent peaks in its mass spectrum at m/z 120 and 105. Compound B, also molecular formula C_9H_{12}, shows prominent peaks at m/z 120 and 91. On vigorous oxidation with chromic acid, both compounds give benzoic acid. From this information, deduce the structural formulas of compounds A and B.

20.20 Compound C shows a molecular ion at m/z 148 and other prominent peaks at m/z 105 and 77. Following are its infrared and ^{1}H-NMR spectra.

(a) Deduce the structural formula of compound C.
(b) Account for the appearance of peaks in its mass spectrum at m/z 105 and 77.

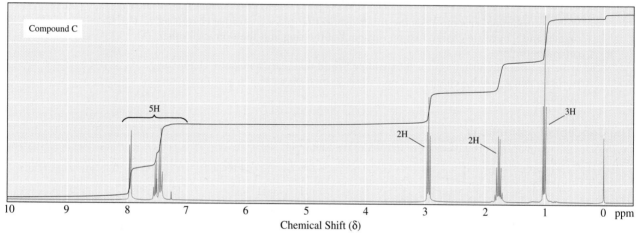

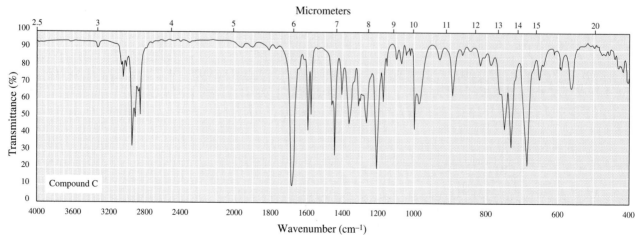

20.21 Following are IR and ^{1}H-NMR spectra of compound D. The mass spectrum of compound D shows a molecular ion peak at m/z 136, a base peak at m/z 107, and other prominent peaks at m/z 118 and 59.

(a) Propose a structural formula for compound D based on this information.

(b) Propose structural formulas for ions in the mass spectrum at m/z 118, 107, and 59.

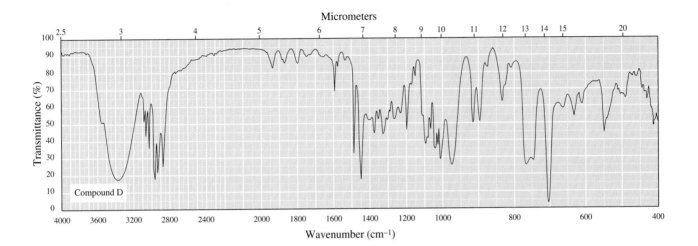

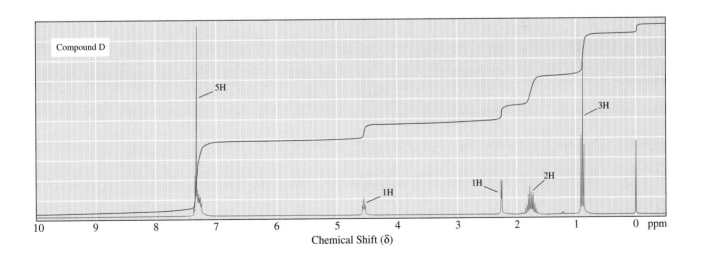

20.22 Compound E is a neutral solid of molecular formula $C_8H_{10}O_2$. Its mass spectrum shows a molecular ion at m/z 138 and prominent peaks at M-1 and M-17. Following are IR and ^{1}H-NMR spectra of compound E. Deduce the structure of compound E.

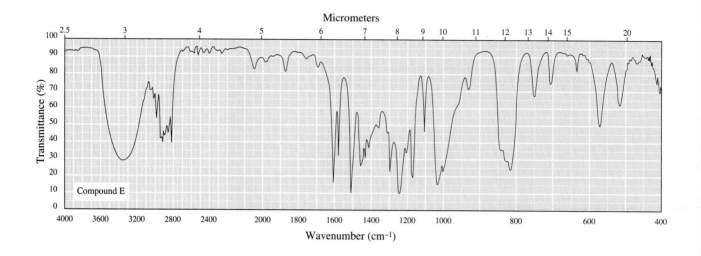

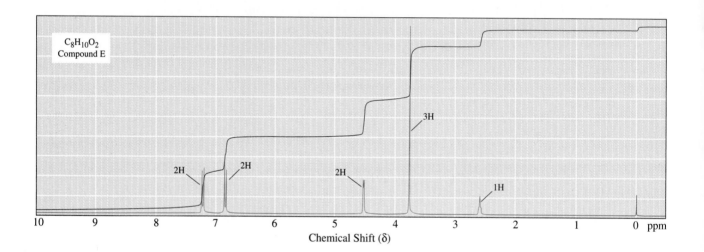

20.23 Following are ^{1}H-NMR and ^{13}C-NMR spectral data for compound F, $C_{12}H_{16}O$. From this information, deduce the structure of compound F.

^{1}H-NMR	^{13}C-NMR	
0.83 (d, 6H)	207.82	50.88
2.11 (m, 1H)	134.24	50.57
2.30 (d, 2H)	129.36	24.43
3.64 (s, 2H)	128.60	22.48
7.2–7.4 (m, 5H)	126.86	

20.24 Following are ^{1}H-NMR and ^{13}C-NMR spectral data for compound G, $C_{10}H_{10}O$. From this information, deduce the structure of compound G.

^{1}H-NMR	^{13}C-NMR	
2.50 (t, 2H)	210.19	126.82
3.05 (t, 2H)	136.64	126.75
3.58 (s, 2H)	133.25	45.02
7.1–7.3 (m, 4H)	128.14	38.11
	127.75	28.34

20.25 Compound H, $C_8H_6O_3$, gives a precipitate when treated with hydroxylamine in aqueous ethanol, and a silver mirror when treated with Tollens' solution. Following is its ^{1}H-NMR spectrum. Deduce the structure of compound H.

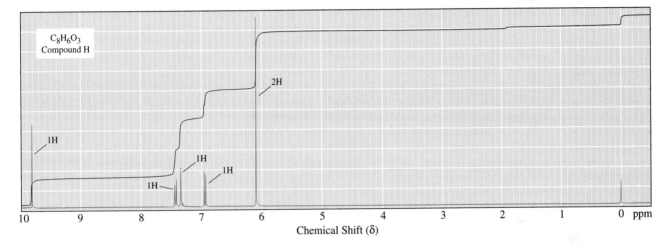

20.26 Compound I, $C_{11}H_{14}O_2$, is insoluble in water, aqueous acid, and aqueous $NaHCO_3$ but dissolves readily in 10% Na_2CO_3 and 10% NaOH. When these alkaline solutions are acidified with 10% HCl, compound I is recovered unchanged. Given this information and its ^{1}H-NMR spectrum, deduce the structure of compound I.

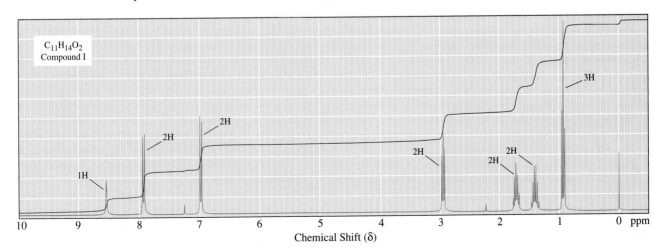

20.27 Propose a structural formula for compound J, $C_{11}H_{14}O_3$, consistent with its ^{1}H-NMR and infrared spectra.

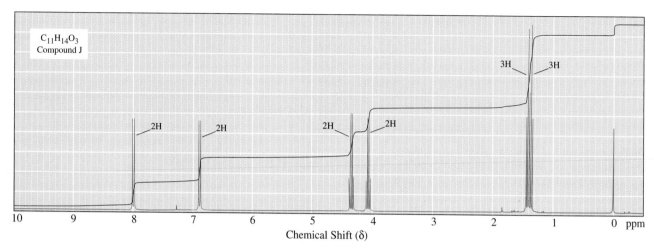

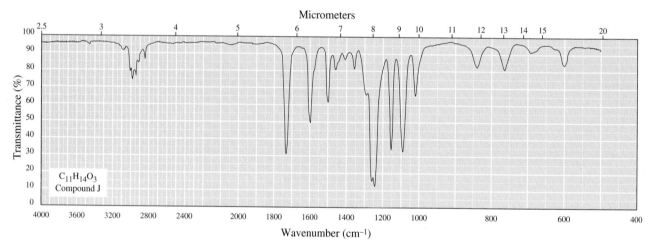

20.28 Propose a structural formula for the analgesic phenacetin, molecular formula $C_{10}H_{13}NO_2$, based on its ^{1}H-NMR spectrum.

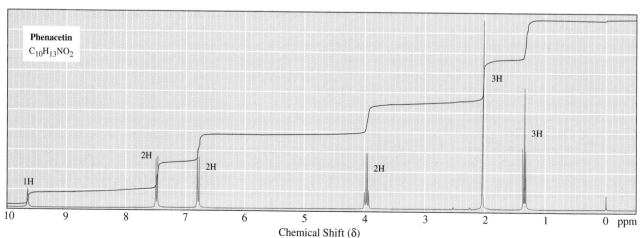

20.29 Compound K, $C_{10}H_{12}O_2$, is insoluble in water, 10% NaOH, and 10% HCl. Given this information and the following ^{1}H-NMR and^{13}C-NMR spectral information, deduce the structural formula of compound K.

^{1}H-NMR	^{13}C-NMR	
2.10 (s, 3H)	206.51	114.17
3.61 (s, 2H)	158.67	55.21
3.77 (s, 3H)	130.33	50.07
6.86 (d, 2H)	126.31	29.03
7.12 (d, 2H)		

20.30 Propose a structural formula for each compound given these NMR data.

(a) $C_9H_9BrO_2$ **(b)** C_8H_9NO **(c)** $C_9H_9NO_3$

^{1}H-NMR	^{13}C-NMR	^{1}H-NMR	^{13}C-NMR	^{1}H-NMR	^{13}C-NMR
1.39 (t, 3H)	165.73	2.06 (s, 3H)	168.14	2.10 (s, 3H)	168.74
4.38 (q, 2H)	131.56	7.01 (t, 1H)	139.24	7.72 (d, 2H)	166.85
7.57 (d, 2H)	131.01	7.30 (m, 2H)	128.51	7.91 (d, 2H)	143.23
7.90 (d, 2H)	129.81	7.59 (d, 2H)	122.83	10.3 (s, 1H)	130.28
	127.81	9.90 (s, 1H)	118.90	12.7 (s, 1H)	124.80
	61.18		23.93		118.09
	14.18				24.09

20.31 Given here are ^{1}H-NMR and ^{13}C-NMR spectral data for two compounds. Each shows strong, sharp absorption between 1700 and 1720 cm^{-1}, and strong, broad absorption over the region 2500–3000 cm^{-1}. Propose a structural formula for each compound.

(a) $C_{10}H_{12}O_3$ **(b)** $C_{10}H_{10}O_2$

^{1}H-NMR	^{13}C-NMR	^{1}H-NMR	^{13}C-NMR
2.49 (t, 2H)	173.89	2.34 (s, 3H)	167.82
2.80 (t, 2H)	157.57	6.38 (d, 1H)	143.82
3.72 (s, 3H)	132.62	7.18 (d, 1H)	139.96
6.78 (d, 2H)	128.99	7.44 (d, 2H)	131.45
7.11 (d, 2H)	113.55	7.56 (d, 2H)	129.37
12.40 (s, 1H)	54.84	12.00 (s, 1H)	127.83
	35.75		111.89
	29.20		21.13

Acidity of Phenols

20.32 Account for the fact that *p*-nitrophenol is a stronger acid than phenol.

$K_a = 1.1 \times 10^{-10}$ $K_a = 7.0 \times 10^{-8}$

20.33 Account for the fact that water-insoluble carboxylic acids (pK_a 4–5) dissolve in 10% aqueous sodium bicarbonate (pH 8.5) with the evolution of a gas but that water-insoluble phenols (pK_a 9.5–10.5) do not dissolve in 10% sodium bicarbonate.

20.34 Match each compound with its appropriate pK_a value.

(a) 4-Nitrobenzoic acid, benzoic acid, 4-chlorobenzoic acid
pK_a = 4.19, 3.98, and 3.41

(b) Benzoic acid, cyclohexanol, phenol
pK_a = 18.0, 9.95, and 4.19

(c) 4-Nitrobenzoic acid, 4-nitrophenol, 4-nitrophenylacetic acid
pK_a = 7.15, 3.85, and 3.41

20.35 Arrange the molecules and ions in each set in order of increasing acidity (from least acidic to most acidic).

(a)

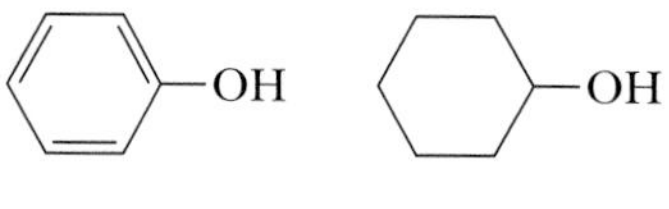

—OH —OH CH_3COOH

(b)

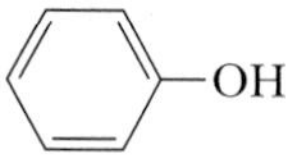

—OH HCO_3^- H_2O

(c)

—C≡CH —OH —CH_2OH

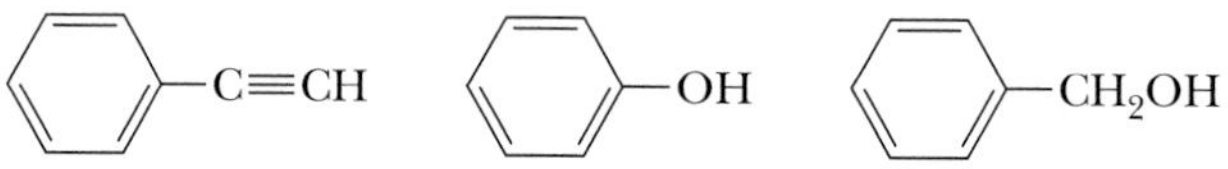

20.36 Explain the trends in the acidity of phenol and the monofluoro derivatives of phenol.

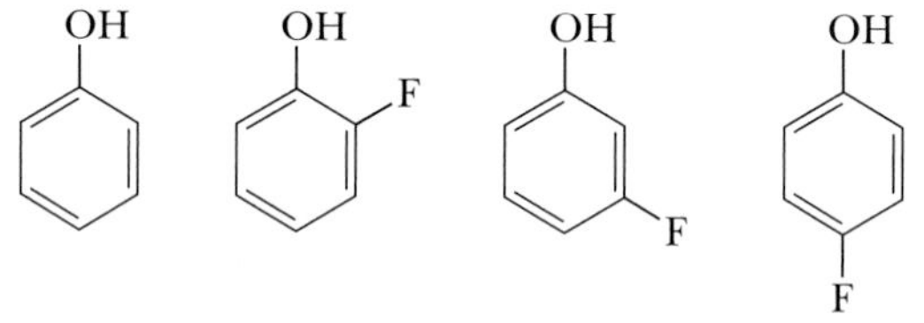

pK_a 10.0 pK_a 8.81 pK_a 9.28 pK_a 9.81

20.37 You wish to determine the inductive effects of a series of functional groups, for example Cl, Br, CN, COOH, and C_6H_5. Is it best to use a series of ortho-, meta-, or para-substituted phenols? Explain your answer.

20.38 From each pair, select the stronger base.

(a) 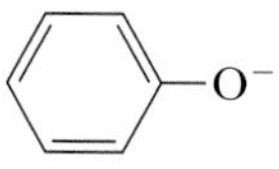 —O^- or OH^- (b) 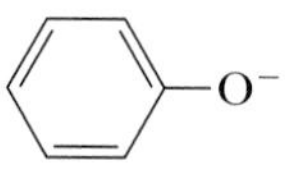 —O^- or —O^-

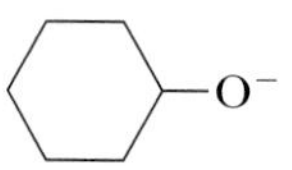

(c) 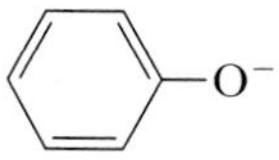 —O^- or HCO_3^- (d) 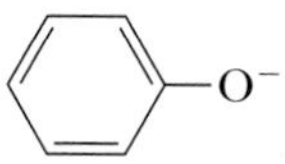 —O^- or CH_3COO^-

20.39 Describe a chemical procedure to separate a mixture of benzyl alcohol and *o*-cresol, and to recover each in pure form.

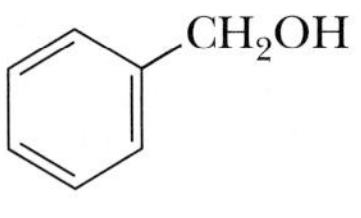

Benzyl alcohol 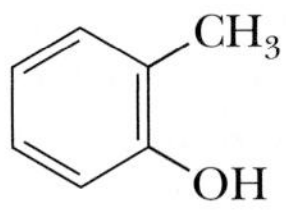   *o*-Cresol

20.40 The compound 2-hydroxypyridine, a derivative of pyridine, is in equilibrium with 2-pyridone. 2-Hydroxypyridine is aromatic. Does 2-pyridone have comparable aromatic character? Explain.

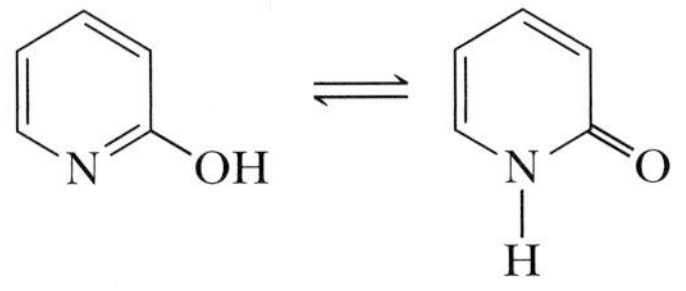

2-Hydroxypyridine   2-Pyridone

Reactions at the Benzylic Position

20.41 Write a balanced equation for the oxidation of *p*-xylene to 1,4-benzenedicarboxylic acid (terephthalic acid) using potassium dichromate in aqueous sulfuric acid. How many milligrams of H_2CrO_4 are required to oxidize 250 mg of *p*-xylene to terephthalic acid?

20.42 Each of the following reactions occurs by a radical chain mechanism. (Consult Appendix 3 for bond dissociation energies.)

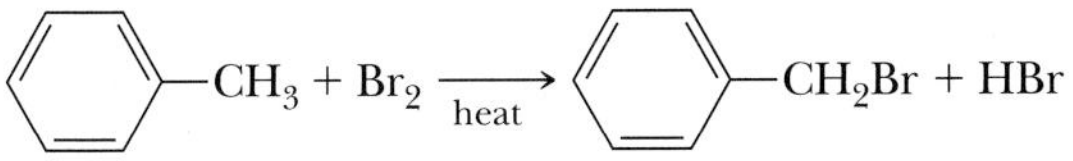

Toluene   Benzyl bromide

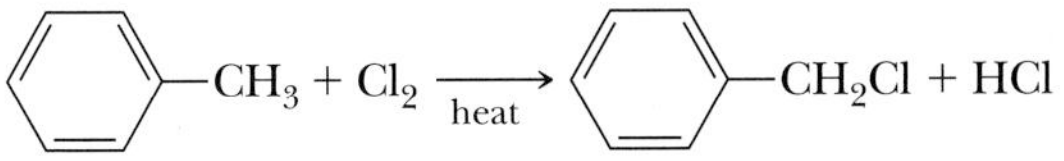

Toluene   Benzyl chloride

(a) Calculate the heat of reaction, ΔH^0, in kilojoules per mole for each reaction.
(b) Write a pair of chain propagation steps for each mechanism, and show that the net result of each pair is the observed reaction.
(c) Calculate ΔH^0 for each chain propagation step, and show that the sum for each pair of steps is identical with the ΔH^0 value calculated in part (a).

20.43 Following is an equation for iodination of toluene.

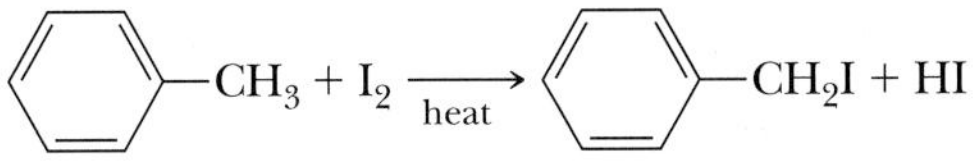

Toluene   Benzyl iodide

This reaction does not take place. All that happens under experimental conditions for the formation of radicals is initiation to form iodine radicals, $I\cdot$, followed by termination to reform I_2. How do you account for these observations?

20.44 Although most alkanes react with chlorine by a radical chain mechanism when reaction is initiated by light or heat, benzene fails to react under the same conditions. Benzene cannot be converted to chlorobenzene by treatment with chlorine in the presence of light or heat.

$$\text{C}_6\text{H}_5-\text{H} + \text{Cl}_2 \xrightarrow[\text{or heat}]{\text{light}} \text{C}_6\text{H}_5-\text{Cl} + \text{HCl}$$

(a) Explain why benzene fails to react under these conditions. Consult Appendix 3 for relevant bond dissociation energies.
(b) Explain why the bond dissociation energy of a C—H bond in benzene is significantly greater than that in alkanes.

20.45 Following is an equation for hydroperoxidation of cumene.

$$\text{C}_6\text{H}_5-\text{CH(CH}_3)_2 + \text{O}_2 \xrightarrow{\text{light}} \text{C}_6\text{H}_5-\overset{\overset{\displaystyle \text{OOH}}{|}}{\text{C}}(\text{CH}_3)_2$$

Cumene   Cumene hydroperoxide

Propose a radical chain mechanism for this reaction. Assume that initiation is by an unspecified radical, R·.

20.46 Para-substituted benzyl halides undergo reaction with methanol by an S_N1 mechanism to give a benzyl ether. Account for the following order of reactivity under these conditions.

$$\text{R}-\text{C}_6\text{H}_4-\text{CH}_2\text{Br} + \text{CH}_3\text{OH} \xrightarrow{\text{methanol}} \text{R}-\text{C}_6\text{H}_4-\text{CH}_2\text{OCH}_3 + \text{HBr}$$

Rate of S_N1 reaction: $R = CH_3O > CH_3 > H > NO_2$

20.47 When warmed in dilute sulfuric acid, 1-phenyl-1,2-propanediol undergoes dehydration and rearrangement to give 2-phenylpropanal.

1-Phenyl-1,2-propanediol   2-Phenylpropanal

(a) Propose a mechanism for this example of a pinacol rearrangement (Section 9.7).
(b) Account for the fact that 2-phenylpropanal is formed rather than its constitutional isomer, 1-phenyl-1-propanone.

20.48 In the chemical synthesis of DNA and RNA, hydroxyl groups are normally converted to triphenylmethyl (trityl) ethers to protect the hydroxyl group from reaction with other reagents.

$$\text{RCH}_2\text{OH} + \text{Ph}_3\text{CCl} \xrightarrow{\text{tertiary amine}} \text{RCH}_2\text{OCPh}_3 + \text{HCl}$$

neutralized by the tertiary amine

Triphenylmethyl chloride   A triphenylmethyl ether
(Trityl chloride)   (A trityl ether)

Triphenylmethyl ethers are stable to aqueous base but are rapidly cleaved in aqueous acid.

$$RCH_2OCPh_3 + H_2O \xrightarrow{H^+} RCH_2OH + Ph_3COH$$

(a) Why are triphenylmethyl ethers so readily hydrolyzed by aqueous acid?

(b) How might the structure of the triphenylmethyl group be modified to increase or decrease its acid sensitivity?

Synthesis

20.49 Using ethylbenzene as the only aromatic starting material, show how to synthesize the following compounds. In addition to ethylbenzene, use any other necessary organic or inorganic chemicals. Note that any compound already synthesized in one part of this problem may then be used to make any other compound in the problem.

20.50 Show how to convert 1-phenylpropane into the following compounds. In addition to this starting material, use any necessary inorganic reagents. Any compound synthesized in one part of this problem may be used to make any other compound in the problem.

20.51 Carbinoxamine is a histamine antagonist, specifically an H_1-antagonist. The maleic acid salt of the levorotatory isomer is sold as the prescription drug Rotoxamine. Given this retrosynthetic analysis, propose a synthesis for carbinoxamine. (*Note:* Aryl bromides form Grignard reagents much more readily than aryl chlorides.)

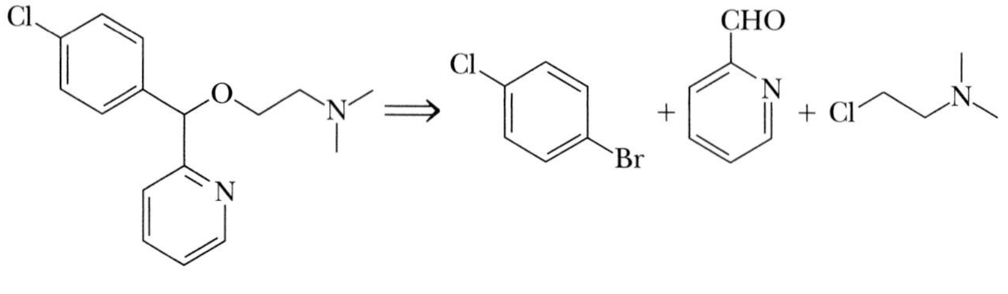

Carbinoxamine

20.52 Cromolyn sodium, developed in the 1960s, is used to prevent allergic reactions primarily affecting the lungs, as for example exercise-induced emphysema. It is thought to block the release of histamine, which prevents the sequence of events leading to swelling, itching, and constriction of bronchial tubes. Cromolyn sodium is synthesized in the following series of steps. Treatment of one mole of epichlorohydrin with two moles of 2,6-dihydroxyacetophenone in the presence of base gives I. Treatment of I with two moles of diethyl oxalate in the presence of sodium ethoxide gives a diester II. Saponification of the diester with aqueous NaOH gives cromolyn sodium.

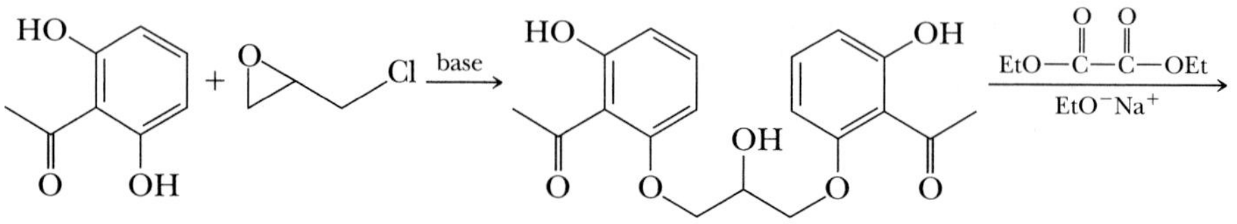

2,6-Dihydroxy-   Epichloro-   I
acetophenone   hydrin

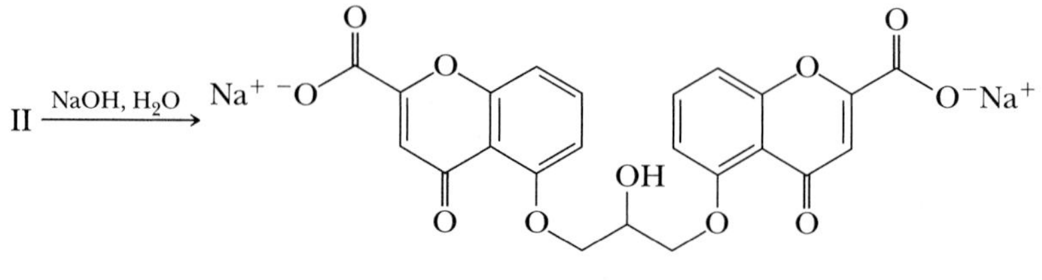

Cromolyn sodium

(a) Propose a mechanism for the formation of compound I.
(b) Propose a structural formula for compound II and a mechanism for its formation.

20.53 The following stereospecific synthesis is part of the scheme used by E. J. Corey of Harvard University in the synthesis of erythronolide B, the precursor of the erythromycin antibiotics. In this remarkably simple set of reactions, the relative configurations of five stereocenters are established. [See E. J. Corey, K. C. Nicolaou, and L. S. Melvin, Jr., *J. Am. Chem. Soc.,* **97:** 654 (1975).]

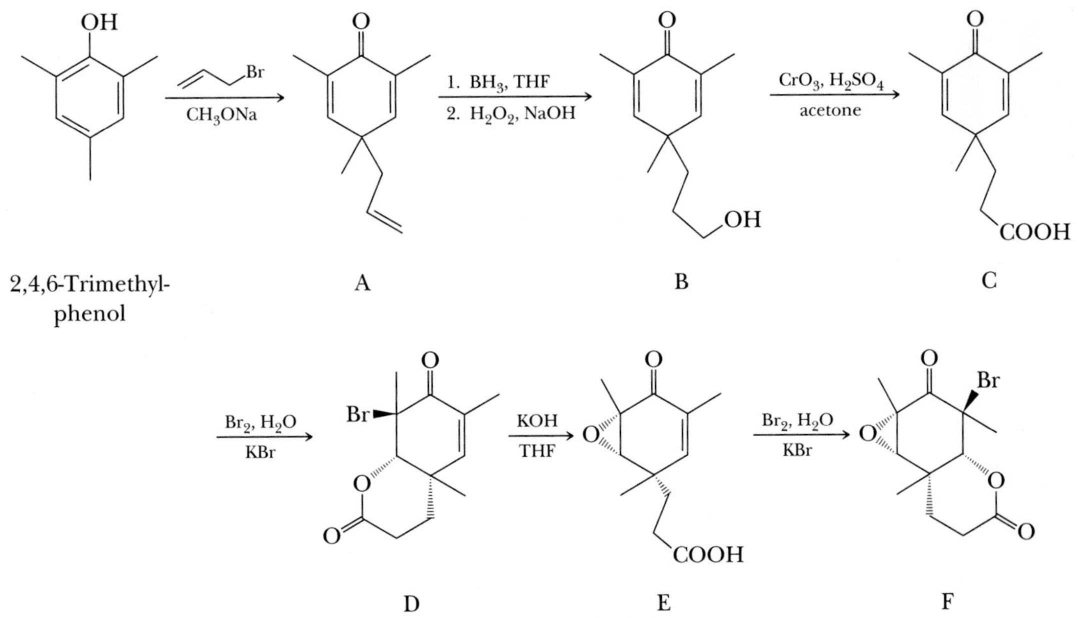

2,4,6-Trimethyl-
phenol

A

B

C

D

E

F

(a) Propose a mechanism for the conversion of 2,4,6-trimethylphenol to compound A.
(b) Account for the stereospecificity and regiospecificity of the three steps in the conversion of compound C to compound F.
(c) Is compound F produced in this synthesis as a single enantiomer or as a racemic mixture? Explain.

21

AROMATICS II: REACTIONS OF BENZENE AND ITS DERIVATIVES

21.1 Electrophilic Aromatic Substitution

21.2 Disubstitution and Polysubstitution

21.3 Nucleophilic Aromatic Substitution

By far the most characteristic reaction of aromatic compounds is substitution at a ring carbon. In this reaction, one of the ring hydrogens is substituted. Some groups that can be introduced directly on the ring are the halogens, the nitro ($-NO_2$) group, the sulfonic acid ($-SO_3H$) group, alkyl ($-R$) groups, and acyl ($RCO-$) groups. Each of these substitution reactions is represented in the following equations.

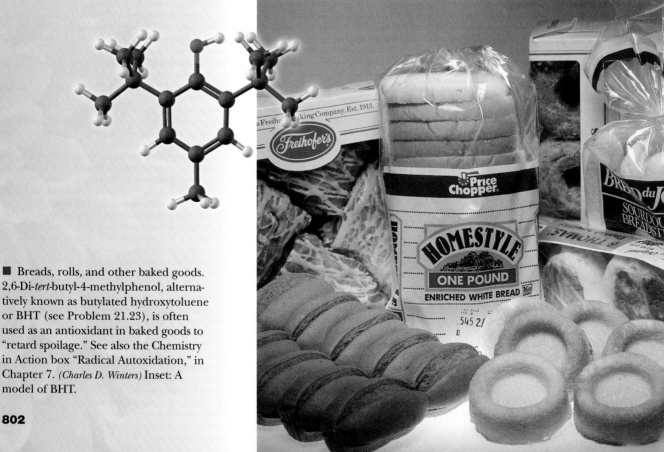

■ Breads, rolls, and other baked goods. 2,6-Di-*tert*-butyl-4-methylphenol, alternatively known as butylated hydroxytoluene or BHT (see Problem 21.23), is often used as an antioxidant in baked goods to "retard spoilage." See also the Chemistry in Action box "Radical Autoxidation," in Chapter 7. *(Charles D. Winters)* Inset: A model of BHT.

Halogenation:

$$\text{C}_6\text{H}_5\text{—H} + \text{Cl}_2 \xrightarrow{\text{FeCl}_3} \text{C}_6\text{H}_5\text{—Cl} + \text{HCl}$$

Chlorobenzene

Nitration:

$$\text{C}_6\text{H}_5\text{—H} + \text{HNO}_3 \xrightarrow{\text{H}_2\text{SO}_4} \text{C}_6\text{H}_5\text{—NO}_2 + \text{H}_2\text{O}$$

Nitrobenzene

Sulfonation:

$$\text{C}_6\text{H}_5\text{—H} + \text{SO}_3 \xrightarrow{\text{H}_2\text{SO}_4} \text{C}_6\text{H}_5\text{—SO}_3\text{H}$$

Benzenesulfonic acid

Alkylation:

$$\text{C}_6\text{H}_5\text{—H} + \text{RX} \xrightarrow{\text{AlX}_3} \text{C}_6\text{H}_5\text{—R} + \text{HX}$$

An alkylbenzene

Acylation:

$$\text{C}_6\text{H}_5\text{—H} + \text{RC}(\text{O})\text{X} \xrightarrow{\text{AlX}_3} \text{C}_6\text{H}_5\text{—C}(\text{O})\text{R} + \text{HX}$$

An acylbenzene

We take these reactions one at a time and examine their common mechanistic theme.

21.1   Electrophilic Aromatic Substitution

Supporting Concepts
Electrophilic Aromatic
Substitution
This section of the CD-ROM
contains an overview of
aromatic substitution
reactions.

In this section, we study several types of **electrophilic aromatic substitution,** that is, reactions in which a hydrogen atom of an aromatic ring is replaced by an electrophile, E^+.

Electrophilic aromatic substitution   A reaction in which there is substitution of an electrophile, E^+, for a hydrogen on an aromatic ring.

$$\text{C}_6\text{H}_5\text{—H} + \text{E}^+ \longrightarrow \text{C}_6\text{H}_5\text{—E} + \text{H}^+$$

We study several common types of electrophiles, how each is generated, and the mechanism by which it replaces hydrogen on an aromatic ring.

A. Chlorination and Bromination

Chlorine alone does not react with benzene, in contrast to its instantaneous addition to cyclohexene. However, in the presence of a Lewis acid catalyst, such as ferric chloride or aluminum chloride (Section 4.5), chlorine reacts to give chlorobenzene and HCl. This reaction begins with interaction of chlorine and the Lewis acid catalyst to give a molecular complex with a positive charge on chlorine and a negative charge on iron. Redistribution of electrons in this complex generates a **chloronium ion,** Cl^+, as part of an ion pair. Reaction of the Cl_2—$FeCl_3$ complex with the pi electron cloud of the aromatic ring forms a resonance-stabilized cation intermediate, here represented as a hybrid of three contributing structures. Proton transfer from the cation intermediate to $FeCl_4^-$ forms HCl, regenerates the Lewis acid catalyst, and gives chlorobenzene.

Mechanisms
Aromatics
Electrophilic aromatic
substitution with Cl

Mechanism   Electrophilic Aromatic Substitution — Chlorination

Step 1: Reaction between chlorine (a Lewis base) and iron(III) chloride (a Lewis acid) gives an ion pair containing a chloronium ion.

Chlorine (a Lewis base) · Ferric chloride (a Lewis acid) · A molecular complex with a positive charge on chlorine and a negative charge on iron · An ion pair containing the chloronium ion

Step 2: Attack of the chloronium ion (a Lewis acid) on the pi system (a Lewis base) of the aromatic ring gives a resonance-stabilized cation intermediate.

Resonance-stabilized cation intermediate; the positive charge is delocalized onto three atoms of the ring

Step 3: Proton transfer from the cation intermediate regenerates the aromatic ring.

Cation intermediate · Chlorobenzene

Treatment of benzene with bromine in the presence of ferric chloride or aluminum chloride gives bromobenzene and HBr. The mechanism of this reaction is the same as that for the chlorination of benzene.

We can write the following general two-step mechanism for electrophilic aromatic substitution. The first and rate-determining step is attack of the electrophile, E^+, on the aromatic ring to give a resonance-stabilized cation intermediate. The second and faster step, loss of H^+ from the cation intermediate, regenerates the aromatic ring and gives the product.

Step 1:

Electrophile   Resonance-stabilized
                                 cation intermediate

Step 2:

The major difference between addition of halogen to an alkene and substitution by halogen on an aromatic ring centers on the fate of the cationic intermediate formed in the first step of each reaction. Recall from Section 6.3D that addition of chlorine to an alkene is a two-step process, the first and slower step of which is formation of a bridged chloronium ion intermediate. This intermediate then reacts with chloride ion to complete the addition. With aromatic compounds, the cationic intermediate loses H^+ to regenerate the aromatic ring and regain the large resonance stabilization. There is no such resonance stabilization to be regained in the case of an alkene. The energy diagram in Figure 21.1 shows both addition and substitution reactions of benzene. Addition causes loss of the aromatic resonance energy and is disfavored except under extreme circumstances.

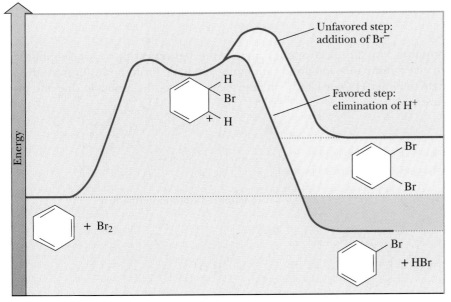

Reaction coordinate

Figure 21.1

Energy diagram for the reaction of benzene with bromine. Reaction of the cationic intermediate to form an addition product results in loss of the resonance stabilization of the aromatic ring. Formation of a substitution product regenerates the resonance-stabilized aromatic ring.

B.   Nitration and Sulfonation

The sequence of steps for nitration and sulfonation of benzene is similar to that for chlorination and bromination. For nitration, the electrophile is the **nitronium ion,** NO_2^+, generated by reaction of nitric acid and sulfuric acid.

Mechanism   Formation of the Nitronium Ion

Step 1:   Proton transfer from sulfuric acid to the OH group of nitric acid gives the conjugate acid of nitric acid.

$$H-\overset{..}{\underset{..}{O}}-NO_2 + H-O-SO_3H \rightleftharpoons H-\overset{\overset{H}{|}}{\underset{..}{\overset{+}{O}}}-NO_2 + HSO_4^-$$

Nitric acid   Conjugate acid
of nitric acid

Step 2:   Loss of water from this conjugate acid gives the nitronium ion.

$$H-\overset{\overset{H}{|}}{\underset{..}{\overset{+}{O}}}-NO_2 \rightleftharpoons H-\overset{\overset{H}{|}}{\underset{..}{O}}: + :\overset{..}{O}=\overset{+}{N}=\overset{..}{O}:$$

The nitronium ion

Example 21.1

Write a stepwise mechanism for the nitration of benzene.

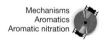

Solution

The nitronium ion (an electrophile) attacks the benzene ring (a nucleophile) in Step 1 to give a resonance-stabilized cation intermediate. Proton transfer from this intermediate to either H_2O or HSO_4^- in Step 2 regenerates the aromatic ring and gives nitrobenzene.

Step 1:

Resonance-stabilized cation intermediate

Step 2: $H_2\overset{..}{O}: +$

Nitrobenzene

A particular value of nitration is that the resulting nitro group can be reduced to a primary amino group, $-NH_2$, by hydrogenation in the presence of a transition metal catalyst such as nickel, palladium, or platinum under fairly mild conditions. This method has the potential disadvantage that other susceptible groups such as carbon-carbon double bonds and aldehyde and ketone carbonyl groups may also be reduced. Note that neither the $-COOH$ nor the aromatic ring is reduced under these conditions.

4-Nitrobenzoic acid   4-Aminobenzoic acid

Alternatively, a nitro group can be reduced to a primary amino group by a metal in acid. The most commonly used metal-reducing agents are iron, zinc, and tin in dilute HCl. When reduced with a metal and hydrochloric acid, the amine is obtained as a salt, which is then treated with strong base to liberate the free amine. The reductant is electrons from the metal.

2,4-Dinitrotoluene   4-Methyl-1,3-benzenediamine
                                            (2,4-Diaminotoluene)

Sulfonation of benzene is carried out using concentrated sulfuric acid containing dissolved sulfur trioxide (fuming sulfuric acid). In the following equation, the sulfonating agent is shown as sulfur trioxide. The electrophile is either SO_3 or HSO_3^+ depending on experimental conditions.

Benzene   Benzenesulfonic acid

Problem 21.1

Write the stepwise mechanism for sulfonation of benzene by hot, concentrated sulfuric acid. In this reaction, the electrophile is SO_3 formed as shown in the following equation.

$$H_2SO_4 \rightleftharpoons SO_3 + H_2O$$

C. Friedel-Crafts Alkylation and Acylation

Alkylation of aromatic hydrocarbons was discovered in 1877 by the French chemist Charles Friedel and a visiting American chemist, James Crafts. They discovered that mixing benzene, an alkyl halide, and $AlCl_3$ results in formation of an alkylbenzene

Friedel-Crafts reaction An electrophilic aromatic substitution in which a hydrogen of an aromatic ring is replaced by an alkyl or acyl group.

and HX. **Friedel-Crafts alkylation** forms a new carbon-carbon bond between benzene and an alkyl group, as illustrated by the reaction of benzene with 2-chloropropane in the presence of aluminum chloride.

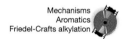

Benzene   2-Chloropropane   Cumene
           (Isopropyl chloride)   (Isopropylbenzene)

Friedel-Crafts alkylation is among the most important methods for forming new carbon-carbon bonds to aromatic rings. It begins with formation of a complex between the alkyl halide and aluminum chloride, in which aluminum has a negative charge, and the halogen of the alkyl halide has a positive charge. The alkyl group can also be written as a carbocation, although it is unlikely that a free carbocation is actually formed, especially in the case of the relatively unstable primary and secondary carbocations. Nonetheless, we very often represent the reactive intermediate as a carbocation to simplify the mechanism. Reaction of an alkyl carbocation with an aromatic ring gives a resonance-stabilized cation intermediate, which then loses a hydrogen to give an alkylbenzene.

Mechanisms
Aromatics
Friedel-Crafts alkylation

Mechanism   Friedel-Crafts Alkylation

Step 1:   Reaction of the alkyl halide (a Lewis base) with aluminum chloride (a Lewis acid) gives an ion pair containing a carbocation.

$$R-\overset{..}{\underset{..}{Cl}}: + \overset{Cl}{\underset{Cl}{\overset{|}{\underset{|}{Al}}}}-Cl \;\rightleftharpoons\; R-\overset{+}{\underset{..}{\overset{..}{Cl}}}-\overset{Cl}{\underset{Cl}{\overset{|}{\underset{|}{\overset{-}{Al}}}}}-Cl \;\rightleftharpoons\; R^+ \; AlCl_4^-$$

A molecular complex   An ion pair
with a positive charge on   containing
chlorine and a negative   a carbocation
charge on aluminum

Step 2:   Reaction of the alkyl carbocation (a Lewis acid) with the pi cloud (a Lewis base) of the aromatic ring gives a resonance-stabilized cation intermediate.

Delocalization of the positive
charge onto three atoms of the ring

Step 3:   Proton transfer regenerates the aromatic character of the ring.

Vinylic and aryl halides do not react under conditions of the Friedel-Crafts alkylation because of the high activation energy required to form vinylic and aryl carbocations.

There are three major limitations on Friedel-Crafts alkylation. The first is the possibility for rearrangement of the alkyl group, which occurs in the following way. Friedel-Crafts alkylation involves the generation of a carbocation, and, as we have already seen in Section 6.3C, carbocations may rearrange to more stable carbocations. Carbocation rearrangements are common in Friedel-Crafts alkylations. For example, reaction of benzene with 1-chloro-2-methylpropane (isobutyl chloride) gives only 2-methyl-2-phenylpropane (*tert*-butylbenzene).

Benzene	1-Chloro-2-methylpropane (Isobutyl chloride)	2-Methyl-2-phenylpropane (*tert*-Butylbenzene)

In this case, the isobutyl chloride/AlCl$_3$ complex rearranges directly to the *tert*-butyl cation/AlCl$_4^-$ ion pair, which is the electrophile in this example of Friedel-Crafts alkylation.

Isobutyl chloride	Isobutyl chloride–aluminum chloride complex	*tert*-Butyl cation/ AlCl$_4^-$ ion pair

In practice, alkylation with primary halides is not useful, and alkylbenzenes containing a primary alkyl group other than —CH$_2$CH$_3$ must be prepared by other means. Alkylation is useful for introducing isopropyl, *tert*-butyl, and other alkyl groups, the cations of which tend not to rearrange.

A second limitation on Friedel-Crafts alkylation is that it fails altogether on benzene rings bearing one or more strongly electron-withdrawing groups. We shall see in the following section that substituents of this type on a benzene ring have a dramatic effect on the ring's reactivity toward further electrophilic aromatic substitution.

When Y Equals Any of These Groups, the Benzene Ring Does Not Undergo Friedel-Crafts Alkylation

---

O‖ —CH	O‖ —CR	O‖ —COH	O‖ —COR	O‖ —CNH$_2$
—SO$_3$H	—C≡N	—NO$_2$	—NR$_3^+$	
—CF$_3$	—CCl$_3$			

The third limitation on Friedel-Crafts alkylation is that it is hard to stop the reaction at monoalkylation because the product is more reactive than the starting material. We will discuss reactivity in detail in Section 21.2, but, in general, alkylated benzenes are more reactive than unsubstituted compounds. This limitation can be overcome if it is feasible to use a large excess of the aromatic compound.

Friedel and Crafts also discovered that treatment of an aromatic hydrocarbon with an acyl halide (Section 18.1A) in the presence of aluminum chloride gives a ketone. An RCO— group is known as an acyl group; hence, reaction of an aromatic hydrocarbon with an acyl halide is known as **Friedel-Crafts acylation,** as illustrated by the reaction of benzene and acetyl chloride in the presence of aluminum chloride to form acetophenone. Because the product is less reactive than the starting material (unreactive in most cases) this overcomes the third limitation of the alkylation.

$$\text{Benzene} \quad + \quad CH_3\overset{\overset{\displaystyle O}{\|}}{C}Cl \quad \xrightarrow{\text{AlCl}_3} \quad \text{Acetophenone} \quad + \text{HCl}$$

Benzene   Acetyl chloride   Acetophenone
        (an acyl halide)   (a ketone)

The following example of electrophilic aromatic substitution involves intramolecular acylation to form a six-membered ring.

4-Phenylbutanoyl   α-Tetralone
chloride

Acylium ion A resonance-stabilized cation with the structure $[RC{=}O]^+$ or $[ArC{=}O]^+$. The positive charge is delocalized over both the carbonyl carbon and the carbonyl oxygen.

Friedel-Crafts acylation begins with donation of a pair of electrons from the halogen of the acyl halide to aluminum chloride to form a molecular complex similar to what we drew for Friedel-Crafts alkylations. In this complex, halogen has a positive formal charge and aluminum has a negative formal charge. Redistribution of electrons of the carbon-chlorine bond gives an ion pair containing an **acylium ion.**

Mechanisms
Aromatics
Friedel-Crafts acylation

Mechanism   Friedel-Crafts Acylation — Generation of an Acylium Ion

Step 1: Reaction between the halogen atom of the acid chloride (a Lewis base) and aluminum chloride (a Lewis acid) gives a molecular complex.

Step 2: Redistribution of valence electrons of this molecular complex gives an ion pair containing an acylium ion.

| An acyl chloride (a Lewis base) | Aluminum chloride (a Lewis acid) | A molecular complex with a positive charge on chlorine and a negative charge on aluminum | An ion pair containing an acylium ion |

Of the two major contributing structures that can be drawn for an acylium ion, the one with complete valence shells for both carbon and oxygen makes the greater contribution to the hybrid.

Both atoms have complete valence shells.

$$R-\overset{+}{C}=\ddot{O}: \longleftrightarrow R-C\equiv\overset{+}{O}:$$

The more important
contributing structure

Friedel-Crafts acylation is free of a second major limitation on Friedel-Crafts alkylations: acyl cations do not undergo rearrangement. Thus, the carbon skeleton of an acyl halide is transferred unchanged to the aromatic ring.

Example 21.2

Write structural formulas for the products from Friedel-Crafts alkylation or acylation of benzene with the following compounds.

(a) $C_6H_5CH_2Cl$ (b) $\overset{O}{\overset{\|}{C_6H_5CCl}}$

Benzyl chloride   Benzoyl chloride

Solution

(a) Benzyl chloride in the presence of a Lewis acid catalyst gives the benzyl cation, which then attacks benzene followed by proton transfer to give diphenylmethane. In this example, the benzyl cation, although primary, cannot rearrange.

Benzyl cation   Diphenylmethane

(b) Treatment of benzoyl chloride with aluminum chloride gives an acylium ion.

Reaction of this cation with the pi electrons of the aromatic ring followed by proton transfer gives benzophenone.

Benzoyl cation   Benzophenone

Problem 21.2

Write structural formulas for the products from Friedel-Crafts alkylation or acylation of benzene with the following compounds.

(a)   (b)   (c) Ph

A special value of Friedel-Crafts acylations in synthesis is the preparation of unrearranged alkylbenzenes, as illustrated by the preparation of isobutylbenzene. Treatment of benzene with 2-methylpropanoyl chloride in the presence of aluminum chloride gives 2-methyl-1-phenyl-1-propanone. Wolff-Kishner or Clemmensen reduction of the carbonyl group to a methylene group (Section 16.14C) gives isobutylbenzene.

2-Methyl-   2-Methyl-1-   Isobutylbenzene
propanoyl   phenyl-1-propanone
chloride

D.   Other Electrophilic Aromatic Alkylations

After the discovery that Friedel-Crafts alkylations and acylations involve cationic electrophiles, it was realized that the same reactions can be accomplished by other combinations of reagents and catalysts. We study two of these: generation of carbocations from alkenes and from alcohols.

As we saw in Section 6.3, treatment of an alkene with a strong acid, most commonly HX, H_2SO_4, H_3PO_4, or HF/BF_3, generates a carbocation. Cumene is synthesized industrially (over 2.8×10^9 kg annually in the United States) by the reaction of benzene with propene in the presence of phosphoric acid as a catalyst.

Benzene   Propene   Cumene

Alkylation with an alkene can also be carried out with a Lewis acid catalyst. Treatment of benzene with cyclohexene in the presence of aluminum chloride gives phenylcyclohexane.

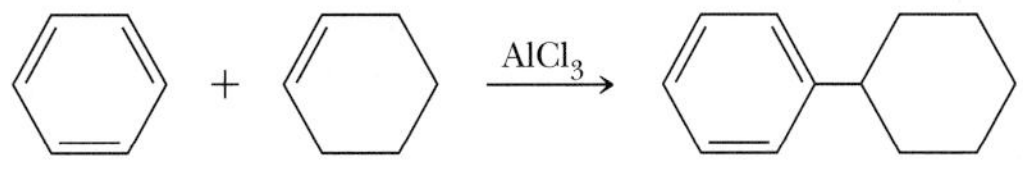

Benzene   Cyclohexene   Phenylcyclohexane

Carbocations can also be generated by treatment of an alcohol with H_2SO_4, H_3PO_4, or HF (Section 9.6).

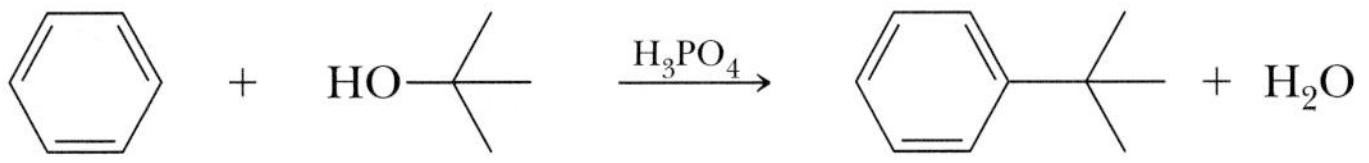

Benzene   2-Methyl-2-propanol   2-Methyl-2-
          (*tert*-Butyl alcohol)   phenylpropane
                                         (*tert*-Butylbenzene)

Example 21.3

Write a mechanism for the formation of isopropylbenzene (cumene) from benzene and propene in the presence of phosphoric acid.

Solution

Step 1:   Proton transfer from phosphoric acid to propene gives the isopropyl cation.

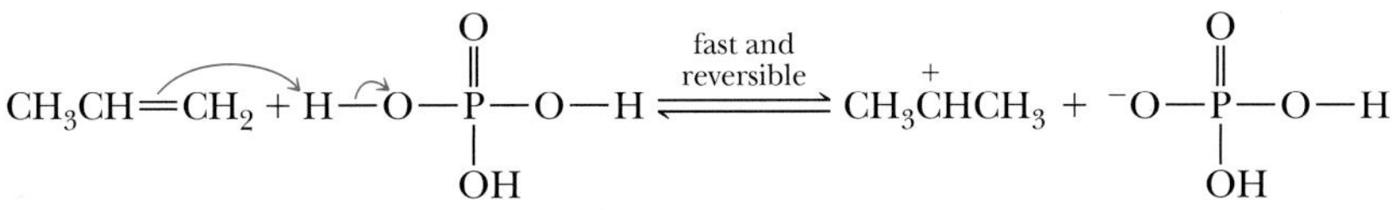

Step 2:   Reaction of the isopropyl cation with the pi electrons of the benzene ring gives a resonance-stabilized carbocation intermediate.

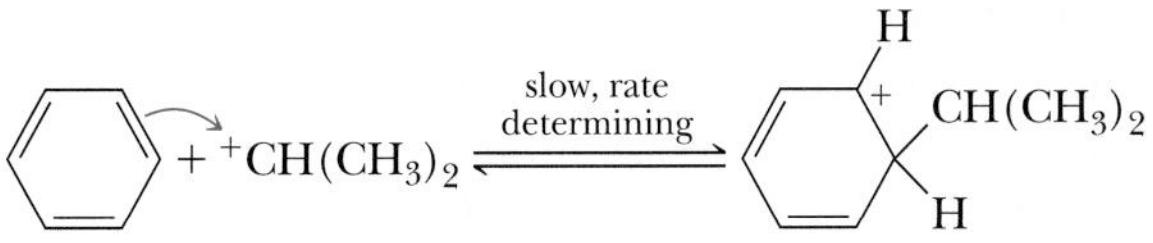

Step 3:   Proton transfer to dihydrogen phosphate ion gives cumene.

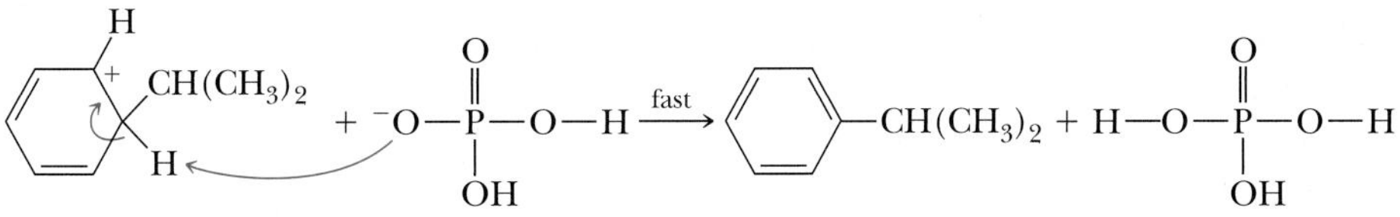

Cumene

Problem 21.3

Write a mechanism for the formation of *tert*-butylbenzene from benzene and *tert*-butyl alcohol in the presence of phosphoric acid.

21.2 Disubstitution and Polysubstitution

A. Effects of a Substituent Group on Further Substitution

See both:
Reactivity Explorer
Aromatics
and
Supporting Concepts
Electrophilic Aromatic Substitution
Orientation

In electrophilic aromatic substitution of a monosubstituted benzene, three products are possible: the new group may become oriented ortho, meta, or para to the existing group. We can make the following generalizations about the manner in which existing groups influence further substitution reactions.

1. Substituents affect the orientation of new groups. Certain substituents direct an incoming group preferentially to the ortho and para positions; other substituents direct it preferentially to the meta position. In other words, substituents on a benzene ring can be classified as **ortho-para directing** or as **meta directing.**
2. Substituents affect the rate of further substitution. Certain substituents cause the rate of a second substitution to be greater than that for benzene itself, whereas other substituents cause the rate of a second substitution to be lower than that for benzene. In other words, groups on a benzene ring can be classified as **activating** or **deactivating** toward further substitution.

Activating group Any substituent on a benzene ring that causes the rate of electrophilic aromatic substitution to be greater than that for benzene.

Deactivating group Any substituent on a benzene ring that causes the rate of electrophilic aromatic substitution to be lower than that for benzene.

These directing and activating-deactivating effects can be seen by comparing the products and rates of bromination of anisole and nitration of nitrobenzene. Bromination of anisole proceeds at a rate considerably greater than that for benzene (the methoxy group is activating), and the product is a mixture of *o*-bromoanisole and *p*-bromoanisole (the methoxy group is ortho-para directing).

$$\text{Anisole} \quad + \text{Br}_2 \xrightarrow[\text{CH}_3\text{COOH}]{} \text{o-Bromoanisole} \quad + \quad \text{p-Bromoanisole} \quad + \quad \text{HBr}$$

Anisole   *o*-Bromoanisole   *p*-Bromoanisole
(4%)   (96%)

Quite another situation is seen in the nitration of nitrobenzene. First, the reaction requires the more reactive fuming nitric acid and a higher temperature than for benzene. Because nitration of nitrobenzene proceeds much more slowly than nitration of benzene itself, we say that a nitro group is strongly deactivating. Second, the product formed consists of approximately 93% of the meta isomer and less than 7% of the ortho and para isomers combined (the nitro group is meta directing).

$$\text{Nitrobenzene} \quad + \quad \text{HNO}_3 \xrightarrow[\text{100°C}]{\text{H}_2\text{SO}_4} \text{m-Dinitrobenzene} \quad + \quad \text{o-Dinitrobenzene} \quad + \quad \text{p-Dinitrobenzene} \quad + \quad \text{H}_2\text{O}$$

Nitrobenzene   *m*-Dinitrobenzene   *o*-Dinitrobenzene   *p*-Dinitrobenzene
(93%)
(less than 7% combined)

Table 21.1   Effects of Substituents on Further Electrophilic Aromatic Substitution

Ortho-Para Directing	Strongly activating	—N̈H₂	—N̈HR	—N̈R₂	—Ö̈H	—Ö̈R		
	Moderately activating	$\overset{O}{\overset{\|}{—N̈HCR}}$	$\overset{O}{\overset{\|}{—N̈HCAr}}$	$\overset{O}{\overset{\|}{—Ö̈CR}}$	$\overset{O}{\overset{\|}{—Ö̈CAr}}$			
	Weakly activating	—R	⟨benzene ring⟩					
	Weakly deactivating	—F̈:	—C̈l:	—B̈r:	—Ï:			
Meta Directing	Moderately deactivating	$\overset{O}{\overset{\|}{—CH}}$	$\overset{O}{\overset{\|}{—CR}}$	$\overset{O}{\overset{\|}{—COH}}$	$\overset{O}{\overset{\|}{—COR}}$	$\overset{O}{\overset{\|}{—CNH_2}}$	$\overset{O}{\underset{O}{\overset{\|}{\underset{\|}{—SOH}}}}$	—C≡N
	Strongly deactivating	—NO₂	—NH₃⁺	—CF₃	—CCl₃			

Relative importance in directing further substitution →

Listed in Table 21.1 are the directing and activating-deactivating effects for the major functional groups with which we are concerned in this text. If we compare these ortho-para and meta directors for structural similarities and differences, we can make the following generalizations.

1. Alkyl groups, phenyl groups, and substituents in which the atom bonded to the ring has an unshared pair of electrons are ortho-para directing. All other substituents are meta directing.
2. All ortho-para directing groups except the halogens are activating. The halogens are weakly deactivating.

The fact that alkyl groups are weakly activating is why it is difficult to stop Friedel-Crafts alkylations at monoalkylation. When a first alkyl group is introduced onto an aromatic ring, the ring is activated toward further alkylation and, unless reaction conditions are very carefully controlled, a mixture of di-, tri-, and polyalkylation products is formed. Friedel-Crafts acylations, on the other hand, never go beyond monoacylation because an acyl group is deactivating toward further substitution.

We can illustrate the usefulness of these generalizations by considering the synthesis of two different disubstituted derivatives of benzene. Suppose we wish to prepare *m*-bromonitrobenzene from benzene. This conversion can be done in two steps: nitration and bromination. If the steps are carried out in just that order, the major product is indeed *m*-bromonitrobenzene. The nitro group is a meta director and, therefore, directs bromination to a meta position.

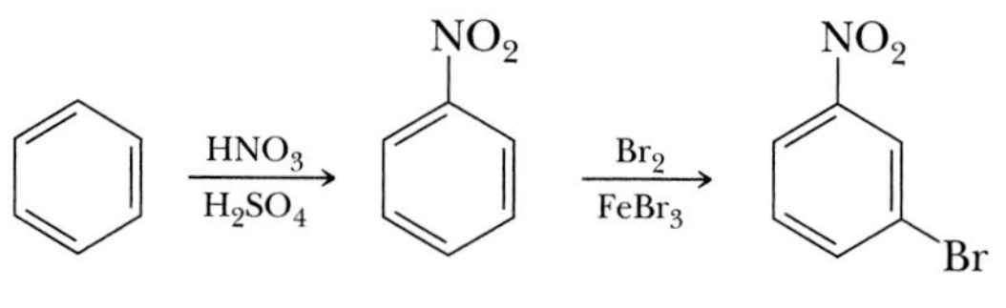

Nitrobenzene   m-Bromonitro-
                   benzene

If, however, we reverse the order of the steps and first form bromobenzene, we now have an ortho-para directing group on the ring, and nitration takes place preferentially at the ortho and para positions.

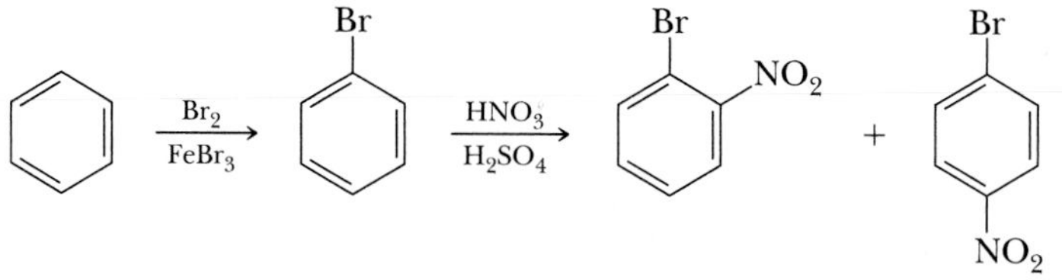

Bromobenzene   o-Bromonitro-   p-Bromonitro-
                        benzene   benzene

As another example of the importance of the order in electrophilic aromatic substitutions, consider the conversion of toluene to *p*-nitrobenzoic acid. The nitro group can be introduced with a nitrating mixture of nitric and sulfuric acids. The carboxyl group can be produced by oxidation of the methyl group of toluene (Section 20.6A).

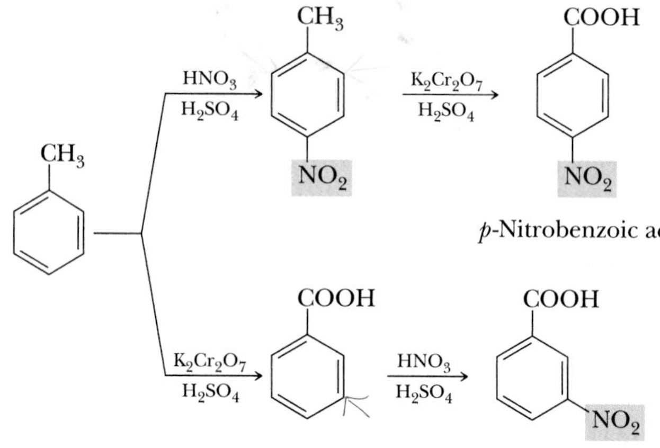

p-Nitrobenzoic acid

m-Nitrobenzoic acid

Nitration of toluene yields a product with the two substituents in the desired para relationship. Nitration of benzoic acid, on the other hand, yields a product with the substituents meta to each other. Again, we see that the order in which the reactions are performed is critical.

Note that, in this last example, we showed nitration of toluene producing only the para isomer. Because methyl is an ortho-para directing group, both the ortho and para isomers are formed; usually the para dominates. In problems of this type in which you are asked to prepare the para isomer, assume that they are both formed but that there are physical methods by which they can be separated and the desired

isomer obtained. The ortho isomer is harder to prepare pure, and special techniques, often using a blocking group, are usually required.

Example 21.4

Complete these electrophilic aromatic substitution reactions. Where you predict
meta substitution, show only the meta product. Where you predict ortho-para substitution, show both products.

(a) [Br-substituted benzene] $+ H_2SO_4$ $\xrightarrow{\text{heat}}$ (b) [SO_3H-substituted benzene] $+ HNO_3$ $\xrightarrow{H_2SO_4}$

Solution

Bromine in (a) is ortho-para directing and weakly deactivating. The sulfonic acid
group in (b) is meta directing and moderately deactivating.

(a) [o-Bromobenzenesulfonic acid structure with Br and SO_3H] + [p-Bromobenzenesulfonic acid structure with Br and SO_3H] (b) [m-Nitrobenzenesulfonic acid structure with SO_3H and NO_2]

o-Bromobenzene- p-Bromobenzene- m-Nitrobenzene-
sulfonic acid sulfonic acid   sulfonic acid

Problem 21.4

Draw structural formulas for the product of nitration of each compound. Where you
predict ortho-para substitution, show both products.

(a) [benzene with $-COCH_3$ group, C=O double bond] (b) [benzene with $-OCCH_3$ group, C=O double bond]

B. Theory of Directing Effects

As we have just seen, a group on a benzene ring exerts a major effect on the pattern
of further substitution. We can account for these patterns by starting with the general
mechanism first presented in Section 21.1 for electrophilic aromatic substitution and
carrying it a step further to consider how groups already present on the ring affect
the energetics of further substitution. Specifically, we need to consider both resonance and inductive effects and the relative importance of each.

Nitration of Anisole

The rate of electrophilic aromatic substitution is limited by the slowest step in the
mechanism. For nitration of anisole, and for almost every other substitution we

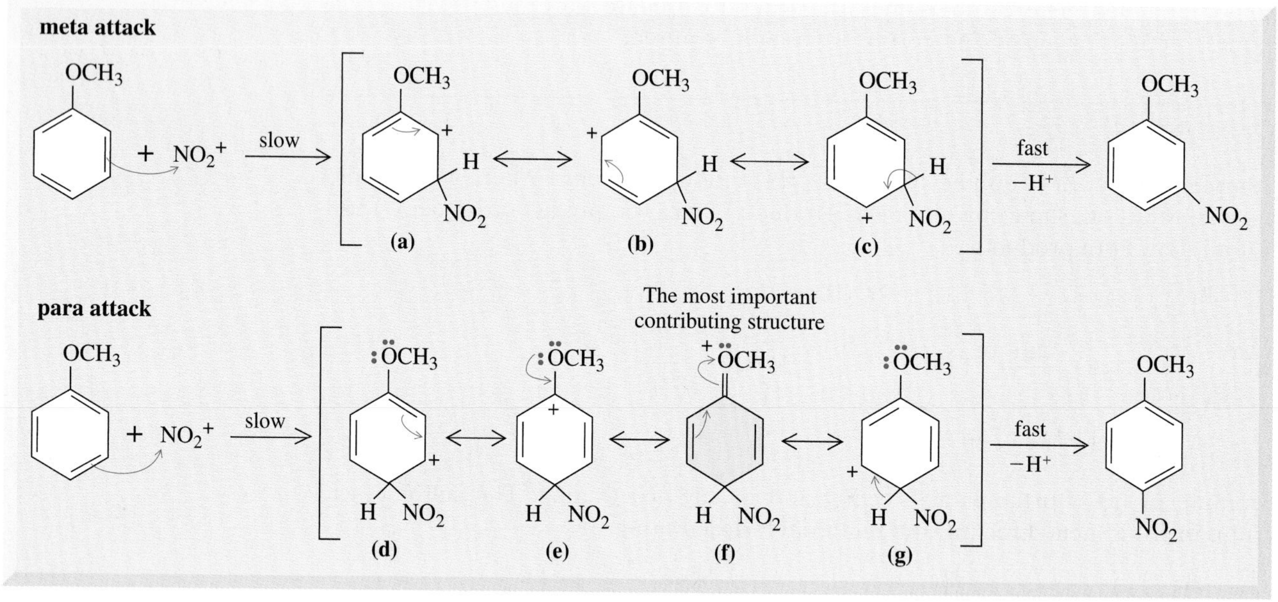

Figure 21.2

Nitration of anisole. Electrophilic attack meta and para to the methoxy group.

consider, the slow and rate-determining step is attack of the electrophile on the aromatic ring. The rate of this step depends on the stability of the cation intermediate formed. The more stable this cation, the faster the rate-determining step and the overall reaction.

Shown in Figure 21.2 is the cation intermediate formed by attack of the nitronium ion meta to the methoxy group. Also shown in Figure 21.2 is the cation intermediate formed by attack para to the methoxy group. Note that in terms of electronic effects, structural formulas for the cation formed by attack ortho to the methoxy group are essentially the same as those for para attack, so, for convenience, we deal only with para attack. The cation intermediate formed by meta attack is a hybrid of contributing structures (a), (b), and (c). The cation intermediate formed by para attack is a hybrid of contributing structures (d), (e), (f), and (g). For each orientation, we can draw three contributing structures that place the positive charge on carbon atoms of the benzene ring. These three structures are the only ones that can be drawn for meta attack. However, for para attack (and for ortho attack as well), a fourth contributing structure, (f), can be drawn that involves an unshared pair of electrons on the oxygen atom of the methoxy group and places a positive charge on this oxygen. Structure (f) contributes more than structures (d), (e), or (g) because, in it, all atoms have complete octets. Because the cation formed by ortho or para attack on anisole has a greater degree of charge delocalization and hence a lower activation energy for its formation, nitration of anisole occurs faster in the ortho and para positions.

Nitration of Nitrobenzene

Shown in Figure 21.3 are resonance-stabilized cation intermediates formed by attack of the nitronium ion meta to the nitro group and then para to it. Each cation in Fig-

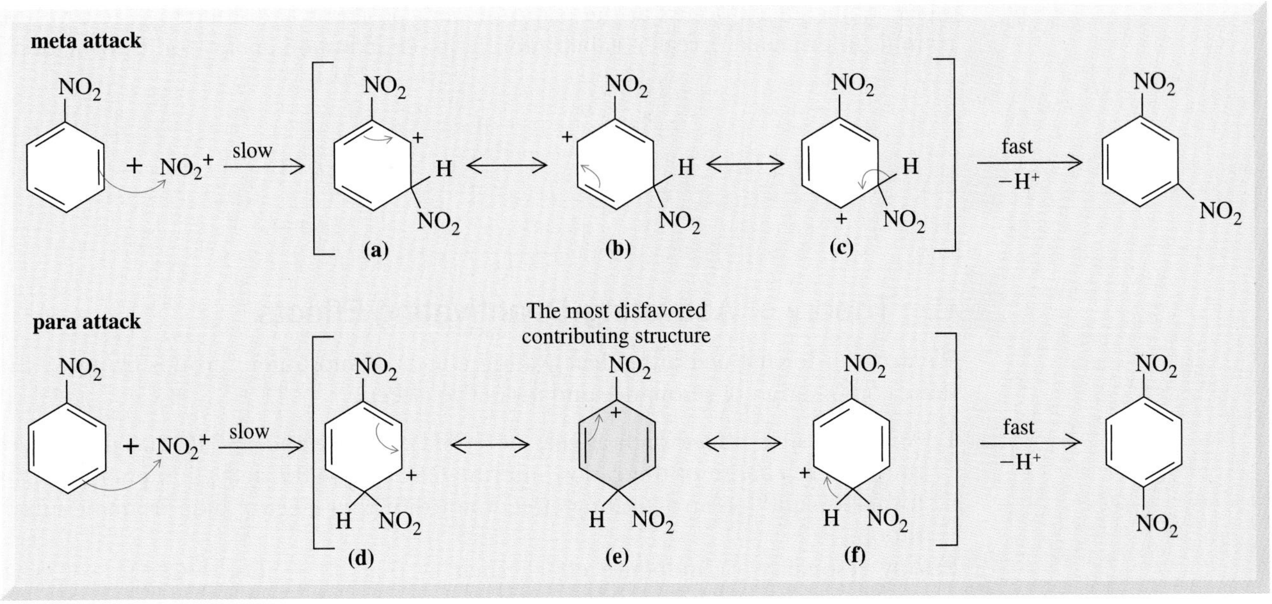

Figure 21.3
Nitration of nitrobenzene. Electrophilic attack meta and para to the nitro group.

ure 21.3 is a hybrid of three contributing structures; no additional ones can be drawn. Now we need to compare the relative resonance stabilization of each hybrid. If we draw a Lewis structure for the nitro group showing the positive charge on nitrogen, we see that contributing structure (e) in Figure 21.3 places positive charges on adjacent atoms. Because of the electrostatic repulsion thus generated, this structure is very unstable and makes only a negligible contribution to the hybrid.

None of the contributing structures for meta attack places positive charges on adjacent atoms. As a consequence, resonance stabilization of the cation for meta attack is greater than that for para (or ortho) attack. Stated alternatively, the activation energy for meta attack is less than that for para attack.

Comparison of the entries in Table 21.1 shows that almost all of the ortho-para directing groups have an unshared pair of electrons on the atom bonded to the aromatic ring. Thus, the directing effect of these groups is due primarily to the ability of the atom bonded to the ring to further delocalize the positive charge of the cation intermediate formed when electrophilic attack occurs at the ortho or para positions.

To account for the fact that alkyl groups are also ortho-para directing, we need to consider their inductive effect. In the case of alkyl groups bonded to an aromatic ring, the carbon of the alkyl group is sp^3 hybridized (25% s-character), whereas the carbon of the aromatic ring to which it is bonded is sp^2 hybridized (33% s-character). Consequently, there is an inductive polarization of electrons from the alkyl substituent toward the aromatic ring. We used the electron-releasing inductive effect of alkyl groups in Section 6.3A to account for the relative stabilities of methyl, primary, secondary, and tertiary carbocations and again in Section 20.5B to account for the decreased acidity of alkylphenols compared with phenol itself.

The net effect of inductive polarization of electrons from alkyl groups is to further delocalize the positive charge of the cation intermediate formed when

delocalizes ⊕ charge of cation intermed.

electrophilic attack occurs at the ortho or para positions and, therefore, to give this cationic intermediate greater stabilization (a lower activation energy for its formation).

A part of the positive charge
on the cation intermediate
is delocalized onto methyl.

C.   Theory of Activating-Deactivating Effects

We account for the activating-deactivating effects of substituent groups by much the same combination of resonance and inductive effects.

1. Any resonance effect, such as that of —NH_2, —OH, and —OR, that delocalizes the positive charge of the cation intermediate lowers the activation energy for its formation and has an activating effect toward further electrophilic aromatic substitution.
2. Any resonance or inductive effect, such as that of —NO_2, —C≡N, —C=O, —SO_2—, and —SO_3H, that decreases electron density on the ring deactivates the ring to further substitution.
3. Any inductive effect, such as that of —CH_3 or another alkyl group, that releases electron density toward the ring activates the ring toward further substitution.
4. Any inductive effect, such as that of halogen, —NR_3^+, —CCl_3, and —CF_3, that decreases electron density on the ring deactivates the ring to further substitution.

The halogens represent an interesting combination of the resonance and inductive effects, the two operating in opposite directions. Recall from Table 21.1 that halogens are ortho-para directing, but, unlike other ortho-para directors listed in that table, the halogens are weakly deactivating. These observations can be accounted for in the following way.

1. The inductive effect of halogens. The halogens are more electronegative than carbon and have an electron-withdrawing inductive effect. Aryl halides, therefore, react more slowly in electrophilic aromatic substitution than benzene.
2. The resonance effect of halogens. When the aromatic ring is attacked by an electrophile to form a cation intermediate, a halogen ortho or para to the site of electrophilic attack can help stabilize the cation intermediate by delocalization of the positive charge.

Thus, the inductive and resonance effects of the halogens are counter to each other but the former is somewhat stronger than the latter. The net effect of this opposition is that the halogens are weakly deactivating but ortho-para directing.

Example 21.5

Predict the major product of each electrophilic aromatic substitution.

(a) [structure: phenol with OH, Br (handwritten), and NO$_2$, with Br$_2$ arrow and handwritten curved arrows]

(b) [structure: phenol with OH, NO$_2$, SO$_3$/H$_2$SO$_4$ arrow, SO$_3$H handwritten]

(c) [structure: toluene with CH$_3$, Br (handwritten), Br$_2$/AlCl$_3$ arrow, SO$_3$H]

Solution

The key to predicting orientation of electrophilic aromatic substitution on each molecule is that ortho-para directing groups activate the ring toward further substitution, whereas meta directing groups deactivate it toward further substitution. Therefore, where there is competition between ortho-para and meta directing groups, ortho-para directing groups win out. For (a) and (b), the next substitution is directed ortho/para to the strongly activating —OH group. In (a), the isomer with bromine between the —OH and —NO$_2$ groups is a very minor product because of steric hindrance. In (c), the next substitution is directed ortho/para to the weakly activating —CH$_3$ group. The major product in each example is that resulting from substitution ortho or para to the activating group.

(a) [structure: phenol with OH, NO$_2$, Br] + [structure: phenol with OH, Br, NO$_2$]

(b) [structure: phenol with OH, NO$_2$, SO$_3$H]

(c) [structure: toluene with CH$_3$, Br, SO$_3$H]

Problem 21.5

Predict the major product(s) of each electrophilic aromatic substitution.

(a) [structure: 1,3-dimethylbenzene with Br$_2$/FeBr$_3$ arrow]

(b) [structure: benzene with Br and NO$_2$, SO$_3$/H$_2$SO$_4$ arrow]

(c) [structure: benzene with COOH and NO$_2$, HNO$_3$/H$_2$SO$_4$ arrow]

21.3 Nucleophilic Aromatic Substitution

One of the important chemical characteristics of aryl halides is that they undergo relatively few reactions involving the carbon-halogen bond. Aryl halides, for example, do not undergo substitution by either of the S$_N$1 or S$_N$2 pathways that are characteristic of nucleophilic aliphatic substitutions. They do, however, undergo **nucleophilic aromatic substitution** under certain conditions but by mechanisms quite different from those for nucleophilic aliphatic substitutions. Nucleophilic aromatic substitution reactions are far less common than electrophilic aromatic substitution reactions and have only limited usefulness in the synthesis of organic compounds. We study these reactions not only for their synthetic usefulness but also for the additional insights they give us into the unique chemical properties of aromatic compounds.

Mechanisms
Aromatics
Nucleophilic Aromatic
Substitution

Nucleophilic aromatic substitution   A reaction in which a nucleophile, most commonly a halogen, on an aromatic ring is replaced by another nucleophile.

A.   Nucleophilic Substitution by Way of a Benzyne Intermediate

An apparent exception to the generalization about the lack of reactivity of aryl halides to nucleophilic substitution is the industrial process developed in 1924 for the synthesis of phenol from chlorobenzene. When heated at 300°C under high pressure with aqueous NaOH, chlorobenzene is converted to sodium phenoxide. Neutralization of this salt with aqueous acid gives phenol.

Chlorobenzene   Sodium phenoxide

In later technological developments, it was discovered that chlorobenzene can be hydrolyzed to phenol by steam under pressure at 500°C. Each of these reactions appears to involve nucleophilic substitution of —OH for —Cl on the benzene ring. However, this reaction is not as simple as it might seem, as can be illustrated by the reaction of substituted halobenzenes with NaOH. For example, *o*-chlorotoluene under these conditions gives a mixture of 2-methylphenol (*o*-cresol) and 3-methylphenol (*m*-cresol).

2-Methylphenol   3-Methylphenol
(*o*-Cresol)   (*m*-Cresol)

The same type of reaction can be brought about by the use of sodium amide in liquid ammonia. Under these conditions, for example, *p*-chlorotoluene gives a mixture of 4-methylaniline (*p*-toluidine) and 3-methylaniline (*m*-toluidine) in approximately equal amounts.

4-Methylaniline   3-Methylaniline
(*p*-Toluidine)   (*m*-Toluidine)

Benzyne intermediate   A reactive intermediate formed by β-elimination from adjacent carbon atoms of a benzene ring and having a triple bond in the benzene ring. The second pi bond of the benzyne triple bond is formed by overlap of coplanar sp^2 orbitals on adjacent carbons.

The difference in this reaction compared with other substitution reactions we have dealt with so far is that the entering group appears not only at the position occupied by the leaving group but also at a position adjacent to it.

To account for these experimental observations, it has been proposed that an elimination of HX occurs to form a **benzyne intermediate** that then undergoes nucleophilic addition to the triple bond to give the products observed.

Mechanism   Nucleophilic Aromatic Substitution via a Benzyne Intermediate

Step 1:   Dehydrohalogenation of the benzene ring gives a benzyne intermediate.

A benzyne
intermediate

Step 2:   Nucleophilic addition of amide ion to a carbon of the benzyne triple bond gives a carbanion intermediate. Addition to both ends of the "triple" bond occurs.

A carbanion intermediate

Step 3:   Proton transfer from ammonia to the carbanion intermediate gives one of the observed substitution products and generates a new amide ion.

3-Methylaniline
(*m*-Toluidine)

The bonding in a benzyne intermediate and also the reason for its extremely reactive nature can be pictured in the following way. According to molecular orbital theory, the benzene ring retains its planarity, pi bonding, and aromatic character. The adjacent sp^2 orbitals formerly bonding to a halogen and a hydrogen atom now overlap to form the second pi bond of the benzyne triple bond. The problem is that the atomic orbitals forming this pi bond are not parallel as in acetylene and unstrained alkynes but rather lie at an angle of 120° to the bond axis connecting them. Consequently, the overlap between these orbitals is reduced. Reduced overlap in turn means a weaker and more reactive pi bond. Therefore, the second pi bond of the benzyne intermediate undergoes addition very readily to form two new and stronger sigma bonds.

pi bonding, but of
reduced strength
because of poor
orbital overlap

A benzyne intermediate

Key Reactions

1. Halogenation (Section 21.1A)

The electrophile is a halonium ion formed as an ion pair by interaction of chlorine or bromine with a Lewis acid. Halogenation of an aromatic ring substituted by strongly activating groups (such as —OH, —OR, and —NH_2) does not require a Lewis acid catalyst.

2. Nitration (Section 21.1B)

The attacking electrophile is the nitronium ion, NO_2^+, formed by interaction of nitric acid and sulfuric acid.

3. Sulfonation (Section 21.1B)

The attacking electrophile is either sulfur trioxide, SO_3, or HSO_3^+ depending on experimental conditions.

4. Friedel-Crafts Alkylation (Section 21.1C)

The attacking electrophile is a carbocation formed as an ion pair by interaction of an alkyl halide with a Lewis acid. Rearrangements from a less stable carbocation to a more stable carbocation are common. Reaction fails with compounds much less reactive than benzene.

5. Friedel-Crafts Acylation (Section 21.1C)

The attacking electrophile is an acyl cation formed as an ion pair by interaction of an acyl halide with a Lewis acid.

6. Alkylation Using an Alkene (Section 21.1D)

The attacking electrophile is a carbocation formed by interaction of the alkene with a Brønsted or Lewis acid.

7. Alkylation Using an Alcohol (Section 21.1D)

The attacking electrophile is a carbocation formed by treatment of the alcohol with a Brønsted or Lewis acid.

8. Nucleophilic Aromatic Substitution: A Benzyne Intermediate (Section 21.3A)

Elimination of HX from an aryl halide by strong base forms a benzyne intermediate, which undergoes nucleophilic addition to give the substitution product(s).

9. Nucleophilic Aromatic Substitution: Addition-Elimination (Section 21.3B)

Addition of the nucleophile to the carbon bearing the leaving group forms a tetrahedral intermediate from which halide ion is ejected to regenerate the aromatic ring. This type of aromatic substitution is made possible by strong electron-withdrawing groups, most commonly nitro groups, located ortho and para to the halogen.

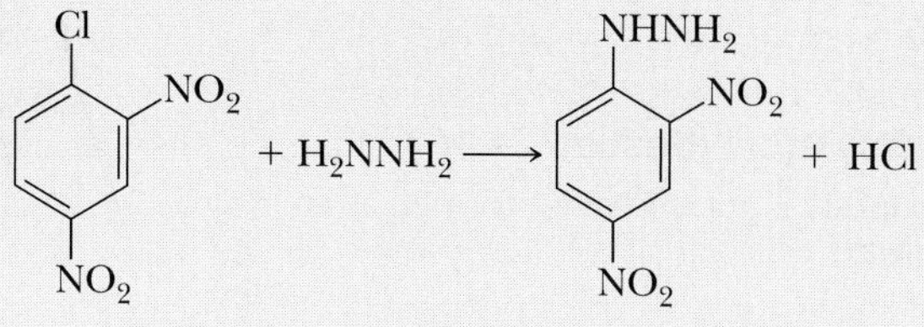

Problems

Electrophilic Aromatic Substitution: Monosubstitution

21.7 Write a stepwise mechanism for each reaction. Use curved arrows to show the flow of electrons in each step.

(a)

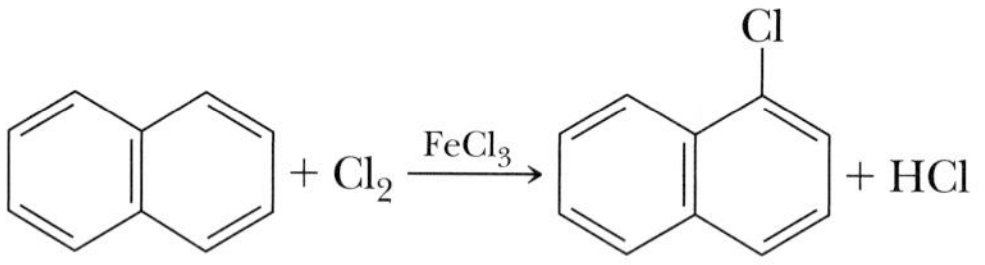

(b)

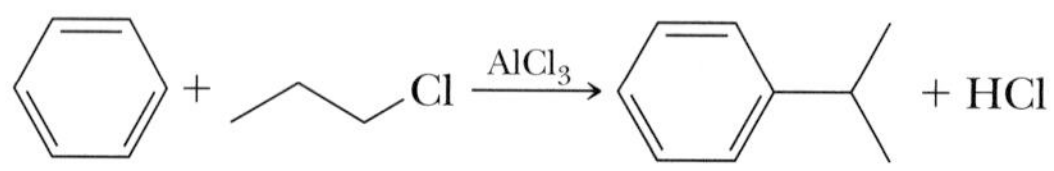

(c)

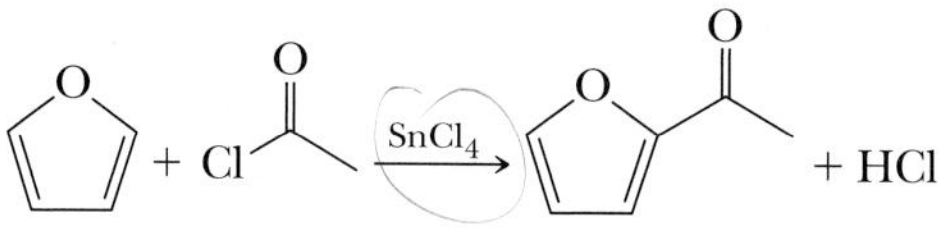

(d)

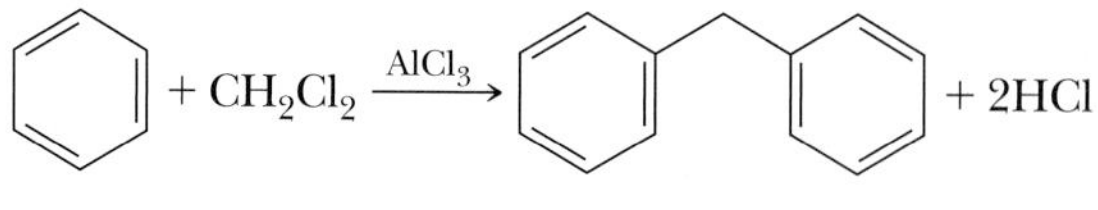

21.8 Pyridine undergoes electrophilic aromatic substitution preferentially at the 3 position as illustrated by the synthesis of 3-nitropyridine. Note that, under these acidic conditions, the species undergoing nitration is not pyridine but its conjugate acid.

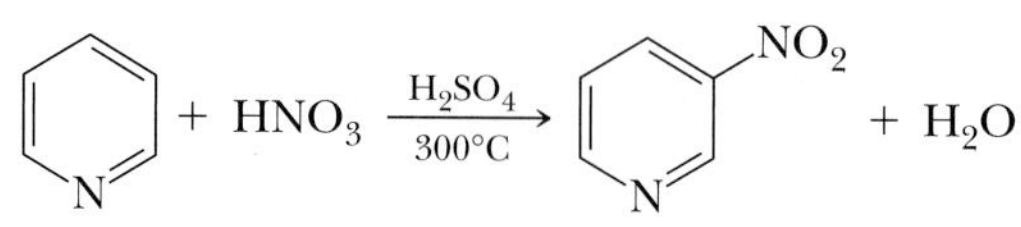

Pyridine   3-Nitropyridine

Write resonance contributing structures for the intermediate formed by attack of NO_2^+ at the 2, 3, and 4 positions of the conjugate acid of pyridine. From examination of these intermediates, offer an explanation for preferential nitration at the 3 position.

21.9 Pyrrole undergoes electrophilic aromatic substitution preferentially at the 2 position as illustrated by the synthesis of 2-nitropyrrole.

Pyrrole   2-Nitropyrrole

Write resonance contributing structures for the intermediate formed by attack of NO_2^+ at the 2 and 3 positions of pyrrole. From examination of these intermediates, offer an explanation for preferential nitration at the 2 position.

21.10 Addition of *m*-xylene to the strongly acidic solvent `HF/SbF_5 at $-45°C$ gives a new species, which shows ^{1}H-NMR resonances at δ 2.88 (3H), 3.00 (3H), 4.67 (2H), 7.93 (1H), 7.83 (1H), and 8.68 (1H). Assign a structure to the species giving this spectrum.

21.11 Addition of *tert*-butylbenzene to the strongly acidic solvent HF/SbF_5 followed by aqueous workup gives benzene. Propose a mechanism for this dealkylation reaction. What is the other product of the reaction?

21.12 What product do you predict from the reaction of SCl_2 with benzene in the presence of $AlCl_3$? What product results if diphenyl ether is treated with SCl_2 and $AlCl_3$?

21.13 Other groups besides H^+ can act as leaving groups in electrophilic aromatic substitution. One of the best of these is the trimethylsilyl group ($Me_3Si—$). For example, treatment of $Me_3SiC_6H_5$ with CF_3COOD rapidly forms DC_6H_5. What are the properties of a silicon-carbon bond that allows you to predict this kind of reactivity?

Disubstitution and Polysubstitution

21.14 The following groups are ortho-para directors. Draw a contributing structure for the resonance-stabilized cation formed during electrophilic aromatic substitution that shows the role of each group in stabilizing the intermediate by further delocalizing its positive charge.

21.15 Predict the major product(s) from treatment of each compound with HNO_3/H_2SO_4.

21.16 How do you account for the fact that *N*-phenylacetamide (acetanilide) is less reactive toward electrophilic aromatic substitution than aniline?

N-Phenylacetamide   Aniline
(Acetanilide)

21.17 Propose an explanation for the fact that the trifluoromethyl group is almost exclusively meta directing.

$$\text{C}_6\text{H}_5\text{CF}_3 + \text{HNO}_3 \xrightarrow{\text{H}_2\text{SO}_4} + \text{H}_2\text{O}$$

21.18 Suggest a reason why the nitroso group, —N=O, is ortho-para directing although the nitro group, —NO₂, is meta directing.

21.19 Arrange the compounds in each set in order of decreasing reactivity (fastest to slowest) toward electrophilic aromatic substitution.

(a)
(A)   (B) —OCCH₃   (C) —COCH₃

(b)
(A) —NO₂   (B) —COOH   (C)

(c)
(A) —CH₃   (B) —CH₂Cl   (C) —CHCl₂

(d)
(A) —Cl   (B) —C≡N   (C) —OCH₂CH₃

(e)
(A) —NH₂   (B) —NHCCH₃   (C) —CNHCH₃

21.20 For each compound, indicate which group on the ring is the more strongly activating and then draw the structural formula of the major product formed by nitration of the compound.

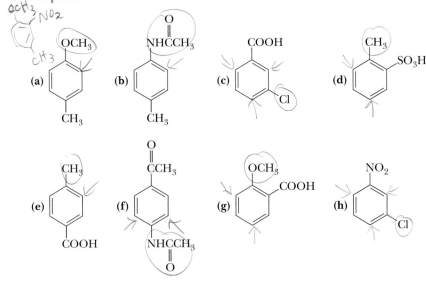

21.21 The following molecules each contain two rings. Which ring in each undergoes electrophilic aromatic substitution more readily? Draw the major product formed on nitration.

(a) [structure: benzene-C(=O)NH-benzene]

(b) O₂N— [biphenyl structure]

(c) [structure: benzene-C(=O)O-benzene]

21.22 Reaction of phenol with acetone in the presence of an acid catalyst gives a compound known as bisphenol A. Bisphenol A is used in the production of epoxy and polycarbonate resins (Section 24.5). Propose a mechanism for the formation of bisphenol A.

$$2 \; \text{C}_6\text{H}_4\text{—OH} + \text{CH}_3\overset{\overset{\displaystyle O}{\|}}{\text{C}}\text{CH}_3 \xrightarrow{\text{H}_3\text{PO}_4} \text{HO—C}_6\text{H}_4\text{—}\underset{\underset{\displaystyle \text{CH}_3}{|}}{\overset{\overset{\displaystyle \text{CH}_3}{|}}{\text{C}}}\text{—C}_6\text{H}_4\text{—OH} + \text{H}_2\text{O}$$

Bisphenol A

21.23 2,6-Di-*tert*-butyl-4-methylphenol, alternatively known as butylated hydroxytoluene (BHT), is used as an antioxidant in foods to "retard spoilage" (see Chemistry in Action box "Radical Autoxidation" in Chapter 7). BHT is synthesized industrially from 4-methylphenol (*p*-cresol) by reaction with 2-methylpropene in the presence of phosphoric acid. Propose a mechanism for this reaction.

4-Methylphenol   2-Methylpropene   2,6-Di-*tert*-butyl-4-methylphenol
                                            "Butylated hydroxytoluene"
                                            (BHT)

21.24 The insecticide DDT is prepared by the following route. Suggest a mechanism for this reaction. The abbreviation DDT is derived from the common name **d**ichloro**d**iphenyl**t**richloroethane.

Chlorobenzene   Trichloro-   DDT
                  acetaldehyde

21.25 Treatment of salicylaldehyde (2-hydroxybenzaldehyde) with bromine in glacial acetic acid at 0°C gives a compound of molecular formula $C_7H_4Br_2O_2$, which is used as a topical fungicide and antibacterial agent. Propose a structural formula for this compound.

21.26 Propose a synthesis for 3,5-dibromo-2-hydroxybenzoic acid (3,5-dibromosalicylic acid) from phenol.

21.27 Treatment of benzene with succinic anhydride in the presence of polyphosphoric acid gives the following γ-ketoacid. Propose a mechanism for this reaction.

Succinic anhydride   4-Oxo-4-phenylbutanoic acid

Nucleophilic Aromatic Substitution

21.28 Following are the final steps in the synthesis of trifluralin B, a pre-emergent herbicide.

Trifluralin B

(a) Account for the orientation of nitration in Step 1.

(b) Propose a mechanism for the substitution reaction in Step 2.

21.29 A problem in dyeing fabrics is the degree of fastness of the dye to the fabric. Many of the early dyes were surface dyes; that is, they do not bond to the fabric with the result that they tend to wash off after repeated laundering. Indigo, for example, which gives the blue color to blue jeans, is a surface dye. Color fastness can be obtained by bonding a dye to the fabric. The first such dyes were the so-called reactive dyes, developed in the 1930s for covalent bonding dyes containing —NH$_2$ groups to cotton, wool, and silk fabrics. In the first stage of the first developed method for reactive dyeing, the dye is treated with cyanuryl chloride, which links the two through the amino group of the dye. The remaining chlorines are then displaced by —OH groups of cotton (cellulose) or —NH$_2$ groups of wool or silk (both proteins).

Cyanuryl chloride   A reactive dye   Dye covalently bonded to cotton

Propose a mechanism for the displacement of a chlorine from cyanuryl chloride by (a) the NH$_2$ group of a dye and (b) by an —OH group of cotton.

Syntheses

21.30 Show how to convert toluene to these compounds.

(a)   (b) Br——CH$_3$

21.31 Show how to prepare compounds (a) and (b) from 1-phenyl-1-propanone.

1-Phenyl-1-propanone   (a)   (b)
(Propiophenone)

21.32 Show how to convert toluene to (a) 2,4-dinitrobenzoic acid and (b) 3,5-dinitrobenzoic acid.

21.33 Show reagents and conditions to bring about the following conversions.

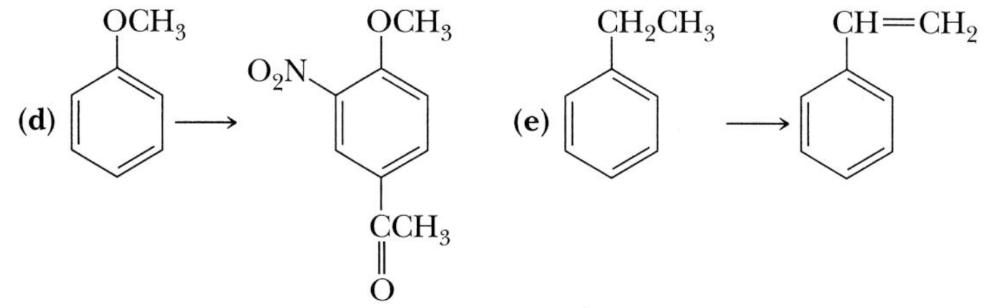

(d) (e)

21.34 Propose a synthesis of triphenylmethane from benzene, as the only source of aromatic rings, and any other necessary reagents.

21.35 Propose a synthesis for each compound from benzene.

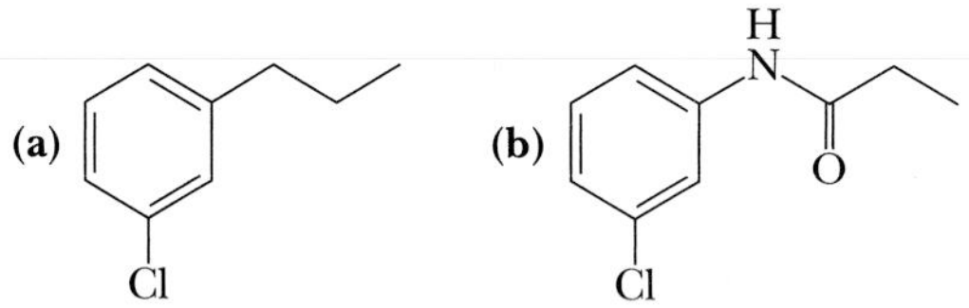

(a) (b)

21.36 The first widely used herbicide for the control of weeds was 2,4-dichlorophenoxyacetic acid (2,4-D). Show how this compound might be synthesized from phenol and chloroacetic acid by way of the given chlorinated phenol intermediate.

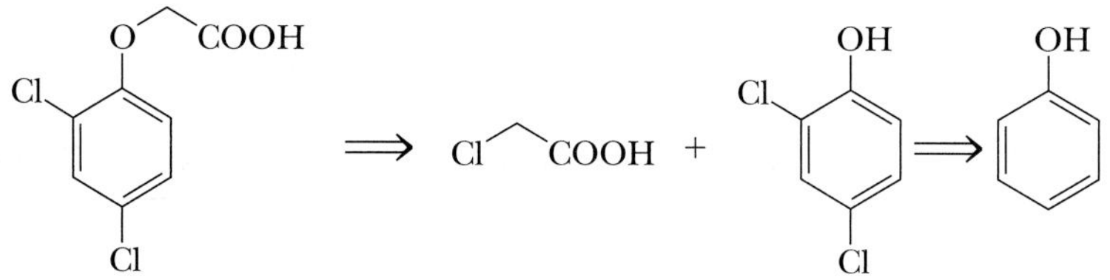

2,4-Dichlorophenoxyacetic acid   Chloroacetic acid   Phenol
(2,4-D)

21.37 Phenol is the starting material for the synthesis of 2,3,4,5,6-pentachlorophenol, known alternatively as pentachlorophenol or more simply as penta. At one time, penta was widely used as a wood preservative for decks, siding, and outdoor wood furniture. Draw the structural formula for pentachlorophenol, and describe its synthesis from phenol.

21.38 Starting with benzene, toluene, or phenol as the only sources of aromatic rings, show how to synthesize the following. Assume in all syntheses that mixtures of ortho-para products can be separated into the desired isomer.

(a) 1-Bromo-3-nitrobenzene   (b) 1-Bromo-4-nitrobenzene
(c) 2,4,6-Trinitrotoluene (TNT)   (d) *m*-Chlorobenzoic acid
(e) *p*-Chlorobenzoic acid   (f) *p*-Dichlorobenzene
(g) *m*-Nitrobenzenesulfonic acid

21.39 3,5-Dibromo-4-hydroxybenzenesulfonic acid is used as a disinfectant. Propose a synthesis of this compound from phenol.

21.40 Propose a synthesis for 3,5-dichloro-2-methoxybenzoic acid starting from phenol.

21.41 The following compound used in perfumery has a violet-like scent. Propose a synthesis of this compound from benzene.

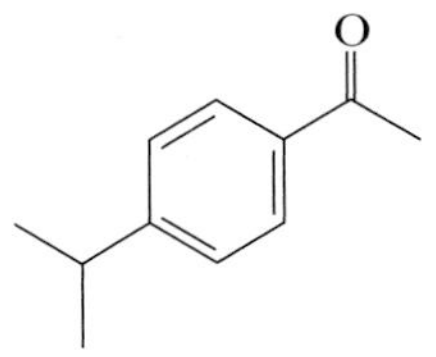

4-Isopropylacetophenone

21.42 Following is the structural formula of musk ambrette, a synthetic musk, essential in perfumes to enhance and retain odor. Propose a synthesis of this compound from *m*-cresol (3-methylphenol).

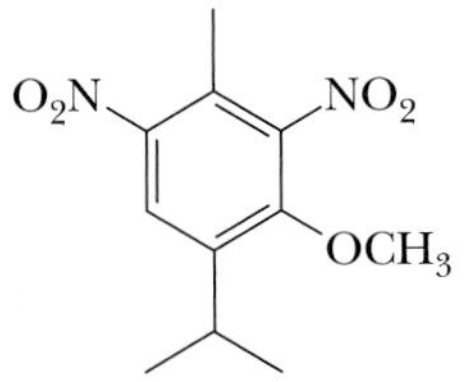

21.43 Propose a synthesis of this compound starting from toluene and phenol.

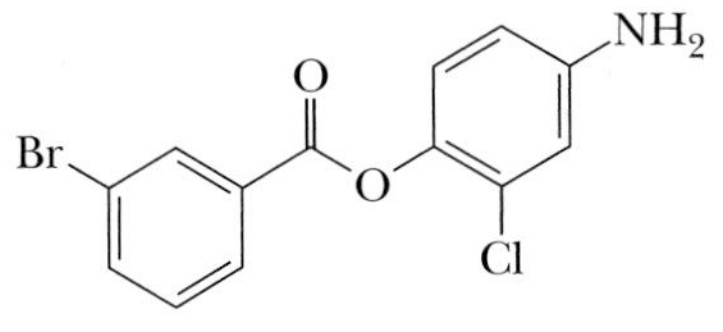

21.44 When certain aromatic compounds are treated with formaldehyde, CH_2O, and HCl, the CH_2Cl group is introduced onto the ring. This reaction is known as chloromethylation.

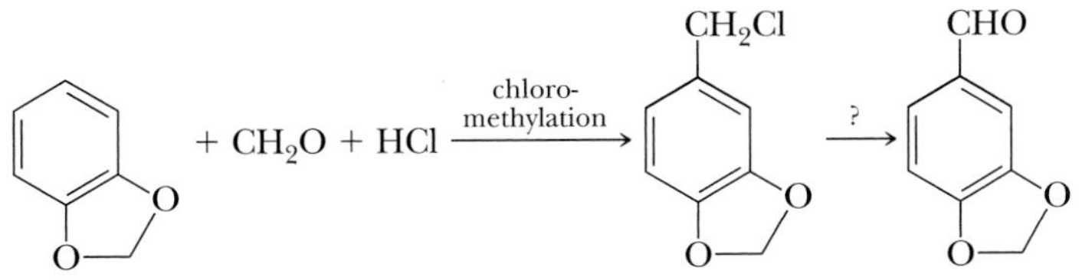

Piperonal

(a) Propose a mechanism for this example of chloromethylation.
(b) The product of this chloromethylation can be converted to piperonal (see *The Merck Index*, 12th ed., #7628), which is used in perfumery and in artificial cherry and vanilla flavors. How might the CH_2Cl group of the chloromethylation product be converted to a CHO group?

21.45 Following is a retrosynthetic analysis for the acaracide (killing mites and ticks) and fungicide dinocap (see *The Merck Index*, 12th ed., #3340). Given this analysis, propose a synthesis for dinocap from phenol and 1-octene.

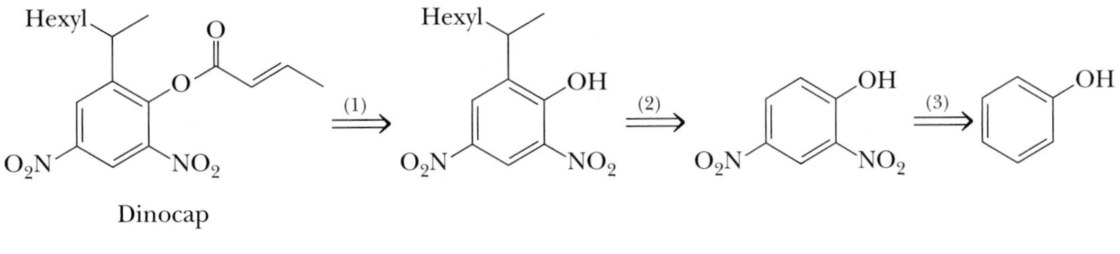

Dinocap

21.46 Following is the structure of miconazole (see *The Merck Index*, 12th ed., #6266), an antifungal agent. It is the active ingredient in a number of over-the-counter preparations, including Monistat which is used to treat vaginal yeast infections. One of the compounds needed for the synthesis of miconazole is the trichloro derivative of toluene shown on the right. Show how this derivative can be synthesized from toluene.

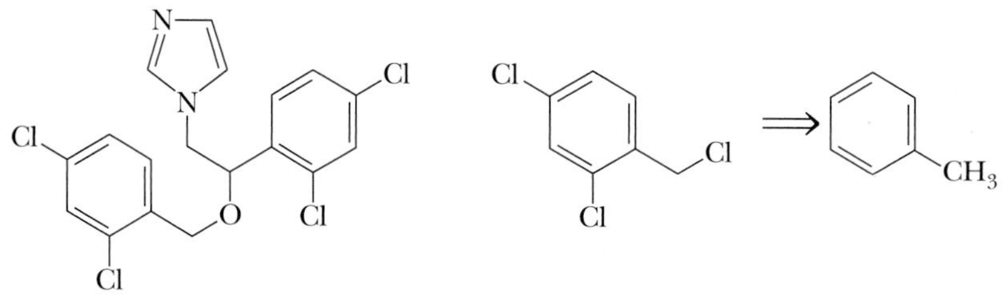

Miconazole   2,4-Dichloro-1-chloromethylbenzene   Toluene

21.47 Following is the structural formula of bupropion, the hydrochloride of which was first marketed in 1985 by Burroughs-Wellcome, now GlaxoWellcome, as an antidepressant under the trade name Wellbutrin. During clinical trials, it was discovered that smokers, after one to two weeks on the drug, reported that their craving for tobacco lessened. Further clinical trials confirmed this finding, and the drug was marketed in 1997 under the trade name Zyban as an aid in smoking cessation. Given this retrosynthetic analysis, propose a synthesis for bupropion.

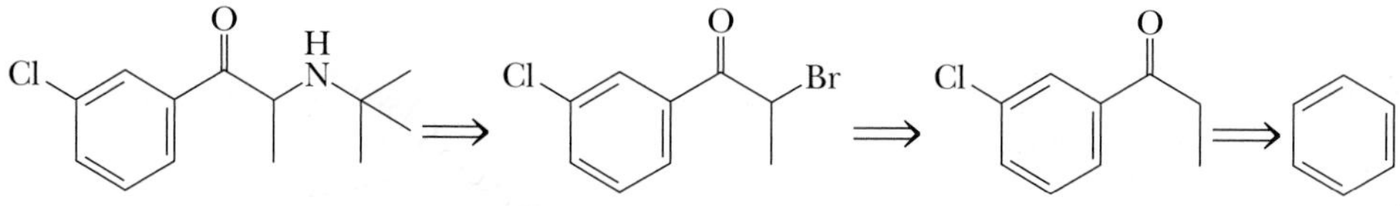

Bupropion

AMINES

Carbon, hydrogen, and oxygen are the three most common elements in organic compounds. Because of the wide distribution of amines in the biological world, nitrogen is the fourth most common element in organic compounds. The most important chemical properties of amines are their basicity and their nucleophilicity.

■ Opium poppies. Morphine, a potent pain killer isolated from the ripe seed heads of the opium poppy, has been a lead drug for chemists in search of potent but less addicting synthetic pain killers. See Problems 22.21 and MC.12 to MC.14. *(Frank Orel/Tony Stone Images)* Inset: A model of morphine.

22.1 Structure and Classification

Amines are derivatives of ammonia in which one or more hydrogens are replaced by alkyl or aryl groups. Amines are classified as primary, secondary, or tertiary, depending on the number of carbon atoms bonded directly to nitrogen (Section 1.3B).

$$CH_3—\ddot{N}H_2 \qquad CH_3—\underset{\underset{CH_3}{|}}{\overset{\cdot\cdot}{N}}H \qquad CH_3—\underset{\underset{CH_3}{|}}{\overset{\overset{CH_3}{|}}{N}}:$$

Methylamine   Dimethylamine   Trimethylamine
(a 1° amine)   (a 2° amine)   (a 3° amine)

Aliphatic amine An amine in which nitrogen is bonded only to alkyl groups.

Aromatic amine An amine in which nitrogen is bonded to one or more aryl groups.

Amines are further divided into aliphatic and aromatic amines. In an **aliphatic amine,** all carbons bonded directly to nitrogen are derived from alkyl groups; in an **aromatic amine,** one or more of the groups bonded to nitrogen are aryl groups.

Aniline   N-Methylaniline   Benzyldimethylamine
(a 1° aromatic amine)   (a 2° aromatic amine)   (a 3° aliphatic amine)

Heterocyclic amine An amine in which nitrogen is one of the atoms of a ring.

Heterocyclic aromatic amine An amine in which nitrogen is one of the atoms of an aromatic ring.

An amine in which the nitrogen atom is part of a ring is classified as a **heterocyclic amine.** When the nitrogen is part of an aromatic ring (Section 20.2D), the amine is classified as a **heterocyclic aromatic amine.** Following are structural formulas for two heterocyclic aliphatic amines and two heterocyclic aromatic amines.

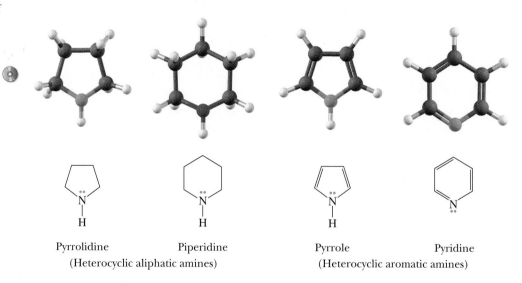

Pyrrolidine   Piperidine   Pyrrole   Pyridine
(Heterocyclic aliphatic amines)   (Heterocyclic aromatic amines)

Alkaloid A basic nitrogen-containing compound of plant origin, many of which are physiologically active when administered to humans.

Example 22.1

Alkaloids are basic nitrogen-containing compounds of plant origin, many of which are physiologically active when administered to humans. Ingestion of coniine, isolated from water hemlock, can cause weakness, labored respiration, paralysis, and eventually death. It is the toxic substance in "poison hemlock" used in the death of Socrates. In small doses nicotine is an addictive stimulant. In larger doses, it causes depression, nausea, and vomiting. In still larger doses, it is a deadly poison. Solutions of nicotine in water are used as insecticides. Cocaine is a central nervous system stimulant obtained from the leaves of the coca plant. Classify each amino group in these alkaloids according to type (primary, secondary, tertiary, aliphatic, aromatic, heterocyclic).

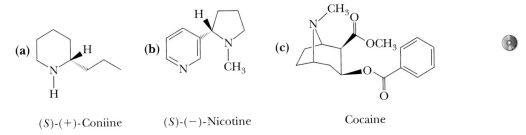

(a) (S)-(+)-Coniine   (b) (S)-(−)-Nicotine   (c) Cocaine

Solution

(a) A secondary heterocyclic aliphatic amine
(b) A tertiary heterocyclic aliphatic amine and a heterocyclic aromatic amine
(c) A tertiary heterocyclic aliphatic amine

Problem 22.1

Identify all carbon stereocenters in coniine, nicotine, and cocaine.

22.2 Nomenclature

A. Systematic Names

Systematic names for aliphatic amines are derived just as they are for alcohols. The suffix -e of the parent alkane is dropped and is replaced by -amine.

$$NH_2$$
$$CH_3CHCH_3$$
2-Propanamine

(S)-1-Phenylethanamine

$$H_2N(CH_2)_6NH_2$$
1,6-Hexanediamine

Example 22.2

Write systematic names for these amines.

(a) $CH_3(CH_2)_5NH_2$ **(b)** H_2N —— NH_2 **(c)**

Example 22.4

Write the IUPAC name and, where possible, a common name for each compound.

(a) $(C_6H_5)_2NH$ (b) [structure with NH₂ and OH on cyclohexane ring]   (c) [structure with H, NH₂ and phenyl group]

Solution

(a) Diphenylamine   (b) *trans*-2-Aminocyclohexanol
(c) Its systematic name is (*S*)-1-phenyl-2-propanamine. Its common name is amphetamine. The dextrorotatory isomer of amphetamine (shown here) is a central nervous system stimulant and is manufactured and sold under several trade names. The salt with sulfuric acid is marketed as Dexedrine sulfate.

Problem 22.4

Write IUPAC and, where possible, common names for these amines.

(a) [structure with O, OH, H, NH₂]   (b) H_2N [structure with O and OH]   (c) [structure with NH₂]

An ion containing a nitrogen atom bonded to any combination of four alkyl or aryl groups is classified as a **quaternary (4°) ammonium ion.** Compounds containing such ions have properties characteristic of salts. Cetylpyridinium chloride (see *The Merck Index*, 12th ed., #2017) is used as a topical antiseptic and disinfectant.

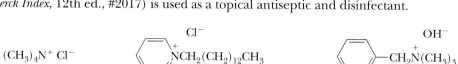

$(CH_3)_4N^+ Cl^-$

Tetramethylammonium chloride

[pyridinium structure] Cl^- $NCH_2(CH_2)_{12}CH_3$

Tetradecylpyridinium chloride (Cetylpyridinium chloride)

[benzyl structure] OH^- $CH_2N(CH_3)_3$

Benzyltrimethylammonium hydroxide

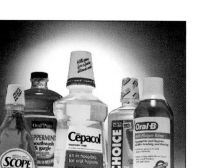

Several over-the-counter mouthwashes contain an *N*-alkylpyridinium chloride as an antibacterial agent. *(Charles D. Winters)*

22.3 Chirality of Amines and Quaternary Ammonium Ions

The geometry of a nitrogen atom bonded to three other atoms or groups of atoms is trigonal pyramidal (Section 1.4). The sp^3-hybridized nitrogen atom is at the apex of the pyramid, and the three groups bonded to it extend downward to form the triangular base of the pyramid. If we consider the unshared pair of electrons on nitrogen as a fourth group, then the arrangement of "groups" around nitrogen is approximately tetrahedral. Because of this geometry, an amine with three different groups bonded to nitrogen is chiral and can exist as a pair of enantiomers, as illustrated by the nonsuperposable mirror images of ethylmethylamine. In assigning configuration to these enantiomers, the group of lowest priority on nitrogen is the unshared pair of electrons.

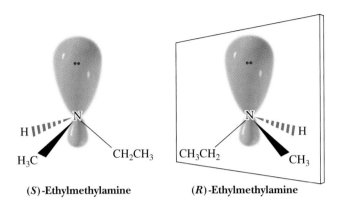

(S)-Ethylmethylamine   (R)-Ethylmethylamine

In principle, a chiral amine can be resolved; that is, it can be separated into a pair of enantiomers. Except for special cases, however, the enantiomers cannot be resolved because they undergo rapid interconversion by a process known as pyramidal inversion. **Pyramidal inversion** is the rapid oscillation of a nitrogen atom from one side of the plane of the three atoms bonded to it to the other side of that plane.

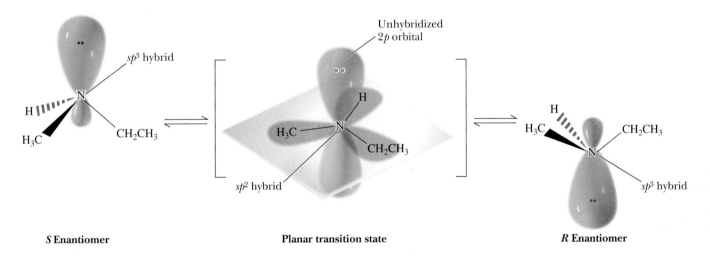

S Enantiomer   Planar transition state   *R* Enantiomer

To visualize this process, imagine the sp^3-hybridized nitrogen atom lying above the plane of the three atoms to which it is bonded. In the transition state for pyramidal inversion, the nitrogen atom and the three groups to which it is bonded become coplanar, and the molecule becomes achiral. In this planar transition state, nitrogen is sp^2 hybridized, and its lone pair of electrons lies in its unhybridized $2p$ orbital. Nitrogen then completes the inversion, becomes sp^3 hybridized again, and now lies below the plane of the three atoms to which it is bonded. As a result of pyramidal inversion, a chiral amine quite literally turns itself inside out, like an umbrella in a stormy wind, and in the process becomes a racemic mixture. The activation energy for pyramidal inversion of simple amines is about 25 kJ (6 kcal)/mol. For ammonia at room temperature, the rate of nitrogen inversion is approximately 2×10^{11} s^{-1}. For simple amines, the rate is less rapid but nonetheless sufficient to make it impossible to resolve them.

Pyramidal inversion is not possible for quaternary ammonium ions, and their salts can be resolved.

R enantiomer   *S* enantiomer

Phosphorus, in the same family as nitrogen, forms trivalent compounds called phosphines, which also have trigonal pyramidal geometry. The activation energy for pyramidal inversion of trivalent phosphorus compounds is considerably greater than it is for trivalent compounds of nitrogen, with the result that a number of chiral phosphines have been resolved.

22.4 Physical Properties

Amines are polar compounds, and both primary and secondary amines form intermolecular hydrogen bonds (Figure 22.1). An N—H----N hydrogen bond is weaker than an O—H----O hydrogen bond because the difference in electronegativity between nitrogen and hydrogen ($3.0 - 2.1 = 0.9$) is less than that between oxygen and hydrogen ($3.5 - 2.1 = 1.4$). The effect of intermolecular hydrogen bonding can be illustrated by comparing the boiling points of methylamine and methanol. Both are polar molecules and interact in the pure liquid by hydrogen bonding. Because hydrogen bonding is stronger in methanol than in methylamine, methanol has the higher boiling point.

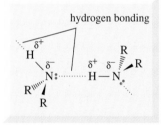

hydrogen bonding

Figure 22.1
Intermolecular association by hydrogen bonding in primary and secondary amines. Nitrogen is approximately tetrahedral in shape with the axis of the hydrogen bond along the fourth position of the tetrahedron.

	CH_3CH_3	CH_3NH_2	CH_3OH
MW (g/mol)	30.1	31.1	32.0
bp (°C)	− 88.6	− 6.3	65.0

All classes of amines form hydrogen bonds with water and are more soluble in water than hydrocarbons of comparable molecular weight. Most low-molecular-weight amines are completely soluble in water (Table 22.1). Higher molecular-weight amines are only moderately soluble or insoluble.

Spectrocopy
Amines

22.5 Spectroscopic Properties

A. Infrared Spectroscopy

The infrared spectroscopy of amines is discussed in Section 12.4G.

B. Nuclear Magnetic Resonance Spectroscopy

The chemical shifts of amine N—H hydrogens, like those of hydroxyl hydrogens (Section 13.14C), are variable and may appear in the region δ 0.5 to δ 5.0, depending on the solvent, the concentration, and the temperature. Furthermore, the rate of in-

Table 22.1 Physical Properties of Selected Amines

Name	Structural Formula	mp (°C)	bp (°C)	Solubility in Water
Ammonia	NH_3	-78	-33	Very soluble
Primary Amines				
Methylamine	CH_3NH_2	-95	-6	Very soluble
Ethylamine	$CH_3CH_2NH_2$	-81	17	Very soluble
Propylamine	$CH_3CH_2CH_2NH_2$	-83	48	Very soluble
Isopropylamine	$(CH_3)_2CHNH_2$	-95	32	Very soluble
Butylamine	$CH_3(CH_2)_3NH_2$	-49	78	Very soluble
Benzylamine	$C_6H_5CH_2NH_2$	—	185	Very soluble
Cyclohexylamine	$C_6H_{11}NH_2$	-17	135	Slightly soluble
Secondary Amines				
Dimethylamine	$(CH_3)_2NH$	-93	7	Very soluble
Diethylamine	$(CH_3CH_2)_2NH$	-48	56	Very soluble
Tertiary Amines				
Trimethylamine	$(CH_3)_3N$	-117	3	Very soluble
Triethylamine	$(CH_3CH_2)_3N$	-114	89	Slightly soluble
Aromatic Amines				
Aniline	$C_6H_5NH_2$	-6	184	Slightly soluble
Heterocyclic Aromatic Amines				
Pyridine	C_5H_5N	-42	116	Very soluble

termolecular exchange of amine hydrogens is sufficiently rapid compared with the time scale of a ^{1}H-NMR measurement that spin-spin splitting between amine hydrogens and hydrogens on an adjacent α-carbon is prevented. Thus, amine hydrogens generally appear as singlets. The amine hydrogens of 1-butanamine, for example, appear as a broad singlet at δ 1.11 (Figure 12.11).

Carbons bonded to nitrogen are deshielded by approximately 20 ppm in a ^{13}C-NMR spectrum relative to their signal in an alkane of comparable structure. Compare, for example, the chemical shift of carbon 1 in 1-butanamine (42.0 ppm) with that of carbon 2 in butane (25.0 ppm). The chemical shift of carbons adjacent to oxygen is in turn approximately 20 ppm greater than for carbons adjacent to nitrogen. Compare, for example, the chemical shift of carbon 1 in 1-butanamine (42.0 ppm) with that of carbon 1 in 1-butanol (62.4 ppm).

Formula	Carbon Atom (δ, ppm)			
	1	2	3	4
$CH_3CH_2CH_2CH_2OH$	62.4	34.9	19.0	13.9
$CH_3CH_2CH_2CH_2NH_2$	42.0	36.1	20.2	13.9
$CH_3CH_2CH_2CH_3$	13.2	25.0	25.0	13.2

CHEMISTRY IN ACTION

The Poison Dart Frogs of South America

The Noanamá and Embrá peoples of the jungles of western Colombia have used poison blow darts for centuries, perhaps millennia. The poisons are obtained from the skin secretions of several brightly colored frogs of the genus *Phyllobates* (*neará* and *kokoi* in the language of the native peoples). A single frog contains enough poison for up to 20 darts. For the most poisonous species (*Phyllobates terribilis*), just rubbing a dart over the frog's back suffices to charge the dart with poison.

Scientists at the National Institutes of Health in the United States became interested in studying these poisons when it was discovered that they act on cellular ion channels, which would make them useful tools in basic research on mechanisms of ion transport. A field station was, therefore, established in western Colombia to collect the relatively common poison dart frogs. From 5000 frogs, 11 mg of two toxins, given the names batrachotoxin and batrachotoxinin A, was isolated. These names are derived from *batrachos*, the Greek word for frog. A combination of NMR spectroscopy, mass spectrometry, and single-crystal x-ray diffraction was used to determine the structures of these compounds.

Poison dart frog, Phyllobates terribilis. (Animals, Animals/© Juan M. Renjifo)

Batrachotoxin and batrachotoxinin A are among the most lethal poisons ever discovered. It is estimated that as little as 200 μg of batrachotoxin is sufficient to induce irreversible cardiac arrest in a human being. It has been determined that they act by causing voltage-gated Na^+ channels in nerve and muscle cells to be blocked in the open position, which leads to a huge influx of Na^+ ions into the affected cell.

Batrachotoxin

Batrachotoxinin A

The batrachotoxin story illustrates several common themes in drug discovery. First, information about the kinds of biologically active compounds and their sources is often obtained from the native peoples of a region. Second, tropical rain forests are a rich source of structurally complex, biologically active substances. Third, the entire ecosystem, not just the plants, provides potential sources of fascinating organic molecules. [See J. W. Daly, *Progress in the Chemistry of Organic Natural Products,* vol. 41, W. Herz, H. Grisebach, and G. W. Kirby (Eds.), Springer-Verlag, Wien, 1982, p. 205.]

22.6 Basicity

Like ammonia, all amines are weak bases, and aqueous solutions of amines are basic. The following acid-base reaction between an amine and water is written using curved arrows to emphasize that, in these proton-transfer reactions, the unshared pair of electrons on nitrogen forms a new covalent bond with hydrogen and displaces hydroxide ion.

$$CH_3-\overset{\overset{\displaystyle H}{|}}{\underset{\underset{\displaystyle H}{|}}{N}}: + H-\overset{\cdot\cdot}{\underset{\cdot\cdot}{O}}-H \rightleftharpoons CH_3-\overset{\overset{\displaystyle H}{|}}{\underset{\underset{\displaystyle H}{|}}{\overset{+}{N}}-H} \quad :\overset{\cdot\cdot}{\underset{\cdot\cdot}{O}}-H$$

| Methylamine | Methylammonium hydroxide |

It is common to discuss the basicity of amines by reference to the acid ionization constant of the corresponding conjugate acid, as illustrated for the ionization of the methylammonium ion.

$$CH_3NH_3{}^+ + H_2O \rightleftharpoons CH_3NH_2 + H_3O^+$$

$$K_a = \frac{[CH_3NH_2][H_3O^+]}{[CH_3NH_3{}^+]} = 2.29 \times 10^{-11} \qquad pK_a = 10.64$$

Values of pK_a for the conjugate acids of selected aliphatic, aromatic, and heterocyclic aromatic amines are given in Table 22.2.

Example 22.5

Predict the position of equilibrium for this acid-base reaction.

$$CH_3NH_2 + CH_3COOH \rightleftharpoons CH_3NH_3{}^+ + CH_3COO^-$$

Solution

Use the approach we developed in Section 4.3 to predict the position of equilibrium in acid-base reactions. Equilibrium favors reaction of the stronger acid with the stronger base to give the weaker acid and weaker base.

$$CH_3NH_2 + CH_3COOH \rightleftharpoons CH_3NH_3{}^+ + CH_3COO^- \qquad \begin{aligned} pK_{eq} &= -5.88 \\ K_{eq} &= 7.6 \times 10^5 \end{aligned}$$

	pK_a 4.76	pK_a 10.64	
(stronger base)	(stronger acid)	(weaker acid)	(weaker base)

Problem 22.5

Predict the position of equilibrium for this acid-base reaction.

$$CH_3NH_3{}^+ + H_2O \rightleftharpoons CH_3NH_2 + H_3O^+$$

A. Aliphatic Amines

All aliphatic amines have about the same base strength, pK_a of conjugate acid 10.0–11.0, and are slightly stronger bases than ammonia. The increase in basicity

Table 22.2   Acid Strengths, pK_a, of the Conjugate Acids of Selected Amines

Amine	Structure	pK_a of Conjugate Acid
Ammonia	NH_3	9.26
Primary Amines		
Methylamine	CH_3NH_2	10.64
Ethylamine	$CH_3CH_2NH_2$	10.81
Cyclohexylamine	$C_6H_{11}NH_2$	10.66
Secondary Amines		
Dimethylamine	$(CH_3)_2NH$	10.73
Diethylamine	$(CH_3CH_2)_2NH$	10.98
Tertiary Amines		
Trimethylamine	$(CH_3)_3N$	9.81
Triethylamine	$(CH_3CH_2)_3N$	10.75
Aromatic Amines		
Aniline	$C_6H_5-NH_2$	4.63
4-Methylaniline	$CH_3-C_6H_4-NH_2$	5.08
4-Chloroaniline	$Cl-C_6H_4-NH_2$	4.15
4-Nitroaniline	$O_2N-C_6H_4-NH_2$	1.0
Heterocyclic Aromatic Amines		
Pyridine	pyridine	5.25
Imidazole	imidazole	6.95

compared with ammonia can be attributed to the greater stability of an alkylammonium ion, as for example $RCH_2NH_3^+$ compared with the ammonium ion, NH_4^+. This greater stability arises from the electron-releasing effect of alkyl groups and the resulting partial delocalization of the positive charge from nitrogen onto carbon in the alkylammonium ion.

Positive charge is partially
delocalized onto the alkyl group.

$$\text{R--CH}_2 \xrightarrow{\delta+} \overset{\overset{\displaystyle H}{|}}{\underset{\displaystyle H}{N}}\overset{\delta+}{\text{--H}}$$

Recall that we invoked a similar argument in Section 6.3A to account for the effect of alkyl groups in stabilizing carbocations.

B. Aromatic Amines

Aromatic amines are considerably weaker bases than aliphatic amines. Compare, for example, values of pK_a for aniline and cyclohexylamine. The ionization constant for the conjugate acid of aniline is larger (the smaller the value of pK_a, the weaker the base) than that for cyclohexylamine by a factor of 10^6.

$$\text{Cyclohexyl}\text{--NH}_2 + \text{H}_2\text{O} \rightleftharpoons \text{Cyclohexyl}\text{--NH}_3^+ \text{ OH}^- \qquad pK_a = 10.66; K_a = 2.19 \times 10^{-11}$$

Cyclohexylamine   Cyclohexylammonium hydroxide

$$\text{Phenyl}\text{--NH}_2 + \text{H}_2\text{O} \rightleftharpoons \text{Phenyl}\text{--NH}_3^+ \text{ OH}^- \qquad pK_a = 4.63; K_a = 2.34 \times 10^{-5}$$

Aniline   Anilinium hydroxide

Aromatic amines are less basic than aliphatic amines because of a combination of two factors. First is the resonance stabilization of the free base form of aromatic amines. For aniline and other arylamines, this resonance stabilization is the result of interaction of the unshared pair on nitrogen with the pi system of the aromatic ring. The resonance energy of benzene is approximately 151 kJ (36 kcal)/mol. For aniline, it is 163 kJ (39 kcal)/mol. Because of this interaction, the electron pair on nitrogen is less available for reaction with acid. No such resonance stabilization is possible for alkylamines. Therefore, the electron pair on the nitrogen of an alkylamine is more available for reaction with an acid; alkylamines are stronger bases than arylamines.

Interaction of the electron pair on
nitrogen with the pi system of the
aromatic ring

No resonance is
possible with
alkylamines.

The second factor contributing to the decreased basicity of aromatic amines is the electron-withdrawing inductive effect of the sp^2-hybridized carbons of the aromatic ring compared with the sp^3-hybridized carbons of aliphatic amines. The unshared pair of electrons on nitrogen in an aromatic amine is pulled toward the ring and, therefore, is less available for protonation to form the conjugate acid of the amine.

Electron-releasing groups (for example, methyl, ethyl, and other alkyl groups) increase the basicity of aromatic amines, whereas electron-withdrawing groups (halogen, nitro, carbonyl) decrease their basicity. The decrease in basicity on halogen substitution is due to the electron-withdrawing inductive effect of the electronegative halogen. The decrease in basicity on nitro substitution is due to a combination of inductive and resonance effects as can be seen by comparing the base ionization constants of 3-nitroaniline and 4-nitroaniline. Note that the conjugate acid of 4-nitroaniline (pK_a 1.0) is a stronger acid than phosphorous acid ($pK_{a1} = 2.0$).

3-Nitroaniline
pK_a 2.47

4-Nitroaniline
pK_b 1.0

The base-decreasing effect of nitro substitution in the 3 position is due almost entirely to its inductive effect, whereas that of nitro substitution in the 4 position is due to both inductive and resonance effects. In the case of para substitution (and ortho substitution as well), delocalization of the lone pair on the amino nitrogen involves not only the carbons of the aromatic ring but also oxygen atoms of the nitro group.

delocalization of the nitrogen lone pair onto the oxygen atoms of the nitro group

C. Heterocyclic Amines

Heterocyclic aromatic amines are weaker bases than heterocyclic aliphatic amines. Compare, for example, the pK_a values for the conjugate acids of piperidine, pyridine, and imidazole.

Piperidine
pK_a 10.75

Pyridine
pK_a 5.25

Imidazole
pK_a 6.95

We discussed the structure and bonding in pyridine and imidazole in Section 20.2D. In accounting for the relative basicities of these and other heterocyclic aromatic amines, it is important to determine first if the unshared pair of electrons on nitrogen is or is not a part of the $(4n + 2)$ pi electrons giving rise to aromaticity. In the case of pyridine, the unshared pair of electrons is not a part of the aromatic sextet. Rather, it lies in an sp^2 hybrid orbital in the plane of the ring and perpendicular to the six $2p$ orbitals containing the aromatic sextet.

Aromaticity is maintained, even when protonated.

This electron pair is not a part of the aromatic sextet.

Pyridine   Pyridinium ion

Proton transfer from water or other acid to pyridine does not involve the electrons of the aromatic sextet. Why, then, is pyridine a considerably weaker base than aliphatic amines? The answer is that the unshared pair of electrons on the pyridine nitrogen lies in a relatively electronegative sp^2 hybrid orbital, whereas in aliphatic amines, the unshared pair lies in an sp^3 hybrid orbital. This effect decreases markedly the basicity of the electron pair on an sp^2-hybridized nitrogen compared with that on an sp^3-hybridized nitrogen.

There are two nitrogen atoms in imidazole, each with an unshared pair of electrons. One unshared pair lies in a $2p$ orbital and is an integral part of the $(4n + 2)$ pi electrons of the aromatic system. The other unshared pair lies in an sp^2 hybrid orbital and is not a part of the aromatic sextet; this pair of electrons functions as the proton acceptor.

This electron pair is not a part of the aromatic sextet.

This electron pair is a part of the aromatic sextet.

Aromaticity is maintained, even when protonated.

Imidazole   Imidazolium ion

As is the case with pyridine, the unshared pair of electrons functioning as the proton acceptor in imidazole lies in an sp^2 hybrid orbital and has markedly decreased basicity compared with an unshared pair of electrons in an sp^3 hybrid orbital. The positive charge on the imidazolium ion is delocalized on both nitrogen atoms of the ring; therefore, imidazole is a stronger base than pyridine.

Example 22.6

Select the stronger base in each pair of amines.

(a) ⟨ ⟩ or ⟨ ⟩   (b) ⟨ ⟩ or ⟨ ⟩   (c) ⟨ ⟩ or ⟨ ⟩

(A)   (B)   (C)   (D)   (E)   (F)

Solution

(a) Morpholine (B) is the stronger base (conjugate acid pK_a 8.2). It has a basicity comparable to that of secondary aliphatic amines. Pyridine (A), a heterocyclic aromatic amine (pK_a 5.25), is considerably less basic than aliphatic amines.

(b) Tetrahydroisoquinoline (C) has a basicity comparable to that of secondary aliphatic amines ($pK_a \sim 10.8$) and is the stronger base. Tetrahydroquinoline (D) has a basicity comparable to that of *N*-substituted anilines ($pK_a \sim 4.4$) and is the weaker base.

(c) Benzylamine (F) is the stronger base (pK_a 9.6). Its basicity is comparable to that of other aliphatic amines. The basicity of *o*-toluidine (E), an aromatic amine, is comparable to that of aniline (pK_a 4.6).

Problem 22.6

Select the stronger acid from each pair of compounds.

(a) O_2N-⟨⟩$-NH_3^+$ or CH_3-⟨⟩$-NH_3^+$ **(b)** ⟨⟩$\overset{+}{N}H$ or ⟨⟩$-NH_3^+$

(A) (B) (C) (D)

D. Guanidine

Guanidine, pK_a 13.6, is the strongest base among neutral organic compounds, almost as basic as hydroxide ion. Alternatively, its conjugate acid is a weaker acid than almost any other protonated amine.

$$H_2N-\overset{\overset{\displaystyle NH}{\|}}{C}-NH_2 + H_2O \rightleftharpoons H_2N-\overset{\overset{\displaystyle NH_2^+}{\|}}{C}-NH_2 + OH^- \quad pK_a = 13.6$$

Guanidine   Guanidinium ion

The remarkable basicity of guanidine is attributed to the fact that the positive charge on the guanidinium ion is delocalized equally over the three nitrogen atoms as shown by these three equivalent contributing structures.

$$H_2\overset{..}{N}-\overset{\overset{\displaystyle \overset{+}{N}H_2}{\|}}{C}-\overset{..}{N}H_2 \longleftrightarrow H_2\overset{+}{N}=\overset{\overset{\displaystyle \overset{..}{N}H_2}{|}}{C}-\overset{..}{N}H_2 \longleftrightarrow H_2\overset{..}{N}-\overset{\overset{\displaystyle \overset{..}{N}H_2}{|}}{C}=\overset{+}{N}H_2$$

Three equivalent contributing structures

22.7 Reactions with Acids

Amines, whether soluble or insoluble in water, react quantitatively with strong acids to form water-soluble salts as illustrated by the reaction of norepinephrine (noradrenaline) with aqueous HCl to form a hydrochloride salt. Norepinephrine, secreted by the medulla of the adrenal gland, is a neurotransmitter. It has been suggested that it acts in those areas of the brain that mediate emotional behavior.

(R)-(−)-Norepinephrine + HCl $\xrightarrow{H_2O}$ (R)-(−)-Norepinephrine hydrochloride

(R)-(−)-Norepinephrine
(only slightly soluble in water)

(R)-(−)-Norepinephrine hydrochloride
(a water-soluble salt)

Example 22.7

Complete each acid-base reaction and name the salt formed.

(a) $(CH_3CH_2)_2NH + HCl \longrightarrow$ (b) $C_6H_5CH_2NH_2 + CH_3COOH \longrightarrow$

Solution

(a) $(CH_3CH_2)_2NH_2^+ Cl^-$ (b) $C_6H_5CH_2NH_3^+ CH_3COO^-$

    Diethylammonium chloride   Benzylammonium acetate

Problem 22.7

Complete each acid-base reaction and name the salt formed.

(a) $(CH_3CH_2)_3N + HCl \longrightarrow$ (b) ⬡NH + $CH_3COOH \longrightarrow$

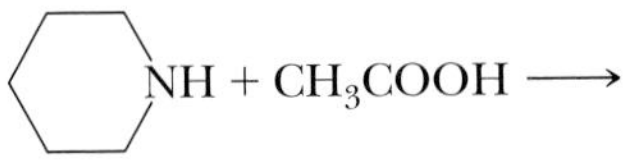

Example 22.8

Following are two structural formulas for alanine (2-aminopropanoic acid), one of the building blocks of proteins (Chapter 27). Is alanine better represented by structural formula (A) or (B)?

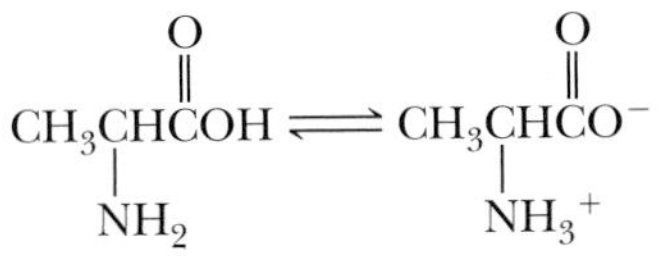

$$CH_3CHCOH \rightleftharpoons CH_3CHCO^-$$

      (with $\overset{O}{\overset{\|}{}}$ and NH_2)   (with $\overset{O}{\overset{\|}{}}$ and NH_3^+)

    (A)   (B)

Solution

Structural formula (A) contains both an amino group (a base) and a carboxyl group (an acid). Proton transfer from the stronger acid (—COOH) to the stronger base (—NH₂) gives an internal salt; therefore, (B) is the better representation for alanine. Within the field of amino acid chemistry, the internal salt represented by (B) is called a **zwitterion** (Section 27.2).

Zwitterion   An internal salt of an amino acid.

Problem 22.8

Following are structural formulas for propanoic acid and the conjugate acids of isopropylamine and alanine, along with pK_a values for each functional group:

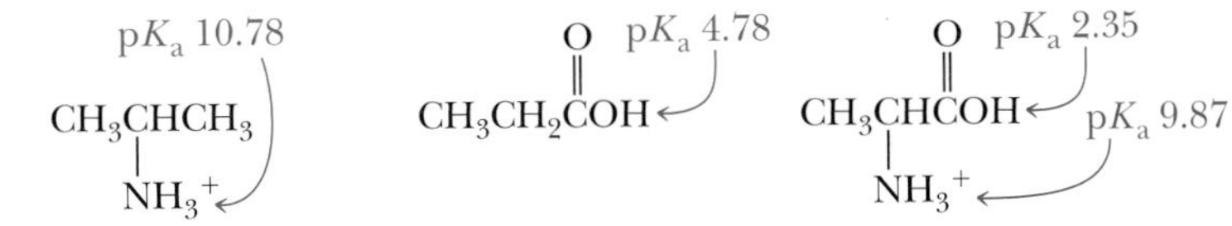

pK_a 10.78

CH_3CHCH_3
|
NH_3^+

Conjugate acid of isopropylamine

O pK_a 4.78

CH_3CH_2COH

Propanoic acid

O pK_a 2.35

CH_3CHCOH — pK_a 9.87
|
NH_3^+

Conjugate acid of alanine

(a) How do you account for the fact that the —NH$_3^+$ group of the conjugate acid of alanine is a stronger acid than the —NH$_3^+$ group of the conjugate acid of isopropylamine?

(b) How do you account for the fact that the —COOH group of the conjugate acid of alanine is a stronger acid than the —COOH group of propanoic acid?

The basicity of amines and the solubility in water of amine salts can be used to separate water-insoluble amines from water-insoluble, nonbasic compounds. Shown in Figure 22.2 is a flow chart for the separation of aniline from acetanilide, a neutral compound.

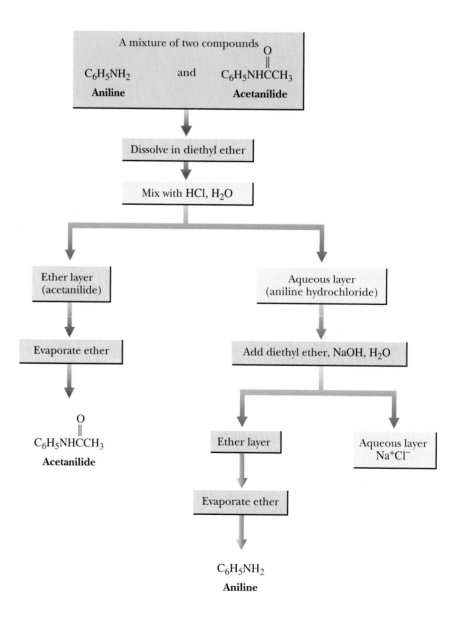

Figure 22.2

Separation and purification of an amine and a neutral compound.

Example 22.9

Here is a flow chart for the separation of a mixture of a primary aliphatic amine (RNH_2, pK_a 10.8), a carboxylic acid (RCOOH, pK_a 5), and a phenol (ArOH, pK_a 10). Assume that each is insoluble in water but soluble in diethyl ether. The mixture is separated into fractions A, B, and C. Which fraction contains the amine, which the carboxylic acid, and which the phenol?

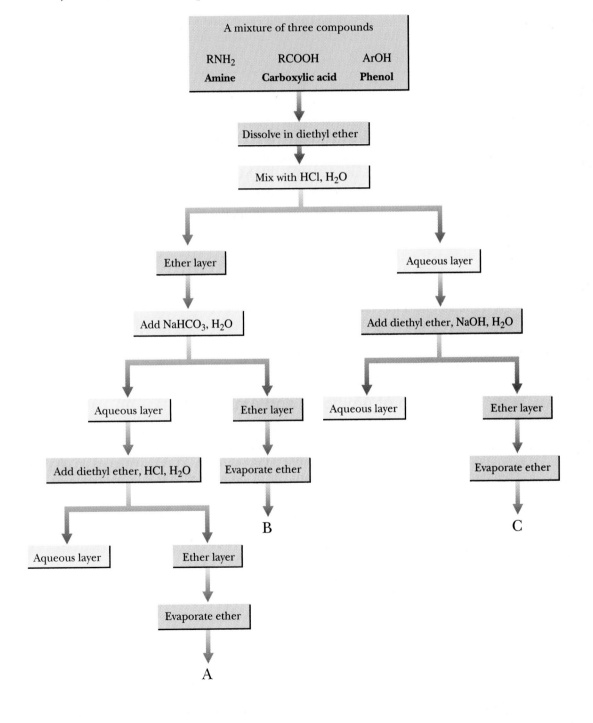

Solution

Fraction C contains RNH_2, fraction B contains ArOH, and fraction A contains RCOOH.

Problem 22.9

In what way(s) might the results of the separation and purification procedure outlined in Example 22.9 be different if the following were to occur?

(a) Aqueous NaOH is used in place of aqueous $NaHCO_3$.
(b) The starting mixture contains an aromatic amine, $ArNH_2$, rather than an aliphatic amine, RNH_2.

Reactivity Explorer
Amines

22.8 Preparation

The synthesis of amines is primarily a problem of how to form a carbon-nitrogen bond and, if the newly formed nitrogen-containing compound is not already an amine, how to convert it into an amine. We have already seen the following methods for the preparation of amines.

1. Nucleophilic ring opening of epoxides by ammonia and amines (Section 11.9B)
2. Addition of nitrogen nucleophiles to the carbonyl group of aldehydes and ketones to form imines (Section 16.10)
3. Reduction of imines to amines (Section 16.10)
4. Reduction of amides by $LiAlH_4$ (Section 18.11B)
5. Reduction of nitriles to primary amines (Section 18.11C)
6. Hofmann rearrangement of amides (Section 18.12)
7. Nitration of an arene followed by reduction of the nitro group to a primary amine (Section 21.1B).

In this chapter, we present two additional methods for the preparation of amines.

A. Alkylation of Ammonia and Amines

Surely one of the most direct synthetic routes to an amine would seem to be treatment of an alkyl halide with ammonia or an amine. Reaction between these two compounds by a second-order nucleophilic substitution (S_N2) reaction gives an alkylammonium salt, as illustrated by treatment of bromomethane with ammonia to give methylammonium bromide.

$$CH_3Br + NH_3 \xrightarrow{S_N2} CH_3NH_3^+ \ Br^-$$

Methylammonium
bromide

Unfortunately, reaction does not stop at this stage but continues to give a complex mixture of products as shown in the following equation.

$$CH_3Br + NH_3 \longrightarrow CH_3NH_3^+Br^- + (CH_3)_2NH_2^+Br^- + (CH_3)_3NH^+Br^- + (CH_3)_4N^+Br^-$$

This mixture is formed in the following way. Proton transfer between ammonia and methylammonium ion gives ammonium ion and methylamine, also a good nucleophile, which then undergoes reaction with bromomethane to give dimethylammonium bromide. A second proton transfer reaction converts the dimethylammonium ion to dimethylamine, yet another good nucleophile, which also participates in nucleophilic substitution, and so on.

$$CH_3NH_3^+ \ Br^- + NH_3 \xrightleftharpoons{\text{proton transfer}} CH_3NH_2 + NH_4^+ \ Br^-$$

Methylammonium bromide   Methylamine

$$CH_3Br + CH_3NH_2 \xrightarrow{S_N2} (CH_3)_2NH_2^+ \ Br^-$$

Dimethylammonium bromide

$$(CH_3)_2NH_2^+ \ Br^- + NH_3 \xrightleftharpoons{\text{proton transfer}} (CH_3)_2NH + NH_4^+ \ Br^-$$

Dimethylamine

The final product from such a series of nucleophilic substitution and proton transfer reactions is a tetraalkylammonium halide. The relative proportions of the various alkylation products depend on the ratio of alkyl halide to ammonia in the reaction mixture. Whatever the starting mixture, however, the product is almost invariably a mixture of alkylated products. For this reason, alkylation of ammonia or amines is not a generally useful laboratory method for the preparation of more complex amines. However, primary amines are easily prepared because ammonia is inexpensive and can be used in large excess. Other amines can also be prepared in this way if the nucleophilic amine is inexpensive enough to be used in large excess.

B.   Alkylation of Azide Ion

As we have just seen, alkylation of ammonia or amines is generally not a useful method for the preparation of amines. One strategy for eliminating the problem of overalkylation is to use a form of nitrogen that can function as a nucleophile but that is no longer an effective nucleophile once it has formed a new carbon-nitrogen bond. One such nucleophilic form of nitrogen is the azide ion, N_3^-. Alkyl azides are easily prepared from sodium or potassium azide and a primary or secondary alkyl halide by an S_N2 reaction. Azides are, in turn, reduced to primary amines by a variety of reducing agents including lithium aluminum hydride.

$$N_3^- \qquad :\overset{-}{\underset{}{N}} = \overset{+}{\underset{}{N}} = \overset{-}{\underset{}{N}}: \qquad\qquad RN_3 \qquad R - \overset{}{\underset{}{N}} = \overset{+}{\underset{}{N}} = \overset{-}{\underset{}{N}}:$$

Azide ion (a good nucleophile)   An alkyl azide

Benzyl chloride   Benzyl azide   Benzylamine

The azide ion can also be used for stereospecific ring opening of epoxides. Reduction of the resulting β-azidoalcohol gives a β-aminoalcohol as illustrated by the conversion of cyclohexene to *trans*-2-aminocyclohexanol. Oxidation of cyclohexene by a peroxyacid (Section 11.8B) gives an epoxide. Stereospecific nucleophilic attack by azide ion anti to the leaving oxygen of the epoxide ring (Section 11.9B) followed by reduction of the azide with lithium aluminum hydride gives *trans*-2-aminocyclohexanol.

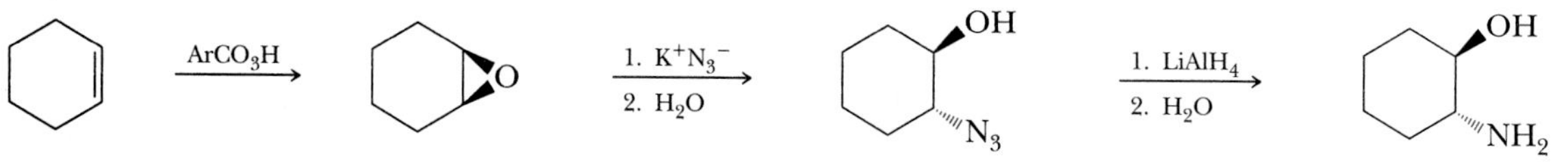

| Cyclohexene | 1,2-Epoxycyclohexane | *trans*-2-Azidocyclohexanol | *trans*-2-Aminocyclohexanol |

Example 22.10

Show how to convert 4-methoxybenzyl chloride to each amine in good yield.

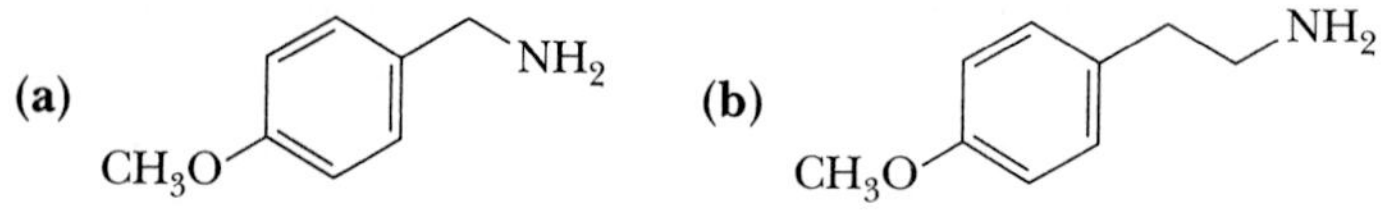

Solution

(a) Two methods might be used: (1) alkylation of NH_3 using a large molar excess of NH_3 to reduce the extent of overalkylation or (2) nucleophilic displacement of chloride using azide ion (from NaN_3) followed by $LiAlH_4$ reduction of the azide. Of these methods, nucleophilic displacement by azide is the more convenient on a laboratory scale.

(b) Nucleophilic displacement of chloride by cyanide ion followed by reduction of the cyano group with lithium aluminum hydride.

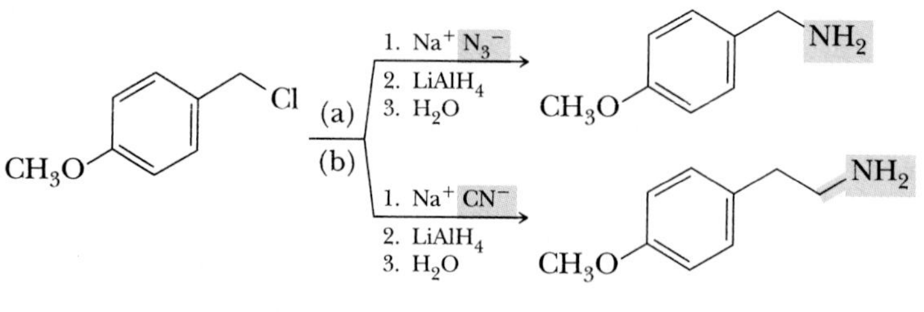

Problem 22.10

Show how to bring about each conversion in good yield. In addition to the given starting material, use any other reagents as necessary.

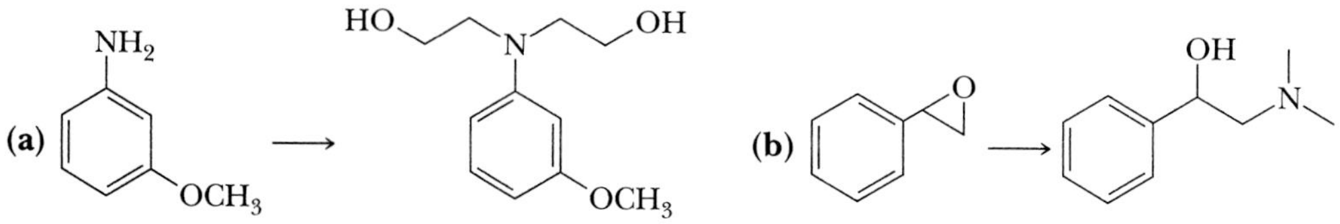

22.9   Reaction with Nitrous Acid

Nitrous acid, HNO_2, is an unstable compound that is prepared by adding sulfuric acid or hydrochloric acid to an aqueous solution of sodium nitrite, $NaNO_2$. Nitrous acid is a weak oxygen acid and ionizes according to the following equation.

$$HNO_2 + H_2O \rightleftharpoons H_3O^+ + NO_2^- \qquad K_a = 4.26 \times 10^{-4}$$
$$pK_a = 3.37$$

Nitrous acid undergoes reaction with amines in different ways, depending on whether the amine is primary, secondary, or tertiary and whether it is aliphatic or aromatic. These reactions are all related by the facts that nitrous acid (1) participates in proton-transfer reactions and (2) is a source of the nitrosyl cation, a weak electrophile.

Mechanism   Formation of the Nitrosyl Cation

Mechanisms
Amines
Nitrosyl Cation, Reaction
with an Amine

Step 1:   Protonation of the OH group of nitrous acid gives an oxonium ion.

Step 2:   Loss of water gives the nitrosyl cation, here represented as a hybrid of two contributing structures.

$$H-\ddot{\underset{..}{O}}-\ddot{N}=\ddot{\underset{..}{O}} + H^+ \overset{(1)}{\rightleftharpoons} H-\overset{+}{\underset{\underset{H}{|}}{\ddot{O}}}-\ddot{N}=\ddot{\underset{..}{O}} \overset{(2)}{\rightleftharpoons} H-\underset{\underset{H}{|}}{\ddot{O}}: \quad + \quad \overset{+}{:}\ddot{N}=\ddot{\underset{..}{O}}: \longleftrightarrow :N\equiv\overset{+}{\ddot{O}}:$$

The nitrosyl cation

A.   Tertiary Aliphatic Amines

When treated with nitrous acid, tertiary aliphatic amines, whether water-soluble or water-insoluble, are protonated to form water-soluble salts. No further reaction occurs beyond salt formation. This reaction is of little practical use.

B.   Tertiary Aromatic Amines

Tertiary aromatic amines are bases and can also form salts with nitrous acid. An alternative pathway, however, is open to tertiary aromatic amines, namely, electrophilic aromatic substitution. The nitrosyl cation is a very weak electrophile and reacts only with aromatic rings containing strongly activating, ortho-para directing groups, such as the hydroxyl and dialkylamino groups. When treated with nitrous acid, these compounds undergo nitrosation, predominantly in the para position to give blue or green aromatic nitroso compounds.

$$(CH_3)_2N-\underset{}{\bigcirc} \quad \xrightarrow[\text{2. NaOH, H}_2\text{O}]{\text{1. NaNO}_2\text{, HCl, 0-5}^\circ\text{C}} \quad (CH_3)_2N-\underset{}{\bigcirc}-N=O$$

N,N-Dimethylaniline   *N,N*-Dimethyl-4-nitrosoaniline

C.   Secondary Aliphatic and Aromatic Amines

Secondary amines, whether aliphatic or aromatic, undergo reaction with nitrous acid to give *N*-nitrosoamines, more commonly called *N*-nitrosamines, as illustrated by the reaction of piperidine with nitrous acid.

$$\text{N—H} + \text{HNO}_2 \longrightarrow \text{N—N}{=}\text{O} + \text{H}_2\text{O}$$

Piperidine   *N*-Nitrosopiperidine

Mechanism   Reaction of a 2° Amine with the Nitrosyl Cation to Give an *N*-Nitrosamine

Step 1:   Reaction of the 2° amine (a Lewis base) with the nitrosyl cation (a Lewis acid) gives an *N*-nitrosammonium ion.

Step 2:   Proton transfer to solvent gives the *N*-nitrosamine.

N-Nitrosamines are of little synthetic or commercial value. They have received considerable attention in recent years, however, because many of them are potent carcinogens. Following are structural formulas of two *N*-nitrosamines, each of which is a known carcinogen.

$$\begin{array}{c} \text{H}_3\text{C} \\ \\ \text{H}_3\text{C} \end{array}\!\!\!\!\!\text{N—N}{=}\text{O} \qquad\qquad \text{N—N}{=}\text{O}$$

N-Nitrosodimethylamine
(found in cigarette smoke
and when bacon "preserved"
with sodium nitrite is fried)

N-Nitrosopyrrolidine
(formed when bacon "preserved"
with sodium nitrite is fried)

It has been common practice within the food industry to add sodium nitrite to processed meats to "retard spoilage," that is, to inhibit the growth of *Clostridium botulinum*, the bacterium responsible for botulism poisoning. Although this practice was well grounded before the days of adequate refrigeration, it is of questionable value today. Sodium nitrite is also added to prevent red meats from turning brown. Controversy over the use of sodium nitrite has been generated by the demonstration that nitrite ion in the presence of acid converts secondary amines to *N*-nitrosamines and that many *N*-nitrosamines are powerful carcinogens. This demonstration led in turn to pressure by consumer groups to force the Food and Drug Administration to ban the use of nitrite additives in foods. The strength of the argument to ban nitrites was weakened with the finding that enzymes in our mouths and intestinal tracts have the ability to catalyze the reduction of nitrate to nitrite. Nitrate ion is normally found in a wide variety of foods and in drinking water. To date, there is no evidence that nitrite as a food additive poses any risk not already present through our existing dietary habits. The FDA has established the current permissible level of sodium nitrite in processed meats as 50 to 125 ppm (that is, 50–125 μg nitrite per gram of cured meat).

D. Primary Aliphatic Amines

Treatment of a primary aliphatic amine with nitrous acid results in loss of nitrogen, N_2, and formation of substitution, elimination, and rearrangement products as illustrated by the treatment of butylamine with nitrous acid.

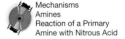

The mechanism by which this mixture of products is formed involves formation of a **diazonium ion.** The conversion of a primary amine to a diazonium ion is called **diazotization.**

Diazonium ion An ArN_2^+ or RN_2^+ ion.

Mechanism   **Reaction of a 1° Amine with Nitrous Acid to Give a Diazonium Ion**

Mechanisms
Amines
Reaction of a Primary
Amine with Nitrous Acid

Step 1: Reaction of the 1° amine with the nitrosonium ion from nitrous acid gives an *N*-nitrosamine, which undergoes keto-enol tautomerism (Section 16.11B) to give a diazotic acid, so named because it has two (di-) nitrogen (-azot-) atoms within its structure.

$$R-\overset{\frown}{\overset{..}{N}H_2} \; + \; \overset{+}{:}N=\overset{..}{\overset{..}{O}} \; \longrightarrow \; R-\overset{\overset{\displaystyle H}{|}}{\underset{..}{N}}-\overset{..}{N}=\overset{..}{\overset{..}{O}} \; \overset{\text{keto-enol}}{\underset{\text{tautomerism}}{\rightleftharpoons}} \; R-\overset{..}{N}=\overset{..}{N}-\overset{..}{\overset{..}{O}}-H$$

A 1° aliphatic   An *N*-nitrosamine   A diazotic acid
amine

Step 2: Protonation of the diazotic acid followed by loss of H_2O gives a diazonium ion.

$$R-\overset{..}{N}=\overset{..}{N}-\overset{..}{\underset{..}{O}}-H \; \overset{H^+}{\rightleftharpoons} \; R-\overset{..}{N}=\overset{..}{N}-\overset{\overset{H}{|}}{\underset{..}{O}}-H \; \overset{-H_2O}{\longrightarrow} \; R-\overset{+}{N}\equiv N: \; \longrightarrow \; R^+ \; + \; :N\equiv N:$$

A diazotic acid   A diazonium acid   A carbo-
                                                                          cation

Aliphatic diazonium ions are unstable, even at 0°C, and immediately lose nitrogen to give carbocations and nitrogen gas. The driving force for this reaction is the fact that N_2 is one of the best leaving groups because it is an extraordinarily weak base. It is removed from the reaction mixture as a gas as it is formed. The carbocation now has open to it the three reactions in the repertoire of aliphatic carbocations: (1) loss of a proton to form an alkene, (2) reaction with a nucleophile to give a substitution product, and (3) rearrangement to a more stable carbocation and then reaction further by (1) or (2).

Because treatment of a primary aliphatic amine with HNO_2 gives a mixture of products, it is generally not a useful reaction. An exception is the **Tiffeneau-Demjanov reaction** in which a cyclic β-aminoalcohol is treated with nitrous acid to form a ring-expanded ketone, with evolution of nitrogen.

A β-aminoalcohol   Cycloheptanone

We can account for this molecular rearrangement as shown in the following mechanism.

Mechanism   The Tiffeneau-Demjanov Reaction

Step 1: Reaction of the 1° amine with nitrous acid gives a diazonium ion.

Step 2: Simultaneous loss of N_2 and rearrangement by a 1,2-shift gives the conjugate acid of the final product as a resonance-stabilized cation.

Step 3: Proton transfer from this cation to solvent completes the reaction.

A diazonium ion

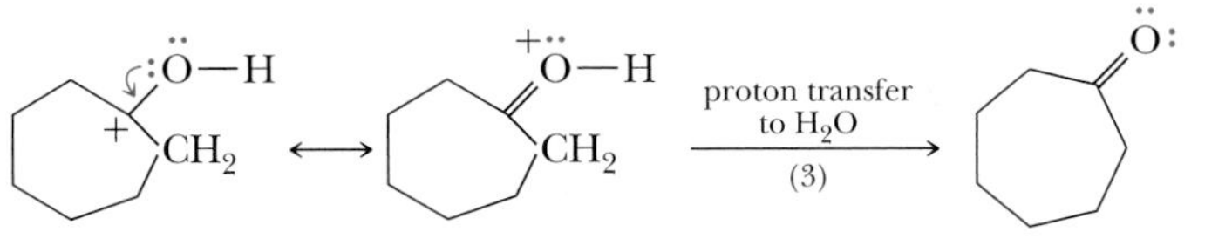

A resonance-stabilized cation   Cycloheptanone

The driving force for this molecular rearrangement is precisely what we already saw for other cation rearrangements: transformation of a less stable cation into a more stable cation. This reaction is analogous to the pinacol rearrangement (Section 9.7) with the leaving group being N_2 rather than H_2O.

Example 22.11

The following sequence of reactions gives cyclooctanone. Propose a structural formula for compound A and a mechanism for its conversion to cyclooctanone.

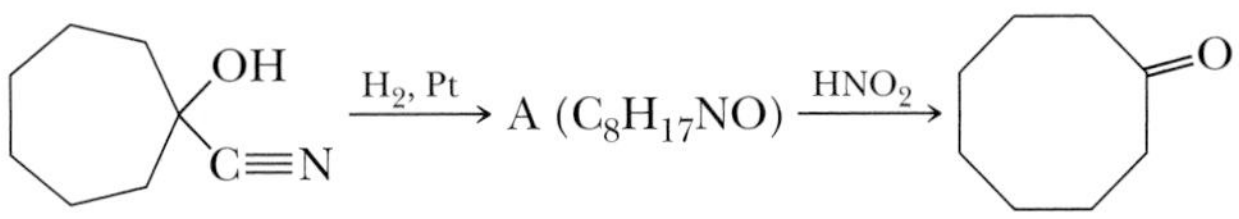

Solution

Catalytic hydrogenation using hydrogen over a platinum catalyst reduces the carbon-nitrogen triple bond to a single bond (Section 18.11C) and gives a β-aminoalcohol. Treatment of the β-aminoalcohol with nitrous acid results in loss of N_2 and expansion of the seven-membered ring to an eight-membered cyclic ketone.

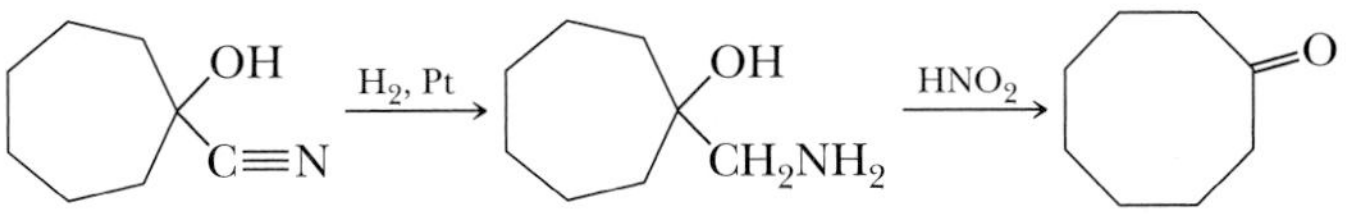

Problem 22.11

How might you bring about this conversion?

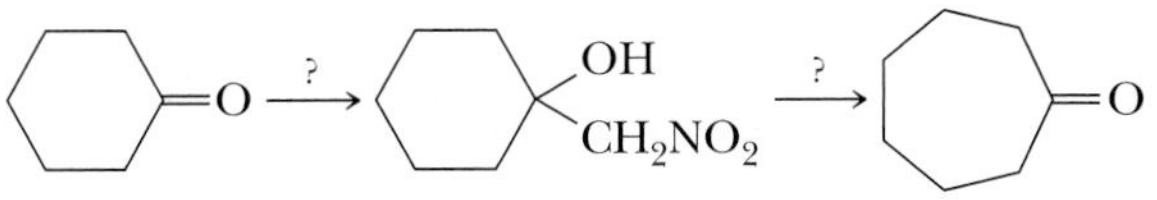

E. Primary Aromatic Amines

Primary aromatic amines, like primary aliphatic amines, undergo reaction with nitrous acid to form arenediazonium salts. Arenediazonium salts, unlike their aliphatic counterparts, are stable at 0°C and can be kept in solution for short periods without decomposition. When an arenediazonium salt is treated with an appropriate reagent, nitrogen is lost and replaced by another atom or functional group. What makes reactions of primary aromatic amines with nitrous acid so valuable is the fact that the amino group can be replaced in a totally regiospecific manner by the groups shown.

Aromatic amines can be converted to phenols by first forming the arenediazonium salt in aqueous sulfuric acid and then heating the solution. In this manner, 2-bromo-4-methylaniline is converted to 2-bromo-4-methylphenol.

2-Bromo-4-methylaniline   2-Bromo-4-methylphenol

The intermediate in the decomposition of an arenediazonium ion in water is an aryl cation, which then undergoes reaction with water to form the phenol. Note that the aryl cation is so unstable that it is only with N_2 as the leaving group that it can be formed.

Benzenediazonium   An aryl cation   Phenol
ion

This reaction of arenediazonium salts represents the main laboratory preparation of phenols.

Example 22.12

What reagents and experimental conditions will bring about each step in the conversion of toluene to 4-hydroxybenzoic acid?

Solution

Step 1:   Nitration of the aromatic ring (Section 21.1B) using HNO_3 in H_2SO_4 followed by separation of the ortho and para isomers gives 4-nitrotoluene.

Step 2: Oxidation at the benzylic carbon (Section 20.6A) using $K_2Cr_2O_7$ in H_2SO_4 gives 4-nitrobenzoic acid.

Step 3: Reduction of the nitro group (Section 21.1B) to an amino group using H_2 in the presence of Ni or other transition metal catalyst gives 4-aminobenzoic acid. Alternatively, reduction of the nitro group can be brought about using Zn, Sn, or Fe in aqueous HCl followed by aqueous NaOH.

Step 4: Reaction of the aromatic amine with $NaNO_2$ in aqueous H_2SO_4 followed by heating gives 4-hydroxybenzoic acid.

Problem 22.12

Show how to convert toluene to 3-hydroxybenzoic acid using the same set of reactions as in Example 22.12, but changing the order in which two or more of the steps are carried out.

The **Schiemann reaction** is the most common method for introduction of fluorine onto an aromatic ring. It is carried out by treatment of a primary aromatic amine with sodium nitrite in aqueous HCl followed by addition of HBF_4 or $NaBF_4$. The diazonium fluoroborate salt precipitates and is collected and dried. Heating the dry salt brings about its decomposition to an aryl fluoride, nitrogen, and boron trifluoride. The Schiemann reaction is also thought to involve an aryl cation intermediate.

A diazonium fluoroborate   Fluorobenzene

Treatment of a primary aromatic amine with nitrous acid followed by heating with HCl/CuCl, HBr/CuBr, or KCN/CuCN results in replacement of the diazonium group by —Cl, —Br, or —CN, respectively, and is known as the **Sandmeyer reaction.** The Sandmeyer reaction fails when attempted with CuI or CuF.

Treatment of an arenediazonium ion with iodide ion, generally from potassium iodide, is the best and most convenient method for introducing iodine onto an aromatic ring.

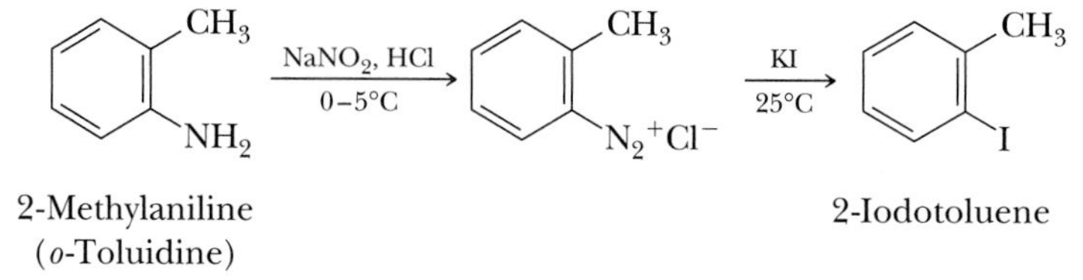

2-Methylaniline
(*o*-Toluidine)

2-Iodotoluene

Treatment of an arenediazonium ion with hypophosphorous acid, H_3PO_2, results in reduction of the diazonium group and its replacement by —H as illustrated by the conversion of aniline to 1,3,5-trichlorobenzene. Recall that —NH_2 is a powerful activating group (Section 21.2A). Treatment of aniline with chlorine requires no catalyst and gives 2,4,6-trichloroaniline. To complete the conversion, the —NH_2 group is removed by treatment with nitrous acid followed by hypophosphorous acid to give 1,3,5-trichlorobenzene.

Aniline   2,4,6-Trichloroaniline   2,4,6-Trichlorobenzene-diazonium chloride   1,3,5-Trichlorobenzene

---

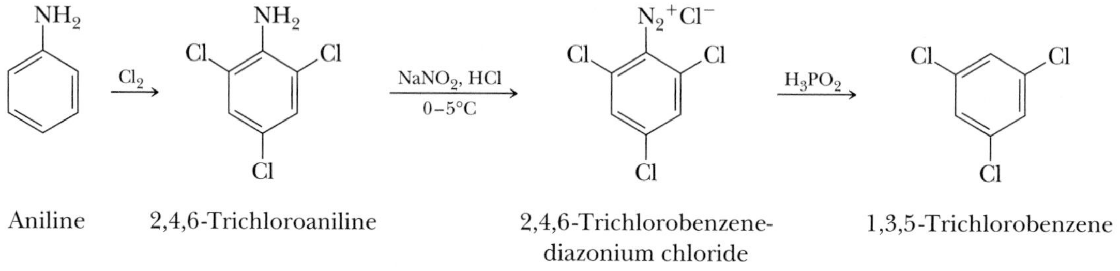

Example 22.13

Show reagents and conditions that will bring about the conversion of toluene to 3-bromo-4-methylphenol.

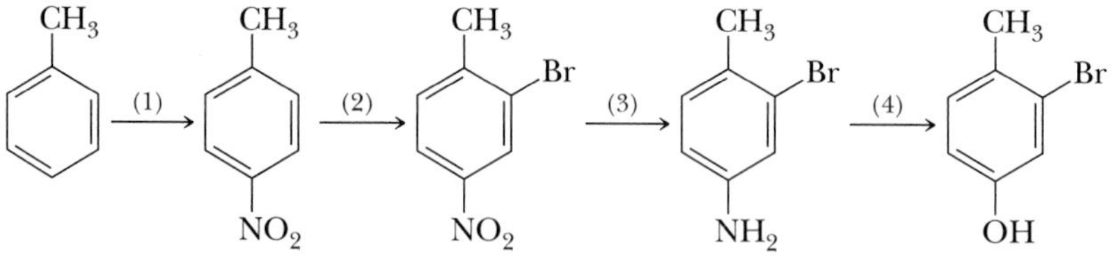

Solution

Step 1: HNO_3 in H_2SO_4. Methyl is ortho-para directing and slightly activating.

Step 2: Treat 4-nitrotoluene with bromine in the presence of $FeCl_3$.

Step 3: Reduce the nitro group either using H_2/Ni or using Sn, Zn, or Fe in aqueous HCl followed by aqueous NaOH.

Step 4: Diazotization of the amine with $NaNO_2$ in aqueous sulfuric acid followed by warming of the solution replaces —N_2^+ by —OH.

Problem 22.13

Starting with 3-nitroaniline, show how to prepare the following compounds.

(a) 3-Nitrophenol   **(b)** 3-Bromoaniline

(c) 1,3-Dihydroxybenzene (resorcinol)   **(d)** 3-Fluoroaniline

(e) 3-Fluorophenol   **(f)** 3-Hydroxybenzonitrile

22.10   Hofmann Elimination

When a quaternary ammonium halide is treated with moist silver oxide (a slurry of Ag_2O in H_2O), silver halide precipitates, leaving a solution of a quaternary ammonium hydroxide.

(Cyclohexylmethyl)trimethyl-   Silver   (Cyclohexylmethyl)trimethyl-
ammonium iodide   oxide   ammonium hydroxide

In the mid-19th century, Augustus Hofmann (for whom the Hofmann rearrangement, Section 18.12, is named) discovered that when a quaternary ammonium hydroxide is heated, it decomposes to an alkene, a tertiary amine, and water. Thermal decomposition of a quaternary ammonium hydroxide to an alkene is known as **Hofmann elimination.**

(Cyclohexylmethyl)trimethyl-   Methylenecyclohexane   Trimethylamine
ammonium hydroxide

Hofmann elimination   When treated with a strong base, a quaternary ammonium halide undergoes β-elimination by an E2 mechanism to give the less-substituted alkene as the major product.

The Hofmann elimination has most of the characteristics of an E2 reaction (Section 8.10B). First, Hofmann eliminations are concerted, meaning that bond-breaking and bond-forming steps occur simultaneously, or nearly so. Second, Hofmann eliminations are stereospecific anti eliminations, meaning that —H and the leaving group must be anti to each other. The following mechanism illustrates the concerted nature of bond forming and bond breaking and the anti arrangement of —H and the trialkylamino group.

Mechanism   The Hofmann Elimination

Simultaneous removal of a β-hydrogen by base, collapse of the electron pair of the C—H bond to become the pi bond of the alkene, and loss of the trialkylamino group occur. The reaction shows anti stereospecificity.

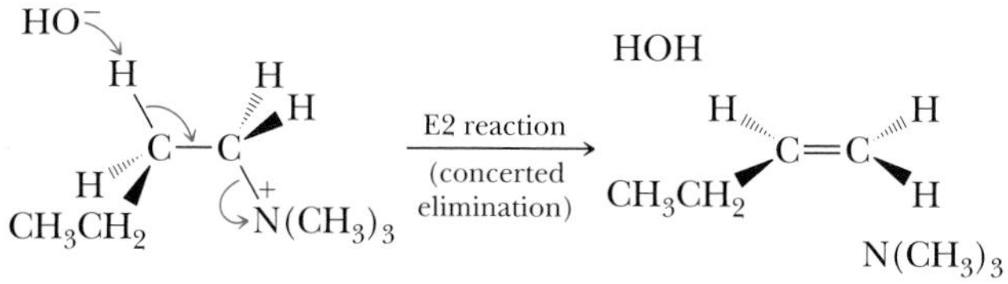

When we studied E2 reactions of alkyl halides in Section 8.10B, we saw that a β-hydrogen must be anti to the leaving group. If only one β-hydrogen meets this requirement, then the double bond is formed in that direction. If, however, two β-hydrogens meet this requirement, then elimination follows Zaitsev's rule: elimination occurs preferentially to form the more substituted double bond.

Thermal decomposition of quaternary ammonium hydroxides is different because elimination occurs preferentially to form the least substituted double bond. Thermal decomposition of *sec*-butyltrimethylammonium hydroxide, for example, gives 1-butene as the major product.

Elimination reactions that give the less substituted alkene as the major product are said to follow the **Hofmann rule.**

The following example further illustrates Hofmann elimination.

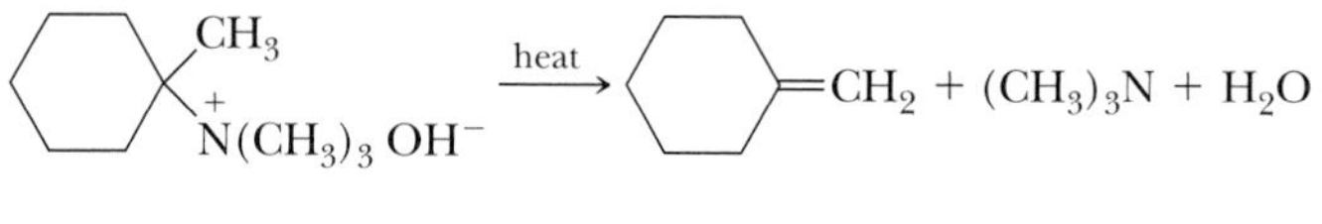

Hofmann rule   Any β-elimination that occurs preferentially to give the less substituted alkene as the major product is said to follow the Hofmann rule.

Example 22.14

Draw a structural formula of the major alkene formed in each β-elimination.

Solution

Thermal decomposition of the quaternary ammonium hydroxide in (a) follows Hofmann elimination and gives 1-octene as the major product. E2 elimination from an alkyl iodide in (b) by sodium methoxide follows Zaitsev's rule and gives 2-octene as the major product.

(a) $CH_3(CH_2)_5CH\!\!=\!\!CH_2 + (CH_3)_3N + H_2O$ **(b)** $CH_3(CH_2)_4CH\!\!=\!\!CHCH_3 + CH_3OH$

   1-Octene   2-Octene

Problem 22.14

The procedure of methylation of amines and thermal decomposition of quaternary ammonium hydroxides was first reported by Hofmann in 1851, but its value as a means of structure determination was not appreciated until 1881 when he published a report of its use in determining the structure of piperidine. Following are the results obtained by Hofmann.

$$C_5H_{11}N \xrightarrow[\substack{\text{1. } CH_3I \text{ (excess), } K_2CO_3 \\ \text{2. } Ag_2O, H_2O \\ \text{3. heat}}]{} C_7H_{15}N \xrightarrow[\substack{\text{4. } CH_3I \text{ (excess), } K_2CO_3 \\ \text{5. } Ag_2O, H_2O \\ \text{6. heat}}]{} CH_2\!\!=\!\!CHCH_2CH\!\!=\!\!CH_2$$

   Piperidine   (A)   1,4-Pentadiene

(a) Show that these results are consistent with the structure of piperidine (Section 22.1).

(b) Propose two additional structural formulas (excluding stereoisomers) for $C_5H_{11}N$ that are also consistent with the results obtained by Hofmann.

In summary, both Hofmann and Zaitsev eliminations are always anti. If only one β-hydrogen is anti to the leaving group, then that is the one removed. If more than one β-hydrogen is anti, then there is competition between Hofmann and Zaitsev elimination.

1. Eliminations involving a negatively charged leaving group, for example Cl^-, Br^-, I^-, and OTs^-, almost always follow Zaitsev's rule, unless a bulky base is used.
2. Eliminations involving a neutral leaving group, for example $N(CH_3)_3$ and $S(CH_3)_2$, almost always follow Hofmann's rule.
3. The bulkier the base, as for example $(CH_3)_3CO^-K^+$ compared with $CH_3O^-Na^+$, the greater the percentage of Hofmann product.

   One of the likeliest explanations for formation of the less stable carbon-carbon double bond is that Hofmann elimination is governed largely by steric factors, namely the bulk of the $-NR_3^+$ group. The hydroxide ion preferentially approaches and removes the least hindered β-hydrogen and gives the least substituted alkene as product. For the same reason, bulky bases, such as $(CH_3)_3CO^-K^+$, give Hofmann elimination also from alkyl halides.

22.11 Cope Elimination

Treatment of a tertiary amine with hydrogen peroxide results in oxidation of the amine to an amine oxide.

An amine oxide

When an amine oxide with at least one β-hydrogen is heated, it undergoes thermal decomposition to form an alkene and an *N,N*-dialkylhydroxylamine. Thermal decomposition of an amine oxide to an alkene is known as a **Cope elimination** after its discoverer, Arthur C. Cope, of the Massachusetts Institute of Technology.

Methylenecyclohexane   *N,N*-Dimethyl-
hydroxylamine

All experimental evidence indicates that the Cope elimination is syn stereoselective and concerted. As shown in the following mechanism, the transition state involves a planar or nearly planar arrangement of the five participating atoms and a cyclic flow of three pairs of electrons.

Mechanism   The Cope Elimination

Redistribution of six electrons in a cyclic transition state gives an alkene. Elimination shows syn stereoselectivity.

Transition state

← an alkene
← *N,N*-dimethyl-
hydroxylamine

If two or more syn β-hydrogens can be removed in a Cope elimination, there is little preference for one over another except when the double bond is conjugated with an aromatic ring. Therefore, as a method of preparation of alkenes, Cope eliminations are best used where only one alkene is possible.

Example 22.15

Following is a formula for 2-dimethylamino-3-phenylbutane. When it is treated with hydrogen peroxide and then made to undergo a Cope elimination, the major alkene formed is 2-phenyl-2-butene.

2-Dimethylamino-   2-Phenyl-2-butene
3-phenylbutane

(a) How many stereoisomers are possible for 2-dimethylamino-3-phenylbutane?

(b) How many stereoisomers are possible for 2-phenyl-2-butene?

(c) Suppose that the starting amine is the $2R,3S$ isomer. What is the configuration of the product?

Solution

(a) There are two stereocenters in the starting amine. Four stereoisomers are possible: two pair of enantiomers.

(b) There is one carbon-carbon double bond about which stereoisomerism is possible. Two stereoisomers are possible: one E,Z pair.

(c) Following is a stereodrawing of the $2R,3S$ stereoisomer showing a syn conformation of the dimethylamino group and the β-hydrogen. Cope elimination on this stereoisomer gives (E)-2-phenyl-2-butene.

(E)-2-Phenyl-2-butene

Problem 22.15

In Example 22.15, you considered the product of Cope elimination from the $2R,3S$ stereoisomer of 2-dimethylamino-3-phenylbutane. What is the product of Cope elimination from the following stereoisomers? What is the product of Hofmann elimination from each of the following?

(a) $2S,3R$ stereoisomer   **(b)** $2S,3S$ stereoisomer

Summary

Amines are classified as primary, secondary, or tertiary, depending on the number of carbon atoms bonded to nitrogen (Section 22.1). In an **aliphatic amine,** all carbon atoms bonded directly to nitrogen are derived from alkyl groups. In an **aromatic amine,** one or more of the groups bonded to nitrogen are aromatic rings. A **heterocyclic amine** is one in which the nitrogen atom is part of a ring. A **heterocyclic aromatic amine** is one in which the nitrogen atom is part of an aromatic ring.

In systematic nomenclature, aliphatic amines are named alkanamines (Section 22.2A). A cation in which a nitrogen is bonded to four alkyl or aryl groups is named as a **quaternary ammonium ion.** In common nomenclature (Section 22.2B), aliphatic amines are named alkylamines; the alkyl groups are listed in alphabetical order in one word ending in the suffix -amine.

Amines in which nitrogen is bonded to three different groups are chiral (Section 22.3) and, in principle, can be resolved. In practice, however, they undergo rapid **pyramidal inversion** with the result that a chiral amine is converted into a racemic mixture. Pyramidal inversion is not possible for quaternary ammonium ions, and chiral quaternary ammonium ions can be resolved. Chiral phosphines invert slowly and have been prepared and resolved.

Amines are polar compounds, and primary and secondary amines associate by intermolecular hydrogen bonding (Section 22.4). An N—H----N hydrogen bond is weaker than an O—H----O hydrogen bond and, therefore, amines have lower boiling points than alcohols of comparable molecular weight and structure. All classes of amines form hydrogen bonds with water and are more soluble in water than hydrocarbons of comparable molecular weight.

Amines are weak bases, and aqueous solutions of amines are basic (Section 22.6). It is common to discuss the acid-base properties of amines by reference to the K_a for their corresponding conjugate acids.

Most aliphatic amines have comparable basicity, and all are slightly stronger bases than ammonia (Section 22.6A). For representative aliphatic amines, values of pK_a are in the range 10–11. Aromatic amines are considerably weaker bases than aliphatic amines. For representative aromatic amines, values of pK_a are in the range 3–5 (Section 22.6B). Unprotonated aromatic amines have resonance stabilization due to interaction of the unshared pair of electrons on nitrogen with the pi system of the aromatic ring. This interaction decreases the availability of the electron pair for protonation.

Heterocyclic aromatic amines are considerably weaker bases than aliphatic amines because the unshared pair of electrons on nitrogen giving rise to basicity lies in an sp^2 hybrid orbital (Section 22.6C). The electron-withdrawing inductive effect of an sp^2-hybridized nitrogen compared with an sp^3-hybridized nitrogen is responsible for the decreased basicity of heterocyclic aromatic amines.

All amines, whether soluble or insoluble in water, react quantitatively with strong acids to form water-soluble salts (Section 22.7). The basicity of amines and the solubility of amine salts in water can be used to separate water-insoluble amines from water-insoluble nonbasic compounds.

Key Reactions

1. Basicity of Aliphatic Amines (Section 22.6A)

Aliphatic amines are slightly stronger bases than ammonia due to the electron-releasing effect of alkyl groups and partial delocalization of positive charge in the alkylammonium ion.

$$CH_3NH_2 + H_2O \rightleftharpoons CH_3NH_3^+ + OH^- \qquad pK_a = 10.64$$

2. Basicity of Aromatic Amines (Section 22.6B)

Aromatic amines are considerably weaker bases than aliphatic amines. Resonance stabilization by interaction of the unshared electron pair on nitrogen with the pi system decreases its availability for protonation.

$$pK_a = 4.63$$

3. Basicity of Heterocyclic Aromatic Amines (Section 22.6C)

Heterocyclic aromatic amines are considerably weaker bases than aliphatic amines.

$$pK_a = 6.95$$

4. Reaction of Amines with Strong Acids (Section 22.7)

All amines react quantitatively with strong acids to form water-soluble salts.

Insoluble in water   A water-soluble salt

5. Alkylation of Ammonia and Amines (Section 22.8A)

This method is seldom used for preparation of pure amines because of overalkylation and the difficulty of separating products.

6. Alkylation of Azide Ion Followed by Reduction (Section 22.8B)

Azides are prepared by treatment of a primary or secondary alkyl halide, or an epoxide with NaN_3 and are reduced to primary amines by a variety of reducing agents including lithium aluminum hydride.

7. Nitrosation of Tertiary Aromatic Amines (Section 22.9B)

The nitrosyl cation is a very weak electrophile and participates in electrophilic aromatic substitution only with highly activated aromatic rings.

8. Formation of *N*-Nitrosamines from Secondary Amines (Section 22.9C)

Reaction of the nitrosyl cation, a Lewis acid, with a 2° amine, a Lewis base, gives an *N*-nitrosamine.

9. Reaction of Primary Aliphatic Amines with Nitrous Acid (Section 22.9D)

Treatment of a primary aliphatic amine with nitrous acid gives an unstable diazonium salt that loses N_2 to give a carbocation. The carbocation may (1) lose a proton to give an alkene, (2) react with a nucleophile, or (3) rearrange, followed by (1) or (2).

$$CH_3CH_2CH_2CH_2NH_2 \xrightarrow[0-5°C]{NaNO_2,\ HCl} [CH_3CH_2CH_2CH_2N_2{}^+] \longrightarrow CH_3CH_2CH_2CH_2{}^+ + N_2$$

10. Reaction of Cyclic *β*-Aminoalcohols with Nitrous Acid (Section 22.9D)

Treatment of a cyclic *β*-aminoalcohol with nitrous acid leads to rearrangement and a ring-expanded ketone.

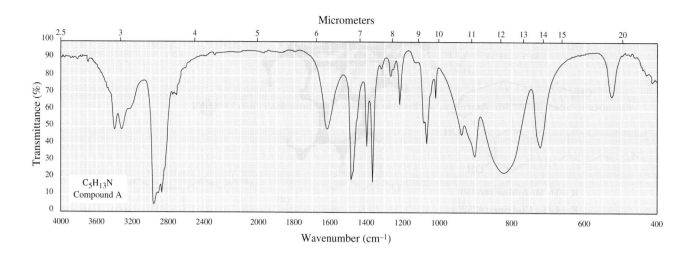

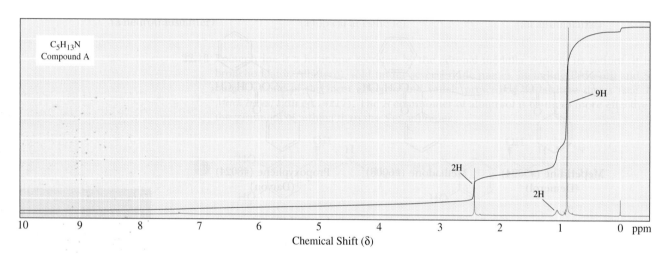

Basicity of Amines

22.24 Select the stronger base from each pair of compounds.

Preparation of Amines

22.32 Propose a synthesis of 1-hexanamine from the following.

    (a) A bromoalkane of six carbon atoms
    (b) A bromoalkane of five carbon atoms

22.33 Show how to convert each starting material into benzylamine in good yield.

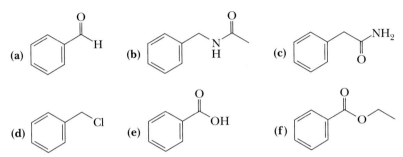

Reactions of Amines

22.34 Treatment of trimethylamine with 2-chloroethyl acetate gives acetylcholine as its chloride. Acetylcholine is a neurotransmitter. Propose a structural formula for this quaternary ammonium salt and a mechanism for its formation.

$$(CH_3)_3N + CH_3\overset{\overset{O}{\|}}{C}OCH_2CH_2Cl \longrightarrow C_7H_{16}ClNO_2$$

$$\text{Acetylcholine chloride}$$

22.35 *N*-Nitrosamines by themselves are not significant carcinogens. However, they are activated in the liver by a class of iron-containing enzymes (members of the cytochrome P-450 family). Activation involves the oxidation of a C—H bond next to the amine nitrogen to a C—OH group. Show how this hydroxylation product can be transformed into an alkyldiazonium ion, an active carcinogen, in the presence of an acid catalyst.

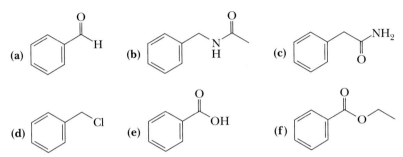

    N-Nitroso-   2-Hydroxy-*N*-   An alkyl diazonium ion
    piperidine   nitrosopiperidine   (a carcinogen)

22.36 Marked similarities exist between the mechanism of nitrous acid deamination of β-aminoalcohols and the pinacol rearrangement. Following are examples of each.

    Nitrous acid
    deamination of
    a β-aminoalcohol:

    Pinacol
    rearrangement:

(a) Analyze the mechanism of each rearrangement, and list their similarities.

(b) Why does the first reaction give ring expansion but not the second?

(c) Suggest a β-aminoalcohol that would give cyclohexanecarbaldehyde as a product.

22.37 Propose a mechanism for this conversion. Your mechanism must account for the fact that there is retention of configuration at the stereocenter. (*S*)-Glutamic acid is one of the 20 amino acid building blocks of polypeptides and proteins (Chapter 27).

(*S*)-Glutamic acid

22.38 The following sequence of methylation and Hofmann elimination was used in the determination of the structure of this bicyclic amine. Compound B is a mixture of two isomers.

(a) Propose structural formulas for compounds A and B.

(b) Suppose that you were given the structural formula of compound B but only the molecular formulas for compound A and the starting bicyclic amine. Given this information, is it possible, working backward, to arrive at an unambiguous structural formula for compound A? For the bicyclic amine?

22.39 Propose a structural formula for the compound $C_{10}H_{16}$, and account for its formation.

22.40 An amine of unknown structure contains one nitrogen and nine carbon atoms. The ^{13}C-NMR spectrum shows only five signals, all between 20 and 60 ppm. Three cycles of Hofmann elimination sequence [(1) CH_3I; (2) Ag_2O, H_2O; (3) heat] give trimethylamine and 1,4,8-nonatriene. Propose a structural formula for the amine.

22.41 The Cope elimination of tertiary amine *N*-oxides involves a planar transition state and cyclic redistribution of $(4n + 2)$ electrons. The pyrolysis of acetic esters to give an alkene and acetic acid is also thought to involve a planar transition state and cyclic redistribution of $(4n + 2)$ electrons. Propose a mechanism for pyrolysis of the following ester.

Butyl acetate   1-Butene   Acetic acid

Synthesis

22.42 Propose steps for the following conversions using a reaction of a diazonium salt in at least one step of each conversion.

(a) Toluene to 4-methylphenol (*p*-cresol)
(b) Nitrobenzene to 3-bromophenol
(c) Toluene to *p*-cyanobenzoic acid
(d) Phenol to *p*-iodoanisole
(e) Acetanilide to *p*-aminobenzylamine
(f) Toluene to 4-fluorobenzoic acid
(g) 3-Methylaniline (*m*-toluidine) to 2,4,6-tribromobenzoic acid

22.43 Starting materials for the synthesis of the herbicide propranil are benzene and propanoic acid. Show how to bring about this synthesis.

Propranil

22.44 Show how to bring about each step in the following synthesis.

22.45 Show how to bring about this synthesis.

22.46 Following are steps in a conversion of phenol to 4-methoxybenzylamine. Show how to bring about each step in good yield.

22.47 Methylparaben (see *The Merck Index*, 12th ed., #6182) is used as a preservative in foods, beverages, and cosmetics. Provide a synthesis of this compound from toluene.

Methyl *p*-hydroxybenzoate
(Methylparaben)

22.48 Show how to synthesize the following tertiary amine from benzene and any necessary reagents.

22.49 *N*-Substituted morpholines are a building block in many drugs. Show how to synthesize *N*-methylmorpholine given this retrosynthetic analysis.

N-Methylmorpholine   Methylamine   Ethylene oxide

22.50 Propose a synthesis for the systemic agricultural fungicide tridemorph (see *The Merck Index*, 12th ed., #9793) from dodecanoic acid (lauric acid) and propene.

$$\Longrightarrow CH_3(CH_2)_{10}COOH + CH_3CH{=}CH_2$$

Tridemorph   Dodecanoic acid   Propene
                       (Lauric acid)

22.51 The Ritter reaction is especially valuable for the synthesis of 3° alkanamines. In fact, there are few alternative routes to them. This reaction is illustrated by the first step in the following sequence. In the second step, the Ritter product is hydrolyzed to the amine.

Ritter product

(a) Propose a mechanism for the Ritter reaction.

(b) What is the product of a Ritter reaction using acetonitrile, CH_3CN, followed by reduction of the Ritter product with lithium aluminum hydride?

22.52 Several diamines are building blocks for the synthesis of pharmaceuticals and agrochemicals. Show how both 1,3-propanediamine and 1,4-butanediamine can be prepared from acrylonitrile.

H_2N⌃⌄NH_2 H_2N⌃⌄⌃NH_2 $CH_2{=}CH{-}C{\equiv}N$

  1,3-Propanediamine   1,4-Butanediamine   Acrylonitrile

22.53 According to the following retrosynthetic analysis, the intravenous anesthetic 2,6-diisopropylphenol (propofol; see *The Merck Index*, 12th ed., #8020) can be synthesized from phenol. Show how this synthesis might be carried out.

    2,6-Diisopropylphenol   Phenol
          (Propofol)

22.54 Following is a retrosynthetic analysis for propoxyphene (see *The Merck Index*, 12th ed., #8024). Darvon is the hydrochloride salt of propoxyphene. Darvon-N is its naphthalenesulfonic acid salt. The configuration of the carbon in Darvon bearing the —OH group is S, and the configuration of the other stereocenter is R. Its enantiomer has no analgesic properties, but is used as a cough suppressant.

    Propoxyphene

                        1-Phenyl-1-propanone
                         (Propiophenone)

Propose a synthesis for propoxyphene from 1-phenyl-1-propanone and any other necessary reagents.

23

CONJUGATED SYSTEMS

In Chapters 5 and 6, we discussed the structure and characteristic reactions of alkenes. We limited this discussion to molecules containing one double bond. In this chapter, we extend our study of alkenes to include molecules that contain two or more conjugated double bonds. As we shall see, although conjugated dienes undergo many of the same reactions characteristic of unconjugated alkenes, they also undergo their own unique set of characteristic reactions.

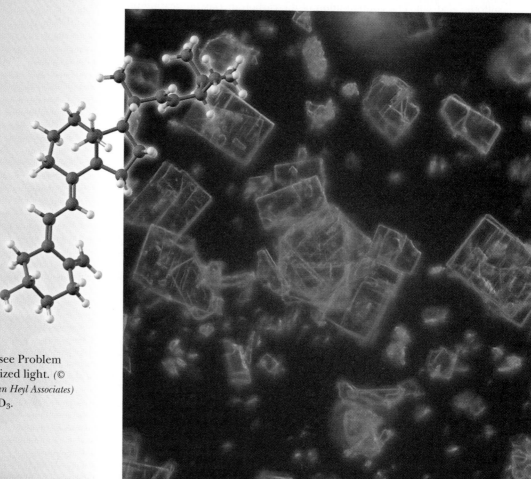

■ Crystals of vitamin D_3 (see Problem 23.41) viewed under polarized light. (© *1997 Herb Charles Ohlmeyer/Fran Heyl Associates*) Inset: A model of vitamin D_3.

23.1 Stability of Conjugated Dienes

Dienes are compounds that contain two carbon-carbon double bonds. Dienes can be divided into three groups: unconjugated, conjugated, and cumulated. An **unconjugated diene** is one in which the double bonds are separated by two or more single bonds. A **conjugated diene** is one in which the double bonds are separated by one single bond. A **cumulated diene** is one in which two double bonds share a carbon.

$$CH_2{=}CHCH_2CH{=}CH_2 \qquad CH_2{=}CHCH{=}CHCH_3 \qquad CH_2{=}C{=}CHCH_2CH_3$$

1,4-Pentadiene   1,3-Pentadiene   1,2-Pentadiene

(an unconjugated diene)   (a conjugated diene)   (a cumulated diene)

Example 23.1

Which of these molecules contain conjugated double bonds?

 (a)   (b)   (c)   (d)

Solution

Compounds (b) and (c) contain conjugated double bonds. The double bonds in compounds (a) and (d) are unconjugated.

Problem 23.1

Which of these terpenes (Section 5.4) contains conjugated double bonds?

(a) Geraniol   (b) Limonene   (c) An aggregating pheromone of bark beetles

---

Given in Table 23.1 are heats of hydrogenation for several alkenes and conjugated dienes. By using these data, we can compare the relative stabilities of conjugated and unconjugated dienes.

The simplest conjugated diene is 1,3-butadiene, but because this molecule has only four carbon atoms, it has no unconjugated constitutional isomer. Nonetheless, we can estimate the effect of conjugation of two double bonds in this molecule in the following way. The heat of hydrogenation of 1-butene is -127 kJ (-30.3 kcal)/mol. A molecule of 1,3-butadiene has two terminal double bonds, each with the same degree of substitution as the one double bond in 1-butene; therefore, we can estimate

Table 23.1 **Heats of Hydrogenation of Several Alkenes and Conjugated Dienes**

Name	Structural Formula	ΔH^0 kJ (kcal)/mol
1-Butene		$-127\ (-30.3)$
1-Pentene		$-126\ (-30.1)$
cis-2-Butene		$-120\ (-28.6)$
trans-2-Butene		$-115\ (-27.6)$
1,3-Butadiene		$-237\ (-56.5)$
trans-1,3-Pentadiene		$-226\ (-54.1)$
1,4-Pentadiene		$-254\ (-60.8)$

that the heat of hydrogenation of 1,3-butadiene should be $2(-127\ \text{kJ/mol})$ or -254 kJ $(-60.6\ \text{kcal})/\text{mol}$. The observed heat of hydrogenation of 1,3-butadiene is -237 kJ $(-56.5\ \text{kcal})/\text{mol}$, a value 17 kJ (4.1 kcal)/mol less than estimated.

$$2 \diagup\!\!\!\!\diagdown\!\!\!\diagup + 2\text{H}_2 \xrightarrow{\text{catalyst}} 2 \diagdown\!\!\!\diagup\!\!\!\diagdown \qquad \Delta H^0 = 2(-127\ \text{kJ/mol})$$
$$= -254\ \text{kJ/mol}$$

$$\diagup\!\!\!\!\diagdown\!\!\!\diagup\!\!\!\diagdown + 2\text{H}_2 \xrightarrow{\text{catalyst}} \diagdown\!\!\!\diagup\!\!\!\diagdown \qquad \Delta H^0 = -237\ \text{kJ/mol}$$

The conclusion from this calculation is that conjugation of two double bonds in 1,3-butadiene gives an extra stability to the molecule of approximately 17 kJ (4.1 kcal)/mol. These energy relationships are displayed graphically in Figure 23.1.

Figure 23.1

Conjugation of double bonds in butadiene gives the molecule an additional stability of approximately 17 kJ (4.1 kcal)/mol.

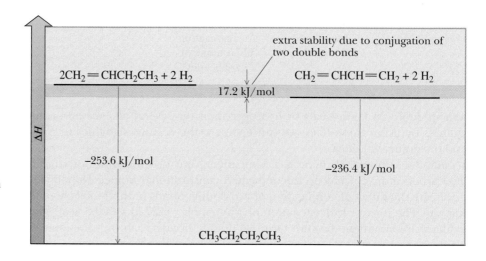

extra stability due to conjugation of two double bonds

$2\text{CH}_2\!=\!\text{CHCH}_2\text{CH}_3 + 2\ \text{H}_2$ 17.2 kJ/mol $\text{CH}_2\!=\!\text{CHCH}\!=\!\text{CH}_2 + 2\ \text{H}_2$

-253.6 kJ/mol -236.4 kJ/mol

ΔH

$\text{CH}_3\text{CH}_2\text{CH}_2\text{CH}_3$

Calculations of this type for other conjugated and unconjugated dienes give similar results. Conjugated dienes are more stable than isomeric unconjugated dienes by approximately 14.5–17.0 kJ (3.5–4.1 kcal)/mol. The effects of conjugation on stability are even more general. Compounds containing conjugated double bonds, not just those in dienes, are more stable than isomeric compounds containing unconjugated double bonds. For example, 2-cyclohexenone is more stable than its isomer 3-cyclohexenone.

2-Cyclohexenone	3-Cyclohexenone
(more stable)	(less stable)

The additional stability of conjugated dienes relative to unconjugated dienes arises from delocalization of electron density in the conjugated diene. In two unconjugated double bonds, each pair of pi electrons is localized between two carbons. In a conjugated diene, however, the four pi electrons are delocalized over the entire set of four parallel $2p$ orbitals.

According to the molecular orbital model, the conjugated system of a diene is described as a set of four pi molecular orbitals arising from combination of four $2p$ atomic orbitals. These MOs have zero, one, two, and three nodes, respectively, as illustrated in Figure 23.2. In the ground state, all four pi electrons lie in pi-bonding MOs. Because the lowest two MOs are at lower energies than that of an isolated pi bond, the net heat given off by filling these orbitals is more than would be the case for two isolated pi bonds.

Example 23.2

Using data from Table 23.1, estimate the extra stability due to the conjugation of double bonds in *trans*-1,3-pentadiene.

Solution

Compare the sum of heats of hydrogenation of 1-pentene and *trans*-2-butene with the heat of hydrogenation of *trans*-1,3-pentadiene. Conjugation of double bonds in *trans*-1,3-pentadiene imparts an added stability of approximately 15 kJ (3.6 kcal)/mol.

Problem 23.2

Estimate the stabilization gained due to conjugation when 1,4-pentadiene is converted to *trans*-1,3-pentadiene. Note that the answer is not as simple as comparing the heats of hydrogenation of 1,4-pentadiene and *trans*-1,3-pentadiene. Although the double bonds are moved from unconjugated to conjugated, the degree of substitution of one of the double bonds is also changed, in this case from a monosubstituted double bond to a trans disubstituted double bond. To answer this question, you must separate the effect due to conjugation from that due to change in degree of substitution.

Figure 23.2
Structure of 1,3-butadiene—molecular orbital model. Combination of four parallel $2p$ atomic orbitals gives two pi-bonding MOs and two pi-antibonding MOs. In the ground state, each pi-bonding MO is filled with two spin-paired electrons. The pi-antibonding MOs are unoccupied.

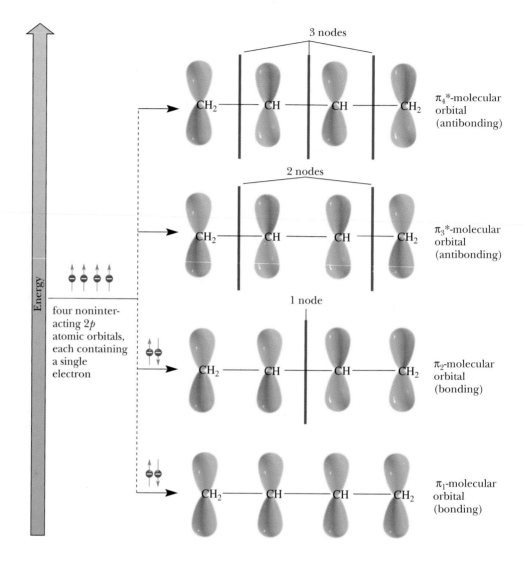

23.2 Electrophilic Addition to Conjugated Dienes

Conjugated dienes undergo two-step electrophilic addition reactions just like simple alkenes (Section 6.3). However, certain features are unique to the reactions of conjugated dienes.

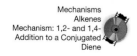

Mechanisms
Alkenes
Mechanism: 1,2- and 1,4-
Addition to a Conjugated
Diene

A. 1,2-Addition and 1,4-Addition

Addition of one mole of HBr to 1,3-butadiene at $-78°C$ gives a mixture of two constitutional isomers, 3-bromo-1-butene and 1-bromo-2-butene.

$$CH_2{=}CH{-}CH{=}CH_2 + \boxed{HBr} \xrightarrow{-78°C} \overset{\boxed{Br}\ \ \boxed{H}}{CH_2{=}CH{-}CH{-}CH_2} + \overset{\boxed{Br}\ \ \ \ \ \ \ \ \boxed{H}}{CH_2{-}CH{=}CH{-}CH_2}$$

1,3-Butadiene   3-Bromo-1-butene   1-Bromo-2-butene
                             (90%)   (10%)
                        (1,2-addition)   (1,4-addition)

The designations "1,2-" and "1,4-" used here to describe additions to conjugated dienes do not refer to IUPAC nomenclature. Rather, they refer to the four-atom system of two conjugated double bonds and indicate that addition takes place at either carbons 1 and 2 or carbons 1 and 4 of the four-atom system. The bromobutenes formed by addition of one mole of HBr to butadiene can, in turn, undergo addition of a second mole of HBr to give a mixture of dibromobutanes. Our concern at this point is only with the products of the first reaction.

Addition of one mole of Br_2 at $-15°C$ also gives a mixture of 1,2-addition and 1,4-addition products.

$$CH_2{=}CH{-}CH{=}CH_2 + \boxed{Br_2} \xrightarrow{-15°C} \overset{\boxed{Br}\ \ \boxed{Br}}{CH_2{-}CH{-}CH{=}CH_2} + \overset{\boxed{Br}\ \ \ \ \ \ \ \ \boxed{Br}}{CH_2{-}CH{=}CH{-}CH_2}$$

1,3-Butadiene   3,4-Dibromo-1-butene   1,4-Dibromo-2-butene
                             (54%)   (46%)
                         (1,2-addition)   (1,4-addition)

We can account for the formation of isomeric products in the addition of HBr and Br_2 by the following mechanism.

Mechanism   1,2- and 1,4-Addition to a Conjugated Diene

Step 1: Electrophilic addition is initiated by reaction of a terminal carbon of one of the double bonds with HBr to give an allylic carbocation intermediate, which can best be represented as a resonance hybrid of two contributing structures. Formation of this cation is the rate-determining step. A similar allylic carbocation (and Br^-) is formed from the addition of Br_2 to the alkene. In this case the bromonium ion opens to the more stable allylic carbocation.

Step 2: Reaction of bromide at one of the carbons bearing partial positive charge gives the 1,2-addition product; reaction at the other gives the 1,4-addition product.

$$CH_2{=}CH{-}CH{=}CH_2 + \boxed{H{-}Br} \longrightarrow \overset{\ \ \ \ \ \ \ \ \ \ \ \boxed{H}}{CH_2{=}CH{-}\overset{+}{CH}{-}CH_2} \longleftrightarrow \overset{\ \ \ \ \ \ \ \ \ \ \ \ \ \ \ \boxed{H}}{\overset{+}{CH_2}{-}CH{=}CH{-}CH_2}$$

A resonance-stabilized allylic carbocation

                  $\downarrow Br^-$ $\downarrow Br^-$

$$\overset{\boxed{Br}\ \ \boxed{H}}{CH_2{=}CH{-}CH{-}CH_2} \qquad \overset{\boxed{Br}\ \ \ \ \ \ \ \ \boxed{H}}{CH_2{-}CH{=}CH{-}CH_2}$$

(1,2-addition)   (1,4-addition)

Example 23.3

Addition of one mole of HBr to 2,4-hexadiene gives a mixture of 4-bromo-2-hexene and 2-bromo-3-hexene. No 5-bromo-2-hexene is formed. Account for the formation of the first two bromoalkenes and for the fact that the third bromoalkene is not formed.

Solution

2,4-Hexadiene is a conjugated diene, and you can expect products from both 1,2- and 1,4-addition. Reaction of the diene with HBr in Step 1, the rate-determining step, gives a resonance-stabilized 2° allylic carbocation intermediate. Reaction of this intermediate in Step 2 at one of the carbons bearing a partial positive charge gives 4-bromo-2-hexene, a 1,2-addition product; reaction at the other gives 2-bromo-3-hexene, a 1,4-addition product.

$$CH_3CH=CH-CH=CHCH_3 \ + \ H-Br \longrightarrow$$

$$CH_3CH=CH-\overset{+}{C}H-\overset{\overset{\displaystyle H}{|}}{C}HCH_3 \longleftrightarrow CH_3\overset{+}{C}H-CH=CH-\overset{\overset{\displaystyle H}{|}}{C}HCH_3$$

(a resonance-stabilized 2° allylic carbocation)

Br⁻ ↓   ↓ Br⁻

$$CH_3CH=CH-\overset{\overset{\displaystyle Br}{|}}{C}H-\overset{\overset{\displaystyle H}{|}}{C}HCH_3 \qquad CH_3\overset{\overset{\displaystyle Br}{|}}{C}H-CH=CH-\overset{\overset{\displaystyle H}{|}}{C}HCH_3$$

4-Bromo-2-hexene	2-Bromo-3-hexene
(1,2-addition)	(1,4-addition)

Formation of 5-bromo-2-hexene requires reaction of the diene with HBr to give a secondary, nonallylic carbocation. The activation energy for formation of this less stable 2° carbocation is considerably greater than that for formation of the resonance-stabilized allylic carbocation; therefore, formation of this carbocation and the resulting product, 5-bromo-2-hexene, does not compete effectively with formation of the observed products.

$$CH_3CH=CH-CH=CHCH_3 \ + \ H-Br \longrightarrow CH_3CH=CH-\overset{\overset{\displaystyle H}{|}}{C}H-\overset{+}{C}HCH_3 \xrightarrow{Br^-} CH_3CH=CH-\overset{\overset{\displaystyle H}{|}}{C}H-\overset{\overset{\displaystyle Br}{|}}{C}HCH_3$$

(2° carbocation, but not allylic)   5-Bromo-2-hexene
                                            (not formed)

Problem 23.3

Predict the product(s) formed by addition of one mole of Br₂ to 2,4-hexadiene.

For a review of electrophilic addition, see: Supporting Concepts (Electrophlic) Addition Reactions

B.   Kinetic Versus Thermodynamic Control of Electrophilic Addition

We saw in the previous section that electrophilic addition to conjugated dienes gives a mixture of 1,2-addition and 1,4-addition products. Following are some additional experimental observations about the products of electrophilic additions to 1,3-butadiene.

1. For addition of HBr at $-78°C$ and addition of Br_2 at $-15°C$, the 1,2-addition product predominates over the 1,4-addition product. Generally at lower temperatures, the 1,2-addition products predominate over 1,4-addition products.
2. For addition of HBr and Br_2 at higher temperatures (generally 40–60°C), the 1,4-addition products predominate.
3. If the products of low-temperature addition are allowed to remain in solution and then are warmed to a higher temperature, the composition of the product changes over time and becomes identical to that obtained when the reaction is carried out at higher temperature. The same result can be accomplished at the higher temperature in a far shorter time by adding a Lewis acid catalyst, such as $FeCl_3$ or $ZnCl_2$, to the mixture of low-temperature addition products. Thus, under these higher temperature conditions, an equilibrium is established between 1,2- and 1,4-addition products in which 1,4-addition products predominate.
4. If either the pure 1,2- or pure 1,4-addition product is dissolved in an inert solvent at the higher temperature and a Lewis acid catalyst added, an equilibrium mixture of 1,2- and 1,4-addition product forms. The same equilibrium mixture is obtained regardless of which isomer is used as the starting material.

Chemists interpret these experimental results using the twin concepts of kinetic control and equilibrium control of reactions. We encountered these concepts earlier in Section 19.2A, where we saw that it is often possible to form alternative enolate anions of ketones, depending on whether experimental conditions are chosen for kinetic control or thermodynamic control.

To review briefly, for reactions under **kinetic (rate) control,** the distribution of products is determined by the relative rates of formation of each. We see the operation of kinetic control in the following way. At lower temperatures, the reaction is essentially irreversible, and no equilibrium is established between 1,2- and 1,4-addition products. The 1,2-addition product predominates under these conditions because the rate of 1,2-addition is greater than that of 1,4-addition.

For reactions under **thermodynamic (equilibrium) control,** the distribution of products is determined by the relative stability of each. We see the operation of thermodynamic control in the following way. At higher temperatures, the reaction is reversible, and an equilibrium is established between 1,2- and 1,4-addition products. The percentage of each product present at equilibrium is in direct relation to the relative thermodynamic stability of that product. The fact that the 1,4-addition product predominates at equilibrium means that it is thermodynamically more stable than the 1,2-addition product.

Relationships between kinetic and thermodynamic control for electrophilic addition of HBr to 1,3-butadiene are illustrated graphically in Figure 23.3. The structure shown in the Gibbs free energy well in the center of Figure 23.3 is the resonance-stabilized allylic cation intermediate formed by proton transfer from HBr to 1,3-butadiene. The dashed lines in this intermediate show the partial double bond character between C2 and C3 and between C3 and C4 in the resonance hybrid. To the left of this intermediate is the activation energy for its reaction with bromide ion to form the less stable 1,2-addition product; to the right is the activation energy for its reaction with bromide ion to form the more stable 1,4-addition product. As shown in Figure 23.3, the activation energy for 1,2-addition is less than that for 1,4-addition, and

Figure 23.3

Kinetic versus thermodynamic control. A plot of Gibbs free energy versus reaction coordinate for Step 2 in the electrophilic addition of HBr to 1,3-butadiene. The resonance-stabilized allylic carbocation intermediate reacts with bromide ion by way of the transition state on the left to give the 1,2-addition product. It reacts with bromide ion by way of the alternative transition state on the right to give the 1,4-addition product.

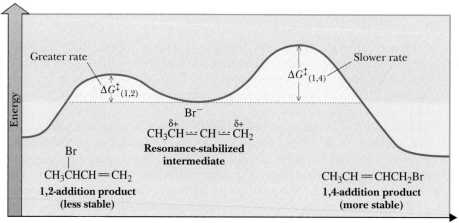

the 1,2-addition product is therefore favored under kinetic control. The 1,4-addition product is the more stable and is favored when the reaction is under thermodynamic control.

To complete our discussion of electrophilic addition to conjugated dienes and of kinetic versus thermodynamic control, we need to ask the following questions.

1. Why is the 1,2-addition product (the less stable product) formed more rapidly at lower temperatures? First, we need to look at the resonance-stabilized allylic carbocation intermediate and determine which Lewis structure makes the greater contribution to the hybrid. We must consider both the degree of substitution of both the positive carbon and the carbon-carbon double bond in each contributing structure.

$$CH_2\text{=}CH\text{—}\overset{+}{CH}\text{—}CH_3 \longleftrightarrow \overset{+}{CH_2}\text{—}CH\text{=}CH\text{—}CH_3$$

Less substituted double bond   More substituted double bond
(secondary carbocation)   (primary carbocation)

A secondary carbocation is more stable than a primary carbocation. If the degree of substitution of the carbon bearing the positive charge were the more important factor, the Lewis structure on the left would make the greater contribution to the hybrid. A more substituted double bond is more stable than a less substituted double bond. If the degree of substitution of the carbon-carbon double bond were the more important factor, the Lewis structure on the right would make the greater contribution to the hybrid.

We know from other experimental evidence that the location of the positive charge in the allylic carbocation is more important than the location of the double bond. Therefore, in the hybrid, the greater fraction of positive charge is on the secondary carbon. Reaction with bromide ion occurs more rapidly at this carbon, giving 1,2-addition, simply because it has a greater density of positive charge.

2. Is the 1,2-addition product also formed more rapidly at higher temperatures, even though the 1,4-addition product predominates under these conditions? The answer is yes. The factors affecting the structure of a resonance-stabilized allylic car-

bocation intermediate and the reaction of this intermediate with a nucleophile are not greatly affected by changes in temperature.

3. Why is the 1,4-addition product the thermodynamically more stable product? The answer to this question has to do with the relative degree of substitution of double bonds. In general, the greater the degree of substitution of a carbon-carbon double bond, the greater the stability of the compound or ion containing it. Following are pairs of 1,2- and 1,4-addition products. In each case, the more stable alkene is the 1,4-addition product.

3-Bromo-1-butene	(E)-1-Bromo-2-butene
(less stable alkene)	(more stable alkene)

3,4-Dibromo-1-butene	(E)-1,4-Dibromo-2-butene
(less stable alkene)	(more stable alkene)

However, there are cases where the 1,2-addition product is more stable and would be the product of thermodynamic control. For example, addition of bromine to 1,4-dimethyl-1,3-cyclohexadiene under conditions of thermodynamic control gives 3,4-dibromo-1,4-dimethylcyclohexene because its trisubstituted double bond is more stable than the disubstituted double bond of the 1,4-addition product.

1,4-Addition product	1,2-Addition product
(less stable)	(more stable)

4. What is the mechanism by which the thermodynamically less stable product is converted to the thermodynamically more stable product at higher temperatures? To answer this question, we must look at the relationships between kinetic energy, potential energy, and activation energy. On collision, a part of the kinetic energy (the energy of motion) is transformed into potential energy. If the increase in potential energy is equal to or greater than the activation energy for reaction, then reaction may occur. At the higher temperatures for electrophilic addition of HBr and Br_2 to conjugated dienes, collisions are sufficiently energetic that ionization of the 1,2-addition product occurs to re-form the resonance-stabilized allylic carbocation intermediate. It then reacts with bromide ion again to form the thermodynamically more stable 1,4-addition product. At lower temperatures, however, the increase in potential energy on collision is not sufficient to overcome the potential energy barrier to bring about this ionization.

5. Is it a general rule that, where two or more products are formed from a common intermediate, the thermodynamically less stable product is formed at a greater rate? The answer is no. Whether the thermodynamically more or less stable product is formed at a greater rate from a common intermediate depends very much on the particular reaction and the reaction conditions.

23.3 The Diels-Alder Reaction

In 1928, Otto Diels and Kurt Alder in Germany discovered another unique reaction of conjugated dienes: They undergo cycloaddition reactions with certain types of carbon-carbon double and triple bonds. For their discovery and subsequent studies of this reaction, Diels and Alder were jointly awarded the 1950 Nobel Prize for chemistry.

> **Dienophile** A compound containing a double bond (consisting of one or two C, N, or O atoms) that can react with a conjugated diene to give a Diels-Alder adduct.

> **Diels-Alder adduct** A cyclohexene resulting from the cycloaddition reaction of a diene and a dienophile.

> **Cycloaddition reaction** A reaction in which two reactants add together in a single step to form a cyclic product. The best known of these is the Diels-Alder reaction.

The compound with the double or triple bond that reacts with the diene in a Diels-Alder reaction is given the special name of **dienophile** (diene-loving), and the product of a Diels-Alder reaction is given the special name of **Diels-Alder adduct.** The designation **cycloaddition** refers to the fact that two reactants add together to give a cyclic product. Following are two examples of Diels-Alder reactions: one with a compound containing a carbon-carbon double bond and the other containing a carbon-carbon triple bond.

1,3-Butadiene
(a diene)

3-Buten-2-one
(a dienophile)

4-Cyclohexenyl methyl ketone
(a Diels-Alder adduct)

1,3-Butadiene
(a diene)

Diethyl 2-butynedioate
(a dienophile)

Diethyl 1,4-cyclohexadiene-
1,2-dicarboxylate
(a Diels-Alder adduct)

Note that the four carbon atoms of the diene and two carbon atoms of the dienophile combine to form a six-membered ring. Note further that there are two more sigma bonds and two fewer pi bonds in the product than in the reactants. This exchange of two (weaker) pi bonds for two (stronger) sigma bonds is a major driving force in Diels-Alder reactions.

We can write a Diels-Alder reaction in the following way, showing only the carbon skeletons of the diene and dienophile. In this representation, curved arrows are used to show that two new sigma bonds are formed, three pi bonds are broken, and one new pi bond is formed. It must be emphasized here that the curved arrows in this diagram are not meant to show a mechanism. Rather they are intended to show which bonds are broken, which new bonds are formed, and how many electrons are involved (six in this case).

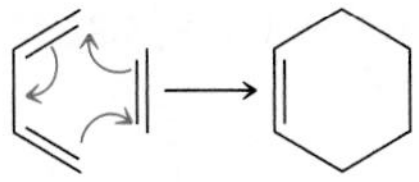

The special values of the reaction discovered by Diels and Alder are that (1) it is one of the simplest reactions that can be used to form six-membered rings, (2) it is one of few reactions that can be used to form two new carbon-carbon bonds at the same time, and, as we will see later in this section, (3) it is stereospecific and regioselective. For these reasons, the Diels-Alder reaction has proved to be enormously valuable in synthetic organic chemistry.

Example 23.4

Draw a structural formula for the Diels-Alder adduct formed by reaction of each diene and dienophile pair.

(a) 1,3-Butadiene and propenal   **(b)** 2,3-Dimethyl-1,3-butadiene and 3-buten-2-one

Solution

First draw the diene and dienophile so that each molecule is properly aligned to form a six-membered ring. Then complete the reaction to form the six-membered ring Diels-Alder adduct.

(a)

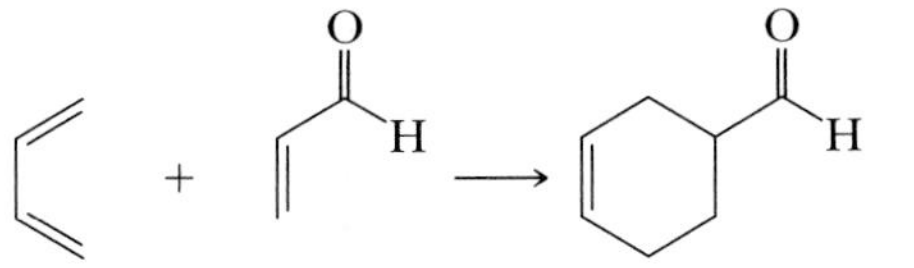

1,3-Butadiene   Propenal
                  (Acrolein)

(b)

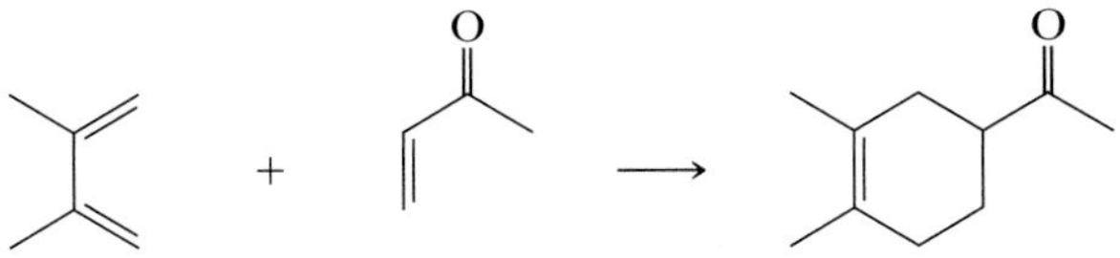

2,3-Dimethyl-   3-Buten-2-one
1,3-butadiene   (Methyl vinyl ketone)

Problem 23.4

What combination of diene and dienophile undergoes Diels-Alder reaction to give each adduct?

(a) 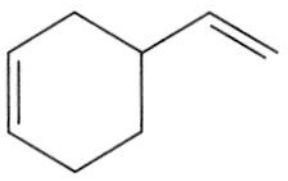   **(b)** 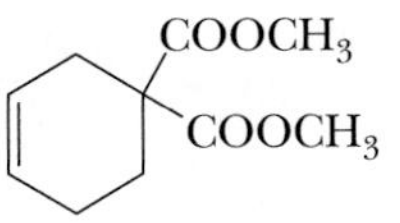 COOCH$_3$ / COOCH$_3$ **(c)** 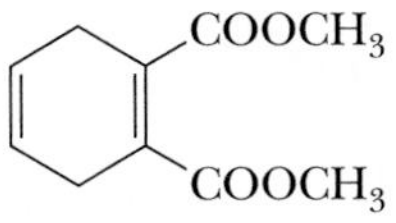 COOCH$_3$ / COOCH$_3$

Now let us look more closely at the scope and limitations, stereochemistry, and mechanism of Diels-Alder reactions.

A. The Diene Must Be Able to Assume an s-Cis Conformation

We can illustrate the significance of conformation of the diene by reference to 1,3-butadiene. For maximum stability of a conjugated diene, overlap of the four unhybridized $2p$ orbitals making up the pi system must be complete, a condition that occurs only when all four carbon atoms of the diene lie in the same plane. It follows then that if the carbon skeleton of 1,3-butadiene is planar, the six atoms bonded to the skeleton of the diene are also contained in the same plane. There are two planar conformations of 1,3-butadiene, referred to as the **s-trans conformation** and the **s-cis conformation** where the designation "s" refers to the carbon-carbon single bond of the diene. Of these, the s-trans conformation is slightly lower in energy and, therefore, slightly more stable.

Although s-*trans*-1,3-butadiene is the more stable conformation, s-*cis*-1,3-butadiene is the reactive conformation in Diels-Alder reactions. In the s-cis conformation, carbon atoms 1 and 4 of the conjugated system are close enough to react with the carbon-carbon double or triple bond of the dienophile and form a six-membered ring. In the s-trans conformation, they are too far apart for this to happen.

The energy barrier for interconversion of the s-trans and s-cis conformations for 1,3-butadiene is low, approximately 11.7 kJ (2.8 kcal)/mol; consequently, 1,3-butadiene is a reactive diene in Diels-Alder reactions.

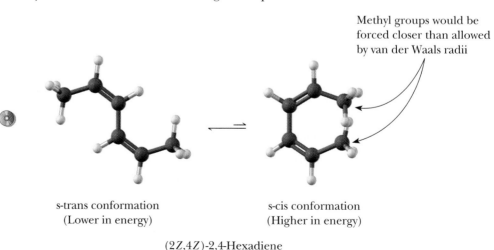

s-trans conformation
(Lower in energy)

s-cis conformation
(Higher in energy)

(2Z,4Z)-2,4-Hexadiene is unreactive in Diels-Alder reactions because it is prevented by steric hindrance from assuming the required s-cis conformation.

Methyl groups would be forced closer than allowed by van der Waals radii

s-trans conformation
(Lower in energy)

s-cis conformation
(Higher in energy)

(2Z,4Z)-2,4-Hexadiene

Example 23.5

Which molecules can function as dienes in Diels-Alder reactions? Explain.

(a) ![structure with CH₂] (b) ![bicyclic structure] (c) ![structure with CH₂ and CH₂]

Solution

The dienes in both (a) and (b) are fixed in the s-trans conformation and, therefore, are not capable of participation in Diels-Alder reactions. The diene in (c) is fixed in the s-cis conformation and, therefore, has the proper orientation to participate in Diels-Alder reactions.

Problem 23.5

Which molecules can function as dienes in Diels-Alder reactions? Explain.

(a) ![cyclohexadiene]   (b) ![cyclohexadiene]   (c) ![cyclopentadiene]

B. The Effect of Substituents on Rate

The simplest example of a Diels-Alder reaction is that between 1,3-butadiene and ethylene, both gases at room temperature. Although this reaction does occur, it is very slow and takes place only if the reactants are heated at 200°C under pressure.

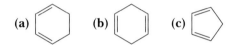

$$\text{1,3-Butadiene} \quad + \quad \text{Ethylene} \xrightarrow[\text{pressure}]{200°C} \text{Cyclohexene}$$

Diels-Alder reactions are facilitated by a combination of electron-withdrawing substituents on one of the reactants and electron-releasing substituents on the other. Most commonly, the dienophile is electron deficient and the diene is electron rich. For example, placing a carbonyl group (electron-withdrawing because of the partial positive charge on its carbon) on the dienophile facilitates the reaction. To illustrate, 1,3-butadiene and 3-buten-2-one form a Diels-Alder adduct when heated at 140°C.

$$\text{1,3-Butadiene} \quad + \quad \text{3-Buten-2-one} \xrightarrow{140°C} \text{(adduct)}$$

Placing electron-releasing methyl groups on the diene further facilitates reaction; 2,3-dimethyl-1,3-butadiene and 3-buten-2-one form a Diels-Alder adduct at 30°C.

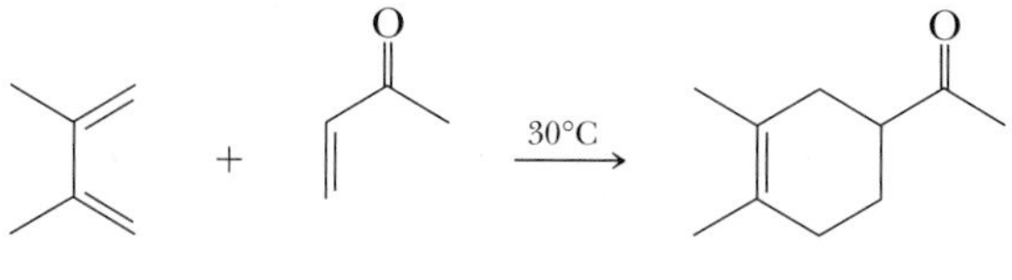

2,3-Dimethyl-   3-Buten-2-one
1,3-butadiene

Several of the electron-releasing and electron-withdrawing groups most commonly encountered in Diels-Alder reactions are given in Table 23.2.

C. Diels-Alder Reactions Can Be Used to Form Bicyclic Systems

Conjugated cyclic dienes, in which the double bonds are of necessity held in an s-cis conformation, are highly reactive in Diels-Alder reactions. Two particularly useful dienes for this purpose are cyclopentadiene and 1,3-cyclohexadiene. In fact, cyclopentadiene is reactive both as a diene and as a dienophile, and, on standing at room temperature, it forms a Diels-Alder adduct known by the common name dicyclopentadiene. When dicyclopentadiene is heated to 170°C, a reverse Diels-Alder reaction takes place, and cyclopentadiene is reformed.

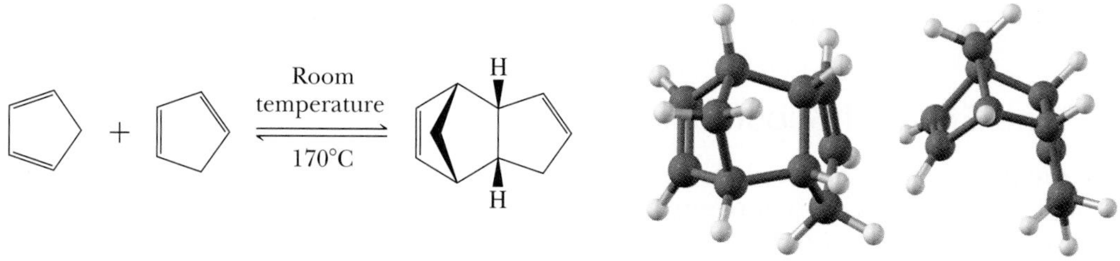

Diene   Dienophile   Dicyclopentadiene   From top   From side
(Endo form)

The terms "endo" and "exo" are used for bicyclic Diels-Alder adducts to describe the orientation of substituents of the dienophile in relation to the two-carbon diene-derived bridge. *Exo* (Greek, outside) substituents are on the opposite side from the diene-derived bridge; *endo* (Greek, within) substituents are on the same side.

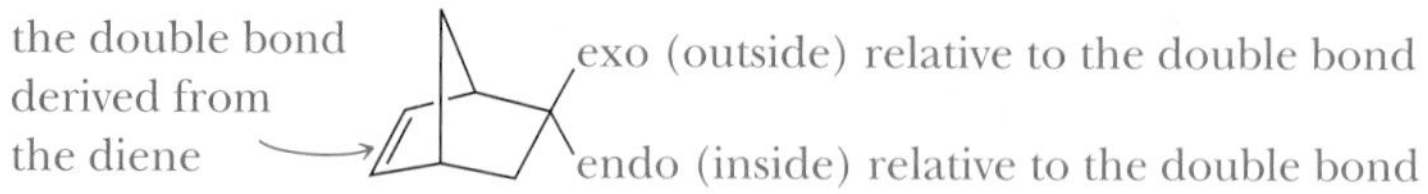

the double bond
derived from   exo (outside) relative to the double bond
the diene   endo (inside) relative to the double bond

For Diels-Alder reactions under kinetic control, the endo orientation of the dienophile is favored. Treatment of cyclopentadiene with methyl propenoate (methyl acrylate) gives the endo adduct almost exclusively. The exo adduct is not formed. Diels-Alder reactions are not always so stereospecific.

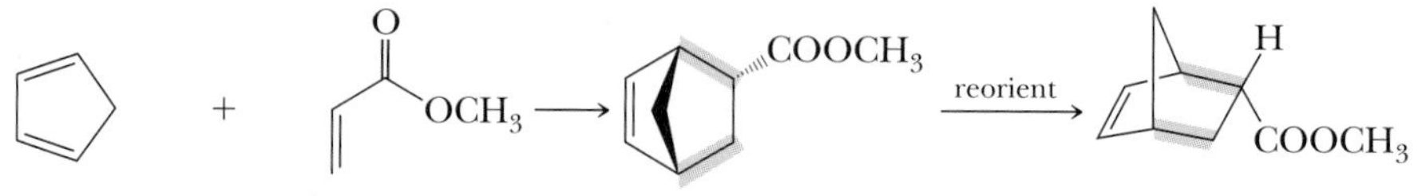

Cyclopentadiene   Methyl   Methyl bicyclo[2.2.1]hept-5-en-
propenoate   endo-2-carboxylate

Table 23.2	Electron-Releasing and Electron-Withdrawing Groups	
Electron-Releasing Groups	**Electron-Withdrawing Groups**
—CH$_3$	$\overset{\text{O}}{\underset{\|}{}}$ —CH (aldehyde)
—CH$_2$CH$_3$	
—CH(CH$_3$)$_2$	$\overset{\text{O}}{\underset{\|}{}}$ —CR (ketone)
—C(CH$_3$)$_3$	
—R (other alkyl groups)	$\overset{\text{O}}{\underset{\|}{}}$ —COH (carboxyl)
—OR (ether)	$\overset{\text{O}}{\underset{\|}{}}$ —COR (ester)
	—NO$_2$ (nitro)
$\overset{\text{O}}{\underset{\|}{}}$ —OCR (ester)	—C≡N (cyano)

D. The Configuration of the Dienophile Is Retained

The reaction is completely stereospecific at the dienophile. If the dienophile is a cis isomer, then the substituents cis to each other in the dienophile are cis in the Diels-Alder adduct. Conversely, if the dienophile is a trans isomer, substituents that are trans in the dienophile are trans in the adduct.

Dimethyl *cis*-2-butenedioate
(a cis dienophile)

Dimethyl *cis*-4-cyclohexene
1,2-dicarboxylate

Dimethyl *trans*-2-butenedioate
(a trans dienophile)

Dimethyl *trans*-4-cyclohexene
1,2-dicarboxylate

E. The Configuration at the Diene Is Retained

The reaction is also completely stereospecific at the diene. Groups on the 1 and 4 position of the diene retain their relative orientation.

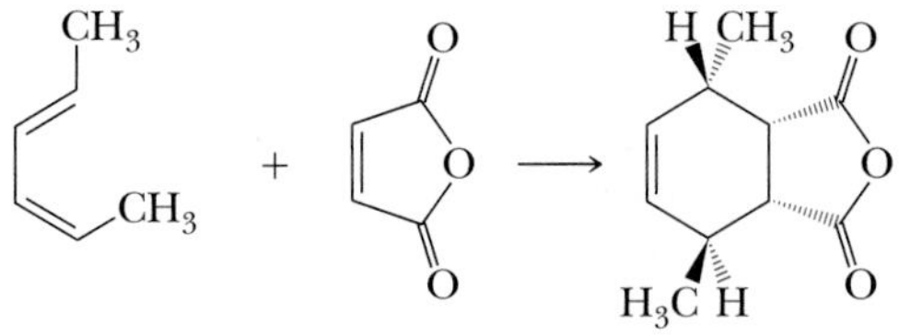

A picture of the transition state will help clarify the reason for this. The groups that are inside on the diene end up on the opposite side from the dienophile.

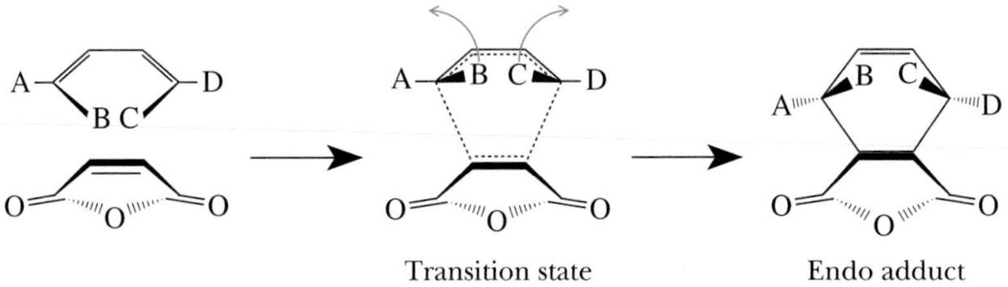

Transition state

Endo adduct

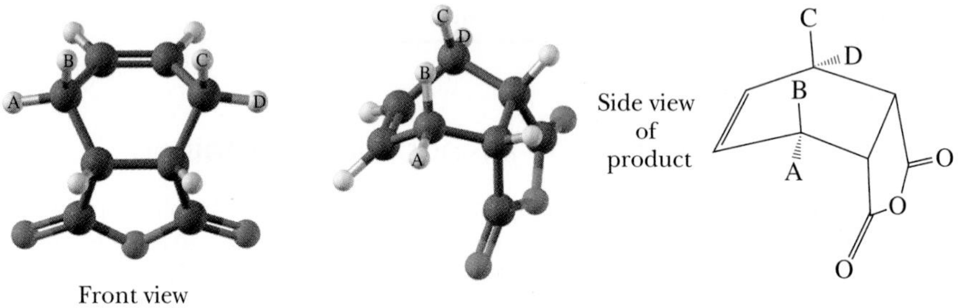

Front view of product

Side view of product

F. Mechanism — The Diels-Alder Reaction Is a Pericyclic Reaction

As chemists probed for details of the Diels-Alder reaction, they found no evidence for participation of either ionic or radical intermediates. Thus, the Diels-Alder reaction is unlike any reaction we have studied thus far. To account for the stereoselectivity of the Diels-Alder reaction and the lack of evidence for either ionic or radical intermediates, chemists have proposed that reaction takes place in a single step during which there is a cyclic redistribution of electrons. During this cyclic redistribution, bond forming and bond breaking are concerted (simultaneous). To use the terminology of organic chemistry, the Diels-Alder reaction is a **pericyclic reaction,** that is, a reaction that takes place in a single step, without intermediates, and involves a cyclic redistribution of bonding electrons. We can envision a Diels-Alder reaction taking place as shown in Figure 23.4.

Pericyclic reaction A reaction that takes place in a single step, without intermediates, and involves a cyclic redistribution of bonding electrons.

Example 23.6

Complete the following Diels-Alder reaction, showing the stereochemistry of the product.

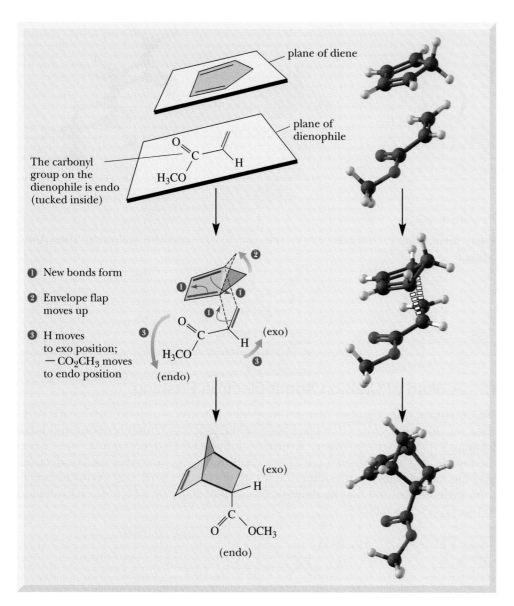

The carbonyl group on the dienophile is endo (tucked inside)

plane of diene

plane of dienophile

❶ New bonds form

❷ Envelope flap moves up

❸ H moves to exo position; —CO₂CH₃ moves to endo position

Figure 23.4
Mechanism of the Diels-Alder reaction. The diene and dienophile approach each other in parallel planes, one above the other, with the substituents on the dienophile endo to the diene. There is overlap of the pi orbitals of each molecule and syn addition of each molecule to the other. As (1) new sigma bonds form in the transition state, (2) the —CH₂— on the diene rotates upward, and (3) the hydrogen atom of the dienophile becomes exo and the ester group becomes endo.

Solution

Reaction of cyclopentadiene with this dienophile forms a disubstituted bicyclo[2.2.1]hept-2-ene. The two ester groups are cis in the dienophile, and, given the stereoselectivity of the Diels-Alder reaction, they are cis and endo in the product.

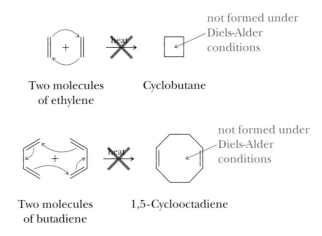

Problem 23.6

What diene and dienophile might you use to prepare the following Diels-Alder adduct?

COOCH$_3$

H H

COOCH$_3$

G.   A Word of Caution About Electron Pushing

We developed a mechanism of the Diels-Alder reaction and used curved arrows to show the flow of electrons that takes place in the process of bond breaking and bond forming. Diels-Alder reactions involve a four-carbon diene and a two-carbon dienophile and are termed [4 + 2] cycloadditions. We can write similar electron-pushing mechanisms for the dimerization of ethylene by a [2 + 2] cycloaddition to form cyclobutane, and for the dimerization of butadiene by a [4 + 4] cycloaddition to form 1,5-cyclooctadiene.

heat

not formed under
Diels-Alder
conditions

Two molecules
of ethylene

Cyclobutane

heat

not formed under
Diels-Alder
conditions

Two molecules
of butadiene

1,5-Cyclooctadiene

Although [2 + 2] and [4 + 4] cycloadditions bear a formal relationship to the Diels-Alder reaction, neither, in fact, takes place under the thermal conditions required for Diels-Alder reactions. These cycloadditions do occur, but only under different (and usually much more vigorous) experimental conditions and by quite different mechanisms. The point of mentioning them here is to add a note of caution. Although electron pushing is a valuable tool in electron bookkeeping, it is not in itself a full description of a reaction mechanism. Reaction mechanisms are often far more complex than the use of curved arrows might suggest.

23.4   Pericyclic Reactions and Transition State Aromaticity

The Diels-Alder reaction is a concerted reaction involving a redistribution of six electrons in a cyclic transition state. The central point here is that there are six electrons and the transition state is cyclic. We can place the transition state for this reaction in a larger context referred to as transition state aromaticity. Recall the Hückel criteria for aromaticity (Section 20.2A): the presence of $(4n + 2)$ pi electrons in a ring that is planar and fully conjugated. Just as aromaticity imparts a stability to certain types of molecules and ions, the presence of $(4n + 2)$ electrons imparts a stability to certain types of transition states. Reactions involving 2, 6, 10, 14, . . . electrons in a cyclic transition state have especially low activation energies and take place particularly readily. In contrast, the reactions in Section 23.3G that involve $4n$ electrons do not take place readily. We now see that, just as the Hückel theory of aromaticity gives us a clearer understanding of the stability of certain types of molecules and ions, it also gives a clearer understanding of certain types of reactions and their transition states. Transition states involving a cyclic redistribution of $4n$ electrons, on the other hand, are antiaromatic and have especially high activation energies. For this reason, the dimerization of ethylene, as shown earlier, to give cyclobutane (a four-electron transition state) and of 1,3-butadiene to give 1,5-cyclooctadiene (an eight-electron transition state) do not occur.

We have seen six examples of reactions that proceed by cyclic, six-electron transition states:

1. The decarboxylation of β-ketoacids and β-dicarboxylic acids (Section 17.9)
2. The Cope elimination of amine *N*-oxides (Section 22.11)
3. The Diels-Alder reaction (Section 23.3)
4. Addition of osmium tetroxide to alkenes (Section 6.5B)
5. Addition of ozone to alkenes (and the resulting fragmentation) (Section 6.5C)
6. Pyrolysis of esters (Problem 22.4)

We now look at two more examples of reactions that proceed by aromatic transition states.

A.   The Claisen Rearrangement

The Claisen rearrangement transforms allyl phenyl ethers to *o*-allylphenols. Heating allyl phenyl ether, for example, the simplest member of this class of compounds, at 200–250°C results in a Claisen rearrangement to form *o*-allyphenol. In this rearrangement, an allyl group migrates from a phenolic oxygen to a carbon atom ortho to it. It has been demonstrated by carbon-14 labeling, here shown in color, that during a Claisen rearrangement, carbon 3 of the allyl group becomes bonded to the ring carbon ortho to the phenolic oxygen.

Allyl phenyl ether   2-Allylphenol

The mechanism of a Claisen rearrangement involves a concerted redistribution of six electrons in a cyclic transition state. The product of this rearrangement is a substituted cyclohexadienone, which undergoes keto-enol tautomerism to reform the aromatic ring.

Mechanism   The Claisen Rearrangement

Mechanisms
Ethers
Claisen Rearrangement

Step 1:   Redistribution of six electrons in a cyclic transition state gives a cyclohexadienone intermediate.

Step 2:   Keto-enol tautomerization restores the aromatic character of the ring.

Allyl phenyl   Transition   A cyclohexadienone   2-Allylphenol
ether   state   intermediate

Thus, we see that the transition state for the Claisen rearrangement bears a close resemblance to that for the Diels-Alder reaction. Both involve a concerted redistribution of six electrons in a cyclic transition state.

Example 23.7

Predict the product of Claisen rearrangement of *trans*-2-butenyl phenyl ether.

trans-2-Butenyl phenyl ether

Solution

In the six-membered transition state for this rearrangement, carbon 3 of the allyl group becomes bonded to the ortho position of the ring.

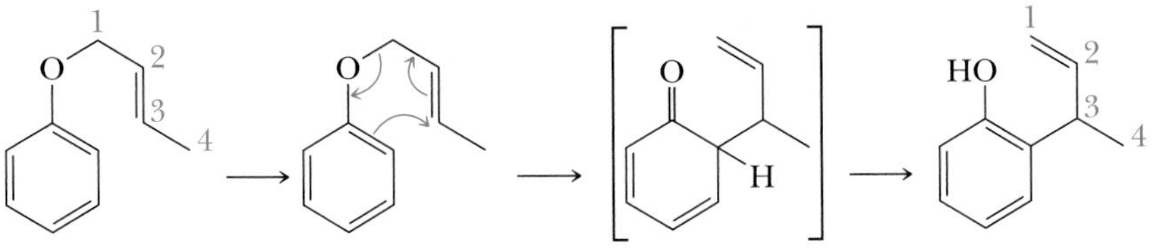

The transition
state

A cyclohexadienone
intermediate

Problem 23.7

Show how to synthesize allyl phenyl ether and 2-butenyl phenyl ether from phenol
and appropriate alkenyl halides.

B.   The Cope Rearrangement

The Cope rearrangement of 1,5-dienes also takes place via a cyclic six-electron transition state. In this example, the product is an equilibrium mixture of isomeric dienes.
The favored product is the diene on the right, which contains the more highly substituted double bonds.

Mechanism   The Cope Rearrangement

Redistribution of six electrons in a cyclic transition state converts a 1,5-diene to an isomeric
1,5-diene.

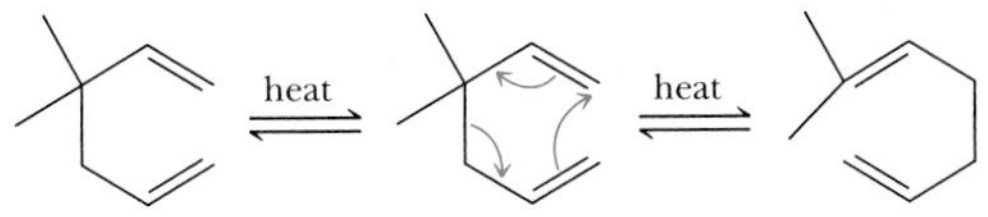

3,3-Dimethyl-
1,5-hexadiene

Transition
state

6-Methyl-1,5-
heptadiene

Example 23.8

Propose a mechanism for the following Cope rearrangement.

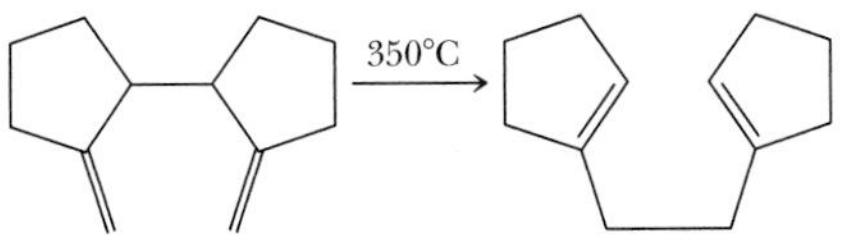

(Example 23.8 continues on p. 910)

C H E M I S T R Y   I N   A C T I O N

Singlet Oxygen

Molecular oxygen differs from most other stable molecules by having two unpaired electrons. This comes about because, during the "aufbau" of molecular oxygen, there are two highest occupied molecular orbitals (π^*) of the same energy and only two electrons to put in them; thus, by Hund's rule, the lowest energy ("ground") state of oxygen has one of these electrons in each orbital and with the same spin. The unpaired electrons cause oxygen to be paramagnetic (attracted by a magnetic field). Most molecules are diamagnetic (repelled by a magnetic field).

The ground state of oxygen is often called a diradical, but in fact oxygen does not really behave as a diradical (it is much less reactive than most radicals). Like protons, electrons have a magnetic moment; two electrons can take on three orientations in a magnetic field, and the ground state of oxygen is

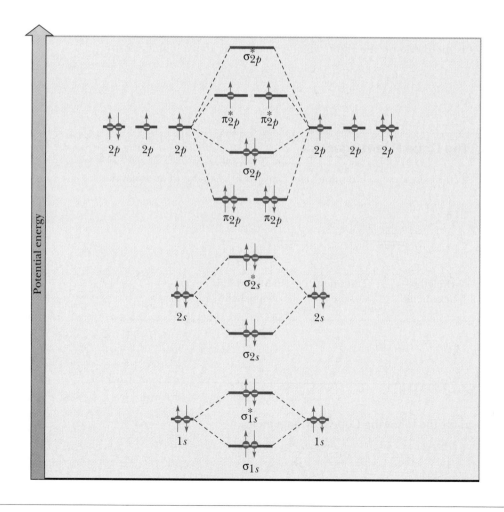

thus called a "triplet" state and given the symbol $^3O_2$. Oxygen is one of an extremely small number of compounds that has a triplet ground state.

However, the triplet state is not the only electronic configuration oxygen can have, just the lowest. There are also two low-energy excited electronic states of oxygen in which its electrons are paired. Only the lower of these, with an energy of 92 kJ (22 kcal)/mol above the ground state, has an appreciable lifetime. Chemists refer to this species as singlet oxygen, $^1O_2$. Its lifetime varies with solvent from about 4 μs to almost 0.1 s, and it has a very different reactivity from oxygen, $^3O_2$.

Singlet oxygen reacts as a dienophile in Diels-Alder reactions. It reacts with cyclohexadiene, for example, to give the Diels-Alder adduct peroxide. This is a general and synthetically useful reaction.

Singlet
oxygen ($^1O_2$)

Singlet oxygen is most often produced photochemically using a dye or other compound called a **photosensitizer** (Sens). Photosensitizers absorb a photon of light ($h\nu$), often in the visible region of the spectrum, and produce an electronically-excited state (Sens*). This state can transfer its energy to $^3O_2$, producing $^1O_2$. This process can be very efficient. The resulting singlet oxygen can then react with dienes and other types of reactive acceptors. 1,4-Dimethylnapthalene, for example, reacts with singlet oxygen by a Diels-Alder reaction to form an endoperoxide. This endoperoxide can be used to store singlet oxygen and then to release it on warming in a reverse Diels-Alder reaction.

1,4-Dimethyl-
naphthalene

An endoperoxide

Singlet oxygen can also react with many biological materials such as unsaturated lipids. This reaction, called **photosensitized oxidation,** can cause damage to or death of an organism. The biological damage caused by photosensitizers, light, and oxygen can be used to selectively kill tumor cells; this **photodynamic therapy** has recently been approved by the FDA. (See C. S. Foote and E. L. Clennan, "Properties and Reactions of Singlet Oxygen," in *Active Oxygen in Chemistry,* C. S. Foote, J. S. Valentine, A. Greenberg, and J. F. Liebman, Eds., Chapman and Hall, London, 1995, pp. 105–140.)

Solution

Redistribution of six electrons in a cyclic transition state gives the observed product.

Transition state

Problem 23.8

Propose a mechanism for the following Cope rearrangement.

23.5   UV-Visible Spectroscopy

An important property of conjugated systems is that they absorb energy in the ultraviolet-visible region of the spectrum as a result of electronic transitions (Table 12.1) In this section, we study the information this absorption gives us about the conjugation of carbon-carbon and carbon-oxygen double bonds and their substitution.

Ultraviolet-visible spectroscopy
A spectroscopic technique in which a compound is irradiated with ultraviolet or visible radiation, absorption of which causes electrons to change from a lower electronic level to a higher one. Ultraviolet spectroscopy is particularly valuable for determining the extent of conjugation in organic molecules containing pi bonds.

A.   Introduction

The region of the electromagnetic spectrum covered by most ultraviolet spectrophotometers runs from 200 to 400 nm, a region commonly referred to as the **near ultraviolet.** Wavelengths shorter than 200 nm require special instrumentation and are not used routinely. The region covered by most visible spectrophotometers runs from 400 nm (violet) to 700 nm (red), with extensions into the (near) IR region to 800 or 1000 nm available on many instruments.

Example 23.9

Calculate the energy of radiation at either end of the near-ultraviolet spectrum, that is, at 200 nm and 400 nm (review Section 12.1).

Solution

Use the relationship $E = hc/\lambda$. Be certain to express the dimension of length in consistent units.

$$E = \frac{hc}{\lambda} = 3.99 \times 10^{-13} \frac{kJ \times s}{mol} \times 3.00 \times 10^8 \frac{m}{s} \times \frac{1}{200 \times 10^{-9} \, m} = 598 \text{ kJ (143 kcal)/mol}$$

By a similar calculation, the energy of radiation of wavelength 400 nm is found to be 299 kJ (71.5 kcal)/mol.

Table 23.3	Wavelengths and Energies of Near Ultraviolet and Visible Radiation	
Region of Spectrum	Wavelength (nm)	Energy [kJ (kcal)/mol]
Ultraviolet	200–400	299–598 (71.5–143)
Visible	400–700	171–299 (40.9–71.5)

Problem 23.9

Wavelengths in ultraviolet-visible spectroscopy are commonly expressed in nanometers; wavelengths in infrared spectroscopy are sometimes expressed in micrometers. Carry out the following conversions.

(a) 2.5 μm to nanometers   **(b)** 200 nm to micrometers

Wavelengths and corresponding energies for near-ultraviolet and visible radiation are summarized in Table 23.3.

Ultraviolet and visible spectral data are recorded as plots of **absorbance (A)** on the vertical axis versus wavelength on the horizontal axis.

$$\text{Absorbance } (A) = \log \frac{I_0}{I}$$

where I_0 is the intensity of radiation incident on the sample, and I is the intensity of the radiation transmitted through the sample.

Absorbance (A)   A quantitative measure of the extent to which a compound absorbs radiation of a particular wavelength. A = log I_0/I where I_0 is the incident radiation and I is the transmitted radiation.

Polyenes

Typically, UV-visible spectra consist of a small number of broad absorption bands, sometimes just one. Figure 23.5 is an ultraviolet absorption spectrum of 2,5-dimethyl-2,4-hexadiene. Absorption of ultraviolet radiation by this conjugated diene begins at

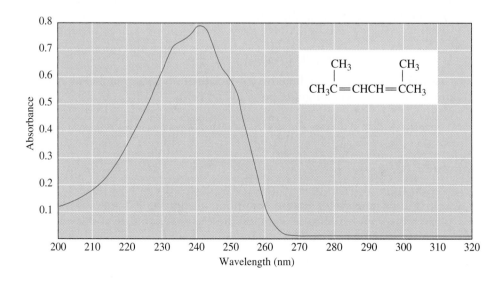

Figure 23.5
Ultraviolet spectrum of 2,5-dimethyl-2,4-hexadiene (in methanol).

wavelengths below 200 nm and continues to almost 270 nm, with maximum absorption at 242 nm. This spectrum is reported as a single absorption peak using the notation λ_{max} 242 nm.

The extent of absorption of ultraviolet-visible radiation is proportional to the number of molecules capable of undergoing the observed electronic transition; therefore, ultraviolet-visible spectroscopy can be used for quantitative analysis of samples. The relationship between absorbance, concentration, and length of the sample cell (cuvette), is known as the **Beer-Lambert law.** The proportionality constant in this equation is given the name **molar absorptivity** (ε) or extinction coefficient.

$$\text{Beer-Lambert Law:} \quad A = \varepsilon c l$$

where A is the absorbance, ε is the molar absorptivity (in liters per mole per centimeter, $M^{-1}cm^{-1}$), c is the concentration of solute (in moles per liter, M), and l is the length of the sample cell, or cuvette (in centimeters, cm).

The molar absorptivity is a characteristic property of a compound and is not affected by its concentration or the length of the light path. Values range from zero to $10^6 \ M^{-1}cm^{-1}$. Values above $10^4 \ M^{-1}cm^{-1}$ correspond to high-intensity absorptions; values below $10^4 \ M^{-1}cm^{-1}$, to low-intensity absorptions. The molar absorptivity of 2,5-dimethyl-2,4-hexadiene, for example, is 13,100 $M^{-1}cm^{-1}$, a high-intensity absorption.

Example 23.10

The molar absorptivity of 2,5-dimethyl-2,4-hexadiene in methanol is 13,100 $M^{-1}cm^{-1}$. What concentration of this diene in methanol is required to give an absorbance of 1.6? Assume a light path of 1.00 cm. Calculate concentration in these units.

(a) Moles per liter   **(b)** Milligrams per milliliter

Solution

Solve the Beer-Lambert equation for concentration, and substitute appropriate values for length, absorbance, and molar absorptivity.

(a) $\quad c = \dfrac{A}{l \times \varepsilon} = \dfrac{1.6}{1.00 \text{ cm} \times 13{,}100 \text{ L·mol}^{-1}\text{·cm}^{-1}} = 1.22 \times 10^{-4} \text{ mol/L}$

(b) The molecular weight of 2,5-dimethyl-2,4-hexadiene is 110 g/mol. The concentration of the sample in milligrams per milliliter is

$$1.22 \times 10^{-4} \frac{\text{mol}}{\text{L}} \times \frac{110 \text{ g}}{\text{mol}} \times \frac{1 \text{ L}}{1000 \text{ mL}} \times \frac{1000 \text{ mg}}{\text{g}} = 1.34 \times 10^{-2} \text{ mg/mL}$$

Problem 23.10

The visible spectrum of β-carotene ($C_{40}H_{56}$, MW 536.89, the orange pigment in carrots) dissolved in hexane shows intense absorption maxima at 463 and 494 nm, both in the blue-green region. Because light of these wavelengths is absorbed by β-carotene, we perceive the color of this compound as that of the complement to blue-green, namely, red-orange.

β-Carotene
$\lambda_{\max}$ 463 nm (log ε 5.10); $\lambda_{\max}$ 494 nm (log ε 4.77)

Calculate the concentration in milligrams per milliliter of β-carotene that gives an absorbance of 1.8 at 463 nm.

Carbonyls

Simple aldehydes and ketones show only weak absorption in the ultraviolet region of the spectrum due to an $n \rightarrow \pi^*$ electronic transition of the carbonyl group (Section 23.5B). If, however, the carbonyl group is conjugated with one or more carbon-carbon double bonds, intense absorption ($\varepsilon = 8{,}000 - 20{,}000 \ M^{-1}cm^{-1}$) occurs due to a $\pi \rightarrow \pi^*$ transition (Section 23.5B); as with polyenes, the position of absorption is shifted to longer wavelengths, and the molar absorptivity, ϵ, of the absorption maximum increases sharply. For the α,β-unsaturated ketone 3-penten-2-one, for example, $\lambda_{\max}$ is 224 nm (log ε 4.10).

2-Pentanone	3-Penten-2-one	Acetophenone
$\lambda_{\max}$ 180 nm (ε 900)	$\lambda_{\max}$ 224 nm (ε 12,590)	$\lambda_{\max}$ 246 nm (ε 9,800)

The greater the extent of conjugation of unsaturated systems with the carbonyl group, the more the absorption maximum is shifted toward the visible region of the spectrum.

Like the carbonyl groups of simple aldehydes and ketones, the carboxyl group shows only weak absorption in the ultraviolet spectrum unless it is conjugated with a carbon-carbon double bond or an aromatic ring.

Aromatic Compounds

Aromatic rings absorb radiation in the ultraviolet region of the spectrum as a result of $\pi \rightarrow \pi^*$ transitions (Section 23.5B). Most commonly, two broad absorptions occur; the first is of high intensity near 205 nm, and the second is a less intense absorption near 270 nm. The presence of these two absorptions in the ultraviolet spectrum is clear evidence for the presence of an aromatic ring. Shown in Table 23.4 are values of $\lambda_{\max}$ for benzene and several of its monosubstituted derivatives.

B. The Origin of Transitions Between Electronic Energy Levels

Absorption of radiation in the ultraviolet-visible spectrum results in promotion of electrons from a lower energy, occupied MO to a higher energy, unoccupied MO.

C H E M I S T R Y   I N   A C T I O N

Ultraviolet Sunscreens and Sunblocks

Ultraviolet (UV) radiation penetrating the earth's ozone layer can be divided into two regions: UVB (290–320 nm) and UVA (320–400 nm). UVB radiation interacts directly with biomolecules of the skin and eyes, causing skin cancer, skin aging, eye damage leading to cataracts, and delayed sunburn that appears 12–24 hours after exposure. UVA radiation, by contrast, causes tanning. It also causes skin damage, though much less efficiently than UVA, and acts directly through reactive oxygen species. Its role in promoting skin cancer is less well understood.

Commercial sunscreen products are rated according to their sun protection factor (SPF), which is defined as the minimum effective dose (MED) of UV radiation in units of joules per square meter (J/m^2) that produces a delayed sunburn on protected skin compared to unprotected skin. In the following equation, protected skin is defined as skin covered by $2 \ mg/cm^2$ of sunscreen.

$$SPF = \frac{MED \ on \ protected \ skin}{MED \ on \ unprotected \ skin}$$

Two types of active ingredients are found in commercial sunblocks and sunscreens. The most common sunblock is titanium dioxide, TiO_2, which reflects and scatters UV radiation. (TiO_2 is also widely used in paints because it covers better than any other inorganic white pigment.) Chemical sunscreens, the second type of active ingredient, absorb UV radiation, and then reradiate it as heat. These compounds are most effective in screening UVB radiation, but they do not screen UVA radiation, thus allowing tanning while preventing UVB damage. Given here are structural formulas for four common chemical UVB-blocking sunscreens along with the name by which each is most commonly listed in the Active Ingredients label on commercial products.

Octyl *p*-methoxycinnamate
(*The Merck Index*, 12th ed., #6864)

Oxybenzone
(*The Merck Index*, 12th ed., #7088)

Homosalate
(*The Merck Index*, 12th ed., #4776)

Padimate A
(*The Merck Index*, 12th ed., #3282)

Table 23.4 Ultraviolet Maxima for Several Aromatic Compounds, C_6H_5—R

R	$\lambda_{\max}$ (nm)	ε ($M^{-1}cm^{-1}$)	$\lambda_{\max}$ (nm)	ε ($M^{-1}cm^{-1}$)
—H	203	7,400	254	204
—CH_3	207	7,000	261	225
—OH	211	6,200	270	1,450
—COOH	230	11,600	273	970

The energy of this radiation is generally insufficient to affect electrons in sigma-bonding molecular orbitals but sufficient to cause $\pi \rightarrow \pi^*$ transitions of electrons, most particularly pi electrons of conjugated systems. Following are three examples of conjugated systems.

$$CH_2{=}CH{-}CH{=}CH_2$$

1,3-Butadiene

$$CH_2{=}CH{-}\overset{\overset{\textstyle O}{\|}}{C}{-}CH_3$$

3-Buten-2-one

Benzaldehyde

As an example of a $\pi \rightarrow \pi^*$ transition, consider ethylene. The double bond in ethylene consists of one sigma bond formed by combination of sp^2 orbitals and one pi bond formed by combination of $2p$ orbitals. The relative energies of the pi-bonding and antibonding molecular orbitals are shown schematically in Figure 23.6. The $\pi \rightarrow \pi^*$ transitions for simple, unconjugated alkenes occur below 200 nm (at 165 nm for ethylene). Because these transitions occur at extremely short wavelengths, they are not observed in conventional ultraviolet spectroscopy and, therefore, are not useful to us for determining molecular structure.

For 1,3-butadiene, the difference in energy between the highest occupied pi molecular orbital and the lowest unoccupied pi-antibonding molecular orbital is less than it is for ethylene with the result that a $\pi \rightarrow \pi^*$ transition for 1,3-butadiene (Figure 23.7) takes less energy (occurs at longer wavelength) than that for ethylene. This transition for 1,3-butadiene occurs at 217 nm.

Electronic excitations are accompanied by changes in vibrational or rotational energy levels. The energy levels for these excitations are considerably smaller than

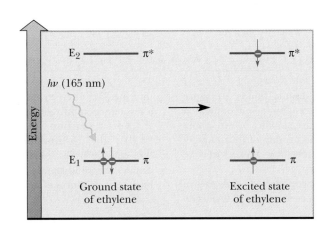

Figure 23.6

A $\pi \rightarrow \pi^*$ transition in excitation of ethylene. Absorption of ultraviolet radiation causes a transition of an electron from a pi-bonding MO in the ground state to a pi-antibonding MO in the excited state. There is no change in electron spin.

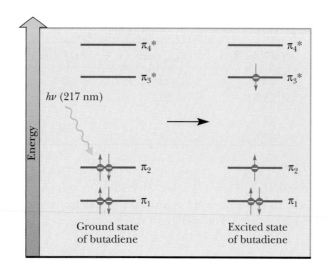

Figure 23.7

Electronic excitation of 1,3-butadiene; a $\pi \rightarrow \pi^*$ transition.

the energy differences between electronic excitations. These transitions are superposed on the electronic excitations, which results in a large number of absorption peaks so closely spaced that the spectrophotometer cannot resolve them. For this reason, UV-visible absorption peaks usually are much broader than IR absorption peaks.

In general, the greater the number of double bonds in conjugation, the longer the wavelength of ultraviolet radiation absorbed. Shown in Table 23.5 are wavelengths and energies required for $\pi \rightarrow \pi^*$ transitions in several conjugated alkenes.

Table 23.5   Wavelengths and Energies Required for $\pi \rightarrow \pi^*$ Transitions of Ethylene and Three Conjugated Polyenes

Name	Structural Formula	λ_{max} (nm)	Energy in kJ (kcal)/mol
Ethylene	$CH_2{=}CH_2$	165	724 (173)
1,3-Butadiene	$CH_2{=}CHCH{=}CH_2$	217	552 (132)
(3E)-1,3,5-Hexatriene	$CH_2{=}CHCH{=}CHCH{=}CH_2$	268	448 (107)
(3E,5E)-1,3,5,7-Octatetraene	$CH_2{=}CH(CH{=}CH)_2CH{=}CH_2$	290	385 (92)

Summary

Compounds containing conjugated double bonds are more stable than isomeric compounds containing unconjugated double bonds (Section 23.1). The extra stability of two conjugated double bonds arises because the overlap of four parallel $2p$ orbitals results in delocalization of electron density over the entire pi framework.

Conjugated dienes undergo electrophilic addition to give 1,2-addition and 1,4-addition products (Section 23.2A). The

intermediate in electrophilic addition to a conjugated diene is a resonance-stabilized allylic carbocation.

For a reaction under **kinetic (rate) control** (Section 23.2B), the distribution of products is determined by the relative rates of formation of each. For a reaction under **thermodynamic (equilibrium) control,** the distribution of products is determined by the relative stabilities of each.

The **Diels-Alder reaction** (Section 23.3) is a **cycloaddition** between a conjugated diene and a dienophile to give a six-membered ring. Dienophiles contain either a double or triple bond. Diels-Alder reactions are facilitated by electron-withdrawing substituents on one of the reactants (either the diene or dienophile) and electron-releasing substituents on the other reactant. The mechanism is described as a **pericyclic reaction;** that is, it takes place in a single step, without intermediates, and involves redistribution of bonding electrons in a cyclic transition state. The Diels-Alder reaction is stereospecific: (1) the configuration of the dienophile is retained [if substituents on the dienophile are cis (or trans), they remain cis (or trans) in the product]; (2) the configuration of the diene is retained; and (3) formation of the endo adduct is favored.

The region of the electromagnetic spectrum covered by most ultraviolet spectrophotometers runs from 200 to 400 nm, a region commonly referred to as the near-ultraviolet (Section 23.5A). Wavelengths and corresponding energies for near-ultraviolet and visible radiation are summarized in Table 23.3. Ultraviolet and visible spectral data are recorded as plots of **absorbance (*A*)** versus wavelength. The relationship between absorbance, concentration, and length of the sample cell is known as the **Beer-Lambert law.**

$$\text{Beer-Lambert law:} \qquad A = \varepsilon c l$$

where A is the absorbance; ε is the molar absorptivity (in liters per mole per centimeter); c is the concentration of solute (in moles per liter), and l is the length of the sample cell (in centimeters).

Absorption of radiation in the near-ultraviolet-visible spectrum is generally sufficient to cause $\pi \rightarrow \pi^*$ transitions of pi electrons of conjugated systems. It is also sufficient to affect $n \rightarrow \pi^*$ transitions of nonbonding electrons associated with carbonyl groups.

Key Reactions

1. Electrophilic Addition to Conjugated Dienes (Section 23.2)

The ratio of 1,2- to 1,4-addition products depends on whether the reaction is under kinetic control or thermodynamic control.

$$CH_2{=}CHCH{=}CH_2 + HBr \longrightarrow CH_3\overset{\overset{\displaystyle Br}{|}}{C}HCH{=}CH_2 + CH_3CH{=}CHCH_2Br$$

Products at −78°C (kinetic control):	90%	10%
Products at 40°C (thermodynamic control):	15%	85%

2. The Diels-Alder Reaction: A Pericyclic Reaction (Section 23.3)

A Diels-Alder reaction takes place in a single step, without intermediates, and involves a redistribution of six pi electrons in a cyclic transition state. The configuration of the diene and dienophile are preserved. Formation of the endo adduct is favored.

3. The Claisen Rearrangement: A Pericyclic Reaction (Section 23.4A)

Rearrangement of an allyl phenyl ether gives an ortho-substituted phenol.

4. The Cope Rearrangement: A Pericyclic Reaction (Section 23.4B)
Rearrangement of a 1,5-diene gives an isomeric 1,5-diene.

Problems

Structure and Stability

23.11 If an electron is added to 1,3-butadiene, into which molecular orbital does it go? If an electron is removed from 1,3-butadiene, from which molecular orbital is it taken?

23.12 Draw an energy diagram (energy versus dihedral angle from 0° to 360°) for rotation about the 2,3 single bond in 1,3-butadiene.

23.13 Draw all important contributing structures for the following allylic carbocations; then rank the structures in order of relative contributions to each resonance hybrid.

(a)   (b) $CH_2=CHCH=CHCH_2^+$ (c) $CH_3\overset{CH_3}{\underset{+}{C}}CH=CH_2$

Electrophilic Addition to Conjugated Dienes

23.14 Predict the structure of the major product formed by 1,2-addition of HCl to 2-methyl-1,3-butadiene (isoprene).

23.15 Predict the major product formed by 1,4-addition of HCl to isoprene.

23.16 Predict the structure of the major 1,2-addition product formed by reaction of one mole of Br_2 with isoprene. Also predict the structure of the major 1,4-addition product formed under these conditions.

23.17 Which of the two molecules shown do you expect to be the major product formed by 1,2-addition of HCl to cyclopentadiene? Explain.

Cyclopentadiene   3-Chlorocyclopentene   4-Chlorocyclopentene

23.18 Predict the major product formed by 1,4-addition of HCl to cyclopentadiene.

23.19 Draw structural formulas for the two constitutional isomers of molecular formula $C_5H_6Br_2$ formed by adding one mole of Br_2 to cyclopentadiene.

23.20 What are the expected kinetic and thermodynamic products from addition of one mole of Br_2 to the following dienes?

(a)   (b)

Diels-Alder Reactions

23.21 Draw structural formulas for the products of reaction of cyclopentadiene with each dienophile.

(a) $CH_2{=}CHCl$ **(b)** $CH_2{=}CHCOCH_3$ (with C=O above)   **(c)** $HC{\equiv}CH$ **(d)** $CH_3OCC{\equiv}CCOCH_3$ (with two C=O above)

23.22 Propose structural formulas for compounds A and B, and specify the configuration of compound B.

$$\text{cyclopentadiene} + CH_2{=}CH_2 \xrightarrow{200°C} C_7H_{10} \xrightarrow[\text{2. }(CH_3)_2S]{\text{1. }O_3} C_7H_{10}O_2$$

$$(A) \qquad\qquad (B)$$

23.23 Under certain conditions, 1,3-butadiene can function both as a diene and a dienophile. Draw a structural formula for the Diels-Alder adduct formed by reaction of 1,3-butadiene with itself.

23.24 1,3-Butadiene is a gas at room temperature and requires a gas-handling apparatus to use in a Diels-Alder reaction. Butadiene sulfone is a convenient substitute for gaseous 1,3-butadiene. This sulfone is a solid at room temperature (mp 66°C) and, when heated above its boiling point of 110°C, decomposes by a reverse Diels-Alder reaction to give s-cis-1,3-butadiene and sulfur dioxide. Draw a Lewis structure for butadiene sulfone, and show by curved arrows the path of this reverse Diels-Alder reaction.

$$\text{Butadiene sulfone} \xrightarrow{140°C} \text{1,3-Butadiene} + \text{Sulfur dioxide} \; (SO_2)$$

23.25 The following trienone undergoes an intramolecular Diels-Alder reaction to give the product shown. Show how the carbon skeleton of the triene must be coiled to give this product, and show by curved arrows the redistribution of electron pairs that takes place to give the product.

$$\xrightarrow{160°C}$$

23.26 The following trienone undergoes an intramolecular Diels-Alder reaction to give a bicyclic product. Propose a structural formula for the product. Account for the observation that the Diels-Alder reaction given in this problem takes place under milder conditions (at lower temperature) than the analogous Diels-Alder reaction shown in Problem 23.25.

$$\xrightarrow{0°C} \text{Diels-Alder adduct}$$

23.27 The following compound undergoes an intramolecular Diels-Alder reaction to give a bi-cyclic product. Propose a structural formula for the product.

$$\xrightarrow{\text{heat}}$$ An intramolecular Diels-Alder adduct

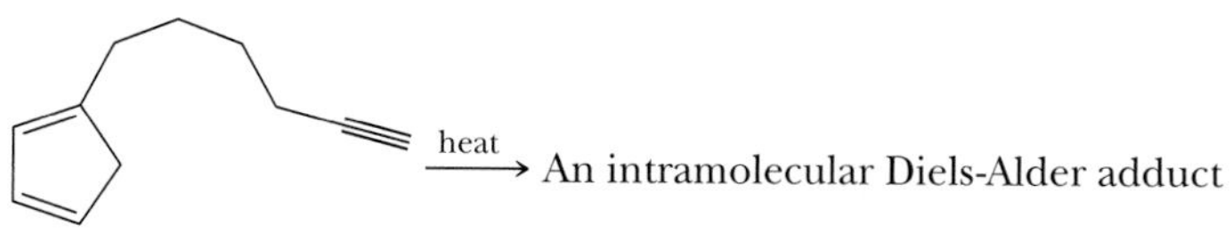

23.28 Draw a structural formula for the product of this Diels-Alder reaction, including the stereochemistry of the product.

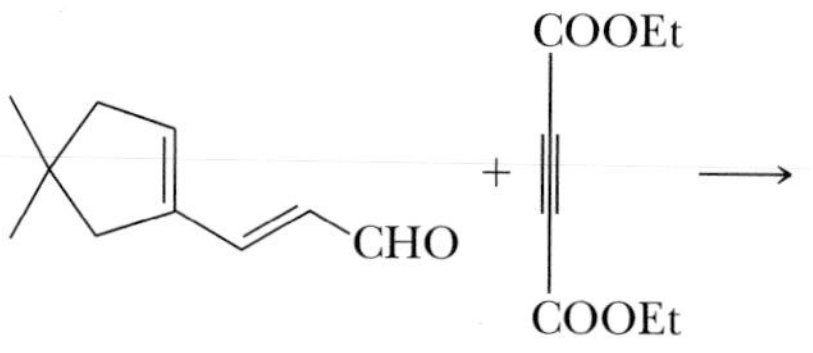

23.29 Following is a retrosynthetic analysis for the dicarboxylic acid shown on the left.

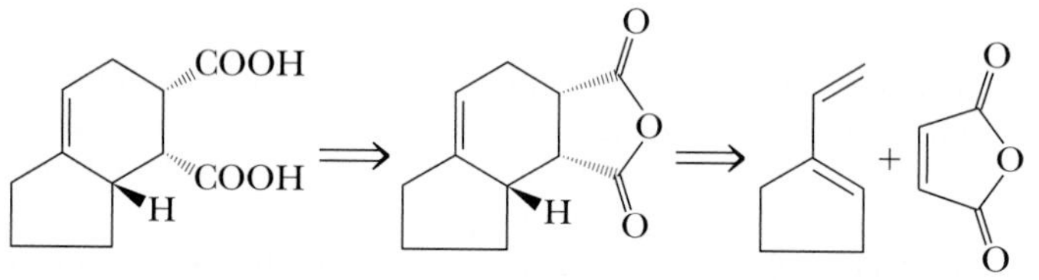

(a) Propose a synthesis of the diene from cyclopentanone and acetylene.
(b) Rationalize the stereochemistry of the target dicarboxylic acid.

23.30 One of the published syntheses of warburganal (Problem 5.28) begins with the follow-ing Diels-Alder reaction. Propose a structure for compound A.

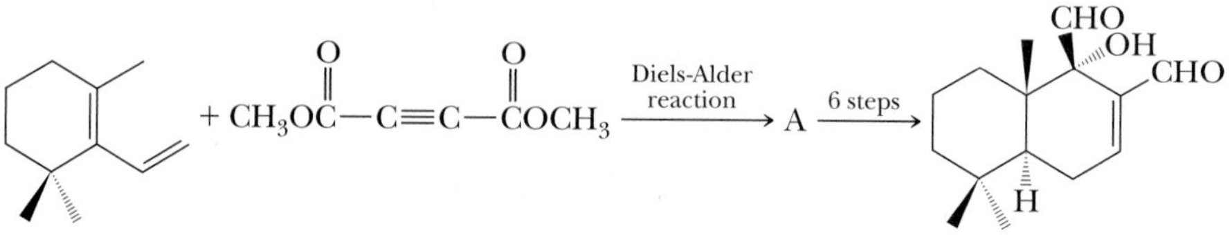

Warburganal

23.31 The Diels-Alder reaction is not limited to making six-membered rings with only carbon atoms. Predict the products of the following reactions that produce rings with atoms other than carbon in them.

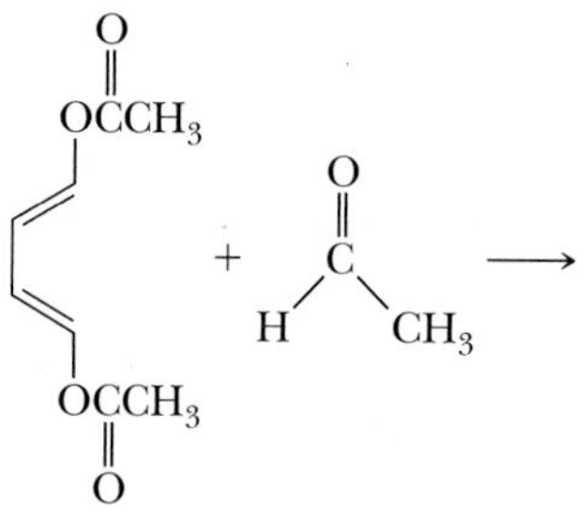

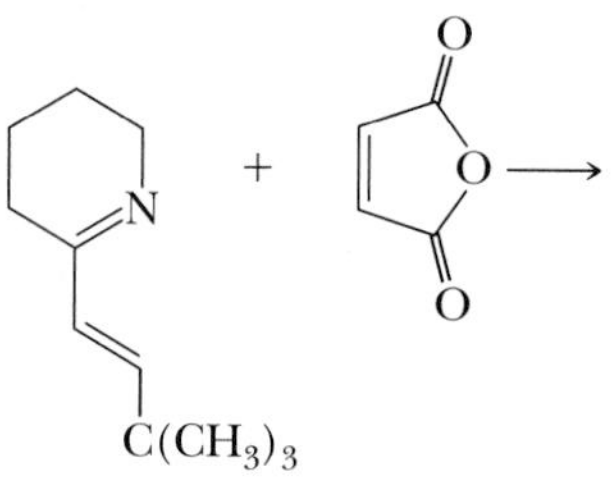

(c) 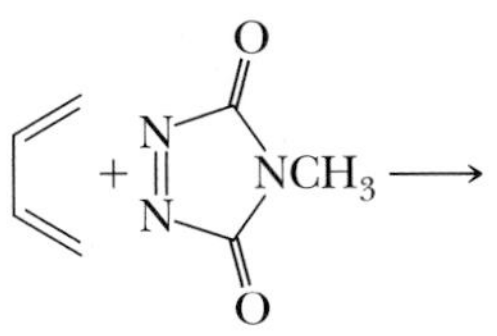 + ⟶   **(d)** 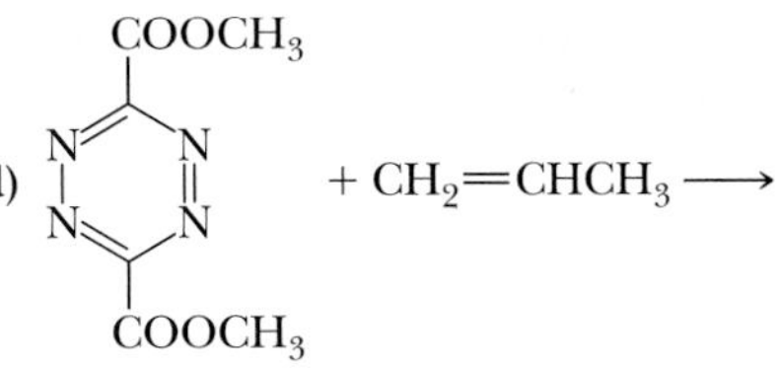 + CH$_2$=CHCH$_3$ ⟶

(e) 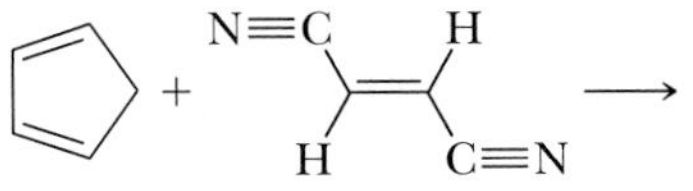 + ⟶

23.32 The first step in a synthesis of dodecahedrane involves a Diels-Alder reaction between the cyclopentadiene derivative (1) and dimethyl acetylenedicarboxylate (2). Show how these two molecules react to form the dodecahedrane synthetic intermediate (3).

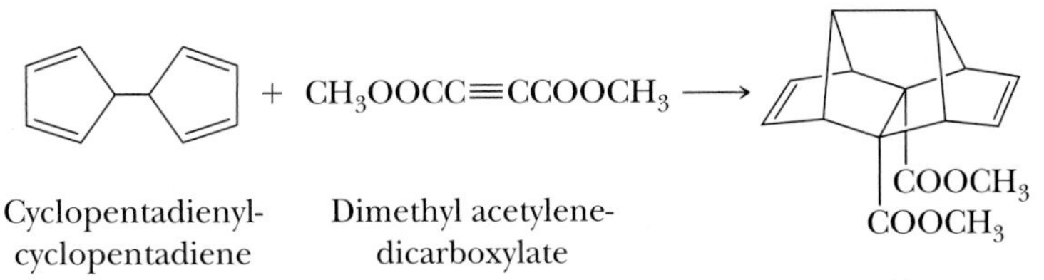

+ CH$_3$OOCC≡CCOOCH$_3$ ⟶

Cyclopentadienyl-   Dimethyl acetylene-
cyclopentadiene   dicarboxylate
(1)   (2)   (3)

23.33 Bicyclo[2.2.1]-2,5-heptadiene can be prepared in two steps from cyclopentadiene and vinyl chloride. Provide a mechanism for each step.

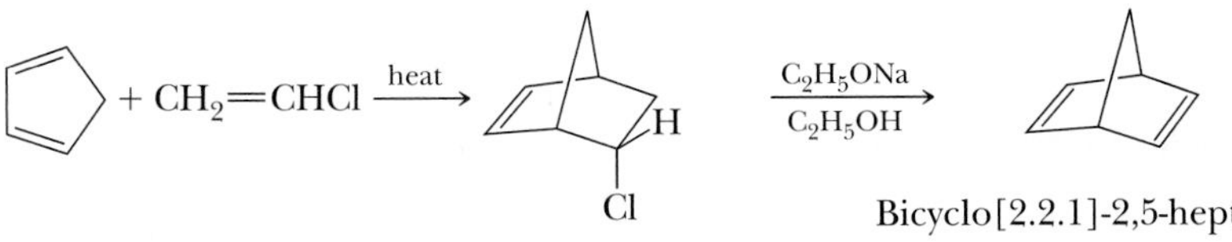

+ CH$_2$=CHCl $\xrightarrow{\text{heat}}$ $\xrightarrow[\text{C}_2\text{H}_5\text{OH}]{\text{C}_2\text{H}_5\text{ONa}}$

Bicyclo[2.2.1]-2,5-heptadiene

23.34 Treatment of anthranilic acid with nitrous acid gives an intermediate, A, that contains a diazonium ion and a carboxylate group. When this intermediate is heated in the presence of furan, a bicyclic compound is formed. Propose a structural formula for compound A and mechanism for formation of the bicyclic product.

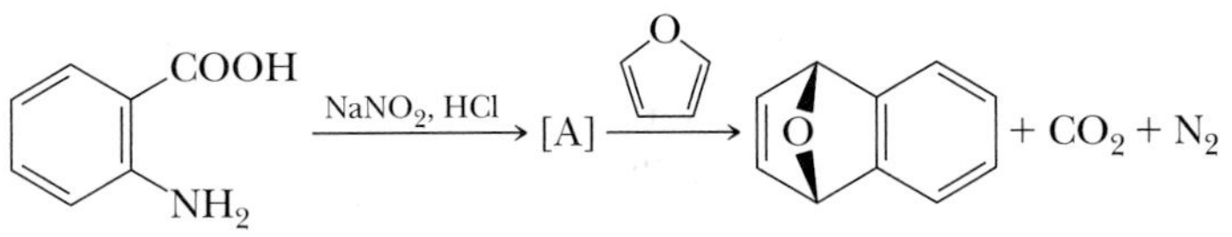

$\xrightarrow{\text{NaNO}_2,\ \text{HCl}}$ [A] $\xrightarrow{\text{furan}}$ + CO$_2$ + N$_2$

Anthranilic
acid

23.35 Propose a mechanism for the following reaction, which is called an allylic rearrangement, or, alternatively, a conjugate addition.

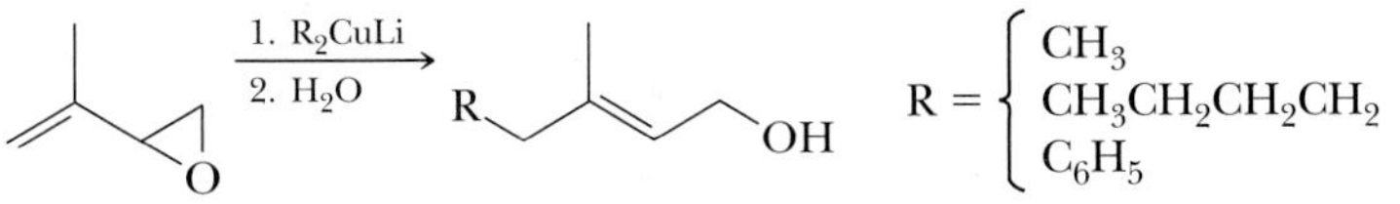

$\xrightarrow[\text{2. H}_2\text{O}]{\text{1. R}_2\text{CuLi}}$ R = $\begin{cases} \text{CH}_3 \\ \text{CH}_3\text{CH}_2\text{CH}_2\text{CH}_2 \\ \text{C}_6\text{H}_5 \end{cases}$

23.36 All attempts to synthesize cyclopentadienone yield only a Diels-Alder adduct dimer. Cycloheptatrienone, however, has been prepared by several methods and is stable.

Cyclopentadienone   Cycloheptatrienone

(a) Draw a structural formula for the Diels-Alder adduct dimer formed by cyclopentadienone.

(b) How do you account for the marked difference in stability of these two ketones? *Hint:* Consider important resonance contributing structures for each compound.

23.37 Following is a retrosynthetic scheme for the tricyclic diene on the left. Show how to accomplish this synthesis from 2-bromopropane, cyclopentadiene, and 2-cyclohexenone.

Other Pericyclic Reactions

23.38 Claisen rearrangement of an allyl phenyl ether with substituent groups in both ortho positions leads to formation of a para-substituted product. Propose a mechanism for the following rearrangement.

23.39 Following are three examples of Cope rearrangements of 1,5-dienes. Show that each product can be formed in a single step by a mechanism involving redistribution of six electrons in a cyclic transition state.

(b)

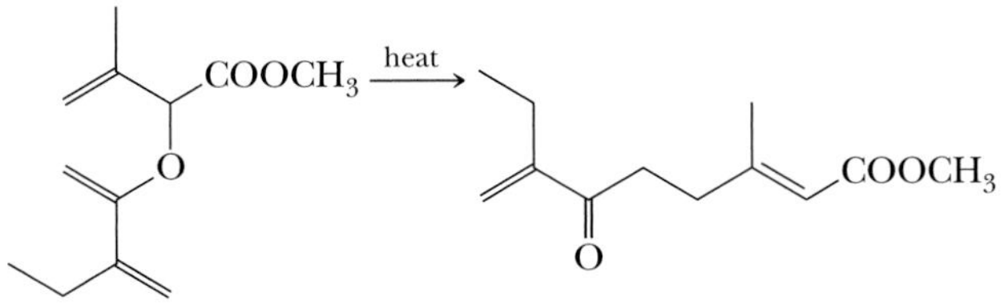

(c)

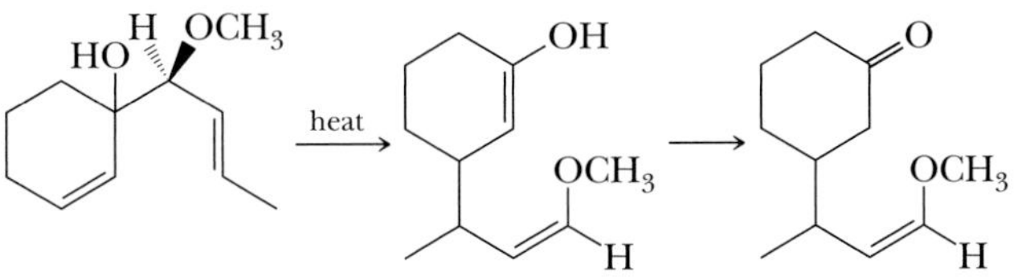

23.40 The following transformation is an example of the Carroll reaction, named after the English chemist, M. F. Carroll, who first reported it. Propose a mechanism for this reaction.

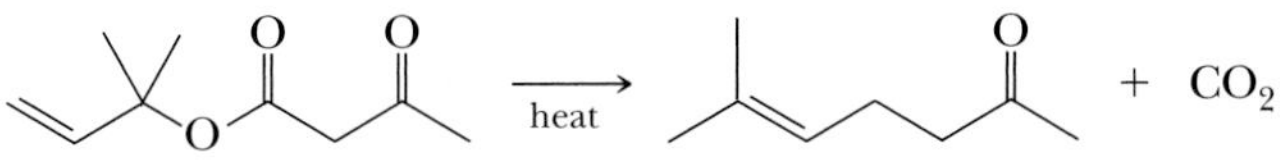

6-Methyl-5-hepten-2-one

23.41 Following are two examples of photoinduced (light-induced) isomerizations. Vitamin D_3 (cholecalciferol) is produced by the action of sunlight on 7-dehydrocholesterol in the skin. First precalciferol is formed, followed by cholecalciferol. Cholecalciferol is shown here in an s-cis conformation. After its formation, it assumes an s-trans conformation. Use curved arrows to show the flow of electrons in these photoisomerizations.

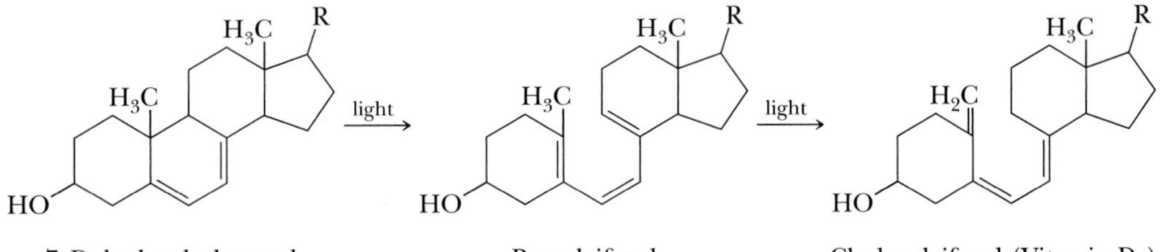

7-Dehydrocholesterol

Precalciferol

Cholecalciferol (Vitamin D_3)
(s-cis conformation)

23.42 Show the product of the following reaction. Include stereochemistry.

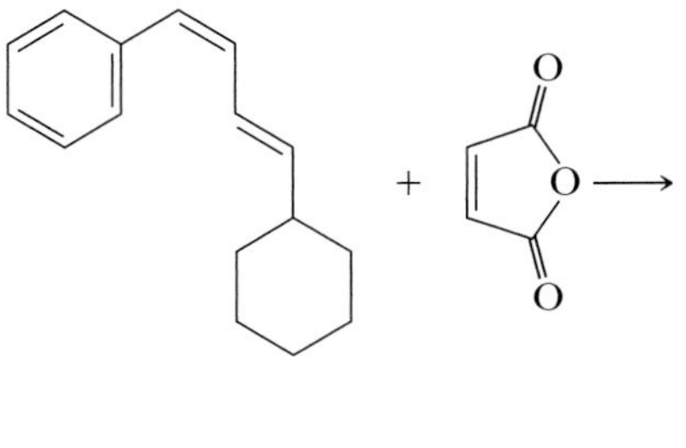

23.43 Following is a synthesis for the antifungal agent tolciclate (see *The Merck Index*, 12th ed., #9647).

4-Bromo-3-iodoanisole   (A)   Tolciclate

(a) Propose a mechanism for formation of (A).

(b) Show how (A) can be converted to tolciclate. Use 3-methyl-*N*-methylaniline as the source of the amine nitrogen, and thiophosgene, $Cl_2C{=}S$, as the source of the $C{=}S$ group.

Ultraviolet-Visible Spectra

23.44 Show how to distinguish between 1,3-cyclohexadiene and 1,4-cyclohexadiene by ultraviolet spectroscopy.

1,3-Cyclohexadiene   1,4-Cyclohexadiene

23.45 Pyridine exhibits a UV transition of the type $n \rightarrow \pi^*$ at 270 nm. In this transition, one of the unshared electrons on nitrogen is promoted from a nonbonding MO to a pi*-antibonding MO. What is the effect on this UV peak if pyridine is protonated?

Pyridine   Pyridinium ion

23.46 The weight of proteins or nucleic acids in solution is commonly determined by UV spectroscopy using the Beer-Lambert law. For example, the ε of double-stranded DNA at 260 nm is 6670 $M^{-1}cm^{-1}$. The formula weight of the repeating unit in DNA (650 Daltons) can be used as the molecular weight. What is the weight of DNA in 2.0 mL of aqueous buffer if the absorbance, measured in a 1-cm cuvette, is 0.75?

23.47 A sample of adenosine triphosphate (ATP) (MW 507, $\varepsilon = 14{,}700$ $M^{-1}cm^{-1}$ at 257 nm) is dissolved in 5.0 mL of buffer. A 250-μL aliquot is removed and placed in a 1-cm cuvette with sufficient buffer to give a total volume of 2.0 mL. The absorbance of the sample at 257 nm is 1.15. Calculate the weight of ATP in the original 5.0-mL sample.

23.48 Biochemical molecules are frequently sold by optical density (OD) units, where one OD unit is the amount of compound that gives an absorbance of 1.0 at its UV maximum in 1.0 mL of solvent in a 1-cm cuvette. If the cost of 10.0 OD units of a DNA polymer ($\varepsilon = 6600$ $M^{-1}cm^{-1}$ at 262 nm) is $51, what is the cost per gram of this biochemical?

A Conversation with . . .

Paul S. Anderson

Paul S. Anderson is Senior Vice President for Chemical and Physical Sciences at DuPont Pharmaceuticals Company in Wilmington, Delaware, a position he has held since 1994. Before that Dr. Anderson spent 30 years at Merck Research Laboratories. To name just two of the many honors he has received in recognition of a fruitful career in pharmaceutical chemistry, he is the 2001 recipient of the American Chemical Society Award in Industrial Chemistry and the 1995 recipient of the Society's E. B. Hershberg Award for Important Discoveries in Medicinally Active Substances.

At Merck, Dr. Anderson was involved in research that ultimately resulted in the creation of Trusopt, a topical medication for glaucoma patients, and Zocor, a cholesterol-lowering medication taken today by millions of people around the world. He and his colleagues also were involved in discovering chemotherapies for fighting AIDS. Two products in that war are the protease inhibitor Crixivan and the reverse transcriptase inhibitor Sustiva.

Like Father, Like Son

Dr. Anderson's interest in chemistry stems from his childhood. His fa-

ther would buy chemicals from the drugstore when he was a boy himself (you could do that in those days) and ran his own experiments, banished by his mother to an abandoned nearby house so that her house wouldn't smell or blow up. Later, as an adult in his own home, Anderson père allowed himself the luxury of running his chemistry experiments at the kitchen sink, with his young son alongside. From these early sessions at the kitchen sink came Paul Anderson's lifelong interest in chemistry.

"We did simple things," he remembered, "generating hydrogen gas or oxygen gas, things like that. What made me realize that I really liked chemistry more than anything else I had to study in school was the

fact that one does experiments and makes things rather than just reading about them. I got a chemistry set as a gift, of course, and began expanding my repertoire to a broader variety of experiments. But the experiments I did with my father were the classic experiments you do when you are beginning to study chemistry. When I think back to that time, I see it was the experimental nature of things, the doing of experiments that really caught my attention and began to trigger my thinking that I'd like to do more of this. And then my high school chemistry course just reinforced the notion."

Dr. Anderson's first plan was to be a teacher, as both his parents were, and when he entered the University of Vermont in 1955, he originally intended to enroll in the College of Education. After a conversation with his father, however, he changed his plans. His father asked the very wise question: "What do you think you'd like to teach?" And when Paul answered, "Chemistry, probably," his father's important advice was that chemistry should be his major. "Get the strongest foundation you can in the subject you want to teach, and then take the education courses you need to get qualified to teach those subjects." Dr. Anderson set out for a teaching career, but four years as a chemistry major told him that experimental chemistry was the life he wanted. After earning his bachelor's degree in chemistry at the University of Vermont, he spent four years at the University of New Hampshire working on this doctorate with the chemist R. E. Lyle.

Designing Molecules

"Pharmaceutical research today is a team effort involving chemists, biologists, and pharmacologists. We all work together to identify target diseases. Then we prioritize, looking for those that are a significant unmet medical need and where we have a good chance of understanding the biomechanism of what's going on in the body of a person suffering from the disease. And so there's a lot of scientific judgment needed in that part of the process. Once we decide on a specific disease we're interested in providing therapy for, we pick a biochemical mechanism that is either the causative factor or a contributor to that disease. Then we look at the chemical reactions in that biochemical mechanism and decide where we want to step in and interrupt the chemistry. Once all these steps have been done, then the medicinal chemistry starts. Now the chemists play the biggest role because it's time to go into the lab and start designing and making molecules that we hope will have the desired properties. All the reactions you studied in school—how to reduce an aldehyde in the presence of a carboxylic acid without reducing the acid, how to oxidize alkenes to glycols— all this knowledge is what you use to make molecules for biological testing.

"Trusopt for glaucoma. Zocor for cholesterol. Crixivan and Sustiva for HIV. All these drugs are enzyme inhibitors. In pharmaceuticals, you have two general classes of drugs— enzyme inhibitors and receptor modulators, which can either activate or inhibit receptors. As it turns out, most of the projects I've been involved in rely on enzyme inhibition to achieve their therapeutic effect. In some cases like Trusopt and Zocor, enzyme inhibition is used to reset the pace of normal enzyme-mediated chemical processes in the body. In other cases like Crixivan and Sustiva, the objective is to shut down a foreign enzyme that has invaded the body.

All the reactions you studied in school—how to reduce an aldehyde in the presence of a carboxylic acid without reducing the acid, how to oxidize alkenes to glycols—all this knowledge is what you use to make molecules for biological testing.

"The whole thing is a very logical, scientific scheme. Once we've selected the disease and selected the biomechanism by which we're going to attack it, then we search for the compound that works by that mechanism. Zocor, for instance, is an inhibitor of an enzyme called HMG-CoA reductase. That enzyme is the rate-limiting step in how the body makes cholesterol, and we knew— from work done by Nobel laureates Brown and Goldstein—that inhibiting HMG-CoA reductase would be useful in patients who have elevated cholesterol levels. At Merck, one of the early important steps on the road to Zocor was our discovery of Mevacor, a natural product that inhibits this enzyme. Then we did our chemistry and synthesized derivatives of this natural product to find a better one as a therapeutic agent, and the result of doing that chemistry was Zocor, which came out in the late 1980s and is still used all around the world.

"The big issue for everyone working on these cholesterol-lowering drugs was safety—would these drugs be safe for long-term therapy? A large number of safety-assessment studies had to be completed, and of course you always worry when you're working on synthetic compounds—or even natural products, for that matter—how those safety studies are going to turn out. Here you are developing a drug that is supposed to keep the body from making cholesterol. One 'minor' detail haunting you is that the body needs cholesterol in every single cell. So if the drug you've made completely inhibits cholesterol synthesis in the body, that could cause unwanted side effects. So we had to figure out just how much reductase inhibition the body will let you get away with, what level we needed to get the desired therapy without any serious side effects. It's a very interesting balancing act, to say the least.

"And then once you think you've got all the balancing done, you start animal studies and keep your fingers crossed. Hopefully, things work out well with the animals and you get to go on to people, which of course was the goal to begin with."

A Batting Average No Major League Could Live with, but I'm Not Complaining

"When you come right down to it, not very many projects succeed in the sense that what you're working on becomes a useful drug. In my own experience, I reckon on an overall success rate of about one out of every seven projects. Which of course means that about six out of every seven projects failed to get to a drug! It's a very high-risk business—pharmaceutical science—and even with good scientists and careful planning and thought, there's still enough unknown so that you fail most of the time. To keep on going, you have to just accept the reality of those numbers—in this business, one out of seven is pretty good. It's a lousy batting average—.142!—but for drug discovery, it's pretty good."

On Being an Administrator

Chemists do chemistry, and when you've been fascinated with chemistry from childhood, it's hard to give up the doing part. Deciding to leave the bench and become an administrator can be a tough decision for a successful scientist in industry, but it's one Dr. Anderson decided to make.

"I had to step away from the bench in 1980 at Merck, when I became an Executive Director. It was a tough decision, though, because I'm an experimentalist at heart. To compensate for having given up the bench, one of the things I've made part of my job at DuPont is spending time with every PhD in the physical sciences, about 90 of them these days. We talk one on one about what they're doing, what they're excited about, and what they think the challenges are in trying to accomplish these things. It helps me because those talks are very much oriented toward experimental science, and so they partially fill the void of not actually being there in the lab.

"I'll always miss being at the bench, but I've got to say that I've had an enormous amount of fun doing what I've done as an administrator. Plus I've been able to do a lot of things I wouldn't have been able to do had I stayed at the bench. I really don't have any regrets. Things just turned out to be different from what I originally planned, that's all."

New Students, Better Training

"Every year, I give several talks on campuses around the country, and I must say that, from what I've seen, students today are better trained than we were as undergraduates—in the sense that the textbooks they have and the teaching professionals that use them are better today than they've ever been. At DuPont, we're of course hiring young chemists all the time, and the ones I see these days are obviously very well trained. The young folk who come here are very committed to using science to achieve goals in medicine and in health care in general.

"If there is a threat to kids in college today, it's that the whole world is becoming better educated, and you have to remember that science is a very competitive endeavor. The rest of the world looks at the United States and realizes that we've achieved such a high standard of living because we are very good at scientific innovation. They realize that science is the best way to achieve this standard of living. So the whole world is working on education, and in no place is it more apparent than in the young folk who are coming from mainland China—they are extremely well educated and very competitive. As long as they stay here, it's fine, but if they all start returning to China, that's going to be a problem for us in the United States. What's eventually going to happen is that innovation in science is not going to be solely a U.S. thing anymore, it's going to be more global. And because I am so very much impressed by the young people I see from other parts of the world, I figure it wouldn't be a bad idea for schools to think about teaching more languages! German is still a very useful language for a chemist because the older German literature in synthetic chemistry is really very good. Today, much of what's done in Germany is published in English, and so in terms of current science, knowing German is not so important. But to be able to look back at work that's already been done, to get a good feel for chemistry as it was done in the first half of the 20th century and before, knowing German is still very useful. Also, I've always thought a fair amount of mathematics is useful, and of course everyone in science needs to know computer science today.

"One thing students preparing today for a career in any branch of the chemical industry have to be prepared for is to work on a team. In today's information-intensive world, where there's just so much information available, it's very difficult for one person to be able to comprehend and project all the potential uses and be able to extract all the value in that information. You need a team working together to sift through all the facts and decide which ones should be focused on.

"This is certainly true in the pharmaceutical industry. Young scientists coming into pharmaceutical drug discovery need to understand they'll be working in a team environment. If they're uncomfortable with that, this may not be the best place for them. Discovering drugs is a team effort. In my judgment, it's an extremely rare situation in medicinal research where one person makes a contribution so significant that the project would have failed without that contribution.

"Now, it usually turns out that it's one person who actually succeeds in making the molecule that's the right one for a given drug, but the long, long process leading up to that situation where one person makes the compound is a very complicated team effort. Different people, different pieces of knowledge all put together is what gets you where you want to go. In my own case, a Merck molecule christened MK-801 is the only one where I was actually the one to make the compound. It didn't become a drug, however. It became a very useful tool, true, but never a drug. If you look in the literature, you'll see thousands of references to MK-801 because it was used around the world to study a certain class of receptors, and I do smile whenever I come across one of those references, but I never forget all the work of all the people that led up to that one day in the lab when my experiment came up with what we were after."

Science and Society — Asking the Right Questions

"One area of chemistry where we've become much more sophisticated is in our understanding of risk assessment, specifically risks to our natural environment. Chemistry gets a bad rap from some quarters these days because of the problems of industrial waste and air pollution, but I firmly believe that the world of chemistry can be quite proud of the way we've expanded our knowledge base on how to deal with these issues. You have to realize that everything we do has risks associated with it—from deciding to go for a bike ride to choosing a family doctor to inventing a new industrial process. As far as assessing the risks associated with the latter, I think the chemical industry has become more sophisticated in the sense that we now have a pretty good idea about what kinds of studies need to be done to do a good job with risk assessment on new products, whether they be agricultural products or pharmaceutical products or anything else that derives from our knowledge of chemistry, and can improve our quality of life.

"The clothes we wear, the ways in which we store food, the ways in which we deal with health-care issues—all these things we've done through the science of chemistry have improved our quality of life. And at one time, it's of course true, we weren't so good at figuring out what risks we were exposing ourselves to as we took these steps toward improving our quality of life. But I think that in the last few decades we've become very sophisticated in how to do risk-assessment studies and how to put things in perspective.

"A good example is DDT—obviously a chemical product—which we used to control mosquito populations and which had a huge positive effect on how we manage malaria around the globe. When this chemical was developed, in the 1940s, we didn't understand that killing mosquitoes and other insects was not the only thing going on. There were unacceptable risks that just weren't known about, but I think that today we are much better at figuring out what those risks are. So if DDT came along today, when we're more sophisticated about assessing risk, we'd know how to frame the right questions. Is saving x number of people from malaria worth the risk DDT could lead to what Rachel Carson famously called a silent spring, silent because all the birds are gone, victims of DDT poisoning? This question shows you the importance of understanding what risk assessment is and how to do it. You have to understand, of course, that you can never prove that something is safe. All you can ever do is define the benefits and the risks and then make a judgment based on the ratio of benefits to risks as to whether or not you should do something.

"Today that judgment is made with the help of regulatory agencies like the EPA, and we have federal and state laws that deal with the risk-assessment issues that need to be addressed whenever a new product is launched. On the personal level, everyone who does science has to be responsible for what she or he does, of course, but in this country the way in which we approach those responsibilities on a communal level is that more than one person is involved in the process of deciding what's going to be safe enough to be acceptable for the benefit that we hope to achieve. It's a team effort, a complicated process involving the input of a lot of scientists from different disciplines, and the thing that matters most is that at the end of the day the right questions are asked and the right experiments are done so that benefits for our quality of life are put into perspective without unacceptable risk for us and our environment."

Interchapter

MEDICINAL CHEMISTRY—PROBLEMS IN ORGANIC SYNTHESIS

W ith the completion of Chapter 23, you have been introduced to the major organic functional groups and their most characteristic reactions. You have also been introduced to the design of organic syntheses. Synthesis is what organic chemists do best—they make things. The goal may be the synthesis of compounds that can be isolated from natural sources but not in quantities sufficient to meet

■ The bark of the Pacific yew contains paclitaxel, a substance that has proven effective in treating certain types of ovarian and breast cancer. *(Tom & Pat Leeson/Photo Researchers, Inc.)* Inset: A model of paclitaxel (Taxol).

929

commercial needs. Alternatively, the goal may be entirely new compounds designed to meet very specific needs, as for example pharmaceuticals, plastics, synthetic fibers, agrochemicals, detergents, paints and other coverings, and elastomers.

In this interchapter section, we concentrate on the synthesis of selected pharmaceuticals. The particular target molecules have been chosen with one criterion in mind, namely that they can be prepared using the chemistry you have encountered in Chapters 1–23. You will find the synthesis of some target molecules quite straightforward. Others you will find challenging. A few you will find very challenging. In a few cases, you will not have studied the specific reactions, but you will be asked to reason by analogy from reactions you do know.

One reason you will find these problems challenging is that they are presented here without a context. The majority of end-of-chapter problems you have done so far have been set within a context. For example, reactions of aldehydes and ketones typically appear at the end of the aldehydes and ketones chapter. Reactions of esters appear at the end of the chapter on functional derivatives of carboxylic acids, and so on. However, in this interchapter section, you must recognize reaction types without the hint of chapter context. In this way, these problems provide an overview of the entire body of organic chemistry you have studied so far. As a help in doing these problems, we suggest that you review the Key Reactions section at the end of each chapter dealing with particular functional groups. With these reactions firmly in your mind, you will be better able to recognize the types of chemical transformations involved in the preparation of each target molecule.

As you work through these problems, keep in mind that there is more than one synthetic pathway by which most target molecules can be prepared. Furthermore, there is often more than one reagent that will bring about a particular step in a synthesis. Your goal should be to recognize the type of reaction involved (oxidation, reduction, nucleophilic displacement, esterification, hydrolysis, and the like) and select an appropriate reagent to bring about that reaction.

A challenge for any synthesis is to begin with readily available starting materials. We have eliminated that problem for you by listing the starting materials for each synthesis.

Finally, we encourage you to discuss your solutions to each synthesis with other members of the class. This interchange of ideas is an invaluable means for developing a deeper understanding of organic chemistry.

The problems are grouped according to the divisions used in the Therapeutic Category and Biological Activity Index of *The Merck Index*.

A note on the naming of drugs in this interchapter section. Chemical names of drugs are not capitalized in the text. Proprietary or trademarked names are always capitalized. It is our practice to list and discuss drugs by their chemical names. Where appropriate, one or more trademarked names may also be given in one of the following formats: ". . . diazepam, better known as Valium . . ." or ". . . the analgesic meperidine (Demerol) . . . ." We do not provide solutions in the Study Guide for the Interchapter Medicinal Chemistry problems. We do, however, provide solutions for them on our Web site for the instructor, who has the option of suppling them to you.

Anxiolytic

MC.1 Both chlorodiazepoxide (see *The Merck Index*, 12th ed., #2132), better known as Librium, and diazepam (see *The Merck Index*, 12th ed., #3042), better known as Valium, belong to a class of central nervous system (CNS) sedative/hypnotics derived from 1,4-

benzodiazepine. As sedatives, benzodiazepines diminish activity and excitement and thereby have a calming effect. As hypnotics, they produce drowsiness and sleep. The benzodiazepines affect neural pathways in the CNS mediated by the neurotransmitter 4-aminobutanoic acid (γ-aminobutyric acid, GABA; see *The Merck Index*, 12th ed., #450). GABA is an inhibitor of neurotransmission. By increasing the affinity of GABA for its neuroreceptors, the benzodiazepines decrease activity in parts of the CNS mediated by GABA with the accompanying reduction in anxiety and alertness.

Librium was introduced in 1960 and was followed soon thereafter by more than two dozen related compounds. Of these, Valium became one of the most widely used. In 1976, for example, based on the number of new and refilled prescriptions processed, it was the most prescribed drug in the United States.

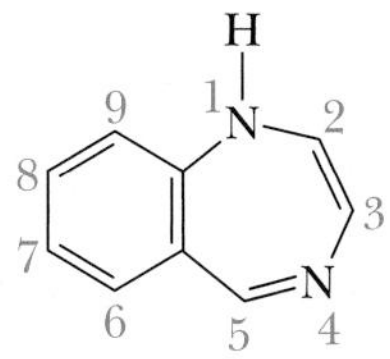

1,4-Benzodiazepine

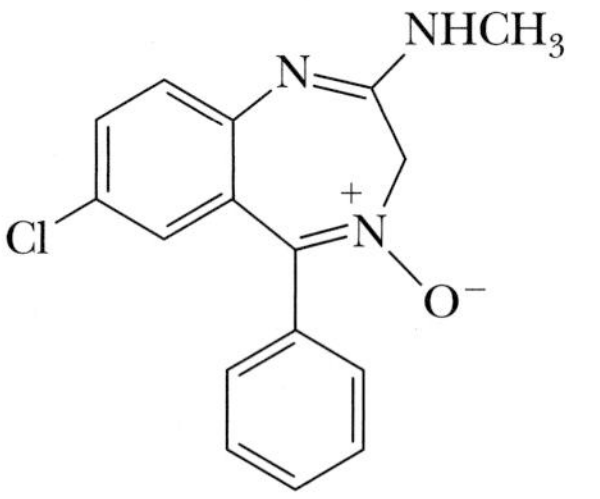

Chlordiazepoxide
(Librium)

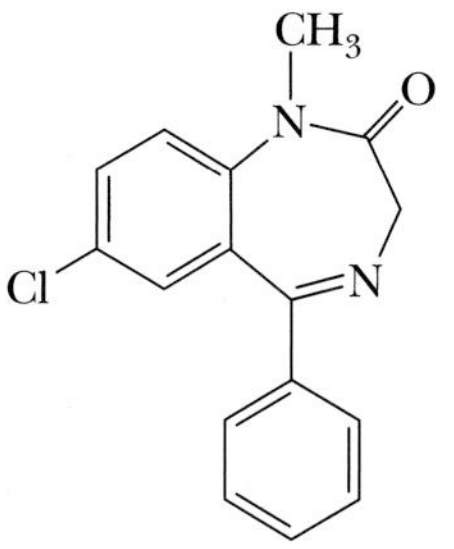

Diazepam
(Valium)

Given the following retrosynthetic analysis, propose a synthesis for diazepam. Note that the formation of compound B involves a Friedel-Crafts acylation. In this reaction it is necessary to protect the 2° amine by prior treatment with acetic anhydride. The acetyl-protecting group is then removed by treatment with aqueous NaOH followed by careful acidification with HCl.

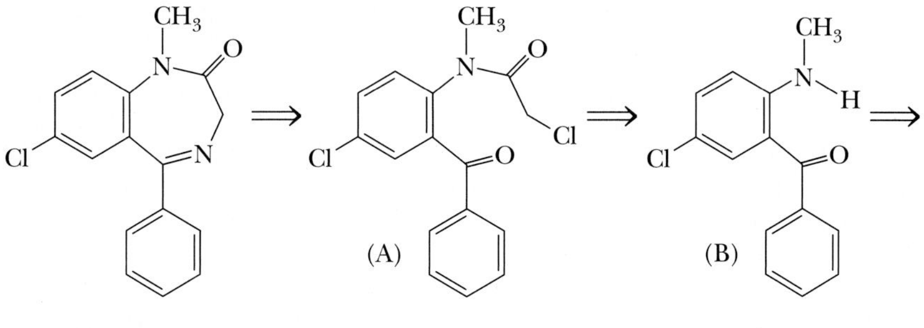

Diazepam

(A)

(B)

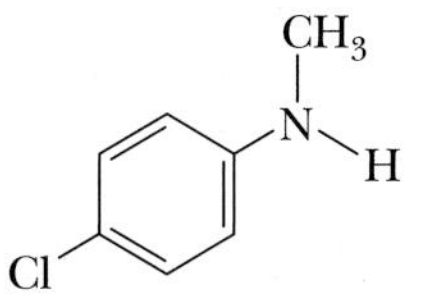

4-Chloro-*N*-methylaniline 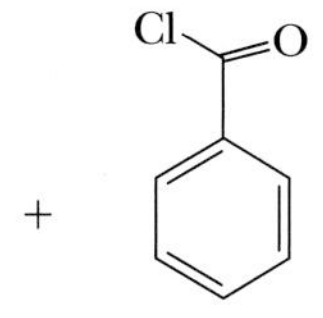   Benzoyl chloride

MC.2 One of the important routes for metabolism of diazepam and related compounds is hydroxylation (a two-electron oxidation) at carbon 3 of the seven-membered ring. This finding raised the possibility that such hydroxylated products might also be tranquilizers. One such drug produced as a result of this reasoning is lorazepam (see *The Merck Index*, 12th ed., #5609). Following is an outline of its synthesis.

(A)   (B)

(C)   (D)   Lorazepam

(a) Propose a mechanism for the formation of (A).
(b) Propose a mechanism for bromination of (A) to give (B).
(c) Propose a mechanism for the conversion of (B) to (C).
(d) Propose reagents for the conversion of (C) to (D).
(e) Propose a mechanism for the cleavage of the methyl ether by boron tribromide. The methyl group of —OCH_3 is converted to CH_3Br.

For the story of illicit and criminal use of a 1,4-benzodiazepine, see D. A. Labianca, "Rohypnol: Profile of the 'Date-Rape Drug,'" *J. Chem. Ed.* **75**(6): 719–722 (1988). The chemical name of this drug is flunitrazepam (see *The Merck Index*, 12th ed., #4181). In addition to its sedative/hypnotic effect, Rohypnol induces amnesia, an effect potentiated by alcohol.

Flunitrazepam

Anticonvulsant

MC.3 Clonazepam (Klonopin; see *The Merck Index*, 12th ed., #2449) is an anticonvulsant used to prevent epileptic seizures. Review your solution to Problem MC.2 and then propose a synthesis for clonazepam from 4-nitroaniline and 2-chlorobenzoic acid.

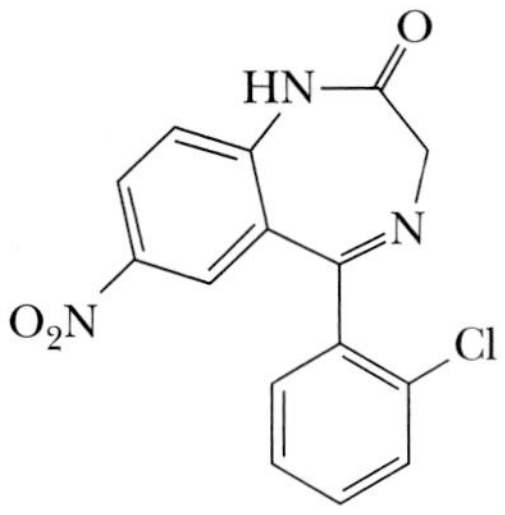

Clonazepam

MC.4 Gabapentin (see *The Merck Index*, 12th ed., #4343), an anticonvulsant used in the treatment of epilepsy, is structurally related to the neurotransmitter 4-aminobutanoic acid (see *The Merck Index*, 12th ed., #450). It was designed specifically to be more lipophilic than GABA and, therefore, more likely to cross the blood-brain barrier, a lipid-like protective membrane that surrounds the capillary system in the brain and prevents hydrophilic compounds from entering the brain by passive diffusion.

Gabapentin

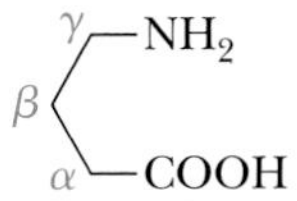

4-Aminobutanoic acid
(γ-Aminobutyric acid, GABA)

Given the following retrosynthetic analysis, propose a synthesis for gabapentin.

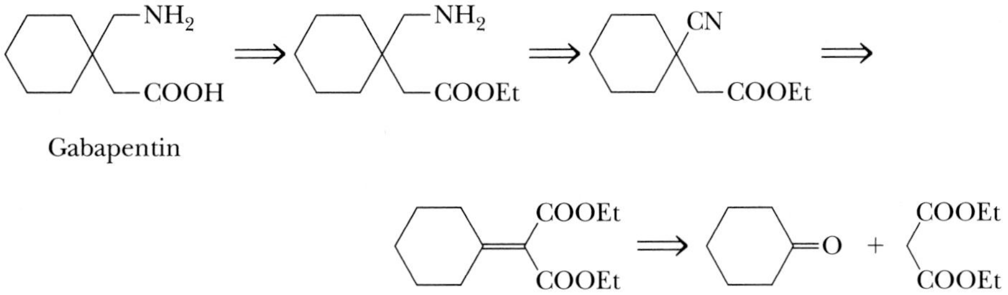

Gabapentin

Cyclohexanone   Diethyl
malonate

MC.5 The following three derivatives of succinimide are anticonvulsants and have found use in the treatment of epilepsy, particularly *petit mal* seizures. Under each name is given its number in *The Merck Index*.

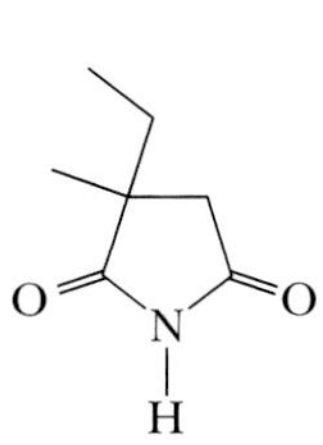

Ethosuximide
(#3794)

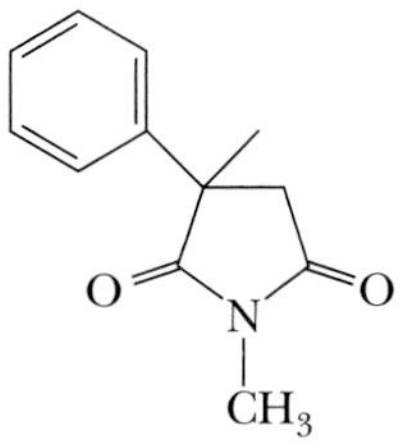

Methsuximide
(#6085)

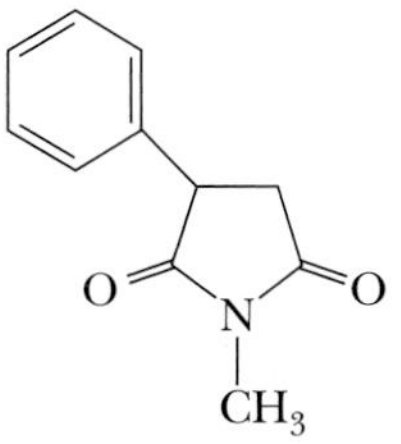

Phensuximide
(#7414)

The synthesis of phensuximide begins with the treatment of benzaldehyde with ethyl cyanoacetate in the presence of sodium ethoxide.

(a) Propose a mechanism for the formation of (A).

(b) Propose a mechanism for the conversion of (A) to (B). What name is given to this type of reaction?

(c) Describe the chemistry involved in the conversion of (B) to (C). You need not present detailed mechanisms. Rather, state what is accomplished by treating (B) first with NaOH and then with HCl followed by heating.

(d) Propose experimental conditions for the conversion of (C) to (D).

(e) Propose a mechanism for the conversion of (D) to phensuximide.

(f) Show how this same synthetic strategy can be used to prepare ethosuximide and methsuximide.

(g) Of these three anticonvulsants, one is considerably more acidic than the other two. Which is the more acidic compound? Estimate its pK_a and account for its acidity. How does its acidity compare with that of phenol? With that of acetic acid?

Anthelmintic (Nematodes)

MC.6 Following is a retrosynthetic analysis for the anthelmintic (against worms) diethylcarbamazine (see *The Merck Index*, 12th ed., #3165). Diethylcarbamazine is used chiefly against nematodes, small cylindrical or slender thread-like worms such as the common round worm, which are parasitic in animals and plants. Given this retrosynthetic analysis, propose a synthesis of diethylcarbamazine from the four named starting materials.

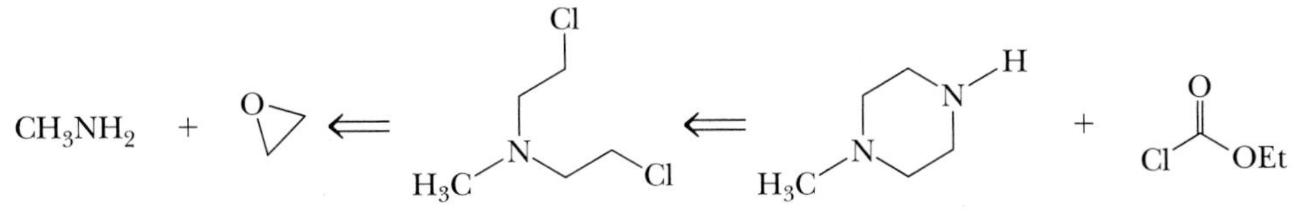

CH$_3$NH$_2$　+　(Ethylene oxide)　⇐　(C)　⇐　(B)　+　Ethyl chloroformate

Methylamine　Ethylene oxide　(C)　(B)　Ethyl chloroformate

Antidepressant

MC.7 The synthesis of chlorpromazine (see *The Merck Index*, 12th ed., #2238) in the 1950s and the discovery soon thereafter of its antipsychotic activity opened the modern era of biochemical investigations of the pharmacology of the central nervous system. One of the compounds prepared in the search for more effective antipsychotics was amitriptyline (see *The Merck Index*, 12th ed., #511). Surprisingly, amitriptyline shows antidepressant activity rather than antipsychotic activity.

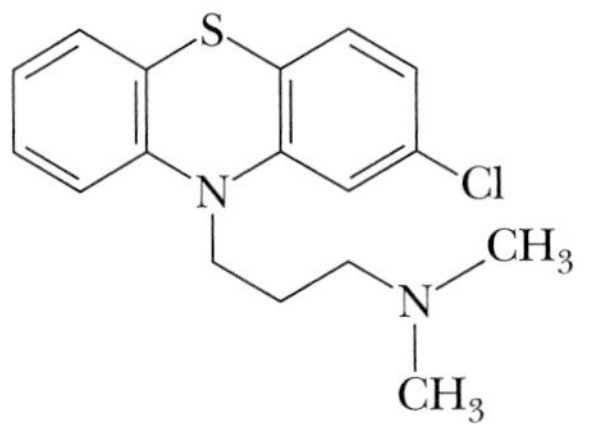

Chlorpromazine 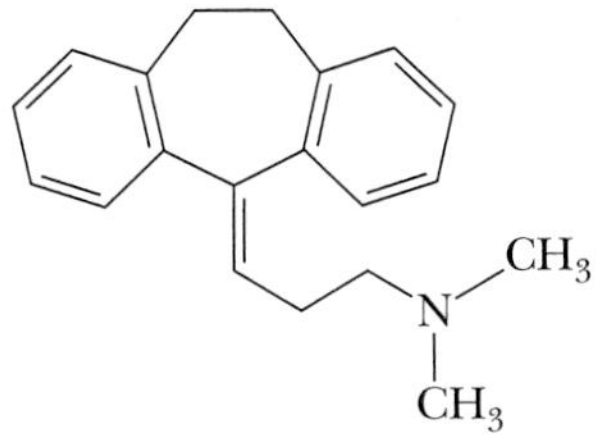　Amitriptyline

It is now known that amitriptyline inhibits the re-uptake of norepinephrine (see *The Merck Index*, 12th ed., #6788) and serotonin (see *The Merck Index*, 12th ed., #8607) from the synaptic cleft. Because the re-uptake of these neurotransmitters is inhibited, their effects are potentiated. They remain available to interact with serotonin and norepinephrine receptor sites longer and continue to cause excitation of serotonin and norepinephrine-mediated neural pathways.

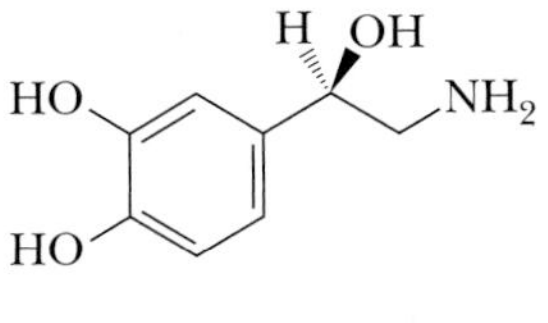

Norepinephrine 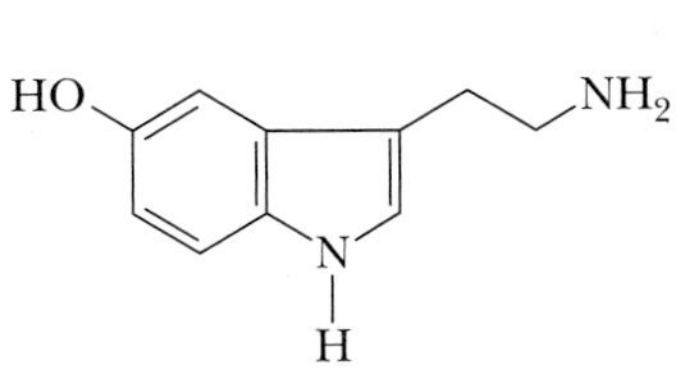　Serotonin

Following is a synthesis for amitriptyline.

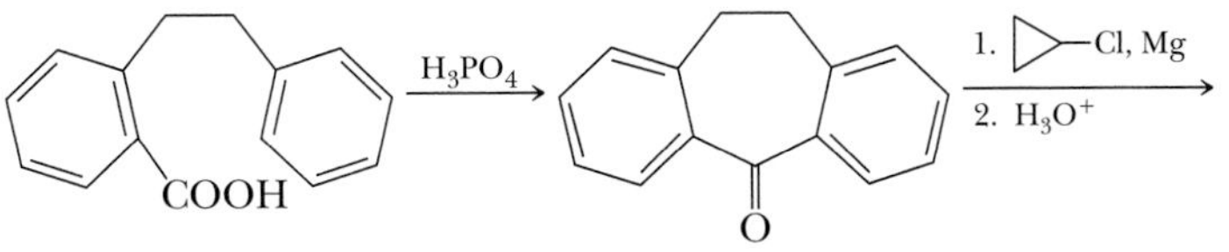

(A)　$\xrightarrow{H_3PO_4}$　(B)　$\xrightarrow[\text{2. } H_3O^+]{\text{1. } \triangleright\!-Cl,\ Mg}$

(A)　(B)

(continued on next page)

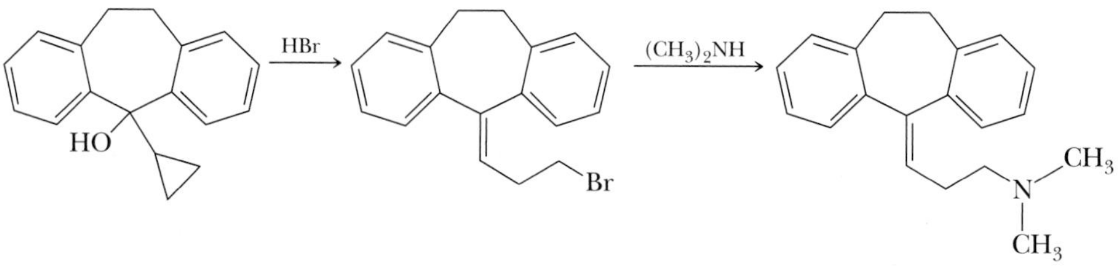

(C) (D) Amitriptyline

(a) Propose a mechanism for the conversion of (A) to (B).

(b) Propose a mechanism for the conversion of (C) to (D). *Note:* It is not acceptable to propose a primary carbocation as an intermediate.

MC.8 Following is an outline of one of the first syntheses of the antidepressant fluoxetine, better known by its trade name Prozac (see *The Merck Index*, 12th ed., #4222). One interesting feature of this synthesis is the use of ethyl chloroformate (ClCOOEt) to bring about *N*-demethylation (removal of a methyl group) of compound E.

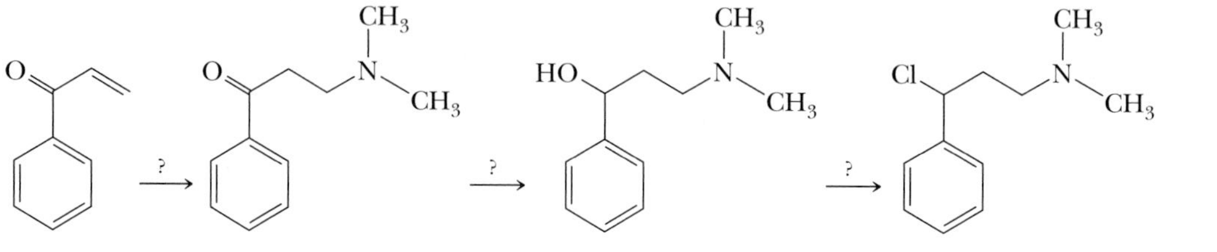

(A) (B) (C) (D)

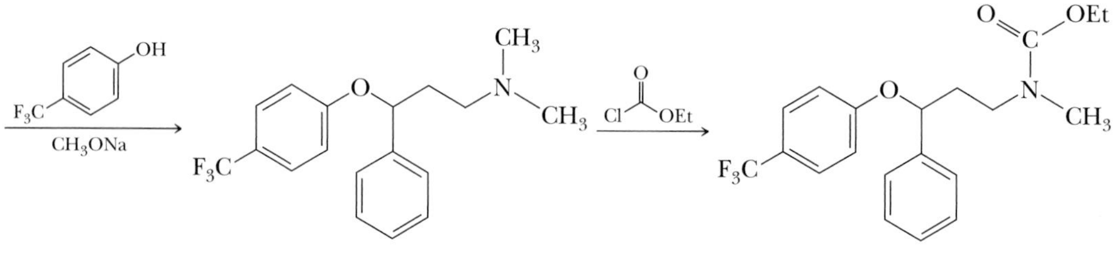

(E) (F)

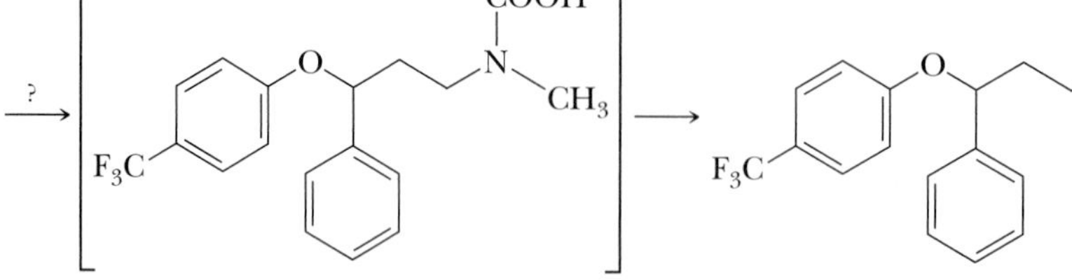

An *N*-substituted carbamic acid

Fluoxetine

(a) Propose a reagent for the conversion of (A) to (B).

(b) Propose a reagent for the conversion of (B) to (C).

(c) Propose a reagent for the conversion of (C) to (D).

(d) Propose a mechanism for the conversion of (E) to (F). The other product of this conversion is chloromethane, CH_3Cl. Your mechanism should show how this compound is formed.

(e) Propose a reagent or reagents to bring about the conversion of (F) to fluoxetine. Note that the bracketed intermediate formed in this step is an *N*-substituted carbamic acid. Such compounds are unstable and break down to carbon dioxide and an amine.

MC.9 Show how the antidepressant venlafaxine (see *The Merck Index*, 12th ed., #10079) can be synthesized from these readily available starting materials.

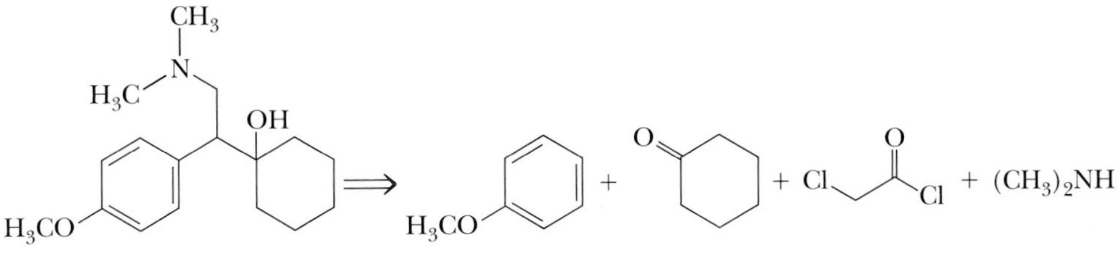

| Venlafaxine | Anisole | Cyclohex-anone | Chloroacetyl chloride | Dimethyl-amine |

MC.10 Propose a synthesis for the antidepressant moclobemide (see *The Merck Index*, 12th ed., #6309) given this retrosynthetic analysis.

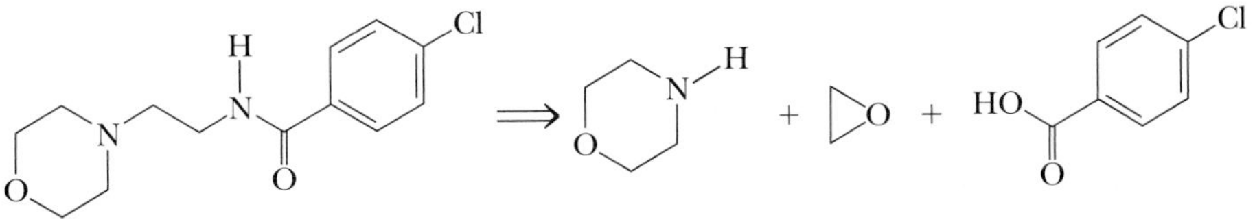

| Moclobemide | Morpholine | Ethylene oxide | *p*-Chlorobenzoic acid |

Nonsteroidal Anti-Inflammatory Drugs (NSAIDs)

MC.11 One of the potential syntheses of the anti-inflammatory and analgesic drug nabumetone (see *The Merck Index*, 12th ed., #6428) is chloromethylation of 2-methoxynaphthalene followed by an acetoacetic ester synthesis. Nabumetone is a prodrug. It has no activity itself but is oxidized in vivo to 6-methoxy-2-naphthylacetic acid, which is the active anti-inflammatory and analgesic drug.

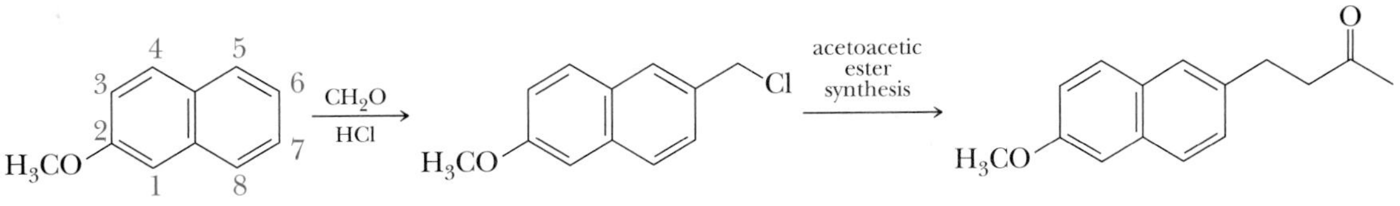

2-Methoxynaphthalene

Nabumetone

(a) Chloromethylation is carried out using formaldehyde in the presence of concentrated HCl. Propose a mechanism for chloromethylation of 2-methoxynaphthalene. Account for the fact that chloromethylation takes place at carbon 6 rather than at carbons 5 or 7.

(b) Show steps in the acetoacetic ester synthesis by which the synthesis of nabumetone is completed.

938 Interchapter Medicinal Chemistry—Problems in Organic Synthesis

CHEMISTRY IN ACTION

From Willow Bark to Aspirin and Beyond

The first drug developed for widespread use was aspirin, one of today's most common pain relievers. Americans alone consume approximately 80 billion tablets of aspirin a year! The story of the development of this modern pain reliever goes back more than 2000 years. In 400 B.C.E., the Greek physician Hippocrates recommended chewing bark of the willow tree to alleviate the pain of childbirth and to treat eye infections.

The active component of willow bark was found to be salicin (see *The Merck Index*, 12th ed., #8476), a compound composed of salicyl alcohol bonded to a unit of β-D-glucose (Section 25.1). Hydrolysis of salicin in aqueous acid gives salicyl alcohol, which can be oxidized to salicylic acid. Salicylic acid proved to be an even more effective reliever of pain, fever, and inflammation than salicin, without its extremely bitter taste. Unfortunately, patients quickly recognized salicylic acid's major side effect: it causes severe irritation of the mucous membrane lining of the stomach.

In the search for less irritating but still effective derivatives of salicylic acid, chemists at the Bayer division of I. G. Farben in Germany in 1883 prepared acetylsalicylic acid and gave it the name aspirin.

Salicin

Salicyl alcohol   Salicylic acid

Salicylic acid   Acetylsalicylic acid (Aspirin)

(c) Draw the structural formula of 6-methoxy-2-naphthylacetic acid.

(d) Compare the structural formulas of this metabolite with those of ibuprofen and naproxen (see the Chemistry in Action box "From Willow Bark to Aspirin and Beyond").

Analgesic, Narcotic

MC.12 The analgesic, soporific, and euphoriant properties of the dried juice obtained from unripe seed pods of the opium poppy *Papaver somniferum* have been known for centuries. By the beginning of the 19th century, the active principle morphine (see *The Merck Index*, 12th ed., #6359) had been isolated and its structure determined.

Even though morphine is one of modern medicine's most effective pain killers, it has two serious disadvantages. First, it is addictive. Second, it depresses the respiratory control center of the central nervous system. Large doses of morphine (or heroin, *N*-acetylmorphine, as well) can lead to death by respiratory failure. For these reasons, chemists have sought to produce pain killers related in structure to morphine, but without these serious disadvantages. One strategy in this on-going research has been to develop an efficient synthesis of the carbon-nitrogen skeleton of morphine, in the

Aspirin proved to be less irritating to the stomach than salicylic acid and also more effective in relieving the pain and inflammation of rheumatoid arthritis. Aspirin, however, is still irritating to the stomach and frequent use can cause duodenal ulcers in susceptible persons.

In the 1960s, in a search for even more effective and less irritating analgesics and anti-inflammatory drugs, chemists at the Boots Pure Drug Company in England synthesized a series of compounds related in structure to salicylic acid. Among them, they discovered an even more potent compound, which they named ibuprofen (see *The Merck Index*, 12th ed., #4925). Soon thereafter, Syntex Corporation in the United States developed naproxen (see *The Merck Index*, 12th ed., #6504), the active ingredient in Aleve. Each compound has one stereocenter and can exist as a pair of enantiomers. For each drug, the physiologically active form is the *S* enantiomer.

In the 1960s, it was discovered that aspirin acts by inhibiting cyclooxygenase (COX), a key enzyme in the conversion of arachidonic acid to prostaglandins (Section 26.3). With this discovery, it became clear why only one enantiomer of ibuprofen and naproxen is active: only the *S* enantiomer of each has the correct handedness to bind to COX and inhibit its activity.

(*S*)-Ibuprofen   (*S*)-Naproxen

Recently it was recognized that there are actually two cyclooxygenases; one is more important for the inflammation pathway, and the other affects the stability of blood vessels. Aspirin and other NSAIDs inhibit both, which is why they can cause gastrointestinal bleeding. New drugs (Celebrex, Vioxx) have been developed that inhibit only the inflammatory enzyme pathway and are remarkably effective for suppression of inflammation (for example, in arthritis) without the gastrointestinal side effects.

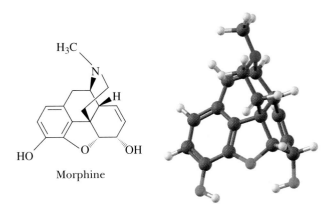

Morphine

hope that it then could be modified to produce medications equally effective but with reduced side effects. One target of this synthetic effort was morphinan (see *The Merck Index*, 12th ed., #6358), the bare morphine skeleton. Following are structural formulas

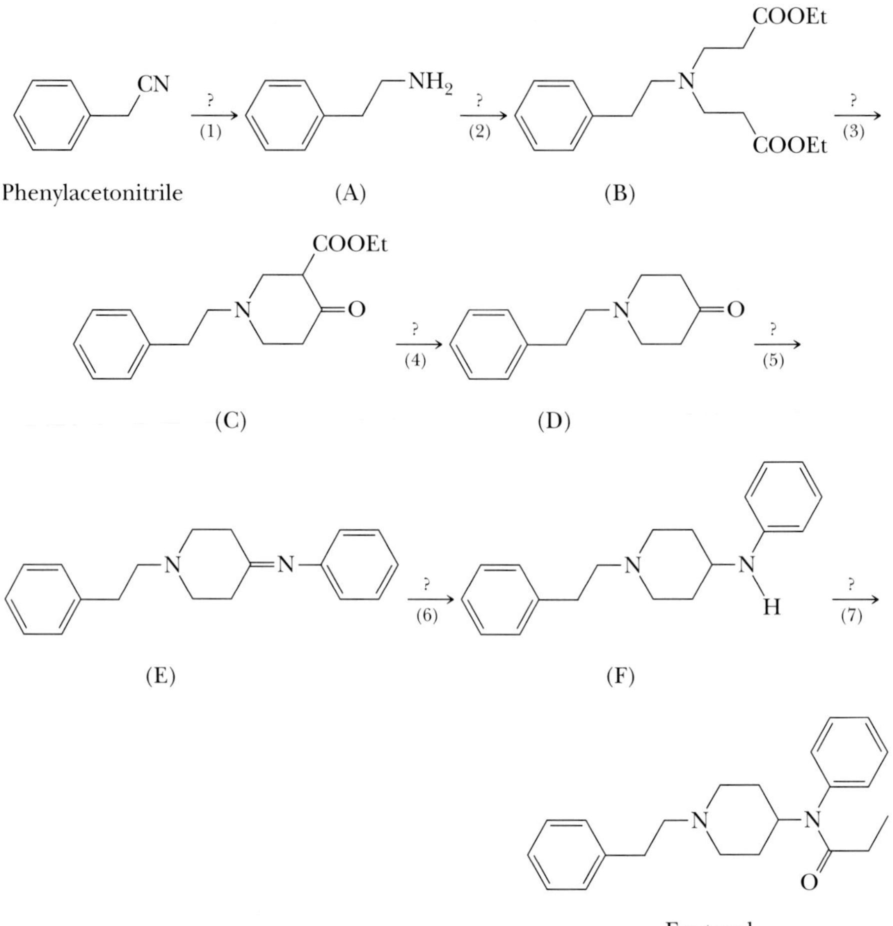

Phenylacetonitrile   (A)   (B)

(C)   (D)

(E)   (F)

Fentanyl

(a) Propose reagents and/or experimental conditions for Steps 1 through 7.

(b) Propose a mechanism for Step 2. By what name is this type of reaction known?

(c) Propose a mechanism for Step 3. By what name is this type of reaction known?

Antiarrhythmic

MC.15 Ibutilide (see *The Merck Index*, 12th ed., #4927) is used to treat cardiac arrhythmia.

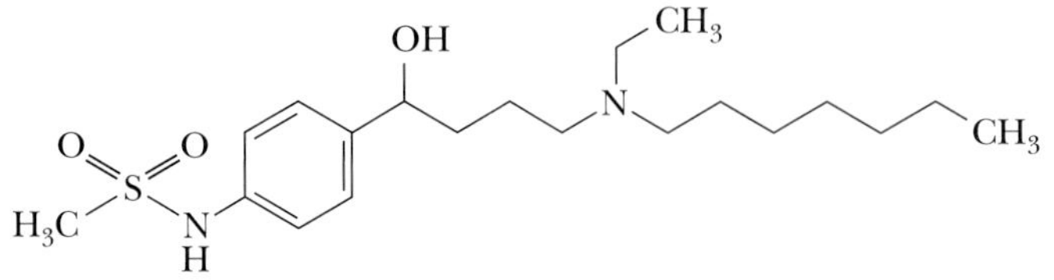

Ibutilide

The lead compounds for the development of this drug were adrenaline and noradrenaline, both of which are intimately associated with the functioning of the sympathetic branch of the autonomic nervous system. These two compounds play a key role in a number of involuntary functions, including regulation of blood pressure, heart rate, and the constriction or dilation of bronchioles. Because of their involvement in these functions, this branch of the autonomic nervous system is also known as the adrenergic system. The adrenergic system is divided into α and β branches. The cardiovascular system falls in the β_1 branch. Ibutilide is a β_1-blocking agent. Following is a retrosynthetic analysis for ibutilide. In this scheme, Hept is an abbreviation for the 1-heptyl group.

Ibutilide   (A)

(B)   (C)

(D)   Aniline

Given this retrosynthetic analysis, propose a synthesis for ibutilide starting with aniline, methanesulfonyl chloride, succinic anhydride, and *N*-ethyl-1-heptananine.

MC.16 Following is a synthesis for the antiarrhythmic drug bidisomide (see *The Merck Index*, 12th ed., #1251). The symbol Bn is an abbreviation for the benzyl group, $C_6H_5CH_2-$.

(A)   (B)   (C)

(D)   (E)   Bidisomide

C H E M I S T R Y   I N   A C T I O N

Ibuprofen — The Evolution of an Industrial Synthesis

A major consideration in any industrial synthesis is atom efficiency; it is most efficient to use only reagents whose atoms appear in the final product. An example of the evolution of syntheses with greater and greater atom efficiency is the industrial synthesis of ibuprofen.

Synthesis I

One of the first industrial syntheses of ibuprofen used the following sequence to introduce a methyl group on the carboxyl side chain of 4-isobutylphenylacetic acid. Fischer esterification of 4-isobutylphenylacetic acid followed by treatment of the ethyl ester with diethyl carbonate in the presence of sodium ethoxide in a crossed Claisen condensation (Section 19.3C) gives the anion of a substituted mal-

onic ester B. Alkylation of this anion (Section 19.7) followed by hydrolysis gives a disubstituted malonic acid C. Decarboxylation (Section 17.9B) of C gives ibuprofen. While this synthesis gives ibuprofen in quite good yield, it is quite wasteful of carbons; there are 19 carbons in intermediate C, only 13 of which appear in ibuprofen.

4-Isobutylphenylacetic acid
($C_{12}H_{16}O_2$)

A
($C_{14}H_{20}O_2$)

B
($C_{18}H_{23}O_4Na$)

C
($C_{19}H_{26}O_4$)

D
($C_{14}H_{18}O_4$)

Ibuprofen
($C_{13}H_{18}O_2$)

Synthesis II

An alternative route with greater atom efficiency starts with 4-isobutylacetophenone. Treatment of this ketone with chloroacetonitrile in the presence of sodium ethoxide gives the epoxynitrile E. Treatment of E with lithium perchlorate, a Lewis acid, gives the α-cyanoketone F. Hydrolysis of F gives ibuprofen.

(a) Propose mechanisms for the conversion of (A) to (B) and for (B) to (C). What is the function of sodium amide in each reaction?

(b) Why is it necessary to incorporate the benzyl group on the chloroamine used to convert (B) to (C)?

(c) Propose a reagent or reagents for the removal of the benzyl group in the conversion of (D) to (E).

(d) Propose a reagent for the conversion of (E) to bidisomide.

The α-cyanoketone group of F is similar to an acid chloride in its chemical properties, and its hydrolysis gives a carboxylic acid and cyanide ion. This synthesis, while more atom-efficient than Synthesis I, uses a cyano group, neither atom of which appears in the final product.

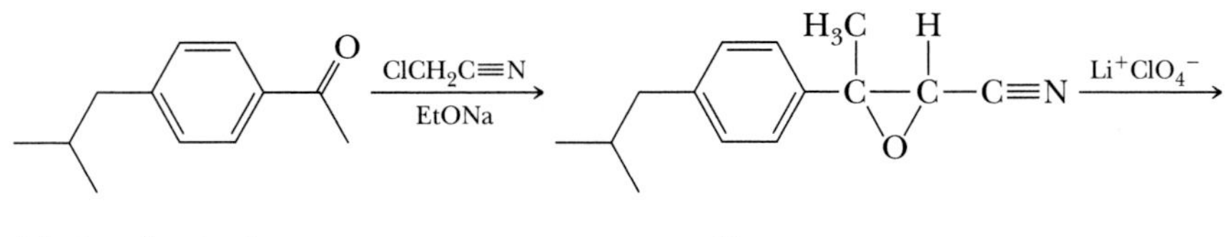

4-Isobutylacetophenone
($C_{12}H_{16}O$)

E
($C_{14}H_{17}NO$)

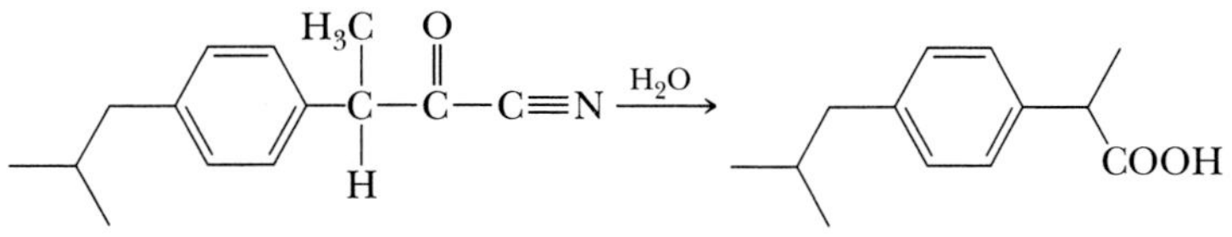

F
($C_{14}H_{17}NO$)

Ibuprofen
($C_{13}H_{18}O_2$)

Synthesis III

The ultimate in atom efficiency in the synthesis of ibuprofen is achieved in the following synthesis. Catalytic reduction of the carbonyl group of 4-isobutylacetophenone gives the alcohol G. Palladium-catalyzed carbonylation of G gives ibuprofen. The one carbon atom introduced in the synthesis appears in the final product!

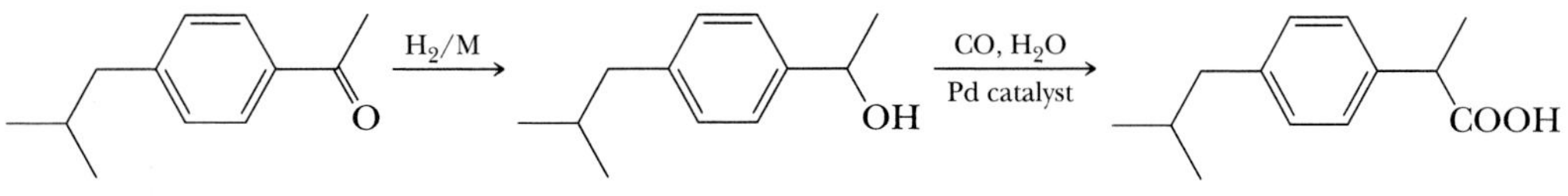

4-Isobutylacetophenone
($C_{12}H_{16}O$)

G
($C_{12}H_{18}O$)

Ibuprofen
($C_{13}H_{18}O_2$)

Antihypertensive

MC.17 Following is the structural formula of the antihypertensive drug labetalol (see *The Merck Index*, 12th ed., #5341), a nonspecific β-adrenergic blocker with vasodilating activity. Members of this class have received enormous clinical attention because of their effectiveness in treating hypertension (high blood pressure), migraine headaches, glaucoma, ischemic heart disease, and certain cardiac arrhythmias. This retrosynthetic analysis involves disconnects to the α-haloketone (F) and the amine (C). Each is in turn derived from a simpler, readily available precursor. Given this retrosynthetic analysis, propose a synthesis for labetalol from salicylic acid and benzyl chloride. *Note:* The conversion of (D) to (E) involves a Friedel-Crafts acylation in which the phenolic —OH must be protected by treatment with acetic anhydride. The protecting group is removed later by treatment with sodium hydroxide.

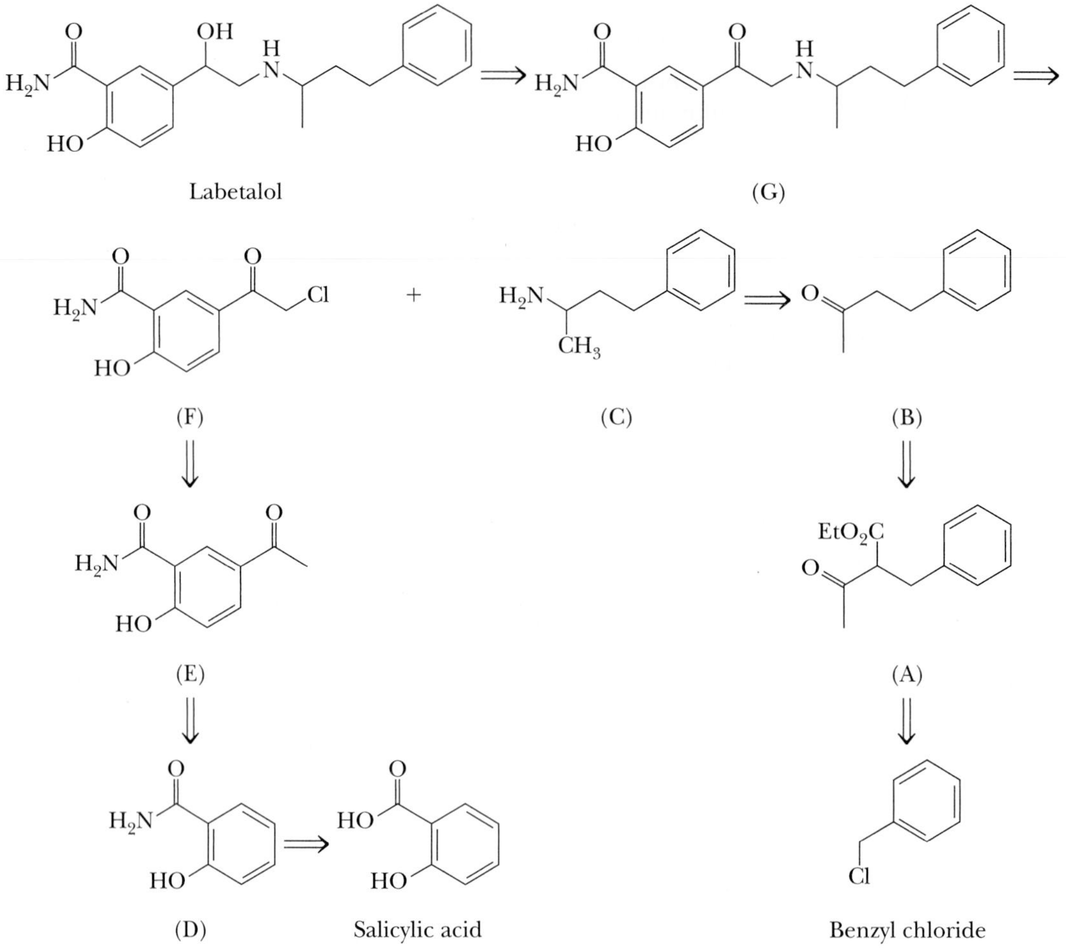

Labetalol

(G)

(F)   +   (C)   (B)

(E)

(A)

(D)   Salicylic acid   Benzyl chloride

Labetalol has two stereocenters and, as produced in this synthesis, is a racemic mixture of the four possible stereoisomers. The active stereoisomer is dilevalol (see *The Merck Index*, 12th ed., #3245). Assign an *R* or *S* configuration to each stereocenter in dilevalol.

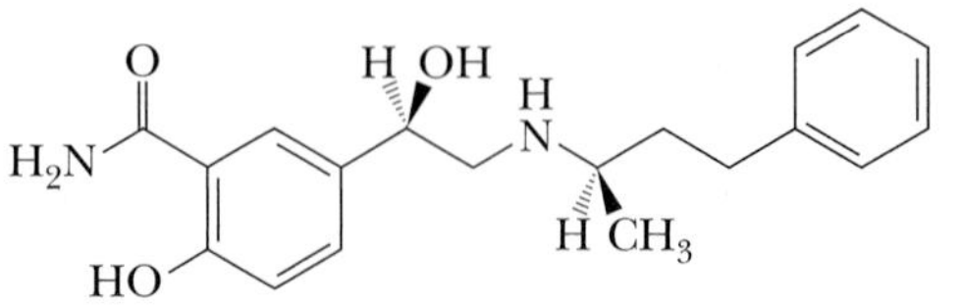

Dilevalol

MC.18 A finding that opened a far easier route to β-blockers was the discovery that β-blocking activity is retained if an oxygen atom is interposed between the aromatic ring and the side chain. To see this difference, compare the structures of labetalol and propranolol. Thus, alkylation of phenoxide ions can be used as a way to introduce this side chain.

The first of this new class of drugs was propranolol (see *The Merck Index*, 12th ed., #8025), prepared as shown here.

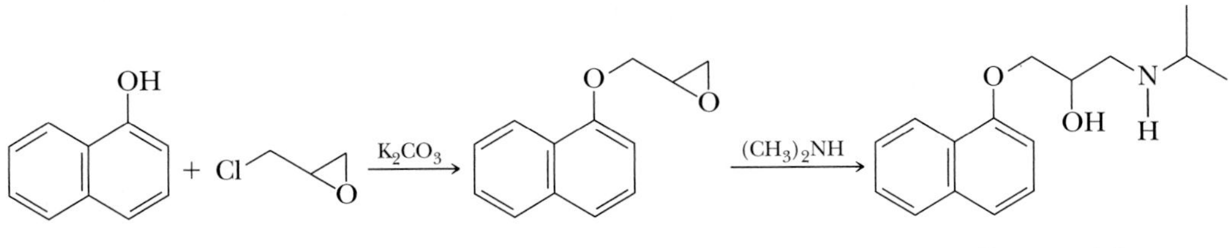

1-Naphthol   Epichloro-   Propranolol
(α-Naphthol)   hydrin

Side effects of propranolol are disturbances of the central nervous system (CNS) such as fatigue, sleep disturbances (including insomnia and nightmares), and depression. Pharmaceutical companies wondered if this drug could be redesigned to eliminate or at least reduce these side effects. Propranolol itself is highly lipophilic (hydrophobic) and readily passes through the blood-brain barrier. Propranolol, it was reasoned, enters the CNS by passive diffusion because of the lipid-like character of its naphthalene ring. The challenge, then, was to design a more hydrophilic drug that does not cross the blood-brain barrier but still retains a β-adrenergic antagonist property. A product of this research is atenolol (see *The Merck Index*, 12th ed., #892), a potent β-adrenergic blocker that is hydrophilic enough that it crosses the blood-brain barrier to only a very limited extent. Atenolol is now one of the most widely used β-blockers.

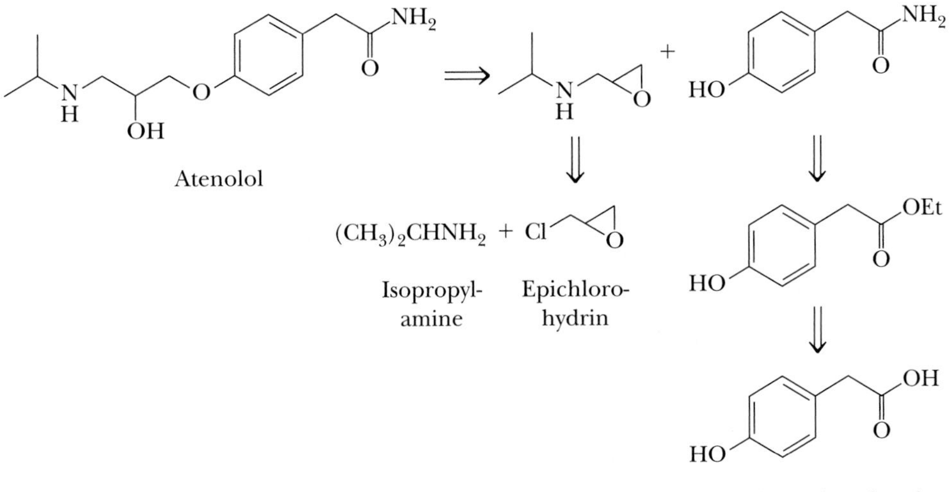

Atenolol

(CH₃)₂CHNH₂ + Cl ...

Isopropyl-   Epichloro-
amine   hydrin

4-Hydroxyphenylacetic
acid

(a) Given this retrosynthetic analysis, propose a synthesis for atenolol from the three given readily available starting materials.
(b) Note that the amide functional group is best made by amination of the ester. Why was this route chosen rather than conversion of the carboxylic acid to its acid chloride and then treatment of the acid chloride with ammonia?

MC.19 In certain clinical situations, there is need for an injectable β-blocker with a short biological half-life. The clue to development of such a drug was taken from the structure of atenolol, whose corresponding carboxylic acid (the product of hydrolysis of its amide) has no β-blocking activity. Substitution of an ester for the amide group and lengthening the carbon side chain by one methylene group resulted in esmolol (see

The Merck Index, 12th ed., #3741). Its ester group is hydrolyzed quite rapidly to a carboxyl group by serum esterases under physiological conditions. This hydrolysis product has no β-blocking activity. Propose a synthesis for esmolol from 4-hydroxycinnamic acid.

Esmolol   4-Hydroxycinnamic   Dimethyl-   Epichloro-
                           acid   amine   hydrin

MC.20 The synthesis of nadolol (see *The Merck Index*, 12th, #6431), another β-adrenergic blocking agent, begins with a dissolving metal (Birch) reduction of 1-naphthol. When an aromatic ring is treated with Li, Na, or K in liquid ammonia in the presence of an alcohol, most commonly methanol or ethanol, the aromatic ring is reduced to a nonconjugated cyclohexadiene. Electron-donating substituents such as alkyl and alkoxyl decrease the rate of reduction; electron-withdrawing substituents such as carboxyl increase the rate of reduction. In the case of Birch reduction of 1-naphthol, reduction occurs by 1,4-addition of hydrogens to the more reactive unsubstituted ring.

(A)   (B)

(C)   (D)   Nadolol

(a) Propose reagents for the conversion of (A) to (D).
(b) Propose a series of steps to convert (D) to nadolol.

Bronchodilator

MC.21 Following is an outline of a synthesis of the bronchodilator carbuterol (see *The Merck Index*, 12th ed., #1882), a β_2 adrenergic blocker with high selectivity for airway smooth muscle receptors.

Carbuterol

(a) Propose reagents to bring about each step.

(b) Why is it necessary to add the benzyl group as a blocking agent in Step 1?

(c) Suggest a structural relationship between carbuterol and ephedrine.

MC.22 Following is a synthesis for albuterol (Proventil; see *The Merck Index*, 12th ed., #217), currently one of the most widely used inhalation bronchodilators.

4-Hydroxybenzaldehyde   (A)   (B)

(C)   (D)   Albuterol

(a) Propose a mechanism for conversion of 4-hydroxybenzaldehyde to (A).

(b) Propose reagents and experimental conditions for conversion of (A) to (B).

(c) Propose a mechanism for the conversion of (B) to (C). *Hint:* Think of trimethylsulfonium iodide as producing a sulfur equivalent of a Wittig reagent. The fate of the adduct is slightly different, however. This reaction is a general one for preparation of epoxides.

(d) Propose reagents and experimental conditions for the conversion of (C) to (D).

(e) Propose reagents and experimental conditions for the conversion of (D) to albuterol.

Antihistaminic

MC.23 Propose a synthesis for the antihistamine *p*-methyldiphenhydramine (see *The Merck Index*, 12th ed., #6130), given this retrosynthetic analysis.

p-Methyldiphenhydramine

MC.24 Propose a synthesis for the antihistamine histapyrrodine (see *The Merck Index*, 12th ed., #4757).

| Histapyrrodine | Pyrrolidine | Ethylene oxide | Aniline | Benzoic acid |

MC.25 Propose a synthesis for diphenhydramine (see *The Merck Index*, 12th ed., #3367) starting from benzene, benzoic acid, and any other organic compounds of two or fewer carbons. The hydrochloride salt of this molecule, best known by its trade name of Benadryl, is an antihistamine.

| Diphenhydramine | Benzophenone | 2-(*N*,*N*-Dimethylamino)-ethanol |

Antiestrogen

MC.26 Estrogens are female sex hormones, the most important of which are estrone, estradiol, and estriol (see *The Merck Index*, 12th ed., #3751, #3747, and #3750). Of these, β-estradiol is the most potent. *Note:* By convention in steroid nomenclature, the designation beta means that a group is toward the reader, on the top side of the molecule; alpha means that it is away from the reader, on the bottom side of the molecule when the rings are written as shown here. As used in β-estradiol, beta refers to the orientation of the —OH group on the five-membered ring.

Estrone   β-Estradiol   Estriol

As soon as these compounds were isolated in the early 1950s and their pharmacology studied, it became clear that they are extremely potent substances. In recent years, there have been intense efforts to design and synthesize molecules that will bind to estrogen receptors. One target of this research has been nonsteroidal estrogen antagonists. (An estrogen antagonist is a compound that interacts with estrogen receptors and blocks the effects of both endogenous and exogenous estrogens.) A feature common to one type of nonsteroidal estrogen antagonist is the presence of a 1,2-diphenylethylene with one of the benzene rings bearing a dialkylaminoethoxyl substituent. The first nonsteroidal estrogen antagonist of this type to achieve clinical importance was tamoxifen (see *The Merck Index*, 12th ed., #9216), now an important drug in the treatment of breast cancer. Tamoxifen has the *Z* configuration as shown here.

R. B. Woodward received the Nobel prize for his achievements in organic synthesis in 1965. As a PhD student, he carried out the synthesis of estrone, a female sex hormone. *(courtesy of Harvard University News Office)*

(A)   (B)

(C)   Tamoxifen

Propose reagents for the conversion of (A) to tamoxifen. *Note:* The final step gives a mixture of *E* and *Z* isomers.

MC.27 Following is a synthesis for toremifene (see *The Merck Index*, 12th ed., #9688), a non-steroidal estrogen antagonist whose structure is closely related to that of tamoxifen (Problem MC.26). Bn is an abbreviation for the benzyl group, $C_6H_5CH_2-$.

(A)   (B)

(C)   (D)   (E)

(F)   Toremifene

(a) This synthesis makes use of two blocking groups, the benzyl (Bn) group and the tetrahydropyranyl (THP) group. Draw a structural formula of each group, and describe the experimental conditions under which it is attached and removed.

(b) Discuss the chemical logic behind the use of each blocking group in this synthesis.

(c) Propose a mechanism for the conversion of (D) to (E).

(d) Propose a mechanism for the conversion of (F) to toremifene.

CHEMISTRY IN ACTION

Taxol — Search and Discovery

In the early 1960s, the National Cancer Institute undertook a program to analyze samples of native plant materials in the hope of discovering substances effective in the fight against cancer. Among the materials tested was an extract of the bark of the Pacific yew, *Taxus brevifolia*, a slow-growing tree found in the old-growth forests of the Pacific Northwest. This extract proved to be remarkably effective in treating certain types of ovarian and breast cancer, even in cases where other forms of chemotherapy failed. The structure of the cancer-fighting component of yew bark was determined in 1962, and the compound was named paclitaxel (Taxol; see *The Merck Index*, 12th ed., #7117).

Paclitaxel (Taxol)

Unfortunately, the bark of a single 100-year-old tree yields only about 1 g of Taxol, not enough for effective treatment of even one cancer patient. Furthermore, getting Taxol means stripping the bark from trees, thus killing them. Chemists succeeded in 1994 in synthesizing Taxol in the laboratory, but its cost was far too high to be economical. Fortunately, an alternative natural source of the drug was found. Researchers in France discovered that the needles of a related plant, *Taxus baccata*, contain a compound that can be converted to Taxol in the laboratory. Because the needles can be gathered without harming the plant, it is not necessary to kill trees to obtain the drug.

Taxol inhibits cell division by acting on microtubules. It does this in two ways: it stimulates microtubule polymerization and stabilizes the resulting structural units. Before cell division can take place, the cell must disassemble these units, and Taxol prevents this disassembly. Because cancer cells are the fastest dividing cells, Taxol effectively controls their spreading.

The remarkable success of Taxol in treatment of breast and ovarian cancer has stimulated research efforts to discover and/or synthesize other substances that work the same way in the body and that may be even more effective anticancer agents than Taxol. One of these is epothilone, isolated from the bacterium *Sorangium cellulosum*. This drug promises to be more effective than Taxol and effective against some Taxol-resistant tumors.

Antianginal

MC.28 Verapamil (see *The Merck Index*, 12th ed., #10083), one of over 30 coronary artery vasodilators listed in *The Merck Index*, is used in the treatment of angina caused by insufficient blood flow to cardiac muscle. Even though its effect on coronary vasculature tone was recognized over 30 years ago, it has only been more recently that its role as a calcium channel blocker has become understood. The following is a retrosynthetic analysis leading to a convergent synthesis; it is convergent because (A) and (B) are made separately and then combined (that is, the route converges) to give the final product.

Convergent syntheses are generally much more efficient than those in which the skeleton is built up stepwise.

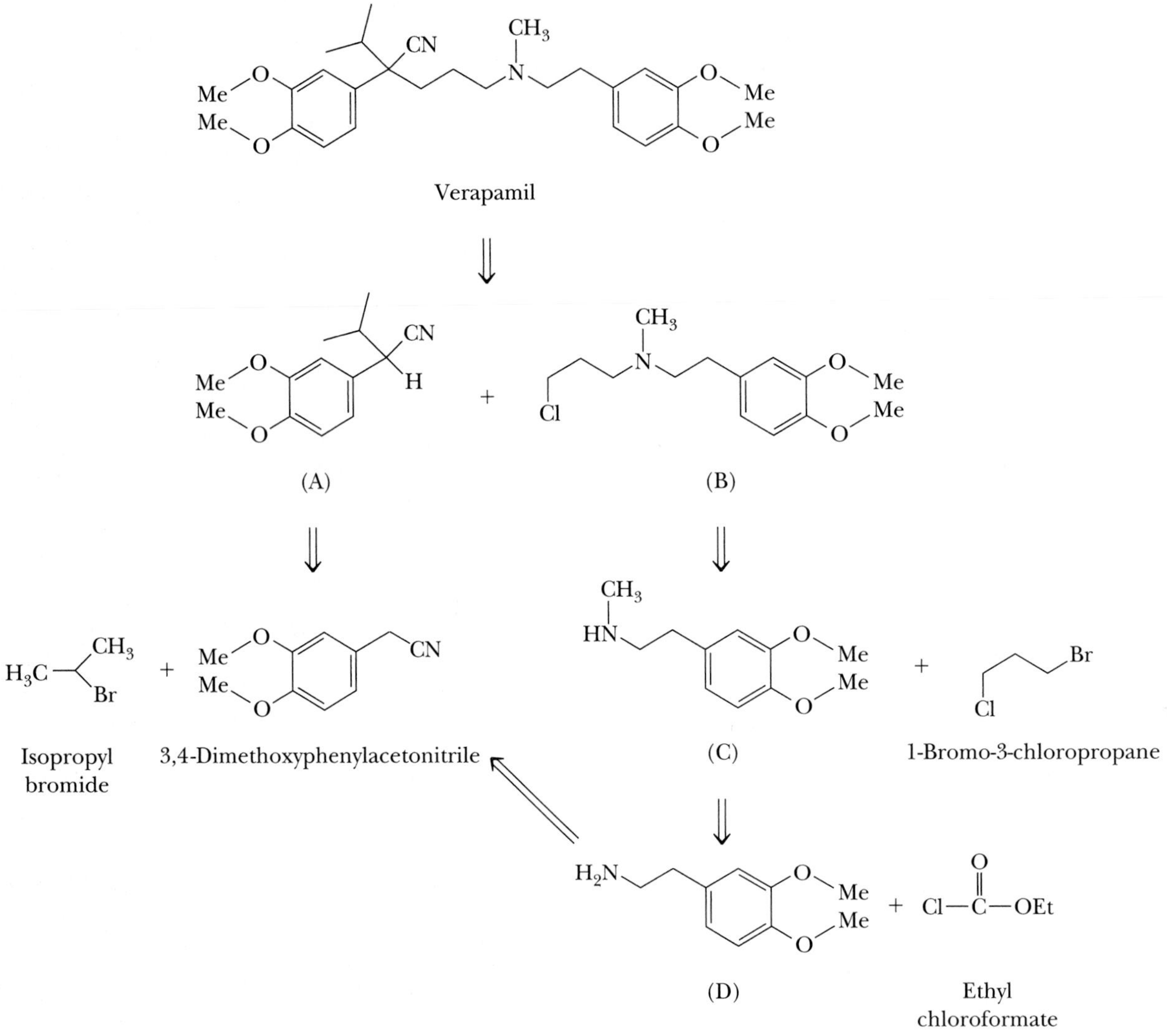

Verapamil

⇓

(A)   +   (B)

⇓   ⇓

Isopropyl   3,4-Dimethoxyphenylacetonitrile   (C)   +   1-Bromo-3-chloropropane
bromide

⇓

(D)   +   Ethyl
chloroformate

(a) Given this retrosynthetic analysis, propose a synthesis for verapamil from the four named starting materials.

(b) It requires two steps to convert (D) to (C). The first is treatment of (D) with ethyl chloroformate. What is the product of this first step? What reagent can be used to convert this product to (C)?

(c) How do you account for the regioselectivity of the nucleophilic displacement involved in converting (C) to (B)?

Vasodilator, Coronary

MC.29 Propose a synthesis for the coronary vasodilator ganglefene (see *The Merck Index*, 12th ed., #4375) from 4-hydroxybenzoic acid and 3-methyl-3-buten-2-one.

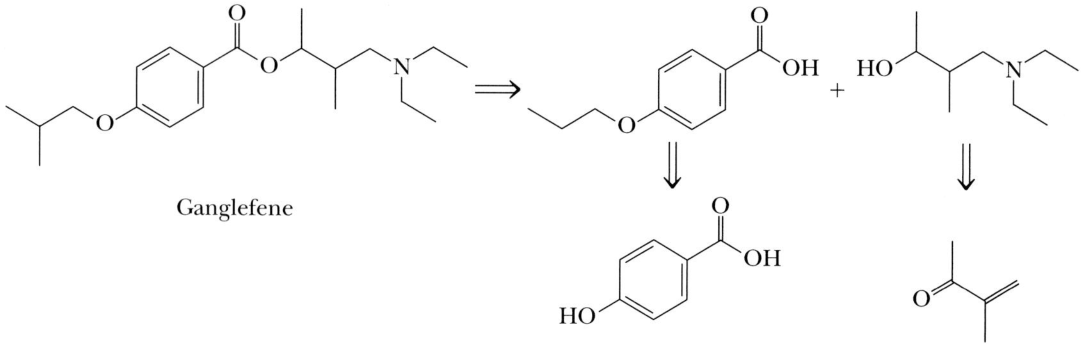

Ganglefene

4-Hydroxybenzoic acid   3-Methyl-3-buten-2-one

Vasodilator, Peripheral

MC.30 Moxisylyte (see *The Merck Index*, 12th ed., #6374), an α-adrenergic blocker, is used as a peripheral vasodilator. Propose a synthesis for this compound from thymol, which occurs in the volatile oils of members of the thyme family. Thymol is made industrially from *m*-cresol.

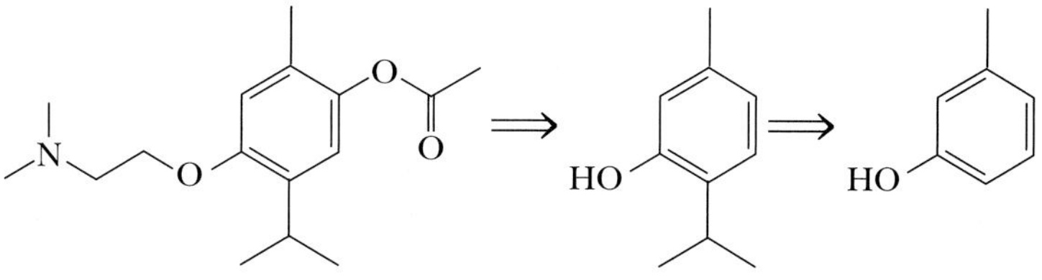

Moxisylyte   Thymol   *m*-Cresol

Anticoagulant

MC.31 Propose a synthesis of the anticoagulant (inhibits blood clotting) diphenadione (see *The Merck Index*, 12th ed., #3363). Because of its anticoagulant activity for blood, this compound is used as a rodenticide. For the story of the discovery of the anticoagulant dicoumarin, see the Chemistry in Action box "From Moldy Clover to a Blood Thinner" in Chapter 18.

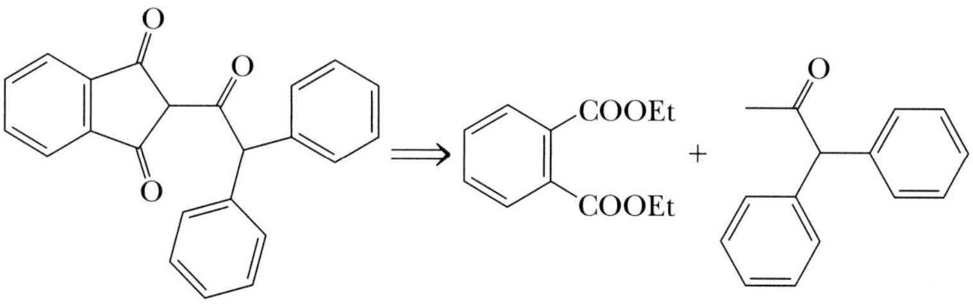

Diphenadione

These drugs are more effective than earlier drugs in improving patient response in the areas of social withdrawal, apathy, memory, comprehension, and judgment. They also produce fewer side effects such as seizures and tardive dyskinesia (involuntary body movements).

Clozapine   Olanzapine

Following is one synthesis for clozapine. [See F. Hunziker, E. Fischer, and J. Schmutz, *Helv. Chem. Acta* **50:** 1588 (1967).] Step 1, an Ulmann coupling, is a type of nucleophilic aromatic substitution that uses a copper catalyst.

2,5-Dichloro-   2-Aminobenzoic acid   (A)
nitrobenzene   (Anthranilic acid)

(B)   (C)   Clozapine

(a) How might you bring about formation of the amide in Step 2?
(b) Propose a reagent for Step 3.
(c) Propose a mechanism for Step 4.

C H E M I S T R Y   I N   A C T I O N

From Cocaine to Procaine and Beyond

Cocaine (see *The Merck Index*, 12th ed., #2517) is an alkaloid present in the leaves of the South American coca plant *Erythroxylon coca*. It was first isolated in 1880, and soon thereafter its property as a local anesthetic was discovered. Cocaine was introduced into medicine and dentistry in 1884 by two young Viennese physicians, Sigmund Freud and Karl Koller. Unfortunately, the use of cocaine can create a dependence, as Freud himself observed when he used it to wean a colleague from morphine and thereby produced one of the first documented cases of cocaine addiction.

Cocaine

Cocaine reduces fatigue, permits greater physical endurance, and gives a feeling of tremendous confidence and power. In some of the Sherlock Holmes stories, the great detective injects himself with a 7% solution of cocaine to overcome boredom.

After determining cocaine's structure, chemists could ask, "How is the structure of cocaine related to its anesthetic effects? Is it possible to separate the anesthetic effects from the habituation?" If these questions could be answered, it might be possible to prepare synthetic drugs with the structural features essential for the anesthetic activity but without those giving rise to the undesirable effects. Chemists focused on three structural features of cocaine: its benzoic ester, its basic nitrogen atom, and something of its carbon skeleton. This search resulted in 1905 in the synthesis of procaine (see *The Merck Index*, 12th ed., #7937), which almost immediately replaced cocaine in dentistry and surgery. Lidocaine (see *The Merck Index*, 12th ed., #5505) was introduced in 1948 and today is one of the most widely used local anesthetics. More recently, other members of the "caine" family of local anesthetics have been introduced, for example etidocaine (see *The Merck Index*, 12th ed., #3907). All these local anesthetics are administered as their water-soluble hydrochloride salts.

Thus, seizing on clues provided by nature, chemists have been able to synthesize drugs far more suitable for a specific function than anything known to be produced by nature itself.

Procaine
(Novocain)

Lidocaine
(Xylocaine)

Etidocaine
(Duranest)

Anesthetic (Local)

MC.36 Proparacaine (see *The Merck Index*, 12th ed., #7991) is one of a class of -caine local anesthetics. See *The Merck Index*, 12th ed., THER-3, for the names of at least three dozen more -caines. Given this retrosynthetic analysis, propose a synthesis of proparacaine from 4-hydroxybenzoic acid. Use diethylamine and ethylene oxide as the source of the carbon atoms of the alkyl portion of the benzoate ester.

Proparacaine

MC.37 Propose a synthesis of the local anesthetic ambucaine (see *The Merck Index*, 12th ed., #402) from 4-nitrosalicylic acid, ethylene oxide, diethylamine, and 1-bromobutane.

Ambucaine   4-Nitrosalicylic acid

Ethylene   Diethylamine
oxide

MC.38 Propose a synthesis for the local anesthetic hexylcaine (see *The Merck Index*, 12th ed., #4746) given this retrosynthetic analysis.

Hexylcaine   Benzoic acid   Propene   Cyclohexylamine

MC.39 Propose a synthesis of the topical anesthetic cyclomethycaine (see *The Merck Index*, 12th ed., #2804) from 4-hydroxybenzoic acid and 2-methylpiperidine and any other necessary reagents.

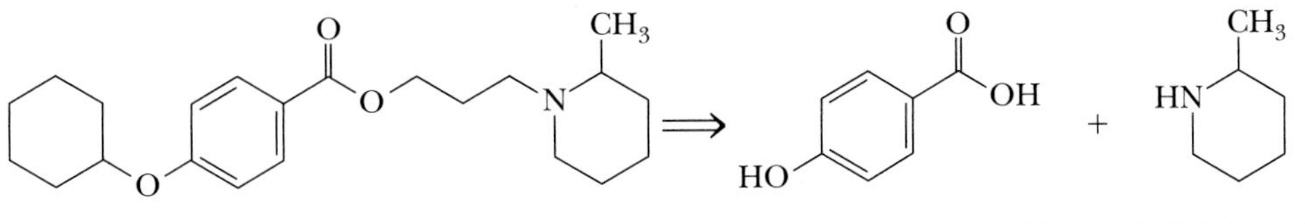

Cyclomethycaine ⟹ 4-Hydroxybenzoic acid + 2-Methylpiperidine

MC.40 Examine a number of the -caine type local anesthetics in *The Merck Index*, 12th ed., THERAP CATS, page THER-13, for structural patterns. You should discover that three of the most common are the following. Under each pattern is a clue to the manner of its synthesis. Name at least three that show each structural pattern.

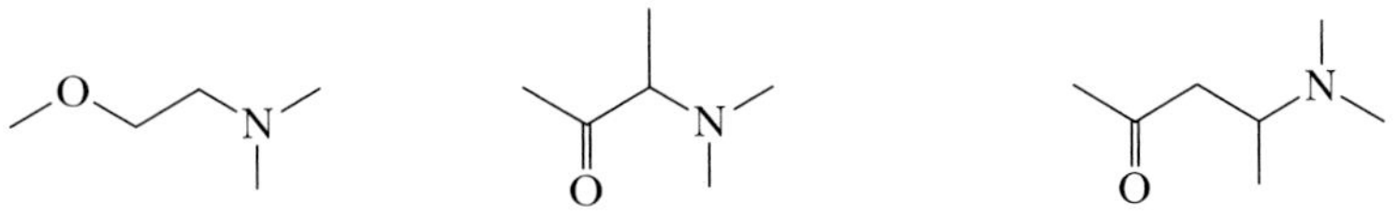

From ethylene oxide   From an α-chloroacid chloride   From a Michael reaction of an α,β-unsaturated carbonyl compound and an amine

Muscle Relaxant (Skeletal)

MC.41 Orphenadrine (see *The Merck Index*, 12th ed., #7007) has been used as a skeletal muscle relaxant. Given this retrosynthetic analysis, propose a synthesis for orphenadrine. *Hint:* Assume that Friedel-Crafts acylation of toluene gives an ortho and Para mixture, and that the ortho isomer can be separated.

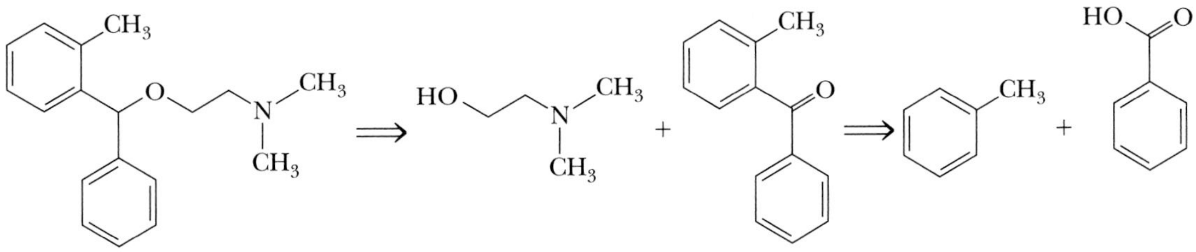

Orphenadrine

Anorexic

MC.42 Following is an outline for a synthesis of the anorexic (appetite suppressant) fenfluramine (see *The Merck Index*, 12th ed., #4015). This compound was one of the two ingredients in Phen-Fen, a weight-loss preparation now banned because of its potential to cause irreversible heart valve damage.

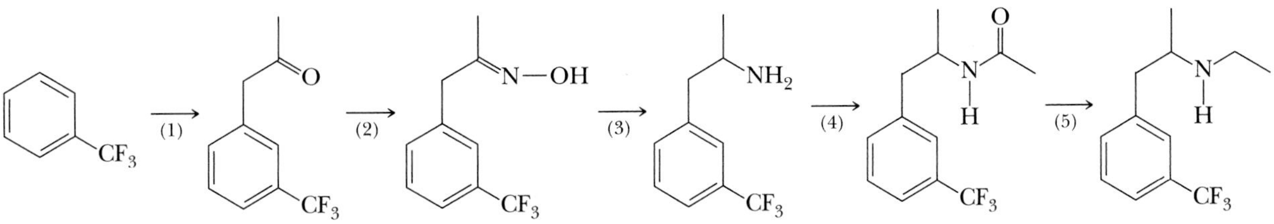

Fenfluramine

(a) Propose reagents and conditions for Step 1. Account for the fact that the CF_3 group is meta directing.

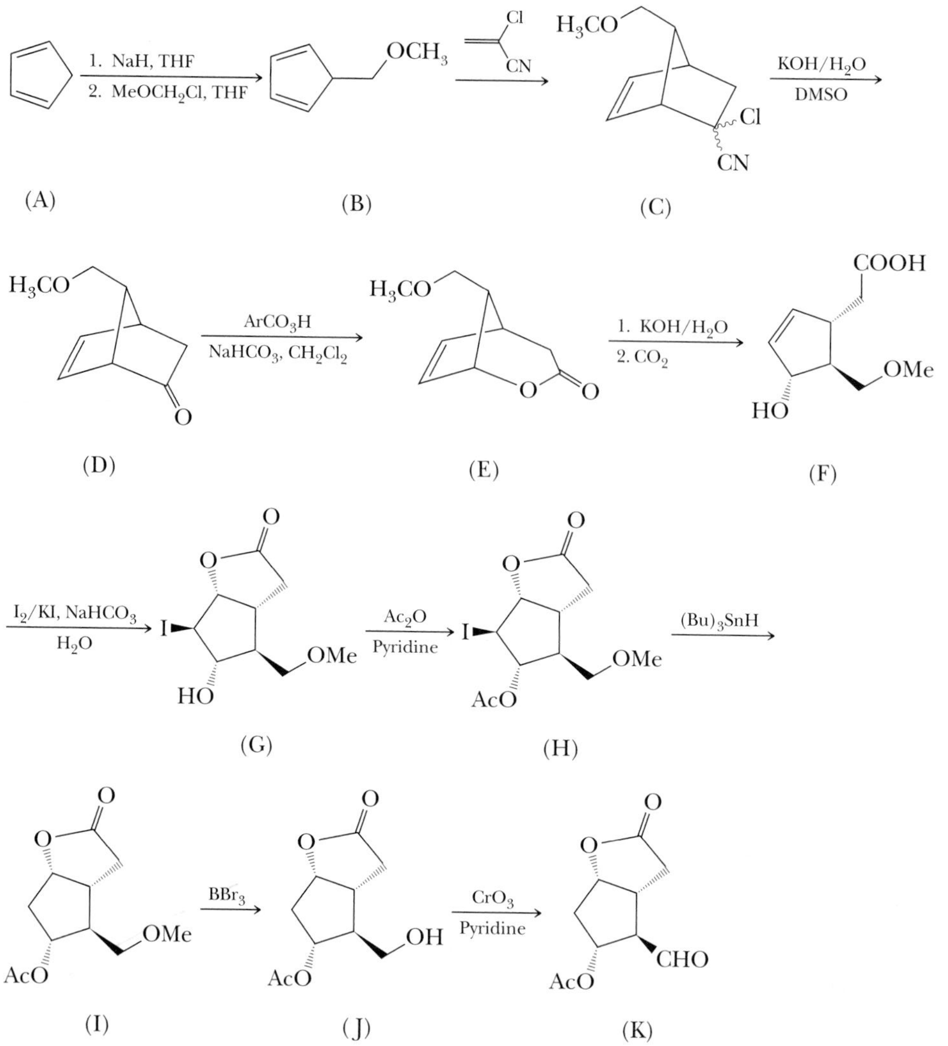

1. NaH, THF
2. MeOCH$_2$Cl, THF

(A)

OCH$_3$

(B)

H$_3$CO

Cl
CN

Cl
CN

(C)

KOH/H$_2$O
DMSO

H$_3$CO

O

(D)

ArCO$_3$H
NaHCO$_3$, CH$_2$Cl$_2$

H$_3$CO

O   O

(E)

1. KOH/H$_2$O
2. CO$_2$

COOH

OMe

HO

(F)

I$_2$/KI, NaHCO$_3$
H$_2$O

O   O

I

OMe

HO

(G)

Ac$_2$O
Pyridine

O   O

I

OMe

AcO

(H)

(Bu)$_3$SnH

O   O

OMe

AcO

(I)

BBr$_3$

O   O

OH

AcO

(J)

CrO$_3$
Pyridine

O   O

CHO

AcO

(K)
(Corey lactone)

shown. In what step or steps in this synthesis is the configuration of each stereocenter determined? Propose a mechanism to account for the observed stereospecificity of the relevant steps.

(f) Compound (F) was resolved using (+)-ephedrine (see *The Merck Index*, 12th ed., #3645). Following is a structural formula for (−)-ephedrine, the naturally occurring stereoisomer. What is meant by "resolution" and what is the rationale for using a chiral, enantiomerically pure amine for the resolution of (F)?

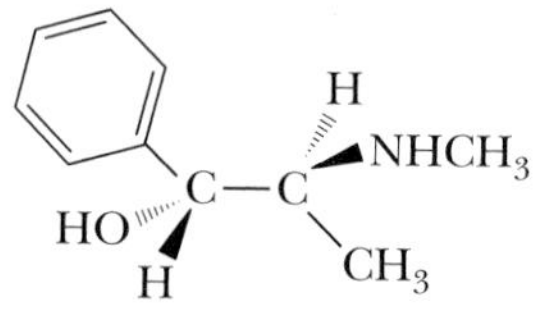

H

NHCH$_3$

C—C

HO

H

CH$_3$

Ephedrine $[\alpha]_D^{21}$ −41°

ORGANIC POLYMER CHEMISTRY

The technological advancement of any society is inextricably tied to the materials available to it. Indeed, historians have used the emergence of new materials as a way of establishing a timeline to mark the development of human civilization. As part of the search to discover new materials, scientists have made increasing use of organic chemistry for the preparation of synthetic polymers. The versatility afforded by these polymers allows for the creation and fabrication of materials with ranges of properties unattainable using such materials as wood, metals, and ceramics. Deceptively simple changes in the chemical

■ Sea of umbrellas on a rainy day in Shanghai, China. *(Gavin Hellier/Tony Stone Images)* Inset: Models of adipic acid and hexamethylenediamine, the two monomer units of nylon 66.

structure of a given polymer, for example, can change its mechanical properties from those of a sandwich bag to those of a bulletproof vest. Furthermore, structural changes can introduce properties never before imagined in organic polymers. For example, using well-defined organic reactions, one type of polymer can be made into an insulator (for example, the rubber sheath that surrounds electrical cords), or, if treated differently, it can be made into an electrical conductor with a conductivity nearly equal to that of metallic copper!

The years since the 1930s have seen extensive research and development in polymer chemistry, and an almost explosive growth in plastics, coatings, and rubber technology has created a worldwide multibillion-dollar industry. A few basic characteristics account for this phenomenal growth. First, the raw materials for synthetic polymers are derived mainly from petroleum. With the development of petroleum-refining processes, raw materials for the synthesis of polymers became generally cheap and plentiful. Second, within broad limits, scientists have learned how to tailor polymers to the requirements of the end use. Third, many consumer products can be fabricated more cheaply from synthetic polymers than from competing materials, such as wood, ceramics, and metals. For example, polymer technology created the water-based (latex) paints that have revolutionized the coatings industry; plastic films and foams have done the same for the packaging industry. The list could go on and on as we think of the manufactured items that are everywhere around us in our daily lives.

24.1 The Architecture of Polymers

Polymers (Greek: *poly* + *meros*, many parts) are long-chain molecules synthesized by linking **monomers** (Greek: *mono* + *meros*, single part) through chemical reactions. The molecular weights of polymers are generally high compared with those of common organic compounds and typically range from 10,000 g/mol to more than 1,000,000 g/mol. The architectures of these macromolecules can also be quite diverse. Types of polymer architecture include linear and branched chains as well as those with comb, ladder, and star structures (Figure 24.1). Additional structural variations can be achieved by introducing covalent cross links between individual polymer chains.

Linear and branched polymers are often soluble in solvents, such as chloroform, benzene, toluene, DMSO, and THF. In addition, many linear and branched polymers can be melted to form highly viscous liquids. In polymer chemistry, the term **plastic** refers to any polymer that can be molded when hot and retains its shape when cooled. **Thermoplastics** are polymers that can be melted and become sufficiently fluid that they can be molded into shapes that are retained when they are cooled. **Thermosetting plastics** or thermosets can be molded when they are first prepared, but once they are cooled, they harden irreversibly and cannot be remelted. Because

Polymer From the Greek: *poly* + *meros*, many parts. Any long-chain molecule synthesized by linking together many single parts called monomers.

Monomer From the Greek: *mono* + *meros*, single part. The simplest nonredundant unit from which a polymer is synthesized.

Plastic A polymer that can be molded when hot and retains its shape when cooled.

Thermoplastic A polymer that can be melted and molded into a shape that is retained when it is cooled.

Thermosetting polymer A polymer that can be molded when it is first prepared, but once cooled, hardens irreversibly and cannot be remelted.

Figure 24.1
Various polymer architectures.

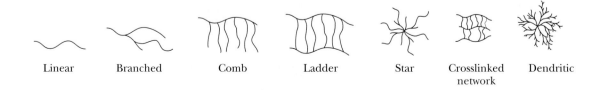

| Linear | Branched | Comb | Ladder | Star | Crosslinked network | Dendritic |

of these very different physical characteristics, thermoplastics and thermosets must be processed differently and are used in very different applications.

The single most important property of polymers at the molecular level is the size and shape of their chains. A good example of the importance of size is a comparison of paraffin wax, a natural polymer, and polyethylene, a synthetic polymer. These two distinct materials have identical repeat units, namely $-CH_2-$, but differ greatly in chain size. Paraffin wax has between 25 and 50 carbon atoms per chain, whereas polyethylene has between 1000 and 3000 carbon atoms per chain. Paraffin wax, such as in birthday candles, is soft and brittle but polyethylene, such as in plastic beverage bottles, is strong, flexible, and tough. These vastly different properties arise directly from the difference in size and molecular architecture of the individual polymer chains.

24.2 Polymer Notation and Nomenclature

We show the structure of a polymer by placing parentheses around the **repeat unit,** which is the smallest molecular fragment that contains all the nonredundant structural features of the chain. Thus, the structure of an entire polymer chain can be reproduced by repeating the enclosed structure in both directions. A subscript n, called the **average degree of polymerization,** is placed outside the parentheses to indicate that this unit is repeated n times.

An exception to this notation are the polymers formed from symmetric monomer units, such as polyethylene, $+CH_2CH_2+_n$, and polytetrafluoroethylene, $+CF_2CF_2+_n$. Although the simplest repeat units are the $-CH_2-$ and $-CF_2-$ groups, respectively, we show two methylene groups and two difluoromethylene groups because they originate from ethylene ($CH_2=CH_2$) and tetrafluoroethylene ($CF_2=CF_2$), the monomer units from which these polymers are derived.

The most common method of naming a polymer is to attach the prefix poly- to the name of the monomer from which the polymer is derived, as for example polyethylene and polystyrene. In the case of a more complex monomer, or where the name of the monomer is more than one word, as for example the monomer vinyl chloride, parentheses are used to enclose the name of the monomer.

Polystyrene   Styrene   Poly(vinyl chloride)   Vinyl chloride
(PVC)

Example 24.1

Given the following structure, determine the polymer's repeat unit, redraw the structure using the simplified parenthetical notation, and name the polymer.

Solution

The repeat unit is —CH_2CF_2— and the polymer is written $\left(CH_2CF_2\right)_n$. The repeat unit is derived from 1,1-difluoroethylene and the polymer is named poly(1,1-difluoroethylene). This polymer is used in microphone diaphragms.

Problem 24.1

Given the following structure, determine the polymer's repeat unit, redraw the structure using the simplified parenthetical notation, and name the polymer.

24.3   Molecular Weights of Polymers

All synthetic polymers and most naturally occurring polymers are mixtures of individual polymer molecules of variable molecular weights. When defining molecular weights in polymer chemistry, the two most common definitions are the number average and weight average molecular weights. The **number average molecular weight, M_n,** is calculated by counting the number of polymer chains of a particular molecular weight, multiplying each number by the molecular weight of its chain, summing these values, and dividing by the total number of polymer chains. The **weight average molecular weight, M_w,** is calculated by recording the total weight of each chain of a particular length, summing these weights, and dividing by the total weight of the sample. Because the larger chains in a sample weigh more than the smaller chains, the weight average molecular weight is skewed to higher values, and M_w is always greater than M_n (Figure 24.2).

Both M_n and M_w are useful values, and their ratio, M_w/M_n, called the **polydispersity index,** provides a measure of the breadth of the molecular-weight distribution. When the M_w/M_n ratio is equal to one, all the polymer molecules in a sample are the same length, and the polymer is said to be **monodisperse.** No synthetic polymers are ever monodisperse unless the individual molecules are carefully fractionated using time-consuming, rigorous separation techniques based on molecular size. On the other hand, natural polymers, such as polypeptides and DNA, that are formed using biological processes are monodisperse polymers.

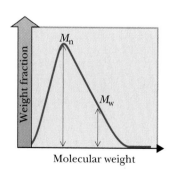

Figure 24.2

The distribution of molecular weights in a given polymer sample.

24.4   Polymer Morphology — Crystalline Versus Amorphous Materials

Polymers, like small organic molecules, tend to crystallize upon precipitation or as they are cooled from a melt. Acting to inhibit this tendency are their very large molecules, which tend to slow diffusion, and their sometimes complicated or irregular structures, which prevent efficient packing of the chains. The result is that polymers in the solid state tend to be composed of both ordered **crystalline domains** (crystallites) and disordered **amorphous domains.** The relative amounts of crystalline and amorphous domains differ from polymer to polymer and often depend upon the manner in which the material is processed.

High degrees of crystallinity are most often found in polymers with regular, compact structures and strong intermolecular forces, such as hydrogen bonding and dipolar interactions. The temperature at which crystallites melt corresponds to the **melt transition (T_m)** of the polymer. As the degree of crystallinity of a polymer increases, its T_m increases, and it becomes more opaque due to scattering of the light by the crystalline domains. There is also a corresponding increase in strength and stiffness with increase in crystallinity. For example, poly(6-aminohexanoic acid) has a T_m of 223°C. At and well above room temperature, this polymer is a hard durable material that does not undergo any appreciable change in properties even on a very hot summer afternoon. Its uses range from textile fibers to shoe heels.

Amorphous domains are characterized by the absence of long-range order. Highly amorphous polymers are sometimes referred to as glassy polymers. Because they lack crystalline domains that scatter light, amorphous polymers are transparent. In addition, they are typically weak polymers, in terms of both their greater flexibility and their smaller mechanical strength. On being heated, amorphous polymers are transformed from a hard glass to a soft, flexible, rubbery state. The temperature at which this transition occurs is called the **glass transition temperature, T_g.** Amorphous polystyrene, for example, has a T_g of 100°C. At room temperature, it is a rigid solid used for drinking cups, foamed packaging materials, disposable medical wares, tape reels, and so forth. If it is placed in boiling water, it becomes soft and rubbery.

This relationship between mechanical properties and the degree of crystallinity can be illustrated by poly(ethylene terephthalate) (PET).

Poly(ethylene terephthalate)

PET can be made with a percent of crystalline domains ranging from 0% to about 55%. Completely amorphous PET is formed by cooling from the melt quickly. By prolonging the cooling time, more molecular diffusion occurs, and crystallites form as the chains become more ordered. The differences in mechanical properties between these forms of PET are great. PET with a low degree of crystallinity is used for plastic beverage bottles, whereas fibers drawn from highly crystalline PET are used for textile fibers and tire cords.

Rubber materials must have low T_g values to behave as **elastomers (elastic polymers).** If the temperature drops below its T_g value, then the material is converted to a rigid glassy solid, and all elastomeric properties are lost. A poor understanding of this

Crystalline domain   An ordered crystalline region in the solid state of a polymer. Also called a crystallite.

Amorphous domain   A disordered, noncrystalline region in the solid state of a polymer.

Melt transition, T_m   The temperature at which crystalline regions of a polymer melt.

Glass transition temperature   The temperature at which a polymer undergoes the transition from a hard glass to a rubbery state.

Elastomer   A material that, when stretched or otherwise distorted, returns to its original shape when the distorting force is released.

behavior of polymers contributed to the *Challenger* spacecraft disaster in 1986. The elastomeric O-rings used to seal the solid booster rockets had a T_g value around 0°C. When the temperature dropped to an unanticipated low on the morning of the *Challenger* launch, the O-ring seals dropped below their T_g value and obediently changed from elastomers to rigid glasses, losing any sealing capabilities. The rest is tragic history. The physicist Richard Feynman sorted this out publicly in a famous televised hearing in which he put a *Challenger*-type O-ring in ice water and showed that its elasticity was lost!

24.5   Step-Growth Polymerizations

Step-growth polymerization   A polymerization in which chain growth occurs in a stepwise manner between difunctional monomers, as for example between adipic acid and hexamethylenediamine to form nylon 66. Also called condensation polymerization.

Condensation polymerization   A polymerization in which chain growth occurs in a stepwise manner between difunctional monomers. Also called step-growth polymerization.

Polymerizations in which chain growth occurs in a stepwise manner are called **step-growth** or **condensation polymerizations.** Step-growth polymers are formed by reaction between difunctional molecules, with each new bond created in a separate step. During polymerization, monomers react with monomers to form dimers, dimers react with dimers to form tetramers, tetramers react with monomers to form pentamers, and so on. This stepwise construction of polymer chains has important consequences for both their molecular weights and molecular-weight distributions. Probability tells us that the most abundant species tend to co-condense. Thus, at the early stages of polymerization, small chains are most likely to react with monomers or other small chains to generate many low-molecular-weight oligomers rather than a small number of high-molecular-weight polymers. This tendency persists until all monomer units are used up. As a result, high-molecular-weight polymers are not produced until very late in the reaction, typically past 99% conversion of monomers to higher molecular-weight chains. Only at this point is there the probability of larger chains reacting with one another to form high-molecular-weight polymer molecules. This restriction points to an important distinction between small-molecule organic reactions and step-growth polymerizations. Although a reaction that typically yields 85% of the desired product is considered "good" in organic synthesis, the same reaction is essentially useless for step-growth polymerizations because high-molecular-weight polymers are rarely formed at such low conversions.

There are two common types of step-growth processes: (1) reaction between A—A and B—B type monomers to give —(A—A—B—B)$_n$— polymers and (2) the self-condensation of A—B monomers to give —(A—B)$_n$— polymers. In each case, an A functional group reacts exclusively with a B functional group, and a B functional group reacts exclusively with an A functional group. New covalent bonds in step-growth polymerizations are generally formed by polar reactions, as for example nucleophilic acyl substitution. In this section, we discuss five types of step-growth polymers: polyamides, polyesters, polycarbonates, polyurethanes, and epoxy resins.

A.   Polyamides

In the years following World War I, a number of chemists recognized the need for developing a basic knowledge of polymer chemistry. One of the most creative of these was Wallace M. Carothers. In the early 1930s, Carothers and his associates at E. I. DuPont de Nemours & Company began fundamental research into the reactions of aliphatic dicarboxylic acids and diols. From adipic acid and ethylene glycol, they obtained a polyester of high-molecular-weight that could be drawn into fibers.

Hexanedioic acid   1,2-Ethanediol   Poly(ethylene adipate)
(Adipic acid)   (Ethylene glycol)

These first polyester fibers had melt transitions (T_m) too low for use as textile fibers, and they were not investigated further. Carothers then turned his attention to the reactions of dicarboxylic acids and diamines to form **polyamides** and, in 1934, synthesized nylon 66, the first purely synthetic fiber. Nylon 66 is so named because it is synthesized from two different monomers, each containing six carbon atoms.

In the synthesis of nylon 66, hexanedioic acid (adipic acid) and 1,6-hexanediamine (hexamethylenediamine) are dissolved in aqueous ethanol where they react to form a one-to-one salt called nylon salt. Nylon salt is then heated in an autoclave to 250°C where the internal pressure rises to about 15 atm. Under these conditions, —COO$^-$ groups from adipic acid and —NH$_3$$^+$ groups from hexamethylenediamine react with loss of H$_2$O to form amide groups.

Polyamide A polymer in which each monomer unit is joined to the next by an amide bond, as for example nylon 66.

Hexanedioic acid   1,6-Hexanediamine
(Adipic acid)   (Hexamethylenediamine)

Nylon salt

heat

Nylon 66

Nylon 66 formed under these conditions has a T_m of 250–260°C and has a molecular-weight range of 10,000 to 20,000 g/mol.

In the first stage of fiber production, crude nylon 66 is melted, spun into fibers, and cooled. Next, the melt-spun fibers are **cold-drawn** (drawn at room temperature) to about four times their original length to increase their degree of crystallinity. As the fibers are drawn, individual polymer molecules become oriented in the direction of the fiber axis, and hydrogen bonds form between carbonyl oxygens of one chain and amide hydrogens of another chain (Figure 24.3). The effects of orientation of polyamide molecules on the physical properties of the fiber are dramatic; both tensile

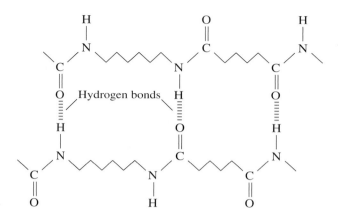

Figure 24.3
The structure of cold-drawn nylon 66. Hydrogen bonds between adjacent polymer chains provide additional tensile strength and stiffness to the fibers.

strength and stiffness are increased markedly. Cold-drawing is an important step in the production of most synthetic fibers.

The current raw material base for the production of nylon 66 is benzene, which is derived almost entirely from catalytic cracking and reforming of petroleum. Catalytic reduction of benzene to cyclohexane followed by catalyzed air oxidation gives a mixture of cyclohexanol and cyclohexanone. Oxidation of this mixture by nitric acid gives adipic acid.

Benzene $\xrightarrow[\text{catalyst}]{3H_2}$ Cyclohexane $\xrightarrow[\text{catalyst}]{O_2}$ [Cyclohexanol + Cyclohexanone] $\xrightarrow{HNO_3}$ Hexanedioic acid (Adipic acid)

Adipic acid, in turn, is a starting material for the synthesis of hexamethylenediamine. Treatment of adipic acid with ammonia gives an ammonium salt, which, when heated, gives adipamide. Catalytic reduction of adipamide gives hexamethylenediamine. Thus, carbon sources for the production of nylon 66 are derived entirely from petroleum, which unfortunately is not a renewable resource.

Ammonium hexanedioate (Ammonium adipate) $\xrightarrow{\text{heat}}$ Hexanediamide (Adipamide) $\xrightarrow[\text{catalyst}]{4H_2}$

1,6-Hexanediamine (Hexamethylenediamine)

The nylons are a family of polymers, the members of which have subtly different properties that suit them to one use or another. The two most widely used members of this family are nylon 66 and nylon 6. Nylon 6 is so named because it is synthesized from caprolactam, a six-carbon monomer. In the synthesis of nylon 6, caprolactam is partially hydrolyzed to 6-aminohexanoic acid and then heated to 250°C to bring about polymerization. Nylon 6 is fabricated into fibers, brush bristles, rope, high-impact moldings, and tire cords.

$$n \text{ Caprolactam} \xrightarrow[\text{2. heat}]{\text{1. partial hydrolysis}} \text{Nylon 6}$$

Caprolactam   Nylon 6

Based on extensive research into relationships between molecular structure and bulk physical properties, scientists at DuPont reasoned that a polyamide containing aromatic rings would be stiffer and stronger than either nylon 66 or nylon 6. In early 1960, DuPont introduced Kevlar, a polyaromatic amide (**aramid**) fiber synthesized from terephthalic acid and *p*-phenylenediamine.

Aramid A poly*aromatic amide*; a polymer in which the monomer units are an aromatic diamine and an aromatic dicarboxylic acid.

$$n\text{HOC} \underset{}{-} \text{COH} + n\text{H}_2\text{N} \underset{}{-} \text{NH}_2 \longrightarrow \text{Kevlar}$$

1,4-Benzenedicarboxylic acid   1,4-Benzenediamine   Kevlar
(Terephthalic acid)   (*p*-Phenylenediamine)

One of the remarkable features of Kevlar is its light weight compared with other materials of similar strength. For example, a 3-in. cable woven of Kevlar has a strength equal to that of a similarly woven 3-in. steel cable. Whereas the steel cable weighs about 20 lb/ft, the Kevlar cable weighs only 4 lb/ft. Kevlar now finds use in such articles as anchor cables for offshore drilling rigs and reinforcement fibers for automobile tires. Kevlar is also woven into a fabric that is so tough that it can be used for bulletproof vests, jackets, and raincoats.

Bulletproof vests have a thick layer of Kevlar. *(Charles D. Winters)*

B. Polyesters

Recall that, in the early 1930s, Carothers and his associates had concluded that **polyester** fibers from aliphatic dicarboxylic acids and ethylene glycol were not suitable for textile use because their melting points are too low. Winfield and Dickson at the Calico Printers Association in England further investigated polyesters in the 1940s and reasoned that a greater resistance to rotation in the polymer backbone would stiffen the polymer, raise its melting point, and thereby lead to a more acceptable polyester fiber. To create stiffness in the polymer chain, they used terephthalic acid. Polymerization of this aromatic dicarboxylic acid with ethylene glycol gives poly(ethylene terephthalate), abbreviated PET (also PETE).

Polyester A polymer in which each monomer unit is joined to the next by an ester bond, as for example poly(ethylene terephthalate).

$$\text{HO} \underset{}{-} \text{OH} + \text{HO} \diagdown \text{OH} \xrightarrow{\text{heat}} \text{Poly(ethylene terephthalate)} + 2n\text{H}_2\text{O}$$

1,4-Benzenedicarboxylic acid   1,2-Ethanediol   Poly(ethylene terephthalate)
(Terephthalic acid)   (Ethylene glycol)   (Dacron, Mylar)

Mylar can be made into extremely strong films. Because the film has very tiny pores, it is used for balloons that can be inflated with helium; the helium atoms diffuse only slowly through the pores of the film. *(Charles D. Winters)*

The crude polyester can be melted, extruded, and then cold-drawn to form the textile fiber Dacron polyester, outstanding features of which are its stiffness (about four times that of nylon 66), very high strength, and remarkable resistance to creasing and wrinkling. Because the early Dacron polyester fibers were harsh to the touch due to their stiffness, they were usually blended with cotton or wool to make acceptable textile fibers. Newly developed fabrication techniques now produce less harsh Dacron polyester textile fibers. PET is also fabricated into Mylar films and recyclable plastic beverage containers.

Ethylene glycol for the synthesis of PET is obtained by air oxidation of ethylene to ethylene oxide (Section 11.8A) followed by hydrolysis to the glycol (Section 11.9A). Ethylene is, in turn, derived entirely from cracking either petroleum or the ethane derived from natural gas (Section 2.10A). Terephthalic acid is obtained by the oxidation of *p*-xylene, an aromatic hydrocarbon obtained along with benzene and toluene from catalytic cracking and reforming of naphtha and other petroleum fractions (Section 2.10B).

$$CH_2{=}CH_2 \xrightarrow[\text{catalyst}]{O_2} \underset{\text{Oxirane}}{CH_2{-}CH_2} \xrightarrow{H^+,\,H_2O} \underset{\text{1,2-Ethanediol}}{HOCH_2CH_2OH}$$

Ethylene   Oxirane   1,2-Ethanediol
(Ethylene oxide)   (Ethylene glycol)

$$H_3C{-}\!\!\left\langle\ \right\rangle\!\!{-}CH_3 \xrightarrow[\text{catalyst}]{O_2} HO\overset{O}{\overset{\|}{C}}{-}\!\!\left\langle\ \right\rangle\!\!{-}\overset{O}{\overset{\|}{C}}OH$$

p-Xylene   Terephthalic acid

C.   Polycarbonates

Polycarbonates, the most familiar of which is Lexan, are a class of commercially important engineering polyesters. In the production of Lexan, an aqueous solution of the disodium salt of bisphenol A (Problem 21.22) is brought into contact with a solution of phosgene dissolved in methylene chloride. The two solutions are immiscible, and no reaction occurs until tetrabutylammonium chloride or other phase-transfer catalyst (Section 8.7) is added. The tetrabutylammonium cation carries the bisphenol A anion into the methylene chloride phase where it reacts smoothly with phosgene to form the polymer. The tetrabutylammonium ion then carries chloride ion back to the aqueous phase.

Disodium salt   Phosgene   Lexan
of bisphenol A   (a polycarbonate)

Lexan is a tough, transparent polymer with high-impact and tensile strengths, and it retains its properties over a wide temperature range. It has found significant use in sporting equipment, such as bicycle, football, motorcycle, and snowmobile helmets as well as hockey and baseball catchers' face masks. In addition, it is used to make light, impact-resistant housings for household appliances and automobile and aircraft equipment and to manufacture safety glass and unbreakable windows.

D.   Polyurethanes

A urethane, or carbamate, is an ester of carbamic acid, H_2NCOOH. Carbamates are most commonly prepared by treatment of an isocyanate with an alcohol.

$$RN{=}C{=}O \ + \ R'OH \ \longrightarrow \ RN\overset{\overset{\displaystyle O}{\displaystyle \|}}{H}COR'$$

An isocyanate   A carbamate

A polycarbonate hockey mask.
(Charles D. Winters)

Polyurethane   A polymer containing the —$NHCO_2$— group as a repeating unit.

Polyurethanes consist of flexible polyester or polyether units (blocks) alternating with rigid urethane units (blocks). The rigid urethane blocks are derived from a diisocyanate, commonly a mixture of 2,4- and 2,6-toluene diisocyanate. The more flexible blocks are derived from low-molecular-weight (MW 1000–4000) polyesters or polyethers with —OH groups at each end of the polymer chain. Polyurethane fibers are fairly soft and elastic and have found use as Spandex and Lycra, the "stretch" fabrics used in bathing suits, leotards, and undergarments.

2,6-Toluene diisocyanate   Low-molecular-weight polyester or polyether   A polyurethane

$+ \ HO{-}polymer{-}OH \ \longrightarrow$

Polyurethane foams for upholstery and insulating materials are made by adding small amounts of water during polymerization. Water reacts with isocyanate groups to produce gaseous carbon dioxide which then acts as the foaming agent.

$$RN{=}C{=}O \ + \ H_2O \ \longrightarrow \ \left[RNH{-}\overset{\overset{\displaystyle O}{\displaystyle \|}}{C}{-}OH \right] \ \longrightarrow \ RNH_2 \ + \ CO_2$$

An isocyanate   A carbamic acid
(unstable)

E.   Epoxy Resins

Epoxy resins are materials prepared by a polymerization in which one monomer contains at least two epoxy groups. Within this range, there are a large number of polymeric materials possible, and epoxy resins are produced in forms ranging from low-viscosity liquids to high-melting solids. The most widely used epoxide monomer is the diepoxide prepared by treatment of one mole of bisphenol A (Problem 21.22) with two moles of epichlorohydrin (Problem 11.40). To prepare the following epoxy resin, the diepoxide monomer is treated with 1,2-ethanediamine (ethylene diamine). This component is usually called the "catalyst" in the two-component formulations that can be bought in any hardware store; it is also the component with the acrid smell. It is not a catalyst but a reagent.

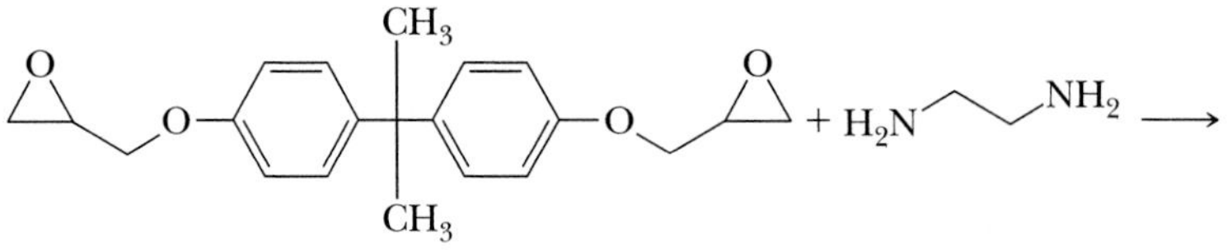

A diepoxide   A diamine

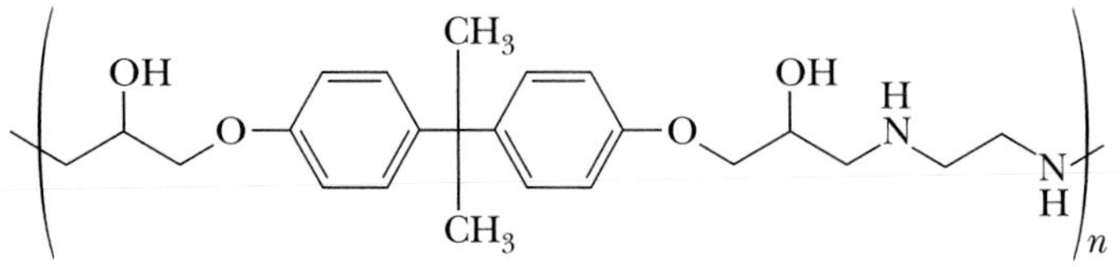

An epoxy resin

Epoxy resins are widely used as adhesives and insulating surface coatings. They have good electrical insulating properties, which leads to their use for encapsulating electrical components ranging from integrated circuit boards to switch coils and insulators for power transmission systems. They are also used as composites with other materials, such as glass fiber, paper, metal foils, and other synthetic fibers to create structural components for jet aircraft, rocket motor casings, and so on.

Example 24.2

Write a mechanism for the acid-catalyzed polymerization of 1,4-benzenediisocyanate and ethylene glycol. To simplify your mechanism, consider only the reaction of one —NCO group with one —OH group.

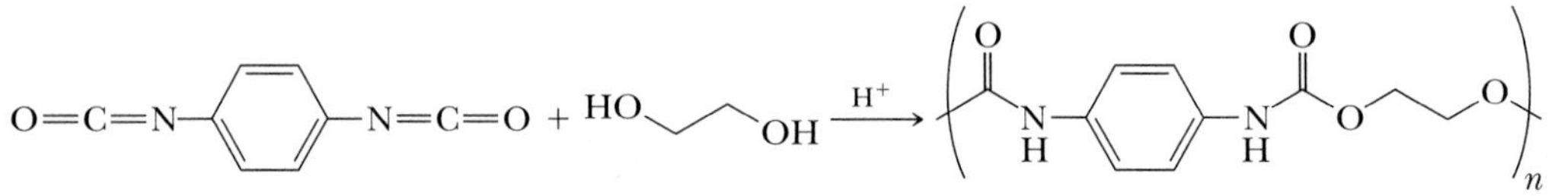

1,4-Benzenediisocyanate   Ethylene glycol   A polyurethane

Solution

A mechanism is shown in three steps. Proton transfer in Step 1 from the acid, HA, to nitrogen followed by addition of ROH to the carbonyl carbon in Step 2 gives an oxonium ion. Proton transfer in Step 3 from the oxonium ion to A^- gives the carbamate ester.

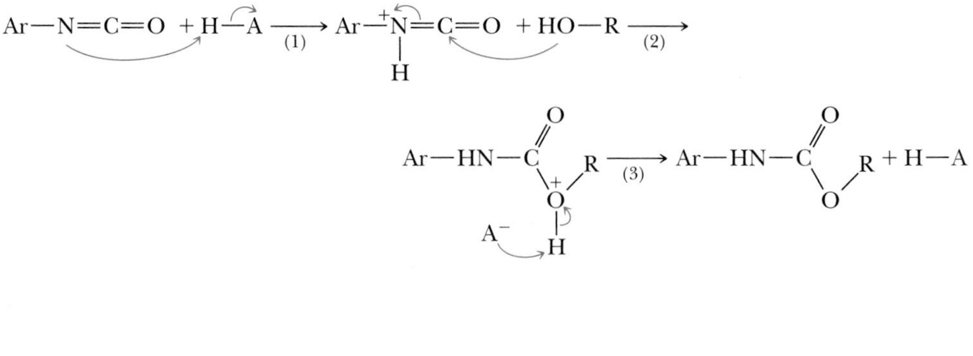

Problem 24.2

Write the repeating unit of the polymer formed from the following reaction, and propose a mechanism for its formation.

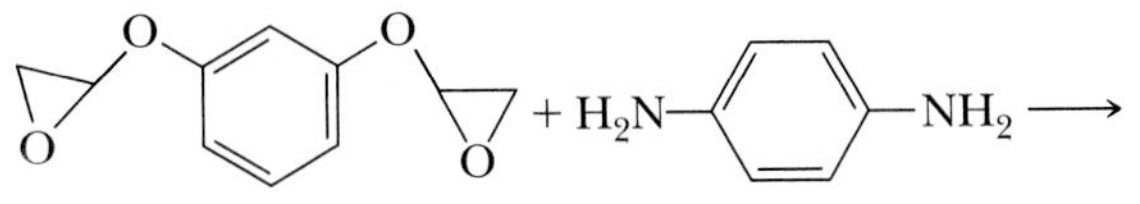

A diepoxide   A diamine

F. Thermosetting Polymers

Thermosetting polymers are composed of long chains that are cross linked by covalent bonds. In effect, a thermosetting polymer is one giant molecule. The first thermosetting polymer was produced by Leo Baekeland (1863–1944) in 1907 by reacting phenol with formaldehyde to form the following three-dimensional structure. The product, known as Bakelite, is a good electrical insulator.

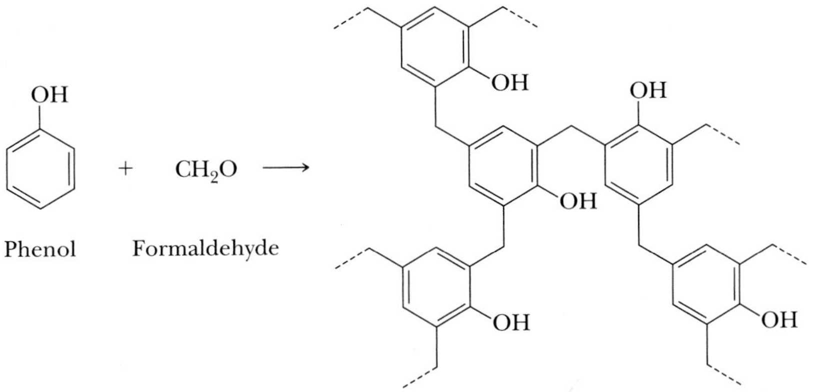

Phenol   Formaldehyde

Bakelite

In the preparation of a thermoset, one of the monomers must be trifunctional. In the case of Bakelite, the trifunctional monomer is phenol. Alkyd thermosets are polyesters of an organic diacid, HOOC—R—COOH, and a trialcohol such as glycerol. Urea-formaldehyde thermosets are polyamides in which one molecule of urea, H_2N—CO—NH_2, can condense with up to four molecules of formaldehyde.

Manufacture of thermosets begins with a fluid mixture of the two monomers. The fluid is first shaped and then polymerized, either by heating or by being mixed with an initiator. The product of the polymerization is a network of covalently bonded atoms that is a solid, even at high temperatures. When heated to high temperatures, thermoset polymers char and decompose, but they do not melt.

CHEMISTRY IN ACTION

Stitches That Dissolve

Medical science has advanced very rapidly in the last few decades. Some procedures considered routine today, such as organ transplantation and the use of lasers in surgery, were unimaginable 60 years ago. As the technological capabilities of medicine have grown, the demand for synthetic materials that can be used inside the body has increased as well. Polymers have many of the characteristics of an ideal biomaterial: they are lightweight and strong, are inert or biodegradable depending on their chemical structure, and have physical properties (softness, rigidity, elasticity) that are easily tailored to match those of natural tissues. Carbon-carbon backbone polymers are degradation resistant and are used widely in permanent organ and tissue replacements.

Whereas most medical uses of polymeric materials require biostability, applications that use the biodegradable nature of some macromolecules have been developed. An example is the use of poly(glycolic acid) and glycolic acid/lactic acid copolymers as absorbable sutures, which go under the trade name Lactomer.

A copolymer of poly(glycolic acid)-poly(lactic acid)

Traditional suture materials such as catgut must be removed by a health care specialist after they have served their purpose. Stitches of Lactomer, however, are hydrolyzed slowly over a period of approximately two weeks, and by the time the torn tissues have fully healed, the stitches have fully degraded, and no suture removal is necessary. Glycolic and lactic acids formed during hydrolysis of the stitches are metabolized and excreted by existing biochemical pathways.

24.6 Chain-Growth Polymerizations

Chain-growth polymerization A polymerization that involves sequential addition reactions either to unsaturated monomers or to monomers possessing other reactive functional groups.

From the perspective of the chemical industry, the single most important reaction of alkenes is **chain-growth polymerization,** a type of polymerization in which monomer units are joined together without loss of atoms. An example is the formation of polyethylene from ethylene.

$$n\,CH_2{=}CH_2 \xrightarrow{\text{catalyst}} \text{---}(CH_2CH_2\text{---})_n$$

Ethylene   Polyethylene

The mechanism of this type of polymerization differs greatly from the mechanism of step-growth polymerizations. In the latter, all monomers plus the polymer endgroups possess equally reactive functional groups, allowing for all possible combinations of reactions to occur, including monomer with monomer, dimer with dimer, and so forth. In contrast, chain-growth polymerizations involve endgroups possessing reactive intermediates that react with a monomer only. The reactive intermediates used in chain-growth polymerizations include radicals, carbanions, carbocations, and organometallic complexes.

The number of monomers that undergo chain-growth polymerizations is large and includes such compounds as alkenes, alkynes, allenes, isocyanates, and cyclic compounds, such as lactones, lactams, ethers, and epoxides. We concentrate on the chain-growth polymerizations of ethylene and substituted ethylenes and show how these compounds can be polymerized by radical, cation, anion, and organometallic-mediated mechanisms.

An alkene

Table 24.1 lists several important polymers derived from ethylene and substituted ethylenes along with their common names and most important uses.

Table 24.1 Polymers Derived from Ethylene and Substituted Ethylenes

Monomer Formula	Common Name	Polymer Name(s) and Common Uses
$CH_2{=}CH_2$	Ethylene	Polyethylene, Polythene; break-resistant containers and packaging materials
$CH_2{=}CHCH_3$	Propylene	Polypropylene, Herculon; textile and carpet fibers
$CH_2{=}CHCl$	Vinyl chloride	Poly(vinyl chloride), PVC; construction tubing
$CH_2{=}CCl_2$	1,1-Dichloroethylene	Poly(1,1-dichloroethylene), Saran; food packaging
$CH_2{=}CHCN$	Acrylonitrile	Polyacrylonitrile, Orlon; acrylics and acrylates
$CF_2{=}CF_2$	Tetrafluoroethylene	Poly(tetrafluoroethylene), PTFE; Teflon, nonstick coatings
$CH_2{=}CHC_6H_5$	Styrene	Polystyrene, Styrofoam; insulating materials
$CH_2{=}CHCOOCH_2CH_3$	Ethyl acrylate	Poly(ethyl acrylate); latex paints
$CH_2{=}CCOOCH_3$ $\mid$ CH_3	Methyl methacrylate	Poly(methyl methacrylate), Lucite, Plexiglas; glass substitutes

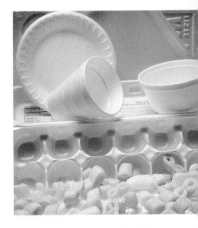

The low thermal conductivity of polystyrene makes it a good insulating material. *(Charles D. Winters)*

A. Radical Chain-Growth Polymerizations

Among the initiators used for radical chain-growth polymerizations are diacyl peroxides, such as dibenzoyl peroxide, which decompose as shown upon heating. In the first step, homolytic cleavage of the weak O—O peroxide bond yields two acyloxy radicals. Each acyloxy radical then decomposes to form an aryl radical and CO_2.

Dibenzoyl peroxide   A benzoyloxy radical   A phenyl radical

Another common class of initiators used in radical polymerizations are azo compounds, such as azoisobutyronitrile (AIBN), which decompose upon heating or by the absorption of UV light to produce alkyl radicals and nitrogen gas.

Azoisobutyronitrile   Alkyl radicals

The chain initiation, propagation, and termination steps for radical polymerization of a substituted ethylene monomer are shown for the monomer $RCH=CH_2$. Dissociation of the initiator produces a radical that reacts with the double bond of a monomer. Once initiated, the chains continue to propagate through successive additions of monomers.

Mechanism   Radical Polymerization of a Substituted Ethylene

Step 1:   Initiation: Radicals form from nonradical compounds.

Step 2:   Propagation: Reaction of a radical and a molecule produces a new radical.

Step 3:   Chain termination: Radicals are destroyed.

Radical reactions with double bonds almost always give the more stable (more substituted) radical. Because additions are biased in this fashion, the polymerizations of vinyl monomers tend to yield polymers with head-to-tail linkages. Vinyl polymers made by radical processes generally have no more than 1–2% head-to-head linkages.

head-to-tail linkages   head-to-head linkage

In radical reactions, the chain termination involves combination of radicals to produce a nonradical molecule or molecules. One common termination step is **radical coupling** to form a new carbon-carbon bond linking two growing polymer chains. This type of termination step is a diffusion-controlled process that occurs without an activation energy barrier. Another common termination process is **disproportionation,** which involves the abstraction of a hydrogen atom from the beta position to the propagating radical of one chain by the radical endgroup of another chain. This process results in two dead chains, one terminated in an alkyl group and the other in an alkenyl group.

Because organic radicals are highly reactive species, it is not surprising that radical polymerizations are often complicated by unwanted side reactions. A frequently observed side reaction is hydrogen abstraction by the radical endgroup from a growing polymer chain, a solvent molecule, or another monomer. These side reactions are called **chain-transfer reactions** because the activity of the endgroup is "transferred" from one chain to another.

Chain transfer is illustrated by radical polymerization of ethylene. Polyethylene formed by radical polymerization exhibits a number of butyl branches on the polymer main chain. These four-carbon branches are generated in a "back-biting" chain-transfer reaction in which the radical endgroup abstracts a hydrogen from the fourth carbon back (the fifth carbon in the chain). Abstraction of this hydrogen is particularly facile because the transition state associated with the process can adopt a conformation like that of a chair cyclohexane. Continued polymerization of monomer from this new radical center leads to branches four carbons long.

> **Disproportionation** A termination process that involves the abstraction of a hydrogen atom from the beta position of the propagating radical of one chain by the radical endgroup of another chain.

> **Chain-transfer reaction** The transfer of reactivity of an endgroup from one chain to another during a polymerization.

A six-membered transition
state leading to
1,5-hydrogen abstraction

As a result of these various abstraction reactions, polymers synthesized by radical processes can have highly branched structures. The number of butyl branches depends on the relative stability of the propagating-radical endgroup and varies depending on the polymer. Polyethylene chains propagate through highly reactive primary radicals, which tend to be susceptible to 1,5-hydrogen abstraction reactions; these polymers typically have 15 to 30 branches per 500 monomer units. In contrast, polystyrene chains propagate through substituted benzyl radicals, which are stabilized by delocalization of the unpaired electron over the aromatic ring. These stabilized radicals are less likely to undergo hydrogen abstraction reactions. Polystyrene typically exhibits only one branch per 4000 to 10,000 monomer units.

The first commercial process for ethylene polymerization used peroxide catalysts at temperatures of 500°C and pressures of 1000 atm and produced a soft, tough polymer known as low-density polyethylene (LDPE). At the molecular level, chains of LDPE are highly branched due to chain-transfer reactions. Because this extensive chain branching prevents polyethylene chains from packing efficiently, LDPE is largely amorphous and transparent, with only a small amount of crystallites of a size too small to scatter light. LDPE has a density between 0.91 and 0.94 g/cm^3 and a melt transition temperature (T_m) of about 108°C. Because its T_m is only slightly above 100°C, it cannot be used for products that will be exposed to boiling water.

CHEMISTRY IN ACTION

Organic Polymers That Conduct Electricity

The influence of chemical structure on the properties of an organic compound is clearly seen in the electrical conducting properties of certain organic polymers. Most organic polymers are insulators. For example, polytetrafluoroethylene with the repeating unit —CF_2CF_2— or polyvinyl chloride with the repeating unit —CH_2CHCl— have conductivities of 10^{-18} S/cm. On the other end of the scale, the conductivity of copper is almost 10^6 S/cm.

Can organic polymers approach the conductivity of copper? Research carried out over the last 20 years shows that the answer is yes. When acetylene is passed through a solution containing certain transition metal catalysts, it can be polymerized to a shiny film of polyacetylene.

Polyacetylene

By itself, polyacetylene is not a conductor. However, by a process called **doping,** which involves introducing small amounts of electron-donating or electron-accepting compounds, it is possible to produce a polyacetylene that shows a conductivity of 1.5×10^5 S/cm.

The purpose of the doping agent is either to remove electrons from the pi system (*p*-doping) or add electrons to the pi system (*n*-doping). A *p*-doped polyacetylene can be represented as a conjugated polyalkene chain containing positively charged carbons at several points along the chain.

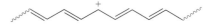

A *p*-doped polyacetylene

We can think of the positive charge as a defect that can move to the left or to the right along the polymer chain, thus giving rise to conductivity.

In crude polyacetylene, the polymer chains are jumbled, pointing in all directions. However, by stretching the film, the chains can be made to line up in a more ordered fashion. The conductivity of doped and oriented polyacetylene chains is greater along the direction of the chain than it is perpendicular to the chain. This result suggests that it is much easier for electrons to travel along a chain than to hop from one chain to the next.

Applications for conducting organic polymers are beginning to be developed. A rechargeable battery with electrodes of *p*-doped and *n*-doped polyacetylene already has been produced. Given the atomic weight of carbon, organic polymer batteries should be lighter than nickel-cadmium or lead-acid batteries. Weight is an important consideration if battery-powered electric cars are ever to be made practical. In addition, many metals used in today's batteries (mercury, nickel, lead) are toxic. If research leads to practical organic batteries, waste disposal problems could be considerably lessened.

Approximately 65% of all low-density polyethylene is used for the manufacture of films. Fabrication of LDPE films is done by a blow-molding technique illustrated in Figure 24.4. A tube of molten LDPE along with a jet of compressed air is forced through an opening and blown into a giant, thin-walled bubble. The film is then cooled and taken up onto a roller. This double-walled film can be slit down the side to give LDPE film, or it can be sealed at points along its length to make LDPE bags. LDPE film is inexpensive, which makes it ideal for trash bags and for packaging such consumer items as baked goods, vegetables, and other produce.

B. Ziegler-Natta Chain-Growth Polymerizations

An alternative method for polymerization of alkenes, which does not involve radicals, was developed by Karl Ziegler of Germany and Giulio Natta of Italy in the 1950s. For their pioneering work, they were awarded the 1963 Nobel Prize in chemistry. The early Ziegler-Natta catalysts were highly active, heterogeneous catalysts composed of a $MgCl_2$ support, a Group 4B transition metal halide, such as $TiCl_4$, and an alkylaluminum compound, such as $Al(CH_2CH_3)_2Cl$. These catalysts bring about polymerization of ethylene and propylene at 1 to 4 atm and at temperatures as low as 60°C. Polymerizations under these conditions do not involve radicals.

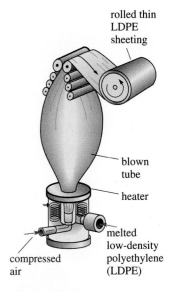

Figure 24.4
Fabrication of an LDPE film.

$$CH_2{=}CH_2 \xrightarrow[MgCl_2]{TiCl_4/Al(CH_2CH_3)_2Cl} \left(\!\!\bigwedge\!\!\right)_n$$

Ethylene   Polyethylene

The active catalyst in a Ziegler-Natta polymerization is thought to be an alkyltitanium compound, which is formed by alkylation of the titanium halide by $Al(CH_2CH_3)_2Cl$ on the surface of a $MgCl_2/TiCl_4$ particle. Once formed, this species repeatedly inserts ethylene into the titanium-carbon bond to yield polyethylene.

Mechanism   Ziegler-Natta Catalysis of Ethylene Polymerization

Step 1:   A titanium-ethyl bond froms.

$MgCl_2/TiCl_4$ particle

Ti—Cl + Cl—Al⟨ ⟶ Ti⌒ + Cl—Al⟨Cl

Diethylaluminum
chloride

Step 2:   Ethylene is inserted into the titanium-carbon bond.

Ti⌒ + $CH_2{=}CH_2$ ⟶ Ti⌒⌒

Polyethylene films are produced by extruding the molten plastic through a ring-like gap and inflating the film into a balloon. *(The Stock Market)*

Over 2.5×10^{11} kg of polyethylene are produced worldwide every year using optimized Ziegler-Natta catalysts, and large-scale reactors can yield up to 1.25×10^5 kg of polyethylene per hour. Production of polymer at this scale is partly due to the mild conditions required for a Ziegler-Natta polymerization and the fact that the polymer obtained has substantially different physical and mechanical properties from those obtained by radical polymerization. Polyethylene from Ziegler-Natta systems, termed high-density polyethylene (HDPE), has a higher density (0.96 g/cm^3) and T_m ($133°C$) than low-density polyethylene, is three to ten times stronger, and is opaque rather than transparent. This added strength and opacity is due to a much lower degree of chain branching and the resulting higher degree of crystallinity of HDPE compared with LDPE.

Approximately 45% of all HDPE used in the United States is blow molded. In blow molding, a short length of HDPE tubing is placed in an open die [Figure 24.5(a)] and the die is closed, sealing the bottom of the tube. Compressed air is then forced into the hot polyethylene/die assembly, and the tubing is literally blown up to take the shape of the mold [Figure 24.5(b)]. After cooling, the die is opened [Figure 24.5(c)], and there is the container!

Even greater improvements in properties of HDPE can be realized through special processing techniques. In the melt state, HDPE chains adopt random coiled conformations similar to those of cooked spaghetti. Engineers have developed special extrusion techniques that force the individual polymer chains of HDPE to uncoil and adopt an extended linear conformation. These extended chains then align with one another to form highly crystalline materials. HDPE processed in this fashion is stiffer

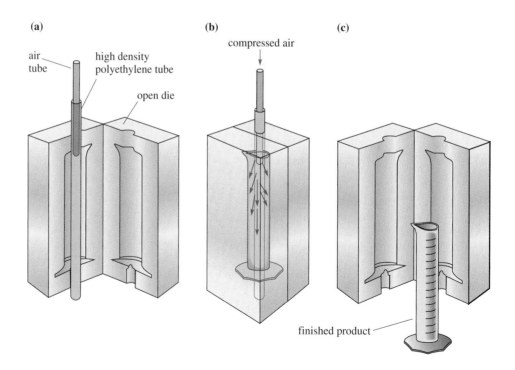

(a)

air tube

high density polyethylene tube

open die

(b)

compressed air

(c)

finished product

Figure 24.5
Blow molding of an HDPE container.

than steel and has approximately four times the tensile strength of steel! Because the density of polyethylene (≈ 1.0 g/cm^3) is considerably less than that of steel (8.0 g/cm^3), these comparisons of strength and stiffness are even more favorable if they are made on a weight basis.

In recent years, there have been several important advances made in catalysts used in Ziegler-Natta type polymerizations. One of the most important has been the discovery of soluble complexes that catalyze the polymerization of ethylene and propylene at extraordinary rates. Because these new homogeneous catalysts are substantially different in structure from the early Ziegler-Natta systems, these polymerizations are referred to as **coordination polymerizations.** Catalysts for coordination polymerizations are frequently formed by allowing bis(cyclopentadienyl)dimethylzirconium, [Cp$_2$Zr(CH$_3$)$_2$], to react with methaluminoxane (MAO). MAO is a complex mixture of methylaluminum oxide oligomers, [—(CH$_3$)AlO—)$_n$], formed by allowing trimethylaluminum to react with small amounts of water. It is thought that MAO activates the zirconium by abstracting a methyl anion to form a zirconium cation that is the active polymerization catalyst.

Mechanism   Homogeneous Catalysis for Ziegler-Natta
Coordination Polymerization

Step 1:   The zirconium catalyst is activated.

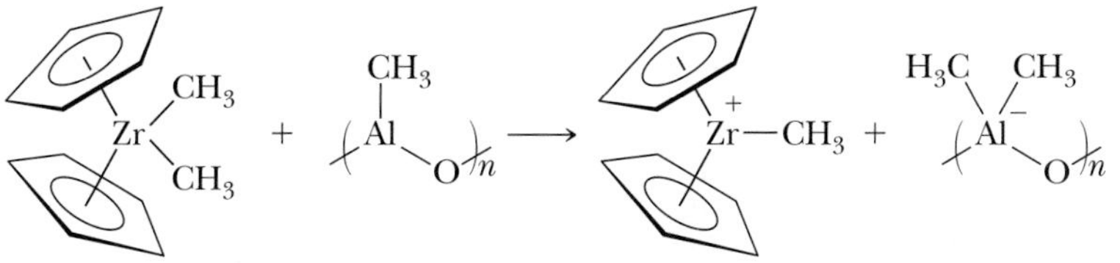

| Bis(cyclopentadienyl)-dimethylzirconium | MAO | A zirconium cation (the active form of the catalyst) |

Step 2:   Ethylene is inserted into the zirconium-carbon bond.

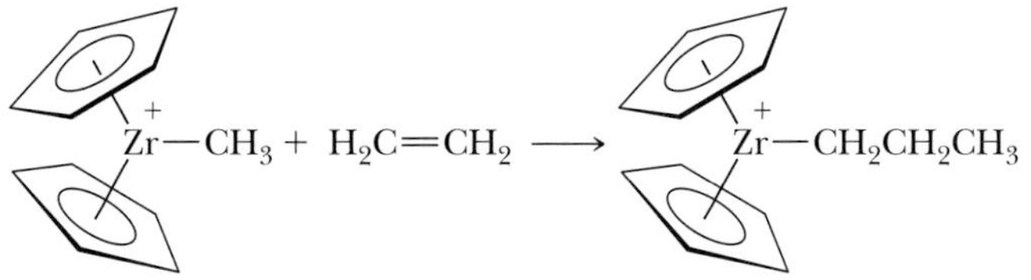

Some of these coordination-polymerization catalysts polymerize up to 20,000 ethylene monomer units per second, a rate reached only by enzyme-catalyzed biological reactions. Another important characteristic of these catalysts is that they show high reactivity toward 1-alkenes, allowing the formation of copolymers, such as that of ethylene and 1-hexene.

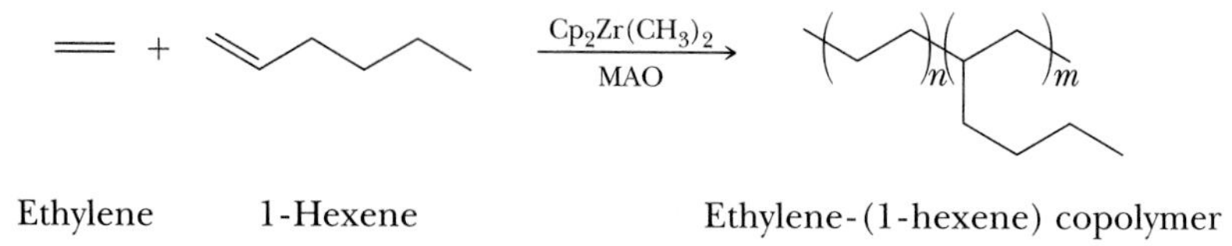

Ethylene   1-Hexene   Ethylene-(1-hexene) copolymer

Copolymers of this type with these moderate length branches (C_4, C_6, and so on) are called linear low-density polyethylene, or LLDPE. These are useful materials because they have many of the properties of LDPE made from radical reactions but are formed at the substantially milder conditions associated with Ziegler-Natta polymerizations.

C.   Stereochemistry and Polymers

Thus far, we have written the formula of a substituted ethylene polymer in the following manner and have not been concerned with the configuration of each stereocenter along the chain.

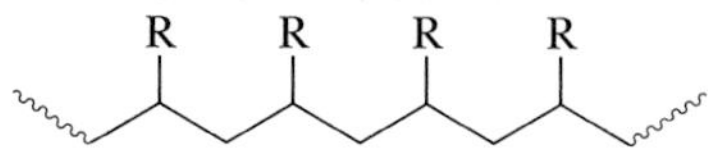

Nevertheless, the relative configurations of these stereocenters are important in determining the properties of a polymer. Polymers with identical configurations at all stereocenters along the chain are called **isotactic polymers.** Those with alternating configurations are called **syndiotactic polymers,** and those with completely random configurations are called **atactic polymers** (Figure 24.6).

In general, the more stereoregular the centers are, that is, the more highly isotactic or highly syndiotactic the polymer is, the more crystalline it is. A random placement of the substituents, such as in atactic materials, results in a polymer that cannot pack well and is usually highly amorphous. Atactic polystyrene, for example, is an amorphous glass, whereas isotactic polystyrene is a crystalline fiber-forming polymer with a high melt transition. The control over the relative configuration, or tacticity, along a polymer backbone is, therefore, an area of considerable interest in modern polymer synthesis.

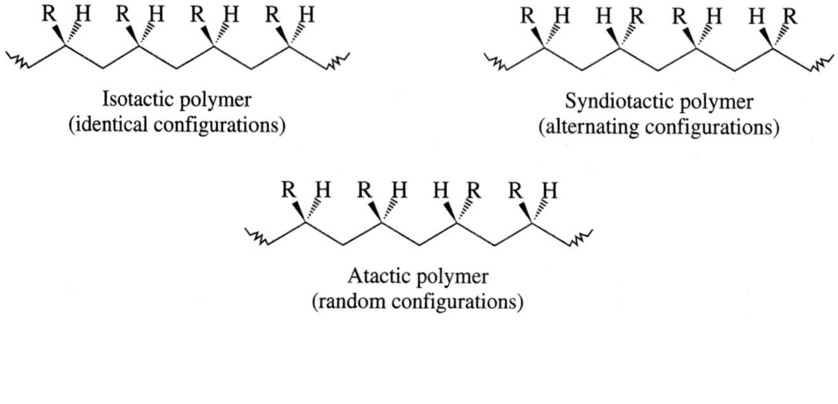

Isotactic polymer   A polymer with identical configurations (either all *R* or all *S*) at all stereocenters along its chain, as for example isotactic polypropylene.

Syndiotactic polymer   A polymer with alternating *R* and *S* configurations at the stereocenters along its chain, as for example syndiotactic polypropylene.

Atactic polymer   A polymer with completely random configurations at the stereocenters along its chain, as for example atactic polypropylene.

Isotactic polymer
(identical configurations)

Syndiotactic polymer
(alternating configurations)

Atactic polymer
(random configurations)

Figure 24.6
Relative configurations of stereocenters in polymers with different tacticities.

Table 24.2   Alkenes Polymerized by Anionic and Cationic Chain-Growth Mechanisms

Anionic polymerizations are most common for monomers subtituted with electron-withdrawing groups.

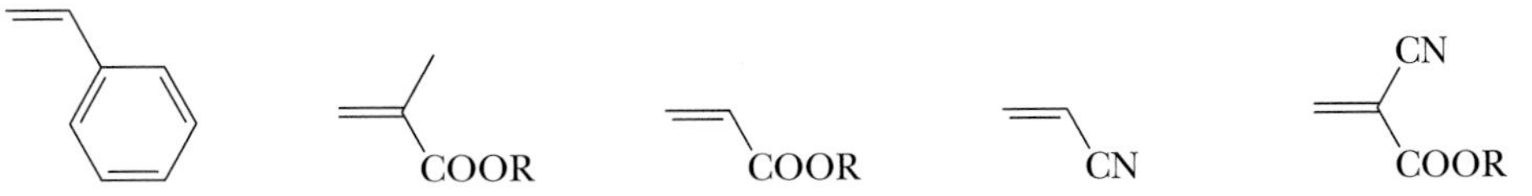

| Styrene | Alkyl methacrylates | Alkyl acrylates | Acrylonitrile | Alkyl cyanoacrylates |

Cationic polymerizations are most common for monomers substituted with electron-donating groups.

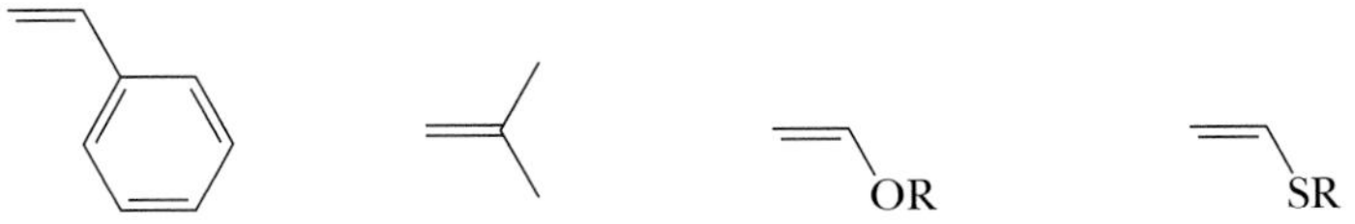

| Styrene | Isobutylene | Vinyl ethers | Vinyl thioethers |

D.   Ionic Chain-Growth Polymerizations

Chain-growth polymers can also be synthesized using reactions that rely on either anionic or cationic species in the propagation steps. The choice of ionic procedure depends greatly on the electronic nature of the monomers to be polymerized. Vinyl monomers with electron-withdrawing groups, which stabilize carbanions, are used in anionic polymerizations, whereas vinyl monomers with electron-donating groups, which stabilize cations, are used in cationic polymerizations (Table 24.2).

Styrene is conspicuous among the monomers given in Table 24.2 because it can be polymerized using either anionic or cationic techniques as well as radical techniques. This characteristic particular to styrene is due to the fact that the phenyl group can stabilize cationic, anionic, and radical benzylic intermediates.

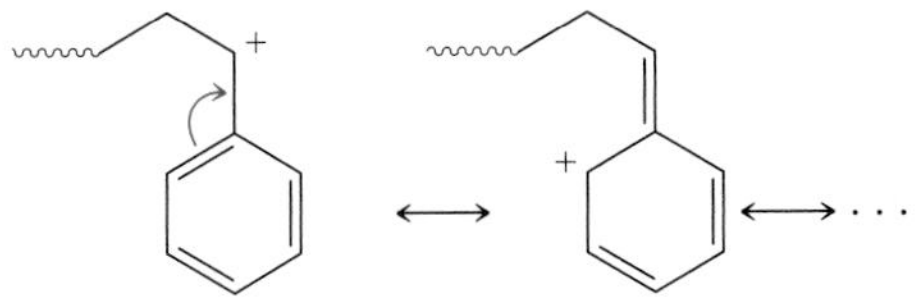

Resonance stabilization of a benzylic cation endgroup

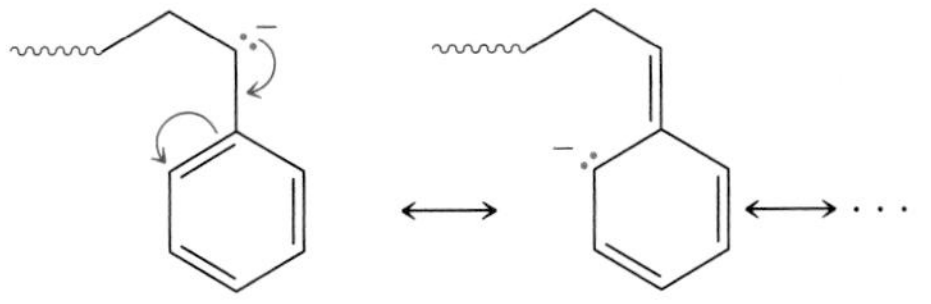

Resonance stabilization of a benzylic anion endgroup

Anionic Polymerizations

Anionic polymerizations can be initiated by addition of a nucleophile to an activated alkene. The most common nucleophiles used for this purpose are metal alkyls, such as methyl or *sec*-butyllithium. The newly formed carbanion then acts as a nucleophile and adds to another monomer unit, and the propagation continues.

Mechanism   Initiation of Anionic Polymerization of Alkenes

Step 1: Polymerization is initiated by addition of nucleophile, shown here as a carbanion derived from an organolithium compound, to an activated carbon-carbon double bond to give a carbanion.

Step 2: This carbanion adds to the activated double bond of a second alkene molecule to give a dimer.

Step 3: Chain growth continues to give a polymer.

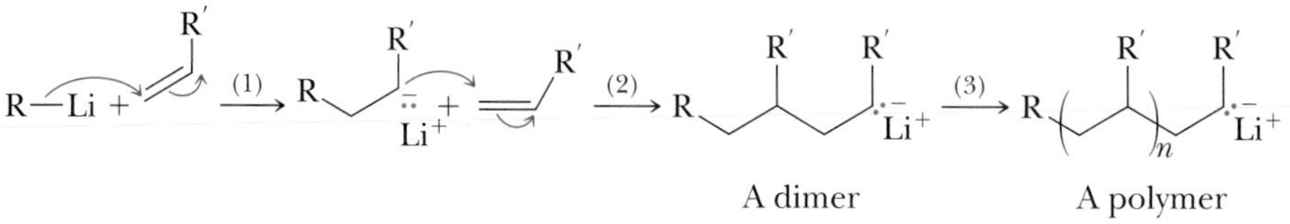

A dimer   A polymer

An alternative method for the initiation of anionic polymerizations involves a one-electron reduction of the monomer by lithium or sodium to form a radical anion. The radical anion thus formed is either further reduced to form a dianion or dimerizes to form a dimer dianion.

Mechanism   Initiation of Anionic Polymerization of Butadiene

Step 1: A one-electron reduction of the diene by lithium metal gives a radical anion.

Step 2: One-electron reduction of this radical anion gives a monomer dianion.

Step 3: Alternatively, radical coupling gives a dimer dianion.

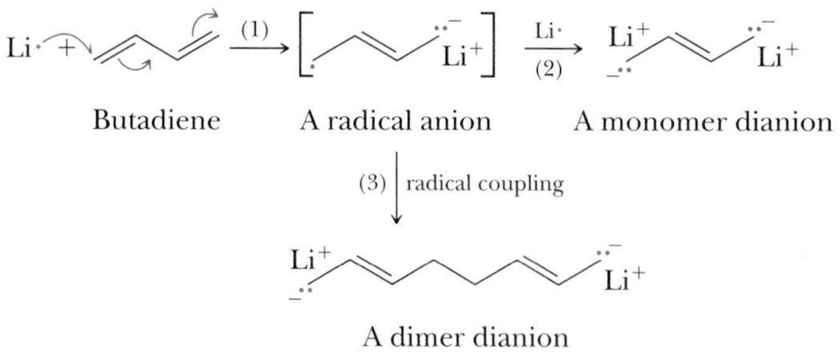

Butadiene   A radical anion   A monomer dianion

(3) radical coupling

A dimer dianion

In either case, a single initiator can now propagate chains from both ends, by virtue of its two active endgroup carbanions. These reactions are heterogeneous and involve transfer of the electron from the surface of the metal. To improve the efficiency of this process, soluble reducing agents such as sodium naphthalide are used. Sodium undergoes electron-transfer reactions with extended aromatic compounds, such as naphthalene, to form soluble radical anions.

Naphthalene   Sodium naphthalide
                       (a radical anion)

The naphthalide radical anion is a powerful reducing agent. For example, styrene undergoes a one-electron reduction to form the styryl radical anion, which couples to form a dianion. The latter then propagates polymerization at both ends, growing chains in two directions simultaneously.

Styrene   A styryl radical anion   A distyryl dianion   Polystyrene

The propagation characteristics of anionic polymerizations are similar to those of radical polymerizations, but with the important difference that many of the chain-transfer and termination reactions that plague radical processes are absent. Furthermore, because the propagating chain ends carry the same charge, bimolecular coupling and disproportionation reactions are also averted. An interesting set of circumstances arises when chain-transfer and chain-termination steps are no longer significant. Under these conditions, polymer chains are initiated and continue to grow until either all the monomer is consumed or some external agent is added to terminate the chains. Polymerizations of this type are called **living polymerizations** because they will restart if more monomer is added after it is initially consumed.

The absence of chain-transfer and chain-termination steps in living polymerizations has far-reaching consequences. One of the most visible of these is in the area of molecular-weight control. The molecular weight of a polymer originating from living polymerizations is determined directly by the monomer-to-initiator ratio. It is, therefore, relatively easy to obtain polymers of a well-defined size simply by controlling the stoichiometry of the reagents. In contrast, the average sizes of polymer chains formed from nonliving, chain-growth processes (radical, Ziegler-Natta, and so on) vary from system to system and are determined by the ratio of the rate of propagation to the rate of termination. In most cases, precise control over the molecular weight of the product obtained in nonliving systems is not possible because it is very difficult to change one of the rates involved without affecting the other.

Living polymer A polymer chain that continues to grow without chain-termination steps until either all the monomer is consumed or some external agent is added to terminate the chain. The polymer chains will continue to grow if more monomer is added.

After consumption of the monomer under living, anionic conditions, electrophilic terminating agents can be added to functionalize the chain ends. Examples of terminating reagents include CO_2 and ethylene oxide, which, after protonation, form carboxylic acid and alcohol-terminated chains, respectively.

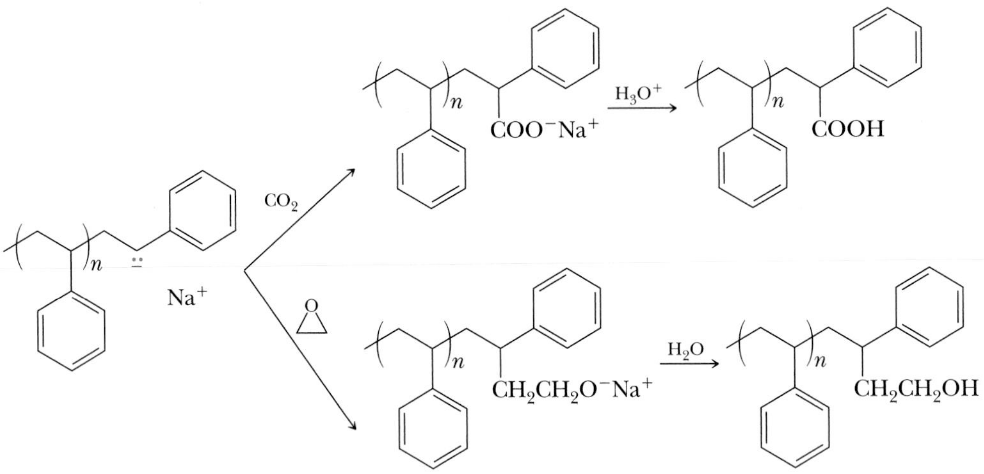

Telechelic polymer A polymer in which its growing chains are terminated by formation of new functional groups at both ends of its chains. These new functional groups are introduced by adding reagents, such as CO_2 or ethylene oxide, to the growing chains.

In a similar fashion, polymer chains with functional groups at both ends, called **telechelic polymers,** can be prepared by addition of these same reagents (CO_2, ethylene oxide, and so on) to solutions of chains with two active ends initiated by sodium naphthalide.

Example 24.3

Show how to prepare polybutadiene that is terminated at both ends with carboxylate groups.

Solution

Form a growing chain with two active endgroups by treatment of butadiene with two moles of lithium metal to form a dianion followed by addition of monomer units and formation of a living polymer. Cap the active endgroups with a carboxylate group by treatment of the living polymer with carbon dioxide.

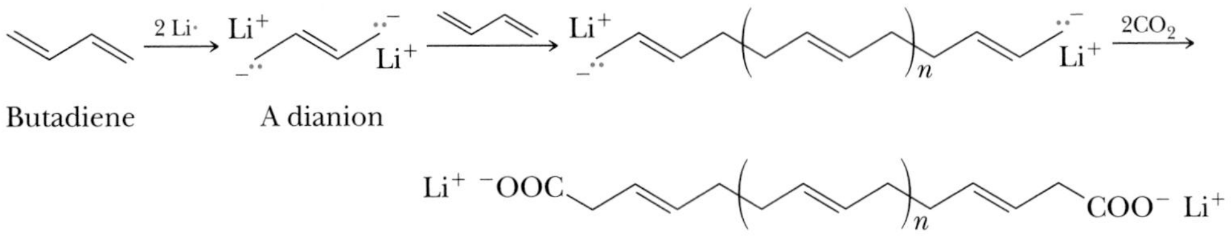

Butadiene   A dianion

Problem 24.3

Show how to prepare polybutadiene that is terminated at both ends with primary alcohol groups.

The Chemistry of Superglue

Used in everything from model planes to passenger planes, Superglue is one of the best-known modern glues. Indeed, anyone who has unwittingly glued his or her fingers together while using the product can acknowledge its remarkable (and insidious) adhesive properties. The curing process that facilitates these properties is a chain polymerization reaction. The ingredient that gives Superglue its adhesive ability is methyl cyanoacrylate. This compound is just one member of a larger family of cyanoacrylates with the following general structure.

$$H_2C=C \begin{cases} C\equiv N \\ C=O \\ \quad\; O \\ \qquad R \end{cases}$$

Contrary to popular understanding, Superglue does not "air dry." In fact, cyanoacrylates cure (convert from liquid to solid) in the presence of weak nucleophiles such as water. Under normal conditions, a thin layer of water is present on almost all surfaces. This accounts for many unintended adhesions involving appendages and/or expensive tools! The curing process involves an anionic chain polymerization reaction, which occurs as follows.

In the chain-initiating step, a weak nucleophile donates an electron pair to a cyanoacrylate monomer. The CH_2 group is highly electropositive as a result of the electron withdrawal by the cyano and ester groups, which makes this Michael reaction very easy. The addition produces a new anion, which is also a weak nucleophile and adds to another monomer, ultimately creating the powerfully adhesive polymer chains of cured Superglue.

Superglue has a rich history. It begins during World War II, when cyanoacrylates were discovered in an attempt to produce optically clear gun sights. Tests with Superglue's main ingredient failed for obvious reasons: the compound stuck to all instruments. Approximately ten years later, researchers at Eastman Kodak rediscovered cyanoacrylates, and Kodak first marketed them in 1958. It is said that researchers discovered superglue's properties when they attempted to take the refractive index of the monomer. A refractometer, the instrument used for this purpose, has two prisms that come together on the liquid whose refractive index is to be determined. The prisms of the instrument were permanently glued together, and a new product was born!

Superglue has been improved since that time with stabilizers and other supplementary ingredients that reinforce its adhesive strength. In recent years, that same adhesive strength has captured attention in new fields. Medical grade Superglues such as 2-octylcyanoacrylate and 2-butylcyanoacrylate are used successfully as sutures in laceration repair. They have also proven effective in skin, bone, and cartilage grafts. Dentists use cyanoacrylates in dental cements and fillings, and paleontologists use them to reorganize fragile fossil specimens. Superglue has many fascinating applications, all facilitated by the anionic chain polymerization reaction.

Based on a Chem30H honors paper by Carrie Brubaker, UCLA.

Cationic Polymerizations

Only alkenes with electron-donating substituents, such as alkyl, aryl, ether, thioether, and amino groups, undergo useful cationic polymerizations. The two most common methods of generating cationic initiators are (1) the reaction of a strong protic acid with an organic monomer and (2) the abstraction of a halide from the organic initiator by a Lewis acid. Cationic chain-growth polymerizations are generally effective only for monomers yielding relatively stable carbocations, that is monomers that form either 3° carbocations or cations stabilized by electron-donating groups, such as ether, thioether, or amino groups.

Initiation by protonation of an alkene requires the use of a strong acid with a nonnucleophilic anion to avoid 1,2-addition across the alkene double bond. Suitable acids with nonnucleophilic anions include HF/AsF_5 and HF/BF_3. In the following general equation, initiation is by proton transfer from $H^+BF_4^-$ to the alkene to form a tertiary carbocation, which then continues the cationic chain growth polymerization.

Mechanism   **Initiation of Cationic Polymerization of an Alkene by HF·BF$_3$**

Step 1:   Proton transfer from the HF·BF$_3$ complex to the alkene gives a carbocation.

Step 2:   Propagation continues the polymerization.

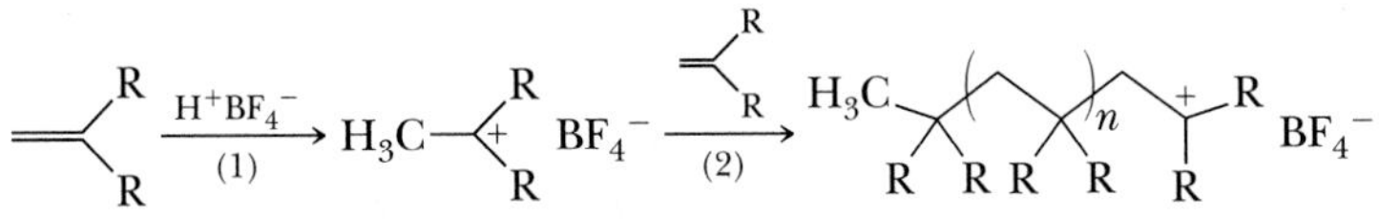

The second common method for generating carbocations involves the reaction between an alkyl halide and a Lewis acid, such as BF_3, $SnCl_4$, $AlCl_3$, $Al(CH_3)_2Cl$, and $ZnCl_2$. When a trace of water is present, the mechanism of initiation using some Lewis acids is thought to involve protonation of the alkene.

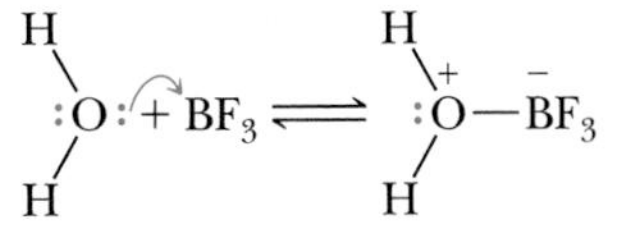

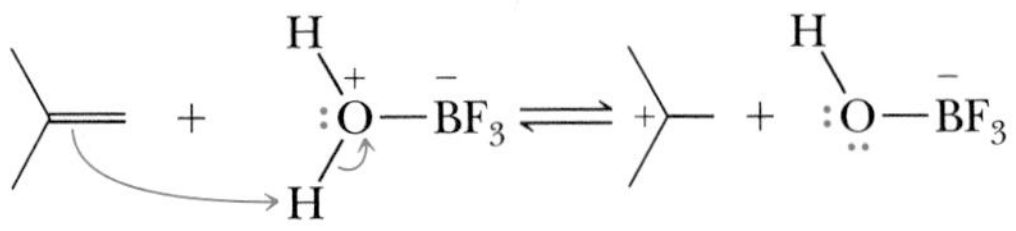

2-Methylpropene
(Isobutylene)

In the absence of water, the Lewis acid removes a halide ion from the alkyl halide to form the initiating carbocation.

Mechanism Initiation of Cationic Polymerization of an Alkene by a Lewis Acid

Step 1: Reaction of the chloroalkane (a Lewis base) with tin(IV) chloride (a Lewis acid) gives a carbocation from which polymerization then proceeds.

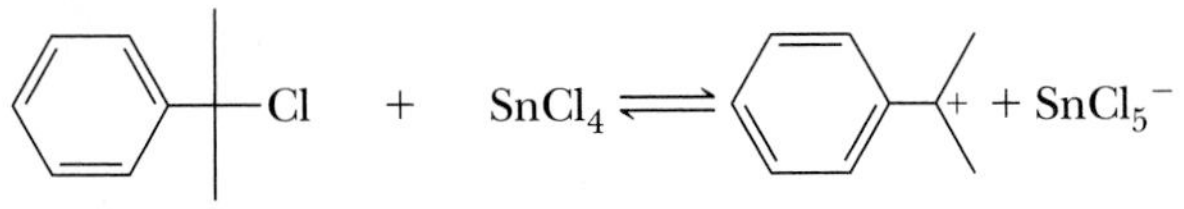

2-Chloro-2-phenylpropane

The polymerization of alkenes then propagates by the electrophilic attack of the carbocation on the double bond of the alkene monomer. The regiochemistry of the addition is determined by the formation of the more stable (the more highly substituted) carbocation.

Example 24.4

Write a mechanism for the polymerization of 2-methylpropene (isobutylene) initiated by treatment of 2-chloro-2-phenylpropane with $SnCl_4$. Label the initiation, propagation, and termination steps.

Solution

Chain initiation: Cations form from nonionic materials.

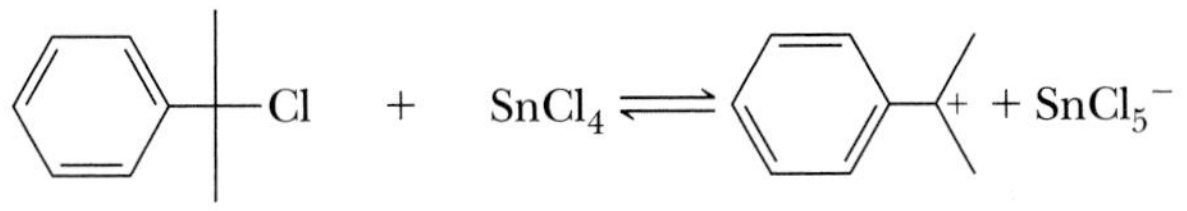

2-Chloro-2-phenylpropane

Chain propagation: A cation and a molecule react to give a new cation.

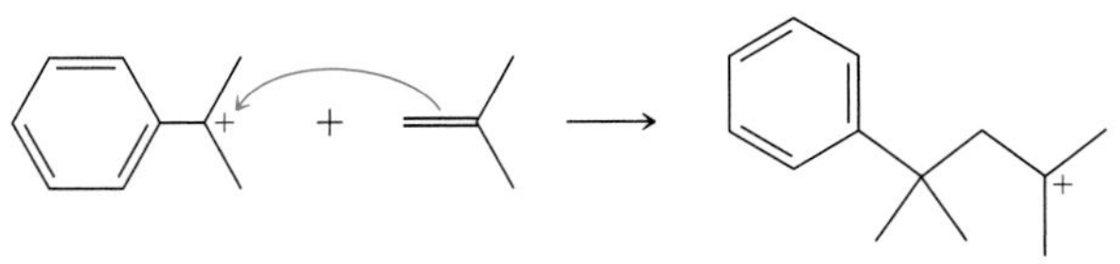

2-Methylpropene

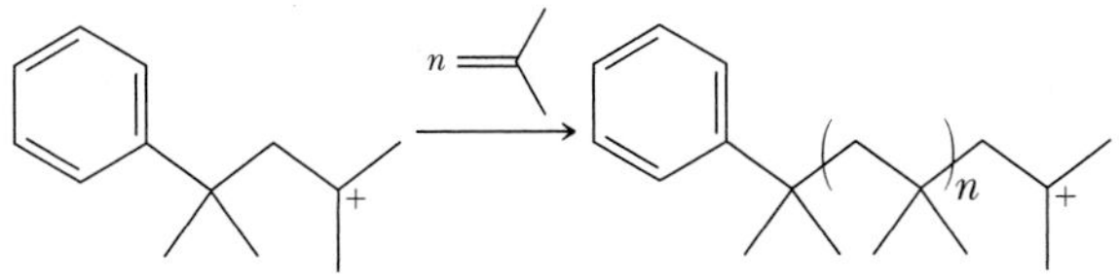

Chain termination: Cations are destroyed.

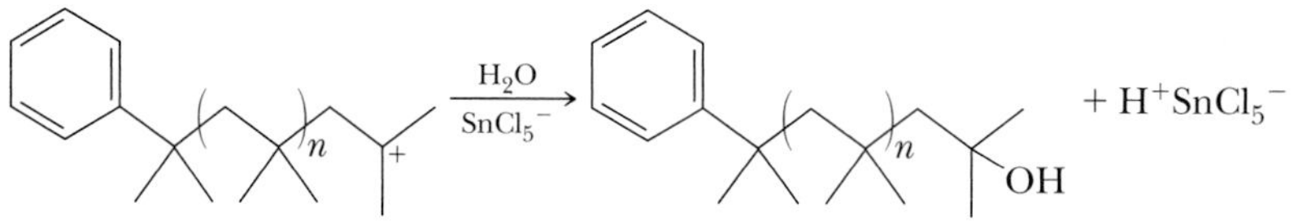

Problem 24.4

Write a mechanism for the polymerization of methyl vinyl ether initiated by 2-chloro-2-phenylpropane and $SnCl_4$. Label the initiation, propagation, and termination steps.

E. Ring-Opening Metathesis Polymerizations

During early investigations into the polymerizations of cycloalkenes by transition metal catalysts such as those used in Ziegler-Natta polymerizations, polymers of unexpected structures were formed that contained the same number of double bonds as originally present in the monomers. This process is illustrated by the polymerization of bicyclo[2.2.1]-2-heptene (norbornene).

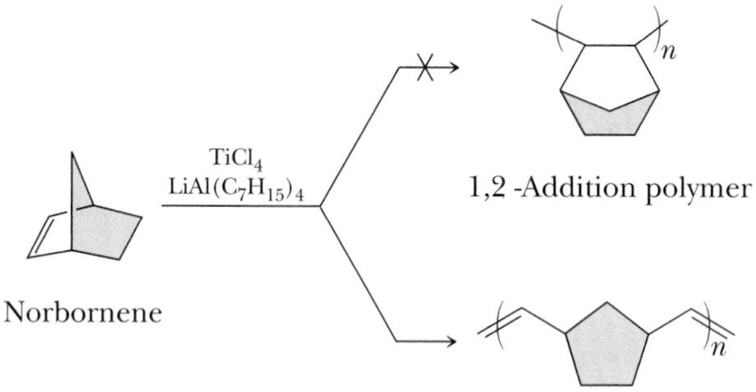

If reaction had proceeded in the same manner as Ziegler-Natta polymerization of ethylene and substituted ethylenes (Section 24.6B), a 1,2-addition polymer would have been formed. What is formed, however, is an unsaturated polymer in which the number of double bonds in the polymer is the same as that in the monomers polymerized. This process is called **ring-opening metathesis polymerization**, or ROMP, after the related process involving reaction of acyclic alkenes and nucleophilic carbenes catalysts described in Section 15.4E.

The fact that ROMP polymers are unsaturated requires that this polymerization proceed by a mechanism substantially different from that involved in polymerization of ethylene and substituted ethylenes by the same catalyst mixtures. Following lengthy and detailed studies, chemists discovered that ROMP polymerizations involve the same metallacyclobutane species as in ring-closing alkene metathesis reactions. The intermediate metallacyclobutane derivative undergoes a ring-opening reaction to give a new substituted carbene. Repetition of these steps leads to the formation of the unsaturated polymer as illustrated here by ROMP polymerization of cyclopentene.

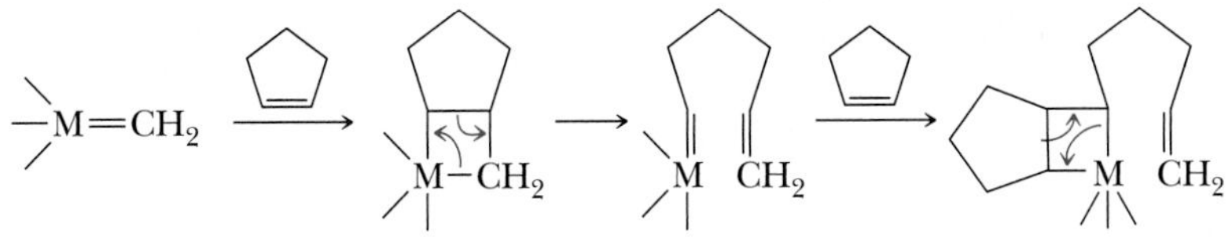

A metal carbene

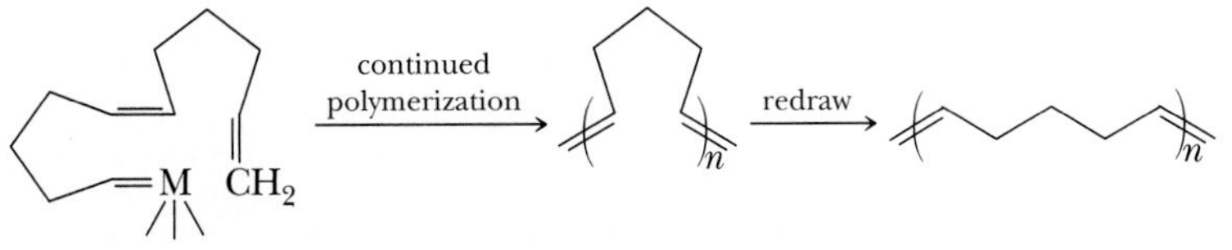

ROMP polymer from cyclopentene

All steps in ROMP polymerization are reversible, and the reaction is driven in the forward direction by the release of ring strain that accompanies the opening of the ring. The reactivity of the following cycloalkenes toward ROMP decreases in this order.

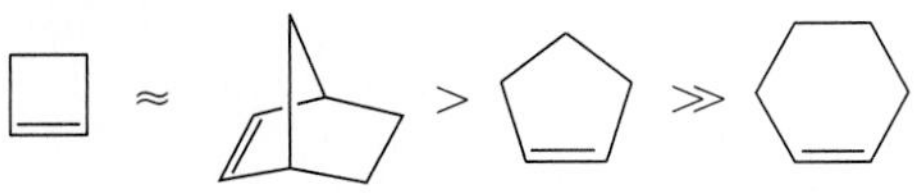

| Ring strain [kJ (kcal)/mol] | 125(29.8) | 113(27) | 24.7(5.9) | 5.9(1.4) |

ROMP reactions are unique in that all the unsaturation present in the monomers is conserved in the polymeric product. This feature makes ROMP techniques especially attractive for the preparation of highly unsaturated, fully conjugated materials. One example is the direct preparation of polyacetylene by the ROMP technique through one of the double bonds of cyclooctatetraene. For further discussion of polyacetylene, see the Chemistry in Action box "Organic Polymers That Conduct Electricity."

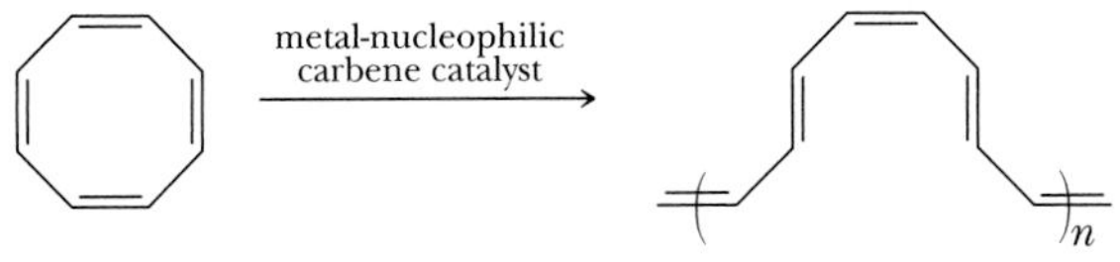

Cyclooctatetraene   Polyacetylene

An important polymer in electro-optical applications is poly(phenylene vinylene) (PPV), which has alternating phenyl and vinyl groups. One of the routes to this polymer starts with a substituted bicyclo[2.2.2]octadiene, which is polymerized using ROMP techniques to form a soluble, processable polymer. Heating the processed polymer results in elimination of two equivalents of acetic acid, which aromatizes the six-membered ring and completes the conjugation.

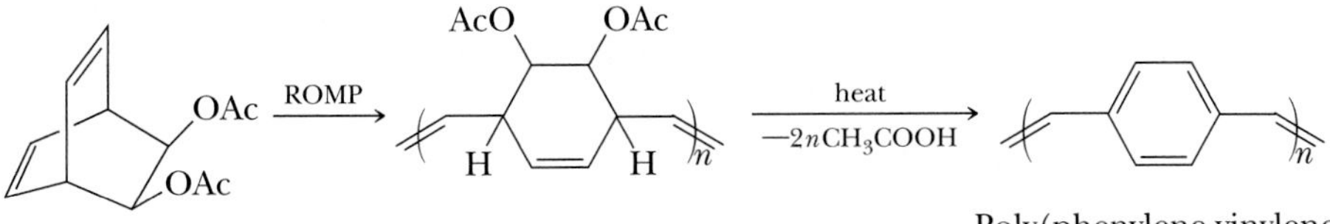

Poly(phenylene vinylene)

C H E M I S T R Y   I N   A C T I O N

Recycling of Plastics

Polymers, in the form of plastics, are materials upon which our society is incredibly dependent. Durable and lightweight, plastics are probably the most versatile synthetic materials in existence; in fact, their current production in the United States exceeds that of steel. Plastics have come under criticism, however, for their role in the garbage crisis. They comprise 21% of the volume and 8% of the weight of solid wastes, most of which is derived from disposable packaging and wrapping. Of the 2.5×10^7 kg of thermoplastic materials produced in 1993 in America, for instance, less than 2% was recycled.

Why aren't more plastics being recycled? The durability and chemical inertness of most plastics make them ideally suited for reuse. The answer to this question has more to do with economics and consumer habits than with technological obstacles. Because curbside pickup and centralized drop-off stations for recyclables are just now becoming common, the amount of used material available for reprocessing has traditionally been small. This limitation, combined with the need for an additional sorting and separation step, rendered the use of recycled plastics in manufacturing expensive compared with virgin materials. Until recently, consumers perceived products made from "used" materials as being inferior to new ones, so the market for recycled products has not been large. However, the increase in environmental concerns over the last few years has resulted in a greater demand for recycled products. As manufacturers adapt to satisfy this new market, plastic recycling will eventually catch up with the recycling of other materials, such as glass and aluminum.

Six types of plastics are commonly used for packaging applications. In 1988, manufacturers adopted recycling code numbers developed by the Society of the Plastics Industry. Because the plastics recycling industry still is not fully developed, only polyethylene terephthalate (PET) and high-density polyethylene are currently being recycled in large quantities, al-

These students are wearing jackets make from recycled PET soda bottles. (Charles D. Winters)

though outlets for the other plastics are being developed. Low-density polyethylene, which accounts for about 40% of plastic trash, has been slow in finding acceptance with recyclers. Facilities for the reprocessing of polyvinyl chloride, polypropylene, and polystyrene exist, but are still rare.

The process for the recycling of most plastics is simple, with separation of the desired plastics from other contaminants the most labor-intensive step. PET soft drink bottles, for example, usually have a paper label, adhesive, and an aluminum cap that must be removed before the PET can be reused. The recycling process begins with hand or machine sorting, after which the bottles are shredded into small chips. An air cyclone removes paper and other lightweight materials, and any remaining labels and adhesives are eliminated with a detergent wash. The PET

Recycling Code	Polymer	Common Uses	Uses of Recycled Polymer
1 PET	Poly(ethylene terephthalate)	Soft drink bottles, household chemical bottles, films, textile fibers	Soft drink bottles, household chemical bottles, films, textile fibers
2 HDPE	High-density polyethylene	Milk and water jugs, grocery bags, bottles	Bottles, molded containers
3 PVC	Poly(vinyl chloride)	Shampoo bottles, pipes, shower curtains, vinyl siding, wire insulation, floor tiles, credit cards	Plastic floor mats
4 LDPE	Low-density polyethylene	Shrink wrap, trash and grocery bags, sandwich bags, squeeze bottles	Trash bags and grocery bags
5 PP	Polypropylene	Plastic lids, clothing fibers, bottle caps, toys, diaper linings	Mixed plastic components
6 PS	Polystyrene	Styrofoam cups, egg cartons, disposable utensils, packaging materials, appliances	Molded items such as cafeteria trays, rulers, Frisbees, trash cans, videocasettes
7	All other plastics and mixed plastics	Various	Plastic lumber, playground equipment, road reflectors

chips are then dried, and aluminum, the final contaminant, is removed electrostatically. The PET produced by this method is 99.9% free of contaminants and sells for about half the price of the virgin material. Unfortunately, plastics with similar densities cannot be separated with this technology, nor can plastics composed of several polymers be broken down into pure components. However, recycled mixed plastics can be molded into plastic lumber that is strong, durable, and graffiti-resistant.

An alternative to this process, which uses only physical methods of purification, is chemical recycling. Eastman Kodak salvages large amounts of its PET film scrap by a transesterification reaction. The scrap is treated with methanol in the presence of an acid catalyst to give ethylene glycol and dimethyl terephthalate.

Poly(ethylene terephthalate)

Ethylene glycol   Dimethyl terephthalate

These monomers are purified by distillation or recrystallization and used as feed stocks for the production of more PET film.

Summary

Polymerization is the process of joining together many small **monomers** into large, high-molecular-weight **polymers** (Section 24.1). The properties of polymeric materials depend on the structure of the repeat unit, molecular weight (Section 24.3), chain architecture, the presence or absence of crystalline phases (Section 24.4), tacticity (Section 24.6C), interchain order and packing, and the materials' morphology. **Step-growth polymerizations** involve the stepwise reaction of difunctional monomers (Section 24.5). Further structural variations, such as cross links and branches, can be introduced into the resulting polymer by the addition of multifunctional monomers to the reaction mixture. The formation of high-molecular-weight polymers from step-growth processes requires the use of reactions that proceed with very high yields. Important commercial polymers synthesized through step-growth processes include polyamides, polyesters, polycarbonates, polyurethanes, and epoxy resins.

Chain-growth polymerization proceeds by the sequential addition of monomer units to an active chain end (Section 24.6). Important mechanisms for chain-growth polymerizations include radical, anionic, cationic, and transition metal-mediated processes. Chain-growth polymerizations involve initiation, propagation, and termination steps. **Chain-transfer steps** terminate one chain but simultaneously initiate the growth of another. Chain polymerizations that proceed without chain-transfer or chain-termination steps are called **living polymerizations.** Among the transition metal-mediated polymerizations, the Ziegler-Natta polymerizations of ethylene and propylene are the most significant (Section 24.6B). These reactions proceed with high specificity to yield polymers that are stereoregular and highly linear. This regularity leads to highly crystalline polymers. When the chains are elongated and oriented through special processing procedures, a polymer with strength and stiffness greater than steel can be obtained. **Ring-opening metathesis polymerization (ROMP)** of strained cycloalkenes and bicycloalkenes is catalyzed by transition-metal nucleophilic-carbene complexes (Section 24.6E). The products of ROMP are unusual in that all the unsaturation present in the monomer is conserved in the resulting polymer. Hence, ROMP reactions are ideal for the formation of unsaturated or fully conjugated materials.

Key Reactions

1. Step-Growth Polymerization of a Dicarboxylic Acid and a Diamine Gives a Polyamide (Section 24.5A)

$$HOC-M-COH + H_2N-M'-NH_2 \xrightarrow{heat} \left[M-N-M'-N \right]_n + 2nH_2O$$

2. Step-Growth Polymerization of a Dicarboxylic Acid and a Diol Gives a Polyester (Section 24.5B)

$$HOC-M-COH + HO-M'-OH \xrightarrow{\substack{acid \\ catalyst}} \left[M-O-M'-O \right]_n + 2nH_2O$$

3. Step-Growth Polymerization of a Diacyl Chloride and a Diol Gives a Polycarbonate (Section 24.5C)

$$Cl-C-Cl + HO-M-OH \longrightarrow \left[O-C-O-M \right]_n + 2nHCl$$

4. Step-Growth Polymerization of a Diisocyanate and a Diol Gives a Polyurethane (Section 24.5D)

$$OCN-M-NCO + HO-M'-OH \longrightarrow$$

5. Step-Growth Polymerization of a Diepoxide and a Diamine Gives an Epoxy Resin (Section 24.5E)

6. Radical Chain-Growth Polymerization of Substituted Ethylenes (Section 24.6A)

7. Titanium-Mediated (Ziegler-Natta) Chain-Growth Polymerization of Ethylene and Substituted Ethylenes (Section 24.6B)

8. Anionic Chain-Growth Polymerization of Substituted Ethylenes (Section 24.6D)

9. Cationic Chain-Growth Polymerization of Substituted Ethylenes (Section 24.6D)

10. Ring-Opening Metathesis Polymerization (ROMP) (Section 24.6E)

Problems

Structure and Nomenclature

24.5 Name the following polymers.

(a) 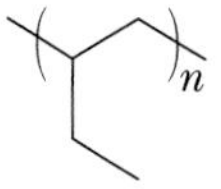   (b) 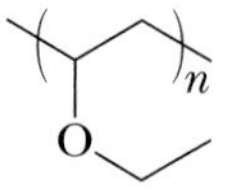   (c) 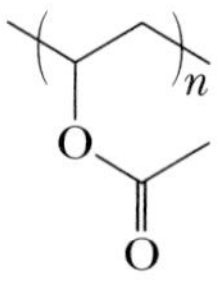   (d)

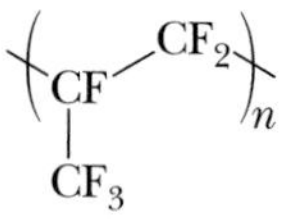

(e) 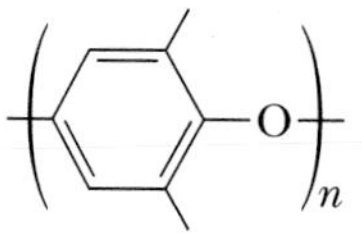   (f) 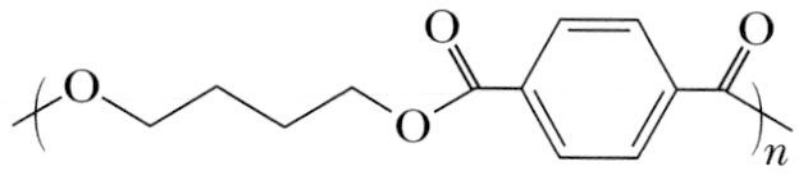   (g)

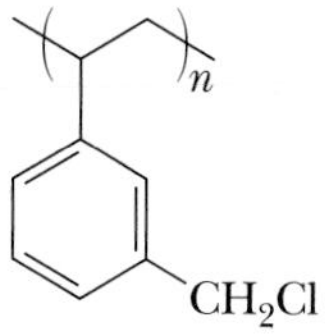

(h)

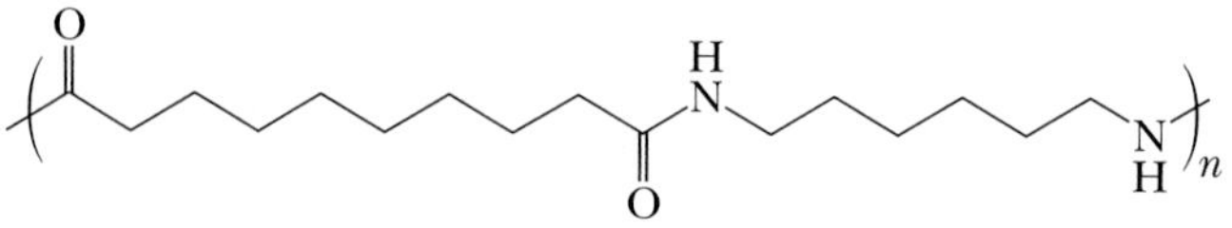

24.6 Draw the structure(s) of the monomer(s) used to make each polymer in Problem 24.5.

Step-Growth Polymerizations

24.7 Draw a structure of the polymer formed in the following reactions.

(a) 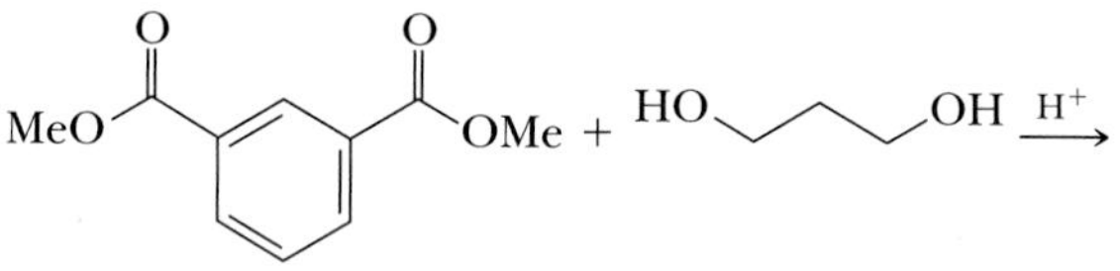 MeO [structure] OMe + HO [structure] OH $\xrightarrow{H^+}$

(b) 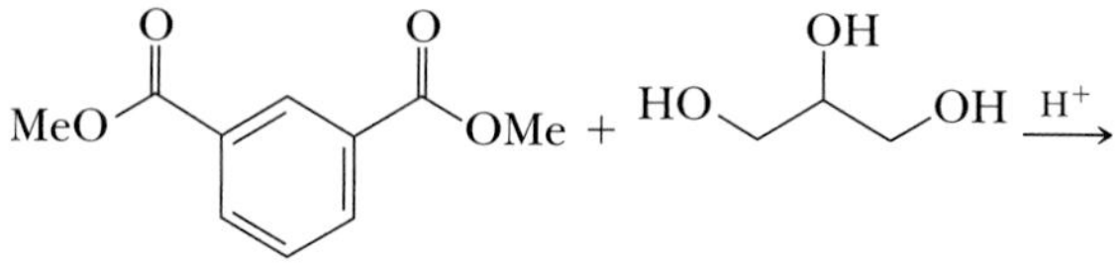 MeO [structure] OMe + HO [structure with OH] OH $\xrightarrow{H^+}$

(c)  [structure] $\xrightarrow{CF_3SO_3H}$ (d) 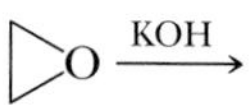 [structure] $\xrightarrow{KOH}$

24.8 At one time, a raw material for the production of hexamethylenediamine was the pentose-based polysaccharides of agricultural wastes, such as oat hulls. Treatment of these wastes with sulfuric acid or hydrochloric acid gives furfural. Decarbonylation of furfural over a zinc-chromium-molybdenum catalyst gives furan. Propose reagents and experimental conditions for the conversion of furan to hexamethylenediamine.

oat hulls, corn cobs, sugar cane stalks, etc. $\xrightarrow[\text{H}_2\text{O}]{\text{H}_2\text{SO}_4}$ [structure] Furfural $\xrightarrow[\text{catalyst}]{\text{Zn—Cr—Mo}}$ [structure] Furan $\xrightarrow{(1)}$ [structure] Tetrahydrofuran (THF) $\xrightarrow{(2)}$

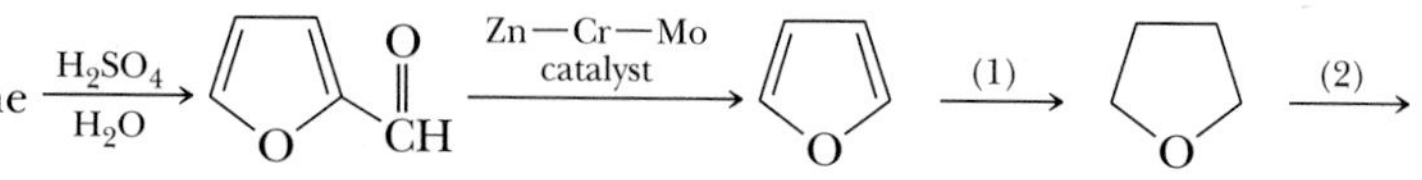

$$Cl(CH_2)_4Cl \xrightarrow{(3)} N{\equiv}C(CH_2)_4C{\equiv}N \xrightarrow{(4)} H_2N(CH_2)_6NH_2$$

| 1,4-Dichloro-butane | Hexanedinitrile (Adiponitrile) | 1,6-Hexanediamine (Hexamethylenediamine) |

24.9 Another raw material for the production of hexamethylenediamine is butadiene derived from thermal and catalytic cracking of petroleum. Propose reagents and experimental conditions for the conversion of butadiene to hexamethylenediamine.

$$CH_2{=}CHCH{=}CH_2 \xrightarrow{(1)} ClCH_2CH{=}CHCH_2Cl \xrightarrow{(2)}$$

| Butadiene | 1,4-Dichloro-2-butene |

$$N{\equiv}CCH_2CH{=}CHCH_2C{\equiv}N \xrightarrow{(3)} H_2N(CH_2)_6NH_2$$

| 3-Hexenedinitrile | 1,6-Hexanediamine (Hexamethylenediamine) |

24.10 Propose reagents and experimental conditions for the conversion of butadiene to adipic acid.

$$\text{1,3-Butadiene} \xrightarrow{?} HOOC{\sim}{\sim}COOH$$

1,3-Butadiene   Hexanedioic acid (Adipic acid)

24.11 Polymerization of 2-chloro-1,3-butadiene under Ziegler-Natta conditions gives a synthetic elastomer called neoprene. All carbon-carbon double bonds in the polymer chain have the trans configuration. Draw the repeat unit in neoprene.

24.12 Poly(ethylene terephthalate) (PET) can be prepared by this reaction. Propose a mechanism for the step-growth reaction in this polymerization.

| Dimethyl terephthalate | Ethylene glycol | Poly(ethylene terephthalate) | Methanol |

24.13 Identify the monomers required for the synthesis of these step-growth polymers.

(a) Kodel (a polyester)   **(b)** Quiana (a polyamide)

24.14 Nomex, another aromatic polyamide (compare aramid) is prepared by polymerization of 1,3-benzenediamine and the diacid chloride of 1,3-benzenedicarboxylic acid. The physical properties of the polymer make it suitable for high-strength, high-temperature applications such as parachute cords and jet aircraft tires. Draw a structural formula for the repeating unit of Nomex.

1,3-Benzenediamine   1,3-Benzene-
                          dicarbonyl chloride

24.15 Caprolactam, the monomer from which nylon 6 is synthesized, is prepared from cyclo-hexanone in two steps. In Step 1, cyclohexanone is treated with hydroxylamine to form cyclohexanone oxime. Treatment of the oxime with concentrated sulfuric acid in Step 2 gives caprolactam by a reaction called a Beckmann rearrangement. Propose a mechanism for the conversion of cyclohexanone oxime to caprolactam.

Cyclohexanone   Cyclohexanone   Caprolactam
                     oxime

24.16 Nylon 6,10 is prepared by polymerization of a diamine and a diacid chloride. Draw a structural formula for each reactant and for the repeat unit in this polymer.

24.17 Polycarbonates (Section 24.5C) are also formed by using a nucleophilic aromatic substitution route (Section 21.3B) involving aromatic difluoro monomers and carbonate ion. Propose a mechanism for this reaction.

An aromatic   Sodium   A polycarbonate
difluoride   carbonate

24.18 Propose a mechanism for the formation of this polyphenylurea. To simplify your presentation of the mechanism, consider the reaction of one —NCO group with one —NH$_2$ group.

1,4-Benzenediisocyanate   1,2-Ethanediamine   Poly(ethylene phenylurea)

24.19 When equal molar amounts of phthalic anhydride and 1,2,3-propanetriol are heated, they form an amorphous polyester. Under these conditions, polymerization is regiose-lective for the primary hydroxyl groups of the triol.

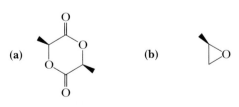

Phthalic anhydride   1,2,3-Propanetriol
(Glycerol)

(a) Draw a structural formula for the repeat unit of this polyester.
(b) Account for the regioselective reaction with the primary hydroxyl groups only.

24.20 The polyester from Problem 24.19 can be mixed with additional phthalic anhydride (0.5 mol of phthalic anhydride for each mole of 1,2,3-propanetriol in the original polyester) to form a liquid resin. When this resin is heated, it forms a hard, insoluble, thermosetting polyester called glyptal.

(a) Propose a structure for the repeat unit in glyptal.
(b) Account for the fact that glyptal is a thermosetting plastic.

24.21 Propose a mechanism for the formation of the following polymer.

24.22 Draw a structural formula of the polymer resulting from base-catalyzed polymerization of each compound. Would you expect the polymers to be optically active? (S)-$(+)$-Lactide is the dilactone formed from two molecules of (S)-$(+)$-lactic acid.

(a)   **(b)**

(S)-$(+)$-Lactide (R)-Propylene oxide

24.23 Poly(3-hydroxybutanoic acid), a biodegradable polyester, is an insoluble, opaque material that is difficult to process into shapes. In contrast, the copolymer of 3-hydroxybutanoic acid and 3-hydroxyoctanoic acid is a transparent polymer that shows good solubility in a number of organic solvents. Explain the difference in properties between these two polymers in terms of their structure.

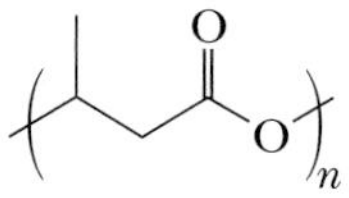

Poly(3-hydroxybutanoic acid) 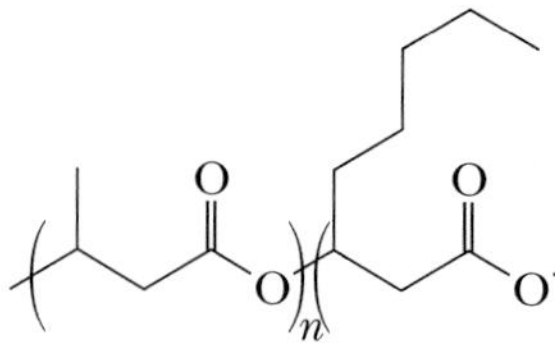   Poly(3-hydroxybutanoic acid-3-hydroxyoctanoic acid) copolymer

Chain-Growth Polymerizations

24.24 How might you determine experimentally if a particular polymerization is propagating by a step-growth or a chain-growth mechanism?

24.25 Draw a structural formula for the polymer formed in the following reactions.

(a) 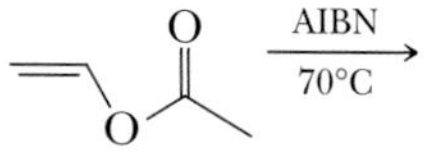 $\xrightarrow[\text{70°C}]{\text{AIBN}}$ **(b)**

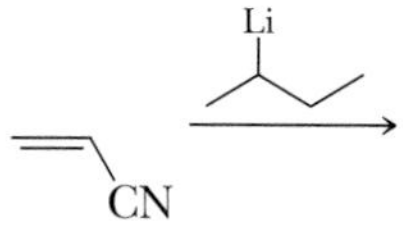

24.26 Select the monomer in each pair that is more reactive toward cationic polymerization.

(a) 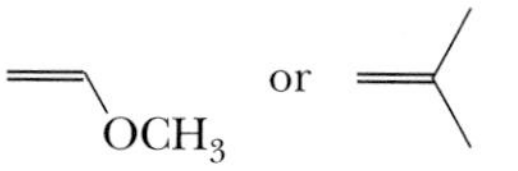 ⟍OCH₃ or   **(b)** 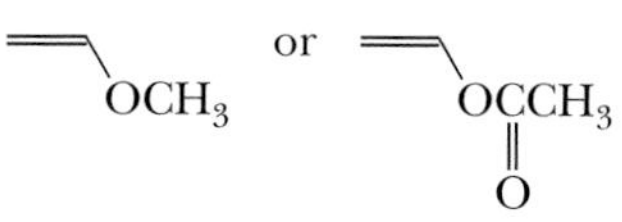 ⟍OCH₃ or ⟍OCCH₃

(c) 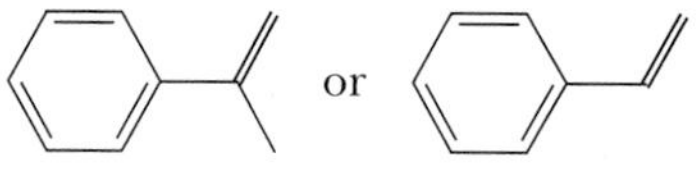 or   **(d)** 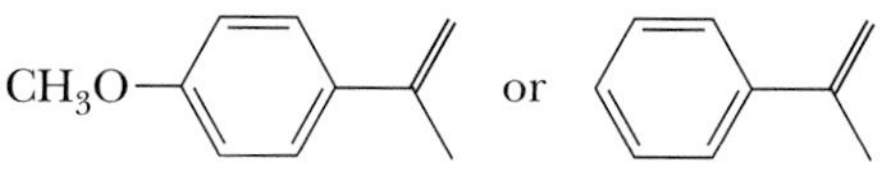 CH₃O— or

24.27 Polymerization of vinyl acetate gives poly(vinyl acetate). Hydrolysis of this polymer in aqueous sodium hydroxide gives the useful water-soluble polymer poly(vinyl alcohol). Draw the repeat units of both poly(vinyl acetate) and poly(vinyl alcohol).

24.28 Benzoquinone can be used to inhibit radical polymerizations. This compound reacts with a radical intermediate, R·, to form a less reactive radical that does not participate in chain propagation steps and, thus, breaks the chain.

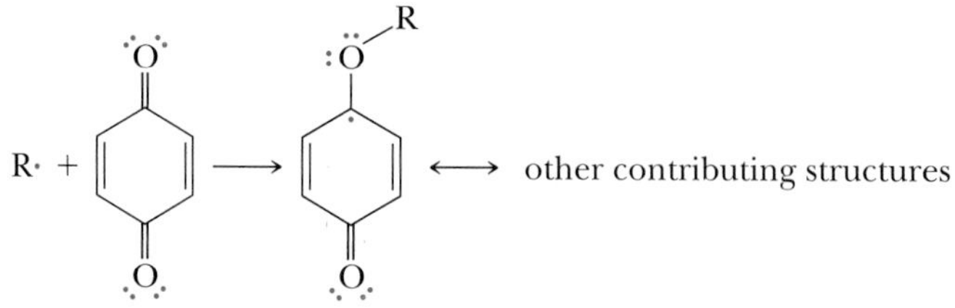

$$R\cdot + \quad \longrightarrow \quad \longleftrightarrow \quad \text{other contributing structures}$$

Draw a series of contributing structures for this less reactive radical, and account for its stability.

24.29 Following is the structural formula of a section of polypropylene derived from three units of propylene monomer.

$$-CH_2CH-CH_2CH-CH_2CH-$$
$$\quad\;\; CH_3 \qquad\; CH_3 \qquad\; CH_3$$

Polypropylene

Draw structural formulas for comparable sections of the following.

(a) Poly(vinyl chloride)
(b) Polytetrafluoroethylene
(c) Poly(methyl methacrylate)
(d) Poly(1,1-dichloroethylene)

24.30 Low-density polyethylene has a higher degree of chain branching than high-density polyethylene. Explain the relationship between chain branching and density.

24.31 We saw how intramolecular chain transfer in radical polymerization of ethylene creates a four-carbon branch on a polyethylene chain. What branch is created by a comparable intramolecular chain transfer during radical polymerization of styrene?

24.32 Compare the densities of low-density polyethylene and high-density polyethylene with the densities of the liquid alkanes listed in Table 2.4. How might you account for the differences between them?

24.33 Natural rubber is the all-cis polymer of 2-methyl-1,3-butadiene (isoprene).

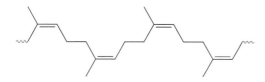

Poly(2-methyl-1,3-butadiene)
(Polyisoprene)

(a) Draw a structural formula for the repeat unit of natural rubber.
(b) Draw a structural formula of the product of oxidation of natural rubber by ozone followed by a workup in the presence of $(CH_3)_2S$. Name each functional group present in this product.
(c) The smog prevalent in many major metropolitan areas contains oxidizing agents, including ozone. Account for the fact that this type of smog attacks natural rubber (automobile tires and the like) but does not attack polyethylene or polyvinyl chloride.
(d) Account for the fact that natural rubber is an elastomer but the synthetic all-trans isomer is not.

24.34 Radical polymerization of styrene gives a linear polymer. Radical polymerization of a mixture of styrene and 1,4-divinylbenzene gives a cross-linked network polymer of the type shown in Figure 24.1. Show by drawing structural formulas how incorporation of a few percent 1,4-divinylbenzene in the polymerization mixture gives a cross-linked polymer.

a copolymer of styrene and divinylbenzene

Styrene   1,4-Divinylbenzene

24.35 One common type of cation exchange resin is prepared by polymerization of a mixture containing styrene and 1,4-divinylbenzene (Problem 24.34). The polymer is then treated with concentrated sulfuric acid to sulfonate a majority of the aromatic rings in the polymer.

(a) Show the product of sulfonation of each benzene ring.
(b) Explain how this sulfonated polymer can act as a cation exchange resin.

24.36 The most widely used synthetic rubber is a copolymer of styrene and butadiene called SB rubber. Ratios of butadiene to styrene used in polymerization vary depending on the end use of the polymer. The ratio used most commonly in the preparation of SB rubber for use in automobile tires is 1 mol styrene to 3 mol butadiene. Draw a structural formula of a section of the polymer formed from this ratio of reactants. Assume that all carbon-carbon double bonds in the polymer chain are in the cis configuration.

24.37 From what two monomer units is the following polymer made?

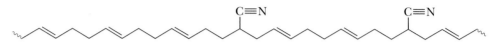

24.38 Draw the structure of the polymer formed from ring-opening metathesis polymerization of each monomer.

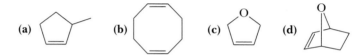

CARBOHYDRATES

Carbohydrates are the most abundant organic compounds in the plant world. They act as storehouses of chemical energy (glucose, starch, glycogen); are components of supportive structures in plants (cellulose), crustacean shells (chitin), and connective tissues in animals (acidic polysaccharides); and are essential components of nucleic acids (D-ribose and 2-deoxy-D-ribose). Carbohydrates make up about three fourths of the dry weight of plants. Animals (including humans) get their carbohydrates by eating plants, but they do not store much of what they consume. Less than 1% of the body weight of animals is made up of carbohydrates.

■ Foxglove (*Digitalis purpurea*), an ornamental flowering plant, is the source of digitoxin and digitalis, medicines widely used in cardiology to reduce pulse rate, regularize heart rhythm, and strengthen heart beat. (*Christy Carter/Grant Heilman Photography, Inc.*) Inset: Digitoxose, a monosaccharide obtained on hydrolysis of digitoxin. See Problem 25.15.

Breads, grains, and pasta are sources of carbohydrates. *(Charles D. Winters)*

The CD-Rom contains molecular models of carbohydrates in both cyclic and open-chain forms.

Carbohydrate A polyhydroxyaldehyde, polyhydroxyketone, or substance that gives these compounds on hydrolysis.

Monosaccharide A carbohydrate that cannot be hydrolyzed to a simpler carbohydrate.

Aldose A monosaccharide containing an aldehyde group.

Ketose A monosaccharide containing a ketone group.

The name "carbohydrate" means hydrate of carbon and derives from the formula $C_n(H_2O)_m$. Following are two examples of carbohydrates with molecular formulas that can be written alternatively as hydrates of carbon.

Glucose (blood sugar): $C_6H_{12}O_6$, or alternatively $C_6(H_2O)_6$

Sucrose (table sugar): $C_{12}H_{22}O_{11}$, or alternatively $C_{12}(H_2O)_{11}$

Not all carbohydrates, however, have this general formula. Some contain too few oxygen atoms to fit this formula, and some others contain too many oxygens. Some also contain nitrogen. The term "carbohydrate" has become so firmly rooted in chemical nomenclature that, although not completely accurate, it persists as the name for this class of compounds.

At the molecular level, most **carbohydrates** are polyhydroxyaldehydes, polyhydroxyketones, or compounds that yield either of these after hydrolysis. Therefore, the chemistry of carbohydrates is essentially the chemistry of hydroxyl groups and carbonyl groups and of the acetal bonds formed between these two functional groups.

The notion that carbohydrates have only two types of functional groups, however, belies the complexity of their chemistry. All but the simplest carbohydrates contain multiple stereocenters. For example, glucose, the most abundant carbohydrate in the biological world, contains one aldehyde group, one primary and four secondary hydroxyl groups, and four stereocenters. Working with molecules of this complexity presents enormous challenges to organic chemists and biochemists alike.

25.1 Monosaccharides

A. Structure and Nomenclature

Monosaccharides have the general formula $C_nH_{2n}O_n$ with one of the carbons being the carbonyl group of either an aldehyde or a ketone. The most common monosaccharides have three to eight carbon atoms. The suffix -ose indicates that a molecule is a carbohydrate, and the prefixes tri-, tetr-, pent-, and so forth indicate the number of carbon atoms in the chain. Monosaccharides containing an aldehyde group are classified as **aldoses;** those containing a ketone group are classified as **ketoses.**

Monosaccharides Classified by Number of Carbon Atoms	
Name	**Formula**
Triose	$C_3H_6O_3$
Tetrose	$C_4H_8O_4$
Pentose	$C_5H_{10}O_5$
Hexose	$C_6H_{12}O_6$
Heptose	$C_7H_{14}O_7$
Octose	$C_8H_{16}O_8$

There are only two trioses: the aldotriose glyceraldehyde and the ketotriose dihydroxyacetone.

$$
\begin{array}{cc}
\text{CHO} & \text{CH}_2\text{OH} \\
| & | \\
\text{CHOH} & \text{C}{=}\text{O} \\
| & | \\
\text{CH}_2\text{OH} & \text{CH}_2\text{OH} \\
\\
\text{Glyceraldehyde} & \text{Dihydroxyacetone} \\
\text{(an aldotriose)} & \text{(a ketotriose)}
\end{array}
$$

Often the designations aldo- and keto- are omitted, and these molecules are referred to simply as trioses, tetroses, and the like.

Glyceraldehyde is a common name; the IUPAC name for this monosaccharide is 2,3-dihydroxypropanal. Similarly, dihydroxyacetone is a common name; its IUPAC name is 1,3-dihydroxypropanone. The common names for these and other monosaccharides, however, are so firmly rooted in the literature of organic chemistry and biochemistry that they are used almost exclusively whenever these compounds are referred to. Therefore, throughout our discussions of the chemistry and biochemistry of carbohydrates, we use the names most common in the literature of chemistry and biochemistry.

B. Fischer Projection Formulas

Glyceraldehyde contains a stereocenter and therefore exists as a pair of enantiomers.

$$
\begin{array}{cc}
\text{CHO} & \text{CHO} \\
H{\blacktriangleright}\text{C}{\blacktriangleleft}\text{OH} & \text{HO}{\blacktriangleright}\text{C}{\blacktriangleleft}\text{H} \\
\text{CH}_2\text{OH} & \text{CH}_2\text{OH} \\
(R)\text{-Glyceraldehyde} & (S)\text{-Glyceraldehyde}
\end{array}
$$

Chemists commonly use two-dimensional representations called **Fischer projections** to show the configuration of carbohydrates. To write a Fischer projection, draw a three-dimensional representation of the molecule oriented so that the vertical bonds from the stereocenter are directed away from you and the horizontal bonds from it are directed toward you. Then write the molecule as a two-dimensional figure with the stereocenter indicated by the point at which the bonds cross.

Fischer projection A two-dimensional representation for showing the configuration of stereocenters; horizontal lines represent bonds projecting forward, and vertical lines represent bonds projecting to the rear.

$$
\begin{array}{ccc}
\text{CHO} & & \text{CHO} \\
H{\blacktriangleright}\text{C}{\blacktriangleleft}\text{OH} & \xrightarrow[\text{Fischer projection}]{\text{convert to a}} & H{-\!\!-}\text{OH} \\
\text{CH}_2\text{OH} & & \text{CH}_2\text{OH} \\
(R)\text{-Glyceraldehyde} & & (R)\text{-Glyceraldehyde} \\
\text{(three-dimensional} & & \text{(Fischer projection)} \\
\text{representation)} & &
\end{array}
$$

The horizontal segments of this Fischer projection represent bonds directed toward you, and the vertical segments represent bonds directed away from you. The only atom in the plane of the paper is the stereocenter.

C.   D- and L-Monosaccharides

Even though the R,S system is widely accepted today as a standard for designating configuration, the configuration of carbohydrates (as well as those of amino acids and many other compounds in biochemistry) is commonly designated by the D,L system proposed by Emil Fischer in 1891. At that time, it was known that one enantiomer of glyceraldehyde has a specific rotation of $+13.5°$; the other has a specific rotation of $-13.5°$. Fischer proposed that these enantiomers be designated D and L (for dextro- and levorotatory), but he had no experimental way to determine which enantiomer has which specific rotation. Fischer, therefore, did the only possible thing—he made an arbitrary assignment. He assigned the dextrorotatory enantiomer an arbitrary configuration as shown here and named it D-glyceraldehyde. He named its enantiomer L-glyceraldehyde. Fischer could have been wrong, but by a stroke of good fortune he was correct. In 1952, his assignment of configuration to the enantiomers of glyceraldehyde was proved correct by a special application of x-ray crystallography.

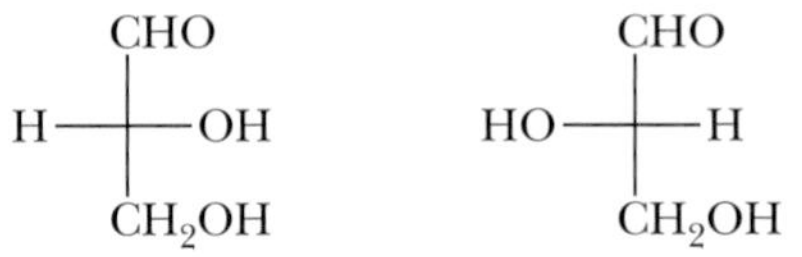

D-Glyceraldehyde
$[\alpha]_D^{25} = +13.5°$

L-Glyceraldehyde
$[\alpha]_D^{25} = -13.5°$

D- and L-glyceraldehyde serve as reference points for the assignment of relative configuration to all other aldoses and ketoses. The reference point is the stereocenter farthest from the carbonyl group. Because this stereocenter is always the next to the last carbon on the chain, it is called the **penultimate carbon.** A **D-monosaccharide** has the same configuration at its penultimate carbon as D-glyceraldehyde (its —OH is on the right when written as a Fischer projection); an **L-monosaccharide** has the same configuration at its penultimate carbon as L-glyceraldehyde (its —OH is on the left).

D-Monosaccharide A monosaccharide that, when written as a Fischer projection, has the —OH on its penultimate carbon to the right.

L-Monosaccharide A monosaccharide that, when written as a Fischer projection, has the —OH on its penultimate carbon to the left.

Tables 25.1 and 25.2 show names and Fischer projections for all D-aldo- and D-2-ketotetroses, pentoses, and hexoses. Each name consists of three parts. The letter D specifies the configuration of the penultimate carbon. Prefixes such as rib-, arabin-, and gluc- specify the configuration of all other stereocenters in the monosaccharide. The suffix -ose shows that the compound is a carbohydrate.

The three most abundant hexoses in the biological world are D-glucose, D-galactose, and D-fructose. The first two are D-aldohexoses; the third is a D-2-ketohexose. Glucose, by far the most common hexose, is also known as dextrose because it is dextrorotatory. Other names for this monosaccharide are grape sugar and blood sugar. Human blood normally contains 65–110 mg of glucose/100 mL of blood. Glucose is synthesized by chlorophyll-containing plants using sunlight as a source of energy. In

25.1 Monosaccharides **1011**

Table 25.1 **Configurational Relationships Among the Isomeric D-Aldotetroses, D-Aldopentoses, and D-Aldohexoses**

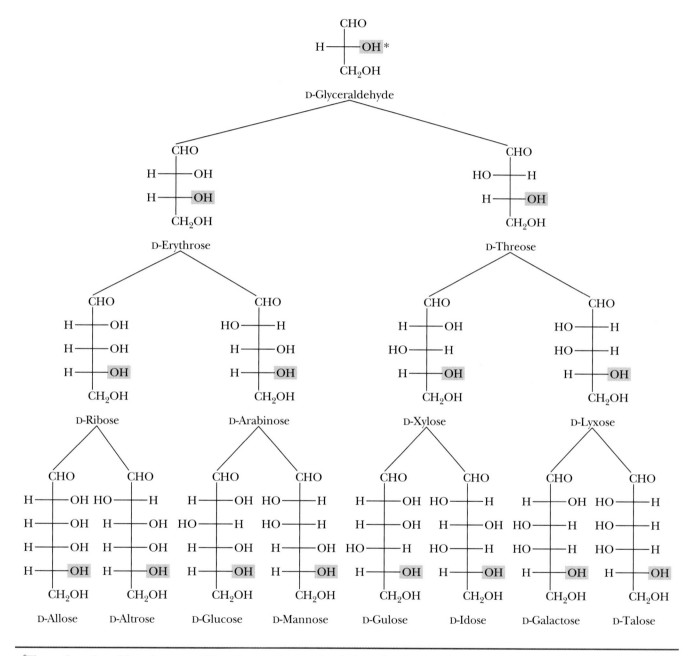

*The configuration of the reference —OH on the penultimate carbon is shown in color.

**Table 25.2   Configurational Relationships Among the Isomeric
D-2-Ketopentoses and D-2-Ketohexoses**

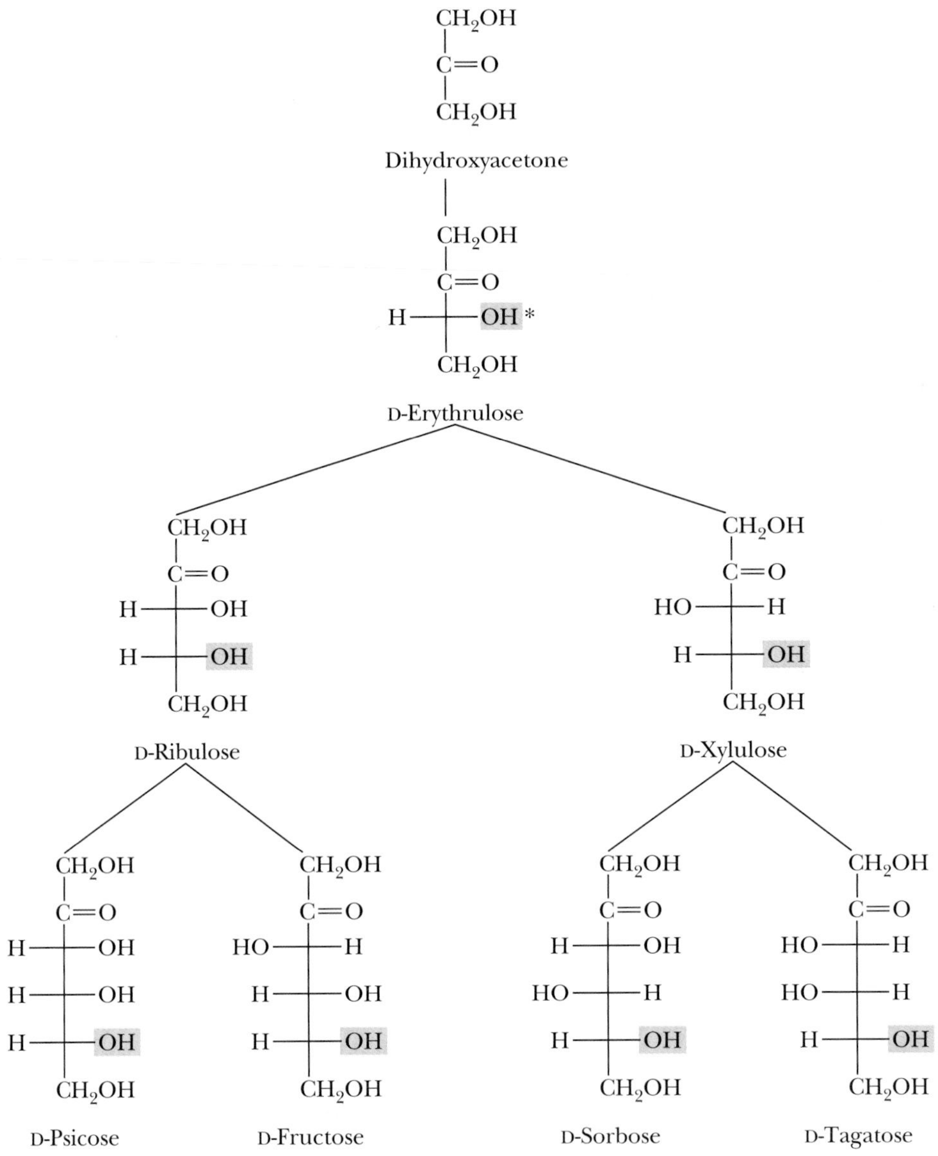

$$
\begin{array}{c}
CH_2OH \\
| \\
C=O \\
| \\
CH_2OH
\end{array}
$$

Dihydroxyacetone

$$
\begin{array}{c}
CH_2OH \\
| \\
C=O \\
H-\!\!\!-OH\ ^* \\
CH_2OH
\end{array}
$$

D-Erythrulose

D-Ribulose		D-Xylulose

$$
\begin{array}{cc}
CH_2OH & \qquad\qquad CH_2OH \\
| & | \\
C=O & C=O \\
H-\!\!\!-OH & HO-\!\!\!-H \\
H-\!\!\!-OH & H-\!\!\!-OH \\
CH_2OH & CH_2OH
\end{array}
$$

D-Ribulose $\qquad\qquad$ D-Xylulose

$$
\begin{array}{cccc}
CH_2OH & CH_2OH & \quad CH_2OH & CH_2OH \\
| & | & | & | \\
C=O & C=O & C=O & C=O \\
H-\!\!\!-OH & HO-\!\!\!-H & H-\!\!\!-OH & HO-\!\!\!-H \\
H-\!\!\!-OH & H-\!\!\!-OH & HO-\!\!\!-H & HO-\!\!\!-H \\
H-\!\!\!-OH & H-\!\!\!-OH & H-\!\!\!-OH & H-\!\!\!-OH \\
CH_2OH & CH_2OH & CH_2OH & CH_2OH
\end{array}
$$

D-Psicose $\qquad$ D-Fructose $\qquad\qquad$ D-Sorbose $\qquad$ D-Tagatose

*The configuration of the reference —OH on the penultimate carbon is shown in color.

the process called photosynthesis, plants convert carbon dioxide from the air and water from the soil to glucose and oxygen.

$$6CO_2 \ + \ 6H_2O \ + \ energy \ \xrightarrow[\text{chlorophyll}]{\text{sunlight}} \ C_6H_{12}O_6 \ + \ 6O_2$$

| Carbon dioxide | Water | | Glucose | Oxygen |

D-Fructose is found combined with glucose in the disaccharide sucrose (table sugar, Section 25.4C). D-Galactose is obtained with glucose in the disaccharide lactose (milk sugar, Section 25.4B).

D-Ribose and 2-deoxy-D-ribose, the most abundant pentoses in the biological world, are essential building blocks of nucleic acids: D-ribose in ribonucleic acids (RNA) and 2-deoxy-D-ribose in deoxyribonucleic acids (DNA).

Example 25.1

Draw Fischer projections for the four aldotetroses. Which are D-monosaccharides, which are L-monosaccharides, and which are enantiomers? Refer to Table 25.1, and write the name of each aldotetrose.

Solution

Following are Fischer projections for the four aldotetroses. The letters D and L refer to the configuration of the penultimate carbon which, in the case of aldotetroses, is carbon 3. In the Fischer projection of a D-aldotetrose, the —OH on carbon 3 is on the right, and, in an L-aldotetrose, it is on the left. The erythroses are each diastereomers of each of the threoses.

one pair of enantiomers   a second pair of enantiomers

D-Erythrose   L-Erythrose   D-Threose   L-Threose

Problem 25.1

Draw Fischer projections for all 2-ketopentoses. Which are D-ketopentoses, which are L-ketopentoses, and which are enantiomers? Refer to Table 25.2, and write the name of each 2-ketopentose.

D. Amino Sugars

Amino sugars contain an —NH_2 group in place of an —OH group. Only three amino sugars are common in nature: D-glucosamine, D-mannosamine, and D-galactosamine.

The structures of amino sugars (Fischer projections):

D-Glucosamine — CHO, H—NH$_2$, HO—H, H—OH, H—OH, CH$_2$OH

D-Mannosamine (C-2 stereoisomer of D-glucosamine) — CHO, H$_2$N—H (position 2), HO—H, H—OH, H—OH, CH$_2$OH

D-Galactosamine (C-4 stereoisomer of D-glucosamine) — CHO, H—NH$_2$, HO—H, HO—H (position 4), H—OH, CH$_2$OH

N-Acetyl-D-glucosamine — CHO, H—NHCCH$_3$ (C=O), HO—H, H—OH, H—OH, CH$_2$OH

N-Acetyl-D-glucosamine, a derivative of D-glucosamine, is a component of many poly-saccharides, including chitin, the hard shell-like exoskeleton of lobsters, crabs, shrimp, and other shellfish. Many other amino sugars are components of naturally occurring antibiotics.

E. Physical Properties

Monosaccharides are colorless, crystalline solids, although they often crystallize with difficulty. Because hydrogen bonding is possible between their polar —OH groups and water, all monosaccharides are very soluble in water. They are only slightly soluble in ethanol and are insoluble in nonpolar solvents such as diethyl ether, chloroform, and benzene.

Although all monosaccharides are sweet to the taste, some are sweeter than others (Table 25.3). D-Fructose tastes the sweetest, even sweeter than sucrose (table sugar, Section 25.4C). The sweet taste of honey is due largely to D-fructose and D-glucose. Lactose (Section 25.4B) has almost no sweetness and is sometimes added to foods as a filler. Some people lack an enzyme that allows them to tolerate lactose well; they should avoid these foods.

Table 25.3 Relative Sweetness of Some Carbohydrates and Artificial Sweetening Agents

Carbohydrate	Sweetness Relative to Sucrose	Artificial Sweetener	Sweetness Relative to Sucrose
Fructose	1.74	Saccharin	450
Invert sugar	1.25	Acesulfame-K	200
Sucrose (table sugar)	1.00	Aspartame	160
Honey	0.97		
Glucose	0.74		
Maltose	0.33		
Galactose	0.32		
Lactose (milk sugar)	0.16		

25.2 The Cyclic Structure of Monosaccharides

We saw in Section 16.8B that aldehydes and ketones react with alcohols to form hemiacetals. We also saw that cyclic hemiacetals form very readily when hydroxyl and carbonyl groups are part of the same molecule and their interaction can form a five- or six-membered ring. For example, 4-hydroxypentanal forms a five-membered cyclic hemiacetal. Note that 4-hydroxypentanal contains one stereocenter and that a second stereocenter is generated at carbon 1 as a result of hemiacetal formation.

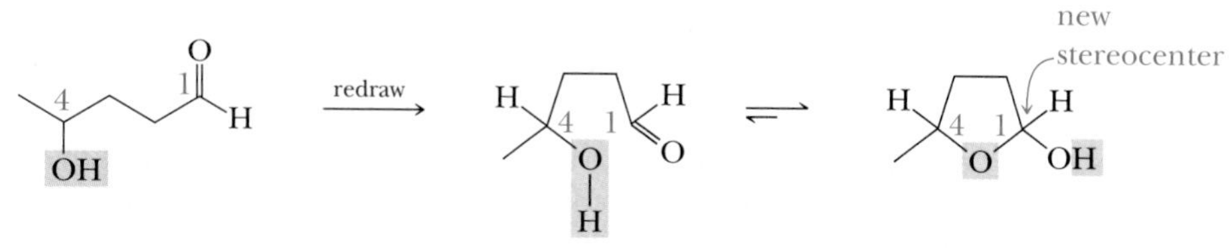

4-Hydroxypentanal (redrawn to show how the cyclic hemiacetal forms) A cyclic hemiacetal

Monosaccharides have hydroxyl and carbonyl groups in the same molecule. As a result, they too exist almost exclusively as five- and six-membered cyclic hemiacetals.

A. Haworth Projections

A common way of representing the cyclic structure of monosaccharides is the **Haworth projection,** named after the English chemist Sir Walter N. Haworth (1937 Nobel Prize for chemistry). In a Haworth projection, a five- or six-membered cyclic hemiacetal is represented as a planar pentagon or hexagon, as the case may be, lying perpendicular to the plane of the paper. Groups attached to the carbons of the ring then lie either above or below the plane of the ring. The new stereocenter created in forming the cyclic structure is called an **anomeric carbon.** Stereoisomers that differ in configuration only at the anomeric carbon are called **anomers.** The anomeric carbon of an aldose is carbon 1; that of the most common ketoses is carbon 2.

Haworth projections are most commonly written with the anomeric carbon to the right and the hemiacetal oxygen to the back (Figure 25.1).

In the terminology of carbohydrate chemistry, the designation β means that the —OH on the anomeric carbon of the cyclic hemiacetal is on the same side of the ring as the terminal —CH_2OH. Conversely, the designation α means that the —OH

Haworth projection A way to view furanose and pyranose forms of monosaccharides. The ring is drawn flat and most commonly viewed through its edge with the anomeric carbon on the right and the oxygen atom of the ring in the rear.

Anomeric carbon A hemiacetal or acetal carbon of the cyclic form of a carbohydrate.

Anomers Carbohydrates that differ in configuration only at their anomeric carbons.

Figure 25.1
Haworth projections for α-D-glucopyranose and β-D-glucopyranose.

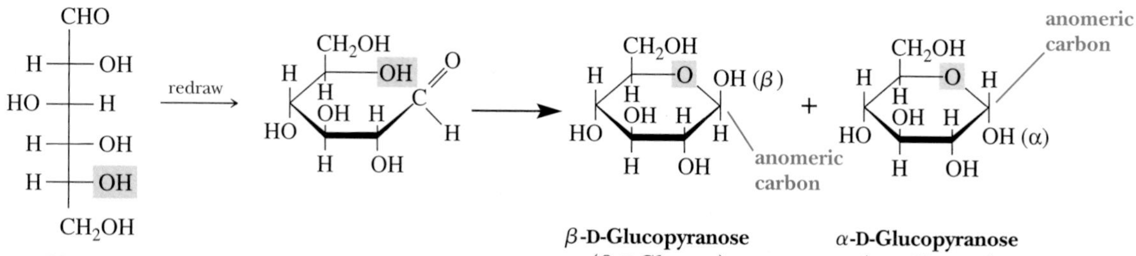

D-Glucose

β-D-Glucopyranose (β-D-Glucose)

α-D-Glucopyranose (α-D-Glucose)

on the anomeric carbon of the cyclic hemiacetal is on the side of the ring opposite from the terminal —CH$_2$OH.

A six-membered hemiacetal ring is indicated by the infix -pyran-, and a five-membered hemiacetal ring is indicated by the infix -furan-. The terms **furanose** and **pyranose** are used because monosaccharide five- and six-membered rings correspond to the heterocyclic compounds furan and pyran.

Furanose A five-membered cyclic form of a monosaccharide.

Pyranose A six-membered cyclic form of a monosaccharide.

Furan   Pyran

Because the α and β forms of glucose are six-membered cyclic hemiacetals, they are named α-D-glucopyranose and β-D-glucopyranose. These infixes are not always used in monosaccharide names, however. Thus, the glucopyranoses, for example, are often named simply α-D-glucose and β-D-glucose.

You would do well to remember the configuration of groups on the Haworth projections of α-D-glucopyranose and β-D-glucopyranose as reference structures. Knowing how the open-chain configuration of any other aldohexoses differs from that of D-glucose, you can then construct its Haworth projection by reference to the Haworth projection of D-glucose.

Example 25.2

Draw Haworth projections for the α and β anomers of D-galactopyranose.

Solution

One way to arrive at these projections is to use the α and β forms of D-glucopyranose as reference and to remember (or discover by looking at Table 25.1) that D-galactose differs from D-glucose only in the configuration at carbon 4. Thus, begin with the Haworth projections shown in Figure 25.1 and then invert the configuration at carbon 4.

Configuration differs from that of D-glucose at C-4.

α-D-Galactopyranose
(α-D-Galactose)

β-D-Galactopyranose
(β-D-Galactose)

Problem 25.2

Mannose exists in aqueous solution as a mixture of α-D-mannopyranose and β-D-mannopyranose. Draw Haworth projections for these molecules.

C H E M I S T R Y   I N   A C T I O N

L-Ascorbic Acid (Vitamin C)

The structure of L-ascorbic acid (vitamin C) resembles that of a monosaccharide. In fact, this vitamin is synthesized both biochemically by plants and some animals and commercially from D-glucose. Humans do not have the enzymes required for this synthesis; therefore, we must obtain it in the food we eat or as a vitamin supplement. Approximately 66 million kilograms of vitamin C are synthesized every year in the United States.

L-Ascorbic acid is very easily oxidized to L-dehydroascorbic acid, a diketone.

Both L-ascorbic acid and L-dehydroascorbic acid are physiologically active and are found together in most body fluids. Ascorbic acid is one of the most important antioxidants (the H in the OH is weakly bonded and easily abstracted by free radicals). One of the most important roles it plays may be to replenish the lipid-soluble antioxidant α-tocopherol by transferring a hydrogen atom to the tocopherol radical, formed by reaction with free radicals in the autoxidation process (see the Chemistry in Action box "Radical Autoxidation" in Chapter 7).

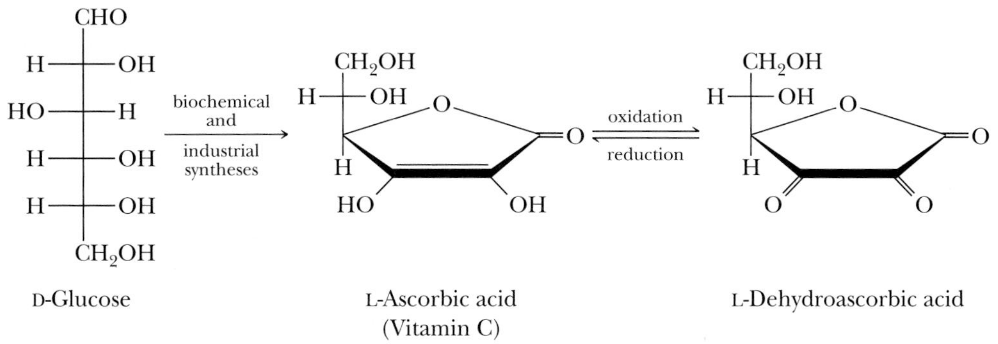

D-Glucose → L-Ascorbic acid (Vitamin C) ⇌ L-Dehydroascorbic acid

Aldopentoses also form cyclic hemiacetals. The most prevalent forms of D-ribose and other pentoses in the biological world are furanoses. Shown in Figure 25.2 are Haworth projections for α-D-ribofuranose (α-D-ribose) and β-2-deoxy-D-ribofuranose (β-2-deoxy-D-ribose). The prefix 2-deoxy indicates the absence of oxygen at carbon 2. Units of D-ribose and 2-deoxy-D-ribose in nucleic acids and most other biological molecules are found almost exclusively in the β configuration.

Other monosaccharides also form five-membered cyclic hemiacetals. Shown in Figure 25.3 are the five-membered cyclic hemiacetals of fructose. The β-D-fructofuranose form is found in the disaccharide sucrose (Section 25.4C).

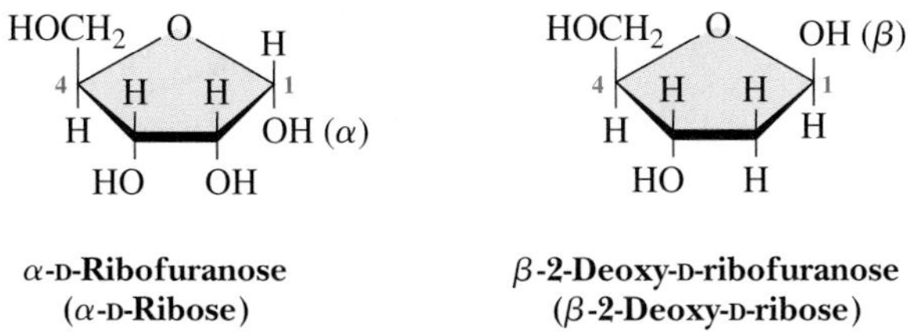

α-D-**Ribofuranose**
(α-D-**Ribose**)

β-**2-Deoxy-D-ribofuranose**
(β-**2-Deoxy-D-ribose**)

Figure 25.2

Haworth projections for two D-ribofuranoses.

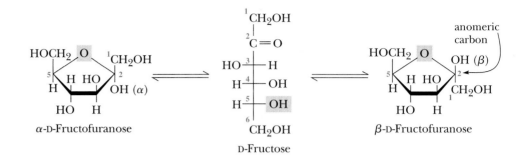

Figure 25.3
Furanose forms of D-fructose at equilibrium in aqueous solution.

B.   Conformation Representations

A five-membered ring is so close to being planar that Haworth projections are adequate representations of furanoses. For pyranoses, however, the six-membered ring is more accurately represented as a chair conformation. Structural formulas for α-D-glucopyranose and β-D-glucopyranose are drawn as chair conformations in Figure 25.4. Also shown is the open-chain or free aldehyde form with which the cyclic hemiacetal forms are in equilibrium in aqueous solution. Notice that each group, including the anomeric —OH, on the chair conformation of β-D-glucopyranose is equatorial. Notice also that the —OH group on the anomeric carbon is axial in α-D-glucopyranose. Because of the equatorial orientation of the —OH on its anomeric carbon, β-D-glucopyranose is more stable and predominates in aqueous solution.

At this point, you should compare the relative orientations of groups on the D-glucopyranose ring in a Haworth projection and a chair conformation. The orientations of groups on carbons 1 through 5 of β-D-glucopyranose, for example, are up, down, up, down, and up in both representations.

Figure 25.4
Chair conformations of β-D-glucopyranose and α-D-glucopyranose.

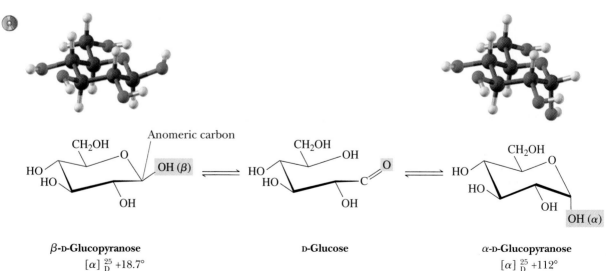

β-D-Glucopyranose
(Haworth projection)

β-D-Glucopyranose
(chair conformation)

Example 25.3

Draw chair conformations for α-D-galactopyranose and β-D-galactopyranose. Label the anomeric carbon in each.

Solution

D-Galactose differs in configuration from D-glucose only at carbon 4. Therefore, draw the α and β forms of D-glucopyranose and then interchange the positions of the —OH and —H groups on carbon 4.

β-D-Galactopyranose
(β-D-Galactose)

D-Galactose

α-D-Galactopyranose
(α-D-Galactose)

Problem 25.3

Draw chair conformations for α-D-mannopyranose and β-D-mannopyranose. Label the anomeric carbon in each.

C. Mutarotation

Mutarotation is the change in specific rotation that accompanies the equilibration of α- and β-anomers in aqueous solution. As an example, a solution prepared by dissolving crystalline α-D-glucopyranose in water has an initial specific rotation of +112°, which gradually decreases to an equilibrium value of +52.7° as α-D-glucopyranose reaches an equilibrium with β-D-glucopyranose. A solution of β-D-glucopyranose also undergoes mutarotation, during which the specific rotation changes from +18.7° to the same equilibrium value of +52.7°. The equilibrium mixture consists of 64% β-D-glucopyranose and 36% α-D-glucopyranose with only a trace (0.003%) of the open-chain form. Mutarotation is common to all carbohydrates that exist in hemiacetal forms.

Mutarotation The change in specific rotation that occurs when an α or β hemiacetal form of a carbohydrate in aqueous solution is converted to an equilibrium mixture of the two forms.

β-D-Glucopyranose
$[\alpha]_D^{25}$ +18.7°

Open-chain form

α-D-Glucopyranose
$[\alpha]_D^{25}$ +112°

25.3   Reactions of Monosaccharides

A.   Formation of Glycosides (Acetals)

We saw in Section 16.8B that treatment of an aldehyde or ketone with one molecule of alcohol gives a hemiacetal and that treatment of the hemiacetal with a molecule of alcohol gives an acetal. Treatment of monosaccharides, all of which exist almost exclusively in a cyclic hemiacetal form, also gives acetals, as illustrated by the reaction of β-D-glucopyranose with methanol.

β-D-Glucopyranose
(β-D-Glucose)

Methyl β-D-glucopyranoside
(Methyl β-D-glucoside)

Methyl α-D-glucopyranoside
(Methyl α-D-glucoside)

Glycoside A carbohydrate in which the —OH on its anomeric carbon is replaced by —OR.

Glycosidic bond The bond from the anomeric carbon of a glycoside to an —OR group.

A cyclic acetal derived from a monosaccharide is called a **glycoside,** and the bond from the anomeric carbon to the —OR group is called a **glycosidic bond.** Mutarotation is not possible in a glycoside because an acetal is no longer in equilibrium with the open-chain carbonyl-containing compound. Glycosides are stable in water and aqueous base, but like other acetals (Section 16.8), they are hydrolyzed in aqueous acid to an alcohol and a monosaccharide.

Glycosides are named by listing the alkyl or aryl group bonded to oxygen followed by the name of the carbohydrate in which the ending -e is replaced by -ide. For example, the glycosides derived from β-D-glucopyranose are named β-D-glucopyranosides; those derived from β-D-ribofuranose are named β-D-ribofuranosides.

Example 25.4

Draw a structural formula for each glycoside. In each, label the anomeric carbon and the glycosidic bond.

(a) Methyl β-D-ribofuranoside (methyl β-D-riboside); draw a Haworth projection.
(b) Methyl α-D-galactopyranoside (methyl α-D-galactoside); draw both a Haworth projection and a chair conformation.

Solution

(a) Methyl β-D-ribofuranoside
(Methyl β-D-riboside)

(b) Methyl α-D-galactopyranoside
(Methyl α-D-galactoside)

Problem 25.4

Draw a structural formula for each glycoside. In each, label the anomeric carbon and the glycosidic bond.

(a) Methyl β-D-fructofuranoside (methyl β-D-fructoside); draw a Haworth projection.
(b) Methyl α-D-mannopyranoside (methyl α-D-mannoside); draw both a Haworth projection and a chair conformation.

Just as the anomeric carbon of a cyclic hemiacetal undergoes reaction with the —OH group of an alcohol to form a glycoside, it also undergoes reaction with the N—H group of an amine to form an *N*-glycoside. Especially important in the biological world are the *N*-glycosides formed between D-ribose and 2-deoxy-D-ribose, each as a furanose, and the heterocyclic aromatic amines uracil, cytosine, thymine, adenine, and guanine (Figure 25.5). *N*-Glycosides of these pyrimidine and purine bases are structural units of nucleic acids (Chapter 28).

Example 25.5

Draw a structural formula for cytidine, the β-*N*-glycoside formed between D-ribofuranose and cytosine.

Solution

Figure 25.5
Structural formulas of the five most important pyrimidine and purine bases found in DNA and RNA. The hydrogen atom shown in color is lost in forming an *N*-glycoside.

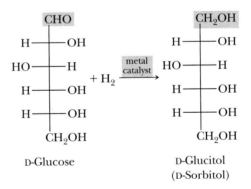

Uracil   Cytosine   Thymine   Adenine   Guanine

Problem 25.5
Draw a structural formula for the *β-N*-glycoside formed between 2-deoxy-D-ribofuranose and adenine.

B.   Reduction to Alditols

The carbonyl group of a monosaccharide can be reduced to a hydroxyl group by a variety of reducing agents, including sodium borohydride and hydrogen in the presence of a transition metal catalyst. The reduction products are known as **alditols.** Reduction of D-glucose gives D-glucitol, more commonly known as D-sorbitol. Note that D-glucose is shown here in the open-chain form. Only a small amount of this form is present in solution, but, as it is reduced, the equilibrium between cyclic hemiacetal forms and the open-chain form shifts to replace it.

Alditol   The product formed when the C=O group of a monosaccharide is reduced to a CHOH group.

Many "sugar-free" products contain sugar alcohols, such as D-sorbitol and xylitol. (*Gregory Smolin*)

Sorbitol is found in the plant world in many berries and in cherries, plums, pears, apples, seaweed, and algae. It is about 60% as sweet as sucrose (table sugar) and is used in the manufacture of candies and as a sugar substitute for diabetics. D-Sorbitol is an important food additive, usually added to prevent dehydration of foods and other materials on exposure to air because it binds water strongly.

Other alditols common in the biological world are erythritol, D-mannitol, and xylitol. Xylitol is used as a sweetening agent in "sugarless" gum, candy, and sweet cereals.

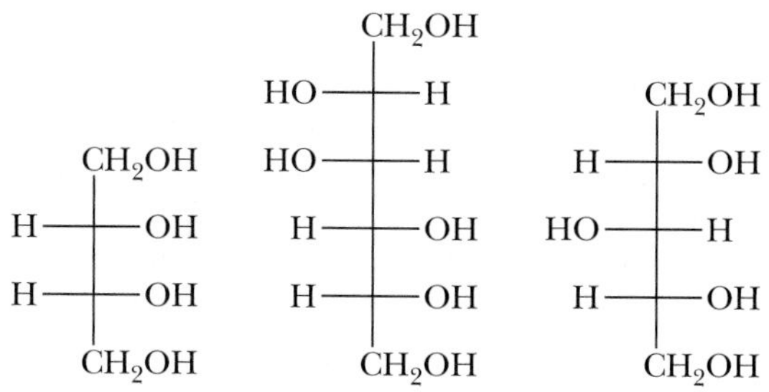

The structural formulas for Erythritol, D-Mannitol, and Xylitol are shown:

Erythritol:
CH₂OH
H—OH
H—OH
CH₂OH

D-Mannitol:
CH₂OH
HO—H
HO—H
H—OH
H—OH
CH₂OH

Xylitol:
CH₂OH
H—OH
HO—H
H—OH
CH₂OH

Erythritol   D-Mannitol   Xylitol

C.   Oxidation to Aldonic Acids—Reducing Sugars

As we saw in Section 16.13A, aldehydes (RCHO) are oxidized to carboxylic acids (RCOOH) by several oxidizing agents, including oxygen, O_2. Similarly, the aldehyde group of an aldose can be oxidized, under basic conditions, to a carboxylate group. Oxidizing agents for this purpose include bromine in aqueous calcium carbonate (Br_2, $CaCO_3$, H_2O) and Tollens' solution [$Ag(NH_3)_2^+$]. Under these conditions, the cyclic form of an aldose is in equilibrium with the open-chain form, which is then oxidized by the mild oxidizing agent. D-Glucose, for example, is oxidized to D-gluconate (the anion of D-gluconic acid).

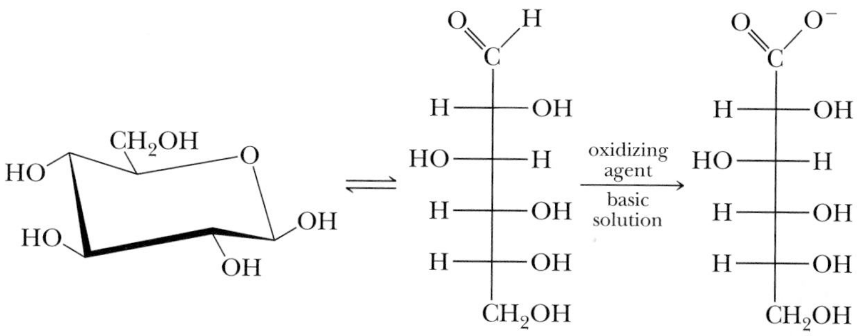

β-D-Glucopyranose (β-D-Glucose) ⇌ D-Glucose —(oxidizing agent, basic solution)→ D-Gluconate

Any carbohydrate that reacts with an oxidizing agent to form an **aldonic acid** is classified as a **reducing sugar** (it reduces the oxidizing agent).

Surprisingly, 2-ketoses are also reducing sugars. Carbon 1 (a CH_2OH group) of a 2-ketose is not oxidized directly. Rather, under the basic conditions of this oxidation, a 2-ketose is in equilibrium with an aldose by way of an enediol intermediate. The aldose is then oxidized by the mild oxidizing agent.

A 2-ketose —(OH⁻)⇌ An enediol —(OH⁻)⇌ An aldose —(oxidizing agent)→ An aldonate

Aldonic acid The product formed when the —CHO group of an aldose is oxidized to a —COOH group.

Reducing sugar A carbohydrate that reacts with an oxidizing agent to form an aldonic acid. In this reaction, the carbohydrate reduces the oxidizing agent.

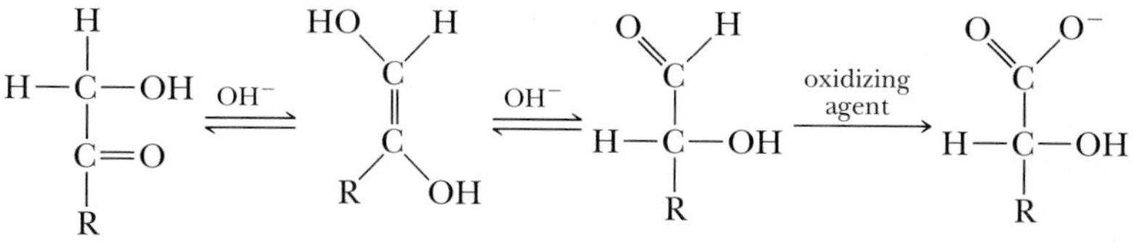

Testing for Glucose

The analytical procedure most often performed in a clinical chemistry laboratory is the determination of glucose in blood, urine, or other biological fluids. This is true because of the high incidence of diabetes mellitus. Approximately two million known diabetics live in the United States, and it is estimated that another one million are undiagnosed.

Diabetes mellitus is characterized by insufficient blood levels of the hormone insulin. If the blood concentration of insulin is too low, muscle and liver cells do not absorb glucose from the blood, which, in turn, leads to increased levels of blood glucose (hyperglycemia), impaired metabolism of fats and proteins, ketosis, and possible diabetic coma. A rapid test for blood glucose levels is critical for early diagnosis and effective management of this disease. In addition to being rapid, a test must also be specific for D-glucose; it must give a positive test for glucose but not react with any other substance normally present in biological fluids.

Blood glucose levels are now measured by an enzyme-based procedure using the enzyme glucose oxidase. This enzyme catalyzes the oxidation of β-D-glucose to D-gluconic acid.

Glucose oxidase is specific for β-D-glucose. Therefore, complete oxidation of any sample containing both β-D-glucose and α-D-glucose requires conversion of the α form to the β form. Fortunately, this interconversion is rapid and complete in the short time required for the test.

$$\beta\text{-D-Glucopyranose} + O_2 + H_2O \xrightarrow{\text{glucose oxidase}}$$

β-D-Glucopyranose
(β-D-Glucose)

$$\text{D-Gluconic acid} + H_2O_2$$

D-Gluconic acid   Hydrogen peroxide

Molecular oxygen, O_2, is the oxidizing agent in this reaction and is reduced to hydrogen peroxide, H_2O_2. In one procedure, hydrogen peroxide formed in the glucose oxidase-catalyzed reaction is used to oxidize colorless o-toluidine to a colored product in a reaction catalyzed by the enzyme peroxidase. The concentration of the colored oxidation product is determined spectrophotometrically and is proportional to the concentration of glucose in the test solution.

$$\text{2-Methylaniline} + H_2O_2 \xrightarrow{\text{peroxidase}} \text{colored product}$$

2-Methylaniline
(o-Toluidine)

Several commercially available test kits use the glucose oxidase reaction for qualitative determination of glucose in urine.

A test kit for the presence of glucose in urine. (Charles D. Winters)

D. Oxidation to Uronic Acids

Enzyme-catalyzed oxidation of the primary hydroxyl group at carbon 6 of a hexose yields a uronic acid. Enzyme-catalyzed oxidation of D-glucose, for example, yields D-glucuronic acid (see *The Merck Index*, 12th ed., #4474), shown here in both its open-chain and cyclic hemiacetal forms.

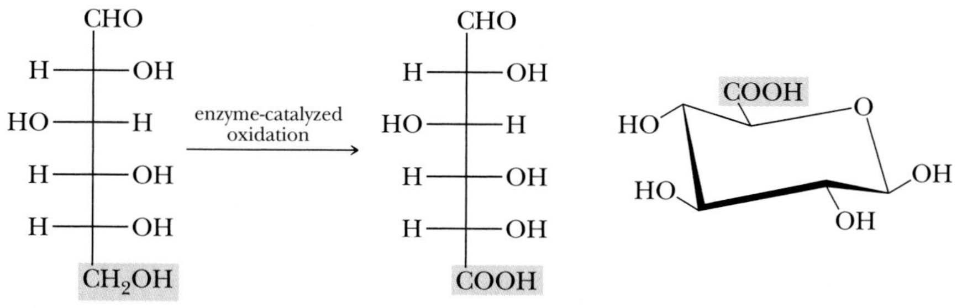

D-Glucose   Fischer projection   Chair conformation

D-Glucuronic acid
(a uronic acid)

D-Glucuronic acid is widely distributed in both the plant and animal world. In humans, it is an important component of the acidic polysaccharides of connective tissues (Section 25.6). It is also used by the body to detoxify foreign hydroxyl-containing compounds, such as phenols and alcohols. In the liver, these compounds are converted to glycosides of glucuronic acid (glucuronides) and excreted in the urine. The intravenous anesthetic propofol (see *The Merck Index*, 12th ed., #8020), for example, is converted to the following glucuronide and excreted in the urine.

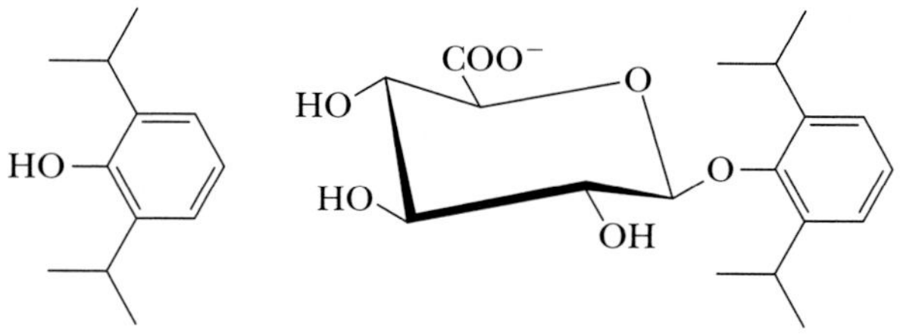

Propofol   A urine-soluble glucuronide

E. Oxidation by Periodic Acid

Oxidation by periodic acid has proven useful in structure determinations of carbohydrates, particularly in determining the size of glycoside rings. Recall from Section 9.8C that periodic acid cleaves the carbon-carbon bond of a glycol in a reaction that proceeds through a cyclic periodic ester. In this reaction, iodine(VII) of periodic acid is reduced to iodine(V) of iodic acid.

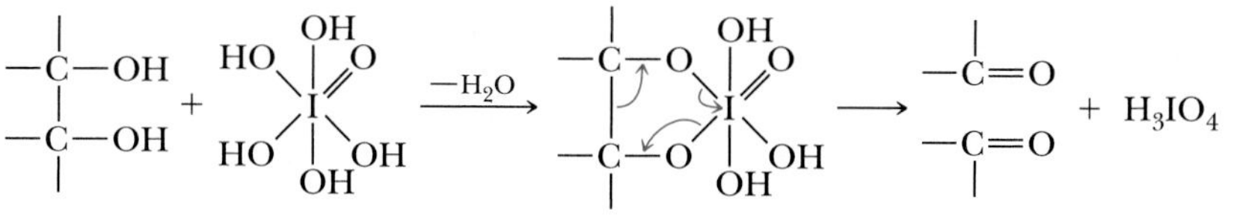

A glycol   Periodic acid   A cyclic periodic
ester   Iodic acid

Periodic acid also cleaves carbon-carbon bonds of α-hydroxyketones and α-hydroxy-aldehydes by a similar mechanism. Following are abbreviated structural formulas for these functional groups and the products of their oxidative cleavage by periodic acid. As a way to help you understand how each set of products is formed, each carbonyl in a starting material is shown as a hydrated intermediate that is then oxidized. In this way, each oxidation can be viewed as analogous to oxidation of a glycol.

α-Hydroxyketone   Hydrated
                         intermediate

α-Hydroxyketone   Hydrated
                         intermediate

α-Hydroxyaldehyde   Hydrated
                         intermediate

As an example of the usefulness of this reaction in carbohydrate chemistry, oxidation of methyl β-D-glucoside consumes two moles of periodic acid and produces one mole of formic acid. This stoichiometry and the formation of formic acid is possible only if —OH groups are on three adjacent carbon atoms.

This is evidence that methyl β-D-glucoside is indeed a pyranoside.

periodic acid cleavage at these two bonds

Methyl β-D-glucopyranoside

$\xrightarrow{2H_5IO_6}$

Methyl β-D-fructoside consumes only one mole of periodic acid and produces neither formaldehyde nor formic acid. Thus, oxidizable groups exist on adjacent carbons only at one site in the molecule. The fructoside, therefore, must be a five-membered ring (a fructofuranoside).

Periodic acid cleaves only this bond.

$\xrightarrow{H_5IO_6}$

Methyl β-D-fructofuranoside

25.4 Disaccharides and Oligosaccharides

Most carbohydrates in nature contain more than one monosaccharide unit. Those that contain two units are called **disaccharides,** those that contain three units are called **trisaccharides,** and so forth. The general term **oligosaccharide** is often used for carbohydrates that contain from four to ten monosaccharide units. Carbohydrates containing larger numbers of monosaccharide units are called **polysaccharides.**

In a disaccharide, two monosaccharide units are joined together by a glycosidic bond between the anomeric carbon of one unit and an —OH of the other. Three important disaccharides are maltose, lactose, and sucrose.

Disaccharide A carbohydrate containing two monosaccharide units joined by a glycosidic bond.

Oligosaccharide A carbohydrate containing four to ten monosaccharide units, each joined to the next by a glycosidic bond.

Polysaccharide A carbohydrate containing a large number of monosaccharide units, each joined to the next by one or more glycosidic bonds.

A. Maltose

Maltose derives its name from its presence in malt, the juice from sprouted barley and other cereal grains (from which beer is brewed). Maltose consists of two molecules of D-glucopyranose joined by an α-1,4-glycosidic bond between carbon 1 (the anomeric carbon) of one unit and carbon 4 of the other unit. Shown in Figure 25.6 is a chair representation for β-maltose, so named because the —OH on the anomeric carbon of the glucose unit on the right is beta. Maltose is a reducing sugar because the hemiacetal group on the right unit of D-glucopyranose is in equilibrium with the free aldehyde and can be oxidized to a carboxylic acid.

B. Lactose

Lactose is the principal sugar present in milk. It makes up about 5–8% of human milk and 4–6% percent of cow's milk. It consists of D-galactopyranose bonded by a β-

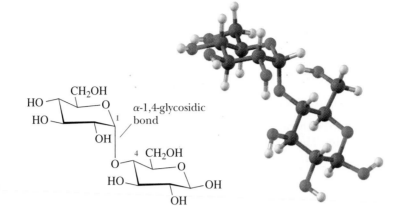

Figure 25.6
β-Maltose.

These products help individuals with lactose intolerance meet their calcium needs.
(Charles D. Winters)

1,4-glycosidic bond to carbon 4 of D-glucopyranose (Figure 25.7). Lactose is a reducing sugar.

C.   Sucrose

Sucrose (table sugar) is the most abundant disaccharide in the biological world (Figure 25.8). It is obtained principally from the juice of sugar cane and sugar beets. In sucrose, carbon 1 of α-D-glucopyranose is joined to carbon 2 of D-fructofuranose by an α-1,2-glycosidic bond. Note that glucose is a six-membered (pyranose) ring, whereas fructose is a five-membered (furanose) ring. Because the anomeric carbons of both the glucopyranose and fructofuranose units are involved in formation of the glycosidic bond, sucrose is a nonreducing sugar.

Example 25.6

Draw a chair conformation for the β anomer of a disaccharide in which two units of D-glucopyranose are joined by an α-1,6-glycosidic bond.

Solution

First draw a chair conformation of α-D-glucopyranose. Then connect the anomeric carbon of this monosaccharide to carbon 6 of a second D-glucopyranose unit by an α-

Figure 25.7
Lactose (milk sugar).

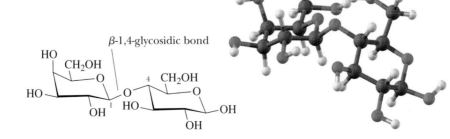

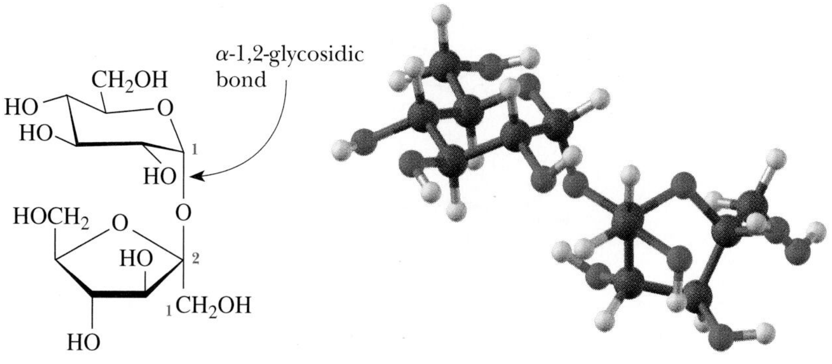

Figure 25.8
Sucrose (table sugar).

glycosidic bond. The resulting molecule is either α or β depending on the orientation of the —OH group on the reducing end of the disaccharide. The disaccharide shown here is β.

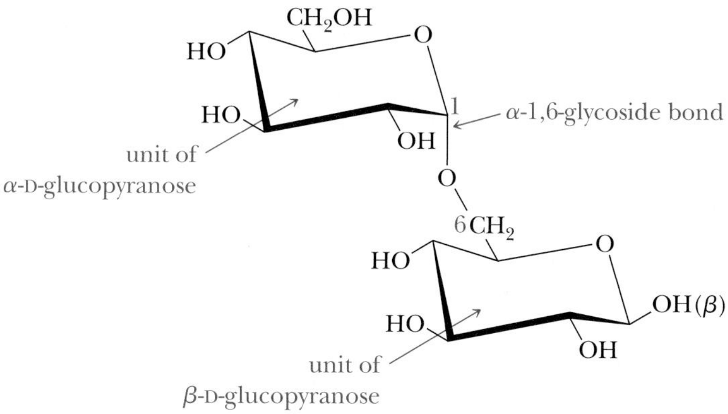

Problem 25.6

Draw a chair conformation for the α form of a disaccharide in which two units of D-glucopyranose are joined by a β-1,3-glycosidic bond.

25.5 Polysaccharides

Polysaccharides consist of large numbers of monosaccharide units bonded together by glycosidic bonds. Three important polysaccharides, all made up of glucose units, are starch, glycogen, and cellulose.

A. Starch—Amylose and Amylopectin

Starch is used for energy storage in plants. It is found in all plant seeds and tubers and is the form in which glucose is stored for later use. Starch can be separated into two principal polysaccharides: amylose and amylopectin. Although the starch from each plant is unique, most starches contain 20–25% amylose and 75–80% amylopectin.

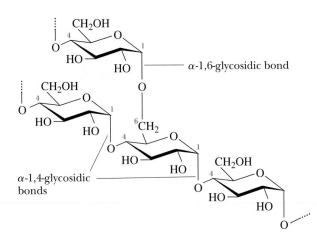

Figure 25.9
Amylopectin is a branched polymer of approximately 10,000 D-glucose units joined by α-1,4-glycosidic bonds. Branches consist of 24–30 D-glucose units started by α-1,6-glycosidic bonds.

Complete hydrolysis of both amylose and amylopectin yields only D-glucose. Amylose is composed of continuous, unbranched chains of up to 4000 D-glucose units joined by α-1,4-glycosidic bonds. Amylopectin contains chains of up to 10,000 D-glucose units also joined by α-1,4-glycosidic bonds. In addition, there is considerable branching from this linear network. At branch points, new chains of 24 to 30 units are started by α-1,6-glycosidic bonds (Figure 25.9).

B. Glycogen

Glycogen is the energy-reserve carbohydrate for animals. Like amylopectin, glycogen is a branched polysaccharide of approximately 10^6 glucose units joined by α-1,4- and α-1,6-glycosidic bonds. The total amount of glycogen in the body of a well-nourished adult human is about 350 g, divided almost equally between liver and muscle.

C. Cellulose

Cellulose, the most widely distributed plant skeletal polysaccharide, constitutes almost half of the cell wall material of wood. Cotton is almost pure cellulose. Cellulose is a linear polysaccharide of D-glucose units joined by β-1,4-glycosidic bonds (Figure 25.10). It has an average molecular weight of 400,000 g/mol, corresponding to approximately 2200 glucose units per molecule. Cellulose molecules act very much like stiff rods, a feature that enables them to align themselves side by side into well-organized water-insoluble fibers in which the OH groups form numerous intermolecular hydrogen bonds. This arrangement of parallel chains in bundles gives cellulose fibers their high mechanical strength. It is also the reason cellulose is insoluble in water. When a piece of cellulose-containing material is placed in water, there are not

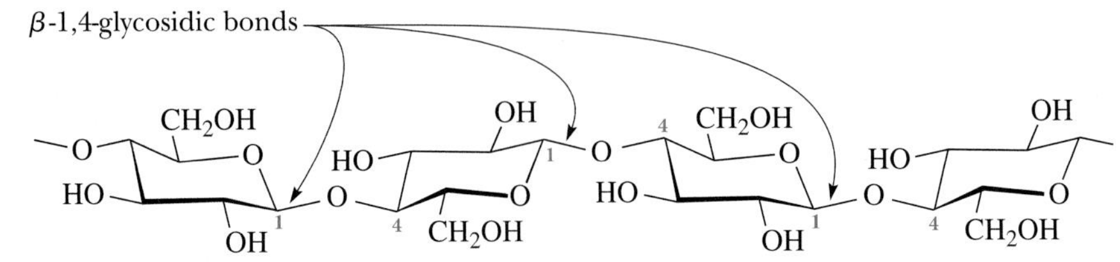

Figure 25.10
Cellulose is a linear polysaccharide of up to 2200 units of D-glucose joined by β-1,4-glycosidic bonds.

enough water molecules on the surface of the fiber to pull individual cellulose molecules away from the strongly hydrogen-bonded fiber.

Humans and other animals cannot use cellulose as food because our digestive systems do not contain β-glucosidases, enzymes that catalyze hydrolysis of β-glucosidic bonds. Instead, we have only α-glucosidases; hence, the polysaccharides we use as sources of glucose are starch and glycogen. On the other hand, many bacteria and microorganisms do contain β-glucosidases and so can digest cellulose. Termites are fortunate (much to our regret) to have such bacteria in their intestines and can use wood as their principal food. Ruminants (cud-chewing animals) and horses can also digest grasses and hay because β-glucosidase-containing microorganisms are present in their alimentary systems.

D.   Textile Fibers from Cellulose

Cotton is almost pure cellulose. Both rayon and acetate rayon are made from chemically modified cellulose and were the first commercially important synthetic textile fibers. In the production of rayon, cellulose-containing materials are treated with carbon disulfide, CS_2, in aqueous sodium hydroxide. In this reaction, some of the —OH groups on a cellulose fiber are converted to the sodium salt of a xanthate ester, which causes the fibers to dissolve in alkali as a viscous colloidal dispersion.

The solution of cellulose xanthate is separated from the alkali-insoluble parts of wood and then forced through a spinneret, a metal disc with many tiny holes, into dilute sulfuric acid to hydrolyze the xanthate ester groups and precipitate regenerated cellulose. Regenerated cellulose extruded as a filament is called viscose rayon thread.

In the industrial synthesis of acetate rayon, cellulose is treated with acetic anhydride. Acetylated cellulose is then dissolved in a suitable solvent, precipitated, and drawn into fibers known as acetate rayon. Today, acetate rayon fibers rank fourth in production in the United States, surpassed only by Dacron polyester, nylon, and rayon.

A glucose unit in   Acetic   A fully acetylated glucose unit
a cellulose fiber   anhydride

25.6 Acidic Polysaccharides

Acidic polysaccharides are a group of polysaccharides that contain carboxyl groups and/or sulfuric ester groups. These compounds play important roles in the structure and function of connective tissues, which form the matrix between organs and cells that provides mechanical strength and also filters the flow of molecular information between cells. Most connective tissues are made up of collagen, a structural protein, in combination with a variety of acidic polysaccharides that interact with collagen to form tight or loose networks.

There is no single general type of connective tissue. Rather, there are a large number of highly specialized forms, such as cartilage, bone, synovial fluid, skin, tendons, blood vessels, intervertebral disks, and cornea.

A. Hyaluronic Acid

Hyaluronic acid is the simplest acidic polysaccharide present in connective tissue. It has a molecular weight of between 10^5 and 10^7 g/mol and contains from 30,000 to 100,000 repeating units, depending on the organ in which it occurs. It is most abundant in embryonic tissues and in specialized connective tissues such as synovial fluid, the lubricant of joints in the body, and the vitreous humor of the eye where it provides a clear, elastic gel that maintains the retina in its proper position.

The repeating disaccharide unit in hyaluronic acid is D-glucuronic acid linked by a β-1,3-glycosidic bond to N-acetyl-D-glucosamine.

The repeating unit of hyaluronic acid

B. Heparin

Heparin is a heterogeneous mixture of variably sulfonated polysaccharide chains, ranging in molecular weight from 6000 to 30,000 g/mol. This acidic polysaccharide is synthesized and stored in mast cells of various tissues, particularly the liver, lungs, and gut. Heparin has many biological functions, the best known and understood of which is its anticoagulant activity. It binds strongly to antithrombin III, a plasma protein involved in terminating the clotting process.

The repeating monosaccharide units of heparin are D-glucosamine, D-glucuronic acid, and L-ioduronic acid bonded by a combination of α-1,4- and β-1,4-glycosidic bonds. Figure 25.11 shows a pentasaccharide unit of heparin that binds to and inhibits the enzymatic activity of antithrombin III.

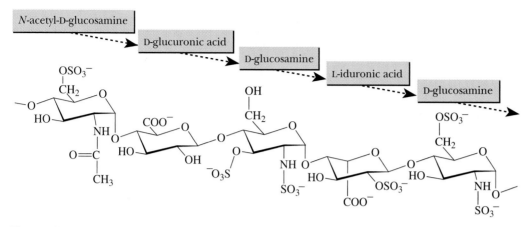

Figure 25.11
A pentasaccharide unit of heparin.

Summary

Monosaccharides (Section 25.1A) are polyhydroxyaldehydes or polyhydroxyketones. They have the general formula $C_nH_{2n}O_n$, where n varies from 3 to 8. Their names contain the suffix -ose. The prefixes tri-, tetr-, pent-, and so on show the number of carbon atoms in the chain. The prefix aldo- shows an aldehyde, and the prefix keto- shows a ketone. In a **Fischer projection** (Section 25.1B) of a monosaccharide, the carbon chain is written vertically with the most highly oxidized carbon toward the top. Horizontal lines show groups projecting above the plane of the page; vertical lines show groups projecting behind the plane of the page.

The **penultimate carbon** is the next to last on the carbon chain of a Fischer projection of a monosaccharide (Section 25.1C). A monosaccharide that has the same configuration at its penultimate carbon as D-glyceraldehyde is called a **D-monosaccharide;** one that has the same configuration at its penultimate carbon as L-glyceraldehyde is called an **L-monosaccharide.**

Monosaccharides exist primarily as cyclic hemiacetals (Section 25.2). The new stereocenter resulting from hemiacetal formation is referred to as the **anomeric carbon,** and the stereoisomers thus formed are called **anomers.** A six-membered cyclic hemiacetal is called a **pyranose;** a five-membered cyclic hemiacetal is called a **furanose.** The symbol β- indicates that the —OH on the anomeric carbon is on the same side of the ring as the terminal —CH$_2$OH. The symbol α- indicates that it is on the opposite side from the terminal —CH$_2$OH. Furanoses and pyranoses can be drawn as **Haworth projections** (Section 25.2A). Pyranoses can also be shown as chair confor-

mations (Section 25.2B). **Mutarotation** (Section 25.2C) is the change in specific rotation that accompanies formation of an equilibrium mixture of α- and β-anomers in aqueous solution.

A **glycoside** (Section 25.3A) is an acetal derived from a monosaccharide. The name of the glycoside is composed of the name of the alkyl or aryl group bonded to the acetal oxygen atom followed by the name of the monosaccharide in which the terminal -e has been replaced by -ide.

An **alditol** (Section 25.3B) is a polyhydroxy compound formed by reduction of the carbonyl group of a monosaccharide to a hydroxyl group. Reduction of D-glucose, for example, gives D-glucitol. An **aldonic acid** (Section 25.3C) is a carboxylic acid formed by oxidation of the aldehyde group of an aldose. Oxidation of D-glucose, for example, gives D-gluconic acid. **Reducing sugars** (Section 25.3C) are oxidized by mild oxidizing agents to aldonic acids.

A **disaccharide** (Section 25.4) contains two monosaccharide units joined by a glycosidic bond. Terms applied to carbohydrates containing larger numbers of monosaccharides are trisaccharide, tetrasaccharide, oligosaccharide, and polysaccharide. Maltose is a disaccharide of two molecules of D-glucose joined by an α-1,4-glycosidic bond. Lactose is a disaccharide consisting of D-galactose joined to D-glucose by a β-1,4-glycosidic bond. Sucrose is a disaccharide containing D-glucose joined to D-fructose by a 1,2-glycosidic bond.

Starch (Section 25.5A) can be separated into two fractions given the names amylose and amylopectin. Amylose is a linear polymer of up to 4000 units of D-glucopyranose joined by α-1,4-glycosidic bonds. Amylopectin is a highly branched polymer of D-glucopyranose joined by α-1,4-glycosidic bonds and, at branch points, by α-1,6-glycosidic bonds. Glycogen (Section 25.5B), the reserve carbohydrate of animals, is a highly branched polymer of D-glucopyranose joined by α-1,4-glycosidic bonds and, at branch points, by α-1,6-glycosidic bonds. Cellulose (Section 25.5C), the skeletal polysaccharide of plants, is a linear polymer of D-glucopyranose joined by β-

1,4-glycosidic bonds. Rayon (Section 25.5D) is made from chemically modified and regenerated cellulose. Acetate rayon is made by acetylation of cellulose.

The carboxyl and sulfate groups of acidic polysaccharides (Section 25.6) are ionized to —COO^- and —SO_3^- at the pH of body fluids, which gives these polysaccharides net negative charges.

Key Reactions

1. Formation of Cyclic Hemiacetals (Section 25.2A)

A monosaccharide existing as a five-membered ring is a furanose; one existing as a six-membered ring is a pyranose. A pyranose is most commonly drawn as either a Haworth projection or a chair conformation.

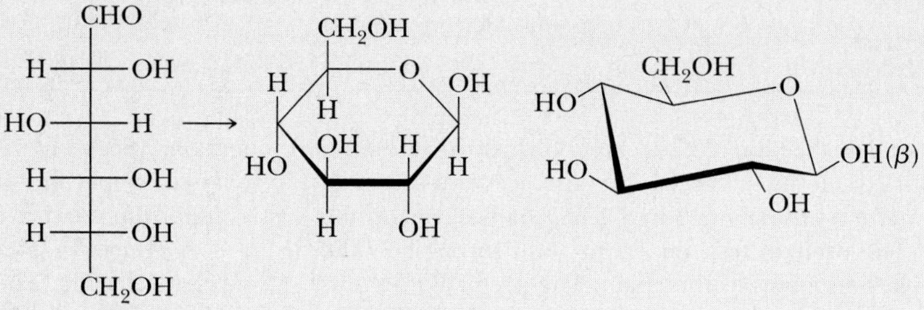

D-Glucose

β-D-Glucopyranose
(β-D-Glucose)

2. Mutarotation (Section 25.2C)

Anomeric forms of a monosaccharide are in equilibrium in aqueous solution. Mutarotation is the change in specific rotation that accompanies this equilibration.

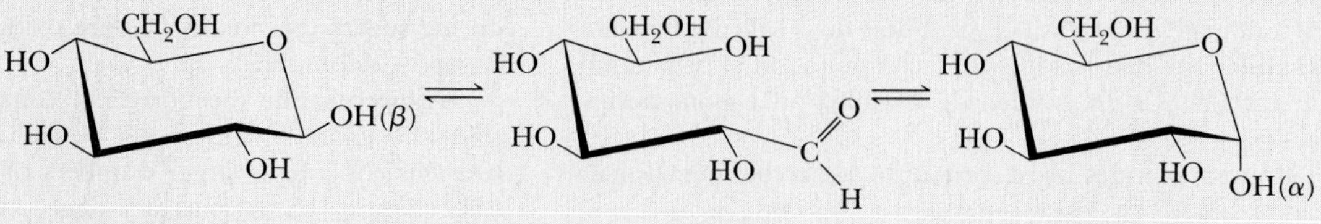

β-D-Glucopyranose
$[\alpha]_D^{25} +18.7°$

Open-chain form

α-D-Glucopyranose
$[\alpha]_D^{25} +112°$

3. Formation of Glycosides (Section 25.3A)

Treatment of a monosaccharide with an alcohol in the presence of an acid catalyst forms a cyclic acetal called a glycoside. The bond to the new —OR group is called a glycosidic bond.

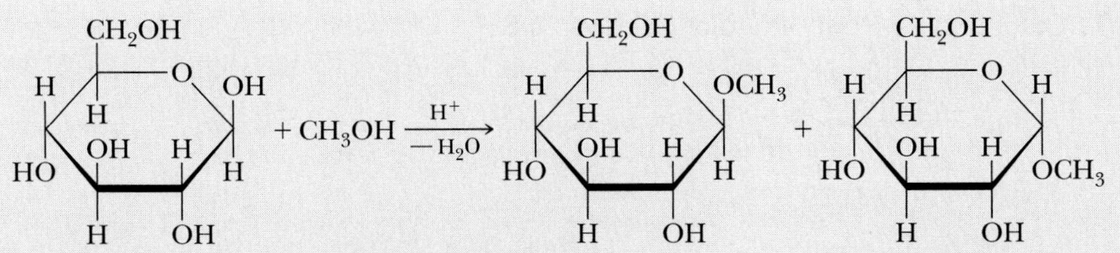

$+ CH_3OH \xrightarrow[- H_2O]{H^+}$

4. Formation of *N*-Glycosides (Section 25.3A)

N-Glycosides formed between a monosaccharide and a heterocyclic aromatic amine are especially important in the biological world.

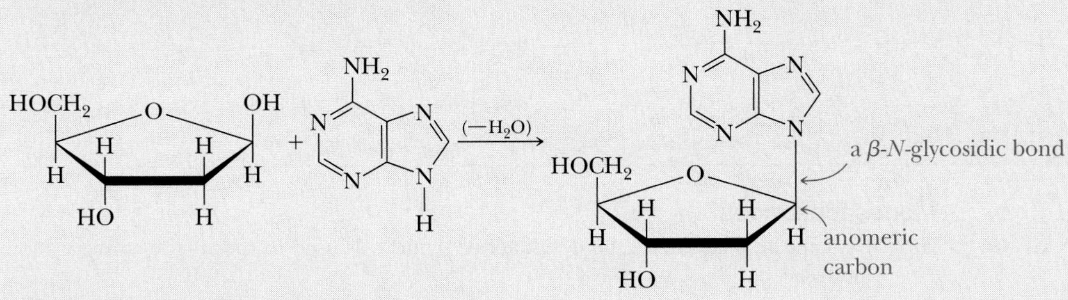

a *β-N*-glycosidic bond

anomeric carbon

5. Reduction to Alditols (Section 25.3B)

Reduction of the carbonyl group of an aldose or ketose to a hydroxyl group yields a polyhydroxy compound called an alditol.

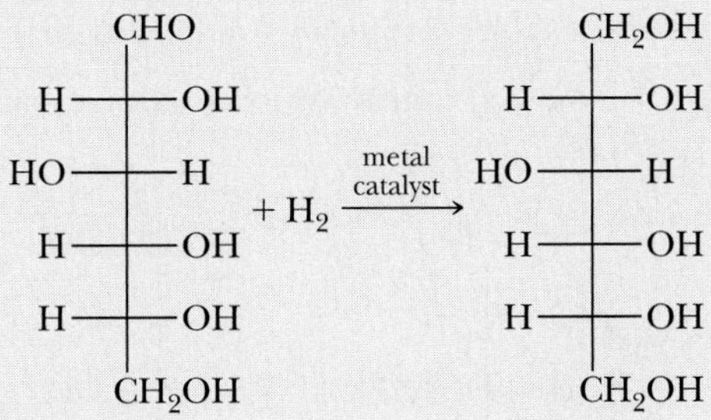

D-Glucose $+ H_2 \xrightarrow[]{\text{metal catalyst}}$ D-Glucitol
(D-Sorbitol)

6. Oxidation to an Aldonic Acid (Section 25.4C)

Oxidation of the aldehyde group of an aldose to a carboxyl group by a mild oxidizing agent gives a polyhydroxycarboxylic acid called an aldonic acid.

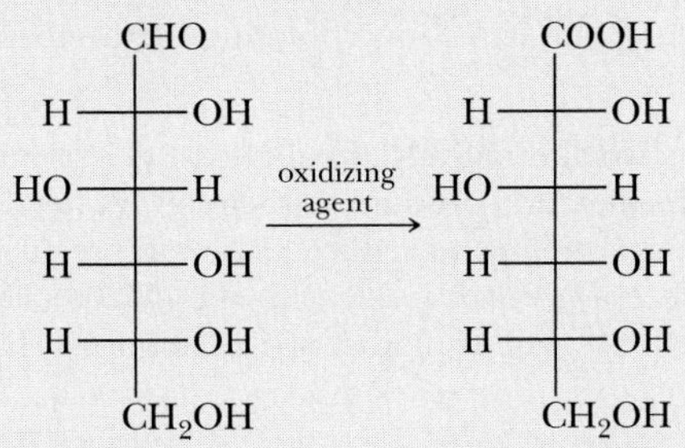

D-Glucose $\xrightarrow[]{\text{oxidizing agent}}$ D-Gluconic acid

7. Oxidation by Periodic Acid (Section 25.3E)

Periodic acid oxidizes and cleaves carbon-carbon bonds of glycol, α-hydroxyketone, and α-hydroxyaldehyde groups.

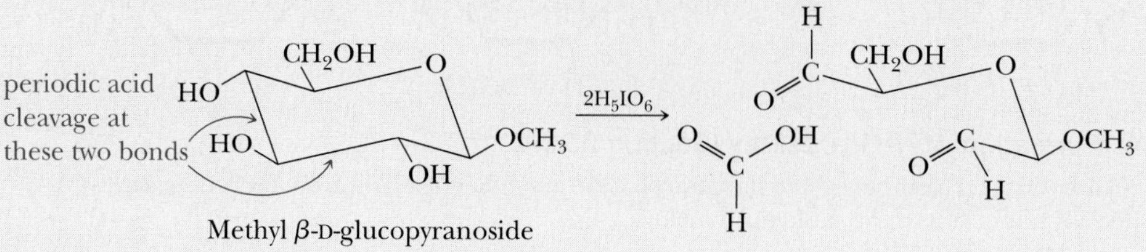

periodic acid
cleavage at
these two bonds

$$\xrightarrow{\text{2H}_5\text{IO}_6}$$

Methyl β-D-glucopyranoside

Problems

Monosaccharides

25.7 Explain the meaning of the designations D and L as used to specify the configuration of monosaccharides.

25.8 How many stereocenters are present in D-glucose? In D-ribose?

25.9 Which carbon of an aldopentose determines whether the pentose has a D or L configuration?

25.10 How many aldooctoses are possible? How many D-aldooctoses are possible?

25.11 Which compounds are D-monosaccharides? Which are L-monosaccharides?

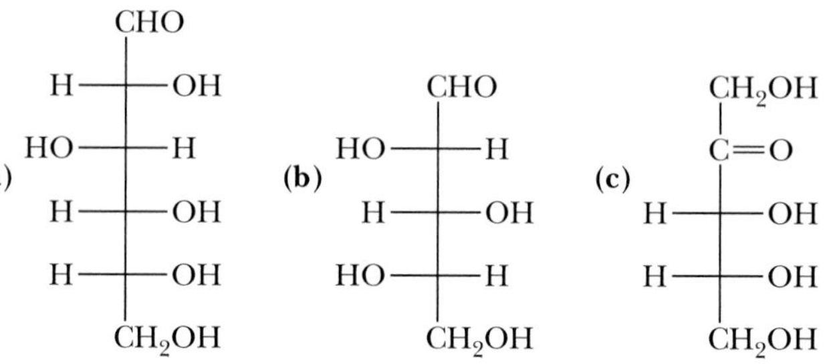

(a)
CHO
H——OH
HO——H
H——OH
H——OH
CH$_2$OH

(b)
CHO
HO——H
H——OH
HO——H
CH$_2$OH

(c)
CH$_2$OH
C=O
H——OH
H——OH
CH$_2$OH

25.12 Write Fischer projections for L-ribose and L-arabinose.

25.13 What is the meaning of the prefix deoxy- as it is used in carbohydrate chemistry?

25.14 Give L-fucose (Problem 25.31) a name incorporating the prefix deoxy- that shows its relationship to galactose.

25.15 2,6-Dideoxy-D-altrose, known alternatively as D-digitoxose, is a monosaccharide obtained on hydrolysis of digitoxin, a natural product extracted from foxglove (*Digitalis purpurea*). Digitoxin is widely use in cardiology to reduce pulse rate, regularize heart rhythm, and strengthen heart beat (see *The Merck Index*, 12th ed., #3206). Draw the structural formula of 2,6-dideoxy-D-altrose. For a model of this monosaccharide, see the chapter opening page.

The Cyclic Structure of Monosaccharides

25.16 Define the term anomeric carbon. In glucose, which carbon is the anomeric carbon?

25.17 Define (a) pyranose and (b) furanose.

25.18 Which is the anomeric carbon in a 2-ketohexose?

25.19 Are α-D-glucose and β-D-glucose enantiomers? Explain.

25.20 Convert each Haworth projection to an open-chain form and then to a Fischer projection. Name the monosaccharide you have drawn.

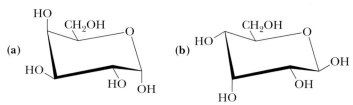

25.21 Convert each chair conformation to an open-chain form and then to a Fischer projection. Name the monosaccharide you have drawn.

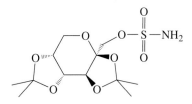

25.22 Explain the phenomenon of mutarotation with reference to carbohydrates. By what means is it detected?

25.23 The specific rotation of α-D-glucose is +112.2°.

(a) What is the specific rotation of α-L-glucose?
(b) When α-D-glucose is dissolved in water, the specific rotation of the solution changes from +112.2° to +52.7°. Does the specific rotation of α-L-glucose also change when it is dissolved in water? If so, to what value does it change?

25.24 Following is a structural formula of the anticonvulsant topiramate (see *The Merck Index*, 12th ed., #9686). See the model on the CD.

Topiramate

(a) Describe the conformation of the six-membered ring in topiramate.
(b) Draw a structural formula of the cyclic structure remaining after hydrolysis of each amide, ester, and cyclic acetal.
(c) The structure remaining is a monosaccharide. Name the monosaccharide and specify whether it belongs to the D series or the L series.

Reactions of Monosaccharides

25.25 Draw Fischer projections for the product(s) formed by reaction of D-galactose with the following. In addition, state whether each product is optically active or inactive.

(continued on next page)

(a) NaBH$_4$ in H$_2$O   (b) H$_2$/Pt   (c) HNO$_3$, warm
(d) Br$_2$/H$_2$O/CaCO$_3$ (e) H$_5$IO$_6$ (f) C$_6$H$_5$NH$_2$

25.26 Repeat Problem 25.25 using D-ribose.

25.27 An important technique for establishing relative configurations among isomeric aldoses and ketoses is to convert both terminal carbon atoms to the same functional group. This can be done either by selective oxidation or reduction. As a specific example, nitric acid oxidation of D-erythrose gives meso-tartaric acid (Table 3.1, Section 3.4). Similar oxidation of D-threose gives (2S,3S)-tartaric acid. Given this information and the fact that D-erythrose and D-threose are diastereomers, draw Fischer projections for D-erythrose and D-threose. Check your answers against Table 25.1.

25.28 There are four D-aldopentoses (Table 25.1). If each is reduced with NaBH$_4$, which yield optically active alditols? Which yield optically inactive alditols?

25.29 Name the two alditols formed by NaBH$_4$ reduction of D-fructose.

25.30 One pathway for the metabolism of glucose 6-phosphate is its enzyme-catalyzed conversion to fructose 6-phosphate. Show that this transformation can be regarded as two enzyme-catalyzed keto-enol tautomerisms.

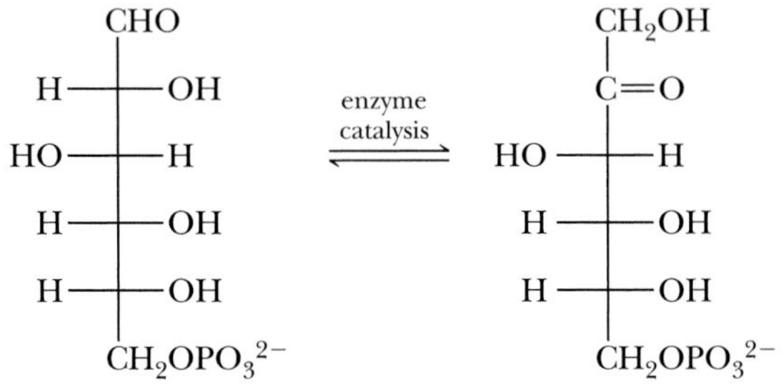

D-Glucose 6-phosphate   D-Fructose 6-phosphate

25.31 L-Fucose, one of several monosaccharides commonly found in the surface polysaccharides of animal cells, is synthesized biochemically from D-mannose in the following eight steps.

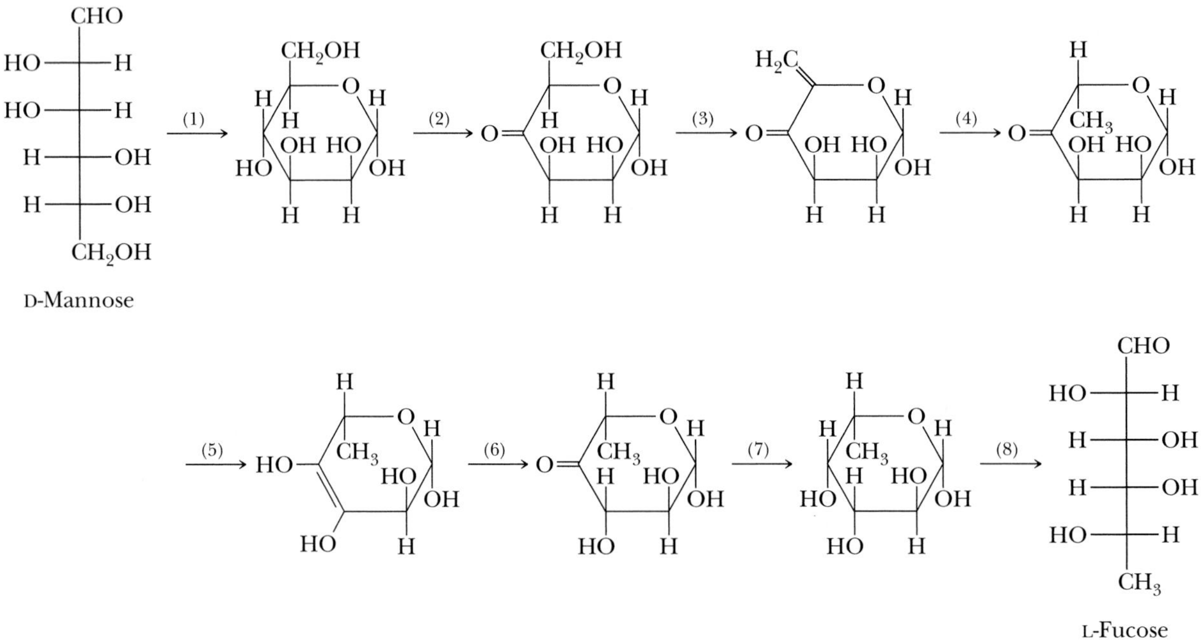

D-Mannose

L-Fucose

(a) Describe the type of reaction (that is, oxidation, reduction, hydration, dehydration, and so on) involved in each step.

(b) Explain why it is that this monosaccharide derived from D-mannose now belongs to the L series.

25.32 What is the difference in meaning between the terms glycosidic bond and glucosidic bond?

25.33 Treatment of methyl β-D-glucopyranoside with benzaldehyde forms a six-membered cyclic acetal. Draw the most stable conformation of this acetal. Identify each new stereocenter in the acetal.

25.34 Vanillin, the principal component of vanilla, occurs in vanilla beans and other natural sources as a β-D-glucopyranoside. Draw a structural formula for this glycoside, showing the D-glucose unit as a chair conformation (see *The Merck Index*, 12th ed., #10069).

Vanillin

25.35 Hot water extracts of ground willow and poplar bark are an effective pain reliever. Unfortunately, the liquid is so bitter that most persons refuse it. The pain reliever in these infusions is salicin, a β-glycoside of D-glucopyranose and the phenolic —OH group of 2-(hydroxymethyl)phenol. Draw a structural formula for salicin, showing the glucose ring as a chain conformation (see *The Merck Index*, 12th ed., #8476).

25.36 Draw structural formulas for the products formed by hydrolysis at pH 7.4 (the pH of blood plasma) of all ester, thioester, amide, anhydride, and glycoside groups in acetyl coenzyme A. Name as many of the products as you can.

Acetyl coenzyme A
(Acetyl-CoA)

Disaccharides and Oligosaccharides

25.37 In making candy or sugar syrups, sucrose is boiled in water with a little acid, such as lemon juice. Why does the product mixture taste sweeter than the starting sucrose solution?

25.38 Trehalose is found in young mushrooms and is the chief carbohydrate in the blood of certain insects. Trehalose is a disaccharide consisting of two D-monosaccharide units, each joined to the other by an α-1,1-glycosidic bond.

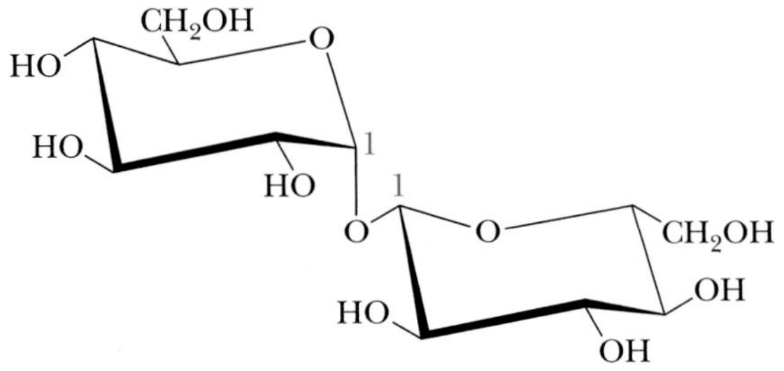

Trehalose

(a) Is trehalose a reducing sugar?
(b) Does trehalose undergo mutarotation?
(c) Name the two monosaccharide units of which trehalose is composed.

25.39 The trisaccharide raffinose occurs principally in cottonseed meal (see *The Merck Index*, 12th ed., #8279).

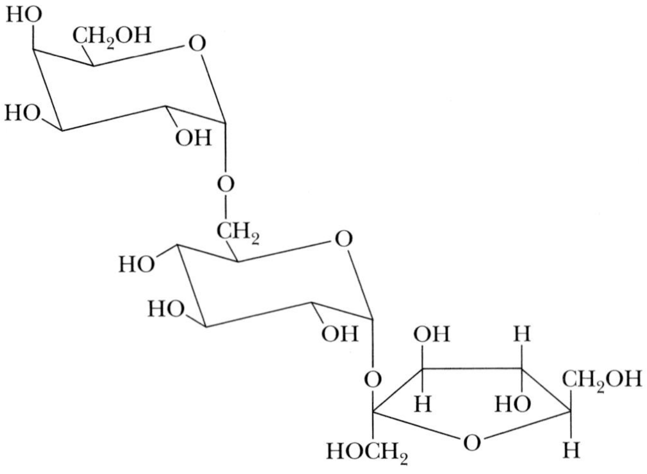

Raffinose

(a) Name the three monosaccharide units in raffinose.
(b) Describe each glycosidic bond in this trisaccharide.
(c) Is raffinose a reducing sugar?
(d) With how many moles of periodic acid will raffinose react?

25.40 Following is the structural formula of laetrile.

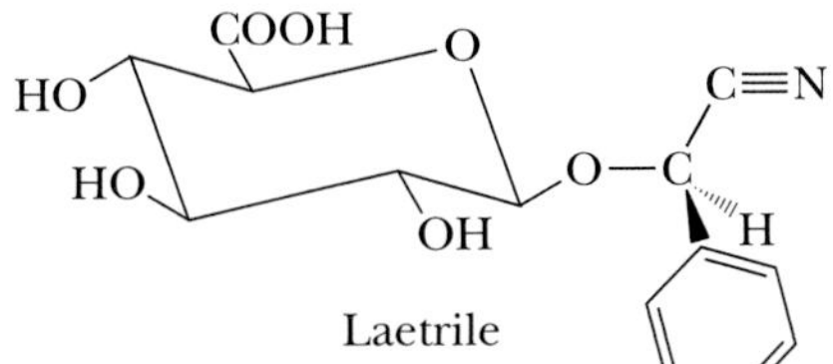

Laetrile

(a) Assign an *R* or *S* configuration to the stereocenter bearing the cyano (—CN) group.

(b) Account for the fact that on hydrolysis in warm aqueous acid, laetrile liberates benzaldehyde and HCN.

Polysaccharides

25.41 What is the difference in structure between oligo- and polysaccharides?

25.42 Why is cellulose insoluble in water?

25.43 Following is the Fischer projection for *N*-acetyl-D-glucosamine.

$$
\begin{array}{c}
\text{CHO}\quad \text{O} \\
\text{H}\!\!-\!\!\!\!-\!\!\text{NHCCH}_3 \\
\text{HO}\!\!-\!\!\!\!-\!\!\text{H} \\
\text{H}\!\!-\!\!\!\!-\!\!\text{OH} \\
\text{H}\!\!-\!\!\!\!-\!\!\text{OH} \\
\text{CH}_2\text{OH}
\end{array}
$$

N-Acetyl-D-glucosamine

(a) Draw a chair conformation for the α- and β-pyranose forms of this monosaccharide.

(b) Draw a chair conformation for the disaccharide formed by joining two units of the pyranose form of *N*-acetyl-D-glucosamine by a β-1,4-glycosidic bond. If you drew this correctly, you have the structural formula for the repeating dimer of chitin, the structural polysaccharide component of the shell of lobster and other crustaceans.

25.44 Propose structural formulas for the following polysaccharides.

(a) Alginic acid, isolated from seaweed, is used as a thickening agent in ice cream and other foods. Alginic acid is a polymer of D-mannuronic acid in the pyranose form joined by β-1,4-glycosidic bonds.

(b) Pectic acid is the main component of pectin, which is responsible for the formation of jellies from fruits and berries. Pectic acid is a polymer of D-galacturonic acid in the pyranose form joined by α-1,4-glycosidic bonds.

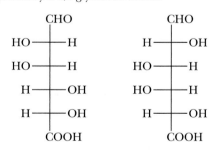

D-Mannuronic acid   D-Galacturonic acid

25.45 Digitalis (see *The Merck Index*, 12th ed., #3201) is a preparation made from the dried seeds and leaves of the purple foxglove (*Digitalis purpurea*), a plant native to southern and central Europe and cultivated in the United States. The preparation is a mixture of several active components, including digitalin. Digitalis is used in medicine to increase the force of myocardial contraction and as a conduction depressant to decrease heart rate (the heart pumps more forcefully but less often).

Digitalin

(a) Describe this glycosidic bond.

(b) Draw an open-chain Fischer projection of this monosaccharide.

(c) Describe this glycosidic bond.

(d) Name this monosaccharide unit.

25.46 Following is the structural formula of ganglioside GM_2, a macromolecular glycolipid (meaning that it contains lipid and monosaccharide units joined by glycosidic bonds). In normal cells, this and other gangliosides are synthesized continuously and degraded by lysosomes, which are cell organelles containing digestive enzymes. If pathways for the degradation of gangliosides are inhibited, the gangliosides accumulate in the central nervous system causing all sorts of life-threatening consequences. In inherited diseases of ganglioside metabolism, death usually occurs at an early age. Diseases of ganglioside metabolism include Gaucher's disease, Niemann-Pick disease, and Tay-Sachs disease. Tay-Sachs disease is a hereditary defect that is transmitted as an autosomal recessive gene. The concentration of ganglioside GM_2 is abnormally high in this disease because the enzyme responsible for catalyzing the hydrolysis of glycosidic bond (b) is absent.

Ganglioside GM_2 or Tay-Sachs ganglioside

(a) Name this monosaccharide unit.

(b) Describe this glycosidic bond (α or β, and between which carbons of each unit).

(c) Name this monosaccharide unit.

(d) Describe this glycosidic bond.

(e) Name this monosaccharide unit.

(f) Describe this glycosidic bond.

(g) This unit is *N*-acetylneuraminic acid, the most abundant member of a family of amino sugars containing nine or more carbons and distributed widely throughout the animal kingdom. Draw the open-chain form of this amino sugar. Do not be concerned with the configuration of the five stereocenters in the open-chain form.

25.47 Hyaluronic acid acts as a lubricant in the synovial fluid of joints. In rheumatoid arthritis, inflammation breaks hyaluronic acid down to smaller molecules. Under these conditions, what happens to the lubricating power of the synovial fluid?

25.48 The anticlotting property of heparin is partly due to the negative charges it carries.

(a) Identify the functional groups that provide the negative charges.

(b) Which type of heparin is a better anticoagulant, one with a high or a low degree of polymerization?

25.49 Keratin sulfate is an important component of the cornea of the eye. Following is the repeating unit of this acidic polysaccharide.

(a) From what monosaccharides or derivatives of monosaccharides is keratin sulfate made?

(b) Describe the glycosidic bond in this repeating disaccharide unit.

(c) What is the net charge on this repeating disaccharide unit at pH 7.0?

26

LIPIDS

Lipids are a heterogeneous group of naturally occurring organic compounds (many related to fats and oils), classified together on the basis of their common solubility properties. Lipids are insoluble in water but soluble in nonpolar aprotic organic solvents, including diethyl ether, methylene chloride, and acetone.

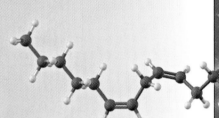

■ Sea lions are marine mammals that require a heavy layer of fat in order to survive in cold waters. *(Doug Perrine/TCL/MasterFile)* Inset: Linoleic acid, a major component of unsaturated triglycerides and phospholipids (Sections 26.1 and 26.5).

Lipids are divided into two main groups. First are those lipids that contain both a relatively large nonpolar hydrophobic region, most commonly aliphatic in nature, and a polar hydrophilic region. Found among this group are fatty acids, triglycerides, phospholipids, prostaglandins, and the fat-soluble vitamins. Second are those lipids that contain the tetracyclic ring system called the steroid nucleus, including cholesterol, steroid hormones, and bile acids. In this chapter, we describe the structures and biological functions of each group of lipids.

26.1   Triglycerides

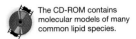

The CD-ROM contains molecular models of many common lipid species.

Animal fats and vegetable oils, the most abundant naturally occurring lipids, are triesters of glycerol and long-chain carboxylic acids. Fats and oils are also referred to as **triglycerides** or **triacylglycerols.** Hydrolysis of a triglyceride in aqueous base followed by acidification gives glycerol and three fatty acids.

Triglyceride (triacylglycerol) An ester of glycerol with three fatty acids.

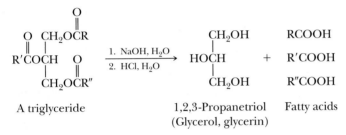

A triglyceride   1,2,3-Propanetriol (Glycerol, glycerin)   Fatty acids

A.   Fatty Acids

More than 500 different **fatty acids** have been isolated from various cells and tissues. Given in Table 26.1 are common names and structural formulas for the most abundant of these. The number of carbons in a fatty acid and the number of carbon-carbon double bonds in its hydrocarbon chain are shown by two numbers separated by a colon. In this notation, for example, linoleic acid is designated as an 18:2 fatty acid; its 18-carbon chain contains two carbon-carbon double bonds. Following are several characteristics of the most abundant fatty acids in higher plants and animals.

Fatty acid A long, unbranched-chain carboxylic acid, most commonly of 12 to 20 carbons, derived from the hydrolysis of animal fats, vegetable oils, or the phospholipids of biological membranes.

1. Nearly all fatty acids have an even number of carbon atoms, most between 12 and 20, in an unbranched chain.
2. The three most abundant fatty acids in nature are palmitic acid (16:0), stearic acid (18:0), and oleic acid (18:1).
3. In most unsaturated fatty acids, the cis isomer predominates; the trans isomer is rare.
4. Unsaturated fatty acids have lower melting points than their saturated counterparts. The greater the degree of unsaturation, the lower the melting point. Compare, for example, the melting points of linoleic acid, a **polyunsaturated fatty acid,** and stearic acid, a saturated fatty acid.

Polyunsaturated fatty acid A fatty acid with two or more carbon-carbon double bonds in its hydrocarbon chain.

Example 26.1

Draw a structural formula of a triglyceride derived from one molecule each of palmitic acid, oleic acid, and stearic acid, the three most abundant fatty acids in the biological world.

Table 26.1 **The Most Abundant Fatty Acids in Animal Fats, Vegetable Oils, and Biological Membranes**

Carbon Atoms/ Double Bonds*	Structure	Common Name	Melting Point (°C)
Saturated Fatty Acids			
12:0	$CH_3(CH_2)_{10}COOH$	Lauric acid	44
14:0	$CH_3(CH_2)_{12}COOH$	Myristic acid	58
16:0	$CH_3(CH_2)_{14}COOH$	Palmitic acid	63
18:0	$CH_3(CH_2)_{16}COOH$	Stearic acid	70
20:0	$CH_3(CH_2)_{18}COOH$	Arachidic acid	77
Unsaturated Fatty Acids			
16:1	$CH_3(CH_2)_5CH{=}CH(CH_2)_7COOH$	Palmitoleic acid	1
18:1	$CH_3(CH_2)_7CH{=}CH(CH_2)_7COOH$	Oleic acid	16
18:2	$CH_3(CH_2)_4(CH{=}CHCH_2)_2(CH_2)_6COOH$	Linoleic acid	−5
18:3	$CH_3CH_2(CH{=}CHCH_2)_3(CH_2)_6COOH$	Linolenic acid	−11
20:4	$CH_3(CH_2)_4(CH{=}CHCH_2)_4(CH_2)_2COOH$	Arachidonic acid	−49

*The first number is the number of carbons in the fatty acid; the second is the number of carbon-carbon double bonds in its hydrocarbon chain.

Solution

In this structure, palmitic acid is esterified at carbon 1 of glycerol; oleic acid, at carbon 2; and stearic acid, at carbon 3.

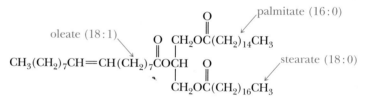

A triglyceride

Problem 26.1

(a) How many constitutional isomers are possible for a triglyceride containing one molecule each of palmitic acid, oleic acid, and stearic acid?

(b) Which of these constitutional isomers are chiral?

Among the components of beeswax is triacontyl palmitate, $CH_3(CH_2)_{14}CO_2(CH_2)_{29}CH_3$, an ester of palmitic acid.
(Charles D. Winters)

B. Physical Properties

The physical properties of a triglyceride depend on its fatty acid components. In general, the melting point of a triglyceride increases as the number of carbons in its hydrocarbon chains increases and as the number of carbon-carbon double bonds decreases. Triglycerides rich in oleic acid, linoleic acid, and other unsaturated fatty

Table 26.2 Grams of Fatty Acid per 100 g of Triglyceride of Several Fats and Oils*

Fat or Oil	Saturated Fatty Acids			Unsaturated Fatty Acids	
	Lauric (12:0)	Palmitic (16:0)	Stearic (18:0)	Oleic (18:1)	Linoleic (18:2)
Human fat	—	24.0	8.4	46.9	10.2
Beef fat	—	27.4	14.1	49.6	2.5
Butter fat	2.5	29.0	9.2	26.7	3.6
Coconut oil	45.4	10.5	2.3	7.5	Trace
Corn oil	—	10.2	3.0	49.6	34.3
Olive oil	—	6.9	2.3	84.4	4.6
Palm oil	—	40.1	5.5	42.7	10.3
Peanut oil	—	8.3	3.1	56.0	26.0
Soybean oil	0.2	9.8	2.4	28.9	50.7

*Only the most abundant fatty acids are given; other fatty acids are present in lesser amounts.

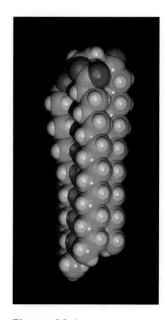

Figure 26.1

Tristearin, a saturated triglyceride.

Oil When used in the context of fats and oils, a mixture of triglycerides that is liquid at room temperature.

Fat A mixture of triglycerides that is semisolid or solid at room temperature.

acids are generally liquids at room temperature and are called **oils,** as for example corn oil and olive oil. Olive oil, which contains mainly the monounsaturated oleic acid, solidifies in the refrigerator, whereas the more unsaturated corn oil will not. Triglycerides rich in palmitic, stearic, and other saturated fatty acids are generally semisolids or solids at room temperature and are called **fats,** as for example human fat and butter fat. Fats of land animals typically contain approximately 40–50% saturated fatty acids by weight (Table 26.2). Most plant oils, on the other hand, contain 20% or less saturated fatty acids and 80% or more unsaturated fatty acids. The notable exceptions to this generalization about plant oils are the **tropical oils** (as for example coconut and palm oils), which are considerably richer in low-molecular-weight saturated fatty acids.

The lower melting points of triglycerides rich in unsaturated fatty acids are related to differences in three-dimensional shape between the hydrocarbon chains of their unsaturated and saturated fatty acid components. Shown in Figure 26.1 is a space-filling model of tristearin, a saturated triglyceride. In this model, the hydrocarbon chains lie parallel to each other, giving the molecule an ordered, compact shape. Because of this compact three-dimensional shape and the resulting strength of the dispersion forces between hydrocarbon chains of adjacent molecules, triglycerides rich in saturated fatty acids have melting points above room temperature.

The three-dimensional shape of an unsaturated fatty acid is quite different from that of a saturated fatty acid. Recall from Section 26.1A that unsaturated fatty acids of higher organisms are predominantly of the cis configuration; trans configurations are rare. Figure 26.2 shows a space-filling model of a **polyunsaturated triglyceride** derived from one molecule each of stearic acid, oleic acid, and linoleic acid. Each double bond in this polyunsaturated triglyceride has the cis configuration.

Polyunsaturated triglycerides have a less ordered structure and do not pack together as closely or as compactly as saturated triglycerides. Intramolecular and intermolecular dispersion forces are weaker, with the result that polyunsaturated triglycerides have lower melting points than their saturated counterparts.

Liquid vegetable oils contain mostly unsaturated fatty acids. *(Charles D. Winters)*

Polyunsaturated triglyceride A triglyceride having several carbon-carbon double bonds in the hydrocarbon chains of its three fatty acids.

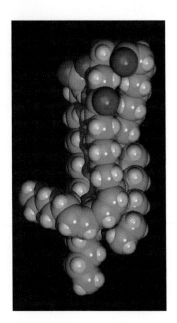

Figure 26.2
A polyunsaturated triglyceride.

C.   Reduction of Fatty Acid Chains

For a variety of reasons, in part convenience and in part dietary preference, conversion of oils to fats has become a major industry. The process is called **hardening** of oils and involves catalytic reduction (Section 6.6A) of some or all carbon-carbon double bonds. In practice, the degree of hardening is carefully controlled to produce fats of a desired consistency. The resulting fats are sold for kitchen use (Crisco, Spry, and others). Margarine and other butter substitutes are produced by partial hydrogenation of polyunsaturated oils derived from corn, cottonseed, peanut, and soybean oils. To the hardened oils are added β-carotene (to give the final product a yellow color and make it look like butter), salt, and about 15% milk by volume to form the final emulsion. Vitamins A and D are also often added. Because the product at this stage is tasteless, acetoin (see *The Merck Index*, 12th ed., #61) and diacetyl (see *The Merck Index*, 12th ed., #3010) are often added. These two compounds mimic the characteristic flavor of butter.

$$CH_3-CH-C-CH_3 \qquad CH_3-C-C-CH_3$$

3-Hydroxy-2-butanone   2,3-Butanedione
(Acetoin)   (Diacetyl)

26.2   Soaps and Detergents

A.   Structure and Preparation of Soaps

Soap   A sodium or potassium salt of a fatty acid.

Natural **soaps** are prepared most commonly from a blend of tallow and coconut oils. In the preparation of tallow, the solid fats of cattle are melted with steam, and the tallow layer formed on the top is removed. The preparation of soaps begins by boiling these triglycerides with sodium hydroxide. The reaction that takes place is called **saponification** (Latin: *saponem*, soap). At the molecular level, saponification corresponds to base-promoted hydrolysis of the ester groups in triglycerides (Section 18.5C). The resulting soaps contain mainly the sodium salts of palmitic, stearic, and oleic acids from tallow and the sodium salts of lauric and myristic acids from coconut oil.

$$\begin{array}{c}
O \quad CH_2OCR \\
RCOCH \quad O \\
CH_2OCR
\end{array} + 3NaOH \xrightarrow{\text{saponification}} \begin{array}{c}
CH_2OH \\
CHOH \\
CH_2OH
\end{array} + 3RCO^-Na^+$$

A triglyceride   1,2,3-Propanetriol   Sodium soaps
                                        (Glycerol; glycerin)

After hydrolysis is complete, sodium chloride is added to precipitate the soap as thick curds. The water layer is then drawn off, and glycerol is recovered by vacuum distillation. The crude soap contains sodium chloride, sodium hydroxide, and other impurities. These are removed by boiling the curd in water and reprecipitating with more sodium chloride. After several purifications, the soap can be used without further processing as an inexpensive industrial soap. Other treatments transform the crude soap into pH-controlled cosmetic soaps, medicated soaps, and the like.

(a)   A soap

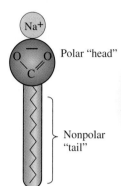

Na⁺

Polar "head"

Nonpolar "tail"

(b)   Cross section of a soap micelle in water

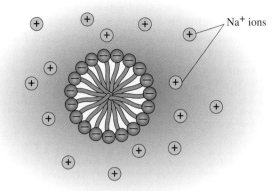

Na⁺ ions

Figure 26.3

Soap micelles. Nonpolar (hydrophobic) hydrocarbon chains are clustered in the interior of the micelle, and polar (hydrophilic) carboxylate groups are on the surface of the micelle. Soap micelles repel each other because of their negative surface charges.

B.   How Soaps Clean

Soap owes its remarkable cleansing properties to its ability to act as an emulsifying agent. Because the long hydrocarbon chains of natural soaps are insoluble in water, they tend to cluster in such a way as to minimize their contact with surrounding water molecules. The polar carboxylate groups, on the other hand, tend to remain in contact with the surrounding water molecules. Thus, in water, soap molecules spontaneously cluster into **micelles** (Figure 26.3).

Most of the things we commonly think of as dirt (such as grease, oil, and fat stains) are nonpolar and insoluble in water. When soap and this type of dirt are mixed together, as in a washing machine, the nonpolar hydrocarbon inner parts of the soap micelles "dissolve" the nonpolar dirt molecules. In effect, new soap micelles are formed, this time with nonpolar dirt molecules in the center (Figure 26.4). In this way, nonpolar organic grease, oil, and fat are "dissolved" and washed away in the polar wash water.

Soaps are not without their disadvantages. Foremost among these, they form insoluble salts when used in water containing Ca(II), Mg(II), or Fe(III) ions (hard water).

Micelle   A spherical arrangement of organic molecules in water solution clustered so that their hydrophobic parts are buried inside the sphere and their hydrophilic parts are on the surface of the sphere and in contact with water.

$$2CH_3(CH_2)_{14}COO^- \ Na^+ + Ca^{2+} \longrightarrow [CH_3(CH_2)_{14}COO^-]_2Ca^{2+} + 2Na^+$$

A sodium soap
(soluble in water as micelles)

Calcium salt of a fatty acid
(insoluble in water)

These calcium, magnesium, and iron salts of fatty acids create problems, including rings around the bathtub, films that spoil the luster of hair, and grayness and roughness that build up on textiles after repeated washings.

C.   Synthetic Detergents

After the cleansing action of soaps was understood, synthetic detergents could be designed. Molecules of a good detergent must have a long hydrocarbon chain, preferably 12 to 20 carbon atoms long, and a polar group at one end of the molecule that does not form insoluble salts with Ca(II), Mg(II), or Fe(III) ions present in hard water. Chemists recognized that these essential characteristics of a soap could be produced in a molecule containing a sulfate or sulfonate group instead of a carboxylate

Soap micelle with "dissolved" grease

Grease

Soap

Figure 26.4

A soap micelle with a "dissolved" oil or grease droplet.

group. Calcium, magnesium, and iron salts of monoalkylsulfuric and sulfonic acids are much more soluble in water than comparable salts of fatty acids.

The most widely used synthetic detergents are the linear alkylbenzenesulfonates (LAS). One of the most common of these is sodium 4-dodecylbenzenesulfonate. To prepare this type of detergent, a linear alkylbenzene is treated with sulfuric acid (Section 21.1B) to form an alkylbenzenesulfonic acid. The sulfonic acid is then neutralized with NaOH, the product is mixed with builders, and spray-dried to give a smooth flowing powder. The most common builder is sodium silicate.

$$CH_3(CH_2)_{10}CH_2 - \text{\textcircled{benzene}} \xrightarrow[\text{2. NaOH}]{\text{1. } H_2SO_4} CH_3(CH_2)_{10}CH_2 - \text{\textcircled{benzene}} - SO_3^- Na^+$$

Dodecylbenzene   Sodium 4-dodecylbenzenesulfonate
(an anionic detergent)

Alkylbenzenesulfonate detergents were introduced in the late 1950s, and today they command close to 90% of the market once held by natural soaps.

Among the most common additives to detergent preparations are foam stabilizers, bleaches, and optical brighteners. A common foam stabilizer added to liquid soaps but not laundry detergents (for obvious reasons: think of a top-loading washing machine with foam spewing out the lid!) is the amide prepared from dodecanoic acid (lauric acid) and 2-aminoethanol (ethanolamine). The most common bleach is sodium perborate tetrahydrate (see *The Merck Index*, 12th ed., #8797), which decomposes at temperatures above 50°C to give hydrogen peroxide, the actual bleaching agent.

$$\overset{\overset{\displaystyle O}{\displaystyle \|}}{CH_3(CH_2)_{10}CNHCH_2CH_2OH} \qquad\qquad O{=}B{-}O{-}O^-Na^+ \cdot 4H_2O$$

N-(2-Hydroxyethyl)dodecanamide   Sodium perborate tetrahydrate
(a foam stabilizer)   (a bleach)

Also added to laundry detergents are optical brighteners, known also as optical bleaches, that are absorbed into fabrics and, after absorbing ambient light, fluoresce with a blue color, offsetting the yellow color caused by fabric aging. Quite literally, these optical brighteners produce a "whiter-than-white" appearance. You most certainly have observed the effects of optical brighteners if you have seen the glow of "white" shirts or blouses when exposed to black light (UV radiation).

26.3 Prostaglandins

Prostaglandin A member of the family of compounds having the 20-carbon skeleton of prostanoic acid.

The **prostaglandins** are a family of compounds all having the 20-carbon skeleton of prostanoic acid.

Prostanoic acid

The story of the discovery and structure determination of these remarkable compounds began in 1930 when gynecologists Raphael Kurzrok and Charles Lieb reported that human seminal fluid stimulates contraction of isolated uterine muscle. A few years later in Sweden, Ulf von Euler confirmed this report and noted that human seminal fluid also produces contraction of intestinal smooth muscle and lowers blood pressure when injected into the bloodstream. Von Euler proposed the name "prostaglandin" for the mysterious substance(s) responsible for these diverse effects because it was believed at the time that they were synthesized in the prostate gland. Although we now know that prostaglandin production is by no means limited to the prostate gland, the name nevertheless has stuck.

Prostaglandins are not stored as such in target tissues. Rather, they are synthesized in response to specific physiological triggers. Starting materials for the biosynthesis of prostaglandins are polyunsaturated fatty acids of 20 carbon atoms, stored until needed as membrane phospholipid esters. In response to a physiological trigger, the ester is hydrolyzed, the fatty acid is released, and the synthesis of prostaglandins is initiated. Figure 26.5 outlines the steps in the synthesis of several prostaglandins from arachidonic acid. A key step in this biosynthesis is the enzyme-catalyzed reaction of arachidonic acid with two molecules of O_2 to form prostaglandin G_2 (PGG_2). The anti-inflammatory and anticlotting effects of aspirin and other nonsteroidal anti-inflammatory drugs (NSAIDs) result from their ability to inhibit the enzyme that catalyzes this step.

Research on the involvement of prostaglandins in reproductive physiology and the inflammatory process has produced several clinically useful prostaglandin derivatives. The observations that $PGF_{2\alpha}$ stimulates contractions of uterine smooth muscle led to a synthetic derivative that is used as a therapeutic abortifacient. A problem with the use of the natural prostaglandins for this purpose is that they are very rapidly degraded within the body. In the search for less rapidly degraded prostaglandins, a number of analogs have been prepared, one of the most effective of which is carboprost (see *The Merck Index*, 12th ed., #1871). This synthetic prostaglandin is 10 to 20 times more potent than the natural $PGF_{2\alpha}$ and is only slowly degraded in the body. The comparison of these two prostaglandins illustrates how a simple change in structure of a drug can make a significant change in its effectiveness.

PGF$_{2\alpha}$

Carboprost
(15S)-15-Methyl-PGF$_{2\alpha}$

The PGEs along with several other PGs suppress gastric ulceration and appear to heal gastric ulcers. The PGE$_1$ analog, misoprostol (see *The Merck Index*, 12th ed., #6297), is currently used primarily for prevention of ulceration associated with aspirin-like NSAIDs (partly caused by their inhibition of clotting).

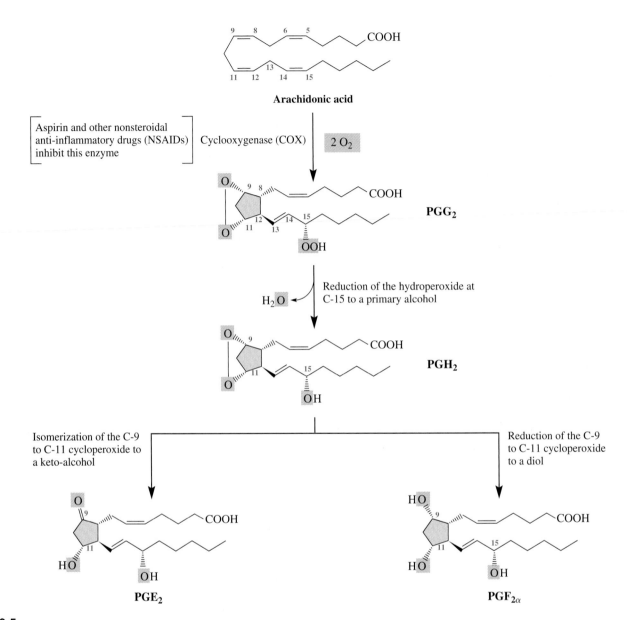

Figure 26.5

Key intermediates in the conversion of arachidonic acid to PGE$_2$ and PGF$_{2\alpha}$. PG stands for prostaglandin. The letters E, F, G, and H are different types of prostaglandins.

Prostaglandins are members of an even larger family of compounds called **eicosanoids.** Eicosanoids contain 20 carbons and are derived from fatty acids. They include not only the prostaglandins but also the leukotrienes, thromboxanes, and prostacyclins. The eicosanoids are extremely widespread, and members of this family of compounds have been isolated from almost every tissue and body fluid.

Leukotriene C_4 (LTC$_4$)
(a smooth muscle constrictor)

Thromboxane A_2
(a potent vasoconstrictor)

Prostacyclin
(a platelet aggregation inhibitor)

Leukotrienes are derived from arachidonic acid and are found primarily in leukocytes (white blood cells). Leukotriene C_4 (LTC$_4$), a typical member of this family, has three conjugated double bonds (hence the suffix -triene) and contains the amino acids L-cysteine, glycine, and L-glutamic acid (Chapter 27). An important physiological action of LTC$_4$ is constriction of smooth muscles, especially those of the lungs. The synthesis and release of LTC$_4$ is prompted by allergic reactions. Drugs that inhibit the synthesis of LTC$_4$ show promise for the treatment of the allergic reactions associated with asthma. Thromboxane A_2 is a very potent vasoconstrictor; its release triggers the irreversible phase of platelet aggregation and constriction of injured blood vessels. It is thought that aspirin and aspirin-like drugs act as mild anticoagulants because they inhibit cyclooxygenase, the enzyme that initiates the synthesis of thromboxane A_2.

26.4 Steroids

Steroids are a group of plant and animal lipids that have the tetracyclic ring system shown in Figure 26.6. The features common to the tetracyclic ring system of most naturally occurring steroids are illustrated in Figure 26.7.

1. The fusion of rings is trans, and each atom or group at a ring junction is axial. Compare, for example, the orientations of —H at carbon 5 and —CH$_3$ at carbon 10.
2. The pattern of atoms or groups along the points of ring fusion (carbons 5 to 10 to 9 to 8 to 14 to 13) is nearly always trans-anti-trans-anti-trans.
3. Because of the trans-anti-trans-anti-trans arrangement of atoms or groups along the points of ring fusion, the tetracyclic steroid ring system is nearly flat and quite rigid.

Steroid A plant or animal lipid having the characteristic tetracyclic ring structure of the steroid nucleus, namely three six-membered rings and one five-membered ring.

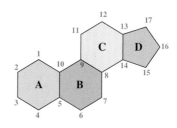

Figure 26.6

The tetracyclic ring system characteristic of steroids.

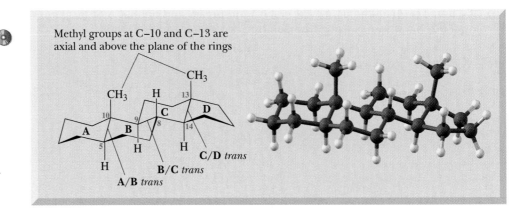

Methyl groups at C–10 and C–13 are axial and above the plane of the rings

Figure 26.7

Features common to the tetracyclic ring system of many steroids.

4. Many steroids have axial methyl groups at carbon 10 and carbon 13 of the tetracyclic ring system.

A.   Structure of the Major Classes of Steroids

Cholesterol

Cholesterol is a white, water-insoluble, waxy solid found in blood plasma and in all animal tissues. This substance is an integral part of human metabolism in two ways: (1) It is an essential component of biological membranes. The body of a healthy adult contains approximately 140 g of cholesterol, about 120 g of which is present in membranes. Membranes of the central and peripheral nervous systems, for example, contain about 10% cholesterol by weight. (2) It is the compound from which sex hormones, adrenocorticoid hormones, bile acids, and vitamin D are synthesized. Thus, cholesterol is, in a sense, the parent steroid.

Cholesterol has eight stereocenters, and a molecule with this many stereocenters can exist as 2^8, or 256, stereoisomers (128 pairs of enantiomers). Only one of these stereoisomers is known to exist in nature: the stereoisomer with the configuration shown in Figure 26.8.

Cholesterol is insoluble in blood plasma but can be transported as a plasma-soluble complex formed with proteins called lipoproteins. **Low-density lipoproteins (LDLs)** transport cholesterol from the site of its synthesis in the liver to the various tissues and cells of the body where it is to be used. It is primarily cholesterol attached

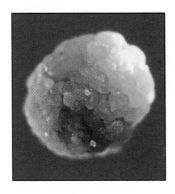

Human gallstones are almost pure cholesterol; this gallstone is about 0.5 cm in diameter. *(© Carolina Biological Supply Company, Phototake, NYC)*

Low-density lipoproteins (LDLs)
Plasma particles, density 1.02–1.06 g/mL, consisting of approximately 26% proteins, 50% cholesterol, 21% phospholipids, and 4% triglycerides.

Figure 26.8

Cholesterol is found in blood plasma and in all animal tissues.

Cholesterol

to LDLs that builds up in atherosclerotic deposits in blood vessels. **High-density lipoproteins (HDLs)** transport excess and unused cholesterol from cells back to the liver for its degradation to bile acids and eventual excretion in the feces. It is thought that HDLs retard or reduce atherosclerotic deposits.

High-density lipoproteins (HDLs) Plasma particles, density 1.06–1.21 g/mL, consisting of approximately 33% proteins, 30% cholesterol, 29% phospholipids, and 8% triglycerides.

Steroid Hormones

Given in Table 26.3 are representations of each major class of steroid hormones, along with the principal functions of each. The two most important female sex hormones, or **estrogens,** are estrone and estradiol. In addition, progesterone, another type of steroid hormone, is essential for preparing the uterus for implantation of a fertilized egg. After the role of progesterone in inhibiting ovulation was understood, its potential as a possible contraceptive was realized. Progesterone itself is relatively ineffective when taken orally. As a result of a massive research program in both industrial and academic laboratories, many synthetic progesterone-mimicking steroids became available in the 1960s. When taken regularly, these drugs prevent ovulation yet allow women to maintain a normal menstrual cycle. Some of the most effective of these preparations contain a progesterone analog, such as norethindrone, combined with a smaller amount of an estrogen-like material to help prevent irregular menstrual flow during prolonged use of contraceptive pills.

Estrogen A steroid hormone, such as estrone and estradiol, that mediates the development of sexual characteristics in females.

"Nor" refers to the absence of a methyl group here. The methyl group is present in ethindrone.

Norethindrone
(a synthetic progesterone analog)

The chief function of testosterone and other **androgens** is to promote normal growth of male reproductive organs (primary sex characteristics) and development of the characteristic deep voice, pattern of body and facial hair, and musculature (secondary sex characteristics). Although testosterone produces these effects, it is not active when taken orally because it is metabolized in the liver to an inactive steroid. A number of oral **anabolic steroids** have been developed for use in rehabilitation medicine, particularly when muscle atrophy occurs during recovery from an injury. Among the synthetic anabolic steroids most widely prescribed for this purpose are methandrostenolone and stanozolol. The structural formula of methandrostenolone differs from that of testosterone by introduction of (1) a methyl group at carbon 17, and (2) an additional carbon-carbon double bond between carbon 1 and carbon 2. In stanozolol, ring A is modified by attachment of a pyrazole ring.

Androgen A steroid hormone, such as testosterone, that mediates the development of sexual characteristics of males.

Anabolic steroid A steroid hormone, such as testosterone, that promotes tissue and muscle growth and development.

Methandrostenolone

Stanozolol

Table 26.3 Selected Steroid Hormones	
Structure	**Source and Major Effects**

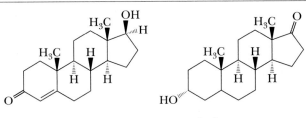

Testosterone   Androsterone

Androgens (male sex hormones) — synthesized in the testes; responsible for development of male secondary sex characteristics

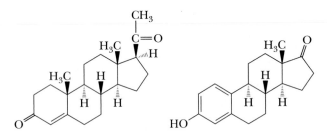

Progesterone   Estrone

Estrogens (female sex hormones) — synthesized in the ovaries; responsible for development of female secondary sex characteristics and control of the menstrual cycle

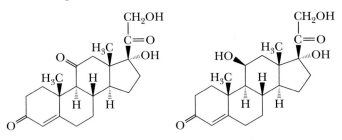

Cortisone   Cortisol

Glucocorticoid hormones — synthesized in the adrenal cortex; regulate metabolism of carbohydrates, decrease inflammation, and are involved in the reaction to stress

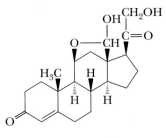

Aldosterone

A mineralocorticoid hormone — synthesized in the adrenal cortex; regulates blood pressure and volume by stimulating the kidneys to absorb Na^+, Cl^-, and HCO_3^-

Among certain athletes, the misuse of anabolic steroids to build muscle mass and strength, particularly for sports that require explosive action, is common. The risks associated with abuse of anabolic steroids for this purpose are enormous: heightened aggressiveness, sterility, impotence, and risk of premature death from complications of diabetes, coronary artery disease, and liver cancer.

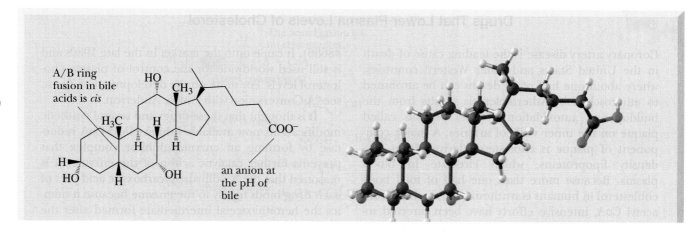

Bile Acids

Shown in Figure 26.9 is a structural formula for cholic acid, a constituent of human bile. The molecule is shown as an anion, as it is ionized in bile and intestinal fluids. **Bile acids,** or more properly, bile salts, are synthesized in the liver, stored in the gallbladder, and secreted into the intestine, where their function is to emulsify dietary fats and thereby aid in their absorption and digestion. Furthermore, bile salts are the end products of the metabolism of cholesterol and, thus, are a principal pathway for the elimination of this substance from the body. A characteristic structural feature of bile salts is a cis fusion of rings A/B.

B. Biosynthesis of Cholesterol

The biosynthesis of cholesterol illustrates a point we first made in our introduction to the structure of terpenes (Section 5.4). In building large molecules, one of the common patterns in the biological world is to begin with one or more smaller subunits, join them by an iterative process, and then chemically modify the completed carbon skeleton by oxidation, reduction, cross linking, addition, elimination, or related processes to give a biomolecule with a unique identity.

The building block from which all carbon atoms of steroids are derived is the two-carbon acetyl group of acetyl-CoA (Problem 25.36). The American biochemist, Konrad Bloch, who shared the 1964 Nobel Prize for medicine and physiology with German biochemist Feodor Lynen for their discoveries concerning the biosynthesis of cholesterol and fatty acids, showed that 15 of the 27 carbon atoms of cholesterol are derived from the methyl group of acetyl-CoA. The remaining 12 carbon atoms are derived from the carbonyl group of acetyl-CoA (Figure 26.10).

A remarkable feature of this synthetic pathway is that the biosynthesis of cholesterol from acetyl-CoA is completely stereoselective; it is synthesized as only one of 256 possible stereoisomers. We cannot duplicate this exquisite degree of stereoselectivity in the laboratory. Cholesterol is, in turn, the key intermediate in the synthesis of most other steroids.

cholesterol ⟶
- bile acids (e.g., cholic acid)
- sex hormones (e.g., testosterone and estrone)
- mineralocorticoid hormones (e.g., aldosterone)
- glucocorticoid hormones (e.g., cortisone)

Figure 26.9
Cholic acid, an important constituent of human bile. Each six-membered ring is in a chair conformation.

Bile acid A cholesterol-derived detergent molecule, such as cholic acid, which is secreted by the gallbladder into the intestine to assist in the absorption of dietary lipids.

CHEMISTRY IN ACTION

Snake Venom Phospholipases

The venoms of certain snakes contain enzymes called phospholipases. These enzymes catalyze the hydrolysis of carboxylic ester bonds of phospholipids. The venom of the eastern diamondback rattlesnake (*Crotalus adamanteus*) and the Indian cobra (*Naja naja*) both contain phospholipase PLA$_2$, which catalyzes the hydrolysis of esters at carbon 2 of phospholipids. The breakdown product of this hydrolysis,

a lysolecithin, acts as a detergent and dissolves the membranes of red blood cells causing them to rupture. Indian cobras kill several thousand people each year.

Milking an Indian cobra for its venom. (Dan McCoy/Rainbow)

$$\begin{array}{c}
CH_2CH_2NH_3{}^+ \\
| \\
O \\
| \\
O{=}P{-}O^- \\
| \\
O \\
| \\
CH_2{-}CH{-}CH_2 \\
| \qquad | \\
O \qquad O \\
| \qquad | \\
O{=}C \;\; O{=}C \\
| \qquad\quad | \\
R_1 \qquad R_2
\end{array} \quad + \quad H_2O \;\xrightarrow{\;PLA_2\;}$$

PLA$_2$ catalyzes hydrolysis of this ester bond.

A phospholipid

$$\begin{array}{c}
CH_2CH_2NH_3{}^+ \\
| \\
O \\
| \\
O{=}P{-}O^- \\
| \\
O \\
| \\
CH_2{-}CH{-}CH_2 \\
| \qquad | \\
O \quad\;\; OH \\
| \\
O{=}C \\
| \\
R_1
\end{array} \quad + \quad R_2{-}\overset{\displaystyle O}{\overset{\|}{C}}{-}O^- + H^+$$

A lysolecithin

are given in Table 26.4. All functional groups in this table and in Figure 26.11 are shown as they are ionized at pH 7.4, the approximate pH of blood plasma and of many other biological fluids. Under these conditions, each phosphate group bears a negative charge and each amino group bears a positive charge.

B. Lipid Bilayers

Figure 26.12 shows a space-filling model of a lecithin (a phosphatidycholine). It and other phospholipids are elongated, almost rodlike molecules, with the nonpolar (hy-

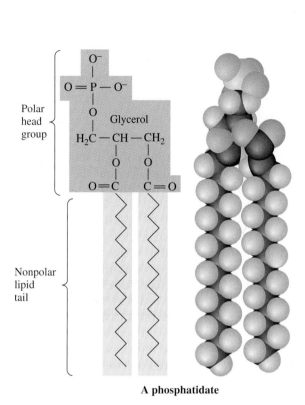

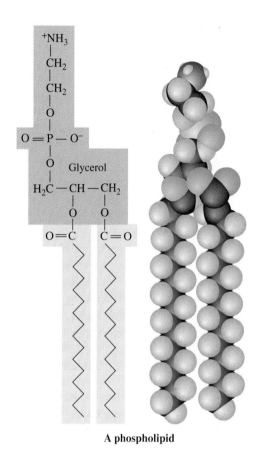

A phosphatidate   A phospholipid

Figure 26.11

In a phosphatidic acid, glycerol is esterified with two molecules of fatty acid and one molecule of phosphoric acid. Further esterification of the phosphoric acid group with a low-molecular-weight alcohol gives a phospholipid.

Table 26.4   Low-Molecular-Weight Alcohols Most Common to Phospholipids

Alcohols Found in Phospholipids

Structural Formula	Name	Name of Phospholipid
$HOCH_2CH_2NH_2$	Ethanolamine	Phosphatidylethanolamine (cephalin)
$HOCH_2CH_2\overset{+}{N}(CH_3)_3$	Choline	Phosphatidylcholine (lecithin)
$HOCH_2\underset{\underset{NH_3^+}{\mid}}{CH}COO^-$	Serine	Phosphatidylserine
(inositol ring structure)	Inositol	Phosphatidylinositol

All of these products contain lecithin. *(Charles D. Winters)*

drophobic) hydrocarbon chains lying roughly parallel to one another and the polar (hydrophilic) phosphoric ester group pointing in the opposite direction.

When placed in aqueous solution, phospholipids spontaneously form a **lipid bilayer** (Figure 26.13) in which polar head groups lie on the surface, giving the bilayer an ionic coating. Nonpolar hydrocarbon chains of fatty acids lie buried within the bilayer. This self-assembly of phospholipids into a bilayer is a spontaneous process, driven by two types of noncovalent forces: (1) hydrophobic effects, which result when nonpolar hydrocarbon chains cluster together and exclude water molecules, and (2) electrostatic interactions, which result when polar head groups interact with water and other polar molecules in the aqueous environment.

Recall from Section 26.2B that formation of soap micelles is driven by these same noncovalent forces; the polar (hydrophilic) carboxylate groups of soap molecules lie on the surface of the micelle and associate with water molecules, and the nonpolar (hydrophobic) hydrocarbon chains cluster within the micelle and are thus removed from contact with water.

The arrangement of hydrocarbon chains in the interior of a phospholipid bilayer varies from rigid to fluid, depending on the degree of unsaturation of the hydrocarbon chains themselves. Saturated hydrocarbon chains tend to lie parallel and closely packed, leading to a rigidity of the bilayer. Unsaturated hydrocarbon chains, on the other hand, have one or more cis double bonds, which cause "kinks" in the chains. As a result, they do not pack as closely and with as great an order as saturated chains. The disordered packing of unsaturated hydrocarbon chains leads to fluidity of the bilayer.

Biological membranes are made of lipid bilayers. The most satisfactory current model for the arrangement of phospholipids, proteins, and cholesterol in plant and animal membranes is the **fluid-mosaic model** proposed in 1972 by S. J. Singer and G. Nicolson (Figure 26.13). The term "mosaic" signifies that the various components in

Lipid bilayer   A back-to-back arrangement of phospholipid monolayers, often forming a closed vesicle or membrane.

Fluid-mosaic model   A biological membrane that consists of a phospholipid bilayer with proteins, carbohydrates, and other lipids on the surface and embedded in the surface of the bilayer.

Figure 26.12
Space-filling model of a lecithin.

Figure 26.13
Fluid-mosaic model of a biological membrane, showing the lipid bilayer and membrane proteins oriented on the inner and outer surfaces of the membrane and penetrating the entire thickness of the membrane.

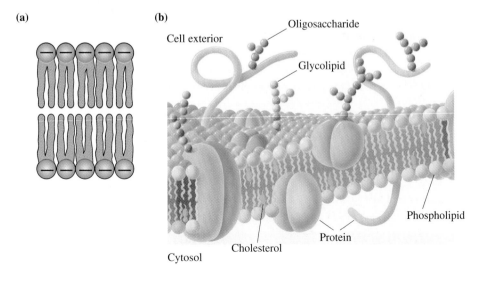

(a)

(b)

the membrane coexist side by side, as discrete units, rather than combining to form new molecules or ions. "Fluid" signifies that the same sort of fluidity exists in membranes that we have already seen for lipid bilayers. Furthermore, the protein components of membranes "float" in the bilayer and can move laterally along the plane of the membrane.

26.6 Fat-Soluble Vitamins

Vitamins are divided into two broad classes on the basis of solubility, those that are fat-soluble (and hence classed as lipids) and those that are water-soluble. The fat-soluble vitamins include A, D, E, and K.

A. Vitamin A

Vitamin A, or retinol, occurs only in the animal world, where the best sources are cod-liver oil and other fish-liver oils, animal liver, and dairy products. Vitamin A in the form of a precursor, or provitamin, is found in the plant world in a group of tetraterpene (C_{40}) pigments called carotenes. The most common of these is β-carotene, abundant in carrots but also found in some other vegetables, particularly yellow and green ones. β-Carotene has activity as an antioxidant; one of its functions in green plants is to quench singlet oxygen, which can be produced as a byproduct of photosynthesis. β-Carotene has no vitamin A activity; however, after ingestion, it is cleaved at the central carbon-carbon double bond followed by reduction of the newly formed aldehyde to give retinol (vitamin A).

Cleavage of this
C=C gives vitamin A.

β-Carotene

enzyme-catalyzed cleavage
and reduction in the liver

Retinol (Vitamin A)

Probably the best understood role of vitamin A is its participation in the visual cycle in rod cells. In a series of enzyme-catalyzed reactions (Figure 26.14), retinol undergoes a two-electron oxidation to all-*trans*-retinal, isomerization about the carbon 11 to carbon 12 double bond to give 11-*cis*-retinal, and formation of an imine (Section 16.10) with the —NH_2 from a lysine unit of the protein, opsin. The product of these reactions is rhodopsin, a highly conjugated pigment that shows intense absorption in the blue-green region of the visual spectrum.

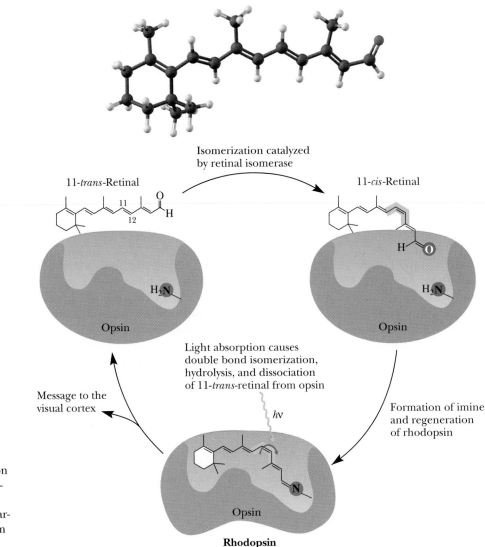

Figure 26.14
The primary chemical reaction of vision in rod cells is absorption of light by rhodopsin followed by isomerization of a carbon-carbon double bond from a cis configuration to a trans configuration.

The primary event in vision is absorption of light by rhodopsin in rod cells of the retina of the eye to produce an electronically excited molecule. Within several picoseconds (1 ps = 10^{-12} s), the excess electronic energy is converted to vibrational and rotational energy, and the 11-cis double bond is isomerized to the more stable 11-trans double bond. This isomerization triggers a conformational change in opsin that causes firing of neurons in the optic nerve and produces a visual image. Coupled with this light-induced change is hydrolysis of rhodopsin to give 11-*trans*-retinal and free opsin. At this point, the visual pigment is bleached and in a refractory period. Rhodopsin is regenerated by a series of enzyme-catalyzed reactions that converts 11-*trans*-retinal to 11-*cis*-retinal and then to rhodopsin. The visual cycle is shown in abbreviated form in Figure 26.14.

B. Vitamin D

Vitamin D is the name for a group of structurally related compounds that play a major role in the regulation of calcium and phosphorus metabolism. A deficiency of vitamin D in childhood is associated with rickets, a mineral-metabolism disease that leads to bone defects that form bowlegs, knock-knees, and enlarged joints. Vitamin D_3, the most abundant form of the vitamin in the circulatory system, is produced in the skin of mammals by the action of ultraviolet radiation on 7-dehydrocholesterol (cholesterol with a double bond between carbons 7 and 8). In the liver, vitamin D_3 undergoes an enzyme-catalyzed, two-electron oxidation at carbon 25 of the side chain to form 25-hydroxyvitamin D_3; the oxidizing agent is molecular oxygen, O_2. 25-Hydroxyvitamin D_3 undergoes further oxidation in the kidneys, also by O_2, to form 1,25-dihydroxyvitamin D_3, the hormonally active form of the vitamin.

7-Dehydrocholesterol

1,25-Dihydroxyvitamin D_3

1. Opening of ring B by UV light
2. Enzyme-catalyzed oxidation by O_2 at C-1 and C-25

C. Vitamin E

Vitamin E was first recognized in 1922 as a dietary factor essential for normal reproduction in rats, hence its name tocopherol from the Greek: *tocos*, birth, and *pherein*, to bring about. Vitamin E is a group of compounds of similar structure, the most active of which is α-tocopherol. This vitamin occurs in fish oil, in other oils such as cottonseed and peanut oil, and in leafy green vegetables. The richest source of vitamin E is wheat germ oil.

4 isoprene units, joined head-to-tail, beginning here and ending at the aromatic ring

Vitamin E (α-Tocopherol)

In the body, vitamin E functions as an antioxidant; it traps peroxy radicals of the type HOO· and ROO· formed as a result of enzyme-catalyzed oxidation by molecular oxygen of the unsaturated hydrocarbon chains in membrane phospholipids (see the Chemistry in Action box "Radical Autoxidation" in Chapter 7). There is speculation that peroxy radicals play a role in the aging process and that vitamin E and other antioxidants may retard that process. Vitamin E is also necessary for the proper development and function of the membranes of red blood cells.

Vitamin K, Blood Clotting, and Basicity

Vitamin K is a fat-soluble vitamin, which must be obtained from the diet. A vitamin K deficiency results in slowed blood clotting, which can be a serious threat to a wounded animal or human. In the process of blood clotting, the natural vitamin, a quinone, is converted to its active hydroquinone form by reduction.

Vitamin K_1 quinone Vitamin K_1 hydroquinone

In the presence of vitamin K hydroquinone, O_2, CO_2, and an enzyme called microsomal carboxylase, side chains of glutamate units in prothrombin, a protein essential for blood clotting, are modified by addition of carboxyl groups (from CO_2) to form γ-carboxyglutamate units. Note that this reaction is the reverse of the decarboxylation of β-dicarboxylic (malonic) acids seen in Section 17.9B. The two carboxyl groups of the chemically modified glutamate now form a tight bidentate ("two teeth") complex with Ca^{2+} during the blood-clotting process. Although there is more to be understood about blood clotting, it is at least clear that if prothrombin is not carboxylated, it does not bind calcium and blood does not clot.

Glutamate side
chain of prothrombin

Carboxylated glutamate
side chain

Carboxylated glutamate side
chain binding calcium ion

These facts have been known for many years. Until quite recently, however, the role of vitamin K_1 in this process remained a mystery. The problem was

this. The anion of vitamin K_1 hydroquinone is a weak base (pK_a of approximately 9) derived from a phenol. To remove a proton from a glutamate side chain requires a very strong base derived from a conjugate acid of pK_a approximately 27. How can molecular oxygen increase the base strength of vitamin K_1 hydroquinone by 18 orders of magnitude?

Recently Paul Dowd discovered that the vitamin K_1 hydroquinone anion reacts with oxygen to give the peroxide anion intermediate 1, which is converted to compound 2. Notice that compound 2 contains a weak O—O bond in a highly strained four-membered ring. Compound 2 then rearranges to vitamin K_1 base, a strong, sterically hindered alkoxide base.

Vitamin K_1
hydroquinone anion
(a weak base)   1

2   Vitamin K_1 base
                         (a very strong base)

The weak O—O bond is, in this way, replaced by a stronger C—O bond in vitamin K_1 base. The extra stability provides the driving force for turning a weak phenoxide base into a strong alkoxide base, which is able to remove a proton from glutamate side chains. This makes possible the addition of CO_2 to form γ-carboxyglutamate side chains, which bind calcium ions during the clotting cascade. Thus, it is now understood why O_2, CO_2, and vitamin K_1 are essential for this phase of blood clotting. Synthetic vitamin K analogs are as effective in this process as naturally occurring vitamin K_1. [See P. Dowd, R. Hershline, S. W. Ham, and S. Naganathan, "Vitamin K and Energy Transduction: A Base Strength Amplification Mechanism," *Science* **269**: 1684 (1995).]

D. Vitamin K

The name of this vitamin comes from the German word *koagulation*, signifying its important role in the blood-clotting process. A vitamin K deficiency results in slowed blood clotting.

Vitamin K₁

Menadione
(a synthetic vitamin K analog)

Natural vitamins of the K family have, for the most part, been replaced in vitamin supplements by synthetic preparations. Menadione, one such synthetic material with vitamin K activity, has only hydrogen in the place of the alkyl chain.

Summary

Lipids are a heterogeneous class of compounds grouped together on the basis of their solubility properties; they are insoluble in water and soluble in diethyl ether, acetone, and methylene chloride. Carbohydrates, amino acids, and proteins are largely insoluble in these organic solvents.

Triglycerides (triacylglycerols), the most abundant lipids, are triesters of glycerol and fatty acids (Section 26.1). **Fatty acids** (Section 26.1A) are long-chain carboxylic acids derived from the hydrolysis of fats, oils, and the phospholipids of biological membranes. The melting point of a triglyceride increases as (1) the length of its hydrocarbon chains increases and (2) its degree of saturation increases. Triglycerides rich in saturated fatty acids are generally solids at room temperature; those rich in unsaturated fatty acids are generally **oils** at room temperature (Section 26.1B).

Soaps are sodium or potassium salts of fatty acids (Section 26.2A). In water, soaps form **micelles,** which "dissolve" nonpolar organic grease and oil. Natural soaps precipitate as water-insoluble salts with Mg^{2+}, Ca^{2+}, and Fe^{3+} ions in hard water. The most common and most widely used synthetic detergents (Section 26.2C) are linear alkylbenzenesulfonates.

Prostaglandins are a group of compounds having the 20-carbon skeleton of prostanoic acid (Section 26.3). They are synthesized, in response to physiological triggers, from phospholipid-bound arachidonic acid and other 20-carbon fatty acids.

Steroids are a group of plant and animal lipids that have a characteristic tetracyclic structure of three six-membered rings and one five-membered ring (Section 26.4). **Cholesterol** is an integral part of animal membranes, and it is the compound from which human sex hormones, adrenocorticoid hormones,

bile acids, and vitamin D are biosynthesized. **Low-density lipoproteins (LDLs)** transport cholesterol from the site of its synthesis in the liver to tissues and cells where it is to be used. **High-density lipoproteins (HDLs)** transport cholesterol from cells back to the liver for its degradation to bile acids and eventual excretion in the feces.

Oral contraceptive pills contain a synthetic progestin, for example norethindrone, which prevents ovulation yet allows women to maintain an otherwise normal menstrual cycle. A variety of synthetic **anabolic steroids** are available for use in rehabilitation medicine where muscle tissue has weakened or deteriorated due to injury. **Bile acids** differ from most other steroids in that they have a cis configuration at the junction of rings A and B.

The carbon skeleton of cholesterol and those of all biomolecules derived from it originate with the acetyl group (a C_2 unit) of acetyl-CoA (Section 26.4B).

Phospholipids (Section 26.5A), the second most abundant group of naturally occurring lipids, are derived from phosphatidic acids, compounds containing glycerol esterified with two molecules of fatty acid and a molecule of phosphoric acid. Further esterification of the phosphoric acid part with a low-molecular-weight alcohol, most commonly ethanolamine, choline, serine, or inositol, gives a phospholipid. When placed in aqueous solution, phospholipids spontaneously form **lipid bilayers** (Section 26.5B).

According to the **fluid-mosaic model** (Section 26.5B), membrane phospholipids form lipid bilayers with membrane proteins associated with the bilayer as both peripheral and integral proteins.

Vitamin A (Section 26.6A) occurs only in the animal world. The carotenes of the plant world are tetraterpenes (C_{40}) and are cleaved, after ingestion, into vitamin A. The best understood role of vitamin A is its participation in the visual cycle.

Vitamin D (Section 26.6B) is synthesized in the skin of mammals by the action of ultraviolet radiation on 7-dehydro-cholesterol. This vitamin plays a major role in the regulation of calcium and phosphorus metabolism. Vitamin E (Section 26.6C) is a group of compounds of similar structure, the most active of which is α-tocopherol. In the body, vitamin E functions as an antioxidant. Vitamin K (Section 26.6D) is required for the clotting of blood.

Problems

Fatty Acids and Triglycerides

26.2 Define the term "hydrophobic."

26.3 Identify the hydrophobic and hydrophilic region(s) of a triglyceride.

26.4 Explain why the melting points of unsaturated fatty acids are lower than those of saturated fatty acids.

26.5 Which would you expect to have the higher melting point, glyceryl trioleate or glyceryl trilinoleate?

26.6 Draw a structural formula for methyl linoleate. Be certain to show the correct configuration of groups about each carbon-carbon double bond.

26.7 Explain why coconut oil is a liquid triglyceride, even though most of its fatty acid components are saturated.

26.8 It is common now to see "contains no tropical oils" on cooking oil labels, meaning that the oil contains no palm or coconut oil. What is the difference between the composition of tropical oils and that of vegetable oils, such as corn oil, soybean oil, and peanut oil?

26.9 What is meant by the term "hardening" as applied to vegetable oils?

26.10 How many moles of H_2 are used in the catalytic hydrogenation of one mole of a triglyceride derived from glycerol, stearic acid, linoleic acid, and arachidonic acid?

26.11 Characterize the structural features necessary to make a good synthetic detergent.

26.12 Following are structural formulas for a cationic detergent and a neutral detergent. Account for the detergent properties of each.

$$\underset{\underset{\underset{CH_2C_6H_5}{|}}{\overset{\overset{CH_3}{|}}{CH_3(CH_2)_6CH_2\overset{+}{N}CH_3}}}{} \quad Cl^-$$

$$\underset{\underset{HOCH_2}{|}}{\overset{\overset{HOCH_2 \quad O}{| \quad \|}}{HOCH_2CCH_2OC(CH_2)_{14}CH_3}}$$

Benzyldimethyloctylammonium chloride (a cationic detergent)   Pentaerythrityl palmitate (a neutral detergent)

26.13 Identify some of the detergents used in shampoos and dish washing liquids. Are they primarily anionic, neutral, or cationic detergents?

26.14 Show how to convert palmitic acid (hexadecanoic acid) into the following.

  (a) Ethyl palmitate   **(b)** Palmitoyl chloride
  (c) 1-Hexadecanol (cetyl alcohol)   **(d)** 1-Hexadecanamine
  (e) *N,N*-Dimethylhexadecanamide

26.15 Palmitic acid (hexadecanoic acid) is the source of the hexadecyl (cetyl) group in the following compounds. Each is a mild surface-acting germicide and fungicide and is used as a topical antiseptic and disinfectant (see *The Merck Index*, 12th ed., #2074 and #2059). They are examples of quaternary ammonium detergents, commonly called "quats."

Cetylpyridinium chloride   Benzylcetyldimethylammonium chloride

(a) Cetylpyridinium chloride is prepared by treating pyridine with 1-chlorohexadecane (cetyl chloride). Show how to convert palmitic acid to cetyl chloride.

(b) Benzylcetyldimethylammonium chloride is prepared by treating benzyl chloride with *N,N*-dimethyl-1-hexadecanamine. Show how this tertiary amine can be prepared from palmitic acid.

26.16 Lipases are enzymes that catalyze the hydrolysis of esters, especially esters of glycerol. Because enzymes are chiral catalysts, they catalyze the hydrolysis of only one enantiomer of a racemic mixture. For example, porcine pancreatic lipase catalyzes the hydrolysis of only one enantiomer of the following racemic epoxyester. Calculate the number of grams of epoxyalcohol that can be obtained from 100 g of racemic epoxyester by this method.

A racemic mixture   This enantiomer is recovered unhydrolyzed.   This epoxyalcohol is obtained in pure form.

Prostaglandins

26.17 Examine the structure of $PGF_{2\alpha}$.

(a) Identify all stereocenters.

(b) Identify all double bonds about which cis,trans isomerism is possible.

(c) State the number of stereoisomers possible for a molecule of this structure.

26.18 Following is the structure of unoprostone (see *The Merck Index*, 12th ed., #9984), a compound patterned after the natural prostaglandins (Section 26.3). Rescula, the isopropyl ester of unoprostone, is an antiglaucoma drug used to treat ocular hypertension. Compare the structural formula of this synthetic prostaglandin with that of $PGF_{2\alpha}$.

Unoprostone
(antiglaucoma)

26.19 Doxaprost, an orally active bronchodilator patterned after the natural prostaglandins (Section 26.3), is synthesized in the following series of reactions starting with ethyl 2-oxocyclopentanecarboxylate. Except for the Nef reaction in Step 8, we have seen examples of all other types of reactions involved in this synthesis.

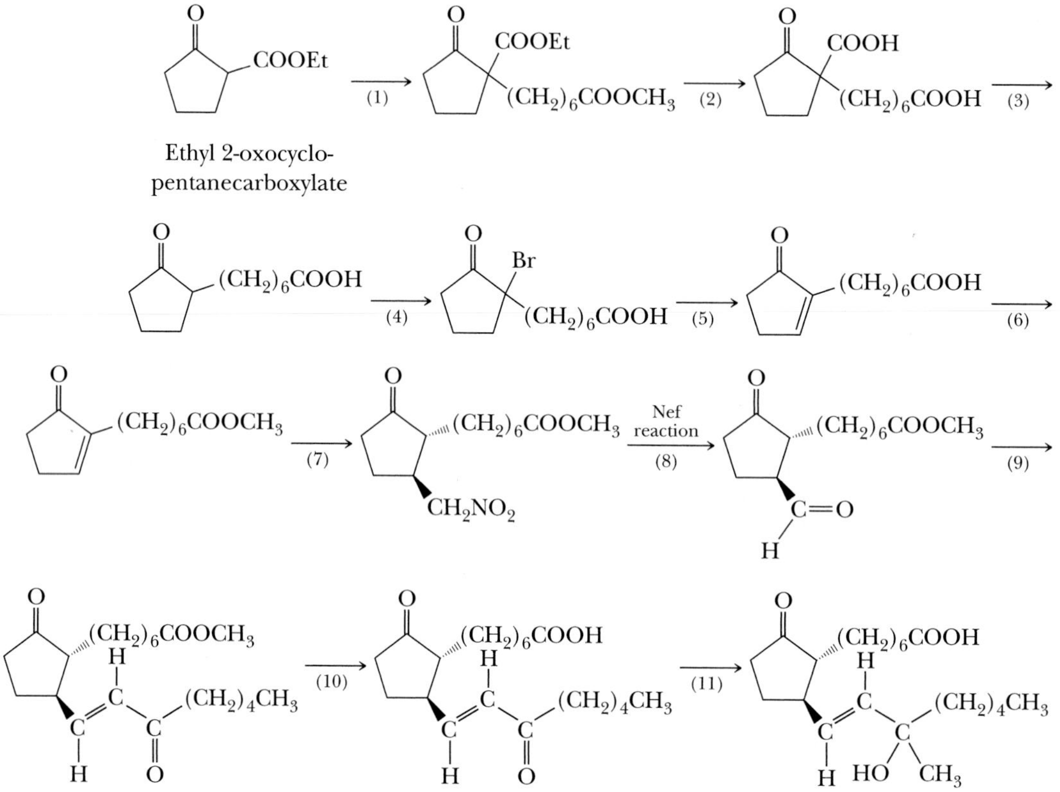

Ethyl 2-oxocyclo-
pentanecarboxylate

Doxaprost
(an orally active
bronchodilator)

(a) Propose a set of experimental conditions to bring about the alkylation in Step 1. Account for the regioselectivity of the alkylation, that is, that it takes place on the carbon between the two carbonyl groups rather than on the other side of the ketone carbonyl.

(b) Propose experimental conditions to bring about Steps 2 and 3.

(c) Propose experimental conditions for bromination of the ring in Step 4 and dehydrobromination in Step 5.

(d) Write equations to show that Step 6 can be brought about using either methanol or diazomethane (CH_2N_2) as a source of the —CH_3 in the methyl ester.

(e) Describe experimental conditions to bring about Step 7 and account for the fact that the trans isomer is formed in this step.

(f) Step 9 is done by a Wittig reaction. Suggest a structural formula for a Wittig reagent that gives the product shown.

(g) Name the type of reaction involved in Step 10.

(h) Step 11 can best be described as a Grignard reaction with methylmagnesium bromide under very carefully controlled conditions. In addition to the observed reaction, what other Grignard reactions might take place in Step 11?

(i) Assuming that the two side chains on the cyclopentanone ring are trans, how many stereoisomers are possible from this synthetic sequence?

Steroids

26.20 Draw the structural formula for the product formed by treatment of cholesterol with the following substances.

(a) H_2/Pd

(b) Br_2

26.21 Both low-density lipoproteins and high-density lipoproteins consist of a core of triacylglycerols and cholesterol esters surrounded by a single phospholipid layer. Draw the structural formula of cholesteryl linoleate, one of the cholesterol esters found in this core.

26.22 Examine the structural formulas of testosterone (a male sex hormone) and progesterone (a female sex hormone). What are the similarities in structure between the two? What are the differences?

26.23 Examine the structural formula of cholic acid and account for the ability of this and other bile salts to emulsify fats and oils and thus aid in their digestion.

26.24 Following is a structural formula for cortisol (hydrocortisone). Draw a stereorepresentation of this molecule showing the conformations of the five- and six-membered rings. See the model of this compound on the CD.

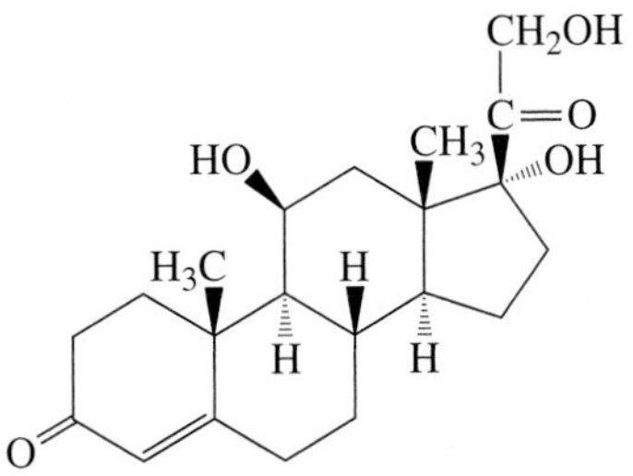

Cortisol
(Hydrocortisone)

26.25 Much of our understanding of conformational analysis has arisen from studies on the reactions of rigid steroid nuclei. For example, the concept of trans-diaxial ring opening of epoxycyclohexanes was proposed to explain the stereoselective reactions seen with steroidal epoxides. Predict the product when each of the following steroidal epoxides is treated with $LiAlH_4$.

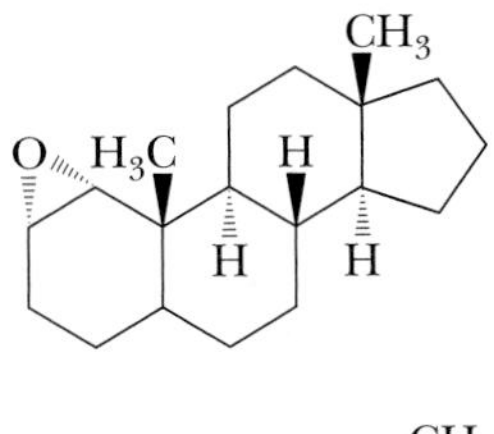

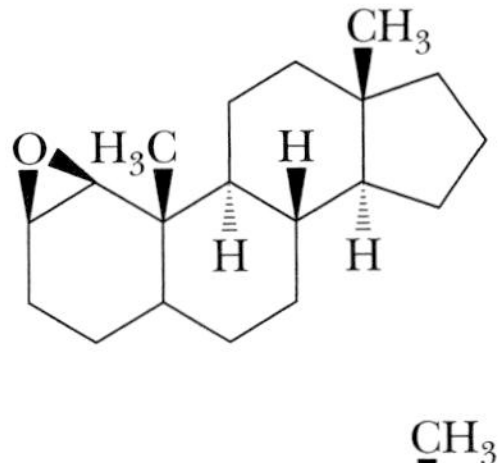

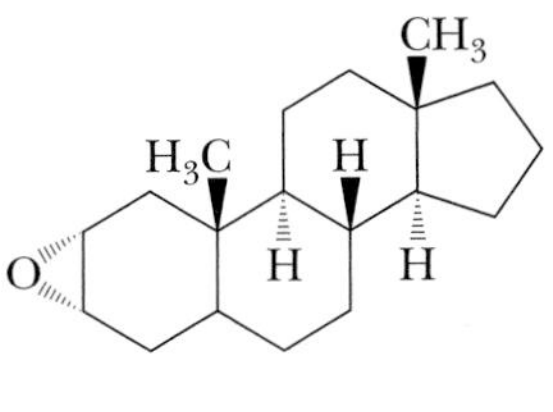

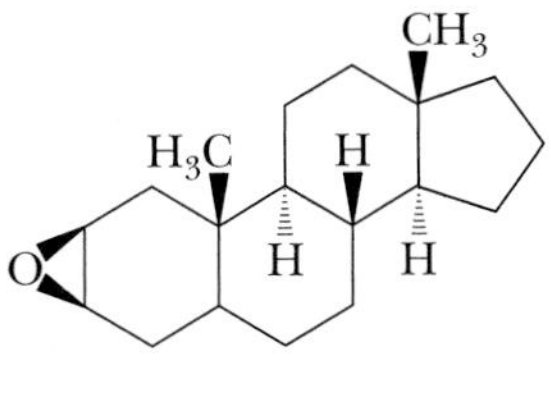

Phospholipids

26.26 Draw the structural formula of a lecithin containing one molecule each of palmitic acid and linoleic acid.

26.27 Identify the hydrophobic and hydrophilic region(s) of a phospholipid.

26.28 The hydrophobic effect is one of the most important noncovalent forces directing the self-assembly of biomolecules in aqueous solution. The hydrophobic effect arises from tendencies of biomolecules (1) to arrange polar groups so that they interact with the aqueous environment by hydrogen bonding and (2) to arrange nonpolar groups so that they are shielded from the aqueous environment. Show how the hydrophobic effect is involved in directing the following.

(a) Formation of micelles by soaps and detergents
(b) Formation of lipid bilayers by phospholipids

26.29 How does the presence of unsaturated fatty acids contribute to the fluidity of biological membranes?

26.30 Lecithins can act as emulsifying agents. The lecithin of egg yolk, for example, is used to make mayonnaise. Identify the hydrophobic part(s) and the hydrophilic part(s) of a lecithin. Which parts interact with the oils used in making mayonnaise? Which parts interact with the water?

Fat-Soluble Vitamins

26.31 Examine the structural formula of vitamin A, and state the number of cis,trans isomers possible for this molecule.

26.32 The form of vitamin A present in many food supplements is vitamin A palmitate. Draw the structural formula of this molecule.

26.33 Examine the structural formulas of vitamin A, 1,25-dihydroxyvitamin-D_3, vitamin E, and vitamin K_1 (Section 26.6). Do you expect them to be more soluble in water or in dichloromethane? Do you expect them to be soluble in blood plasma?

27

AMINO ACIDS AND PROTEINS

W e begin this chapter with a study of amino acids, compounds whose chemistry is built on amines (Chapter 22) and carboxylic acids (Chapter 17). We concentrate in particular on the acid-base properties of amino acids because these properties are so important in determining many of the properties of proteins, including the catalytic functions of enzymes. With this understanding of the chemistry of amino acids, we then examine the structure of proteins themselves.

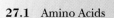

■ Spider silk is a fibrous protein that exhibits unmatched strength and toughness. *(Stone)* Inset: D-Alanine and Glycine, which are major components of the fibrous protein of spider silk. See the Chemistry in Action box "Spider Silk."

1073

The CD-ROM contains molecular models of examples of proteins and enzymes.
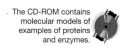

27.1 Amino Acids

A. Structure

An **amino acid** is a compound that contains both a carboxyl group and an amino group. Although many types of amino acids are known, the **α-amino acids** are the most significant in the biological world because they are the monomers from which proteins are constructed. A general structural formula of an α-amino acid is shown in Figure 27.1.

Although Figure 27.1(a) is a common way of writing structural formulas for amino acids, it is not accurate because it shows an acid (—COOH) and a base (—NH₂) within the same molecule. These acidic and basic groups react with each other to form a dipolar ion or internal salt [Figure 27.1(b)]. The internal salt of an amino acid is given the special name **zwitterion.** Note that a zwitterion has no net charge; it contains one positive charge and one negative charge.

Because they exist as zwitterions, amino acids have many of the properties associated with salts. They are crystalline solids with high melting points and are fairly soluble in water but insoluble in nonpolar organic solvents such as ether and hydrocarbon solvents.

Amino acid A compound that contains both an amino group and a carboxyl group.

α-Amino acid An amino acid in which the amino group is on the carbon adjacent to the carboxyl group.

Zwitterion An internal salt of an amino acid.

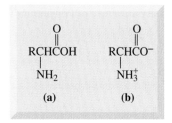

Figure 27.1

An α-amino acid. (a) Un-ionized form and (b) internal salt (zwitterion) form.

B. Chirality

With the exception of glycine, H₂NCH₂COOH, all protein-derived amino acids have at least one stereocenter and, therefore, are chiral. Figure 27.2 shows Fischer projection formulas for the enantiomers of alanine. The vast majority of carbohydrates in the biological world are of the D series, whereas the vast majority of α-amino acids in the biological world are of the L series.

The alternative *R,S* convention is also used to specify the configurations of amino acids. According to this convention, L-alanine is designated (*S*)-alanine. Because D- and L- are used more commonly to describe the configuration of amino acids, we use this convention throughout the remainder of the chapter.

C. Protein-Derived Amino Acids

Table 27.1 gives common names, structural formulas, and standard three-letter and one-letter abbreviations for the 20 common L-amino acids found in proteins. The amino acids in this table are divided into four categories: those with nonpolar side

Figure 27.2

The enantiomers of alanine. The vast majority of α-amino acids in the biological world have the L configuration at the α-carbon.

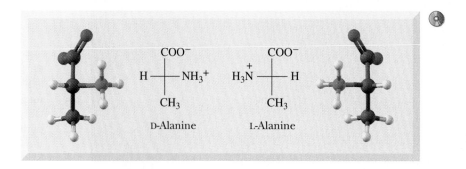

Table 27.1   The 20 Common Amino Acids Found in Proteins*

Nonpolar Side Chains

Alanine
(Ala, A)

Glycine
(Gly, G)

Isoleucine
(Ile, I)

Leucine
(Leu, L)

Methionine
(Met, M)

Phenylalanine
(Phe, F)

Proline
(Pro, P)

Tryptophan
(Trp, W)

Valine
(Val, V)

Polar Side Chains

Asparagine
(Asn, N)

Glutamine
(Gln, Q)

Serine
(Ser, S)

Threonine
(Thr, T)

Acidic Side Chains

Aspartic acid
(Asp, D)

Glutamic acid
(Glu, E)

Cysteine
(Cys, C)

Tyrosine
(Tyr, Y)

Basic Side Chains

Arginine
(Arg, R)

Histidine
(His, H)

Lysine
(Lys, K)

* Each ionizable group is shown in the form present in highest concentration at pH 7.0.

chains, polar but un-ionized side chains, acidic side chains, and basic side chains. The following structural features of these amino acids should be noted.

1. All 20 of these protein-derived amino acids are α-amino acids, meaning that the amino group is located on the carbon alpha to the carboxyl group.

2. For 19 of the 20 amino acids, the α-amino group is primary. Proline is different; its α-amino group is secondary.
3. With the exception of glycine, the α-carbon of each amino acid is a stereocenter. Although not shown in this table, all 19 chiral amino acids have the same relative configuration at the α-carbon. In the D,L convention, all are L-amino acids.
4. Isoleucine and threonine contain a second stereocenter. Four stereoisomers are possible for each amino acid, but only one is found in proteins.
5. The sulfhydryl group of cysteine, the imidazole group of histidine, and the phenolic hydroxyl of tyrosine are partially ionized at pH 7.0, but the ionic form is not the major form present at this pH.

Example 27.1

Of the 20 protein-derived amino acids shown in Table 27.1, how many contain the following?
(a) Aromatic rings   **(b)** Side-chain hydroxyl groups
(c) Phenolic —OH groups   **(d)** Sulfur

Solution

(a) Phenylalanine, tryptophan, tyrosine, and histidine contain aromatic rings.
(b) Serine and threonine contain side-chain hydroxyl groups.
(c) Tyrosine contains a phenolic —OH group.
(d) Methionine and cysteine contain sulfur.

Problem 27.1

Of the 20 protein-derived amino acids shown in Table 27.1, which contain the following?
(a) No stereocenter   **(b)** Two stereocenters

D.   Some Other Common L-Amino Acids

Although the vast majority of plant and animal proteins are constructed from just these 20 α-amino acids, many other amino acids are also found in nature. Ornithine and citrulline, for example, are found predominantly in the liver and are an integral part of the urea cycle, the metabolic pathway that converts ammonia to urea.

L-Ornithine   L-Citrulline

Thyroxine and triiodothyronine, two of several hormones derived from the amino acid tyrosine, are found in thyroid tissue. Their principal function is to stimulate metabolism in other cells and tissues.

L-Thyroxine, T_4

L-Triiodothyronine, T_3

4-Aminobutanoic acid is found in high concentration (0.8 mM) in the brain but in no significant amounts in any other mammalian tissue. It is synthesized in neural tissue by decarboxylation of the α-carboxyl group of glutamic acid and is a neurotransmitter in the central nervous system of invertebrates and in humans as well.

Glutamic acid

4-Aminobutanoic acid
(γ-Aminobutyric acid, GABA)

Only L-amino acids are found in proteins, and only rarely are D-amino acids a part of the metabolism of higher organisms. Several D-amino acids, however, along with their L-enantiomers, are found in lower forms of life. D-Alanine and D-glutamic acid, for example, are structural components of the cell walls of certain bacteria. Several D-amino acids are also found in peptide antibiotics.

27.2 Acid-Base Properties of Amino Acids

A. Acidic and Basic Groups of Amino Acids

Among the most important chemical properties of amino acids are their acid-base properties; all are weak polyprotic acids because of their —COOH and —NH$_3{}^+$ groups. Given in Table 27.2 are pK_a values for each ionizable group of the 20 protein-derived amino acids.

Acidity of α-Carboxyl Groups

The average value of pK_a for an α-carboxyl group of a protonated amino acid is 2.19. Thus, the α-carboxyl group is a considerably stronger acid than acetic acid (pK_a 4.76) and other low-molecular-weight aliphatic carboxylic acids. This greater acidity is accounted for by the electron-withdrawing inductive effect of the adjacent —NH$_3{}^+$ group. Recall that we used similar reasoning in Section 17.4A to account for the relative acidities of acetic acid and its mono-, di-, and trichloroderivatives.

The ammonium group has an electron-withdrawing inductive effect.

$$RCHCOOH + H_2O \rightleftharpoons RCHCOO^- + H_3O^+ \qquad pK_a = 2.19$$
$$\quad | \qquad\qquad\qquad\qquad | $$
$$NH_3{}^+ \qquad\qquad\qquad NH_3{}^+$$

Table 27.2 pK$_a$ Values for Ionizable Groups of Amino Acids				
Amino Acid	**pK$_a$ of α-COOH**	**pK$_a$ of α-NH$_3^+$**	**pK$_a$ of Side Chain**	**Isoelectric Point (pI)**
Alanine	2.35	9.87	—	6.11
Arginine	2.01	9.04	12.48	10.76
Asparagine	2.02	8.80	—	5.41
Aspartic acid	2.10	9.82	3.86	2.98
Cysteine	2.05	10.25	8.00	5.02
Glutamic acid	2.10	9.47	4.07	3.08
Glutamine	2.17	9.13	—	5.65
Glycine	2.35	9.78	—	6.06
Histidine	1.77	9.18	6.10	7.64
Isoleucine	2.32	9.76	—	6.04
Leucine	2.33	9.74	—	6.04
Lysine	2.18	8.95	10.53	9.74
Methionine	2.28	9.21	—	5.74
Phenylalanine	2.58	9.24	—	5.91
Proline	2.00	10.60	—	6.30
Serine	2.21	9.15	—	5.68
Threonine	2.09	9.10	—	5.60
Tryptophan	2.38	9.39	—	5.88
Tyrosine	2.20	9.11	10.07	5.63
Valine	2.29	9.72	—	6.00

Acidity of Side-Chain Carboxyl Groups

Due to the electron-withdrawing inductive effect of the α-NH$_3^+$ group, the side-chain carboxyl groups of protonated aspartic and glutamic acids are also stronger acids than acetic acid (pK$_a$ 4.76). Notice that this acid-strengthening inductive effect decreases with increasing distance of the —COOH from the α-NH$_3^+$. Compare the acidities of the α-COOH of alanine (pK$_a$ 2.35), the β-COOH of aspartic acid (pK$_a$ 3.86), and the γ-COOH of glutamic acid (pK$_a$ 4.07).

Acidity of α-Ammonium Groups

The average value of pK$_a$ for an α-ammonium group, α-NH$_3^+$, is 9.47 compared with an average value of 10.60 for primary aliphatic ammonium ions (Section 22.6A). Thus, the α-ammonium group of an amino acid is a slightly stronger acid than a primary aliphatic ammonium ion. Conversely, an α-amino group is a slightly weaker base than a primary aliphatic amine.

$$\underset{\underset{NH_3^+}{|}}{RCHCOO^-} + H_2O \rightleftharpoons \underset{\underset{NH_2}{|}}{RCHCOO^-} + H_3O^+ \qquad pK_a = 9.47$$

$$\underset{\underset{NH_3^+}{|}}{CH_3CHCH_3} + H_2O \rightleftharpoons \underset{\underset{NH_2}{|}}{CH_3CHCH_3} + H_3O^+ \qquad pK_a = 10.60$$

Basicity of the Guanidine Group of Arginine

The side-chain guanidine group of arginine is a considerably stronger base than an aliphatic amine. As we saw in Section 22.6D, guanidine (pK_a 13.6) is the strongest base of any neutral compound. The remarkable basicity of the guanidine group of arginine is attributed to the large resonance stabilization of the protonated form relative to the neutral form.

The guanidinium ion side chain of arginine is a hybrid of three contributing structures

No resonance stabilization without charge separation

$pK_a = 12.48$

Basicity of the Imidazole Group of Histidine

Because the imidazole group on the side chain of histidine contains six π electrons in a planar, fully conjugated ring, imidazole is classified as a heterocyclic aromatic amine (Section 20.2D). The unshared pair of electrons on one nitrogen is a part of the aromatic sextet, whereas that on the other nitrogen is not. The pair of electrons that is not part of the aromatic sextet is responsible for the basic properties of the imidazole ring. Protonation of this nitrogen produces a resonance-stabilized cation.

This lone pair is not a part of the aromatic sextet; it is the proton acceptor.

$pK_a = 6.10$

Resonance-stabilized imidazolium cation

B. Titration of Amino Acids

Values of pK_a for the ionizable groups of amino acids are most commonly obtained by acid-base titration and by measurement of the pH of the solution as a function of added base (or added acid, depending on how the titration is done). To illustrate this experimental procedure, consider a solution containing 1.00 mol of glycine to which has been added enough strong acid that both the amino and carboxyl groups are fully protonated. Next, this solution is titrated with 1.00 M NaOH; the volume of base added and the pH of the resulting solution are recorded and then plotted as shown in Figure 27.3. The most acidic group and the one to react first with added sodium hydroxide is the carboxyl group. When exactly 0.50 mol of NaOH has been added, the carboxyl group is half neutralized. At this point, the concentration of the zwitterion equals that of the positively charged ion, and the pH of 2.35 equals the pK_a of the carboxyl group (pK_{a1}).

At pH = pK_{a1} $[\overset{+}{H_3N}CH_2COOH] = [\overset{+}{H_3N}CH_2COO^-]$

Positive ion   Zwitterion

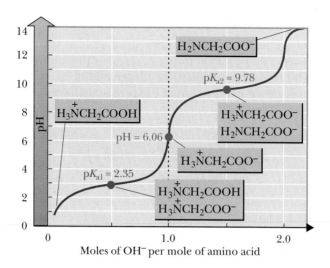

Figure 27.3
Titration of glycine with sodium hydroxide.

The end point of the first part of the titration is reached when 1.00 mol of NaOH has been added. At this point, the predominant species present is the zwitterion, and the observed pH of the solution is 6.06.

The next section of the curve represents titration of the —NH_3^+ group. When another 0.50 mol of NaOH has been added (bringing the total to 1.50 mol), half of the —NH_3^+ groups are neutralized and converted to —NH_2. At this point, the concentrations of the zwitterion and negatively charged ion are equal, and the observed pH is 9.78, the pK_a of the amino group of glycine (pK_{a2}).

$$\text{At pH} = pK_{a2} \quad [\overset{+}{H_3}NCH_2COO^-] = [H_2NCH_2COO^-]$$

Zwitterion   Negative ion

The second end point of the titration is reached when a total of 2.00 mol of NaOH have been added and glycine is converted entirely to an anion.

C.   Isoelectric Point

Titration curves such as that for glycine permit us to determine pK_a values for the ionizable groups of an amino acid. They also permit us to determine another important property: isoelectric point. **Isoelectric point, pI,** for an amino acid is the pH at which the majority of molecules in solution have a net charge of zero (they are zwitterions). By examining the titration curve, you can see that the isoelectric point for glycine falls half way between the pK_a values for the carboxyl and amino groups.

$$pI = \tfrac{1}{2}(pK_a \,\alpha - COOH + pK_a \,\alpha - NH_3^+)$$

$$= \tfrac{1}{2}(2.35 + 9.78) = 6.06$$

At pH 6.06, the predominant form of glycine molecules is the dipolar ion; furthermore, at this pH, the concentration of positively charged glycine molecules equals the concentration of negatively charged glycine molecules.

Given a value for the isoelectric point of an amino acid, it is possible to estimate the charge on that amino acid at any pH. For example, the charge on tyrosine at pH

5.63, its isoelectric point, is zero. A small fraction of tyrosine molecules are positively charged at pH 5.00 (0.63 unit less than its pI), and virtually all are positively charged at pH 3.63 (2.00 units less than its pI). As another example, the net charge on lysine is zero at pH 9.74. At pH values smaller than 9.74, an increasing fraction of lysine molecules are positively charged.

D. Electrophoresis

Electrophoresis, a process of separating compounds on the basis of their electric charges, is used to separate and identify mixtures of amino acids and proteins. Electrophoretic separations can be carried out using paper, starch, agar, certain plastics, and cellulose acetate as solid supports. In paper electrophoresis, a paper strip saturated with an aqueous buffer of predetermined pH serves as a bridge between two electrode vessels (Figure 27.4). Next, a sample of amino acids is applied as a spot on the paper strip. When an electrical potential is then applied to the electrode vessels, amino acids migrate toward the electrode carrying the charge opposite to their own. Molecules having a high charge density move more rapidly than those with a lower charge density. Any molecule already at its isoelectric point remains at the origin. After separation is complete, the strip is dried and sprayed with a dye to make the separated components visible.

Electrophoresis The process of separating compounds on the basis of their electric charge.

A dye commonly used to detect amino acids is ninhydrin (1,2,3-indanetrione monohydrate). Ninhydrin reacts with α-amino acids to produce an aldehyde, carbon dioxide, and a purple-colored anion. This reaction is used very commonly in both qualitative and quantitative analysis of amino acids.

$$
\underset{\substack{\text{An } \alpha\text{-amino} \\ \text{acid}}}{\text{RCHCO}^-} + 2\ \underset{\text{Ninhydrin}}{\text{(indanetrione)}} \longrightarrow \underset{\text{Purple-colored anion}}{\text{(anion)}} + \text{RCH} + CO_2 + H_3O^+
$$

Nineteen of the 20 protein-derived α-amino acids have primary amino groups and give the same purple-colored ninhydrin-derived anion. Proline, a secondary amine, gives a different, orange-colored compound.

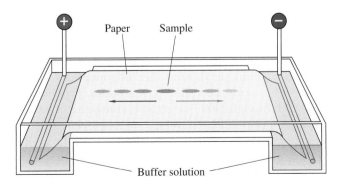

Figure 27.4
An apparatus for electrophoresis of a mixture of amino acids. Those with a negative charge move toward the positive electrode; those with a positive charge move toward the negative electrode; those with no charge remain at the origin.

Example 27.2

The isoelectric point of tyrosine is 5.63. Toward which electrode does tyrosine migrate on paper electrophoresis at pH 7.0?

Solution

On paper electrophoresis at pH 7.0 (more basic than its isoelectric point), tyrosine has a net negative charge and migrates toward the positive electrode.

Problem 27.2

The isoelectric point of histidine is 7.64. Toward which electrode does histidine migrate on paper electrophoresis at pH 7.0?

Example 27.3

Electrophoresis of a mixture of lysine, histidine, and cysteine is carried out at pH 7.64. Describe the behavior of each amino acid under these conditions.

Solution

The isoelectric point of histidine is 7.64. At this pH, histidine has a net charge of zero and does not move from the origin. The pI of cysteine is 5.02; at pH 7.64 (more basic than its isoelectric point), cysteine has a net negative charge and moves toward the positive electrode. The pI of lysine is 9.74; at pH 7.64 (more acidic than its isoelectric point), lysine has a net positive charge and moves toward the negative electrode.

Problem 27.3

Describe the behavior of a mixture of glutamic acid, arginine, and valine on paper electrophoresis at pH 6.0.

27.3   Polypeptides and Proteins

In 1902, Emil Fischer proposed that proteins are long chains of amino acids joined together by amide bonds between the α-carboxyl group of one amino acid and the α-amino group of another. For these amide bonds, Fischer proposed the special name **peptide bond.** Figure 27.5 shows the peptide bond formed between serine and alanine in the dipeptide serylalanine.

Peptide is the name given to a short polymer of amino acids. Peptides are classified by the number of amino acid units in the chain. A molecule containing 2 amino acids joined by an amide bond is called a **dipeptide.** Those containing 3 to 10 amino acids are called **tripeptides, tetrapeptides, pentapeptides,** and so on. Molecules containing more than 10 but fewer than 20 amino acids are called **oligopeptides.** Those containing several dozen or more amino acids are called **polypeptides. Proteins** are biological macromolecules of molecular weight 5000 or greater, consisting of one or more polypeptide chains. The distinctions in this terminology are not precise.

By convention, polypeptides are written from the left, beginning with the amino acid having the free $-NH_3^+$ group and proceeding to the right toward the amino acid with the free $-COO^-$ group. The amino acid with the free $-NH_3^+$ group is called the **N-terminal amino acid** and that with the free $-COO^-$ group is called the

Peptide bond   The special name given to the amide bond formed between the α-amino group of one amino acid and the α-carboxyl group of another amino acid.

Dipeptide   A molecule containing two amino acid units joined by a peptide bond.

Polypeptide   A macromolecule containing many amino acid units, each joined to the next by a peptide bond.

N-Terminal amino acid   The amino acid at the end of a polypeptide chain having the free $-NH_2$ group.

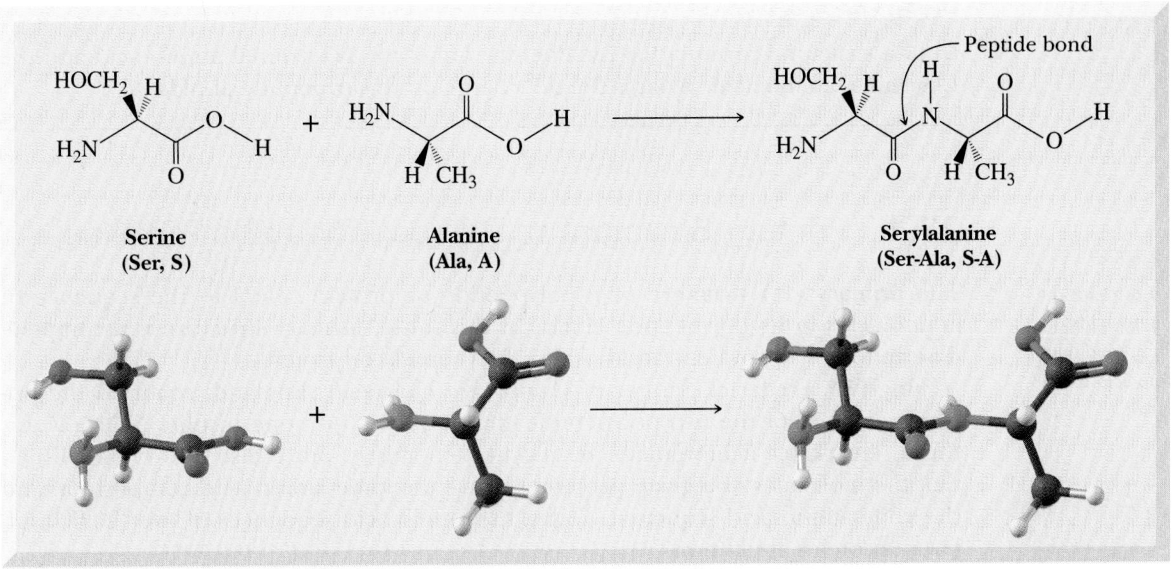

Figure 27.5
The peptide bond in serylalanine.

C-terminal amino acid. Notice the repeating pattern in the peptide chain of N—α-carbon—carbonyl, and so on.

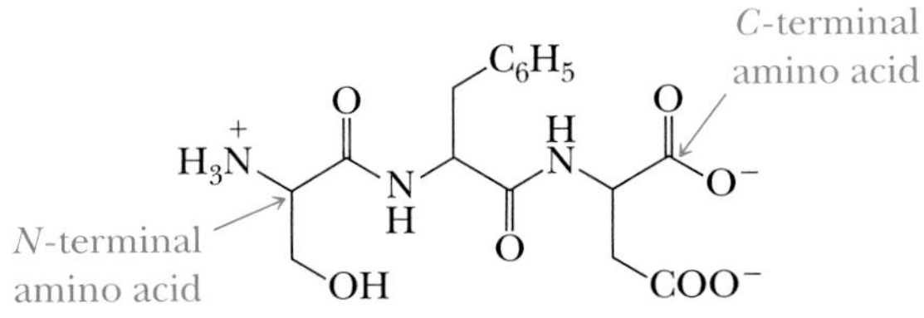

Ser-Phe-Asp

C-Terminal amino acid   The amino acid at the end of a polypeptide chain having the free —COOH group.

Example 27.4

Draw a structural formula for Cys-Arg-Met-Asn. Label the *N*-terminal amino acid and the *C*-terminal amino acid. What is the net charge on this tetrapeptide at pH 6.0?

Solution

The backbone of this tetrapeptide is a repeating sequence of nitrogen—α-carbon—carbonyl. The net charge on this tetrapeptide at pH 6.0 is +1.

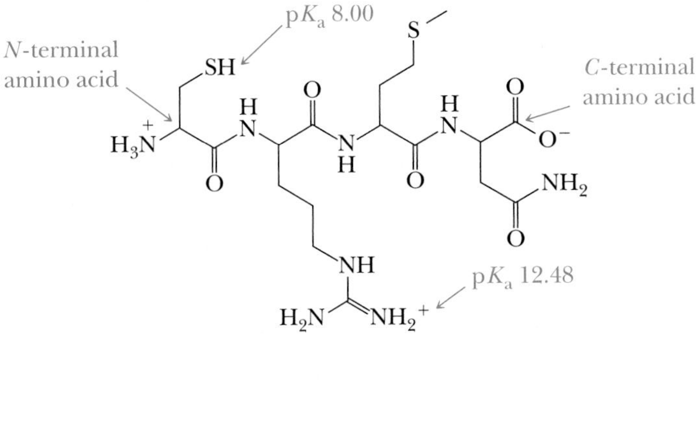

Problem 27.4

Draw a structural formula for Lys-Phe-Ala. Label the *N*-terminal amino acid and the *C*-terminal amino acid. What is the net charge on this tripeptide at pH 6.0?

27.4 Primary Structure of Polypeptides and Proteins

Primary structure of proteins
The sequence of amino acids in the polypeptide chain, read from the *N*-terminal amino acid to the *C*-terminal amino acid.

The **primary (1°) structure** of a polypeptide or protein refers to the sequence of amino acids in its polypeptide chain. In this sense, primary structure is a complete description of all covalent bonding in a polypeptide or protein.

In 1953, Frederick Sanger of Cambridge University, England, reported the primary structure of the two polypeptide chains of the hormone insulin. Not only was this a remarkable achievement in analytical chemistry, but it also clearly established that the molecules of a given protein all have the same amino acid composition and the same amino acid sequence. Today, the amino acid sequences of over 20,000 different proteins are known.

A. Amino Acid Analysis

The first step for determining the primary structure of a polypeptide is hydrolysis and quantitative analysis of its amino acid composition. Recall from Section 18.5D that amide bonds are very resistant to hydrolysis. Typically, samples of protein are hydrolyzed in 6 *M* HCl in sealed glass vials at 110°C for 24 to 72 hours. This hydrolysis can be done in a microwave oven in a shorter time. After the polypeptide is hydrolyzed, the resulting mixture of amino acids is analyzed by ion-exchange chromatography. Amino acids are detected as they emerge from the column by reaction with ninhydrin (Section 27.2D) followed by absorption spectroscopy. Current procedures for hydrolysis of polypeptides and analysis of amino acid mixtures have been refined to the point where it is possible to obtain amino acid composition from as little as 50 nanomoles (50×10^{-9} mol) of polypeptide. Figure 27.6 shows the analysis of a polypeptide hydrolysate by ion-exchange chromatography. Note that during hydrolysis, the side-chain amide groups of asparagine and glutamine are hydrolyzed, and these amino acids are detected as aspartic acid and glutamic acid. For each glutamine or asparagine hydrolyzed, an equivalent amount of ammonium chloride is formed.

B. Sequence Analysis

After the amino acid composition of a polypeptide has been determined, the next step is to determine the order in which the amino acids are joined in the polypeptide chain. The most common sequencing strategy is to cleave the polypeptide at specific peptide bonds (using, for example, cyanogen bromide or certain proteolytic enzymes), determine the sequence of each fragment (using, for example, the Edman degradation), and then match overlapping fragments to arrive at the sequence of the polypeptide.

Cyanogen Bromide

Cyanogen bromide (BrCN) is specific for cleavage of peptide bonds formed by the carboxyl group of methionine (Figure 27.7). The products of this cleavage are a sub-

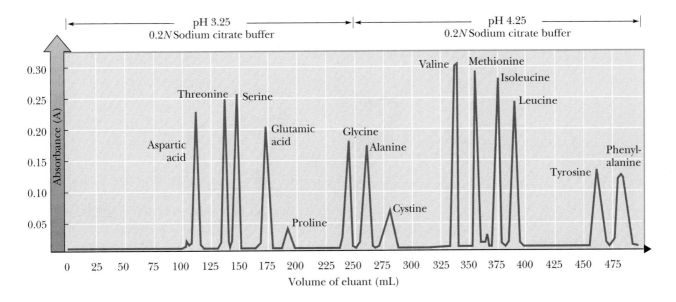

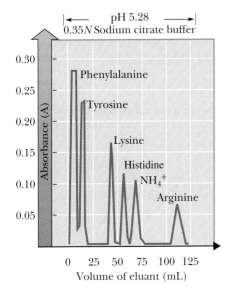

Figure 27.6

Analysis of a mixture of amino acids by ion-exchange chromatography using Amberlite IR-120, a sulfonated polystyrene resin. The resin contains phenyl-$SO_3^-Na^+$ groups. The amino acid mixture is applied to the column at low pH (3.25) under which conditions the acidic amino acids (Asp, Glu) are weakly bound to the resin and the basic amino acids (Lys, His, Arg) are tightly bound. Sodium citrate buffers at two different concentrations and three different values of pH are used to elute the amino acids from the column. Cysteine is determined as cystine, Cys-S-S-Cys, the disulfide of cysteine.

stituted γ-lactone (Section 18.1C) derived from the *N*-terminal portion of the polypeptide and a second fragment containing the *C*-terminal portion of the polypeptide.

A three-step mechanism can be written for this reaction. The strategy for cyanogen bromide cleavage depends on chemical manipulation of the leaving ability of the sulfur atom of methionine. Because CH_3S^- is the anion of a weak acid, it is a very poor leaving group, just as OH^- is a poor leaving group (Section 8.4F). Yet, just as the oxygen atom of an alcohol can be transformed into a better leaving group by converting it into an oxonium ion (by protonation), so too can the sulfur atom of methionine be transformed into a better leaving group by converting it into a sulfonium ion.

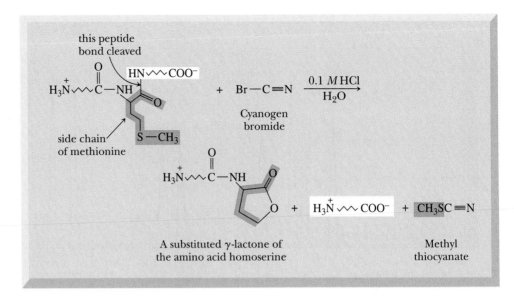

Figure 27.7
Cleavage by cyanogen bromide, BrCN, of a peptide bond formed by the carboxyl group of methionine.

Mechanism   Cleavage of a Peptide Bond at Methionine by Cyanogen Bromide

Step 1:   Reaction is initiated by nucleophilic attack of the divalent sulfur atom of methionine on the carbon of cyanogen bromide displacing bromide ion. The product of this nucleophilic displacement is a sulfonium ion.

Step 2:   An internal S_N2 reaction in which the oxygen of the methionine carbonyl group attacks the γ-carbon and displaces methyl thiocyanate gives a five-membered ring. Note that the oxygen of a carbonyl group is at best a weak nucleophile. This displacement is facilitated, however, because the sulfonium ion is a very good leaving group and because of the ease with which a five-membered ring is formed.

Step 3:   Hydrolysis of the imino group gives a γ-lactone derived from the *N*-terminal end of the original polypeptide.

A substituted γ-lactone of
the amino acid homoserine

Enzyme-Catalyzed Hydrolysis of Peptide Bonds

A group of proteolytic enzymes, among them trypsin and chymotrypsin, can be used to catalyze the hydrolysis of specific peptide bonds. Trypsin catalyzes the hydrolysis of peptide bonds formed by the carboxyl groups of arginine and lysine; chymotrypsin catalyzes the hydrolysis of peptide bonds formed by the carboxyl groups of phenylalanine, tyrosine, and tryptophan (Table 27.3).

Example 27.5

Which of these dipeptides are hydrolyzed by trypsin? By chymotrypsin?
(a) Arg-Glu-Ser   **(b)** Phe-Gly-Lys

Solution

(a) Trypsin catalyzes the hydrolysis of peptide bonds formed by the carboxyl groups of lysine and arginine. Therefore, the peptide bond between arginine and glutamic acid is hydrolyzed in the presence of trypsin.

$$\text{Arg-Glu-Ser} + H_2O \xrightarrow{\text{trypsin}} \text{Arg} + \text{Glu-Ser}$$

Chymotrypsin catalyzes the hydrolysis of peptide bonds formed by the carboxyl groups of phenylalanine, tyrosine, and tryptophan. Because none of these three aromatic amino acids is present, tripeptide (a) is not affected by chymotrypsin.

(b) Tripeptide (b) is not affected by trypsin. Although lysine is present, its carboxyl group is at the *C*-terminal end and not involved in peptide bond formation. Tripeptide (b) is hydrolyzed in the presence of chymotrypsin.

$$\text{Phe-Gly-Lys} + H_2O \xrightarrow{\text{chymotrypsin}} \text{Phe} + \text{Gly-Lys}$$

Table 27.3 **Cleavage of Specific Peptide Bonds Catalyzed by Trypsin and Chymotrypsin**	
Enzyme	**Catalyzes Hydrolysis of Peptide Bond Formed by Carboxyl Group of**
Trypsin	Arginine, lysine
Chymotrypsin	Phenylalanine, tyrosine, tryptophan

Problem 27.5

Which of these tripeptides are hydrolyzed by trypsin? By chymotrypsin?

(a) Tyr-Gln-Val   (b) Thr-Phe-Ser   (c) Thr-Ser-Phe

Edman Degradation

Edman degradation A method for selectively cleaving and identifying the N-terminal amino acid of a polypeptide chain.

Of the various chemical methods developed for determining the amino acid sequence of a polypeptide, the one most widely used today is the **Edman degradation,** introduced in 1950 by Pehr Edman of the University of Lund, Sweden. In this procedure, a polypeptide is treated with phenyl isothiocyanate, C_6H_5N=C=S, and then with acid. The effect of Edman degradation is to remove the N-terminal amino acid selectively as a substituted phenylthiohydantoin (Figure 27.8), which is then separated and identified.

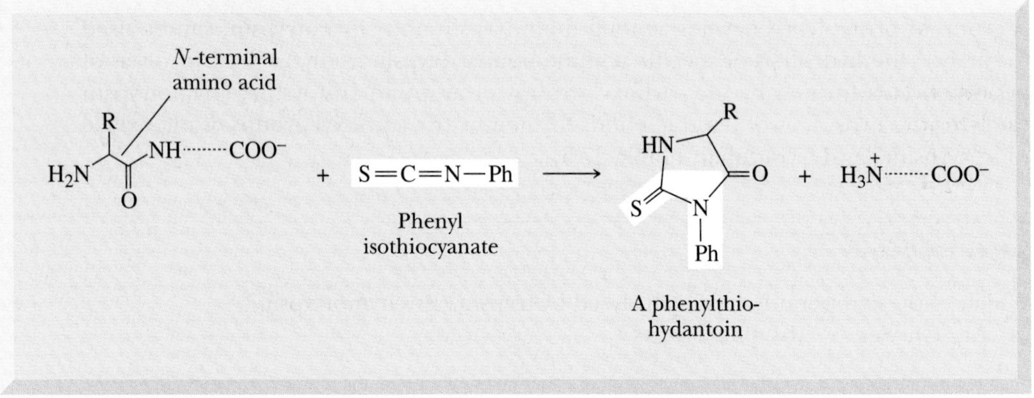

Figure 27.8
Edman degradation. Treatment of a polypeptide with phenyl isothiocyanate followed by acid selectively cleaves the N-terminal amino acid as a substituted phenylthiohydantoin.

The key feature of the Edman degradation is successive C=N, C=O, and C=O addition reactions.

Mechanism   Edman Degradation—Cleavage of an
N-Terminal Amino Acid

Step 1:   Nucleophilic addition of the N-terminal amino group to the C=N bond of phenyl isothiocyanate gives a derivative of N-phenylthiourea.

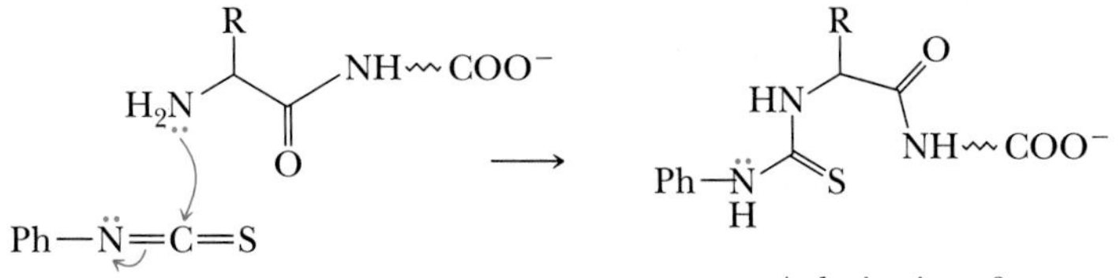

A derivative of
N-phenylthiourea

Step 2: Heating the derivatized polypeptide in HCl at 100°C results in nucleophilic addition of sulfur to the carbonyl of the adjacent amide group to give a tetrahedral carbonyl addition intermediate, which collapses to give a thiazolinone ring derived from the *N*-terminal amino acid.

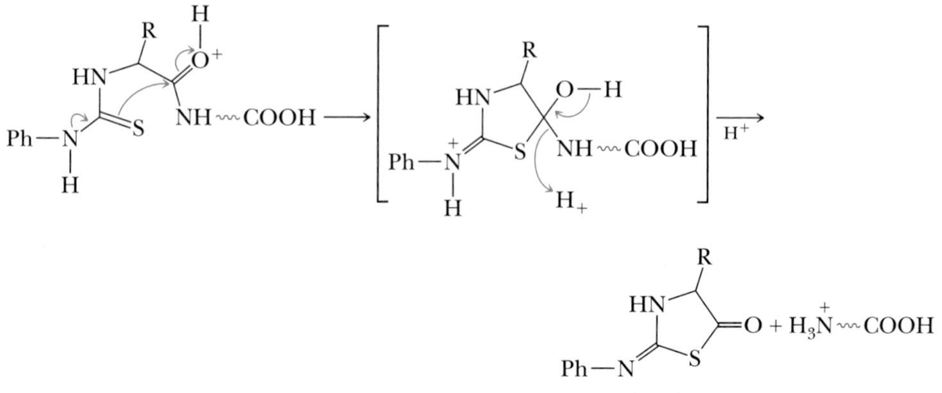

A thiazolinone

Step 3: The thiazolinone ring undergoes isomerization by ring opening followed by reclosure to give a more stable phenylthiohydantoin, which is separated and identified by chromatography.

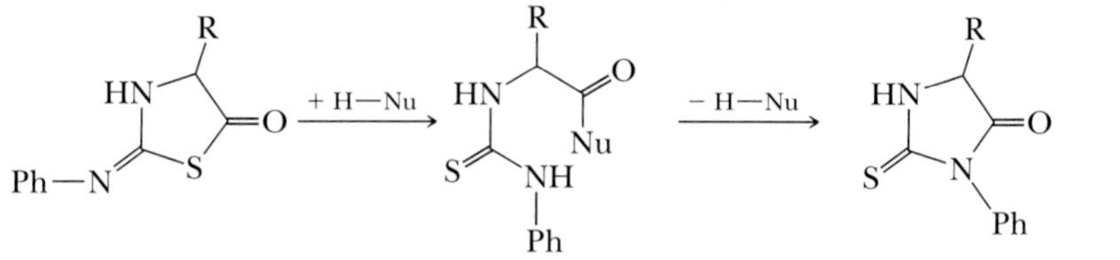

A thiazolinone

A phenylthiohydantoin

The special value of the Edman degradation is that it cleaves the *N*-terminal amino acid from a polypeptide without affecting any other bonds in the chain. Furthermore, Edman degradation can be repeated on the shortened polypeptide, causing the next amino acid in the sequence to be cleaved and identified. In practice, it is now possible to sequence as many as the first 20 to 30 amino acids in a polypeptide by this method using only a few milligrams of material.

Most polypeptides in nature are longer than 20 to 30 amino acids, the practical limit to the number of amino acids that can be sequenced by repetitive Edman degradation. The special value of cleavage with cyanogen bromide, trypsin, and chymotrypsin is that a long polypeptide chain can be cleaved at specific peptide bonds into smaller polypeptide fragments, and each fragment can then be sequenced separately.

Example 27.6

Deduce the amino acid sequence of a pentapeptide from the following experimental results. Note that under the column Amino Acid Composition, the amino acids are listed in alphabetical order. In no way does this listing give any information about primary structure.

Experimental Procedure	Amino Acid Composition
Pentapeptide	**Arg, Glu, His, Phe, Ser**
Edman degradation	Glu
Hydrolysis catalyzed by chymotrypsin	
Fragment A	Glu, His, Phe
Fragment B	Arg, Ser
Hydrolysis catalyzed by trypsin	
Fragment C	Arg, Glu, His, Phe
Fragment D	Ser

Solution

Edman degradation cleaves Glu from the pentapeptide; therefore, glutamic acid must be the *N*-terminal amino acid.

Glu-(Arg, His, Phe, Ser)

Fragment A from chymotrypsin-catalyzed hydrolysis contains Phe. Because of the specificity of chymotrypsin, Phe must be the *C*-terminal amino acid of fragment A. Fragment A also contains Glu, which we already know is the *N*-terminal amino acid. From these observations, conclude that the first three amino acids in the chain must be Glu-His-Phe, and now write the following partial sequence.

Glu-His-Phe-(Arg, Ser)

The fact that trypsin cleaves the pentapeptide means that Arg must be within the pentapeptide chain; it cannot be the *C*-terminal amino acid. Therefore, the complete sequence must be the following.

Glu-His-Phe-Arg-Ser

Problem 27.6

Deduce the amino acid sequence of an undecapeptide (11 amino acids) from the experimental results shown in the table.

Experimental Procedure	Amino Acid Composition
Undecapeptide	**Ala, Arg, Glu, Lys$_2$, Met, Phe, Ser, Thr, Trp, Val**
Edman degradation	Ala
Trypsin-catalyzed hydrolysis	
Fragment E	Ala, Glu, Arg
Fragment F	Thr, Phe, Lys
Fragment G	Lys
Fragment H	Met, Ser, Trp, Val
Chymotrypsin-catalyzed hydrolysis	
Fragment I	Ala, Arg, Glu, Phe, Thr
Fragment J	Lys$_2$, Met, Ser, Trp, Val
Treatment with cyanogen bromide	
Fragment K	Ala, Arg, Glu, Lys$_2$, Met, Phe, Thr, Val
Fragment L	Trp, Ser

Sequencing by Mass Spectrometry

As a result of enormous instrumental advances made during the last few years, mass spectrometry (MS) is becoming the method of choice for sequencing peptides. It is now possible using special MS techniques to obtain complete amino acid sequences on polypeptides of substantial length. The mass spectrometer cleaves polypeptides into fragments. Although there are many cleavage modes possible, the main cleavage is at peptide bonds, as shown in the following diagram. Both fragments (from the *N*-terminal and *C*-terminal part) can usually be identified. A mass analyzer determines the mass of each fragment. Because each amino acid has a slightly different mass (except for Leu and Ile), the exact amino acid composition of the peptide can be determined. Using powerful computers, it is possible to align overlapping fragments and determine the exact sequence.

$$H_2N-\underset{H}{\overset{R}{C}}-\overset{O}{C}+N-\underset{H}{\overset{R}{C}}-\overset{O}{C}+N-\underset{H}{\overset{R}{C}}-\overset{O}{C}+N-\underset{H}{\overset{R}{C}}-\overset{O}{C}+N-\underset{H}{\overset{R}{C}}-\overset{O}{C}OH$$

For larger proteins, the mass spectrometric method is usually preceded by enzymatic cleavage. However, in the most advanced systems, which should come into routine use in the near future, the individual cleavage fragments in the digest do not need to be separated. They are injected into a tandem MS (often called an MS-MS); the first segment of the instrument separates the individual digest fragments by mass, then the second segment fragments each peptide and analyzes it separately. The tandem MS can also be used to separate, and then further fragment, the primary ions obtained in the first stage to obtain sequence information directly. Thus enormous proteins can be sequenced on a picomole or lower concentration level. This method is particularly helpful because it can be used on mixtures (and ultimately, it is believed, on whole cell contents). Enormous computing resources are required for this task. This emerging field is called proteomics; the ambitious goal is to locate all proteins made from each gene in an organism and characterize their post-translational modification.

Sequencing Proteins from the Coding Nucleotide Sequence

As it has become easier to determine nucleotide sequences (see Section 28.5), it is now often easier to sequence the nucleotide that codes for a protein than to sequence the protein itself. In some cases, this has lead to discovery of new proteins of unknown function. Often comparison of the revealed protein sequences with those of known proteins from simpler organisms discloses sequence homologies that suggest the function of the new proteins. Comparisons with yeast, whose genome was one of the first to be sequenced, and for which most of the proteins coded for have known functions, have been particularly fruitful.

27.5 Synthesis of Polypeptides

A. The Problems

The problem in peptide synthesis is to join the carboxyl group of amino acid 1 (aa$_1$) by an amide (peptide) bond to the amino group of amino acid 2 (aa$_2$).

$$\underset{aa_1}{\overset{+}{H_3}NCH\overset{O}{C}O^-} + \underset{aa_2}{\overset{+}{H_3}NCH\overset{O}{C}O^-} \xrightarrow{?} \underset{aa_1}{\overset{+}{H_3}NCH\overset{O}{C}}NH\underset{aa_2}{CH\overset{O}{C}O^-} + H_2O$$

B.   The Strategy

A rational strategy for the synthesis of peptide bonds and polypeptides requires three steps.

1. Protect the α-amino group of amino acid aa$_1$ to reduce its nucleophilicity so that it does not participate in nucleophilic addition to the carboxyl group of either aa$_1$ or aa$_2$.
2. Protect the α-carboxyl group of amino acid aa$_2$ so that it is not susceptible to nucleophilic attack by the α-amino group of another molecule of aa$_2$.
3. Activate the α-carboxyl group of amino acid aa$_1$ so that it is susceptible to nucleophilic attack by the α-amino group of aa$_2$.

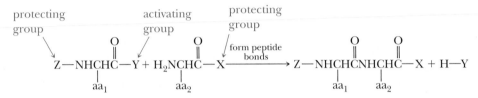

Once dipeptide aa$_1$—aa$_2$ has been formed, the protecting group Z can be removed, and chain growth can be continued from the *N*-terminal end of the dipeptide. Alternatively, the protecting group X can be removed, and chain growth can be continued from the *C*-terminal end. The range of protecting groups and activating groups is large, and experimental conditions have been found to attach and remove them as desired.

C.   Amino-Protecting Groups

The most common strategy for protecting amino groups and reducing their nucleophilicity is to convert them to amides. The reagents most commonly used for this purpose are benzyloxycarbonyl chloride and di-*tert*-butyl dicarbonate. In the terminology adopted by the IUPAC, the benzyloxycarbonyl group is given the symbol Z—, and the *tert*-butoxycarbonyl group is given the symbol BOC—.

$$\text{PhCH}_2\text{OCCl} \qquad \text{PhCH}_2\text{OC}— \qquad (\text{CH}_3)_3\text{COCOCOC}(\text{CH}_3)_3 \qquad (\text{CH}_3)_3\text{COC}—$$

Benzyloxycarbonyl   Benzyloxycarbonyl   Di-*tert*-butyl dicarbonate   *tert*-Butoxycarbonyl
chloride   (Z—) group   (BOC—) group

Treatment of an amino group with either of these reagents forms a new functional group called a carbamate. A **carbamate** is an ester of carbamic acid; that is, it is an ester of the monoamide of carbonic acid.

$$\text{PhCH}_2\text{OCCl} \;+\; \overset{+}{\text{H}_3}\text{NCHCO}^- \xrightarrow[\text{2. HCl, H}_2\text{O}]{\text{1. NaOH}} \text{PhCH}_2\text{OCNHCHCOH}$$

Benzyloxycarbonyl   Alanine   *N*-Benzyloxycarbonylalanine
chloride   (Z—Ala)
(Z—Cl)   A carbamate

The special advantage of the carbamate group is that it is stable to dilute base but can be removed by treatment with HBr in acetic acid.

$$\underset{\text{A Z-protected peptide}}{PhCH_2O\overset{\overset{\displaystyle O}{\|}}{C}NH\text{—peptide}} \quad\xrightarrow[\text{CH}_3\text{COOH}]{\text{HBr}}\quad \underset{\text{Benzyl bromide}}{PhCH_2Br} \;+\; CO_2 \;+\; \underset{\text{Unprotected peptide}}{\overset{+}{H_3N}\text{—peptide}}$$

A study of the mechanism for removal of this protecting group has shown that the reaction is first order in $[H^+]$ and involves formation of a carbocation and a carbamic acid. A carbamic acid spontaneously loses carbon dioxide to form the free amine. The carbocation may lose a proton to form an alkene or react with an available nucleophile such as halide ion to form an alkyl halide.

Mechanism   **Acid-Catalyzed Removal of a Benzyloxycarbonyl Protecting Group**

A carbamic acid

Note that because acid-catalyzed removal of this protecting group is carried out in nonaqueous media, there is no danger of simultaneous acid-catalyzed hydrolysis of peptide (amide) bonds within the newly synthesized polypeptide. This relation exists because water is required for hydrolysis of a peptide bond.

The benzyloxycarbonyl group can also be removed by treatment with H_2 in the presence of a transition metal catalyst (hydrogenolysis, Section 20.6C). In hydrogenolysis of a Z-protecting group, one product is toluene. The other is a carbamic acid, which undergoes spontaneous decarboxylation to give carbon dioxide and the unprotected peptide.

$$\underset{\text{A Z-protected peptide}}{PhCH_2O\overset{\overset{\displaystyle O}{\|}}{C}NH\text{—peptide}} + H_2 \;\xrightarrow{\;Pd\;}\; \underset{\text{Toluene}}{PhCH_3} \;+\; CO_2 \;+\; \underset{\text{Unprotected peptide}}{H_2N\text{—peptide}}$$

D. Carboxyl-Protecting Groups

Carboxyl groups are most often protected by conversion to methyl, ethyl, or benzyl esters. Methyl and ethyl esters are prepared by Fischer esterification (Section 17.7A) and are removed by hydrolysis in aqueous base (Section 18.5C) under mild conditions. Benzyl esters are conveniently removed by hydrogenolysis with H_2 over a palladium or platinum catalyst (Section 20.6C). Benzyl groups can also be removed by treatment with HBr in acetic acid.

E. Peptide Bond–Forming Reactions

The reagent most commonly used to bring about peptide bond formation is 1,3-dicyclohexylcarbodiimide (DCC). This reagent is the anhydride of a disubstituted urea, and, when treated with water, it is converted to N,N'-dicyclohexylurea (DCU).

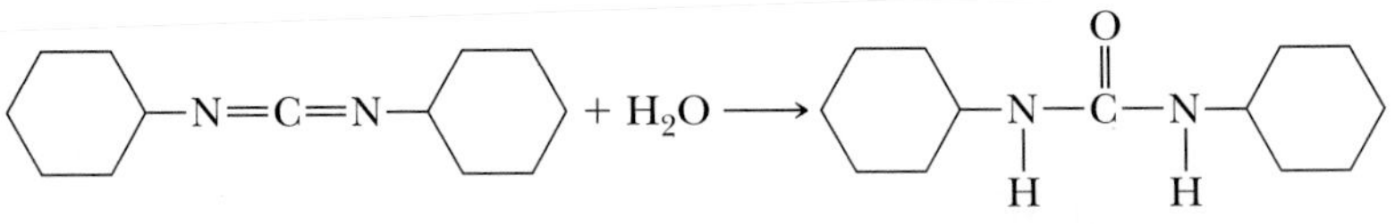

1,3-Dicyclohexylcarbodiimide
(DCC)

N,N'-Dicyclohexylurea
(DCU)

When an amino-protected aa$_1$ and a carboxyl-protected aa$_2$ are treated with DCC, this reagent acts as a dehydrating agent; it removes —OH from the carboxyl group and —H from the amino group to form an amide bond. More specifically, DCC activates the α-carboxyl group of aa$_1$ toward nucleophilic acyl substitution by converting its —OH group into a better leaving group.

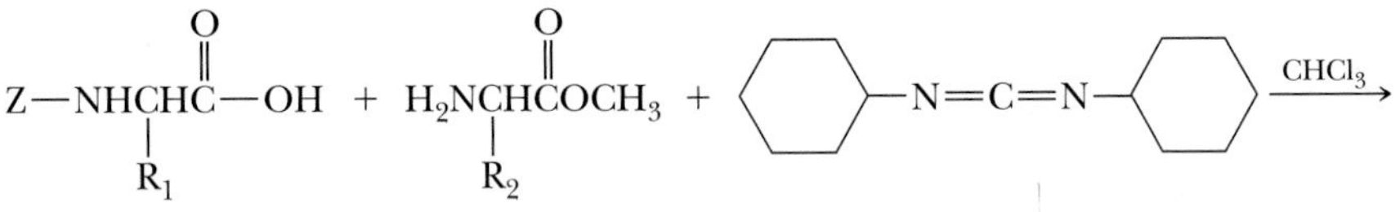

Amino protected
aa$_1$

Carboxyl protected
aa$_2$

1,3-Dicyclohexylcarbodiimide
(DCC)

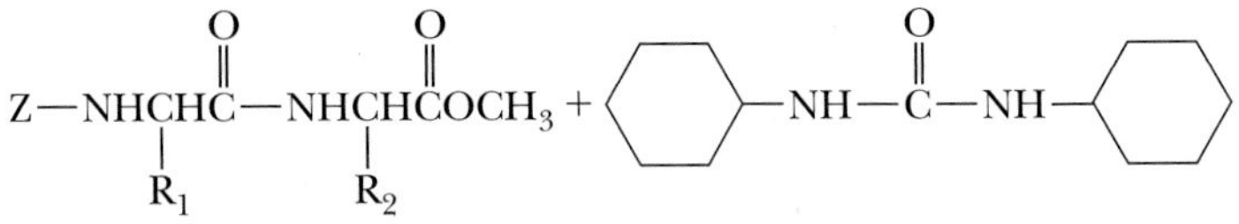

Amino and carboxyl
protected dipeptide

N,N'-Dicyclohexylurea
(DCU)

An abbreviated mechanism for this intermolecular dehydration is shown in Figure 27.9. An acid-base reaction in Step 1 between the carboxyl group of aa$_1$ and a nitrogen of DCC followed in Step 2 by addition of the carboxylate anion to the C=N double bond results in electrophilic addition to a C=N double bond. The O-acylisourea formed is the nitrogen analog of a mixed anhydride. Nucleophilic addition of the amino group of aa$_2$ to the carbonyl group of the O-acylisourea in Step 3 generates a tetrahedral carbonyl addition intermediate that collapses in Step 4 to give a dipeptide and DCU.

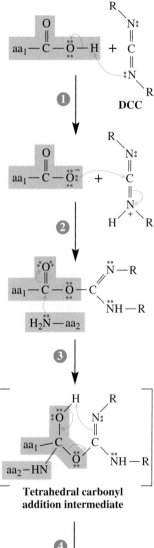

Figure 27.9

The role of 1,3-dicyclohexylcarbodiimide (DCC) in the formation of a peptide bond between an amino-protected amino acid (aa_1) and a carboxyl-protected amino acid (aa_2).

F. Solid-Phase Synthesis

A major problem associated with polypeptide synthesis is purification of intermediates after each protection, activation, coupling, and deprotection step. If unreacted starting materials are not removed after each step, the final product is contaminated by polypeptides missing one or more amino acids. The required purification steps are not only laborious and time-consuming, but they also inevitably result in some loss of the desired product. These losses become especially severe in the synthesis of larger polypeptides.

A major advance in polypeptide synthesis came in 1962 when R. Bruce Merrifield of Rockefeller University described a solid-phase synthesis (alternatively called polymer-supported synthesis) of the tetrapeptide Leu-Ala-Gly-Ala by a technique that now bears his name. Merrifield was awarded the 1984 Nobel Prize for chemistry for his work in developing the solid-phase method for peptide synthesis.

The solid support used by Merrifield was a type of polystyrene in which about 5% of the phenyl groups carry a chloromethyl ($-CH_2Cl$) group in their para positions (Figure 27.10). These chloromethyl groups, like all benzylic halides, are particularly reactive in nucleophilic substitution reactions.

In the Merrifield method, the *C*-terminal amino acid is joined as a benzyl ester to the solid polymer support, and then the polypeptide chain is extended one amino acid at a time from the *N*-terminal end. The advantage of polypeptide synthesis on a solid support is that the polymer beads with the peptide chains anchored on them are completely insoluble in the solvents used in the synthesis. Furthermore, excess reagents (for example, DCC) and byproducts (for example, DCU) are removed after each step simply by washing the polymer beads. When synthesis is completed, the polypeptide is released from the polymer beads by cleavage of the benzyl ester. The steps in solid-phase synthesis of a polypeptide are summarized in Figure 27.11.

A dramatic illustration of the power of the solid-phase method was the synthesis of the enzyme ribonuclease by Merrifield in 1969. The synthesis involved 369 chemical reactions and 11,931 operations, all of which were performed by an automated machine and without any intermediate isolation stages. Each of the 124 amino acids was added as an *N-tert*-butoxycarbonyl derivative and coupled using DCC. Cleavage from the resin and removal of all protective groups gave a mixture that was purified by ion-exchange chromatography. The specific activity of the synthetic enzyme was 13–24% of that of the natural enzyme. The fact that the specific activity of the synthetic enzyme was lower than

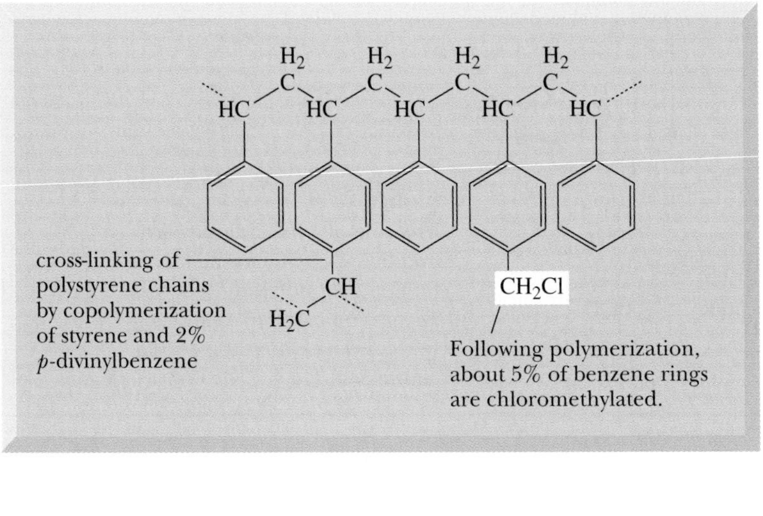

Figure 27.10

The support used for the Merrifield solid phase synthesis is a chloromethylated polystyrene resin.

cross-linking of polystyrene chains by copolymerization of styrene and 2% *p*-divinylbenzene

Following polymerization, about 5% of benzene rings are chloromethylated.

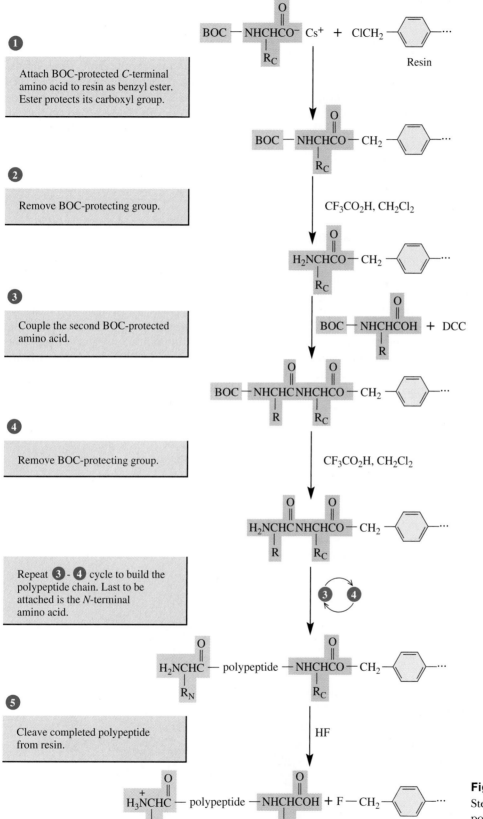

1 Attach BOC-protected *C*-terminal amino acid to resin as benzyl ester. Ester protects its carboxyl group.

2 Remove BOC-protecting group.

CF₃CO₂H, CH₂Cl₂

3 Couple the second BOC-protected amino acid.

4 Remove BOC-protecting group.

CF₃CO₂H, CH₂Cl₂

Repeat **3** - **4** cycle to build the polypeptide chain. Last to be attached is the *N*-terminal amino acid.

5 Cleave completed polypeptide from resin.

HF

Figure 27.11
Steps in the Merrifield solid phase polypeptide synthesis.

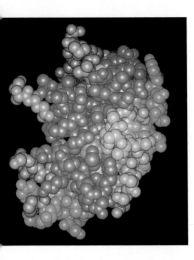

A model of the protein ribonuclease A. (*Charles Grisham*)

that of the natural enzyme was probably due to the presence of polypeptide byproducts closely related to but not identical to the natural enzyme. Synthesizing ribonuclease (124 amino acids) requires forming 123 peptide bonds. If each peptide bond is formed in 99% yield, the yield of homogeneous polypeptide is $0.99^{123} = 29\%$. If each peptide bond is formed in 98% yield, the yield is 8%. Thus, even with yields as high as 99% in each peptide bond-forming step, a large portion of the synthetic polypeptides have one or more sequence defects. Many of these, nonetheless, may be fully or partially active.

27.6 Three-Dimensional Shapes of Polypeptides and Proteins

A. Geometry of a Peptide Bond

In the late 1930s, Linus Pauling began a series of studies to determine the geometry of a peptide bond. One of his first and most important discoveries was that a peptide bond itself is planar. As shown in Figure 27.12, the four atoms of a peptide bond and the two α-carbons joined to it all lie in the same plane. Had you been asked in Chapter 1 to describe the geometry of a peptide bond, you probably would have predicted bond angles of 120° about the carbonyl carbon and 109.5° about the amide nitrogen. This prediction agrees with the observed bond angles of approximately 120° about the carbonyl carbon. Bond angles of 120° about the amide nitrogen, however, are unexpected. To account for this observed geometry, Pauling proposed that a peptide bond is more accurately represented as a resonance hybrid of these two contributing structures.

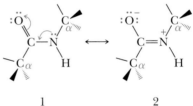

Contributing structure 1 shows a carbon-oxygen double bond, and structure 2 shows a carbon-nitrogen double bond. The hybrid, of course, is neither of these; in the real

Figure 27.12

Planarity of a peptide bond. Bond angles about the carbonyl carbon and the amide nitrogen are approximately 120°.

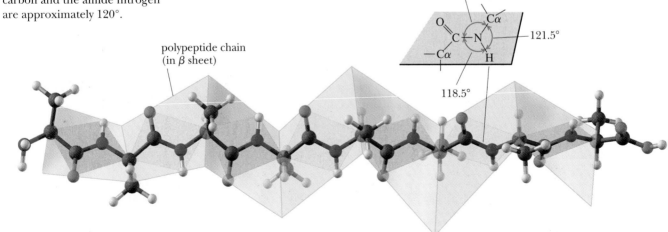

polypeptide chain (in β sheet)

structure, the carbon-nitrogen bond has considerable double-bond character. Accordingly, in the hybrid, the six-atom group is planar.

Two configurations are possible for the atoms of a planar peptide bond. In one, the two α-carbons are cis to each other; in the other, they are trans to each other. The trans configuration is more favorable because the α-carbons with the bulky groups bonded to them are farther from each other than they are in the cis configuration. Virtually all peptide bonds in naturally occurring proteins studied to date have the trans configuration.

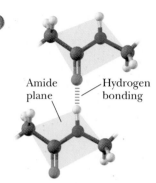

Figure 27.13
Hydrogen bonding between amide groups.

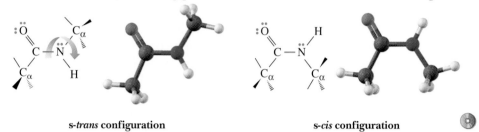

s-*trans* configuration   **s-*cis* configuration**

B. Secondary Structure

Secondary (2°) structure refers to ordered arrangements (conformations) of amino acids in localized regions of a polypeptide or protein molecule. The first studies of polypeptide conformations were carried out by Linus Pauling and Robert Corey, beginning in 1939. They assumed that in conformations of greatest stability, all atoms in a peptide bond lie in the same plane, and there is hydrogen bonded between the N—H of one peptide bond and the C═O of another, as shown in Figure 27.13. On the basis of model building, Pauling proposed that two types of secondary structure should be particularly stable: the α-helix and the antiparallel β-pleated sheet. X-ray crystallography has vindicated this prediction completely.

Secondary structure of proteins The ordered arrangements (conformations) of amino acids in localized regions of a polypeptide or protein.

The α-Helix

In an **α-helix** pattern, shown in Figure 27.14, a polypeptide chain is coiled in a spiral. As you study this section of α-helix, note the following:

α-Helix A type of secondary structure in which a section of polypeptide chain coils into a spiral, most commonly a right-handed spiral.

1. The helix is coiled in a clockwise, or right-handed, manner. Right-handed means that if you turn the helix clockwise, it twists away from you. In this sense, a right-handed helix is analogous to the right-handed thread of a common wood or machine screw.
2. There are 3.6 amino acids per turn of the helix.
3. Each peptide bond is trans and planar.

Figure 27.14
An α-helix. The peptide chain is repeating units of L-alanine.

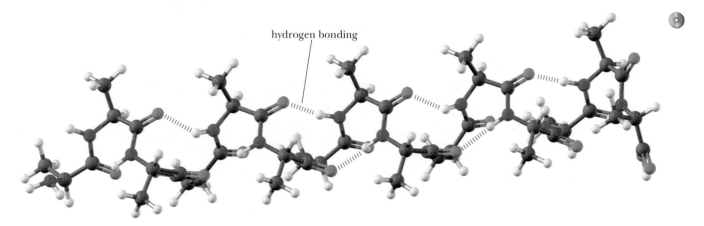

hydrogen bonding

4. The N—H group of each peptide bond points roughly downward, parallel to the axis of the helix, and the C═O of each peptide bond points roughly upward, also parallel to the axis of the helix.
5. The carbonyl group of each peptide bond is hydrogen-bonded to the N—H group of the peptide bond four amino acid units away from it. Hydrogen bonds are shown as dotted lines.
6. All R— groups point outward from the helix.

Almost immediately after Pauling proposed the α-helix conformation, other researchers proved the presence of α-helix conformations in keratin, the protein of hair and wool. It soon became obvious that the α-helix is one of the fundamental folding patterns of polypeptide chains.

The β-Pleated Sheet

β-Pleated sheet   A types of secondary structure in which sections of polypeptide chains are aligned parallel or antiparallel to one another.

An antiparallel **β-pleated sheet** consists of extended polypeptide chains with neighboring chains running in opposite (antiparallel) directions. In a parallel β-pleated sheet, the polypeptide chains run in the same direction. Unlike the α-helix arrangement, N—H and C═O groups lie in the plane of the sheet and are roughly perpendicular to the long axis of the sheet. The C═O group of each peptide bond is hydrogen-bonded to the N—H group of a peptide bond of a neighboring chain (Figure 27.15).

As you study this section of β-pleated sheet, note the following:

1. The three polypeptide chains lie adjacent to each other and run in opposite (antiparallel) directions.
2. Each peptide bond is planar, and the α-carbons are trans to each other.
3. The C═O and N—H groups of peptide bonds from adjacent chains point at each other and are in the same plane so that hydrogen bonding is possible between adjacent polypeptide chains.
4. The R— groups on any one chain alternate, first above, then below the plane of the sheet, and so on.

The β-pleated sheet conformation is stabilized by hydrogen bonding between N—H groups of one chain and C═O groups of an adjacent chain. By comparison, the α-helix is stabilized by hydrogen bonding between N—H and C═O groups within the same polypeptide chain.

Figure 27.15

β-Pleated sheet conformation with three polypeptide chains running in opposite (antiparallel) directions. Hydrogen bonding between chains is indicated by dashed lines.

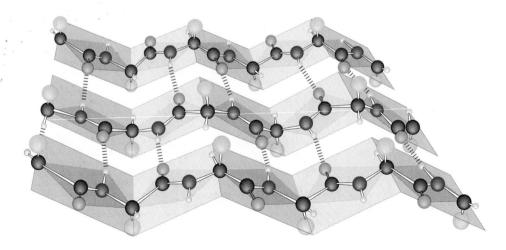

CHEMISTRY IN ACTION

Spider Silk

Spider silk has some remarkable properties. Research is currently concentrated on the strong dragline silk that forms the spokes of a web of the golden orb weaver (*Nephila clavipes*). This silk has three times the impact strength of Kevlar and is 30% more flexible than nylon. The commercial application of spider silk is not a novel concept. Eighteenth century French entrepreneur Bon de Saint-Hilaire attempted to mass-produce silk in his high-density spider farms but failed because of cannibalism among his territorial arachnid workers. In contrast, native New Guineans continue to successfully collect and utilize spider silk for a wide range of applications including bags and fishing nets. Today, the only way to obtain large amounts of silk is to extract it from the abdomens of immobilized spiders, but scientific advances make the mass production and industrial application of spider silk increasingly possible.

Biologically produced dragline silk is a combination of two liquid proteins, Spidroin 1 and 2, which become oriented and solidify as they travel through a complex duct system in the spider's abdomen. These proteins are composed largely of alanine and glycine, the two smallest amino acids. Although glycine comprises almost 42% of each protein, the short, 5 to 10 peptide chains of alanine, which account for 25% of each protein's composition, are more important for the properties. Nuclear magnetic resonance (NMR) techniques have vastly improved the level of understanding of spider silk's structure, which was originally determined by x-ray crystallography. NMR data of spidroins containing deuterium-tagged alanine have shown that all alanines are configured into β-pleated sheets. Furthermore, the NMR data suggest that 40% of the alanine β-sheets are highly structured while the other 60% are less oriented, forming fingers that reach out from each individual strand. These fingers are believed to join the oriented alanine β-sheets and the glycine-

rich, amorphous "background" sectors of the polypeptide.

Currently, genetically modified *Escherichia coli* is used to mass-produce Spidroin 1 and 2. However, DNA redundancy initially caused synthesis problems when the spider genes were transposed into the bacteria. The *E. coli* did not transcribe some of the codons in the same way that spider cells would, forcing scientists to modify the DNA. When the proteins could be synthesized, it was necessary to develop a system to mimic the natural production of spider silk while preventing the silk from contacting the air and subsequently hardening. After the two proteins are separated from the *E. coli*, they are drawn together into methanol through separate needles. Another approach is to dissolve the silk in formic acid or to add codons for hydrophilic amino acids, in this case histidine and arginine, to keep the artificial silk pliable. The industrial and practical applications of spider silk will not be fully known until it can be abiotically synthesized in large quantities.

Based on a Chem30H honors paper by Paul Celestre, UCLA.

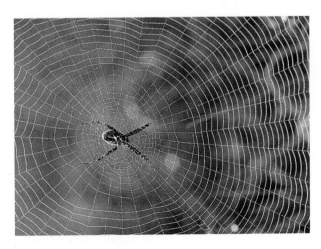

Golden orb weaver. (Tom Bean/Stone)

C. Tertiary Structure

Tertiary structure of proteins
The three-dimensional arrangement in space of all atoms in a single polypeptide chain.

Tertiary (3°) structure refers to the overall folding pattern and arrangement in space of all atoms in a single polypeptide chain. No sharp dividing line exists between secondary and tertiary structures. Secondary structure refers to the spatial arrangement of amino acids close to one another on a polypeptide chain, whereas tertiary structure refers to the three-dimensional arrangement of all atoms of a polypeptide chain. Among the most important factors in maintaining 3° structure are disulfide bonds, hydrophobic interactions, hydrogen bonding, and salt linkages.

Disulfide bonds (Section 9.9F) play an important role in maintaining tertiary structure. Disulfide bonds are formed between side chains of two cysteine units by oxidation of their thiol groups (—SH) to form a disulfide bond. Treatment of a disulfide bond with a reducing agent regenerates the thiol groups.

Figure 27.16 shows the amino acid sequence of human insulin. This protein consists of two polypeptide chains: an A chain of 21 amino acids and a B chain of 30 amino acids. The A chain is bonded to the B chain by two interchain disulfide bonds. An intrachain disulfide bond also connects the cysteine units at positions 6 and 11 of the A chain.

As an example of 2° and 3° structure, let us look at the three-dimensional structure of myoglobin—a protein found in skeletal muscle and particularly abundant in diving mammals, such as seals, whales, and porpoises. Myoglobin and its structural relative, hemoglobin, are the oxygen transport and storage molecules of vertebrates. Hemoglobin binds molecular oxygen in the lungs and transports it to myoglobin in muscles. Myoglobin stores molecular oxygen until it is required for metabolic oxidation.

Myoglobin consists of a single polypeptide chain of 153 amino acids. Myoglobin also contains a single heme unit. Heme consists of one Fe^{2+} ion coordinated in a square planar array with the four nitrogen atoms of a molecule of porphyrin (Figure 27.17).

Determination of the three-dimensional structure of myoglobin represented a milestone in the study of molecular architecture. For their contribution to this research, John C. Kendrew and Max F. Perutz, both of Britain, shared the 1962 Nobel Prize for chemistry. The secondary and tertiary structures of myoglobin are shown in

Figure 27.16

Human insulin. The A chain of 21 amino acids and B chain of 30 amino acids are connected by interchain disulfide bonds between A7 and B7 and between A20 and B19. In addition, a single intrachain disulfide bond occurs between A6 and A11.

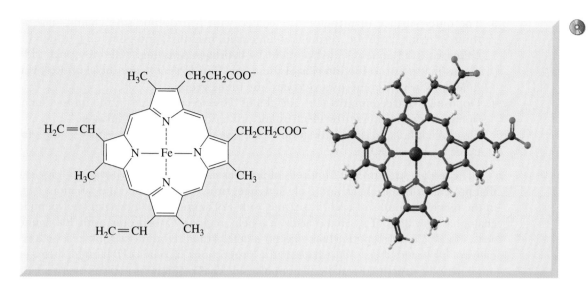

Figure 27.18. The single polypeptide chain is folded into a complex, almost boxlike shape.

Following are important structural features of the three-dimensional shape of myoglobin.

1. The backbone consists of eight relatively straight sections of α-helix, each separated by a bend in the polypeptide chain. The longest section of α-helix has 24 amino acids; the shortest has 7. Some 75% of the amino acids are found in these eight regions of α-helix.
2. Hydrophobic side chains of phenylalanine, alanine, valine, leucine, isoleucine, and methionine are clustered in the interior of the molecule where they are shielded from contact with water. **Hydrophobic interactions** are a major factor in directing the folding of the polypeptide chain of myoglobin into this compact, three-dimensional shape.

Figure 27.18

Ribbon model of myoglobin. The polypeptide chain is shown in yellow; the heme ligand, in red; and the Fe atom, as a white sphere.

3. The outer surface of myoglobin is coated with hydrophilic side chains, such as those of lysine, arginine, serine, glutamic acid, histidine, and glutamine, which interact with the aqueous environment by **hydrogen bonding.** The only polar side chains that point to the interior of the myoglobin molecule are those of two histidine units, which point inward toward the heme group.

4. Oppositely charged amino acid side chains close to each other in the three-dimensional structure interact by electrostatic attractions called **salt linkages.** An example of a salt linkage is the attraction of the side chains of lysine ($-NH_3^+$) and glutamic acid ($-COO^-$).

The tertiary structures of hundreds of proteins have also been determined. It is clear that proteins contain α-helix and β-pleated sheet structures, but that wide variations exist in the relative amounts of each. Lysozyme, with 129 amino acids in a single polypeptide chain, has only 25% of its amino acids in α-helix regions. Cytochrome, with 104 amino acids in a single polypeptide chain, has no α-helix structure but does contain several regions of β-pleated sheet. Yet, whatever the proportions of α-helix, β-pleated sheet, or other periodic structure, virtually all nonpolar side chains of water-soluble proteins are directed toward the interior of the molecule, whereas polar side chains are on the surface of the molecule and in contact with the aqueous environment. Note that this arrangement of polar and nonpolar groups in water-soluble proteins very much resembles the arrangement of polar and nonpolar groups of soap molecules in micelles (Figure 26.3). It also resembles the arrangement of phospholipids in lipid bilayers (Figure 26.13).

Example 27.7

With which of the following amino acid side chains can the side chain of threonine form hydrogen bonds?

(a) Valine   (b) Asparagine   (c) Phenylalanine
(d) Histidine   (e) Tyrosine   (f) Alanine

Solution

The side chain of threonine contains a hydroxyl group that can participate in hydrogen bonding in two ways: its oxygen has a partial negative charge and can function as a hydrogen bond acceptor, and its hydrogen has a partial positive charge and can function as a hydrogen bond donor. Therefore, the side chain of threonine can form hydrogen bonds with the side chains of tyrosine, asparagine, and histidine.

Problem 27.7

At pH 7.4, with what amino acid side chains can the side chain of lysine form salt linkages?

D. Quaternary Structure

Quaternary structure   The arrangement of polypeptide monomers into a noncovalently bonded aggregate.

Most proteins of molecular weight greater than 50,000 consist of two or more noncovalently linked polypeptide chains. The arrangement of protein monomers into an aggregation is known as **quaternary (4°) structure.** A good example is hemoglobin, a

Figure 27.19
Ribbon model of hemoglobin.
The α-chains are shown in
purple; the β-chains, in yellow; the heme ligands, in red;
and the Fe atoms, as white
spheres.

protein that consists of four separate polypeptide chains: two α-chains of 141 amino
acids each and two β-chains of 146 amino acids each. The quaternary structure of hemoglobin is shown in Figure 27.19.

A major factor stabilizing the aggregation of protein subunits is the **hydrophobic
effect.** When separate polypeptide chains fold into compact three-dimensional shapes
to expose polar side chains to the aqueous environment and shield nonpolar side
chains from water, hydrophobic "patches" still appear on the surface, in contact with
water. These patches can be shielded from water if two or more monomers assemble
so that their hydrophobic patches are in contact. The numbers of subunits of several
proteins of known quaternary structure are shown in Table 27.4. Other important
factors include correctly located complementary hydrogen bonding and charged
sites on different subunits. The formation of aggregates of well-defined structure
based on specific structural units on the subunits is being explored in the recent field
of molecular recognition.

Hydrophobic effect The tendency of nonpolar groups to cluster to shield themselves from contact with an aqueous
environment.

Table 27.4	**Quaternary Structure of Selected Proteins**
Protein	**Number of Subunits**
Alcohol dehydrogenase	2
Aldolase	4
Hemoglobin	4
Lactate dehydrogenase	4
Insulin	6
Glutamine synthetase	12
Tobacco mosaic virus protein disc	17

Summary

Amino acids are compounds that contain both an amino group and a carboxyl group (Section 27.1A). A **zwitterion** is an internal salt of an amino acid. With the exception of glycine, all protein-derived amino acids are chiral (Section 27.1B). In the D,L convention, all are L-amino acids. In the R,S convention, 18 are S-amino acids. Although cysteine has the same absolute configuration, it is an R-amino acid because of the manner in which priorities are assigned about the tetrahedral stereocenter. Isoleucine and threonine contain a second stereocenter. The 20 protein-derived amino acids are commonly divided into four categories (Section 27.1C): nine with nonpolar side chains, four with polar but un-ionized side chains, four with acidic side chains, and three with basic side chains.

The **isoelectric point, pI,** of an amino acid, polypeptide, or protein is the pH at which it has no net charge (Section 27.2C). **Electrophoresis** is the process of separating compounds on the basis of their electric charge (Section 27.2D). Compounds having a high charge density move more rapidly than those with a lower charge density. Any amino acid or protein in a solution with a pH that equals the pI of the compound remains at the origin.

A **peptide bond** is the special name given to the amide bond formed between α-amino acids (Section 27.3). A **polypeptide** is a biological macromolecule containing many amino acids, each joined to the next by a peptide bond. By convention, the sequence of amino acids in a polypeptide is written beginning with the **N-terminal amino acid** toward the **C-terminal amino acid**. **Primary (1°) structure** of a polypeptide is the sequence of amino acids in the polypeptide chain (Section 27.4).

In solid-phase synthesis (Section 27.5F), or polymer-supported synthesis of polypeptides, the C-terminal amino acid is joined to a chloromethylated polystyrene resin as a benzyl ester. The polypeptide chain is then extended one amino acid at a time from the N-terminal end. When synthesis is completed, the polypeptide chain is released from the solid support by cleavage of the benzyl ester.

A peptide bond is planar (Section 27.6A), that is, the four atoms of the amide and the two α-carbons of a peptide bond lie in the same plane. Bond angles about the amide nitrogen and the amide carbonyl carbon are 120°. **Secondary (2°) structure** (Section 26.7B) refers to the ordered arrangement (conformations) of amino acids in localized regions of a polypeptide or protein. Two types of secondary structure are the α-helix and the β-pleated sheet. **Tertiary (3°) structure** (Section 27.6C) refers to the overall folding pattern and arrangement in space of all atoms in a single polypeptide chain. **Quaternary (4°) structure** (Section 27.6D) is the arrangement of polypeptide monomers into a noncovalently bonded aggregate.

Key Reactions

1. Acidity of an α-Carboxyl Group (Section 27.2A)

An α-COOH (pK_a approximately 2.19) of a protonated amino acid is a considerably stronger acid than acetic acid (pK_a 4.76) or other low-molecular-weight aliphatic carboxylic acid due to the electron-withdrawing inductive effect of the α-NH$_3^+$ group.

$$\underset{\underset{NH_3^+}{|}}{RCHCOOH} + H_2O \rightleftharpoons \underset{\underset{NH_3^+}{|}}{RCHCOO^-} + H_3O^+ \qquad pK_a = 2.19$$

2. Acidity of an α-Ammonium Group (Section 27.2A)

An α-NH$_3^+$ group (pK_a approximately 9.47) is a slightly stronger acid than a primary aliphatic ammonium ion (pK_a approximately 10.76).

$$\underset{\underset{NH_3^+}{|}}{RCHCOO^-} + H_2O \rightleftharpoons \underset{\underset{NH_2}{|}}{RCHCOO^-} + H_3O^+ \qquad pK_a = 9.47$$

3. Reaction of an α-Amino Acid with Ninhydrin (Section 27.2D)

Treatment of an α-amino acid with ninhydrin gives a purple-colored solution. Treatment of proline with ninhydrin gives an orange-colored solution.

An α-amino   Ninhydrin   Purple-colored anion
acid

4. Cleavage of a Peptide Bond by Cyanogen Bromide (Section 27.4B)

Cleavage is regioselective for a peptide bond formed by the carboxyl group of methionine.

A substituted γ-lactone of
the amino acid homoserine

5. Edman Degradation (Section 27.4B)

Treatment with phenyl isothiocyanate followed by acid removes the *N*-terminal amino acid
as a substituted phenylthiohydantoin, which is then separated and identified.

Phenyl   A phenylthiohydantoin
isothiocyanate

6. The Benzyloxycarbonyl (Z—) Protecting Group (Section 27.5C)

Prepared by treatment of an unprotected α-NH$_2$ group with benzyloxycarbonyl chloride.
Removed by treatment with HBr in acetic acid or by hydrogenolysis.

Benzyloxycarbonyl   Alanine   *N*-Benzyloxycarbonylalanine
chloride   (Z—Ala)
(Z—Cl)

7. Peptide Bond Formation Using 1,3-Dicyclohexylcarbodiimide (Section 27.5E)

This substituted carbodiimide is a dehydrating agent and is converted to a disubstituted urea. The reaction is efficient, and yields are generally very high.

$$Z{-}NHCHC{-}OH \ + \ H_2NCHCOCH_3 \ + \ \bigcirc{-}N{=}C{=}N{-}\bigcirc \xrightarrow{CHCl_3}$$

Amino protected aa$_1$	Carboxyl protected aa$_2$	1,3-Dicyclohexylcarbo-diimide (DCC)

$$Z{-}NHCHC{-}NHCHCOCH_3 \ + \ \bigcirc{-}NH{-}C{-}NH{-}\bigcirc$$

Amino and carboxyl protected dipeptide	N,N'-Dicyclohexylurea (DCU)

Problems

Amino Acids

27.8 What amino acid does each abbreviation stand for?

(a) Phe (b) Ser (c) Asp (d) Gln (e) His (f) Gly (g) Tyr

27.9 Configuration of the stereocenter in α-amino acids is most commonly specified using the D,L convention. It can also be identified using the R,S convention (Section 3.3). Does the stereocenter in L-serine have the R or the S configuration?

27.10 Assign an R or S configuration to the stereocenter in each amino acid.

(a) L-Phenylalanine (b) L-Glutamic acid (c) L-Methionine

27.11 The amino acid threonine has two stereocenters. The stereoisomer found in proteins has the configuration $2S,3R$ about the two stereocenters. Draw the following:

(a) A Fischer projection of this stereoisomer
(b) A three-dimensional representation

27.12 Define the term zwitterion.

27.13 Draw zwitterion forms of these amino acids.

(a) Valine (b) Phenylalanine (c) Glutamine

27.14 Why are Glu and Asp often referred to as acidic amino acids?

27.15 Why is Arg often referred to as a basic amino acid? Which two other amino acids are also basic amino acids?

27.16 What is the meaning of the alpha as it is used in α-amino acid?

27.17 Several β-amino acids exist. There is a unit of β-alanine, for example, contained within the structure of coenzyme A (Problem 25.36). Write the structural formula of β-alanine.

27.18 Although only L-amino acids occur in proteins, D-amino acids are often a part of the metabolism of lower organisms. The antibiotic actinomycin D, for example, contains a unit of D-valine, and the antibiotic bacitracin A contains units of D-asparagine and D-glutamic acid. Draw Fischer projections and three-dimensional representations for these three D-amino acids.

27.19 Histamine (see *The Merck Index*, 12th ed., #4756) is synthesized from one of the 20 protein-derived amino acids. Suggest which amino acid is its biochemical precursor and the type of organic reaction(s) involved in its biosynthesis (for example, oxidation, reduction, decarboxylation, nucleophilic substitution).

Histamine

27.20 As discussed in the Chemistry in Action box "Vitamin K, Blood Clotting, and Basicity" (Chapter 26), vitamin K participates in carboxylation of glutamic acid residues of the blood-clotting protein prothrombin.

(a) Write a structural formula for γ-carboxyglutamic acid.

(b) Account for the fact that the presence of γ-carboxyglutamic acid escaped detection for many years; on routine amino acid analyses, only glutamic acid was detected.

27.21 Both norepinephrine (see *The Merck Index*, 12th ed., #6788) and epinephrine (see *The Merck Index*, 12th ed., #3656) are synthesized from the same protein-derived amino acid. From which amino acid are they synthesized and what types of reactions are involved in their biosynthesis?

(a) Norepinephrine

(b) Epinephrine (Adrenaline)

27.22 From which amino acid are serotonin (see *The Merck Index*, 12th ed., #8607) and melatonin (see *The Merck Index*, 12th ed., #5857) synthesized and what types of reactions are involved in their biosynthesis?

(a) Serotonin

(b) Melatonin

Acid-Base Behavior of Amino Acids

27.23 Draw a structural formula for the form of each amino acid most prevalent at pH 1.0.

(a) Threonine   **(b)** Arginine   **(c)** Methionine   **(d)** Tyrosine

27.24 Draw a structural formula for the form of each amino most prevalent at pH 10.0.

  (a) Leucine   **(b)** Valine   **(c)** Proline   **(d)** Aspartic acid

27.25 Write the zwitterion form of alanine and show its reaction with the following.

  (a) 1 mol NaOH   **(b)** 1 mol HCl

27.26 Write the form of lysine most prevalent at pH 1.0, and then show its reaction with the following. Consult Table 27.2 for pK_a values of the ionizable groups in lysine.

  (a) 1 mol NaOH   **(b)** 2 mol NaOH   **(c)** 3 mol NaOH

27.27 Write the form of aspartic acid most prevalent at pH 1.0, and then show its reaction with the following. Consult Table 27.2 for pK_a values of the ionizable groups in aspartic acid.

  (a) 1 mol NaOH   **(b)** 2 mol NaOH   **(c)** 3 mol NaOH

27.28 Given pK_a values for ionizable groups from Table 27.2, sketch curves for the titration of **(a)** glutamic acid with NaOH and **(b)** histidine with NaOH.

27.29 Draw a structural formula for the product formed when alanine is treated with the following reagents.

  (a) Aqueous NaOH   **(b)** Aqueous HCl
  (c) CH_3CH_2OH, H_2SO_4 **(d)** $(CH_3CO)_2O$, CH_3COONa

27.30 For lysine and arginine, the isoelectric point, pI, occurs at a pH where the net charge on the nitrogen-containing groups is $+1$ and balances the charge of -1 on the α-carboxyl group. Calculate pI for these amino acids.

27.31 For aspartic and glutamic acids, the isoelectric point occurs at a pH where the net charge on the two carboxyl groups is -1 and balances the charge of $+1$ on the α-amino group. Calculate pI for these amino acids.

27.32 Account for the fact that the isoelectric point of glutamine (pI 5.65) is higher than the isoelectric point of glutamic acid (pI 3.08).

27.33 Enzyme-catalyzed decarboxylation of glutamic acid gives 4-aminobutanoic acid (Section 27.1D). Estimate the pI of 4-aminobutanoic acid.

27.34 Guanidine and the guanidino group present in arginine are two of the strongest organic bases known. Account for their basicity.

27.35 At pH 7.4, the pH of blood plasma, do the majority of protein-derived amino acids bear a net negative charge or a net positive charge? Explain.

27.36 Do the following compounds migrate to the cathode or to the anode on electrophoresis at the specified pH?

  (a) Histidine at pH 6.8   **(b)** Lysine at pH 6.8
  (c) Glutamic acid at pH 4.0   **(d)** Glutamine at pH 4.0
  (e) Glu-Ile-Val at pH 6.0   **(f)** Lys-Gln-Tyr at pH 6.0

27.37 At what pH would you carry out an electrophoresis to separate the amino acids in each mixture?

  (a) Ala, His, Lys   **(b)** Glu, Gln, Asp   **(c)** Lys, Leu, Tyr

27.38 Examine the amino acid sequence of human insulin (Figure 27.16), and list each Asp, Glu, His, Lys, and Arg in this molecule. Do you expect human insulin to have an isoelectric point nearer that of the acidic amino acids (pI 2.0–3.0), the neutral amino acids (pI 5.5–6.5), or the basic amino acids (pI 9.5–11.0)?

27.39 A chemically modified guanidino group is present in cimetidine (Tagamet), a widely prescribed drug for the control of gastric acidity and peptic ulcers. Cimetidine reduces gastric acid secretion by inhibiting the interaction of histamine with gastric H_2 receptors. In the development of this drug, a cyano group was added to the substituted guanidino group to alter its basicity significantly. Do you expect this modified guanidino group to be more basic or less basic than the guanidino group of arginine? Explain.

$$
\underset{\substack{\\ \text{HN} \quad \text{N}}}{\overset{\substack{\text{N}-\text{CN} \\ \|}}{\underset{\text{H}_3\text{C} \qquad \text{CH}_2\text{SCH}_2\text{CH}_2\text{NHCNHCH}_3}{}}}
$$

Cimetidine
(Tagamet)

27.40 Draw a structural formula for the product formed when alanine is treated with the following reagents.

(a) $\langle\ \rangle$—CCl, $(CH_3CH_2)_3N$ (b) (indane-1,3-dione-2,2-diol structure)

(c) $\langle\ \rangle$—CH$_2$OCCl, NaOH   (d) $(CH_3)_3COCOCOC(CH_3)_3$, NaOH

(e) Product (c) + L-alanine ethyl ester + DCC
(f) Product (d) + L-alanine ethyl ester + DCC

Primary Structure of Polypeptides and Proteins

27.41 If a protein contains four different SH groups, how many different disulfide bonds are possible if only a single disulfide bond is formed? How many different disulfides are possible if two disulfide bonds are formed?

27.42 How many different tetrapeptides can be made under the following conditions?

(a) The tetrapeptide contains one unit each of Asp, Glu, Pro, and Phe.
(b) All 20 amino acids can be used, but each only once.

27.43 A decapeptide has the following amino acid composition:

Ala$_2$, Arg, Cys, Glu, Gly, Leu, Lys, Phe, Val

Partial hydrolysis yields the following tripeptides:

Cys-Glu-Leu + Gly-Arg-Cys + Leu-Ala-Ala + Lys-Val-Phe + Val-Phe-Gly

One round of Edman degradation yields a lysine phenylthiohydantoin. From this information, deduce the primary structure of this decapeptide.

27.44 Following is the primary structure of glucagon, a polypeptide hormone of 29 amino acids. Glucagon is produced in the α-cells of the pancreas and helps maintain blood glucose levels in a normal concentration range.

1   5   10   15   20   25   29
His-Ser-Glu-Gly-Thr-Phe-Thr-Ser-Asp-Tyr-Ser-Lys-Tyr-Leu-Asp-Ser-Arg-Arg-Ala-Gln-Asp-Phe-Val-Gln-Trp-Leu-Met-Asn-Thr

Which peptide bonds are hydrolyzed when this polypeptide is treated with each reagent?

(a) Phenyl isothiocyanate   (b) Chymotrypsin   (c) Trypsin   (d) Br—CN

27.53 The side-chain carboxyl groups of aspartic acid and glutamic acid are often protected as benzyl esters.

$$
\underset{\text{BOC as amino-}\atop\text{protecting group}}{\text{Me}_3\text{COCNHCHCOCH}_3}
$$

O   O
‖   ‖
Me₃COCNHCHCOCH₃
          |
BOC as amino-   CH₂
protecting group |
          CH₂   benzyl ester as carboxyl-
          | protecting group
          C=O
          |
          OCH₂Ph

(a) Show how to convert the side-chain carboxyl group to a benzyl ester using benzyl chloride as a source of the benzyl group.

(b) How do you deprotect the side-chain carboxyl under mild conditions without removing the BOC protecting group at the same time?

Three-Dimensional Shapes of Polypeptides and Proteins

27.54 Examine the α-helix conformation. Are amino acid side chains arranged all inside the helix, all outside the helix, or randomly?

27.55 Distinguish between intermolecular and intramolecular hydrogen bonding between the backbone groups on polypeptide chains. In what type of secondary structure do you find intermolecular hydrogen bonds? In what type do you find intramolecular hydrogen bonding?

27.56 Many plasma proteins found in an aqueous environment are globular in shape. Which amino acid side chains would you expect to find on the surface of a globular protein and in contact with the aqueous environment? Which would you expect to find inside, shielded from the aqueous environment? Explain.

(a) Leu   (b) Arg   (c) Ser   (d) Lys   (e) Phe

27.57 Denaturation of a protein is a physical change, the most readily observable result of which is loss of biological activity. Denaturation stems from changes in secondary, tertiary, and quaternary structure through disruption of noncovalent interactions including hydrogen bonding and hydrophobic interactions. Three common denaturing agents are sodium dodecyl sulfate (SDS), urea, and heat. What kinds of noncovalent interactions might each reagent disrupt?

NUCLEIC ACIDS

The organization, maintenance, and regulation of cellular function require a tremendous amount of information, all of which must be processed each time a cell is replicated. With very few exceptions, genetic information is stored and transmitted from one generation to the next in the form of deoxyribonucleic acids (DNA). **Genes,** the hereditary units of chromosomes, are long stretches of double-stranded DNA. If the DNA in a human chromosome in a single cell were uncoiled, it would be approximately 1.8 meters long!

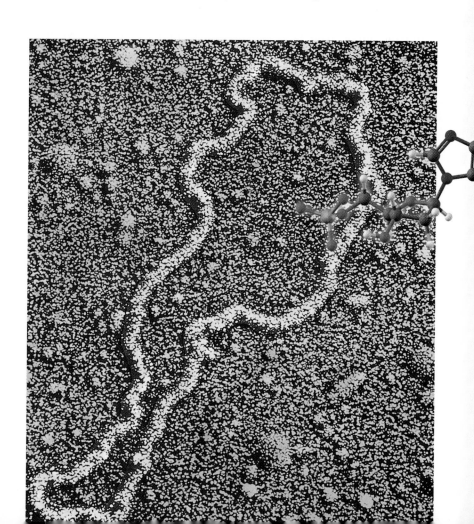

■ False-colored transmission electron micrograph of plasmid of bacterial DNA. If the cell wall of a bacterium such as *Escherichia coli* is partially digested and the cell then osmotically shocked by dilution with water, its contents are extruded to the exterior. Shown here is the bacterial chromosome surrounding the cell. *(Professor Stanley Cohen/Science Photo Library/Photo Researchers, Inc.)* Inset: 2-Deoxyadenosine 5′-monophosphate (dAMP), a building block of DNA (Section 28.2).

Genetic information is expressed in two stages: transcription from DNA to ribonucleic acids (RNA) and then translation for the synthesis of proteins.

$$DNA \xrightarrow{\text{transcription}} RNA \xrightarrow{\text{translation}} proteins$$

Thus, DNA is the repository of genetic information in cells, whereas RNA serves in the transcription and translation of this information, which is then expressed through the synthesis of proteins.

In this chapter, we examine the structure of nucleosides and nucleotides and the manner in which these monomers are covalently bonded to form **nucleic acids.** Then, we examine the manner in which genetic information is encoded on molecules of DNA, the function of the three types of ribonucleic acids, and finally how the primary structure of a DNA molecule is determined.

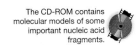
The CD-ROM contains molecular models of some important nucleic acid fragments.

28.1   Nucleosides and Nucleotides

Controlled hydrolysis of nucleic acids yields three components: heterocyclic aromatic amine bases, the monosaccharides D-ribose or 2-deoxy-D-ribose (Section 25.1A), and phosphate ions. The five heterocyclic aromatic amine bases most common to nucleic acids are shown in Figure 28.1. Uracil, cytosine, and thymine are referred to as pyrimidine bases after the name of the parent base; adenine and guanine are referred to as purine bases.

A **nucleoside** is a compound containing D-ribose or 2-deoxy-D-ribose bonded to a heterocyclic aromatic amine base by a β-N-glycosidic bond (Section 25.3A). The monosaccharide component of DNA is 2-deoxy-D-ribose, whereas that of RNA is D-ribose. The glycosidic bond is between C-1′ (the anomeric carbon) of ribose or 2-deoxyribose and N-1 of a pyrimidine base or N-9 of a purine base. Figure 28.2 shows a structural formula for uridine, a nucleoside derived from ribose and uracil.

A **nucleotide** is a nucleoside in which a molecule of phosphoric acid is esterified with a free hydroxyl of the monosaccharide, most commonly either the 3′-hydroxyl or the 5′-hydroxyl. A nucleotide is named by giving the name of the parent nucleo-

Nucleoside   A building block of nucleic acids, consisting of D-ribose or 2-deoxy-D-ribose bonded to a heterocyclic aromatic amine base by a β-N-glycoside bond.

Nucleotide   A nucleoside in which a molecule of phosphoric acid is esterified with an —OH of the monosaccharide, most commonly either the 3′-OH or the 5′-OH.

Figure 28.1

Names and one-letter abbreviations for the heterocyclic aromatic amine bases most common to DNA and RNA. Bases are numbered according to the patterns of the parent compounds, pyrimidine and purine.

| Pyrimidine | Uracil (U) | Cytosine (C) | Thymine (T) |

| Purine | Adenine (A) | Guanine (G) |

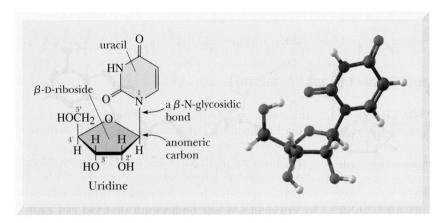

Figure 28.2
Uridine, a nucleoside. Atom numbers on the monosaccharide rings are primed to distinguish them from atom numbers on the heterocyclic aromatic amine bases.

side followed by the word "monophosphate." The position of the phosphoric ester is specified by the number of the carbon to which it is bonded. Figure 28.3 shows a structural formula of 5′-adenosine monophosphate (AMP). Monophosphoric esters are diprotic acids with pK_a values of approximately 1 and 6. Therefore, at pH 7, the two hydrogens of a phosphoric monoester are fully ionized giving a nucleotide a charge of -2.

Nucleoside monophosphates can be further phosphorylated to form nucleoside diphosphates and nucleoside triphosphates. Shown in Figure 28.4 is a structural formula for adenosine 5′-triphosphate (ATP). Nucleoside diphosphates and triphosphates are also polyprotic acids and are extensively ionized at pH 7.0. The pK_a values of the first three ionization steps for adenosine triphosphate are less than 5.0. The value of pK_{a4} is approximately 7.0. Therefore, at pH 7.0, approximately 50% of adenosine triphosphate is present as ATP^{4-}, and 50% is present as ATP^{3-}.

Example 28.1

Draw a structural formula for each nucleotide.

(a) 2′-Deoxycytidine 5′-diphosphate   **(b)** 2′-Deoxyguanosine 3′-monophosphate

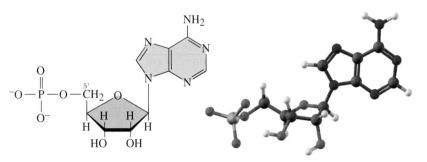

Figure 28.3
Adenosine 5′-monophosphate, AMP. The phosphate group is fully ionized at pH 7.0 giving this nucleotide a charge of -2.

three levels of structural complexity in nucleic acids. Although these levels are somewhat comparable to those in polypeptides and proteins, they also differ in significant ways.

A. Primary Structure — The Covalent Backbone

Primary structure of nucleic acids The sequence of bases along the pentose-phosphodiester backbone of a DNA or RNA molecule read from the 5′ end to the 3′ end.

Deoxyribonucleic acids consist of a backbone of alternating units of deoxyribose and phosphate in which the 3′-hydroxyl of one deoxyribose unit is joined by a phosphodiester bond to the 5′-hydroxyl of another deoxyribose unit (Figure 28.5). This pentose-phosphodiester backbone is constant throughout an entire DNA molecule. A heterocyclic aromatic amine base—adenine, guanine, thymine, or cytosine—is bonded to each deoxyribose unit by a β-N-glycosidic bond. **Primary structure** of DNA refers to the order of heterocyclic bases along the pentose-phosphodiester backbone. The sequence of bases is read from the 5′ end to the 3′ end.

Example 28.2

Draw a structural formula for the DNA dinucleotide TG that is phosphorylated at the 5′ end only.

Solution

Problem 28.2

Draw a structural formula for the section of DNA that contains the base sequence CTG and is phosphorylated at the 3′ end only.

B. Secondary Structure — The Double Helix

By the early 1950s, it was clear that DNA molecules consist of chains of alternating units of deoxyribose and phosphate joined by 3′,5′-phosphodiester bonds with a base attached to each deoxyribose unit by a β-N-glycosidic bond. In 1953, the American

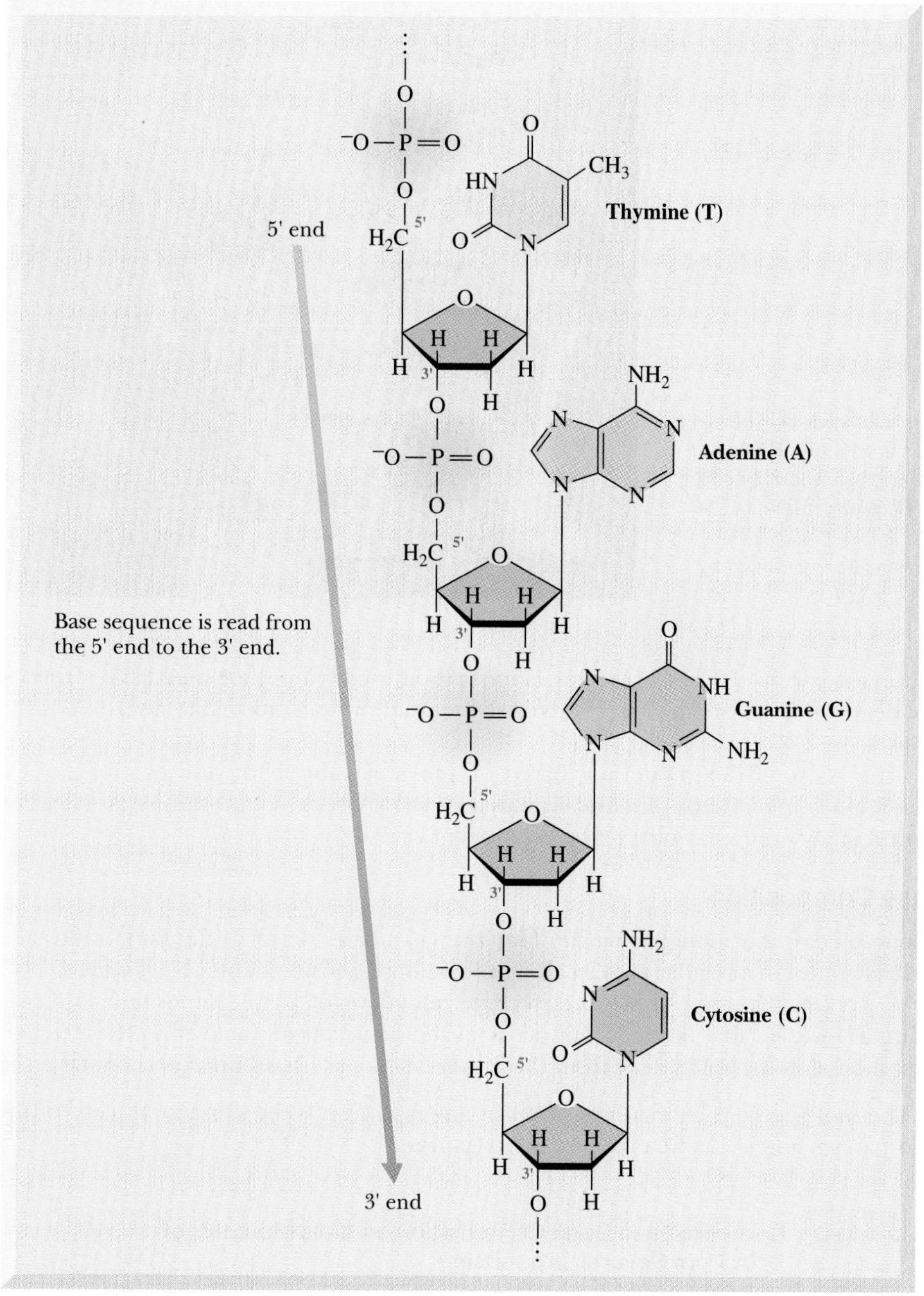

5' end

Base sequence is read from the 5' end to the 3' end.

3' end

Thymine (T)

Adenine (A)

Guanine (G)

Cytosine (C)

Figure 28.5
A tetranucleotide section of a single-stranded DNA.

biologist James D. Watson and the British physicist Francis H. C. Crick proposed a double-helix model for the **secondary structure** for DNA. Watson, Crick, and Maurice Wilkins shared the 1962 Nobel Prize for physiology and medicine for "their discoveries concerning the molecular structure of nucleic acids, and its significance for information transfer in living material." Although Rosalind Franklin also played an

Secondary structure of nucleic acids The ordered arrangement of nucleic acid strands.

Watson and Crick with their model of a DNA molecule. *(From M. M. Jones et al., Chemistry, Man, and Society, 4th ed. Philadelphia: Saunders College Publishing, 1983, p. 256)*

Rosalind Franklin (1920–1958). In 1951, she joined the Biophysical Laboratory at King's College, London, where she began her studies on the application of x-ray diffraction methods to the study of DNA. She is credited with discoveries that established the density of DNA, its helical conformation, and other significant aspects. Her work was, thus, important to the model of DNA developed by Watson and Crick. She died in 1958 at the age of 37 and, because the Nobel Prize is never awarded posthumously, she did not share in the 1962 Nobel Prize for physiology and medicine with Watson, Crick, and Wilkins. Although the relation between Watson, Crick, and Franklin was initially strained, Watson later said that "we later came to appreciate . . . the struggles the intelligent woman faces to be accepted by the scientific world which often regards women as mere diversions from serious thinking." *(Vittorio Luzzati/Centre Nationale de Génétique Moléculaire)*

important part in this research, her name was omitted from the Nobel list because of her death in 1958 at age 37. The Nobel foundation does not make awards posthumously.

The Watson-Crick model was based on molecular modeling and two lines of experimental observations: chemical analyses of DNA base compositions and mathematical analyses of x-ray diffraction patterns of crystals of DNA.

Base Composition

At one time, it was thought that the four principal bases occur in the same ratios and perhaps repeat in a regular pattern along the pentose-phosphodiester backbone of DNA for all species. However, more precise determinations of base composition by Erwin Chargaff revealed that bases do not occur in the same ratios (Table 28.1). Researchers drew the following conclusions from this and related data. To within experimental error,

1. The mole-percent base composition in any organism is the same in all cells of the organism and is characteristic of the organism.

Table 28.1 **Comparison in Base Composition, in Mole-Percent, of DNA from Several Organisms**

Organism	Purines		Pyrimidines		A/T	G/C	Purines/ Pyrimidines
	A	G	C	T			
Human	30.4	19.9	19.9	30.1	1.01	1.00	1.01
Sheep	29.3	21.4	21.0	28.3	1.04	1.02	1.03
Yeast	31.7	18.3	17.4	32.6	0.97	1.05	1.00
E. coli	26.0	24.9	25.2	23.9	1.09	0.99	1.04

2. The mole-percents of adenine (a purine base) and thymine (a pyrimidine base) are equal. The mole-percents of guanine (a purine base) and cytosine (a pyrimidine base) are also equal.
3. The mole-percents of purine bases (A + G) and pyrimidine bases (C + T) are equal.

Analyses of X-Ray Diffraction Patterns

Additional information about the structure of DNA emerged when x-ray diffraction photographs taken by Rosalind Franklin and Maurice Wilkins were analyzed. These diffraction patterns revealed that, even though the base composition of DNA isolated from different organisms varies, DNA molecules themselves are remarkably uniform in thickness. They are long and fairly straight, with an outside diameter of approximately 2000 pm, and not more than a dozen atoms thick. Furthermore, the crystallographic pattern repeats every 3400 pm. Herein lay one of the chief problems to be solved. How could the molecular dimensions of DNA be so regular even though the relative percentages of the various bases differ so widely? With this accumulated information, the stage was set for the development of a hypothesis about DNA structure.

Figure 28.6
A DNA double helix has a chirality associated with the helix. Right-handed and left-handed double helices of otherwise identical DNA chains are non-superposable mirror images.

The Watson-Crick Double Helix

The heart of the **Watson-Crick model** is the postulate that a molecule of DNA is a complementary **double helix.** It consists of two antiparallel polynucleotide strands coiled in a right-handed manner about the same axis to form a double helix. As illustrated in the ribbon models in Figure 28.6, chirality is associated with a double helix; left-handed and right-handed double helices are related by reflection just as enantiomers are related by reflection.

   To account for the observed base ratios and uniform thickness of DNA, Watson and Crick postulated that purine and pyrimidine bases project inward toward the axis of the helix and are always paired in a very specific manner. According to scale models, the dimensions of an adenine-thymine base pair are almost identical to the dimensions of a guanine-cytosine base pair, and the length of each pair is consistent with the core thickness of a DNA strand (Figure 28.7). Thus, if the purine base in one strand is adenine, then its complement in the antiparallel strand must be thymine. Similarly, if the purine in one strand is guanine, its complement in the anti-parallel strand must be cytosine. The "fits" between the TA pair and between the CG pair are remarkable and represent another important example of molecular

Watson-Crick model A double-helix model for the secondary structure of a DNA molecule.

Double helix A type of secondary structure of DNA molecules in which two antiparallel polynucleotide strands are coiled in a right-handed manner about the same axis.

Thymine = Adenine

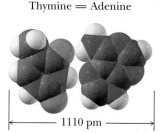

|← 1110 pm →|

Cytosine ≡ Guanine

|← 1080 pm →|

Figure 28.7
Base-pairing between adenine and thymine (A-T) and between guanine and cytosine (G-C). An A-T base pair is held by two hydrogen bonds, whereas a G-C base pair is held by three hydrogen bonds.

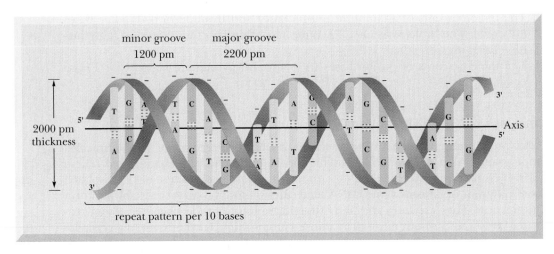

Figure 28.8

Ribbon model of double-stranded B-DNA. Each ribbon shows the pentose-phosphodiester backbone of a single-stranded DNA molecule. The strands are antiparallel; one strand runs to the left from the 5′ end to the 3′ end, the other runs to the right from the 5′ end to the 3′ end. Hydrogen bonds are shown by three dotted lines between each G-C base pair and two dotted lines between each A-T base pair.

recognition and supramolecular complex formation. The resulting hydrogen bonding holds the two strands together very strongly.

A significant feature of Watson and Crick's model is that no other base-pairing is consistent with the observed thickness of a DNA molecule. A pair of pyrimidine bases is too small to account for the observed thickness, whereas a pair of purine bases is too large. Thus, according to the Watson-Crick model, the repeating units in a double-stranded DNA molecule are not single bases of differing dimensions but rather base pairs of almost identical dimensions.

To account for the periodicity observed from x-ray data, Watson and Crick postulated that base pairs are stacked one on top of the other with a distance of 340 pm between base pairs and with ten base pairs in one complete turn of the helix. There is one complete turn of the helix every 3400 pm. Shown in Figure 28.8 is a ribbon model of double-stranded **B-DNA,** the predominant form of DNA in dilute aqueous solution and thought to be the most common form in nature.

In the double helix, the bases in each base pair are not directly opposite one another across the diameter of the helix but rather are slightly displaced. This displacement and the relative orientation of the glycosidic bonds linking each base to the sugar-phosphate backbone leads to two differently sized grooves, a major groove and a minor groove (Figure 28.8). Each groove runs along the length of the cylindrical column of the double helix. The major groove is approximately 2200 pm wide; the minor groove is approximately 1200 pm wide.

Figure 28.9 shows more detail of an idealized B-DNA double helix. The major and minor grooves are clearly recognizable in this model.

Other forms of secondary structure are known that differ in the distance between stacked base pairs and in the number of base pairs per turn of the helix. One of the

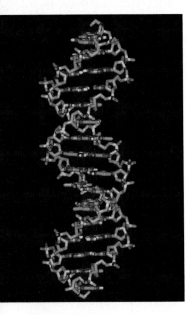

Figure 28.9
An idealized model of B-DNA.

most common of these, **A-DNA,** also a right-handed helix, is thicker than B-DNA, and has a repeat distance of only 2900 pm. There are ten base pairs per turn of the helix with a spacing of 290 pm between base pairs.

Example 28.3

One strand of a DNA molecule contains the base sequence 5′-ACTTGCCA-3′. Write its complementary base sequence.

Solution

Remember that the base sequence is always written from the 5′ end of the strand to the 3′ end, that A pairs with T, and that G pairs with C. In double-stranded DNA, the two strands run in opposite (antiparallel) directions so that the 5′ end of one strand is associated with the 3′ end of the other strand.

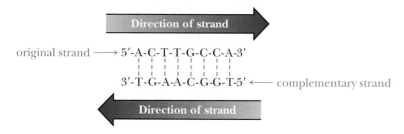

original strand ⟶ 5′-A-C-T-T-G-C-C-A-3′

3′-T-G-A-A-C-G-G-T-5′ ⟵ complementary strand

Written from the 5′ end, the complementary strand is 5′-TGGCAAGT-3′.

Problem 28.3

Write the complementary DNA base sequence for 5′-CCGTACGA-3′.

C.   Tertiary Structure — Supercoiled DNA

The length of a DNA molecule is enormously greater than its diameter, and the extended molecule is quite flexible. A DNA molecule is said to be relaxed if it has no twists other than those imposed by its secondary structure. Said another way, relaxed DNA does not have a clearly defined tertiary structure. We consider two types of **tertiary structure,** one type induced by perturbations in circular DNA and a second type introduced by coordination of DNA with nuclear proteins called histones. Tertiary structure, whatever the type, is referred to as **supercoiling.**

Supercoiling of Circular DNA

Circular DNA is a type of double-stranded DNA in which the two ends of each strand are joined by phosphodiester bonds [Figure 28.10(a)]. This type of DNA, the most prominent form in bacteria and viruses, is also referred to as circular duplex (because it is double-stranded) DNA. One strand of circular DNA may be opened, partially unwound, and then rejoined. The unwound section introduces a strain into the molecule because the nonhelical gap is less stable than hydrogen-bonded,

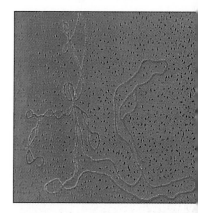

Supercoiled DNA from a mitochondrion. *(Fran Heyl Associates)*

Tertiary structure of nucleic acids The three-dimensional arrangement of all atoms of a nucleic acid, commonly referred to as supercoiling.

Circular DNA A type of double-stranded DNA in which the 5′- and 3′-ends of each strand are joined by phosphodiester groups.

C H E M I S T R Y   I N   A C T I O N

Mustard Gases and the Treatment of Neoplastic Diseases

Sulfur mustard (Section 8.5) is a highly toxic gas used in World War I. Autopsies of soldiers killed by sulfur mustard revealed, among other things, very low white blood cell counts and defects in bone marrow development. From these observations, it was realized that sulfur mustards have profound effects on rapidly dividing cells. This became a lead observation in the search for less toxic alkylating agents for use in clinical medicine. Attention turned to the less reactive nitrogen mustards. One of the first compounds tested was mechlorethamine (see *The Merck Index*, 12th ed., #5815). In reactions of this and other nitrogen mustards, assisted ionization of halogen forms a cyclic aziridinium ion intermediate. This is

followed by attack of a nucleophile on a carbon atom of the three-membered ring to give an alkylated product (this type of neighboring group participation was introduced in Section 8.5).

Mechlorethamine undergoes very rapid reaction with water (hydrolysis) and with other nucleophiles, so much so that within minutes after injection into the body, it has completely reacted. The problem for the chemist, then, was to find a way to decrease the nucleophilicity of nitrogen while maintaining a reasonable water solubility. Substitution of phenyl for methyl reduced the nucleophilicity, but the resulting compound was not sufficiently soluble in water for intravenous injection. The solubility problem was solved by adding a carboxyl group. When the carboxyl group was added directly to the aromatic ring,

Mechlorethamine

(An aziridinium ion intermediate)

The nucleophilicity of nitrogen is acceptable, but the compound is too insoluble in water for intravenous injection.

The solubility in water is acceptable, but the nucleophilicity of nitrogen is reduced so much that the compound is unreactive.

base-paired helical sections. The strain can be localized in the nonhelical gap. Alternatively, it may be spread uniformly over the entire circular DNA by introduction of **superhelical twists,** one twist for each turn of a helix unwound. The circular DNA shown in Figure 28.10(b) has been unwound by four complete turns of the helix. The strain introduced by this unwinding is spread uniformly over the entire molecule by introduction of four superhelical twists [Figure 28.10(c)]. Interconversion of re-

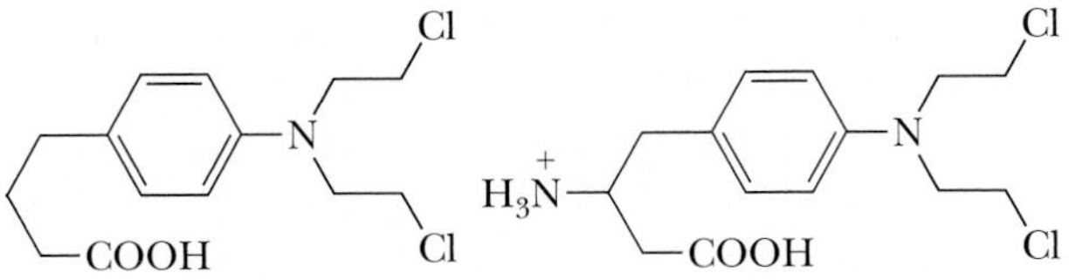

Chlorambucil

Melphalan

however, the resulting compound was too stable and, therefore, not biologically active.

Adding a propyl bridge (chlorambucil; see *The Merck Index*, 12th ed., #2116) or an aminoethyl bridge (melphalan; see *The Merck Index*, 12th ed., #5871) between the aromatic ring and the carboxyl group solved both the solubility problem and the reactivity problem. Note that melphalan is chiral. It has been demonstrated that the *R* and *S* enantiomers have approximately equal therapeutic potency.

The clinical value of the nitrogen mustards lies in the fact that they undergo reaction with certain nucleophilic sites on the heterocyclic aromatic amine bases in DNA. For DNA, the most reactive nucleophilic site is N-7 of guanine. Next in reactivity is N-3 of adenine, followed by N-3 of cytosine.

The nitrogen mustards are bifunctional alkylating agents; one molecule of nitrogen mustard undergoes reaction with two molecules of nucleophile. When guanine is the target nucleophile, N-7 of each guanine is converted to an ammonium ion. This in turn increases the acidity of guanine and shifts the keto-enol equilibrium from the keto form to the enol form. In the keto form, guanine forms a base pair with cytosine. In the enol form, however, it forms a base pair with thymine, which leads to miscoding during DNA replication. The therapeutic value of the nitrogen mustards, then, lies in their ability to form covalent cross links on DNA strands and to disrupt normal base pairing.

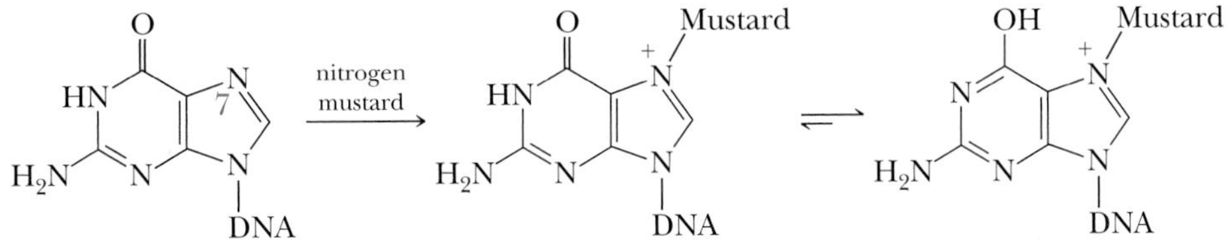

Guanine

Keto form

Enol form

laxed and supercoiled DNA is catalyzed by groups of enzymes called topoisomerases and gyrases.

Supercoiling of Linear DNA

Supercoiling of linear DNA in plants and animals takes another form and is driven by interaction between negatively charged DNA molecules and a group of positively

C H E M I S T R Y   I N   A C T I O N

The Fountain of Youth

In 1997, a sheep named Dolly became the first mammal to be born through cloning. By the time she was three years old, however, her genetic material was aging at the rate of the six-year-old sheep from which she was cloned. It turned out that Dolly had shortened telomeres, and—as a consequence—her genetic structure was considerably "older" than Dolly herself.

Telomeres are the physical ends of eukaryotic chromosomes, and they surrender a bit of themselves during each cell division because the DNA-replicating machinery normally fails to copy the DNA at a chromosome's tips. This process limits the total number of divisions a cell can undergo because the telomeres are ultimately consumed. In egg and cancer cells, among others, an enzyme called telomerase rebuilds the telomere cap after each division, making such cells effectively immortal. Consequently, telomerase may be the fountain of youth because it is responsible for the extension of telomeres in most species. If telomerase activity is limited, as it is in normal cells, the telomeres will shorten, and cells will age like Dolly's.

Researchers are trying to alter the integrity of telomeres by controlling the expression and location of telomerase. Preserving telomere caps may rejuvenate aging organisms by enabling cells to survive additional divisions. Conversely, doing away with telomerase activity in tumors may cause

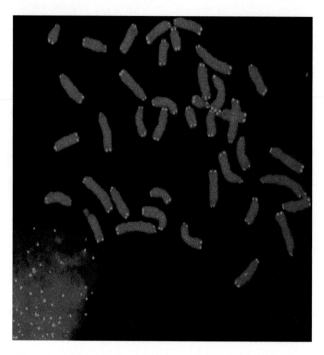

Telomeres are the repeating DNA strings (TTAGGC in vertebrates) that cap chromosomes. (from the laboratory of doctors Jerry W. Shay and Woodring E. Wright)

cancer cells to cease their endless replications and die.

Based on a Chem30H honors paper by James Stinebaugh, UCLA.

Histone A protein, particularly rich in the basic amino acids lysine and arginine, that is found associated with DNA molecules.

charged proteins called **histones.** Histones are particularly rich in lysine and arginine; therefore, at the pH of most body fluid they have an abundance of positively charged sites along their length. The complex between negatively charged DNA and positively charged histones is called **chromatin.** Histones associate to form core particles about which double-stranded DNA then wraps. Further coiling of DNA produces the chromatin found in cell nuclei.

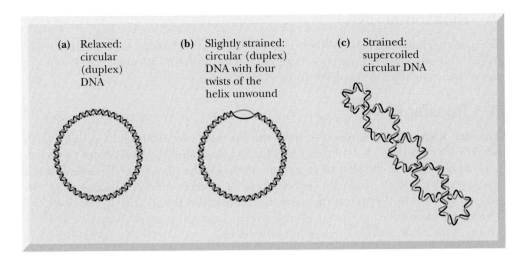

(a) Relaxed: circular (duplex) DNA

(b) Slightly strained: circular (duplex) DNA with four twists of the helix unwound

(c) Strained: supercoiled circular DNA

Figure 28.10
Relaxed and supercoiled DNA. (a) Circular DNA is relaxed. (b) One strand is broken, unwound by four turns, and the ends then rejoined. The strain of unwinding is localized in the nonhelical gap. (c) Supercoiling by four twists distributes the strain of unwinding uniformly over the entire molecule of circular DNA.

28.3 Ribonucleic Acids

Ribonucleic acids (RNA) are similar to deoxyribonucleic acids (DNA) in that they, too, consist of long, unbranched chains of nucleotides joined by phosphodiester groups between the 3′-hydroxyl of one pentose and the 5′-hydroxyl of the next. There are, however, three major differences in structure between RNA and DNA.

1. The pentose unit in RNA is β-D-ribose rather than β-2-deoxy-D-ribose.
2. The pyrimidine bases in RNA are uracil and cytosine rather than thymine and cytosine (Figure 28.1).
3. RNA is single-stranded rather than double-stranded.

Cells contain up to eight times as much RNA as DNA; in contrast to DNA, RNA occurs in different forms and in multiple copies of each form. RNA molecules are classified, according to their structure and function, into three major types: ribosomal RNA, transfer RNA, and messenger RNA. The molecular weight, number of nucleotides, and percent cellular abundance of these types in cells of *Escherichia coli* are summarized in Table 28.2.

Table 28.2   Types of RNA Found in Cells of _E. coli_

Type	Molecular-Weight Range (g/mol)	Number of Nucleotides	Percentage of Cell RNA
mRNA	25,000–1,000,000	75–3000	2
tRNA	23,000–30,000	73–94	16
rRNA	35,000–1,100,000	120–2904	82

A. Ribosomal RNA

Ribosomal RNA (rRNA) A ribonucleic acid found in ribosomes, the sites of protein synthesis.

The bulk of **ribosomal RNA (rRNA)** is found in the cytoplasm in subcellular particles called ribosomes, which contain about 60% RNA and 40% protein. Ribosomes are the sites in cells at which protein synthesis takes place.

B. Transfer RNA

Transfer RNA (tRNA) A ribonucleic acid that carries a specific amino acid to the site of protein synthesis on ribosomes.

Transfer RNA (tRNA) molecules have the lowest molecular weight of all nucleic acids. They consist of 73–94 nucleotides in a single chain. The function of tRNA is to carry amino acids to the sites of protein synthesis on the ribosomes. Each amino acid has at least one tRNA dedicated specifically to this purpose. Several amino acids have more than one. In the transfer process, the amino acid is joined to its specific tRNA by an ester bond between the α-carboxyl group of the amino acid and the $3'$ hydroxyl group of the ribose unit at the $3'$ end of the tRNA.

C. Messenger RNA

Messenger RNA (mRNA) A ribonucleic acid that carries coded genetic information from DNA to the ribosomes for the synthesis of proteins.

Messenger RNAs (mRNA) are present in cells in relatively small amounts and are very short-lived. They are single-stranded, and their synthesis is directed by information encoded on DNA molecules. Double-stranded DNA is unwound, and a complementary strand of mRNA is synthesized along one strand of the DNA template, beginning from the $3'$ end. The synthesis of mRNA from a DNA template is called transcription, because genetic information contained in a sequence of bases of DNA is transcribed into a complementary sequence of bases on mRNA. The name "messenger" is derived from the function of this type of RNA, which is to carry coded genetic information from DNA to the ribosomes for the synthesis of proteins.

Example 28.4

Following is a base sequence from a portion of DNA. Write the sequence of bases of the mRNA synthesized using this section of DNA as a template.

$$3'\text{-A-G-C-C-A-T-G-T-G-A-C-C-}5'$$

Solution

RNA synthesis begins at the $3'$ end of the DNA template and proceeds toward the $5'$ end. The complementary mRNA strand is formed using the bases C, G, A, and U. Uracil (U) is the complement of adenine (A) on the DNA template.

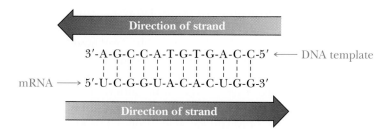

Reading from the 5′ end, the sequence of mRNA is 5′-UCGGUACACUGG-3′.

Problem 28.4

Here is a portion of the nucleotide sequence in phenylalanine tRNA.

3′-ACCACCUGCUCAGGCCUU-5′

Write the nucleotide sequence of its DNA complement.

28.4 The Genetic Code

A. Triplet Nature of the Code

It was clear by the early 1950s that the sequence of bases in DNA molecules constitutes the store of genetic information and directs the synthesis of messenger RNA, which, in turn, directs the synthesis of proteins. However, the statement that "the sequence of bases in DNA directs the synthesis of proteins" presents the following problem. How can a molecule containing only four variable units (adenine, cytosine, guanine, and thymine) direct the synthesis of molecules containing up to 20 variable units (the protein-derived amino acids)? How can an alphabet of only four letters code for the order of letters in the 20-letter alphabet that occurs in proteins?

An obvious answer is that there is not one base but rather a combination of bases coding for each amino acid. If the code consists of nucleotide pairs, there are $4^2 = 16$ combinations; this code is more extensive, but it is still not extensive enough to code for 20 amino acids. If the code consists of nucleotides in groups of three, there are $4^3 = 64$ combinations, which is more than enough to code for the primary structure of a protein. This answer appears to be a very simple solution to a system that must have taken eons of evolutionary trial and error to develop. Yet proof now exists, from comparison of gene (nucleic acid) and protein (amino acid) sequences, that nature does indeed use this simple three-letter or triplet code to store genetic information. A triplet of nucleotides is called a **codon.**

Codon A triplet of nucleotides on mRNA that directs incorporation of a specific amino acid into a polypeptide sequence.

B. Deciphering the Genetic Code

The next question is, which of the 64 triplets code for which amino acid? In 1961, Marshall Nirenberg provided a simple experimental approach to the problem, based on the observation that synthetic polynucleotides direct polypeptide synthesis in much the same manner as do natural mRNAs. Nirenberg incubated ribosomes, amino acids, tRNAs, and appropriate protein-synthesizing enzymes. With only these components, there was no polypeptide synthesis. However, when he added synthetic polyuridylic acid (poly U), a polypeptide of high molecular weight was synthesized. What was more important, the synthetic polypeptide contained only phenylalanine. With this discovery, the first element of the genetic code was deciphered: the triplet UUU codes for phenylalanine.

Similar experiments were carried out with different synthetic polyribonucleotides. It was found, for example, that polyadenylic acid (poly A) leads to the synthesis of polylysine, and that polycytidylic acid (poly C) leads to the synthesis of polyproline. By 1964, all 64 codons had been deciphered (Table 28.3).

C. Properties of the Genetic Code

Several features of the genetic code are evident from a study of Table 28.3.

1. Only 61 triplets code for amino acids. The remaining three (UAA, UAG, and UGA) are signals for chain termination; they signal to the protein-synthesizing machinery of the cell that the primary sequence of the protein is complete. The three chain termination triplets are indicated in Table 28.3 by "Stop."

Table 28.3 The Genetic Code — mRNA Codons and the Amino Acid Each Codon Directs

First Position (5'-end)	Second Position								Third Position (3'-end)
	U		C		A		G		
U	UUU	Phe	UCU	Ser	UAU	Tyr	UGU	Cys	U
	UUC	Phe	UCC	Ser	UAC	Tyr	UGC	Cys	C
	UUA	Leu	UCA	Ser	UAA	Stop	UGA	Stop	A
	UUG	Leu	UCG	Ser	UAG	Stop	UGG	Trp	G
C	CUU	Leu	CCU	Pro	CAU	His	CGU	Arg	U
	CUC	Leu	CCC	Pro	CAC	His	CGC	Arg	C
	CUA	Leu	CCA	Pro	CAA	Gln	CGA	Arg	A
	CUG	Leu	CCG	Pro	CAG	Gln	CGG	Arg	G
A	AUU	Ile	ACU	Thr	AAU	Asn	AGU	Ser	U
	AUC	Ile	ACC	Thr	AAC	Asn	AGC	Ser	C
	AUA	Ile	ACA	Thr	AAA	Lys	AGA	Arg	A
	AUG*	Met	ACG	Thr	AAG	Lys	AGG	Arg	G
G	GUU	Val	GCU	Ala	GAU	Asp	GGU	Gly	U
	GUC	Val	GCC	Ala	GAC	Asp	GGC	Gly	C
	GUA	Val	GCA	Ala	GAA	Glu	GGA	Gly	A
	GUG	Val	GCG	Ala	GAG	Glu	GGG	Gly	G

*AUG also serves as the principal initiation codon.

2. The code is degenerate, which means that several amino acids are coded for by more than one triplet. Only methionine and tryptophan are coded for by just one triplet. Leucine, serine, and arginine are coded for by six triplets, and the remaining amino acids are coded for by two, three, or four triplets.
3. For the 15 amino acids coded for by two, three, or four triplets, only the third letter of the code varies. For example, glycine is coded for by the triplets GGA, GGG, GGC, and GGU.
4. There is no ambiguity in the code, meaning that each triplet codes for only one amino acid.

Finally, we must ask one last question about the genetic code. Is the code universal, that is, is it the same for all organisms? Every bit of experimental evidence available today from the study of viruses, bacteria, and higher animals, including humans, indicates that the code is universal. Furthermore, the fact that it is the same for all these organisms means that it has been the same over millions of years of evolution.

Example 28.5

During transcription, a portion of mRNA is synthesized with the following base sequence:

5'-AUG-GUA-CCA-CAU-UUG-UGA-3'

(a) Write the nucleotide sequence of the DNA from which this portion of mRNA was synthesized.
(b) Write the primary structure of the polypeptide coded for by this section of mRNA.

Solution

(a) During transcription, mRNA is synthesized from a DNA strand, beginning from the 3' end of the DNA template. The DNA strand must be the complement of the newly synthesized mRNA strand.

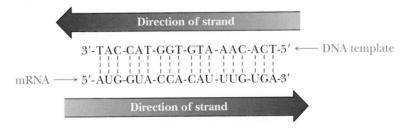

Note that the codon UGA codes for termination of the growing polypeptide chain; therefore, the sequence given in this problem codes for a pentapeptide only.

(b) The sequence of amino acids is shown in the following mRNA strand.

5'-AUG-GUA-CCA-CAU-UUG-UGA-3'

met—val—pro—his—leu—stop

Problem 28.5

The following section of DNA codes for oxytocin, a polypeptide hormone.

3′ACG-ATA-TAA-GTT-TTA-ACG-GGA-GAA-CCA-ACT-5′

(a) Write the base sequence of the mRNA synthesized from this section of DNA.

(b) Given the sequence of bases in part (a), write the primary structure of oxytocin.

28.5   Sequencing Nucleic Acids

As recently as 1975, the task of determining the primary structure of a nucleic acid was thought to be far more difficult than determining the primary structure of a protein. Nucleic acids, it was reasoned, contain only four different units, whereas proteins contain 20 different units. With only four different units, there are fewer specific sites for selective cleavage, distinctive sequences are more difficult to recognize, and there is greater chance of ambiguity in the assignment of sequence. Two breakthroughs reversed this situation. First was the development of a type of electrophoresis called polyacrylamide gel electrophoresis, a technique so sensitive that it is possible to separate nucleic acid fragments that differ from one another in only a single nucleotide. The second breakthrough was the discovery of a class of enzymes called restriction endonucleases, isolated chiefly from bacteria.

A.   Restriction Endonucleases

Restriction endonuclease An enzyme that catalyzes hydrolysis of a particular phosphodiester bond within a DNA strand.

A **restriction endonuclease** recognizes a set pattern of four to eight nucleotides and cleaves a DNA strand by hydrolysis of the linking phosphodiester bonds at any site that contains that particular sequence. Close to 1000 restriction endonucleases have been isolated and their specificities characterized; each cleaves DNA at a different site and produces a different set of restriction fragments. *E. coli*, for example, has a restriction endonuclease, EcoRI, that recognizes the hexanucleotide sequence, GAATTC, and cleaves it between G and A.

cleavage here

5′---G-A-A-T-T-C---3′ $\xrightarrow{\text{EcoRI}}$ 5′---G + 5′-A-A-T-T-C---3′

Note that the action of restriction endonucleases is analogous to the action of trypsin (Section 27.4B), which catalyzes hydrolysis of amide bonds formed by the carboxyl groups of Lys and Arg, and of chymotrypsin, which catalyzes cleavage of amide bonds formed by the carboxyl groups of Phe, Tyr, and Trp.

Example 28.6

The following is a section of the gene coding for bovine rhodopsin along with several restriction endonucleases, their recognition sequences, and their hydrolysis sites. Which endonucleases will catalyze cleavage of this section of DNA?

5′-GCCGTCTACAACCCGGTCATCTACTATCATGATCAACAAGCAGTTCCGGAACT-3′

Enzyme	Recognition Sequence	Enzyme	Recognition Sequence
A*lu*I	AG ↓ CT	H*pa*II	C ↓ CGG
B*al*I	TGG ↓ CCA	M*bo*I	↓ GATC
F*nu*DII	CG ↓ CG	N*ot*I	GC ↓ GGCCGC
H*ea*III	GG ↓ CC	S*ac*I	GAGCT ↓ C

Solution

Only restriction endonucleases H*pa*II and M*bo*I catalyze cleavage of this polynucleotide: H*pa*II at two sites and M*bo*I at one site.

Problem 28.6

The following is another section of the bovine rhodopsin gene. Which of the endonucleases given in Example 28.6 will catalyze cleavage of this section?

5′-ACGTCGGGTCGTCGTCCTCTCGCGGTGGTGAGTCTTCCGGCTCTTCT-3′

B. Methods for Sequencing Nucleic Acids

Any sequencing of DNA begins with site-specific cleavage of double-stranded DNA by one or more restriction endonucleases into smaller fragments called **restriction fragments.** Each restriction fragment is then sequenced separately, overlapping base sequences are identified, and the entire sequence of bases is then deduced.

Two methods for sequencing restriction fragments have been developed. The first of these, developed by Allan Maxam and Walter Gilbert and known as the **Maxam-Gilbert method,** depends on base-specific chemical cleavage. The second method, developed by Frederick Sanger and known as the **chain termination** or **dideoxy method,** depends on interruption of DNA-polymerase catalyzed synthesis. Sanger and Gilbert shared the 1980 Nobel Prize for biochemistry for their "development of chemical and biochemical analysis of DNA structure." Sanger's dideoxy method is currently more widely used, and it is on this method that we concentrate.

Sanger dideoxy method A method developed by Frederick Sanger for sequencing DNA molecules.

C. DNA Replication in Vitro

To appreciate the rationale for the dideoxy method, we must first understand certain aspects of the biochemistry of DNA replication. During replication, the sequence of nucleotides in one strand is copied as a complementary strand to form the second strand of a double-stranded DNA molecule. Synthesis of the complementary strand is catalyzed by the enzyme DNA polymerase. DNA polymerase will also carry out this synthesis in vitro using single-stranded DNA as a template, provided that both the four deoxynucleotide triphosphate (dNTP) monomers and a primer are present. A **primer** is an oligonucleotide capable of forming a short section of double-stranded DNA

Figure 28.11

DNA polymerase catalyzes the synthesis in vitro using single-stranded DNA as a template provided that both the four deoxynucleotide triphosphate (dNTP) monomers and a primer are present. The primer provides a short stretch of double-stranded DNA by base-pairing with its complement on the single-stranded DNA.

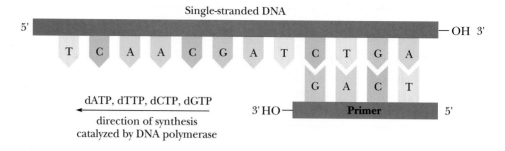

(dsDNA) by base-pairing with its complement on a single-stranded DNA (ssDNA). Because a new DNA strand grows from its 5′ to 3′ end, the primer must have a free 3′-OH group to which the first nucleotide of the growing chain is added (Figure 28. 11).

D.   The Chain Termination or Dideoxy Method

The key to the chain termination method is the addition to the synthesizing medium of a 2′,3′-dideoxynucleoside triphosphate (ddNTP). Because a ddNTP has no —OH group at the 3′ position, it cannot serve as an acceptor for the next nucleotide to be added to the growing polynucleotide chain. Thus, chain synthesis is terminated at any point where a ddNTP becomes incorporated; hence the designation chain termination method.

$$\text{^-O}-\overset{\overset{\displaystyle O}{\|}}{\underset{\underset{\displaystyle O^-}{|}}{P}}-O-\overset{\overset{\displaystyle O}{\|}}{\underset{\underset{\displaystyle O^-}{|}}{P}}-O-\overset{\overset{\displaystyle O}{\|}}{\underset{\underset{\displaystyle O^-}{|}}{P}}-O-CH_2$$

A 2′,3′-dideoxynucleoside triphosphate
(ddNTP)

In the chain termination method, a single-stranded DNA of unknown sequence is mixed with primer and divided into four separate reaction mixtures. To each reaction mixture are added all four deoxynucleoside triphosphates (dNTPs), one of which is labeled in the 5′ phosphoryl group with ^{32}P so that the newly synthesized fragments can be visualized by autoradiography.

$$^{32}_{15}\text{P} \longrightarrow\ ^{32}_{16}\text{S} + \text{Beta particle} + \text{Gamma rays}$$

Also added to each reaction mixture are DNA polymerase and one of the four ddNTPs. The ratio of dNTPs to ddNTP in each reaction mixture is adjusted so that incorporation of a ddNTP takes place infrequently. In each reaction mixture, DNA synthesis takes place but, in a given population of molecules, synthesis is interrupted at every possible site (Figure 28.12).

When gel electrophoresis of each reaction mixture is completed, a piece of x-ray film is placed over the gel, and gamma rays released by radioactive decay of ^{32}P darken the film and create a pattern on it that is an image of the resolved oligonucleotides. The base sequence of the complement to the original single-stranded template is then read directly from bottom to top of the developed film.

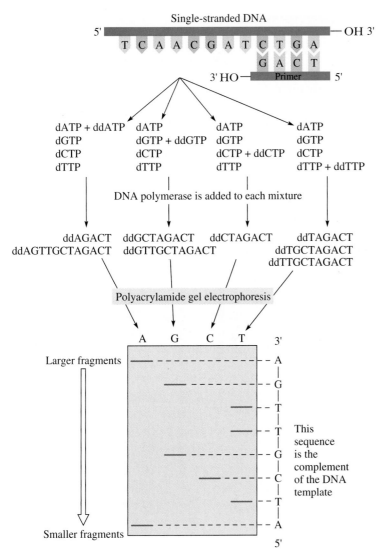

Figure 28.12

The chain termination or dideoxy method of DNA sequencing. The primer-DNA template is divided into four separate reaction mixtures. To each is added the four dNTPs, DNA polymerase, and one of the four ddNTPs. Synthesis is interrupted at every possible site. The mixtures of oligonucleotides are separated by polyacrylamide gel electrophoresis. The base sequence of the DNA complement is read from the bottom to the top (from the 5′ end to the 3′ end) of the developed gel.

A variation on this method is to use a single reaction mixture with each of the four ddNTPs labeled with a different fluorescent indicator. Each label is then detected by its characteristic spectrum. Automated DNA sequencing machines using this variation are capable of sequencing up to 10,000 base pairs per day.

E. Sequencing the Human Genome

As nearly everyone knows, the sequencing of the human genome was announced in the spring of 2000 by two competing groups, the so-called Human Genome Project, a loosely linked consortium of publicly funded groups, and a private company called

CHEMISTRY IN ACTION

DNA Fingerprinting

Each human being has a genetic makeup consisting of approximately 3 billion base pairs of nucleotides, and, except for identical twins, the base sequence of DNA in one individual is different from that of every other individual. As a result, each person has a unique DNA fingerprint. To determine a DNA fingerprint, a sample of DNA from a trace of blood, skin, or other tissue is treated with a set of restriction endonucleases, and the 5′ end of each restriction fragment is labeled with phosphorus-32. The resulting ^{32}P-labeled restriction fragments are then separated by polyacrylamide gel electrophoresis and visualized by placing a photographic plate over the developed gel.

In the DNA fingerprint patterns shown in the figure, lanes 1, 5, and 9 represent internal standards or control lanes. They contain the DNA fingerprint pattern of a standard virus treated with a standard set of restriction endonucleases. Lanes 2, 3, and 4 were used in a paternity suit. The DNA fingerprint of the mother in lane 4 contains five bands, which match with five of the six bands in the DNA fingerprint of the child in lane 3. The DNA fingerprint of the alleged father in lane 2 contains six bands, three of which match with bands in the DNA fingerprint of the child. Because the child inherits only half of its genes from the father, only half of the child's and father's DNA fingerprints are expected to match. In this instance, the paternity suit was won on the basis of the DNA fingerprint matching.

Lanes 6, 7, and 8 contain DNA fingerprint patterns used as evidence in a rape case. Lanes 7 and 8 are DNA fingerprints of semen obtained from the rape victim. Lane 6 is the DNA fingerprint pattern of the alleged rapist. The DNA fingerprint patterns of the semen do not match that of the alleged rapist and excluded the suspect from the case.

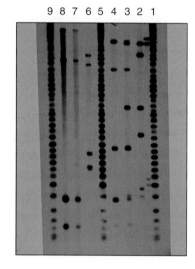

A DNA "fingerprint." (Courtesy of Dr. Lawrence Kobilinsky)

Celera. Actually, this milestone doesn't represent a complete sequence, but a so-called rough draft, comprising about 85% of the entire genome. The methodology used is based on a refinement of the techniques described earlier using massively parallel separations of fragments by electrophoresis in capillary tubes. The Celera approach used some 300 of the fastest sequencing machines in parallel, each operating on many parallel DNA fragments. Supercomputers were used to assemble and compare millions of overlapping sequences.

It is now much easier to obtain nucleic acid sequences than protein sequences. Sequencing a gene immediately gives the investigator the sequence of the protein produced by that gene because the code is known. Because many simpler organisms have been sequenced and the functions of many of the coded proteins are known, it is possible to take the sequence of a protein of unknown function and determine sequence homologies with the vast number of proteins of known function in the Protein Data Bank, and thereby make an educated guess about the function of a protein the new gene codes for. This technology is about to produce a revolution in chemistry, biology, and medicine.

This achievement represents the beginning of a new era of molecular medicine, in which specific genetic deficiencies leading to inherited diseases will be understood on a molecular basis, and new therapies targeted at shutting down undesired genes or turning on desired ones will be developed.

Summary

Nucleic acids are composed of three types of monomer units: heterocyclic aromatic amine bases derived from purine and pyrimidine, the monosaccharides D-ribose or 2-deoxy-D-ribose, and phosphate ions (Section 28.1). A **nucleoside** is a compound containing D-ribose or 2-deoxy-D-ribose bonded to a heterocyclic aromatic amine base by a β-N-glycosidic bond. A **nucleotide** is a nucleoside in which a molecule of phosphoric acid is esterified with an —OH of the monosaccharide, most commonly either the 3′-OH or the 5′-OH. Nucleoside mono-, di-, and triphosphates are strong polyprotic acids and are extensively ionized at pH 7.0. At pH 7.0, adenosine triphosphate, for example, is a 50:50 mixture of ATP^{3-} and ATP^{4-}.

The **primary structure of deoxyribonucleic acids** consists of units of 2-deoxyribose bonded by 3′,5′-phosphodiester bonds (Section 28.2A). A heterocyclic aromatic amine base is attached to each deoxyribose unit by a β-N-glycosidic bond. The sequence of bases is read from the 5′ end of the polynucleotide strand to the 3′ end.

The heart of the **Watson-Crick model** is the postulate that a molecule of DNA consists of two antiparallel polynucleotide strands coiled in a right-handed manner about the same axis to form a **double helix** (Section 28.2B). Purine and pyrimidine bases point inward toward the axis of the helix and are always paired G-C and A-T. In **B-DNA**, base pairs are stacked one on top of another with a spacing of 340 pm and ten base pairs per 3400-pm helical repeat. In **A-DNA**, bases are stacked with a spacing of 290 pm between base pairs and ten base pairs per 2900-pm helical repeat.

The **tertiary structure** of DNA is commonly referred to as **supercoiling** (Section 28.2C). **Circular DNA** is a type of double-stranded DNA in which the ends of each strand are joined by phosphodiester groups. Opening of one strand followed by partial unwinding and rejoining the ends introduces strain in the nonhelical gap. This strain can be spread over the entire molecule of circular DNA by introduction of **superhelical twists**. **Histones** are particularly rich in lysine and arginine and, therefore, have an abundance of positive charges. The association of DNA and histones produces a pigment called **chromatin.**

There are two important differences between the primary structure of ribonucleic acids and DNA (Section 28.3). (1) The monosaccharide unit in RNA is D-ribose. (2) Both RNA and DNA contain the purine bases adenine (A) and guanine (G), and the pyrimidine base cytosine (C). As the fourth base, however, RNA contains uracil (U), whereas DNA contains thymine (T).

The genetic code (Section 28.4) consists of nucleosides in groups of three; that is, it is a triplet code. Only 61 triplets code for amino acids; the remaining three code for termination of polypeptide synthesis.

Restriction endonucleases recognize a set pattern of four to eight nucleotides and cleave a DNA strand by hydrolysis of the linking phosphodiester bonds at any site that contains that particular sequence (Section 28.5A). In the **chain termination** or **dideoxy method** of DNA sequencing developed by Frederick Sanger (Section 28.5D), a primer-DNA template is divided into four separate reaction mixtures. To each is added the four dNTPs, one of which is labeled with ^{32}P. Also added are DNA polymerase and one of the four ddNTPs. Synthesis is interrupted at every possible site. The mixtures of newly synthesized oligonucleotides are separated by polyacrylamide gel electrophoresis and visualized by autoradiography. The base sequence of the DNA complement to the original DNA template is read from the bottom to the top (from the 5′ end to the 3′ end) of the developed photographic plate.

Problems

Nucleosides and Nucleotides

28.7 A pioneer in designing and synthesizing antimetabolites that could destroy cancer cells was George Hitchings at Burroughs Wellcome Co. In 1942 he initiated a program to discover DNA antimetabolites, and in 1948 he and Gertrude Elion synthesized 6-mercaptopurine (see *The Merck Index*, 12th ed., #5919), a successful drug for treating acute leukemia. Another DNA antimetabolite synthesized by Hitchings and Elion was 6-thioguanine (see *The Merck Index*, 12th ed., #9473). Hitchings and Elion along with Sir James W. Black won the 1988 Nobel Prize for physiology or medicine for their discoveries of "important principles of drug treatment." In each drug, the oxygen at carbon 6 of the parent molecule is replaced by divalent sulfur. Draw structural formulas for the enethiol (the sulfur equivalent of an enol) forms of 6-mercaptopurine and 6-thioguanine.

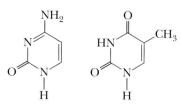

6-Mercaptopurine   6-Thioguanine

28.8 Following are structural formulas for cytosine and thymine. Draw two additional tautomeric forms for cytosine and three additional tautomeric forms for thymine.

Cytosine (C)   Thymine (T)

28.9 Draw a structural formula for a nucleoside composed of the following.

  (a) β-D-Ribose and adenine   **(b)** β-2-Deoxy-D-ribose and cytosine

28.10 Nucleosides are stable in water and in dilute base. In dilute acid, however, the glycosidic bond of a nucleoside undergoes hydrolysis to give a pentose and a heterocyclic aromatic amine base. Propose a mechanism for this acid-catalyzed hydrolysis.

28.11 Explain the difference in structure between a nucleoside and a nucleotide.

28.12 Draw a structural formula for each nucleotide, and estimate its net charge at pH 7.4, the pH of blood plasma.

  (a) 2′-Deoxyadenosine 5′-triphosphate (dATP)
  (b) Guanosine 3′-monophosphate (GMP)
  (c) 2′-Deoxyguanosine 5′-diphosphate (dGDP)

28.13 Cyclic-AMP, first isolated in 1959, is involved in many diverse biological processes as a regulator of metabolic and physiological activity. In it, a single phosphate group is esterified with both the 3′ and 5′ hydroxyls of adenosine. Draw a structural formula of cyclic-AMP.

The Structure of DNA

28.14 Why are deoxyribonucleic acids called acids? What are the acidic groups in their structure?

28.15 Human DNA contains approximately 30.4% A. Estimate the percentages of G, C, and T and compare them with the values presented in Table 28.1.

28.16 Draw a structural formula of the DNA tetranucleotide 5'-A-G-C-T-3'. Estimate the net charge on this tetranucleotide at pH 7.0. What is the complementary tetranucleotide to this sequence?

28.17 List the postulates of the Watson-Crick model of DNA secondary structure.

28.18 The Watson-Crick model is based on certain experimental observations of base composition and molecular dimensions. Describe these observations and show how the Watson-Crick model accounts for each.

28.19 If you read J. D. Watson's account of the discovery of the structure of DNA, *The Double Helix*, you will find that for a time in their model-building studies, he and Crick were using alternative (and incorrect, at least in terms of their final model of the double helix) tautomeric structures for some of the heterocyclic bases.
(a) Write at least one alternative tautomeric structure for adenine.
(b) Would this structure still base-pair with thymine, or would it now base-pair more efficiently with a different base? If so, identify that base.

28.20 Compare the α-helix of proteins and the double helix of DNA in the following ways.
(a) The units that repeat in the backbone of the polymer chain
(b) The projection in space of substituents along the backbone (the R groups in the case of amino acids; purine and pyrimidine bases in the case of double-stranded DNA) relative to the axis of the helix

28.21 Discuss the role of the hydrophobic interactions in stabilizing the following.
(a) Double-stranded DNA   (b) Lipid bilayers   (c) Soap micelles

28.22 Name the type of covalent bond(s) joining monomers in these biopolymers.
(a) Polysaccharides   (b) Polypeptides   (c) Nucleic acids

28.23 In terms of hydrogen bonding, which is more stable, an A-T base pair or a G-C base pair?

28.24 At elevated temperatures, nucleic acids become denatured; that is, they unwind into single-stranded DNA. Account for the observation that the higher the G-C content of a nucleic acid, the higher the temperature required for its thermal denaturation.

28.25 Write the DNA complement for 5'-ACCGTTAAT-3'. Be certain to label which is the 5' end and which is the 3' end of the complement strand.

28.26 Write the DNA complement for 5'-TCAACGAT-3'.

Ribonucleic Acids

28.27 Compare the degree of hydrogen bonding in the base pair A-T found in DNA with that in the base pair A-U found in RNA.

28.28 Compare DNA and RNA is these ways.
(a) Monosaccharide units   (b) Principal purine and pyrimidine bases
(c) Primary structure   (d) Location in the cell
(e) Function in the cell

28.29 What type of RNA has the shortest lifetime in cells?

28.30 Write the mRNA complement for 5′-ACCGTTAAT-3′. Be certain to label which is the 5′ end and which is the 3′ end of the mRNA strand.

28.31 Write the mRNA complement for 5′-TCAACGAT-3′.

The Genetic Code

28.32 What does it mean to say that the genetic code is degenerate?

28.33 Write the mRNA codons for:

(a) Valine   (b) Histidine   (c) Glycine

28.34 Aspartic acid and glutamic acid have carboxyl groups on their side chains and are called acidic amino acids. Compare the codons for these two amino acids.

28.35 Compare the structural formulas of the aromatic amino acids phenylalanine and tyrosine. Compare also the codons for these two amino acids.

28.36 Glycine, alanine, and valine are classified as nonpolar amino acids. Compare their codons. What similarities do you find? What differences do you find?

28.37 Codons in the set CUU, CUC, CUA, and CUG all code for the amino acid leucine. In this set, the first and second bases are identical, and the identity of the third base is irrelevant. For what other sets of codons is the third base also irrelevant, and for what amino acid(s) does each set code?

28.38 Compare the codons with a pyrimidine, either U or C, as the second base. Do the majority of the amino acids specified by these codons have hydrophobic or hydrophilic side chains?

28.39 Compare the codons with a purine, either A or G, as the second base. Do the majority of the amino acids specified by these codons have hydrophilic or hydrophobic side chains?

28.40 What polypeptide is coded for by this mRNA sequence?

5′-GCU-GAA-GUC-GAG-GUG-UGG-3′

28.41 The alpha chain of human hemoglobin has 141 amino acids in a single polypeptide chain. Calculate the minimum number of bases on DNA necessary to code for the alpha chain. Include in your calculation the bases necessary for specifying termination of polypeptide synthesis.

28.42 In HbS, the human hemoglobin found in individuals with sickle-cell anemia, glutamic acid at position 6 in the beta chain is replaced by valine.

(a) List the two codons for glutamic acid and the four codons for valine.
(b) Show that one of the glutamic acid codons can be converted to a valine codon by a single substitution mutation, that is by changing one letter in one codon.

29

THE ORGANIC CHEMISTRY OF METABOLISM

W e have now studied the structure and typical reactions of the major types of organic functional groups. Further, we have studied the structure of carbohydrates, lipids, amino acids and proteins, and nucleic acids. Now let us see how this background can be applied to the study of the organic chemistry of metabolism. In this chapter, we study two key metabolic pathways, namely β-oxidation of

■ These U.S. runners have just won the gold medal in the 4 × 400 relay at the 1996 Olympic games. *(RB-GM-J.O. Atlanta/Liason Agency)* Inset: A model of lactic acid. The build-up of lactic acid can cause severe muscle pain (Section 29.7A).

fatty acids and glycolysis. The first of these is a pathway by which the hydrocarbon chains of fatty acids are degraded, two carbons at a time, to acetyl coenzyme A. The second is a pathway by which glucose is converted to pyruvate and then to acetyl coenzyme A.

Those of you who go on to courses in biochemistry will undoubtedly study these metabolic pathways in considerable detail, including their role in energy production and conservation, their regulation, and the diseases associated with errors of particular metabolic steps. Our concern in this chapter is more limited. It is our purpose to show that reactions of these pathways are biochemical equivalents of organic functional group reactions we have already studied in detail. In these pathways, we find examples of keto-enol tautomerism; oxidation of an aldehyde to a carboxylic acid; oxidation of a secondary alcohol to a ketone; a reverse aldol reaction; a reverse Claisen condensation; and formation and hydrolysis of esters, imines, thioesters, and mixed anhydrides. In this chapter, we use the mechanisms we have studied earlier to give us insights into the mechanisms of these reactions, all of which are enzyme-catalyzed.

29.1   Five Key Participants in Glycolysis and β-Oxidation

To understand what happens in β-oxidation of fatty acids and glycolysis, we first need to introduce five of the principal compounds participating in these and a great many other metabolic pathways. Three of these compounds (ATP, ADP, and AMP) are central to the storage and transfer of phosphate groups. The other two, $NAD^+/NADH$ and $FAD/FADH_2$, are **coenzymes** involved in the oxidation/reduction of metabolic intermediates.

Coenzyme   A low-molecular-weight, nonprotein molecule or ion that binds reversibly to an enzyme, functions as a second substrate for the enzyme, and is regenerated by further reaction.

A.   ATP, ADP, and AMP — Agents for Storage and Transfer of Phosphate Groups

Following is a structural formula of adenosine triphosphate (Section 28.1), the most important of the compounds involved in the storage and transport of phosphate groups. A building block for ATP is adenosine, which consists of a unit of adenine bonded to a unit of D-ribofuranose by a β-N-glycosidic bond. Bonded to the terminal $-CH_2OH$ of ribose are three units of phosphate: one joined by a phosphoric ester bond and the remaining two joined by phosphoric anhydride bonds.

Adenosine triphosphate (ATP)

Hydrolysis of the terminal phosphate group of ATP gives ADP. In the following abbreviated structural formulas, adenosine and its single phosphoric ester group are represented by the symbol AMP.

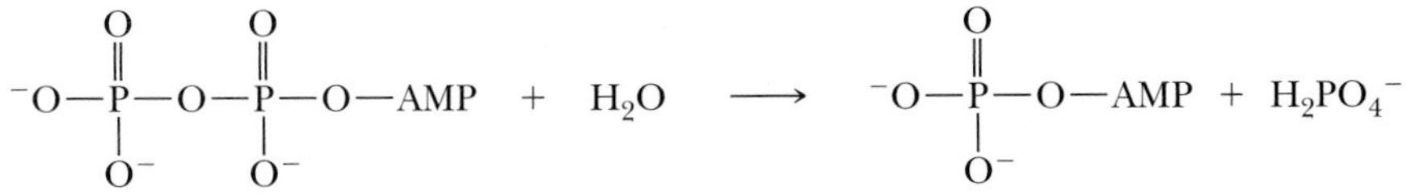

Adenosine triphosphate   Water   Adenosine diphosphate
(ATP)   (a phosphate   (ADP)
acceptor)

The reaction shown is hydrolysis of a phosphoric anhydride; the phosphate acceptor is water. In the first two reactions of glycolysis, the phosphate acceptors are —OH groups of glucose and fructose, respectively, to form phosphoric esters of these molecules. In glycolysis are two reactions in which the phosphate acceptor is ADP, which is converted to ATP.

B. NAD⁺/NADH — Agents for Electron Transfer in Biological Oxidation/Reductions

Nicotinamide adenine dinucleotide (NAD⁺) is one of the central agents for the transfer of electrons in metabolic oxidations and reductions. NAD⁺ is constructed of a unit of ADP joined by a phosphoric ester bond to the terminal —CH_2OH of β-D-ribofuranose, which is in turn joined to the pyridine ring of nicotinamide by a β-N-glycosidic bond.

Nicotinamide adenine dinucleotide (NAD⁺) A biological oxidizing agent. When acting as an oxidizing agent, NAD⁺ is reduced to NADH.

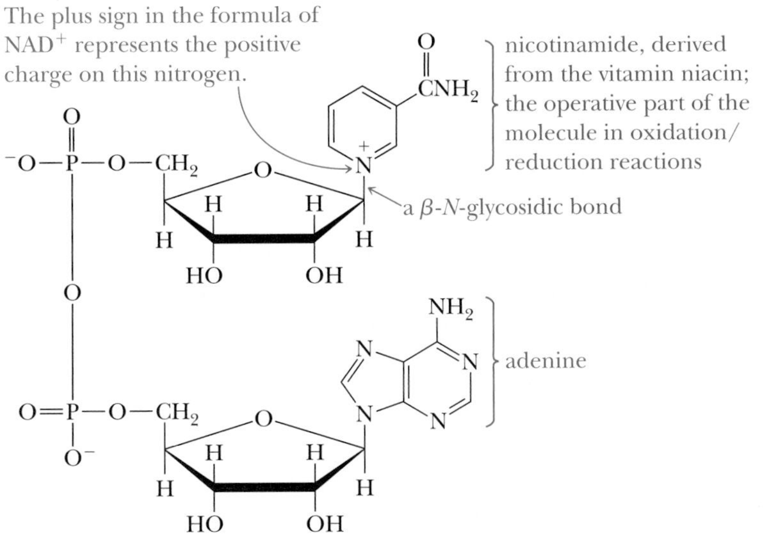

The plus sign in the formula of NAD⁺ represents the positive charge on this nitrogen.

nicotinamide, derived from the vitamin niacin; the operative part of the molecule in oxidation/reduction reactions

a β-N-glycosidic bond

adenine

Nicotinamide adenine dinucleotide
(NAD⁺)

NAD⁺ is a two-electron oxidizing agent, as seen in the following balanced half-reaction, and is reduced to NADH. NADH is, in turn, a two-electron reducing agent and is oxidized to NAD⁺. In these abbreviated structural formulas, the adenine dinucleotide part of each molecule is represented by the symbol Ad.

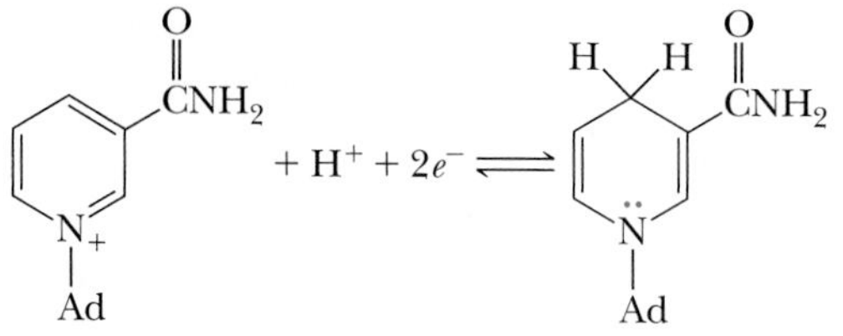

NAD$^+$
(oxidized form)

NADH
(reduced form)

NAD$^+$ is involved in a variety of enzyme-catalyzed oxidation/reduction reactions. The two types of oxidations we deal with in this chapter are oxidation of a secondary alcohol to a ketone and oxidation of an aldehyde to a carboxylic acid. Each is a two-electron oxidation.

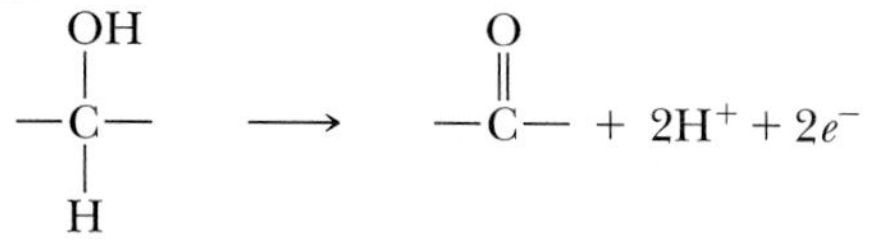

A secondary alcohol   A ketone

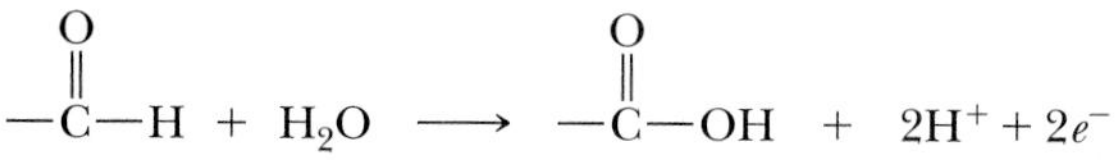

An aldehyde   A carboxylic acid

Oxidation of each functional group involves transfer of a hydride ion to NAD$^+$. This reaction can proceed in either direction depending on whether NAD$^+$ or NADH is in excess.

Mechanism Oxidation of an Alcohol by NAD$^+$

Although this oxidation is probably concerted, we show individual steps for clarity.

Step 1: A basic group, B$^-$, on the surface of the enzyme removes H$^+$ from the —OH group.

Step 2: Electrons of the H—O sigma bond become the pi electrons of the C=O bond.

Step 3: A hydride ion is transferred from carbon to NAD$^+$ to create a new C—H bond.

Step 4: Electrons within the ring flow to the positively charged nitrogen.

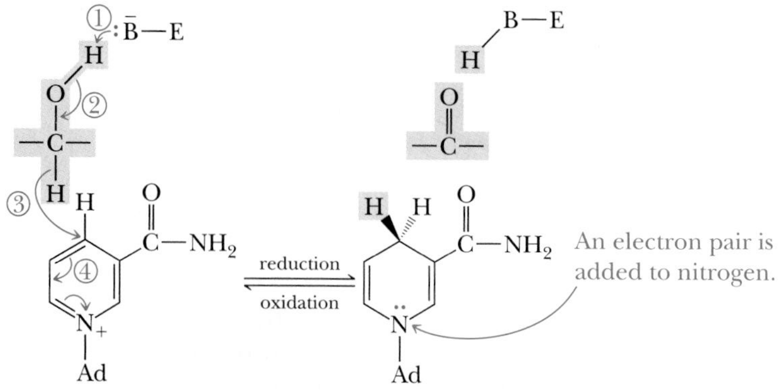

NAD$^+$ NADH

An electron pair is added to nitrogen.

The hydride ion, H:$^-$, which is transferred from the secondary alcohol to NAD$^+$, contains two electrons, and, thus, NAD$^+$ and NADH function exclusively in two-electron oxidations and two-electron reductions.

C.   FAD/FADH$_2$ — Agents for Electron Transfer in Biological Oxidation/Reductions

Flavin adenine dinucleotide (FAD) is also a central component in the transfer of electrons in metabolic oxidations and reductions. In FAD, flavin is bonded to the five-carbon monosaccharide ribitol which is, in turn, bonded to the terminal phosphate group of ADP.

Flavin adenine dinucleotide (FAD) A biological oxidizing agent. When acting as an oxidizing agent, FAD is reduced to FADH$_2$.

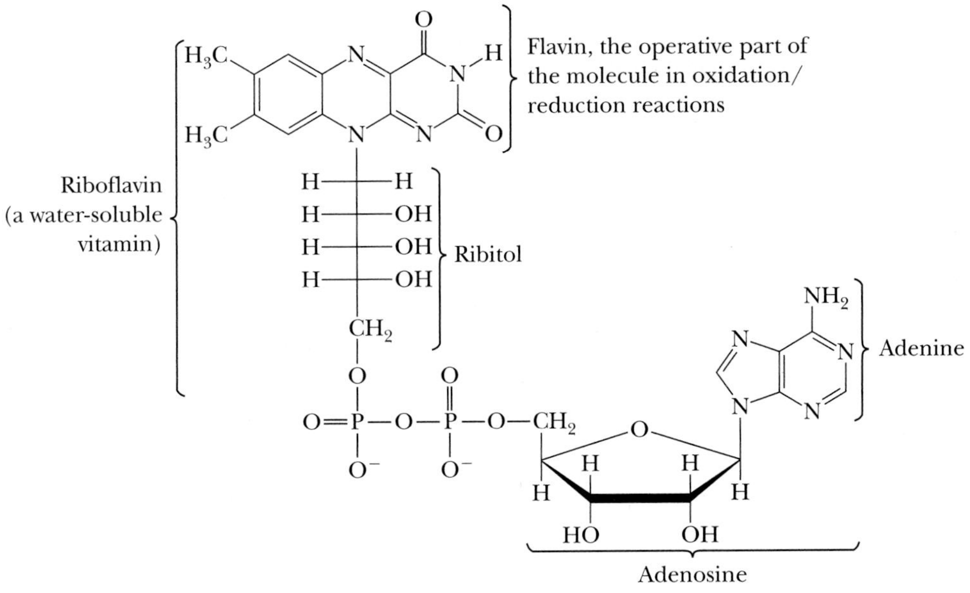

Flavin adenine dinucleotide
(FAD)

FAD participates in several types of enzyme-catalyzed oxidation/reduction reactions. Our concern in this chapter is with its role in the oxidation of a carbon-carbon single bond in the hydrocarbon chain of a fatty acid to a carbon-carbon double bond. As seen from balanced half-reactions, the two-electron oxidation of the hydrocarbon chain is coupled with the two-electron reduction of FAD.

Balanced half-reactions:

Oxidation of the
hydrocarbon chain: $-CH_2-CH_2- \longrightarrow -CH=CH- + 2H^+ + 2e^-$

Reduction of FAD: $FAD + 2H^+ + 2e^- \longrightarrow FADH_2$

The mechanism by which FAD oxidizes $-CH_2-CH_2-$ to $-CH=CH-$ involves transfer of a hydride ion from the hydrocarbon chain of the fatty acid to FAD, as shown in the following diagram. The individual curved arrows in this figure are numbered 1–6 to help you follow the flow of electrons in this transformation.

Mechanism   Oxidation of a Fatty Acid $-CH_2-CH_2-$
                   to $-CH=CH-$ by FAD

Step 1:   A basic group on the surface of the enzyme removes a hydrogen from the carbon alpha to the carboxyl group (shown here as R_2).

Step 2:   Electrons from this C—H sigma bond become the pi electrons of the new C—C double bond.

Step 3:   A hydride ion is transferred from the carbon beta to the carboxyl group to a nitrogen atom of flavin.

Step 4:   The pi electrons within the flavin are redistributed.

Step 5:   Electrons of the C=N bond remove a proton from the enzyme.

Step 6:   A new basic group is created on the enzyme.

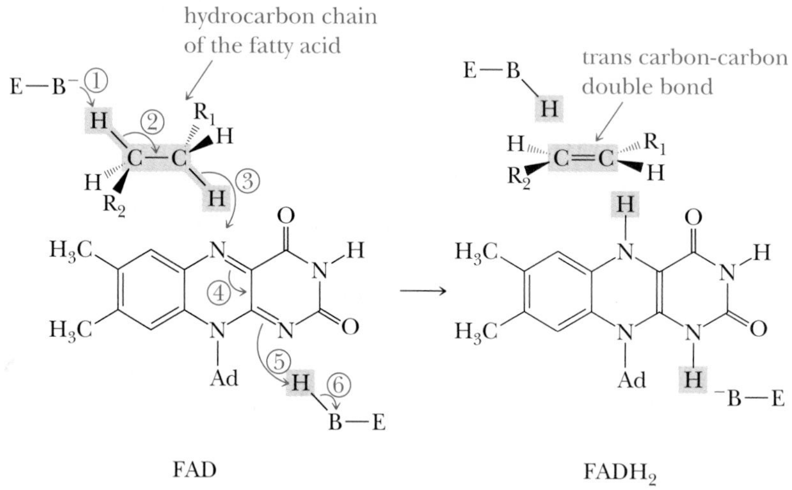

FAD $FADH_2$

Note that one of the hydrogen atoms added to FAD to produce $FADH_2$ comes from the hydrocarbon chain; the other comes from an acidic group on the surface of the enzyme catalyzing this oxidation. Also note that one group on the enzyme functions as a proton acceptor and that another functions as a proton donor.

29.2   Fatty Acids as a Source of Energy

Fatty acids in the form of triglycerides are the principal storage form of energy for most organisms. The principal advantage of storing energy in this form is that the hydrocarbon chains of fatty acids, consisting mostly of CH_2 groups, are a more highly reduced form of carbon than the oxygenated chains of carbohydrates; therefore, the energy yield per gram of fatty acid oxidized is greater than that per gram of carbohydrate oxidized. Complete oxidation of 1 g of palmitic acid, for example, yields almost 2.5 times the energy obtained from 1 g of glucose.

	Yield of Energy	
	(kJ/mol)	**(kJ/g)**
$C_6H_{12}O_6 + 6O_2 \longrightarrow 6CO_2 + 6H_2O$ Glucose	−2870	−15.9
$CH_3(CH_2)_{14}COOH + 23O_2 \longrightarrow 16CO_2 + 16H_2O$ Palmitic acid	−9791	−38.9

29.3 β-Oxidation of Fatty Acids

The first phase in the catabolism of fatty acids involves their release from triglyc-erides, either those stored in adipose tissue or from the diet, by hydrolysis catalyzed by a group of enzymes called lipases. The free fatty acids then pass into the blood-stream, where they are carried to sites of utilization. There are two major stages in **β-oxidation of fatty acids:** (1) activation of a free fatty acid in the cytoplasm and its transport across the inner mitochondrial membrane followed by (2) β-oxidation, a repeated sequence of four reactions.

β-Oxidation A series of four en-zyme-catalyzed reactions that cleaves carbon atoms, two at a time, from the carboxyl end of a fatty acid by way of intermediates that are oxidized at the β-carbon.

A. Activation of Fatty Acids — Formation of a Thioester with Coenzyme A

The process of β-oxidation begins in the cytoplasm with the formation of a **thioester** between the carboxyl group of a fatty acid and the sulfhydryl group of coenzyme A (Problem 25.36). Formation of this acyl-CoA derivative is coupled with hydrolysis of ATP to AMP and pyrophosphate ion. It is common in writing biochemical reactions to show some reactants and products by a curved arrow set over or under the main reaction arrow. We use this convention here to show ATP as a reactant and AMP and pyrophosphate ion as products.

Thioester An ester in which one atom of oxygen in the carboxylate group is replaced by an atom of sulfur.

$$
\underset{\substack{\text{Fatty acid} \\ \text{(as anion)}}}{R-\overset{\overset{\textstyle O}{\|}}{C}-O^-} + \underset{\text{Coenzyme A}}{HS-CoA} \xrightarrow[]{\;\;\overset{ATP\;\;\;AMP + P_2O_7^{4-}}{\curvearrowright}\;\;} \underset{\substack{\text{An acyl-CoA} \\ \text{derivative}}}{R-\overset{\overset{\textstyle O}{\|}}{C}-S-CoA} + OH^-
$$

The mechanism of this reaction involves attack by the fatty acid carboxylate an-ion on P=O of a phosphoric anhydride group of ATP to form an intermediate anal-ogous to the tetrahedral carbonyl addition intermediate formed in C=O chemistry. In the intermediate formed in the fatty acid-ATP reaction, the phosphorus attacked by the carboxylate anion becomes bonded to five groups. This intermediate then col-lapses to give an acyl-AMP, which is a highly reactive mixed anhydride of the carboxyl group of the fatty acid and the phosphate group of AMP.

$$
\underset{\substack{\text{Fatty acid} \\ \text{(as anion)}}}{R-\overset{\overset{\textstyle O}{\|}}{C}-O^-} + \underset{\text{ATP}}{Ad-O-\overset{\overset{\textstyle O}{\|}}{\underset{\underset{\textstyle O^-}{|}}{P}}-O-\overset{\overset{\textstyle O}{\|}}{\underset{\underset{\textstyle O^-}{|}}{P}}-O-\overset{\overset{\textstyle O}{\|}}{\underset{\underset{\textstyle O^-}{|}}{P}}-O^-} \rightleftharpoons \begin{bmatrix}\text{intermediate}\\\text{with one}\\\text{phosphorus}\\\text{bonded to}\\\text{five groups}\end{bmatrix} \longrightarrow \underset{\substack{\text{An acyl-AMP} \\ \text{(a mixed anhydride)}}}{R-\overset{\overset{\textstyle O}{\|}}{C}-O-\overset{\overset{\textstyle O}{\|}}{\underset{\underset{\textstyle O^-}{|}}{P}}-O-Ad} + \underset{\substack{\text{Pyrophosphate} \\ \text{ion}}}{{}^-O-\overset{\overset{\textstyle O}{\|}}{\underset{\underset{\textstyle O^-}{|}}{P}}-O-\overset{\overset{\textstyle O}{\|}}{\underset{\underset{\textstyle O^-}{|}}{P}}-O^-}
$$

This mixed anhydride then undergoes a carbonyl addition reaction with the sulfhydryl group of coenzyme A to form a tetrahedral carbonyl addition intermediate, which collapses to give AMP and an acyl-CoA (a fatty acid thioester of coenzyme A).

| Coenzyme A | An acyl-AMP | | An acyl-CoA | AMP |

At this point, the activated fatty acid is transported into the mitochondrion where its carbon chain is degraded by the reactions of β-oxidation.

B.   The Four Reactions of β-Oxidation

Reaction 1: Oxidation of a Carbon-Carbon Single Bond to a Double Bond

In the first reaction of β-oxidation, the carbon chain is oxidized, and a double bond is formed between the alpha- and beta-carbons of the fatty acid chain. The oxidizing agent is FAD, which is reduced to $FADH_2$.

An acyl-CoA   A *trans*-enoyl-CoA

Reaction 2: Hydration of the Carbon-Carbon Double Bond

The second reaction of β-oxidation is enzyme-catalyzed hydration of the carbon-carbon double bond to give an (R)-β-hydroxyacyl-CoA. The reaction is regiospecific in that —OH is added to carbon 3 of the chain. It is stereospecific in that only the R enantiomer is formed.

A *trans*-enoyl-CoA (R)-β-Hydroxyacyl-CoA

Reaction 3: Oxidation of the β-Hydroxy Group to a Carbonyl Group

In the second oxidation step of β-oxidation, the secondary alcohol group is oxidized to a ketone group. The oxidizing agent is NAD^+, which is reduced to NADH.

(R)-β-Hydroxyacyl-CoA β-Ketoacyl-CoA

Reaction 4: Cleavage of the Carbon Chain by a Reverse Claisen Condensation

The final step of β-oxidation is a reverse Claisen condensation, which results in cleavage between carbons 2 and 3 of the chain to form a molecule of acetyl coenzyme A and a new acyl-CoA, the hydrocarbon chain of which is now shortened by two carbon atoms.

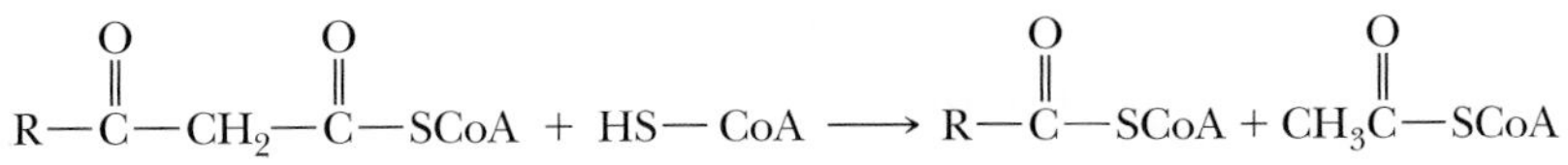

$$
\underset{\text{β-Ketoacyl-CoA}}{R-\overset{\overset{\displaystyle O}{\|}}{C}-CH_2-\overset{\overset{\displaystyle O}{\|}}{C}-SCoA} \;+\; \underset{\text{Coenzyme A}}{HS-CoA} \longrightarrow \underset{\text{An acyl-CoA}}{R-\overset{\overset{\displaystyle O}{\|}}{C}-SCoA} + \underset{\text{Acetyl-CoA}}{CH_3\overset{\overset{\displaystyle O}{\|}}{C}-SCoA}
$$

Mechanism   **A Reverse Claisen Condensation in β-Oxidation of Fatty Acids**

Step 1:   Attack of a sulfhydryl group of the enzyme thiolase on the carbonyl carbon of the ketone gives a tetrahedral carbonyl addition intermediate.

Step 2:   Collapse of this addition intermediate gives an enzyme-bound thioester, which is now shortened by two carbons.

Step 3:   Reaction of the enolate anion with a proton donor gives acetyl-CoA.

Step 4:   Reaction of the enzyme-thioester intermediate with a molecule of coenzyme A regenerates the sulfhydryl group on the surface of the enzyme and liberates the fatty acyl-CoA, now shortened by two carbon atoms.

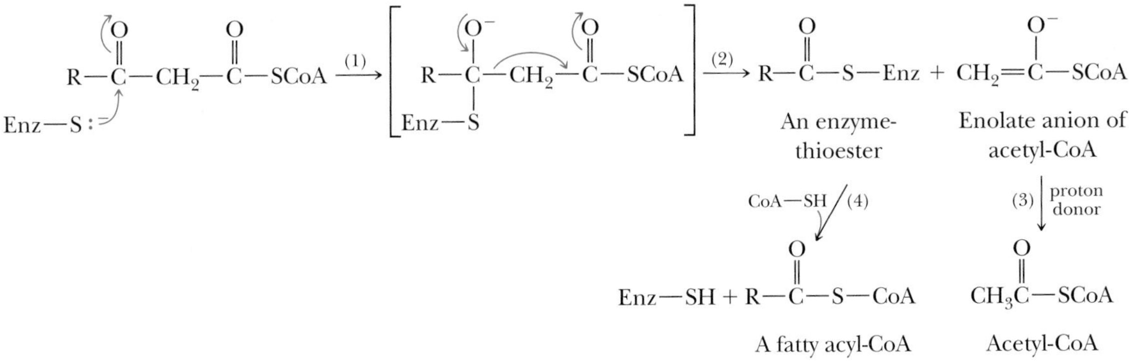

If Steps 1–3 of this mechanism are read in reverse, it is seen as an example of a **Claisen condensation** (Section 19.3A): the attack of the enolate anion of acetyl-CoA on the carbonyl group of a thioester to form a tetrahedral carbonyl addition intermediate, followed by its collapse to give a β-ketothioester.

The four steps in β-oxidation are summarized in Figure 29.1.

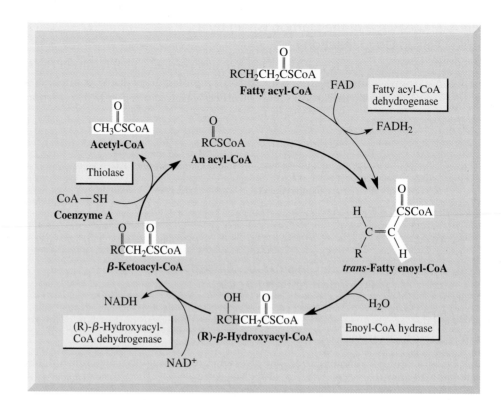

Figure 29.1

The four reactions of β-oxidation. The steps of β-oxidation are called a spiral because, after each series of four reactions, the carbon chain is shortened by two carbon atoms.

C. Repetition of the β-Oxidation Spiral Yields Additional Acetate Units

This series of four reactions is then repeated on the shortened fatty acyl-CoA chain and continues until the entire fatty acid chain is degraded to acetyl-CoA. Seven cycles of β-oxidation of palmitic acid, for example, give eight molecules of acetyl-CoA and involve seven oxidations by FAD and seven oxidations by NAD^+.

$$CH_3(CH_2)_{14}\overset{\overset{\displaystyle O}{\|}}{C}OH + 8CoA{-}SH + 7NAD^+ + 7FAD \xrightarrow{\quad ATP \quad AMP + P_2O_7{}^{4-}\quad}$$

Hexadecanoic acid
(Palmitic acid)

$$8CH_3\overset{\overset{\displaystyle O}{\|}}{C}SCoA + 7NADH + 7FADH_2$$

Acetyl coenzyme A

29.4   Digestion and Absorption of Carbohydrates

The main function of dietary carbohydrates is as a source of energy. In the typical American diet, carbohydrates provide about 50–60% of daily energy needs. The remainder is supplied by fats and proteins. During digestion of carbohydrates, glycosidic (acetal) bonds of disaccharides and polysaccharides are hydrolyzed to the hemi-acetal and hydroxyl groups of monosaccharides. Hydrolysis is catalyzed by a group of enzymes called **glycosidases,** each given a specific common name indicating the carbohydrate whose hydrolysis it catalyzes.

$$\text{starch} + n\text{H}_2\text{O} \xrightarrow[\substack{\text{(mouth and} \\ \text{intestine)}}]{\alpha\text{-amylase}} n\text{ maltose}$$

$$\text{maltose} + \text{H}_2\text{O} \xrightarrow[\text{(intestine)}]{\text{maltase}} \text{glucose} + \text{glucose}$$

$$\text{sucrose} + \text{H}_2\text{O} \xrightarrow[\text{(intestine)}]{\text{sucrase}} \text{glucose} + \text{fructose}$$

$$\text{lactose} + \text{H}_2\text{O} \xrightarrow[\text{(intestine)}]{\text{lactase}} \text{glucose} + \text{galactose}$$

Hydrolysis of starch begins in the mouth, catalyzed by α-amylase, a component of saliva. There, it is hydrolyzed to smaller polysaccharides and the disaccharide maltose. Hydrolysis of sucrose, lactose, maltose, and remaining polysaccharides is completed in the small intestine.

29.5 Glycolysis

Nearly every living cell carries out glycolysis. Living things first appeared in an environment lacking O_2, and glycolysis was an early and important pathway for extracting energy from nutrient molecules; its steps occur with no requirement of oxygen. Glycolysis played a central role in anaerobic metabolic processes for the first billion or so years of biological evolution on earth. Modern organisms still employ it to provide precursor molecules for aerobic pathways, such as the tricarboxylic acid cycle, and as a short-term energy source when the supply of oxygen is limited.

Glycolysis is a series of ten enzyme-catalyzed reactions that brings about the oxidation of glucose to two molecules of pyruvate.

Glycolysis From the Greek: *glyko*, sweet, and *lysis*, splitting. A series of ten enzyme-catalyzed reactions by which glucose is oxidized to two molecules of pyruvate.

$$\underset{\text{Glucose}}{\text{C}_6\text{H}_{12}\text{O}_6} \xrightarrow[\substack{\text{10 enzyme-} \\ \text{catalyzed steps}}]{\text{glycolysis}} \underset{\text{Pyruvate}}{2\text{CH}_3\overset{\overset{\text{O}}{\|}}{\text{C}}\text{COO}^- + 6\text{H}^+}$$

Example 29.1

Show by a balanced half-reaction that the net reaction of glycolysis involves oxidation. Do not worry at this point what the oxidizing agent is; we will come to that later in this chapter.

Solution

The balanced half-reaction requires four electrons on the right; therefore, the net reaction of glycolysis is a four-electron oxidation. As we will see, this occurs in two separate two-electron oxidation steps, each requiring NAD^+ as the oxidizing agent.

$$\underset{\text{Glucose}}{\text{C}_6\text{H}_{12}\text{O}_6} \xrightarrow{\text{glycolysis}} \underset{\text{Pyruvate}}{2\text{CH}_3\overset{\overset{\text{O}}{\|}}{\text{C}}\text{COO}^- + 6\text{H}^+ + 4e^-}$$

Problem 29.1

Under anaerobic (without oxygen) conditions, glucose is converted to lactate by a metabolic pathway called anaerobic glycolysis or, alternatively, lactate fermentation. Is anaerobic glycolysis a net oxidation, a net reduction, or neither?

$$C_6H_{12}O_6 \xrightarrow[\text{glycolysis}]{\text{anaerobic}} 2CH_3\overset{\overset{\displaystyle OH}{|}}{C}HCOO^- + 2H^+$$

Glucose   Lactate

29.6   The Ten Reactions of Glycolysis

Although writing the net reaction of glycolysis is simple, it took several decades of patient, intensive work by scores of scientists to discover the separate reactions by which glucose is converted to pyruvate. Glycolysis is frequently called the Embden-Meyerhof pathway, in honor of the two German biochemists, Gustav Embden and Otto Meyerhof, who contributed so greatly to our present knowledge of it. The ten reactions of glycolysis are summarized in Figure 29.2.

Reaction 1: Phosphorylation of α-D-Glucose

Transfer of a phosphate group from ATP to glucose to give α-D-glucose 6-phosphate is an example of reaction of an anhydride with an alcohol to form an ester (Section 18.6B), in this case reaction of a phosphoric anhydride with the primary alcohol group of glucose to form a phosphoric ester. In Section 18.10, we saw how a more reactive carboxyl functional group can be transformed into a less reactive carboxyl functional group. The same principle applies to functional derivatives of phosphoric acid. In Reaction 1 of glycolysis, a phosphoric anhydride, a more reactive functional group, is converted to a phosphoric ester, a less reactive functional group.

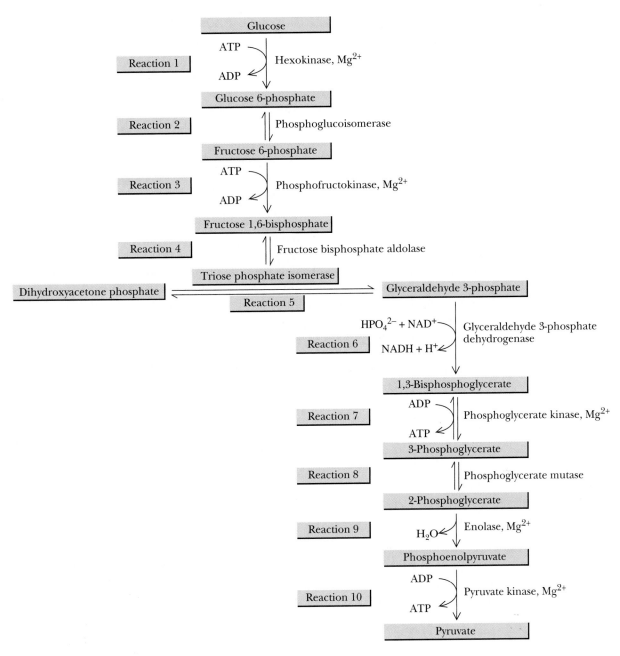

Figure 29.2
The ten reactions of glycolysis.

This enzyme-catalyzed reaction requires divalent magnesium ion, Mg^{2+}, whose function is to coordinate with two negatively charged oxygens of the terminal phosphate group of ATP and to facilitate attack by the —OH group of glucose on the phosphorus atom of the P=O group.

Reaction 2: Isomerization of Glucose 6-Phosphate to Fructose 6-Phosphate

In this reaction, glucose 6-phosphate, an aldohexose, is converted to fructose 6-phosphate, a 2-ketohexose.

α-D-Glucose 6-phosphate   α-D-Fructose 6-phosphate

Although both glucose 6-phosphate and fructose 6-phosphate exist almost entirely in cyclic hemiacetal forms, the chemistry involved in this isomerization is most easily seen by considering the open-chain forms of these two monosaccharides. This transformation is an example of keto-enol tautomerism to form an enediol, which then forms the ketone carbonyl group in fructose 6-phosphate.

Glucose 6-phosphate   An enediol   Fructose 6-phosphate
(an aldohexose phosphate)   (a 2-ketohexose phosphate)

Reaction 3: Phosphorylation of Fructose 6-Phosphate

In the third reaction, a second mole of ATP is used to convert fructose 6-phosphate to fructose 1,6-bisphosphate.

Fructose 6-phosphate   Fructose 1,6-bisphosphate

Reaction 4: Cleavage of Fructose 1,6-Bisphosphate to Two Triose Phosphates

In the fourth reaction, fructose 1,6-bisphosphate is cleaved to dihydroxyacetone phosphate and glyceraldehyde 3-phosphate by a reaction that is the reverse of an aldol reaction. Recall that an aldol reaction takes place between the α-carbon of one carbonyl-containing compound and the carbonyl carbon of another and that the functional group of the product of an aldol reaction is a β-hydroxyaldehyde or ketone (Section 19.1).

The characteristic structural features of the product of an aldol reaction are:

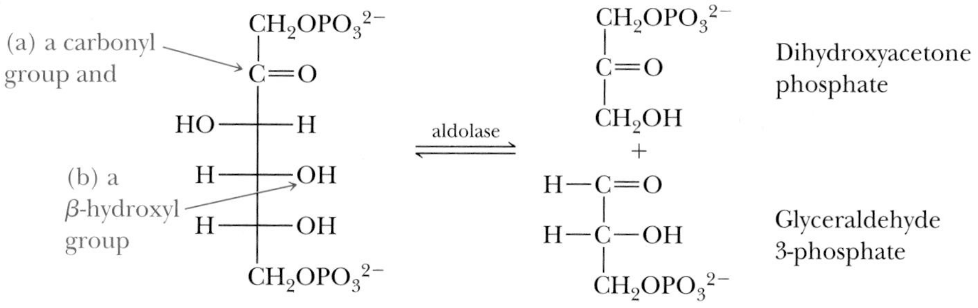

(a) a carbonyl group and

(b) a β-hydroxyl group

Fructose 1,6-bisphosphate

Dihydroxyacetone phosphate

Glyceraldehyde 3-phosphate

A key intermediate in this enzyme-catalyzed reverse aldol reaction is an imine (Section 16.10) formed between the carbonyl group of fructose 1,6-bisphosphate and the side-chain amino group of lysine (Section 27.1A), one of the amino acid building blocks of the enzyme catalyzing this reaction. In the following formulas, the symbol B represents the group on the enzyme involved in proton transfer. As you study this sequence, note the formation and hydrolysis of the imine and the concerted nature of bond breaking and bond forming in the reverse aldol reaction. Also note that the enzyme is regenerated unchanged after its catalytic role is accomplished.

Mechanism   A Reverse Aldol Reaction in Step 4 of Glycolysis

Step 1:   Reaction of an ammonium group on the surface of the enzyme with the carbonyl group of fructose 1,6-bisphosphate gives a protonated imine.

Step 2:   An enzyme-catalyzed reverse aldol reaction cleaves the six-carbon monosaccharide into two three-carbon units.

Step 3:   Proton transfer from the enzyme accompanied by a redistribution of valence electrons gives a new protonated imine.

Step 4: Hydrolysis of the new protonated imine gives dihydroxyacetone phosphate and regenerates the active enzyme, which is now ready to catalyze the cleavage of another molecule of fructose 1,6-bisphosphate.

Fructose 1,6-bisphosphate   Protonated imine

Glyceraldehyde
3-phosphate

Protonated imine   Dihydroxyacetone
phosphate

Reaction 5: Isomerization of Dihydroxyacetone Phosphate to Glyceraldehyde 3-Phosphate

This interconversion of triose phosphates occurs by the same type of keto-enol tautomerism and enediol intermediate we have already seen in the isomerization of glucose 6-phosphate to fructose 6-phosphate.

Dihydroxyacetone   An enediol   Glyceraldehyde
phosphate   intermediate   3-phosphate

Reaction 6: Oxidation of the Aldehyde Group of Glyceraldehyde 3-Phosphate

To simplify structural formulas in Reaction 6, glyceraldehyde 3-phosphate is abbreviated G—CHO. Two changes occur in this molecule. First, the aldehyde group is oxidized to a carboxyl group, which is, in turn, converted to a mixed anhydride. Oxidation of the aldehyde group is a two-electron oxidation. The oxidizing agent is NAD^+, which is reduced to NADH.

$$G-\overset{\overset{\displaystyle O}{\|}}{C}-H \ + \ H_2O \ \longrightarrow \ G-\overset{\overset{\displaystyle O}{\|}}{C}-OH \ + \ 2H^+ \ + \ 2e^- \qquad \text{A two-electron oxidation}$$

$$NAD^+ \ + \ H^+ \ + \ 2e^- \ \longrightarrow \ NADH \qquad\qquad\qquad \text{A two-electron reduction}$$

The reaction is considerably more complicated than might appear from combination of the balanced half-reactions. As shown in the mechanism box, it involves (1) formation of a thiohemiacetal, (2) hydride ion transfer to form a thioester, and (3) conversion of a thioester to a mixed anhydride.

Mechanism   Oxidation of Glyceraldehyde 3-Phosphate to 1,3-Bisphosphoglycerate

Step 1:   Reaction between glyceraldehyde 3-phosphate and a sulfhydryl group of the enzyme gives a thiohemiacetal (Section 16.8B).

$$G-\overset{\overset{\displaystyle O}{\|}}{C}-H \ + \ HS-Enz \ \rightleftharpoons \ G-\overset{\overset{\displaystyle OH}{|}}{\underset{\underset{\displaystyle H}{|}}{C}}-S-Enz$$

Glyceraldehyde 3-phosphate   A thiohemiacetal

Step 2:   Oxidation occurs by transfer of a hydride ion from the thiohemiacetal to NAD^+.

Step 3:   Reaction of the thioester with phosphate ion gives a tetrahedral carbonyl addition intermediate, which then collapses to regenerate the enzyme and give a mixed anhydride of phosphoric acid and glyceric acid.

1,3-Bisphosphoglycerate
(a mixed anhydride)

Reaction 7: Transfer of a Phosphate Group from 1,3-Bisphosphoglycerate to ADP

Transfer of a phosphate group in this reaction involves exchange of one anhydride group for another, namely the mixed anhydride of 1,3-bisphosphoglycerate for the new phosphoric anhydride in ATP.

$$
\underset{\substack{\text{1,3-Bisphospho-}\\\text{glycerate}}}{\underset{\displaystyle CH_2OPO_3{}^{2-}}{\overset{\displaystyle \overset{O}{\|}}{\underset{\displaystyle H-C-OH}{C-OPO_3{}^{2-}}}}}
\;+\;
\underset{\text{ADP}}{\overset{\displaystyle \overset{O}{\|}}{\underset{\displaystyle O^-}{{}^{-}O-P-O-AMP}}}
\;\underset{Mg^{2+}}{\overset{\substack{\text{phospho-}\\\text{glycerate kinase}}}{\rightleftharpoons}}\;
\underset{\substack{\text{3-Phosphoglycerate}}}{\underset{\displaystyle CH_2OPO_3{}^{2-}}{\overset{\displaystyle \overset{O}{\|}}{\underset{\displaystyle H-C-OH}{C-O^-}}}}
\;+\;
\underset{\text{ATP}}{\overset{\displaystyle \overset{O}{\|}\quad\overset{O}{\|}}{\underset{\displaystyle O^-\quad O^-}{{}^{-}O-P-O-P-O-AMP}}}
$$

Reaction 8: Isomerization of 3-Phosphoglycerate to 2-Phosphoglycerate

A phosphate group is transferred from the primary —OH group on carbon 3 to the secondary —OH group on carbon 2.

$$
\underset{\text{3-Phosphoglycerate}}{\underset{\displaystyle CH_2OPO_3{}^{2-}}{\overset{\displaystyle COO^-}{\underset{\displaystyle H-C-OH}{\vphantom{|}}}}}
\;\underset{}{\overset{\substack{\text{phosphoglycerate}\\\text{mutase}}}{\rightleftharpoons}}\;
\underset{\text{2-Phosphoglycerate}}{\underset{\displaystyle CH_2OH}{\overset{\displaystyle COO^-}{\underset{\displaystyle H-C-OPO_3{}^{2-}}{\vphantom{|}}}}}
$$

Reaction 9: Dehydration of 2-Phosphoglycerate

Dehydration of the primary alcohol (Section 8.4E) gives phosphoenolpyruvate, which is the ester of phosphoric acid and the enol form of pyruvic acid.

$$
\underset{\text{2-Phosphoglycerate}}{\underset{\displaystyle CH_2OH}{\overset{\displaystyle COO^-}{\underset{\displaystyle H-C-OPO_3{}^{2-}}{\vphantom{|}}}}}
\;\underset{Mg^{2+}}{\overset{\text{enolase}}{\longrightarrow}}\;
\underset{\text{Phosphoenolpyruvate}}{\underset{\displaystyle CH_2}{\overset{\displaystyle COO^-}{\underset{\displaystyle C-OPO_3{}^{2-}}{\vphantom{|}}}}}
\;+\;
H_2O
$$

Reaction 10: Transfer of a Phosphate Group from Phosphoenolpyruvate to ADP

Reaction 10 is divided into two steps: transfer of a phosphate group to ADP to produce ATP and conversion of the enol form of pyruvate to its keto form by keto-enol tautomerism (Section 16.11B).

$$
\underset{\substack{\text{Phosphoenol-}\\\text{pyruvate}}}{\underset{\displaystyle CH_2}{\overset{\displaystyle COO^-}{\underset{\displaystyle C-OPO_3{}^{2-}}{\vphantom{|}}}}}
\;\underset{\underset{\displaystyle \text{ADP}\quad\text{ATP}}{\searrow}}{\overset{\substack{\text{pyruvate kinase}\\ Mg^{2+}}}{\longrightarrow}}\;
\underset{\substack{\text{Enol of}\\\text{pyruvate}}}{\underset{\displaystyle CH_2}{\overset{\displaystyle COO^-}{\underset{\displaystyle C-OH}{\vphantom{|}}}}}
\;\rightleftharpoons\;
\underset{\text{Pyruvate}}{\underset{\displaystyle CH_3}{\overset{\displaystyle COO^-}{\underset{\displaystyle C=O}{\vphantom{|}}}}}
$$

Summing these ten reactions gives a balanced equation for the net reaction of glycolysis:

$$C_6H_{12}O_6 + 2NAD^+ + 2HPO_4{}^{2-} + 2ADP \xrightarrow{\text{glycolysis}}$$

Glucose

$$\underset{\text{Pyruvate}}{2CH_3\overset{O}{\overset{\|}{C}}COO^-} + 2NADH + 2ATP + 2H_2O$$

29.7 The Fates of Pyruvate

Pyruvate does not accumulate in cells but rather undergoes one of three possible enzyme-catalyzed reactions, depending on the state of oxygenation and the type of cell in which it is produced. A key to understanding the biochemical logic responsible for two of the possible fates of pyruvate is to recognize that it is produced by the oxidation of glucose through the reactions of glycolysis. NAD^+ is the oxidizing agent and is reduced to NADH. Glycolysis requires a continuing supply of NAD^+; therefore, under anaerobic conditions (where there is no oxygen present for the reoxidation of NADH), two of the metabolic pathways we describe use pyruvate in ways that regenerate NAD^+.

A. Reduction to Lactate — Lactate Fermentation

In vertebrates, the most important pathway for regeneration of NAD^+ under anaerobic conditions is reduction of pyruvate to lactate, catalyzed by the enzyme lactate dehydrogenase.

$$\underset{\text{Pyruvate}}{CH_3\overset{O}{\overset{\|}{C}}COO^-} + NADH + H_3O^+ \underset{\text{dehydrogenase}}{\overset{\text{lactate}}{\rightleftharpoons}} \underset{\text{Lactate}}{CH_3\overset{OH}{\overset{\|}{C}HCOO^-}} + NAD^+ + H_2O$$

Even though **lactate fermentation** allows glycolysis to continue in the absence of oxygen, it also brings about an increase in the concentration of lactate, and, perhaps more importantly, it increases the concentration of hydrogen ion, H_3O^+, in muscle tissue and in the bloodstream. This build-up of lactate and H_3O^+ is associated with muscle fatigue. When blood lactate reaches a concentration of about 0.4 mg/100 mL, muscle tissue becomes almost completely exhausted.

Lactate fermentation A metabolic pathway that converts glucose to two molecules of lactate.

Example 29.2

Show by writing a balanced equation that glycolysis followed by reduction of pyruvate to lactate (lactate fermentation) leads to an increase in the hydrogen ion concentration in the bloodstream.

Solution

Lactate fermentation produces lactic acid, which is completely ionized at pH 7.4, the normal pH of blood plasma. Therefore, the hydrogen ion concentration increases.

$$C_6H_{12}O_6 + 2H_2O \xrightarrow[\text{fermentation}]{\text{lactate}} 2CH_3\overset{\overset{\displaystyle OH}{|}}{C}HCOO^- + 2H_3O^+$$

Glucose   Lactate

Problem 29.2

Does lactate fermentation result in an increase or decrease in blood pH?

B. Reduction to Ethanol — Alcoholic Fermentation

Yeast and several other organisms have developed an alternative pathway to regenerate NAD^+ under anaerobic conditions. In the first step of this pathway, pyruvate undergoes enzyme-catalyzed decarboxylation to give acetaldehyde.

$$CH_3\overset{\overset{\displaystyle O}{\|}}{C}COO^- + H_3O^+ \xrightarrow[\text{decarboxylase}]{\text{pyruvate}} CH_3\overset{\overset{\displaystyle O}{\|}}{C}H + CO_2 + H_2O$$

Pyruvate   Acetaldehyde

The carbon dioxide produced in this reaction is responsible for the foam on beer and the carbonation of naturally fermented wines and champagnes. In a second step, acetaldehyde is reduced by NADH to ethanol.

$$CH_3\overset{\overset{\displaystyle O}{\|}}{C}H + NADH + H_3O^+ \xrightarrow[\text{dehydrogenase}]{\text{alcohol}} CH_3CH_2OH + NAD^+ + H_2O$$

Acetaldehyde   Ethanol

Adding the reactions for decarboxylation of pyruvate and reduction of acetaldehyde to the net reaction of glycolysis gives the overall reaction of **alcoholic fermentation.**

Alcoholic fermentation A metabolic pathway that converts glucose to two molecules of ethanol and two molecules of CO_2.

$$C_6H_{12}O_6 + 2HPO_4{}^{2-} + 2ADP + 2H^+ \xrightarrow[\text{fermentation}]{\text{alcoholic}} 2CH_3CH_2OH + 2CO_2 + 2ATP$$

Glucose   Ethanol

C. Oxidation and Decarboxylation to Acetyl-CoA

Under aerobic conditions, pyruvate undergoes oxidative decarboxylation. The carboxylate group is converted to carbon dioxide, and the remaining two carbons are converted to the acetyl group of acetyl-CoA.

$$CH_3\overset{\overset{\displaystyle O}{\|}}{C}COO^- + NAD^+ + CoASH \xrightarrow[\text{decarboxylation}]{\text{oxidative}} CH_3\overset{\overset{\displaystyle O}{\|}}{C}SCoA + CO_2 + NADH$$

Pyruvate   Acetyl-CoA

The oxidative decarboxylation of pyruvate is considerably more complex than is suggested by the preceding equation. In addition to NAD^+ and coenzyme A, this transformation also requires FAD, thiamine pyrophosphate, which is derived from thiamine (vitamin B_1), and lipoic acid.

Thiamine pyrophosphate

Lipoic acid
(as the carboxylate anion)

Acetyl coenzyme A then becomes a fuel for the tricarboxylic acid cycle, which results in oxidation of the two-carbon chain of the acetyl group to CO_2 with the production of NADH and $FADH_2$. These reduced coenzymes are, in turn, oxidized to NAD^+ and FAD during respiration with O_2 as the oxidizing agent.

Summary

ATP, ADP, and AMP (Section 29.1A) are agents for the storage and transport of phosphate groups. **Nicotinamide adenine dinucleotide (NAD^+)** (Section 29.1B) and **flavin adenine dinucleotide (FAD)** (Section 29.1C) are agents for the storage and transport of electrons in metabolic oxidations and reductions. NAD^+ is a two-electron oxidizing agent and is reduced to NADH. NADH is, in turn, a two-electron reducing agent and is oxidized to NAD^+. In the reactions of FAD involved in β-oxidation of fatty acids, it is a two-electron oxidizing agent and is reduced to $FADH_2$.

Fatty acids in the form of triglycerides are the principal storage form of energy for most organisms (Section 29.2). The hydrocarbon chains of fatty acids are a more highly reduced form of carbon than the oxygenated chains of carbohydrates. The energy yield per gram of fatty acid oxidized is greater than that per gram of carbohydrate.

There are two major stages in the metabolism of fatty acids (Section 29.3): (1) activation of free fatty acids in the cytoplasm by formation of thioesters with coenzyme A and transport of the activated fatty acids across the inner mitochondrial membrane followed by (2) β-oxidation. **β-Oxidation of fatty acids** (Section 29.3) is a series of four enzyme-catalyzed reactions by which a fatty acid is degraded to acetyl-CoA.

During digestion of carbohydrates, hydrolysis of glycosidic bonds is catalyzed by a group of enzymes called **glycosidases** (Section 29.4). **Glycolysis** (Section 29.5) is a series of ten enzyme-catalyzed reactions that brings about the oxidation of glucose to two molecules of pyruvate. The ten reactions of glycolysis (Section 29.6) can be grouped in the following way:

- Transfer of a phosphate group from ATP to an —OH group of a monosaccharide to form a phosphoric ester (Reactions 1 and 3).
- Interconversion of constitutional isomers by keto-enol tautomerism (Reactions 2 and 5).
- Reverse aldol reaction (Reaction 4).
- Oxidation of an aldehyde group to the mixed anhydride of a carboxylic acid and phosphoric acid (Reaction 6).
- Transfer of a phosphate group from a monosaccharide intermediate to ADP to form ATP (Reactions 7 and 10).
- Transfer of a phosphate group from a 1° alcohol to a 2° alcohol (Reaction 8).
- Dehydration of a 1° alcohol to form a carbon-carbon double bond (Reaction 9).

Pyruvate, the product of anaerobic glycolysis, does not accumulate in cells but rather undergoes one of three possible enzyme-catalyzed reactions, depending on the state of oxygenation and the type of cell in which it is produced (Section 29.7). In **lactate fermentation,** pyruvate is reduced to lactate by NADH. In **alcoholic fermentation,** pyruvate is converted to acetaldehyde, which is reduced to ethanol by NADH. Under aerobic conditions, pyruvate is oxidized to acetyl coenzyme A by NAD^+.

Key Reactions

1. β-Oxidation of Fatty Acids (Section 29.3)

A series of four enzyme-catalyzed reactions occurs, after each set of which the carbon chain of a fatty acid is shortened by two carbon atoms.

$$CH_3(CH_2)_{14}\overset{\overset{\displaystyle O}{\|}}{C}OH + 8CoA\!-\!SH + 7NAD^+ + 7FAD \xrightarrow{\;\;ATP\;\;AMP + P_2O_7^{4-}\;\;}$$

Hexadecanoic acid
(Palmitic acid)

$$8CH_3\overset{\overset{\displaystyle O}{\|}}{C}SCoA \;\; + \;\; 7NADH + 7FADH_2$$

Acetyl coenzyme A

2. Glycolysis (Section 29.5)

A series of ten enzyme-catalyzed reactions converts glucose to pyruvate.

$$C_6H_{12}O_6 \xrightarrow[\substack{\text{ten enzyme-}\\\text{catalyzed steps}}]{\text{glycolysis}} 2CH_3\overset{\overset{\displaystyle O}{\|}}{C}COO^- + 2H^+$$

Glucose   Pyruvate

3. Reduction of Pyruvate to Lactate — Lactate Fermentation (Section 29.7A)

$$CH_3\overset{\overset{\displaystyle O}{\|}}{C}COO^- + NADH + H_3O^+ \underset{}{\overset{\text{lactate}\atop\text{dehydrogenase}}{\rightleftharpoons}} CH_3\overset{\overset{\displaystyle OH}{|}}{C}HCOO^- + NAD^+ + H_2O$$

Pyruvate   Lactate

4. Reduction of Pyruvate to Ethanol — Alcohol Fermentation (Section 29.7B)

The carbon dioxide formed in this reaction is responsible for the foam on beer and the carbonation of naturally fermented wines and champagnes.

$$CH_3\overset{\overset{\displaystyle O}{\|}}{C}COO^- + NADH + 2H_3O^+ \longrightarrow CH_3CH_2OH + CO_2 + NAD^+ + 2H_2O$$

Pyruvate   Ethanol   Carbon
dioxide

5. Oxidative Decarboxylation of Pyruvate to Acetyl-CoA (Section 29.7C)

$$CH_3\overset{\overset{\displaystyle O}{\|}}{C}COO^- + NAD^+ + CoASH \xrightarrow{\substack{\text{oxidative}\\\text{decarboxylation}}} CH_3\overset{\overset{\displaystyle O}{\|}}{C}SCoA + CO_2 + NADH$$

Pyruvate   Acetyl-CoA

Problems

β-Oxidation

29.3 Write structural formulas for palmitic, oleic, and stearic acids, the three most abundant fatty acids.

29.4 A fatty acid must be activated before it can be metabolized in cells. Write a balanced equation for the activation of palmitic acid.

29.5 Name three coenzymes necessary for β-oxidation of fatty acids. From what vitamin is each derived?

29.6 We have examined β-oxidation of saturated fatty acids, such as palmitic acid and stearic acid. Oleic acid, an unsaturated fatty acid, is also a common component of dietary fats and oils. This unsaturated fatty acid is degraded by β-oxidation but, at one stage in its degradation, requires an additional enzyme named enoyl-CoA isomerase. Why is this enzyme necessary, and what isomerization does it catalyze?

Glycolysis

29.7 Name one coenzyme required for glycolysis. From what vitamin is it derived?

29.8 Number the carbon atoms of glucose 1 through 6 and show from which carbon atom of glucose the carboxyl group of each molecule of pyruvate is derived.

29.9 How many moles of lactate are produced from three moles of glucose?

29.10 Although glucose is the principal source of carbohydrates for glycolysis, fructose and galactose are also metabolized for energy.

  (a) What is the main dietary source of fructose? Of galactose?
  (b) Propose a series of reactions by which fructose might enter glycolysis.
  (c) Propose a series of reactions by which galactose might enter glycolysis.

29.11 How many moles of ethanol are produced per mole of sucrose through the reactions of glycolysis and alcoholic fermentation? How many moles of CO_2 are produced?

29.12 Glycerol derived from hydrolysis of triglycerides and phospholipids is also metabolized for energy. Propose a series of reactions by which the carbon skeleton of glycerol might enter glycolysis and be oxidized to pyruvate.

29.13 Ethanol is oxidized in the liver to acetate ion by NAD^+.

  (a) Write a balanced equation for this oxidation.
  (b) Do you expect the pH of blood plasma to increase, decrease, or remain the same as a result of metabolism of a significant amount of ethanol?

29.14 Write a mechanism to show the role of NADH in the reduction of acetaldehyde to ethanol.

29.15 When pyruvate is reduced to lactate by NADH, two hydrogens are added to pyruvate: one to the carbonyl carbon, the other to the carbonyl oxygen. Which of these hydrogens is derived from NADH?

29.16 Review the oxidation reactions of glycolysis and β-oxidation and compare the types of functional groups oxidized by NAD^+ with those oxidized by FAD.

29.17 Why is glycolysis called an anaerobic pathway?

29.18 Which carbons of glucose end up in CO_2 as a result of alcoholic fermentation?

29.19 Which steps in glycolysis require ATP? Which steps produce ATP?

29.20 The respiratory quotient (RQ) is used in studies of energy metabolism and exercise physiology. It is defined as the ratio of the volume of carbon dioxide produced to the volume of oxygen used:

$$RQ = \frac{\text{Volume } CO_2}{\text{Volume } O_2}$$

(a) Show that RQ for glucose is 1.00. (*Hint:* Look at the balanced equation for complete oxidation of glucose to carbon dioxide and water.)

(b) Calculate RQ for triolein, a triglyceride of molecular formula $C_{57}H_{104}O_6$.

(c) For an individual on a normal diet, RQ is approximately 0.85. Would this value increase or decrease if ethanol were to supply an appreciable portion of caloric needs?

29.21 Acetoacetate, β-hydroxybutyrate, and acetone are commonly known within the health sciences as ketone bodies, in spite of the fact that one of them is not a ketone at all. They are products of human metabolism and are always present in blood plasma. Most tissues, with the notable exception of the brain, have the enzyme systems necessary to use them as energy sources. Synthesis of ketone bodies occurs by the following enzyme-catalyzed reactions. Enzyme names are (1) thiolase, (2) β-hydroxy-β-methylglutaryl-CoA synthase, (3) β-hydroxy-β-methylglutaryl-CoA lyase, and (5) β-hydroxybutyrate dehydrogenase. Reaction (4) is spontaneous and uncatalyzed.

Describe the type of reaction involved in each step.

29.22 A connecting point between anaerobic glycolysis and β-oxidation is formation of acetyl-CoA. Which carbon atoms of glucose appear as methyl groups of acetyl-CoA? Which carbon atoms of palmitic acid appear as methyl groups of acetyl-CoA?

THERMODYNAMICS AND THE EQUILIBRIUM CONSTANT

For the equilibrium $A \rightleftharpoons B$ $K_{eq} = \dfrac{[B]}{[A]}$

$$\Delta G^0 = -2.303 \, RT \log K_{eq}$$

R = molar gas constant = 8.3145 J (1.987 cal) $\cdot$ mol$^{-1}\cdot$K^{-1}

T = degrees Kelvin = K

$$\% B = \dfrac{B}{A + B} \times 100$$

K_{eq}	ΔG^0 (kcal/mol)	log K_{eq}	% B in Mixture
1	0.00	0.00	50.00
2	−0.41	0.30	66.67
5	−0.95	0.70	83.33
10	−1.36	1.00	90.91
20	−1.77	1.30	95.24
100	−2.73	2.00	99.01
1,000	−4.09	3.00	99.90
10,000	−5.49	4.00	99.99

MAJOR CLASSES OF ORGANIC ACIDS

Class and Example	Typical pK_a	Class and Example	Typical pK_a
Sulfonic acid	0–1	β-Ketoester	11
		Water $HO-H$	15.7
Carboxylic acid	3–5	Alcohol CH_3CH_2O-H	15–19
Arylammonium ion	4–5	Amide	15–19
Imide	8–9	Cyclopentadiene	16
		α-Hydrogen of an aldehyde or ketone	18–20
Thiol CH_3CH_2S-H	8–12	α-Hydrogen of an ester	23–25
Phenol	9–10	Alkyne $HC\equiv C-H$	25
Ammonium ion NH_3-H^+	9.24	Ammonia NH_2-H	38
β-Diketone	10	Amine $[(CH_3)_2CH]_2N-H$	40
Nitroalkane $H-CH_2NO_2$	10	Alkene $CH_2{=}CH-H$	44
Alkylammonium ion $(CH_3CH_2)_3\overset{+}{N}-H$	10–12	Alkane CH_3CH_2-H	51

BOND DISSOCIATION ENERGIES

Bond dissociation energy (BDE) is defined as the amount of energy required to break a bond homolytically into two radicals in the gas phase at 25°C.

$$A \overset{\frown}{\frown} B \longrightarrow A\cdot + B\cdot \qquad \Delta H^0 [\text{kJ (kcal)/mol}]$$

Bond	ΔH^0	Bond	ΔH^0	Bond	ΔH^0
H—H bonds		**C—C multiple bonds**		**C—Br bonds**	
H—H	435 (104)	$CH_2{=}CH_2$	720 (172)	CH_3—Br	293 (70)
D—D	444 (106)	$HC{\equiv}CH$	837 (200)	C_2H_5—Br	285 (68)
				$(CH_3)_2CH$—Br	285 (68)
X—X bonds		**C—H bonds**		$(CH_3)_3C$—Br	285 (68)
F—F	159 (38)	CH_3—H	439 (105)	$CH_2{=}CHCH_2$—Br	230 (55)
Cl—Cl	247 (59)	C_2H_5—H	418 (100)	C_6H_5—Br	339 (81)
Br—Br	192 (46)	$(CH_3)_2CH$—H	402 (96)	$C_6H_5CH_2$—Br	243 (58)
I—I	151 (36)	$(CH_3)_3C$—H	389 (93)		
		$CH_2{=}CH$—H	444 (106)	**C—I bonds**	
H—X bonds		$CH_2{=}CHCH_2$—H	360 (86)	CH_3—I	238 (57)
H—F	569 (136)	C_6H_5—H	464 (111)	C_2H_5—I	222 (53)
H—Cl	431 (103)	$C_6H_5CH_2$—H	368 (88)	$(CH_3)_2CH$—I	226 (54)
H—Br	368 (88)	$HC{\equiv}C$—H	552 (132)	$(CH_3)_3C$—I	213 (51)
H—I	297 (71)			$CH_2{=}CHCH_2$—I	172 (41)
		C—F bonds		C_6H_5—I	272 (65)
O—H bonds		CH_3—F	452 (108)	$C_6H_5CH_2$—I	201 (48)
HO—H	498 (119)	C_2H_5—F	444 (106)		
CH_3O—H	435 (104)	$(CH_3)_2CH$—F	448 (107)	**C—N single bonds**	
		C_6H_5—F	527 (126)	CH_3—NH_2	356 (85)
O—O bonds				C_6H_5—NH_2	427 (102)
HO—OH	213 (51)	**C—Cl bonds**			
CH_3O—OCH_3	159 (38)	CH_3—Cl	356 (85)	**C—O single bonds**	
$(CH_3)_3CO$—$OC(CH_3)_3$	159 (38)	C_2H_5—Cl	335 (80)	CH_3—OH	385 (92)
		$(CH_3)_2CH$—Cl	339 (81)	C_6H_5—OH	464 (111)
C—C single bonds		$(CH_3)_3C$—Cl	331 (79)		
CH_3—CH_3	377 (90)	$CH_2{=}CHCH_2$—Cl	289 (69)		
C_2H_5—CH_3	368 (88)	C_6H_5—Cl	402 (96)		
$CH_2{=}CH$—CH_3	427 (102)	$C_6H_5CH_2$—Cl	301 (72)		
$CH_2{=}CHCH_2$—CH_3	305 (73)				
C_6H_5—CH_3	423 (101)				
$C_6H_5CH_2$—CH_3	314 (75)				

CHARACTERISTIC ^{1}H-NMR CHEMICAL SHIFTS

Type of Hydrogen (R = alkyl, Ar = aryl)	Chemical Shift (δ)*	Type of Hydrogen (R = alkyl, Ar = aryl)	Chemical Shift (δ)*
$(CH_3)_4Si$	0 (by definition)	$\underset{\displaystyle \ \ \ \ \parallel}{O}$ RCOCH$_3$	3.7–3.9
RCH$_3$	0.8–1.0		
RCH$_2$R	1.2–1.4	$\underset{\displaystyle \ \ \ \ \parallel}{O}$ RCOCH$_2$R	4.1–4.7
R$_3$CH	1.4–1.7		
R$_2$C=CRCHR$_2$	1.6–2.6	RCH$_2$I	3.1–3.3
RC≡CH	2.0–3.0	RCH$_2$Br	3.4–3.6
ArCH$_3$	2.2–2.5	RCH$_2$Cl	3.6–3.8
ArCH$_2$R	2.3–2.8	RCH$_2$F	4.4–4.5
ROH	0.5–6.0	ArOH	4.5–4.7
RCH$_2$OH	3.4–4.0	R$_2$C=CH$_2$	4.6–5.0
RCH$_2$OR	3.3–4.0	R$_2$C=CHR	5.0–5.7
R$_2$NH	0.5–5.0	ArH	6.5–8.5
$\underset{\displaystyle \ \ \ \ \parallel}{O}$ RCCH$_3$	2.1–2.3	$\underset{\displaystyle \ \ \ \parallel}{O}$ RCH	9.5–10.1
$\underset{\displaystyle \ \ \ \ \parallel}{O}$ RCCH$_2$R	2.2–2.6	$\underset{\displaystyle \ \ \ \ \parallel}{O}$ RCOH	10–13

*Values are relative to tetramethylsilane. Other atoms within the molecule may cause the signal to appear outside these ranges.

CHARACTERISTIC ^{13}C-NMR CHEMICAL SHIFTS

Type of Carbon	Chemical Shift (δ)	Type of Carbon	Chemical Shift (δ)
RCH$_3$	10–40	C—R (aromatic)	110–160
RCH$_2$R	15–55		
R$_3$CH	20–60	$\underset{\text{RCOR}}{\overset{\text{O}}{\parallel}}$	160–180
RCH$_2$I	0–40		
RCH$_2$Br	25–65	$\underset{\text{RCNR}_2}{\overset{\text{O}}{\parallel}}$	165–180
RCH$_2$Cl	35–80		
R$_3$COH	40–80	$\underset{\text{RCOH}}{\overset{\text{O}}{\parallel}}$	165–185
R$_3$COR	40–80		
RC≡CR	65–85	$\underset{\text{RCH, RCR}}{\overset{\text{O}\quad\text{O}}{\parallel\quad\parallel}}$	180–215
R$_2$C=CR$_2$	100–150		

CHARACTERISTIC INFRARED ABSORPTION FREQUENCIES

Bonding		Frequency (cm^{-1})	Intensity*	Type of Vibration (Stretching unless noted
C—H	Alkane	2850–3000	w–m	
	—CH$_3$	1375 and 1450	w–m	Bending
	—CH$_2$—	1450	m	Bending
	Alkene	3000–3100	w–m	
		650–1000	s	Out-of-plane bending
	Alkyne	~3300	s	
	Aromatic	3000–3100	s	
		690–900	s	Out-of-plane bending
	Aldehyde	2700–2800	w	
		2800–2900	w	
C≕C	Alkene	1600–1680	w–m	
	Aromatic	1450 and 1600	w–m	
C≡C	Alkyne	2100–2250	w–m	
C—O	Alcohol, ether, ester, carboxylic acid, anhydride	1000–1100 (sp^3 C—O)	s	
			s	
		1200–1250 (sp^3 C—O)	s	
C=O	Amide	1630–1680	s	
	Carboxylic acid	1700–1725	s	
	Ketone	1705–1780	s	
	Aldehyde	1705–1740	s	
	Ester	1735–1800	s	
	Anhydride	1760 and 1810	s	
	Acid chloride	1800	s	
O—H	Alcohol, phenol			
	Free	3600–3650	m	
	Hydrogen bonded	3200–3500	m	
	Carboxylic acid	2500–3300	m	
N—H	Amine and amide	3100–3550	m–s	
C≡N	Nitrile	2200–2250	m	

*m = medium, s = strong, w = weak

REAGENTS AND THEIR USES

Aluminum chloride, $AlCl_3$

    With arenes; catalyst for chlorination and bromination (Section 21.1A)

    With arenes; catalyst for Friedel-Crafts alkylations and acylations (Section 21.1C)

Amalgamated zinc/hydrochloric acid, $Zn(Hg)/HCl$

    With aldehydes and ketones; Clemmensen reduction of $C{=}O$ to CH_2 (Section 16.14C)

Amine, primary, RNH_2

    With acid anhydrides; formation of amides (Section 18.7B)

    With acid chlorides; formation of amides (Section 18.7A)

    With aldehydes and ketones; formation of imines (Section 16.10A)

    With epoxides; regioselective and stereoselective ring opening (Section 11.9B)

    With esters; formation of amides (Section 18.7C)

Amine, secondary, R_2NH

    With acid anhydrides; formation of amides (Section 18.7B)

    With acid chlorides; formation of amides (Section 18.7A)

    With aldehydes and ketones; formation of enamines (Section 16.10B)

    With epoxides; regioselective and stereoselective ring opening (Section 11.9B)

    With esters; formation of amides (Section 18.7C)

Ammonia, NH_3

    With acid anhydrides; formation of amides (Section 18.7B)

    With acid chlorides; formation of amides (Section 18.7A)

    With aldehydes and ketones; formation of imines (Section 16.10A)

    With epoxides; regioselective and stereoselective ring opening (Section 11.9B)

    With esters; formation of amides (Section 18.7C)

Azobisisobutyronitrile (AIBN)

    With substituted ethylenes; initiation of radical chain-growth polymerizations (Section 24.6A)

2,2-Bis(diphenylphosphanyl)-1,1′-binaphthyl (BINAP)

    With alkenes and $H_2 + RuCl_2$; enantioselective reduction of alkenes (Section 6.7B)

Boron trifluoride, BF_3

    With substituted ethylenes; initiation of cationic chain-growth polymerizations (Section 24.6D)

Bromine, Br_2

    With aldoses; oxidation to aldonic acids (Section 25.3D)

With alkanes; regioselective bromination (Section 7.4)

With alkenes + water; regiospecific and stereospecific formation of bromohydrins (Section 6.3E)

With alkenes; allylic bromination (Section 7.6)

With alkenes; stereospecific addition (Section 6.3D)

With alkylarenes; benzylic bromination (Section 20.6B)

With alkynes; stereoselective addition (Section 10.9A)

With arenes and a Lewis acid catalyst; bromination (Section 21.1A)

With ketones; formation of α-bromoketones (Section 16.12C)

With methyl ketones; the haloform reaction (Section 16.12C)

With primary amides; Hofmann rearrangement (Section 18.12)

N-Bromosuccinimide (NBS)

With alkenes; radical allylic bromination (Section 7.6)

With alkylarenes; radical benzylic bromination (Section 20.6B)

Butyllithium, BuLi

With alkyltriphenylphosphonium salts; formation of Wittig reagents (Section 16.7)

Carbon dioxide, CO_2

With Grignard reagents; formation of carboxylic acids (Section 17.5A)

With organolithium reagents; formation of carboxylic acids (Section 17.5A)

With phenoxide ions; Kolbe carboxylation (Section 20.5E)

Chlorine, Cl_2

With alkanes; regioselective chlorination (Sections 7.4 and 7.5)

With alkenes + water; regiospecific and stereospecific formation of chlorohydrins (Section 6.3E)

With alkenes; allylic chlorination (Section 7.6)

With alkenes; stereospecific addition (Section 6.3D)

With alkylarenes; benzylic chlorination (Section 20.6B)

With alkynes; stereoselective addition (Section 10.9A)

With arenes and a Lewis acid catalyst; chlorination (Section 21.1A)

With ketones; formation of α-chloroketones (Section 16.12C)

With methyl ketones; the haloform reaction (Section 16.12C)

Chromic acid, H_2CrO_4

With 1° alcohols; oxidation to carboxylic acids (Section 9.8A)

With 2° alcohols; oxidation to ketones (Section 9.8A)

With aldehydes; oxidation to carboxylic acids (Section 16.13A)

With alkylarenes; oxidation to arenecarboxylic acids (Section 20.6A)

With phenols; oxidation to *p*-quinones (Section 20.5F)

Copper(I) iodide, CuI

With RLi; formation of lithium diorganocopper (Gilman) reagents (Section 15.2A)

Cyanogen bromide, BrCN

With polypeptides; cleavage of peptide bonds at the carboxyl group of methionine (Section 27.4B)

Di(*sec*-isoamyl)borane, $(sia)_2BH$

With terminal alkynes; regioselective and stereoselective hydroboration (Section 10.8)

Diazomethane, CH_2N_2

With carboxylic acids; formation of methyl esters (Section 17.7C)

Diborane, B_2H_6

With alkenes; regioselective and stereospecific hydroboration (Section 6.4)

With internal alkynes; stereoselective hydroboration (Section 10.8)

Dicyclohexylcarbodiimide (DCC)

With carboxyl and amino groups; formation of peptide bonds (Section 27.5E)

Dihydropyran (DHP)

With alcohols; formation of tetrahydropyranyl (THP) protecting group (Section 16.8C)

Diiodomethane (methylene iodide), CH_2I_2

With Zn(Cu); formation of Simmons-Smith reagent, ICH_2ZnI (Section 15.4D)

Diisobutylaluminum hydride (DIBALH)

With esters; reduction to aldehydes (Section 18.11A)

Dimethyl sulfate, $(CH_3)_2SO_4$

With alcohols and $NaNH_2$ or NaH; formation of methyl ethers (Section 25.4B)

With phenols; formation of methyl ethers (Section 20.5D)

1,2-Ethanediol (ethylene glycol), $HO(CH_2)_2OH$

With aldehydes and ketones; formation of a carbonyl protecting group (Section 16.8)

Hydrazine, N_2H_4

With aldehydes and ketones and KOH; Wolff-Kishner reduction C=O to CH_2 (Section 16.14C)

Hydrogen bromide, HBr

With 1°, 2°, and 3° alcohols; formation of alkyl bromides (Section 9.5A)

With alkenes; regioselective addition (Section 6.3A)

With alkynes; regioselective addition (Section 10.9B)

With dialkyl and alkyl-aryl ethers; cleavage of the ether (Section 11.5A)

Hydrogen chloride, HCl

With 2° and 3° alcohols; formation of alkyl chlorides (Section 9.5A)

With alkenes; regioselective addition (Section 6.3A)

With alkynes; regioselective addition (Section 10.9B)

Hydrogen cyanide/potassium cyanide, HCN/KCN

With aldehydes and ketones; formation of cyanohydrins (Section 16.6D)

Hydrogen iodide, HI

With 1°, 2°, and 3° alcohols; formation of alkyl iodides (Section 9.5A)

With alkenes; regioselective addition (Section 6.3A)

With dialkyl and alkyl-aryl ethers; ether cleavage (Section 11.5A)

Hydrogen peroxide, H_2O_2

With 3° amines; oxidation to amine *N*-oxides (Section 22.11)

With aldehydes; oxidation to carboxylic acids (Section 16.13A)

With trialkylboranes; oxidation to alcohols (Section 6.4)

Hydrogen, H_2

With aldehydes and ketones; catalytic reduction to alcohols (Section 16.14A)

With aldoses and ketoses; catalytic reduction (Section 25.3C)

With alkenes + BINAP; enantioselective reduction to alkanes (Section 6.7B)

With alkenes; catalytic reduction to alkanes (Section 6.6A)

With alkynes and Lindlar catalyst; stereoselective reduction to E alkenes (Section 10.7A)

With alkynes; catalytic reduction to alkanes (Section 10.7A)

With benzyl ethers; catalytic hydrogenolysis (Section 20.6C)

With enamines; catalytic reduction to amines (Section 16.10A)

Iron(III) chloride (ferric chloride), $FeCl_3$

Lewis acid catalyst for chlorination and bromination of arenes (Section 21.1A)

Lithium, Li

With alcohols; formation of lithium alkoxides (Section 9.4)

With alkyl and aryl halides; formation of organolithium reagents (Section 15.2)

With Cu(I) and alkyl and alkyl halides; formation of Gilman reagents (Section 15.2)

Lithium aluminum hydride (LAH), $LiAlH_4$

With aldehydes and ketones; reduction to alcohols (Section 16.14B)

With amides; reduction to amines (Section 18.11B)

With carboxylic acids; reduction to primary alcohols (Section 17.6A)

With epoxides; regioselective reduction to alcohols (Section 11.9B)

With esters; reduction to two alcohols (Section 18.11A)

With nitriles; reduction to 1° amines (Section 18.11C)

Lithium diorganocopper(I); alternatively, a Gilman reagent, R_2CuLi

With acid chlorides; formation of ketones (Section 18.9C)

With alkyl, vinylic, and aryl halides; stereoselective coupling (Section 15.2B)

With epoxides; regioselective nucleophilic ring opening (Section 15.2C)

With α,β-unsaturated carbonyl compounds; conjugate addition (Section 19.8B)

Lithium diisopropylamide (LDA)

And directed aldol reactions (Section 19.2B)

With aldehydes, ketones, and esters; formation of lithium enolates (Section 19.2A)

Magnesium, Mg

With alkyl and aryl halides; formation of Grignard reagents (Section 15.1A)

Mercury(II) acetate (mercuric acetate) $Hg(OAc)_2$

With alkenes; regioselective oxymercuration (Section 6.3F)

Mercury(II) sulfate (mercuric sulfate), $HgSO_4$

With alkynes and H_2SO_4; hydration to acetaldehyde or ketones (Section 10.9C)

Methanesulfonyl chloride (MsCl), CH_3SO_2Cl

With alcohols; formation of methanesulfonate esters (Section 9.5D)

Nitric acid, HNO_3

With arenes; nitration (Section 21.1B)

Nitrous acid, HNO_2

See sodium nitrite, $NaNO_2/HCl$

Organolithium reagent, RLi

With aldehydes and ketones; formation of alcohols (Section 16.6B)

With carbon dioxide; formation of carboxylic acids (Section 17.5A)

With CuI; formation of lithium diorganocopper (Gilman) reagents (Section 15.2A)

With epoxides; formation of alcohols (Section 15.1C)

With esters; formation of alcohols (Section 18.9B)

With protic acids: formation of alkanes or arenes (Section 15.1B)

Organomagnesium reagent (Grignard reagent), RMgX

With aldehydes and ketones; formation of alcohols (Section 16.6A)

With carbon dioxide; formation of carboxylic acids (Section 17.5A)

With epoxides formation of alcohols (Section 15.1C)

With esters; formation of alcohols (Section 18.9A)

With protic acids: formation of alkanes or arenes (Section 15.1B)

Osmium tetroxide, OsO_4

With alkenes; syn stereoselective oxidation to glycols (Section 6.5B)

Oxygen, O_2

With aldehydes; oxidation to carboxylic acids (Section 16.13A)

With ethers; oxidation to hydroperoxides (Section 11.5B)

With thiols; oxidation to disulfides (Section 9.9F)

Ozone, O_3

With alkenes; oxidative cleavage to two carbonyl-containing compounds (Section 6.5C)

Palladium, Pd

With alkenes + H_2; catalyst for reduction to alkanes (Section 6.6A)

With alkynes + H_2; catalyst for reduction to alkanes (Section 10.7A)

With aldehydes and ketones + H_2; catalyst for reduction to alcohols (Section 16.14A)

With a haloalkane or haloarene and alkenes; catalyst for the Heck reaction (Section 15.3)

Periodic acid, HIO_4

With glycols; oxidative cleavage to two carbonyl-containing compounds (Section 9.8C)

With α-hydroxyaldehydes and α-hydroxyketones; oxidative cleavage (Section 25.3F)

Peroxycarboxylic acid (a peracid), $ArCO_3H$ or RCO_3H

With alkenes; stereoselective oxidation to epoxides (Section 11.8B)

Phenyl isothiocyanate, C_6H_5NCS

With polypeptides; Edman degradation of *N*-terminal amino acid (Section 27.4B)

Phosphoric acid, H_3PO_4

With alcohols and arenes; catalyst for alkylation of arenes (Section 21.1D)

With alcohols; catalyst for intermolecular dehydration to ethers (Section 11.4B)

With alcohols; catalyst for regioselective dehydration (Section 9.6)

With alkenes and arenes; catalyst for alkylation of arenes (Section 21.1D)

Phosphorus tribromide, PBr_3

With alcohols; formation of alkyl bromides (Section 9.5B)

Potassium, K

With alcohols; formation of potassium alkoxides (Section 9.4)

1,3-Propanedithiol, $HS(CH_2)_3SH$

With aldehydes; formation of 1,3-dithianes (Section 16.9)

Pyridinium chlorochromate (PCC)

With 1° alcohols; oxidation to aldehydes (Section 9.8B)

With 2° alcohols; oxidation to ketones (Section 9.8B)

Ruthenium-nucleophilic carbene complexes

With alkenes; catalyst for alkene metathesis reactions (Section 15.4E)

Silver oxide, Ag_2O

With aldehydes; oxidation to carboxylic acids (Section 16.13A)

Sodium, Na

With alcohols; formation of sodium alkoxides (Section 9.4)

With $NH_3(l)$; stereoselective reduction of alkynes to Z alkenes (Section 10.7B)

Sodium amide, $NaNH_2$

With alkyltriphenylphosphonium salts; formation of Wittig reagents (Section 16.7)

With haloarenes; formation of benzyne intermediates (Section 21.3A)

With terminal alkynes; formation of alkyne anions (Section 10.4)

Sodium borohydride, $NaBH_4$

With products of oxymercuration; reduction to alcohols (Section 6.4)

With aldehydes and ketones; reduction to alcohols (Section 16.14B)

Sodium nitrite, $NaNO_2/HCl$

With 1° alkylamines; formation of unstable alkanediazonium salts (Section 22.9D)

With 1° arylamines; formation of arenediazonium salts (Section 22.9E)

With 2° amines; formation of *N*-nitrosoamines (Section 22.9C)

With β-aminoalcohols; Tiffeneau-Demjanov rearrangement (Section 22.9D)

With 3° aromatic amines; formation of nitrosoarenes (Section 22.9B)

Sulfuric acid, H_2SO_4

With alcohols; catalyst for intermolecular dehydration to ethers (Section 11.4B)

With alcohols; catalyst for regioselective dehydration (Section 9.6)

With alkenes; catalyst for regioselective hydration (Section 6.3B)

With alkynes and $HgSO_4$; catalysts for regioselective hydration (Section 10.9C)

With arenes; sulfonation (Section 21.1B)

With glycols; catalyst for pinacol rearrangements (Section 9.7)

Tetrabutylammonium chloride, Bu_4NCl

In nucleophilic substitutions; a phase-transfer catalyst (Section 8.7)

Tetrabutylammonium fluoride, Bu_4NF

With trimethylsilyl (TMS) ethers; cleavage of the silyl ether (Section 11.6B)

Thionyl chloride, $SOCl_2$

With alcohols; formation of alkyl chlorides (Section 9.5C)

With carboxylic acids; formation of acid chlorides (Section 17.8)

Tin(IV) chloride (stannic chloride), $SnCl_4$

Lewis acid catalyst for Friedel-Crafts acylations (Section 21.1C)

Tollens' reagent, $Ag(NH_3)_2^+$

With aldehydes; oxidation to carboxylic acids (Section 16.13A)

With aldoses and ketones; oxidation to aldonic acids (Section 25.3D)

p-Toluenesulfonyl chloride (TsCl), *p*-$CH_3C_6H_4SO_2Cl$

With alcohols; formation of *p*-toluenesulfonate esters (Section 9.5D)

Tribromomethane (bromoform), $CHBr_3$

With strong base; formation of dibromocarbene, CBr_2 (Section 15.4B)

Trichloromethane (chloroform), $CHCl_3$

With strong base; formation of dichlorocarbene, CCl_2 (Section 15.4B)

Trimethylsilyl chloride (TMSCl), $(CH_3)_3SiCl$

With alcohols; formation of trimethylsilyl ethers as alcohol-protecting groups (Section 11.6B)

Triphenylphosphine, $(C_6H_5)_3P$

With alkyl halides; triphenylphosphonium salts and Wittig reagents (Section 16.7)

Zinc/copper couple, $Zn(Cu)$

With CH_2I_2; formation of the Simmons-Smith reagent, $IZnCH_2I$ (Section 15.4C)

SUMMARY OF METHODS FOR THE SYNTHESIS OF FUNCTIONAL GROUPS

Acetals

1. Reaction of an aldehyde or ketone with an alcohol (Section 16.8B)

Acyl halides

1. Reaction of a carboxylic acid with $SOCl_2$ (Section 17.8)

Alcohols

1. Acid-catalyzed hydration of an alkene (Section 6.3B)
2. Oxymercuration-reduction of an alkene (Section 6.3F)
3. Hydroboration-oxidation of an alkene (Section 6.4)
4. Reaction of an alkyl halide with OH^- (Section 8.3)
5. Reduction of an epoxide by $LiAlH_4$ (Section 11.9B)
6. Reaction of a Grignard reagent with an epoxide (Section 15.1C)
7. Reaction of an organolithium reagent with an epoxide (Section 15.1C)
8. Reaction of Grignard reagent with an aldehyde or ketone (Section 16.6A)
9. Reaction of an organolithium reagent with an aldehyde or ketone (Section 16.6B)
10. Reaction of the anion of a terminal alkyne with an aldehyde or ketone (Section 16.6C)
11. Catalytic reduction of an aldehyde or ketone (Section 16.14A)
12. Metal hydride reduction of an aldehyde or ketone (Section 16.14B)
13. Reduction of an ester with lithium aluminum hydride (Section 17.6A)
14. Reaction of an ester with a Grignard or organolithium reagent (Section 18.9A)

Aldehydes

1. Oxidation of an alkene by ozone followed by reductive workup (Section 6.5C)
2. Pinacol rearrangement of a glycol (Section 9.7)
3. Oxidation of a primary alcohol with PPC (Section 9.8)
4. Oxidative cleavage of a glycol with HIO_4 (Section 9.8C)
5. Hydroboration-oxidation of a terminal alkyne using $(sia)_2BH$ (Section 10.8)
6. Reduction of an ester using diisobutylaluminum hydride (Section 18.11A)

Alkanes

1. Catalytic reduction of an alkene (Section 6.6)
2. Enantioselective reduction of an alkene (Section 6.7)
3. Catalytic reduction of an alkyne (Section 10.7A)

4. Reaction of a Gilman reagent with an alkyl halide (Section 15.3B)

5. Clemmensen reduction of an aldehyde or ketone (Section 16.14C)

6. Wolff-Kishner reduction of an aldehyde or ketone (Section 16.14C)

Alkenes

1. Elimination of HX from an alkyl halide (Section 8.9)

2. Acid-catalyzed dehydration of an alcohol (Section 9.6)

3. Reduction of an alkyne by H_2 and Lindlar catalyst to a cis alkene (Section 10.7A)

4. Reduction of an alkyne by Na or Li in NH_3(l) to a trans alkene (Section 10.7B)

5. Wittig reaction (Section 16.7)

6. Hofmann elimination of a quaternary ammonium hydroxide (Section 22.10)

7. Cope elimination by pyrolysis of a tertiary amine oxide (Section 22.11)

Alkyl halide

1. Addition of HX to an alkene (Section 6.3A)

2. Addition of bromine and chlorine to an alkene (Section 6.3D)

3. Chlorination and bromination of an alkane (Section 7.4)

4. Bromination or chlorination at an allylic position (Section 7.6)

5. Reaction of an alcohol with HCl, HBr, and HI (Section 9.5A)

6. Reaction of an alcohol with PBr_3 (Section 9.5B)

7. Reaction of an alcohol with $SOCl_2$ (Section 9.5C)

8. Acid-catalyzed cleavage of a dialkyl or alkyl aryl ether (Section 11.5A)

9. Bromination or chlorination at a benzylic position (Section 20.6B)

Alkynes

1. Double dehydrohalogenation of a geminal- or vecinal-dihalide (Section 10.6)

2. Alkylation of an acetylide anion (Section 10.5)

Amides

1. Reaction of acid chloride with ammonia or a 1° or 2° amine (Section 18.7A)

2. Reaction of an acid anhydride with ammonia or a 1° or 2° amine (Section 18.7B)

3. Reaction of an ester with ammonia or a 1° or 2° amine (Section 18.7C)

Amines

1. Reaction of an alkyl halide with NH_3, RNH_2, or R_2NH (Section 8.3)

2. Reaction of an epoxide with ammonia or a 1° or 2° amine (Section 11.9B)

3. Reduction of an imine (Section 16.10A)

4. Reduction of an amide (Section 18.11B)

5. Reduction of a nitrile (Section 18.11C)

6. Hofmann rearrangement of a primary amide (Section 18.12)

7. Alkylation of azide ion followed by reduction (Section 22.8B)

8. Reduction of a nitroarene (Section 21.1B)

Anhydrides

1. Reaction of an acid chloride with a carboxylic salt (Section 18.8)

Aryl halides

1. Reaction of an arene with Cl_2 or Br_2 and a Lewis acid catalyst (Section 21.1A)

Carboxylic acids

1. Oxidation of a primary alcohol (Section 9.8)

2. Haloform reaction of methyl ketones (Section 16.12B)

3. Reaction of a Grignard or organolithium reagent with CO_2 (Section 17.5A)
4. Hydrolysis of an acid chloride, anhydride, ester, or amide (Sections 18.5A–18.5D)
5. Hydrolysis of a nitrile (Section 18.5E)
6. Malonic ester synthesis (Section 19.7)
7. Kolbe synthesis: carboxylation of a phenol (Section 20.5E)
8. Oxidation at a benzylic position (Section 20.6A)

Cyanohydrins

1. Reaction of an aldehyde or ketone with HCN/KCN (Section 16.6D)

1,2-Diols

1. Oxidation of an alkene by OsO_4 (Section 6.5B)
2. Acid-catalyzed hydrolysis of an epoxide (Section 11.9A)

Disulfides

1. Oxidation of a thiol (Section 9.9F)

Enamines

1. Reaction of an aldehyde or ketone with a 2° amine (Section 16.10A)

Epoxides

1. Oxidation of an alkene by a peroxycarboxylic acid (Section 11.8B)
2. Reaction of a halohydrin with base (Section 11.8C)
3. Sharpless asymmetric epoxidation of an alkene (Section 11.8D)
4. Reaction of an aldehyde or ketone with a sulfonium ylide (Problem 16.28)

Esters

1. Reaction of an alkyl halide with a carboxylic ion (Section 8.3)
2. Fischer esterification (Section 17.7A)
3. Reaction of a carboxylic acid with diazomethane (Section 17.7C)
4. Reaction of an acid chloride with an alcohol (Section 18.6A)
5. Reaction of an acid anhydride with an alcohol (Section 18.6B)

Ethers

1. Williamson ether synthesis (Section 11.4A)
2. Acid-catalyzed dehydration of an alcohol (Section 11.4B)
3. Acid-catalyzed addition of an alcohol to an alkene (Section 11.4C)

Halohydrins

1. Addition of Cl_2/H_2O or Br_2/H_2O to an alkene (Section 6.3E)

α-Haloketones

1. Reaction of a ketone with Cl_2 or Br_2 (Section 16.12B)

Imines

1. Reaction of an aldehyde or ketone with NH_2 or RNH_2 (Section 16.10A)

Ketones

1. Oxidation of an alkene by ozone (Section 6.5C)
2. Pinacol rearrangement of a glycol (Section 9.7)
3. Oxidation of a secondary alcohol (Section 9.8)
4. Oxidative cleavage of a glycol (Section 9.8C)
5. Hydroboration-oxidation of an alkyne (Section 10.8)
6. Acid-catalyzed hydration of an alkyne (Section 10.9C)
7. Reaction of an acid chloride with a Gilman reagent (Section 18.9C)

8. Acetoacetic ester synthesis (Section 19.6)
9. Friedel-Crafts acylation of an arene (Section 21.1C)

New carbon-carbon bonds

1. Alkylation of an acetylide anion (Section 10.5)
2. Reaction of a Gilman reagent with an alkyl, aryl, or alkenyl halide (Section 15.2B)
3. Heck reaction (Section 15.3)
4. Reaction of dichloro- or dibromocarbene with an alkene (Section 15.4B)
5. Simmons-Smith reaction (Section 15.4C)
6. Alkene metathesis (Section 15.5)
7. Reaction of Grignard reagent with an aldehyde or ketone (Section 16.6A)
8. Reaction of an organolithium reagent with an aldehyde or ketone (Section 16.6B)
9. Reaction of the anion of a terminal alkyne with an aldehyde or ketone (Section 16.6C)
10. Reaction of an aldehyde or ketone with HCN/KCN (Section 16.6D)
11. Wittig reaction (Section 16.7)
12. Alkylation of the anion derived from an aldehyde 1,3-dithiane (Section 16.9)
13. Reaction of an acid chloride with a Gilman reagent (Section 18.9C)
14. Aldol reaction (Section 19.1)
15. Directed aldol reaction (Section 19.2B)
16. Claisen condensation (Section 19.3A)
17. Dieckmann condensation (Section 19.3B)
18. Crossed Claisen condensation (Section 19.3C)
19. Alkylation of an enamine (Section 19.5A)
20. Acylation of an enamine (Section 19.5B)
21. Acetoacetic ester synthesis (Section 19.6)
22. Malonic ester synthesis (Section 19.7)
23. Michael reaction (Section 19.8A)
24. Robinson annulation (Section 19.8A)
25. Addition of Gilman reagent to an α,β-unsaturated carbonyl compound (Section 19.8B)
26. Kolbe synthesis: carboxylation of a phenol (Section 20.5E)
27. Friedel-Crafts alkylation of an arene (Section 21.1C)
28. Friedel-Crafts acylation of an arene (Section 21.C)
29. Alkylation of an arene using an alkene (Section 21.1D)
30. Alkylation of an arene using an alcohol (Section 21.1D)
31. Sandmeyer reaction using CuCN (Section 22.9E)
32. Diels-Alder reaction (Section 23.3)
33. Claisen rearrangement of an allyl phenyl ether (Section 23.4A)
34. Cope rearrangement of a 1,5-diene (Section 23.4B)

Nitriles

1. Reaction of an alkyl halide with CN^- (Section 8.3)
2. Reaction of an aldehyde or ketone with HCN/KCN (Section 16.6D)

Sulfides

1. Reaction of an alkyl halide with RS^- (Section 8.3)

Thiols

1. Reaction of an alkyl halide with HS^- (Section 8.3)

Appendix 9

ELECTROSTATIC POTENTIAL MAPS

The term **electronic structure** refers to the distribution of electron density in a molecule. According to the laws of quantum mechanics, electrons have no definite locations. Instead, they collectively produce a negatively charged region around a nucleus, measured by electron density in units of $e/Å^3$ (electrons per cubic Angstrom). For an atom, density is high near the nucleus and vanishingly small far from the nucleus. Electron density maps are now easily computed for small molecules using desktop computers and various software packages. Those in this text were done using MacSpartan (Wavefunction, Inc.) then modified by the artists (Woolsey Associates) for better visibility.

Electrostatic potential (elpot) maps provide a way to visualize the distribution of electron density in a molecule. **Electrostatic potential** is defined as the potential energy that a positively charged particle would experience in a molecule's presence. The electrostatic potential is made up of two parts.

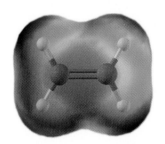

1. The repulsive component (positive potential, repulsion) exerted by the positively charged nuclei
2. The attractive component (negative potential, attraction) exerted by the negatively charged electrons cloud

Thus, electrostatic potential contains information about the entire electron distribution.

Electrostatic potential maps are color-coded. By convention, the most negative potential is red and the most positive potential is blue. Intermediate potentials are coded accordingly (red-orange-yellow-green-blue). While any surface might be chosen to display an electrostatic potential map, the most common is $0.002 \ e/Å^3$. Nearly all of a molecule's electron density lies within this surface, which corresponds almost exactly to how closely another molecule can approach without running into severe steric repulsive forces; that is, the surface corresponds almost exactly to the van der Waals surface of the molecule.

Figure 1 shows an electrostatic potential map for ethylene (top and side views). You can see areas of high electron density to which electrophiles will be attracted (red) over the pi orbitals. There are four blue patches, one over each hydrogen; these regions are relatively electron-poor.

Figure 1
Electrostatic potential map for ethylene.

A.17

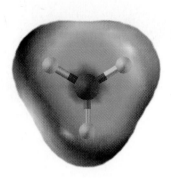

Figure 2

Electrostatic potential map for
methyl carbocation

The methyl carbocation provides an even more dramatic visualization of the electrostatic potential. Figure 2 shows that the central atom (the C^+) is blue, or positively charged, whereas the hydrogen nuclei are green, or somewhat positive. Only at the outside do they become red, or negative.

GLOSSARY

Absorbance (A) (Section 23.5A) A quantitative measure of the extent to which a compound absorbs radiation of a particular wavelength. $A = \log I_0/I$ where I_0 is the incident radiation and I is the transmitted radiation.

Acetal (Section 16.8B) A molecule containing two —OR or —OAr groups bonded to the same carbon.

Aceto group (Section 17.2B) A CH_3CO— group; also called an acetyl group.

Achiral (Section 3.2) An object that lacks chirality; an object that has no handedness.

Activating group (Section 21.1A) Any substituent on a benzene ring that causes the rate of electrophilic aromatic substitution to be greater than that for benzene.

Activation energy (Section 6.2A) The difference in energy between reactants and the transition state.

Acyl group (Section 18.1A) An RCO— or ArCO— group.

Acylation (Section 19.5) The process of introducing an acyl group, RCO— or ArCO—, onto an organic molecule.

Acylium ion (Section 21.1C) A resonance-stabilized cation with the structure $[RC{=}O]^+$ or $[ArC{=}O]^+$. The positive charge is delocalized over both the carbonyl carbon and the carbonyl oxygen.

Addition (Section 6.1) A reaction in which two atoms or ions react with a double bond, forming a compound with the two new groups bonded to the carbons of the original double bond.

Alcohol (Section 1.3A) A compound containing an —OH (hydroxyl) group bonded to an sp^3-hybridized carbon.

Alcoholic fermentation (Section 29.7B) A metabolic pathway that converts glucose to two molecules of ethanol and two molecules of CO_2.

Aldehyde (Section 1.3C) A compound containing a —CHO group.

Alditol (Section 25.3B) The product formed when the $C{=}O$ group of a monosaccharide is reduced to a CHOH group.

Aldonic acid (Section 25.3C) The product formed when the —CHO group of an aldose is oxidized to a —COOH group.

Aldose (Section 25.1A) A monosaccharide containing an aldehyde group.

Aliphatic amine (Section 22.1) An amine in which nitrogen is bonded only to alkyl groups.

Aliphatic hydrocarbon (Section 2.1) An alternative term to describe an alkane.

Alkaloid (Section 22.1) A basic nitrogen-containing compound of plant origin, many of which are physiologically active when administered to humans.

Alkane (Section 2.1) A saturated hydrocarbon whose carbon atoms are arranged in an open chain.

Alkoxy group (Section 11.2) An —OR group, where R is an alkyl group.

Alkyl group (Section 2.3A) A group derived by removing a hydrogen from an alkane.

Alkylation (Section 10.5) Any reaction in which a new carbon-carbon bond to an alkyl group is formed.

Alkyne (Chapter 10 introduction) An unsaturated hydrocarbon that contains one or more carbon-carbon triple bonds.

Allene (Section 10.6) The compound $CH_2{=}C{=}CH_2$. Any compound that contains adjacent carbon-carbon double bonds; that is, the arrangement $C{=}C{=}C$.

Allylic (Section 8.4B) Next to a carbon-carbon double bond.

Allylic carbocation (Section 8.4E) A carbocation in which an allylic carbon bears the positive charge.

Allylic carbon (Section 7.6) A carbon adjacent to a carbon-carbon double bond.

Allylic substitution (Section 7.6) Any reaction in which an atom or group of atoms is substituted for another atom or group of atoms at an allylic carbon.

Amino acid (Section 27.1A) A compound that contains both an amino group and a carboxyl group.

α-Amino acid (Section 27.1A) An amino acid in which the amino group is on the carbon adjacent to the carboxyl group.

Amino group (Section 1.3B) A compound containing an sp^3-hybridized nitrogen atom bonded to one, two, or three carbon atoms.

Amorphous domain (Section 24.4) A disordered, noncrystalline region in the solid state of a polymer.

Anabolic steroid (Section 26.4A) A steroid hormone, such as testosterone, that promotes tissue and muscle growth and development.

Androgen (Section 26.4A) A steroid hormone, such as testosterone, that mediates the development of sexual characteristics of males.

Angle strain (Section 2.6B) The strain that arises when a bond angle is either compressed or expanded compared to its normal value.

Anion (Section 1.2B) An atom or group of atoms bearing a negative charge.

Annulene (Section 20.2B) A cyclic hydrocarbon with a continuous alternation of single and double bonds.

Anomeric carbon (Section 25.2A) The hemiacetal or acetal carbon of the cyclic form of a carbohydrate.

Anomers (Section 25.2A) Carbohydrates that differ in configuration only at their anomeric carbons.

Anti conformation (Section 2.6A) A conformation about a single bond of an alkane in which the groups lie at a dihedral angle of 180°.

Anti stereoselectivity (Section 6.3D) The addition of atoms or groups of atoms to opposite faces of a carbon-carbon double bond.

Antiaromatic compound (Section 20.2C) A monocyclic compound that is planar or nearly so, has one p orbital on each atom of the ring, and has $4n$ pi electrons in the cyclic arrangement of overlapping p orbitals, where n is an integer. Antiaromatic compounds are especially unstable.

Antibonding molecular orbital (Section 1.8A) A molecular orbital in which electrons have a higher energy than they would in isolated atomic orbitals.

Aprotic acid (Section 4.5) An acid that is not a proton donor; an acid that is an electron pair acceptor in a Lewis acid-base reaction.

Aprotic solvent (Section 8.2) A solvent that cannot serve as a hydrogen bond donor; nowhere in the molecule is a hydrogen bonded to an atom of high electronegativity. Common aprotic solvents are dichloromethane, diethyl ether, and dimethylsulfoxide.

Aramid (Section 24.5A) A poly*aromatic amide*; a polymer in which the monomer units are an aromatic diamine and an aromatic dicarboxylic acid.

Arene (Chapter 5 introduction) A term used to classify benzene and its derivatives.

Aromatic amine (Section 22.1) An amine in which nitrogen is bonded to one or more aryl groups.

Aromatic compound (Chapter 20 introduction) A term used initially to classify benzene and its derivatives. More accurately, it is used to classify any compound that meets the Hückel criteria for aromaticity (Section 20.2A).

Aryl group (Ar—) (Chapter 5 introduction) A group derived from an arene by removal of an H. Given the symbol Ar—.

Atactic polymer (Section 24.6C) A polymer with completely random configurations at the stereocenters along its chain, as for example atactic polypropylene.

Aufbau principle (Section 1.1A) Orbitals fill in order of increasing energy, from lowest to highest.

Axial position (Section 2.6B) A position on a chair conformation of a cyclohexane ring that extends from the ring parallel to the imaginary axis of the ring.

Azeotrope (Section 16.8B) A liquid mixture of constant composition with a boiling point that is different from that of any of its components.

Base peak (Section 14.1) The peak due to the most abundant ion in a mass spectrum; the most intense peak. It is assigned an arbitrary intensity of 100.

Basicity (Section 8.4B) An equilibrium property measured by the position of equilibrium in an acid-base reaction, as for example the acid-base reaction between ammonia and water.

Benzyl group $C_6H_5CH_2$— (Section 20.3A) The group derived from toluene by removing a hydrogen from its methyl group.

Benzylic position (Section 20.6) An sp^3-hybridized carbon attached to a benzene ring.

Benzyne intermediate (Section 21.3A) A reactive intermediate formed by β-elimination from adjacent carbon atoms of a benzene ring and having a triple bond in the benzene ring. The second pi bond of the benzyne triple bond is formed by overlap of coplanar sp^2 orbitals on adjacent carbons.

Betaine (Section 16.7) A neutral molecule with non-adjacent positive and negative charges. An example of a betaine is the intermediate formed by addition of a Wittig reagent to an aldehyde or ketone.

Bicycloalkane (Section 2.4B) An alkane containing two rings that share two carbons.

Bile acid (Section 26.4A) A cholesterol-derived detergent molecule, such as cholic acid, which is secreted by the gallbladder into the intestine to assist in the absorption of dietary lipids.

Bimolecular reaction (Section 8.3) A reaction in which two species are involved in the rate-determining step.

Boat conformation (Section 2.6B) A nonplanar conformation of a cyclohexane ring in which carbons 1 and 4 of the ring are bent toward each other.

Bond dipole moment (μ) (Section 1.2C) A measure of the polarity of a covalent bond. The product of the charge on either atom of a polar bond times the distance between the atoms.

Bond length (Section 1.2C) The distance between atoms in a covalent bond; now most commonly given in picometers (pm; $1 \text{ pm} = 10^{-12} \text{ m}$).

Bonding electrons (Section 1.2D) Valence electrons involved in forming a covalent bond (i.e., shared electrons).

Bonding molecular orbital (Section 1.8A) A molecular orbital in which electrons have a lower energy than they would in isolated atomic orbitals.

Brønsted-Lowry acid (Section 4.1) A proton donor.

Brønsted-Lowry base (Section 4.1) A proton acceptor.

Carbanion (Section 16.6A) An anion in which carbon has an unshared pair of electrons and bears a negative charge.

Carbene (Section 15.4) A neutral molecule that contains a carbon atom surrounded by only six valence electrons (R_2C:).

Carbenoid (Section 15.4) A compound that delivers the elements of a carbene without actually producing a free carbene.

Carbocation (Section 6.3A) A species in which a carbon atom has only six electrons in its valence shell and bears a positive charge.

Carbohydrate (Chapter 25 introduction) A polyhydroxyaldehyde or polyhydroxyketone, or a substance that gives these compounds on hydrolysis.

α-Carbon (Section 16.11A) A carbon atom adjacent to a carbonyl group.

Carbonyl group (Section 1.3C) A C=O group.

Carboxyl group (Section 1.3D) A —COOH group.

Carboxylic acid (Section 1.3D) A compound containing a carboxyl, —COOH, group.

Carboxylic ester (Section 1.3E) A derivative of a carboxylic acid in which H of the carboxyl group is replaced by a carbon group.

Cation (Section 1.2B) An atom or group of atoms bearing a positive charge.

Center of symmetry (Section 3.2) A point so situated that identical components of an object are located on opposite sides and equidistant from that point along any axis passing through it.

Chain initiation (Section 7.5B) A step in a chain reaction characterized by the formation of reactive intermediates (radicals, anions, or cations) from nonradical or noncharged molecules.

Chain length (Section 7.5B) The number of times the cycle of chain propagation steps repeats in a chain reaction.

Chain propagation (Section 7.5B) A step in a chain reaction characterized by the reaction of a reactive intermediate and a molecule to give a new reactive intermediate and a new molecule.

Chain termination (Section 7.5B) A step in a chain reaction that involves destruction of reactive intermediates.

Chain-growth polymerization (Section 24.6) A polymerization that involves sequential addition reactions, either to unsaturated monomers or to monomers possessing other reactive functional groups.

Chain-transfer reaction (Section 24.6A) The transfer of reactivity of an endgroup from one chain to another during a polymerization.

Chair conformation (Section 2.6B) The most stable nonplanar conformation of a cyclohexane ring; all bond angles are approximately 109.5°, and all bonds on adjacent carbons are staggered.

Chemical shift, δ (Section 13.3) The shift in parts per million (ppm) of an NMR signal from the signal of TMS.

Chiral (Section 3.2) From the Greek, *cheir*, hand; an

object that is not superposable on its mirror image; an object that has handedness.

Circular DNA (Section 28.2C) A type of double-stranded DNA in which the $5'$ and $3'$ ends of each strand are joined by phosphodiester groups.

Cis (Section 2.7A) A prefix meaning on the same side.

Cis,trans isomerism Isomers that have the same order of attachment of their atoms but a different arrangement of their atoms in space due to the presence of either a ring (Section 2.7A) or a carbon-carbon double bond (Section 5.1C).

Clemmensen reduction (Section 16.14C) Reduction of the $C=O$ group of an aldehyde or ketone to a CH_2 group using $Zn(Hg)$ and HCl.

Codon (Section 28.4A) A triplet of nucleotides on mRNA that directs incorporation of a specific amino acid into a polypeptide sequence.

Coenzyme (Section 29.1) A low-molecular-weight, nonprotein molecule or ion that binds reversibly to an enzyme, functions as a second substrate for the enzyme, and is regenerated by further reaction.

Condensation polymerization (Section 24.5) A polymerization in which chain growth occurs in a stepwise manner between difunctional monomers. Also called step-growth polymerization.

Conformation (Section 2.6A) Any three-dimensional arrangement of atoms in a molecule that results by rotation about a single bond.

Conjugate acid (Section 4.1) The species formed from a base when it accepts a proton from an acid.

Conjugate addition (Section 19.8) Addition of a nucleophile to the β-carbon of an α,β-unsaturated carbonyl compound.

Conjugate base (Section 4.1) The species formed from an acid when it donates a proton to a base.

Conjugation (Section 12.4H) A situation in which two multiple bonds are separated by a single bond. Alternatively, a series of overlapping p orbitals. 1,3-Butadiene, for example, is a conjugated diene, and 3-buten-3-one is a conjugated enone.

Constitutional isomers (Section 2.2) Compounds with the same molecular formula but a different connectivity (order of attachment) of their atoms.

Contributing structures (Section 1.6A) Representations of a molecule, ion, or radical that differ only in the distribution of valence electrons by resonance.

Coupling (Section 13.9) The magnetic interaction of the nuclear spins of nearby atoms.

Coupling constant (J) (Section 13.10) The distance between peaks in a split signal, expressed in hertz. The value of J is a quantitative measure of the magnetic interaction of nuclei whose spins are coupled.

Covalent bond (Section 1.2C) A chemical bond formed between two atoms by sharing one or more pairs of electrons.

Crown ether (Section 11.10) A cyclic polyether derived from ethylene glycol and substituted ethylene glycols. Crown ethers are excellent phase-transfer catalysts.

Crystalline domain (Section 24.4) An ordered crystalline region in the solid state of a polymer; also called a crystallite.

Curved arrow (Section 1.6B) A symbol used to show the redistribution of valence electrons in resonance contributing structures or reactions.

Cyanohydrin (Section 16.6D) A molecule containing an —OH group and a —CN group bonded to the same carbon.

Cycloaddition reaction (Section 23.3) A reaction in which two reactants add together in a single step to form a cyclic product. The best known of these is the Diels-Alder reaction.

Cycloalkane (Section 2.4A) A saturated hydrocarbon that contains carbons joined to form a ring.

Deactivating group (Section 21.1A) Any substituent on a benzene ring that causes the rate of electrophilic aromatic substitution to be lower than that for benzene.

Decarboxylation (Section 17.10) Loss of CO_2 from a carboxyl group.

Dehydration (Section 9.6) Elimination of water.

Dehydrohalogenation (Section 8.8B) Removal of —H and —X from adjacent carbons; a type of β-elimination.

DEPT-NMR (Section 13.13) Distortionless Enhancement by Polarization Transfer. A spectroscopic technique for distinguishing among ^{13}C signals for CH_3, CH_2, CH, and quaternary carbons in ^{13}C-NMR.

Deshielding (Section 13.3) An effect produced when electron density is decreased around a nucleus, causing it to absorb toward the left (downfield) on the chart paper.

Dextrorotatory (Section 3.7B) Refers to a substance that rotates the plane of polarized light to the right.

Diamagnetic current (Section 13.3) The circulation of

electrons about a nucleus in an applied field. The resulting nuclear shielding resulting is called diamagnetic shielding.

Diastereomers (Section 3.4A) Stereoisomers that are not mirror images of each other; refers to relationships among two or more objects.

Diastereotopic groups (Section 13.11) Atoms or groups on an atom that is bonded to two nonidentical groups, one of which contains a stereocenter. When one of the atoms or groups is replaced by another group, a new stereocenter is created, and a set of diastereomers results. The hydrogens of the CH_2 group of 2-butanol, for example, are diastereotopic. Diastereotopic groups have different chemical shifts under all conditions.

Diaxial interactions (Section 2.6B) Nonbonded interactions between atoms or groups in axial positions on the same side of a chair conformation of a cyclohexane ring.

Diazonium ion (Section 22.9D) An ArN_2^+ or RN_2^+ ion.

Dielectric constant (Section 8.2) A measure of a solvent's ability to insulate opposite charges from one another.

Diels-Alder adduct (Section 23.3) A cyclohexene produced by a cycloaddition reaction between a diene and a dienophile.

Dienophile (Section 23.3) A compound containing a double bond (consisting of one or two C, N, or O atoms) that can react with a conjugated diene to give a Diels-Alder adduct.

Dihedral angle (Section 2.6A) The angle created by two intersecting planes.

Diol (Section 9.1B) A compound containing two hydroxyl groups.

Dipeptide (Section 27.3) A molecule containing two amino acid units joined by a peptide bond.

Dipole moment (μ) (Section 1.5) The vector sum of individual bond dipole moments in a molecule. Reported in a unit called the debye (D).

Dipole-dipole interaction (Section 9.2) The attraction between the positive end of one dipole and the negative end of another.

Disaccharide (Section 25.4) A carbohydrate containing two monosaccharide units joined by a glycosidic bond.

Dispersion forces (Section 2.8) Very weak intermolecular coulombic forces of attraction.

Disproportionation (Section 24.6A) A termination process that involves the abstraction of a hydrogen atom from the beta position of the propagating radical of one chain by the radical endgroup of another chain.

Disulfide (Section 11.11) A molecule containing an —S—S— group.

Double helix (Section 28.2B) A type of secondary structure of DNA molecules in which two antiparallel polynucleotide strands are coiled in a right-handed manner about the same axis.

Double-headed arrow (Section 1.6A) A symbol used to connect resonance contributing structures.

Downfield (Section 13.4) The shift of an NMR signal to the left on the chart paper.

E (Section 5.2C) From the German, *entgegen*, opposite. Specifies that groups of higher priority on the carbons of a double bond are on opposite sides.

E-Z system (Section 5.2C) A system to specify the configuration of groups about a carbon-carbon double bond.

Eclipsed conformation (Section 2.6A) A conformation about a carbon-carbon single bond in which the atoms or groups on one carbon are as close as possible to the atoms or groups on an adjacent carbon.

Edman degradation (Section 27.4B) A method for selectively cleaving and identifying the *N*-terminal amino acid of a polypeptide chain.

Elastomer (Section 24.4) A material that, when stretched or otherwise distorted, returns to its original shape when the distorting force is released.

Electromagnetic radiation (Section 12.1) Light and other forms of radiant energy.

Electronegativity (Section 1.2C) A measure of the force of an atom's attraction for electrons it shares with another atom in a chemical bond.

Electrophile (Section 6.3A) From the Greek meaning electron loving. Any species that can accept a pair of electrons to form a new covalent bond; a Lewis acid.

Electrophilic aromatic substitution (Section 21.1) A reaction in which there is substitution of an electrophile, E^+, for a hydrogen on an aromatic ring.

Electrophoresis (Section 27.2D) The process of separating compounds on the basis of their electric charge.

β-Elimination (Chapter 8 introduction) A reaction in which a molecule, such as HCl, HBr, HI, or HOH, is split out or eliminated from adjacent carbons.

Enamine   (Section 16.10A) An unsaturated compound derived by the reaction of an aldehyde or ketone and a secondary amine followed by loss of H_2O; $R_2C=CR—NR_2$.

Enantiomeric excess (ee)   (Section 3.7D) The percentage difference in the number of moles of each enantiomer in a mixture compared with the total number of moles of both.

Enantiomers   (Section 3.2) Stereoisomers that are nonsuperposable mirror images of each other; refers to a relationship between pairs of objects.

Enantiotopic groups   (Section 13.11) Atoms or groups on an atom that give a stereocenter when one of the atoms or groups is replaced by another group. A pair of enantiomers results. The hydrogens of a CH_2 group of ethanol, for example, are enantiotopic. Replacing one of them by deuterium gives (R)-1-deuteroethanol; replacing the other by deuterium gives (S)-1-deuteroethanol. Enantiotopic groups have identical chemical shifts in achiral environments. In chiral environments, they have different chemical shifts.

Endergonic reaction   (Section 6.2) A reaction in which the Gibbs free energy of the products is higher than that of the reactants.

Endothermic reaction   (Section 6.2A) A reaction in which the enthalpy of the products is higher than the enthalpy of the reactants; a reaction in which heat is absorbed.

Energy diagram   (Section 6.2A) A graph showing the changes in energy that occur during a chemical reaction; energy is plotted on the vertical axis, and reaction progress is plotted on the horizontal axis.

Enol   (Section 10.8) A compound containing a hydroxyl group bonded to a doubly bonded carbon atom.

Enolate anion   (Section 16.11A) An anion derived by loss of a hydrogen from a carbon alpha to a carbonyl group; the anion of an enol.

Epoxide   (Section 11.7) A cyclic ether in which oxygen is one atom of a three-membered ring.

Equatorial position   (Section 2.6B) A position on a chair conformation of a cyclohexane ring that extends from the ring roughly perpendicular to the imaginary axis of the ring.

Equivalent hydrogens   (Section 13.5) Hydrogens that have the same chemical environment.

Estrogen   (Section 26.4A) A steroid hormone, such as estrone and estradiol, that mediates the development of sexual characteristics in females.

Ether   (Section 11.1) A compound containing an oxygen atom bonded to two carbon atoms.

Exergonic reaction   (Section 6.2) A reaction in which the Gibbs free energy of the products is lower than that of the reactants.

Exothermic reaction   (Section 6.2A) A reaction in which the enthalpy of the products is lower than that of the reactants; a reaction in which heat is liberated.

Fat   (Section 26.1B) A mixture of triglycerides that is semisolid or solid at room temperature.

Fatty acid   (Section 26.1A) A long, unbranched-chain carboxylic acid, most commonly of 12 to 20 carbons, derived from the hydrolysis of animal fats, vegetable oils, or the phospholipids of biological membranes.

Fingerprint region   (Section 12.3D) The portion of the vibrational infrared region that extends from 1000 to 400 cm^{-1}.

Fischer esterification   (Section 17.8A) The process of forming an ester by refluxing a carboxylic acid and an alcohol in the presence of an acid catalyst, commonly H_2SO_4 or HCl.

Fischer projection   (Section 25.1B) A two-dimensional representation for showing the configuration of stereocenters; horizontal lines represent bonds projecting forward and vertical lines represent bonds projecting to the rear.

Fishhook arrow   (Section 7.5A) A barbed curved arrow used to show the change in position of a single electron.

Flavin adenine dinucleotide (FAD)   (Section 29.1C) A biological oxidizing agent. When acting as an oxidizing agent, FAD is reduced to $FADH_2$.

Fluid-mosaic model   (Section 26.5B) A model wherein a biological membrane consists of a phospholipid bilayer with proteins, carbohydrates, and other lipids on the surface and embedded in the surface of the bilayer.

Formal charge   (Section 1.2E) The charge on an atom in a polyatomic ion or molecule.

Frequency   (Section 12.1) The number of full cycles of a wave that pass a fixed point in a second; given the symbol ν (Greek nu) and reported in hertz (Hz), which has the units s^{-1}.

Friedel-Crafts reaction   (Section 21.1C) An electrophilic aromatic substitution in which a hydrogen of an aromatic ring is replaced by an alkyl or acyl group.

Frost circle (Section 20.2A) A graphic method for determining the relative energies of pi MOs for planar, fully conjugated, monocyclic compounds.

Functional group (Section 1.3) An atom or group of atoms within a molecule that shows a characteristic set of physical and chemical properties.

Furanose (Section 25.2A) A five-membered cyclic form of a monosaccharide.

Gauche conformation (Section 2.6A) A conformation about a single bond of an alkane in which two groups lie at a dihedral angle of 60°.

Geminal (gem) dihalide (Section 10.6) Geminal, from the Latin: *geminatus*, twin. A compound containing two halogens on the same carbon atom.

Gibbs free energy, ΔG (Section 6.2A) A thermodynamic function relating enthalpy, entropy, and temperature, given by the equation $\Delta G = \Delta H - T\Delta S$. If $\Delta G < 0$ for a reaction, the reaction is spontaneous. If $\Delta G > 0$, the reaction is nonspontaneous.

Glass transition temperature (Section 24.4) The temperature at which a polymer undergoes the transition from a hard glass to a rubbery state.

Glycol (Section 6.5B) A compound with two hydroxyl (—OH) groups on adjacent carbons.

Glycolysis (Section 29.5) From the Greek: *glyko*, sweet, and *lysis*, splitting. A series of ten enzyme-catalyzed reactions by which glucose is oxidized to two molecules of pyruvate.

Glycoside (Section 25.3A) A carbohydrate in which the —OH on its anomeric carbon is replaced by —OR.

Glycosidic bond (Section 25.3A) The bond from the anomeric carbon of a glycoside to an —OR group.

Ground-state electron configuration (Section 1.1A) The electron configuration of lowest energy for an atom, molecule, or ion.

Haloalkane (alkyl halide) (Section 7.1) A compound containing a halogen atom covalently bonded to an sp^3-hybridized carbon atom and given the symbol R—X.

Haloalkene (vinylic halide) (Section 7.1) A compound containing a halogen bonded to one of the carbons of a carbon-carbon double bond.

Haloarene (aryl halide) (Section 7.1) A compound containing a halogen atom bonded to a benzene ring and given the symbol Ar—X.

Haloform (Section 7.2B) A compound of the type CHX_3 where X is a halogen.

Halohydrin (Section 6.3E) A compound containing a halogen atom and a hydroxyl group on adjacent carbons; those containing Br and OH are bromohydrins, and those containing Cl and OH are chlorohydrins.

Hammond's postulate (Section 7.5D) The structure of the transition state for an exothermic step looks more like the reactants of that step. Conversely, the structure of the transition state for an endothermic step looks more like the products of that step.

Haworth projection (Section 25.2A) A way to view furanose and pyranose forms of monosaccharides. The ring is drawn flat and viewed through its edge with the anomeric carbon on the right and the oxygen atom of the ring in the rear.

Heat of combustion (Section 2.9C) Standard heat of combustion is the heat released when one mole of a substance in its standard state (gas, liquid, solid) is oxidized completely to CO_2 and H_2O and is given the symbol ΔH^0.

***α*-Helix** (Section 27.6B) A type of secondary structure in which a section of polypeptide chain coils into a spiral, most commonly a right-handed spiral.

Hemiacetal (Section 16.8B) A molecule containing an —OH and an —OR or —OAr group bonded to the same carbon.

Hertz (Hz) (Section 12.1) The unit in which frequency is measured; s^{-1} (read "per second").

Heterocycle (Section 11.2) A cyclic compound whose ring contains more than one kind of atom. Ethylene oxide, for example, is a heterocycle whose ring contains two carbon atoms and one oxygen atom.

Heterocyclic amine (Section 22.1) An amine in which nitrogen is one of the atoms of a ring.

Heterocyclic aromatic amine (Section 22.1) An amine in which nitrogen is one of the atoms of an aromatic ring.

High-density lipoprotein (HDL) (Section 26.4A) Plasma particles, density 1.06–1.21 g/mL, consisting of approximately 33% proteins, 30% cholesterol, 29% phospholipids, and 8% triglycerides.

High-resolution mass spectrometry (Section 14.2A) Use of instrumentation that gives data capable of distinguishing between ions that differ in mass by as little as 0.0001 amu.

Histone (Section 28.2C) A protein, particularly rich in

the basic amino acids lysine and arginine, that is found associated with DNA molecules.

Hofmann elimination (Section 22.10) When treated with a strong base, a quaternary ammonium halide undergoes β-elimination by an E2 mechanism to give the less-substituted alkene as the major product.

Hofmann rule (Section 22.10) Any β-elimination that occurs preferentially to give the less-substituted alkene as the major product is said to follow the Hofmann rule.

Homotopic groups (Section 13.11) Atoms or groups on an atom that give an achiral molecule when one of the groups is replaced by another group. The hydrogens of the CH_2 group of propane, for example, are homotopic. Replacing either one of them by deuterium gives 2-deuteropropane, which is achiral. Homotopic atoms or groups have identical chemical shifts under all conditions.

Hückel criteria for aromaticity (Section 20.2A) To be aromatic, a monocyclic compound must have one p orbital on each atom of the ring, be planar or nearly so, and have $(4n + 2)$ pi electrons in the cyclic arrangement of p orbitals.

Hund's rule (Section 1.1A) When orbitals of equivalent energy are available but there are not enough electrons to fill all of them completely, then one electron is added to each equivalent orbital before a second electron is added to any one of them.

Hybrid orbital (Section 1.8B) An orbital formed by the combination of two or more atomic orbitals.

Hybridization (Section 1.8B) The combination of atomic orbitals of different types.

Hydration (Section 6.3B) Addition of water.

Hydride ion (Section 16.14B) A hydrogen atom with two electrons in its valence shell; $H:^-$.

Hydroboration-oxidation (Section 6.4) A method for converting an alkene to an alcohol. The alkene is treated with borane (BH_3) to give a trialkylborane, which is then oxidized with aqueous sodium hydroxide to give the alcohol.

Hydrocarbon (Section 2.1) A compound composed of only carbon and hydrogen.

α-Hydrogen (Section 16.11A) A hydrogen on a carbon alpha to a carbonyl group.

Hydrogen bonding (Section 9.2) The attractive interaction between a hydrogen atom bonded to an atom of high electronegativity (most commonly F, O, or N), and a lone pair of electrons on another atom of high electronegativity (again, most commonly F, O, or N).

Hydrogenolysis (Section 20.5C) Cleavage of a single bond by H_2, most commonly accomplished by treatment of a compound with H_2 in the presence of a transition metal catalyst.

Hydroperoxide (Section 11.5B) A compound containing an —OOH group.

Hydrophilic (Section 8.7) From the Greek, meaning water-loving.

Hydrophobic (Section 8.7) From the Greek, meaning water-fearing.

Hydrophobic effect (Section 27.6D) The tendency of nonpolar groups to cluster so as to shield themselves from contact with an aqueous environment.

Hydroxyl group (Section 1.3A) An —OH group.

Hyperconjugation (Section 6.3A) Interaction of electrons in a sigma-bonding orbital with the va-cant $2p$ orbital of an adjacent positively charged carbon.

Imide (Section 18.1D) A functional group in which two acyl groups, RCO— or ArCO—, are bonded to a nitrogen atom.

Imine (Section 16.10A) A compound containing a carbon-nitrogen double bond, $R_2C{=}NR'$; also called a Schiff base.

Inductive effect (Section 4.3D) The polarization of the electron density of a covalent bond due to the electronegativity of a nearby atom.

Infrared (IR) spectroscopy (Chapter 12 introduction) A spectroscopic technique in which a compound is irradiated with infrared radiation, absorption of which causes covalent bonds to change from a lower vibrational energy level to a higher one. Infrared spectroscopy is particularly valuable for determining the kinds of functional groups present in an organic molecule.

Ionic bond (Section 1.2C) A chemical bond resulting from the electrostatic attraction of an anion and a cation.

Ionization potential (IP) (Section 14.1) The minimum energy required to remove an electron from an atom or molecule to a distance where there is no electrostatic interaction between the resulting ion and electron.

Isoelectric point (pI) (Section 27.2C) The pH at which an amino acid, polypeptide, or protein has no net charge.

Isomers (Section 3.1) Different compounds with the same molecular formula. They may be stereoisomers, constitutional isomers, or conformational isomers.

Isotactic polymer (Section 24.6C) A polymer with identical configurations (either all *R* or all *S*) at all stereocenters along its chain, as, for example, isotactic polypropylene.

Keto-enol tautomerism (Section 10.8) A type of isomerism involving keto (from ketone) and enol tautomers.

Ketone (Section 1.3C) A compound containing a carbonyl group bonded to two carbons.

Ketose (Section 25.1A) A monosaccharide containing a ketone group.

Kinetic control (Section 19.2) Experimental conditions under which the composition of the product mixture is determined by the relative rates of formation of each product.

Lactam (Section 18.1D) A cyclic amide.

Lactate fermentation (Section 29.7A) A metabolic pathway that converts glucose to two molecules of lactate.

Lactone (Section 18.1C) A cyclic ester.

Levorotatory (Section 3.7B) Refers to a substance that rotates the plane of polarized light to the left.

Lewis acid (Section 4.5) Any molecule or ion that can form a new covalent bond by accepting a pair of electrons.

Lewis base (Section 4.5) Any molecule or ion that can form a new covalent bond by donating a pair of electrons.

Lewis structure of an atom (Section 1.1B) The symbol of an element surrounded by a number of dots equal to the number of electrons in the valence shell of the atom.

Ligand (*L*) (Chapter 15 introduction) A Lewis base bonded to a metal atom in a coordination compound. It may bond strongly or weakly. Refer to your general chemistry text for a more detailed description of types of ligand and coordination compounds.

Lindlar catalyst (Section 10.7A) Finely powdered palladium metal deposited on solid calcium carbonate that has been specially modified with lead salts. Its particular use is as a catalyst for the reduction of an alkyne to a cis alkene.

Line-angle drawing (Section 2.1) An abbreviated way to draw structural formulas in which each angle and line ending represents a carbon and each line represents a bond.

Lipid (Chapter 26 introduction) A biomolecule isolated from plant or animal sources by extraction with nonpolar organic solvents, such as diethyl ether and hexane.

Lipid bilayer (Section 26.5B) A back-to-back arrangement of phospholipid monolayers, often forming a closed vesicle or membrane.

Living polymer (Section 24.6D) A polymer chain that continues to grow without chain-termination steps until either all the monomer is consumed or some external agent is added to terminate the chain. The polymer chains will continue to grow if more monomer is added.

Low-density lipoprotein (LDL) (Section 26.4A) Plasma particles, density 1.02–1.06 g/mL, consisting of approximately 26% proteins, 50% cholesterol, 21% phospholipids, and 4% triglycerides.

Low-resolution mass spectrometry (Section 14.2A) Use of instrumentation that is capable of distinguishing only between ions that differ by 1 amu or more.

Markovnikov's rule (Section 6.3A) In the addition of HX, H_2O, or ROH to an alkene, hydrogen adds to the carbon of the double bond having the greater number of hydrogens.

Mass spectrometry (Chapter 14 introduction) An analytical technique for measuring the mass-to-charge ratio (m/z) of ions.

Mass spectrum (Section 14.1) A plot of the relative abundance of ions versus their mass-to-charge ratio.

Melt transition, T_m (Section 24.4) The temperature at which crystalline regions of a polymer melt.

Mercaptan (Section 9.9B) A common name for a thiol; that is, any compound containing an —SH (sulfhydryl) group.

Meso compound (Section 3.4B) An achiral compound possessing two or more stereocenters.

Messenger RNA (mRNA) (Section 28.3C) A ribonucleic acid that carries coded genetic information from DNA to the ribosomes for the synthesis of proteins.

Meta (*m*) (Section 20.3B) Refers to groups occupying 1,3- positions on a benzene ring.

Micelle (Section 26.2B) A spherical arrangement of organic molecules in water solution clustered so that their hydrophobic parts are buried inside the sphere and their hydrophilic parts are on the surface of the sphere and in contact with water.

Molecular ion (M^+) (Section 14.1) The cation formed by removal of a single electron from a parent molecule in a mass spectrometer.

Molecular orbital theory (Section 1.8) A theory of chemical bonding in which electrons in molecules occupy molecular orbitals formed by the combination of the atomic orbitals that make up the molecule.

Molecular spectroscopy (Section 12.2) The study of which frequencies of electromagnetic radiation are absorbed or emitted by substances and the correlation between these frequencies and specific types of molecular structure.

Monomer (Section 24.1) From the Greek: *mono + meros*, single part. The simplest nonredundant unit from which a polymer is synthesized.

Monosaccharide (Section 25.1A) A carbohydrate that cannot be hydrolyzed to a simpler carbohydrate.

D-Monosaccharide (Section 25.1C) A monosaccharide that, when written as a Fischer projection, has the —OH on its penultimate carbon to the right.

L-Monosaccharide (Section 25.1C) A monosaccharide that, when written as a Fischer projection, has the —OH on its penultimate carbon to the left.

Mutarotation (Section 25.2C) The change in specific rotation that occurs when an α or β hemiacetal form of a carbohydrate in aqueous solution is converted to an equilibrium mixture of the two forms.

(n + 1) Rule (Section 13.8) The ^{1}H-NMR signal of a hydrogen or set of equivalent hydrogens is split into (n + 1) peaks by a nonequivalent set of n equivalent neighboring hydrogens.

Newman projection (Section 2.6A) A way to view a molecule by looking along a carbon-carbon single bond.

Nicotinamide adenine dinucleotide (NAD$^+$) (Section 29.1B) A biological oxidizing agent. When acting as an oxidizing agent, NAD$^+$ is reduced to NADH.

Nitrile (Section 18.1E) A compound containing a —C≡N (cyano) group bonded to a carbon atom.

Nitrogen rule (Section 14.3) A rule stating that a compound with an odd number of nitrogen atoms has an odd m/z ratio; if it has zero or an even number of nitrogen atoms, the molecular ion has an even m/z ratio.

Node (Section 1.7A) Any point in space where the value of a wave function is zero.

Nonbonded interaction strain (Section 2.6A) The strain that arises when atoms not bonded to each other are forced abnormally close to one another.

Nonbonding electrons (Section 1.2D) Valence electrons not involved in forming covalent bonds; also called unshared pairs or lone pairs.

Nonpolar covalent bond (Section 1.2C) A covalent bond between atoms whose difference in electronegativity is less than approximately 0.5.

Nuclear magnetic resonance (NMR) spectroscopy (Chapter 13 introduction) A spectroscopic technique that gives us information about the number and types of atoms in a molecule, for example, about hydrogens (^{1}H-NMR) and carbons (^{13}C-NMR).

Nucleic acid (Chapter 28 introduction) A biopolymer containing three types of monomer units: heterocyclic aromatic amine bases derived from purine and pyrimidine, the monosaccharides D-ribose or 2-deoxy-D-ribose, and phosphoric acid.

Nucleophile (Chapter 8 introduction) From the Greek, meaning nucleus-loving. A molecule or ion that donates a pair of electrons to another atom or ion to form a new covalent bond; a Lewis base.

Nucleophilic acyl substitution (Section 18.4) A reaction in which a nucleophile bonded to the carbon of an acyl group is replaced by another nucleophile.

Nucleophilic aromatic substitution (Section 21.3B) A reaction in which a nucleophile, most commonly a halogen, on an aromatic ring is replaced by another nucleophile.

Nucleophilic substitution (Chapter 8 introduction) Any reaction in which one nucleophile is substituted for another at a tetravalent carbon atom.

Nucleophilicity (Section 8.4B) A kinetic property measured by the rate at which a nucleophile causes nucleophilic substitution on a reference compound under a standardized set of experimental conditions.

Nucleoside (Section 28.1) A building block of nucleic acids, consisting of D-ribose or 2-deoxy-D-ribose bonded to a heterocyclic aromatic amine base by a β-N-glycosidic bond.

Nucleotide (Section 28.1) A nucleoside in which a molecule of phosphoric acid is esterified with an —OH of the monosaccharide, most commonly either the 3′—OH or the 5′—OH.

Observed rotation (Section 3.7B) The number of degrees through which a compound rotates the plane of polarized light.

Octane rating (Section 2.10B) The percentage of isooctane in a test mixture of isooctane and heptane that has equivalent antiknock properties to a gasoline being tested.

Octet rule (Section 1.2A) Group 1A–7A elements react to achieve an outer shell of eight valence electrons.

Oil   (Section 26.1B) When used in the context of fats and oils, a mixture of triglycerides that is liquid at room temperature.

Oligosaccharide   (Section 25.4) A carbohydrate containing four to ten monosaccharide units, each joined to the next by a glycosidic bond.

Optical purity   (Section 3.7D) The specific rotation of a mixture of enantiomers divided by the specific rotation of the enantiomerically pure substance.

Optically active   (Section 3.7) Refers to a compound that rotates the plane of polarized light.

Orbital   (Section 1.1) A region of space where there is a high probability of finding electrons.

Order of precedence of functions   (Section 16.2B) A ranking of functional groups in order of priority for the purposes of IUPAC nomenclature.

Organic synthesis   (Section 10.10) A series of reactions by which a set of organic starting materials is converted to a more complicated structure.

Organometallic compound   (Chapter 15 introduction) A compound that contains a carbon-metal bond.

Ortho (o)   (Section 20.3B) Refers to groups occupying 1,2-positions on a benzene ring.

Oxidation   (Section 6.5A) The loss of electrons; alternatively, either the loss of hydrogens, the gain of oxygens, or both.

β-Oxidation   (Section 29.3) A series of four enzyme-catalyzed reactions that cleaves carbon atoms, two at a time, from the carboxyl end of a fatty acid by way of intermediates that are oxidized at the β-carbon.

Oxonium ion   (Section 6.3B) An ion in which oxygen bears a positive charge.

Oxymercuration-reduction   (Section 6.3F) A method for converting an alkene to an alcohol. The alkene is treated with mercury(II) acetate followed by reduction with sodium borohydride.

Para (p)   (Section 20.3B) Refers to groups occupying 1,4-positions on a benzene ring.

Pauli exclusion principle   (Section 1.1A) No more than two electrons may be present in an orbital. If two electrons are present, their spins must be paired.

Peak   (Section 13.8) The units into which an NMR signal is split: two in a doublet, three in a triplet, four in a quartet, and so on.

Peptide bond   (Section 27.3) The special name given to the amide bond formed between the α-amino group of one amino acid and the α-carboxyl group of another amino acid.

Pericyclic reaction   (Section 23.2F) A reaction that takes place in a single step, without intermediates, and involves a cyclic redistribution of bonding electrons.

Phase-transfer catalyst   (Section 8.7) A substance that transfers ions from an aqueous phase into an organic phase and vice versa.

Phenol   (Section 20.5A) A compound that contains an —OH bonded to a benzene ring; a benzenol.

Phenyl group, C_6H_5—   (Chapter 5 introduction) An aryl group derived by removing an H from benzene; abbreviated C_6H_5— or Ph—.

Phospholipid   (Section 26.5A) A lipid containing glycerol esterified with two molecules of fatty acid and one molecule of phosphoric acid.

Pi (π) bond   (Section 1.8D) A covalent bond formed by the overlap of parallel p orbitals.

Pi (π) bonding molecular orbital   (Section 1.8D) A molecular orbital formed by overlap of parallel p orbitals on adjacent atoms; its electron density lies above and below the line connecting the atoms.

Plane of symmetry   (Section 3.2) An imaginary plane passing through an object dividing it so that one half is the mirror image of the other half.

Plane-polarized light   (Section 3.7A) Light vibrating in only parallel planes.

Plastic   (Section 24.1) A polymer that can be molded when hot and retains its shape when cooled.

β-Pleated sheet   (Section 27.6B) A type of secondary structure in which sections of polypeptide chains are aligned parallel or antiparallel to one another.

Polar covalent bond   (Section 1.2C) A covalent bond between atoms whose difference in electronegativity is between approximately 0.5 and 1.9.

Polarimeter   (Section 3.7B) An instrument for measuring the ability of a compound to rotate the plane of polarized light.

Polarizability   (Section 7.3B) A measure of the ease of distortion of the distribution of electron density about an atom or group in response to interaction with other molecules or ions. Fluorine, which has a high electronegativity, holds its electrons tightly and has a very low polarizability. Iodine, which has a lower electronegativity and holds its electrons less tightly, has a very high polarizability.

Polyamide   (Section 24.5A) A polymer in which each monomer unit is joined to the next by an amide bond, as, for example, nylon 66.

Polycarbonate   (Section 24.5C) A polyester in which the carboxyl groups are derived from carbonic acid.

Polyester   (Section 24.5B) A polymer in which each monomer unit is joined to the next by an ester bond, as, for example, poly(ethylene terephthalate).

Polymer   (Section 24.1) From the Greek: *poly + meros*, many parts. Any long-chain molecule synthesized by linking together many single parts called monomers.

Polynuclear aromatic hydrocarbon (PAH)   (Section 20.3B) A hydrocarbon containing two or more fused aromatic rings.

Polypeptide   (Section 27.3) A macromolecule containing many amino acid units, each joined to the next by a peptide bond.

Polysaccharide   (Section 25.5) A carbohydrate containing a large number of monosaccharide units, each joined to the next by one or more glycosidic bonds.

Polyunsaturated fatty acid   (Section 26.1A) A fatty acid with two or more carbon-carbon double bonds in its hydrocarbon chain.

Polyunsaturated triglyceride   (Section 26.1B) A triglyceride having several carbon-carbon double bonds in the hydrocarbon chains of its three fatty acids.

Polyurethane   (Section 24.5D) A polymer containing the —NHCO$_2$— group as a repeating unit.

Primary (1°) alcohol   (Section 1.3A) An alcohol in which the —OH group is bonded to a primary carbon.

Primary (1°) amine   (Section 1.3B) An amine in which nitrogen is bonded to only one carbon atom.

Primary (1°) carbon   (Section 2.3C) A carbon bonded to one other carbon.

Primary structure of nucleic acids   (Section 28.2A) The sequence of bases along the pentose-phosphodiester backbone of a DNA or RNA molecule read from the 5′ end to the 3′ end.

Primary structure of proteins   (Section 27.4) The sequence of amino acids in the polypeptide chain, read from the *N*-terminal amino acid to the *C*-terminal amino acid.

Principle of microscopic reversibility   (Section 9.6) A principle stating that the sequence of transition states and reactive intermediates in the mechanism of any reversible reaction must be the same, but in reverse order, for the backward reaction as for the forward reaction.

Prochiral hydrogens   (Section 13.11) Refers to two hydrogens bonded to a carbon atom. When one or the other of them is replaced by a different atom, the carbon atom becomes a stereocenter. The hydrogens of the CH$_2$ group of ethanol, for example, are prochi-

ral. Replacing one of them by deuterium gives (*R*)-1-deuteroethanol; replacing the other by deuterium gives (*S*)-1-deuteroethanol.

Pro-R-hydrogen   (Section 13.11) Replacing this hydrogen by deuterium gives a stereocenter with an *R* configuration.

Pro-S-hydrogen   (Section 13.11) Replacing this hydrogen by deuterium gives a stereocenter with an *S* configuration.

Prostaglandin   (Section 26.3) A member of the family of compounds having the 20-carbon skeleton of prostanoic acid.

Protic acid   (Section 4.5) An acid that is a proton donor in an acid-base reaction.

Protic solvent   (Section 8.2) A solvent that is a hydrogen bond donor; the most common protic solvents contain —OH groups. Common protic solvents are water and low-molecular-weight alcohols such as ethanol.

Pyranose   (Section 25.2A) A six-membered cyclic form of a monosaccharide.

Quantum mechanics   (Section 1.7A) The branch of science that studies particles and their associated waves.

Quaternary (4°) ammonium ion   (Section 22.1) An ion in which nitrogen is bonded to four carbons and bears a positive charge.

Quaternary (4°) carbon   (Section 2.3C) A carbon bonded to four other carbons.

Quaternary structure   (Section 27.6D) The arrangement of polypeptide monomers into a noncovalently bonded aggregate.

R   (Section 2.3A) A symbol used to represent an alkyl group.

R   (Section 3.3) From the Latin: *rectus*, straight, correct; used in the *R,S* convention to show that the order of priority of groups on a stereocenter is clockwise.

R,S system   (Section 3.3) A set of rules for specifying configuration about a stereocenter; also called the Cahn-Ingold-Prelog system.

Racemic mixture   (Section 3.7C) A mixture of equal amounts of two enantiomers.

Radical   (Section 7.5) Any chemical species that contains one or more unpaired electrons.

Rate-determining step   (Section 6.2A) The step in a multistep reaction sequence that crosses the highest energy barrier.

Reaction coordinate   (Section 6.2A) A measure of the

change in the positions of atoms during a reaction; plotted on the horizontal axis in a reaction energy diagram.

Reaction intermediate (Section 6.2A) A species, formed between two successive reaction steps, that lies in an energy minimum between the two transition states.

Reaction mechanism (Section 6.2A) A step-by-step description of how a chemical reaction occurs.

Rearrangement (Section 6.3C) A change in connectivity of the atoms in a product compared with the connectivity of the same atoms in the starting material.

Reducing sugar (Section 23.4D) A carbohydrate that reduces Ag(I) to Ag or Cu(II) to Cu(I)

Reduction (Section 6.5A) The gain of electrons; alternatively, either the gain of hydrogen, loss of oxygens, or both.

Reductive amination (Section 16.10A) A method for preparation of amines by treating an aldehyde or ketone with an amine in the presence of a reducing agent.

Regioselective reaction (Section 6.3A) A reaction in which one direction of bond forming or bond breaking occurs in preference to all other directions.

Regiospecific reaction (Section 6.3) A reaction in which one direction of bond forming or breaking occurs to the exclusion of all other directions of bond forming or breaking.

Resolution (Section 3.8) Separation of a racemic mixture into its enantiomers. (Section14.2A) In mass spectrometry, a measure of how well a mass spectrometer separates ions of different mass.

Resonance energy (Section 20.1C) The difference in energy between a resonance hybrid and the most stable of its hypothetical contributing structures in which electrons are localized on particular atoms and in particular bonds.

Resonance hybrid (Section 1.6A) A molecule, ion, or radical described as a composite of a number of contributing structures.

Resonance in NMR spectroscopy (Section 13.3) The absorption of electromagnetic radiation by a precessing nucleus and the resulting "flip" of its nuclear spin from a lower energy state to a higher energy state.

Restriction endonuclease (Section 28.5A) An enzyme that catalyzes hydrolysis of a particular phosphodiester bond within a DNA strand.

Retrosynthesis (Section 10.10) A process of reasoning backwards from a target molecule to a suitable set of starting materials.

Ribosomal RNA (rRNA) (Section 28.3A) A ribonucleic acid found in ribosomes, the sites of protein synthesis.

S (Section 3.3) From the Latin: *sinister*, left; used in the *R,S* convention to show that the order of priority of groups on a stereocenter is counterclockwise.

Sanger dideoxy method (Section 28.5B) A method developed by Frederick Sanger for sequencing DNA molecules.

Saponification (Section 18.5C) Hydrolysis of an ester in aqueous NaOH or KOH to an alcohol and the sodium or potassium salt of a carboxylic acid.

Saturated hydrocarbon (Section 2.1) A hydrocarbon containing only carbon-carbon single bonds.

Schiff base (Section 16.10A) An alternative name for an imine.

Secondary (2°) alcohol (Section 1.3A) An alcohol in which the —OH group is bonded to a secondary carbon.

Secondary (2°) amine (Section 1.3B) An amine in which nitrogen is bonded to two carbon atoms.

Secondary (2°) carbon (Section 2.3C) A carbon bonded to two other carbons.

Secondary structure of nucleic acids (Section 28.2B) The ordered arrangement of nucleic acid strands.

Secondary structure of proteins (Section 27.6A) The ordered arrangements (conformations) of amino acids in localized regions of a polypeptide or protein.

Shell (Section 1.1) A region of space around a nucleus in which electrons are found.

Shielding (Section 13.3) An effect produced when electron density is increased about a nucleus, causing it to absorb toward the right (upfield) on the chart paper.

1,2-Shift (Section 6.3C) A type of rearrangement in which an atom or group of atoms with its bonding electrons moves from one atom to an adjacent electron-deficient atom.

Sigma (σ) bonding molecular orbital (Section 1.8A) A molecular orbital in which electron density is concentrated between two nuclei and along the axis joining them.

Signal (Section 13.3) A recording in an NMR spectrum of a nuclear magnetic resonance.

Signal splitting (Section 13.8) Splitting of an NMR

signal into a set of peaks by the influence of non-equivalent nuclei on the same or adjacent atom(s).

S_N1 reaction (Section 8.3) A unimolecular nucleophilic substitution reaction.

S_N2 reaction (Section 8.3) A bimolecular nucleophilic substitution reaction.

Soap (Section 26.2A) A sodium or potassium salt of a fatty acid.

Solvolysis (Section 8.3) A nucleophilic substitution in which the solvent is also the nucleophile.

***sp* Hybrid orbital** (Section 1.8E) A hybrid atomic orbital formed by the combination of one *s* atomic orbital and one *p* atomic orbital.

***sp*² **Hybrid orbital** (Section 1.8D) A hybrid atomic orbital formed by the combination of one *s* atomic orbital and two *p* atomic orbitals.

***sp*³ **Hybrid orbital** (Section 1.8C) A hybrid atomic orbital formed by the combination of one *s* atomic orbital and three *p* atomic orbitals.

Specific rotation (Section 3.7B) Observed rotation of the plane of polarized light when a sample is placed in a tube 1.0 dm in length and at a concentration of 1 g/mL.

Staggered conformation (Section 2.6A) A conformation about a carbon-carbon single bond in which the atoms or groups on one carbon are as far apart as possible from atoms or groups on an adjacent carbon.

Step-growth polymerization (Section 24.5) A polymerization in which chain growth occurs in a stepwise manner between difunctional monomers, as, for example, between adipic acid and hexamethylenediamine to form nylon 66; also called condensation polymerization.

Stereocenter (Section 3.2) An atom that has four different atoms or groups of atoms attached to it; also called a stereogenic center.

Stereoisomers (Section 3.1) Isomers that have the same molecular formula and the same connectivity but a different orientation of their atoms in space.

Stereoselective reaction (Section 6.3D) A reaction in which one stereoisomer is formed or destroyed in preference to all others.

Stereospecific reaction (Section 6.3D) A reaction in which one stereoisomer is formed or destroyed to the exclusion of all others.

Steric hindrance (Section 8.4D) The ability of groups, because of their size, to hinder access to a reaction site within a molecule.

Steroid (Section 26.4A) A plant or animal lipid having the characteristic tetracyclic ring structure of the steroid nucleus, namely three six-membered rings and one five-membered ring.

Substitution (Section 7.4) A reaction in which an atom or group of atoms in a compound is replaced by another atom or group of atoms.

Sulfide (Section 11.11) The sulfur analog of an ether; a molecule containing a sulfur atom bonded to two carbon atoms.

Syn stereospecific (Section 6.4) The addition of atoms or groups of atoms to the same face of a carbon-carbon double bond.

Syndiotactic polymer (Section 24.6C) A polymer with alternating *R* and *S* configurations at the stereocenters along its chain, as for example syndiotactic polypropylene.

Tautomers (Section 10.8) Constitutional isomers in equilibrium with each other that differ in the location of a hydrogen atom and a double bond relative to a heteroatom, most commonly O, N, or S.

Telechelic polymer (Section 24.6D) A polymer in which its growing chains are terminated by formation of new functional groups at both ends of its chains. These new functional groups are introduced by adding reagents, such as CO_2 or ethylene oxide, to the growing chains.

***C*-Terminal amino acid** (Section 27.3) The amino acid at the end of a polypeptide chain having the free —COOH group.

***N*-Terminal amino acid** (Section 27.3) The amino acid at the end of a polypeptide chain having the free —NH₂ group.

Terpene (Section 5.4) A compound whose carbon skeleton can be divided into two or more units identical with the carbon skeleton of isoprene.

Tertiary (3°) alcohol (Section 1.3A) An alcohol in which the —OH group is bonded to a tertiary carbon.

Tertiary (3°) amine (Section 1.3B) An amine in which nitrogen is bonded to three carbon atoms.

Tertiary (3°) carbon (Section 2.3C) A carbon bonded to three other carbon atoms.

Tertiary structure of nucleic acids (Section 28.2C) The three-dimensional arrangement of all atoms of a nucleic acid; commonly referred to as supercoiling.

Tertiary structure of proteins (Section 27.6C) The three-dimensional arrangement in space of all atoms in a single polypeptide chain.

Thermodynamic control   (Section 19.2) Experimental conditions that permit the establishment of equilibrium between two or more products of a reaction. The composition of the product mixture is determined by the relative stabilities of the products.

Thermoplastic   (Section 24.1) A polymer that can be melted and molded into a shape that is retained when it is cooled.

Thermoset plastic   (Section 24.1) A polymer that can be molded when it is first prepared but, once cooled, hardens irreversibly and cannot be remelted.

Thioester   (Section 29.3A) An ester in which one atom of oxygen in the carboxylate group is replaced by an atom of sulfur.

Thiol   (Chapter 9 introduction) A compound containing an —SH (sulfhydryl) group bonded to an sp^3-hybridized carbon.

Tollens' reagent   (Section 16.13A) A solution prepared by dissolving Ag_2O in aqueous ammonia; used for selective oxidation of an aldehyde to a carboxylic acid.

Torsional strain   (Section 2.6A) The force that opposes the rotation of one part of a molecule about a bond, while the other part of the molecule is held stationary.

Trans   (Section 2.7A) A prefix meaning across from.

Transesterification   (Section 18.6C) Exchange of the —OR or —OAr group of an ester for another —OR or —OAr group.

Transfer RNA (tRNA)   (Section 28.3B) A ribonucleic acid that carries a specific amino acid to the site of protein synthesis on ribosomes.

Transition state   (Section 6.2A) An unstable species of maximum energy formed during the course of a reaction; a maximum on an energy diagram.

Triglyceride (triacylglycerol)   (Section 26.1) An ester of glycerol with three fatty acids.

Triol   (Section 9.1B) A compound containing three hydroxyl groups.

Tripeptide   (Section 27.3) A molecule containing three amino acid units, each joined to the next by a peptide bond.

Twist-boat conformation   (Section 2.6B) A nonplanar conformation of a cyclohexane ring that is twisted from and slightly more stable than a boat conformation.

Ultraviolet-visible spectroscopy   (Section 23.5) A spectroscopic technique in which a compound is irradiated with ultraviolet or visible radiation, absorption of which causes electrons to change from a lower electronic level to a higher one. Ultraviolet spectroscopy is particularly valuable for determining the extent of conjugation in organic molecules containing pi bonds.

Unimolecular reaction   (Section 8.3) A reaction in which only one species is involved in the rate-determining step.

Unsaturated hydrocarbon   (Chapter 5 introduction) A hydrocarbon containing one or more carbon-carbon double or triple bonds. The three classes of unsaturated hydrocarbons are alkenes, alkynes, and arenes.

Upfield   (Section 13.4) The shift of an NMR signal to the right on the chart paper.

Valence electrons   (Section 1.1B) Electrons in the valence (outermost) shell of an atom.

Valence shell   (Section 1.1B) The outermost electron shell of an atom.

Valence-shell electron-pair repulsion (VSEPR) model   (Section 1.4) A method for predicting bond angles in a molecule or ion based on the idea that regions of electron density about an atom repel each other so that each is as far away from the others as possible.

Van der Waals forces   (Section 7.3B) A group of intermolecular attractive forces including dipole-dipole, dipole–induced dipole, and induced dipole–induced dipole (dispersion) forces.

Vibrational infrared   (Section 12.3A) The portion of the infrared region that extends from 4000 to 400 cm^{-1}.

Vicinal (vic) dihalide   (Section 10.6) Vicinal from the Latin: *vicinalis*, neighbor. A compound containing two halogen atoms on adjacent carbon atoms.

Vinylic carbocation   (Section 10.9B) A carbocation in which the positive charge is on one of the carbons of a carbon-carbon double bond.

Watson-Crick model   (Section 28.2B) A double-helix model for the secondary structure of a DNA molecule.

Wave function   (Section 1.7A) A solution to a set of equations that defines the energy of an electron in an atom.

Wavelength (λ)   (Section 12.1) The distance between consecutive peaks on a wave.

Wavenumber ($\bar{\nu}$)   (Section 12.3A) The frequency of electromagnetic radiation expressed as the number of waves per centimeter.

Williamson ether synthesis   (Section 11.4A) A general method for the synthesis of dialkyl ethers by an S_N2 reaction between an alkyl halide and an alkoxide ion.

Wolff-Kishner reduction   (Section 16.14C) Reduction of the C=O group of an aldehyde or ketone to a CH_2 group using hydrazine and base.

Ylide   (Section 16.7) A neutral molecule with positive and negative charges on adjacent atoms.

Z   (Section 5.3C) From the German: *zusammen,* together. Specifies that groups of higher priority on the carbons of a double bond are on the same side.

Zaitsev's rule   (Section 8.8) A rule stating that the major product of a β-elimination reaction is the most stable alkene; that is, it is the alkene with the greatest number of substituents on the carbon-carbon double bond.

Zwitterion   (Section 27.1A) An internal salt of an amino acid.

INDEX